ROBOTICS AND MANUFACTURING

RECENT TRENDS IN RESEARCH, EDUCATION, AND APPLICATIONS

VOLUME 3

ASME PRESS SERIES: PROCEEDINGS OF THE INTERNATIONAL SYMPOSIA ON ROBOTICS AND MANUFACTURING ISSN 1052-4150

Volume 1, Recent Trends in Robotics: Modeling, Control, and Education, edited by M. Jamshidi, J.Y.S. Luh, and M. Shahinpoor, North-Holland, New York, 1986

Volume 2, Robotics and Manufacturing: Recent Trends in Research, Education, and Applications, edited by M. Jamshidi, J.Y.S. Luh, H. Seraji, and G.P. Starr, ASME Press, New York, 1988

Volume 3, Robotics and Manufacturing: Recent Trends in Research, Education, and Applications, edited by M. Jamshidi and M. Saif, ASME Press, New York, 1990

ASME PRESS SERIES

ROBOTICS AND MANUFACTURING

RECENT TRENDS IN RESEARCH, EDUCATION, AND APPLICATIONS

VOLUME 3

Proceedings of the Third International Symposium on Robotics and Manufacturing: Research, Education, and Applications (ISRAM '90), held July 18-20, 1990, in Burnaby, British Columbia, Canada

EDITORS

Mohammad Jamshidi
AT&T Professor and Director
CAD Laboratory for Systems/Robotics
Department of Electrical and Computer Engineering
University of New Mexico
Albuquerque, NM 87131 USA

Mehrdad Saif
School of Engineering Science
Simon Fraser University
Burnaby, BC, Canada V5A 1S6

School of Engineering Science
Simon Fraser University

Advanced Systems Institute
of British Columbia

Centre for Systems Science
Simon Fraser University

Natural Science Research
Council of Canada

CAD Laboratory for Systems/Robotics
University of New Mexico

ASME PRESS NEW YORK 1990

Library of Congress Cataloging-in-Publication Data

International Symposium on Robotics and Manufacturing--Research, Education, and Applications (3rd : 1990 : Burnaby, B.C.)
Robotics and manufacturing : recent trends in research, education, and application : proceedings of the Third International Symposium on Robotics and Manufacturing : Research, Education, and Applications (ISRAM '90), held July 18-20, 1990, in Burnaby, British Columbia, Canada / editors, Mohammad Jamshidi, Mehrdad Saif.
p. cm. -- (ASME Press series--proceedings of the international symposia on robotics and manufacturing : v. 3)
Includes bibliographical references and index.
ISBN 0-7918-0016-4
1. Robots, Industrial--Congresses. 2. Manufactruing processes--Automation--Congresses. 3. Robotics--Congresses. I. Jamshidi, Mohammad. II. Saif, Mehrdad. III. Title. IV. Series.
TS191.8.I58 1990
670.42'72--dc20 90-14453
CIP

THIS VOLUME IS DEDICATED

TO :

PROFESSOR FAZLOLLAH REZA

FOR

FIFTY YEARS OF CONTRIBUTIONS TO CIRCUIT AND NETWORK THEORIES, CONTROL SYSTEMS, SYSTEM THEORY, INFORMATION THEORY, CYBERNETICS, APPLIED MATHEMATICS,
AND
PERSIAN POETRY AND LITERATURE.

Fazlollah Reza, a prominent scientist, engineer and Persian literary personality, received his B.S. degree in engineering from University of Tehran in 1940, M.S. from Columbia University, in 1946 and Doctor of Electrical Engineering from the Polytechnic Institute of New York in 1950. He has served as a member of faculty of MIT and Syracuse University. He has been guest professor at ETH (Zürich), Royal Technical University (Copenhagen), University of Colorado (Boulder), University of Paris, and Carleton University (Ottawa). He has numerous publications including the first book on information theory published by McGraw Hall in 1961. His administrative posts include the Chancellorships of University of Tehran, Tehran University of Technology, Ambassador of Iran to UNESCO (1969-1974) and to Canada (1974-1978). He is currently an honorary professor at McGill University and visiting professor at Concordia University in Montreal.

TABLE OF CONTENTS

PREFACE

Robotics and Manufacturing remain as two highly interdisciplinary and critical technological areas for the future economic, social and even political well-being of many nations around the globe. Automated manufacturing with or without robots stands to be of immense critical value in fast-emerging peaceful world around us.

The object of this volume is to report on the technical presentations of ISRAM '90 — The Third International Symposium on Robotics and Manufacturing, held July 18-20, 1990 in Burnaby, B.C., Canada. ISRAM, which began in 1986 in Albuquerque, New Mexico, USA, is considered as a relatively small, highly interactive, and international gathering of robotics and manufacturing engineers, educators, researchers, and administrators. This most recent ISRAM was attended by over 200 individuals with some 160 papers and 7 keynote speeches delivered. The distinguished group of keynote speakers and their presentations were:

1. Dr. Nick Cercone, Director Centre for System Science on "The CSS Research Interstructure; One Example is Human Interface"

2. Dr. Melvin Cohen, Vice President AT&T Bell Laboratories on "Time as a Competitive Tool for Product Realization"

3. Dr. John M. Hollerbach, NSERC Professor of McGill University on "Robotics Research in Canada and at McGill"

4. Mr. Ronald McCollough, Vice President SPAR Aerospace on "Advances in Robotics in Hazardous Environments - How Space Drives the Way"

5. Mr. Mick Kahan, Director of CIM Integration Systems, IBM-Canada, Ltd. on "Manufacturing Excellence in the 90's

6. Dr. Margarette Eastwood, Director of Factory Automation, Motorola Corporation on "U.S. Electronic Manufacturing Technology"

7. Dr. Lotfi A. Zadeh, Professor, University of California - Berkeley on "Fuzzy Logic and Control"

The presentations of the technical papers of ISRAM '90 in this volume have been similar to other ISRAMs and have been grouped in accordance to specialized areas of robotics and manufacturing for easy access of the readers. In order to access the technical papers based on the location or the author, the subject and author indices at the end of the volume can be consulted as well.

The editors would like to take this opportunity to thank every one of the keynote speakers and authors for their contributions to robotics and manufacturing in general, and to this volume in particular. We wish to express our gratitude to the program and organizing team members of ISRAM '90. Numerous colleagues have gracefully taken interest in this meeting and have put a great deal of effort in organizing sessions. We would like to thank: M. Beckerman, D. Bosomworth, R.L. Carroll, C.H. Dagli, P. Dauchez, A. Goldenberg, W. Hamel, S. Hayati, J.N. Herndon, R.M. Inigo, D.C. Krupka, R. Lumia, S.J. Mendonca, J.H. Mullins, C.C. Nguyen, S. Ollendorf, T. Salcudean, H. Seraji, A. Steiger-Garcao, J. Vozenilek, C.R. Weisbin and J. Yuh. Without their efforts, the Symposium's technical content would not have been the same. Thanks are also due to all the Chairs and Co-chairs of the sessions.

We would like to express our appreciation to several members of Simon Fraser University and the University of New Mexico's administrations. Foremost, we appreciate the financial sponsorships of a number of Canadian institutions including: Simon Fraser University's School of Engineering Science, SFU's Centre for System Science, Natural Science, Research Council of Canada (NSERC), and Advanced Systems Institute of British Columbia. We thank President W. Saywell, Dr. Marilyn Bowman, VP for Research and Information Systems, SFU, Dean Donald George, and Professor Nick Cercone of SFU for their valuable help and en-couragements. We thank Dean James E. Thompson of UNM's College of Engineering for his continuous support. We appreciate encouragements of Professor Nasir Ahmed, Chair of EECE Department at UNM. We sincerely thank the fine efforts of Professors Andy Goldenberg (Toronto), Rajini Patel (Concordia), Joe Mullins (UNM), Dr. Doug Bosomworth (Motorola) and Dr. Steve Mendonca (IBM) for playing major roles in inviting and arranging for the participation of our keynote speakers.

We wish to express our most sincere gratitude to Mrs. Janet Watts of UNM's CAD Laboratory for Systems/Robotics for her tireless assistance to the editors. Our task was made very simple thanks to her dedicated upkeep of the affairs of yet another Symposium. We appreciate the valuable and dedicated editorial assistance of Mrs. Janet S. Jafari and Mrs. Jila S. Jamshidi for organizing the two indices and much much more. We appreciate the collaborations of Ms. Caryl Dreiblatt and Dr. Ken Metzner of ASME Press and their staff during the printing of this volume.

Mo Jamshidi
Albuquerque, NM, USA

Mehrdad Saif
Vancouver, BC, Canada

SECTION 1 : KINEMATICS AND DYNAMICS

An Algorithmic Approach to the Direct-Kinematics of Special Stewart-Platform Manipulators

Ou Ma   Jorge Angeles

Department of Mechanical Engineering and
McGill Research Centre for Intelligent Machines, McGill University
817 Sherbrooke Street W, Montreal, Canada H3A 2K6

Abstract

In this paper, the direct displacement problem of a special type of Stewart-platform manipulators is solved using an invariant approach. The method is derived based on a virtual-cut technique and the well-developed technique of spatial four-bar linkage analysis. Such an approach leads to a computationally efficient algorithm and also provides a simple criterion for singularity identification. In deriving the method, it is also verified that the direct displacement problem of Stewart-platform manipulators leads to a sixteenth-order polynomial equation.

1. Introduction

While the kinematics of manipulators with serial kinematic chains has been well developed in the past decade, that of manipulators with closed kinematic chains is still the subject of current research. The six-degree-of-freedom Stewart-platform mechanism (Stewart 1965), like that shown in Fig. 1, can be considered as a typical architecture of parallel manipulators with six parallel closed kinematic loops. The end effector of the manipulator, namely, the six-degree-of-freedom movable platform, is driven by six translational actuators fixed to the six legs. Since this type of mechanism was first designed as a manipulator by Stewart in 1965, extensive research work on its kinematics has been reported in the literature—see, for example, (Hunt 1983, Yang and Lee 1984, Fichter 1986, Do and Yang 1988, Merlet 1987, 1989). However, until 1987 no report on the direct kinematics of this type of manipulators appeared in the literature. Merlet (1987) first proposed a method of direct kinematics for the special Stewart-platform mechanism shown in Fig. 2. In that manipulator, every two legs merge at one point connecting to the platform with a spherical joint (S). The joint connecting each leg to the base is either a spherical joint or a universal joint (U). With Merlet's method, several simultaneous nonlinear equations must be solved numerically. Nanua and Waldron (1989) presented a method of direct kinematics for the same mechanism. In their method, these nonlinear equations are algebraically manipulated such that only one final polynomial equation with a single unknown is derived; a quite similar method was independently developed by Charentus and Renaud (1989).

In the method proposed here a virtual-cut technique and RSSR linkage analysis are applied, which yields the kinematics model of three-legged Stewart platforms in the form of one equation in a single variable. Based on the equivalence of this platform with triangularly-connected six-legged platforms, the results of the former are directly applicable to the latter. Using an invariant approach in this method, an algorithm-oriented kinematics model is derived, which reduces the problem to solving one nonlinear and two linear equations. Thus, an algorithm designed based on this method will be computationally efficient, which is an essential requirement for robotics applications. On the other hand, the method provides a simple criterion to identify singularities. Such a simple criterion can be easily implemented in a direct-kinematics algorithm being capable of handling or avoiding singular configurations. The geometric representation of a singular configuration can be easily found using this singularity criterion.

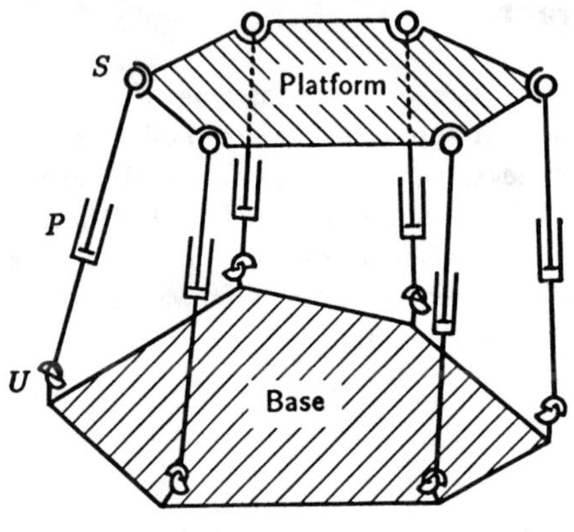

Fig. 1 A general Stewart mechanism

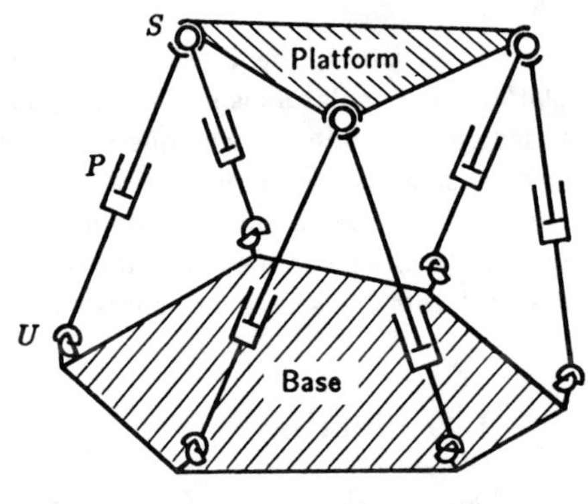

Fig. 2 A special Stewart mechanism

2. Direct Kinematics of Three-Legged Stewart Platforms

The Stewart-platform manipulator under study has three legs, each of which is connected to the movable platform with a spherical pair (S) and to the base with a cylindrical pair (C), as shown in Fig. 3. The platform undergoes a six-degree-of-freedom motion which is controlled by six linear actuators mounted on the C and P (prismatical) joints.

2.1 Formulation

The direct kinematics problem is defined as: *Given the displacements of the six linear actuators, namely,* $q_1, \cdots, q_6$, *determine the corresponding pose (position and orientation) of the end effector.* Since the pose of the end effector can be defined by the positions of spherical joints S_1, S_2 and S_3, the problem is equivalent to finding the following functions:

$$\mathbf{s}_i = \mathbf{s}_i(q_1, \cdots, q_6), \qquad \text{for } i = 1,\ 2,\ 3 \tag{1}$$

where $\mathbf{s}_i$ denotes the 3-dimensional position vector of the center of the spherical joint S_i. As one can see from Fig. 3, the coordinates of the centers of joints S_1, S_2

and S_3 can be readily computed if angles ψ_1, ψ_2 and ψ_3 are known. Thus, the problem reduces to solving ψ_1, ψ_2 and ψ_3. Kinematic relations between these three angles can be readily formulated if we consider the mechanism being composed of three $RSSR$ spatial four-bar linkages. In this way, the relations can be described by the three input-output functions (IOF) of the three associated linkages, namely,

$$\begin{cases} \phi_1(\psi_1, \psi_2) = 0 & \longleftarrow \quad \text{IOF of 4-bar linkage } C_1S_1S_2C_2; \\ \phi_2(\psi_2, \psi_3) = 0 & \longleftarrow \quad \text{IOF of 4-bar linkage } C_2S_2S_3C_3; \\ \phi_3(\psi_3, \psi_1) = 0 & \longleftarrow \quad \text{IOF of 4-bar linkage } C_3S_3S_1C_1. \end{cases} \tag{2}$$

The parameters of eqs. (2) depend only on linkage dimensions (Hartenberg and Denavit 1964). Moreover, in this problem these parameters are constant because the dimensions of all links, as well as the displacements q_1 through q_6, are fixed.

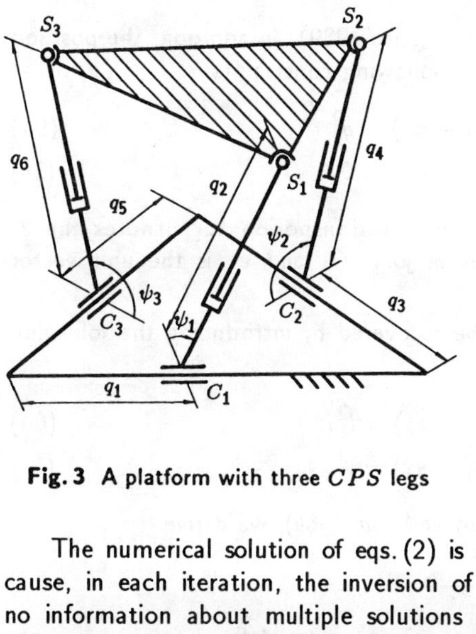

Fig. 3 A platform with three CPS legs

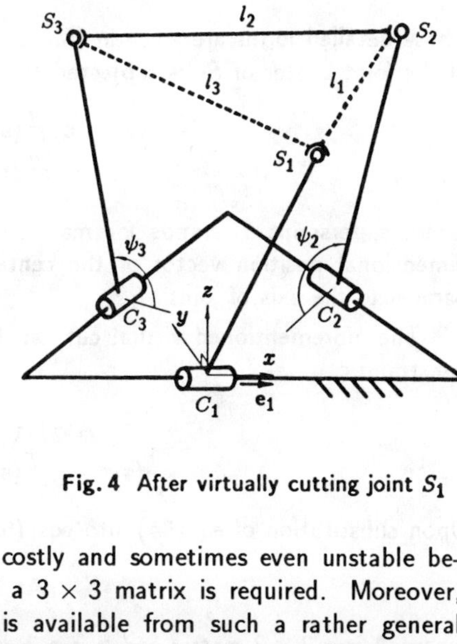

Fig. 4 After virtually cutting joint S_1

The numerical solution of eqs. (2) is costly and sometimes even unstable because, in each iteration, the inversion of a 3×3 matrix is required. Moreover, no information about multiple solutions is available from such a rather general numerical approach. Equations (2) can be algebraically manipulated using trigonometric identities such that a final polynomial equation with only one unknown, say $\tan(\psi_1/2)$, is obtained (Nanua and Waldron 1989, Charentus and Renaud 1989), namely,

$$f\left(\tan \frac{\psi_1}{2}\right) = 0. \tag{3}$$

Many numerical methods are available to find the roots of this polynomial. Once these roots are known, the other unknowns can be readily found. The advantage of this method is that only one equation is solved numerically, which reduces the computational cost substantially. Moreover, because the problem is reduced to a single polynomial equation, all of its roots can be found. Consequently, all of the possible solutions of the direct-kinematics problem can be obtained. To apply this method, however, one requires significant algebraic skill since the method is based on a tedious algebraic manipulation of eqs. (2) and some trigonometric identities.

Moreover, singularity conditions are not available with such a method. It is well known that a simple singularity-identification criterion is desirable for both singularity analysis and singularity avoidance in trajectory planning.

As an alternative, we propose a method which applies a virtual-cut technique and a method of four-bar linkage analysis, as described next. One of the spherical joints, say S_1, is virtually cut off, as shown in Fig. 4, which causes the remaining kinematic chain $C_2S_2S_3C_3$ to become a one-degree-of-freedom $RSSR$ linkage, where ψ_2 and ψ_3 are input and output angles, respectively. Based on four-bar linkage analysis, the input-output equation of the linkage can be solved for ψ_3 in terms of ψ_2. Thus, the positions of the centers of both S_2 and S_3 can be expressed as functions of the input angle of the $RSSR$ linkage only, i.e.,

$$\mathbf{s}_2 = \mathbf{s}_2(\psi_2) \tag{4a}$$
$$\mathbf{s}_3 = \mathbf{s}_3(\psi_2) \tag{4b}$$

whose detailed forms are given in (Ma and Angeles 1989). In addition, the position of the joint center of S_1 is subjected to the following constraints:

$$(\mathbf{s}_1 - \mathbf{c}_1)^T(\mathbf{s}_1 - \mathbf{c}_1) = q_2^2 \tag{5a}$$
$$\mathbf{e}_1^T(\mathbf{s}_1 - \mathbf{c}_1) = 0 \tag{5b}$$

where superscript T stands for matrix or vector transposion; $\mathbf{c}_1$ denotes the 3-dimensional position vector of the center of joint C_1 and $\mathbf{e}_1$ is the unit vector parallel to the axis of joint C_1.

The aforementioned virtual cut can be recovered by introducing the following constraints:

$$(\mathbf{s}_1 - \mathbf{s}_2)^T(\mathbf{s}_1 - \mathbf{s}_2) = l_1^2, \tag{6a}$$
$$(\mathbf{s}_1 - \mathbf{s}_3)^T(\mathbf{s}_1 - \mathbf{s}_3) = l_3^2. \tag{6b}$$

Upon substitution of eq. (5a) into eqs. (5b) and (6a & 6b), we derive

$$\mathbf{A}\mathbf{s}_1 = \mathbf{b} \tag{7}$$

where $\mathbf{A}$ is a 3×3 matrix and $\mathbf{b}$ is a 3-dimensional vector defined as

$$\mathbf{A} \equiv [\mathbf{e}_1,\ \mathbf{s}_2 - \mathbf{c}_1,\ \mathbf{s}_3 - \mathbf{c}_1]^T = \mathbf{A}(\psi_2), \tag{8}$$

$$\mathbf{b} \equiv \frac{1}{2}\begin{bmatrix} \mathbf{e}_1^T\mathbf{c}_1 \\ \mathbf{s}_2^T\mathbf{s}_2 - \mathbf{c}_1^T\mathbf{c}_1 + q_2^2 - l_1^2 \\ \mathbf{s}_3^T\mathbf{s}_3 - \mathbf{c}_1^T\mathbf{c}_1 + q_2^2 - l_3^2 \end{bmatrix} = \mathbf{b}(\psi_2). \tag{9}$$

Substituting eq. (7) back into eq. (5a), we derive

$$(\mathbf{A}^{-1}\mathbf{b} - \mathbf{c}_1)^T(\mathbf{A}^{-1}\mathbf{b} - \mathbf{c}_1) = q_2^2. \tag{10}$$

We define now a single-variable function:

$$f(\psi_2) \equiv (\mathbf{A}^{-1}\mathbf{b} - \mathbf{c}_1)^T(\mathbf{A}^{-1}\mathbf{b} - \mathbf{c}_1) - q_2^2. \tag{11}$$

Thus, the problem is reduced to one equation:

$$f(\psi_2) = 0 \tag{12}$$

which is a polynomial equation in either $\sin\psi_2$ and $\cos\psi_2$ or $\tan(\psi_2/2)$. After numerically solving this equation for ψ_2, one can immediately compute the following:

$$\begin{cases} \mathbf{s}_1 = \mathbf{A}^{-1}\mathbf{b} \\ \mathbf{s}_2 = \mathbf{s}_2(\psi_2) \\ \mathbf{s}_3 = \mathbf{s}_3(\psi_2) \end{cases} \tag{13}$$

With no loss of generalarity, we define a coordinate frame, shown in Fig. 4, as follows: The x axis is parallel to the axis of joint C_1; the y and z axes are parallel and perpendicular to the base plane, respectively. Moreover, the origin of this frame is located at the center of joint C_1. Thus,

$$\mathbf{c}_1 = [0,\ 0,\ 0]^T \quad \text{and} \quad \mathbf{e}_1 = [1,\ 0,\ 0]^T, \tag{14}$$

which leads to

$$f(\psi_2) = \mathbf{b}^T(\mathbf{A}\mathbf{A}^T)^{-1}\mathbf{b} - q_2^2 = 0 \tag{15}$$

and

$$\mathbf{A} = [\mathbf{e}_1,\ \mathbf{s}_2,\ \mathbf{s}_3]^T = \begin{bmatrix} 1 & 0 & 0 \\ x_2 & y_2 & z_2 \\ x_3 & y_3 & z_3 \end{bmatrix}, \quad \mathbf{b} = \frac{1}{2}\begin{bmatrix} 0 \\ \mathbf{s}_2^T\mathbf{s}_2 + q_2^2 - l_1^2 \\ \mathbf{s}_3^T\mathbf{s}_3 + q_2^2 - l_3^2 \end{bmatrix} \tag{16}$$

where x_i, y_i and z_i are the coordinates of the center of joint S_i, for $i = 1, 2, 3$. Equations (16) can also be written in block form, namely,

$$\mathbf{A} \equiv \begin{bmatrix} 1 & 0 \ 0 \\ x_2 & \mathbf{A}' \\ x_3 & \end{bmatrix} \quad \text{and} \quad \mathbf{b} \equiv \begin{bmatrix} 0 \\ \mathbf{b}' \end{bmatrix} \tag{17}$$

where $\mathbf{A}'$ is a 2×2 submatrix of $\mathbf{A}$ and $\mathbf{b}'$ is a 2-dimensional subvector of $\mathbf{b}$. Thus, the solution of vector $\mathbf{s}_1$ is reduced to the following trivial form:

$$x_1 = 0 \quad \text{and} \quad \begin{bmatrix} y_1 \\ z_1 \end{bmatrix} = (\mathbf{A}')^{-1}\mathbf{b}'. \tag{18}$$

2.2 Multiple Solutions

An interesting theoretical issue of direct kinematics is how many solutions a given direct-kinematics problem may have. Merlet (1987) formulated the problem as a sixteenth-order polynomial equation and hence, he claimed that there can be up to sixteen solutions, and so did Charentus and Renaud in 1989. However, Nanua and Waldron (1989) claimed that there can be up to twenty-four solutions for they formulated the problem as a twenty-fourth-order polynomial equation. With our method, the order and the solutions are discussed in the following: From the definition of $\mathbf{A}$, as given in eq. (8), it can be readily shown that matrix $\mathbf{A}$ is $O(1)$ in

$\sin\psi_2$, $\cos\psi_2$, $\sin\psi_3$, and $\cos\psi_3$, which leads to $\mathbf{AA}^T$ being $O(2)$ and its inverse being $O(2)/O(4)$ in $\sin\psi_2$, $\cos\psi_2$, $\sin\psi_3$, and $\cos\psi_3$. Moreover, vector $\mathbf{b}$ is $O(1)$ in $\cos\psi_2$ and $\cos\psi_3$ (Ma and Angeles 1989). Thus, the order of the polynomial equation (12) will be:

$$f = \mathbf{b}^T(\mathbf{AA}^T)^{-1}\mathbf{b} - q_2^2 = 0 \longleftarrow O(4) \text{ in } \sin\psi_2,\ \cos\psi_2,\ \sin\psi_3,\ \cos\psi_3.$$

Furthermore, from the input-output equation of the four-bar linkage $C_2S_2S_3C_3$ (Ma and Angeles 1989), $\sin\psi_3$ and $\cos\psi_3$ can be expressed as quadratic functions of $\sin\psi_2$ and $\cos\psi_2$. It follows that

$$f = f(\sin\psi_2, \cos\psi_2) = 0 \quad\longleftarrow\quad O(8) \text{ in } \sin\psi_2 \text{ and } \cos\psi_2$$

which can be expressed in terms of $\tan(\psi_2/2)$. Thus, we finally find the order of the problem:

$$f = f(\tan\frac{\psi_2}{2}) = 0 \longleftarrow O(16) \text{ in } \tan(\frac{\psi_2}{2}).$$

This result verifies that this direct-kinematics problem leads to a sixteenth-order polynomial equation in one single variable. This result has been verified from the symbolic derivation of eq. (12) using MACSYMA. Since the result generated by MACSYMA is very lengthy, it is not presented here due to space limitations. It is available in (Ma and Angeles 1989). Because a sixteenth-order polynomial equation has at most sixteen real roots, there are at most sixteen solutions in this direct-kinematics problem.

2.3 Singularity Criterion

There are two types of failure associated with the method presented here. One is that there are no real solutions of the polynomial equation (12). This happens if the given set of displacements $q_1, \cdots, q_6$ lie outside the mobility range (domain of the input variables) of the mechanism. In other words, the given displacements violate the necessary kinematic constraints of the manipulator and hence, the corresponding direct-kinematics problem is not feasible. The other type of failure happens if the 3×3 matrix $\mathbf{A}$ is singular. There are two possible singularities associated with this type of failure. One is termed *structural singularity*, while the other is called *formulation singularity*. The structural singularity is caused by a special kinematic structure of the manipulator and hence, it is difficult to handle. The treatment of such a singularity is beyond the scope of this paper; an extensive study of this issue has been presented by Merlet (1989). On the contrary, the formulation singularity is caused by the formulation itself and hence, can be readily handled. In fact, such a singularity happens if eq. (5b) and eqs. (6a & 6b) are dependent. This kind of dependency may not occur if the S_2 or the S_3 joint were virtually cut instead of the S_1 joint. Therefore, a formulation singularity can be eliminated simply by switching the numbering of joints. However, joint-number switching cannot eliminate a structural singularity. A more detailed discussion on singularities will be given in an oncoming paper. From the foregoing discussion, we know that both types of singularity are characterized by

$$\det(\mathbf{A}) = 0 \tag{19}$$

which tells that a singularity happens if the three vectors $\mathbf{e}_1$, $\mathbf{s}_2-\mathbf{c}_1$ and $\mathbf{s}_3-\mathbf{c}_1$ are coplanar. In other words, if the axis of joint C_1 lies on the plane formed by joints C_1, S_2 and S_3, the manipulator is in either a formulation or a structural singularity.

If the x-y-z frame is defined as shown in Fig. 4, then the singularity condition given by eq. (19) reduces to

$$y_2z_3 - y_3z_2 = 0 \tag{20}$$

which is so simple that no additional computation is required to detect it in the context of the direct-kinematics algorithm presented here.

3. Direct Kinematics of Six-Legged Stewart Platforms

Given a six-legged Stewart-platform manipulator whose movable platform is triangularly connected, like the one shown in Fig. 2, a kinematically equivalent three-legged manipulator can be found (Nanua and Waldron 1989). In fact, for such a manipulator, the first pair of legs, namely, legs $U_1P_1S_1$ and $U_2P_2S_1$, can be replaced by a new leg, $C_1P_1'S_1$, as shown in Fig. 5. Similar replacements can be made to the other two pairs of legs, as shown in the same figure. With such replacements, the three-legged equivalent Stewart-platform manipulator is defined; the latter is exactly the one we studied in the previous section. Thus, the aforementioned direct-kinematics method for three-legged Stewart-platform manipulators is also applicable to the six-legged ones, as long as their movable platforms are triangularly connected. Since such a replacement is uniquely defined without any associated conditions for this direct-kinematics problem, the results on multiple solutions and singularities for three-legged Stewart-platform manipulators are also applicable here.

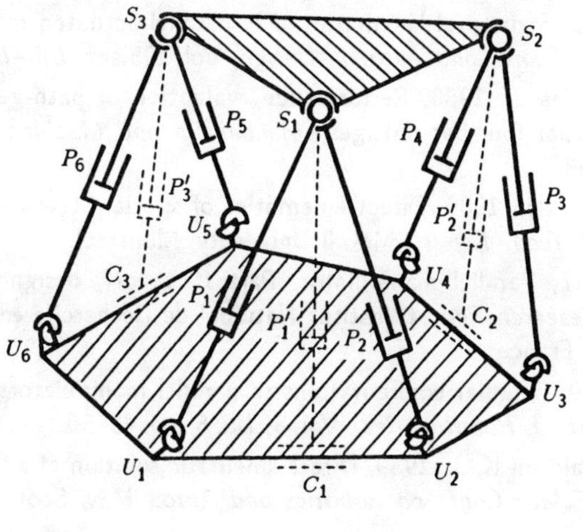

Fig. 5 A six-legged platform and its equivalent three-legged mechanism

4. Conclusions

Based on the method of four-bar linkage analysis and a virtual-cut technique, a new method for the direct displacement problem of special Stewart-platform manipulators was presented. An invariant approach is used throughout the whole procedure of formula derivation, which leads to an algorithm with computational efficiency. The method provides a simple singularity criterion which can be readily implemented for identifying singularities without involving additional computations. Using this singularity criterion, the geometric representation of a singular configuration can be easily found. Moreover, with this approach, it has been verified that the direct-displacement problem can be defined by a sixteenth-order polynomial equation in a single unknown.

Acknowledgements

The research work reported here was possible under NSERC Grant No. A4532 and FCAR Grant No. 88AS25A.

References

Charentus, S. and Renaud, M., 1989, Modelling and control of a modular, redundant robot manipulator, *The First Int. Symp. on Experimental Robotics*, June, Montreal, pp. 508–527.

Do, D. Q. W. and Yang, D. C. H., 1988, Inverse dynamic analysis of a platform type of robot, *J. Robotic Syst.*, Vol. 5, No. 3, pp. 210–227.

Fichter, E. F., 1986, A Stewart platform-based manipulator: general theory and practical construction, *The Int. J. Robotics Research*, Vol. 5, No. 2, pp. 157–182.

Hartenberg, R. S. and Denavit, J., 1964, *Kinematic Synthesis of Linkages*, McGraw-Hill, New York.

Hunt, K. H., 1983, Structural kinematics of in-parallel-actuated robot-arms, *Trans. ASME, J. Mech., Trans., and Auto. in Design*, Vol. 105, pp. 705–712.

Ma, O. and Angeles J., 1988, Performance evaluation of path-generating planar, spherical and spatial four-bar linkages, *Mechanism and Machine Theory*, vol. 23, No. 1, pp. 257–268.

Ma, O. and Angeles J., 1989, Direct kinematics of special Stewart-platform manipulators, *McRCIM Tech. Report*, McGill University, Montreal.

Merlet, J-P., 1987, Parallel manipulator, Part 1: theory, design, kinematics and control, *INRIA Research Report*, Institut National de Recherche en Informatique et en Automatique, France.

Merlet, J-P., 1989, Singular configurations of parallel manipulators and Grassmann geometry, *The Int. J. Robotics Res.*, Vol. 8, No. 5, pp. 45–56.

Nanua, P. and Waldron K. J., 1989, Direct kinematic solution of a Stewart platform, *Proc. 1989 IEEE Int. Conf. on Robotics and Auto.*, May, Scottsdale, Vol. 1, pp. 431–437.

Stewart, D., 1965, A platform with six degrees of freedom, *Proc. of the Institution of Mechanical Engineers*, Vol. 180, No. 1, pp. 371–386.

Yang, D. C. H. and Lee, T. W., 1984, Feasibility study of a platform type of robotic manipulators from a kinematic viewpoint, *Trans. ASME, J. Mech., Trans., and Auto. in Design*, 106: 191–198.

Direct Forward Kinematic Solution for General Stewart Platforms *

Zheng Geng *Leonard S. Haynes*
Intelligent Automation, Inc., 1370 Piccard Drive
Rockville, MD 20850, USA

Robert L. Carroll
Department of Electrical Engineering and Computer Science
Goerge Washington University, Washington, DC 20052,USA

ABSTRACT

A direct forward kinematic solution is obtained for a general class of Stewart Platform-type parallel link robotic manipulators of six degree of freedom. A formula of a sixteenth order polynomial equation corresponding to the forward kinematic problem is derived which will give all possible global solutions of configuration of manipulator given a set of six leg length.

1. Introduction

The Stewart Platform-type of parallel link robotic manipulators[7] have recently received considerable research interests due to their successful applications and potential advantages over the conventional serial link manipulators. A general Stewart Platform is shown in Figure 1, which has six legs connecting upper plate and base in the fashion that three pairs of legs meet at three point, R, S,and T, in the upper plate and the other ends of the six legs are put arbitrarily on six points , P_1, P_2, P_3, P_4, P_5, and P_6, in the base, which are not necessarily coplanar. The length of legs can be controlled to obtain desired configuration (position and orientation) of the upper plate. One of the reason we consider this type of general Stewart Platform is the practical consideration of singularity. Some singularity will be avoided by employing a set of non-coplanar P_1, P_2, P_3, P_4, P_5, and P_6.

The forward kinematics problem for the general Stewart Platform is addressed as following: Let R, S, and T be fixed on the upper plate, and P_1, P_2, P_3, P_4, P_5, and P_6be fixed on the static base of the platform. Determine the position, x, y, z, and orientation, α, β, γ, of the moving upper plate relative to the base, given a set of six leg length , namely L_1, L_2, L_3, L_4, L_5, and L_6. Notice that in general the solution for the forward kinematic problem is not unique.

Unlike its counterpart in conventional serial link robotic manipulators, the forward kinematic problem for parallel link mechanism is difficult while the inverse problem is comparatively simple. The approaches to forward kinematic problem proposed in literatures so far can be catalogued into two categories: *iterative methods* and *direct solutions*. Among iterative approaches, Dieudonne[1] proposed a treatment based on Newton-Raphson's method.This method consists of first defining a function of the unknowns and find its Jacobian matrix, then iteratively applying the recursive algorithm until convergence criterion is met. Lansberger[5] suggested a search method based on geometrical interpretation. The inverse kinematic problem has to be solved for a series of candidate configuration then find one which satisfies the given leg length constraints. The common drawbacks shared by iterative approaches are: The performance and the result of the algorithms heavily depend on the selection of initial condition, because they can only find local solution. There is not a systematic way in this category which is able to give all possible kinematic solutions (pose of upper plate) for a given set of leg length.

* This work is supported by National Science Foundation Grant # ISI 8821640

There are relatively very few methods available in the category of direct solution. Nanua and Waldron[6] presented a solution for a special case of Stewart Platform shown in Figure 1 in which P_1, P_2, P_3, P_4, P_5, and P_6 have to be coplanar. They formulated the problem as a 24-th order polynomial equation. Griffis and Duffy[4] recently gave a solution for an ever deteriorated case of Stewart Platform shown in Figure 1 in which P_1, P_2, P_3, P_4, P_5, and P_6 are coplanar and meet in pair-wise fashion at three point on the base. They derived a 16-th order polynomial equation for the problem based on the results for spherical four-bar linkage.

In this paper, a solution for the general Stewart Platform of Figure 1 is presented. The forward kinematic problem is converted through a series of symbolic calculation into the problem of solving a 16-th order polynomial equation. All possible solution for the pose of upper plate can be found given a set of leg length.

2. Geometrical Interpretation of the Forward Kinematic Problem

If we take off the upper plate RST from the general Stewart Platform shown in Fig 1, the mechanism then becomes three independent triangle link pairs (L_1,L_2) ---> P_1RP_2, (L_3,L_4) ---> P_3SP_4, (L_5,L_6) --- > P_5TP_6. For given set of the length L_i, i=1,2,...6, the shapes of these three triangles are fixed. We can calculate the positions of O_r,O_s,O_t and the values of m_r, m_s,m_t . Since P_1 and P_2 are fixed in the base of the Stewart Platform and R has some freedom to move, the locus of point R must be a circle with centre located on O_r and radius m_r. This circle is a curve in a plane perpendicular to line P_1P_2 . Similarly, point S will move along a circle with a centre located on O_s and a radius m_s. Point T will move along a circle with a centre located on O_t and a radius m_t. Therefore, given leg length, three independent variables θ_r,θ_s, and θ_t determine the positions of points R, S, and T.

It can be easily verified that

$$|P_1P_2| = r_1+r_2 \quad , \quad |P_3P_4| = s_1+s_2 \quad , \quad |P_5P_6| = t_1+t_2$$

and (1)

$$m_r = \sqrt{(L_1^2 - r_1^2)} \qquad m_s = \sqrt{(L_3^2 - s_1^2)} \qquad m_t = \sqrt{(L_5^2 - s_1^2)}$$

where

$$r_1 = \frac{|P_1P_2|^2 + L_1^2 - L_2^2}{2\,|P_1P_2|} \qquad s_1 = \frac{|P_3P_4|^2 + L_3^2 - L_4^2}{2\,|P_3P_4|} \quad . \quad t_1 = \frac{|P_5P_6|^2 + L_5^2 - L_6^2}{2\,|P_5P_6|}$$

The coordinates of points O_r, O_s, and O_t can be calculated as following:

$$O_r=P_1+(P_2-P_1)r_1/|P_2-P_1|,\ O_s=P_3+(P_4-P_3)s_1/|P_4-P_3|,\ O_t=P_5+(P_6-P_5)t_1/|P_6-P_5| \qquad (2)$$

Given the position of point P_i, i=1,2,...6, the rotation angles β and α can be calculated as:

$$\beta_r = \cos^{-1}\left[\frac{(P_2 - P_1)_x}{\sqrt{(P_2-P_1)_x^2 + (P_2-P_1)_y^2}}\right] \qquad \alpha_r = \cos^{-1}\left[\frac{\sqrt{(P_2 - P_1)_x^2 + (P_2 - P_1)\ldots}}{|P_2 - P_1|}\right.$$

$$\beta_s = \cos^{-1}\left[\frac{(P_4 - P_3)_x}{\sqrt{(P_4-P_3)_x^2 + (P_4-P_3)_y^2}}\right] \qquad \alpha_s = \cos^{-1}\left[\frac{\sqrt{(P_4 - P_3)_x^2 + (P_4 - P_3\ldots}}{|P_4 - P_3|}\right.$$

$$\beta_t = \cos^{-1}\left[\frac{(P_6 - P_5)_x}{\sqrt{(P_6-P_5)_x^2 + (P_6-P_5)_y^2}}\right] \qquad \alpha_t = \cos^{-1}\left[\frac{\sqrt{(P_6 - P_5)_x^2 + (P_6 - P_5)_y^2}}{|P_6 - P_5|}\right]$$

Therefore the position vectors v_r, v_s, and v_t can be represented as following:

$$v_r = m_r(\cos\beta_r \sin\alpha_r \sin\theta_r - \sin\beta_r \cos\theta_r)\mathbf{i} + m_r(\sin\beta_r \sin\alpha_r \sin\theta_r + \cos\beta_r \cos\theta_r)\mathbf{j} + m_r(\cos\alpha_r \sin\theta_r)\mathbf{k}$$

$$v_s = m_s(\cos\beta_s \sin\alpha_s \sin\theta_s - \sin\beta_s \cos\theta_s)\mathbf{i} + m_s(\sin\beta_s \sin\alpha_s \sin\theta_s + \cos\beta_s \cos\theta_s)\mathbf{j} + m_s(\cos\alpha_s \sin\theta_s)\mathbf{k}$$

$$v_t = m_t(\cos\beta_t \sin\alpha_t \sin\theta_t - \sin\beta_t \cos\theta_t)\mathbf{i} + m_t(\sin\beta_t \sin\alpha_t \sin\theta_t + \cos\beta_t \cos\theta_t)\mathbf{j} + m_t(\cos\alpha_t \sin\theta_t)\mathbf{k} \qquad (4)$$

The position vector of points R, S, and T becomes

$$R = O_r + v_r, \quad S = O_s + v_s, \quad T = O_t + v_t \qquad (5)$$

Define G_1, G_2, and G_3 as the length of vertice RS, ST and RT, respectively. G_1, G_2, and G_3 are constants for a given platform. Thus the additional physical constraints to the positions of points R, S, and T are the distance between R and S should equal to the length of the vertice G1, that of S and T equal to G2, and that of R and T equal to G3. These conditions can be expressed as:

$$|R - S| = G1, \; |R - T| = G2, \; |S - T| = G3$$

or

$$(x_R - x_S)^2 + (y_R - y_S)^2 + (z_R - z_S)^2 = G1^2 \qquad (6)$$

$$(x_R - x_T)^2 + (y_R - y_T)^2 + (z_R - z_T)^2 = G2^2 \qquad (7)$$

$$(x_S - x_T)^2 + (y_S - y_T)^2 + (z_S - z_T)^2 = G3^2 \qquad (8)$$

3. Direct Forward Kinematic Solution

3.1 Equations

Substitute equation (4) into equation (6) gives:

$$[(O_r - O_s)_x + m_r \cos\beta_r \sin\alpha_r \sin\theta_r - m_r \sin\beta_r \cos\theta_r - m_s \cos\beta_s \sin\alpha_s \sin\theta_s + m_s \sin\beta_s \cos\theta_s]^2 + [(O_r - O_s)_y + m_r \sin\beta_r \sin\alpha_r \sin\theta_r + m_r \cos\beta_r \cos\theta_r - m_s \sin\beta_s \sin\alpha_s \sin\theta_s - m_s \cos\beta_s \cos\theta_s]^2 + [(O_r - O_s)_z + m_r \cos\alpha_r \sin\theta_r - m_s \cos\alpha_s \sin\theta_s]^2 = G1^2 \qquad (9)$$

Through some manipulation, we have

$$2(O_r - O_s)_x m_r \cos\beta_r \sin\alpha_r \sin\theta_r - 2(O_r - O_s)_x m_r \sin\beta_r \cos\theta_r - 2(O_r - O_s)_x m_s \cos\beta_s \sin\alpha_s \sin\theta_s + 2(O_r - O_s)_x m_s \sin\beta_s \cos\theta_s - 2 m_r m_s \cos\beta_r \sin\alpha_r \sin\theta_r \cos\beta_s \sin\alpha_s \sin\theta_s + 2 m_r m_s \cos\beta_r \sin\alpha_r \sin\theta_r \sin\beta_s \cos\theta_s + 2 m_r m_s \sin\beta_r \cos\theta_r \cos\beta_s \sin\alpha_s \sin\theta_s - 2 m_r m_s \sin\beta_r \cos\theta_r \sin\beta_s \cos\theta_s + 2(O_r - O_s)_y m_r \sin\beta_r \sin\alpha_r \sin\theta_r + 2(O_r - O_s)_y m_r \cos\beta_r \cos\theta_r - 2(O_r - O_s)_y m_s \sin\beta_s \sin\alpha_s \sin\theta_s - 2(O_r - O_s)_y m_s \cos\beta_s \cos\theta_s - 2 m_r m_s \sin\beta_r \sin\alpha_r \sin\theta_r \sin\beta_s \sin\alpha_s \sin\theta_s - 2 m_r m_s \sin\beta_r \sin\alpha_r \sin\theta_r \cos\beta_s \cos\theta_s - 2 m_r m_s \cos\beta_r \cos\theta_r \sin\beta_s \sin\alpha_s \sin\theta_s - 2 m_r m_s \cos\beta_r \cos\theta_r \cos\beta_s \cos\theta_s + 2(O_r - O_s)_z m_r \cos\alpha_r \sin\theta_r - 2(O_r - O_s)_z m_s \cos\alpha_s \sin\theta_s - 2 m_r m_s \cos\alpha_r \sin\theta_r \cos\alpha_s \sin\theta_s = G1^2 - (O_r - O_s)_x^2 - (O_r - O_s)_y^2 - (O_r - O_s)_z^2 - m_r^2 - m_s^2 \qquad (10)$$

or we can express equation (10) as

$$A_1 \sin\theta_r + A_2 \cos\theta_r + A_3 \sin\theta_s + A_4 \cos\theta_s + A_5 \sin\theta_r \sin\theta_s + A_6 \sin\theta_r \cos\theta_s + A_7 \cos\theta_r \sin\theta_s + A_8 \cos\theta_r \cos\theta_s = A_0 \qquad (11)$$

where

$$A_1 = 2(O_r - O_s)_x m_r \cos\beta_r \sin\alpha_r + 2(O_r - O_s)_y m_r \sin\beta_r \sin\alpha_r + 2(O_r - O_s)_z m_r \cos\alpha_r$$

$$A_2 = -2(O_r - O_s)_x m_r \sin\beta_r + 2(O_r - O_s)_y m_r \cos\beta_r$$

$$A_3 = -2(O_r - O_s)_x m_s \cos\beta_s \sin\alpha_s - 2(O_r - O_s)_y m_s \sin\beta_s \sin\alpha_s - 2(O_r - O_s)_z m_s \cos\alpha_s$$

$A_4 = +2(O_r - O_s)_x m_s \sin\beta_s - 2(O_r - O_s)_y m_s \cos\beta_s$
$A_5 = -2 m_r m_s \sin\alpha_r \sin\alpha_s \cos(\beta_s - \beta_r) - 2 m_r m_s \cos\alpha_r \cos\alpha_s$
$A_6 = 2 m_r m_s \sin\alpha_r \sin(\beta_s - \beta_r)$, $\quad A_7 = 2 m_r m_s \sin\alpha_s \sin(\beta_s - \beta_r)$
$A_8 = -2 m_r m_s \cos(\beta_s - \beta_r)$, $A_0 = G1^2 - (O_r - O_s)_x^2 - (O_r - O_s)_y^2 - (O_r - O_s)_z^2 - m_r^2 - m_s^2$

Equation (11) contains both sin and cos of θ_s and θ_r. To make the expression concise, the tan-half-angle relationship is used:

$$u_r = \tan(\frac{\theta_r}{2}), \quad u_s = \tan(\frac{\theta_s}{2}), \quad u_t = \tan(\frac{\theta_t}{2}) \tag{12}$$

Equation (11) then becomes
$2A_1 u_r(1+u_s^2) + A_2(1-u_r^2)(1+u_s^2) + 2A_3(1+u_r^2)u_s + A_4(1+u_r^2)(1-u_s^2) + 4A_5 u_r u_s + 2A_6 u_r(1-u_s^2) + 2A_7(1-u_r^2)u_s + A_8(1-u_r^2)(1-u_s^2) = A_0(1+u_r^2)(1+u_s^2)$
or

$$B_1 u_r^2 u_s^2 + B_2 u_r^2 u_s + B_3 u_r u_s^2 + B_4 u_r u_s + B_5 u_r^2 + B_6 u_s^2 + B_7 u_r + B_8 u_s + B_9 = 0 \tag{13}$$

where
$B_1 = -A_2 - A_4 + A_8 - A_0$, $B_2 = +2A_3 - 2A_7$, $B_3 = 2A_1 - 2A_6$, $B_4 = 4A_5$, $B_5 = -A_2 + A_4 - A_8 - A_0$, $B_6 = A_2 - A_4 - A_8 - A_0$, $B_7 = 2A_1 + 2A_6$, $B_8 = +2A_3 + 2A_7$, $B_9 = +A_2 + A_4 + A_8 - A_0$

In the similar fashion, substitute equation (4) into equation (7) gives:

$$[(O_r - O_t)_x + m_r \cos\beta_r \sin\alpha_r \sin\theta_r - m_r \sin\beta_r \cos\theta_r - m_t \cos\beta_t \sin\alpha_t \sin\theta_t + m_t \sin\beta_t \cos\theta_t]^2$$
$$+[(O_r - O_t)_y + m_r \sin\beta_r \sin\alpha_r \sin\theta_r - m_r \cos\beta_r \cos\theta_r - m_t \sin\beta_t \sin\alpha_t \sin\theta_t - m_t \cos\beta_t \cos\theta_t]^2$$
$$+[(O_r - O_t)_z + m_r \cos\beta_r \sin\theta_r - m_t \cos\beta_t \sin\theta_t]^2 = G2^2 \tag{14}$$

Upon simplification and variable transformation, the following equations are obtained:

$$C_1 \sin\theta_r + C_2 \cos\theta_r + C_3 \sin\theta_t + C_4 \cos\theta_t + C_5 \sin\theta_r \sin\theta_t + C_6 \sin\theta_r \cos\theta_t + C_7 \cos\theta_r \sin\theta_t + C_8 \cos\theta_r \cos\theta_t = C_0 \tag{15}$$

$$D_1 u_r^2 u_t^2 + D_2 u_r^2 u_t + D_3 u_r u_t^2 + D_4 u_r u_t + D_5 u_r^2 + D_6 u_t^2 + D_7 u_r + D_8 u_t + D_9 = 0 \tag{16}$$

and

$$[(O_s - O_t)_x + m_s \cos\beta_s \sin\alpha_s \sin\theta_s - m_s \sin\beta_s \cos\theta_s - m_t \cos\beta_t \sin\alpha_t \sin\theta_t + m_t \sin\beta_t \cos\theta_t]^2$$
$$+[(O_s - O_t)_y + m_s \sin\beta_s \sin\alpha_s \sin\theta_s - m_s \cos\beta_s \cos\theta_s - m_t \sin\beta_t \sin\alpha_t \sin\theta_t - m_t \cos\beta_t \cos\theta_t]^2 + [(O_s - O_t)_z + m_s \cos\beta_s \sin\theta_s - m_t \cos\beta_t \sin\theta_t]^2 = G3^2 \tag{17}$$

$$E_1 \sin\theta_s + E_2 \cos\theta_s + E_3 \sin\theta_t + E_4 \cos\theta_t + E_5 \sin\theta_s \sin\theta_t + E_6 \sin\theta_s \cos\theta_t + E_7 \cos\theta_s \sin\theta_t + E_8 \cos\theta_s \cos\theta_t = E_0 \tag{18}$$

$$F_1 u_s^2 u_t^2 + F_2 u_s^2 u_t + F_3 u_s u_t^2 + F_4 u_s u_t + F_5 u_s^2 + F_6 u_t^2 + F_7 u_s + F_8 u_t + F_9 = 0 \tag{19}$$

where C_i, D_i, E_i, and F_i are all constants. Now the forward kinematics problem has been converted into the algebraic problem of solving simultaneously three equations (13), (16), and (19).

3.2 Solving Method

Express equation (16) and (19) as the quadratic equations in terms of u_s and u_t:

$$a_1(u_r)u_s^2 - a_2(u_r)u_s - a_3(u_r) = 0 \tag{20}$$

where

$$a_1 = (B_1u_r^2+B_3u_r+B_6),\quad a_2 = -(B_2u_r^2+B_4u_r+B_8),\ a_3 = -(B_5u_r^2+B_7u_r+B_9) \tag{21}$$

$$b_1(u_r)u_t^2 - b_2(u_r)u_t - b_3(u_r) = 0 \tag{22}$$

where

$$b_1 = (D_1u_r^2+D_3u_r+D_6),\ b_2 = -(D_2u_r^2+D_4u_r+D_8),\quad b_3 = -(D_5u_r^2+D_7u_r+D_9) \tag{23}$$

$a_i(u_r)$ and $b_i(u_r)$, (i = 1, 2, 3) , are second order polynomials of u_r. Substitute (20) and (22) into (19) gives:

$$F_1(a_2u_s + a_3)(b_2u_t+ b_3)+ b_1F_2u_t(a_2u_s+ a_3)+a_1F_3u_s(b_2u_t+ b_3)+a_1b_1F_4u_su_t+ b_1F_5(a_2u_s + a_3)+a_1F_6(b_2u_t+ b_3)+a_1b_1F_7u_s+a_1b_1F_8u_t+a_1b_1F_9= 0$$

or

$$c_1u_su_t+ c_2u_s+ c_3u_t+ c_4= 0 \tag{24}$$

where

$$\begin{aligned} c_1 &= F_1 a_2 b_2 + F_2 a_2 b_1+ F_3 a_1 b_2 + F_4 a_1 b_1\\ c_2 &= F_1 a_2 b_3+ F_3 a_1 b_3+ F_5 a_2 b_1+F_7 a_1 b_1\\ c_3 &= F_1 a_3 b_2+ F_2 a_3 b_1+ F_6 a_1 b_2+ F_8 a_1 b_1\\ c_4 &= F_1 a_3 b_3+ F_5 a_3 b_1+ F_6 a_1 b_3 + F_9 a_1 b_1 \end{aligned} \tag{25}$$

c_i , i = 1,2,3,4 , are the forth order polynomials of u_r.

Define : $a_4 = +-(a_2^2+ 4 a_1a_3)^{1/2}$, $\quad b_4 = +-(b_2^2+ 4 b_1b_3)^{1/2}$ (26)

we then have the relationships (solution for a quadratic equation):

$$2 a_1u_s = a_2 + a_4 ,\qquad 2 b_1u_s = b_2 +b_4 \tag{27}$$

Substitute equation (27) into (24), we have

$$d_1 a_4b_4+ d_2 a_4+ d_3 b_4+ d_4 = 0 \tag{28}$$

$$d_1 = c_1,\qquad d_2 = b_2c_1+ 2 b_1c_2,\qquad d_3 = a_2c_1+ 2 a_1c_3$$

$$d_4 = a_2b_2c_1 + 2 a_2b_1c_2+ 2 a_1b_2c_3+ 4 a_1b_1c_4 \qquad \text{(Eighth order polynomial of } u_r\text{)}$$

Squaring the equation (28), we have

$$\mathbf{2 (d_1d_4- d_2 d_3) a_4b_4 = - d_1^2 a_4^2 b_4^2+ d_2^2 a_4^2 + d_3^2 b_4^2 - d_4^2} \tag{29}$$

Based on equation (25), the left hand side of equation (29) is

$$\mathbf{2 (d_1d_4- d_2 d_3) a_4b_4 = 8 a_1b_1(c_1c_4- c_2c_3) a_4b_4}$$

Further simplification generates

$$\mathbf{2 (d_1d_4- d_2 d_3) a_4b_4 = 8 a_1^2b_1^2}[(F_4F_9- F_7F_8) a_1b_1+ (F_3F_9- F_6F_7)a_1b_2+ (F_4F_6- F_3F_8) a_1b_3+ (F_2F_9 - F_5F_8) a_2 b_1+ (F_1F_9- F_5F_6)a_2 b_2+ (F_2F_6 - F_1F_8)a_2b_3+ (F_4F_5 - F_2F_7)a_3 b_1+(F_3F_5 - F_1F_7)a_3 b_2+(F_1F_4- F_2F_3)a_3 b_3]\mathbf{a_4b_4} \tag{30}$$

Canceling the common factor $8 a_1^2 b_1^2$ make the equation (29) an eighth order polynomial of u_r :

$$[(F_4F_9- F_7F_8) a_1b_1+ (F_3F_9- F_6F_7)a_1b_2+ (F_4F_6- F_3F_8) a_1b_3+ (F_2F_9 - F_5F_8) a_2 b_1+ (F_1F_9- F_5F_6)a_2 b_2+ (F_2F_6 - F_1F_8)a_2b_3+ (F_4F_5 - F_2F_7)a_3 b_1+(F_3F_5 - F_1F_7)a_3 b_2+(F_1F_4- F_2F_3)a_3 b_3]\mathbf{a_4b_4}$$

$$=(F_3F_5 + F_1F_7)a_2a_3 b_2^2 + (F_4F_5 + F_2F_7) a_2 a_3 b_1b_2+ 2 (F_3F_7- F_1F_9-F_5F_6)a_1a_3 b_2^2 +2(F_3^2- 2F_1F_6) a_1a_3b_3^2+ 2 (F_7^2 -2F_5F_9)a_1a_3 b_1^2+2 (F_1F_3+ F_1F_7)a_2a_3b_3^2+4 (F_3F_7+F_2F_8 -0.5 F_4^2 - F_5F_6- F_1F_9) a_1a_3b_1b_3 +2 F_5F_7a_2a_3b_1^2 +2 (F_2^2-2 F_1F_5)a_3^2b_1b_3+2 (F_8^2 - 2F_6F_9)a_1^2 b_1b_3 +2 F_1F_2a_3^2b_2b_3 + 2(F_1F_8+F_2F_6-F_3F_4)a_1a_3 b_2b_3 +2F_6F_8a_1^2b_2b_3 - F_6^2 a_1^2 b_3^2 +(F_2F_6 + F_1F_8)a_2^2b_2 b_3+ (F_4F_6 + F_3F_8)a_1a_2 b_2b_3 + 2(F_4F_8-F_3F_6 - F_3F_9 - F_6F_7)a_1a_2 b_1b_3+2(F_4F_7 -$$

$$F_5F_8)a_1a_3\,b_1\,b_2 - 2\,F_2F_4a_2a_3b_1b_3 - (F_1F_4 + F_2F_3)\,a_2\,a_3\,b_2b_3 - (F_1F_9 + F_5F_6 + F_2F_9 + F_5F_8)$$
$$a_2^{\,2}b_2^{\,2} - (F_3F_9 + F_6F_7)\,a_1\,a_2b_2^{\,2} - (F_4F_9 + F_7F_8)a_1\,a_2b_1b_2 - 2\,F_1F_6a_2^{\,2}b_3^{\,2} - 2(F_1F_9 + F_5F_6 - F_2F_8$$
$$)a_2^{\,2}\,b_1b_3 - 2F_5F_9\,a_2^{\,2}\,b_1^{\,2} - 2F_7F_9\,a_1a_2b_1^{\,2} - 2\,F_1F_5a_3^{\,2}b_2^{\,2} - 2F_2F_5a_3^{\,2}\,b_1b_2 - 2\,F_2F_9\,a_1\,a_3\,b_1b_2 - 2$$
$$F_6F_9\,a_1^{\,2}\,b_2^{\,2} - 2\,F_8F_9a_1^{\,2}\,b_1b_2 - 2F_1^{\,2}\,a_3^{\,2}b_3^{\,2} - 2\,F_5^{\,2}\,a_3^{\,2}\,b_1^{\,2} - 2\,F_9^{\,2}\,a_1^{\,2}\,b_1^{\,2} \tag{31}$$

Equation (31) still contains the square root terms $a_4\,b_4$. Squaring both sides of (31), a sixteenth order polynomial of u_r is obtained. Solve the resulted equation, we are able to find out all possible solutions of the forward kinematic problem.

Remark: If the points P_1, P_2, P_3, P_4, P_5, and P_6 are arranged to be coplanar and pair-wise, it can be shown that the coefficients of the odd-order terms in the 16-th polynomial equation are all zeros. Therefore the equation can be reduced to a 8-th order polynomial equation by a simple variable transform $w = u_r^{\,2}$.

5. Conclusion

In this paper, a solution for the general Stewart Platform is presented. The forward kinematic problem is converted through a series of symbolic calculation into the problem of solving a 16-th order polynomial equation. All possible solution for the pose of upper plate can be found given a set of leg length.

References

[1] J.E. Dieudonne, et al "An actuator extension transformation for a motion simulator and an inverse transformation applying Newton-Raphson's method",NASA Tech report D-7067, 1972

[2] E.F.Fichter, "A stewart Platform - Based Manipulator : General Theory and Practical Construction", Int. J. of Robotics Research, Vol.5, No.2, 1986, p157

[3] Z.Geng, J.Lee, R.Carroll and L.Haynes,"Learning control system design for a parallel link robotic manipulator", IEEE Int. Conf. Robotics and Automation, 1990

[4] M.Griffis and J.Duffy,"A forward displacement analysis of a class of Stewart Platforms", Int. J. of Robotics Research, Vol.6, No.6, 1989, p703

[5] S.E.Landsberger, " Design and construction of a cable-controlled parallel link manipulator", Master thesis, MIT 1984

[6] P.Nanua and K.Waldron,"Direct kinematic solution of a Stewart Platform", IEEE Int. Conf. Robotics and Automation, 1989, P 431

[7] D. Stewart, "A platform with six degree of freedom", Proc. Inst. Mech Eng., London, Vol. 180, 1965, p 371

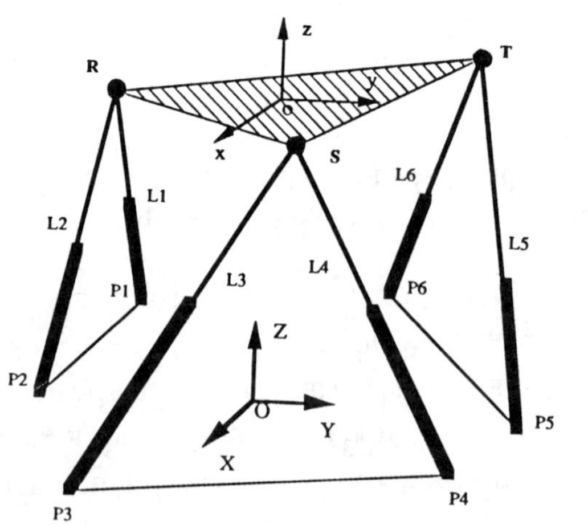

Figure 1 **General Stewart Platform: P_1, P_2, P_3, P_4, P_5, and P_6 are not coplanar**

ON THE COMPUTATIONAL EFFICIENCY OF FOUR NUMERICAL TECHNIQUES FOR SOLVING THE PRACTICAL INVERSE KINEMATICS PROBLEM OF REDUNDANT MANIPULATORS

Tarek M. Abdel-Rahman
Mechanical Engineering Department
Qatar University, P.O. Box 2713, Doha, Qatar.

ABSTRACT

In this paper we compare the computational efficiency of four methods for solving the practical inverse kinematics problem of redundant manipulators. Two methods use recursive approaches for sequential determination of feasible joint rates while the other two methods derive complete joint rate solutions at first and then check the joint kinematical constraints. The methods were implemented on a 386 microcomputer. Issues of improving efficiency by changing velocity reference frames and workspace decomposition were examined.

INTRODUCTION

Non-redundant manipulators have limited dexterous workspaces because of joint kinematical constraints and manipulator kinematical singularities. Dexterity can be improved by employing redundant manipulators. The redundant degrees of freedom can be used for singularity, obstacle or joint constraint avoidance or for optimization of a selected criterion for resolving the kinematical redundancy in the inverse velocity problem. Algorithms for redundancy resolution differ in their computational complexity. As the complexity increases, the computational cycle time increases and errors due to velocity linearization also increase. The effect of these increases on the performance is evaluated in this paper.

Most of the algorithms proposed in the literature either ignore the joint kinematical constraints or consider them indirectly via weighting matrices in quadratic joint rate functions. The later approach has been evaluated and compared with the direct consideration of the kinematical constraints in the solution procedure. Four solution methods were applied to solve the practical inverse kinematics problem of the 10 DOF manipulator shown in Fig. 1. The methods were the LU factorization [1], the Singular Value Decomposition (SVD) method [2], the Most-Effective-Direction (MED) method [3] and the

Recursive-Orthogonal-Motion-Resolution (ROMR) method [4]. The first two methods considered the kinematical constraints via the weighting matrices approach while the last two methods considered the joint rate constraints recursively in the solution procedures.

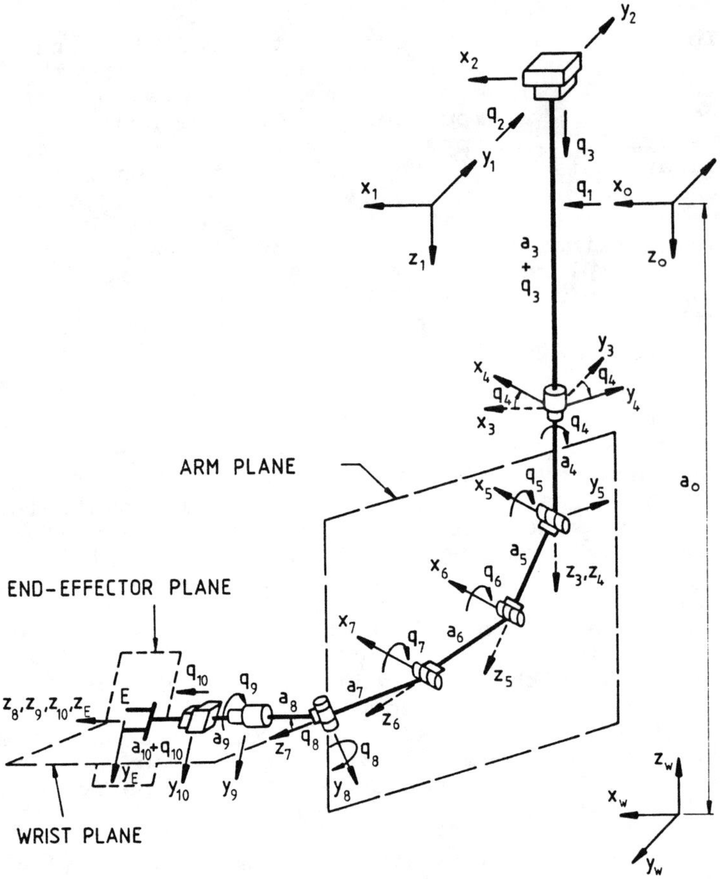

Fig. 1 A Ten-Degrees-of-Freedom Manipulator

The first three methods: the LU, SVD and MED were implemented on a Wang VS100 computer in [3]. In this paper, the programs of [3] were run on a 16MHz-386 Compaque microcomputer and the results were compared with recent ones determined by applying the ROMR method and the workspace decomposition approach [4].

THE PRACTICAL INVERSE KINEMATICS PROBLEM

The inverse kinematics problem can be formulated as follows: Let $\underline{x}$ denote a vector of dimension m containing the desired positions and orientation of the end-effector in workspace coordinates referred to an inertial base frame, and let $\underline{q}$ denote a vector of dimension n containing the joint positions. The relation between the joint rates $\dot{\underline{q}}$ and the end-effector velocity $\dot{\underline{x}}$ is given by :

$$\dot{\underline{x}} = J(q)\dot{\underline{q}} \tag{1}$$

where J(q) is the Jacobian matrix of dimension mxn. If n>m, the manipulator is kinematically redundant. In practice, each joint variable q_i has position, rate and acceleration constraints that can be described by:

$$\begin{aligned} q_{i,min} &< q_i < q_{i,max} \\ \dot{q}_{i,min} &< \dot{q}_i < \dot{q}_{i,max} \qquad i=1,2,...,n \\ \ddot{q}_{i,min} &< \ddot{q}_i < \ddot{q}_{i,max} \end{aligned} \tag{2}$$

The position boundaries in (2) are constants for most designs while the rate and acceleration bounds are function of links and payload inertia, the manipulator configuration and maximum motor torques. If $\underline{T}$ denote an n-element torque vector of the joint actuator, the torque constraints can be defined by :

$$\underline{T}_{min} < \underline{T} < \underline{T}_{max} \tag{3}$$

The practical inverse kinematics problem aims at finding a $\underline{q}$ solution of (1) for a given $\underline{x}$ such that the constraints (2) and (3) are satisfied.

By sampling the motion at various stages, we can map the position and dynamics constraints onto the joint rate space [4], hence, (2) and (3) yield:

$$\dot{\underline{q}}^k_{inf} < \dot{\underline{q}}^k < \dot{\underline{q}}^k_{sup} \tag{4}$$

at the time stage k. In this paper, we address the solution of the practical inverse kinematics problem with joint rate constraints in the form (4) and assume that they represent (2) and (3).

THE SOLUTION METHODS AND PERFORMANCE INDEX

Redundancy resolution methods vary in their approaches and hence provide diversified solutions for the local inverse kinematics problem. For comparison, we apply a constant velocity command $\dot{\underline{x}}_c$ for a period of time and compute the performance index from :

$$\text{Performance Index} = \frac{\text{Average End-Effector Speed}}{\text{End-Effector Speed Command}}$$

Four redundancy resolution methods were considered. The first two, the LU factorization and the SVD methods, resolve all the joint rates simultaneously. Each of the two methods determines, at first, a generalized inverse of the Jacobian matrix and then computes a solution for the given velocity command. When one of the computed joint rates exceeds its kinematical constraints, all the joint rates are scaled down proportionally to obtain the largest feasible solution. The LU solution minimizes a quadratic function of weighted joint rates, while the SVD solution minimizes the norm of the joint rates.

The operation counts of the LU and SVD methods are:

$$\text{LU} \quad : \quad m^2\left(\frac{3n-m}{6}\right) + (k^2+k+1)\,(n+m) + \frac{1}{3}k^2$$

$$\text{SVD} \quad : \quad 2mn^2 + f(n)$$

where an operation is a multiply or division plus an add, k is the rank of the Jacobian matrix and f(n) is a term that accounts for the iterative phase of the SVD method. If normal convergence occurs: $f(n) \cong 4n^3$. The last two methods, the MED and the ROMR, determine feasible joint rates recursively. The methods compute the contribution of each joint rate to the end-effector velocity and then determine the residual motion to be provided by the remaining joints. The ROMR method follows the same recursive resolution approach of the MED method but the resolution occurs in orthogonalized directions. The computation of the orthogonalized resolution directions requires a relatively small additional computational time but is essential for fast convergence to a solution. In many cases, the orthogonalized resolution direction could be derived analytically and hence reduces the computational time. In such cases the number of operations becomes mn^2+2n^2+mn for the two recursive methods.

A CASE STUDY

The manipulator shown in Fig. 1 was used in evaluating the computational efficiency of the four methods. The Jacobian matrix of the manipulator was a 6x10 matrix [3]. To facilitate the computation of the recursive methods, the Jacobian matrix was decomposed in the joint space. The end-effector velocity equation (1) became :

$$\underline{\dot{x}} = J_{j1}[\ \dot{q}_1\ \dot{q}_2\ \dot{q}_3\ ]^T + J_{j2}[\ \dot{q}_4\ \cdot\cdot\ \dot{q}_{10}\ ]^T \qquad (5)$$

where the subscripts j1 and j2 refer to the first and second decomposed joint spaces. Velocity resolution was done, at first, in the joint space, j2, of the distal joints with the first three joint rates set equal to zero. If some linear components remained unresolved the errors were compensated by the first three joints.

Experience in using the MED method [4] showed difficulty in convergence if the Jacobian matrix columns were not nearly orthogonal. To improve the performance of the MED method the reference frame of the end-effector velocity was changed to the fourth frame instead of the initial frame. The performance was then evaluated by applying six velocity commands v_1, $v_2, \ldots, v_6$ of unity magnitudes in the various six dimension directions, e.g. $v_1 = (1 \ 0 \ 0 \ 0 \ 0 \ 0)^T$. Each command was applied for 10 secs.

Computer programs for resolving the kinematics problem were written in Fortran and implemented on a 16MHz-386 Compaque microcomputer. All methods yielded satisfactory results if no computational time constraints were introduced. Many runs were done to determine appropriate cycle times for the four methods. The smallest appropriate time cycles were 1, .5, and .25 sec. for the SVD, LU and recursive methods respectively. When the cycle times were considered we obtained the results shown in Fig. 2. Fig. 2-a shows the performance index of the LU method, while Figs. 2-b and 2-c show the performance index of the SVD and the two recursive methods. The Figures illustrate that the performance degrades as the cycle time increases and that small index values occur for the v_5 and v_6 angular velocity commands. This was attributed to the kinematical constraints associated with the few number of joints that contribute to those commands. The results reported of the MED method has shown improvement than those of [2]. The improvement was due to the selection of an intermediate reference frame that enhanced the structure of the associated Jacobian matrix. The recursive approaches yielded the best performances. This is attributed to smaller cycle times and consideration of the joint kinematical constraints directly in the solution procedure.

To further improve the performance of the v_5 and v_6 command cases, the workspace was decomposed into two sub-workspaces. The first sub-workspace comprised the joints of major contribution to motion in the directions of the v_5 and v_6 commands and the second sub-workspace comprised the remaining joints. The velocity was resolved at first in the first sub-workspace then in the second sub-workspace. The results are shown in Fig. 3. In this figure, the increase of the v_5 and v_6 performance indeces illustrates the advantages of the workspace decomposition.

(a)

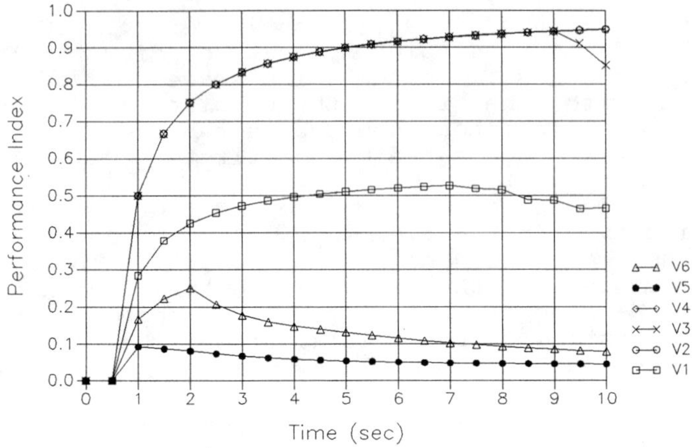

(b)

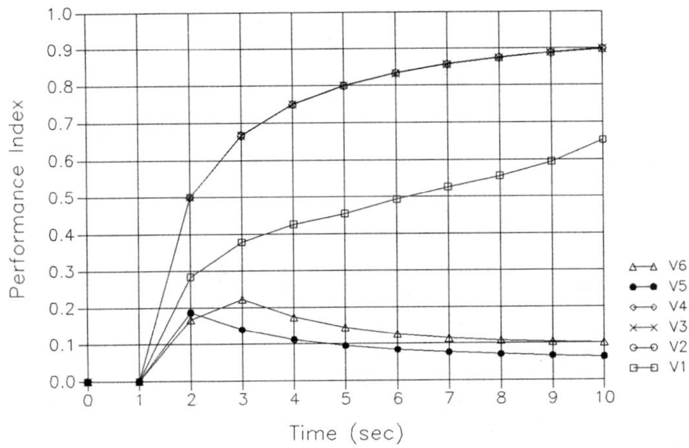

(c)

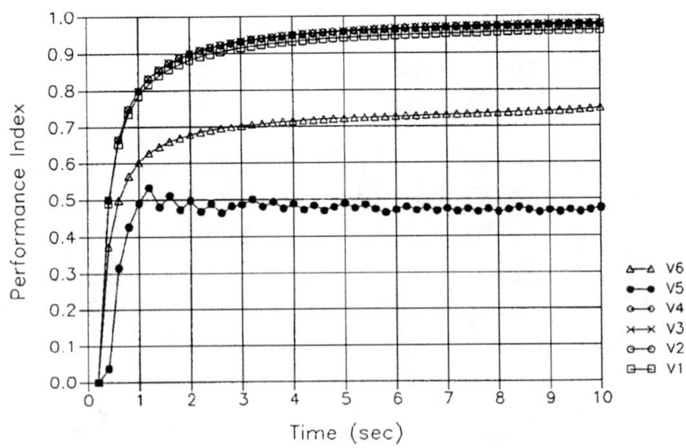

FIG. 2 Performance Index Variations:
(a) LU method, (b) SVD method
(c) MED and ROMR methods

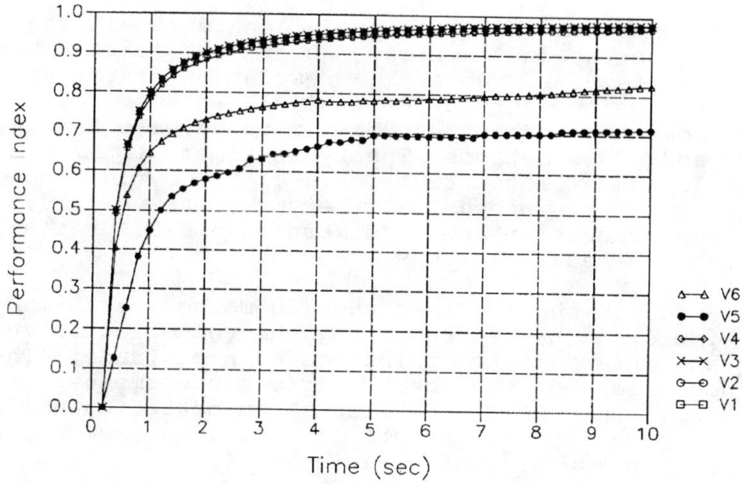

(a) MED method

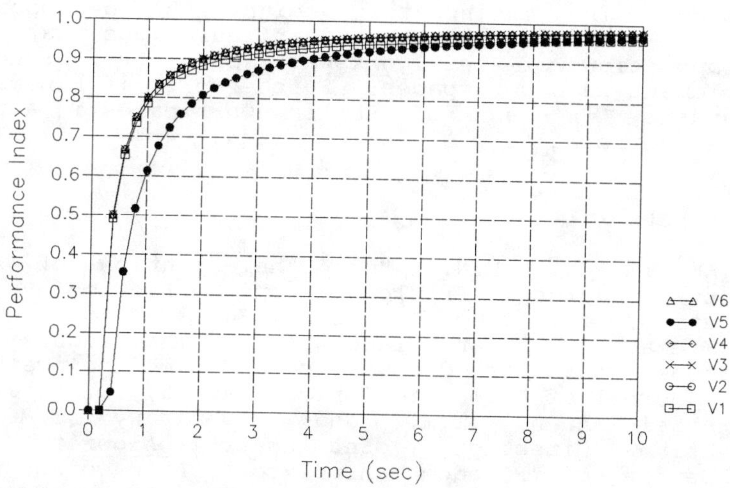

(b) ROMR Method

FIG. 3 Performance Index Variation for the Recursive Decomposed Workspace Methods

CONCLUSION

Two approaches for solving the practical inverse problem of redundant manipulators were evaluated. The first approach, simultaneously considers all the joints while the second approach uses the joint kinematics constraints for sequential resolution of the velocity. Four solution methods were implemented on a 16MHz-386 microcomputer. The methods evaluated were the LU, SVD, MED and ROMR methods. The cycle times of the recursive methods were less than half of the cycle time of the non recursive methods. Furthermore, the velocity errors in the recursive methods were much smaller than those of the non recursive ones.

Results showed that the MED method performance can be improved by representing the end-effector velocity in an intermediate joint reference frame. The ROMR method yielded the best performance amongst the four methods. The ROMR has also shown better capability in determining robust solutions and identifying singular and near singular configurations.

The recursive methods performance can be improved by decomposing the workspace into two sub-workspaces and solving two smaller-dimension inverse kinematics problem. In manipulator design, the decomposition implies selecting structural configurations that allow decomposition of the workspace into two orthogonal complement sub-workspaces. Furthermore, if kinematical redundancy in each of the sub-workspaces is achieved better performance would be expected.

REFERENCES

1. Abdelmalek, N.N. "On the Solution of Linear Least Squares Problems and Pseudo-Inverses", Computing, Vol. 13, pp.215-228, 1974.
2. Klema, V.C., and Laub, A.J. "The Singular Value Decomposition: Its Computation and Some Applications", IEEE Journal of Automatic Control, AC-25, No. 2, 1980.
3. Abdel-Rahman, T.M. and Elghamrawy, T.M."Most-Effective Direction Joint Control Algorithm of a Ten-Degree of Freedom Manipulator", Proceedings of the IASTED International Symposium on Robotics and Automation, Lugano, Switzerland, June 24-26, pp. 175-179, 1985.
4. Abdel-Rahman, T.M., and Takriti, M.S., "On-line Computational Schemes for the Inverse Problem in Robotics", Proceedings of the IASTED International Symposium on Robatics and Artificial Intelligence, Toulouse, France, June 18-20, pp. 249-267, 1986.
5. Abdel-Rahman, T.M. "A Generalized Practical Method for Analytical Solution of the Constrained Inverse-Kinematics Problems of Redundant Manipulators", To appear in the International Journal of Robotics Research.

STATIC FORCE ANALYSIS OF A THREE-CYLINDRIC ROBOT USING THE THEORY OF SCREWS

G.R. Pennock
Associate Professor

B. A. Oncu
Teaching Assistant

School of Mechanical Engineering
Purdue University
West Lafayette, IN 47907

ABSTRACT

Screw theory provides geometric insight which is important in the kinematic and dynamic analysis of spatial mechanisms, in general, and multi-body open chains, in particular. The screw approach also provides a compact notation which considerably simplifies symbolic and algebraic manipulation. In this paper, screw theory is applied to the inverse static force problem of a three-cylindric robot. The symmetry between the inverse static force solution and the inverse velocity solution is discussed. Then the dual actuator force applied at each joint and the power required by each rotary-linear actuator to support a specified external wrench acting on the end-effector are presented. The paper applies the results to a kinematically simple three-cylindric robot and presents a numerical example for illustrative purposes.

1. Introduction

Ball [1] defined a screw as a straight line associated with a definite linear magnitude called the pitch. Some of the earliest applications of screw theory to spatial kinematics are in the work of Dimentberg [2] who combined vector calculus and dual numbers and introduced the calculus of screws. Yang [19] applied quaternion algebra to the analysis of spatial mechanisms and Woo and Freudenstein [18] used line geometry to develop a numerical approach for the kinematic and dynamic analysis of spatial mechanisms. Keler [6] used screw theory and dual-vectors in the kinematics and statics including friction of single-loop mechanisms. Hunt [5] studies special configurations of robot arms using screw theory. A method for the solution of joint rates of open and closed-chains, based on an orthogonal screw basis for the screw system formed by individual screws along the joint axes, was introduced by Sugimoto and Duffy [14].

The simplification in algebraic manipulation, that comes from the compact notation of screw theory, is especially important in computer controlled multi-degree-of-freedom mechanisms. The simplification may represent a considerable saving in computational time. Many researchers have applied screw theory to robotics and obtained concise and elegant expressions for the inverse kinematics and dynamics problems. For example, Hunt [3] showed that the Jacobian matrix of a serial manipulator [16] can be obtained from inspection of the joint screws. He applied this geometric insight to solve the inverse velocity problem of commercially available robots. Pennock and Yang [10,11] used screw matrices to obtain closed-form solutions for the inverse kinematics and dynamics of a multi-rigid body open-chain system. Pennock and Vierstra [12] introduced the inverse of a dual-Jacobian matrix in the solution of the inverse kinematic problems of a three-cylindric robot.

In this paper, a geometric interpretation of screw theory is used to determine the joint forces and torques required for the static equilibrium of a three-cylindric robot. Since the robot joints are cylindric, each actuator is modeled as a rotary-linear actuator [17]. The paper discusses the symmetry between the inverse static force solution and the inverse velocity solution when the gravitational and inertial effects of the robot links are neglected. It is common practice to neglect gravity and the effects of inertia when a robot is operating at a low speed and/or when there is a large external wrench acting on the end-effector. The screw approach to the inverse kinematic and static problems is computationally efficient, since an understanding of the geometry of the screws eliminates many of the steps required in the solution of both problems. The results may be applied to a three-axis robot with any combination of revolute, prismatic, helical, and cylindric joints. A numerical example of a kinematically-simple three-cylindric robot is presented in Section 5 for illustrative purposes.

2. Notation

A schematic diagram of a three-link open-chain with cylindric joints, henceforth referred to as the three-cylindric robot, is shown in Fig. 1. Associated with link i (i = 0, 1, 2, 3) is a body frame F_i defined by the $\overline{Z}_{i+1}$ axis, directed along the axis of joint i + 1, and the $\overline{X}_i$ axis, directed along the common perpendicular between $\overline{Z}_i$ and $\overline{Z}_{i+1}$. Link 0; i.e., the ground, is regarded as the fixed base frame F_0. The terminal link 3 is the end-effector with origin of the end-effector frame F_3 denoted by W. A frame defined by the $\overline{X}_i$ axis and the $\overline{Z}_i$ axis is referred to as the ith joint frame and denoted by J_i. The origin of the ith joint frame is denoted by O_i. The position of O_{i+1} relative to O_i and the position of W relative to O_i are denoted by the vectors $\overline{R}_{i+1,\,i}$ and $\overline{R}_{w,\,i,}$ respectively. The location of the mass center of link i, relative to O_i, is denoted by $\overline{G}_i$, and the acceleration vector due to gravity is denoted by $\overline{g}$.

The structure of link i may be completely described by the dual angle $\hat{\alpha}_i = \alpha_i + \varepsilon l_i$, where the dual operator is defined as $\varepsilon^n = 0$ with the integer n > 1. The real part α_i is the twist angle between the joint axes $\overline{Z}_i$ and $\overline{Z}_{i+1}$, measured about the common perpendicular $\overline{X}_i$; and the dual part li is the link length, which is the length of the common perpendicular between the joint axes $\overline{Z}_i$ and $\overline{Z}_{i+1}$. The position and displacement of link i relative to link i - 1 is defined by the dual angle $\hat{\Theta}_i = \Theta_i + \varepsilon\, d_i$. The real part Θ_i is the angular joint displacement, defined by the rotation of the X_i axis relative to the $\overline{X}_{i-1}$ axis measured about Z_i; and the dual part di is the linear joint displacement along $\overline{Z}_i$ defined by the distance between the common perpendiculars $\overline{X}_{i-1}$ and $\overline{X}_i$. The trigonometric identifies of a dual angle may be obtained from a Taylor series expansion [6,9]. For example, the sine and cosine of $\hat{\Theta}_i$ (= Θi + $\varepsilon\, d_i$) are

$$\sin\hat{\theta}_i = \sin\theta_i + \varepsilon\, d_i \cos\theta_i \qquad \text{and} \qquad \cos\hat{\theta}_i = \cos\theta_i - \varepsilon\, d_i \sin\theta_i$$

Screw matrices are convenient for both forward and backward transformations between Cartesian reference frames [11]. For example, the forward transformations from F_{i-1} to J_i and from J_i to F_i can be expressed, respectively, by the screw matrices

$$[\,\hat{\theta}_i\,]_{Z_i} = \begin{bmatrix} C\hat{\theta}_i & S\hat{\theta}_i & 0 \\ -S\hat{\theta}_i & C\hat{\theta}_i & 0 \\ 0 & 0 & 1 \end{bmatrix} \quad \text{and} \quad [\,\hat{\alpha}_i\,]_{X_i} = \begin{bmatrix} 1 & 0 & 0 \\ 0 & C\hat{\alpha}_i & S\hat{\alpha}_i \\ 0 & -S\hat{\alpha}_i & C\hat{\alpha}_i \end{bmatrix} \; ; \; (\, i = 1,2,3\,)$$

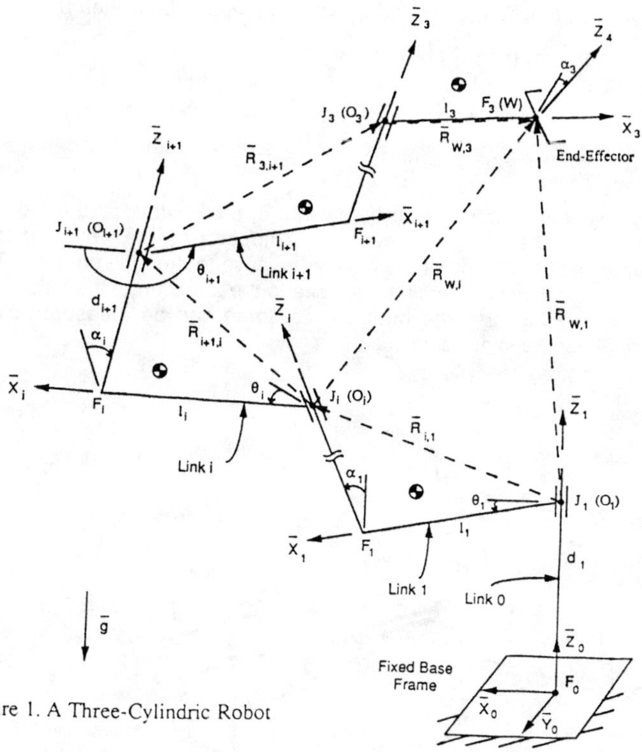

Figure 1. A Three-Cylindric Robot

where sine and cosine are abbreviated throughout the paper by S and C, respectively, Screw matrices are orthogonal and are computationally more efficient than the more well-known but highly redundant (4x4) homogeneous matrices.

3. The Static Force Analysis

Consider the case where the three-cylindric robot is in static equilibrium due to a resultant force vector, F, and a resultant moment vector about point W, $\overline{M}_W$, acting on the end-effector. The resultant force and moment vectors can be expressed, in the end-effector frame F_3, by the external wrench

$$\{\hat{F}_W\}_{F_3} = \overline{F} + \varepsilon \overline{M}_W = \hat{\Gamma}\,\{\hat{T}_W\}_{F_3} \tag{1}$$

where $\hat{\Gamma}$ denotes the dual modulus of the wrench, and $\hat{T}_W$ denotes the unit screw along the axis of the wrench, referred to point w. In general, the external wrench acting on the end-effector will depend upon: (i) the weight of the payload, (ii) the moment about point W caused by the weight of the payload, (iii) the contact force $\overline{F}_C$ between the payload and the environment, and (iv) the moment $\overline{M}_F$ about point W caused by the contact force. The contact force, see Fig. 2, could be the result of such things as; coulomb friction, a spring force, an impact force, etc.

The external wrench, referred to the end-effector frame, can be written as

$$\{\hat{F}_W\}_{F_3} = \{\hat{F}_P\}_{F_3} + \{\hat{F}_C\}_{F_3}$$

$$\{\hat{F}_P\}_{F_3} = m_P\,\bar{g} + \varepsilon\,(\bar{r} \times m_P\,\bar{g}) \quad \text{and} \quad \{\hat{F}_C\}_{F_3} = \bar{F}_C + \varepsilon\,\overline{M}_F$$

are the dual gravitational force due to the weight of the payload and the dual contact force, respectively. The dual contact force is applied by the environment on the payload and opposes the motion of the payload relative to the environment. Therefore, the direction of the dual contact force is opposite to the direction of the velocity screw of the end-effector, see Fig. 2. The dual contact force can be measured by sensors placed at the origin of the joint frame J_3.

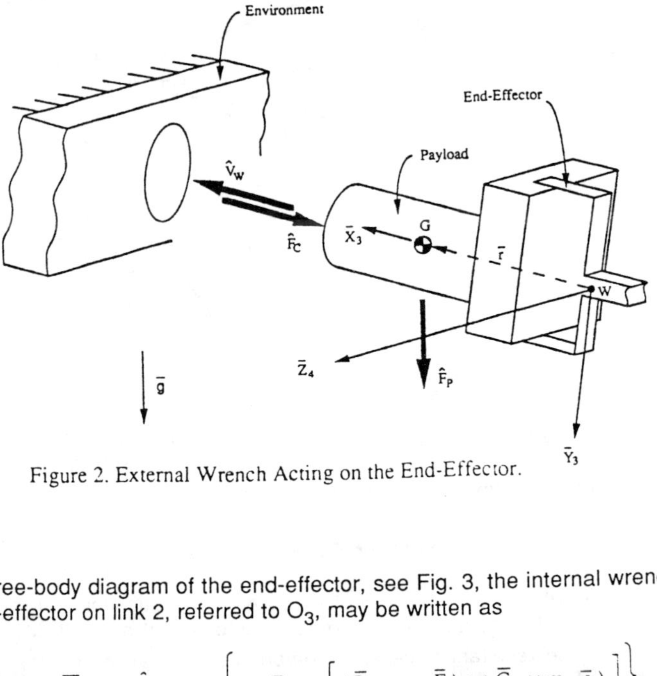

Figure 2. External Wrench Acting on the End-Effector.

From the free-body diagram of the end-effector, see Fig. 3, the internal wrench applied by the end-effector on link 2, referred to O_3, may be written as

$$\hat{F}_{3,2} = \bar{F}_{3,2} + \varepsilon\,\overline{M}_{3,2} = \{\hat{F}_W\}_{F_3} + \left\{ m_3\,\bar{g} + \varepsilon\left[(\bar{R}_{W,3} \times \bar{F}) + (\bar{G}_3 \times m_3\,\bar{g}) \right] \right\} \tag{2}$$

The real part of Eq. (2) satisfies the force equilibrium condition and the dual part satisfies the moment equilibrium condition.

From the free-body diagram of link i, see Fig. 4, the internal wrench applied by link i on link i - 1, referred to O_i, may be written as

$$\hat{F}_{i,i-1} = \hat{F}_{i+1,i} + \left\{ m_i\,\bar{g} + \varepsilon\left[(\bar{R}_{i+1,i} \times \bar{F}_{i+1,i}) + (\bar{G}_i \times m_i\,\bar{g}) \right] \right\}; \quad (i = 2, 1) \tag{3}$$

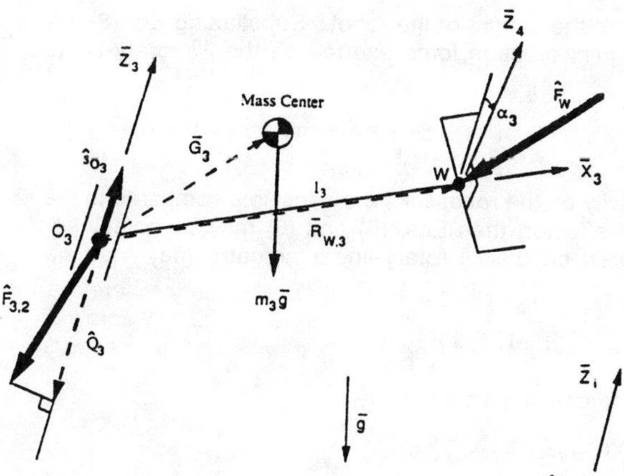

Figure 3. Free-Body Diagram of the End-Effector.

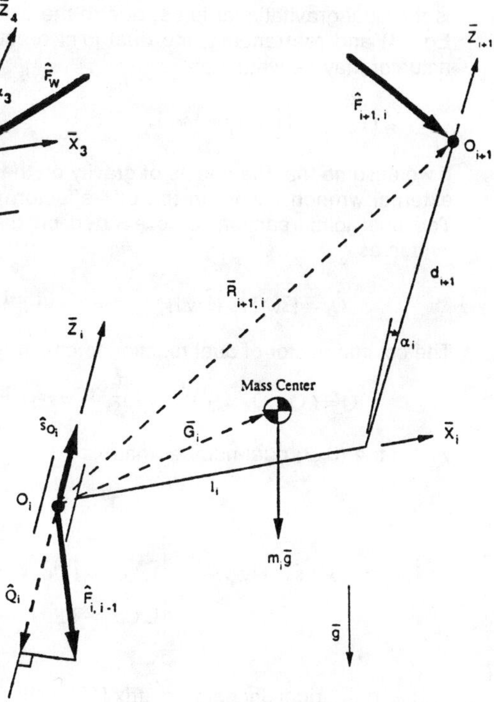

Figure 4. Free-Body Diagram of Link i.

The projection of the internal wrench $\hat{F}_{i,\,i-1}$ unto the ith joint axis is the dual joint reaction force exerted on the ith actuator; i.e.,

$$\hat{Q}_i = Q_i + \varepsilon\, M_i = \{\hat{s}_{O_i}\}_{J_i} \cdot \hat{F}_{i,\,i-1} \qquad ; \qquad (\, i = 3,\, 2,\, 1\,) \tag{4}$$

where the real part Q_i is the joint reaction force and the dual part M_i is the joint reaction moment; $\{\hat{s}_{Oi}\}_{Ji}$ $(= \bar{s}_i + \varepsilon\, \bar{s}_{Oi})$ is the unit screw along the ith joint axis expressed in the J_i frame (i.e., the column vector of dual direction cosines of the ith joint screw referred to the J_i frame). Since the unit screw contains point O_i, the dual part is zero and the unit screw reduces to a unit vector. Substituting Eq. (2) into Eq. (4) the dual joint reaction force exerted on the terminal rotary-linear actuator may be written as

$$\hat{Q}_3 = \{\hat{s}_{W_3}\}_{F_3} \cdot \{\hat{F}_W\}_{F_3} + \{\hat{s}_{O_3}\}_{J_3} \cdot \{\hat{F}_{G3}\} \tag{5}$$

where $\{\hat{s}_{W3}\}_{F3}$ is the unit screw along the 3rd joint axis expressed in the end-effector frame F_3, and

$$\{\hat{F}_{G_i}\} = \sum_{k=i}^{3} [\, m_k\, \bar{g} + \varepsilon\, (\, \bar{G}_k \times m_k\, \bar{g}\,)\,]$$

is the dual gravitational force, due to the weight of the robot. Substituting Eq. (3) into Eq. (4) and rearranging, the dual joint reaction force exerted on the ith rotary-linear actuator may be written as

$$\hat{Q}_i = \{\hat{s}_{W_i}\}_{F_3} \cdot \{\hat{F}_W\}_{F_3} + \{\hat{s}_{O_i}\}_{J_i} \cdot \{\hat{F}_{G_i}\} + \{\hat{s}_{O_i}\}_{J_i} \cdot [\sum_{k=i+1}^{3} (\bar{R}_{k,i} \times m_k \bar{g})] \quad ; \quad (i = 2, 1) \qquad (6)$$

If we assume that the effects of gravity on the robot links are negligible compared to the external wrench acting on the end-effector, then Eqs. (5) and (6) may be combined. The dual joint reaction force exerted on the ith rotary-linear actuator may then be written as

$$\hat{Q}_i = \{\hat{s}_{W_i}\}^T_{F_3} \{\hat{F}_W\}_{F_3} \quad ; \quad (i = 1, 2, 3) \qquad (7)$$

The column vector of dual reaction forces may be written as

$$\hat{Q} = (\hat{Q}_1 \ \hat{Q}_2 \ \hat{Q}_3)^T = [\hat{J}]^T_{F_3} \{\hat{F}_W\}_{F_3} \qquad (8)$$

where the (3x3) dual-number matrix

$$[\hat{J}]_{F_3} = [\hat{s}_{W_1} \ \hat{s}_{W_2} \ \hat{s}_{W_3}]_{F_3} = \begin{bmatrix} \hat{E} & S\hat{\theta}_3 S\hat{\alpha}_2 & 0 \\ \hat{L} S\hat{\alpha}_3 + \hat{I} C\hat{\alpha}_3 & \hat{M} & S\hat{\alpha}_3 \\ \hat{L} C\hat{\alpha}_3 - \hat{I} S\hat{\alpha}_3 & \hat{N} & C\hat{\alpha}_3 \end{bmatrix} \qquad (9)$$

is called the dual-Jacobian matrix [12], with

$$\hat{E} = \hat{K} S\hat{\theta}_3 + C\hat{\theta}_3 S\hat{\theta}_2 S\hat{\alpha}_1 \quad ; \quad \hat{L} = C\hat{\alpha}_2 C\hat{\alpha}_1 - S\hat{\alpha}_2 S\hat{\alpha}_1 C\hat{\theta}_2$$

$$\hat{I} = \hat{K} C\hat{\theta}_3 - S\hat{\theta}_3 S\hat{\theta}_2 S\hat{\alpha}_1 \quad ; \quad \hat{K} = S\hat{\alpha}_2 C\hat{\alpha}_1 + C\hat{\alpha}_2 S\hat{\alpha}_1 C\hat{\theta}_2$$

$$\hat{M} = S\hat{\alpha}_3 C\hat{\alpha}_2 + C\hat{\alpha}_3 S\hat{\alpha}_2 C\hat{\theta}_3 \quad \text{and} \quad \hat{N} = C\hat{\alpha}_3 C\hat{\alpha}_2 - S\hat{\alpha}_3 S\hat{\alpha}_2 C\hat{\theta}_3$$

An important observation, from Eq. (8), is that a symmetry exists between the inverse static force formulation and the inverse velocity formulation [7,9,15]. The inverse of the dual-Jacobian matrix plays the same role in the inverse velocity problem as the transpose of the dual-Jacobian matrix plays in the inverse static force problem. The two screw systems; i.e., the velocity screw and the external wrench, are interrelated by the theorem of reciprocity presented by Ball [1] and discussed in the work of Hunt [4].

This relationship has also been used in the kinematic and static force analyses of closed-loop manipulators and platform-type manipulators [13].

Since the dual actuator force at joint i is equal and opposite to the dual joint reaction force exerted on the ith rotary-linear actuator, we have

$$\hat{f}_i = f_i + \varepsilon t_i = -\hat{Q}_i \quad ; \quad (i = 1, 2, 3) \qquad (10)$$

where f_i and t_i are the actuator force and the actuator torque, respectively.

4. The Actuator Power

In many practical robot applications, the gravitational and inertial effects are negligible compared to the external wrench acting on the end-effector [13]. In this case, the dual actuator force obtained from a static force analysis can be used to calculate the power required by each rotary-linear actuator. The dot product of the column vector of dual actuator forces with the column vector of dual joint rates may be written as

$$\hat{f} \cdot \dot{\hat{\Theta}} = \hat{f}^T \dot{\hat{\Theta}} = \dot{\hat{\Theta}}^T \hat{f} = \{\hat{F}_W\}_{F_3} \cdot \{\hat{V}_W\}_{F_3} \tag{11}$$

where

$$\{\hat{V}_W\}_{F_3} = \overline{\omega} + \varepsilon \, \overline{V}_W = \hat{\Omega} \, \{\hat{s}_W\}_{F_3} = [\, \hat{J} \,]_{F_3} \, \dot{\hat{\Theta}} \tag{12}$$

is called the velocity screw of the end-effector; and Ω is called the dual modulus of the velocity screw. For the inverse velocity problem, the velocity screw of the end-effector is specified, and the dual joint rates are obtained from

$$\dot{\hat{\Theta}} = [\, \hat{J} \,]_{F_3}^{-1} \, \{\hat{V}_W\}_{F_3} \tag{13}$$

Substituting Eqs. (1) and (12) into the right-hand side of Eq. (11), gives

$$\hat{f}^T \dot{\hat{\Theta}} = \overline{F} \cdot \overline{\omega} + \varepsilon \, P \tag{14}$$

where the dual part

$$P = \overline{F} \cdot \overline{V}_W + \overline{M}_W \cdot \overline{\omega} = \sum_{i=1}^{3} P_i = \sum_{i=1}^{3} (\, f_i \, \dot{d}_i + t_i \, \dot{\theta}_i \,) \tag{15}$$

is the net power required by the actuators to support a specified external wrench acting on the end-effector, and P_i is the power required by the ith actuator to support the internal wrench applied by link i on link i - 1. There are three possible cases: (i) the net power is zero (i.e., the external wrench does not net work on the robot). The external wrench and the velocity screw (of the end-effector) are said to be reciprocal screws [1,4,8]; (ii) the net power is a positive value; i.e., power is going into the system, then the two screws are said to be repelling screws [8]. This implies that the end-effector has motion along the positive direction of the external wrench; and (iii) the net power is a negative value, i.e., power is removed from the system, then the two screws are said to be contrary screws [8]. This implies that the end-effector has motion along the negative direction of the external wrench.

5. A Numerical Example

In this section, we consider a kinematically simple three-cylindric robot with an external wrench acting at point W of the end effector, see Fig. 5. The twist angles of the robot links are specified as: α_1 = -90°, α_2 = 90° and α_3 = 0°. From an inspection of the three joint screws, the dual-Jacobian matrix expressed in the end-effector frame is

$$[\, \hat{J} \,]_{F_3} = \begin{bmatrix} -S\theta_2 C\theta_3 & S\theta_3 & 0 \\ S\theta_2 S\theta_3 & C\theta_3 & 0 \\ C\theta_2 & 0 & 1 \end{bmatrix} + \varepsilon \begin{bmatrix} q\, S\theta_3 - d_2\, C\theta_2\, C\theta_3 & d_3\, C\theta_3 & 0 \\ q\, C\theta_3 + (d_2\, S\theta_3 + l_3)\, C\theta_2 & -d_3\, S\theta_3 & l_3 \\ -(d_2 + l_3\, S\theta_3)\, S\theta_2 & -(l_2 + l_3\, C\theta_3) & 0 \end{bmatrix}$$

where q (= $1_1 + 1_2\, C\Theta_2 + d_3\, S\Theta_2$) is the component of the vector from O_1 to O_3 in the X_1 direction. The dual-Jacobian matrix may also be obtained analytically by substituting the specified twist angles into Eq. (9).

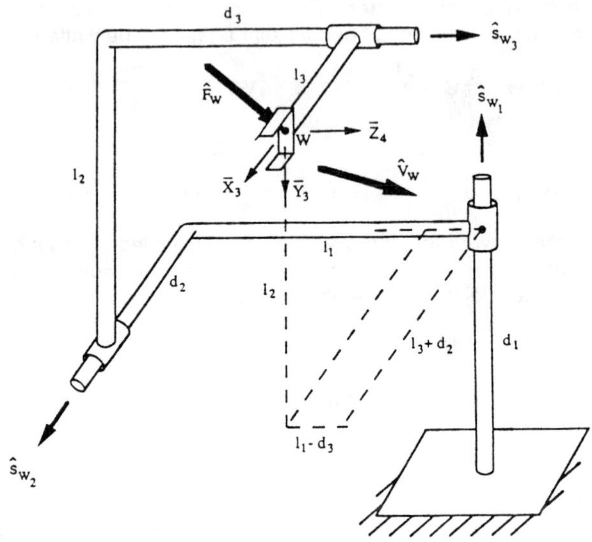

Figure 5. The Kinematically Simple Three-Cylindric Robot.

To illustrate the procedure for the inverse static force analysis (and at the same time simplify the calculations), the angular joint displacements at the instant under consideration are taken to be: $\Theta_2 = 270^o$ and $\Theta_3 = 90^o$. Therefore, the dual-Jacobian matrix reduces to

$$[\,\hat{J}\,]_{F_3} = \begin{bmatrix} 0 & 1 & 0 \\ -1 & 0 & 0 \\ 0 & 0 & 1 \end{bmatrix} + \varepsilon \begin{bmatrix} l_1 - d_3 & 0 & 0 \\ 0 & -d_3 & l_3 \\ l_3 + d_2 & -l_2 & 0 \end{bmatrix} \tag{16}$$

For convenience, we will assume that the link lengths are: $1_1 = 1_2 = 1_3 = 0.30$ m; and the linear joint displacements, at the instant under consideration, are: d_2 - 0.10 m and d_3 - 0.40 m. Substituting these values into Eq. (16), the dual-Jacobian matrix is

$$[\,\hat{J}\,]_{F_3} = \begin{bmatrix} 0 & 1 & 0 \\ -1 & 0 & 0 \\ 0 & 0 & 1 \end{bmatrix} + \varepsilon \begin{bmatrix} -0.10 & 0 & 0 \\ 0 & -0.40 & 0.30 \\ 0.40 & -0.30 & 0 \end{bmatrix} \quad (\text{rad} + \varepsilon\ \text{m}) \tag{17}$$

The inverse of the dual-Jacobian matrix is

$$[\,\hat{J}\,]_{F_3}^{-1} = \begin{bmatrix} 0 & -1 & 0 \\ 1 & 0 & 0 \\ 0 & 0 & 1 \end{bmatrix} + \varepsilon \begin{bmatrix} -0.40 & 0 & 0.30 \\ 0 & -0.10 & 0 \\ 0.30 & 0.40 & 0 \end{bmatrix} \quad (\text{rad} + \varepsilon\ \text{m}) \tag{18}$$

For a complete discussion on the inverse of the dual-Jacobian matrix the reader is referred to Pennock and Vierstra [12] and Oncu [9]. The velocity screw of the end-effector and the external wrench acting on the end-effector are specified, respectively, as

$$\{\hat{V}_W\}_{F_3} = \begin{Bmatrix} 2 + \varepsilon\ 0.15 \\ 0 + \varepsilon\ 0.10 \\ 1 - \varepsilon\ 0.20 \end{Bmatrix} \text{(rad/s + } \varepsilon \text{ m/s) and } \{\hat{F}_W\}_{F_3} = \begin{Bmatrix} 40 - \varepsilon\ 5 \\ 80 + \varepsilon\ 6 \\ 30 + \varepsilon\ 2 \end{Bmatrix} \text{(N + } \varepsilon \text{ N m)} \quad (19)$$

Therefore, substituting Eqs. (18) and (19) into Eq. (13), the dual joint rates are

$$\begin{Bmatrix} \dot{\hat{\theta}}_1 \\ \dot{\hat{\theta}}_2 \\ \dot{\hat{\theta}}_3 \end{Bmatrix} = \begin{Bmatrix} 0 - \varepsilon\ 0.60 \\ 2 + \varepsilon\ 0.15 \\ 1 + \varepsilon\ 0.40 \end{Bmatrix} \text{(rad/s + } \varepsilon \text{ m/s)} \quad (20)$$

Similarly, substituting Eqs. (17) and (19) into Eq. (8), the dual joint reaction forces are

$$\begin{Bmatrix} \hat{Q}_1 \\ \hat{Q}_2 \\ \hat{Q}_3 \end{Bmatrix} = \begin{Bmatrix} -80 + \varepsilon\ 2 \\ 40 - \varepsilon\ 46 \\ 30 + \varepsilon\ 26 \end{Bmatrix} \text{(N + } \varepsilon \text{ N m)} \quad (21)$$

Therefore, from Eq. (10), the dual actuator forces are

$$\begin{Bmatrix} \hat{f}_1 \\ \hat{f}_2 \\ \hat{f}_3 \end{Bmatrix} = \begin{Bmatrix} 80 - \varepsilon\ 2 \\ -40 + \varepsilon\ 46 \\ -30 - \varepsilon\ 26 \end{Bmatrix} \text{(N + } \varepsilon \text{ N m)} \quad (22)$$

Substituting Eqs. (20) and (22) into Eq. (15), the net power required by the actuators to support the specified external wrench acting on the end-effector is

$$P = P_1 + P_2 + P_3 = -48 + 86 - 38 = 0 \text{ N m} \quad (23)$$

Since the net power is zero, the velocity screw of the end-effector and the external wrench acting on the end-effector must be reciprocal screws. The power required by each rotary-linear actuator, however, to hold the robot in static equilibrium are not zero.

6. Conclusions

In this paper, the theory of screws is applied to the inverse static force problem of a three-cylindric robot. Emphasis is placed on the geometric interpretation of the (3x3) dual-Jacobian matrix. In this way, concise expressions for the efficient computational programming of the inverse kinematics and inverse statics problems may be obtained. We believe that an understanding of the kinematic geometry provides the greatest contribution to the efficient kinematic and dynamic analysis of robot manipulators. The paper also demonstrates that the duality between the velocity screw and the static force screw is a basic property of screw systems. The inverse velocity and static force solutions of a kinematically simple three-cylindric robot are presented. Finally, a numerical example of the kinematically simple robot is included in the paper for the purpose of illustration.

REFERENCES

1. Ball, R.S., A Treatise on the Theory of Screws, Cambridge University Press, Cambridge, 1900.
2. Dimentberg, F.M., The Screw Calculus and Its Application in Mechanics, Izdat, Nauka, 1965.
3. Hunt, K.H., "Robot Kinematics - A Compact Analytic Inverse Solution for Velocities," *ASME J. of Mech., Trans., and Auto. in Design,* Vol. 109, No. 1, March 1987, pp. 42-49
4. Hunt, K.H., Screw Systems in Spatial Kinematics, MMERS Report 3, Department of Mechanical Engineering, Monash University, Melbourne, Australia, 1970.
5. Hunt, K.H., "Special Configurations of Robot-Arms Via Screw Theory, Part 1, The Jacobian and Its Matrix of Cofactors," *Robotics,* Vol. 4, Part 3, July-Sept. 1986, pp. 171-180.
6. Keler, M.L., "Kinematics and Statics Including Friction in Single-Loop Mechanisms by Screw Calculus and Dual Vectors," *ASME J. of Eng. for Industry,* Vol. 95, No. 2, May 1973, pp. 471-480.
7. Lipkin, H., and Duffy, J., "Analysis of Industrial Robots Via the Theory of Screws," Proc. of 12th International Symposium on Industrial Robots, Paris, 1982, pp. 359-370
8. Ohwovoriole, M.S., An Extension of Screw Theory and Its Application to the Automation of Industrial Assemblies, Ph.D. dissertation, Stanford University, California, 1980, 173 p.
9. Oncu, B.A., A Geometric Interpretation of Screw Theory: Application to Rigid Body Dynamics and Robotics, Master of Engineering Science Thesis, Purdue University, May 1990, 148 p.
10. Pennock, G.R., and Yang, A.T., "Dynamic Analysis of a Multi-Rigid-Body Open-Chain System," *ASME J. of Mech., Trans., and Auto. in Design,* Vol. 105, No. 1, March 1983, pp. 28-34.
11. Pennock, G.R., and Yang, A.T., "Application of Dual-Number Matrices to the Inverse Kinematics Problem of Robot Manipulators," *ASME J. of Mech., Trans., and Auto. in Design,* Vol. 107, No. 2, June 1985, pp. 201-208.
12. Pennock, G.R., and Vierstra, B.C., "The Inverse Kinematics of a Three-Cylindric Robot," *Proc. of lst National Appl. Mech. and Robotics Conf.*, Cincinnati, Ohio, Nov. 5-8, 1989, Vol. 1, pp. 1-8.
13. Raghavan, M., and Roth, B., "On the Design of Manipulators for Applying Wrenches," Proc. of 1989 *IEEE Inter. Conf. on Robotics and Auto.*, Scottsdale, Arizona, Vol. 1, May 14-19, 1989, pp. 438-443.
14. Sugimoto, K., and Duffy, J., "Application of Linear Algebra to Screw Systems," Mechanism and Machine Theory, Vol. 17, No. 1, 1982, pp. 73-83.
15. Waldron, K.J. and Hunt, K.H., "Series-Parallel Dualities in Actively Coordinated Mechanisms," Proc. of *4th Inter. Symposium on Robotics Research,* Santa Cruz, CA., 1987, pp. 175-181.
16. Waldron, K.J., Wang, S.L., and Bolin, S.J., "A Study of the Jacobian Matrix of Serial Manipulators," *ASME J. of Mech., Trans., and Auto. in Design,* Vol. 107, No. 2, June 1985, pp. 230-238.
17. Weng, T.C., Sandor, G.N., Xu, Y., and Kohli, D., "Analytical Method of Determining the Workspace of Closed-Loop Manipulators with Ground-Mounted Rotary-Linear Actuators and Infinitesimal Platform," *Proc. 1987 ASME Design Tech. Conf.,* Boston, MA, 2, Sept. 1987, pp. 55-61.
18. Woo, L.S. and Freudenstein, F., "Dynamic Analysis of Mechanisms Using Screw Coordinates," ASME J. of Eng. for Industry, Vol. 93, No.1, Feb. 1971, pp. 273-276.
19. Yang, A.T., Application ofQuaternion Algebra and Dual Numbers to the Analysis of Spatial Mechanisms, Ph.D. Dissertation, Columbia University, New York, NY, 1963, 241 p.

Optimum Kinematic Design of Planar Three-Degree-of-Freedom Dextrous Manipulators

Clément M. Gosselin
Département de Génie Mécanique
Université Laval
Ste-Foy, Québec, Canada, G1K 7P4

Abstract

In this paper, planar three-degree-of-freedom manipulators are optimized using the concept of kinematic dexterity. The dexterity index used here is based on the condition number of a matrix representing the mapping between the joint velocities of the manipulator and a set of Cartesian velocities used to describe the motion of the end effector. The Cartesian coordinates are chosen in such a way that the dexterity index is frame invariant. Hence, it is not affected by a scaling of the manipulator, as opposed to existing indices. A comparison of the results obtained with existing indices and with the new index is given and an isotropic design is found using the new dexterity index. Finally, an example of application to a practical optimization problem is presented.

1 Introduction

The mechanical design of robotic manipulators is a very important issue in the context of robotic system optimization. Indeed, with the development of precise manipulators and articulated hands, the need for inherently accurate kinematic structures has become quite obvious. Some performance indices have been defined in order to characterize the dexterity of robotic manipulators or systems. The condition number of the Jacobian matrix has been used by Salibury and Craig [1] as an optimization criterion to obtain ideal dimensions for the fingers of the Stanford/JPL articulated hand and by Angeles and Rojas [2] to optimize the design of three-degree-of-freedom spatial and spherical manipulators. On the other hand, Yoshikawa [3] used the determinant of the same matrix as a manipulability index whereas Stoughton and Kokkinis [4] used the minimum singular value and the condition number. A good review of the dexterity measures is presented in [5] where redundant manipulators are treated. In the latter reference, it is pointed out that, although a vanishing determinant indicates the presence of a singularity, its actual value cannot be used as a practical measure of the ill-conditioning. Instead, numerical analysts recommend the use of the condition number [6].

Other results of optimization based on the condition number are given in [7] where a six-degree-of-freedom isotropic manipulator is obtained and in [8-9] for parallel manipulators. Moreover, a global dexterity index based on the condition number was defined in [10] to characterize the kinematic accuracy of a manipulator over its whole workspace. A similar index is also discussed in [11].

However, when both the position and the orientation of the end effector are included in the kinematic equations, it can be readily seen that the condition number of the Jacobian matrix will not be invariant under a scaling of the dimensions of the manipulator. Hence, the need for a formulation of the kinematic equations that would lead to a Jacobian matrix in which each of the entries would bear the same physical units so that the condition number would not be affected by a scaling of the manipulator. Dexterity indices based on such a formulation of the kinematic equations are presented in [12] for both planar and spatial manipulators.

In this paper, one of the dexterity indices derived in the latter reference is applied to the optimum design of planar three-degree-of-freedom manipulators. An isotropic design is found, leading to the maximum possible kinematic accuracy. The results are compared with those obtained with another dexterity index. Finally, an example of application to a particular design problem is presented.

2 Dexterity indices for planar manipulators

The dexterity index used here was derived in [12] where both planar and spatial cases are treated. Two dexterity indices for planar manipulators are proposed in the latter reference. The derivation of one of these indices will now be summarized.

Planar motion is concerned with the position and orientation of rigid bodies on a plane. Therefore, the Cartesian coordinates usually associated with the end effector of a planar manipulator consist of two positioning variables, e.g., the Cartesian components x and y of one point of the end effector, and of a rotation angle ϕ.The end effector of an arbitrary planar manipulator is shown in Fig. 1 where the reference frame $\mathcal{R}$ (O_{xy}) is fixed to the base of the manipulator while frame $\mathcal{R}'$ $(O'_{x'y'})$ is attached to the end effector. In general, the Cartesian coordinates and velocities used for the end effector are given by vectors $\mathbf{c}$ and $\dot{\mathbf{c}}$ which are defined as:

$$\mathbf{c} = [x,\, y,\, \phi]^T, \qquad \dot{\mathbf{c}} = [\dot{x},\, \dot{y},\, \dot{\phi}]^T \tag{1}$$

where x and y are the components of the position vector of point O' and ϕ is the angle of orientation.

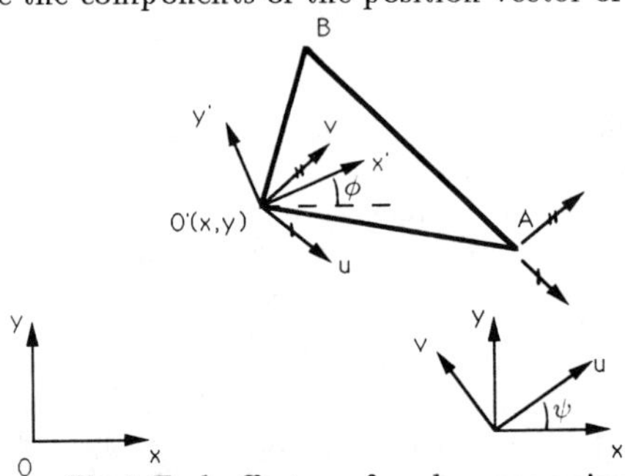

Fig.1 End effector of a planar manipulator.

On the other hand, the joint coordinate vector $\boldsymbol{\theta}$, and the joint velocity vector $\dot{\boldsymbol{\theta}}$ are written as

$$\boldsymbol{\theta} = [\theta_1,\, \theta_2,\, \ldots \theta_n]^T, \qquad \dot{\boldsymbol{\theta}} = [\dot{\theta}_1,\, \dot{\theta}_2,\, \ldots \dot{\theta}_n]^T \tag{2}$$

where n is the number of joints. The velocity relationship is then written as

$$\mathbf{J}\dot{\boldsymbol{\theta}} = \dot{\mathbf{c}} \tag{3}$$

where $\mathbf{J}$ is the Jacobian matrix of the manipulator. The dexterity index ν based on the condition number is then written

$$\nu = \kappa(\mathbf{J}) \tag{4}$$

where $\kappa(\mathbf{J})$ is the condition number of the Jacobian matrix and is defined as

$$\kappa(\mathbf{A}) = ||\mathbf{A}||\,||\mathbf{A}^{-1}||, \qquad \text{with} \qquad ||\mathbf{A}|| = \sqrt{\mathrm{tr}(\mathbf{AWA}^T)} \tag{5}$$

$\mathbf{W}$ being defined as $w\mathbf{1}$, with $w = 1/n$, where n is the dimension of the square matrix $\mathbf{A}$ and $\mathbf{1}$ is the $n \times n$ identity matrix. When $\mathbf{A}$ is not square, an alternative definition of the condition number is used [5].

However, because the Cartesian velocity vector includes components that are of a different nature, namely, two linear velocities and one angular velocity, the entries of matrix $\mathbf{J}$ will bear different physical units and hence, scaling the dimensions of the manipulator will affect the value of the condition number.

One way to avoid the problem of having mixed dimensions in the Jacobian matrix is to express the Cartesian velocity of the end effector differently. Instead of using the components shown in eq.(1), we can use the velocity of two points on the end effector. Let these two points be noted A and B (Fig. 1). The Cartesian velocity vector becomes

$$\dot{\mathbf{c}}' = [v_{ax},\, v_{ay},\, v_{bx},\, v_{by}]^T \tag{6}$$

where $\mathbf{v}_a = [v_{ax}, v_{ay}]^T$ and $\mathbf{v}_b = [v_{bx}, v_{by}]^T$ are the velocity vectors of point A and B, respectively. It is pointed out that this four-dimensional vector contains redundant information since the velocity vectors of two points on a rigid body are not independent. In fact, the velocity component of each of the points on the line connecting the two points has to be the same for the rigidity condition to be respected. Therefore, it is possible to reduce the number of kinematic parameters by choosing only three of the four velocity components. To this end, we will use a coordinate frame attached to the end effector, whose origin is located at point O' and which is oriented in such a way that the line passing through points A and B is along one of its axes. This coordinate frame will be noted $\mathcal{R}''$ (O'_{uv}). The orientation of frame $\mathcal{R}''$ with respect to $\mathcal{R}$ is given by angle ψ, as shown in Fig. 1. According to the constraint on the velocities of two points of a rigid body stated above, the component of the velocity of points A and B on axis u will be the same. Therefore, the independent velocity components will be

$$\dot{\mathbf{c}}' = [v_{au},\, v_{av},\, v_{bv}]^T \tag{7}$$

What needs to be established now is the relationship between these velocity components in the local frame $\mathcal{R}''$ and the velocity components given in eq.(1). This is written as

$$\dot{\mathbf{c}}' = \mathbf{R}\dot{\mathbf{c}} \tag{8}$$

Matrix $\mathbf{R}$ is obtained by writing the velocity vectors of points A and B, denoted as $\mathbf{v}_a$ and $\mathbf{v}_b$ in frame $\mathcal{R}''$, in the fixed frame [12]. We find:

$$\mathbf{R} = \begin{bmatrix} \cos\psi & \sin\psi & -y_a \\ -\sin\psi & \cos\psi & x_a \\ -\sin\psi & \cos\psi & x_b \end{bmatrix} \tag{9}$$

Now, if we premultiply each member of eq.(3) by $\mathbf{R}$, we obtain

$$\mathbf{K}\dot{\boldsymbol{\theta}} = \dot{\mathbf{c}}', \quad \text{with} \quad \mathbf{K} = \mathbf{RJ} \tag{10}$$

and the new dexterity index is defined as

$$\nu' = \kappa(\mathbf{K}) \tag{11}$$

An important property of matrix $\mathbf{K}$ is that all its entries bear the same physical units, namely the units of a length and hence ν' will be invariant under a scaling of the manipulator since the condition number of a matrix remains unchanged when the matrix is multiplied by an arbitrary nonzero real number. This is clearly seen in eq.(5). Moreover, we can readily verify that the premultiplication of matrix $\mathbf{J}$ by $\mathbf{R}$ does not introduce spurious singularities. Indeed, we can assume, without loss of generality, that points A and O' are coincident. The determinant of matrix $\mathbf{R}$ then becomes

$$\det(\mathbf{R}) = x_b \tag{12}$$

Hence, matrix $\mathbf{R}$ is nonsingular as long as x_b is different than zero, or, in other words, as long as points A and B are distinct.

Finally, it is noted that angles ϕ and ψ can be easily related because they both represent the orientation of the end effector with respect to the fixed frame.

3 Optimization of planar three-degree-of-freedom manipulators

A planar three-degree-of-freedom manipulator is shown in Fig. 2. This manipulator will now be optimized using the dexterity index presented in Section 2. First, we write the joint and Cartesian velocity vectors as in eqs.(1-2), where $\dot{x}$ and $\dot{y}$ are the velocity components in x and y of point O'. The Jacobian matrix is defined as in eq.(3) and is written as

$$\mathbf{J} = \begin{bmatrix} (-l_1 s_1 - l_2 s_{12} - l_3 s_{123}) & (-l_2 s_{12} - l_3 s_{123}) & -l_3 s_{123} \\ (l_1 c_1 + l_2 c_{12} + l_3 c_{123}) & (l_2 c_{12} + l_3 c_{123}) & l_3 c_{123} \\ 1 & 1 & 1 \end{bmatrix} \tag{13}$$

where s_{ij} stands for $\sin(\theta_i + \theta_j)$ and c_{ij} stands for $\cos(\theta_i + \theta_j)$. Hence, the condition number of the Jacobian matrix is written as

$$\nu = \kappa(\mathbf{J}) = ||\mathbf{J}||\,||\mathbf{J}^{-1}|| = \frac{\sqrt{N_1 N_2}}{3 l_1 l_2 s_2} \tag{14}$$

with:

$$N_1 = 3 + l_1^2 + 2l_2^2 + 3l_3^2 + 2l_1l_2c_2 + 2l_1l_3c_2 + 4l_2l_3c_3 \tag{15}$$

$$\begin{aligned} N_2 &= 2l_1^2 + 2l_2^2 + 2l_1l_2c_2 + l_1^2l_2^2s_2^2 + 2l_1^2l_3^2s_{23}^2 \\ &+ 2l_2^2l_3^2s_3^2 + 2l_1l_2l_3^2s_3s_{23} + 2l_1^2l_2l_3s_2s_{23} \end{aligned} \tag{16}$$

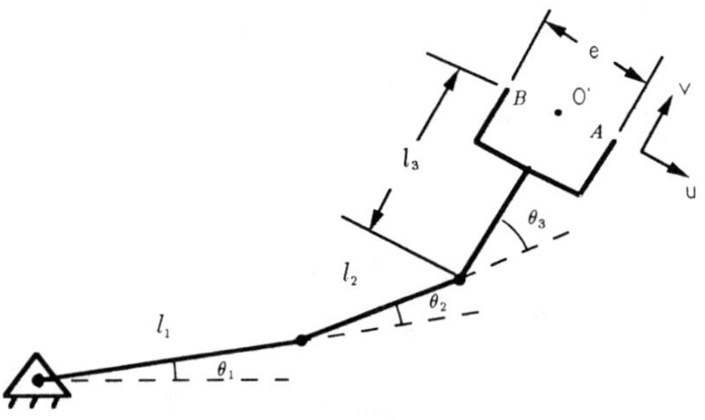

Fig.2 Planar 3-DOF serial manipulator.

It is clear, from eqs.(14–16) that the condition number of $\mathbf{J}$ is not invariant under a scaling of the manipulator.

A minimization procedure was used to find a design that would lead to the smallest possible condition number for $\mathbf{J}$. The results obtained are shown in Table 1 where the dimensions of a manipulator having a minimum condition number of 1.12 are given. However, if we apply a scaling factor to this manipulator the condition number of $\mathbf{J}$ will vary, as shown in Fig. 3, where the reciprocal of the condition number is plotted as a function of the scaling factor, all other variables remaining unchanged. Hence, the meaning of the isotropic design obtained here is not clear.

l_1	l_2	l_3	θ_2(rad)	θ_3(rad)	ν
2.15	2.37	1.23	2.14	2.91	1.117

Table 1: Dimensions of the manipulator leading to a minimum condition number of the Jacobian matrix.

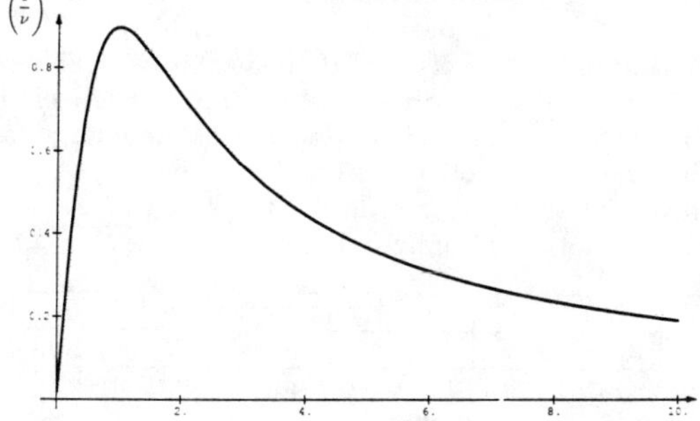

Fig. 3 Reciprocal of the condition number of matrix $\mathbf{J}$ as a function of the scaling factor η for the dimensions of table 1.

Now, in order to apply the dexterity index of Section 2, we write the new Cartesian velocity vector as

$$\dot{\mathbf{c}}' = [v_{au},\, v_{av},\, v_{bv}]^T \tag{17}$$

and we recall eq.(8). Referring to Fig. 2, the origin of coordinate frame $\mathcal{R}''$ is located in O' and the distance between point O' and either one of points A and B is equal to $e/2$. Hence, vectors $\mathbf{p}''_a$ and $\mathbf{p}''_b$ are given by

$$\mathbf{p}''_a = [e/2\,,\, 0]^T, \qquad \text{and} \qquad \mathbf{p}''_b = [-e/2\,,\, 0]^T \tag{18}$$

Moreover, from the same figure, the relationship between angles ϕ and ψ is written as

$$\phi = \psi + \pi/2 = \theta_1 + \theta_2 + \theta_3 \tag{19}$$

Therefore, matrix $\mathbf{R}$ is found to be equal to

$$\mathbf{R} = \begin{bmatrix} \cos\psi & \sin\psi & 0 \\ -\sin\psi & \cos\psi & e/2 \\ -\sin\psi & \cos\psi & -e/2 \end{bmatrix} = \begin{bmatrix} s_{123} & -c_{123} & 0 \\ c_{123} & s_{123} & e/2 \\ c_{123} & s_{123} & -e/2 \end{bmatrix} \tag{20}$$

The new kinematic relationship is written as eq.(10), from which we obtain an expression for the dexterity index as

$$\nu' = \kappa(\mathbf{K}) = \kappa(\mathbf{RJ}) = ||(\mathbf{RJ})||\,||(\mathbf{J}^{-1}\mathbf{R}^{-1})|| \tag{21}$$

Explicit expressions for the matrices involved have been obtained with the help of a symbolic manipulator. After simplifying the expressions, the dexterity index ν' is written as

$$\nu' = \frac{1}{3}\sqrt{V_1 V_2} \tag{22}$$

with

$$\begin{aligned} V_1 &= 3/2e^2 + l_1^2 + 2l_2^2 + 3l_3^2 + 4l_2l_3c_3 + 2l_1l_3c_{23} \\ &+ 2l_1l_2c_2 + 2l_2^2s_3^2 + 2l_1l_2s_3s_{23} + l_1^2s_{23}^2 \end{aligned} \tag{23}$$

$$\begin{aligned} V_2 &= \frac{1}{e^2l_1^2l_2^2s_2^2}(2l_1^2l_2^2s_2^2 + l_2^2e^2 + l_1l_2e^2c_2 + l_1^2e^2 + l_2^2e^2s_3^2 + 4l_2^2l_3^2s_3^2 \\ &+ 4l_1^2l_2l_3s_{23} + l_1l_2e^2s_3s_{23} + 4l_1l_2l_3^2s_3s_{23} + l_1^2e^2s_{23}^2 + 4l_1^2l_3^2s_{23}^2) \end{aligned} \tag{24}$$

Clearly, ν' is invariant under a scaling of the manipulator. Moreover, it is pointed out that neither ν nor ν' depend on θ_1 because this angle doesn't play any role in the kinematic accuracy.

Although the partial derivatives of ν' with respect to the kinematic and design variables are not simple, they have been obtained—with the help of the symbolic manipulator—and the minimization of ν' with respect to variables $l_1, l_2, l_3, e, \theta_2, \theta_3$ has been pursued. This allowed us to obtain an isotropic design of the manipulator, i.e., a design for which a set of configurations, located in the manipulator's workspace lead to a dexterity index of 1, which corresponds to the lowest value that can be attained. The dimensions of the isotropic design are given in Table 2 together with the joint angles giving the isotropic configurations.

l_1	l_2	l_3	e	θ_2(rad)	θ_3(rad)	ν
1.55	2.05	1.27	1.27	1.96	2.75	1.00

Table 2: Kinematic parameters of the isotropic design for the planar 3-DOF manipulator.

4 Example

Planar manipulators have several potential and existing applications[13]. Moreover, the development of articulated hands may lead to fingers of a planar architecture. An example of application of the dexterity index derived above is now illustrated:

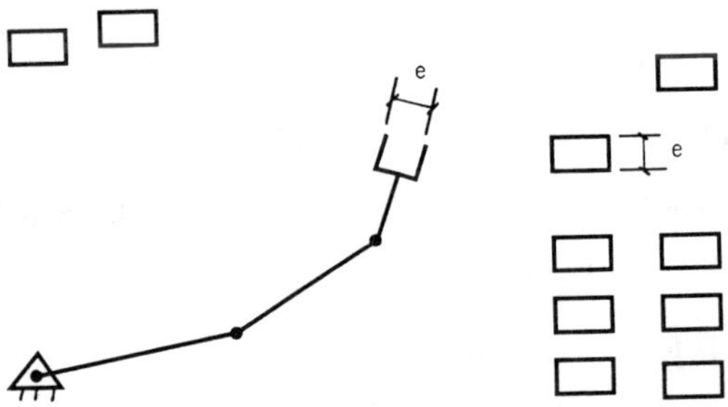

Fig. 4 Planar 3-DOF manipulator in a given workcell.

A planar three-degree-of-freedom manipulator is used to handle objects of a given size in a given environment as illustrated in Fig. 4. A workspace analysis has shown that, in order for the manipulator to be able to conveniently reach each portion of the working cell, the dimensions of the links should be chosen as

$$l_1 = l_2 = 1, \qquad 0 \leq l_3 \leq .7 \tag{25}$$

The problem is now to choose the value of l_3 that will lead to the best dexterity while satisfying the above constraints.

First, e is chosen to match the size of the objects to be handled. Hence, in this case, we will have

$$e = 0.2 \tag{26}$$

Then, the dexterity of the manipulator is rewritten as a function of l_3, θ_2, θ_3 only. We can plot the dexterity of the manipulator over its workspace for a few values of l_3 as shown in Fig. 5, where the reciprocal of ν' is used. Since we have a representation of the condition number over the workspace, we can either choose to optimize the local dexterity or the global dexterity, the latter being an average of the dexterity over the workspace [10].

The minimum local dexterity for a few values of l_3 are shown in Table 3. In this example, we obtain the optimum local dexterity with a value of l_3 of 0.5 whereas the best overall performance (global dexterity) is obtained with $l_3 = 0$.

l_3	$\nu'_{\min}$	θ_2^*(rad)	θ_3^*(rad)
0.0	2.94	2.82	3.15
0.2	2.02	2.80	2.84
0.5	1.72	2.65	2.89
0.7	1.91	1.89	3.68

Table 3: The minimum local dexterity for a few values of l_3.

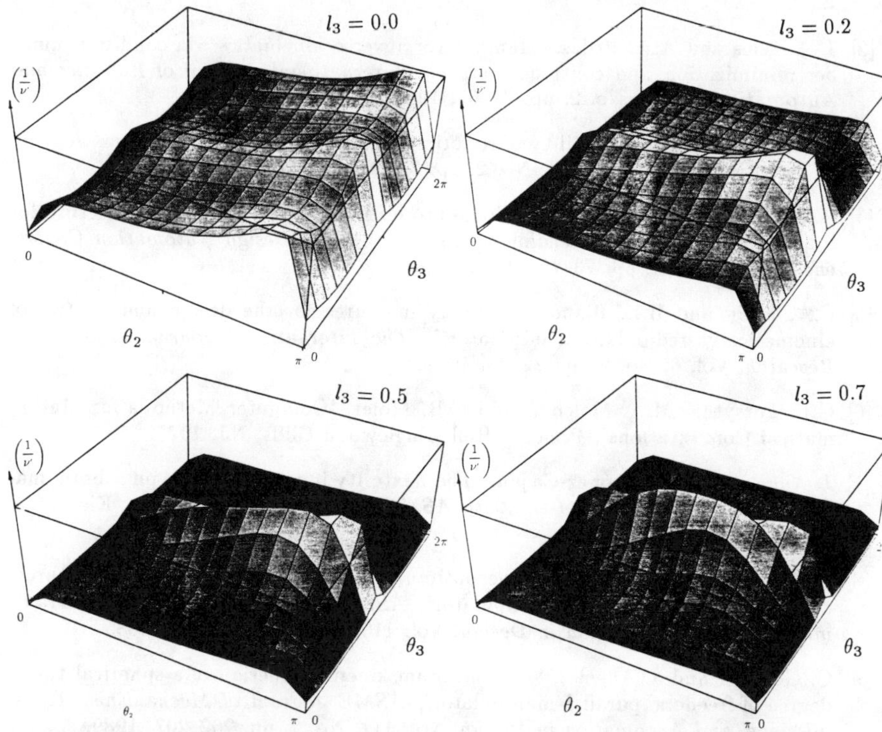

Fig. 5 Reciprocal of ν' as a function of θ_2 and θ_3 for different values of l_3.

5 Conclusion

An invariant dexterity index has been used to optimize a planar three-degree-of-freedom manipulator. As opposed to other existing dexterity indices, this one is invariant under a change of coordinate or a scaling of the manipulator even when both positioning and orientation are considered. This was illustrated in the example where the dexterity index based on the sole Jacobian was shown to vary with the scaling factor whereas the index used here was constant. An isotropic design was found for the planar manipulator based on the new dexterity index. Finally, an example of application to a constrained design problem was solved. The approach used here is also applicable to any other planar manipulator whose joints are all of the same type.

6 Acknowledgements

This work was completed under a research grant from the Faculté des Sciences et de Génie, Université Laval.

References

[1] J.K. Salisbury and J.J. Craig, 'Articulated hands: force control and kinematic issues', *The International Journal of Robotics Research*, Vol. 1, No. 1, pp. 4–17, 1982.

[2] J. Angeles and A.A. Rojas, 'Manipulator inverse kinematics via condition number minimization and continuation', *The International Journal of Robotics and Automation*, Vol. 2, No. 2, pp. 61–69, 1987.

[3] T. Yoshikawa, 'Manipulability of robotic mechanisms', *The International Journal of Robotics Research*, Vol. 4, No. 2, pp. 3–9, 1985.

[4] R. Stoughton and T. Kokkinis, 'Some properties of a new kinematic structure for robot manipulators', *Proceedings of the 13th ASME Design Automation Conference*, Boston, Ma., pp. 73–79, 1987.

[5] C.A. Klein and B.E. Blaho, 'Dexterity measures for the design and control of kinematically redundant manipulators', *The International Journal of Robotics Research*, Vol. 6, No. 2, pp. 72–83, 1987.

[6] G.E. Forsythe, M.A. Malcom, and C.B. Moler, 'Computer Methods for Mathematical Computations', Prentice-Hall, Englewood Cliffs, NJ, 1977.

[7] J. Angeles and C.S. López-Cajún, 'The dexterity index of serial-type robotic manipulators', *Proceedings of the 20th ASME Mechanisms Conference*, Kissimme, Florida, pp. 74–84, 1988.

[8] C. Gosselin and J. Angeles, 'The optimum kinematic design of a planar three-degree-of-freedom parallel manipulator', *ASME Journal of Mechanisms, Transmissions, and Automation in Design*, Vol. 110, No. 1, pp. 35–41, 1988.

[9] C. Gosselin and J. Angeles, 'The optimum kinematic design of a spherical three-degree-of-freedom parallel manipulator', *ASME Journal of Mechanisms, Transmissions, and Automation in Design*, Vol. 111, No. 2, pp. 202–207, 1989.

[10] C. Gosselin and J. Angeles, 'A new performance index for the kinematic optimization of robotic manipulators', *Proceedings of the 20th ASME Mechanisms Conference*, Kissimmee, Florida, pp. 441–447, 1988.

[11] T. Kokkinis and R. Stoughton, 'Optimal parallel actuation linkage for 3 DOF elbow manipulators', *Proceedings of the 20th ASME Mechanisms Conference*, Kissimmee, Florida, pp. 465–472, 1988.

[12] C. Gosselin, 'Dexterity indices for planar and spatial robotic manipulators', *Proceedings of the IEEE International Conference on Robotics and Automation*, 1990.

[13] T.W. Nye, D.J. LeBlanc and R.J. Cipra, 'Design and Modeling of a computer-controlled planar manipulator', *The International Journal of Robotics Research*, Vol. 6, No. 1, pp. 85–95, 1987.

Assembly modes and direct kinematics of parallel manipulators

J-P. Merlet
INRIA Centre de Sophia-Antipolis
2004 Route des Lucioles
06565 Valbonne Cedex , France

Abstract: Direct kinematics is known to be one of the most difficult problem during the study of parallel manipulators although a solution is needed for most of the practical applications of parallel manipulators.

We consider here a parallel manipulator with an hexagonal base, a triangular mobile plate and time-varying links lengths. We first show that the direct kinematics problem can be reduced to solve a polynomial in one variable which degree will give an upper-bound of maximum number of assembly mode (UBAM for brief) of the manipulator. We establish theoretically an upper-bound of the degree of this polynomial and then we calculate the polynomial. We present an example where the theoretical maximum number of assembly mode is reached.

1 Introduction

The direct kinematics of closed-loop mechanism is known to be one of the most challenging problem when one deal with this kind of mechanism. We consider here specific closed-loop mechanisms called "parallel manipulator" where a mobile plate is connected by articulated links to a fixed plate. Each link has one extremity linked to the base, the other being linked to the mobile plate. The motion of the mobile plate is ensured through one actuator in each link: for example an electric ram enables to change the link's length. The positions of the actuators are called the articular coordinates. It is in general easy to determine the articular coordinates from the position and orientation of the base (i.e. the inverse kinematics problem) but this yield to a set of non-linear equations.

Solving a system of non-linear equations is a difficult task and a numerical resolution is tedious. Furthermore we have no a-priori information about the uniqueness of the solution (except in the case of singular configurations [1]). Nanua and Waldron [2] have initiated a new approach to this problem. They reduce the resolution of the system of non-linear equations to the one of a polynomial in one variable. The number of assembly-modes of the manipulator (i.e. the number of way one can assemble the manipulator with fixed articular coordinates) is clearly related to the degree of this polynomial: it cannot be greater than this degree.

In the case of the manipulator called the TSSM [4], [5], [3] (Triangular Symmetric Simplified Manipulator, see figure 1) these authors show that the direct kinematics problem may be expressed as solution of a polynomial in one variable, which degree is 24. Charentus and Renaud [6] have studied the same manipulator, in the case where the mobile plate is an equilateral triangle. They have shown that the degree of the polynomial can be reduced to 16. Hunt [7] has proposed a conjecture for the same manipulator which states that the number of assembly-modes cannot be greater than 16.

2 The TSSM

2.1 Equivalent mechanism

The TSSM (figure 1) is a 6 d.o.f. parallel manipulator in which a mobile plate is connected to a fixed base through 6 articulated links, each link being connected both at the base and the mobile plate through ball-and-socket and universal joints. By controlling the links lengths we are able to control the position and orientation of the mobile plate [4], [8], [9], [10].

For fixed links lengths the articulation points B_1, B_3, B_5 of the mobile plate are able to describe circles centered in O_{12}, O_{34}, O_{56} whose radius are r_{12}, r_{34}, r_{56}. The characteristics of these circles can be determined using only the knowledge of the links lengths. Thus the TSSM is equivalent to a mechanism constituted of three links articulated on revolute joints and connected to the mobile plate (figure 1) for which the articular coordinates are the angles p_{12}, p_{34}, p_{56}. This mechanism is called the *equivalent mechanism* of the TSSM.

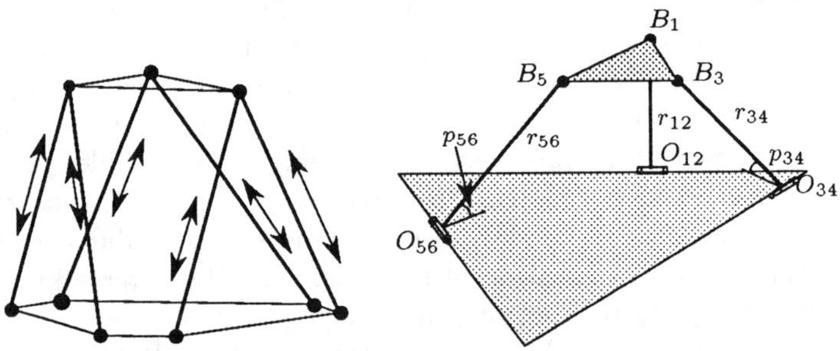

Figure 1: The TSSM and its equivalent mechanism

2.2 Minimal degree of the TSSM polynomial

Hunt [7] has conjectured that an UBAM of the TSSM is 16 by using the following method: if we dismantle one of the link of the equivalent mechanism of the TSSM we get a RSSR mechanism.

It is known [11] that a point of the coupler of this mechanism describes a sixteenth order surface, the *RSSR spin surface*. In order to find the possible configurations of mobile plate we have to intersect this surface with the circle described by the extremity of the dismantled link. A sixteenth order surface is intersected by a circle in no more than 32 points. Working on the conjecture that the RSSR spin-surface contains the imaginary spherical circle eight times Hunt deduces that at least 16 points are imaginary, and therefore there is at most 16 assembly-modes for the TSSM.

Thus to demonstrate this conjecture we have to determine the circularity of the RSSR spin-surface. In the following part we will use the notation defined in figure 2.

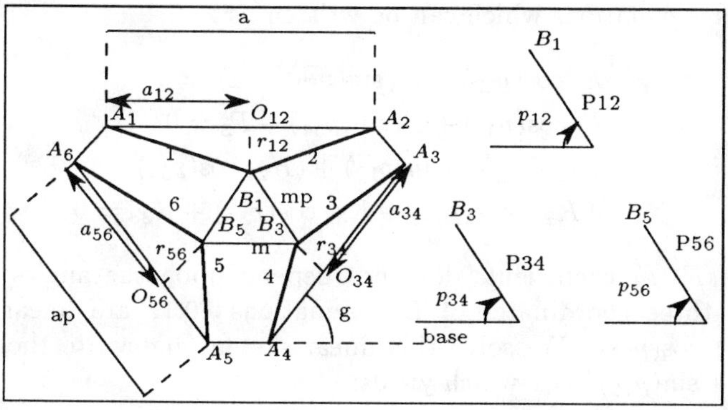

Figure 2: Notation (the TSSM is represented in top view)

Let X, Y, Z denote the coordinates of B, a point on the coupler of the RSSR. The coordinates of the articulation points B_1, B_3 can be expressed as a function of the unknown angles p_{12}, p_{34}. Thus we are able to write three equations relating the known distance between the articulation points to the unknowns p_{12}, p_{34}, X, Y, Z. We have:

$$a_{12} = \frac{\rho_1^2 + a^2 - \rho_2^2}{2a} \qquad a = 2xa_2 \tag{1}$$

$$xO_{12} = -xa_2 + a_{12} = \frac{\rho_1^2 - \rho_2^2}{4xa_2} \quad yO_{12} = ya_2 \quad r_{12}^2 = \rho_1^2 - a_{12}^2 \tag{2}$$

If n_{ij} denote the unit vector between O_{ij} and the corresponding articulation points we get :

$$n_{12} = -\cos(p_{12})\mathbf{j} + \sin(p_{12})\mathbf{k} \tag{3}$$

In the same way:

$$a_{34} = \frac{\rho_3^2 + ap^2 - \rho_4^2}{2ap} \tag{4}$$

$$xO_{34} = xa_3 - a_{34}\cos(g) \quad yO_{34} = ya_3 - a_{34}\sin(g) \quad r_{34}^2 = \rho_3^2 - a_{34}^2 \tag{5}$$

$$n_{34} = -\cos(p_{34})\sin(g)\mathbf{i} + \cos(p_{34})\cos(g)\mathbf{j} + \sin(p_{34})\mathbf{k} \tag{6}$$

We may write then

$$\mathbf{OB_1} = \mathbf{OO_{12}} + r_{12}\mathbf{n_{12}} \tag{7}$$

$$\mathbf{OB_3} = \mathbf{OO_{34}} + r_{34}\mathbf{n_{34}} \tag{8}$$

We are able to express the norm of the vectors $\mathbf{B_1B_3}$, $\mathbf{B_1B}$, $\mathbf{B_3B}$ i.e. the distances between the articulation points of the mobile whose values are mp and m, m. This yields to the following three equations :

$$||B_1B_3||^2 - mp^2 = 0 \quad ||B_1B||^2 - m^2 = 0 \quad ||B_3B||^2 - m^2 = 0 \tag{9}$$

We get three equations which can be written as :

$$E_1\cos(p_{12}) + E_2\sin(p_{12}) + E_3 = 0 \tag{10}$$
$$F_1\cos(p_{34}) + F_2\sin(p_{34}) + F_3 = 0 \tag{11}$$
$$K_{11}\sin(p_{34})\sin(p_{12}) + (K_{21}\cos(p_{34}) + K_{22})\cos(p_{12}) + K_{32}\cos(p_{34}) + K_{33} = 0 \tag{12}$$

where the E_i, F_j coefficients does not depend upon the angles but only upon the three coordinates of B. Equations 10,12 are linear in term of $\sin(p_{12}), \cos(p_{12})$. We solve this linear system and write the equation $\cos(p_{12})^2 + \sin(p_{12})^2 = 1$ which yields:

$$(N_1 - N_2)\cos(p_{34})^2 + N_3\sin(p_{34})\cos(p_{34}) + N_4\sin(p_{34}) + N_5\cos(p_{34}) + N_6 + N_2 = 0 \tag{13}$$

Then $\sin(p_{34})$ is determined using equation 11. If we put this value in equation 13 and write $\sin(p_{34})^2 + \cos(p_{34})^2 = 1$ we get two equations:

$$I_1\cos(p_{34})^2 + I_2\cos(p_{34}) + I_3 = 0 \tag{14}$$
$$H_1\cos(p_{34})^2 + H_2\cos(p_{34}) + H_3 = 0 \tag{15}$$

where the coefficients of I_i, H_j are function only of the coordinates of B. The orders of these coefficients are 4, 4, 4, 2, 3, 4. The equations 14, 15 yield to :

$$\begin{vmatrix} |I_1H_2| & |I_1H_3| \\ |I_1H_3| & |I_2H_3| \end{vmatrix} = 0 \tag{16}$$

where

$$|I_iH_j| = I_iH_j - I_jH_i \tag{17}$$

Using this method we get a sixteenth order polynomial. Its higher degree term is:

$$F_{21}^4(Y^2 + X^2 + Z^2)^8(N_{13} - N_{21})^2 \tag{18}$$

If $(N_{13} - N_{21})$ is equal to zero the mobile plate is reduced to a line. As for F_{21} it cannot be equal to zero. Therefore the circularity of the RSSR spin-surface is 8 and the conjecture of Hunt is verified. Thus there is at most 16 assembly-modes for the TSSM.

By expressing the coordinates of point B as a function of the angle p_{56} we are able to determine a sixteenth order polynomial for the direct kinematics problem of the TSSM. This polynomial has only even power (this mean that for a given configuration of the mobile the symetrical configuration with respect to the base is also a solution of the polynomial). A numerical resolution has shown that there may be effectively up to sixteen solutions to this polynomial : figure 3 shows one of the eight over-the-base assembly mode of a given TSSM.

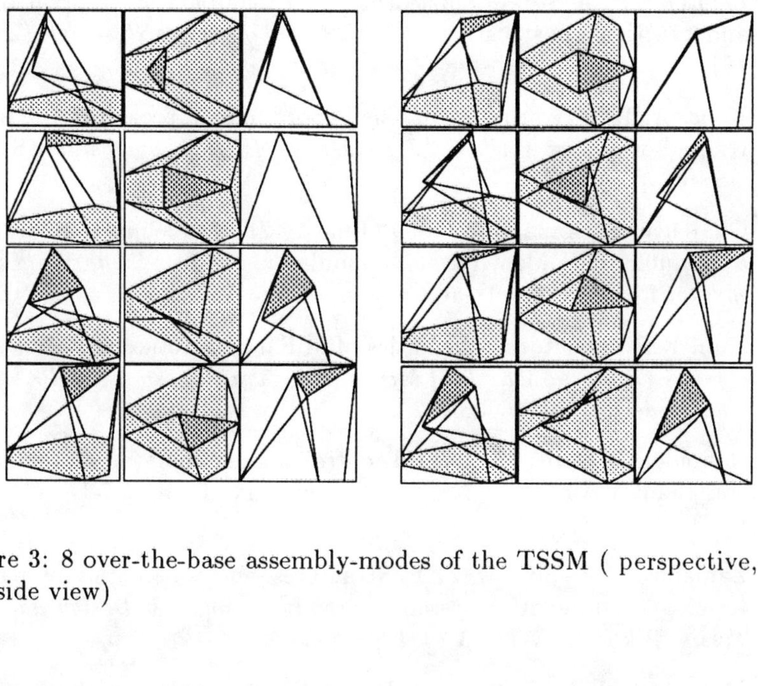

Figure 3: 8 over-the-base assembly-modes of the TSSM (perspective, top and side view)

3 Conclusion

We have shown that for the TSSM there will not be possible to find a closed-form for the solution of the direct kinematics problem. An algorithm has been proposed for finding all the assembly modes of a TSSM (although all may not be reachable without dismantling the manipulator) for a given set of links lengths. A generalization of this algorithm has been proposed in [12] for various kind of parallel manipulators.

References

[1] Merlet J-P. 1989 (October). Singular configurations of parallel manipulators and Grassmann geometry. The Int. J. of Robotics Research, Vol 8, n°5,pp. 45-56

[2] Nanua P., Waldron K.J. 1989 (May 14-19). Direct kinematic Solution of a Stewart Platform. *IEEE Int. Conf. on Robotics and Automation*, Scotsdale, Arizona, pp. 431-437.

[3] Gosselin. C. 1988. Kinematic analysis, optimization and programming of parallel robotic manipulators. Ph.D. thesis, McGill University, Montréal, Québec, Canada.

[4] Fichter E.F. A Stewart platform based manipulator: general theory and practical construction.*The Int. J. of Robotics Research*5(2):157-181.

[5] Koliskor, 1986 (April 22-25). The l-coordinate approach to the industrial robot design. *V IFAC/IFIP/IMACS/IFORS Symposium.*, Suzdal, USSR, pp. 108-115.

[6] Charentus S., Renaud M. 1989 (June 19-21). Modeling and control of a modular, redundant robot manipulator. *1st Int. Symp. on Experimental Robotics*, Montréal, Canada.

[7] Hunt K.H. Structural kinematics of in Parallel Actuated Robot Arms . *Trans. of the ASME, J. of Mechanisms,Transmissions, and Automation in design* (105): 705-712.

[8] Reboulet C., Robert A. Hybrid control of a manipulator with an active compliant wrist. *Proc 3th ISRR*, Gouvieux, France, 7-11 Oct.1985, pp.76-80.

[9] Zamanov V.B, Sotirov Z.M. Structures and kinematics of parallel topology manipulating systems. *Proc.Int. Symp. on Design and Synthesis*, Tokyo, July 11-13 1984, pp.453-458.

[10] Mohamed M.G., Duffy J. A Direct Determination of the Instantaneous Kinematics of Fully Parallel Robot Manipulators. *Trans. of the ASME, J. of Mechanisms, Transmissions, and Automation in design*, Vol 107:226-229.

[11] Hunt K.H. 1978. *Kinematic geometry of mechanisms.* Oxford: Clarendon Press.

[12] Merlet J-P. 1989 (December). Manipulateurs parallèles, 4eme partie: Mode d'assemblage et cinématique directe sous forme polynomiale. INRIA Research Report, n°1135.

DESIGN, MOBILITY ANALYSIS & ANIMATION OF A DOUBLE EQUILATERAL TETRAHEDRAL MECHANISM

ALEXANDER HYDER and P.J. ZSOMBOR-MURRAY
Robotic Mechanical Systems, McRCIM
McGill University, Mechanical Engineering
817 Sherbrooke St. o., Montréal, QC
Canada H3A 2K6 tel. (514)398-6311

ABSTRACT

Input-output kinematics of a novel, exoskeletal manipulator mechanism has been analysed. The abstract model is two equilateral tetrahedral wire frames which move relatively so that their edges remain in intersecting, parallel or coincident pairs. A mechanical, rigid body model was constructed and a computer animation was performed.

INTRODUCTION

The mobility of two equilateral tetrahedral wire frames, assembled so that the edges are constrained to form six intersecting pairs, was discovered in 1982 by Louis Tompos, a student of Ernö Rubik at the Hungarian Academy of Craft and Design. Tompos took the model to the Hungarian Institute for Building Science and showed it to Tibor Tarnai who together with E. Makai, a mathematician of the Hungarian Academy of Science, proceeded to demonstrate the nature of possible motions of the equilateral set[1,2] and, recently, of more general, similar sets[3,4,5]. Stachel, at the Institut für Geometrie, Technical University, Vienna, derived independently a geometric proof that a typical vertex of the moving tetrahedron traces, under rotations, a suface of parabolic-hyperboloidal form[6]. Apparently, Stachel and the Hungarians regarded the tetrahedral pair only as a curious toy and a challenging exercise in mobility analysis. They did not consider any serious physical implementation or possible application of such a mechanism.

In this work we take a mechanistic approach. (1) A perfectly equivalent mechanism, composed of rigid bodies and exhibiting much of the mobility of the theoretical wire frames was designed and built and will be described. (2) An analysis is carried out and a computer animation of the wire frame set is described. It should be noted that this algorithmic procedure would be equally well suited to commanding displacements of a powered mechanism model. (3) Certain limited motions are described in terms of the simplest possible geometric invariants and parameters so that direct and inverse kinematics may be presented. The motion input-output is shown to be easily inverted and readily differentiable with respect to time. (4) Some possible applications and avenues of further research are outlined. Considerable detail concerning (1) and (2), above, was presented by Hyder[7].

RIGID BODY MODEL MECHANISM

The mobile rigid body assembly of two tetrahedral frames shown in Fig. 1 is believed to be the only existing one. It was built to help visualize the motions, investigate design, fabrication and assembly problems and to plan the ultimate motorization and control of this mechanism.

In any physical implementation it is difficult to satify the requirement that the frames be exactly congruent but one be located an infinitessimal distance "inside" the other. The device in Fig. 1 overcomes this problem. Eight

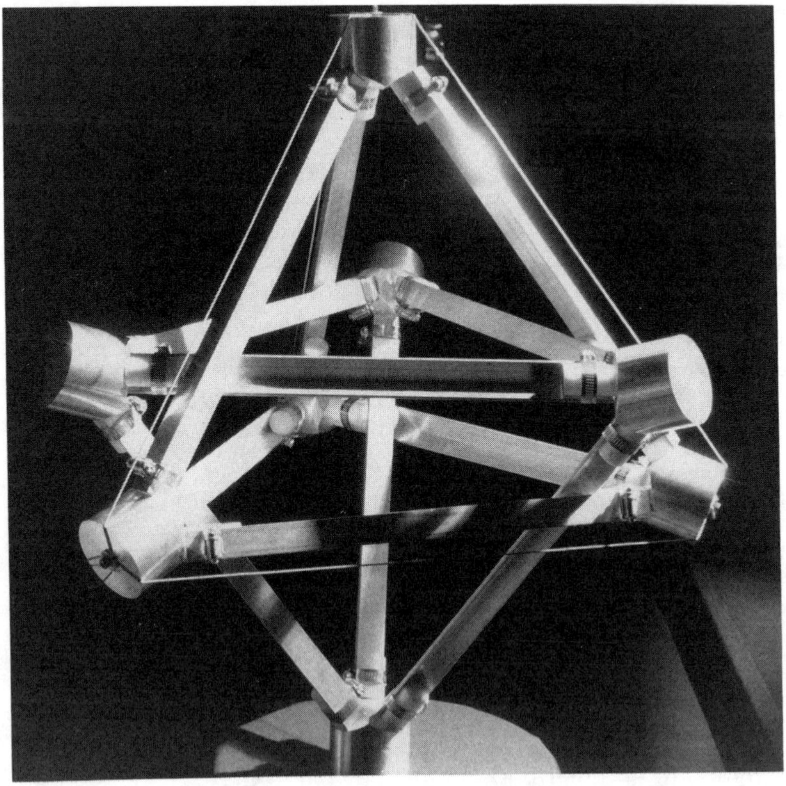

FIGURE 1
Double Tetrahedral Mechanism

aluminum vertex blocks can be seen there. The cylinder has three symmetrical flats milled on it such that the three normals make an angle $\beta = \boldsymbol{tan}^{-1}(1/\sqrt{2})$ with the cylinder axis. A hole is drilled normal to each flat such that the three hole axes intersect on the cylinder axis. The edges consist of $1x1x^{1}/_{8}$THK aluminum angle. Three are seen to converge on each vertex block. Also visible there, mounted at each end of every edge angle, in the *vee* concavity with a metallic tangent screw garter, is an edge axle pin. This 1ø plastic cylinder has a portion of its length reduced to $^{3}/_{8}$ø which protrudes into one of the vertex block holes. Thus, when the garters are tightened upon assembly, vertex blocks are held together with three adjacent edges by the axle pins in a "three fingered bowler's grip". Edges are free to rotate about vertex block hole/edge axle pin axes. These axes comprise the virtual edges of the abstract wire frame model. Finally, the sliding of four edge pair intersections is effected by arbitrarily assigning an inside/outside parity to the pair of assembled tetrahedra. The inside partner has its edge angles turned concavely outward while the outer tetrahedron's angles' *vee* faces in. Although not evident in Fig. 1, each edge pair intersection cusp embraces a 1ø plastic ball girdled by a plastic washer to keep the angle lips parallel to the plane of intersection of the virtual edge axes.

This is an easy to build design but it is by no means ideal. The "bowler's grip" is not positive; notice the guy wires wrapping the outer member's vertices to control axial play at each axle pin. Sliding, rather than rolling, at the plastic balls introduces undue friction; motion is stiff, not smooth. Furthermore the vertex block design precludes coincidence of vertex pairs; a condition required for some theoretically valid poses of this mechanism. Nevertheless it should be quite easy to adapt three tractive, *i.e.*, tensile, and three rotary actuators on each vertex block so as to motorize the sliding balls and to turn the edge angles. The preliminary model admirably serves to help visualize motion and as an exploratrory basis upon which to found better designs.

MOBILITY AND SINGULARITY

The description of mechanism mobility starts from a key symmetrical singular configuration called the "basic position"(BP)[1,2]. Here the vertices of the two frames occupy the eight corners of a cube to form a *stella octangula* or 8-pointed 3-D star. All edges are bisected at intersection with their pair mate as shown in Fig. 2. Each edge occupies a cube face diagonal. Mobility can be reduced to three types of rotation of one frame, called the moveable one, with respect to the other, arbitrarily fixed one. The rotations are classified according to axial direction as follows:-

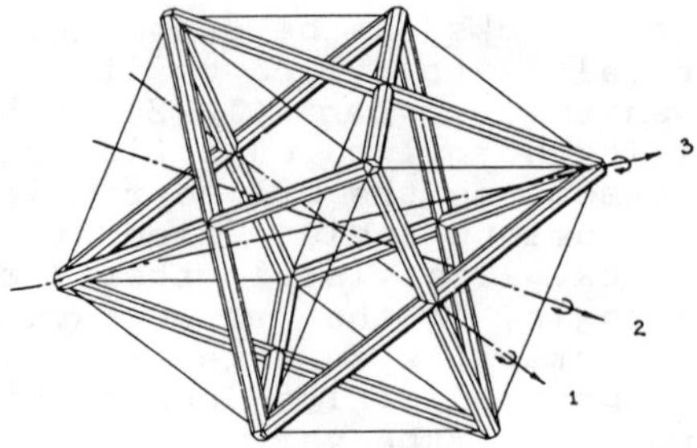

FIGURE 2
Basic Position/Conceptual Cube

1. Perpendicular to a face of the conceptual cube(CC); there are three such directions,
2. Parallel to an edge or face diagonal; there are six such directions and
3. Parallel to cubic body diagonal; there are four.

Rotation of type 1. may proceed continuously through ±90° until the assembly assumes one of six immediately adjacent singularities called "coincident positions" (CP)[1,2]. There are twelve such singularities and the other six are adjacent to the "counter-basic position"(CBP)[1,2]. This other *stella octangular* singularity occurs when a type 1. rotation continues, from (BP), through ±180°. During a type 1. rotation an opposite edge pair of each frame remains bisected though, obviously, not orthogonal. The frame centroids remain coincident.

A type 2. rotation may proceed continuously to a maximum angle $\phi = \tan^{-1}\sqrt{2} \approx 54.7°$, the angle between a tetrahedral edge and face. This results an a "terminal vertex singularity"(TV) wherein a single pair of vertices coincide. There are 24 of these, 12 immediately attainable via a rotation of $\pm\phi$ from (BP), the other 12 similarly adjacent to (CBP). During a type 2. rotation two edge pairs remain orthogonal and one edge on each frame remains bisected. Frame centroids do not remain coincident.

Type 3. rotation has a range of ±60°, ending in one of eight "terminal face singularities"(TF) adjacent to (BP). Another eight occur at 60° from (CBP). At (TF) three vertex pairs and a face pair become coincident and the assembly assumes an equilateral hexahedral form. The centroid of the moving frame translates in a direction parallel to the type 3. axis of rotation. At (TF) centroids are separated by $g=\sqrt{3}p/2$ where p is the edge length of (CC), not the tetrahedral edge length, $s=\sqrt{2}p$. During type 3. rotation, no orthogonal or bisected edge pairs survive.

The three rotation types are motions of one degree of freedom, *i.e.*, can be sustained by only one of the 24 (12 linear/tractive ball and 12 rotary edge angle) possible actuation modes. Although, in principle, one actuator can continue the frame rotation from (BP) or (CBP) to any immediately adjacent (CP), (TV) or (TF), a minimum of three actuators are required to initiate an excursion from or through a basic singularity and along a deliberately selected branch of motion.

GEOMETRIC SOLUTION: TRIANGULAR "RING" ON TRIPOD MODEL

Consider the tripod of three edges converging at a vertex of the fixed frame and the triangular face or "ring" of the moveable frame whose edges intersect the tripod legs as shown in Fig. 3. Distances along the tripod legs to the points intersection with ring edges are seen to be a, b and c. p, q and r are the sides of triangle formed by the three points of intersection. p lies between legs, length a and b, while q spans legs of length b and c. Normalizing (CC) by setting p=1 (note that p is not p) it may be seen that

$p=\sqrt{[(1-a)^2+(1-b)^2-(1-a)(1-b)]}$, now defining

$k_1=\sqrt{3}/2$, $k_2=-q\sin(\pi/3+\phi_r)$, $k_3=q\cos(\pi/3+\phi_r)-p$ where

$\phi_r=\cos^{-1}[(p^2+q^2-r^2)/2pq]$ we obtain

$$\cos\theta_q=\{-2k_1k_2\pm\sqrt{[4k_1^2k_2^2-4(k_2^2+k_3^2)(k_1^2-k_3^2)]}\}/[2(k_2^2+k_3^2)]$$

and similarly, the angles ϕ_p, ϕ_q, θ_p and θ_r. This is adequate to compute all positions for rotations of the three types just described.

Input parameters a, b and c, for the three rotative types of motion, are constrained to be

1.	$a=1/2$	$b=1-c$	$0<c<1$
2.	$\sqrt{6}-2<a<1$	$b=a$	$c=[1-a+\sqrt{(a^2+4a-2)}]/2$
3.	$0<a<1/2$	$b=a$	$c=a$

These constitute the limited motions refered to as (3) in INTRODUCTION, above. Note that the relationships among θ, ϕ, a, b and c are invertable and differentiable.

Motion with Two Degrees of Freedom

The nature of rotational motion with two degrees of freedom can be understood with the help of Fig. 4 which shows at the centre the (CC) representing the *stella octangula* of a (BP) or (CBP). Notice the three directions of type 1. rotative axes are parallel to the three cartesian axes while six directions parallel to type 2. axes appear as

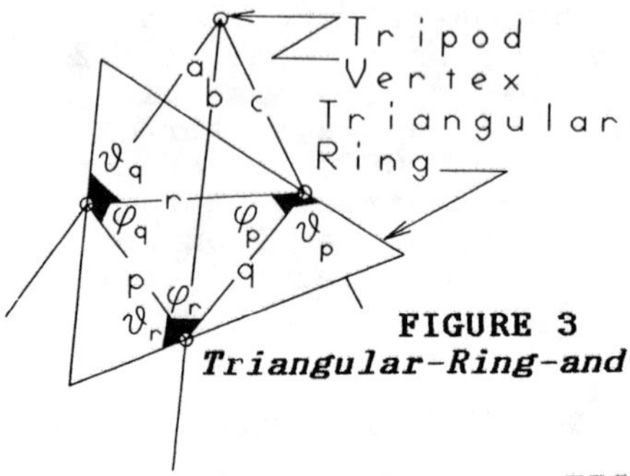

FIGURE 3
Triangular-Ring-and-Tripod

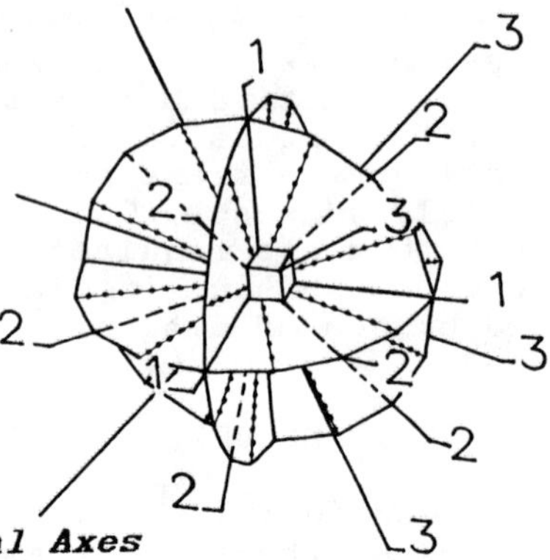

FIGURE 4
Planes of Rotational Axes

angular bisectors, the broken lines, to the cartesian axes. Four type 3. axes are parallel to the four lines emerging from opposite corners of the (CC). Each line is along a (CC) body diagonal. Rotation of the moveable frame may take place, however, about any axial direction parallel to one of the principal cartesian planes, shown as circular segments in Fig. 4. These axes are directed to lie between type 1. and type 2. axial directions. In fact, this so-called *intermediate* type of motion, type *I*., is characterized by precession of the axis of rotation. Axes in directions midway between types 1. and 2. direction are shown in Fig. 4 as beaded lines. But, of course, along any typical type *I*. motion trajectory, the axis of rotation changes direction continuously.

Animation

A type *I*. rotation may be animated by successive computations of compatible values of {*abc*} under rotation of prescribed magnitude which takes place about a prescribed type *I*. axial direction. Γ is the rotation axis angle measured from a type 1. axial direction. Ω is the desired anle of rotation about the axis inclined at Γ. $\Gamma=0$ for type 1. rotations and $\Gamma=\pi/4$ for type 2. motions. A computational solution begins by considering a typical tripod of AB, AC, AD fanning from vertex A of the moveable frame and intersecting the ring of edges FG, GE, EF of the fixed frame, respectively. With the frames in a (BP), define two cartesian coordinate systems, one attached to ABCD, the other to EFGH. A rotation of (Γ,Ω) is imposed on ABCD and the tripod leg vectors

$$[A]=\begin{bmatrix} AB_x AC_x AD_x \\ AB_y AC_y AD_y \\ AB_z AC_z AD_z \end{bmatrix} \quad \text{become} \quad [A']=\begin{bmatrix} AB_1' AC_1' AD_1' \\ AB_2' AC_2' AD_2' \\ AB_3' AC_3' AD_3' \end{bmatrix} \quad \text{by means of } [R],$$

the rotational transformation in $[A']=[R][A]$ where

$$[R]=\begin{bmatrix} l_1 m_1 n_1 \\ l_2 m_2 n_2 \\ l_3 m_3 n_3 \end{bmatrix}=\{\mathbf{lmn}\},$$ the direction cosines of the rotated axes with respect to the fixed ones.

$$\mathbf{l}=\begin{bmatrix} n_3\cos^2\Gamma+\sin^2\Gamma \\ (1-n_3)\sin\cos\Gamma \\ \cos\Gamma\sin\Omega \end{bmatrix},\quad \mathbf{m}=\begin{bmatrix} l_2 \\ n_3\sin^2\Gamma+\cos^2\Gamma \\ -\sin\Gamma\cos\Omega \end{bmatrix},\quad \mathbf{n}=\begin{bmatrix} -l_3 \\ -m_3 \\ \cos\Omega \end{bmatrix}$$

A set of three mutually perpendicular unit vectors oriented to the typical fixed triangular ring EFG is

$$[N]=\begin{bmatrix} \mathbf{N}_1 \\ \mathbf{N}_2 \\ \mathbf{N}_3 \end{bmatrix}=\begin{bmatrix} -1/\surd 6 & 1/\surd 6 & -\surd(2/3) \\ -1/\surd 2 & -1/\surd 2 & 0 \\ -1/\surd 3 & 1/\surd 3 & 1/\surd 3 \end{bmatrix}$$

This defines a coordinate system, natural to the ring, in which the projection of [A′] is given by

$$[A'']=[\mathbf{AB}'\mathbf{AC}'\mathbf{AD}']\begin{bmatrix} \mathbf{N}_1 \\ \mathbf{N}_2 \\ \mathbf{N}_3 \end{bmatrix}=\begin{bmatrix} AB_1''AC_1''AD_1'' \\ AB_2''AC_2''AD_2'' \\ AB_3''AC_3''AD_3'' \end{bmatrix}$$

Now a plane parallel to EFG and at a distance h is defined where $h=\mathbf{min}(AB_3'',AC_3'',AD_3'')/2$. The portions of AB′, AC′ and AD′, external to this plane, are, respectively,

$$a^*=h/AB_3'',\quad b^*=h/AC_3'',\quad c^*=h/AD_3'' \text{ and } a^*/a=b^*/b=c^*/c$$

Spanning a^*b^* is p^*, b^*c^* is q^* and c^*a^* is r^*, in the same proportions as abc and pqr in Fig. 3. The vectors of these two scalene triangle sides, expressed in the planar cartesian system of the ring EFG are

$$\mathbf{p}^*=\begin{bmatrix} p_1^* \\ p_2^* \end{bmatrix}=\begin{bmatrix} c^*AD_2''-b^*AC_2'' \\ b^*AC_1''-c^*AD_1'' \end{bmatrix},\quad \mathbf{q}^*=\begin{bmatrix} q_1^* \\ q_2^* \end{bmatrix}=\begin{bmatrix} a^*AB_2''-c^*AD_2'' \\ a^*AB_1''-c^*AD_1'' \end{bmatrix}$$

and the magnitudes are $p^*=\surd(p_1^{*2}+p_2^{*2})$ and $q^*=\surd(q_1^{*2}+q_2^{*2})$

Circumscribing the appropriate equilateral triangle about the scalene one with sides pqr requires the angles

$$\theta_r=\mathbf{cos}^{-1}(p_1^*/p^*) \text{ and } \pi-(\theta_r+\phi_r)=\mathbf{cos}^{-1}(q_1^*/q^*)$$

Sides of the inappropriate equilateral triangle, from which protrudes the tripod $\{a^*b^*c^*\}$, are of length, s^*, where

$$s^*=(p^*/(\surd 3/2))\,\mathbf{sin}(2\pi/3-\theta_r)+(q^*/(\surd 3/2))\,\mathbf{sin}(\theta_r+\phi_r-\pi/3)$$

Then any valid solution is computed as $\{abc\}=(1/s^*)\{a^*b^*c^*\}$

CONCLUSIONS

Any attempt at design improvement must address the

problem of devising a mechanism which can assume (BP) and (CBP). Aside from the obvious problem to contrive virtual vetices which may coincide there is also the question of how to resolve the ambiguity of intersection between a pair of coincident, parallel lines. The practical consideration involves the redistribution of the plastic ball-and-washer joints, *i.e.*, how to have two emerge from a vertex where only one is available and the complementary problem of disposing of an extra one from some vertices when the mechanism moves away from (BP). Another interesting proposition to enhance the mechanism is the addition of two more degrees of freedom with a third frame.

As regards applications, the suitability of this device as a wrist or as a universal joint may be investigated. A possible, revolutionary "vehicle", based on this model, might also be considered. A (BC) is vertex stable, *i.e.*, in this position the assembly rests on any of three adjacent vertices. Its mass centre is symmetrical; monopolar. At the end of a type 3. rotation it assumes an octahedral (TF) configuration which is absolutely convex, hence face stable. The mass centres of the two frames are at extreme displacement, a dipolar distribution. It would appear that the assembly might be made to tumble in prescribed direction over irregular terrain in an alternating sequence of face and vertex support stances.

Acknowlegement

This research is supported by Natural Sciences and Engineering Research Canada grant #A4219.

References

1. Tarnai, T. & Makai, E., "Physically Inadmissable Motions of a Pair of Tetrahedra", Proc. 3rd Int. Conf. Eng. Graph. & Desc. Geom. 2, 264-271, edited by Stachel & Slaby (Technical Univ., Vienna July 1988).
2. ________________, "A Moveable Pair of Tetrahedra", Proc. R. Soc. Lon. A **423**, 423-442 (June 9, 1989).
3. ________________, "Kinematical Indeterminacy of a Pair of Tetrahedral Frames", *in preparation*, (1989).
4. ________________, "Moveability of a Pair of Tetrahedral Frames", *in preparation*, (1989).
5. Makai, E. & Tarnai, T., "Generalized Forms of an Overconstrained Sliding Mechanism Consisting of Two Equal Tetrahedra", *in preparation*, (1989).
6. Stachel, H., "Ein Bewegliches Tetraederpaar", Elemente der Mathematik 43(3), 65-75 (1988).
7. Hyder, A. & Zsombor-Murray, P.J., "Design, Mobility Analysis and Animation of a Double Equilateral Tetrahedral Mechanism", CIM-89-15, 57pp. (McRCIM Int. Rpt. McGill Univ., Montréal, H3A 2K6, September 15, 1989).

AN ALTERNATIVE APPROACH FOR MODELLING A CLASS OF HYDRAULICALLY-POWERED MANIPULATORS

N. Sepehri and F. Sassani
Department of Mechanical Engineering
P.D. Lawrence
Department of Electrical Engineering
The University of British Columbia
Vancouver, B.C., Canada

ABSTRACT

Modelling and simulation of a class of hydraulic systems used in heavy-duty manipulators is investigated. Different techniques are explored through rearrangement of hydraulic equations in terms of different state-space variables.

Two techniques are specifically studied. The first one uses fluid pressure as state-space along with steady-state equations as a solution to stiff equations. The second approach combines the steady-state equations with the hydraulic lines' differential equations and fluid flow is chosen as a new state-space variable. Both techniques are exemplified with a two-link hydraulic manipultor in a computer simulation program.

1 INTRODUCTION

Hydraulic power has proven to be reliable and flexible in design for a variety of mechanical manipulations, including heavy-duty tasks. However, they are nonlinear and could become complex and difficult to analyze. An example is the hydraulic system for machines used in primary industries, such as excavators and forest harvesters. This paper addresses the problem involved in modelling and simulation of such hydraulic systems. The emphasis is on the development of a methodology and discovering the effect of changing the hydraulic state-space variables on the computational efficiency. Two modelling techniques are discussed. It is shown that the way a hydraulic system is modelled has a direct effect on the resulting efficiency.

A hydraulic system can be considered as a large-scale system consisting of standard components such as valves, lines and pumps with known models [1]. A systematic approach for analyzing large-scale systems was first introduced through Bond-Graph Theory. It was then applied to model [2] and to simulate [3] hydraulic systems. A similar approach to [3], in concept, was further developed [4]. The development of a general purpose hydraulic simulation package has also been investigated [5].

These studies indicate that fluid power systems are mathematically stiff. Computer techniques for analyzing stiff systems have already been reported [6]. The origin of the stiffness in hydraulic systems was also studied. and special solutions such as Pseudo-Analytic Techniques [7] were applied to certain configurations.

In this paper a basic hydraulic circuit used in heavy-duty manipulation is studied. The initial mathematical model is found to be stiff. Fast response states are located and the corresponding equations are redefined to form a set of steady-state ones. Although numerical iteration is required in the arrangement, it proves to be more efficient.

The possibility of rearranging the hydraulic equations in terms of different state-spaces is then considered. Steady-state equations are combined with the hydraulic lines' differential equations; fluid-flow rate is chosen as a state-space variable rather than the conventional fluid pressure. The equations are rearranged and new differential equations are derived. This eliminates the iteration process. The number of state-space variables remains the same as well as the integration time-interval. The new arrangement, in spite of its merits, does not contain the strength of the first approach in the presence of certain discontinuities. Special solutions should be provided to meet the requirement for such cases.

This study is exemplified with a single and a two-link manipulator in a computer simulation.

2 FORMULATION

2.1 Single Actuation

Figure 1 shows a typical single actuated system. A constant flow, Q, is distributed between the actuator and the tank. Pressures across the lines as well as flows are shown in the diagram. Orifices a_i, a_o and a_e which control the flow distribution, belong to an open-center valve as shown in Figure 2.

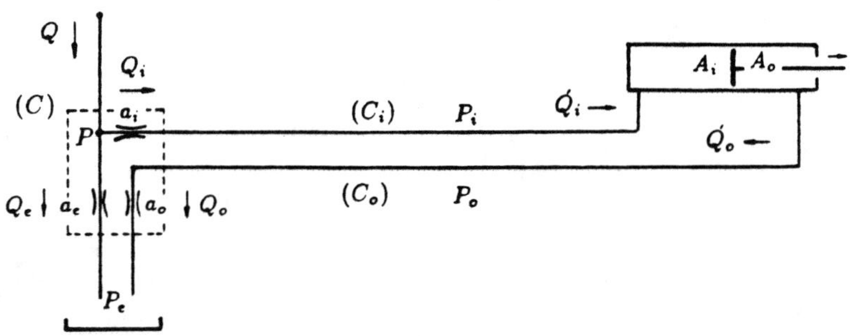

Fig. 1- A single link actuating system.

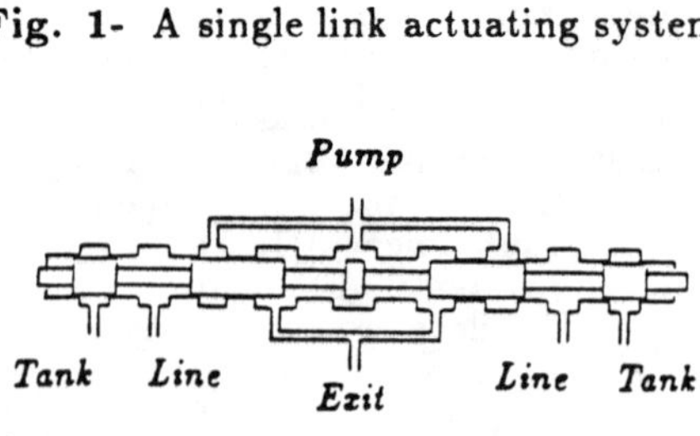

Fig. 2- An open-center valve used in heavy-duty machines.

Equations describing the hydraulic dynamics are written as follows:
(*To the actuator*)

$$Q_i = Ka_i\sqrt{P - P_i} \tag{1a}$$
$$Q_e = Ka_e\sqrt{P - P_e} \tag{1b}$$
$$\frac{dP}{dt} = \frac{1}{C}(Q - Q_i - Q_e) \tag{1c}$$
$$\frac{dP_i}{dt} = \frac{1}{C_i}(Q_i - \acute{Q}_i) \tag{1d}$$

(*Out of actuator*)

$$Q_o = Ka_o\sqrt{Po - P_e} \tag{2a}$$
$$\frac{dP_o}{dt} = \frac{1}{C_o}(\acute{Q}_o - Q_o) \tag{2b}$$

where K is the discharge coefficient. $\acute{Q}_i$ and $\acute{Q}_o$ are related to the actuator's velocity. The compliance of the pump pipe, C, is small compared with C_i and C_o which belong to hoses connecting the valve to the actuator. Equation (1*c*) is thus a stiff equation which determines the integration time-interval in the simulation. It can be replaced with the following

$$C\frac{dP}{dt} \approx \emptyset = (Q - Q_i - Q_e) \tag{1\acute{c}}$$

The smallest time constant has, in this way, been eliminated which allows a larger step size [6]. Also, the number of states to be integrated has been reduced. However, the way these equations are set up requires some iterations.

Equations (1) can be rearranged in terms of different states in a similar manner as in [7]. Equations (1*a*), (1*b*) and (1*ć*) are combined and rewritten as follows:

$$\left(\frac{Q_i}{Ka_i}\right)^2 = P - P_i \tag{3a}$$
$$\left(\frac{Q - Q_i}{Ka_e}\right)^2 = P - P_e \tag{3b}$$

when differentiated with respect to time,

$$\left(\frac{2Q_i}{(Ka_i)^2}\right)\frac{dQ_i}{dt} = \frac{dP}{dt} - \frac{dP_i}{dt} \tag{4a}$$
$$-\left(\frac{2(Q - Q_i)}{(Ka_e)^2}\right)\frac{dQ_i}{dt} = \frac{dP}{dt} \tag{4b}$$

It is noted that the pump flow, Q, and the tank pressure, P_e, are constant. Substituting from (4*b*) and (1*d*) into (4*a*) a new differential equation in terms of Q_i is derived:

$$\frac{dQ_i}{dt} = \frac{(\acute{Q}_i - Q_i)}{2C_i\left(\frac{Q_i}{(Ka_i)^2} + \frac{Q - Q_i}{(Ka_e)^2}\right)} \tag{5}$$

which is used to update Q_i during the simulation whose value can be used to determine other flow and pressure states recursively as follows:

$$Q_e = Q - Q_i \tag{6a}$$

$$P = \left(\frac{Q_e}{K a_e}\right)^2 + P_e \tag{6b}$$

$$P_i = P - \left(\frac{Q_i}{K a_i}\right)^2 \tag{6c}$$

Equations (5) and (6) replace equations (1) and (2). Iteration is no longer needed. The number of differential equations is the same.

2.2 Multi Actuation

Figure 3 shows a hydraulic actuation system with priority action. Two open-center valves are connected in series, a case similar to a *Caterpillar 215B* excavator's hydraulic system. In this configuration, one actuator has priority over the other one. The connection between the two valves has a very low compliance. Flexible hoses delivers fluid to the actuators. Referring to Figure 3, the following equations are applied for the power side:

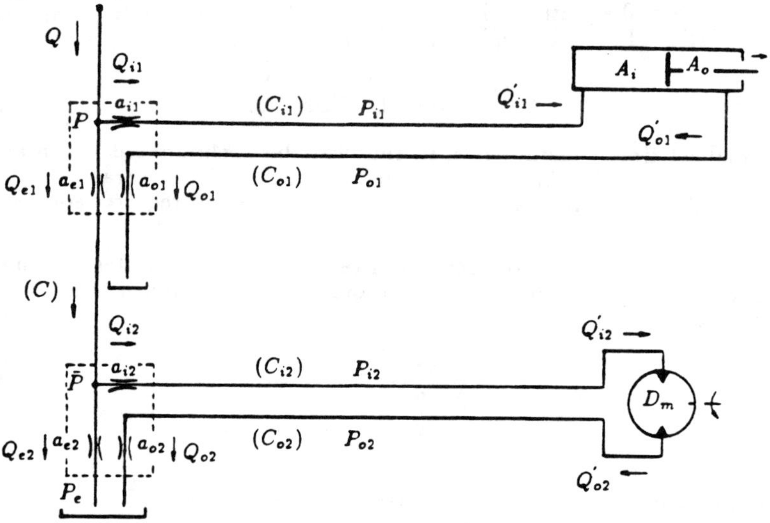

Fig. 3- A two-link actuating system with priority action.

$$Q = Q_{i1} + Q_{e1} \tag{7a}$$

$$Q_{i1} = K a_{i1} \sqrt{P - P_{i1}} \tag{7b}$$

$$Q_{e1} = K a_{e1} \sqrt{P - \bar{P}} \tag{7c}$$

$$Q_{e1} = Q_{i2} + Q_{e2} \tag{8a}$$

$$Q_{i2} = K a_{i2} \sqrt{\bar{P} - P_{i2}} \tag{8b}$$

$$Q_{e2} = K a_{e2} \sqrt{\bar{P} - P_e} \tag{8c}$$

$$\frac{dP_{i1}}{dt} = \frac{1}{C_{i1}}(Q_{i1} - \acute{Q}_{i1}) \tag{9a}$$

$$\frac{dP_{i2}}{dt} = \frac{1}{C_{i2}}(Q_{i2} - \acute{Q}_{i2}) \tag{9b}$$

Equations (7) and (8) are to be solved iteratively.

As an alternative solution to (9), fluid flows Q_{i1} and Q_{i2} are taken as state variables. Following the same approach as before, equations (7) and (8) are differentiated with respect to time. After some manipulation and substitutions, the following equation are derived:

$$\alpha_1 \frac{dQ_{i1}}{dt} + \beta_1 \frac{dQ_{i2}}{dt} = \gamma_1 \tag{10a}$$

$$\alpha_2 \frac{dQ_{i1}}{dt} + \beta_2 \frac{dQ_{i2}}{dt} = \gamma_2 \tag{10b}$$

where

$$\alpha_1 = \frac{2Q_{i1}}{(Ka_{i1})^2} + \frac{2(Q - Q_{i1})}{(Ka_{e1})^2} + \frac{2(Q - Q_{i1} - Q_{i2})}{(Ka_{e2})^2}$$

$$\alpha_2 = \frac{2(Q - Q_{i1} - Q_{i2})}{(Ka_{e2})^2}$$

$$\beta_1 = \frac{2(Q - Q_{i1} - Q_{i2})}{(Ka_{e2})^2}$$

$$\beta_2 = \frac{2Q_{i2}}{(Ka_{i2})^2} + \frac{2(Q - Q_{i1} - Q_{i2})}{(Ka_{e2})^2}$$

$$\gamma_1 = -\left(\frac{Q_{i1} - \acute{Q}_{i1}}{C_{i1}}\right)$$

$$\gamma_2 = -\left(\frac{Q_{i2} - \acute{Q}_{i2}}{C_{i2}}\right)$$

which can be easily solved for $\frac{dQ_{i1}}{dt}$ and $\frac{dQ_{i2}}{dt}$. Other states are then calculated recursively as follows:

$$Q_{e2} = Q - Q_{i1} - Q_{i2} \tag{11a}$$

$$Q_{e1} = Q - Q_{i1} \tag{11b}$$

$$\bar{P} = \left(\frac{Q_{e2}}{Ka_{e2}}\right)^2 + Pe \tag{11c}$$

$$P = \left(\frac{Q_{e1}}{Ka_{e1}}\right)^2 + \bar{P} \tag{11d}$$

$$P_{i2} = \bar{P} - \left(\frac{Q_{i2}}{Ka_{i2}}\right)^2 \tag{11e}$$

$$P_{i1} = P - \left(\frac{Q_{i1}}{Ka_{i1}}\right)^2 \tag{11f}$$

3 SIMULATION RESULTS

Figure 4 shows the simulation result of a single link motion due to a step in spool displacement. Both modelling techniques were used and the results were compared. The forth order Runge Kutta integration routine with the same integration time interval was used for both models. Figure 5 shows a similar result for a two link manipulator. Table 1 compares the computation time required for both models to simulate 4*sec* of motion.

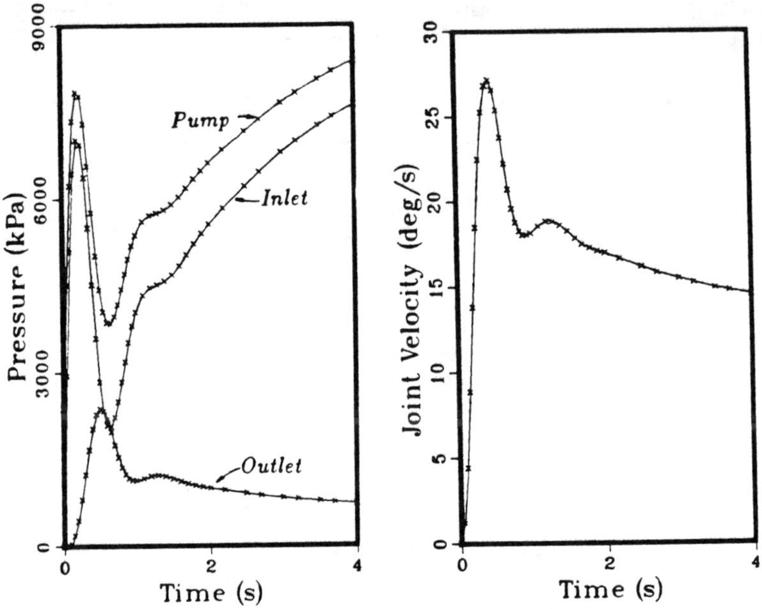

Fig. 4- Response to step input (single link)
— *original method*, × *alternative solution*

4 CONCLUDING REMARKS

Using fluid-flow as an alternative state-space variable can reduce the computing time; however, compared with the conventional method which considers pressure as the cause and flow as the effect (result), great care "should" be taken during the programming when the alternative state-space variable is used. Equations describing pressure as a function of flow are inversely related to the orifices areas. The equations fail to characterize the pressure properly when the orifices areas are approaching zero. However, this is not a problem when flow is calculated based on the pressure.

The differential equations describing the fluid flow pattern change for different orifice arrangement, or, when some orifices are closed (cases when the valve is fully open, $a_e = \emptyset$, or when it passes through its dead-band, $a_i and a_o = \emptyset$). It is thus necessary to redefine the differential equations during the execution. The inclusion of nonlinearities in the form of discontinuities such as a check valve or relief valve action, which is trivial for the conventional method, is much more difficult when the new technique is applied.

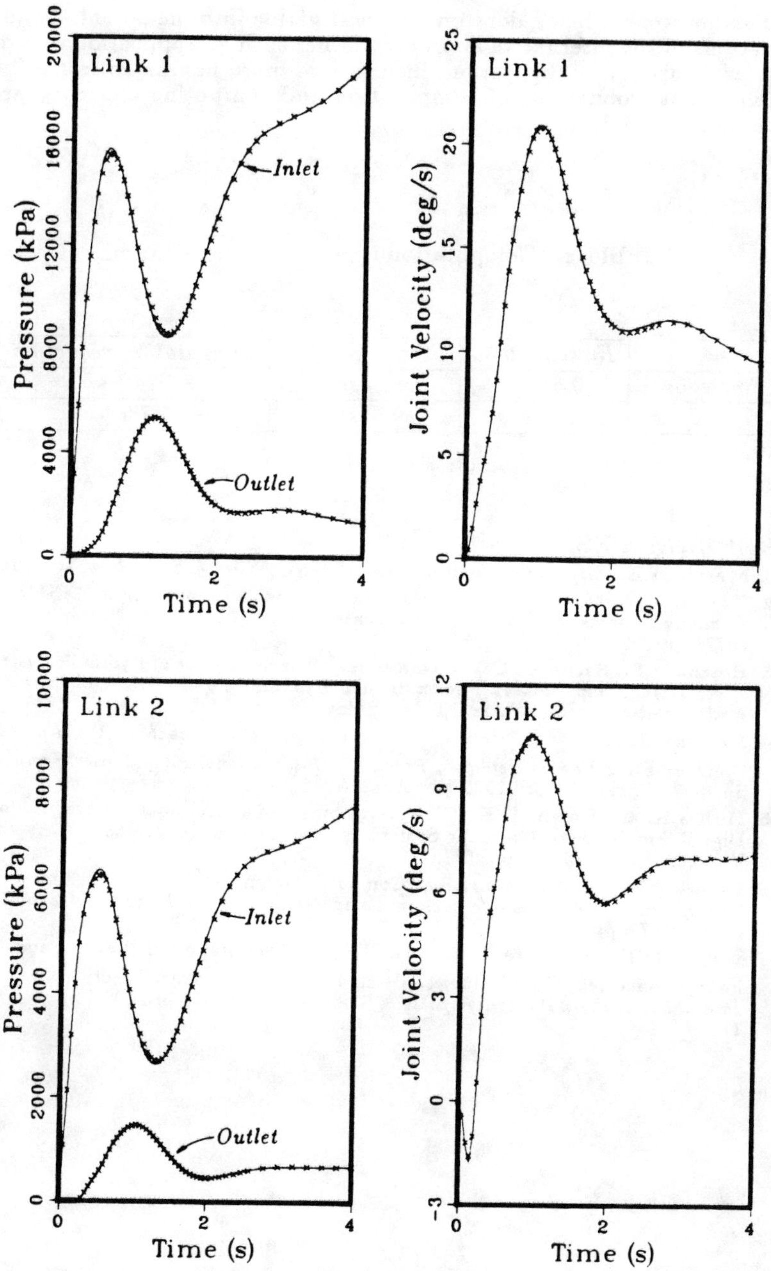

Fig. 5- Response to step input (two link)
— *original method*, × *alternative solution*

Further work is being done on an investigation into the potential and the limitations of the alternative approach. Issues such as application to a more complex hydraulic configuration, inclusion of more nonlinearities, software complications, robustness of computation and computing efficiency are of concerns.

Table 1- Computation time for 4*sec* simulation.

	First Method	*Second Method*	*computation reduction*
Single Link	9.51*sec*	8.34*sec*	14%
Two Link	10.05*sec*	9.82*sec*	7%

5 REFERENCES

1 **Merrit H.E.**, "*Hydraulic Control Systems*", John Wiley & Sons, New York, 1967.

2 **Karnopp D.**, "Bond-Graph Models Fluid Dynamic Systems", *Trans. ASME, J. of Dynamic Systems, Measurement and Control*, Series G., Vol. 94, No. 3, September 1972, pp. 222–229.

3 **Bowns D.E.**, **Richards C.W.**, **Caney K.**, "The Simulation of Hydraulic Systems", *Proc. IFAC Symposium on Pneumatic and Hydraulic Components and Instruments in Automatic Control*, Warsaw, Poland, May 1980, pp. 3–8.

4 **Iyengar S.K.R.** and **Fitch E.C.**, "A systematic Approach to the Analysis of Complex Fluid Power Systems", *Proc.* 4th *International Fluid Power Symposium*, Sheffield, England, April 1975, pp. A219-A234.

5 **Hull S.R.** and **Bowns D.E.**, "The Development of an Automatic Procedure for the Digital Simulation of Hydraulic Systems", *Proc. Institute of Mechanical Engineers*, Vol. 199, No. B1, 1985, pp. 51–57.

6 **Hronsky P.** and **Martens H.R.**, "Computer Techniques for Stiff Differential Equations", *Proc. Summer Conference in Computer Simulation*, Montreal, Canada, Vol. 1, July 1973, pp. 131–139.

7 **Bowns D.E.** and **Rolfe A.C.**, "The Digital Computation of Pressure and Flows in Interconnected Fluid Volumes, Using Lumped Parameter Theory", *Proc.* 4th *International Fluid Power Symposium*, Sheffield, England, April 1975, pp. A1-1–A1-17.

ON THE EQUIVALENCE OF NEWTON-EULER FORMULATION AND TREE SELECTION

Jack C. K. Chou
Program in Engineering Science
Computer-Aided Design and Systems Analysis Laboratory
Erik Jonsson School of Engineering and Computer Science
The University of Texas at Dallas, P. O. Box 830688, MP 33
Richardson, Texas 75083-0688, USA

ABSTRACT

The GraphNE formulation is compared with the recursive Newton-Euler formulation. We can show that the Newton-Euler equations derived from the two formulations are equivalent, and they correspond to two different trees chosen from one system graph. A tree selection scheme designed for a given system graph determines its corresponding formulation scheme and the form of its mathematical model.

INTRODUCTION

The well-known recursive Newton-Euler equations became very popular after Luh, Walker, and Paul (1980) [1] developed their systematic computation method based on moving coordinate systems for efficient robot control computations. However, the recursive NE equations may not be directly applicable to forward dynamics simulation of a robot manipulator. $O(n)$ algorithm for the complete calculation of robot forward dynamics. In 1987, Chou, Baciu, and Kesavan [2,3] presented a graph-theoretic Newton-Euler model and an efficient solution method for robot manipulators. The equations in this model are different in form but they retain the recursive property and provide a matrix structure applicable to robust implicit integrations. Further work in developing an efficient computation method for an n-body pendulum (a robot-like mechanism) by exploiting the recursive matrix structure was accomplished by Chou in 1989 [4].

In this paper, a graph-theoretic Newton-Euler formulation (GraphNE) is compared with the recursive Newton-Euler formulation. Modeling robot manipulators with graphs provides a general representation that can be utilized to develop various formulation schemes. We can show that the Newton-Euler equations derived from the two formulations are equivalent, and they correspond to two different trees chosen from one system graph.

TREE SELECTION

Consider an n-link robot manipulator. The construction of the graph-theoretic model (GTM) of this mechanical system involves, first, identifying the components of the system and, then, modeling the components and the system. The GTM's of major components such as a rigid body, rigid arms, and prismatic and revolute joints were presented in [2]. Using the component models, we can draw the system graphs directly by connecting the terminal graphs of the components. A complete GTM of a 3D robot-manipulator system consists of one translational graph, one rotational graph, and a set of terminal equations characterizing the edges of the graphs. The system graphs for translation and rotation are illustrated in Figures 1 and 2, respectively.

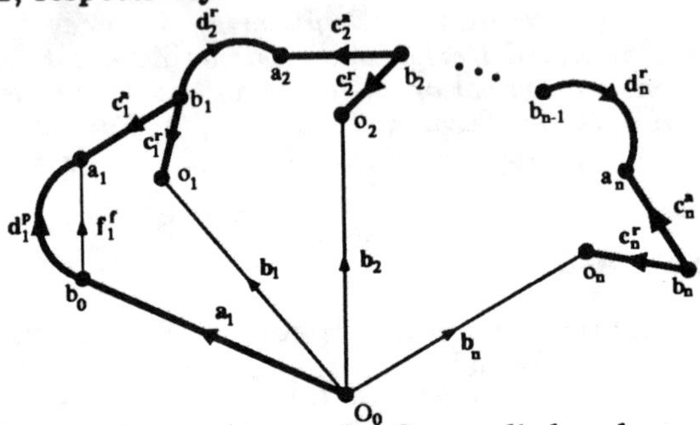

Figure 1: Translational system graph of an n-link robot arm. A tree in thicker edges is chosen for the GraphNE formulation.

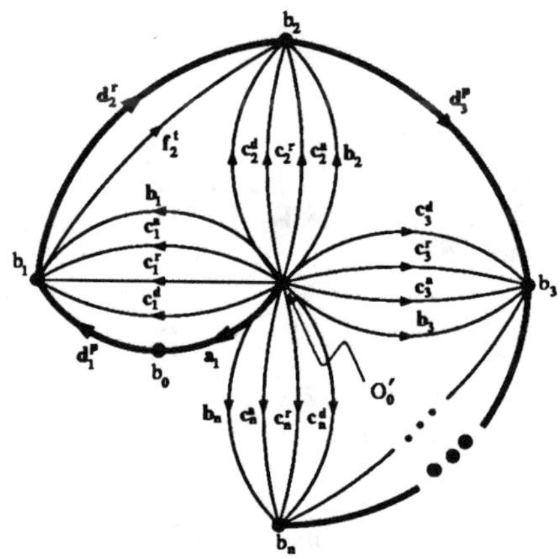

Figure 2: Rotational system graph of an n-link robot arm. A tree in thicker edges is chosen for the GraphNE formulation.

The GraphNE formulation is a Newton-Euler formulation which incorporates with GTM's. The first step in deriving the equations of motion for a robot arm is to select a formulation tree from each system graph. Two trees, as thicker edges, selected from the translational and rotational graphs are illustrated in Figures 1 and 2. However, we can choose another set of trees as those illustrated in Figures 3 and 4. We are going to show that we can use the second set of trees to derive the recursive NE equations, and the equations derived using the first set of trees are equivalent to the recursive NE equations.

KINEMATIC EQUATIONS OF ROTATION

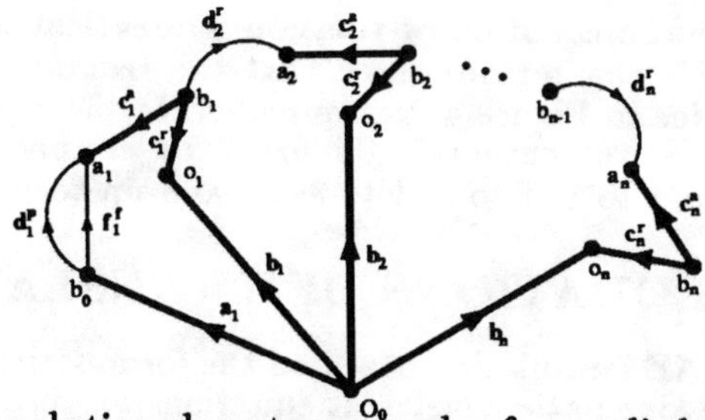

Figure 3: Translational system graph of an n-link robot arm. A tree in thicker edges is chosen for the recursive NE formulation.

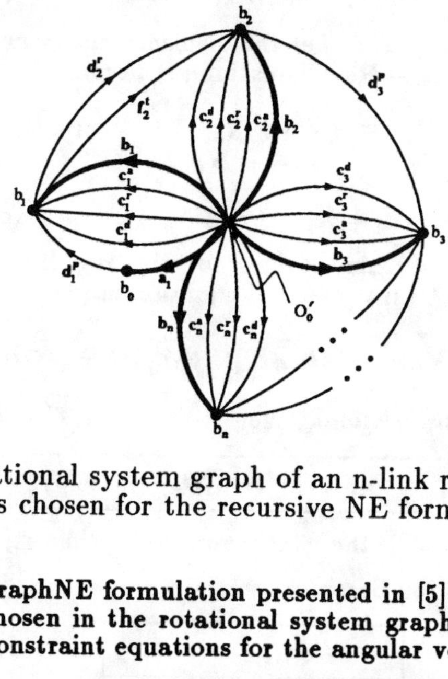

Figure 4: Rotational system graph of an n-link robot arm. A tree in thicker edges is chosen for the recursive NE formulation.

Applying the GraphNE formulation presented in [5] according to the formulation tree chosen in the rotational system graph in Figure 4, we can derive a set of constraint equations for the angular velocities as follows:

$$\mathbf{\Omega}_{d_i^j} - \mathbf{\Omega}_{b_i} + \mathbf{\Omega}_{b_{i-1}} = \mathbf{0} \qquad (1)$$

for $i = 1, ..., n$ and j = p or r. Let $\mathbf{\Omega}_i$ and $\mathbf{\Omega}_{i-1}$ be, respectively, the angular velocities of coordinate systems b_i and b_{i-1} with reference to the inertial system O_0, and let $\mathbf{\Omega}_{i,i-1}$ be the angular velocity of the system b_i with reference to the system b_{i-1}. Then, $\mathbf{\Omega}_i \equiv \mathbf{\Omega}_{b_i}$, $\mathbf{\Omega}_{i-1} \equiv \mathbf{\Omega}_{b_{i-1}}$, and $\mathbf{\Omega}_{d_i^j} \equiv \mathbf{\Omega}_{i,i-1}$, and (1) becomes $\mathbf{\Omega}_i = \mathbf{\Omega}_{i-1} + \mathbf{\Omega}_{i,i-1}$. If the ith joint is translational, $\mathbf{\Omega}_{i,i-1}$ is zero. If the joint is rotational, $\mathbf{\Omega}_{i,i-1}$ is not zero. In terms of Denavit-Hartenberg parameters, we have $\mathbf{\Omega}_{i,i-1}^{i-1} = \dot{\theta}_i \mathbf{u}_i$; that is, $\mathbf{\Omega}_{i,i-1} = \dot{\theta}_i \mathbf{U}_i$ where $\mathbf{U}_i = \boldsymbol{A}_{i-1}\mathbf{u}_i$. Now, substituting it back, we reach

$$\mathbf{\Omega}_i = \begin{cases} \mathbf{\Omega}_{i-1} + \dot{\theta}_i \mathbf{U}_i, & \text{if joint } i \text{ is rotational} \\ \mathbf{\Omega}_{i-1}, & \text{if joint } i \text{ is translational} \end{cases} \qquad (2)$$

This recursive kinematic equation of rotation is identical to the equation (15) given in [1]. We can further show that the recursive equations (1) derived due to the tree in Figure 4 are equivalent to the equations derived due to an alternative tree chosen in Figure 2 by simply by adding the equations in (1) from 1 to 1, 1 to 2, 1 to 3, ..., and up to 1 to n.

KINEMATIC EQUATIONS OF TRANSLATION

Applying the GraphNE formulation based on the formulation tree in Figure 3, we can derive the kinematic constraint equations for linear velocities, as follows:

$$\mathbf{V}_{d_i^j} - \mathbf{V}_{c_i^a} + \mathbf{V}_{c_i^r} - \mathbf{V}_{b_i} + \mathbf{V}_{b_{i-1}} - \mathbf{V}_{c_{i-1}^r} = \mathbf{0} \tag{3}$$

for $i = 1, ..., n$ and $j = \mathrm{p}$ or r. Let the displacement from O_0 to the point b_i be $\mathbf{R}_i$; then $\mathbf{R}_i = \mathbf{R}_{b_i} - \mathbf{R}_{c_i^r}$. Equation (3) becomes

$$\begin{cases} \mathbf{V}_i = (\mathbf{V}_{d_i^j} - \mathbf{V}_{c_i^a}) + \mathbf{V}_{i-1} \\ \mathbf{V}_{b_i} = \mathbf{V}_{c_i^r} + \mathbf{V}_i \end{cases} \tag{4}$$

Substituting $\mathbf{V}_{d_i^j} = \dot{\boldsymbol{A}}_{i-1}\mathbf{d}_i + \boldsymbol{A}_{i-1}\dot{\mathbf{d}}$ and $\mathbf{V}_{c_i^a} = \dot{\boldsymbol{A}}_i\mathbf{a}_i$, applying the Poisson's equation $\tilde{\boldsymbol{\Omega}} = \dot{\boldsymbol{A}}\boldsymbol{A}^T$, and further substituting $\mathbf{d}_i = [0, 0, d_i]^T$, $\mathbf{U}_i = \boldsymbol{A}_{i-1}\mathbf{u}_i$, and $\mathbf{R}_{d_i^j} = d_i\mathbf{U}_i$ into the first equation in (4) gives

$$\mathbf{V}_i = \mathbf{V}_{i-1} + \tilde{\boldsymbol{\Omega}}_{i-1}\mathbf{R}_{d_i^j} + \dot{d}_i\mathbf{U}_i - \tilde{\boldsymbol{\Omega}}_i\mathbf{R}_{c_i^a}$$

If the ith joint is *translational*, then $\boldsymbol{\Omega}_i = \boldsymbol{\Omega}_{i-1}$. The equation becomes

$$\boxed{\mathbf{V}_i = \mathbf{V}_{i-1} + \tilde{\boldsymbol{\Omega}}_{i-1}\mathbf{R}_{c_i^w} + \dot{d}_i\mathbf{U}_i} \tag{5}$$

where $\mathbf{R}_{c_i^w} = \mathbf{R}_{d_i^j} - \mathbf{R}_{c_i^a}$. If the joint is *rotational*, then $\boldsymbol{\Omega}_i = \boldsymbol{\Omega}_{i-1} + \boldsymbol{\Omega}_{i,i-1}$ and $\dot{d}_i = 0$. The equation becomes

$$\boxed{\mathbf{V}_i = \mathbf{V}_{i-1} + \tilde{\boldsymbol{\Omega}}_i\mathbf{R}_{c_i^w}} \tag{6}$$

where $\tilde{\boldsymbol{\Omega}}_{i,i-1}\mathbf{R}_{c_i^d} = \dot{\theta}_i d_i\tilde{\mathbf{U}}_i\mathbf{U}_i = \mathbf{0}$ was substituted. Equations (5) and (6) form a set of recursive kinematic equations for the velocity of point b_i and are identical to the recursive NE equation (20) given in [1].

The second equation in (4) gives the linear velocity of the mass center. Since $\mathbf{V}_{c_i^r} = \tilde{\boldsymbol{\Omega}}_i\mathbf{R}_{c_i^r}$, the equation becomes

$$\boxed{\mathbf{V}_{b_i} = \tilde{\boldsymbol{\Omega}}_i\mathbf{R}_{c_i^r} + \mathbf{V}_i} \tag{7}$$

This recursive equation describes the linear velocity of the center of mass of link i and is identical to the equation (24) given in [1].

We can show that these equations are equivalent to the equations derived from the tree in Figure 1 by taking the first set of equations in (4) and adding the equations from 1 to 1, 1 to 2, 1 to 3, ..., and up to 1 to n, and then substituting them back into the second set of equations in (4).

FORCE-BALANCE EQUATIONS

Directly using the translational system graph and the tree chosen in Figure 3, we can derive a set of cutset equations defined by the edges b_i ($i = 1, ..., n$), as follows:

$$\mathbf{F}_{b_i} - (\mathbf{F}_{d^j_{i+1}} + \mathbf{F}_{f^f_{i+1}}) + (\mathbf{F}_{d^j_i} + \mathbf{F}_{f^f_i}) = \mathbf{0} \quad (8)$$

for $i = 1, ..., n$ and $j =$ p or r. If a joint is revolute, then $\mathbf{F}_{f^f_i} = \mathbf{0}$. In [1], the force $(\mathbf{F}_{d^j_i} + \mathbf{F}_{f^f_i})$ was defined as a combined force $\mathbf{F}_i$ of the reaction force and the applied force. Therefore, (8) becomes

$$\boxed{\mathbf{F}_i = \mathbf{F}_{i+1} - \mathbf{F}_{b_i}} \quad (9)$$

This recursive equation is identical to the equation (28) given in [1].

Taking the general equation (8) and summing up from 1 to n, 2 to n, ..., and n to n, we can show that (8) is equivalent to the equations derived directly from the translational graph in Figure 1.

TORQUE-BALANCE EQUATIONS

Directly using the rotational system graph and the tree chosen in Figure 4, we can derive a set of cutset equations, which are defined by the tree edges b_i (i.e., bodies), as follows:

$$\mathbf{T}_{b_i} + (\mathbf{T}_{d^j_i} + \mathbf{T}_{f^t_i}) - (\mathbf{T}_{d^j_{i+1}} + \mathbf{T}_{f^t_{i+1}}) + \mathbf{T}_{c^a_i} + \mathbf{T}_{c^d_i} + \mathbf{T}_{c^r_i} = \mathbf{0} \quad (10)$$

for $i = 1, ..., n$ and $j =$ p or r. If a joint is prismatic, the actuator applied torque $\mathbf{T}_{f^t_i}$ is zero. In [1], the reaction and applied torques in a joint were combined to a torque $\mathbf{T}_i \equiv \mathbf{T}_{d^j_i} + \mathbf{T}_{f^t_i}$. Therefore, (10) becomes

$$\mathbf{T}_{b_i} + \mathbf{T}_i - \mathbf{T}_{i+1} + \mathbf{T}_{c^a_i} + \mathbf{T}_{c^d_i} + \mathbf{T}_{c^r_i} = \mathbf{0} \quad (11)$$

Substituting the terminal equations for the a-arm, d-arm, and r-arm, we get

$$\mathbf{T}_{b_i} + \mathbf{T}_i - \mathbf{T}_{i+1} - \mathbf{R}_{c^a_i} \times \mathbf{F}_{c^a_i} - \mathbf{R}_{c^d_i} \times \mathbf{F}_{c^d_i} - \mathbf{R}_{c^r_i} \times \mathbf{F}_{c^r_i} = \mathbf{0}$$

From the translational system graph in Figure 3 and its particular tree, we can get the following cutset equations:

$$\begin{cases} \mathbf{F}_{c^a_i} = -(\mathbf{F}_{d^j_i} + \mathbf{F}_{f^t_i}) \\ \mathbf{F}_{c^r_i} = (\mathbf{F}_{d^j_i} + \mathbf{F}_{f^t_i}) - (\mathbf{F}_{d^j_{i+1}} + \mathbf{F}_{f^t_{i+1}}) \end{cases}$$

and they can be further written as

$$\begin{cases} \mathbf{F}_{c^a_i} = -\mathbf{F}_{c^d_i} = -\mathbf{F}_i = \mathbf{F}_{b_i} - \mathbf{F}_{i+1} \\ \mathbf{F}_{c^r_i} = \mathbf{F}_{c^d_i} - \mathbf{F}_{c^d_{i+1}} = \mathbf{F}_i - \mathbf{F}_{i+1} = -\mathbf{F}_{b_i} \end{cases}$$

Substituting them into (11), we get

$$\boxed{\mathbf{T}_i = \mathbf{T}_{i+1} - \mathbf{T}_{b_i} + \mathbf{R}_{c_i^w} \times \mathbf{F}_{i+1} - (\mathbf{R}_{c_i^r} + \mathbf{R}_{c_i^w}) \times \mathbf{F}_{b_i}} \tag{12}$$

where $\mathbf{R}_{c_i^w} = \mathbf{R}_{d_i^j} - \mathbf{R}_{c_i^a} = \mathbf{R}_{c_i^d} - \mathbf{R}_{c_i^a}$. Equation (12) is identical to the equation (29) given in [1].

Taking the general equations (10) and summing them up from 1 to n, 2 to n, 3 to n, ..., and n to n, we can show that (10) is equivalent to the equations derived directly from the rotational system graph in Figure 2.

CONCLUSIONS

The recursive Newton-Euler equations were derived for facilitating the computation of inverse dynamics for robot real-time control. For the forward dynamics of a robot manipulator, it usually involves numerical integrations and simultaneous equations. The matrix structure generated by the recursive NE equations may not be appropriate for this purpose.

In this paper, an alternative system of equations are introduced for the forward dynamics of a robot manipulator. The equations are called the GraphNE equations and are derived using a graph-theoretic approach given in [2]. The GraphNE equations retain the recursive nature and provide an appropriate matrix structure for numerical computations [3].

Modeling robot manipulators with graphs provides a general representation that can be utilized to develop various formulation schemes. Two particular schemes are given in this paper. We show that the well-known recursive NE equations and the GraphNE equations are equivalent, and that they are derived from the same graph-theoretic approach using two particularly chosen trees.

References

[1] J. Y. S. Luh, M. W. Walker, and R. P. C. Paul. On-line computational scheme for mechanical manipulators. *ASME Journal of Dynamic Systems, Measurement and Control*, 102:69–76, June 1980.

[2] J. C. K. Chou, G. Baciu, and H. K. Kesavan. Graph-theoretic models for simulating robot manipulators. *Proceedings of 1987 IEEE International Conference on Robotics and Automation*, 2:953–959, Raleigh, North Carolina, March 31–April 3, 1987.

[3] J. C. K. Chou, G. Baciu, and H. K. Kesavan. Computational scheme for simulating robot manipulators. *Proceedings of 1987 IEEE International Conference on Robotics and Automation*, 2:961–966, Raleigh, North Carolina, March 31–April 3, 1987.

[4] Jack C. K. Chou. A nearly-linear computational-cost scheme for the forward dynamics of an n-body pendulum. *Proceedings of the 3rd Annual Conference on Aerospace Computational Control*, JPL 89-45, 1:88–101, Ed. D.E. bernard and G. K. Man, Pasadena, CA, December 15, 1989.

[5] Jack C. K. Chou. Forward dynamics simulation of robot manipulators: Part I – GraphNE formulation and DAE model. *IEEE Transactions on Robotics and Automation*, 1990, submitted.

AN EFFECTIVE CONTROL SCHEME FOR FLUID POWERED SERVOMECHANISMS

JOSEPH C. K. WONG and JOHN P. T. MO
City Polytechnic of Hong Kong

ABSTRACT

This paper addresses to the philosophy in the design of the control scheme and its implementation in two examples of pneumatic and hydraulic servomechanisms. Analysis on the impact of the control algorithm was attempted and the mathematical models were referenced to determine the effectiveness of the control schemes. The controller configuration and the control software are also described to illustrate the universal applicability of the scheme.

NOTATION

- A_j Cross-sectional area of cylinder chamber j, j=1,2
- B Bulk modulus of oil
- C_{ij} Valve flow gain by spool displacement in cylinder chamber j, j=1,2
- C_{pj} Valve flow gain by pressure variation in cylinder chamber j, j=1,2
- K Spring stiffness
- K_L Leakage flow coefficient
- M Mass of load
- P_j Pressure in cylinder chamber j, j=1,2
- Q_j Volume flow rate from/to cylinder chamber j, j=1,2
- R Universal gas constant
- s Laplace transform operator
- V_j Volume of cylinder chamber j, j=1,2
- x_o Load displacement
- x_s Spool displacement
- γ Ratio of specific heats of gas

INTRODUCTION

Fluid powered systems including hydraulic or pneumatic actuation are traditionally used as logic controlled positioning systems in which the devices are operated in binary states. There are fundamental difficulties to extend its use to programmable servomechanism because of the non-linearity inherent in the media. Initial investigation by Bell et al [1] and Shearer [2] showed that the non-linear flow characteristics of the components in the system induced the requirement for better valve, stable feedback devices and more sophisticated controller.

This paper discusses the research by applying modern digital techniques to tackle the problem. Several prototypes of hydraulic and pneumatic powered linear axes were built. More specifically, a specially designed proportional valve was used for pneumatic systems and a commercially available servovalve was used for hydraulic systems. Both valves provide an approximately linear signal-spool displacement characteristics which is predictable within a narrow range of operation. The feedback path was provided by optical incremental encoders mounted on plastic, pre-stressed rack and pinion arrangement are incorporated as the position feedback devices.

With the advent of microprocessor, an effective variable positional gain algorithm has been designed on intelligent controller based on Motorola 68000 CPU to control both pneumatic and hydraulic systems. Slight modification on the actual implementation on the hydraulic system was required due to the difference in stiffness, but the basic philosophy is identical.

THEORETICAL ANALYSIS ON FLUID POWERED SYSTEMS

The analysis on fluid powered systems has been performed as two separate streams due the difference in compressibility of the working medium of hydraulic and pneumatic systems.

A detail analysis on hydraulic devices was made by Leaney [3] in a research to develop a configuration software for hydraulic sizing and selection. For an asymmetric cylinder load situation (Fig. 1), the transfer function is given by :

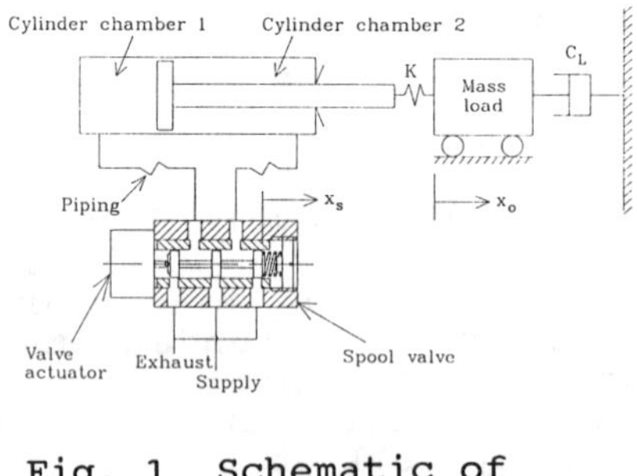

Fig. 1 Schematic of servomechanism

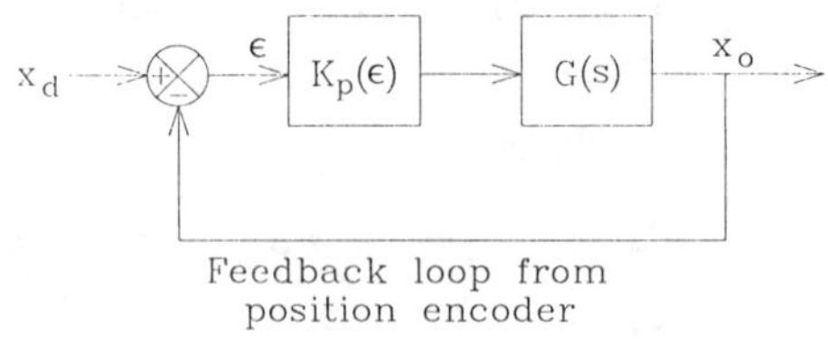

Fig. 2 Position loop control

$$\frac{\delta x_o}{\delta x_s} = \frac{sB_1 + B_2}{s^4B_3 + s^3B_4 + s^2B_5 + sB_6} \qquad (1)$$

where $B_1 = [A_1C_{i1}V_2 + A_2C_{i2}V_1]/B$

$B_2 = A_1C_{i1}K_{L2} + A_2C_{i2}K_{L1}$

$B_3 = MV_1V_2/B^2 + M(V_2A_1{}^2+V_1A_2{}^2)/BK$

$B_4 = M(V_1K_{12}+V_2K_{L1})/B + C_LV_1V_2/B^2 + A_1{}^2(MK_{L2}+C_LV_2/B)/K + A_2{}^2(MK_{L1}+C_LV_1/B)/K$

$B_5 = MK_{L1}K_{L2} + C_L(V_1K_{L2}+V_2K_{L1})/B + (A_1{}^2V_2+A_2{}^2V_1)/B + C_L(A_1{}^2K_{L2}+A_2{}^2K_{L1})/K$

$B_6 = C_LK_{L1}K_{L2} + A_1{}^2K_{L2} + A_2{}^2K_{L1}$

$K_{L1} = K_L[1+(A_2/A_1)^2] + C_{p1}$

$K_{L2} = K_L[1+(A_1/A_2)^2] + C_{p2}$

In case of pneumatics, the involvement of temperature as one of the deciding factors complicates the equations required to describe the behavior of the devices. By applying small perturbation analysis, the same format of the transfer function in (1) was derived with the values of the coefficients represented by:

$B_1 = A_1C_{i1}K_3 + A_2C_{i2}K_{12}$

$B_2 = A_1C_{i1}P_2K_1 + A_2C_{i2}P_1K_1 - A_1C_{i1}C_{p2} - A_2C_{i2}C_{p1} - A_1C_{i2}P_2K_1 - A_2C_{i1}P_1K_1$

$B_3 = MK_2K_3$

$B_4 = M[K_3(P_1K_1-C_{p1})+K_2(P_2K_1-C_{p2})] + C_LK_2V_3$

$B_5 = C_L[K_3(P_1K_1-C_{p1})+K_2(P_2K_1-C_{p2})] + M[(P_1K_1-C_{p1})(P_2K_1-C_{p2})-P_1P_2K_1{}^2] + A_1{}^2\tau K_2K_3P_1/V_1 + A_2{}^2\tau K_2K_3P_2/V_2$

$B_6 = C_L[(P_1K_1-C_{p1})(P_2K_1-C_{p2})-P_1P_2K_1{}^2] - A_1A_2\tau K_1[K_2P_1{}^2/V_1 + K_3P_2{}^2/V_2 + A_1{}^2\tau P_1K_2(P_2K_1-C_{p2})/V_1+A_2{}^2\tau P_2K_3(P_1K_1-C_{p1})/V_2$

and $K_1 = K_L/RT_1$; $K_2 = V_1/RT_1$; $K_3 = V_2/RT_2$

DESIGN OF THE CONTROL ALGORITHM

From the theoretical formulae, it is apparent that the coefficients are each functions of load position. In case of pneumatics, the complication of pressures and temperatures are involved. It is therefore necessary to derive a control scheme which will adapt dynamically to changes of these parameters.

The approach adopted in this research was to incorporate versatility at the position feedback loop (Fig. 2). Since there is no information as regards to the pressures and temperatures variations, the current attempt was to provide the best control algorithm achieveable on the minimum system. The position loop gain is adjusted by the digital controller according to position error ϵ. The question here is to find a feedback gain function $K_p(\epsilon)$ whereby the stability criteria can be ascertained.

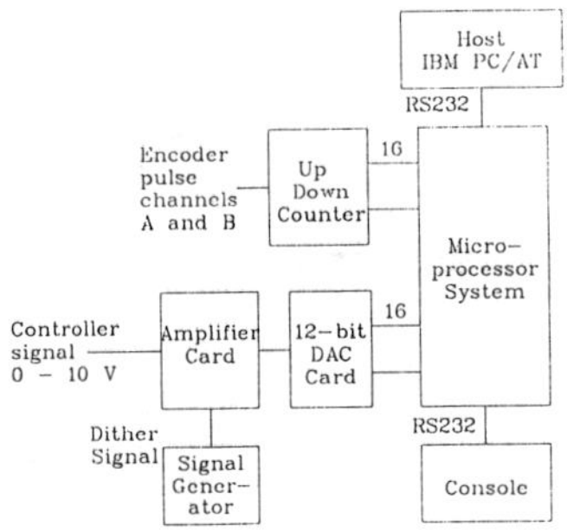

Fig. 3 Microprocessor controller

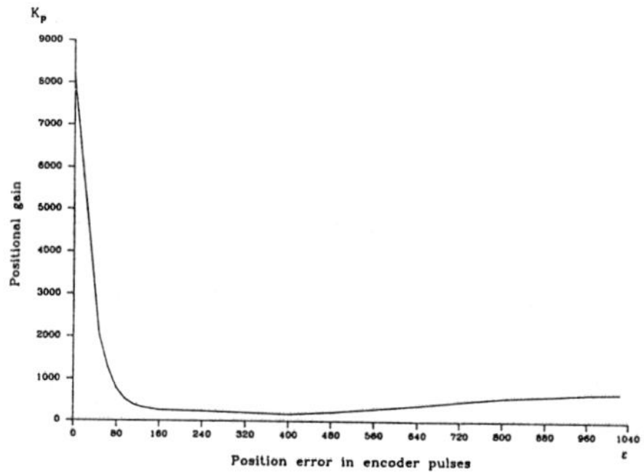

Fig. 4 Control gain plot

IMPLEMENTATION OF THE CONTROL SCHEME

The implementation of the variable positional gain algorithm was on a Motorola VME microprocessor system with custom-made input/output interfaces for the encoder feedback and digital-to-analogue converter solenoid output (Fig. 3). The software was written in assembly language with features allowing data transfer to SUN workstations for analysis purposes [4]. The position gain function was represented by an array of 256 words which was expanded to cover $|\epsilon|$ within the range of 0 to 1024. The sampling time of the microprocessor was 1.3 ms.

Pneumatic Servomechanism

Due to the high frictions relative to all other

forces in pneumatic systems, the control algorithm was designed such that significant gain values were computed at low position errors so as to minimise the dead band errors. At higher position errors, the gain values were reduced to induce damping into the system when it approached the set value. It is not important at high errors where the phenomenon of choked flow through the valve may exist. Fig. 4 shows the control algorithm graphically. Fig. 5 shows a typical transient plot.

<u>Hydraulic Servomechanism</u>

Extending the concept of variable position gain function to hydraulic servomechanism, a hydraulic test rig was built on a smaller Motorola microprocessor system to reduce the total costs involved. The software has been modified slightly to suit the new environment which was essentially the same as the pneumatic test rig schematically.

However, it was found that the system started to drift after a number of moves (typically within 5-10). Furthermore, the drift was towards the rod side where the cross-section of the hydraulic pressure projection area is smaller. Referring back to the model, the volume flow rates of the two chambers are given by:

$$\delta Q_1 = s(A_1 \delta x_o + V_1 \delta P_1 / B) + K_L(P_1 - P_2) \quad (2)$$

$$\delta Q_2 = s(A_2 \delta x_o + V_2 \delta P_2 / B) + K_L(P_1 - P_2) \quad (3)$$

The difference in flow and hence pressure accumulated at the two chambers of the hydraulic cylinder could induce unbalanced flows which resulted in increase of one side of the fluid volume.

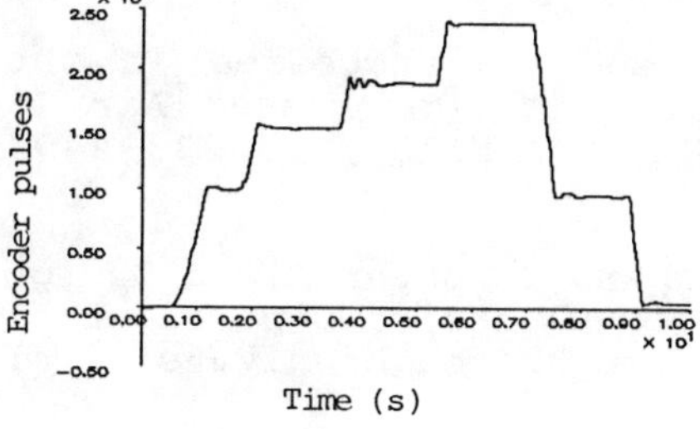

Fig. 5 Typical transient plot

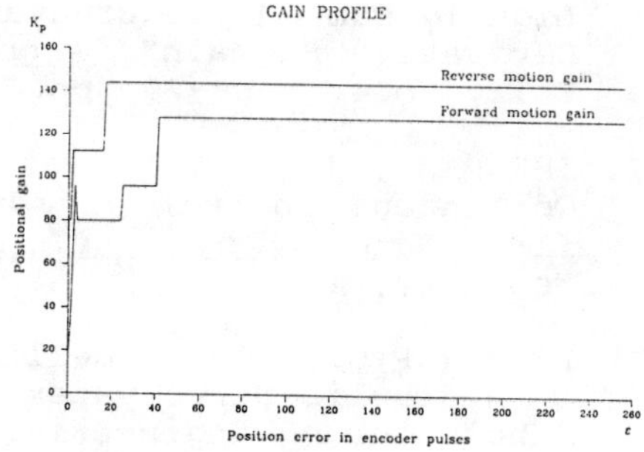

Fig. 6 Dual gain functions for hydraulic servomechanism

Subsequently, the position gain function K_ϵ was splitted into the forward and backward gains. The forward position gain function produced a smaller solenoid analogue signal so that the asymmetricity could be balanced artificially by the microprocessor. The two position gain functions are shown in Fig. 6. As the whole system was treated as two different closed loop systems, the tuning of one part did not affect the other part and hence allowed tuning to be done on a wider working range as the load increased. In addition, since the working medium was basically incompressible, the dual function control scheme possessed the capability to separately control the speed of the test rig. This characteristics is particularly important when several such axes are put together in future to form a modular robot. The amount of energy change calculated from the multiple axes Lagrangian dynamics is highly interactive due to inertia effect and this can be tuned accurately under the dual gain function scheme.

CONCLUSION

An effective control alogrithm was designed to work on fluid powered servomechanisms to cater for the non-linearity of the media. The control scheme was implemented in the form of a position gain function array which could be tuned on the microprocessor systems to suit the characteristics of individual hardware. The versatility of the control scheme was proved particularly sucessful with the modification of dual gain functions for hydraulic systems.

REFERENCES

1. BELL R. and PENNINGTON A. de, "Active Compensation of Lightly Damped Electrohydraulic Cylinder Drives Using Derivative Signals", Proc Instn Mech Engrs, Vol.184, Pt.1, No.4, 1969-70, pp.83-98

2. SHEARER J.L., "Study of Pneumatic Processes in the Continuous Control of Motion with Compressed Air I-II", Trans. of the ASME, Vol.78, pp.233-249, February, 1956

3. LEANEY P.G., "The Modelling and Computer Aided Design of Hydraulic Servosystems", PhD Thesis, Department of Manufacturing Engineering, Loughborough University of Technology, U.K., 1986

4. MO P.T.J., "Control Software for Pneumatic Servomechanism Analysis", International Conference on Computers in Engineering Education, 24-25 October, 1988, Singapore

Recursive Dynamics of Multiarm Robotic Systems in Loose Grasp of Articulated Task Objects

A. Jain and G. Rodriguez
Jet Propulsion Laboratory/California Institute of Technology

K. Kreutz–Delgado
University of California, San Diego

Abstract

This paper develops the spatial algebra dynamics formalism for modeling and analysis of multiple parallel manipulators grasping an articulated object. The held object, may be in contact with the external environment. The grasps (i.e. the contacts between the manipulators and the held object) may be fully constrained ("rigid") or only partially constrained ("loose"), where holonomic and nonholonomic grasp constraints are allowed. The objects handled by the manipulators may be articulated, meaning that the object is viewed as a hinged ensemble of rigid bodies. The forward dynamics algorithm is of complexity $O(\mathcal{N})$ + $O(\ell)$, where $\mathcal{N}$ is the number of motion dofs for the whole system, and ℓ is the number of manipulators.

1 Introduction

This paper develops the dynamics formalism for modeling and analysis of multiple parallel manipulators grasping an articulated object. The held object, may be in contact with the external environment. The grasps (i.e. the contacts between the manipulators and the held object) may be fully constrained ("rigid") or only partially constrained ("loose"), where holonomic and nonholonomic grasp constraints are allowed. The objects handled by the manipulators may be non–articulated (i.e. a single rigid body) or articulated, meaning that the object is viewed as a hinged ensemble of rigid bodies. The recently developed *spatial operator algebra* framework [1, 2] is used to model the problem discussed above, and develop an efficient forward dynamics algorithm for the combined multiarm/task system.

The modeling is based upon the natural partitioning between a multi–arm system and a task system which it manipulates. The multi–arm system is typically relatively time–invariant in its structure and topology. The task system represents the part of the overall system which is most time dependent due to a variety of reasons: a) there may be a tool change; b) there may be a change of task object; c) the nature of the grasp between the multi–arm system and the tool or task object may change; d) the nature of the contact between the tool and the task object may change. Note that when the arms grasp the tool or task object closed kinematic chains may be formed. A major rationale for a partitioning into the multiarm and task systems is that in the forward dynamics algorithm we develop here, a major portion of the dynamics algorithm remains independent of the particular nature of the task object and the constraints. Thus the time–invariant parts of the algorithm can be optimized. The partitioning also allows the use of the previously reported efficient dynamics algorithms for unconstrained single manipulators [2], and their integration into the task scenarios discussed here. Some previous work on the dynamics of a system of multiple parallel manipulators can be found in [3]–[7].

2 Equations of Motion

We use coordinate–free *spatial notation* (see [2]) throughout this paper. Thus V denotes the *spatial velocity*, α the *spatial acceleration*, f the *spatial force* and $\phi(.,.) \in \mathcal{R}^{6\times 6}$ the *rigid body transformation operator*, and M the *spatial inertia* of a rigid body. The equations of motion of a single manipulator are expressed compactly using spatial operators. In this notation, the $V(k)$'s, $\alpha(k)$'s etc are viewed as components of vectors V, α etc. Also $M \triangleq diag[M(1) \cdots M(n)]$, $H = [H(1) \cdots H(n)]$, and $B^* \triangleq [\phi^*(1,0),\ 0,\ \cdots\ 0]$.

$$\phi \triangleq = \begin{pmatrix} I & 0 & \dots & 0 \\ \phi(2,1) & I & \dots & 0 \\ \vdots & \vdots & \ddots & \vdots \\ \phi(n,1) & \phi(n,2) & \dots & I \end{pmatrix}, \quad \text{where} \quad ,\phi(i,j) \triangleq \phi(i,i-1) \cdots \phi(j+1,j) \tag{2.1}$$

With β denoting the motion generalized coordinates for the joints, the equations of motion for a single manipulator are

$$\begin{aligned} V &= \phi^* H^* \beta, \qquad \alpha = \phi^*(H^*\dot{\beta} + a) \\ f &= \phi(M\alpha + b) = \phi M \phi^* H^* \dot{\beta} + \phi(M\phi^* a + b + Bf(0)) \\ T &= Hf = H\phi M \phi^* H^* \dot{\beta} + H\phi(M\phi^* a + b + Bf(0)) \\ &= \mathcal{M}\dot{\beta} + \mathcal{C} + Bf(0), \quad \text{where} \quad \mathcal{M} \triangleq H\phi M \phi^* H^*, \quad \text{and} \quad \mathcal{C} \triangleq H\phi(M\phi^* a + b) \end{aligned} \tag{2.2}$$

$\mathcal{M} \in \mathcal{R}^{\mathcal{N}\times\mathcal{N}}$ is the *mass matrix* of the serial chain and $\mathcal{C} \in \mathcal{R}^{\mathcal{N}}$ consists of the Coriolis, centrifugal and gyroscopic joint forces. Since the tip velocity, $V(0)$, is given by $V(0) = B^*V$, it follows that $V(0) = J\beta$, where $J = B^*\phi^*H^*$ is the Jacobian matrix which maps the joint rates β to the tip spatial velocity.

Equations of Motion of the (Primary) Manipulators System

Multiarm robotic systems may be partitioned as follows: (a) The *primary* system consisting of all the manipulators. This may be regarded as the part of the overall system that varies the least. When the arm system is fixed and has a fixed end–effector, the primary system is time–invariant; (b) The *secondary* (or task) system consisting of articulated tools and/or task–object(s). It will vary from task to task; (c) The *closure constraints* between the primary and secondary systems and their interactions with the world. These will change with time during the execution of a task as there are changes in the grasp, mating, tool operation etc.

The *primary system* is defined as the collection of serial manipulators mounted on a fixed base. Thus n_j and $\mathcal{N}_j$ denote the number of links and the motion dofs respectively for the jth manipulator. The total number of links n_P, and the total motion dofs $\mathcal{N}_P$ for the primary system are given by

$n_P \triangleq \sum_{j=1}^{\ell} n_j$ and $\mathcal{N}_P \triangleq \sum_{j=1}^{\ell} \mathcal{N}_j$. Furthermore, for the jth manipulator, in the stacked notation we have the quantities V_j, M_j, $\mathcal{E}_{\phi_j}$, ϕ_j etc.. We use double indexing to identify an individual link/joint in the system. Thus a link/joint is identified by the index of the manipulator it is on, plus its location within the manipulator. For instance, $V(k_j)$ (or more accurately $V_j(k)$) denotes the spatial velocity of the the kth link of the jth manipulator at its inboard joint location.. The overall stacked spatial velocity, acceleration etc. vectors are denoted simply as V_P, α_P, f_P etc. with $V_P \triangleq [V_1^* \cdots V_\ell^*]^*$ etc.. We extend in a straightforward manner the dynamics of a single manipulator to a multiple manipulator context. For that we define

$$\phi_P \triangleq diag(\phi_j), \quad \mathcal{E}_{\phi_P} \triangleq diag(\mathcal{E}_{\phi_j}), \quad M_P = diag(M_j), \quad \text{etc.}$$

The dynamics of the multiarm system are then given by Eq. (2.2) with the quantities therein being replaced by the corresponding quantities with the subscript "P". The collective Jacobians of all the manipulators is given by

$$J_P = diag(J_j) = B_P^* \phi_P^* H_P^*$$

Let $\hat{V}_P \in \mathcal{R}^{6\ell}$ denote the vector of spatial velocities at the tips (grasp points) for the manipulators. Thus,

$$\hat{V}_P = J_P \beta_P$$

For the task system, referred to as the *secondary system*, we will carry over the notation used for the primary system but with the use of the subscript S. Thus the secondary system contains n_S links with $\mathcal{N}_S$ motion dofs and its internal joint forces are given by T_S. Also $\mathcal{M}_S \in \mathcal{R}^{\mathcal{N}_S \times \mathcal{N}_S}$ denotes the mass matrix of the secondary system, $\beta_S \in \mathcal{R}^{\mathcal{N}_S}$ its motion generalized coordinates, $T_S \in \mathcal{R}^{\mathcal{N}_S}$ its generalized joint forces, and $\hat{T}_S$ the "bias" free generalized joint forces. J_S denotes the Jacobian matrix which maps the β_S into spatial velocities $\hat{V}_S$ at the points of closure, i.e., $\hat{V}_S = J_S \beta_S$. The points of closure may either be due to grasping by a manipulator, or due to interactions between the tool/task–object and the environment.

The kinematical constraints due to the interactions between the task–object and the environment leads to constraint equations of the form:

$$Q_S \hat{V}_S = 0 \tag{2.3}$$

The grasping of the tool/task-object by the manipulators leads to constraint equations of the form:

$$\hat{Q}_P \hat{V}_P + \hat{Q}_S \hat{V}_S = 0 \tag{2.4}$$

These constraint equations characterize constraints including those that arise from a "loose grasp". In contrast with "rigid" grasps/constraints, such "loose" grasps/constraints, permit relative motion along certain directions between the the task object and the manipulators/environment. Defining

$$A_P \triangleq \begin{pmatrix} \hat{Q}_P \\ 0 \end{pmatrix}, \quad A_S \triangleq \begin{pmatrix} \hat{Q}_S \\ Q_S \end{pmatrix}, \quad \text{and} \quad A \triangleq [A_P \;\; A_S]$$

the closure constraints can be collectively expressed in the form:

$$A \begin{pmatrix} \hat{V}_P \\ \hat{V}_S \end{pmatrix} = [A_P \;\; A_S] \begin{pmatrix} J_P & 0 \\ 0 & J_S \end{pmatrix} \begin{pmatrix} \beta_P \\ \beta_S \end{pmatrix} = [A_P J_P \;\; A_S J_S] \begin{pmatrix} \beta_P \\ \beta_S \end{pmatrix} = 0 \tag{2.5}$$

It is assumed that $[A_P J_P \;\; A_S J_S]$ is of full row rank $\mathcal{N}_E$, or steps have been taken to discard the dependent constraint equations. With this setup, as the grasps change, and contacts are made and broken, only the updating of the A_P and A_S matrices is necessary.

Closed–Chain System Dynamics

Based on the principle of virtual work, Eq. (2.5) implies that the closure constraint joint forces are of the form

$$\begin{pmatrix} J^* A_P^* \\ J_S^* A_S^* \end{pmatrix} \hat{f}$$

for some $\hat{f} \in \mathcal{R}^{\mathcal{N}_E}$. $\hat{f}$ may be regarded as the Lagrange multipliers associated with the closure constraints. When constraints at the contact points between the primary and secondary systems are on the relative spatial velocity, the corresponding components of $\hat{f}$ correspond to the forces that the primary and secondary systems exert on each other at the point of contact. The overall equations of motion can be put in the form

$$\begin{pmatrix} \mathcal{M}_P & 0 & J_P^* A_P^* \\ 0 & \mathcal{M}_S & J_S^* A_S^* \\ 0 & 0 & -[A_P \Lambda_P A_P^* + A_S \Lambda_S A_S^*] \end{pmatrix} \begin{pmatrix} \dot{\beta}_P \\ \dot{\beta}_S \\ \hat{f} \end{pmatrix} = \begin{pmatrix} \hat{T}_P \\ \hat{T}_S \\ \hat{U} \end{pmatrix} \tag{2.6}$$

where

$$U \triangleq -[(\dot{A_P} J_P) \;\; (\dot{A_S} J_S)], \quad \text{and} \quad \hat{U} \triangleq U - [A_P J_P \mathcal{M}_P^{-1} \hat{T}_P + A_S J_S \mathcal{M}_S^{-1} \hat{T}_S]$$

Also $\Lambda_P \triangleq J_P \mathcal{M}_P^{-1} J_P^*$ and $\Lambda_S \triangleq J_S \mathcal{M}_S^{-1} J_S^*$. Note that $\Lambda_P = diag(\Lambda_j)$ where $\Lambda_j \triangleq J_j \mathcal{M}_j^{-1} J_j^*$. In the above, J_j and $\mathcal{M}_j$ are the Jacobian and mass matrices for the jth manipulator. Λ_P and Λ_S are the effective "admittances" of the primary and secondary systems reflected to the points of closure, and are closely related to Khatib's Operational Space Inertia Matrices.

3 Recursive Forward Dynamics Algorithms

For the equations of motion of the closed chain system given by Eq. (2.6), the solution of the forward dynamics problem can be solved by the following sequence of steps:

(A) Solve $\mathcal{M}_P \dot{\beta}_P^f = \hat{T}_P$ for $\dot{\beta}_P^f$ — Solve $\mathcal{M}_S \dot{\beta}_S^f = \hat{T}_S$ for $\dot{\beta}_S^f$

(B) Compute $\hat{\alpha}_P^f = \hat{J}_P \dot{\beta}_P^f$ — Compute $\hat{\alpha}_S^f = \hat{J}_S \dot{\beta}_S^f$

(C) Compute $\Lambda_P = \hat{J}_P \mathcal{M}_P^{-1} \hat{J}_P^*$ — Compute $\Lambda_S = \hat{J}_S \mathcal{M}_S^{-1} \hat{J}_S^*$

(D) Solve $[A_P \Lambda_P A_P^* + A_S \Lambda_S A_S^*] \hat{f} = (A_P \hat{\alpha}_P^f + A_S \hat{\alpha}_S^f) - U$ for $\hat{f}$

(E) Solve $\mathcal{M}_P \dot{\beta}_P^\delta = -\hat{J}_P^* A_P^* \hat{f}$ for $\dot{\beta}_P^\delta$ — Solve $\mathcal{M}_S \dot{\beta}_S^\delta = -\hat{J}_S^* A_S^* \hat{f}$ for $\dot{\beta}_S^\delta$

(F) $\dot{\beta}_P = \dot{\beta}_P^f + \dot{\beta}_P^\delta$ — $\dot{\beta}_S = \dot{\beta}_S^f + \dot{\beta}_S^\delta$

STEP (A) solves for the "free" joint accelerations $\dot{\beta}_P^f$ and $\dot{\beta}_S^f$ of the primary and secondary systems assuming that there are no closure constraints on the system. STEP (B) computes the corresponding "free" tip spatial accelerations α_P^f and α_S^f for the systems. In STEP (C) and STEP (D), the constraint forces at the points of closure are computed, and they are used in STEP (E) to compute the "correction" joint accelerations $\dot{\beta}_P^\delta$ and $\dot{\beta}_S^\delta$ for the two systems. Combining the free and correction joint accelerations in STEP (F) gives the true joint accelerations for the two systems.

Changes in the grasp of the task–object, tool mating operation, handover etc. during the execution of a task lead to changes in the closure constraints on the overall system. This requires the updating of the A closure constraints matrix, and in the forward dynamics procedure described above, only STEP (D) is effected. Changes in the task being performed lead to changes in the tool/task–object, and this requires updating the secondary system. The impact on the forward dynamics algorithm is confined to the steps in the column on the right above. In a given environment, the manipulators do not change structurally, and thus neither do the steps in the left column, and as a consequence optimized algorithms to carry them out can be implemented.

We summarize below spatial operator expressions for the computation of the various steps in the above table. Due to space limitations, derivations, proofs and the equivalent recursive algorithms are not included here, but they closely resemble those for the simpler cases studied in [1, 2]. We restrict our attention to the steps in the first column in the above table (since the ideas directly carry over to the task system as well), and so drop the use of the subscript "P" for notational convenience (except for STEP (D), where the primary and secondary systems come together).

STEP (A) $\dot{\beta}^f = \mathcal{M}^{-1}\hat{T} = [I - H\psi K]^* D^{-1}[I - H\psi K]\hat{T}$

STEP (B) $\hat{\alpha}^f = B^*\tilde{\alpha}^f$, where $\tilde{\alpha}^f = \psi^* H^* D^{-1}[I - H\psi K]\hat{T}$

STEP (C) $\Lambda = B^*\psi^* H^* D^{-1} H\psi B = B^*\Omega B$, where $\Omega \triangleq \psi^* H^* D^{-1} H\psi$

STEP (D) Solve $[A_P \Lambda_P A_P^* + A_S \Lambda_S A_S^*]\hat{f} = (A_P \hat{\alpha}_P^f + A_S \hat{\alpha}_S^f) - U$ for $\hat{f}$

STEP (E) $\dot{\beta}^\delta = -[I - H\psi K]^* D^{-1} H\psi B A^* \hat{f}$

Exploiting further simplifications when none of the arms is at a kinematic singularity, the overall complexity of this recursive forward dynamics algorithm is $O(\mathcal{N}_P + \mathcal{N}_S) + O(\mathcal{N}_S^3) + O(\ell)$. Typically $\mathcal{N}_S$ is small resulting in a particularly efficient and recursive algorithm for the forward dynamics of the closed chain system. Note that as a consequence of the topological partitioning employed here, changes in the grasp etc, constraints or in the task–object require only small changes in the overall algorithm.

4 Conclusions

In this paper we have presented a very general framework for modeling systems consisting of arbitrarily complex articulated–body tools in contact with task objects, and where a multiarm system manipulates and grasps these tools via rigid or loose grasps. We partition the system into a primary and a secondary system and a set of constraints for the purposes of modeling the kinematical and dynamical behavior of the system. The spatial operator algebra is then used to derive efficient forward dynamics algorithms of complexity $O(\mathcal{N})+O(\ell)$ for these systems. The algorithms are structured so as to be adaptable to time-variations in the system arising from reconfiguration, changes in the task object or the in the constraints themselves. The algorithms are recursive in nature, and only localized changes corresponding to the changes in the system are needed.

5 Acknowledgement

The research described in this paper was performed at the Jet Propulsion Laboratory, California Institute of Technology, under contract with the National Aeronautics and Space Administration.

References

[1] G. Rodriguez and K. Kreutz, "Recursive Mass Matrix Factorization and Inversion: An Operator Approach to Manipulator Forward Dynamics," *IEEE Transactions on Robotics and Automation*, 1990. Accepted for publication. (see also JPL Publication 88-11, 1988).

[2] G. Rodriguez, K. Kreutz, and A. Jain, "A Spatial Operator Algebra for Manipulator Modeling and Control," *The International Journal of Robotics Research*, 1990. Accepted for publication. (Also at the 1989 IEEE Conf. on Robotics and Automation).

[3] O. Khatib, "Object Manipulation in a Multi-Effector System," in *4th International Symposium Robotics Research*, 1988.

[4] J. Luh and Y. Zheng, "Constrained Relations between Two Coordinated Industrial Robots for Motion Control," *The International Journal of Robotics Research*, vol. 6, pp. 60–70, Fall 1987.

[5] Y. Nakamura, K. Nagai, and T. Yoshikawa, "Mechanics of Coordinated Manipulation by Multiple Robotic Systems," in *IEEE International Conference on Robotics and Automation*, (Raleigh, NC), 1987.

[6] K. Kreutz and A. Lokshin, "Load Balancing and Closed-Chain Multiple Arm Control," in *1988 American Control Conference*, (Atlanta, GA), June 1988.

[7] Z. Li, J. Canny, and S. Sastry, "On Motion Planning for Dextrous Manipulation, Part I: the Problem Formulated," in *IEEE International Conference on Robotics and Automation*, (Scottsdale, AZ), May 1989.

Iterative Formulation of Operational Space Control

K. Kreutz–Delgado
Applied Mechanics and Engineering Sciences
University of California, San Diego
La Jolla, California 92093–0411

A. Jain, G. Rodriguez
Jet Propulsion Laboratory/California Institute of Technology
4800 Oak Grove Drive, Mail Stop 198–219
Pasadena, California 91109

Abstract

A Spatial Operator Algebra is used to derive an O(n) iterative algorithm which implements Operational Space Control of an n–link manipulator. Algorithms for the Operational Space mass matrix and the coriolis and gravity terms are also derived.

1. Introduction and Background

The Operational Space of a manipulator is defined by the configuration space of the end–effector. Operational Space Control (OSC) is an approach to manipulator control that focuses on the dynamical behavior of a serial rigid link manipulator as seen at the end–effector as it evolves in its Operational Space [1]. Let $\beta(0) \in R^6$ denote a generalized velocity for the end–effector (the Operational Space velocity). The argument "0" denotes the fact that our interest is with some point fixed with respect to the "tip" of the manipulator. The associated Operational Space dynamics is

$$\Lambda(0)\dot{\beta}(0) + c(0) + g(0) = F(0) \qquad (1.1)$$

where the relationship between the end–effector generalized spatial force, $F(0) \in R^6$, and the joint–level force, T, is given by $T = J^* F(0)$. The term $c(0)$ gives the coriolis/centrifugal forces and $g(0)$ is the gravity loading term. Now take $T = J^* F_c(0)$ where $F_c(0)$ is given by

$$F_c(0) = \Lambda(0)u + c(0) + g(0) \ . \qquad (1.2)$$

Assuming that $\Lambda(0)$ is nonsingular, this results in decoupled and (almost) linearized end–effector dynamics of the form $\dot{\beta}(0) = u$. It is now straightforward to perform stable control by the use of standard control techniques. In this paper the Spatial Operator Algebra of [2]–[4] is used to develop an O(n) iterative algorithm which implements OSC control, for a given control signal u, where n is the number of manipulator links. To do this, we will show how $\Lambda(0)$, $c(0)$, and $g(0)$ can each be generated by separate O(n) algorithms.

The joint space dynamical equation of motion for an n–link manipulator is

$$\mathcal{M}\ddot{\theta} + \mathcal{C} + \mathcal{G} = T \qquad (1.3)$$

where θ denotes the joint space configuration variable. $\mathcal{C}$ is the coriolis/centrifugal forces term and $\mathcal{G}$ is the gravity loading term. Each joint has a single degree–of–freedom (dof) so that $\ddot{\theta} \in R^6$ and $T \in R^6$. Despite this restriction to an n dof manipulator, the results developed in this paper easily extend to the multiple–dof joint case.

Generally, it is convenient to work with the end–effector generalized velocity $\beta(0)^* = [\omega^*(0), \dot{x}^*(0)]$, where $\omega(0)$ and $\dot{x}(0)$ are the angular and linear velocity of the end–effector. The relationship between this choice of $\beta(0)$ and the joint velocity $\dot{\theta}$ is given by $\beta(0) = J\dot{\theta}$, where the corresponding Jacobian, J is referred to as the natural or standard Jacobian [1] . The Operational Space quantities $\Lambda(0)$, $c(0)$, and $g(0)$ of eq. (1.1) can be expressed in terms of the joint level quantities $\mathcal{M}$, $\mathcal{C}$ and $\mathcal{G}$ of eq. (1.3), and the standard Jacobian, J, as follows [1]:

$$\Lambda(0) = \Omega(0)^{-1}; \quad \Omega(0) = J\mathcal{M}^{-1}J^* \tag{1.4}$$

$$c(0) = \Lambda(0)(J\mathcal{M}^{-1}\mathcal{C} - \dot{J}\dot{\theta}) \tag{1.5}$$

$$g(0) = \Lambda(0)J\mathcal{M}^{-1}\mathcal{G} \; . \tag{1.6}$$

2. A Spatial Operator Algebra

Fundamental linear spatial operators associated with a manipulator are H^*, M, and $\mathcal{E}_\phi$. They are defined as follows:

$$H^* \triangleq \text{diag}[H^*(1), \cdots, H^*(n)] \in R^{6n \times n} \; , \tag{2.1}$$

where $H^*(k) \in R^6$ is the joint axis of the kth link (which is along the joint connecting links k and $k+1$);

$$M \triangleq \text{diag}[M(1), \cdots, M(n)] \in R^{6n \times 6n} \; , \tag{2.2}$$

where $M(k) \in R^{6 \times 6}$ is the spatial inertia of the kth link; and

$$\mathcal{E}_\phi \triangleq \begin{pmatrix} 0 & 0 & 0 & 0 & 0 \\ \phi(2,1) & 0 & \cdots & 0 & 0 \\ 0 & \phi(3,2) & \cdots & 0 & 0 \\ \vdots & \vdots & \ddots & \vdots & \vdots \\ 0 & 0 & \cdots & \phi(n, n-1) & 0 \end{pmatrix} \in R^{6n \times 6n} \; , \tag{2.3}$$

where $\phi(k, k-1) \in R^{6 \times 6}$ is the rigid–body force transformation jacobian that transforms a spatial force acting on link k at joint $k-1$ to the equivalent force acting on link k at joint k. Note that $\mathcal{E}_\phi$ only has nonzero block entries along its first block subdiagonal. Two additional fundamental spatial operators associated with the manipulator are

$$B^* \triangleq [\phi^*(1,0), 0, \cdots, 0] \in R^{6 \times 6n} \tag{2.4}$$

$$E^* \triangleq [0, \cdots, 0, \phi(n+1, n)] \in R^{6 \times 6n} \; . \tag{2.5}$$

It should be noted that the link numbering scheme used here increases from the tip of the link to the base, so that link 1 is the outboard–most link and link $n+1$ denotes the base. Point "0" the end effector location.

From the fundamental quantities (2.1)–(2.5), additional linear spatial operators associated with the manipulator can be derived. The important additional spatial operators ϕ, P, D, G, K, $\mathcal{E}_\psi$, and ψ are obtained as follows:

$$\begin{aligned}
\phi &= (I-\mathcal{E}_\phi)^{-1} \in R^{6n\times 6n} && (2.6)\\
P &= \mathcal{E}_\psi P \mathcal{E}_\psi^* + M = \text{diag}[P(1),\cdots,P(n)] \in R^{6n\times 6n} && (2.7)\\
D &= HPH^* = \text{diag}[D(1),\cdots,D(n)] \in R^{n\times n} && (2.8)\\
G &= PH^*D^{-1} = \text{diag}[G(1),\cdots,G(n)] \in R^{6n\times n} && (2.9)\\
K &= \mathcal{E}_\phi G \in R^{6n\times n} && (2.10)\\
\mathcal{E}_\psi &= \mathcal{E}_\phi(I-GH) \in R^{6n\times 6n} && (2.11)\\
\psi &= (I-\mathcal{E}_\psi)^{-1} \in R^{6n\times 6n}\ . && (2.12)
\end{aligned}$$

Note that P, D, and G are block diagonal. $P(k)$ is the articulated body inertia of the kth link while the nonzero scalar $D(k) = H(k)P(k)H^*(k)$ is the projection of $P(k)$ onto the kth joint axis. Also note that $G(k) = P(k)H^*(k)D^{-1}(k) \in R^6$. Since $\mathcal{E}_\phi$ and $\mathcal{E}_\psi$ are both nilpotent operators, the operators ϕ and ψ given by (2.6) and (2.12) are both lower block triangular with 6×6 block elements. Similar to $\mathcal{E}_\phi$, K only has non–zero block entries, $\{K(2,1),\cdots,K(n,n-1)\}$, along its first block subdiagonal where $K(k,k-1) = \phi(k,k-1)G(k-1) \in R^6$. $\mathcal{E}_\psi$ also only has non–zero block entries, $\{\psi(2,1),\cdots,\psi(n,n-1)\}$, along its first block subdiagonal where $\psi(k,k-1) = \phi(k,k-1)[I-G(k-1)H(k-1)]$. The quantities (2.7)–(2.11) can be computed by a tip–to–base discrete–step riccati–like iterative processing of the fundamental quantities (2.1)–(2.3). This iteration is implicit in equation (2.7) and is given by

$$P(k) = \psi(k,k-1)P(k-1)\psi^*(k,k-1) + M(k)\ , \qquad (2.13)$$

for $k = 1\cdots n$, where $P(0) = 0$ and $\psi(1,0) = 0$. Note that $D(k-1)$, $G(k-1)$, $\psi(k,k-1)$, and $K(k,k-1)$, $k \geq 2$, can be computed as soon as $P(k-1)$ is available so that the iteration (2.13) depends upon available quantities.

The dynamic expressions describing multibody systems can be expressed in terms of the linear spatial operators. The spatial operators show how physical quantities such as forces, velocities, and accelerations propagate between bodies. For example, the standard jacobian can be written as

$$J^* = H\phi B\ , \qquad (2.14)$$

a factorization into spatial operators which shows how the effect of a manipulator tip force propagates from the tip to the base and then projects onto the joint axes. This reflects the fact that the action of the lower block triangular operator ϕ is equivalent to a tip–to–base propagation of link interaction forces. Similarly, it can be shown that

$$\begin{aligned}
\mathcal{M} &= H\phi M\phi^* H^* && (2.15)\\
\mathcal{C} &= H\phi(M\phi^* a + b) && (2.16)\\
\mathcal{G} &= H\phi M\phi^* E a_g && (2.17)
\end{aligned}$$

where $a_g = \text{col}(0, 9.8\cdot e_g)$. The vectors a and b are quadratic in the link velocities and are readily computed given these velocities. It is also true that

$$\dot{J}\dot{\theta} = B^*\phi^* a + a(0) \qquad (2.18)$$

where $a(0)$ also depends quadratically on the link velocities. The quantities a, $a(0)$, and b are computable once the manipulator link velocities V have been found from the O(n) base–to–tip iteration given by $V = J\dot{\theta} = B^*\phi^*H^*\dot{\theta}$. The manipulator mass operator factorization, eq. (2.15), is referred to as the Newton–Euler factorization and reflects the equivalence between the Newton–Euler and the Lagrangian formulations of multibody dynamics. An alternative mass operator factorization is given by the Innovations factorization

$$\mathcal{M} = [I + H\phi K]D[I + H\phi K]^* \; . \tag{2.19}$$

Since it is also true that $[I+H\phi K]^{-1} = [I-H\psi K]$, $\mathcal{M}$ has the operator inversion

$$\mathcal{M}^{-1} = [I - H\psi K]^* D^{-1}[I - H\psi K] \; . \tag{2.20}$$

We now give some useful identities. An important identity is

$$\phi K H\psi = \psi K H\phi = \phi - \psi \tag{2.21}$$

which can be easily used to show that

$$[I - H\psi K]H\phi = H\psi \; . \tag{2.22}$$

It is also true that

$$\psi M\phi^* = P + \tilde{\psi}P + P\tilde{\phi}^* = \psi P + P\tilde{\phi}^* \; , \tag{2.23}$$

where $\tilde{\phi} \triangleq \phi - I$, $\tilde{\psi} \triangleq \psi - I$, and P is given by the tip–to–base iteration (2.7) (namely (2.13)). Note from (2.6) that it is easy to show

$$\tilde{\phi} \triangleq \phi - I = \phi\mathcal{E}_\phi = \mathcal{E}_\phi\phi \; . \tag{2.24}$$

A related result to (2.23) is

$$\psi^* H^* D^{-1} H\psi \; = \; \Omega + \tilde{\psi}^*\Omega + \Omega\tilde{\psi} \tag{2.25}$$

$$\Omega \; = \; \mathcal{E}_\psi^*\Omega\mathcal{E}_\psi + H^* D^{-1} H \; . \tag{2.26}$$

Note that $H^*D^{-1}H$ is block diagonal. As a conseqence $\Omega = \text{diag}[\Omega(1), \cdots, \Omega(n)]$ is block diagonal. Ω can be computed via a base–to–tip iteration that is implied by eq. (2.26). For an n–link manipulators, n iterates are required. Note also that given the fundamental spatial operators (2.1)–(2.5), two iterations are required to do this since P, D, and $\mathcal{E}_\psi$ must be computed before (2.26) can be performed.

3. Iterative Operational Space Control

The operator factorization of the jacobian, eq. (2.14), tell us how to iteratively compute $T = J^*F_c(0)$ given the end effector force (1.2) via a tip–to–base iteration. It remains to show how to iteratively compute $\Lambda(0)$, $c(0)$, and $g(0)$. Turning first to $\Lambda(0)$, note that (1.4), (2.14), (2.20), and (2.22) yield

$$\Omega(0) = B^*\psi^* H^* D^{-1} H\psi B$$

which with (2.25) immediately gives

$$\Omega(0) = B^*\Omega B \; , \tag{3.1}$$

since $B^*\tilde{\psi}^*\Omega B = 0$. The complete iterative algorithm for the computation of $\Lambda(0)$ is given by the combination of equations (2.26), (3.1), and (1.4):

$$\Lambda(0) = \Omega(0)^{-1} \ , \quad \Omega(0) = B^*\Omega B \ , \quad \Omega = \mathcal{E}_\psi^*\Omega\mathcal{E}_\psi + H^*D^{-1}H \ . \qquad (3.2)$$

Since Ω is block diagonal, $\Omega(0) = \phi^*(1,0)\Omega(1)\phi(1,0)$. Computing $\Lambda(0) = \Omega(0)^{-1}$ from (3.2) requires $O(n)$ computations since n iterations are needed to produce $\Omega(0)$ (see the comments following (2.26)) and there is a flat cost associated with inverting the 6×6 matrix $\Omega(0)$. Thus the computation of $\Lambda(0)$ requires a single $O(n)$ tip–to–base interative sweep.

Focusing now on the development of a $c(0)$–algorithm, note that equations (1.5), (2.16), (2.14), (2.20), (2.18), and (2.22) taken together result in

$$c(0) = \Lambda(0)\left\{B^*\psi^*H^*D^{-1}H\psi(M\phi^*a + b) - B^*\phi^*a - a(0)\right\} \ . \qquad (3.3)$$

From (2.23), (2.24), (2.10), and (2.21),

$$\begin{aligned} \psi^*H^*D^{-1}H\psi M\phi^* &= \psi^*H^*D^{-1}H(\psi P + P\tilde{\phi}^*) \\ &= \psi^*H^*D^{-1}H\psi P + \psi^*H^*G^*\tilde{\phi}^* \\ &= \psi^*H^*D^{-1}H\psi P + \psi^*H^*K^*\phi^* \\ &= \psi^*H^*D^{-1}H\psi P + \phi^* - \psi^* \ . \end{aligned} \qquad (3.4)$$

Equations (3.3) and (3.4) together yield

$$c(0) = \Lambda(0)\left\{B^*\psi^*[H^*D^{-1}H\psi(Pa + b) - a] - a(0)\right\} \ . \qquad (3.5)$$

It is numerically preferable to work with the equivalent form

$$\Omega(0)c(0) = B^*\psi^*[H^*D^{-1}H\psi(Pa + b) - a] - a(0) \ . \qquad (3.6)$$

During the $O(n)$ tip–to–base iterative sweep in (3.6) represented by ψ, the quantities P and D can be computed. After the subsequent $O(n)$ base–to–tip sweep given by ψ^* and ϕ^* is done, the right hand side of (3.6) and $\Omega(0)$ can be computed and $c(0)$ can then be solved for using a flat–cost efficient method. Thus to compute $c(0)$ (given the velocity dependent terms a, $a(0)$, and b) by the algorithm (3.6) requires two $O(n)$ sweeps along the manipulator. Another base–to–tip $O(n)$ iterative sweep is needed to compute the velocity dependent quantities a, $a(0)$, and b before algorithm (3.6) can be applied. Thus, most generally, three iterative $O(n)$ sweeps across the manipulator are needed to compute $c(0)$. Note however that by using slightly delayed velocity dependent terms, only two sweeps are needed since the velocity dependent terms can be computed during the second sweep of (3.6) and then used during the first sweep at the next invocation of algorithm (3.6).

Similarly to the development for the $c(0)$ algorithm, (1.6), (2.17), (2.14), (2.20), and (2.22) result in

$$g(0) = \Lambda(0)B^*\psi^*H^*D^{-1}H\psi M\phi^*Ea_g \ . \qquad (3.7)$$

Equations (3.4) and (3.7) together give

$$g(0) = \Lambda(0)B^*\left\{\psi^*(H^*D^{-1}H\psi HP - I) + \phi^*\right\}Ea_g \ . \qquad (3.8)$$

An equivalent statement is

$$\Omega(0)g(0) = B^* \left\{ \psi^*(H^* D^{-1} H \psi H P - I) + \phi^* \right\} E a_g \ . \tag{3.9}$$

Algorithm (3.9) is very similar to (3.6) and the same implementation comments apply, except that there is no need to process velocity dependent terms. To compute $g(0)$, then, requires two O(n) iterative sweeps along the manipulator.

4. Conclusions

Algorithm (3.2) enables $\Lambda(0)$ to be computed via two iterative sweeps of the manipulator. Algorithm (3.9) allows $g(0)$ to also be computed via two O(n) iterative sweeps across the manipulator. Assuming the availability of the velocity dependent terms a, $a(0)$, and b, algorithm (3.6) allows $c(0)$ to be computed via two O(n) sweeps of the manipulator. These algorithms can be run simultanously to produce $\Lambda(0)$, $c(0)$ $g(0)$, and $F_c(0)$ of (1.2) as the output of a two–sweep O(n) iterative procedure. As discussed above, generally an additional O(n) iterative sweep is needed to compute the velocity dependent terms but if a slight time delay can be accepted in the values of these terms, the need for an additional distinct sweep can be avoided. Once $F_c(0)$ is available a final O(n) tip–to–base iterative sweep is needed to compute $T = J^* F_c(0)$.

Acknowledgement

This work has been partially performed at the Jet Propulsion Laboratory, California Institute of Technology, under a contract with the National Aeronautics and Space Administration. K. Kreutz–Delgado is also supported by an NSF Presidential Young Investigator Award.

References

[1] Khatib, O., "A Unified Approach for Motion and Force Control of Robot Manipulators: the Operational Space Formulation", *IEEE J. Robotics and Automation*, Vol. RA–3, pp. 43–53.

[2] Rodriguez, G., and Kreutz, K., "Recursive Mass Matrix Factorization and Inversion: An Operator Approach to Open– and Closed–Chain Multibody Dynamics", JPL Publication 88–11, Jet Propulsion Laboratory, Pasadena, CA, 1988.

[3] Rodriguez, G., Kreutz, K., and Jain, A., "A Spatial Operator Algebra for Manipulator Modeling and Control", *1989 IEEE Conf. Robotics and Automation*, Scottsdale, AZ, May 1989.

[4] Rodriguez, G., Jain, A., and Kreutz, K., "Spatial Operator Algebra Framework for Multibody System Dynamics", 3rd *Annual Conf. Aerospace Computational Control*, Oxnard, CA, Aug. 1989.

SECTION 2 : PATH AND TASK PLANNING

Fast Planning of Collision-Free Paths For A Point Robot

Charles E. Campbell, Jr.
NASA, Goddard Space Flight Center

Abstract

A safe trajectory for a point robot is determined by specifying a sequence of edges which are alternately traversed (*slide*) and then linked (*jump*) by line segment paths, referred to herein as *slide-jump trajectories*. A slide-jump trajectory exists if any safe trajectory exists. Hence, if no slide-jump trajectories exist, then there is no solution. An $o(n^3)$ algorithm (in path safety checks) is developed herein, where n is the number of edges in the environment. Heuristic algorithms are presented which relax the slide-jump trajectory towards a sequence of locally optimal but not necessarily safe trajectories. Safety of the relaxed trajectory is maintained throughout the planning process.

Introduction

In this article, a *path* will refer to a single straight line segment and a *trajectory* will be a sequence of paths. Consider a two-dimensional environment with obstacles made up of edges and a point robot with two degrees of freedom. a start, and a goal. Then, the shortest safe trajectory for the robot between the start and goal, assuming one exists, will be a simply connected set of paths. Furthermore, each path will terminate upon a vertex, start point, or goal point.

In three dimensions, and with a three DOF point robot, some simply connected set of paths will comprise the shortest safe trajectory, but with the trajectory's paths terminating on edges (and the start/goal points). An algorithm was developed in [5] which determined the shortest distance trajectory given a sequence of edges to be reached (optimal in the sense of the sum of the squared lengths of paths).

One may find a path which achieves the shortest distance between two edges by considering at most three types: a path which is orthogonal to both edges, a path which terminates on one edge and is orthogonal to the other, or a path which terminates on vertices of both edges; such a path will be termed a *jump*. In the slide-jump algorithm, trajectories are generated which alternate between slides along edges and jumps between edges. In [1], it has been shown that if any safe trajectory exists between a start and a goal, then a slide-jump trajectory will also exist, where jumps include both the shortest distance path between the two edges and the four possible endpoint to endpoint connecting paths.

The Slide-Jump Algorithm

Since slides connect jumps, one need only specify the coordinates of jumps; slides are thereby implicitly defined. Assume that edge e_i is specified by two endpoints, P_{A_i} and P_{B_i}, and that points along the edge are specified by $P_i = \alpha(P_{A_i} - P_{B_i}) + P_{B_i}$, $\alpha \in [0,1]$. Then, a pair of αs will specify a jump between edges. Algorithm 1 uses these properties to find a path which achieves the shortest distance between two edges, taking into account parallel edges, "edges" which are really points, and skew edges. Note that the resulting αs will be in the interval $[0,1]$.

Algorithm 1 *Determine a shortest distance path between two edges.*

1. Given: P_{A_i}, P_{B_i}, P_{A_j}, and P_{B_j}
2. Let $P_{AB_i} = P_{A_i} - P_{B_i}$, $P_{AB_j} = P_{A_j} - P_{B_j}$, and $den = \|P_{AB_i}\|^2\|P_{AB_j}\|^2 - (P_{AB_i}{}'P_{AB_j})^2$
3. If $|den| > \epsilon$: (ϵ is a small positive constant)
   (a) $r_i = P_{AB_i}{}'(P_{B_j} - P_{B_i})/den$
   (b) $r_j = P_{AB_j}{}'(P_{B_i} - P_{B_j})/den$
   (c) $\alpha_i = r_i\|P_{AB_j}\|^2 + r_j(P_{AB_i}{}'P_{AB_j})$
   (d) $\alpha_j = r_j\|P_{AB_i}\|^2 + r_i(P_{AB_i}{}'P_{AB_j})$
4. Else if $|den| \leq \epsilon$, then:
   Let ℓ_i be true if $\|P_{AB_i}\|^2 < \epsilon^2$, false otherwise.
   Let ℓ_j be true if $\|P_{AB_j}\|^2 < \epsilon^2$, false otherwise.
   (a) If $(\ell_i \wedge \neg\ell_j)$: set $\alpha_i = 0$, $\alpha_j = (P_{B_i} - P_{B_j})'P_{AB_j}/\|P_{AB_j}\|^2$
   (b) Else if $(\neg\ell_i \wedge \ell_j))$: set $\alpha_j = 0$, $\alpha_i = (P_{B_j} - P_{B_i})'P_{AB_i}/\|P_{AB_i}\|^2$
   (c) Else if $(\neg\ell_i)$: set both $\alpha_i = \alpha_j = -1$.
   (d) Else set both $\alpha_i = \alpha_j = 0$.
5. (a) If ($\|P_{AB_j}\|^2 > \epsilon^2$):
      i. If $(\alpha_i < -\epsilon)$: set $\alpha_i = 0$, $\alpha_j = (P_{B_i} - P_{B_j})'P_{AB_j}/\|P_{AB_j}\|^2$
      ii. Else if $(\alpha_i > 1+\epsilon)$: set $\alpha_i = 1$, $\alpha_j = (P_{A_i} - P_{B_j})'P_{AB_j}/\|P_{AB_j}\|^2$
   (b) Else $\alpha_j = 0$
6. (a) If ($\|P_{AB_i}\|^2 > \epsilon^2$):
      i. If $(\alpha_j < -\epsilon)$: set $\alpha_j = 0$, $\alpha_i = (P_{B_j} - P_{B_i})'P_{AB_i}/\|P_{AB_i}\|^2$
      ii. Else if $\alpha_j > 1+\epsilon$: $\alpha_2 = 1$, $\alpha_i = (P_{A_j} - P_{B_i})'P_{AB_i}/\|P_{AB_i}\|^2$

(b) Else $\alpha_i = 0$

7. (a) If $\alpha_i < 0$: set $\alpha_i = 0$
   (b) Else if $\alpha_i > 1$: set $\alpha_i = 1$
   (c) If $\alpha_j < 0$: set $\alpha_j = 0$
   (d) Else if $\alpha_j > 1$: set $\alpha_j = 1$

The slide-jump algorithm constructs both a list of edges and viable trajectories. Initially, there are two sets: 1) $AEL_0 = \{s\}$, where s is the start point and AEL_0 is the 0th Attained Edge List, and 2) NAEL = { all other edges, and g} (the Not Attained Edge List). The algorithm proceeds by constructing AEL_1, the set of all edges that can be arrived at via safe slide-jumps from s and removing those edges from NAEL. Then, it constructs AEL_2, the set of all edges not in AEL_1 or AEL_0 which can be reached via a safe slide-jump from some edge in AEL_1. This process continues until the goal g is reached or no further edges in NAEL can be reached. In the latter case, no safe trajectory to the goal exists for the 3DOF point robot.

This algorithm can be modified: one may choose to continue constructing new AEL_is even after the goal has been attained in the hopes of finding a shorter trajectory, and one may have several points or edges comprising the "goal". The algorithm as shown below is already modified to determine all edges attainable via slide-jump trajectories; another set, the "Goal List" GL_i, is introduced which is comprised of those edges in AEL_i which safely connect to the goal set via a safe slide-jump path pair.

Trajectory information is assumed to be retained at each attained edge. In the algorithm below, the sum of slide and jump distances needed to reach each edge and a pointer to the edge in the previous AEL_{k-1} from which it was reached is kept. In the frequent case that an edge may be attained from several edges in the previous AEL_{k-1}, only the trajectory information yielding the shortest trajectory is kept.

Algorithm 2 *Constructing Slide-Jump Trajectories*

1. Set $i = 0$, $AEL_i = \{s\}$, and NAEL = { all edges except s}, and $GL_i = \{\emptyset\}$.
2. Determine if the path $s \rightarrow g$ is safe; if so, terminate the algorithm with $GL_0 = \{s\}$.
3. While $AEL_i \neq \{\emptyset\}$:
   (a) Set $i = i + 1$: $GL_i = \{\emptyset\}$
   (b) For each edge $\bar{e}$ in NAEL, do:
   (c) For each edge e in AEL_{i-1}, do:
      i. Use Algorithm 1 to determine the αs for the jump path between $\bar{e}$ and e.

ii. Determine if the algorithm 1 jump or any of the endpoint to endpoint connecting paths between the two edges is safe.

iii. If edge $\bar{e}$ has not yet been attained, or if the slide+jump path distances + the slide-jump trajectory distance to e is less than that stored with $\bar{e}$: enter a pointer to e and the total slide-jump trajectory distance to $\bar{e}$ into $\bar{e}$'s trajectory information.

iv. If $\bar{e} = g$, then remove e from AEL_{i-1} and append it to GL_{i-1}.

(d) If $\bar{e}$ has been attained, and $\bar{e} \notin \{g\}$, then move $\bar{e}$ from NAEL to AEL_i.

Slide-Jump Trajectory Relaxation

The slide-jump algorithm produces safe and feasible trajectories for a point robot whenever any safe trajectory exists. By backtracking from the non-empty GL_i, using the trajectory information retained with the edges in the AEL_is, usually quite a few trajectories are available. These trajectories are not the shortest ones possible, and it is sometimes desirable to compute shorter trajectories. One approach is to use the slide-jump trajectories as starting information for optimal control formulations which include dynamics [2, 3, 4]. Alternatively, one may first use the heuristic relaxation algorithms discussed below, possibly in order to identify topologically distinct trajectories prior to using optimal control methods on them.

The first relaxation strategy forms trajectories solely composed of jumps between edges. Associated with each edge in a slide-jump trajectory sequence is an α^a (arrival) and α^j (jump-off). Relaxation of the slide-jump trajectory between e_i and e_{i+1} occurs by letting α^a_{i+1} approach α^j_{i+1}. This is called single-pivot relaxation, as the jump from e_i to e_{i+1} appears to pivot about $e_i : \alpha^j_i$ and sweep along e_{i+1}. Note that the single-pivot operation sweeps out a triangular region; if some edge intersects the triangular region, then it "blocks" the relaxation. One then continues to pivot from the latest blocking edge until $\alpha = \alpha^j_{i+1}$. This strategy is illustrated in Figure 1 with one blocking edge and detailed below.

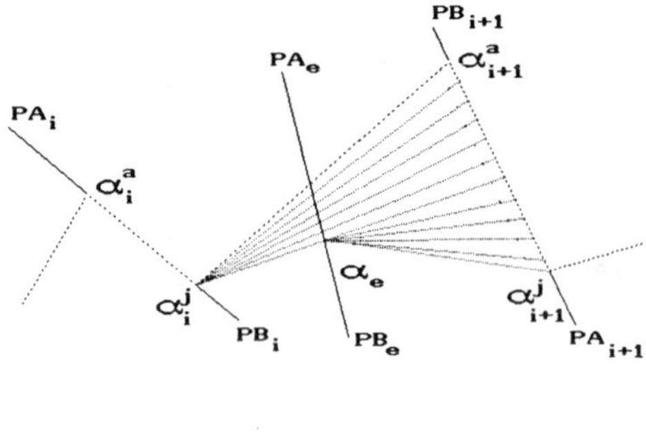

Figure 1: Slide-Jump Trajectory Relaxation

An edge which "blocks" the sweeping of the path from α^a_{i+1} to α^j_{i+1} has a common point $\text{P} = \alpha_e \text{P}_{\text{AB}_e} + \text{P}_{\text{B}_e} = \Delta \text{R}_{\text{BA}} + \text{R}_{\text{A}}$, where $\text{R}_{\text{A}} = \alpha^j_i \text{P}_{\text{AB}_i} + \text{P}_{\text{B}_i}$,

$R_B = \alpha^a_{i+1}P_{AB_{i+1}} + P_{B_{i+1}}$, and $R_{BA} = R_B - R_A$ (R_A and R_B specify the end-points of the jump from edge e_i to e_{i+1}). For skew edges, this observation can be used to obtain the following equation.

$$\Delta\alpha_{i+1}P_{AB_{i+1}} + \Delta(P_{B_{i+1}} - R_A) - \alpha_e P_{AB_e} = P_{B_e} - R_A$$

This constitutes three equations in the three unknowns $\Delta\alpha_{i+1}$, Δ, and α_e. Note that $\Delta = 0$ implies an edge e in contact with the beginning of the jump, and since such an edge can't block the sweep, then such edges can be ignored. Thus, one can obtain α_{i+1} for potentially blocking edges. A sweep is blocked, of course, only when $\Delta \in [0,1]$, $\alpha_{i+1} \in [\alpha^a_{i+1}, \alpha^j_{i+1}]$, and $\alpha_e \in [0,1]$.

When the determinant of the 3x3 matrix is zero, then the three vectors $P_{AB_{i+1}}$, $P_{B_{i+1}} - R_A$, and $-P_{AB_e}$ are coplanar, and so the edges e, e_{i+1}, and the point R_A all lie in the same plane. In this case, letting $V_1 = P_{AB_{i+1}} \times (P - R_A)$ and $V_2 = (P_{B_{i+1}} - R_A) \times (P - R_A)$, where $P = P_{A_e}$ and then P_{B_e} for each edge e, one can obtain α_{i+1}.

$$\alpha^2_{i+1}\|V_1\|^2 + 2\alpha_{i+1}V_1'V_2 + \|V_2\|^2 = 0$$

When this equation is needed, then α_e and Δ are:

$$\alpha_e = \begin{cases} 1 & \text{if } P = P_{Ae} \\ 0 & \text{otherwise} \end{cases}$$

$$\Delta = \frac{(P - R_A)'(\alpha_{i+1}P_{AB_{i+1}} + P_{B_{i+1}} - R_A)}{\|\alpha_{i+1}P_{AB_{i+1}} + P_{B_{i+1}} - R_A\|^2}$$

The relaxation process either sweeps the jump to α^j_{i+1} or is blocked by the edge e which minimizes $\left|\alpha_{i+1} - \alpha^a_{i+1}\right|$ and satisfies the boundary constraints on Δ and α_e. The point P on edge e which lies at the blockage takes the place of e_i:α^j_i as the new pivot. The path is then swept onwards until either α_{i+1} equals α^j_{i+1} or with edges blocking the sweep and specifying new pivots. Since $|\alpha_{i+1} - \alpha^j_{i+1}|$ is bounded from below by zero, the number of edges is finite, and the area of the triangle of the pivot, P, and α_{i+1}:e_{i+1} is approaching zero, this process must terminate in finite time. Note that new edges are inserted into the trajectory's edge sequence, associated with their α_es.

A second relaxation strategy uses the algorithm in [5] which generates optimal α^*s for a specific sequence of edges. The optimal α^*s are not guaranteed to produce safe trajectories, however, so they act as a direction for relaxation to proceed. In Figure 2, three edges are shown with the current jump-jump αs (no superscript), sequence-optimal α^*s, and the result of applying single relaxation pivots until edges first blocked the sweeps (α^-, α^+).

For each blocked path sweep, a number $f = (\alpha^\pm_i - \alpha^*_i)/(\alpha_i - \alpha^*_i)$ is determined. Choose $f = \max_i f_i$ for all blocked sweeps ($f = 0$ if there are no blocked sweeps) and set α_i for all edges in the sequence to $f(\alpha_i - \alpha^*_i) + \alpha^*_i$. Note

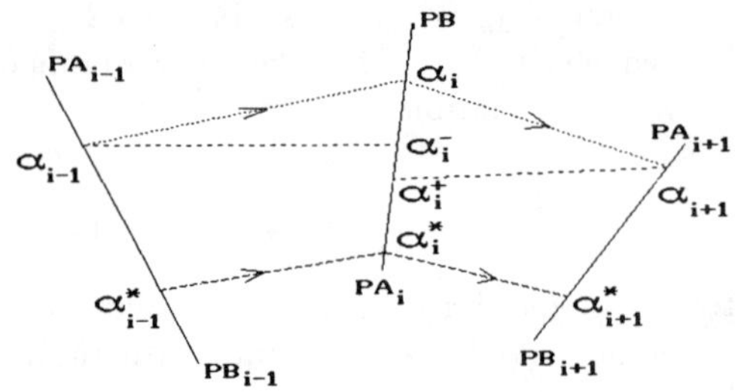

Figure 2: Jump-Jump Trajectory Relaxation

that $f \in [0,1]$ and, at $f = 1$, the original jump-jump α sequence applies and at $f = 0$, the sequence-optimal α^* applies. The edge for the selected f is inserted appropriately into the edge sequence along with the associated α_e.

A third relaxation strategy determines if edges can be safely and beneficially removed from the safe jump-jump trajectory. The shortest vector from point $e_i{:}\alpha_i$ to the path directly connecting $e_{i-1}{:}\alpha_{i-1}$ to $e_{i+1}{:}\alpha_{i+1}$ is used as the endpoint of sweeps originating from edges on either side of edge e_i in the trajectory's edge sequence. If the trajectories can be safely swept in that direction, then edge e_i is dropped from the sequence and any blocking edges appropriately inserted.

The slide-jump trajectory relaxation is performed once to form a jump-jump trajectory. The jump-jump relaxation techniques retain the jump-jump form and should be applied until no more relaxation occurs. Each jump-jump relaxation application will shorten the trajectory.

References

1. C. E. Campbell, Jr. *Analysis and Planning of Optimal and Collision-Free Paths for Industrial Robots with a Prismatic Joint.* PhD thesis, Purdue Univ., 1985.
2. E. G. Gilbert and D. W. Johnson. Distance functions and their application to robot path planning in the presence of obstacles. In *Proceedings Eighteenth Ann. Conf. Information Sciences and Systems*, Mar 1984.
3. E. G. Gilbert and D. W. Johnson. Distance functions and their application to robot path planning in the presence of obstacles. *IEEE J. of Robotics and Automation*, RA-1:21–30, Mar 1985.
4. D. W. Johnson and E. G. Gilbert. Minimum time robot path planing in the presence of obstacles. In *Proceedings of 24th Conference on Decision and Control*, pages 1748–1753, Dec 1985.
5. J. Luh and C. E. Campbell, Jr. Minimum distance collision-free path planning for industrial robots with a prismatic joint. *IEEE Trans. Auto. Ctrls*, AC-29:675–680, Aug 1984.

ENSURING TASK MOTION FEASIBILITY USING ALTERNATIVE INSTANTANEOUS MOTION PLANNING

Ron P. Podhorodeski, Assistant Professor
Dept. of Mechanical Engineering, University of Victoria
Victoria, B.C., CANADA, V8W 2Y2

ABSTRACT

For a desired end effector task space velocity to be possible, corresponding feasible joint motions must exist. For manipulators with insufficient joint degrees of freedom, or for manipulators operating at or near a degenerate configuration arbitrary task space motion is not possible.

In this paper, an approach for the on-line planning of optimal feasible task space velocities for the above cases is considered. The approach is based upon using reciprocal screw quantities and joint rate truncation to constrain the planned motion to be feasible. Subject to these feasibility constraints optimization is used to find the alternative motion best satisfying priority components of the original motion.

Explicit expressions are derived for planning alternative motions when translational velocity is considered a priority over angular velocity. The introduced concepts and formulations are demonstrated with a kinematic simulation of a Puma-type manipulator executing a motion trajectory near its wrist degeneracy.

1. INTRODUCTION

1.1 Preamble

For a desired robot end effector instantaneous motion (velocity) to be possible, the corresponding joint motions must be feasible. As an example where this is not the case, consider a manipulator with five joint axes. Use of such a device for a task requiring six independent motion degrees of freedom (d.o.f.) is not possible. Similarly, the encounter of configurations involving degenerate alignments of the joint axes for a nominally full d.o.f. manipulator will lead to the loss of task motion d.o.f.. Near these degenerate configurations, infeasibly high joint rates can be required to provide a desired task space motion.

For some applications, the task can be carefully preplanned such that the capabilities of the robot are never exceeded. Complete preplanning, however, can be difficult due to the prevalence of certain degenerate configurations [1]. Programmers of industrial robots can attest to the difficultly caused by the wrist degeneracy, in the planning of tasks involving continuous end effector motions. Furthermore, complete task preplanning is impossible for applications involving end effector motion specifications based on sensor input. This autonomy in motion specification is present in many industrial and telerobotic applications.

In this paper, a new approach for ensuring feasible task motion specification is addressed.

1.2 Background

An end effector velocity can be described by the screw quantity $\mathbf{M} = \{\mathbf{m}; \mathbf{m}_o\}^T$. The primary 3×1 vector $\mathbf{m}$ is comprised of components of the end effector's angular velocity about a set of Cartesian reference axes. The secondary 3×1 vector $\mathbf{m}_o$ is the translational velocity that the reference frame origin would have if attached to the end effector.

If the n joint axes, $\$_i$, $i = 1, n$, of an open chain serial manipulator are described by the screw coordinates of the axes, $\$_i = \{\mathbf{s}_i; \mathbf{s}_{o_i}\}^T$, †1 the end effector velocity resulting from joint rates $\dot{q}_i$, $i = 1, n$, about the axes is

$$\mathbf{M} = \sum_{i=1}^{n} \$_i \dot{q}_i = [\$_1 \ \$_2 \ \cdots \ \$_n]\{\dot{q}_1 \ \dot{q}_2 \ \cdots \ \dot{q}_n\}^T \equiv [\mathbf{J}]\dot{\mathbf{q}} \tag{1-1}$$

The joint screw system Ψ_{jt} is comprised of all linear combinations of the $\$_i$, and has an order equal to the number of linearly independent $\$_i$. The right hand side of (1-1) indicates the equivalence of the screw representation and the matrix commonly referred to as the manipulator's Jacobian (see also Hunt [3]). While equivalent, it is important to note that each column of the Jacobian is comprised of two vectors having different units.

If the joint axes provide insufficient task space d.o.f. the desired motion, $\mathbf{M}_d$, will not belong to Ψ_{jt} and the linear system of (1-1) becomes overdetermined. In this case, the forward velocity solution can be expressed as

$$\mathbf{M}_d = \mathbf{M}_{err} + \sum_{i=1}^{n} \dot{q}_i \$_i \tag{1-2}$$

where $\mathbf{M}_{err}$ is the error of the motion provided by $\dot{q}_i$, $i = 1, n$.

1.3 Previous Approaches

Pseudo-inverse based solutions have been suggested for overdetermined inverse kinematic problems, i.e.,

$$\{\dot{q}\} = [\mathbf{J}]^+ \mathbf{M}_d = ([\mathbf{J}]^T [\mathbf{J}])^{-1} [\mathbf{J}]^T \mathbf{M}_d \tag{1-3}$$

The joint rates of (1-3) yield a least-squares error solution. That is,

$$\mathbf{M}_{err} \cdot \mathbf{M}_{err} = \mathbf{m}_{err} \cdot \mathbf{m}_{err} + \mathbf{m}_{o_{err}} \cdot \mathbf{m}_{o_{err}} \tag{1-4}$$

is minimal. The product of (1-4) is the orthogonal product of screws (Sugimoto and Duffy [4]). This product is unit inconsistent, and it is not invariant of the location of the frame of reference or of the measure of length (see also Lipkin and Duffy [5]). As such, the solution of (1-3) and its associated motion error, will depend on the choice of reference and of measure of length. Further, the minimization of (1-4) does not have a physical significance in manipulation.

Damped least-squares methods have been recommended for the solution of joint rates near degenerate configurations (Nakumura and Hanafusa [6], Wampler [7]). These solutions are based on minimizing the Euclidean norm of the motion error (equation (1-4)) summed with a weighting of the norm of the joint rate demand. The weighting determines the emphasis placed on the joint rate demand and was based on a dexterity measure in [6] and a constant weighting in [7]. The earlier comments on the variance and lack of physical significance apply to the damped solutions. As well, the damped solutions will lead to unnecessary errors for feasible motions near degeneracies, and can not ensure the satisfaction of hard constraints on the joint rates.

†1 $\mathbf{s}_i = \mathbf{L}_i$; $\mathbf{s}_{o_i} = \mathbf{L}_{o_i} + p_L \mathbf{s}_i$; where $\{\mathbf{L}; \mathbf{L}_o\}$ denote the line coordinates (direction and moment about the reference origin) of the joint axes i, and p_{L_i} is the linear pitch associated with the axis (infinite ≡ prismatic, 0 ≡ revolute). See Hunt [2] for screw system definitions and terminology.

Maciejewski and Klein [8], using the damped solution, employed numerical filtering to eliminate the induced errors for feasible motions, and formulated a technique for the selection of the joint demand weighting to ensure the satisfaction of hard joint rate limits. Again, however, the solution is based on the least-square of the motion error discussed earlier.

Aboaf and Paul [9] employed velocity truncation and Jacobian reduction and reformulation, to ensure the satisfaction of joint rate limits and translational velocity components. for spherical wristed manipulators operating near the wrist degeneracy, The philosophy of this approach is the closest to that of this work, but is limited to a specific application and motion priority objective, and is burdened by the need to reformulate a reduced Jacobian with respect to a frame located at the end effector tip.

1.4 The Presented Approach

The approach presented in this work is based upon the planning of à feasible alternative instantaneous motion satisfying both the motion capabilities of the robot and priority components of the original desired motion. Substitution of the planned alternative motion for the original desired motion specification, allows feasible joint velocities to be resolved. The novelity of the approach stems from the finding of a best feasible alternative motion based on optimization subject to capability constraints. The objective function used in the optimization can be chosen such as to reflect the requirements of the task.

Once the optimal alternative motion, $\mathbf{M}_{alt} \in \Psi_{jt}$, has been found, the required joint velocities are resolved by finding $\dot{q}_i$, $i = 1, n$, satisfying

$$\mathbf{M}_{alt} = \sum_{i=1}^{n} \dot{q}_i \$_i \tag{1-5}$$

The inverse instantaneous kinematic problem of (1-5) can be resolved using techniques such as screw decomposition [10].

The remainder of this paper is organized as follows. In Section 2, reciprocity conditions defining the feasibility of the specified motion are presented for overdetermined systems, and are extended to near degenerate systems. In Section 3, an alternative velocity expression, optimally satisfying the desired translational velocity at the expense of a minimal angular velocity error, is derived. In Section 4, the introduced approach is demonstrated through application to spherical wristed manipulators operating near the wrist degeneracy. Further consideration and conclusions are addressed in Sections 5 and 6.

2. ENFORCING MOTION FEASIBILITY

2.1 Overdetermined Cases

An exact inverse velocity solution will exist for a desired end effector motion , $\mathbf{M}_d$, if $\mathbf{M}_d \in \Psi_{jt}$. If $\mathbf{M}_d$ is an arbitrary screw quantity belonging to the screw system Ψ_m, an inverse velocity solution will exist only if $\Psi_m \subset \Psi_{jt}$. For an arbitrary velocity in three dimensional space Ψ_m corresponds to the six-system of all screw quantities [2]. Hence, to provide arbitrary motion requirements in three dimensional space Ψ_{jt} must also be the six-system of all screws. For the remainder of this paper it will be assumed that $\mathbf{M}_d$ belongs to the six-system.

The manipulator joint axes screws will form an r-system, where $r \le 6$. For $r < 6$ there is a $(6 - r)$-system, Γ_{jt}, which is reciprocal Ψ_{jt}. Let $\mathbf{W}_i = \{\mathbf{w}_i; \mathbf{w}_{o_i}\}^T$, $i = 1, (6 - r)$, denote linearly independent screw quantities forming a basis for screw system Γ_{jt}. Since Ψ_{jt} and Γ_{jt} are reciprocal,

$$\mathbf{W}_i \ \chi \ \$_j = \mathbf{w}_i \cdot \$_{o_j} + \mathbf{w}_{o_i} \cdot \$_j = 0, \quad i = 1, (6 - r), \ j = 1, n \tag{2-1}$$

where χ denotes the reciprocal product of the screw quantities.

The instantaneous motion at the end effector is $\mathbf{M} = \sum_{i=1}^{n} \dot{q}_i \, \$_i$. Therefore, to be feasible, a specified motion must be reciprocal to Γ_{jt}, i.e.,

$$\mathbf{M} \, \chi \, \mathbf{W}_i = 0, \quad i = 1, (6 - r) \tag{2.2}$$

Satisfaction of the reciprocal conditions of (2-2), ensures that $\mathbf{M} \in \Psi_{jt}$. If Γ_{jt} is considered to be a screw system representing a force system, satisfaction of the constraints of (2-2) can be interpreted as forcing **M** to be incapable of performing work which the manipulator end effector is unable to produce.

Let $\mathbf{M}_{alt}^{end} = \{\omega_x', \omega_y', \omega_z'; v_v', v_y', v_z'\}^T$, represent an arbitrary alternative velocity specification with respect to an end effector tip located frame. Assuming that the screws defining Γ_{jt} are known with respect to a frame, *ref*, enforcement of (2-2) yields

$$a_i \omega_x' + b_i \omega_y' + c_i \omega_z' + d_i v_x' + e_i v_y' + f_i v_z' = 0, \quad i = 1, (6 - r) \tag{2-3}$$

where $\{d, e, f, a, b, c\}_i^T = [\mathrm{T}]_{end \to ref} \mathbf{W}_i^{ref}$, $i = 1, (6 - r)$, with $[\mathrm{T}]_{end \to ref}$ denoting the screw transformation[2] from the *end* to *ref* frames. †2 The conditions of (2-3) define the instantaneous motions belonging to Ψ_{jt}, forming $6 - r$ constraints on the alternative velocity components.

Alternative instantaneous motion planning involves finding an $\mathbf{M}_{alt} \in \Psi_{jt}$ which optimally satisfies priority components of the original desired motion at the sacrifice of secondary motion components. In Section 3, expressions are developed for finding alternative motions satisfying translational velocity components at the sacrifice of a minimum angular velocity error.

2.2 Near Degenerate Configurations

Near degenerate configurations the joint velocities required to generate a desired end effector velocity can be infeasibly high. Velocity limits, $|\dot{q}_i| \le \dot{q}_{i_{\lim}}$, are assumed to exist for each of the joints. If $\mathbf{M}_d$ is found to require a joint velocity $\dot{q}_j$ where $|\dot{q}_j| > \dot{q}_{j_{\lim}}$ then $\dot{q}_j$ is truncated to $\dot{q}_{j_{\lim}}$. The remaining desired velocity screw, after removing the contribution from $\dot{q}_{j_{\lim}}$ is

$$\mathbf{M}_{d_{rem}} = \mathbf{M}_d - (\dot{q}_j / |\dot{q}_j|) \dot{q}_{j_{\lim}} \$_j \tag{2-4}$$

From $\mathbf{M}_{d_{rem}}$, the remaining desired translational, $\mathbf{v}_{d_{rem}}$, and angular, $\omega_{d_{rem}}$, velocities of the end effector can be found. $\mathbf{M}_{alt}$ is sought such that it optimally satisfies priority conditions based on the remaining desired motion components, while requiring no further velocity of joint j.

$\mathbf{M}_{alt}$ requires no further input from joint j, if

$$\mathbf{W} \, \chi \, \mathbf{M}_{alt} = 0 \tag{2-5}$$

where $\mathbf{W} \, \chi \, \$_i = 0$, $i \ne j$. That is, the joint screws of the manipulator after velocity truncation can be treated as effectively forming a five-system. This yields a constraint equation of the form of (2-3), in terms of the translational, $\mathbf{v}'$, and angular, ω', components of the alternative velocity.

In Section 4, alternative instantaneous motion planning is applied to spherical wristed manipulators operating near the wrist degeneracy.

3. TRANSLATIONAL VELOCITY A PRIORITY OVER ANGULAR VELOCITY

In this section, alternative velocity components are derived for the case when translational velocity is considered a priority over angular velocity. Specificly, the minimization of the Euclidean norm of the end effector translational error is considered to be the primary objective, with minimization of the

†2 For implementation it is more efficient to transform $\mathbf{M}_{alt}$ if $r < 5$.

angular velocity error norm a secondary priority, i.e.,

$$\text{1st priority: } \| \mathbf{v}_{err} \| = \text{minimum}, \quad \text{2nd priority: } \| \boldsymbol{\omega}_{err} \| = \text{minimum} \tag{3-1}$$

The alternative velocity solution will be derived for the case when the screw system reciprocal to the joint screws is a one-system. This reciprocal screw system may be due to insufficient joint d.o.f., or may be due to the approach or encounter of a degenerate configuration.

Requiring that $\mathbf{W} \chi \mathbf{M}_{alt} = 0$, where $\mathbf{W}$ defines the reciprocal one-system, yields

$$a\omega_x' + b\omega_y' + c\omega_z' + dv_x' + ev_y' + fv_z' = 0 \tag{3-2}$$

where $a \ldots f$ are described in (2-3). If $a, b,$ and c are not all zero, it is possible to satisfy the desired translational velocity while minimizing the norm of the error in the angular velocity.

Normalizing the reciprocity condition of (3-2), with respect to the largest magnitude of a, b and c, yields

$$\omega_A' = -(B\omega_B' + C\omega_C' + Dv_D' + Ev_E' + Fv_F') \tag{3-3}$$

In (3-3), $B, \ldots, F$ are $b, \ldots, f$ normalized with respect to a, where for the discussion a has been assumed to be the largest of a, b and c. The first objective of minimum translational velocity error is perfectly satisfied by letting $\mathbf{v}'$ be identical to the desired velocity, $\mathbf{v}_d$. The secondary objective requires $\| \boldsymbol{\omega}_d - \boldsymbol{\omega}' \|$ to be minimal, where $\boldsymbol{\omega}_d$ is the desired end effector angular velocity. This is equivalent to the following minimization problem

$$\text{minimize} \quad (\{\omega_{A_d} - \omega_A', \omega_{B_d} - \omega_B', \omega_{C_d} - \omega_C'\}^T)^2 \tag{3-4}$$

subject to (3-3).

Substituting (3-3) into (3-4), and rearranging, yields

$$\text{minimize} \quad \left(\begin{Bmatrix} \omega_{A_d} \\ \omega_{B_d} \\ \omega_{C_d} \end{Bmatrix} + \begin{Bmatrix} Dv_{D_d} + Ev_{E_d} + Fv_{F_d} \\ 0 \\ 0 \end{Bmatrix} - \begin{bmatrix} -B & -C \\ 1 & 0 \\ 0 & 1 \end{bmatrix} \begin{Bmatrix} \omega_B' \\ \omega_C' \end{Bmatrix} \right)^2 \tag{3-5}$$

where $\mathbf{v}_d$ has been substituted for $\mathbf{v}'$. Equating the derivative of (3-5) with respect to $\{\omega_B', \omega_C'\}^T$ to zero, and solving for $\{\omega_B', \omega_C'\}^T_{opt}$ yields

$$\begin{Bmatrix} \omega_B' \\ \omega_C' \end{Bmatrix}_{opt} = \left(\begin{bmatrix} -B & 1 & 0 \\ -C & 0 & 1 \end{bmatrix} \begin{bmatrix} -B & -C \\ 1 & 0 \\ 0 & 1 \end{bmatrix} \right)^{-1} \begin{bmatrix} -B & 1 & 0 \\ -C & 0 & 1 \end{bmatrix} \begin{Bmatrix} \omega_{A_d} + Dv_{D_d} + Ev_{E_d} + Fv_{F_d} \\ \omega_{B_d} \\ \omega_{C_d} \end{Bmatrix} \tag{3-6}$$

This is equivalent to the pseudo-inverse solution of the overdetermined system

$$\begin{bmatrix} -B & -C \\ 1 & 0 \\ 0 & 1 \end{bmatrix} \begin{Bmatrix} \omega_B' \\ \omega_C' \end{Bmatrix} = \{\omega_{A\,d} + Dv_{D_d} + Ev_{E_d} + Fv_{z\,d}, \omega_{B\,d}, \omega_{C\,d}\}^T \tag{3-7}$$

Substitution of the values found for $\omega_{B\,opt}'$ and $\omega_{C\,opt}'$ into (3-3) yields $\omega_{A\,opt}'$, completing the solution of the optimal alternative velocity components. From $\omega_{A\,opt}'$, $\omega_{B\,opt}'$, $\omega_{C\,opt}'$, and $\mathbf{v}_d$, the alternative screw $\mathbf{M}_{alt}$ can be found.

As stated above, an alternative instantaneous motion exactly satisfying the translational velocity requirement, can be found if all of a, b, and c, are not zero. Due to the "end" frame of reference of the components in (2-3), the coefficients a, b, and c, are the elements of the secondary vector of $\mathbf{W}^{end}$, i.e., $\{a, b, c\}^T = \mathbf{w}_o^{end}$. For $\mathbf{w}_o^{end}$ to be null, $\mathbf{W}$ must be a zero pitch screw passing through the end effector tip. For this case, where the translational velocity cannot be exactly satisfied, the norm of the translational velocity error, $\| \mathbf{v}_d - \mathbf{v}' \|$ is to be minimized. From $\mathbf{W} \chi \mathbf{M}_{alt} = 0$, a constraint of the form

$$v_D' = -(Ev_E' + Fv_G') \tag{3-8}$$

is found. The norm of the error in the translational velocity can be minimized by pseudo-inverse solution of the overdetermined system,

$$\begin{bmatrix} -E & -F \\ 1 & 0 \\ 0 & 1 \end{bmatrix} \begin{Bmatrix} v_F' \\ v_G' \end{Bmatrix} = \{v_{Dd}, v_{Ed}, v_{Fd}\}^T \tag{3-9}$$

4. APPLICATION TO SPHERICAL WRIST MANIPULATORS NEAR THE WRIST DEGENERACY

Spherical wrists commonly consist of three revolute joints having a common point of intersection. A design found on many commercial robots has the three axes successively perpendicular as illustrated on Figure 1. Three revolute joints intersecting at a point are degenerate when all three axes lie in a common plane ($\theta_5 = 0$ for the illustrated wrist). High counter rotating velocities of joints four and six can occur near this degeneracy.

The joint screw coordinates for a six d.o.f. manipulator, with respect to a reference frame having its origin at the wrist intersection, are of the form

$$[\$_1 \; \$_2 \; \$_3 \; \$_4 \; \$_5 \; \$_6] = \begin{bmatrix} \$_1 & \$_2 & \$_3 & \$_4 & \$_5 & \$_6 \\ \$_{o_1} & \$_{o_2} & \$_{o_3} & 0 & 0 & 0 \end{bmatrix} \tag{4-1}$$

Assuming that the joint with the excessive velocity corresponds to $\$_4$ [†3], a screw quantity $\mathbf{W} = \{\mathbf{w}; \mathbf{w}_o\}^T$ reciprocal to $\$_1$, $\$_2$, $\$_3$, $\$_5$ and $\$_6$ is required.

Enforcing reciprocity of $\mathbf{W}$ with $\$_5$ and $\$_6$, yields

$$[\$_5^T \;\; \$_6^T \;]^T \; \mathbf{w}_o = 0 \tag{4-2}$$

Since the directions of joints 5 and 6 are not the same, $\mathbf{w}_o$ must be

$$\mathbf{w}_o = c\,(\$_5 \times \$_6) \tag{4-3}$$

where $\times$ denotes the vector cross product, and c is an arbitrary constant ($c \neq 0$).

Enforcing reciprocity of $\mathbf{W}$ with $\$_1$, $\$_2$ and $\$_3$, yields

$$[\$_{o_1} \;\; \$_{o_2} \;\; \$_{o_3} \;]^T \, \mathbf{w} + [\$_1 \;\; \$_2 \;\; \$_3 \;]^T \, \mathbf{w}_o = 0 \tag{4-4}$$

---

†3 The derivation is valid for truncation of any of the wrist velocities if the indices of the fourth and the truncated joint are interchanged.

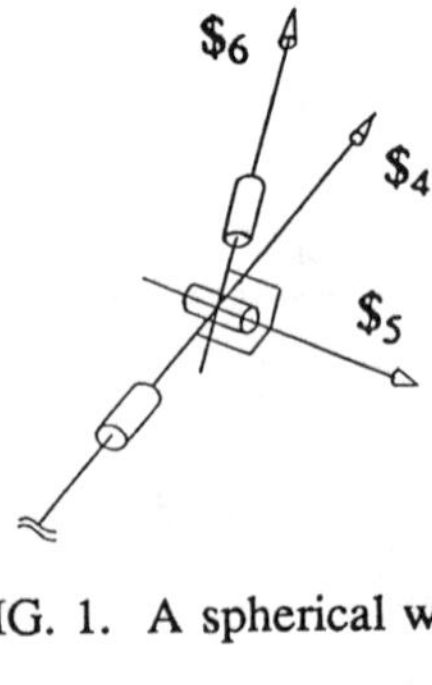

FIG. 1. A spherical wrist

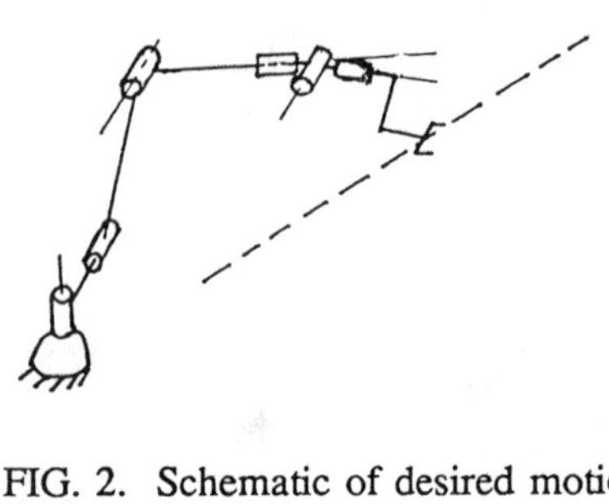

FIG. 2. Schematic of desired motion

If the manipulator is not near any degenerate configuration other than that of the wrist singularity, the 3×3 matrix in (4-4) consisting of $\$_{o_1}$, $s\$o2$, and $\$_{o_3}$ will be well-conditioned, i.e., invertible. Therefore, $\mathbf{w}$ can be found as

$$\mathbf{w} = -([\$_{o_1}{}^T \;\; \$_{o_2}{}^T \;\; \$_{o_3}{}^T]^T)^{-1}[\$_1{}^T \;\; \$_2{}^T \;\; \$_3{}^T]^T \; \mathbf{w}_o \tag{4-5}$$

which in combination with (4-3) yields $\mathbf{W}$, hence defining Γ_{jt}.

EXAMPLE → A Puma-type manipulator was simulated for a straight-line, constant orientation end effector motion as depicted schematically in Figure 2. The velocity specification was based upon a desired velocity along the line and a motion error feedback term. The translational velocity priority solution of Section 3 was used for the simulation. The motion brought the joint displacements near the wrist degeneracy. Truncation of the velocity of joint four occurred as shown on Figure 3(a). Orientation error during the motion was as depicted on Figure 3(b). Translational position error (zero) for the motion was as shown on Figure 3(c).

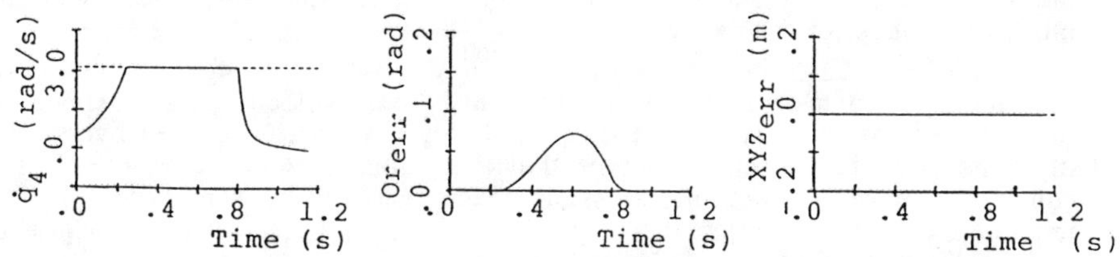

(a) Velocity of Joint 4 (b) Orientation Error (c) Positioning Error

FIG. 3. Alternative Instantaneous Motion Planning Results

5. FURTHER CONSIDERATIONS

It must be emphasized that alternative instantaneous motion planning is based on selecting a best motion from the existing feasible motions. As such, the motion priorities must be chosen such that this selection is plausible. For example, in Section 3 an objective having translational velocity a priority at the expense of minimal angular velocity error was considered. It was demonstrated that exact satisfaction of the translational velocity could not be ensured if the reciprocal screw system could be characterized by a zero pitch screw passing through the end effector tip. A motion priority based on exact satisfaction of the translational velocity would fail in this case. However, a properly chosen motion priority objective, such as the least-squares objectives of Section 3, ensures satisfaction of the translational velocity when possible, and the selection of a best solution for other cases.

Further details must be considered for successful implementation of the alternative instantaneous motion planning scheme for manipulators operating near degenerate configurations. In some applications attempting to optimally satisfy the remaining desired end effector velocity after truncation of a joint velocity value, may lead to infeasible motion requirements at the other joints. For example, this could occur if wrist axes were required to provide a tip translational velocity after truncation of the velocity of a joint of the main-arm subassembly. The feasibility of application must be considered a priori to implementation.

For the wrist application of Section 4, truncation of the velocity at the first or third joints of the wrist will lead to an acceptable velocity requirement at the other joint axes. The loss of end effector translational velocity caused by this truncation is easily compensated for by joints of the main-arm subassembly.

If the displacement of the second wrist axis (the pitch of the wrist) passes through the singularity during the alternative motion planning, i.e., switches from positive to negative or vice versa, the required counter rotating joint rates will also switch sign. This causes an infeasible discontinuity in the required joint rates. For the implementation considered in this paper the change in joint rates was monitored and limited. That is, the maximum joint accelerations were limited. This allows, through further sacrifice of the angular velocity accuracy, a feasible transition from maximal positive and negative joint rates.

6. CONCLUSIONS

The basic concept of alternative instantaneous motion planning comes from the simple idea that a specified motion for a manipulator should be a feasible motion. Constraints, based on reciprocal screw quantities, are used to ensure that the alternative end effector motion is selected from the existing feasible motions (Section 2). The motion planning (selection) of the alternative motion is based on finding the "best" motion amongst the feasible motions. The optimization objective function for this planning can be chosen such that it represents the motion priorities for the required task.

To demonstrate the method, a solution was developed for the alternative velocity required to satisfy end effector translational velocity at the expense of minimal angular velocity error, for manipulators having a (effective) five-system of joint axes. Exact satisfaction of the translational velocity components was concluded to be theoretically possible, except when the screw defining the reciprocal one-system instantaneously corresponded to a zero pitch screw passing through the end effector tip. (Section 3).

Through a process of joint velocity truncation and finding an effective reciprocal screw system, the alternative instantaneous motion planning method was adapted for application to manipulators executing motion near degenerate configurations (Section 2.2). This was applied to spherical wristed manipulators and demonstrated to be effective with a kinematic simulation of a Puma-type manipulator executing a motion near its wrist degeneracy (Section 4).

REFERENCES

[1] Paul, R.P., Stevenson, C.N., "Kinematics of Robot Wrists", *Int. J. of Rob. Res.*, Vol. 2, No. 1, pp. 31-38, Spring 1983.

[2] Hunt, K.H., *Kinematic Geometry of Mechanisms*, Clarendon Press, Oxford, 1978.

[3] Hunt, K.H., "Robot Kinematics - A Compact Analytic Inverse Solution for Velocities", *ASME Trans. J. of Mech., Trans., and Aut. in Des.*, Vol. 109, March, pp.42-49, 1987.

[4] Sugimoto, K., Duffy, J., "Application of Linear Algebra to Screw Systems", *Mech. and Mach. Theory*, Vol. 17, No. 1, pp. 73-83, 1982.

[5] Lipkin, H., Duffy, J., "On the Geometry of Orthogonal and Reciprocal Screws", *1984 IFToMM Sym. on the Theory and Pract. of Rob. and Man.*, pp. 47-55, 1985.

[6] Nakumura, Y., Hanafusa, H., "Inverse Kinematic Solutions With Singularity Robustness for Robot Manipulator Control", *ASME Trans. J. of Dyn. Sys., Meas., and Cont.*, Vol. 108, pp. 163-171, Sept. 1986.

[7] Wampler II, C.W., "Manipulator Inverse Kinematic Solutions Based on Vector Formulations and Damped Least-squares Methods", *IEEE Trans. on Sys., Man and Cyber.*, SMC-16, pp. 93-101, Jan./Feb. 1986.

[8] Maciejewski, A.A., Klein, C.A., "Numerical Filtering for the Operation of Robotic Manipulators through Kinematically Singular Configurations", *J. of Rob. Sys.*, Vol. 5, No. 6, pp. 527-552, 1988.

[9] Aboaf, E.W., Paul, R.P., "Living With the Singularity of Robot Wrists", *1987 IEEE Int. Conf. on Rob. and Auto.*, pp. 1713-1717, 1987.

[10] Podhorodeski, R.P., Goldenberg, A.A., Fenton, R.G., "A Screw Based Decomposition Method for the Inverse Solution of the Instantaneous Kinematics of Manipulators", *Int. Meet: Adv. in Rob. Kin.*, Ljubljana, Yugoslavia, pp. 103-112, Sept. 1988.

Planning Tactile Recognition Paths in Three Dimensions

R. E. Ellis
Computing and Information Science
Queen's University at Kingston

ABSTRACT

A general problem in model-based object recognition is how the system should proceed when the available data are ambiguous, i.e., when either the object or its location in space are not uniquely identified. In previous work, we have shown that the problem can be solved in two dimensions, given noisy data and polygonal models, in polynomial time. It was also shown that in three dimensions the problem becomes inherently nonlinear and extremely difficult.

Here we present a description of the problem, a discussion of possible approaches, a taxonomy that subsumes previous attempts at solution, and simulation results that describe the efficacy of our approach. We show that the heuristic of attempting to decompose the three-dimensional, four-degree-of-freedom, nonlinear problem into a pair of two-degree-of-freedom, linear problems produces perfect paths about half of the time, and thus is the most efficient complete decomposition method that has been implemented.

1. Introduction

A topic of considerable recent interest is *active sensing*, in which a sensor can be actively moved in order to acquire new data. Here we are interested in how a tactile sensor, providing estimates of the position and local surface normal of contact, can be moved in linear paths by a robot to reduce or eliminate the existing of multiple false interpretations of the data in a model-based recognition system. The problem thus devolves to the computational geometric equivalent of finding a line in space that pierces model elements according to given constraints.

We have shown previously, in [Ellis *et al.*,1986], [Ellis, 1987], and [Ellis, 1990], that this can be done very effectively in two dimensions. We were able to solve this problem, and incorporate data errors and the propagated effects [Ellis, 1989], in polynomial time.

The three-dimensional problem, however, is much harder. The parameter space is non-convex, and so the path planning is inherently much more difficult. We consider three alternatives to this: optimization, linearizing the problem, and attempting to decompose the problem into several simpler problems. The first two are very unlikely to yield feasible algorithms, and the latter can take a variety of forms which can be discussed in the form of a taxonomy of possible parametric decompositions.

2. Finding Sensing Paths in Three Dimensions

The three-dimensional path-planning problem is much more difficult to solve computationally than is the two-dimensional problem. The principal reason is that the non-linearities of projection, which we were able to circumvent

---

This research was supported in part by a grant from the Natural Sciences and Engineering Research Council of Canada.

in two dimensions, cannot be eliminated in a three-dimensional projection. We will now describe analytically the three-dimensional problem, and examine three general approaches to its solution: a direct approach of optimization of the path parameters; heuristic linearization; and heuristic decomposition of the problem into simpler forms.

2.1 Analytic Description

In three dimensions, a line (which is how a sensing path is represented in our paradigm) has four degrees of freedom. Following our paradigm, we seek the representation of all of the lines passing through a face – a polygon in real three-space – in a four-dimensional manifold whose points directly encode the lines' parameters. This is very complicated, because the projection of a polygon is a hypervolume bounded by curved hypersurfaces, and to find the intersection of an arbitrary number of such hypervolumes is, to say the least, non-trivial.

The fundamental reason for the increase in complexity is that the boundary of a polygonal face in three-space is a line, whereas the boundary of an edge face in two-space is a point. In both cases, the projection into the position–direction parameter space of the path is non-linear, but the boundary conditions in two-space permit the problem to be simplified by considering only the endpoints of a line segment. In three-space, the non-linear projection cannot be reduced in such a manner.

The nature of this complexity increase can be explained by examining the parameters of the set of all lines that contact a given line segment. Consider a three-space line segment $L(\lambda)$ that we wish to contact when starting from the **X-Y** plane. The parameters defining the sensing path in this construction are the two coordinates of the starting point on the **X-Y** plane, the slope of the path from the **Z** axis in the **X** direction (which we will call α), and the slope of the path from the **Z** axis in the **Y** direction (which we will call β).

A line $L(\lambda) = P + \lambda D$ is projected into the **X-Y-A-B** manifold of piercing-line parameters by the formula

$$\begin{aligned} \mathbf{Q_L}(\lambda, \alpha, \beta) &= (L_X(\lambda) - \alpha \cdot L_Z(\lambda)\,,\; L_Y - \beta \cdot L_Z(\lambda)\,,\; \alpha\,, \beta) \\ &= (P_X + \lambda \cdot D_X - \alpha \cdot P_Z - \lambda \cdot \alpha \cdot D_Z\,, \\ &\quad\; P_Y + \lambda \cdot D_Y - \beta \cdot P_Z - \lambda \cdot \beta \cdot D_Z\,, \\ &\quad\; \alpha\,, \beta) \end{aligned} \tag{1}$$

where the subscripts indicate the components in the original three-space Cartesian coordinates.

If $D_Z \neq 0$, i.e., if the line is not parallel to the **X-Y** plane, then even if we hold one of the parameters constant – say, β – the locus in the **X-Y-A** subspace described by this projection is a ruled hyperbolic paraboloid. Thus, even when we simply seek to intersect a line when holding one of the path directions constant, the problem of finding the path parameters is non-linear.

This difficulty is compounded when we wish to pierce a polygon in **X-Y-Z** space, because a polygon is bounded by a set of line segments. The space that describes the locus of paths that pierces such a polygon is a three-dimensional volume in **X-Y-A** space that is bounded by ruled quadric surfaces. Piercing a candidate set of polygons requires finding the intersection of a number of such volumes, which is computationally very expensive.

If we now allow β to vary, a given polygon projects to the hypervolume bounded by the ruled hypersurfaces formed by the projection of the polygon's bounding line segments into the four-dimensional manifold. The path parameters of all of the lines that pierces a set of polygons are the coordinates of the points that are within the hypervolume that is formed from the intersection of the hypervolumes of projection of all of the original three-space polygons.

Clearly, this intersection hypervolume is in general very complicated, and a full description of its boundary in terms of implicit parametric equations is not a computationally feasible approach to planning a sensing path. Remembering, however, that what we seek is a *single point* in the interior of this four-dimensional structure, there are several possible attacks we might mount. Also, the solution might become simpler if we seek not always a perfect path or even an optimal path, but are content with just a good path, one that reduces the ambiguity considerably but can be found with reasonable computational expense.

2.2 Optimization

One approach to solving the problem of piercing a set of polygonal faces is to construct the projections of the faces, and then try to apply an optimization procedure over the projection parameters to find the point in projection space that provides the "best" path through the faces. Such an approach is possible, but it is also very costly and is not likely to be practical.

In [Ellis, 1990], we have shown that there is an optimization construction that permits solution for convex faces by means of quadratic programming. Unfortunately, the complexity increases exponentially if we allow non-convex polyhedra. We can summarize this approach as computationally tractable but expensive, and one in which the uncertainty in the position of the object faces deeply affects the efficacy of the optimization.

2.3 Linearization

Another approach to efficiently finding sensing paths for multiple interpretations of three-dimensional objects is to try to eliminate the nonlinearities of the projection equation of a line segment. If the projection function can be made linear, then linear programming or some similar rapid search procedure might be applied to the problem.

As we see from Equation 1, the nonlinearities arise from the product of the projection direction, the line parameter, and the component of the line direction in the **Z** direction. It is not possible to eliminate either of the first two multipliers, but we might try to set D_Z to zero. Geometrically, this is equivalent to approximating a tilted polygon by one parallel to the **X-Y** plane at some arbitrary height.

The approximation of one polygon by another introduces error. We have shown, in [Ellis, 1990], that this error effectively makes this approach infeasible. For a polygon of N sides, maximum diameter W, and tilted from the **X-Y** plane by an angle of Ω, then the total error E_P in approximation is

$$0 \leq E_P \leq \frac{2\pi N W \tan^3 \Omega \sin \Omega}{3}$$

This is a very poor upper bound. For example, if $N = 9$, then in the worst case we observe that for $\Omega \geq 16.5°$, i.e., a polygon tilted by more than 16.5°, the error is more than the area of the original polygon. Our experience with planning sensing paths in two dimensions indicates that values of Ω up to 60° are often needed; if this experience is at all applicable to planning in three dimensions, linearization is not likely to prove an effective approach.

2.4 Decomposition

The three-dimensional sensing-path problem can, as we have seen, be attacked by a costly direct optimization or by a linearizing approximation. If we are willing to trade optimal information-gathering power for rapidity of calculation, another attack can be mounted: we might try to decompose the four-degree-of-freedom problem into simpler linear problems.

2.4.1 A Taxonomy of Decompositions

A simple, exhaustive taxonomy of decompositions can easily be formed by considering all of the possible subsets of the four path parameters. Rather than attempting to optimize all four parameters simultaneously, we may instead optimize a *pair* of parameters, and then optimize the other pair. In such a paradigm there are only a few combinations of pairs; we will represent the decomposition as $(a, b) \times (c, d)$ to denote that we first optimize (a, b) and **then** optimize (c, d). In addition to the four parameters, we will also note that there is a variant representation of a line segment as a pair of endpoints, i.e., as $\{P, P'\}$. We can now categorize a decomposition as belonging to one of the following classes:

1. $(X, \beta) \times (Y, \alpha)$ or $(Y, \alpha) \times (X, \beta)$ or $(X, Y') \times (Y, X')$ or $(Y, X') \times (X, Y')$.
2. $(X, Y) \times (\alpha, \beta)$.
3. $(\alpha, \beta) \times (X, Y)$.
4. $(X', Y') \times (X, Y)$ or $(X, Y) \times (X', Y')$.
5. $(X, \alpha) \times (Y, \beta)$ or $(Y, \beta) \times (X, \alpha)$ or $(X, X') \times (Y, Y')$ or $(Y, Y') \times (X, X')$.

This exhaustive decomposition can be described thusly:

1. A poor breakdown of the problem, in that there is no interaction between the parameters in a pair.
2. An attempt to find a good location, and optimize the path direction. This was used in [Grimson, 1986], assuming perspective projection.
3. An attempt to find a good direction, and optimize the starting position. This was used in [Grimson, 1986], assuming orthographic projection.
4. An attempt to find a starting point and a point to aim at. This was described in [Schneiter, 1986].
5. An attempt to find the **X** parameters and then the **Y** parameters, or *vice versa*. This takes advantage of the orthogonality of the projection equations, and we will call this the **decoupling** approach.

2.4.2 Decoupling

The final approach in our taxonomy attempts heuristically to decouple the four–dimensional search into a pair of two–dimensional searches. This decoupling is an attempt to sacrifice guaranteed optimality for rapid computation of acceptable paths.

In decoupling, the path parameters are broken into orthogonal pairs. Without loss of generality, let us suppose that we attempt to find first the $\{X, \alpha\}$ pair and then the $\{Y, \beta\}$ pair. The first effect of this procedure is that we must approximate each face in a candidate set by some line segment on the **X-Z** plane, and use this linear approximation as an edge in our 2–D path planner.

Two alternatives are to be conservative and plan only guaranteed paths and potentially miss many good paths, or to be liberal and potentially plan poor paths. The advantage of the latter is, extrapolating from our experience with planar objects, a sub-optimal path often provides a good deal of information. Pursuing this route, we would approximate a polygon that is orthographically projected onto the **X-Z** plane as a line whose endpoints are chosen as a maximally distant pair. When all faces of the candidate set have been approximated as edges, our 2–D algorithm is capable of piercing the edges so as to maximize the information gained.

3. Experimental Assessment of Decoupling

We have attempted to assess the efficacy of the decoupling approach by implementing it, and conducting several hundred simulations. The simulations used the minimum amount of data – 3 points and normals – and the objects shown in Figure 1. These objects (which are approximations of parts of a robot in our laboratory) generate a total of 32 distinguishable interpretations, which is a rich test environment.

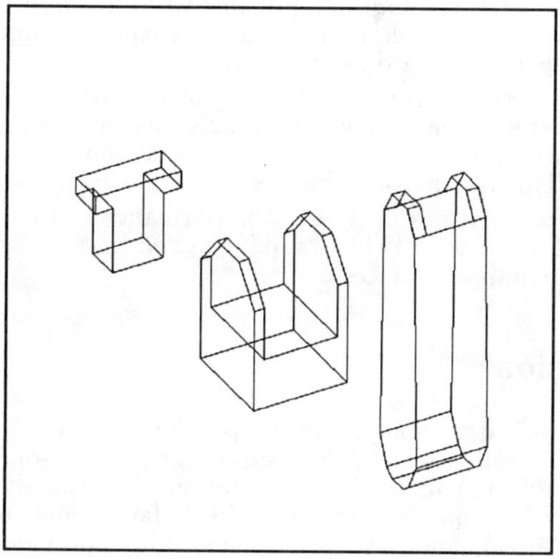

Figure 1: Objects used in the simulations.

The method we employed was to select various pairs of interpretations and plan decoupled paths. Because the calculation is a cubic function of the number of faces, we restricted these tests to cases of 2, 3, 4, or 5 ambiguities. For each set, we then planned 20 paths, which are the result of rotating each set from 0° to 90° in increments of 10°. At each stage paths were planned in the **X-Z** plane then the **Y-Z** plane, and *vice versa*, to test the sensitivity of decoupling to rotations and order of operation. The termination criterion was that the decoupled paths identified at least two projected interpretations.

The results, shown in Table 1, indicate that the heuristic is very effective but by no means perfect. We can see that for the examples used, roughly half of the time we are able to distinguish *all* of the interpretations. But, a large number of times, we touch none or touch them so as to provide no distinguishing information.

Table 1: Path Identification Success.

# of Interps	Number of distinguishing hits					
	0	1	2	3	4	5
2	20.4%	35.2%	44.4%			
3	19.4%	27.8%	1.4%	51.4%		
4	11.1%	18.8%	9.7%	8.3%	52.8%	
5	12.5%	30.6%	1.4%	11.1%	11.1%	33.3%

The reason for the relatively large number of useless paths seems to be that in combining the orthogonal views, it not uncommonly happens that the path passes through a concavity in the object. This circumstance would of course not occur for convex objects, but realistically this is the result of the decoupling heuristic.

One obvious improvement that is possible is to be more intelligent about the way in which faces are projected and then approximated. Presently, we use the first maximal pair of vertices as the approximation, and apply no constraints to the path. It might be possible to substantially increase the effectiveness by choosing a good "horizontal" approximation and then preferring those generated paths that are most vertical.

The second observation is that the quality of approximation varies as the local coordinate system rotates about the **Z** axis. We observed that by planning paths for 9 orientations we were invariably able, at some rotation, to generate a perfect path. This is expensive, but seems to be effective experimentally.

This heuristic, then, has fairly good performance in that it found perfect paths about half of the time (which is quite a good hit rate). Combinatorially, it is $O(n^3)$ in the number of faces.

4. Conclusions

We have presented an analysis of the problem of finding sensing paths in three dimensions, and discussed five approaches to its solution. The most rigourous approach is a direct implementation of the analytical description by optimization of the parameters; this is likely far too expensive to be worth implementing in a real-time system, although it might provide some interesting insights into the frequency of easy *vs.* hard cases.

The second approach is a linearization of the projection equations, which can lead to the formulation of the problem of finding a sensing path as a problem in linear programming. This is not a promising heuristic, because of the nature of the approximation function that derives from this procedure.

Two other approaches fall under our taxonomy of decomposition, the approaches being due to Grimson and Schneiter. The first does not attempt to optimize all of the parameters, but excellently incorporates uncertainty into the analysis; the second attempts a direct attack on the problem of optimizing the parameters, but does so at considerable computational expense.

Our final approach attempts to decompose the problem into two simpler ones and then combine the results into a path that distinguishes the ambiguous interpretations. This decomposition is effective in a large variety of circumstances, although it is very difficult to analyse and predict those cases in which it is likely to fail.

We should note that the approaches described, particularly our decomposition taxonomy, are not the only ones possible. These are the approaches that result from fairly direct analysis of the recognition paradigm in which we operate, but there are many other ways of recognizing and modelling objects. Other paradigms, no doubt, have approaches that could effectively solve the general problem we pose; they might also suggest methods that could be used within our paradigm. Finding ways to gather information efficiently and effectively to aid the recognition of three-dimensional objects is demonstrably a hard problem, but one whose solution is important in robotics.

REFERENCES

Ellis, R.E., Riseman, E.M., & Hanson, A.R.: 1986. "Tactile recognition by probing: Identifying a polygon on a plane", *Proceedings of the AAAI Conference (1986)*, pp. 632–637.

Ellis, R.E.: 1987. "Acquiring tactile data for the recognition of planar objects", *Proceedings of the IEEE Conference on Robotics and Automation (1987)*, pp. 1799–1805.

Ellis, R.E.: 1989. "Uncertainty estimates for polyhedral object recognition". *Proceedings of the IEEE Conference on Robotics and Automation (1989)*, pp. 348–353.

Ellis, R.E.: 1990. "Planning tactile recognition paths in three dimensions, *Technical Report 90-277*, Department of Computing and Information Science, Queen's University at Kingston.

Grimson, W.E.L.: 1984. "The combinatorics of local constraints in model-based recognition and localization from sparse data". *AI Memo 763*, Artificial Intelligence Laboratory, Massachusetts Institute of Technology.

Grimson, W.E.L.: 1986. "Sensing strategies for disambiguating among multiple objects in known poses". *IEEE Journal of Robotics and Automation*, RA-2(4)196–213.

Grimson, W.E.L., and Lozano-Pérez, T.: 1984. "Model-based recognition and localization from sparse range or tactile data". *International Journal of Robotics Research*, 3(3):3–35.

Schneiter, J.L.: 1986. "An objective tactile sensing strategy for object recognition and localization". *Proceedings of the IEEE Conference on Robotics and Automation (1986)*, pp. (2)1262–1267.

PATH PLANNING AND STEERING CONTROL OF MOBILE ROBOTS *

Carlos Resendes(1) Ricardo Cabral(2) M. Isabel Ribeiro(3)

(1) *LRPI/ITEC, Av.Rovisco Pais, 1, P-1096 Lisboa Codex, Portugal*

(2) *Arthur Anderson Co., Av.Eng.Duarte Pacheco, Torre 1,* 15°, *P-1200 Lisboa, Portugal*

(3) *CAPS/INIC and Instituto Superior Técnico, Av.Rovisco Pais 1, P-1096 Lisboa Codex, Portugal*

ABSTRACT

This paper addresses the trajectory control for Autonomous Mobile Robots (AMR) in dynamic environments. The approach taken in this paper implements the problem solution with an hierarchical structure that integrates the functions of global planning, local navigation and collision avoidance. The emphasis of this paper is on the first component of this structure. It is solved off-line by a path planner algorithm that evaluates a set of critical points along a globally optimized geometric trajectory that accounts for the trajectory length and for the obstacle proximity. This evaluation uses a quadtree representation of the free space, which proved to be efficient and to have a low computational cost. The local navigation and the collision avoidance are executed by a steering control algorithm. Simulation results are presented illustrating the advantages of the quadtree representation and of the problem decomposition proposed in this paper.

1. INTRODUCTION

The trajectory control is fundamental in many problems of mobile robotics. In non trivial applications involving Autonomous Mobile Robots (AMR), these vehicles exhibit some degree of intelligence in order to perform a set of planned tasks in dynamic environments. Therefore trajectory control highly determines the vehicle performance.

The development of the trajectory control problem has been the subject of intensive research in recent years [1,2,4,5,7] and many approaches, decomposing the problem in hierarchical levels, have been proposed to solve it. In [1,7] the path planning is solved by applying artificial intelligence techniques to search a path on a state-space based on the a priori knowledge of the environment map. In [7] the minimal length path is found by modeling the obstacles by a set of convex polygons, while in [1] the free space is represented by a set of overlapping corridors, the paths being defined on the middle of

*Work partially supported by JNICT research program n° 23/INF/87.

the corridors. In these algorithms any change on the given map leads to the computation of a new trajectory. On the contrary, an on-line approach is presented in [2,4] addressing the steering control of AMRs by defining subgoals in a local map in order to bypass immediate obstacles. In [5], a method is presented for combining the path planning and the dynamic steering control. The path planner covers the workspace by a regular grid of points and uses path relaxation to find a minimum cost trajectory. The flexibility of the cost functional (accounting for length traveled, obstacle proximity, and unmapped areas) is achieved through an high computational burden which increases for high grid resolution.

The algorithm proposed in this paper takes this previous approach and combines a path planner procedure, solved off-line, with a steering control module implemented on-line. These two modules integrate the functions of global planning, local navigation and collision avoidance, represented as different levels in the hierarchical structure displayed in Fig. 1. In this structure,

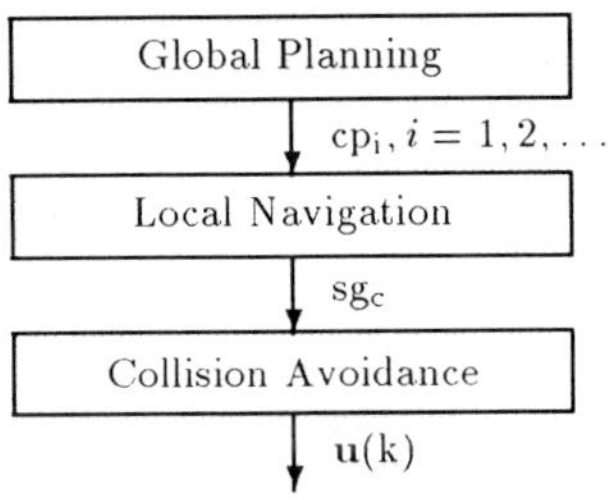

Figure 1: Levels of competence of the trajectory control decomposition.

the lower level implements the simplest actions (collision avoidance) while the higher levels are responsible for more complex tasks (planning, navigation) and impose a set of constraints to the lower levels. For the efficiency of this trajectory control system, a good performance is required on each level, some degree of redundancy between them being required.

The path planner uses a quadtree representation to model the free space and the A* algorithm to search, at a low computational cost, a set of free cells extending from the initial to the final point in a desired trajectory. The central points of these cells, named as critical points, cp_i, define a geometric trajectory that will be optimized by the on-line steering control algorithm. Given the set of critical points, this local optimization procedure derives a dynamic trajectory (e.g., velocity and acceleration) instead of driving the AMR directly through each cp_i ignoring the vehicle dynamics. The steering algorithm, modifies the Generalized Potential Field(GPF) technique, [4], used on the Feedback Steering Control Algorithm, [6], and guides the AMR through a smooth trajectory.

The paper organization is the following. In Section 2 the path planning algorithm is presented, and the advantages of the quadtree representation and of the global optimization criterion are discussed. The steering control problem is addressed in Section 3. Relevant simulation results are presented in Section 4, displaying the combination of the two modules of the proposed algorithm. A discussion of the trade-off between a tolerance around each critical point and the AMR motion smoothness is carried out. Finally Section 5 concludes the paper.

2. PATH PLANNER

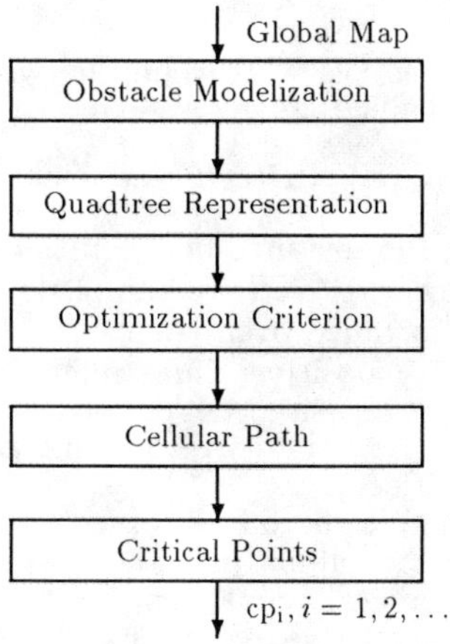

Figure 2: Block diagram of the path planner.

The path planning algorithm presented in this section computes a geometric trajectory, from an initial point to a goal, according to the block diagram of Fig. 2. Given a global map of a static environment, the planner defines a collision-free geometric trajectory as a set of critical points $cp_i, i = 1, 2, \ldots$ obtained from a global optimization procedure that accounts for the trajectory length and for the obstacle proximity. The advantages of the quadtree representation of the free space used by the path planner as well as the global optimization criterion are discussed in this section.

The algorithm starts from the specification of the static environment, where the mobile robot will move. The obstacles are modeled by a set of non strictly convex polygons which are extended to account for the dimensions of the AMR, [7]. The search of an optimal path in such a state-space, when accomplished by artificial intelligence techniques, requires an efficient representation of the free space. This is done by transforming the obstacle modelization into another representation where the free space is modeled by a quadtree, [3]. This representation, recursively decomposes the entire space into square region cells of decreasing area. For each region, the decomposition process stops whenever the corresponding cell, taken as a node of the tree, contains no obstacles. When this is not the case, the decomposition proceeds until the minimum size cell is obtained. See Fig. 3 a) in Section 4 for a quadtree decomposition of a particular plant.

Alternative representations of the free space are provided in [1,5]. In [1] the free space is represented by a set of overlapping corridors and a geometric trajectory is defined in the middle of the corridors. The approach taken in [5] relies on a rectangular grid and uses path relaxation to find a minimum cost trajectory. The flexibility of the cost functional (accounting for traveled length, for obstacle proximity and for unmapped areas) leads to an high computational burden which increases for high grid resolutions. The quadtree representation proposed in this paper combines the advantages of the Free Space and of the Grid modeling presented in, [1,5] by representing large free regions as single nodes of the tree. Therefore, the flexibility is maintained without increasing the computational cost.

Given the quadtree representation, the planner proceeds by using the A^* algorithm, [9], in a global optimization procedure that minimizes the cost functional

$$F = \sum_{i=1}^{n-1} C(cell_i, cell_{i+1}), \tag{1}$$

where $C(cell_i, cell_{i+1})$ is the cost of traveling between the adjacent cells, $cell_i$ and $cell_{i+1}$. This cost is defined as

$$C(cell_i, cell_j) = D(cell_i, cell_j) + \alpha P(cell_j), \tag{2}$$

with $D(cell_i, cell_j)$ being the distance between the central points of $cell_i$ and $cell_j$. The proximity of the free $cell_j$ to the obstacles, $P(cell_j)$, is computed as in [10] by taking the distance from the central point of $cell_j$ to the nearest obstacle. In (2), α is a positive scalar referred to as the approximation factor, which determines the desired degree of proximity of the trajectory to the obstacles. If α equals zero the corresponding path is the minimal length collision-free geometric trajectory.

The minimization of (1) leads to a set of free cells from the initial point to the goal, defining an optimal cellular path, $cell_i, i = 1, \ldots, n$. Finally, the path planner derives a set of critical points, $\mathrm{cp_i}, i = 1, \ldots, n$, by taking the final goal and the central point of the free cells in the cellular path. Associated with each $\mathrm{cp_i}$, let tol_i be a tolerance value, given by $P(cell_i)$, which represents the proximity to the obstacles.

3. STEERING CONTROL

The mobile robot steering control is guided by the set of critical points, $\mathrm{cp_i}$, determined by the path planner and derives the control strategy (e.g., velocity, acceleration) using these points as trajectory local subgoals. The steering control implements two levels of competence, the local navigation and the collision avoidance, as showed in Fig. 1. The local navigation is accomplished by the Subgoal Selection Algorithm (SSA), [6], which defines subgoals, $\mathrm{sg_i}$, in a local map in order to bypass immediate obstacles. This module is assisted by the AMR sensing system, in order to update the local map in non static environments. The collision avoidance is based on the modification of the Generalized Potential Field (GPF) technique proposed in [8]. See [8] for the details of the steering control algorithm, namely for the modification of the GPF technique.

4. SIMULATION RESULTS

In this section several simulation examples are presented to illustrate the combination of the path planning and of the steering control algorithms. The role played by the tolerance, tol_i, around each critical point is discussed, trading the optimality of the executed trajectory with its execution time and the robot motion smoothness. The plant used for the experiments is the one presented in [2].

In Fig. 3 a) the set of critical points obtained by the path planner is displayed. The white square areas are the free cells that correspond to the lower level nodes on the quadtree decomposition of the free space. The dotted cells represent the cellular path obtained from the optimization procedure, while the asterisks located at the center of the cells stand for the critical points defining the geometric trajectory. This trajectory is obtained with $\alpha = 0.0$ (i.e., no account is made for the obstacle proximity), and thus it has the minimum length, 424.1 m. In Fig. 3 b), the steering control module takes the

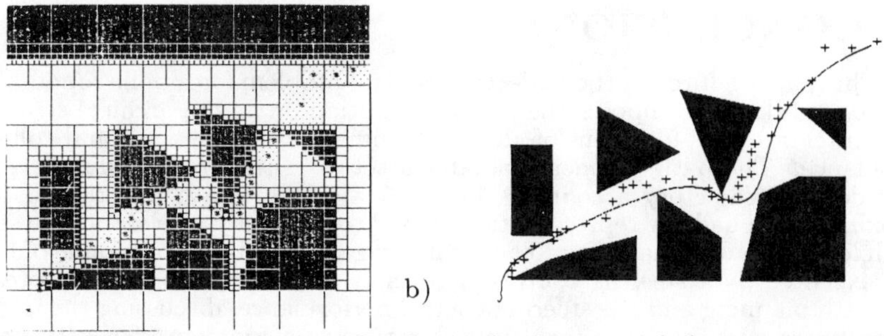

Figure 3: Trajectories obtained by the path planner and the steering control with $\alpha = 0.0$.

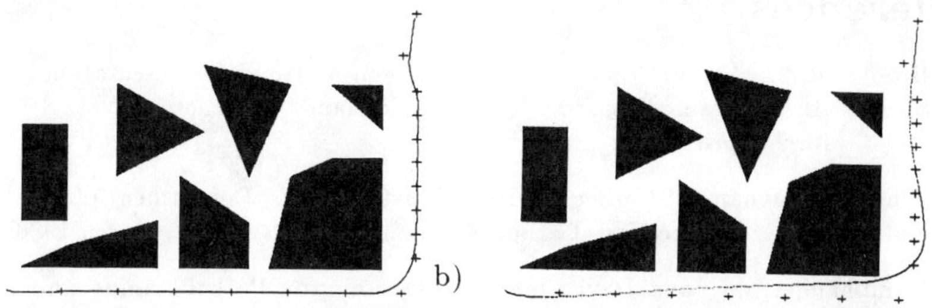

Figure 4: Trajectories obtained by the steering control algorithm with $\alpha = 0.35$ and different tolerances around each critical point.

critical points indicated in Fig. 3 a) (now represented by '+') and steers the vehicle along the trajectory illustrated by the dotted line. In this figure it is clear the local optimization process achieved on-line by the steering control, leading to a smooth trajectory that is executed in 59.6 sec. This simulated trajectory is obtained by ignoring the tolerance value, tol_i, around each critical point and by assuming that it has an infinite value. As pointed out before, the critical points act as local trajectory subgoals. Therefore, with high values of tolerance, the steering control updates the critical points as trajectory subgoals soon before reaching them, allowing the AMR to travel further away from these points.

In the Figures 4 a) and b) the critical points were computed by the path planner with $\alpha = 0.35$, leading to a trajectory length of 486.7 m, greater than the length obtained with $\alpha = 0.0$ (Fig. 3). The difference between the dynamic trajectories in these figures is due to the difference on the tolerance considered around each critical point. The tolerance values used in Fig. 4 a) are provided by the path planner while Fig. 4 b) assumes an infinite value for them. With low values of the tolerance, the dynamic trajectory is constrained to remain in a narrow band defined along the critical points, while high values of tol_i may lead to AMR motion far apart from the critical points. The robot trajectory at the lower right corner of the plant (Fig. 4) illustrates that there is a trade--off between the tolerance around each critical point and the AMR motion smoothness. The great disparity of values on the execution times (104.7 sec in Fig. 4 a) and 51.7 sec in Fig. 4 b)) is justified by the fact that in Fig. 4 a) the robot moves very slowly because the critical regions, [2], degenerate in a line.

Finally, by comparing the execution times obtained for different values of α (51.7 sec in Fig. 4 b) and 59.6 sec in Fig. 3 b)) we conclude that a longer geometric trajectory may be executed in a shorter time interval.

5. CONCLUSIONS

This paper addresses the trajectory control problem for mobile robots. The approach taken decomposes the problem solution on an hierarchical structure that integrates the functions of global planning, local navigation and collision avoidance. The path planner evaluates a set of critical points along a globally desired trajectory accounting for the trajectory length and the obstacle proximity. A quadtree representation is used to efficiently model the free space with low computational costs. The local navigation and the collision avoidance are executed by a steering control algorithm. The simulation results present the path planning and the steering control performance, discussing the motion smoothness and the trajectory time duration as a function of the tolerance allowed around each critical point.

References

[1] Brooks, R.A., "Solving the Find-Path Problem by Good Representation of Free Space," IEEE Transactions on Systems Man and Cybernetics, Vol. SMC-13, No. 3, March-April 1983.

[2] Feng, D., "Dynamic Steering Control," Master Thesis, Department of Electrical and Computer Engineering, Carnegie-Mellon University, USA, March 1986.

[3] Kambhampati, S., and Davis, L.S., "Multiresolution Path Planning for Mobile Robots," IEEE Transactions on Robotics and Automation, Vol. 2, No. 3, September 1986.

[4] Krogh, B.H., "A Generalized Potential Field Approach to Obstacle Avoidance Control," Proceedings Robotics Research the Next Five Years and Beyond, International Robotics Research Conference, Bethlehem, USA, August 1984.

[5] Krogh, B.H., and Thorpe, C.E., "Integrated Path Planning and Dynamic Steering Control for Autonomous Vehicles," Proceedings of International Conference on Robotics and Automation, pp. 1664-1669, 1985.

[6] Krogh, B.H., Feng, D., "Dynamic Generation of Subgoals for Autonomous Mobile Robots Using Local Feedback Information," IEEE Transactions on Automatic Control, Vol. 34, No. 5, pp. 483-493, May 1989.

[7] Lozano-Pérez, T., and Wesley, M.A., "An Algorithm for Planning Collision-Free Paths Among Polyhedral Obstacles," Communications of the ACM, Vol. 22, No. 10, pp. 560-570, October 1979.

[8] Resendes, C., Cabral, R., and Ribeiro, M.I., "An Algorithm for Steering Control of Mobile Robots," IROS'90 IEEE International Workshop on Intelligent Robots and Systems'90, Japan, July 1990.

[9] Rich, E., Artificial Intelligence, McGraw-Hill 1986.

[10] Samet, H., "Distance Transform for Images Represented by Quadtrees," IEEE Transactions on Pattern Analysis and Machine Intelligence, Vol. PAMI-4, No. 3, May 1982.

Research on the Collision Avoidance of Manipulators Based on the Global Subgoals and a Heuristic Graph Search

Yuuki Inoue
Graduate School of Science and Technology, Kobe University, Rokkodai, Kobe 657, Japan

Toshihide Yoshimura and Shinzo Kitamura
Department of Instrumentation Engineering, Faculty of Engineering, Kobe University

ABSTRACT

A collision avoidance algorithm based on a heuristic graph search and the setting of subgoals on desired path is presented. The evaluation function for a heuristic search is defined by the sum of the distance between the links of a manipulator and the middle planes among obstables and the distance between the end-effector and the subgoals on desired trajectory. Such a subgoal method reduces the combinatorial explosion in the search space. Its effectiveness has been verified by simulation studies.

1. Introduction

Collision avoidance problem has been an important issue for the task planning of robot manipulators. A number of collision avoidance methods have been proposed, which are classified into two types; "potential method"[1] and "configuration space method"[2]. Since the potential method is based on the minimization of a potential function, the dead-lock problem at the local minima may exist and general solution for the dead-lock problems have never found. The configuration space method can remove the dead-lock problem. However, in the case of high d.o.f. manipulators, the number of nodes for the configuration space grows exponentially. Therefore, it is impossible to generate the entire configuration space even for the 6 d.o.f. manipulators. Lozano-Perez[2] and Hasegawa[3] proposed methods to decrease the dimension of configuration space, based on the idea that only few joints have an important effect on the gross motion. But the solutions of this method are a subset of general solution. Kondo [3] applied a heuristic graph search method to this problem. But, because the heuristic function is designed for finding the shortest path in the configuration space, the links and end-effector may move near obstacles and the method to modify the path is applied after the shortest path is found.

In this paper, we propose a heuristic graph search algorithm with the subgoals on a global path of end-effector. The heuristic evaluation function can give the local information necessary for path searching. We defined a evaluation function as a sum of two functions: a potential function for end-effector to reach the goal configuration and that to avoid the collision between the links and the obstacles. Further, employing the global path, the combinational explosion can be reduced in the search procedure. If no global path is used, heavy back tracking may occur and the computational cost for searching may explode quickly, because the heuristic function does not work efficiently in the back tracking.

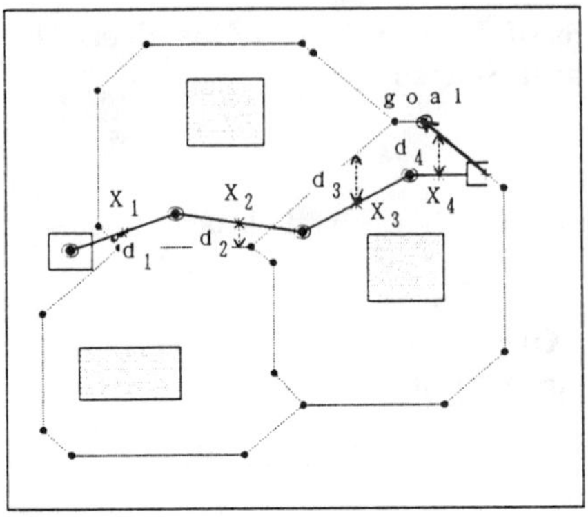

Fig. 1 Safety First Planes in 2 dimensional workspace and distances d_k

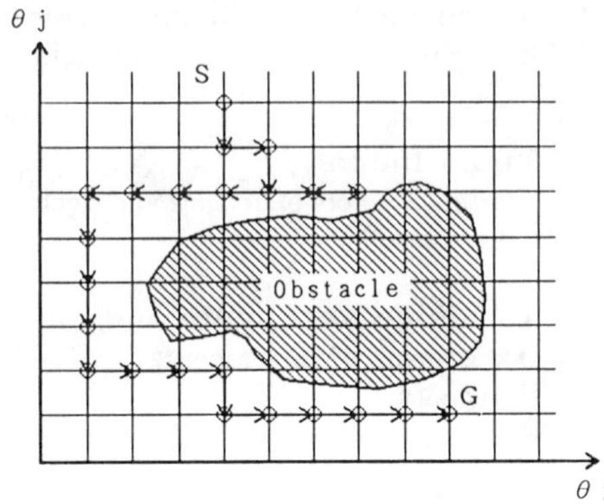

Fig. 2 The process of the expansion of the node of the graph

2. A heuristic function

The graph for configuration space is defined as a multi-dimensional grid in the joint parameter space. A closed set of collision free configuration is called "free space". A heuristic function is used to find the free space including initial and final postures, and yet not being close to obstacles. The trajectory of manipulator is defined as the shortest path in the searched free space.

We define the heuristic function as a quadratic form

$$ev = \frac{1}{2}(X_e - X_G)^T W_e (X_e - X_G) + \sum_{k=1}^{L} w_{ok} \left[\frac{1}{2} d(X_k)^2 \right] \tag{1}$$

where X_e and X_G are 9×1 configuration vectors in the Cartesian coordinates, and W_e is a 9×9 weighting matrix, and w_{ok} is a scalar weighting factor. This function takes the same form defined for an algorithm based on the potential method[5]. The first term of the right hand side is an error between the current configuration of end-effector X_e and a desired configuration X_G, and the second term is a sum of distances between the center point of the k-th link X_k and the nearest middle plane among the obstacles. Sets of middle plane are uniquely defined from geometrical data of the workspace. We have called it the 'Safety First Plane'. Figure 1 shows an example of a Safety First Plane in 2 dimensional workspace and of distances $d_k = d(X_k)$ appearing in the second term of eq. (1).

3. A heuristic graph search algorithm based on local information

We first apply the heuristic graph search algorithm, which does not take into account the global path of end-effector, to the collision free path finding. Its capability was studied by simulation.
In order to realize the dynamic memory allocation for the free space, the data of free space was represented as a list structure. Parameters for each node of the free space are defined as

NODE	**th**	:parameter list of joint angle
	ev	:value of heuristic function
	pc	:path cost from start node
	pp	:pointer to the parent node

Figure 2 illustrates a process of the node expansion during a searching. We propose an algorithm:

HGSL (Heuristic graph search based on local information for collision avoidance)

Step 1 Compute a set of middle planes from given geometrical data for working environment.

Step 2 Search a node by "the best first search algorithm[6]" from eq. (1). We call the node **n**.

Step 3 If the node **n** is a final configuration, a trajectory is computed by tracing the parent pointer **pp** from **n** to **s**, and the searching process ends successfully.

Step 4 Compute a set of the nearest neighbor of the node **n** and determine the descendant of **n** in the searching tree. Make a list $\mathcal{M}$ which stores the collision free descendant of **n**.

Step 5 Find the elements of $\mathcal{M}$ which do not belong to the searched free space, and let **n** be their parent node. In the remainder of $\mathcal{M}$, if **n** becomes the shorter path from **s** than the previous searching, let **n** be their parent node.

Step 6 Go to the step 2

Figure 3 shows the simulation results by the *HGSL*. A 4 d.o.f. articulated manipulator in the 2 dimensional workspace is assumed. Every axis of the configuration space was equally quantized into 360 divisions. Figure 4 shows an example of the environment for which the *HGSL* cannot apply, because heavy back-trackings occured and the configuration space became too large to store.

For the motion generation proposed above, we have the following problems;

- The end-effector tends to move near the obstacles.
- Back-track works efficiently, there is a possibility of the explosion of calculation cost.

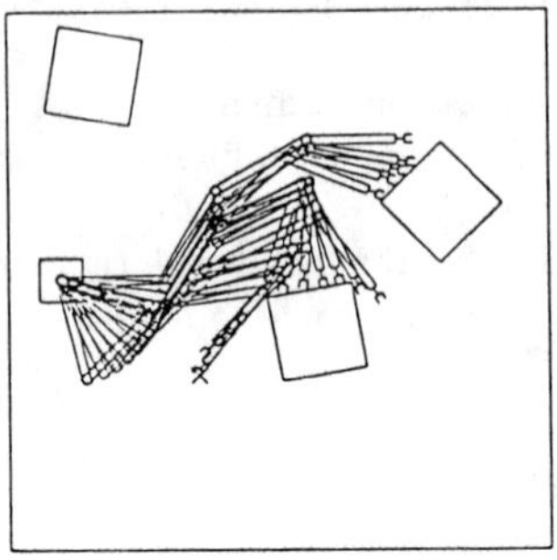

Fig. 3 A simulation of the *HGSL*.

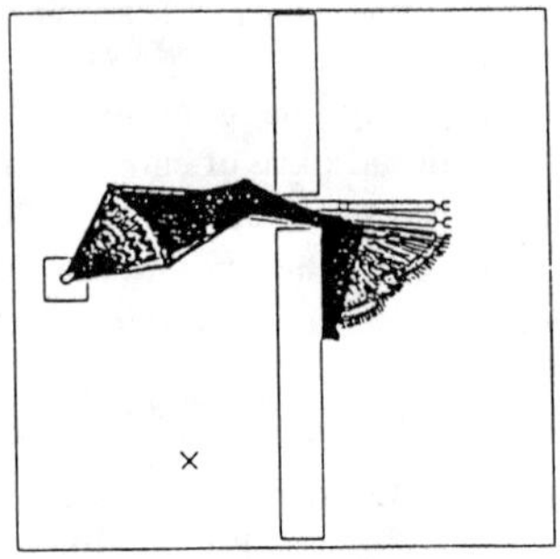

Fig. 4 An example of the environment for which the *HGSL* could not apply.

4. A heuristic search algorithm based on local and global information

Since the path planning *HGSL* is based on local information only, the generated motion may not be safe as shown above. To cope with this problem, we take the global information of the working environment into consideration.

4.1 Global path

The global information for collision avoidance is given as the safety path of end-effector. We use a graph called the 'Safety first graph' as a candidate of desired path. It is defined as a graph whose arcs are lines of intersection of any two safety first planes and whose nodes are points of intersection of two arcs. Subgoals are defined as the nodes on the shortest path in the graph between the initial and final points of end-effector. Since the subgoals lie in the middle of the obstacles, the end-effector is expected to approach safely to the subgoal. Figure 5 shows an example of such subgoals.

HGSG (Heuristic graph search based on global and local information for the collision avoidance)

Step 1 Compute the safety first planes and the safety first graph from geometric model for the working environment, and let $i = 1$.

Step 2 Execute the *HGSL* algorithm from step 2 to step 8 to find a trajectory from the current posture to a subgoal SG_i.

Step 3 If $i \neq N$, let the trajectory put into P_i, and $i \leftarrow i + 1$. If $i = N$, connect all trajectories from P_1 to P_{N-1} and the searching process ends in success.

Step 4 Go to the step 2

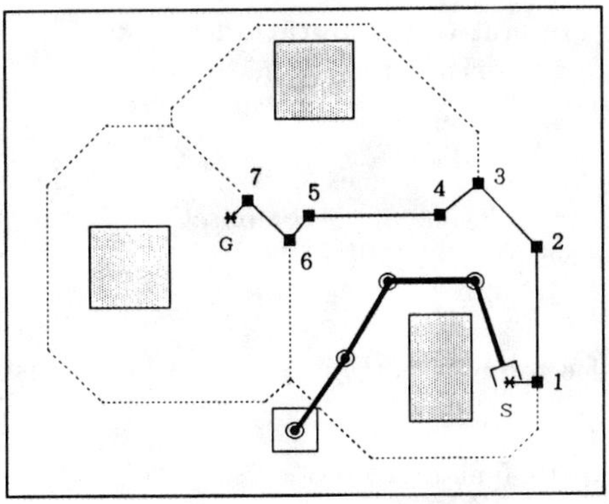

Fig. 5 Subgoals on the desired trajectory along the Safty First Graph.

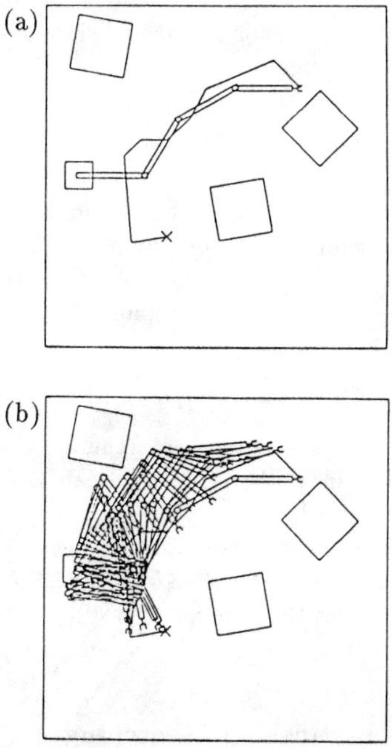

Fig. 6 A simulation of the *HGSG*. (a) Initial posture of manipulator and a goal point. (b) Planned motion.

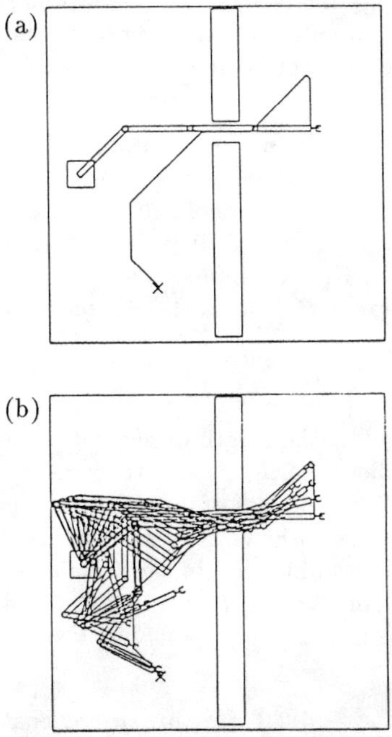

Fig. 7 A simulation of *HGSG* in the environment for which the *HGSL* cannot apply . (a) Initial posture of manipulator and a goal point. (b) Planned motion.

4.2 Simulation result of *HGSG*

Figure 6 shows an example of planned motion using the *HGSG*. The environment for experiment is the same as for the case of the *HGSL*. The path of end-effector has improved from the viewpoint of the safety as compared with that by *HGSL*. Figure 7 shows another example for which the *HGSL* could not apply.

5. Estimation of the amount of computation and the efficiency of heuristic function

For collision avoidance problems, the amount of computation becomes an important issue. Since the dimension of the search space becomes large, memory explosion or computational explosion may happen. Here, the amount of computation of a heuristic graph search are estimated by the following factors[6].

- The number of nodes to be expanded to find a path (i.e., efficiency of heuristic graph search) .
- The amount of computation of heuristic function.

We estimated the first factor from several experimental results, and the second factor is computed by summing up the amount of computation in the collision detection, computation of heuristic function and data management.

5.1 The efficiency measure of heuristic function

The performance of heuristic function of graph search algorithm depends on application domains, and it is difficult to define the generalized performance of searching process. Here we applied a measure the "penetrance"[6] as an efficiency measure. The penetrance P denotes the sharpness of the algorithm, and it is defined as :

$$P = L/T \tag{2}$$

where L is the length of generated path, and T is the number of generated node as a candidate of the path. We examined it by the simulation studies of the *HGSL* and *HGSG* for 10 cases of different workspace.

Figure 8 shows the value of penetrance in experiments. We could not observe an explicit difference in the searching performance between *HGSL* and *HGSG*. In the almost of all cases, P is close to 1.0 and this indicates that the number of trivial search was less.

5.2 Computational cost for an expansion of searching

We estimated the growth of computation in respect to the complexity of workspace and the number of links. If we assume that the length of the planned path is constant, the computational cost depends on the cost to search for one node. Therefore we examined the cost for one node expansion. The factors we considered are the degree of freedom of manipulator d , the number of edge to describe the environment model e, the number of face of environment model f, the number of links for calculation of distance from the middle plane in a heuristic function l, and the depth of search tree k. Computational order is a sum of the order of collision detection, that of calculation for eq. (1) and that of node sorting. They are estimated as

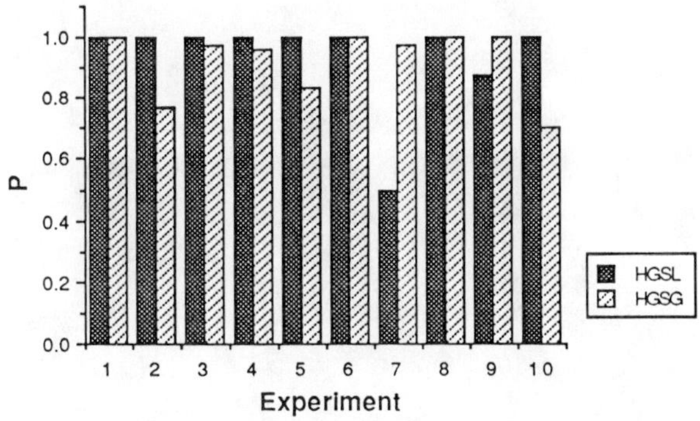

Fig. 8 Value of penetrance of each experiment

$$t = 2d \cdot O(e \cdot f) + 2d \cdot O(l \cdot f^2) + O(d^2 \cdot k) \quad (3)$$

From this estimation, we conclude that d and f become main factors. Since d is at most 6 or 7, the number of faces practically becomes important factor.

6. Conclusion

A collision avoidance algorithm using the heuristic graph search and the global subgoals was presented. As the value of penetrance in the result of experiments showed, a trajectory can be searched efficiently. However, when the *HGSL* method was applied, the manipulator tended to move near the obstacles. We coped with this problem to use the subgoals on a global path for end-effector. We could verify the fact that it reduces the depth of back-tracking in the search process and saves the computational cost. This enables us to apply the algorithm to any kind of manipulator with higher degree of freedom. Such results will be presented elsewhere.

Referance

1. O. Khatib,and J.F.Le Maitre, *"Dynamic control of manipulators operation in a complex environment"*,Int. *Proc. of the 3rd CISM-IFToMM*,267-282 (1978)
2. T. Lozano-Perez, *"Automatic planning of manipulator transfer movement,"*, IEEE Trans. Syst. Man Cybern., SMC-11(10), 681-698 (1981)
3. T. Hasegawa ,*"Collision avoidance using characterized description of free space"*,Proc. '85 I.C.A.R. , 69-76 (1985)
4. K. Kondo, *"A simple motion planning algorithm using heuristic free space enumeration"*, Proc. 1988 IEEE Int. Workshop on Intel. Robots and Systems, 751-756 (1988)
5. H. Hirukawa, and S. Kitamura *"A collision avoidance algorithm for robot manipulators using the potential method and Safety First Graph"*,Proc. of JAPAN-USA Symposium on Flexible Automation,JAACE-ASME, 99-102 (1986)
6. Nils J. Nilson, *"Principles of artificial intelligence"*,Springer-Verlag (1982)

Toward Practical Motion Planners: Experiments with A Sequential Search Strategy

Kamal Kant Gupta*
School of Engineering Science
Simon Fraser University, Burnaby, B.C. V5A 1S6.

Abstract

We have implemented a motion planner based on a novel sequential strategy to plan collision-free motions for a manipulator arm (such as PUMA 260) among 3-D polyhedral obstacles. The basic idea behind our approach is to successively plan the motion of each link starting from the base link. Suppose that the motion of links till link i (including link i) has been planned. This already determines the path of one end (the proximal end) of link $i+1$. The motion of link $i+1$ is now planned along this path by controlling the degree of freedom associated with it – a 2-dimensional motion planning problem. This strategy results in one 1-dimensional (the first link is degenerate) and $n-1$ 2-dimensional planning problems instead of one n-dimensional problem for an n link manipulator arm. The search strategy leads to fast and efficient algorithms, and is especially suited for highly redundant arms. However, the penalty is that the strategy may miss certain paths. At the very least, it will form a front end to a more complete planner. We believe that in most reasonable cases, the strategy will quickly find a path. We further present some interesting examples on which the strategy succeeds admirably.

1 Introduction

Motion planning for manipulator arms has been a much researched topic [Udupa 1977, Lozano-Perez 1981, 1987, Schwartz and Sharir 1982, Brooks 1983, Hopcroft *et al.* 1985]. As generally stated in the literature, the motion planning problem is to find a path from a given initial robot configuration to a given final configuration that avoids collisions with a known set of stationary obstacles in its workspace. Many of these algorithms are based on the use of configuration space (Cspace) – the space of degrees of freedom of the robot. This permits explicit characterization of the constraints that the physical obstacles impose on the robot's motion. However, there are two major difficulties with this approach for practical manipulator arms: (i) high dimensionality of the Cspace, and (ii) complexity of the Cspace obstacles. It is well known that the worst case time bound for motion planning for an arm is exponential in the dimensionality of its configuration space, [Schwartz and Sharir 1982, and Hopcroft *et al.* 1985]. In practice, for a six degree of freedom arm, building and searching the six dimensional Cspace, although possible, is extremely cumbersome and slow. [Donald 1984] implemented a full six degree of freedom

*This research was supported by an NSERC operating grant and SFU Centre for Systems Science grant.

(three rotational and three translational) planner for a polyhedral body moving among polyhedral obstacles. His implementation took several minutes on a Lisp machine to plan a path. The problem for a manipulator arm, such as PUMA 260, would be computationally more intensive, since all the six degrees of freedom are rotational. Furthermore, with increasing emphasis on redundant arms, this problem of dimensionality is going to be worse. Motion planning algorithms, to be useful, must run in reasonable time. Thus, researchers have suggested heuristics and search strategies to cope with this problem of dimensionality (See [Gupta 1990] for more details).

2 The Sequential Search Strategy

In this paper, we have implemented a sequential search strategy for fast collision-free motion planning of manipulator arms. The strategy is based on a key observation that in many instances of human arm motions, *the motion of a proximal link does not depend on the motion of a distal link.* Other researchers have pointed out this sequential nature of human arm motion in a different context. For example, [Korein 1985] mentions that to reach and pick up an object, humans use the proximal joints first and the distal ones later.

This simple observation, applied to motion planning, motivates our algorithm as follows. A simple manipulator arm (like a human arm) has an inherent ordering of links, from the base link to the last link. Consider a 3 link (planar manipulator) shown in Figure 1. The links have been labeled in increasing order from the base, i.e., the base link is labeled 1, the next link is labeled 2, and the last (free) link is labeled 3. Suppose, the motion of link 1 has been planned. Now, as it moves link 2 can be treated as a moving robot with another degree of freedom (pure rotation for revolute joints). Thus, its motion (rotation) can be planned to avoid the obstacles. Once the motion of link two has been planned, now the motion of link three can be planned, and so on... *Notice that the problem of planning in an n-dimensional Cspace (for an n-link arm) has been converted into n much simpler single link problems.* This is shown in Figure 1.

Assuming that we can solve the single link problem (see Section 3), we have an algorithm for planning the motion of the whole arm. We have implemented this algorithm for planar arms as well as 3-dimensional arms. Two examples are shown here. Figure 2 shows an example for a planar four link arm polygonal obstacles is shown in Figure 2. This example is similar to the one in [Lozano-Perez 1987], and our algorithm finds a path in this fairly cluttered situation. Note that the motion is quite complex. Link 1 has to back up to allow link 2 to rotate out of the narrow region; similarly, while these links are moving, link 3 also has to backup to avoid colliding with the obstacles. Figure 3 shows another difficult example for a three link arm. Note that the only solution in this case is to "shrink" the arm and then "stretch" it back, and our algorithm finds it.

These examples are a clear indication that our sequential strategy should find collision-free motions in a fairly large class of situations. This sequential strategy, however, is not complete, i.e., in very cluttered environments it may not find a collision-free path even if one such path exists. We discuss these issues in Section 5.

3 Planning Motion of a Single Link Along a Given Path

The motion of one end of the link is already known (the motion of previous links having been already planned. As the link moves along this path, the problem now is to adjust its angle (or equivalently its rotational velocity) to avoid the obstacles in its way. Now, suppose we parameterize the path of the link endpoint by parameter t. The physical obstacles give rise to constraints on the angle θ of the link as a function of t. These constraints are approximated (at a given degree of resolution) by methods similar to [Lozano-Perez 1987], and further approximated as polygonal regions in $t \times \theta$ space. Having characterized the forbidden regions in $t \times \theta$ space, we may now use 2-dimensional path planning approaches e.g., visibility graph or free-space based approaches [Kant and Zucker 1986]. The basic steps are illustrated in Figure 4. Given that we have solved the problem for a single link, we can now state the full algorithm.

4 The Complete Algorithm

The manipulator joint parameters are represented by q_i, the corresponding links by l_i, $i = 1, 2, \ldots, n$, l_1 being the most proximal link, and l_n the most distal link. Let $\mathbf{r_i}$ be the ref. vertex for link l_i. The base of the first link is assumed to be fixed at $\mathbf{r_1}$. Let $\mathbf{r_i}(t)$ denote the trajectory of the reference vertex $\mathbf{r_i}$. The initial and goal configurations of the arm are given as q_i^s, and q_i^g, $i = 1, n$. The algorithm is as follows:

Algorithm

Find the ranges of legal values of q_1.
if ($[q_1^s, q_1^g]$ is a sub-interval of a legal range)
  Parameterize $[q_i^s, q_i^g]$ as $q_1(t), t \in [0, T]$
  Discretize $q_1(t)$ to get $q_1(k\Delta T) \stackrel{\text{def}}{=} q_1(k)$, $k = 0, \ldots, M-1$; ΔT being the resolution.
  $i = 2$.
  while $(i \leq n)$
    for $k = 0, M-1$
      Given $q_1(k), q_2(k), \ldots q_{i-1}(k)$, compute $\mathbf{r_i}(k)$.
      Compute the forbidden ranges (slices) of $q_i(k)$ for each $\mathbf{r_i}(k)$.
    end for
    Group $q_i(k)$, $k = 0, M-1$ slices into regions.
    Group regions into polygonal obstacles.
    Build visibility graph with $I = (0, q_i^s)$, $F = (1, q_i^g)$.
    Search the graph to get a collision-free piecewise linear path for q_i, $q_i(t)$ from I to F.
    if (success)
      Discretize $q_i(t)$, i.e., compute $q_i(k)$, $k = 0, M-1$ along the collision-free path.
      Increment i.
    else return failure
  end while
  return success
else

return no path exists

The $t \times \theta$ space for each link in example shown in Figure 2 are shown in Figure 5.

4.1 Backup Movements Outside the Start to Goal Interval

Consider Figure 6. Suppose we do not allow the links to cross over. With this constraint, the only way for the arm to reach the final configuration is to allow link 1 to move beyond its final position so that link 2 can move and then allow link 1 to move back to its final position.

A straightforward modification as shown in Figure 6 leads to such a mechanism. In the initial parameterization for link 1, the algorithm considers the sub-interval from start value q_1^s to the goal value q_1^g, i.e., $[q_1^s, q_1^g]$ for the first link. In general, this sub-interval will be a subset of a whole legal range. We parameterize the whole interval (instead of only the sub-interval from start position to goal position) and then plan the path for the second link. Note that the beginning and the end of motion of link 1 now does not correspond to $t = 0$ and $t = T$, but to some intermediate values, $t = t_1$ and $t = t_2$. The start and the goal points for link 2 are not at $t = 0$ and $t = T$, but are at $t = t_1$ and $t = t_2$. Thus the path for link 2 can go beyond $[t_1, t_2]$ interval. Such motions correspond to back up movements beyond the start to goal interval. Hence link 1 can now back up away from its goal position and then move back to its goal position. With this modification, the two link example is easily solved. The motion is shown in Figure 7. Furthermore, it is easy to see that with this modification, the algorithm is complete for a two link arm (up to the resolution of discretization), i.e., it will find a collision-free path if there exists one. However, this property does not hold for more than two links.

For the first link there is a natural mechanism for the link to move beyond its final (or initial) configuration – by considering the whole free-interval in which the initial and the final configuration values for the first joint lie. There seems no natural way of extending such "beyond the final (or initial) configuration motion" for subsequent links. Consider the example shown in Figure 3. In this case, there does not exist a collision-free motion for the last link along the path determined for the previous links, hence the algorithm fails. Note that the only solution in this case is to "shrink" the arm and then "stretch" it back. Thus, every link and not just the first link, has to back away from its initial position and move farther than its final position in order to avoid the obstacles. We have incorporated such motion to some degree for link i in $t \times \theta$ space by appending segments of vgraph to the path in $t \times \theta$ space before the initial point and after the final point, and then building the $t \times \theta$ space for link $i+1$ along this augmented path. With this modification, the algorithm does find a collision-free path as shown in Figure 3.

5 Discussion

Undoubtedly, our strategy is going to lead to significant speed-ups in computation simply because, for an n link arm, it solves one 1-dimensional and $n-1$ 2-dimensional planning problems rather than one n-dimensional problem. It is well known that the complexity of the motion planning problem is exponential in n. The strategy, therefore breaks this exponential barrier of complexity. But the price paid is that it is incomplete. A main question, therefore, is in which situations will this sequential planning fail? The first thought that all situations where the motion of more than one link has to be planned simultaneously will be missed, is not true. Because, although the planning is sequential, the planning of a link depends upon all the previous links. As stated earlier, for a two link arm, algorithm B (non-temporal) is complete up to the resolution of discretization. However, this property does not hold for more than two links. If along the path (determined by the previous $1, \ldots i$ links), no motion exists for $i + 1^{th}$ link, our sequential strategy will fail. With the mechanism for each joint to move beyond its start to goal interval, we believe that the algorithm will work in most reasonable cases. Nevertheless, in extreme cases, one may have to search the full n-dimensional Cspace. At the very least, our strategy will form a front end to a more complete planner.

References

[1] Donald, B. 1984. Motion Planning with six degrees of freedom. MIT AI Tech. Rep. 791. 504-512

[2] Faverjon, B. 1984. Obstacle avoidance using an octree in the configuration space of a manipulator. *IEEE Conference on Robotics and Automation.* pp. 504-512

[3] Gouzenes, Laurent. 1984. Strategies for solving collision-free trajectories for mobile and manipulator robots. *Int. Journal of Robotics Research.* 3(4). MIT Press.

[4] Gupta, Kamal Kant, and Yalpani, Kuros. 1989. Motion planning with Velocity and Acceleration Constraints. CSS-IS TR 90-02. Centre for Systems Science, Simon Fraser University.

[5] Hopcroft, Joseph, and Whitesides, S.,. 1985. On the movement of robot arms in 2-dimensional bounded regions. *SIAM Journal of Computing,* 14(2).

[6] Korein, J. 1985. *A geometric investigation of reach.* MIT Press.

[7] (Gupta) Kant, Kamal, and Zucker, S.W. 1986. Toward efficient trajectory planning: the path-velocity decomposition. *International Journal of Robotics Research.* 5(3).

[8] (Gupta) Kant, Kamal, and Zucker, S.W. 1988. Planning smooth collision-free trajectories: path, velocity, and splines in free-space. *International Journal of Robotics and Automation.* 2(3).

[9] Lozano-Perez, T. June 1987. A simple motion planning algorithm for general robot manipulators. *IEEE Journal of Robotics and Automation.* 3(3).

[10] Schwartz, Jacob T. and Sharir, Micha. October 1981. On the 'Piano Movers' problem I. The case of a two-dimensional rigid polygonal body moving amidst polygonal barriers. Report No. 39. Dept. of Comp. Sc., Courant Institute of Math. Sc., New York.

[11] Schwartz, Jacob T. and Sharir, Micha. October 1982. On the 'Piano Movers' problem II. General techniques for computing topological properties of real algebraic manifolds. Report No. 41. Dept. of Comp. Sc., Courant Institute of Math. Sc., New York.

[12] Udupa, S. 1977. Collision detection and avoidance in computer controlled manipulators. *5th Int. Joint Conf. on Artificial Intelligence.* Cambridge.

[13] Xing, D. M. 1987. Collision-free trajectory generation for a general robot manipulator. Tech. Report. CRIIF, Laboratoire de Robotique. Paris.

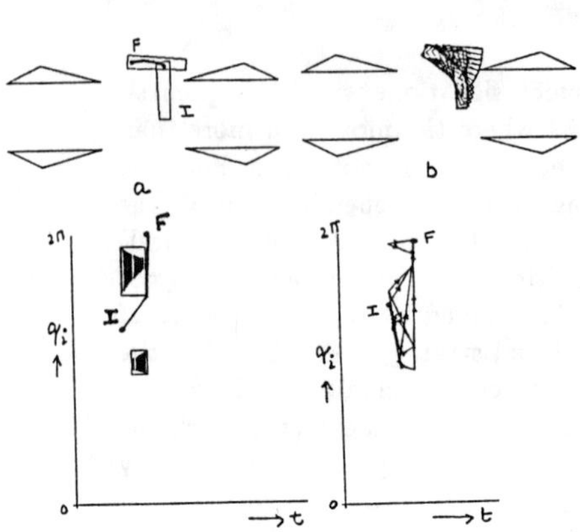

Figure 4:

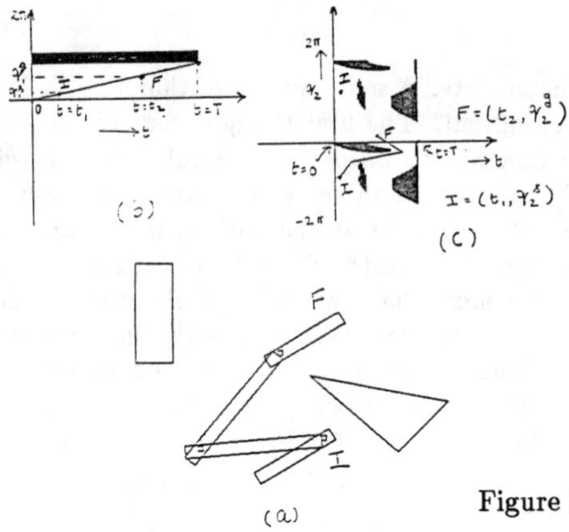

Figure 6:

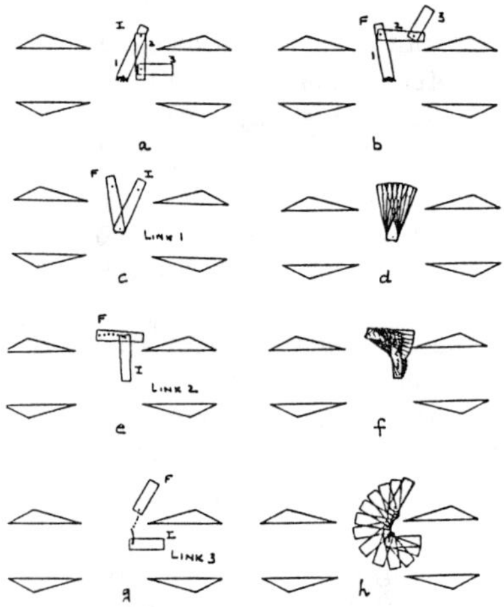

Figure 1:

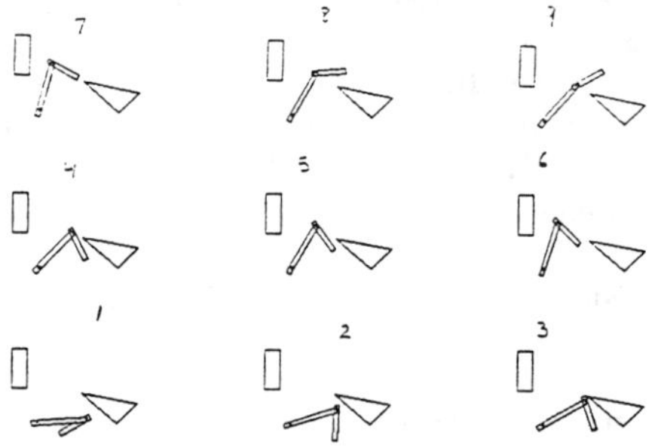

Figure 7:

Figure 2:

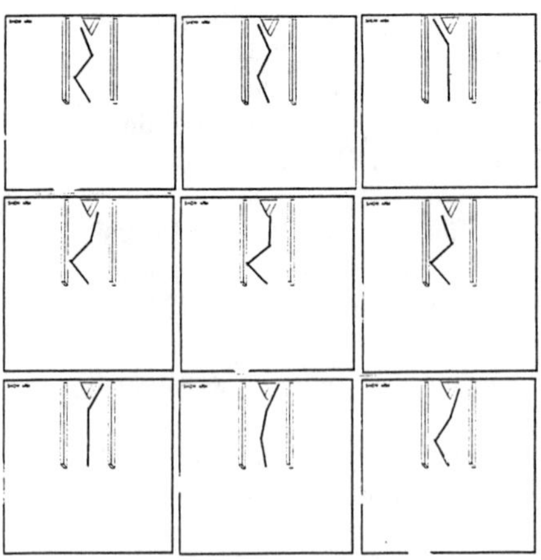

Figure 3:

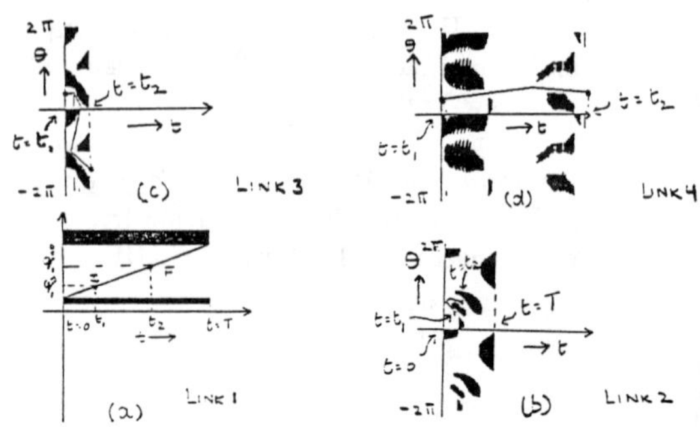

Figure 5:

DEVELOPMENT OF A HEAVY LOAD HANDLING ROBOT AND ITS PNEUMATIC POSITION SERVO CONTROLLER

W.K. CHUNG* and Y. YOUM*

* Professor, Department of Mechanical Engineering
Pohang Institute of Science and Technology and
Research Scientist, Division of Mechatronics
Research Institute of Industrial Science and Technology
Pohang, Korea

S.D. PARK**

** Research Associate, Division of Mechatronics, RIST

ABSTRACT

A number of industrial robots have been used in various work tasks from small parts assembly jobs to heavy load handling ones. To handle heavy loads a manipulator becomes heavy because the payload to weight(p/w) ratios of the conventional robots are very low. In this project, a manipulator whose p/w ratio is near to one was designed and fabricated. A pneumatic proportional plus on/off position controller was also developed to achieve fine positioning accuracy and overshoot-free response for various payload conditions. The robot's controller is developed and implemented by both simulation and experiment.

INTRODUCTION

Although industrial robots have become widely used in various industrial application fields, its use in heavy load handling jobs has been restricted because of the payload limits of conventional robots. To handle over a 50kg payload, the weight of the robot itself easily exceeds 1 ton to satisfy the desired motion. For the conventional robot the payload to weight ratio is very low. For example, these ratios are 1/20 and 1/48 for the PUMA - 560 and 760 robots, respectively. In this paper, the developed manipulator is described in two parts, 1) design of the manipulator whose payload/weight ratio is near 1 and 2) design of a pneumatic position controller which has no overshoot, even with 40kg payload.

Design of the Manipulator

The manipulator has a double-parallelogram mechanism as shown in Fig. 1 schematically. It has 3 - degrees-of-freedom to give positioning motion by base rotation, and horizontal and vertical movements through the corresponding slots. The kinematics of the robot is derived for the planar 2-d.o.f. joints to obtain the decoupled motion between the horizontal and vertical motions. If the input translation vector of the slot joints is defined by $s = [x, y]^T$, the displacement of revolute joints by $q =$

$[\theta_1, \theta_2]^T$ and the output translation vector of the end effector by $\underset{\sim}{h} = [u, v]^T$, then the time derivatives of these vectors have the following relationship.[1]

$$\dot{\underset{\sim}{h}} = J_1 \dot{\underset{\sim}{q}}, \quad \dot{\underset{\sim}{s}} = J_2 \dot{\underset{\sim}{q}} \tag{1}$$

The complete equations of J_1 and J_2 are described in detail in Ref.[1]. In the case of $\det(J_2) \neq 0$ there exists a nonsingular manipulator Jacobian (J) such that $\underset{\sim}{h} = J \underset{\sim}{s}$, where J can be obtained from $J_1 J_2^{-1}$. Through the careful observation of J, we can diagonalize the Jacobian by selecting k as $k = l_5/l_2 = l_4/l_2$ and $l_1/l_3 = 1 + k$, where k represents the arm ratio. This is the decoupling condition and the Jacobian is now expressed as a function of k only:

$$\begin{bmatrix} \delta u \\ \delta v \end{bmatrix} = \begin{bmatrix} -k & 0 \\ 0 & k+1 \end{bmatrix} \begin{bmatrix} \delta x \\ \delta y \end{bmatrix} . \tag{2}$$

When we consider the real time control aspects, this kinematic relation is very simple to use in controlling the robot.

In addition to these kinematic considerations, the manipulator must be designed to have a balanced configuration irrespective of robot position and payload variations. In Fig.2, a free body diagram of the manipulator is shown for the force analysis. By using the force and moment equilibrium conditions, the balancing equation can be obtained as follows:

$$F_p = \frac{(l_3 + l_4)\cos\theta_1 + (l_2 + l_5)\cos\theta_2}{l_3\cos\theta_1 + l_2\cos\theta_2} W_L + \frac{A\cos\theta_1 + B\cos\theta_2 + C}{l_3\cos\theta_1 + l_2\cos\theta_2} \tag{3}$$

Here, A,B and C are complex functions of masses of links, link lengths and payloads.[1] To obtain configuration-independent characteristics, F_p, which is the force exerted by the vertical cylinder, should be almost constant for all robot configurations i.e., F_p is independent of θ_1 and θ_2. These configuration-independent characteristics can be obtained from the Eq.(3) if the following relations are satisfied:

$$(l_3 + l_4)l_3 = (l_2 + l_5)/l_2 \ , \ A/l_3 = B/l_3 \text{ and C is negligible.} \tag{4}$$

The manipulator, whose arm ratio (k) is 3.5, was designed and manufactured based on the decoupling and balancing conditions obtained above. This manipulator can handle up to 150kg while the weight of the manipulator itself is near 150kg. For this heavy load handling robot, the overshoot response can damage both the workpiece and the robot, so the first goal of our position controller was to prevent overshoot response. The pneumatic servo controller for the manipulator will be explained in following section.

Design of a Pneumatic Position Control System

Since the pneumatic system is characterized by nonlinear flow phenomena, the classical pneumatic position control systems have been mainly used for simple end-point control using on/off signals which can produce considerable overshoot. To reduce the overshoot of the system, large deadband is needed. In this robot development, a flow type E-P (Electro-Pneumatic) proportional valve was used to pass the weight flow rate proportional to the position error through the valve. To control the direction of the weight flow rate, which is fed into a cylinder, two 3/2 way directional valves were used.[2] The resulting pneumatic position control system can have a damping effect and can reduce the minimum stable deadband and overshoot by governing the speed of load properly. The mathematical modelling for theoretical analysis of this system was made as follows.

Fig. 3 shows the schematics of the electro-pneumatic servo system. In the figure the voltage V_e which is the difference between the feedback voltage, V_f, and reference input voltage, V_r, makes the actuating voltage, V_a, of the E-P proportional valve as follows:

$$V_{act} = V_{min} + K_p|V_e|. \tag{5}$$

where V_{min} is the minimum actuating voltage of the E-P proportional valve, and K_p is the voltage gain of the E-P proportional valve. The V_{act} in Eq.(5) changes the effective orifice area, A_{es}, of E-P proportional valve and the weight flow rate, W_i. In this case W_i can be modeled as[3]:

$$W_i = C_2 A_{es} \frac{P_{sup}}{\sqrt{T_i}} f_{re}\left(\frac{P_i}{P_{sup}}\right) h\,(t-\tau) \tag{6}$$

where C_2 = 0.3968($\sqrt{{}^{\circ}K}$/sec) (constant), P_i, the pressure of the inlet chamber(kg_f/cm^2), $f_{re}(P_i/P_{sup})$, the flow function[3], P_{sup}, the pressure of the air supply(kg_f/cm^2), T_i, the temperature of the air supply(°K), and τ, the time delay of the servo valve (sec). When the piston ram begins to move, the outlet weight flow rate, W_o, which is exhausted to the atmosphere through an exhaust orifice valve, can be represented as follows.

$$W_o = C_2 A_{eo} \frac{P_o}{\sqrt{T_i}} f_{re}\left(\frac{P_{atm}}{P_o}\right) \tag{7}$$

where A_{eo} is the effective orifice area of the exhaust valve (cm^2), P_{atm} is the atmospheric pressure (kg_f/cm^2) and P_o is the pressure in outlet chamber (kg_f/cm^2). The rate of pressure change in both chambers in the cylinder can be described as:

$$\begin{aligned} \dot{P}_i &= \frac{\kappa R T_i}{g(\upsilon_i + A_c x)} W_i - \frac{\kappa A_c}{\upsilon_i + A_c x} P_i \dot{x}, \\ \dot{P}_o &= -\frac{\kappa R T_i}{g(\upsilon_i - A_c x)} W_o + \frac{\kappa A_c}{\upsilon_i - A_c x} \dot{P}_o \dot{x}. \end{aligned} \tag{8}$$

where κ is the specific heat ratio of air (1.4), R, the gas constant of air (cm^2/s^2 °K), υ_i, the initial volume in chamber (cm^3), A_c, the ram area (cm^2), x, the position of the piston ram (cm), $\dot{x}$, the velocity of the piston ram (cm/s), and g, the gravitational acceleration(981 (cm/s^2)).

The dynamic equation of the system is written as:

$$m_{eff}\ddot{x} = A_c (P_i - P_o) - F_f. \quad (9)$$

where m_{eff} is the effective mass of the system(kg) and the friction force, F_f, is modeled as:

$$\begin{aligned} F_f &= f_s && \text{for } \dot{x} = 0 \\ &= f_v\dot{x} + \text{sgn}(\dot{x})\, f_c && \text{for } \dot{x} \neq 0 \end{aligned} \quad (10)$$

where f_s is the static friction force (kg_f), f_c is the Coulomb friction force (kg_f) and f_v is the viscous friction coefficient (kg_f·sec/cm). The values of f_s, f_c and f_v were obtained through extensive experiments[2] and determined as 4.2, 2.5, and 0.2, respectively.

Simulations and Experiments

Computer simulations were carried out to investigate the validity of the mathematical modelling and to predict the performances of the designed control system. An experimental testbed was made to test the performances. The resulting position response by both simulation and experiment on a testbed is shown in Fig. 4 for the operating conditions: $V_{min} = 0.74$, $V_{max} = 0.84$, $A_{eo} = 8 \times 10^{-4}$ (cm^2) and $K_p = 0.025$. In this figure, it can be shown that the response obtained from the simulation coincides very well with that from the experiment which demonstrates the validity of the modelling of the system.

The effects of system parameters, such as K_p, V_{min}, V_{max}, which will be used in theoretical and experimental analysis, were also investigated. The results can be summarized as follows:

1) For the same K_p, the response with the larger V_{min} and V_{max} is faster than that with the smaller V_{min} and V_{max} as shown in Fig. 5.
2) For the same V_{min} and V_{max}, the response with the larger K_p is faster than that with the smaller K_p. But, if K_p is too large, then large overshoot and instable system response can result.
3) At the same V_{min}, V_{max} and K_p, the response with the larger exhaust orifice area, A_{eo}, is faster than that with the smaller A_{eo}.

Using these results from the testbed, the pneumatic position control system was applied to the designed manipulator and the experiments were made for various payload conditions. The designed manipulator is shown in Fig. 6 with the pneumatic controller. The end effector was commanded to move 35.875cm horizontally and the position of a horizontal slot was measured and fedback to the position controller.

Through experiments, we obtained the relation between the payload and proportional gain, K_p, which produces no overshoot characteristics as shown in Fig. 7.

If a K_p is selected according to this curve, the manipulator can move to any position without overshoot. In the implementation point of view, this can be done by measuring the pressure in the vertical cylinder which is an indicator for payloads, and a K_p can be selected through the table-look-up method using the measured pressure. The response for the payloads of 0 - 40kg are shown in Fig. 8, where no overshoot characteristics were found. The response differs only in the rise time, which is a function of system inertia(payloads). The dynamic analysis of the system is presently under investigation.

CONCLUSION

A heavy load handling robot whose payload/weight ratio is near to 1 was developed using a double-parallelogram mechanism. The robot is designed by kinematic decoupling scheme and force balancing condition. A new controller with proportional plus on/off controller was proposed to control the heavy load handling robot. The performance of the control system was examined through both simulation and experiments. The main advantage of the system is that it can produce overshoot-free response for the payload up to 40kg.

Acknowledgements : This paper was supported in part by research fund for Junior Scholars, Korea Research Foundation and RIST.

REFERENCES

1. Y. Youm, W.K. Chung and W.J. Chung, "Design of Heavy Payload Manipulator with Balanced Double-Parallelogram", KOREA-U.S.A. Design Engineering Seminar, Oct. 13-19, 1988. Seoul, Korea. pp.303-320.
2. S.D Park, "Development of Pneumatic Proportional Plus On/Off Servo System for the Industrial Robots", M.S. Thesis, Dept. of Mech. Engg., POSTECH, 1990.
3. F.D. Ezekiel and J.L. Shearer, "Pressure-Flow Characteristics of Pneumatic Valves", Trans. of ASME, Oct. 1957, pp.1577-1590.

Figures

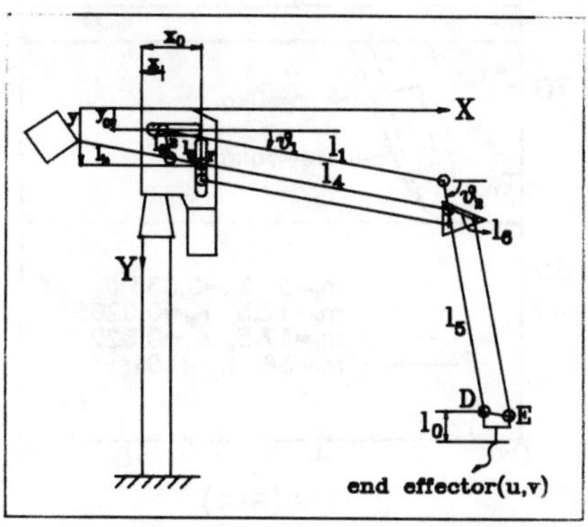

Fig. 1. Schematic diagram of the manipulator

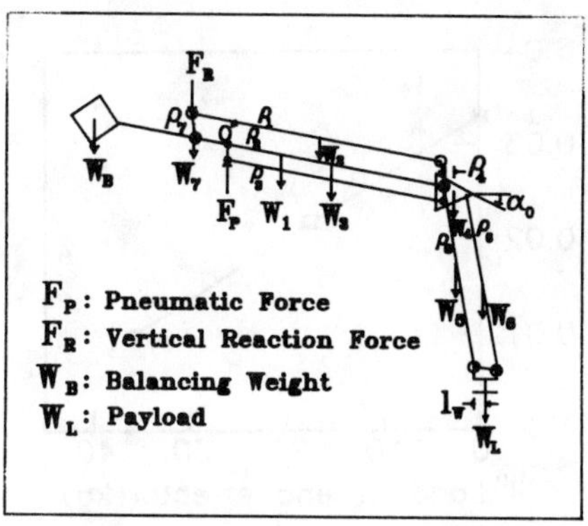

Fig. 2. Free body diagram of the designed manipulator

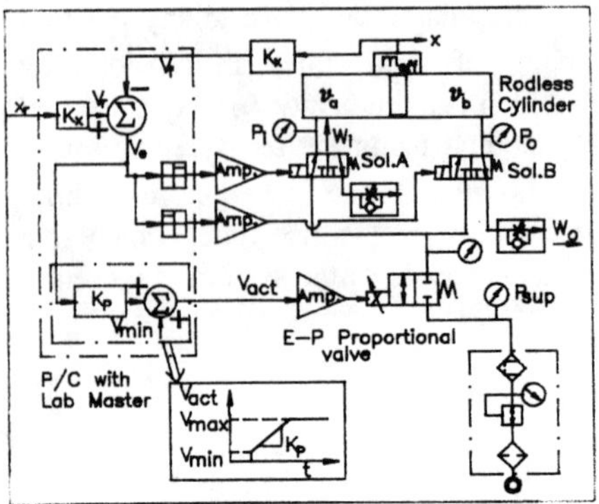

Fig. 3. Schematic diagram of the pneumatic position servo controller

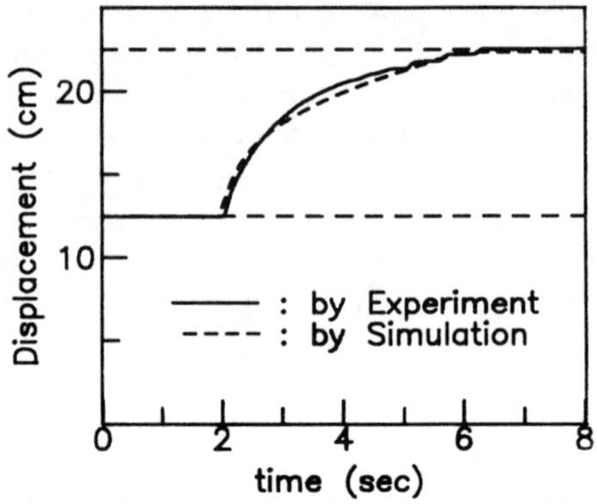

Fig. 4. Comparison of experimental response with computed result at testbed experiment

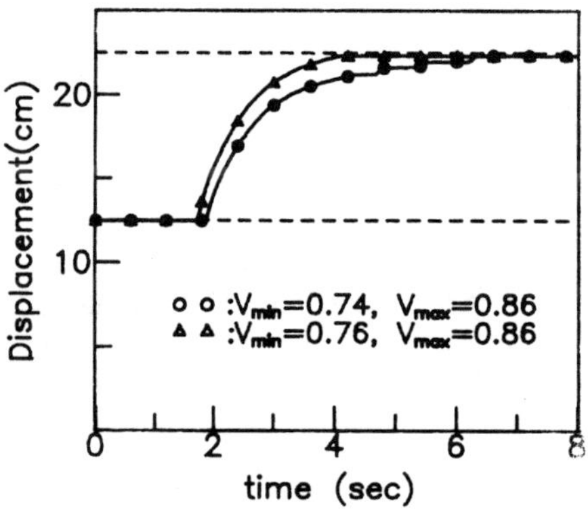

Fig. 5. Experimental response w.r.t. step input at testbed (Effect of V_{max}, V_{min} change)

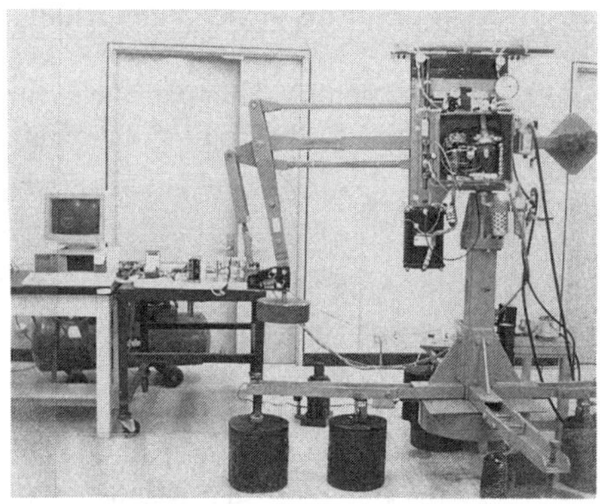

Fig. 6. The designed manipulator with pneumatic position controller

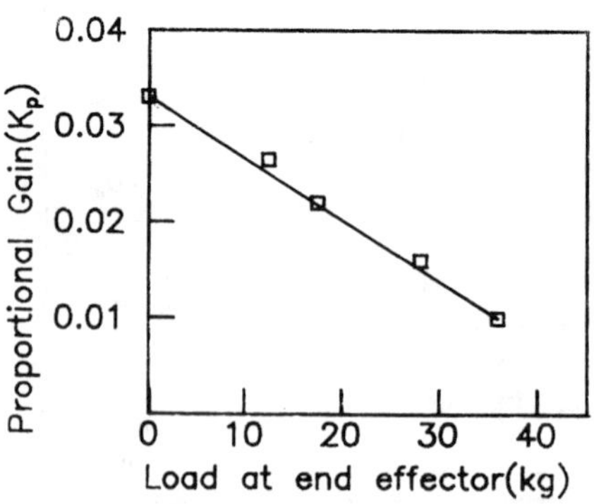

Fig. 7. Relation between the payload and K_p

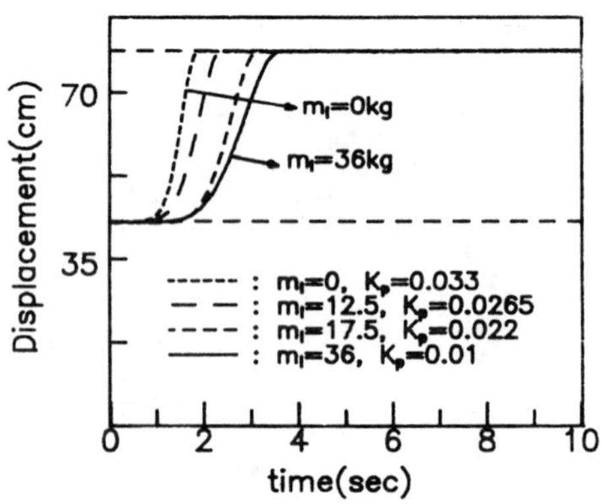

Fig. 8. The critical responses w.r.t. various loads

An Implementation of a Novel Method for Concurrent Process Control in Robot Programming

S. Marin, M. Ristic

Imperial College of Science, Technology and Medicine
Department of Mechanical Engineering
Exhibition Road, London SW7, U.K.

Z. Sojat

RDP Technology, London, U.K.

Abstract

During the development of complex robotic applications (such as assembly) there is a strong need for the use of concurrent processes. Existing robot languages which allow concurrent programming realize concurrent process control through statements embedded within the application. This programming structure does not lend itself to easy analysis, prediction and debugging of conflicting situations (e.g. deadlocks).

A real-time Robot Control Language (RCL) has been developed as a part of the robot research programme at Imperial College. The new approach for concurrent process control adopted in RCL is based on a strict division between user's *application programme modules* (concurrent processes associated with execution of distinct tasks) and the *control module* (supervisor).

Key Words : application programming module, control module,
independent activity, process, task, supervision, supervisor

1. Introduction

During the development of complex robot applications (such as assembly operations), there is a strong demand for the use of concurrent processes. The current robot programming languages do not offer a satisfactory concurrent process controll, they realize this control through statements embedded within the applications. Programmers are not able to check possible dead-lock or starvation situations until the robot program is executed in a real application environment.

To overcome these limitations the robot programming language should offer the facilities to examine user specified process control before its execution, allowing the programmer to survey the control patterns and to correct them where necessary.

This paper addresses the new method for concurrent process control employed in RCL language. The new approach is based on a strict division between user's *application programming modules* and program *control module* (supervisor).

Each application module accommodates programming statements related to a specific task (i.e. robot devices, sensors, conveyors, etc.), with no reference to the global performance of the robot program, while process control statements are excluded from these modules.

The control module groups together only process control statements which are related to the overall behaviour of all the devices engaged within the robot working environment. It has the role of real time supervision of the various autonomous activities (processes). Communication between supervisor and processes is established through the notion of a *transfer* parameter. Process control statements are taken out from the application modules and are grouped together within the supervisor. Only the transfer parameters suggest the possible existence of the process control.

Based on these ideas, a robot program which is programmed to handle concurrent activities will consist of: set of independent processes (activities), global data area (data exchange) and of the control module (supervision and control). These relations are shown in Figure 1.

Figure 1

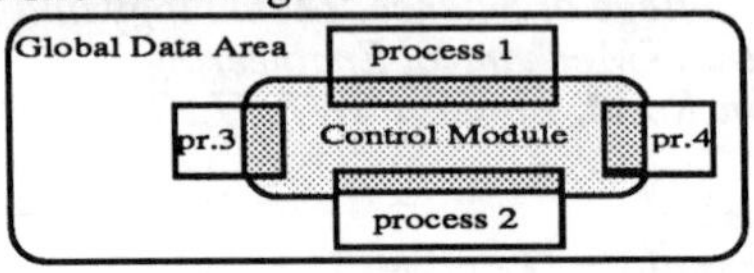

The first version of the RCL language offered standard program-based process control statements built on the semaphore concept such as: start, kill, resume, suspend, wait, signal statements [1], [2]. It was soon realised that the coordination of more than three processes is complicated and difficult to accomplish and the amount of programming code dedicated to process control grows substantially, leading to the loss of the initial program concept. It became difficult to understand the programs and to predict the system behaviour. These problems initiated a new approach to the philosophy of the control, and to the process definition itself.

2. Formal Definition of a New Method for Concurrent Process Control

2.1. Preliminaries

An independent activity or a *process* is defined as a finite set of all possible states in which process can be during its lifetime,

$\mathcal{P} = \{p_i \mid 0 \leq i \leq m\}, m$ = maximum number of process states,

p_o = initial process state,

p_{end} = final process state. Process is terminated.

Each process state is one processor instruction which is executed at time $\mathcal{T}$ and it is a function of time.

$p_i = f(\mathcal{T}_i)$, where for $i = 0 \dashrightarrow p_i = p_o$ and $i = \mathcal{T}_{end} \dashrightarrow p_i = p_{end}$

We define the characteristic process states as a finite set of all *transfer* sections declared within the process:

$\mathcal{P}_t = \{p_{t_i} \mid 0 \leq i \leq k, k$ = number of transfer sections in process $\mathcal{P}\}$,

where $\mathcal{P}_t \subseteq \mathcal{P}$ and $p_o \in \mathcal{P}_t$. A system of processes Q is defined as a set product of all defined processes $\mathcal{P}$:

$$Q = \mathop{\times}_{i=0}^{n} \mathcal{P}_i \quad (n = \text{maximum number of defined processes})$$

A characteristic set which consists of all processes which at time $\mathcal{T}$ have reached their transfer states is:

$$Q_t = \mathop{\times}_{i=0}^{j} \mathcal{P}_{t_i} \quad \text{where } Q_t \subseteq Q \text{ and } j = \text{number of processes at the transfer state}$$

Initial state of the system Q is n-tuple:

$$q_o = (p_{o_1}, p_{o_2}, \ldots, p_{o_i}, \ldots, p_{o_n}), \text{ where } p_{o_i} \in \mathcal{P}_i,\ i = 1,2,3,\ldots,n$$

Final state of the system Q is n-tuple:

$$q_{end} = (p_1, p_2, ..., p_i, ..., p_n), \text{ where } p_i \in \mathcal{P}_i,\ i = 1,2,3,...,n$$

A supervisor S is a pair $S = (\mathcal{R}, f)$, where $\mathcal{R}$ is a set of the rules which are given by following:

$$\mathcal{R} = \{\delta_{iy} \mid \forall i \exists \chi \mid \delta_{i\chi} : (p_{t_\chi} \times Q) \text{ ---> } Q \text{ iff } \chi = y\} \quad \text{-----} \quad (1)$$

A function f is a *state transition function* which maps state of the system q_v at time T_v into new state: q_{v+1} : $f : q_v \text{ ---> } q_{v+1}$, where f is a partial function, defined only at some elements of q_v.

Formally supervisor S is equivalent to a directed graph with node set Q and an edge $q_v \text{ ---> } q_{v+1}$ labeled t for each triple (t, q_v, q_{v+1}) such that $q_{v+1} = f(t, q_v)$. Such an edge, or state transition is called an *event*. Result of an event is a new system state q_{v+1} .

We define a generator $\mathcal{G}$ to be the action of the supervisor S over system Q. It is a set of all possible finite sequences of new system states which can be generated:

$$\mathcal{G} \equiv (S/Q) = \{(q_0, q_1, ..., q_s)_i \mid i = 1,2,..k\},$$

s = last generated system state, k = final generated sequence.

2.2 Controllability

System Q is controllable if there exist such a generator $\mathcal{G}_c$:

$$q_0 \xrightarrow{\mathcal{G}_c} {}^{*} q_{end}$$

$$\mathcal{G}_c = \{(q_0, q_1, ..., q_v, ..., q_s)_i \mid i = 1,2,..k, \text{ and } (\exists q_v = q_{end})\},$$

and $\mathcal{G}_c \subseteq \mathcal{G}$,

meaning that the system Q will reach its final state q_{end} when there exists a state transfer function f which will map state q_v into the final state q_{end}.

System Q is not controllable if there exists state q_v such that :

$$q_0 \xrightarrow{\mathcal{G}} {}^{*} q_v,$$

$q_v \in (q_0, q_1, ..., q_s)$ and $q_v \in Q_t$, (system is in dead-lock state)

It is not known if the system Q is controllable if there exists an external input to the system Q , where the next process state is a function of time:

$$p = f(T, \text{external event}) .$$

3. RCL Language Structures for Concurrent Programming

3.1. Process Declaration

A process is a sequence of operations carried out one at a time. These computational operations must be carried out in a certain order of precedence to ensure that the final result of a particular operation can be used by others. This type of computation is called a *sequential process*. It consists of a set of operations which are totally ordered in time.

$$\mathcal{P} = \{punit_i \mid i = 1, 2, \ldots, m\}, \text{ Figure 2.}$$

Process $\mathcal{P}$ consists of a sequence of time ordered basic operations. Process unit operations A,B,C do not overlap in time and they are not interdependent.

Processes are *concurrent* if their executions overlap in time (Figure 3). Processes A,B,C are concurrent because the first operation of one process is started before the last operation of another process is completed. In order to accomplish the whole task "R", the constituent concurrent processes have to be synchronized.

Looking at the problem more practically, we distinguish two cases: the first, when process is associated with the processor, and second where all processes are within the scope of one processor. If we have an installation where several processors work simultaneously, the machine instruction of concurrent processes can overlap in time. But if one processor is multiplexed among concurrent processes, the machine instructions of these processes can only be interleaved in time. The logical problems of process control (synchronization) turn out to be the same in both cases; they are caused by our ignorance of the relative speeds of concurrent processes [5].

Figure 2

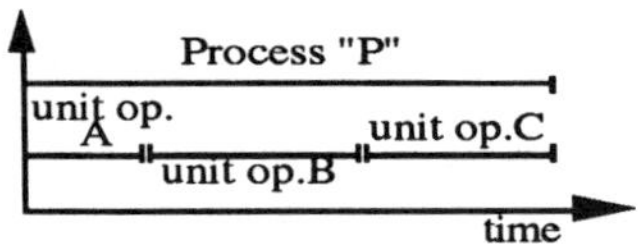

Figure 3

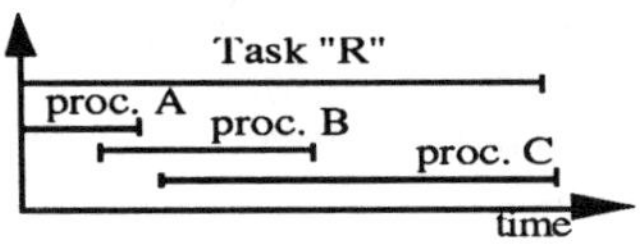

3.2 Definition of RCL concurrent structures

An RCL *task* is a set of coordinated, independent activities which are ordered in time. Control of the task realization is performed by the *supervisor*. An RCL *process* is an independent activity and it doesn't communicate with another, process-defined activities. A process can also use programming system resources without explicit concern for other processes. RCL processes and tasks are characterized by their state in time.

Task states are recognized as :

a) initial state ..... q_o, initial state of all defined processes,

b) final state ..... q_{end}, final state of all defined processes,

c) working state ..... q_i, all states which originate during state transfer from initial to final state,

Process states are: a) *activated* process which is defined and started,but it is immediately blocked before p_{oi}. b) *running* process which is active (i.e. it uses processor cpu_time) and its state is: $p_i \in (\mathcal{P} - \mathcal{P}_t,)$.

c) *blocked* process state is, $p_i \in \mathcal{P}_t$.

Declaration of the concurrent structures in the RCL:
1) Declaration of RCL task formation:
TASK activity1, activity2, activity3 ...... (3)
2) Declaration of RCL initial task state:
INITIAL STATE activity1[+], activity2[-], activity3[+] **
3) Declaration of RCL final task state:
FINAL STATE activity1[+], activity2[-], activity3[-]
4) Declaration of RCL independent activity or a process

The application programming module is an RCL module associated with the sequential computation actions performed on the single device (robot device, sensing device, conveyers, etc.). Statement (3) introduced procedures "activity1-3" to be declared as processes, and the new rules which will apply at these procedures:

a) computational and physical actions declared within the procedure are related only to an unique device. Consequently no other process can be declared from that procedure. The procedure itself is a single and only allowable process.

b) communication with the control module or a supervisor is formulated through the transfer parameter declared within the procedure.

c) existence of the supervisor and procedure declaration as a process implies that the beginning and the end of the procedure are also beginning and the end of process.

** Note: Process states are symbolised with +/- sign where "+" means a running process, and "-" blocked process.

3.3 Definition of RCL control module (supervisor)

The supervisor is a device which takes initial task state as an entry and performs task state transitions until final task state is not reached. Each supervisor event is initiated by a singular request from the task element process. This will occur when currently running process reaches its *transfer* section, and its state changes from running to blocked (Figure 4).

Figure 4

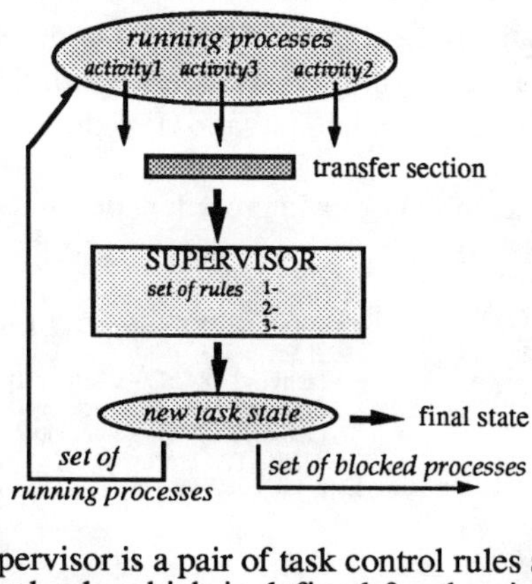

By definition the supervisor is a pair of task control rules and state transition functions. The task control rule which is defined for the single process being blocked on an unique transfer parameter, launches state transfer function *f*. At the time of request, task state is q_r where:

$q_r = (p_{i_1}, p_{i_2}, ..., p_{i_j})$, where j = number of defined processes and i = process state index,

We distinguish two characteristic subsets of the task state q_r:

a) all processes which are in the running state, and can not be affected by the supervisor action.

$$\mathcal{PR}_r = \{p_i \mid p_i \in \mathcal{P} \text{ and } p_i \notin \mathcal{P}_t\},$$

b) all processes which are blocked and their status could be influenced by the supervisor action.

$$\mathcal{PR}_b = \{p_i \mid p_i \in \mathcal{P}_t\},$$

Special element of the $\mathcal{PR}_b$ set is that process state p_{t_x} which has inaugurated the request for the supervision.

A generator $\mathcal{G}$ is defined as:

$$\mathcal{G} = \{(q_0, q_1, \ldots, q_{v-1}, q_v, q_{v+1}, \ldots, q_s)_i \mid q_v = f_t(t, q_{v-1}), i \text{ and s as defined at (2)}\},$$

t ≡ process transfer section,

x ≡ process which has initiated request,

q_{v-1} ≡ task status before request for the supervisor action,

State transfer function f_t is defined as: $f_t : (p_{t_x} \times \mathcal{R}_x) \times Q \text{ --> } Q$, where set product $(p_{t_x} \times \mathcal{R}_x)$ represents a rule which will be applied on the task state q_{v-1}. There exists a single rule for process state

p_{t_x}: $(\exists r_{t_x}) \in (p_{t_x} \times \mathcal{R}_x) \mid r_{t_x} \in \mathcal{R})$, where $\mathcal{R}$ is defined at (1). If we mark all the task states which can be reached from q_{v-1} to q_v as:

$Q_m = \{q_v \mid q_v = f_m(t, q_{v-1})\}$, where transfer function f_m is defined as:

$f_m : (p_{t_x} \times \mathcal{P}_{t_x}) \times Q \text{ --> } Q$ and $\mathcal{P}_{t_x} = \{t_i \mid 0 \leq i \leq n\},$

and n is number of transfer parameters defined at process x, then we can extend the meaning of the controllability by:

Task Q is *not controllable* if, for the event $e \equiv q_{v-1} \xrightarrow{\mathcal{G}} q_v$ applies that, $\exists q_v \mid q_{v-1} \in Q_m$ and $(q_v \in Q_m)$ and $(q_v \# q_{end})$, and task state q_v doesn't depend on any external input $p_x \# f(\text{external input})$. For such q_v, system will go into the continuous closed loop (starvation), without success to complete a final goal q_{end}.

Example of the RCL supervisor and procedure declaration:

```
PROCEDURE activity1;
VAR ...
CONST ...
TRANSFER c1, c2
BEGIN
 ...
 ...
 c1;
 ...
 c2;
 ...
END activity;
```

```
SUPERVISOR
  TASK activity1, activity2
  INITIAL STATE activity1[+], activity2[-]
  RULES
     (activity1 ON c1) --> activity1[-], activity2[+]
     (activity1 ON c2) --> activity1[+], activity2[+]
     (activity2 ON c1) --> activity2[-], activity1[+]
  FINAL STATE activity1[-], activity2[+]
 END_SUPERVISOR;
```

The role of the supervisor is to establish proper non-blocking and finite control of the robot task which will lead toward the user specified final state. Control rules formulated by the programmer might not be correct to ultimately settle the system in the final task state. They might lead towards a system dead-lock or starvation state.

To avoid these problems it is necessary to investigate user specified rules before the robot program influences the real working environment.
Such an investigation *(correctness algorithm)* should be done during the robot program compilation stage. The result of that analysis is the controllability report, which was defined before. This will be the subject of our next articles.

4. Conclusion

The new method presented in this paper applies a new approach to concurrent control problem. The difficulty which has existed before at the application program level (i.e. communication, synchronization and data exchange between various processes) has been transferred to the supervisor level. This resulted with considerably simplified control whose suitability is examined during the robot program development. It also addressed the problem of communications. Various devices defined within the certain automated system can not communicate between themselves without the knowledge of the supervisor.
Advantages of this proposed division between user application's modules and the task control module are:
- application module consists of the expressions related only to the description of an unique device activity . Two device activities which will be synchronized cannot be declared within the single application module.
- application module is liberated from complex process control statements such as: start, wait, signal, suspend, resume, etc.
- communication and control statements are declared only within the supervisor. Change of the supervisor implies different relations between controlled processes.
- division of the robot program into two parts achieved an easy identification of most conflicting situations (i.e. dead-lock, starvation) during the compilation time.
- a significant benefit of this approach is that application development is made much easier and programs more readable.

5. References

1. M.Ristic and S.Marin, "Robot Programming Language Functionality for use in Off-Line Programming", Proceedings from 21st International Symposium on Automotive Technology & Automation, Volume I, Wiesbaden 6th-10th November 1989.
2. S.Marin, "Off-Line Robot Programming, Robot Control Language - Report", Unpublished report, Imperial College, Department Of Mechanical Engineering, Exhibition Road, London SW7, January 1989.
3. S.Marin, R.Thorpe and Z.Sojat, "A Modular Robot Programming System", Internal Report on the Robot Project Development, RDP Technology, London, U.K. , January 1990.
4. P.J.Ramadge,W.M. Wonham, " Supervisory Control of a Class of Discrete Event Processes", SIAM Control and Optimization, Vol.25,No.1, January 1987.
5. P.B.Hansen, "Operation System Principles", Prentice-Hall, Inc.,EngleWood Cliffs,N.J 1973.

A STEP based Information System for Robot Assembly Planning Applications

R. DILLMANN and S. SCHNEIDER
Institute for Realtime Computer Systems and Robotics
Prof.Dr.-Ing. U. Rembold, Prof. Dr.-Ing. R. Dillmann
IPR, University of Karlsruhe

ABSTRACT

The STEP/ISO (Standard for the Exchange of product model data) initiative is of great importance for the development of integrated robot planning systems.The model driven approach of STEP allows the integration of standardized data schemata of geometry, topology, form features etc. with not (yet) standardized data schemata describing robot assembly cells and their components.

A flexible architecture for a robot information system using the STEP specification language EXPRESS is presented in this paper.

INTRODUCTION

The future ISO/STEP standard overcomes restrictions of other CAD data exchange standards (IGES, VDA-FS, SET) concerning a fixed data schema. These previous standards have been restricted to fixed data schema as for example to boundary representations of solids (FEMGEN) or to free form shapes (VDA-FS). The use of the PASCAL like specification language EXPRESS [1] within STEP, allows the extension of standardized data schema with information related to robotics. This feature of STEP can be used to exchange any data of interest in the area of robotics.

The exchange of robotics data with the help of neutral STEP files (STEP level 1,2) will not be dicussed in this paper. The main topic is the presentation of the STEP oriented information system IBASE. IBASE is used to integrate different components of a robot assembly planning system. For this purpose a common database and a common data schema described with help the specification language EXPRESS is used. An example of a data schema for the representation of structured robot trajectories will show how a robot offline programming system can use the IBASE system to achieve the robot trajectories.

IBASE : A STEP ORIENTED INFORMATION SYSTEM

The development of the IBASE (IMPPACT database) system is topic of research within the ESPRIT project no. 2165 IMPPACT. IBASE is an information system which is capable to manage instances of STEP/EXPRESS entities. This management includes consistency checks concerning the restrictions and constraints which are defined by the EXPRESS data schema of IBASE.

The general architecture [2] of IBASE is illustrated in figure 1. The specification of the required common data schema is given by a standardized EXPRESS description or by a private EXPRESS description. A DBMS generator is using this specification to generate an individual IBASE system which is able to manage the specified classes of objects (EXPRESS entities). The DBMS generator is using a NSCDD (Neutral STEP Common Data Dictionary) to represent the meta knowledge together with additional information about the location and the physical representation of the data.

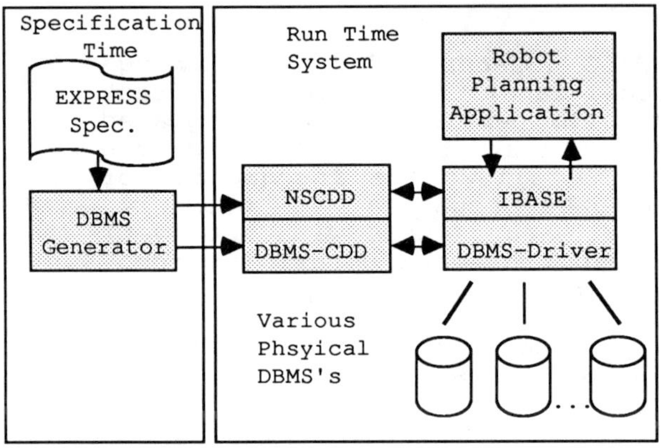

IBASE	IMPPACT database
CDD	Common Data Dictionary
DBMS	Database Management System
DBMS-CDD	DBMS specific CDD
NSCDD :	Neutral STEP CDD

Fig 1.: Generation of a STEP oriented information system

The application interface of the IBASE run time system is a procedural interface which offers the following operations for every class of objects (entities):

- store_object - delete_object
- get_object - get_object_list

This uniform application interface has to be able to use lower DBMS drivers which are supporting different DBMS models like relational, hierarchical or object-oriented models. The prototype implementation of IBASE will be based on a relational SQL DBMS. The generation of the required SQL database definitions has been discussed by Eggers [3] and Kleinert [4,5]. The access of IBASE to the relational DBMS is performed by automatically generated SQL queries [6]. This set of queries is controlled by the DBMS driver which is specific for each DBMS product. The query generator itself uses meta knowledge of the NSCDD (Neutral STEP CDD) which is equivalent to the specification of the data schema specified in EXPRESS.

The compatibility of IBASE to the STEP standard is assured by the common specification language EXPRESS.

Any standardized data schema of STEP can be used by IBASE to generate a multiuser information system. The IBASE system can be seen as a proposal for a STEP level 3 implementation. This STEP level 3 which is covering shared databases is not yet standardized.

OBJECT-ORIENTED DATA MODELLING FOR ROBOTICS APPLICATIONS

The IBASE information system has a procedural interface for the programming language C. Other languages like FORTRAN or PASCAL can also be supported easily. The final goal is the integration of IBASE in an object-oriented (o-o) environment. Such an environment will facilitate the development of future robot assembly planning systems. The powerful o-o programming paradigms as data encapsulation by methods, message passing, inheritance and late binding are supporting rapid design and prototyping of applications in the robotics area. Some o-o paradigms as inheritance of attributes and data encapsulation are also supported by the STEP specification language EXPRESS. This is the reason why to combine a STEP based information system with an o-o environment. The remaining problem is the transformation of an o-o data schema which consists of objects and relations into the STEP/EXPRESS data model. This transformation process from o-o CRL (Carnegie Representaion Language) data schemata into STEP/EXPRESS specifications has been discussed in detail by Sieth [7]. The online integration of the information system with the o-o environment of the CRL language is the remaining task. This integration will be performed with the help the procedural interface of IBASE as a set of methods in the o-o CRL. The interaction between the o-o environment of CRL and the IBASE system will be shortly

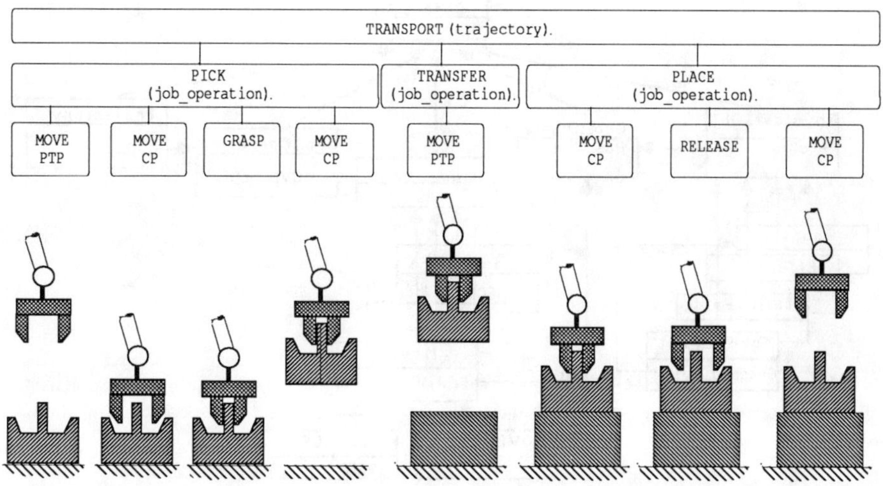

Fig 2.: Example of a structured robot trajectory (Pick & Place operation)

discussed. A prototype implementation of a database which is able to present structured robot trajectories in relation to the components of a robot assembly cell is shown.

AN EXAMPLE : REPRESENTATION OF STRUCTURED ROBOT TASKS

A common data schema for the neutral representation of structured robot tasks has been developped within the ESPRIT project no. 623 [8,9] at the University of Karlsruhe and by an industrial partner. The goal of this work has been to develop a common database and a common data schema which should be able to achieve and to exchange complete descriptions of robot tasks including the description of the associated cell layout. Pre- and postprocessors have been developped to map the robot tasks given in the native robot offline languages with the neutral representation of the common database.

A representative robot task may be a pick and place operation as shown by figure 2. The robot task is an object of type 'trajectory'. Objects of type 'trajectory' are consisting of a list of 'job_operation's. This are the three robot subtasks PICK, TRANSFER and PLACE. Every 'job_operation' itself may consist of a list of objects with type 'move_seq'. The object class ´move_seq´ has various specializations to represent the different move commands of a robot.

The relations between the most important object classes themself and the taxonomy of the object classes can be seen by figure 3. Figure 3 is based on an o-o representation as it is used within the o-o programming enivronment of KC (Knowledge Craft).

The CRL (Carnegie Representation Language) data

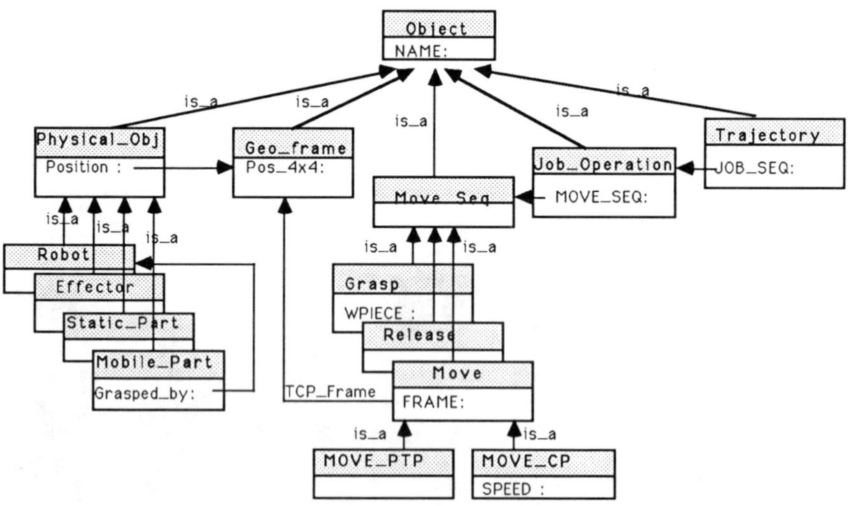

Fig 3.: Object-oriented data schmema of a structured robot trajectory. Taxonomy of object classes.

schemata of KC can be used to generate an equivalent EXPRESS schema [7]. A small part of the generated EXPRESS schema is represented in figure 4. The EXPRESS entity 'move_cp in figure 4 is used to describe continuous path motions of robots. The parameter 'speed' is indicating the robot TCP (tool center point) in terms of percents of the maximum speed. The parameter is controlled by local rules which are assuring that the value of this parameter will have a range between 0 and 100. The attribute 'frame' for the final position of the TCP is inherited by a more general entity 'move'. The keywords SUBTYPE and SUPERTYPE are used to indicate this inheritance path.

This EXPRESS schema is the required information to build post- and preprocessors for neutral STEP files. Another posibility is the generation of a shared database, which is able to achieve the required robot tasks. An automatically generated IBASE system for the representation of robot tasks is offering the following functions to manipulate robot trajectories :

```
get_trajectory (name, &pointer),
store_trajectory(&pointer),
delete_trajectory(name),
get_trajectory_list(&pointer_to_list).
```

The IBASE system is now able to manage robot tasks of the type 'trajectory' including the relevant context of the related object classes 'job_operation', 'move_seq', and 'geo_frame'.

A given robot task, as shown in figure 2, can be achieved easily. The object TRANSPORT of type 'trajectory' has to be stored by use of the procedural interface of IBASE. The whole context of the object TRANSPORT is achieved automatically. This are the 3 objects of type 'job_operation' (PICK, TRANS, PLACE), 8 objects of type 'move_seq' and various geometric frames and cell components. Not yet existing objects will be generated automatically and modified objects will be updated. Violations concerning bad use of relations are indicated and refused by the IBASE system. Local rules as given within the entity 'move_cp' are not yet controlled.

```
SCHEMA esprit_demod;
...
ENTITY move
  SUPERTYPE OF (move_ptp XOR
           move_cp);
  frame : geo_frame;
END_ENTITY

ENTITY move_cp
  SUBTYPE OF (move);
  speed_profil : speed_prof;
  speed : REAL;
 WHERE
  0 <= speed;
  100 >= speed;
END_ENTITY
...
END_SCHEMA
```

Fig 4.:EXPRESS specification for a special move command

CONCLUSION

The IBASE system is a DBMS which is using the specification language of the STEP standard to generate shared databases for CIM applications. STEP standard objects [10] and user defined objects can be integrated easily by the use of the EXPRESS language. IBASE is able to manage EXPRESS entities including their semantic context.
The IBASE system is providing a procedural application interface to communicate with applications written in conventional programming languages. An interface for a more sophisticated o-o environment is also going to be implemented. The use of the complete language EXPRESS, including local and global rules is a topic of research at the IPR. The presented architecture can be seen as a proposal for STEP level 3 implementation.

ACKNOWLEDGEMENT

This work is based on two CEC sponsored ESPRIT projects (Project No. 623 and 2165 IMPPACT : Integrated Modelling of Products and processes using advanced Computer Technologies). The work was performed at the Institute of Real-Time Computer Systems and Robotics (Prof.Dr.-Ing. U. Rembold and Prof.Dr.-Ing. R. Dillmann) at the University of Karlsruhe, Germany.

REFERENCES

1. Schenck D.; Information Modeling Language EXPRESS, Language Reference Manual; ISO TC184/SC4/WG1 N362
2. Schneider S. ; Integration of "STEP" Data Models in Object-oriented Robot Assembly Planning Systems; MSET 21; May 1990, Tokyo
3. Eggers J.A.; Implementing EXPRESS in SQL ; ISO TC184/SC4/WG1 N292
4. Schneider S.; Kleinert R. ; An EXPRESS-SQL Compiler; ESPRIT project No. 2165 IMPPACT, Modelling tools workshop 11.1989 held in Delft (NL)
5. Kleinert R.; Generation of an EXPRESS-SQL Compiler (in German) ; Studienarbeit 12.1989; Fac. for Computer Sc.; Univ. of Karlsruhe
6. Schlaich D.; Generation of SQL queries for STEP based relational DBMS (in German); Studienarbeit; 1.1990; Faculty for Computer Science; University of Karlsruhe
7. Sieth T. ; Translating CRL into EXPRESS (in German); Studienarbeit 2.90; Fac. for Comp. Sc.; Univ. Karlsruhe
8. Duffau B. ; Planning and Interactive Off-Line Robot Programming Applic. in the Automotive Industry; ISATA'89 Proc. Vol. III p.1955 6th-10th Nov. 89; Wiesbaden FRG
9. Sewell P.; Design of a robot trajectory database for the robot simulation system ROSI (in German); Diplomarbeit; 5.89; Fac. for Comp. Sc.; Univ. Karlsruhe
10. Wilson P. R., Kennicot P.R.; ISO STEP Baseline Requirements Document N284 ISOTC184/SC4/WG1; 10.1988

SECTION 3 : SENSORS, MAGNETIC LEVITATION AND CONTROL ARCHITECTURES

DEVELOPMENT OF A PROTOTYPE INTELLIGENT ROBOT TASK PLANNING ALGORITHM USING SENSOR FUSION

T. Sarkodie-Gyan and M. Jamshidi
CAD Laboratory for Systems/Robotics
Dept. of Electrical and Computer Engineering
University of New Mexico, Albuquerque, NM 87131 USA

ABSTRACT

Single sensor systems work well in situations where the environment is structured and objects are well known. These single-sensor single-algorithm systems are, therefore, severely limited in their ability to resolve ambiguities especialy in unknown or partially known environments, to identify spurious information and to detect error or failure.

If robot systems are to operate in unstructured or uncertain environments, they must make use of multiple sources of sensory information. The combination of information from multiple sensors and sensor types makes it possible to reduce the uncertainty and ambiguity inherent in making decisions based on only a single information source, thus, increasing the accuracy and resolving ambiguities in the knowledge about the environment.

The goal of sensor fusion in robot guidance and task planning is aimed at considering the problem of integrating information from different sensory sources; how to organize the distributed sensing systems, how to integrate their (the sensors) diverse observations, how to coordinate and guide the decisions made by each sensor, and how to control devices toward improving sensor system performance.

1. INTRODUCTION

In recent years, there has been an increase in importance and interest in sensor fusion as limitations in the performance of single function sensors become apparent. For operations in unstructured and/or hazardous environments, sensor systems have to be developed to dynamically interpret observations of the environment for the guidance and task planning of the robot, accounting for uncertainties and refining the robot's model of the world.

A multi-sensor robot system comprises many diverse sources of information, and therefore, receives information about changes in the environment through this array of sensors for its assessment, guidance and task planning executions. The sensors are characterized by the variety of information that they can provide and by the complexity of their operation.Some sensors are specifically suited for detecting objects but are poor to recognized or identify them. As a result, various sensors are implemented to detect different aspects (features, attributes) of the objects and some intelligence is then required to match these aspects together in order to identify an object confidently.

To be able to plan and guide the execution of tasks, the robot system has to coordinate the sensors by directing them to view areas of interest and integrating the resulting observations into a consistent consensus view of the environment.

The overall objective of sensor fusion in this research paper is, therefore, aimed at planning and monitoring the robot system in guidance and task planning executions which involve input from multiple disparate sources, whereby sensor data fusion forms the fundamental component. Without fusion it is difficult to see how an accurate perception of the situation being monitored can evolve consistently.

A prototype robot guidance and task planning algorithm system consisting of vision, tactile, proximity and force/torque sensors will be considered.. A fuzzy logic algorithm for combining the sensor data information is developed as the expert system for high-level decision making. The use of fuzzy algorithms permits the representation of uncertainty in the functions involved in the fusion process since it cannot be guaranteed that the sensors are perfect in every aspect.

2. GUIDANCE AND TASK PLANNING OF THE ROBOT SYSTEM

The guidance, task planning and also the assessment of a robot system consists of a hierarchy of navigation and control phases that are simultaneously combined together with search and recognition strategies and the processing from the environment [1]. The decomposition of the robot as a mobile system, as shown below, gives the components **Display**, the **Information Processing Unit**, and the **Motion Unit.**

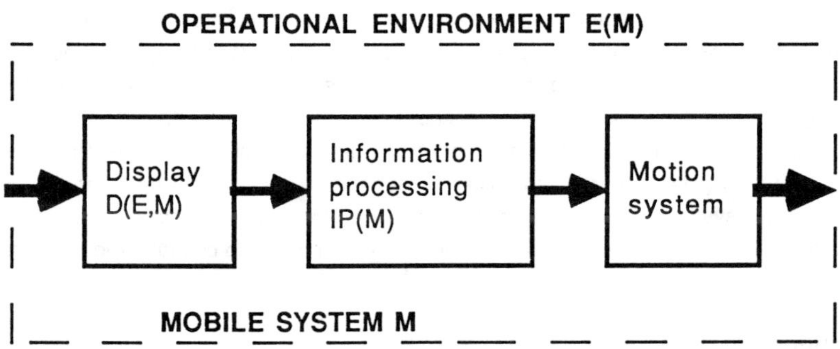

The Display may be viewed as the sensory interface of the mobile system (robot), whereas the environment may be considered as consisting of a finite configuration of objects which might be as well defined by their temporal and spatial rates. For the guidance and task planning of a robot system three separate components are involved: these are the sensors with which the relevant parts of the whole pattern of the environmental clusters are filtered out; the information processing unit is used to convert the sensor data into action commands; and the motion interface where the action commands are converted into task actions. Fuzzy set theory is used as basis of the information processing [1].

2.1. SENSOR FUSION

The fusion of multiple sensor data is defined as the technology to integrate data from multiple sensors by means of algorithms. The aim is to develop multi-sensor systems, fuse their information data together so that they behave as if they were fabricated on a single chip. Sensor fusion relies on supervisory control, because supervisory tasks require knowledge and expertise to interpret data in an intelligent manner.

An intelligent controller which requires two interface blocks which tie the fuzzy controller to the physical system so as to make the fuzzy controller work in the real-world is applied. These two blocks are known in the literature as the Fuzzifier (F) and the Defuzzifier (DF) respectively. Purpose of the fuzzifier is to translate the numerical (digital) information coming from the system into a format acceptable by the fuzzy controller developed on the basis of some linguistic information.The defuzzifier transforms the fuzzy sets of the control into a simple numerical (digital) quantity that is required to control the physical plant. The result of the attribute fusion is then an improved estimate of the identity of the object.

2.2. INFORMATION PROCESSING USING FUZZY SETS THEORY

2.2.1.OBJECTIVES OF THE THEORY OF FUZZY SETS

An important form of information processing with high significance for the treatment of robotic guidance and planning tasks at the complexity with the real environment is the recognition. The aim of recognition could be, in general, objects (symbols, signals, etc.), situations (relationships between the symbols), or also perceptions (changes in situations). Upon these basis, the robot guidance and planning tasks behaviour is modelled to adapt to the workstation (environment). The objective of recognition lies in the relation with control under uncertainties and learning.

The recognition has the task to coordinate external situations that describe the perceptions (classes). Such classes have certain common characteristic situations and are, therefore, similar. This similarity derives itself from the projection of the environmental conditions through patterns of the features that describe the perception. Thus, the use of the terminology "pattern recognition." From the character, it follows that the features to be classified through the available perceptions are of fuzzy mappings. The features belong to a stochastic environment. In this way, recognition can be understood as cases of decisions under uncertainties [1,2].

3. FUZZY PATTERN RECOGNITION

In order to identify the object, information is required about changes in the environment through an array of sensors.These sensors have different operating characteristics and capabilities. Some sensors may be very good at detecting objects but poor in recognizing or identifying them. Because of their difference, these sensors detect different aspects of the objects and some intelligence is needed to tie these aspects together to confidently identify an object.

To detect an object, the plane pattern of the environmental clusters may be made available in matrix notation for further processing thus indicating the fuzzy relationship in the two-dimensional X x Y object-plane indicating the strength of the relations between $x_i \in X$ and $y_j \in Y$.

In order to obtain maximum conformity between two comparable fuzzy sets we need to shift them mutually. Hence, a shifting operator q^k for the fuzzy set $\underline{v}_o$ is introduced to cause a right-hand shift for k->0:

$$\underline{v}_k^T = q^k \underline{v}_o^T \qquad (1)$$

Therefore, the reference set $\underline{V}$ takes on the shape

$\underline{V}$ = Shift ($\underline{v}_o$)

The fuzzy set $\underline{r}^*_i$ following from

$$(\underline{r}^*_i)^T = \underline{u}_i^T * \text{Shift}(\underline{v}_o) \qquad (2)$$

corresponds to the desired fuzzy shift which is needed for maximum conformity between $\underline{u}_i$ and $\underline{v}_k$. Obviously the equation (2) is the cross correlation of discrete sequences, and is therefore, named fuzzy correlation which may complement the notion of fuzzy convolution.

It is well known in classical detection theory that the cross correlation is the optimum detector. The fuzzy detector's methodology inherently has the power to eliminate noises to certain extents in the preprocessing stage by using maximum and minimum operations operated on the shifted data. Classical methods lack this distinction since they assume that noise cannot be eliminated from signals [3].

The composition of matrices $\underline{U}$ and $\underline{Shift}(\underline{v}_o)$ representing fuzzy relations in X x Y and in Y xZ gives a matrix $\underline{R}$

$$\underline{R} = \underline{U} * \underline{Shift}(\underline{v}_o) \qquad (3)$$

of which row vectors $\underline{r}_i^T$ of $\underline{R}$ indicate the fuzzy shifts needed to attain maximum conformity between the corresponding row vectors $\underline{u}_i^T$ of $\underline{U}$ and $\underline{v}_o$ (i = 1,...,n). Therefore, an equivalent representation of equation (2) is:

$$\underline{R} = \underline{U} \overset{hor}{*} \underline{Cyl}(\underline{v}^T{}_o) \qquad (4)$$

where $\underline{Cyl}(\underline{v}^T{}_o)$ is the cylindrical extension of the fuzzy set $\underline{v}^T{}_o$ from Y to X x Y; this means that $\underline{R}$ follows from a row-wise comparision of $\underline{U}$ and $\underline{Cyl}(\underline{v}_o{}^T)$ by means of horizontal shifting of $\underline{Cyl}(\underline{v}_o{}^T)$, relative to $\underline{U}$.

Here, $\underline{U}$ and $\underline{Cyl}(\underline{v}_o{}^T)$ are fuzzy relations in X x Y, while the number of columns in $\underline{R}$ depends on the number of shifting steps.

3.1. THE FUZZY LOGIC ALGORITHM

For a fuzzy logic control it is assumed that all of the knowledge about a control policy, , can be structured by a set of **"if...then"** rules. To implement a real-world fuzzy logic controller, it follows that the fuzzy controls contributing to the condition and the action part are defined on a finite universe of discourse. Assuming a single-input, single-output controller for simplicity, let X be the input space of the fuzzy controller, and U be the control space),

$$X = \{x_1, x_2, \ldots, x_n\}; \qquad (5)$$

$$U = \{u_1, u_2, \ldots, u_n\};$$

Then, the control rules are of the form:

$$\text{If } X_k \text{ then } U_k, \quad k = 1,2,\ldots,K \qquad (6)$$

where X_k describe the various conditions of the system output specified on the input space X_k where X belongs to the interval [0,1] of the controller. Similarly, U_k is a corresponding fuzzy control (action).

Knowing all the control rules (which by default, are considered to be consistent and complete), the fuzzy logic controller is designed by developing a fuzzy relation R_k conveying knowledge covered by an individual k-th, i.e.

$$R_k = g(X_k, U_k) \qquad (7)$$

whereby in most designs, a mapping "g" is viewed as the cartesian product of X_k and U_k,

$$R_k = (x_i, u_j) = X_k(x_i) \wedge U_k(u_j), \qquad (8)$$

and then aggregating all the rules k = 1,2,...,K and taking a union of R_k's

$$R = \bigcup_k R_k \qquad (9)$$

From Equations (8) and (9), it follows that

$$R_k(x_i,u_j) = \bigcup_k [X_k(x_i \;\wedge\; U_k(u_j)]. \qquad (10)$$

This form of aggregation reflects that the rules are being combined by "or" conjunctions thus If X_{k1} then U_{k1} or If X_{k2} then U_{k2} etc. (11)

These two steps complete the process of knowledge aggregation during modelling. The control signal is generated now by taking the max-min composition (interpreted as a matching of two fuzzy sets) of the input to the fuzzy logic controller X, and the fuzzy relation R. This yields a fuzzy control U given by

$$U = X \circ R \qquad (12)$$

or equivalentlty, $U(u_j) = \bigcup_i [(x_i) \;\wedge\; R(x_i,u_j)]$.

By making use of the form of composition of the R_k, U can be written in a more convenient fashion that makes use of a relation that displays a method for matching X with a relevant condition part of the rule. Thus, combining equation (6) and equation (7) we obtain

$$V(u_j) = \bigcup_k [\; Poss(x|X_u) \;\wedge\; U_u(u_j)] \qquad (13)$$

where $Poss(x|X_u) = Poss(X_u|x)$ denotes a measure of matching between X_k and X,

$$Poss(x|X_k) = \bigcup_i [x(x_i) \;\wedge\; X_k(x_i)]. \qquad (14)$$

More precisely the possibility measure specifies a degree of overlap of the two fuzzy sets being considered.Because of the fact that a fuzzy logic controller presents a set of algorithms modelling the control policy realized by a human expert, and therefore of operators with fuzzy sets, while a real-world system requires simple numerical values, a stage of defuzzification is necessary. In other words, in order to create a closed loop for a real system, a defuzzification procedure is required [3].

The cross-correlation of the weight, distance and the visual-perception is required to work synergestically to display the object. The workcell tactile array supplements the vision system by providing additional information with respect to the weight of the object, whereas the finger tactile sensor is mounted on the end-effector of the robot to provide it with the sense of touch. The proximity sensor adds to the determination of the distance from the robot manipulator to the recognized object. The strain-gage sensor which requires both the touch and stress sensing with a degree of accuracy measures the contact pressure of the jaws and a weight transducer determines the weight.

4. CONCLUSION

In this paper a model of an intelligent robot guidance and task planning system consisting of vision, tactile, proximity and force/torque sensors is introduced. Fuzzy logic algorithm is used as the high-level decision making tool for the fusion of the sensor data.

REFERENCES

1. T. Sarkodie-Gyan(1988):"Erkennungs-und Entscheidungsalgorithmen bei der Fuehrung von Kraftfahrzeugen auf der Grundlage unscharfer Systeme," Research paper, German Research Foundations (DFG).

2. T. Sarkodie-Gyan (1987): "A fuzzy cross-correlation method to model driver behaviour in lane-keeping tasks," Research report, Techincal Univ. Berlin, West Germany.
3. M.M.Gupta;W. Pedrycz (1989): "Cognitive and Fuzzy Logic Controllers: A retrospective and perspecticve," 1989 Acc, Vol. 3.
4. J.A.Bernard (1988): "Use of a rule-based system for process control,"IEEE Control Systems, vol.8, No.5.

DEPTH PERCEPTION USING BLURRING AND ITS APPLICATION IN VLSI WAFER PROBING*

R.V. Dantu, N.J. Dimopoulos†, R.V. Patel and A.J. Al-Khalili

Department of Electrical & Computer Engineering,
Concordia University Montreal, Canada H3G 1M8

ABSTRACT

In this paper, we present a technique for measuring the amount of blur of an edge and using this information to determine the distance of a micro-manipulator probe from a wafer surface in VLSI wafer probing. In this application, a soft and reliable touch of the probe with a metal pad in the wafer is a sensitive operation. The wafer is focussed and several images of the probe while approaching the wafer are analyzed. In our theory, the amount of blur is calculated from the step height of the edge and the slope of the intensity profile at the zero-crossing. Hence, our formula is simple and easy to use. We estimate the distance of the probe from the surface of the wafer and obtain a robust measure i.e., one which is valid even in the presence of significant noise in the images. In order to validate our method, we have experimented with various VLSI patterns as backgrounds.

1. INTRODUCTION

Recent developments in VLSI wafer design indicate a shift towards design for testability [1]. The semiconductor parameters obtained through tests during fabrication improve the monitoring of the ongoing process quality and therefore the yield. The need for such process control and its yield benefits are discussed in [2] and [3]. In wafer testing, metallic probes are used to inject test vectors or to check for continuity [4]. Input and output or test pads are used for such probings. The test pads are intermediate test points provided during layout design to increase testability. These test pads have linear dimensions of the order of 10 microns or less. Also, several probings are required for testing each parameter. The accurate and sensitive operation of touching a probe onto a pad has to be performed repeatedly with good resolution, requiring highly skilled labour. So far, to our knowledge, no work has been reported on the automatic probing of the wafers for semiconductor parameters during manufacturing. We are presently developing a micro-manipulator system for automating such high resolution tasks like wafer probing [5], [6]. The system is being automated with visual information used for feedback.

Our set-up consists of a VLSI wafer probing station complete with a microscope, probe platform, probes, wafer chuck, and a 256*256 CCD camera. Through-the-microscope images can be captured, displayed and transferred to back-end processors for further processing. In this set-up, the probe tips are fixed at an angle to the optical axis. Also for

---

* This research was supported by the Natural Sciences and Engineering Research Council of Canada.

† Department of Electrical & Computer Engineering, University of Victoria, Victoria, Canada V8W 2Y2

our experimental work, the wafer is fixed and the probe is lowered using the vertical drive provided by the probe-platform of the VLSI test station. The probes are currently under manual control, and can be interfaced to a variety of analyzing equipment for signature or parameter analysis. Currently, our main research effort is in processing the acquired images and extracting parameters related to wafer orientation, pad and probe identification. We are also using vision to determine the distance of the probe from the wafer so that accurate guidance of the probe can be achieved. In this paper, we discuss a method for estimating the distance. In the following sections we describe the measurement of the degree of blur of the edges and relate this to the distance of the object. In Section 2 we review the previous work done in perception of depth using a single view. A theoretical relation for estimation of blur near the edges is derived in Section 3. In Section 4, an algorithm for implementation of this relation is described. Finally, the algorithm is applied to experimental data, and the results are described in Section 5.

2. MEASUREMENT OF DEPTH USING BLURRING

A recent approach to depth perception is to measure the degree of blur in the image. To our knowledge, two researchers, Pentland [7], and Subbarao [8] have used edge information in measurement of depth of objects. Pentland [7] used the second derivative of the intensity profile for calculating the blur in an image. The blurred image of a point object is considered as the point spread function of the optical system. This function is usually modeled by a two-dimensional Gaussian $G(r,\sigma)$ with a spatial constant σ and radial distance r [7]. Pentland [7] used the Laplacian operator of the intensity profiles near the edges for the calculation of σ. He derived the following expression for the calculation of the spread parameter σ of the Gaussian distribution as a linear regression of x^2.

$$ln\frac{\delta}{\sqrt{2\pi\sigma^3}} - \frac{x^2}{2\sigma^2} = ln\left|\frac{C(x,y)}{x}\right| \quad (2.1)$$

where x is the variable perpendicular to the edge, δ is the step height of the edge and $C(x,y)$ is the Laplacian around each edge point.

Subbarao [8] considered the first derivative of the intensity profiles perpendicular to each edge. He computed the first derivative g_x along the intensity gradient by taking the difference of gray values of adjacent pixels (perpendicular to the edge direction). A line spread function $\theta(i)$ is computed using the following equation:

$$\theta(i) = \frac{g_{x}(i)}{\sum_{i=o}^{N} g_x(i)} \quad (2.2)$$

where g_x is the first derivate of the intensity function near the edge and N is the number of pixels considered. The spread σ of the line spread function is computed using the following expression..

$$\sigma = \pm\left[\sum_{i=0}^{N}(i-\bar{i})^2\,\theta(i)\right]^{1/2} \quad (2.3)$$

where $\bar{i}$ is the edge location. Although the height of the edge is considered in this approach, the first derivative in a noisy environment results in erroneous depth values.

The two methods described above attempt to estimate the spread σ of the point spread function which is modeled as a Gaussian distribution. They assume that the two surfaces separated by the edge are homogeneous and smooth. We found it difficult to use the above methods in our application as we have a mix of blurred and focussed regions. Also the object of interest (i.e. the probe) is at an angle to the optical axis, and therefore the focus varies along its edges. This can be observed in Figure 2.1. In addition, in our environment, we do not have smoothed regions or edges for estimation of proximity. The task is made even more difficult by the fact that we have to work with unconstrained illumination, background and hidden surfaces. We have derived a workable solution to this problem by dividing the region around an edge into three sections. The first and third sections belong to the background and the blurred object respectively while the second one belongs to the edge itself. We analyze these three parts of the edge separately and calculate the spread σ of the point spread function contributing to the blurring of the edge. By doing so, we arrived at a robust procedure for estimating σ and hence the distance from the tip of the probe to the focal plane i.e., the target pad. We have applied our algorithm to images of different backgrounds such as those shown in Figure 2.1. The following sections describe the theoretical background to our approach, the details of the algorithm, and the results obtained in our application.

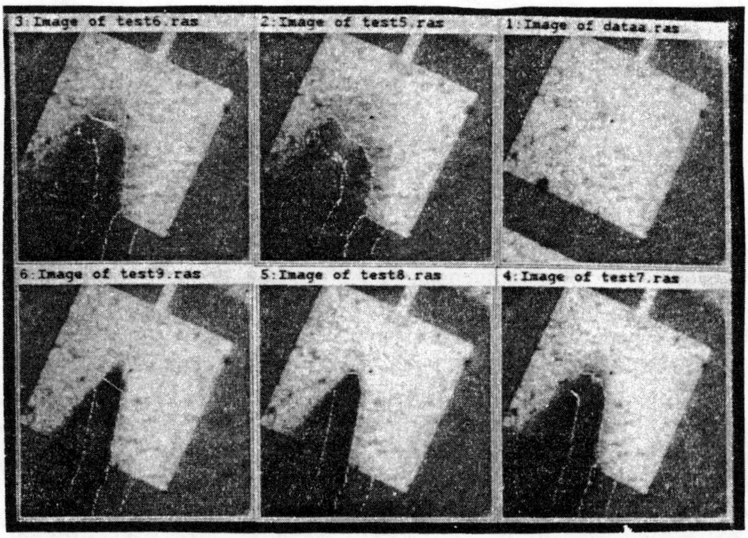

Figure 2.1 Images of the probe while approaching an Input/Output pad.

3. THEORETICAL DEVELOPMENT

We consider a step edge $f(x)$ with magnitude δ, i.e.

$$f(x) = \begin{cases} k & \text{if } x<0 \\ k+\delta & \text{if } x \geq 0 \end{cases} \tag{3.1}$$

The line spread function of the camera with the lens system is modeled as a Gaussian distribution

$$g(x,\sigma) = \frac{1}{\sqrt{2\pi}\sigma} e^{\frac{-x^2}{2\sigma^2}} \tag{3.2}$$

The blurred edge is represented by the following convolution function.

$$h(x,\sigma) = \int_{-\infty}^{\infty} f(\xi)\, g(x-\xi,\sigma)\, d\xi \tag{3.3}$$

The above integral can be expanded as

$$h(x,\sigma) = k \int_{-\infty}^{\infty} g(x-\xi,\sigma)\, d\xi + \delta \int_{0}^{\infty} g(x-\xi,\sigma)\, d\xi \tag{3.4}$$

By evaluating the above integrals [5], we get

$$h(x,\sigma) = k + \frac{\delta}{2} + \delta \int_{0}^{x} \frac{1}{\sigma\sqrt{2\pi}} e^{\frac{-\tau^2}{2\sigma^2}}\, d\tau \tag{3.5}$$

Differentiating the above equation with respect to x, we get

$$\frac{\partial h(x,\sigma)}{\partial x}\Big|_{x=0} = \frac{\delta}{\sigma\sqrt{2\pi}} \tag{3.6}$$

In order to calculate σ, we calculate the derivative of the blurred edge at its zero crossing and also the height of the step edge δ. In our case, we approximate the blurred edge by a least-squares fit of the piece-wise linear segments. Then, we differentiate this function with respect to x and evaluate the result at $x=0$ (say a_1). From equation 3.6, we obtain an expression for σ as

$$\sigma = \frac{\delta}{a_1\sqrt{2\pi}} \tag{3.7}$$

The value of the σ (which from now onwards we shall call "the blur parameter") of the Gaussian distribution allows us to determine the distance of the object [7], [8].

4. ALGORITHM

In this section we describe the different steps in evaluating (3.7). These processing steps include the following: i) Obtaining an edge-map of the probe, ii) Using this estimate to find the actual zero-crossing and hence the "blur parameter" σ along each edge, and iii) Correlating σ with the measured distance of the probe from the surface of the pad. A detailed description of the algorithm is given in the following sections.

4.1 Generation of an Edge Map of the Probe

For analyzing the neighborhood of a blurred edge we need to determine the location of the edge. The procedure for obtaining the edge-map of the probe is described here. The image of the VLSI wafer consists of pads, probes and layout patterns. Due to bright through-the-lens illumination and different reflection characteristics of the materials of the wafer (e.g. substrate, metal and poly-silicon) the image is corrupted by a considerable amount of noise. Hence, for the extraction of edges, we smooth the image using a neighborhood of 3x3 pixels. The probe, pad and background have distinct grey levels; we use these levels as threshold values for creating their images. The edge-map of the probe is generated from the zero-crossings of the Laplacian of the probe-image.

4.2 Interpolation of the Edge Map

Due to the presence of noise, the edge-map obtained in the previous section is not continuous, but contains gaps within the neighboring edges. These gaps are "filled in" with linear segments as follows: If (x_1, y_1), and (x_2, y_2) are the end points of two adjacent regions, then the new point (x_{new}, y_{new}) at a distance d from (x_1, y_1) can be calculated and the procedure is given in [5].

4.3 Chaining of Edges

The probe is a conical object and the projection of this on the image plane has a triangular shape. We can observe this in Figure 2.1. The edges in the image plane represent the boundaries of the conical surface of the probe. Also, the probe is at an angle to the optical axis of the microscope. The end point of the tip is considered as the first element of the edge-map. The next element of the edge-map is located by searching the neighborhood and then linking to the first one. This chaining is continued upto the mount of the probe, i.e. until the farthest end is reached. This procedure results in a linked list of the edges starting from the tip to the mount.

4.4 Determination of Pixels Perpendicular to Edges

The blurring around each edge in the edge-map is analyzed. More specifically, we have to analyze the pixels perpendicular to each edge. This section describes the procedure for determining these pixels.

Each probe pixel in the image plane is the vertical projection of the three dimensional surface of the probe. Also, the grey levels of the pixels depend on the blurring due to defocusing. So the grey level profile around all the edges is considered for the calculation of the blur parameter σ. As per our assumption in (3.1), a profile of the edge is perpendicular to the edge. This profile is a set of image pixels lying normal to the edge under consideration. For each edge at (x,y) with slope m, we calculate a point (x_i, y_i) in the direction perpendicular to the edge and at a distance d_i from (x,y). Sixty pixels (i=1,2,3...60) in the perpendicular direction are considered for each edge. This number is determined experimentally to give a sufficiently detailed profile for most of the edges in our images. The edges and the region of pixels in the perpendicular direction are highlighted in Figure 2.1.

4.5 Extraction of the Three Regions in an Edge Profile

In this section we describe the procedure for determining the three regions in the edge profile. The vertical cross-section (profile of the edges) obtained with the above procedure (previous section) constitutes the blurred edge. The degree of the blur σ at each edge point can be calculated from the slope of the edge at its zero-crossing, and the height of the edge (see (3.7)). In our application, the height of the edge is the grey level difference between the two regions, i.e., the body of the probe and the background. The edge location computed using a Laplacian (described in Section 4.1) may not be the true zero-crossing of the edge profile. Hence we re-calculate the zero-crossing of the Laplacian using the following procedure. All the grey levels along the perpendicular direction are fitted with a polynomial. A zero-crossing of the second order derivative of this polynomial is taken as the true edge. Let us assume that the points p_1 and p_3 for every edge profile are obtained by detecting the zero-crossings of the third order derivative of the edge profile. Then, the grey values of the points with coordinates in the range from the first point on the perpendicular direction to p_1 are considered to belong to the body of the probe. The grey levels of points with coordinates in the range (p_1, p_3) are considered to belong to the edge. Finally, the grey levels of points with coordinates from p_3 to the last point on the perpendicular direction (60 in our case) are considered to belong to the background. These three sets of grey levels are used in the next step to calculate the blur parameter σ .

4.6 Piecewise Plane Fitting Around the Surface of the Edge

To determine the accurate slope of the edge profile (even for noisy profiles) we have used the neighboring edge profiles. This operation is valid since all the edge profiles are formed by a single object (probe) and the background. Thus, in order to evaluate the slope of the zero-crossing for an edge, we obtain the best fit plane for the set of points lying near the zero-crossing. This is additionally parallel to the edge. Similarly, the sets of points in the range $(-\infty, p_2)$ and $(p_3, +\infty)$ are also fitted with two planes and represent the probe and the background. Their difference yields an estimate of the height of the edge. In practice, the grey level value of the probe and the background cannot be accurately evaluated using the above mentioned fitting of two planes. This is because the grey levels of the probe in the image are corrupted by the high intensity reflection of the background. Moreover, the true height of the step edge is the grey level difference between the two ends of the edge profile. Since we have a single object (probe) and a unique background (metal pad), the grey level difference between these two ends is a constant for a selected lamp setting. This value was determined experimentally.

5. APPLICATION OF THE ALGORITHM TO IMAGES OF VLSI WAFERS

The real challenge for any new technique is the ability to solve problems in the real world. In fact, all our efforts are directed towards the successful guidance of the probe to the target surface on the wafer. In this section we describe the results obtained by applying the formula (3.7) to experimentally obtained images of the probe. The algorithm described in Section 3 is used to estimate the blur σ in the image of the probe. The algorithm is applied to images containing different backgrounds in VLSI wafers. The deviation in the calculated value of σ for ten sets of experimental data is discussed in Section 5.1. Finally, with this value of σ, the actual distance can also be evaluated as described in [7] and [8]; however we require the optical constants of the microscope and the camera.

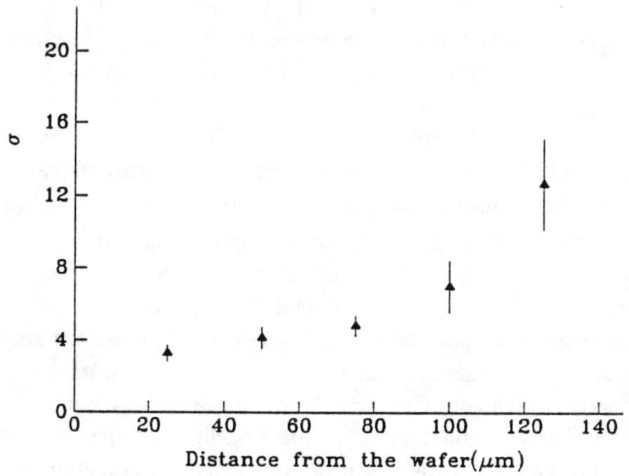

Figure 2.2 Deviation from mean value of σ.

5.1 Metal Pad as Background

The final objective for the measurement of σ is the controlled guidance of the probe to the target pad. The degree of confidence in the measurement of σ and subsequently the estimation of distance is an important consideration in any new technique, and ours is no exception. We describe below some experimental results to meet this consideration. One useful observation in our application is that only metal pads are used as contact points in any VLSI design. Therefore, we have considered 10 sets of images, where each set contains images of the probe while approaching different metal pads. For each image of the probe, the value of σ is calculated as described in the previous paragraph. The mean values of all 10 sets is plotted against the distance and are shown in Figure 2.2. The deviation of σ for each distance is also shown. By observing the figure, we can conclude that for any calculated value of σ it is possible to estimate the distance of the probe and hence the required Z-drive to the probe.

6. CONCLUSIONS

We have developed a method to estimate the depth of an object using information concerning blurring around the edges. The methods suggested by Pentland [7], [9] and Subbarao [8] use second and first derivatives of the intensity profiles perpendicular to the edge. Because of the derivative calculations along the entire cross-section, these methods are sensitive to noise and are not suitable for some uncontrolled environments, like the application considered in this thesis, where the image is corrupted by noise due to incorrect illumination, motion etc. In our method, we need to find the slope of the edge-profile at one point, i.e. at zero-crossing. In order to estimate the slope of an edge, even in the presence of noise, we fit a plane surface in the neighborhood of the zero-crossing. The slope of this plane is used in calculating σ. In contrast to other methods, the measured slope in our procedure is less susceptible to noise.

In order to confirm our theory, as well as to test it in a real environment, we have applied it to determine the distance of a probe from the surface of a VLSI wafer. We have considered 10 sets of images of different VLSI wafers belonging to different batches in the fabrication process. The parameter which is uniformly varied in all these images is the blur parameter σ. The deviation in the measurement of σ can be reduced by considering a higher number of grey levels (e.g. 256) than in the present set-up (64 levels). Another concern is in the measurement of δ (the height of the step edge in (3.7)). The values of δ need to be calibrated for different settings of the lamp illumination. This will further improve the measurement. Due to the above mentioned limitations, we use this method for coarse measurement of distance of the probe from the wafer.

Another interesting observation is that a smoother variation of σ is observed for a uniform background, e.g. a single layer of the background material than for non-uniform backgrounds, e.g. one having several layers. Hence a blurred object on a uniform background gives a more accurate measure of the distance from the object to the background. Also, we observed that the slope of the plane near zero-crossings, for different backgrounds, varies very significantly and agrees with common experience that a blurred object looks sharper against a high contrast background than against a low contrast one. This type of information may also be used for identifying boundaries between different regions. Our human visual system can neither estimate nor identify the vertical orientation of the probe by viewing along the optical axis (see Figure 2.1). But it is possible through machine vision to obtain this information by measuring the blur parameter of the blurred object. This interesting result may be useful in other applications of machine perception.

REFERENCES

[1] T. W. Williams, *VLSI Testing*. Amsterdam: North-Holland Publishing Company, 1986.

[2] K. L. Harris, P. Sandland, and R. M. Singleton, "Automated inspection of wafer patterns with applications in stepping, projection and direct-write lithography," *Solid State Technology*, pp. 159–179, February 1984.

[3] W. L. Engl, *Process and Device Modelling*. Amsterdam: North-Holland Publishing Company, 1986.

[4] C. Beddoe-Stephens, "Semiconductor wafer probing," *Test and Measurement World*, pp. 33–35, November 1982.

[5] R. V. Dantu, N. J. Dimopoulos, R. V. Patel, and A. J. Al-Khalili, "Depth perception using blurring and its application to VLSI wafer probing," *International Journal of Machine Vision and Applications*, 1990. Accepted for publication.

[6] R. V. Dantu, N. J. Dimopoulos, R. V. Patel, and A. J. Al-Khalili, "Micromanipulator vision for wafer probing," *IEEE Transactions on Semiconductor Manufacturing*, vol. 2, pp. 114–117, August 1989.

[7] A. P. Pentland, "A new sense for depth of field," *IEEE Transactions on Pattern Analysis and Machine Intelligence*, vol. 9, pp. 523–531, July 1987.

[8] M. Subbarao and G. Natarajan, "Depth recovery from blurred edges," in *IEEE Computer Society International Conference on Computer Vision and Pattern Recognition*, (Ann Arbor, Michigan), pp. 498–503, June 1988.

[9] A. P. Pentland, T. Darrel, M. Turk, and W. Huang, "A simple real-time range camera," in *IEEE Computer Society International Conference on Computer Vision and Pattern Recognition*, (San Diego, California), June 1989.

Optimal Motion for Active Sensing*

Nicola J. Ferrier James J. Clark

Division of Applied Sciences
Harvard University
Cambridge, MA 02138

Abstract

Active sensing can be defined as the process that controls the motion of the sensor in order to improve the performance of sensory information processing tasks. For low level vision tasks multiple images can be obtained via motion of the visual sensor (camera) and these additional images can be used to improve the performance of the vision system[1]. We consider the manner in which the additional sensory information obtained via sensor motion is used and determine which motions "best" improve the task at hand. We present a methodology for the determination of optimal motions in an active vision system based on the stability, or robustness, of low level vision algorithms.

1 Introduction

Active sensing controls the motion of the sensor to improve performance of sensory information processing tasks. Motion can be seen to aid sensing tasks by the addition of new sensory information which varies temporally and spatially from the initial sensed data. Three main applications of active sensing are: (1) regularization of sensory processes, (2) exploration or search for new information, and (3) temporal algorithm adaptation. *regularization* of sensory processes is the process by which an ill-posed problem is made well-posed through the addition of a stabilization process[11]. Aloimonos *et al*[1] propose methods to make ill-posed vision tasks well posed with the addition of images from different viewpoints and times since many computational vision tasks have no unique solution to the sensory information task or the solution is unstable with respect to small perturbations in the input data given a single image. The additional images provide necessary constraints to make problems well-posed. *Exploration*, or search, allows the sensor to move in order to sense portions of the environment that were previously out of range of the sensing elements. Exploration may involve arbitrary motion or be guided in the sense that the sensor moves in a manner to extract desired information from the environment. For example, movement can expose objects which were occluded from the initial vantage point. Active vision systems of Bajcsy[3] and Burt[5] use movement about the world in order to obtain information about the world. *Adaptation* of an algorithm involves altering the algorithm based on the consistency of the output of an algorithm as the sensor moves[7]. The position of the sensors can be moved then the solution obtained by one sensor position is compared with the solution obtained at another sensor position to check for consistency. If the solutions are inconsistent then the constraints can be adapted in a way to make subsequent sensory data more consistent as the sensors move.

Each of the above applications of active sensing require the sensors to move. In robotic systems these motions can be performed through the use of sensors affixed to movable links of the robot or on a mobile robot, or by systems designed especially to perform active sensory data acquistion[3, 4, 6]. We consider the problem of *specification of motion* and concentrate on determining the motions which are most effective at improving the sensory task or providing the sensory information sought. We present a methodology for the determination of optimal motions in active sensing systems. This approach is based on an examination of the stability, or robustness, of active sensing algorithms. We derive strategies for motion of the sensing systems based on optimization of an optimality criterion dependent on solution stability and accuracy.

*** This research was supported in part by ARO Brown/Harvard/MIT Center for Intelligent Control Systems under grant DAA103-86-K-0170 and by the Harvard/University of Maryland NSF Systems Research Center under grant CDR-85-00108.**

Two problems addressed specifically are *object mapping* and *shape from shading*. In either problem, any motion would provide new sensory data when used in conjunction with the original data will provide well-posed solutions to these low level vision tasks. Any motion is better than static sensing since it still produces different viewpoints in time and space. However, it is easily seen that some motions are more effective than others for a given active sensing process. In the next two sections we will examine the robustness of the algorithms as well as the uncertainty in the solution with respect to the motion used. In each case we show that an optimal motion can be found using a robustness criterion.

2 Optimal Motions for Active Object Mapping

In this section we look at the problem of optimally mapping an object with a mobile, noisy, depth measurement system. A depth measurement system gives the distances to the surface of the object along rays normal to the sensor. Typically one would like to use the depth information to build a *representation* of the object as the depth values alone do not transform in a simple way when the object moves (see [9], chapter 16 for a discussion on object representations). In this problem the camera must move in order to obtain data from all sides of the object as the object will occlude itself. Combining the measured depth from different viewpoints around an object, one can recover some representation of the object. In general, mapping can be achieved with views at fixed intervals around the object (such as viewing in one degree "steps", for example). However, this will produce redundancy and computational overhead in fusing the data. We would like to obtain an accurate object mapping while minimizing the computational overhead. The criterion of optimality in this case is made up of a term reflecting the expected error in the object surface measurement (i.e. the surface mapping) and a term expressing the number of measurements made. The motions (of the ranging device) produced by our approach to optimal active vision will be such as to obtain a mapping of the surface that jointly minimizes the number of measurements made and the error in the resulting representation of the object.

The algorithm produces a finite set of vantage points, at which depth measurements are made and the depth information is combined to produce a mapping of the object. The number of views and the viewpoints required to accurately map an object are object dependent. Without knowing the shape of the object, the viewpoints cannot be determined ***a priori***. Instead, we minimize a "cost" function which embodies our optimality criterion. These resulting viewpoints of the object correspond to what is know in the literature as the edges of the *aspect graph* of the object[8, 12, 10].

In the following section we discuss object mapping in a simple two dimensional example. Using a one dimensional ranging device we determine the optimal motions to map an object in the plane.

2.1 One Dimensional Object Mapping

Consider the following configuration shown in figure 1. A camera system is mounted on a mobile robot acting in a *plane*. The camera system is restrained to move along a circle of radius R but is free to rotate about this circle (position measured by ϕ from some fixed reference point) and the camera system can pivot around its center point (measured by θ). This planar camera system gives data in the form of a one dimensional array of "depth" values, $y(x)$, where x is displacement along this array and y is the sensed depth at that point. The object positioned within the camera system can be described in polar coordinates by $f(\psi)$, $0 \leq \psi < 2\pi$, where the coordinate system is centred within the camera system. This coordinate frame will be called the *object* frame. The data supplied by the camera system will be in the *camera* frame, which will vary with respect to the object frame as a function of θ and ϕ. We would like to "reconstruct" the object, $\tilde{f}(\psi)$, using the data obtained from various viewing positions. The values of (ϕ, θ) which give the "best" views and how many views are required to reconstruct the object are determined by minimizing a robustness criterion. We now determine what this criterion should be by analyzing the mapping algorithm.

Let (X_o, Y_o) be the object centred coordinates and (X_c, Y_c) be the coordinates of the camera reference frame (see figure1). The conversion between coordinates is given by the operator: ${}^c_oT$ (ie. $(x_c, y_c)^T = {}^c_oT(x_o, y_o)^T$). We assume that the sensor array is infinite, the depth value is infinite for regions extending beyond the object, and orthographic projection is used.

From our "sensed" data $(\hat{x}, \hat{y})$ (in camera coordinates), we compute $(\tilde{x}, \tilde{y}) = {}^c_oT^{-1}(\hat{x}, \hat{y})$. $(\tilde{x}, \tilde{y})$ are now in the object coordinate frame. A "piece" of the estimated function $\tilde{f}_i$ is given by:

$$\tilde{f}_i(\psi) = \sqrt{\tilde{x}^2 + \tilde{y}^2} \quad \psi = \tan^{-1}(\frac{\tilde{y}}{\tilde{x}}).$$

If the data is defined for $x_L \leq x \leq x_H$ then $\tilde{f}_i$ is defined for $\tan^{-1}(\frac{y(x_L)}{x_L}) \leq \psi \leq \tan^{-1}(\frac{y(x_H)}{x_H})$

We must use data from various viewpoints to ensure we cover $0 \leq \psi \leq 2\pi$. The depth data from these viewpoints gives us a set of functions $\{f_1, f_2, ..., f_n\}$ and their corresponding intervals over which they are defined. The pieces f_i can be combined to give the estimated function $\tilde{f}$. The accuracy of this estimate depends on the error in the depth measurement. Error arises from error in the sensor, error in combining the data, and from error caused by occlusions at a viewpoint. In the case of occlusions, some viewpoints can be seen to be "better" than others. That is, we can move to remove, or minimize, the occlusions.

We now reformulate the problem as a regularization problem. Consider the problem of finding the function, f, which minimizes the energy function:

$$E(f) = \int (f - \tilde{f})^2 + \lambda(\frac{d\tilde{f}}{d\psi})^2 d\psi$$

The first term is a "data" term, forcing f to match the data, $\tilde{f}$, and the second term forces smoothness. We cannot obtain $\tilde{f}$ with one view so we can break up the energy function on intervals:

$$E_{(\phi,\theta)}(f) = \sum_i E_{(\phi_i,\theta_i)}(f), \quad E_{(\phi_i,\theta_i)}(f) = \int_{\psi_{1i}}^{\psi_{2i}} (f - \tilde{f}_i(\psi))^2 + \lambda(\frac{d\tilde{f}_i}{dx_i}(\psi))^2 d\psi$$

where the sum is taken over intervals $[\psi_{1i}, \psi_{2i}]$ which cover $[0, 2\pi]$. We could also add a specific cost for each measurement made (or consider that the overlap will "naturally" add cost to the energy function as overlapping regions are "counted twice"). Now we minimize $E_{(\phi,\theta)}(f)$ over all $\{f, \phi, \theta\}$. This is computationally very difficult to solve in closed form. However, locally we can compute $\Delta E_{(\phi_i,\theta_i)}(f)$ and decide the direction of motion which would minimize the energy function. As we move about, we can update the estimate of f incorporating new views and recording values of (ϕ, θ) which produced low energy measures. Hence the robot performs a form of *active regularization*. The result is a set of views which produce smooth, accurate recovery of the object depth in order to map the object surface.

Figure 2 shows the results from an experiment using the simple 2-D ranging device to map a polygon. The locations (ϕ, θ) which minimized the corresponding energy function are shown. Also shown is a cross-section of their corresponding energy functions (given here for all values of ϕ and $\theta = 0$). Observe that the vantage points for the star correspond to unstable or

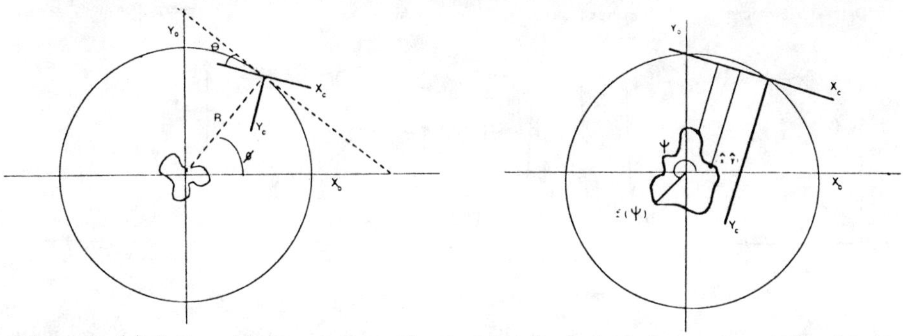

Figure 1: (a) Active object mapping with a moving depth (range) sensor, and (b) object mapping coordinate conventions.

accidental viewpoints in that a small perturbation from that viewpoint will change the aspect since one of the edges of the star will be occluded as we vary our viewpoint. Such viewpoints correspond to edges in the aspect graph of the object, as these points are transitions between two aspects.

This section has demonstrated how one can analyze a given active sensing algorithm (in this case that of obtaining object mapping) and determine the sensor motions that should be made in order for the result of the active sensing algorithm to be as accurate as possible. In the next section apply this process another active sensing task (that of shape from shading).

3 Optimal Motions for Shape from Shading

In the shape from shading problem, one tries to recover the shape of an object from the intensity image (see [9]). We can measure the intensity of light reflecting off an object with a photosensor array (camera). There exists a relationship between the surface normal vector (of the object surface) and the observed intensity. This relationship depends on the surface normal vector and the lighting. Even with known lighting conditions, there is not enough information in a single image to uniquely recover the shape (surface normal vectors). Aloimonos and Basu[2] show how additional information can come from the optical flow induced in the image by the motion of the camera relative to the object surface. Images of points on the surface move in directions that are functions of the surface normal. Knowing how the images of points move (the optical flow), and the lighting, along with the observed intensity, the object shape (surface normal vectors) can be uniquely determined.

In this section we determine the *optimal* directions of movement to be taken for a camera in using Aloimonos' active vision solution to the shape from shading problem[2]. In our approach, the motion of the camera is not arbitrary, but is taken so as to maximize the stability of the computation of the surface normal components. The algorithm for determining the optimal camera movement is seen to be object dependent. We therefore examine two cases. In the first, we assume that we know the object shape beforehand. In this way we can compute what the "correct" camera motion should be. In the second case we make no assumptions about the object shape. The camera motion is therefore adapted to as the active shape from shading algorithm proceeds.

3.1 Problem Formulation

The algorithm for determining shape from shading will be analyzed to determine the relationship between the motion of the camera and the object shape. From this, a criterion for the sensitivity of the solution (surface normal vectors) as a function of the motion will be derived. Minimization of this criterion results in the optimal motion with respect to solution robustness.

The motion of the camera can be described by a rotation matrix ${}^2_1R$ and a translation

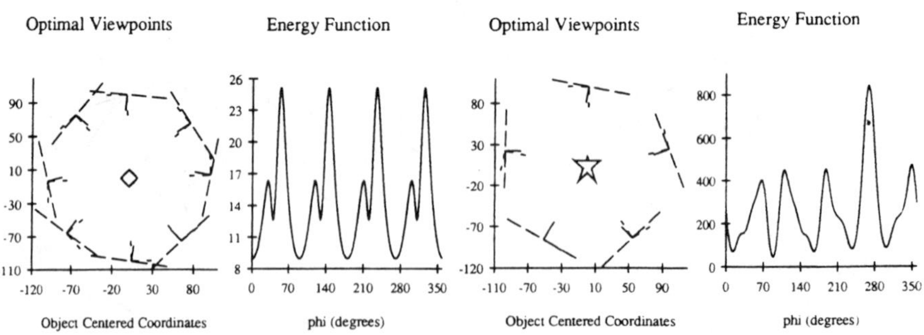

Figure 2: Polygonal objects shown with their optimal viewing positions, (ϕ, θ), which minimize the energy function $E_{(\phi,\theta)}(f)$.

vector ${}^2t_1$ where ${}^2_1R$ is the 3X3 orthogonal matrix describing the rotation from the coordinate frame assigned to the first camera position to the coordinate frame assigned to the second camera position and ${}^2t_1$ gives the translation from the origin of the first coordinate frame to that of the second. Using *orthographic projection* for the imaging, the following relation holds between the image flow and the camera motion:

$$(R1) \qquad \begin{bmatrix} u(x,y) \\ v(x,y) \end{bmatrix} = \begin{bmatrix} 1 & 0 & 0 \\ 0 & 1 & 0 \end{bmatrix} \left[\quad {}^2_1R - I_3 \quad \right] \begin{bmatrix} x \\ y \\ z \end{bmatrix} + \left[\; {}^2t_1 \; \right]$$

where I_3 is the 3X3 identity matrix. Observe that the flow depends on both the transformation (ie. the camera motion) and on the object (ie. the depth $z(x,y)$).

As observed earlier, multiple images make the shape from shading problem well-posed through the additional constraints provided in the additional images. In particular Aloimonos and Basu[2] utilize two constraints: (1) the *shape motion* constraint

$$Ap^2 + Bq^2 - 2Cpq + D = 0 \qquad (C1)$$

where

$$A = {u_y}^2 + {v_y}^2 + 2v_y \;,\; B = {u_x}^2 + {v_x}^2 + 2u_x C = u_y + u_y u_x + v_x + v_x v_y \;,\; D = C^2 - AB$$

where $u(x,y)$ and $v(x,y)$ are the image flow vector fields and $u_x = u(x+1,y) - u(x,y)$, $u_y = u(x,y+1) - u(x,y)$, $v_x = v(x+1,y) - v(x,y)$, and $v_y = v(x,y+1) - v(x,y)$. and (2) the *image irradiance* constraint: $I(x,y) = \rho f(\mathbf{n} \cdot \mathbf{l})$, where ρ is the albedo, $\mathbf{n}$ is the surface normal and $\mathbf{l}$ is the direction of the light source. For unit albedo and Lambertian image this constraint simplifies to

$$I(x,y) = (-p, -q, 1) \cdot (-l_1, -l_2, 1)^T \;\; / \;\; \sqrt{p^2+q^2+1}\sqrt{l_1^2+l_2^2+1}. \qquad (C2)$$

The p and q variables represent the slant and tilt of the object surface, and vary over the surface. They are related to the components of the surface normal.

The above two constraints, $C1$ and $C2$, yield two equations for the shape $(p(x,y), q(x,y))$ in terms of the variation in the image flow, the lighting position, $\mathbf{l}$, and the image intensity, $I(x,y)$. Aloimonos and Basu[2] proposed that by moving the camera and hence obtaining two images, then solving the equations, $C1$ and $C2$, will yield the shape (i.e. the p and q values over the image of the object).

In our approach, the camera motion is not arbitrary, but is taken so as to minimize the effect of errors in the measurement of the flow vectors and of the image intensity values. The determination of this optimal motion proceeds as follows.

$(R1)$ gives us a relationship between the flow and the camera motion. Thus we can choose the camera motion which will give us the solution to maximize the stability of the solution found. In general, solving the system $(C1)$ and $(C2)$ is difficult. However, by differentiating $(C1)$ and $(C2)$ with respect to the change in the flow we get the following system of equations which are linear in the quantities we want to minimize, the sensitivities of p and q with respect to the uncertainty in the measured variables:

$$\alpha_\delta \frac{\partial p}{\partial \delta} + \beta_\delta \frac{\partial q}{\partial \delta} + \gamma_\delta = 0 \quad \kappa_\delta \frac{\partial p}{\partial \delta} + \lambda_\delta \frac{\partial q}{\partial \delta} + \nu_\delta = 0 \quad \text{or} \quad \begin{bmatrix} \alpha & \beta \\ \kappa & \lambda \end{bmatrix} \begin{bmatrix} \frac{\partial p}{\partial \delta} \\ \frac{\partial q}{\partial \delta} \end{bmatrix} = \begin{bmatrix} 0 \\ -\nu_\delta \end{bmatrix}$$

Where α, β, γ, κ, λ, ν are functions of I, $\mathbf{l}$, p, q, u_x, u_y, v_x, and v_y and $\delta \in \{u_x, u_y, v_x, v_y\}$. These equations are linear (which is much easier to solve than the conic equations). Solving this yields

$$\begin{bmatrix} P_\delta \\ Q_\delta \end{bmatrix} = \begin{bmatrix} \frac{\partial p}{\partial \delta} \\ \frac{\partial q}{\partial \delta} \end{bmatrix} = \begin{bmatrix} \alpha & \beta \\ \kappa & \lambda \end{bmatrix}^{-1} \begin{bmatrix} 0 \\ -\nu_\delta \end{bmatrix} = \frac{1}{\alpha\lambda - \beta\kappa} \begin{bmatrix} \beta\nu_\delta \\ -\alpha\nu_\delta \end{bmatrix}.$$

These terms give us the *sensitivity* of the shape to the flow parameters. The optimal motion is the one that provides a solution that is most stable to small changes in the data. Starting with an initial estimate of the shape (p,q) we can compute

$$T(p,q) = \min_{{}^2_1T} \left[\textstyle\sum_\delta P_\delta^2 + Q_\delta^2 \right] = \min_{{}^2_1T} \left[\frac{\alpha^2+\beta^2}{\alpha\lambda - \beta\kappa} (\nu_{ux}^2 + \nu_{uy}^2 + \nu_{vx}^2 + \nu_{vy}^2) \right]$$

where the minimization occurs over all possible transformations ${}^2_1T = ({}^2_1R, {}^2t_1)$. Note that camera transformations will induce a flow on the image and it is the sensitivity to that change in flow which we are computing (see equation (R1) for the relationship between image flow and camera motion). The sensitivity, $T(p,q) = T(p,q,x,y,1,I)$ is a complicated expression and varies at each point. Thus the tranformation which yields the minimum will be the camera motion that *locally* is optimum. Finding a solution $(\hat{u}_x, \hat{u}_y, \hat{v}_x, \hat{v}_y)$ that minimizes the above expression gives us the desired image flow from which we can determine the optimal camera motion (which induces this desired image flow).

If we know the shape *a priori* then we can substitute the known (p,q) into the above formulae and determine exactly the transformation that minimizes the above expression. In general, the object shape is not known (as it is what we are trying to find!). Therefore we must begin with an arbitrary motion and obtain an estimate of the shape. Based on this estimate we can compute a new motion and repeat the experiment. We can combine the results of the two experiments using a least squares form of estimation to provide new estimates for the p and q values. Given this new estimate we can compute a new direction of motion and obtain a new solution which can be again combined with the previous results.

This process can be iterated, with the sensor moving along an optimal trajectory, and the error in the computed shape will drop as we move along this trajectory, integrating new solutions as we go.

4 Summary

This paper has presented two examples of how one can improve the performance of an active sensing task by optimizing the motions made by the sensor. The principle applied was to choose the direction of motion of the sensor such that the error in the estimate of whatever parameter we are trying to recover from the sensory measurements is minimized. The first example demonstrated how to determine the sensor motions such that the results of the active sensing algorithm, shape from shading, are as accurate as possible. In the second example, the motions of the range sensor are chosen to minimize an energy function in order to accurately map the surface of an object.

References

[1] Aloimonos, Y., Weiss, I., Bandyopadhyay, A., "Active vision", Proceedings of the 1st IEEE Conference on Computer Vision, London, 1987, pp35-54.

[2] Aloimonos, Y., Basu, A., "Combining Information in Low-Level Vision", Proceedings of the Image Understanding Workshop, Cambridge, MA, 1988, pp862-906.

[3] Bajcsy, R., "Active perception vs. passive perception", Proceedings 3rd IEEE Workshop on Computer Vision, Bellaire, pp. 55-59, 1985.

[4] Brown, C.M., "Progress in image understanding at the University of Rochester", in the Proceedings of the 1988 Darpa Image Understanding Workshop, pp 73-77

[5] Burt, P., "Algorithms and architectures for smart sensing", in the Proceedings of the 1988 Darpa Image Understanding Workshop, pp 139-153

[6] Clark, J.J., Ferrier, N.J., "Modal Control of Attentive Vision", Proceedings of the 2nd International Conference on Computer Vision, Florida, December, 1988.

[7] Clark, J.J., Yuille, A., *Data Fusion For Sensory Information Processing Systems*, Kluwer Academic Publishers, Boston, in press.

[8] Gigus, Z., Canny, J., Seidel, R., "Efficiently Computing and Representing Aspect Graphs of Polyhedral Objects", Proceedings of the 2nd International Conference on Computer Vision, Florida, December, 1988.

[9] Horn, B.K.P., *Robot Vision*, MIT Press, Cambridge, MA, 1986.

[10] Koenderink, J., van Doorn, A.J., "The Singularities of the Visual Mapping", Biological Cybernetics, 1976, 24, pp. 51-59.

[11] Marroquin, J., Mitter, S., Poggio, T., "Probabilistic Solution of Ill-Posed Problems in Computational Vision", Journal of the American Statistical Association, March 1987, Vol. 82, No. 397, pp. 76-89.

[12] Plantinga, W.H., Dyer, C.R., "An algorithm for constructing the aspect graph", Proceedings of the 27th Symp. on the Foundation of Computer Science, IEEE, N.Y., 1986, pp. 123-131.

Illumination Independent Imaging for Industrial Robotics Applications

M. Vaezi and B. Bavarian

University of California, Irvine
Department of Electrical and Computer Engineering
Irvine, California 91727

ABSTRACT

A nonlinear filter for image processing is presented which extracts reflectance map ratio of adjacent pixels regardless of the lighting condition. We present this filter as the basis for illumination independent imaging system for industrial machine vision applications. Images under varying lighting conditions are tested and presented.

1 Introduction

In most robotics applications of vision guided systems special lighting or illumination is required for the proper operation. In general this is a severe limitation since it reduces the flexibility of the automated systems. The problem is more severe for an Automated Guided Vehicle (AGV) which relies on computer vision for navigation and obstacle avoidance.

The fundamental problem in automated vision guided systems lies on the variations of illumination level. For example, a vision guided system which has to undergo a variety of lighting conditions such as indoor, outdoor lighting, shadows, etc., requires continious recalibration of the imaging system [1-2]. Also a proper contrast detection requires a high level of illumination against the background scene to detect fine details[3]. However, in some applications where the products are sensitive to illumination level[4], exposure to high illumination might be hazardous to the product.

In view of the above constraints and limitations, we propose the use of a nonlinear filter as a preprocessor for the robot's vision system which removes the effect of illumination.

In an earlier work, we have introduced a contrast dependent filter [8] which was later formally extended into a class of nonlinear space dependent filters [9]. Here we concentrate on the feasibility of the industrial application of these class of filters.

In the next section, the characteristics of the nonlinear filter is described. Section III illustrates the simulation results and concluding remarks are presented in section IV.

2 Space-Dependent Spread Filters

The human eye is the perfect example of a vision system working under variety of lighting condition from bright sunshine to dim star light. Vision researchers state that the human visual system is a reflectance based processor rather than an intensity one [5-6], i.e; the images are processed as independent of local illumination as possible. This idea is embedded in some popular filtering schemes such as the retinex theory[6] and Intensity-Dependent Spread (IDS) filters[7]. The latter filter has gained popularity among vision researchers due to some interesting properties which resembles the human visual system calculating the weber ratio across edge boundaries. However, this filter suffers from some side effects such as blurring and missing edges if they are closely spaced. We have studied the limitations of the IDS filter [8-9] and proposed a Space-Dependent Spread (SDS) filter. The SDS filter has the following properties:

- It minimizes the image noise.
- It produces (reflectance map) weber ratio across each edge boundary under a variety of lighting conditions, i,e; uinform or nonuniform, low or high illumination.
- It is a nonlinear bandpass filter hence the filter performs edge enhancement.

The spread of the filter is chosen to be

$$\lambda(x,y) = Z(x,y)\left[1 + \frac{Z(x,y)Z'(x,y)}{1+Z''(x,y)}\right] \tag{1}$$

where

$$\lambda(x,y) = \frac{1}{\sigma(x,y)} \tag{2}$$

and Z is the input image, noisy or noise free, and $Z' and Z''$ denote the first and second derivatives of image intensity Z, respectively, are defined as

$$Z'(x,y) = \sqrt{\left(\frac{\partial Z}{\partial x}\right)^2 + \left(\frac{\partial Z}{\partial y}\right)^2} \tag{3}$$

and

$$Z''(x,y) = \frac{\partial^2 Z}{\partial x^2} + \frac{\partial^2 Z}{\partial y^2} \tag{4}$$

The output of the SDS filter is calculated according to

$$O(x,y) = \int H(x,y:\zeta,\eta)\,d\zeta d\eta \tag{5}$$

where H is defined as, for example, a Gaussian spread

$$H = \frac{\lambda(x,y)}{2\pi} e^{-\frac{1}{2}\lambda^2(x,y)(x^2+y^2)} \tag{6}$$

In the next section, the result of combining equatuions (1) through (6) and applying them to 1D and 2D images is illustrated.

3 Simulation results

In Figure 1, a staircase, in 1D, with constant weber ratio across each edge boundary at the input is applied to the filter to demonstrate the edge localization capability of the SDS filter. The output intensity has sharp resolution with constant peak-to-peak trough values at all steps regardless of the intensity value at each step.

In Fig. 2, a real image of mechanical tools at high and low intensity is presented. The output of the filter is shown in Figs. 2b and 2c correspoding to the high and low intensity images, respectively. The property of the SDS filter being independent of illumination is demonstrated in the output images. The Weber ratio across edge boundaries in low and high illumination is also measures correctly at the output of the filter.

4 Conclusion

An optimum nonlinear filter to extract Weber ratio across edge boundaries is described. The one- and two-dimensional simulation under high and low intensity values are demonstrated. It is verified that the Weber ratio is accurately calculated across the edges as it is at the input.

Currently, we are using B-splines and series expansion to approximate spread function for real-time implementation.

5 Acknowledgement

This research was supported in part by the IBM Almaden Research Center in San Jose, California and in part by UCMICRO Project.

6 References

1. M.A. Turk, D.G. Morgenthaler, K.D. Gremban, and M. Marra,"Video road-following for the autonomous land vehicle", in Proc. of IEEE International Conference on Robotics and Automation, pp. 273-280, North Carolina, March 31-April 3, 1987.

2. K. Krishen, R.J.P. de Figueirdo, and O. Graham, "Robotic vision/senseing for space applications", in Proc. of IEEE International Conference on Robotics and Automation, pp. 138-150, North Carolina, March31-April 3, 1987.

3. W.E. Blanz, J.L.C. Sanz, and E.B. Hinkle,"Image analysis methods for solder ball inspection in integerated circuit manufacturing", in the Proc. of IEEE International Conference on Robotics and Automation, pp. 509-512, North Carolina, March 31-April 3, 1987.

4. F.U. King, G.V. Puskorius, F. Yuan, R.C. Meier, V. Jeyabalan, and L.A. Feldkamp, "Vision guided robots for automated assembly", in Proc. of

IEEE Internatinal Conference on Robotics and Automation, pp. 1611-1616, Philadelphia, Pensylvania, April 24-29, 1988.

5. B.K.P. Horn,"Determining lightness from images", Computer Vision and Image Processing, Vol. 3, No.1, pp. 277-299, 1974.

6. E.H. Land and J. Mc Cann,"Lightness and retinex theory", Journal of Optical Society of America 61, pp. 1-11, Jan 1971.

7. T.N. Cornsweet and J.I. Yellot,"Intensity-dependent spatial summation", Journal of Optial Society of America-A, Vol. 2, No. 10, pp. 1769-1786, 1985.

8. M. Vaezi, and B. Bavarian, "Contrast-dependent spread filters", in Proc. of SPIE/SPSE symposium on Electronic Imaging Conference, Feb. 11-16, Santa Clara, California 1990.

9. M. Vaezi and B. Bavarian,"Space-dependent spread filters", submitted to Journal of Optical Society of America-A, Feb. 1990.

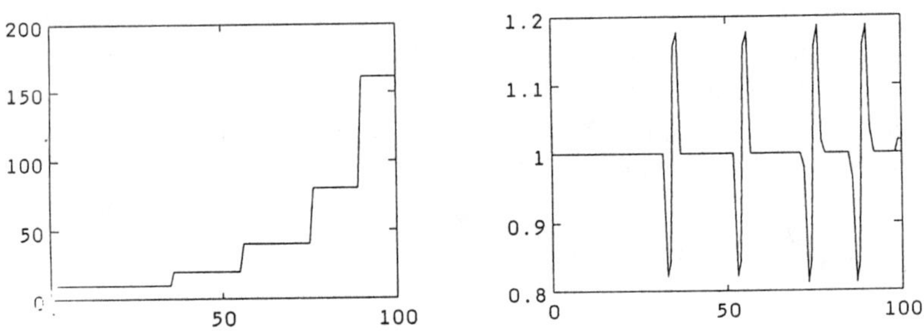

(a) (b)

Figure 1. The output of the SDS filter for 1D signal with staircase intensity and constant Weber ratio; a)input and b)output.

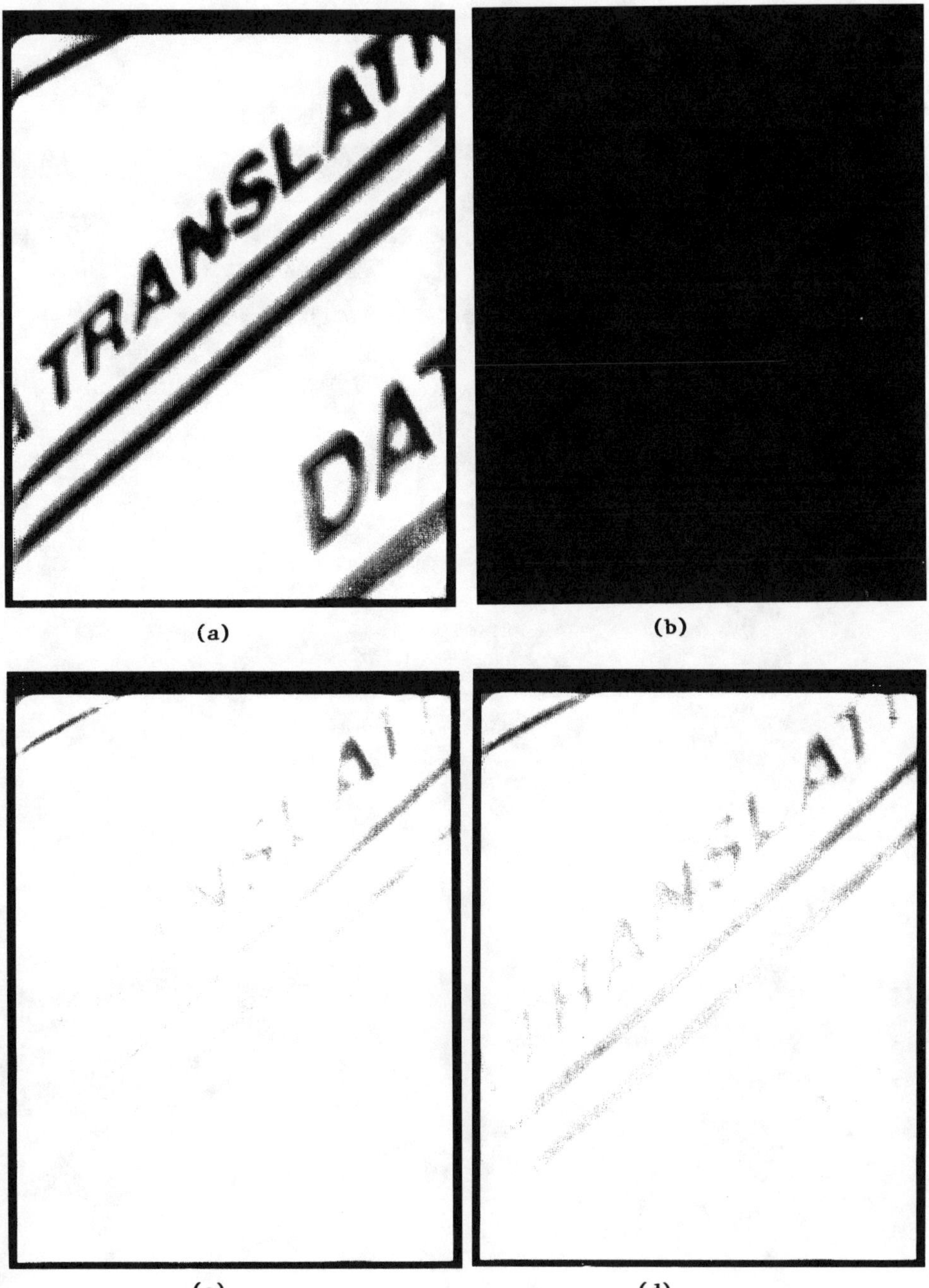

Figure 2. The real image experiment; a)high intensity, b)low intensity
c)and d) the corresponding reconstructed output.

3-D SURFACE RECONSTRUCTION USING CROSS-STRIPE STRUCTURED-LIGHT SYSTEM

Whoi-Yul (Yura) Kim
Erik Jonsson School of Engineering and Computer Science
The University of Texas at Dallas,
P.O. Box 830688, MP 33
Richardson, TX 75083-0688, USA

ABSTRACT

In this paper, we present a method of reconstructing surfaces with a sparse set of range data using a computer aided-design technique -- an indispensable step when the large size of the surface is called for. Most active range sensors gather depth information by sweeping the light source in one direction. Subsequent scanning of the scene yields a set of range data. The spatial sampling rate of range data along the stripe is different from that between stripes unless the distance between stripes is the same as the distance between points lying on a single stripe, which can be achieved by dense scanning. In our method, two sets of range data along the meshed lines are obtained by cross-structured light range sensor. The interpolation technique utilizes *Coon's patch* formalism that reconstructs the surface solely from the boundary information of the patch and results in a visual surface of a solid object. Simulation and experimental results of our method are presented.

1. Introduction

In this paper we show how to incorporate the 3-D computer-aided design technique to the 3-D computer vision research. Surface interpolation, or construction of composite surfaces, has been used in computer-aided methods for the design and manufacture of industrial parts whose geometry involves smoothly curved surfaces. Only a few researchers have employed those methods for application to their computer vision systems [1, 7, 8, 9, 10], and the related literature can be found in [6].

During the past decade, a number of range sensors have been developed for rapid and accurate measurement of the range information. The review of the recent development in range sensors can be found elsewhere [3]. Among them, we will focus on the structured light scanner type, because it shares the basic principle with many other types of range sensors.

Since its early development in 1970s, the stripe structured light scanner, has been used successfully in various applications. Here, such structured light scanner is referred to as *single stripe structured light scanner (SSSLS)* to differentiate from our *cross stripe structured light scanner (CSSLS)*. Also, for simplicity in this exposition, unless otherwise described, we use the term *stripe* to refer the single pixel wide stripe image after the thinning process in SSSLS. For the time-of-flight-type sensor or synchronized scanner, for example, the sweeping of the laser beam can be considered as a stripe and the lateral movement by galvanometer for rotating the scanner can be considered as the distance between the stripes.

One of the drawbacks of SSSLS is that the spatial sampling rate of range data along the stripes is different from that between stripes unless the distance between stripes is the same as the distance between points that lying on a single stripe. Most range sensors have similar characteristics in this respect, for example, with the time-of-flight type sensor the resolution along the stripe is dense while the resolution of between the stripes is very sparse. Because of this different sampling rate, a number of useful algorithms developed for intensity image analysis are difficult to use in the analysis of range data obtained by SSSLS, for example, algorithm which employs window operation.

When the sampling rates are not the same, the range data has to be either undersampled by discarding some part of range information along the stripes, or oversampled by interpolation between the stripes to make the sampling rates the same. In an effort to mimic the intensity image format to make the sampling interval even on the *xy* plane, some sort of interpolation technique has been used to determine the quantized value between two adjacent points [5]. One of the typical treatments for such problem was either to increase the number of projections in the lateral direction or decrease the sampling rate along the stripes, or preprocess *xyz* format in order to make an array format. However, as a result of keeping the same rate in both directions, the amount of the range data and the time for range data acquisition greatly increases, and the computation time for feature extraction process increases as well. Especially when the size of the surface is large, the number of data points may be computationally prohibitive.

In Section 2 we discuss our approach of *Cross Scanning Structured Light Scanner* (CSSLS) for the range data acquisition. The *Coons Patch* is described in Section 4, and the results are discussed in Section 5.

2. Cross Stripe Structured Light Scanner

The actual scanner is composed of two orthogonal stripe projectors and a camera as illustrated in Fig. 1. As the conveyor belt moves, each projector is turned on alternatively, resulting in stripes in left and right diagonal directions, referred as *left stripe (LS)* and *right stripe (RS)*, respectively. This situation is illustrated in Fig. 2 when there is no object present on the conveyor belt. By combining the camera image of successive projections, a mesh network of light stripes is obtained. The range values (x, y, z) are computed via calibration procedure along all the mesh lines in both the right stripes (RS's) and left stripes (LS's) directions. Thus, as shown in Fig. 3, we have a mesh network that defines a number of topologically rectangular patches, each bounded by four continuous curves, with known range values along their boundaries. The interior of the cell needs to be 'filled in' using two-dimensional interpolation [4]. In other words, the range map inside of the patch will be generated solely from the boundary curves of the rectangular patch. Since each patch has range information along four boundaries but no other information is available regarding inner surface of the patch, the *Coons patch* is the best candidate for the inner surface reconstruction. Implementation details and the necessary formulations for the *Coons patch* are discussed in the following sections.

3. Coons Patch

With the four three dimensional boundary curves $\mathbf{r}(u,0)$, $\mathbf{r}(u,1)$, $\mathbf{r}(0,v)$, $\mathbf{r}(1,v)$ obtained from CSSLS, $\mathbf{P}(u,v)$ is computed by

$$\mathbf{P}(u,v) = \mathbf{r}(u,0)(1-v) + \mathbf{r}(u,1)v + \mathbf{r}(0,v)(1-u) + \mathbf{r}(1,v)u \quad (1)$$

$$-\mathbf{r}(0,0)(1-u)(1-v)-\mathbf{r}(0,1)(1-u)v-\mathbf{r}(1,0)u(1-v)-\mathbf{r}(1,1)uv$$

where **r** is the vector obtained from known or input measured data on a surface, and **P** is the vector defining the surface constructed from the input data **r**. In compact matrix form

$$\mathbf{P}(u,v)=\begin{bmatrix}1-u & u & 1\end{bmatrix}\begin{bmatrix}-\mathbf{r}(0,0) & -\mathbf{r}(0,1) & -\mathbf{r}(0,v)\\ -\mathbf{r}(1,0) & -\mathbf{r}(1,1) & -\mathbf{r}(1,v)\\ \mathbf{r}(u,0) & \mathbf{r}(u,1) & 0\end{bmatrix}\begin{bmatrix}1-v\\ v\\ 1\end{bmatrix} \qquad (2)$$

Notice that in Eq. (1) the first two items and the second two items are ruled surfaces in the u and v directions, respectively. The remaining four items are simply a bilinear surface of four corner points of a patch. Therefore Eq. (1) can be rewritten in terms of two basic surfaces,

$$\mathbf{P}(u,v)=p_{ru}(u,v)+p_{rv}(u,v)-p_b(u,v)$$

where p_{ru}, p_{rv} stand for the ruled surfaces in the u and v direction, respectively, while p_b is for the bilinear surface. The functions $1-u$, u, $1-v$ and v are called blending functions, because their effect is to blend together four separate boundary curves to give a single well-defined surface, and provide C^0 continuity at the boundary of the patches. These blending functions are the simplest linear interpolation functions.

Patch Construction

Although the concept of CSSLS is very simple, the actual implementation requires several phases. Prior to the surface interpolation, first we must ensure that the range data is as accurate as possible so that two curves meeting at a corner of a patch intersect each other. In CSSLS, since each scan with RS and LS is similar to the SSSLS, and since the range data from RS and LS is obtained by a similar procedure, the two curves (corresponding to RS and LS), may not meet in 3 space at the corner of a patch. This is primarily because of the global errors from the calibration matrices for RS and LS that we used in our experiment. Therefore, we introduce the virtual intersection points of two curves, RS and LS. This can be done by locating the intersection points of a mesh network of range data from RS and LS, on the projection plane, say, xy plane. In other words, we first find the intersection points on the projection of the curves. Then the intersection points are back projected to the curve resulting two corner points on each curve. The virtual intersection point is determined as the average of these two points.

Shifting Right and Left Stripe Range Data at the Virtual Intersection

Now, we need to have 4 boundary curves to define a patch. In order to identify and index each patch boundary from the jumble of range data so that each patch has 4 boundary curve segments, each set of (x,y,z) coordinates from each scan should be subdivided into boundary segments. The virtual intersection points obtained using the above method do not belong to either of the plane curves. Therefore, we need to compensate for these global errors from the two systems by shifting each curve segment by half of the deviation at each virtual intersection point as shown in Fig. 5. The shifted curve along RS would then be

$$\mathbf{X}_s(s)=\mathbf{X}(s)-\frac{1}{2}(s\Delta E_{i,j+1}+(1-s)\Delta E_{i,j})$$

and along LS, be

$$\mathbf{X}_s(t) = \mathbf{X}(t) - \frac{1}{2}(t\,\Delta E_{i+1,j} + (1-t)\Delta E_{i,j}).$$

Here, $X_s(\cdot)$ denotes the shifted curve, and $\Delta E_{i,j}$ is the deviation between two curves, i^{th} RS and j^{th} LS, at a virtual intersection point. A linear function is used to shift each segment, however, a more complex function which satisfies the blending function conditions can be used.

4. Experimental Results

In order to simulate our CSSLS experiment, 12 stripes (6 stripes in each diagonal direction) were placed on a 64×64 synthesized image composed of spherical surface on top of a cylindrical surface. Fig. 6 a shows the 3-D profile of the image and Fig. 6 b shows the corresponding range map. Fig. 6 c shows the stripes overlaid on the range map and the gray value of the stripes indicates the height or the range value of the object. Notice that all stripes do not necessarily have to be parallel in each direction, and the stripes do not cover the whole region of the object (a stripe on the lower left corner short). In this experiment, at the boundary of the object, stripes make open triangular, rectangular and pentagonal shapes as illustrated in Fig. 6 d. For these regions, the range information on the missing boundary was obtained by using the linear interpolation between two open end points of the polyhedra. The resulting output of our system is shown in Fig. 6 e and the range map is shown in Fig. 6 f, which recovers the original shape successfully.

For a real world scene, the scanner sweeps a stair-like object where each surface is a planar region as shown in Fig. 7 a. Figure 7 b shows the overlapped raw stripe images and Fig. 7 c shows the normalized range data. The 3D profile of the reconstructed range map is shown in Fig. 7 d. In this experiment, we did not attempt to reconstruct the incomplete patches.

5. References

[1] H. Ailisto, M. Pietikainen, "Ferguson Patch Based Method for Representation of 3-D Vision," *Proceedings of 1987 IEEE International Conference on Robotics and Automation*, Raleigh, North Carolina, Mar. 31 - Apr. 3, 1987.

[3] Paul J. Besl, "Range Imaging Sensors," *Advances in Machine Vision: Applications and Architectures*, J. Sanz, Editor, Springer-Verlag, New York, 1988

[4] S.A. Coons, "Surfaces for Computer-aided Design of Space Figures," M.I.T. ESL 9442-M-139, January 1964.

[5] J-L Jezouin, P. Saint-Marc and G. Medioni, "Build an Accurate Range Finder with Off the Shelf Components," *Proceedings of International Conference on Computer Vision*, June 5-9, 1988

[6] Whoi-Yul Kim, "Efficient Strategies for Reducing the Size of Search Space in 3-D Object Recognition and Pose Estimation", *Ph.D. Thesis*, School of Electrical Engineering, Purdue University, W.L., IN.

[7] M. Potmesil, "Generation of 3D Surface Descriptions from Images of Pattern-Illuminated Objects," *IEEE Computer Society Conference on Pattern Recognition and Image Processing*, 1979.

[8] M. Potmesil, "Generating Models of Solid Objects by Matching 3D Surface Segments," *Proc. of the 8th Int. Joint Conf. on Artificial Intel.*, 1983.

[9] B.W. York, A.R. Hanson, E.M. Riseman, "A Surface Representation for Computer Vision," *Proc. of IEEE workshop on Picture Data Description and Management* Asilomar, Ca., August 1980, 124-129.

[10] B.W. York, A.R. Hanson, "3D Object Representation and Matching with B-Splines and Surface Patches," *Proc. of the 7th Int. Joint Conf. on Artificial Intelligence,* 1981.

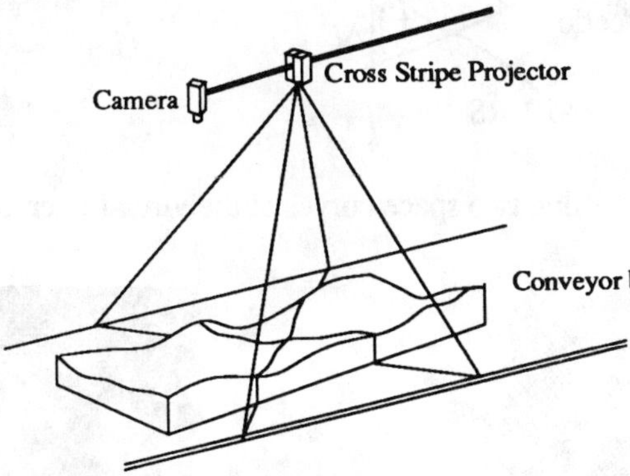

Figure 1 A typical application of CSSLS on a conveyor belt

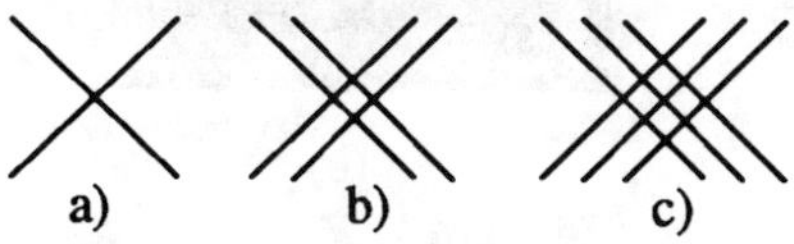

Figure 2 Overlap of "X" Patterns.

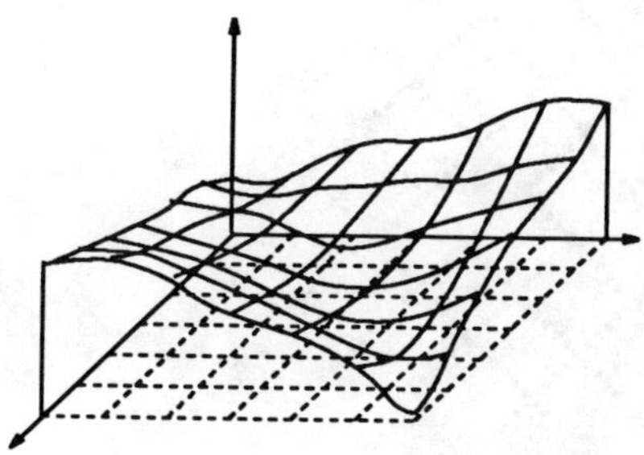

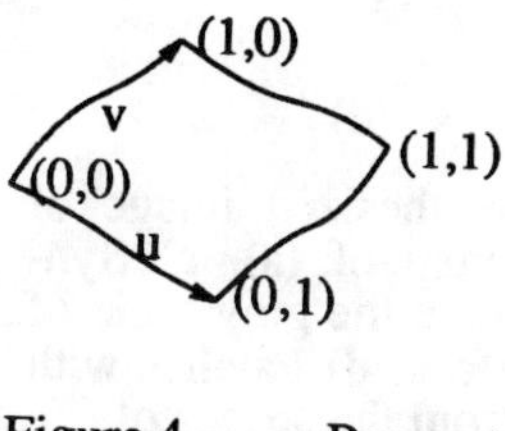

Figure 3 A mesh network of range data.

Figure 4 Parametric curve and surface.

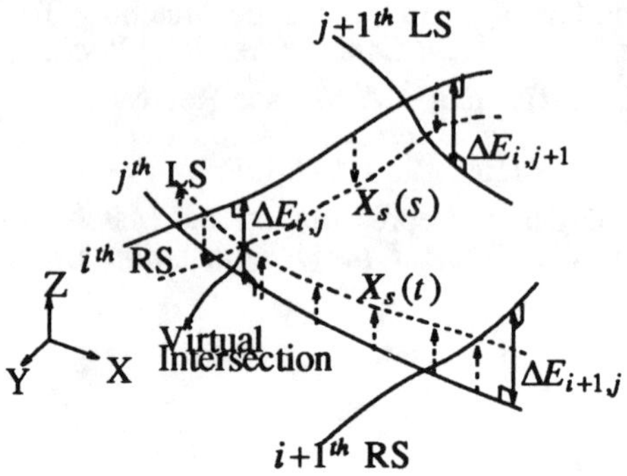

Figure 5 Shifting two space curves at the virtual intersection point.

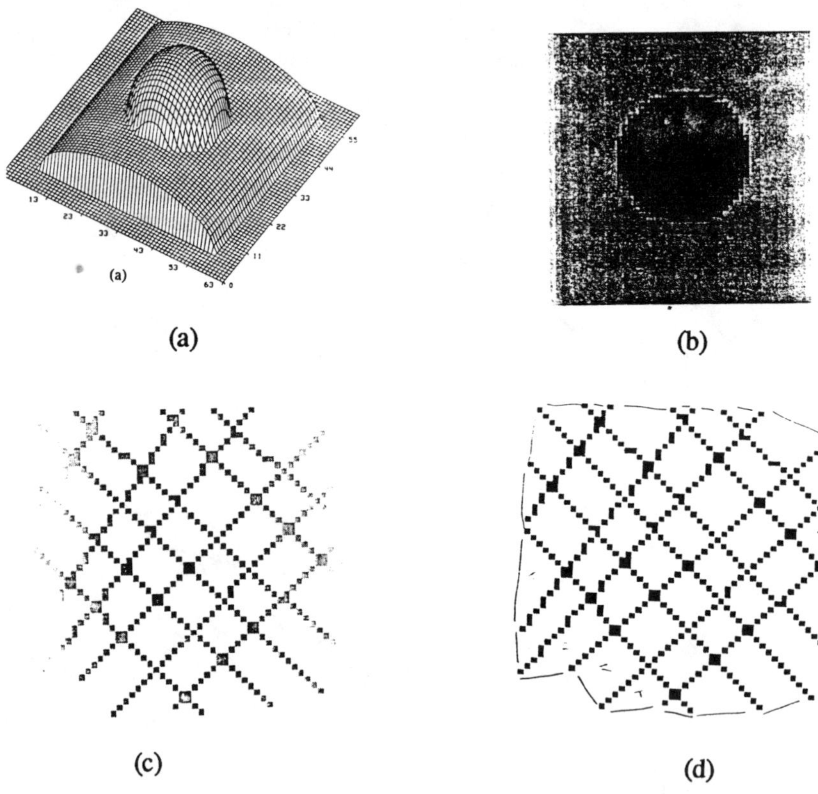

(a) (b)

(c) (d)

Figure 6 a) Three dimensional profile of a 64x64 synthesized image of cylindrical and spherical surface. b) Range map of (a). c) Synthesized stripes placed on the surface of (a) where the gray scale of the stripes indicates the range value of the surface. d) Patches with missing boundaries. e) Reconstructed surface from the stripes of (c). f) Range map of the reconstructed surface.

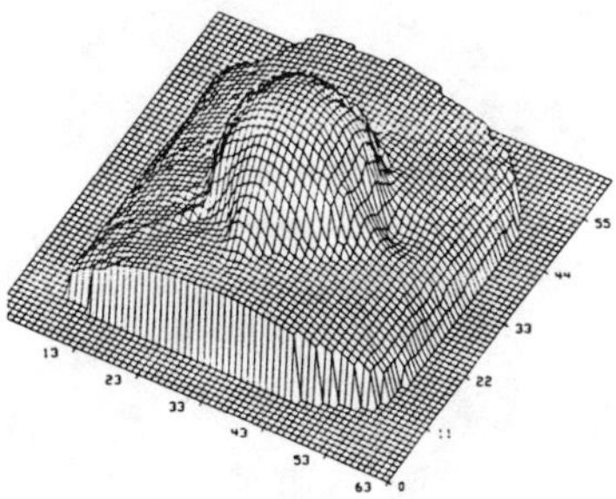

(e)

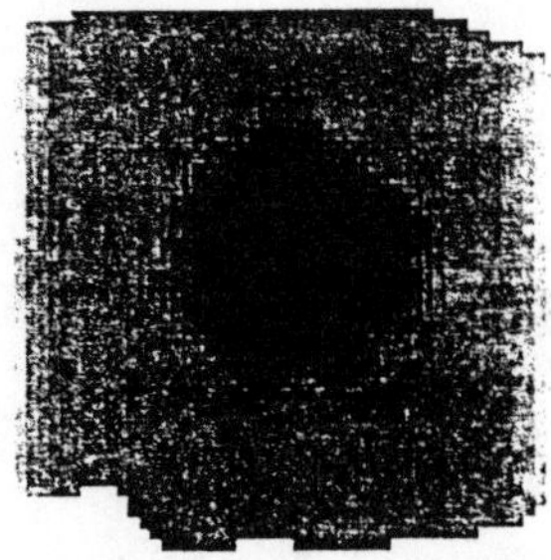

(f)

Figure 6 Continued

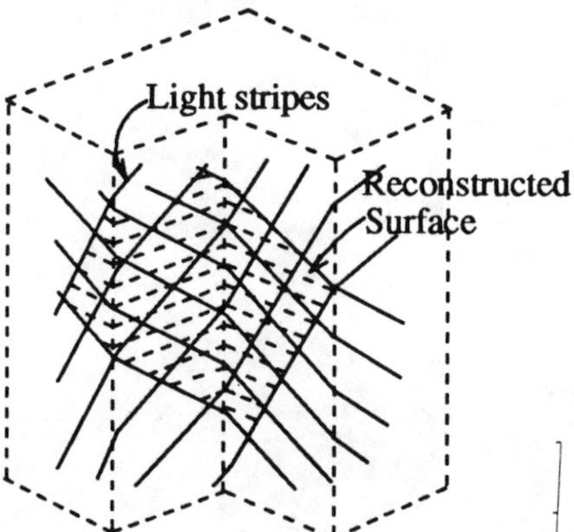

(a)

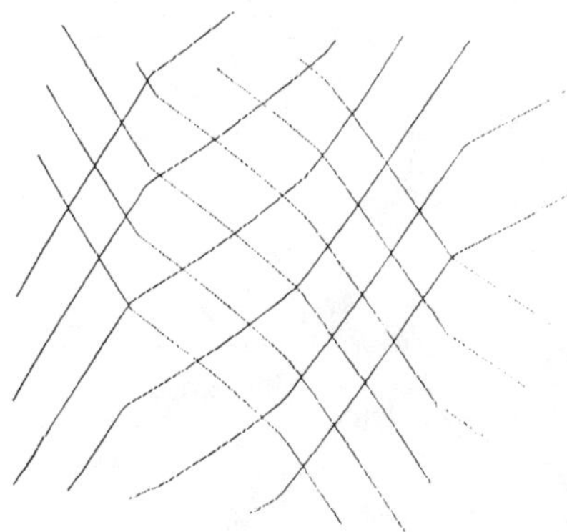

(b)

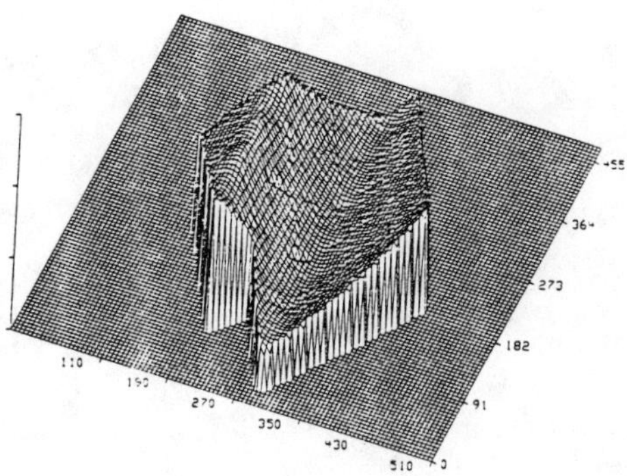

(c)

Figure 7 a) Line drawing of a scene of the stair case with the stripes. b) Overlap of the stripe images from the camera's point of view. c) Three dimensional profile of the reconstructed surface.

TRI-MODE COLLISION AVOIDANCE SKIN FOR ROBOT ARMS IN SPACE

John M. Vranish
NASA/Goddard Space Flight Center
Greenbelt, Maryland 20771

D. S. Chauhan

ABSTRACT

Safety is a primary concern for robots operating in space. The tri-mode skin sensor, being developed in-house at the NASA/Goddard Space flight center, addresses this concern by employing a collision avoidance/management skin around the robot arms. This capacitive sensor is at present dual mode, proximity and tactile. The proximity mode permits the robot to sense an intruding object (including a human being), to range the object (in time to permit the robot to react and to detect the edges of the object (with sufficient directionality and resolution to permit the robot to go around it). The tactile mode permits the robot to sense when it has contracted an object, where on the arm contact has been made, and provides a three dimensional image of the shape of the contact impression. A third, pyroelectric, mode will be added to permit the robot arms to detect the proximity and location of a hot object and to add sensing redundancy to the other two modes.

The sensing system will perform despite sun glint, temperature extremes and vacuum (conditions which predominate in space). It will not disturb other instruments, and since its excitation field is frequency modulated and stabilized by a phase locked loop, EMI from nearby systems will not disturb the sensor.

In this paper the capacitive modes of the sensing skin are described along with the fundamental operating principles, design particulars and test results.

INTRODUCTION

The objective is to develop a sensing skin that will prevent robot arms from colliding with objects in space; particularly human beings. It is essential that this sensing system be able to function reliably in the extreme environment of space and that it not disturb or be disturbed by neighboring NASA instruments. To date, this has been an unsolved problem. However, our evidence indicates that an approach based on an array of capacitors is promising. This approach will satisfy both the proximity and the tactile modes. A thin sheet of polyvinylidine fluoride (PVDF) will be added later to provide a temperature sensing capability and complete the third mode.

FUNCTIONAL REQUIREMENTS

At this point the functional requirements are general and address providing the suitability of the sensing principle for space use as a collision-avoidance sensor rather than addressing the details (such as out-gassing materials etc.). These are as follows:

1. The sensor should be able to detect a human at ranges sufficient to permit the robot to stop (est. 12 in. (30.5 cm)).

2. The sensor should be able to detect the near-range location of an object sufficient to avoid collisions and to find the object edges. (est. edge detection to +/- 1 in. (2.5 cm) at 3 in. (7.6 cm) range. Est. range determination to +/- 1 inc. (2.5 cm) at 3 in. (7.6 cm) nominal range).

3. The sensor should be able to detect contact and where and how hard the contact is (Est. location to +/- 1 in. (2.5 cm) and force to +/- 0.25 lbf (1.1N)).

4. The sensor should be able to detect the presence, direction and temperature of hot objects (to be determined).

5. The sensor should not interfere with other instruments nor be adversely effected by them (To be determined).

PRINCIPLES OF OPERATION

The proximity sensing drive circuit consists of two oscillators, one fixed frequency (FFO) and one whose frequency is controlled by a tuning capacitor which can be termed a capacitive controlled oscillator (CCO). The phase outputs of these two oscillators are compared and a beat frequency results. The capacitive coupling between the skin plates shown in fig. 1 and an approaching object changes the tuning of the CCO resulting in a change in the output beat frequency (signal). The nearer the object, the greater the change in the beat frequency[1,2]. The columns and rows of the skin plates are individually connected to the CCO by a multiplexer arrangement so the robot can determine the direction of the approaching object. It can also estimate the object edges. Thus the proximity mode permits locating (range and direction) an approaching object and determining its edges provided it is a dielectric and/or a conductor. This information permits the robot to avoid collisions.

For tactile sensing, if the object touches a sensing element, the outer capacitive plate is pushed nearer the grounded robot arm and we observe a change in the potential on the thin conductor near the arm. We interpret this change as tactile sensing[3,4]. The skin plates, the thin conductor near the robot arm, and the arm itself provide a system of capacitors in series. Pressure on a plate changes the capacitance between the skin plate and the thin conductor while the capacitance between the thin conductor and the grounded robot arm is unchanged. This system, then, acts as a pressure-dependent voltage divider. Since we can pre-select which skin plates are energized, we can determine location of tactile contact and edges of the object as well. Fig. 2 shows equivalent circuit details.[2]

EXPERIMENTAL RESULTS

The purpose of the measurements was to determine the approximate performance of the proximity mode of the skin, specifically range and edge detection. To accomplish this, a capacitive strip 0.25 in. (0.6 cm) wide and 13 in. (33 cm) long was used as the capacitive drive plate. When mounted vertically (see fig. 3), it acted as a column of the skin array and when mounted horizontally it acted as a row. Of particular interest was its ability to detect a human being.

Graph 1 provides an indication of the range at which the sensor can detect a human. It also gives an indication of the sensor's directionality. For this graph, five of the capacitive strips were shorted together and mounted in the vertical orientation. They were separated by 0.25 in. (0.6 cm) except at the point where they were shorted together. In this configuration we would expect relatively long range detection and graph I bears this out. It shows the frequency shift in the CCO caused by the change in its tuning capacitance caused by a human hand approaching the conductive strips acting as a sensor array column. Measurements were taken at different ranges and at different angular orientations to the column. The curves are essentially hyperbolic as predicted. The directionality at the close in ranges is also as predicted. The irregularities in the curves are primarily due to the hand having a relatively large area, thus blurring the directionality, due to signal jitter, and due to the fact that the hand was unsupported. Further managements were made using a human hand approaching a single capacitive strip acting as a sensor array column. The range of the sensor was reduced; but its directionality was improved. Measurements using a steady, supported aluminum rod as a probe; yielded a clearer indication of the directionality of a sensor column. The directionality was much improved with the signal dropping to near zero for an approach angle angle of 45 deg.

The strip was then mounted horizontally; thus acting as a row (see fig. 3b) and a human hand was used for the measurements. Again the curves were essentially hyperbolic and again uncertainties were caused by using an unsupported human hand. The directionality was apparent; but the range was slightly reduced from the column configuration because the simulated robot arm is a cylinder that favors the vertical column arrangement.

Although no tactile measurements were taken, the skin proved to be very sensitive to touch and this will be an easy mode to incorporate.

SUMMARY AND CONCLUSIONS

The results to date on the development of a tri-mode sensing skin are described. The ranging and edge detection mode is examined in detail. Its operational concept is explained and modeled and a description is given of the sensor hardware. Also, experimental results are given to measure the sensor's range and directionality against a human hand. Though optimization is yet to be seriously employed, we can already see that the sensor shows promise as a collision avoidance sensor array for robots in space.

REFERENCES

1. Hayt, William H. Jr., "Engineering Electromagnetics," McGraw-Hill, Inc., Copyright 1981, 1974, 1967, 1958.
2. Fischer, C.R. "Build the Digital Theremin", 1989 PE HOBBYISTS HANDBOOK, pp. 142,143,144,152.
3. Pugh, E.M. and E. W. Pugh, "Principles of Electricity and Magnetism". 2nd ed. Addison-Wesley Publishing Co. Reading, Mass 1970, Chapter IV.
4. Berec, A., Soykan, O. and Neuman, M.R., "A Multi-elemental Capacitive Force Sensor", Proc. Annu. Conf. Eng. Med. Biol, 29:101, 1987.

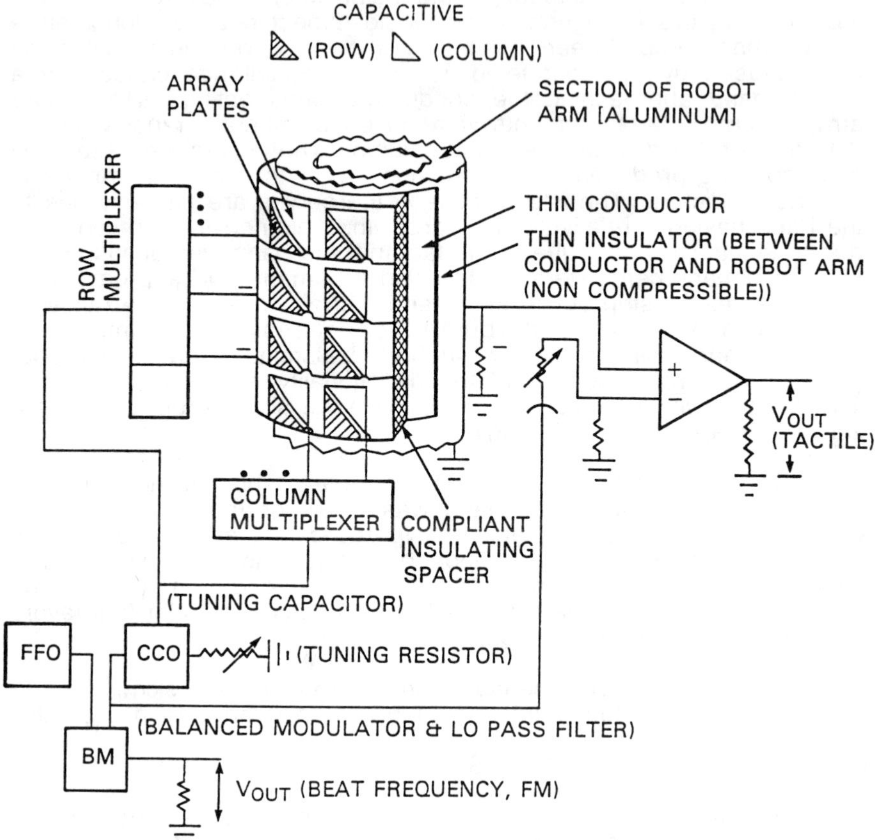

FIG. 1. SENSOR CONCEPT (PROXIMITY/TACTILE)

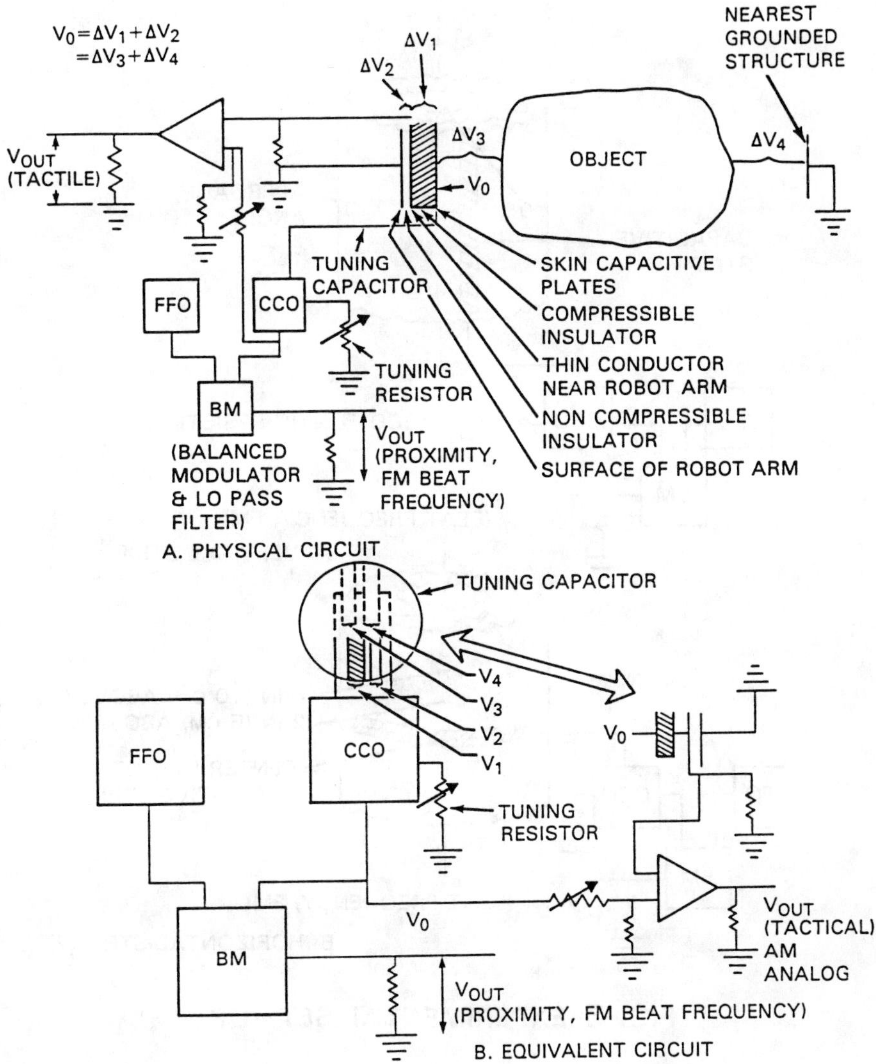

FIG. 2 DEVELOPMENT OF EQUIVALENT CIRCUIT

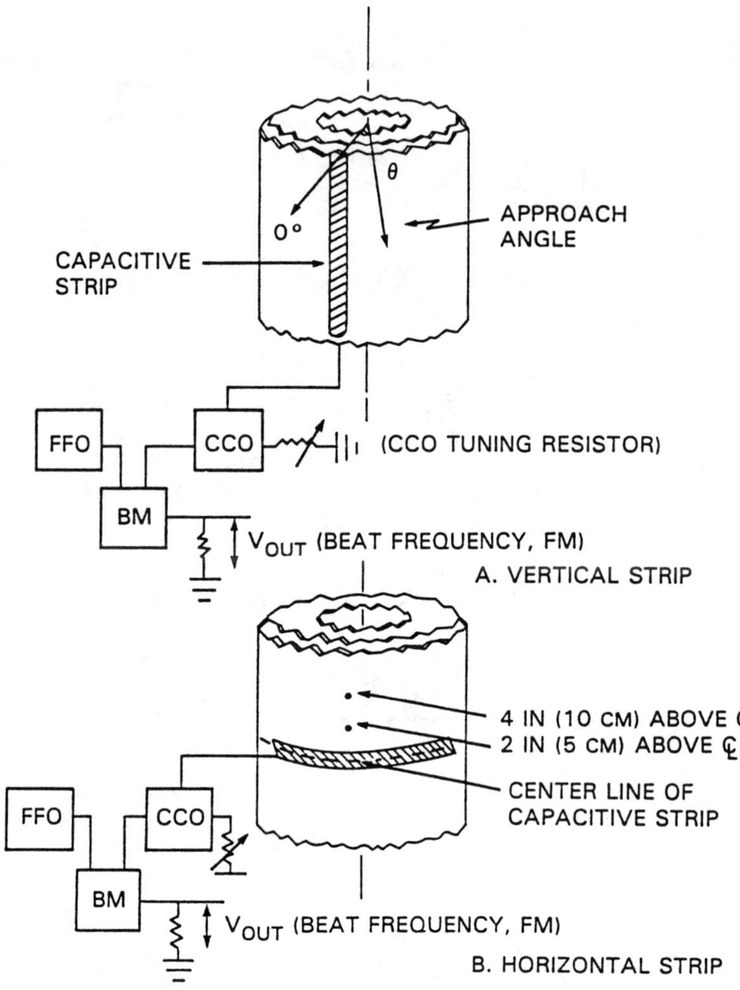

FIG. 3 EXPERIMENTAL SET—UP

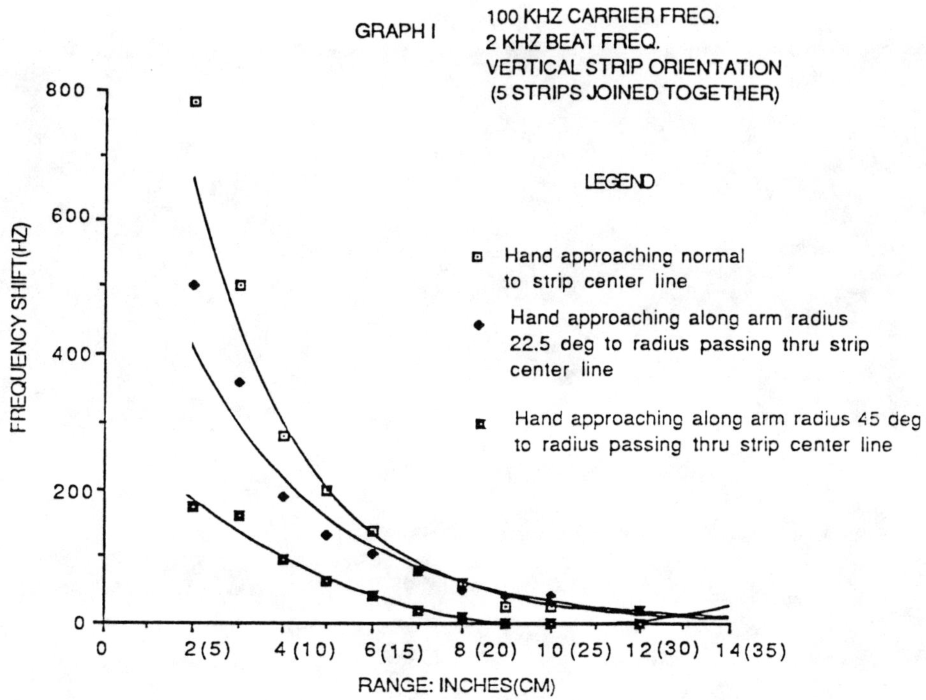
GRAPH I
100 KHZ CARRIER FREQ.
2 KHZ BEAT FREQ.
VERTICAL STRIP ORIENTATION
(5 STRIPS JOINED TOGETHER)
LEGEND
Hand approaching normal to strip center line
Hand approaching along arm radius 22.5 deg to radius passing thru strip center line
Hand approaching along arm radius 45 deg to radius passing thru strip center line
FREQUENCY SHIFT(HZ)
800
600
400
200
0
0
2(5)
4(10)
6(15)
8(20)
10(25)
12(30)
14(35)
RANGE: INCHES(CM)

Self-Calibration of a Laser Tracking Coordinate Measuring Machine

Hanqi Zhuang, Baiyuan Li and Zvi S. Roth
Robotics Center, College of Engineering, Florida Atlantic University
Boca Raton, FL 33431

Fumio Hamano
Electrical Engineering, California State University at Long Beach
Long Beach, CA 90840

Abstract

Two new methods for self-calibration of a laser tracking coordinate measuring machine are reported. In the first method, an extra laser beam is added to a three-beam laser tracking system to provide measurements necessary for self-calibration. The other method uses an external precision planar surface to constrain the target motion during a calibration process. Iterative optimization algorithms with simulation results are also presented.

1 Introduction

Systems that combine laser interferometers, optics and electro-mechanical mirror-positioning devices to perform coordinate measuring have been developed by several researchers [1]. Since intereferometer reading is incremental and provides only relative displacement information of a moving target, there must exist a reference point from which relative displacements are measured. When the coordinates of such point are not explicitly available, measurements need to be taken from which such coordinates may be deduced. Furthermore, in the case of multiple trackers the relative distances of the tracking mirrors should be determined very accurately in advance. These tasks are referred to as the "calibration of the laser tracking system". This paper focuses on self-calibration of a multiple-beam target tracking system, developed at Florida Atlantic University [2].

Two new self-calibration methods are proposed. The first method is based on a four-trackers system. An extra laser beam is added to a three-beam laser tracking system to provide measurements necessary for self-calibration. The other method uses an external precision planar surface to constrain the target motion during the self-calibration process. Iterative optimization algorithms with simulation results are presented for the estimation of the geometric parameters of the laser tracking system and its reference point. The self-calibration ideas presented here were motivated by the early work on self-calibration of robot manipulators as presented in [3].

2 Preliminaries

2.1 Operation of a three-beam Laser Tracking System

A schematic diagram of the laser tracking mirror positioning servo mechanism used in the Florida Atlantic University machine is shown in Fig. 2.1. Two orthogonally polarized beams at narrowly spaced frequencies, referred to as the reference beam having a frequency of f_2 and measuring beam having a frequency of f_1, are emitted by the laser head. The reference beam is diverted to the receiver by the polarizing beamsplitter located at the single beam interferometer. The measuring beam proceeds through the interferometer and the 50% beamsplitter and is directed by the tracking mirror to the retroreflector located on the moving target. The returning beam from the retroreflecter goes parallel to the incoming beam and enters the 50% beamsplitter from the opposite direction. Half of the beam when passes through the 50% beamsplitter is reflected by the polarizing beamsplitter and emerges into the receiver. If the movable retroreflector changes position, a Doppler frequency shift of $\pm \Delta f$ occurs, which is then translated through interferometry to a relative displacement reading accurate to within a fraction of a wavelength resolution. The other half of the beam reflected by the 50% beamsplitter is transmitted to a four quadrant detector, which detects the deviations of the returning beam from the center. Error signals drive the servo system to adjust the angles of the two degrees-of-freedom mirror gimbal.

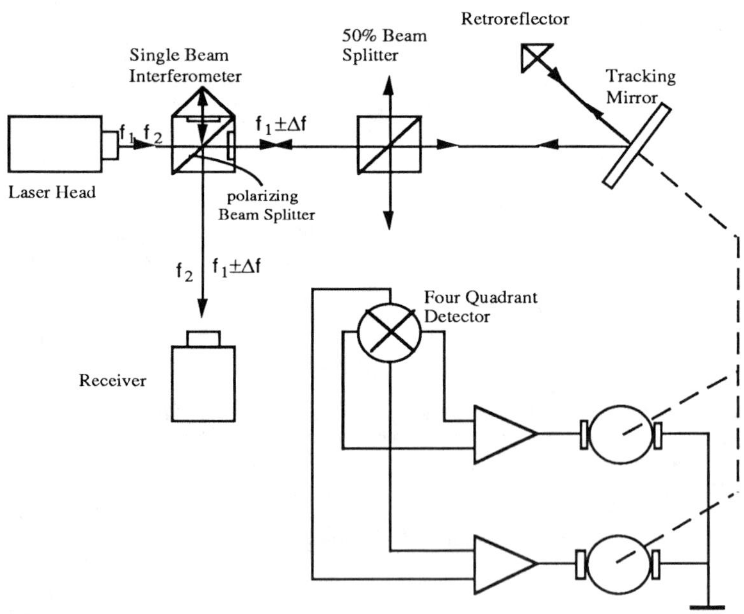

Fig 2.1 Schematic of a laser tracking system

A basic assumption in this paper is that measurements are taken at steady state when the returning beams are perfectly centered at their tracking mirrors. Error sources include misalignment of the gimbal-axes, mirror center offsets from the intersection of the dual axes and servo motor dead zones. These issues go beyond the scope of this paper.

2.2 Basic Equations for 3-D Measurements

A minimum of three trackers is needed for 3-D measuring using distance measurements only. The target coordinates are obtained from the distance measurements through triangulation, based on the knowledge of the locations of the three trackers (i.e., the relative distances among the mirror centers) and a reference point in the world coordinate frame. Denote by Points *A*, *B*, *C* the centers of each mirror, respectively. Denote by *P* the reference point. Assume that the locations of Points *A*, *B* and *C*, represented respectively in the reference frame by $\boldsymbol{r}_{s1}$, $\boldsymbol{r}_{s2}$ and $\boldsymbol{r}_{s3}$, are known. Also assume that the relative distances from Points *A*, *B* and *C* to Point *P*, denoted by l_1, l_2 and l_3 respectively, are measured. The location of Point *P*, represented in the reference frame by $\boldsymbol{r}_p$, can then be calculated from the following equations,

$$(\boldsymbol{r}_p - \boldsymbol{r}_{sj})\cdot(\boldsymbol{r}_p - \boldsymbol{r}_{sj}) = l_j^2 \qquad j = 1, 2, 3 \qquad (2.1)$$

Without loss of generality, let Point *A* be at the origin of the reference coordinate frame. Thus $\boldsymbol{r}_{s1} = \boldsymbol{0}$. Also let Point *B* be located on the x axis and Point *C* on the x-y plane. From (2.1),

$$r_{p,x} = 0.5(l_1^2 - l_2^2 + r_{s2,x}^2)/\, r_{s2,x} \qquad (2.2a)$$

$$r_{p,y} = 0.5(l_1^2 - l_3^2 + \boldsymbol{r}_{s3}\cdot\boldsymbol{r}_{s3} - 2\, r_{p,x}\, r_{s3,x})/\, r_{s3,y} \qquad (2.2b)$$

$$r_{p,z} = \pm\,(l_1^2 - r_{p,x}^2 - r_{p,y}^2)^{0.5} \qquad (2.2c)$$

The sign of $r_{p,z}$ is often uniquely determined taking into account physical constraints of the system configuration.

3 Four-Tracker Calibration Method

3.1 Problem Formulation

Three length measurements of a single target provided by three trackers uniquely determine the location of the target point in 3-D space. Adding one more length measurement of the test point creates an overdetermined system. A least-squares technique can be applied to estimate the location of the point. The fitting error is primarily contributed by the uncertainty in the locations of the mirror centers and the reference point. If a sufficient number of points in 3-D space are measured using the four-tracker system, the geometric parameters of the laser system as well as the target locations can be found by least squares techniques. The basic equations for the four-tracker system are as follows

$$(\boldsymbol{r}_i - \boldsymbol{r}_{sj})\cdot(\boldsymbol{r}_i - \boldsymbol{r}_{sj}) = (l_{0j} + \Delta l_{ji})^2 \qquad j = 1, 2, 3, 4 \quad i = 1, 2, \cdots, m \qquad (3.1)$$

where m is the number of measurements; $\boldsymbol{r}_i$ denotes the coordinates of the ith point; l_{0j} is the distance between the jth tracker's mirror and the reference point; and Δl_{ji} is the relative displacement measured by the jth interferometer for the ith point; see Fig. 3.1.

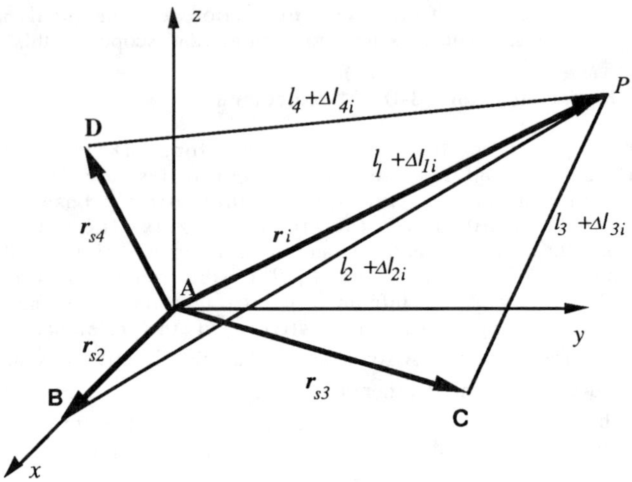

Fig. 3.1 Parameters of a four-tracker laser measurement system

Four trackers can provide a set of four solutions of $\boldsymbol{r}_i$ as there are four different combinations of three trackers; i.e., Combinations {A,B,C}, {A,B,D}, {A,C,D} and {B,C,D}. Denote by $\boldsymbol{x}^4$, $\boldsymbol{x}^3$, $\boldsymbol{x}^2$ and $\boldsymbol{x}^1$ the geometric parameter vectors of Combinations {A,B,C}, {A,B,D}, {A,C,D} and {B,C,D}, respectively. Clearly $\boldsymbol{r}_{sj}$ is excluded from $\boldsymbol{x}^j$; i.e.,

$$\boldsymbol{x}^1 = [\, r_{s2,x}, r_{s3,x}, r_{s3,y}, \boldsymbol{r}_{s4}^T, l_{02}, l_{03}, l_{04}]^T \tag{3.2a}$$
$$\boldsymbol{x}^2 = [r_{s3,x}, r_{s3,y}, \boldsymbol{r}_{s4}^T, l_{01}, l_{03}, l_{04}]^T \tag{3.2b}$$
$$\boldsymbol{x}^3 = [\, r_{s2,x}, \boldsymbol{r}_{s4}^T, l_{01}, l_{02}, l_{04}]^T \tag{3.2c}$$
$$\boldsymbol{x}^4 = [\, r_{s2,x}, r_{s3,x}, r_{s3,y}, l_{01}, l_{02}, l_{03}]^T \tag{3.2d}$$

Let $\boldsymbol{x}$ be the aggregated geometric parameter vector of the four-trackers system, $\boldsymbol{x} = [x_1, x_2, ..., x_{10}]^T = [\, r_{s2,x}, r_{s3,x}, r_{s3,y}, \boldsymbol{r}_{s4}^T, l_{01}, l_{02}, l_{03}, l_{04}]^T$. That is $\boldsymbol{x} = \boldsymbol{x}^1 \cup \boldsymbol{x}^2 \cup \boldsymbol{x}^3 \cup \boldsymbol{x}^4$. Let

$$h_{ij}(\boldsymbol{x}^k) = (\boldsymbol{r}_i(\boldsymbol{x}^k) - \boldsymbol{r}_{sj})\cdot(\boldsymbol{r}_i(\boldsymbol{x}^k) - \boldsymbol{r}_{sj}) - (l_{0j} + \Delta l_{ji})^2 \quad k = 1, 2, 3, 4;$$
$$j = 1, 2, 3, 4; \; i = 1, 2, \cdots, m \tag{3.3}$$

If $\boldsymbol{x}$ is known exactly, $h_{ij}(\boldsymbol{x}^k) \equiv 0$. Uncertainties in $\boldsymbol{x}$ cause $h_{ij}(\boldsymbol{x}^k)$ in general to be nonzero for k = j. The optimization problem is to choose $\boldsymbol{x}$ such that the cost function J is minimized, where

$$J(\boldsymbol{x}) = \sum_{k=1}^{4} \sum_{j=1}^{4} \sum_{i=1}^{m} h_{ij}(\boldsymbol{x}^k)^2 \tag{3.4}$$

3.2 An Optimization Algorithm

The problem formulated in (3.4) is a classical nonlinear least squares one. An algorithm is proposed next to solve this problem.

Step 1. x^k, k = 1, 2, 3, 4, are used to estimate $r_i(x^k)$. An iterative algorithm may be adopted to perform the task for k = 1, 2, 3, in which cases the error model, obtained by linearizing (3.3) in r_i, is

$$2(r_i(x^k) - r_{sj})\cdot dr_i - \{(l_{0j} + \Delta l_{ji})^2 - (r_i(x^k) - r_{sj})\cdot(r_i(x^k) - r_{sj})\} = 0$$
$$j \in \{1, 2, 3, 4\} \text{ and } j \neq k \qquad (3.5)$$

$r_i(x^4)$ is used as the initial value of $r_i(x^k)$. Section 2 provides the closed form solution for $r_i(x^4)$.

Step 2. The resulting $r_i(x^k)$, i = 1, 2, ···, m and k = 1, 2, 3, 4, are used to update the geometric parameter vector x by applying a nonlinear optimization algorithm [4]. The error model, obtained by linearizing (3.3) with respect to x, takes the following form

$$2(r_i(x^k) - r_{sj})^T \nabla r_i\, dx - 2(r_i(x^k) - r_{sj})\cdot dr_{sj} - 2(l_{0j} + \Delta l_{ji})\cdot dl_{0j}$$
$$- \{(l_{0j} + \Delta l_{ji})^2 - (r_i(x^k) - r_{sj})\cdot(r_i(x^k) - r_{sj})\} = 0$$
$$k = 1, 2, 3, 4 \quad j = 1, 2, 3, 4 \quad i = 1, 2, \cdots, m \qquad (3.6)$$

where "∇" denotes a gradient operation. Here dr_{s1}, $dr_{s2,y}$, $dr_{s2,z}$ and $dr_{s3,z}$ are zero; i. e., $dx = [\, dr_{s2,x}, dr_{s3,x}, dr_{s3,y}, dr_{s3}{}^T, dl_{01}, dl_{02}, dl_{03}, dl_{04}]^T$. The Jacobian matrix is made up of the coefficients in (3.6).

The above two steps are iteratively solved until the difference of two consecutive solutions is within a preset limit.

4 Planar Calibration Method

4.1 Problem Formulation

In this method, the target is constrained to move on a precision planar surface during the calibration process. If three different locations of the target are measured, the plane equation can be uniquely determined. More than three measurements provide an overdetermined system, in which case the plane equation can be estimated using least squares techniques. The fitting errors are mainly contributed by the uncertainty of the geometric parameters of the laser system as both the precision planar surface, assumed very smooth, and the relative measurements provided by the interferometry, assumed very accurate, are assumed error free. The geometric parameters of the laser system and the precision plane can be estimated together by minimizing the fitting errors.

Let n be a normal to a plane L and let the distance between the origin of the reference frame and Plane L be a (see Fig. 4.1). The plane equation is

$$n\cdot r_i - a = 0 \qquad i = 1, 2, \cdots, m \quad (4.1)$$

where r_i is the coordinate vector of a random target point P_i lying on Plane L, and m is the number of the measurements. Let l_{0j} be the distance between the jth tracker's mirror and the reference point. r_i is a function of the parameter vector x, where $x = [x_1, x_2, ..., x_6]^T = [\, r_{s2,x}, r_{s3,x}, r_{s3,y}, l_{01}, l_{02}, l_{03}]^T$. Let

$$h_i(\boldsymbol{x},\boldsymbol{n},a) = \boldsymbol{n}\cdot\boldsymbol{r}_i(\boldsymbol{x}) - a \qquad i = 1, 2, \cdots, m \quad (4.2)$$

The problem is to choose $\boldsymbol{x}$, $\boldsymbol{n}$ and a such that the cost function J defined below is minimized,

$$J(\boldsymbol{x},\boldsymbol{n},a) = \sum_{i=1}^{n} h_i(\boldsymbol{x},\boldsymbol{n},a)^2 \qquad (4.3)$$

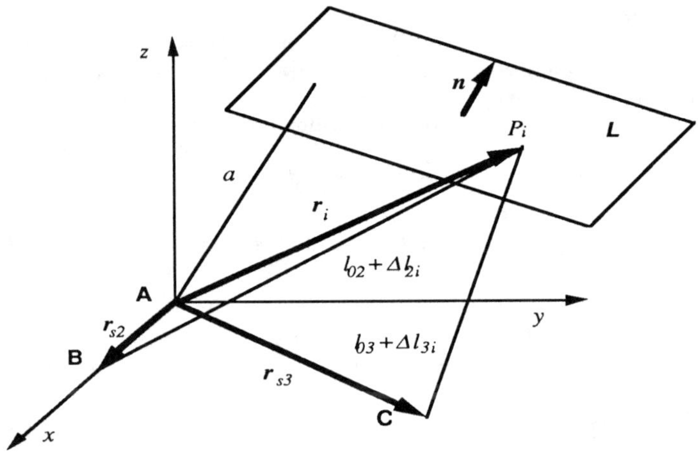

Fig. 4.1 Parameters of a three-tracker laser measurement system with a precision planar surface

4.2 An Optimization Algorithm

It is assumed that all the three trackers lie on the x-y plane. If this is not the case, slight modifications of the algorithm are required. The procedure is:

Step 1. The solution formulas for the basic measurement equations given in Section 2 are used to compute $\boldsymbol{r}_i$, i = 1, 2, ···, m, for a given initial $\boldsymbol{x}$.

Step 2. The resulting $\boldsymbol{r}_i$ is used to update the geometric parameter vector consisting of $\boldsymbol{x}$, $\boldsymbol{n}$ and a by applying a nonlinear optimization algorithm [4]. A linearized error model, obtained by linearizing (4.2) with respect to $\boldsymbol{x}$, $\boldsymbol{n}$ and a, is

$$\boldsymbol{n}\cdot\boldsymbol{r}_i(\boldsymbol{x}) - a + \boldsymbol{n}^T\nabla\boldsymbol{r}_i d\boldsymbol{x} + \boldsymbol{r}_i(\boldsymbol{x})\cdot d\boldsymbol{n} - da = 0 \qquad i = 1, 2, \cdots, m \qquad (4.4)$$

The Jacobian matrix is made up of the coefficients in (4.4).

The above two steps need to be iteratively solved until the difference of two consecutive solutions is within a preset limit. Note that among the three elements of $\boldsymbol{n}$, only two are independent. A condition handling technique should be applied to eliminate one redundant element of $\boldsymbol{n}$ while maintaining the vector with unit magnitude in the minimization process.

To increase the robustness of the algorithm, the precision planar surface may be placed at two or three different places with different orientations. It is important to maintain the continuity of target tracking when the target is switched from one plane to another. In this case, (4.1) is replaced by

$$n_k \cdot r_{ik}(x) - a_k = 0 \qquad k = 1, 2, \cdots, p \quad i = 1, 2, \cdots, m_k \quad (4.5)$$

where p is the number of planes, and m_k is the number of measurements sampled on the kth plane. The cost function in (4.2) and the algorithm also needs to be modified slightly.

5 Simulation Results and Summary

A modified version of the algorithm given in Section 3 were implemented to test the four tracker scheme. About 2% of uncertainties were added to both the geometric parameters and the reference point. Different number of test points were tried. The algorithm converged after a few hundred iterations with 15, 25 and 50 measurements. Using more than 50 measurements did not significantly improve the estimation results.

To test the planar scheme, about 2% uncertainties were added to both the parameters of the planes and the geometric parameters of the laser tracking system. Both the single-plane and three-plane methods were tried. The single-plane algorithm converged after a few iterations with about 20 sampling points. The three-plane algorithm converged after also a few iterations with about 10 points from each plane.

The planar methods are more efficient than the four-tracker scheme in terms of computational cost. On the other hand, the four-tracker scheme allows the laser system to calibrate itself without the help of any external objects. Among the single-plane and three-plane schemes, the latter is more robust when facing process and measurement noises. All of the methods however do not take into account measurement noise in the interferometers, which may produce large deviations between simulations results and actual experimental results. This is still an open research topic.

References

[1] Lau, K., N. Dagalakis and D. Myers, "Testing," *International Encyclopedia of Robotics, Application and Automation,* R. C. Dorf and S. Y. Nof, John Wiley & Sons, Inc., 1988, pp. 1753-1769.

[2] Zhuang, H., *Kinematic Modeling, Identification and Compensation of Robot Manipulators*, Ph. D. Dissertation, Florida Atlantic University, Dec., 1989.

[3] Bennett, D. J. and J. M. Hollerbach, "Self-calibration of Single-loop Closed Kinematic Chain Formed by Dual or Redundant Manipulators," *Proc. 27th IEEE Int. Conf. Decision & Control*, Austin, Texas, December, pp. 627-629.

[4] Lawson, C. L. and R. J. Hanson, *Solving Least-Sqares Problems*, Prentice-Hall, Inc., 1974.

ON THE DESIGN OF A CONCURRENT IMAGE ANALYSIS SYSTEM*

JUDSON P. JONES
Center for Engineering Systems Advanced Research
Oak Ridge National Laboratory
Oak Ridge, TN 37831-6364 USA
jonesjp@ornl.gov

ABSTRACT

Because programming parallel computers is challenging, their use has been limited. To make the technology more accessible it is necessary to develop programming environments which hide the details. An environment for image analysis on concurrent computers is described which provides fundamental services for parallelism. The system is organized in a tripartite abstraction hierarchy. The topmost level can be programmed as though it were an ordinary sequential computer. Distributed memory and message passing appear only at lower levels. An operational taxonomy consisting of mathematical transformations, memory management, domain decomposition, and input/output facilities for relevant data structures form the nucleus of the environment. A collection of applications are enumerated; advantages and drawbacks are discussed.

INTRODUCTION

In robotics there is a strong motivation to minimize the execution time of image processing and analysis so that the results can be used to control robot motion. Two common strategies for reducing execution time are first, to construct special purpose hardware, and second, to use some form of parallel processing -- shared memory, common bus, SIMD, MIMD, etc. -- and distribute the workload over the many processors. These two strategies are complimentary, not contradictory, since obviously a parallel machine can be constructed from multiple special purpose machines.

Programming parallel computers is challenging. Consequently, the use of parallel computers has been limited to date to a relatively small group of experts. Adoption of parallel computing technology in a broad spectrum of applications awaits the development of methods which insulate non-experts from the details.

---

* The submitted manuscript has been authored by a contractor of the U.S. Government under contract DE-AC05-84OR21400. Accordingly, the U.S. Government retains a nonexclusive, royalty-free license to publish or reproduce the published form of this contribution, or allow others to do so, for U.S. Government purposes. The author gratefully acknowledges the support of the U.S Department of Energy Office of Technology Support Programs and Office of Basic Energy Sciences.

In this paper we describe an image processing and analysis environment (Ipac) for distributed-memory concurrent (MIMD) multicomputers. The primary design goal was to develop a reasonably friendly environment which allows programmers to exploit the power of parallel computers for image processing and analysis operations without necessarily becoming expert in parallel processing. To be useful, the system had to be accessible from a conventional programming language and provide a broad range of services without compromising performance.

AN ABSTRACTION HIERARCHY

Ipac is organized into a tripartite abstraction hierarchy. At the topmost level, we employ the concept of a "virtual sequential machine", which can be programmed as though it were an ordinary sequential computer. An intermediate level contains image processing and analysis utilities. The existence of distributed memory and the requirement for message passing are known at this level, but knowledge about the specific communication topology, mechanisms for message passing, and collective communications algorithms are hidden at the lowest, machine specific level.

The virtual sequential machine. Perhaps the most profound conceptual barrier facing programmers of parallel computers is the lack of an appropriate conceptual model for the machine. Whereas the operation of a von-Neumann style computer can be conveniently thought of as a sequence of arithmetic and logical operations, the concept of a single instruction stream does not apply to parallel computers.

Image processing and analysis tasks are characterized by voluminous data, and an appropriate model for parallel computing is the "data parallel" model. In this model, the entire task to be performed is decomposed into a sequence of sub-tasks, each involving all of the data, and algorithms are written which use the whole parallel machine for the solution of each subtask.

A profound advantage of the data parallel model, in contrast with the "pipeline" or "task decomposition" models, is that at a certain level, the familiar abstraction of a single instruction sequence is formally preserved and can be usefully manipulated to solve problems. From the perspective of the applications programmer, the whole parallel computer can be thought of as a single CPU, for which libraries have been conveniently written to support specific types of applications. We call this concept the "virtual sequential machine". Applications programmers who wish only to write sequential programs can do so at this level without regard for distributed memory or message passing.

The Data Parallel Computer. Generic image processing and analysis routines are written in a manner which ignores the specific topology and message passing mechanisms of the concurrent computer. Certain data structures are recognized to be distributed over the memories of the many CPUs, and programs must therefore make explicit calls to collective communications routines. These calls specify generic communications requirements, such as perimeter communication or global summation, rather than specific source and destination nodes. The abstraction is therefore one of a fully-connected distributed-memory multicomputer whose CPUs communicate using an unspecified mechanism. Any real machine deviates from this abstraction, but the abstraction nevertheless serves two useful purposes. First, the development of efficient parallel programs is isolated from the context of entire applications, so that problems are small enough to be solved individually. Second, it frees the developer from considering simultaneously methods for parallelizing a particular problem and methods for implementing the solution on a machine with a given topology.

Specific real machines. Routines responsible for I/O, domain decomposition, and collective communications form the lowest level of the hierarchy. These routines are developed in response to the requirements of the routines at the intermediate level.

The topology of the interconnection network and the mechanisms for interacting with the message passing system are taken into account. In particular, mechanisms for distributing data structures are implemented at this level, so that efficient input/output routines can be developed [1].

DESCRIPTION OF IPAC ENVIRONMENT

Ipac operations are conceived as mathematical operations which transform one data structure into another, possibly under the influence of a third. To describe the environment as seen from the level of the virtual sequential machine, it is therefore necessary to describe only the available data structures and operations, although the environment as seen from lower levels involves some consideration of concurrency. In the sections which follow, brief descriptions are given for the data structures, domain decompositions, operations, input/output, typing, function nomenclature and behavior, code generation, and portability.

Data Structures. Ipac data structures are classified as scalar, vector, neighborhood, image, and pyramidal. These data structures are dynamically allocated and deallocated using routines from the lowest level of the hierarchy, so that optimal assignment of sub-structures to processors can be made on a machine specific basis. Information regarding the location of parts of distributed data structures are encapsulated into records which are allocated and deallocated along with the data itself. This information is used by automatic type conversion routines and by communications, including input/output routines, to determine type casting and communications requirements for various operations. Multidimensional data structures are allocated using an "edge pointer" method, so that references to array elements are accomplished using standard array indexing.

Domain decompositions. In general, image and multiresolution pyramidal data structures are distributed, whereas scalar, vector, and neighborhood data structures are replicated (one copy on each CPU in the network). In most applications, images and pyramidal data structures are far larger than the vectors and neighborhoods. Parallelism is exploited in operations on these data structures. Neighborhoods and vectors (e.g. histograms and feature lists) are applied to or derived from entire images, and it is therefore most convenient to replicate these entities.

Domain decompositions for images and pyramids [2,3] are typically subimages where neighboring subimages are assigned to neighboring processors (grid). This decomposition scheme is not necessarily optimal, since optimal domain decompositions are only achieved in the context of specific algorithms, with their associated communication requirements. Grid decompositions are optimized for calculations involving neighborhood operations, i.e. having the same algorithmic structure as linear convolutions, and in some cases [4], are also optimal for non-neighborhood operations involving perimeter communication. However, some algorithms enjoy better performance if strip decompositions are used [5]. Accordingly, Ipac provides both strip and grid decompositions, and facilities for converting between one domain decomposition and another [1].

Operations. A taxonomy of functions have been developed based on the identities of the data structures and the various operations which can be performed on them. The taxonomy includes categories for operations involving scalars and images, images and images, scalars and pyramids, pyramids and pyramids, images and neighborhoods, pyramids and neighborhoods, images and pyramids, with some facilities included for the manipulation of vectors.

Ipac supports a full complement of arithmetic and logical operations. These operations use either scalar variables applied to each pixel in an image or a pyramid, or pixels from one image or pyramid applied to the corresponding pixels of another. The available arithmetic and logical operations are assignment, negation, addition, subtraction, multiplication, division, modulus, relational and logical operators, binary arithmetic operators and transcendental functions. Function names are composed of string fragments indicating the identity of the operative data structures and the identity of the operation.

Neighborhood operations include both linear and morphological operations, the latter for both binary and grey-level images. In each case, it is necessary to specify the image to be operated on, and the neighborhood surrounding a pixel over which the desired operation is to be applied. Neighborhood operators include linear convolution, greylevel maximum, minimum and median, and binary dilation and erosion. From these operations and mechanisms for specifying neighborhoods of various size and shape a large variety of linear and non-linear filters can be constructed, including detectors for various features. Other operations include a small number of vector operations, including histogramming and feature list accumulation, geometric transformations such as component labelling and hough transforms, creation of multiresolution representations, and particular multiresolution feature detectors.

Transparent I/O. Processors in a distributed memory multiprocessors communicate only with other processors through message passing. Usually a dedicated processor, not part of the topology of the network and usually called the host, is responsible for executing conventional I/O operations to devices like terminals, disk drives, and image acquisition and display devices. [6]. Ipac supports a subset of the standard Unix I/O system [7] augmented with functions specialized for I/O of multidimensional data structures. These I/O routines are classified as belonging to the lowest machine specific level of the abstraction hierarchy, and are therefore accessible at both superior levels.

Transparent typing. Different image analysis applications require different storage classes for pixels. Some require only 8-bit unsigned integers, whereas others [8] require floating point or double precision. A single application may require different storage classes at different stages of the calculation. In Ipac, storage class assignments for pixels in images and pyramids at the level of the virtual sequential machine for all arithmetic and logical operations are implicit. Applications programmers specify the desired storage class at allocation time. Thereafter, operations on images and pyramids are conducted without reference to storage class. Arithmetic type conversion is handled automatically according to the definition the "C" programming language uses for scalars [7]. For image-image operations, we cast the storage format having less precision into the storage format having greater precision, execute the operation using the arithmetic with greater precision, and cast the result to the desired precision.

Function Nomenclature and Behavior. A systematic nomenclature for function naming, and consistent behavior of functions permits applications to be written quickly and conveniently with a minimum of bug-producing side effects. Function names are composed by concatenating strings representing the data structures to be operated upon and the operation to be performed, in that order. The data structures are abbreviated by the single character strings s (scalar), v (vector), i (image), p (pyramid), and n (neighborhood). Operators have a similarly convenient system for nomenclature. For example, the function which adds two images is iiadd(.,.), and the function which computes the median over a given neighborhood is inmed(.,.).

All functions take arguments with the destination given first and return pointers to the results. This permits function nesting. As an extreme example, it is possible to allocate memory for an image, read an image from the current image source, execute a median filter over an 8 neighborhood, convolve it with a kernel over an 8-neighborhood, and threshold the result using

```
bin = inconv(inmed(imagein(imalloc(high,wide,"char")),n8),n8);
```

The resulting image is then referred to as "bin".

Code Generation. Efficient operation on all storage classes implies a different function for each operation for each storage class. The large number of operations in Ipac implies a correspondingly large number of routines. For example, an operation on two images of arbitrary storage class implies 10 functions, one for each storage class. Implementation of Ipac was accomplished using an indirect method for code generation. "Abstracts" of certain function were written which perform abstract operations on a data structures having an abstract storage class. These functions were then automatically instantiated a large number of times for specific operations on specific storage classes using a hierarchy of macro expansions defined using the m4 macro preprocessor. Abstract functions were written describing the operation of scalars on images and the operation of neighborhoods on images and used to generate all of the individual functions which could be generated by the use of different operators.

Portability. The isolation of machine dependencies to a restricted set of routines enhances portability. Only memory allocation, input/output, and communications routines are affected. To further enhance portability, communications routines are written using generic message passing primitives for which specific interfaces for a variety of computers have been implemented [9]. This guarantees that the resulting code is portable across a wide variety of parallel computers, and assists in the generation of performance visualization [10].

APPLICATIONS

A number of applications have been developed under Ipac, which has been implemented on four NCUBE, Inc. (Beaverton, OR) hypercubes. Two of these hypercubes are installed on-board mobile robots [11,12]. Two are standalone installations. Most of the Ipac applications have used an earlier version of the system than the one described here. These applications have included multi-spectral texture analysis [13], dynamic scene analysis [14], object recognition for mobile robots [11], multiresolution and spatio-temporal image analysis [3,15], development of concurrent algorithms [11,16], advanced representational methods [17], multisensor integration [18], and image restoration [8].

DISCUSSION

The conceptual model around which this image processing and analysis facility is constructed has proven to be sufficiently powerful to support a number of projects in research and development, either in concurrent algorithms proper, or in applications where concurrency is not the main issue.

Current limitations on the system is that feature lists which are not constructed through semi-group operations on pixel coordinates are not handled very well. For example, the construction of a chain code for a contour is not well balanced.

Finally, we note that although using a general purpose parallel computer to solve image processing and analysis problems does have a favorable cost/benefit ratio when compared to traditional sequential machines (at least once the code generation problem has been solved), these machines cannot currently compete on a cost/performance basis with dedicated hardware, such as dedicated convolvers. Although parallel computers can achieve raw performance superior to dedicated hardware (which is typically bound to a 30 Hz frame rate), they are usually much more expensive. This limits their application in robotics to situations in which there are additional constraints on system configuration, such as limits on the space, weight, or power supply (as in on-board computing for mobile robotics), or time-sharing the parallel computer with other computationally demanding tasks having different algorithmic structure.

REFERENCES

1. H. Embrechts, J.P. Jones (1990) "An input-output algorithm for M-dimensional rectangular domain decompositions on N-dimensional hypercube multicomputers." Proc. 5th Conference on Distributed Memory Concurrent Computers, *in press.*
2. Q.F. Stout (1987) "Hypercubes and pyramids." In: *Pyramidal Systems for Computer Vision*, ed. V. Cantonini & S. Levialdi, NATO ASI Series F, Vol. 25, Springer-Verlag, New York.
3. H.I.Christensen, J.P. Jones (1989) "Concurrent multi-resolution image analysis." Proc. 4th Conference on Hypercubes, Concurrent Computers, and Applications, pp. 1031-1038.
4. H. Embrechts, D. Roose (1989) "Efficiency and load balancing issues for a parallel component labelling algorithm." Proc. 4th Conference on Hypercubes, Concurrent Computers, and Applications, pp. 1005-1008.
5. R. Miller, S.E. Miller (1987) "Using hypercube multiprocessors to determine geometric properties of digitized pictures." Proc. 1987 Int. Conf. Parallel Processing, ed. S.K. Sahni, pp. 638-640.
6. J. Salmon (1987) "CUBIX: Programming hypercubes without programming hosts." In: *Hypercube Multiprocessors 1987*, ed. M.T. Heath, SIAM, Philadelphia, PA.
7. B.W. Kernighan, D.M. Ritchie (1978) *The C Programming Language*. Prentice-Hall, Englewood Cliffs, NJ
8. H. Hiriyanniah, G.L. Bilbro, W.E. Snyder, R.C. Mann (1989) "Resoration of piecewise constant images via mean field annealing." J. Opt. Soc. Am. A (6) 1901-1912.
9. G.A. Geist, M.T. Heath, B.W. Peyton, P.H. Worley (1989) "A machine-independent communication library." Proc. 4th Conference on Hypercubes, Concurrent Computers, and Applications, pp. 565-568.
10. M.T. Heath (1990) "Visual animation of parallel algorithms for matrix computation." Proc. 5th Conference on Distributed Memory Concurrent Computers, *in press.*
11. J.P. Jones (1988) "A concurrent on-board vision system for a mobile robot." Proc. 3rd Conference on Hypercube Concurrent Computers and Applications, G. Fox, ed., pp.1022-1032.
12. C.R. Weisbin, B.L. Burks, J.R. Einstein, R.R. Feezel, W.W. Manges, D.H. Thompson (1990) "Hermies-III: A step towards autonomous mobility, manipulation, and perception." *Robotica* (8) 7-12.
13. B.A. Huntsberger, T.L. Huntsberger (1989) "Hypercube algorithms for multi-spectral texture analysis." Proc. 4th Conference on Hypercubes, Concurrent Computers, and Applications, pp. 1009-1012.
14. Y. Soh, T.L. Huntsberger (1989) "Hypercube algorithms for dynamic scene analysis." Proc. 4th Conference on Hypercubes, Concurrent Computers, and Applications, pp. 1061-1068.
15. H.I. Christensen (1989) "Concurrent spatio-temporal image analysis." Proc. 4th Conference on Hypercubes, Concurrent Computers, and Applications, pp. 991-994.
16. K.W. Bowyer, J.P. Jones, C.H. Lake (1989) "Computing the Hough transform on an MIMD concurrent computer." Proc. 6th Scandanavian Conference on Image Analysis, ed. M. Pietikainen & J. Roning, pp.1172-1181.
17. T.L. Huntsberger, B.A. Huntsberger (1990) "Hypercube algorithms for image decomposition and analysis in the wavelet representation." Proc. 5th Conference on Distributed Memory Concurrent Computers, *in press.*
18. J.P. Jones, M. Beckerman, R.C. Mann (1989) "Design and implementation of two concurrent multi-sensor integration algorithms for mobile robots." In: Sensor Fusion II: Human and Machine Strategies, P.S. Schenker, ed., Proc. SPIE 1198, pp. 301-312.

SENSOR ENVIRONMENT FOR PROGNOSIS AND MONITORING SYSTEMS SUPPORT

M. Martins-Barata; A. Steiger-Garção
Universidade Nova de Lisboa
Faculdade de Ciências e Tecnologia
Departamento de Informática
Quinta da Torre
2825 Monte de Caparica
Portugal

Abstract

Previous work [1] done at the Intelligent Robotics Group launched a four layer based conceptual architecture to support prognosis and monitoring environments. These four layers can be supported by an heterogeneous system, based on multiprocessors running specialized and cooperating process. Logical and procedimental programming paradigms specifically integrated in a transputer based multiprocessor system were made available. This paper presents a preliminary monitoring and prognosis experiment on a CNC LATH machine, running on top of the previous described system. A set of sensors to monitor the machine working conditions (forces, noise, vibrations, etc.) is discussed as well their low level processing and integration. Sensed information is processed and analyzed according to existing models stored in a proper constructed Knowledge Base.

1- Introduction

Flexible automation system's service quality is assured, or evaluated, by monitoring. Its successive refinements leads to more powerful requirements: Besides giving information about the actual system's work quality and inform about successful/unsuccessful events, the monitoring system should also make predictions about possible events that can happen in a forseable future: Prognostics.

Diagnosis is supported by almost every monitoring systems, however it only can inform more specifically about the detected malfunction at system breakdown time. Prognosis works in the reverse way, it informs about possible malfunctions and associated occurrence probability, as early as possible. It is an important improvement that can be made within a monitoring system, because besides giving information about the actual system's work quality and successful / unsuccessful actions, it can also make predictions about possible undesirable events that might happen.

Because systems to be monitored can be very complex, an appropriated environment should be available. It must support several programming paradigms and processing architectures as well. Such environment is presented in [2] and proposes a four layer Conceptual Architecture. The first level concerns sensor and low level data processing. The next one supports the sensor integration and feeds the System's Knowledge Base at the third level. User interaction is supported by the fourth layer. Associated with this conceptual architecture exists a Reference System Model, that defines the several relationships between the process controller, process monitor and prognostic processor. Based on the requirements demanded a transputer based multiprocessor system was developed in order to provide the necessary processing power. This system can run in parallel several tasks that can be programmed using procedimental or declarative / logic languages.

This paper presents the work that is being done at UNL Intelligent Robotics Group, about the sensor processing environment and its interaction with the upper levels.

2- Sensor Environment

Sensors give the Prognostic and monitoring System the capability to be informed about the process status and its behavior as well. We are concerned about a sensorial environment that can be used in a broad range of applications, therefore it should be as versatile as possible. The versatility of our environment is presented in this paragraph.

2.1 Sensor data acquisition

In our environment sensorial data acquisition can be done following one of two strategies: i) using existing standard hardware or ii) developing special acquisition boards based on transputers. Both options require previous work; in case i) the existing hardware must be integrated into the transputer network while case ii) needs the development and test of the specific boards. At the beginning we started with integrating the existing acquisition hardware. We will develop specific boards only for those situations where the existing hardware can't be employed, e.g. sampling rates at high speeds.

The integration of existing acquisition hardware was done following the service--server strategy, fig(1). It assumes that an existing laboratory computer hosting the acquisition hardware, can be adapted to communicate with the transputer network following a specific logical protocol. The way how an IBM PC/AT can be integrated in the transputer network is presented in [1]. Basically the IBM PC/AT hosts an InMos Link Adapter board (specially developed inhouse) and executes a specific service-server program. The Link Adapter board is general purpose hardware, however the service server program is specific for each kind of server; it must execute the service and support also the respective protocol.

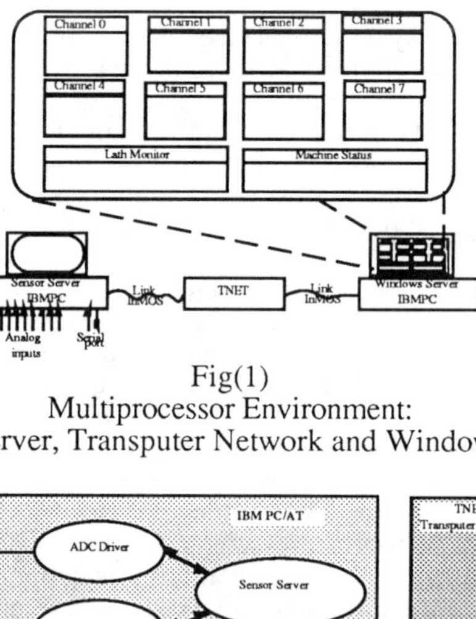

Fig(1)
Multiprocessor Environment:
Sensor Server, Transputer Network and Window Server

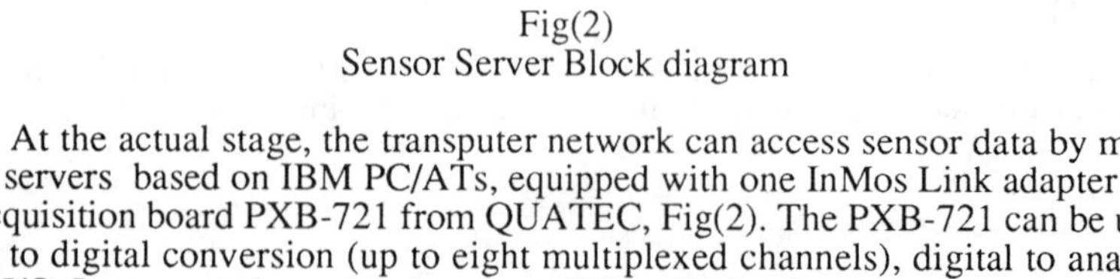

Fig(2)
Sensor Server Block diagram

At the actual stage, the transputer network can access sensor data by means of sensor servers based on IBM PC/ATs, equipped with one InMos Link adapter and the data acquisition board PXB-721 from QUATEC, Fig(2). The PXB-721 can be used for analog to digital conversion (up to eight multiplexed channels), digital to analog and digital I/O. It can sample analog data from DC to 30Khz, but in practice the maximum data acquisition rate is about 17Khz (if using AT compatible operating at 10 Mhz). That is because the sampled data can only be stored in memory by means of CPU I/O instructions (doesn't support DMA). It is also possible to access the RS 232 ports available on the PC/AT, and it can be used to communicate with other available sensors that use this communication interface. Three basic modules compose the SENSOR SERVER program: i) Service's Protocol Interpreter/Executor, ii) PXB-721 board driver and iii) Serial ports driver. The module i) receives acquisition/configuration commands from the transputer network, execute them and send back the result; it is written in Microsoft C. Module ii) does the interface to the PXB-721 board; it does the local data acquisition service required from module i). This module is all written in assembly language because it should execute as fast as possible to avoid CPU time waste. The sampling interval's accuracy is controlled by the real time clock interrupt. Module iii) is also interrupt driven and it collects serial incoming data in a local FIFO as well as the data to be transmitted. It is controlled by commands received from module i).

Vision is another sensor that is now being integrated in our transputer network. Following the same strategy an existing vision processing machine will be used as an image processing server. The existing hardware is a MAGISCAN II from Joyce Loebl[7]: it is based on a bit slice processor to support frame grabbing and image low level processing like histogram correction, convolution, threshold etc. It works controlled by an IBM PC/AT that supports the user interface and the sequences of image processing operations. Working as an image server, this system will provide a set of low level image processing functions to the transputer network tasks.

2.2 Sensor processing

We mean by sensor processing the application of a specific computation algorithm on the sampled data function **f**(t) in order to extract specific features. This processing algorithm is composed by a chain of sub-algorithms of which some are well known. One of them is filtering: It can enhance the sensed signals according to specific aspects, for example corrupting noise removal. Normally this operation is done by the sensor electronics and it does not require computation loading. Other filtering actions are sometimes necessary to be done by a computer like non-linear filtering.

In a sensorial environment it is difficult to define in advance the sensor processing algorithms, because we probably do not know exactly what kind of sensors we are going to use. A good approach to this problem is to enable sensor integration each time a new sensor is necessary. However we can construct a sensor processing library to store the already developed algorithms and to include the future ones. The primitives available in this library should be executed as transputer tasks in the network.

We started this library by writing the necessary signal processing routines to permit frequency domain analysis. A Fast Fourier Transform algorithm is now available. It can be executed by one (more than one in the near future) T800 transputer. Continuously it requests a buffer of signal samples from the dedicated sensor server, then does the transformation to the frequency domain and sends the results to the subsequent task. Hartley transform is also available and can be used as an alternative for time domain transformation.

Thus it is possible to develop basic transformation algorithms and their concatenation in order to get a low level sensor parallel algorithm and it's connection to the upper levels as well. The power of this approach is presented in the next paragraph.

3- CNC Machine application

Computerized numerical controlled machine tools play an important role in FMS systems. Their productivity and work quality can be increased by using a proper monitoring and prognostic system. We found out that productivity and work quality are contextual parameters and, of course, they render the statement of optimal criteria more difficult. That means the sensorial data interpretation and resulting actions should be done having in mind the specific working conditions: machine to be used, machined material type, desired accuracy, tool wear permitted, etc.

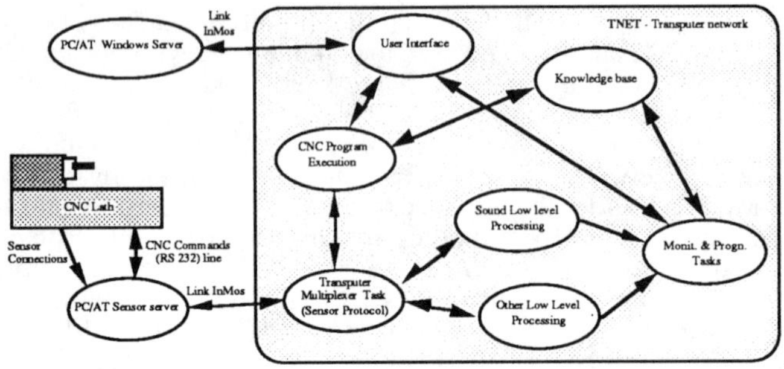

Fig(3)
CNC Lath Monitoring and Prognostic System

The monitoring system structure is presented in fig(3). Each module is implemented as one or more transputer tasks running in parallel on one or more transputers. They cooperate by means of communications channels supported by the TNET environment [2]. The machine used is a small LATH (1/2hp DC spindle motor)[6] with automatic tool exchange. Because the machine's CNC instructions are also controlled by a network task, it is possible to keep the Knowledge Base updated about what the machine is doing at any time. This permits the sensor data to be contextually interpreted by the monitoring system.

3.1 Machine sensors

The sensors to be used should provide information about the i) machine's working conditions and the ii) resulting work (machined piece evolution). Related to situation i) we can sense motors' power consumption and temperature, cutting forces, vibrations, tool's temperature and cutting status, and working resulting noise. For case ii) we need real time measuring equipment based on optical principles: laser, infrared and video cameras.

Because these sensing needs are new, CNC machines don't provide these kind of sensors, and their integration must be done: Sensors should be developed calibrated and integrated according to the machines' specifics. Fortunately machine working sound is easy to sense, it only requires the installation of a microphone. So while we are developing the other sensors, we decided to do preliminary experiments based only on working sound sensing, in order to start developing the prognostic and monitoring processing architecture.

The sound sensor, low level processing and feature extraction is presented in fig(4). The signal is captured by a microphone. It drives the input of an amplifier followed by a low pass (7Khz) antialiasing filter. The filter output is connected to one analog input of the transputer network sensor server. Low level sensor processing is done by transputer tasks. Sampled data is first translated from time to frequency domain by the FFT transform task. Then the frequency domain power spectrum data is analyzed and normalized in the following. Low power components are removed to cancel corrupting noise. The remaining spectral components constitute the sensor signal frequency domain description; by this way a list of frequency features is extracted from each sampled data buffer.

T800 performance doing FFT transform is presented in fig(5) b). Some examples of extracted frequency domain features are presented in fig(5) a). The background spectrum corresponds to the machine's idle status; in this case the noise detected originates from the cooling fans.

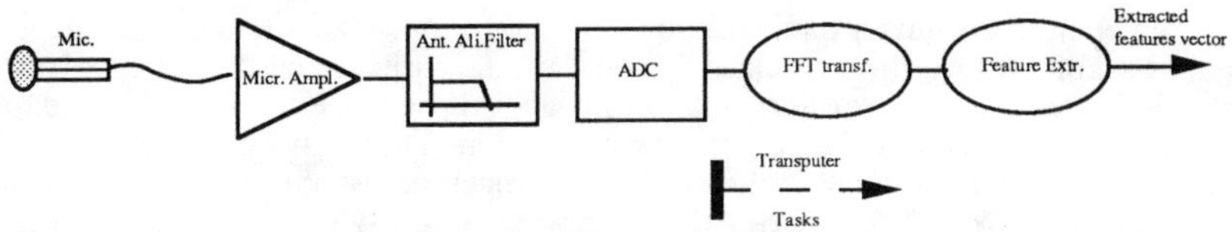

Fig(4)
Sound Sensing and low level processing

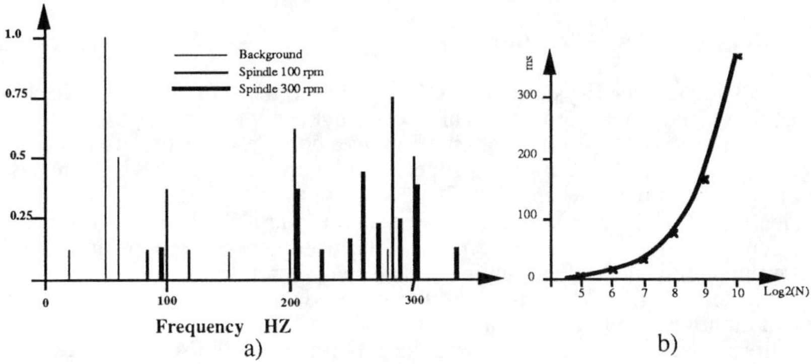

Fig(5)
a) Spectral features for different situations
b) FFT times using one T800 (3LC used)

3.2 Knowledge Base

In a monitoring and prognostic system the knowledge base handles all the information needed. The enumeration of all its aspects is without scope of this paper, but we present how context sensorial data interpretation can be done by using **KB** information. The **KB** is supported by one or more transputers running a Prolog-based interpreted (compiled in a near future) frame engine program. The frames support inherited slots, methods and demons. Other network tasks can consult/transform the information stored in the **KB** by sending proper messages to the frame engine processor and receiving back the result.

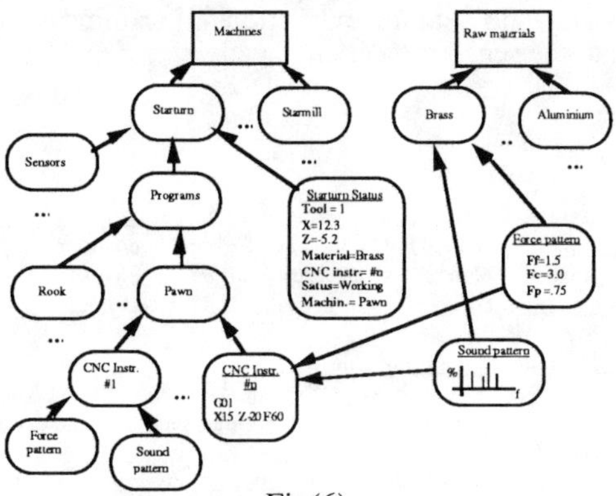

Fig(6)
KB Frame structure

We want to do sensor data verbalization, i.e for each parts that can be machined, we must be able at any time to classify sensor data with verbal quantifiers (normal, abnormal, increasing, decreasing, etc.). To do this we need to associate patterns known as normal behavior to each working state. That can be done using CNC instructions as references because they define the machine working status. This strategy is illustrated in Fig(6), it is a partial **KB** frame structure view. Some slots from the machine status frame are presented, as well as the relation between it and the CNC program under execution. The sensor patterns expected during this status are also exemplified.

3.3 Monitoring and Training

The described system, before being able to monitor a specific machine work, must first be trained accordingly. This means that every time a new program or material is added to the **KB**, a training phase should be done in order to determine the normal behavior patterns. The training is supervised by an expert user while the system gets experience with the new situation. During this phase the monitoring tasks are replaced by learning tasks. New frames are then added to the **KB.** After having been trained, the sensed patterns frames are ready to be used by the monitoring tasks. In our environment it is possible to run in parallel several monitoring/training tasks each for a specific sensor type. At the moment the system is working using only one task to do sound monitoring/training, but as soon as new sensors are available the corresponding monitoring tasks will be added, and the **KB** frames will also be refined to store the data.

Prognosis tasks can also be added, however it requires a different kind of information. The system must be able to analyze sensed data evolution, and of course the **KB** must also contain data related with the past behavior. Because we have started recently and all the necessary sensors aren't yet integrated, we don't have such historical data. We hope do some experiments about prognosis in a close future.

4- Conclusions and Future work

We have presented a sensorial environment and it's integration into the transputer network TNET. The sensor server concept was presented as well as an application example on a laboratory CNC lath machine. Sensor low level processing methods and feature extraction is illustrated. A global **KB** able to support the monitoring environment information and it's access by any task is also presented.

The existing system will be enriched with new sensors and dedicated transputer based acquisition modules, in order to do experiments in a wider sensorial environment. More signal processing and feature extraction algorithms will be developed and integrated according to the sensor processing needs.

References

[1] Steiger-Garção, A.; Barata, M.M.; Gomes, Luís F.S.; Integrated environment for prognosis and monitoring systems support, UNIDO 1st Workshop, Lisbon, September 89.

[2] Manuel M. Barata, José C. Cunha, A. Steiger Garção; Transputer Environment to support heterogeneous systems in robotics, paper presented in Transputer Applications 90 conference, Southampton July 10-13, 1990.

[3] InMos Limited, The Transputer Data Book, InMos, 1988.

[4] Barata M. M., Transputer Based Multiprocessor System, UNL DI 49/88, Lisbon, Portugal.

[5] N. Thrane, The Discrete Fourier Transform and FFT Analysers, Technical Review No. 1- 1979, Brüel & Kjær

[6] DENDFORD Conputerised Machines and Systems, Starturn Programnming Instruction & Maintenance Manual, UK 1988.

[7] Joyce Loebl, Joyce-Loebl Mx Vision System, UK 1986.

[8] 3L Ltd, Parallel C User Guide, Peel House, Scotland.

HAND - EYE COORDINATION FOR FINE ROBOT MOTION

Anatole Lokshin

Jet Propulsion Laboratory
California Institute of Technology
4800 Oak Grove, Pasadena, Ca. 91109

I. Introduction

The future success and wide use of robotic manipulators depends heavily on their ability to operate in a changing and poorly structured environment. This ability can become feasible only if robotic systems can be guided by sensors, such as vision, force/torque, proximity, etc., instead of relying on open loop motion based on a prestored world model.

While all types of real-time sensory information are important for success of intelligent manipulators, effective eye-hand coordination promises the largest pay-off in terms of robot performance and versatility. Despite an enormous interest in developing such facilities, there does not seem to be any practical and effective solution to this problem due mainly to an enormous computational and algorithmic complexity required for real time sensor processing and model matching.

Setting aside a problem of object recognition, which is purely a problem in machine vision, one still has to map object coordinates obtained from vision into manipulator joint angles which define the robot configuration space. This is usually called the hand-eye coordination problem. The usual approach to the problem is to calibrate separately the vision and manipulation system in a common "absolute" frame of reference and then to use the nominal inverse kinematics of the manipulator for vision induced motions. This approach has two major difficulties. First, a vision system has to be calibrated over the total work space. Second, the absolute frame, which is usually quite arbitrary, has to be the same with a high degree of precision for both Robot and Vision subsystem calibrations.

This paper describes work done at the JPL which achieved an acceptable level of hand-eye coordination for fine robot motion without high requirements for vision/manipulation calibration.

II. System Setup Description

The JPL Telerobot System, where the described work has been done, includes several subsystems working cooperatively: a low level mechanization subsystem - Manipulation and Control Mechanization (MCM); a Run Time Control (RTC) subsystem; an Operator Control Station (OCS); an AI Task Planner; and a Sensing and Perception (S&P) subsystem. More detailed description of the Telerobot system and individual subsystems can be found in [Matij], [Kan 1-2], [Tso], [Stone]. Robot Vision is provided by the S&P subsystem, which determines object 3–D positions and orientations by matching a pair of stereo images with pre-stored polyhedral models. A detailed description of the algorithms and special purpose hardware used by the S&P is given in [Gennery].

The Telerobot system has five video cameras which are available for the Sensing and Perception subsystem (S&P). Three cameras are stationary and provide wide angle (F = 12.5 mm) view of the total workspace. Two cameras are mounted on a wrist of a Puma-560 manipulator, called Camera Arm, and can be used to provide a close view stereo (F = 25.0 mm). The stationary cameras are used mainly to acquire and track large objects such as satellite replicas, and do not provide enough resolution for fine motion manipulation. Only the movable close view cameras provide positional information for the objects in a task space. Here, the concern is only with the Puma arm mounted stereo cameras.

III. Extension of the Vision Work Space

The camera model used by the Sensing and Perception Subsystem is composed of four 3-vectors: c, a, h, and v, which are known as the center, axis, horizontal, and vertical vectors, respectively (Fig. 1). These vectors describe the relationship between the 3D coordinates of the Telerobot Reference Frame (used as Absolute Frame) and the 2D image camera coordinates. Vector values are derived from a rather extensive and time-consuming calibration process. As long as the camera is not moved from its calibrated position, then these vectors are sufficient. However, arm mounted cameras need to be positioned anywhere in the working space to get a close view on the particular fine motions. As a result, these four vectors need to be altered to reflect the movement.

For each model, an arbitrarily placed camera reference point which is rigidly attached to the camera is defined. While this additional coordinate frame can be set anywhere, to simplify computations it was chosen to coincide with the wrist origin of the Camera Arm. If this point is located in the Telerobot Reference Frame at position p (a 3-vector) and orientation q (a quaternion) before the camera is moved, then the camera model (c, a, h, v) can be transformed to a new camera model (c', a', h', v') at location (p', q') after the move by the following:

$$\begin{aligned} c' &= p' + R(c - p) \\ a' &= Ra \\ h' &= Rh \\ v' &= Rv \end{aligned} \tag{1}$$

where

$$R = r(q')r^T(q)$$

and $r(q)$ is a function which transforms a quaternion q into a rotation matrix (Fig. 1).

However, a practical application of this algorithm leads to a significant mismatch between the real and S&P perceived absolute object positions of about 2 – 4 cm (Fig.2). In addition, the manipulator has its own absolute positioning errors which are about 0.5 – 1.0 cm. These errors come from several sources: imperfection of manipulator nominal kinematics; uncertainty in the shoulder transform; and errors in the deduced camera models. Some of these errors could be significantly reduced by using more detailed models for arm kinematics [Hayati] and by much more elaborate measurements for the Task Data Base. However, it is our opinion, that an application of computer vision in robotics should lead to inherently robust algorithms which work with nominally known systems operating in an unstructured world.

The next section describes two closely related methods which allow employment of moving cameras by : (a) using vision to determine only relative motions of manipulators toward the target; and (b) calibrating vision system directly in the manipulator's nominal kinematic space.

IV. Relative Motion

Assume that the Vision System has been calibrated in an absolute space once. Then, after moving cameras into position P, an object's position S derived via eq. (1) will have an error $D_{cam}(P, S)$:

$$T^{P}_{cam}(S) = T_{abs}(S)D_{cam}(P, S) \tag{2}$$

We want to position the End Effector at a location A relative to an object located at S_{obj}.

$$T_{abs}(S_{des}) = T_{abs}(S_{obj})A \tag{3}$$

Suppose that a motion based on an *a priori* CAD Data Base brought EE close to the desired position. Suppose that the cameras can be located in such a way that both End Effector and an object can be seen simultaneously. Then, to position an EE at P_{des}, the manipulator should make a relative motion B such that

$$T_{abs}(S_{EE})B = T_{abs}(S_{obj})A \tag{4}$$

Since small relative motions can be usually performed very accurately, the problem of hand-eye coordination will be solved in this case by accurately estimating B from vision information.

A vision based estimate for corrective motion B is

$$\hat{B} = T^{-1}_{cam}(S_{EE})T_{cam}(S_{obj})A \tag{5}$$

or by substituting (2) and (3) into (5)

$$\hat{B} = D^{-1}T^{-1}_{abs}(S_{EE})T_{abs}(S_{obj})DA = D^{-1}BA^{-1}DA \tag{6}$$

since when cameras are not moving

$$D_{cam}(P, S_{EE}) = D_{cam}(P, S_{obj}) = D$$

Eq. (6) shows that this method is robust to translational errors in the camera model but still sensitive to orientation errors.

Indeed, since A is a known transformation we can group BA^{-1} together and denote as X. Then from (6)

$$\hat{X} = D^{-1}XD \tag{7}$$

where X and D are homogeneous transformations. Let X have a rotational part R and translational part p and D has W and d correspondingly. Then [Paul]

$$\hat{R} = WRW^{-1} \; ; \; \hat{p} = WpW^{-1} + W^{-1}d - d \tag{8}$$

Eq.(8) shows, that if $W = I$ then $\hat{R} = R$ and $\hat{p} = p$. In practice p, R, W are small and the method works satisfactory even when the orientational error W is not unity.

V. Camera Calibration in Manipulator Space

By analyzing the sources of errors, one may see that an Absolute Frame used by both Vision and Manipulation Systems is not required for hand-eye coordination. Ultimately we are interested in a mapping between Manipulation configuration space and a cartesian product of two 2-D visual mappings and (p, q) reference point

$$F : R^6 \times R^2 \times R^2 \rightarrow R^6$$

or any known function of those. For example, it seems reasonable to look for a mapping between the manipulators nominal kinematic and vision model of a 3-D object.

As was stated above, an Absolute Frame is used only as a common reference for Vision and Manipulation calibration. Our ultimate goal is to find a direct mapping between manipulator and vision space representation. For this reason, one can use a manipulator itself as a measuring device by setting a shoulder transform to unity $Z = I$ and for each 6-D point assigning its coordinate values via nominal forward kinematics. First, a Vision System is calibrated using a target rigidly attached to an End Effector (grasped). Suppose it is done at a Camera Arm at a position P_0.

$$T^0_{cam}(S_{obj}) = T_{nom\_kin}(S_{obj}) \tag{9}$$

for any object S. Then, at any other point P : $T^P_{cam}(S_{obj})$ is computed via (1) and has an error depending on the camera position

$$T^P_{cam}(S_{obj}) = T_{nom\_kin}(S_{obj})D(P) \tag{10}$$

For a set of reference points $\{P^{ref}_k\}$ (different positions of Camera Arm), let the errors D be measured by comparing T_{cam} and T_{nom_kin} for a known target held by the manipulator according to (10)

$$D(P^{ref}_k) = T^{-1}_{nom\_kin}(S_{obj})T^{P^{ref}_k}_{cam}(S_{obj}) \tag{11}$$

Then, for any other camera position P, object position in nominal kinematic space can be estimated as

$$\hat{T}_{nom\_kin}(S) = T^P_{cam}(S)D^{-1}(P^{ref}_m) \tag{12}$$

where P^{ref}_m is the nearest point in configuration space to P from $\{P^{ref}_k\}$. Substituting (10) into (12) we have

$$\hat{T}_{nom\_kin}(S) = T_{nom\_kin}(S)D(P)D^{-1}(P^{ref}_m) \tag{13}$$

which gives a perfect estimate if $D(P)$ and $D(P^{ref}_m)$ are equal. The proposed technique works very well since it is much easier to determine $D(P^{ref})$ than to perform a full camera calibration. Therefore, a set of $D(P^{ref})$ is determined off-line along a working space volume.

VI. Summary

The described techniques were used at the JPL Telerobot system laboratory. They allowed a reduction in errors in hand-eye coordination by an order of magnitude without any complicated calibration fixtures and sophisticated methods for inverse kinematics solutions. Together with Operator Assistant Vision and Compliant Motion, these methods allow robust autonomous operations in an unstructured environment in the presence of significant CAD Data Base errors. Vision Based Relative Motion together with teleoperation is a connecting link between autonomous free motions and specific task oriented macros based on complaint motion, proximity sensors, and CAD models. A right combination of these technologies allows performance of sophisticated assembly and servicing operations in an unstructured world by using easily calibrated systems and a limited number of robotics primitives.

Acknowledgment.

This work was done at the Jet Propulsion Laboratory, California Institute of Technology under a contract with the NASA.

References

[Paul] Paul R., *Robot Manipulators: Mathematics, Programming, and Control*, MIT Press, Cambridge, Mass., 1981.

[Hayati] Hayati S. and Mirmirani M., "Improving the Absolute Positioning Accuracy of Robot Manipulators", *Journal of Robotic Systems*, vol. II, No. IV, 1985.

[Matij] J. Matijevic, W. Zimmerman, S. Dolinsky, "The NASA–OAST Telerobot Testbed Architecture", *Proc. NASA Conf. on Space Telerobotics*, JPL Publ. 89-7, Pasadena, Jan. 31 – Feb. 2 , 1989.

[Kan 1] E. Kan, *et al.* "The JPL Telerobot Operator Control Station : Part I – Hardware". *Proc. NASA Conf. on Space Telerobotics.* JPL Publ. 89-7, Pasadena, Jan 31 – Feb. 2 , 1989.

[Kan 2] E.Kan, *et al.*, "The JPL Telerobot Operator Control Station : Part II – Software". *Proc. NASA Conf. on Space Telerobotics.* JPL Publ. 89-7, Pasadena, Jan 31 – Feb. 2 , 1989.

[Stone] H.Stone, *et al.*, "Experiences with the JPL Telerobot Testbed : Issues and Insights." *Proc. NASA Conf. on Space Telerobotics.* JPL Publ. 89-7, Pasadena, Jan. 31 – Feb. 2 , 1989.

[Tso] K.S.Tso, P.G.Backes, T.S.Lee, S.Hayati, "A Multi-Arm Tele/Autonomous Executive System", *Proc.* 3rd *ISRM, July 1990, British Columbia, Canada.*

[Gennery] Gennery et al., "Sensing and Perception Research for Space Telerobotics at JPL." *IEEE Proc. on ICRA,* March 1987, Raleigh, NC.

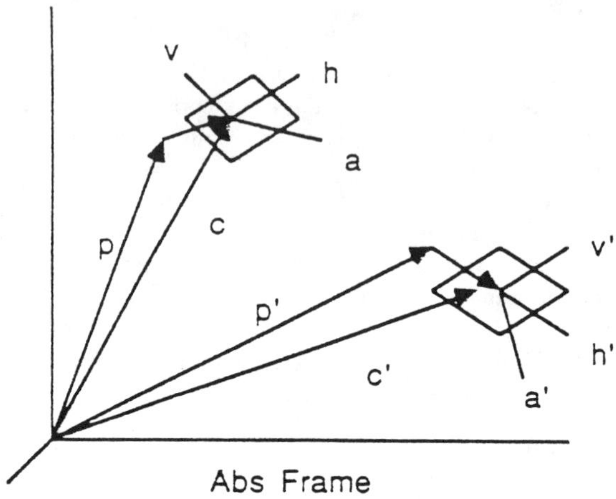

Figure 1

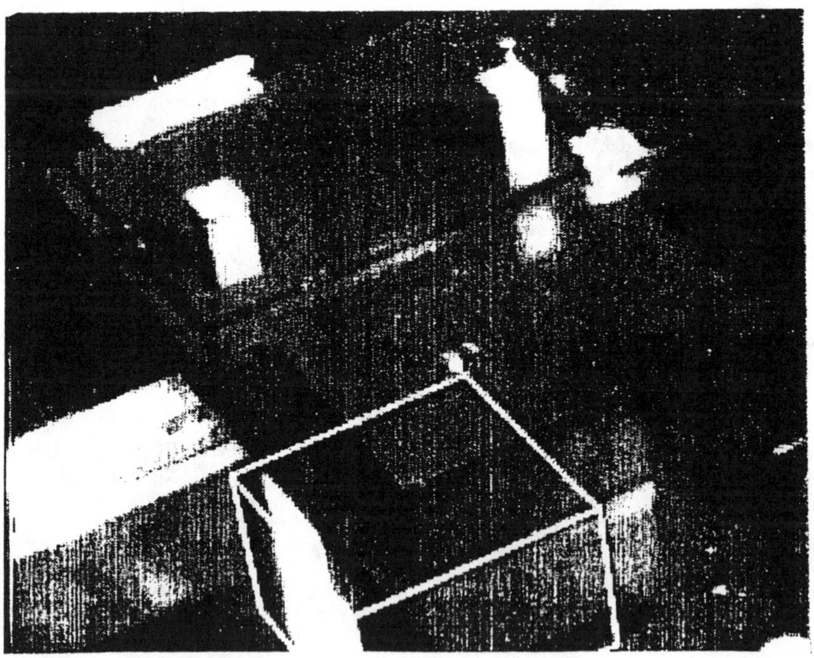

Figure 2

ROBOTIC ARM CONTROL SYSTEM

A. Śmiarowski, Jr. J. N. Anderson
Center for Manufacturing Research and Technology Utilization
Tennessee Technological University
Cookeville, Tennessee

ABSTRACT

Fast digital controllers are essential for robotics control research. This paper proposes a practical computer architecture based on off-the-shelf parts. Although the controller was designed to implement the Distributed Sampling Control concept as applied to model-based control, it may readily be used for any other control method. Because of incorporating a force/torque feedback, the Robotic Arm Control System (RACS) is a flexible testbed for further research in the area of force/position control. The fast computing rate allows testing complex control schemes, like the Computed-Voltage method, in a real-time. The RACS has been implemented in a Unimation VAL-II MkIII controller replacing the existing low-level servo controller.

INTRODUCTION

Many robotic controllers have been proposed but few implemented. Quite frequently, proposed architectures exploring the depths of possible parallelism and/or pipelining, were simply too expensive. But technological advances have expanded the choice of computers that can provide enough power to handle complex control algorithms. The proposed architecture is built around two digital signal processors (DSPs). These third generation, fast floating-point devices have hardware intensive solutions allowing unprecedented computer power on a single chip.

Three goals are achieved by the controller described in this paper. First, the controller can implement the Distributed Sampling Control (DSC) method. As applied to model-based control, this method uses the Newton-Euler recursive equations to compute the model, and hence, processes input and output data at different instants of time [1,2]. Second, the controller has to be flexible enough to provide a testbed for further research, particularly in the area of Reflex-Action Position (RAP) control and multiple arm control [3]. Speed is of prime concern, so that more complex schemes like Computed-Voltage control [4] can be tested. Third, the controller must be simple and inexpensive so that it may be used in an industrial environment. Once a control technique is developed and checked, the system may be stripped of R&D nonessentials and supplied as a plug-in package replacing the existing low-level servo controller.

The paper is organized as follows. The RACS architecture is outlined first followed by more detailed description of the system's modules. Then the choice of the processors driving two main controller units is explained. The paper is wrapped up with calculations of the control sampling rate.

ARCHITECTURE OUTLINE

The basic architecture of the DSC controller was presented in [1]. This architecture is adapted here to the target system; in our case, to the VAL-II MkIII controller. In order to be general, the design should assume some level of independence from the target system. This requirement is achieved by introducing a private internal bus, the T-bus, as shown in Figure 1.

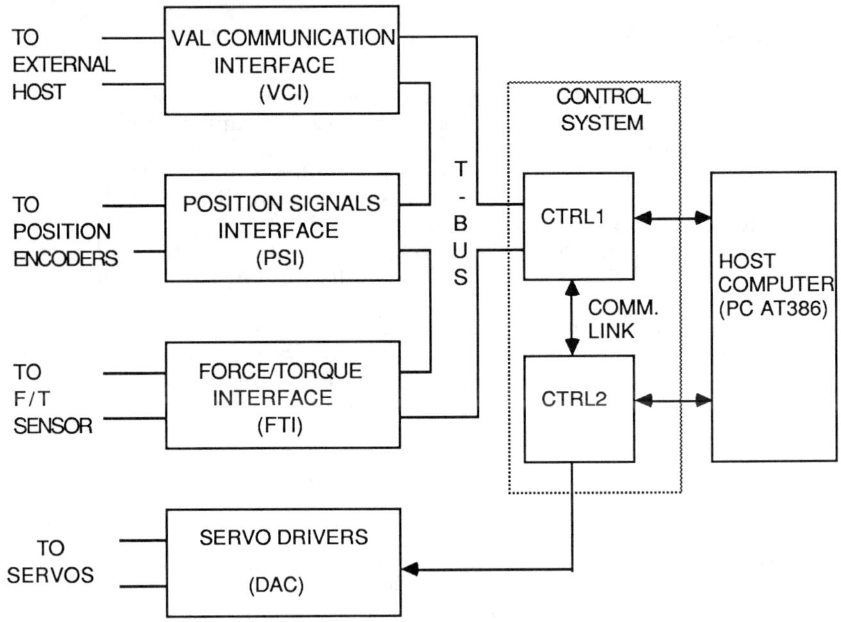

Figure 1. RACS Block Diagram.

The main communication artery of the RACS is the T-bus (a simple, non-standard design), which connects the control system to real world interfaces. This modular design makes the system's core virtually independent of the host computer and external sensors. Command signals driving the target plant are output as 16-bit words to the DAC Interface, which in turn provides signals compatible with the PUMA 562 servo boards. Thus the main control unit may be transported to control any other system that can supply its inputs in accordance with the T-bus protocol and where the outputs can be adapted to its own servo drivers.

The control system is a master on the T-bus and all the interfaces are in a slave (actually a passive) mode. Any communication with the external world is via these interfaces or via dedicated communication ports. The T-bus is used as a memory extension, thus decreasing communications overhead. Higher level commands and feedback from sensors are provided to bus buffers, which are accessed by the control system whenever relevant data is needed. This allows a simple bus operation (no need for handshaking) and synchronization of all the operating components.

The RACS may operate in two modes: VAL and independent. In the VAL mode the MkIII controller provides the commands generated by the higher level control algorithms of VAL. In the independent mode the higher level controlalgorithms are implemented on the host computer. This feature makes the RACS a very flexible research tool.

DESCRIPTION OF THE FUNCTIONAL BLOCKS

The RACS has 5 basic modules: 4 interfaces and a main control system (see Figure 1). The three interfaces VCI, PSI, and FTI adapt the real world signals to the internal T-bus structure. The fourth interface, the DAC, converts command digital signals to analog voltages driving the power amplifiers. The servo amplifiers of the Unimation controller [5] are used to interface the system's analog outputs with plant.

VCI (VAL Communication Interface)

The VCI module (Figure 2) contains the handler for communication with the higher control level. This handler accepts interrupts, arbitrates them, reads/writes data from/to the external bus (J-Bus for VAL). It loads setpoint data and commands to the appropriate output registers on the T-bus. This module also communicates with the main control system, passing data and (arbitrated) interrupts. Optionally, it may also run the interpolation chores for the control system.

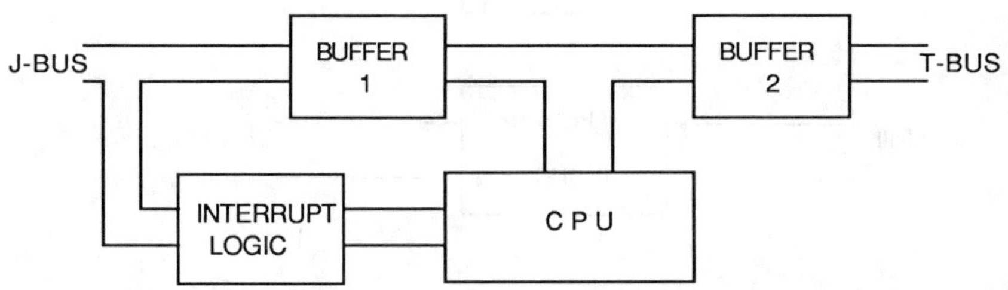

Figure 2. VAL Communication Interface.

PSI (Position Signals Interface)

The PSI module (Figure 3) provides the position feedback data to the control system. The main components of the PSI module are the absolute position counters (in the Encoder, Counter, Buffer block) connected to the six incremental encoders located on the joint motor shafts. They are reset from the control system at the predefined zero position and, afterwards, they keep track of the absolute position. This relieves the control system software from tracking absolute positions via relevant operations on the index registers. Internal logic secures also the proper count by testing the index pulses and detecting encoder errors.

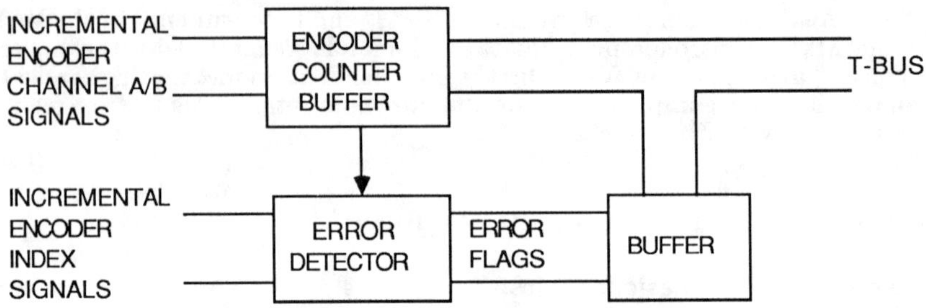

Figure 3. Position Signal Interface.

FTI (Force/Torque Interface)

The FTI module (Figure 4) provides the force/torque feedback data required by the control system (used for load-decoupling in RAP control). This interface allows proper communication between the control system and force/torque sensor hardware via providing the handshaking required by the force/torque sensor manufacturer.

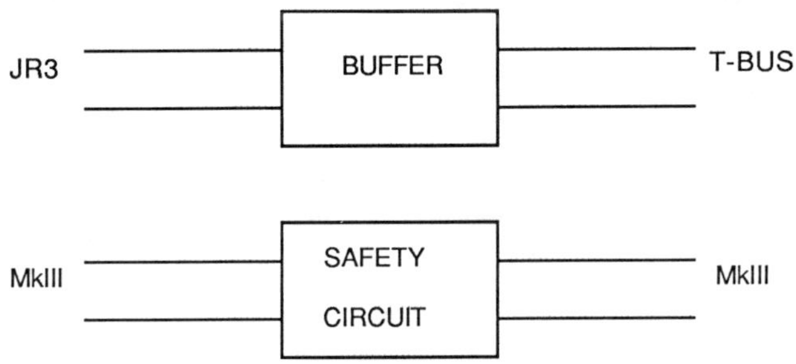

Figure 4. Force/Torque Interface.

The FTI module also hosts a safety circuit to turn on/off the power for the arm motors. There are voltage comparators equipped with adjustable reference sources that can be compared with the signals coming from the joint potentiometers (equivalent to the joint positions). They allow the user to define the safe working space. When working with the VCI (VAL mode) this feature parallels one already provided with the MkIII controller. In this case the power may be turned on only when both the MkIII and the FTI safety circuit allow it. However, when working in an independent mode, the safety circuit takes over the control of the power on/off function.

Control System

The Control System module consists of two identical controllers — Spirit-

30 application development systems from Sonitech International, Inc. (see details and figures in [6]). Controller 1 drives the T-bus, accepts host commands, and communicates data to the host via the VCI or EHI interface or via the host computer communication port. It preprocesses raw data and passes prepared information to Controller 2. This Controller runs the selected control scheme and generates the control command signals for the target plant. These signals are output to the DAC board, which in turn drives the servo amplifier boards. If the selected control scheme is model-based control, the plant model resides in the Controller 2 memory.

Controllers 1 and 2 implement two stages of a conceptual architecture of the DSC [1]. The output stage (signal conversion), i.e., the actual torque generation and application, occurs outside, in the servo boards and actuators.

DAC (Digital to Analog Converters)

The DAC module (Figure 5) accepts the 16-bit words computed by the Control System and converts them to analog voltages. These voltages must be compatible with the Unimation servo boards. In essence, this module emulates the six separate output stages located on the Unimation's Digital Servo Controller cards. The DAC module, however, has better precision, since Unimation's design uses 12-bit DACs.

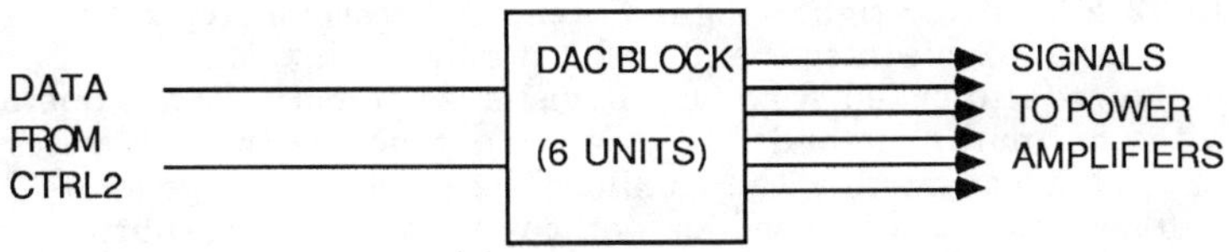

Figure 5. DAC Module.

CHOICE OF THE PROCESSORS

In view of the stiff timing requirements of a real-time robotic controller, speed of computation is of prime concern. Time performance of the RACS depends on the speed of the main control system, assuming that the T-bus allows fast enough access to external resources to be considered transparent. Therefore the selection of processors used for this block is the most critical choice in the design process.

Among many advanced computer technologies available to a system designer, the third generation DSPs offer great advantages in robotics applications. The DSPs are specifically designed to accomodate a sum-of-products type of operation — the most frequent operation in digital signal processing. It happens that many modern robotic control methods also use algorithms which depend heavily on this kind of operation.

Choice of the TMS320C30

There are many factors to look at when trying to decide which DSP to use for a design. From the hardware and system integration side, the most important ones are speed, precision, power consumption, package size, and cost. Considering the choice from a research and development point of view, some important factors are ease of programming, availability and cost of development tools, portability, and vendor's technical assistance. After comparing available chips it was decided to base the design upon the Texas Instruments' TMS320C30 processor — currently the best supported DSP.

The TMS320C30 is a third-generation, high performance, 32-bit CMOS machine in the TMS320 family of single-chip DSPs [7]. Its 60 nsec cycle time, with one instruction per cycle operation and internal parallelism, allows an optimized code to achieve up to 33 MFLOPS (millions floating-point operations per second).

Many of the functions of the TMS320C30 are implemented in hardware, which, combined with the large address space and peripheral interfaces, lends itself to fast general-purpose applications. Internal busing boosts processor performance because of the parallelism. There are separate program buses, data buses, and Direct Memory Access (DMA) buses which allow for concurrent program fetches, data fetches, and DMA access. The TMS320C30 also supports two external interfaces: the primary bus and the expansion bus — both consisting of 32-bit data lines and a set of control signals. There are also internal and external interrupts, a reset signal and I/O programmable flags.

The CPU has a floating-point/integer multiplier in parallel with an arithmetic logic unit (ALU) so that one instruction can execute a multiply and an ALU operation in a single cycle. The ALU includes a barrel shifter that can shift up to 32 bits left or right. Eight Extended Precision Registers support operations on both 32-bit integer and 40-bit floating-point numbers. Two Address Generators (also called Auxiliary Register Arithmetic Units) operate in parallel with the multiplier and ALU, and can generate two addresses in a single cycle. They can modify the Auxiliary Registers which generate 24-bit addresses. They may also be used as loop counters or general-purpose registers. The remaining 12 Control Registers of the CPU may perform a variety of functions like data page pointer, stack pointer, index register, interrupt and I/O registers, or repeat counter. Thus the CPU may readily support a high-level language through this register-based architecture.

The TMS320C30 has a total address space of 16 million 32-bit words. Data, program, and I/O ports are mapped into this space. On-chip memory has two RAM blocks ($1K \times 32$-bits each) and one ROM block ($4K \times 32$-bits). Each of these blocks supports two accesses in a single cycle. This solution together with multiple buses allows for concurrent program fetches, data read/writes, and DMA operations. A small 64×32-bit instruction cache stores frequently used blocks of code so that slower external memories can be used without seriously compromising performance.

The DMA has access to the entire memory space without interfering with the CPU. The DMA's Address Generators keep track of source, destination, and transfer count. The Control Registers control the state of the DMA controller and monitor its status. The DMA releases the CPU from input/output functions and allows for interfacing the processor to slow external memories and peripherals.

All the on-chip peripherals communicate with the CPU through a dedicated peripheral bus. They are controlled through memory mapped registers. The two independent Serial Ports can be configured to transfer 8, 16, 24, or 32 bits words with either an internal or external clock. A special handshake mode guarantees synchronization. The two timers can signal either specified intervals or count external events and interrupt the CPU after a specified number of events.

CONTROLLER PERFORMANCE

The TMS320C30 processor is a powerful machine designed for computationally intensive applications. It is still general-purpose enough so that it can readily be used to drive external peripherals. The Spirit-30 boards (Controllers 1 and 2) may be used on a variety of host computers. In this project an IBM PC-AT hosts both boards. Software support consists of Sonitech's debugger running under MS-DOS. Application programs are written in "C" or TMS320C30 assembly language. The "C" compiler and the assembler/linker are both available from Texas Instruments.

TABLE 1. TIMING CHART FOR DISTRIBUTED SAMPLING.

J_j - calculation of the position control law, rotational matrix and joint variable derivatives for joint j;
F/T - force/torque data acquisition;
F_j - forward recursion computations for joint j;
B_j - backward recursion computations for joint j;
u_j - generating voltage for actuator of joint j.

TIME		$\longrightarrow$												
CTRL1	J_1	J_2	J_3	J_4	J_5	J_6	F/T							
CTRL2		F_1	F_2	F_3	F_4	F_5	F_6	B_6	B_5	B_4	B_3	B_2	B_1	
OUTPUT									u_6	u_5	u_4	u_3	u_2	u_1

Performance of the RACS is evaluated for the RAP controller using DSC. The sampling period is broken down in Table 1. Calculations J_j (Controller 1) consist of numerical differentiation of the joint j variable and compution of the PD control law and terms of the rotational matrices. The actual position is fetched from an appropriate absolute position counter at the beginning of the subinterval and outputs are sent to Controller 2 towards the end of the subinterval. During the F/T subinterval, Controller 1 acquires the current readings from the wrist force/torque sensor and passes them down to Controller 2. Controller 2 computes Newton-Euler equations of motion — F_j a forward recursion and B_j a backward recursion.

TABLE 2. RACS TIMING (in μs).

TIME		$\longrightarrow$												
CTRL1	10	10	10	10	10	10	10							
CTRL2		15	15	15	15	15	15	15	15	15	15	15	15	
OUTPUT									10	10	10	10	10	10

Table 2 shows estimated timing of the subintervals shown in Table 1 for a general-purpose 6 DOF robot including communications delays. A liberal margin of 50% is added for computation overhead. The total time comes from summing up the longer subintervals of the parallel operations. In this case it is $12 \times 15\mu sec + 10\mu sec + 10\mu sec = 200\mu sec$. Serial link communication occurs in parallel with other processes but at a relatively slow rate (around 8 Mbits/sec). It may eventually be supplemented with a direct access from Controller 1 to the memory of Controller 2 with the appropriate arbitration. On the other hand, the programs may be optimized to take a maximum advantage of the TMS320C30 internal parallelism. Also the Newton-Euler equations can be customized for the target robot.

CONCLUSIONS

A fast testbed for robotic control research is presented. Because of the feedback from the force/torque sensor, many force and hybrid position/force control methods may be tested. In particular, the RACS readily implements RAP control.

Performance evaluation data given in Table 2 estimates the total sampling period of the 6 DOF general-purpose robot to be 200 μsec. Since the RACS structure is based on the DSC concept, the computational delay for the joint 6 is 55 μsec (the overall sample period remains the same for all joints).

One of the advantages of the RACS is that it uses off-the-shelf processors and hence is widely available. All the RACS modules may be put on plug-in boards which replace the existing controller for the Puma 562. Once the control method is established, the boards may be packaged with only the necessary memory and peripherals so that the RACS can be used in an industrial environment. For the research community it is important that the RACS has a full software support. TI's "C" compiler and assembler/linker package provide this support.

REFERENCES

1. Anderson, J. N. and A. Śmiarowski, "A Specialized Parallel Architecture for Robotic Controller," in Robotics and Manufacturing, (M. Jamshidi et al. eds.), ASME Press, 1988.
2. Anderson, J. N. and A. Śmiarowski, "A Computer Architecture for Robotic Controllers," in Proc. 31st Midwest Symposium on Circuits and Systems, August 1988, pp. 656–659.
3. Anderson, J. N. and M. E. Pittelkau, "Reflex-Action Position Control and Two-Arm Load-Sharing Force Control for Robotic Manipulators," Technical Report No. MCTR 1-88-1, Tennessee Technological University, Cookeville, Tn, January 1988.
4. Anderson, J. N. and A. Śmiarowski, "Model-Based Control of Robots Including Actuator Dynamics," in Proc. of the 22nd Southeastern Symposium on Systems Theory, Tallahassee, FL, March 1989.
5. PUMA 552/562 Equipment Manual 398H1A, Unimation Inc., January 1987.
6. Spirit-30 System Technical Reference Manual, Sonitech International Inc., 1989.
7. Third-Generation TMS320C30 User's Guide, Texas Instruments, 1988.

A Real–time Control Architecture for a Robotic Filament Winding System

Edwin Castro, Alan Desrochers, Sanjeev Seereeram, Jaidev Singh, John Wen

Department of Electrical, Computer and Systems Engineering, Rensselaer Polytechnic Institute, Troy, NY 12181

Keywords. Real Time Control, Robotic Filament Winding, Hierarchical Control, Trajectory Planning, Multi–Processing

Abstract. This paper describes an architecture that was designed and implemented for the path–planning and real–time control of a robotic filament winding system. A Unimate 6000 robotic system and a filament winder are synchronized and coordinated by a multi–processor hierarchical control system based on RCS–II which was designed by the National Institute of Standards and Technology (NIST). The performance of this system significantly improves over a previous implementation where the robot and winder are not closely synchronized.

1. Introduction

Filament winding is a manufacturing term given to a fabrication technique for forming reinforce plastic parts of high strength and light weight. It is made possible by exploiting the remarkable strength of continuous fibers or filaments encased in a matrix of either organic or inorganic resinous material [1].

Conventional filament winding, where the filament is wrapped around a rotating mandrel, has been very successfully used and matured to a high level of sophistication. Generation of complex winding patterns and precise filament placement by using winding machines having several degrees of freedom has increased a great deal over the past 25 years [2]. However, the inherent limitation of the basic filament winding technique as it is commonly used today is in its ability to produce axisymmetric structures. Any structure that does not possess a longitudinal axis of symmetry, like L–shaped and T–shaped pipe fittings, cannot be wound using the conventional technique.

Robotic filament winding is one solution to this problem. This is a fairly new idea that is still in its early stages of development. Unlike conventional filament winding where the mandrel rotates and the feeding head traverses the length of the mandrel laying down the fiber, in robotic filament winding the concept of a stationary mandrel is used. In this technique, as the name implies, a robotic system is used to position and orient the specially designed end effector containing the payout eye while the mandrel remains stationary in the task space. The winding pattern for the mandrel is transformed into a trajectory for the robot, and since the mandrel is not rotating, the entire burden of the correct placement of the filament and its wrapping lies on the robotic system. Extremely complex shapes can be wound using this method, with no limitations being placed on the complexity of the structure being fabricated, except that the generation of the winding pattern for the robot be feasible.

The basic principle of robotic filament winding has been applied by the Automated dynamics Corporation (ADC), a company located in Troy, NY. ADC designs and produces advanced robotic filament winding systems used for manufacturing complex–shaped composite structures. The physical system that performs the actual winding of the structures consists of two main components — a robotic system and a patented winder. The winder is attached to the end of the robot arm, which provides a mobile base for it. The robot moves the winder to the desired position and orientation, while the winder rotates in order to wrap the filament around the mandrel. This system is described in detail in [3].

ADC was having problems with the synchronization between the robot and winder motions and was operating the filament winder at speeds well below the theoretical capability of such a system. Fig. 1 shows the original organization of the ADC filament winding system. Path planning is done separately for the winder and the robot. The execution of the path control program is independent from each other. Once the robot (or winder) reaches the set point, it comes to a complete halt and sends a message to the PC indicating this condition, and wait for the arrival of the other mechanism. As a result of the lack of synchronization between the two mechanisms, there is jerky motion of the system which affects both the overall winding speed and the winding quality. Furthermore, due to the potential buildup of tension in the filament, constant operator supervision is required.

This paper presents a solution to the above problem. The modified system is called the Robotic Filament Winding System (RFWS). The robot and the winder are now treated as a single mechanism and the path planning is done for the position and orientation of the payout eye of the winder. Tighter synchronization is also achieved by sending set points to the robot and the winder every 28ms. High level path planning and error checking are also incorporated into this multi–processing architecture to achieve a self–contained and robust system. This system is implemented on the RCS–II hierarchical real time control designed by NIST [4].

In this paper, we will present the RFWS architecture in Section 2. In section 3, we will discuss the path interpolation strategy and high level winding path generation. Winding for the T–shaped mandrel will be highlighted; as far as we know, this is the first complete geodesic solution to this problem. The inverse kinematic for the 5–DOF Unimate 6000 and the 1–DOF winder will be discussed in Section 4.

2. Architecture

The architecture of the Robotic Filament Winding System (RFWS) shown in Fig. 2 follows the basic NIST concept of hierarchical control with some modification to tailor it to the filament winding system. The main part RFWS is hierarchical in nature. The user interface communicates with the PC which performs high level winding path generation. The gross winding path description is sent to the path interpolator which generates a position/orientation set point for the winder payout eye every 28ms. This set point is then converted into a winder set point and a robot end flange position/orientation set point through the kinematics routine. The low level communication routine then sends the set points to the appropriate servo controllers. The gross winding path can also be learned via teach pendant positioning. Error checking is performed in all levels.

The RFWS hierarchy is partitioned into the following levels: TASK, TRAJECTORY, KINEMATICS and COMMUNICATION. The role of each level is briefly explained below:

(1) The TASK level serves as the top level planner and interfaces with the operator and possibly other computers. It decomposes high level commands and translates them into gross trajectory specification. The output of this level is a sequence of position and orientation set points for the payout eye of the winder with specification of the traversal time or velocity between consecutive points

and the amount of blending (see the Trajectory Planning Section) between motion segments. This sequence of set points is called the gross knot points.

(2) The TRAJECTORY level takes the gross knot points and convert them into 28ms position and orientation set points via a straight line interpolation algorithm and parabolic blending.

(3) The KINEMATICS level converts each position and orientation set point for the payout eye to set points for the winder angle and position and orientation for the robot end plate, by using an analytic inverse kinematics algorithm (see the KINEMATICS Section).

(4) The COMMUNICATION level talks to the winder servo controller and the VAL II controller via serial lines. It ensures the proper protocol are followed in each case, and also keeps track of the timing. The external alter feature in VAL II is used to directly command the robot in its tool coordinate.

In the robot control application with RCS–II at NIST, the slave interface is used to command VAL–II to following the specified *joint level* path. Since the task specification is typically in the tool frame, this necessitates an additional level of inverse kinematics done to the robot joint angles. We elected to use the external alter feature which requires the tool set points for the robot end plate. Therefore, in effect we are using the LSI–11 processor in the UNIMATE controller to perform the inverse kinematics. This saves development time and further enhances the parallelism in our architecture.

Note the last three level operate in real–time in the sense that they must execute within 28ms. The top level is performed off–line and is generally the most time consuming portion and it is highly dependent on the complexity of the mandrel geometry. We are currently transferring this level to a PC which will compute the winding path off–line based on a CAD model of the mandrel and also provides visualization before the actual winding.

In contrast to the standard hierarchical control architecture such as RCS–II, there are also two multi–level functions in RFWS. The first is the error checking and correction function which has the following roles:

(1) Gross knot points are checked to see if the trajectory is feasible in terms of robot capability (joint stops, velocity limit, acceleration limit) and possible collision with the environment (such as with robot itself, mandrel, mandrel support etc.).

(2) Error in communication with robot through external alter and with the winder controller.

VAL–II itself also performs error checking, for example, if the tracking error exceeds the tolerance in 28ms, then an error flag is sent to RCS–II and the robot will stop. Recovery of off–line error checking basically prompts the operator to replan the path to conform with the constraints. Real–time error will cause the winding to immediately stop with possible restart from the same point later.

The second multi–level function is the TEACH function. For a complex shaped mandrel, frequently all or part of the winding path needs to be taught by manually positioning the arm to a desired configuration. The actual positioning is done by the RCS–II teach pendant which contains all the Unimate teach pendant features and also controls the winder. Upon positioning the payout eye at a desired configuration, the position and orientation of the robot end plate and the winder angle are stored as a gross knot point.

The heart of the RFWS is the RCS–II multi–processing system. RCS–II is a hierarchically structured system for the real–time control of automated systems. It supports multiple 8086/8087 single board floating point computers on the MULTIBUS. The operating system is based on a compiled version of the FORTH programming language. This operating system supports multiple processors, inter–processor communication, background tasking, and a set of mid–level high speed compiled macros called SMACROS. The RFWS is implemented on three processors with communication through the shared memory.

Comparing the RFWS architecture with the previous ADC architecture, we note the following improvement of the new scheme:

(1) In the previous implementation, path planning is done separately for the robot and the winder. In RFWS, this is done directly for the payout eye.

(2) In the old implementation, gross set points generated are for motions for the robot and winder to be completed within a certain time period (usually several seconds). Low level VAL controller and winder servo controller operate independently in this interval to reach their respective set points. Hence, coordination is only done infrequently at these gross set points. In RFWS, synchronization is done every 28ms (this is limited by the external alter feature in VAL II), which is 200–800 times more often than before.

Due to this improvement, RFWS results in greater control of the filament and produces smoother, faster, and more coordinated motion of the entire system.

3. Planning Algorithm

The planner consists of two levels. The top level planner generates a path for the surface of the mandrel. The criterion is purely based on the mandrel geometry and coverage requirement. The bottom level planner converts the gross path from the top level into a real time trajectory and transmit the set point to the robot and the winder every 28ms. This section describes the algorithms that are used for each of these two levels.

For a stable wind, ideally the winding path should follow the geodesic (shortest path between two points on a curved surface) on the mandrel surface. For a cylinder, this is simply a helix. For more complex axisymmetric parts, calculus of variation can be used to find the geodesic paths [**5**]. When a part is not axisymmetric, such as an L–shaped or T–shaped part, the problem is more challenging. For an L with a smooth bend, this problem can still be intuitively solved by using the cylinder algorithm (i.e., helix) with a bent center axis. This idea unfortunately does not apply to T's. Indeed, automated winding of T's has never been successfully implemented before. We have recently proposed a stable algorithm which is currently being applied for the winding of a T–shaped pipe fitting. The basic idea is now briefly described. A T consists of the adjoining of two cylinders for which we know the geodesic path as helix. If two geodesic paths, one on each cylinder, can be smoothly (same tangent at the interface) connected together, then the resulting path remains a geodesic for the T. Hence, a simple geodesic winding algorithm can be constructed by winding helix for the top of the T and when the helix intersect the stem of the T, it is splined onto a helix on the stem cylinder, which satisfies the tangency matching condition. This approach can also be generalized to parts that are the union of several axisymmetric parts.

After the top level planner generates a coarse set of set points (called the gross knot points) based on the mandrel geometry. A real time path planner then generates the fine knot points for the robot and the winder every 28 ms, by interpolating between the gross knot points. We used the standard straight–line algorithm with parabolic blend [**6**]. Either the length of time for the segment or the velocity can be specified by the user.

For a smooth operation in between the straight line segments, a parabolic path is constructed to smoothly (i.e., continuously in both position and velocity) join the segments together. The amount of blending is user specifiable. The beginning and ending segments are smoothly blended into a zero velocity segment.

4. Kinematics

The external alter feature in VAL II only accepts set points in the Cartesian coordinate for the end plate of the 5–DOF Unimate 6000 robot arm. Since the trajectory planner provides the position and orientation (in terms of the unit quaternion) set points for the payout eye, an inverse kinematics calculation needs to be performed to generate the winder angle, defined as θ_6, and the position and orientation for the robot end plate. The algorithm for this computation is presented in this section.

We use the standard Denavit–Hartenberg convention for the link coordinate frame assignment (see Fig. 3). The interlink homogeneous transformation is given by ${}^{i-1}A_i$:

$$ {}^{i-1}A_i = \begin{bmatrix} \cos\theta_i & -\sin\theta_i \cos\alpha_i & \sin\theta_i \sin\alpha_i & a_i \cos\theta_i \\ \sin\theta_i & \cos\theta_i \cos\alpha_i & -\cos\theta_i \sin\alpha_i & a_i \sin\theta_i \\ 0 & \sin\alpha_i & \cos\alpha_i & d_i \\ 0 & 0 & 0 & 1 \end{bmatrix} $$

where θ_i, α_i, a_i, d_i are the standard Denavit–Hartenberg parameters and $i = 1, 2, \ldots, 6$.

The homogeneous transformation between frame $i-1$ to the last frame, frame six (at the payout eye), is

$$ {}^{i-1}T_6 = {}^{i-1}A_i \, {}^{i}A_{i+1} \ldots {}^{5}A_6 \quad . $$

In this convention, the base to payout eye transformation is ${}^{0}T_6$. Every 28 ms, the trajectory planner provides a set point which can be converted into ${}^{0}T_6$; write it in the form of

$$ {}^{0}T_6 = \begin{bmatrix} n_x & s_x & a_x & p_x \\ n_y & s_y & a_y & p_y \\ n_z & s_z & a_z & p_z \\ 0 & 0 & 0 & 1 \end{bmatrix} $$

Then from the inverse kinematics, the winder angle can be solved [7]

$$ \theta_6 = \arctan2(s_x, -n_x) \quad . $$

The set point for the robot can be extracted from ${}^{0}T_5$ which can be obtained from ${}^{0}T_5 = {}^{0}T_6 \, {}^{6}A_5$, where ${}^{6}A_5$ is given by

$$ {}^{6}A_5 = \begin{bmatrix} \cos\theta_6 & \sin\theta_6 & 0 & -a_6 \cos\theta_6 \\ -\sin\theta_6 & \cos\theta_6 & 0 & -a_6 \sin\theta_6 \\ 0 & 0 & 1 & -d_6 \\ 0 & 0 & 0 & 1 \end{bmatrix} $$

The last step is to convert ${}^{0}T_5$ to a form required by VAL II which is (x, y, z, O, A, T), where (O, A, T) are the Euler angles [8].

5. Conclusion

This paper presented a real time control architecture for robotic filament winding. The heart of the system is the multi–processing RCS–II designed by NIST. The winding operation is partitioned into high level winding path generation, trajectory interpolation, inverse kinematics, and low level communication. We have successfully integrate all the components together for an integrated control of the robot and the winder to achieve smooth motion at the payout eye. At the present, several cylinders have been wound, and the all–geodesic algorithm for T's described in the trajectory planning section is under implementation. Future goal includes a front end CAD based user interface and the incorporation of collision avoidance in the winding path planning stage.

Acknowledgment

This project was supported by the National Institute of Standard and Technology through the North East Manufacturing Technology Center at Rensselaer Polytechnic Institute. The authors would like to thank Ron Brandow, Albert Santos, and Gary Wager for helpful technical discussion and assistance.

Bibilography

[1] Rosato, D.V. and C.S. Grove, Jr., *Filament Winding: Its Development, Manufacture, Applications, and Design*, Interscience Publishers, New York, NY, 1964.

[2] Wood, A.S., "The Filament–Wound Car Frame and Other Practical Miracles," Modern Plastics, July, 1984. pp.62–65.

[3] Bubeck, K.B., "Robotic Winding: Continuous Fibers In Complex Shapes," SME's Quaterly on Composite Technology, Vol.3, No.4, 1987. pp.2–5.

[4] Leake, S.A. and R.D. Kilmer, "The NBS Real–Time Control System User's Reference Manual," National Institute of Standarrds and Technology, Gaithersburg, MD, Oct., 1988.

[5] Soanes, R.W., "Mathematical Aspects of the Off–Line Programming of Filament Winding Machines for General Surfaces of Revolution," ARCCB–TR–88036, US Army ARDEC, Benet Laboratories, Watervaliet, NY, Sept., 1988.

[6] Taylor, R.H., "Planning and Execution of Straight Line Manipulator Trajectories," IBM J. Res. & Dev., 1979. pp.253–264.

[7] Singh, J., *A Hierarchical Architecture in Computer Integrated Manufacturing for Real–Time Control of a Robotic Filament Winding System*, M.S. These, ECSE Dept, Rensselaer Polytechnic Institute, Troy, NY, March, 1990.

[8] "Unimate Industrial Robot 6000 Series Equipment Manual for VAL II Operating Systems," Unimation Inc., Danbury, CT, Dec., 1986.

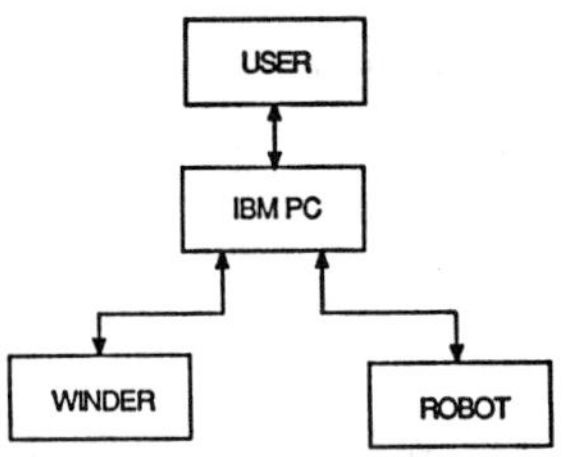

Fig. 1 Existing ADC Architecture

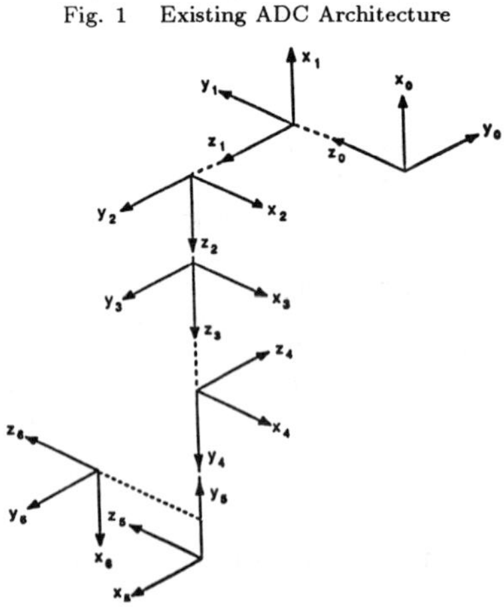

Fig. 3 RFWS Frame Assignment

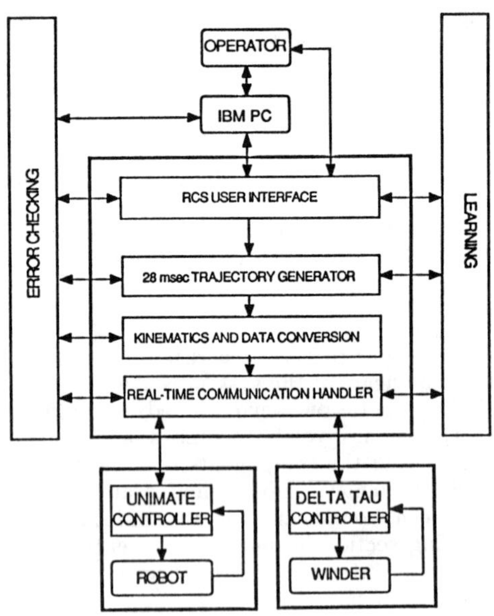

Fig. 2 RFWS Architecture

Real Time Control Under UNIX for RCCL

John Lloyd and Mike Parker
McGill University, Research Center for Intelligent Machines

Gaylord Holder
University of Pennsylvania

Abstract

RCI (Real-time Control Interface) is a C/UNIX package which gives a programmer the ability to develop real-time control applications within a UNIX environment. It allows functions linked into a standard UNIX program to be executed as real-time "tasks", either on the UNIX (host) CPU or on an auxiliary CPU connected to the host by shared memory. RCI has been implemented on MicroVAX-II and Sun-4 systems, and is used as a platform for the robot programming library RCCL. The paper outlines the system's design and discusses aspects of its implementation.

1. Introduction

This paper describes RCI (Real-time Control Interface), a software package for creating simple real-time tasks in a UNIX environment. It was developed over the last few years at McGill University and the General Electric Advanced Technology Laboratory.

RCI was designed as a substrate for implementing RCCL (Robot Control C Library), a software package for doing robot control from UNIX [Lloyd, et al. 88, Hayward and Paul 86]. In RCCL, robot actions are requested from a UNIX process, and the trajectories to implement these actions are computed by a real-time control task, which communicates with the UNIX process using shared memory. The control task typically runs at 50 to 100 Hz., which is beyond the real-time capabilities of most current stock UNIX systems.

The RCI package makes it possible to create such tasks. Depending on the system, RCI tasks can be run either on the UNIX CPU itself (with some hooks introduced into UNIX to make this possible), or on auxiliary CPUs which are attached to the system backplane and utilize a real-time kernel.

Versions of RCI capable of running tasks on the UNIX CPU have been implemented on MicroVAX-II systems running 4.3 BSD, and Sun SPARC systems running SUNOS 3.2 and 4.0.3. A version capable of running tasks on auxiliary CPUs has been implemented for the MicroVAX-II system, using KA630 or KA620 CPU boards attached to the system Q-bus and running a special kernel written by Mike Parker. A version capable of using auxiliary SPARC CPUs attached to a Sun system and running the VxWorks real-time system is currently under development.

2. RCI Functionality and an Example

From a software point of view, an RCI task looks very much like a signal handler: a particular *control* function is designated to be executed either on a particular hardware interrupt, or off of a system clock tick. The control function runs to completion, and communicates with the parent UNIX process (known as the *planning* task) through shared memory. RCI tasks which run on the UNIX CPU are executed directly at interrupt level (see section 4), in order to achieve the necessary real-time response. RCI tasks running on an auxiliary CPU execute as tasks under the real-time kernel for that CPU. In either case, control functions do not have access to normal UNIX system services, although this is not a great loss since such services are typically time consuming. RCI tasks can directly access all or part of the external system bus through mapped memory (an idea also used in the **/dev/bus** project at the University of Pennsylvania [Holder 89]).

Space limitations prevent a detailed description of the RCI primitives [Lloyd 88]. Instead, we will give an example RCI program in which a task running at 100 Hz. computes a running average of data collected from a bus register. The averaged value is read back by the planning task.

The planning and control tasks can be compiled together as one program, although the control level portion needs to be placed in a separate module(s).

```
/* FILE plan.c: */

#include <rci.h>                          /* RCI definitions, etc. */

main()
{
        extern taskFxn();                 /* function used by task */

        RCI_DESC *td;                     /* task descriptor */
        float *avg;                       /* pointer to shared mem */
        int t;

        td = rciCreate ("task1", 0);
        avg = (float*)allocMem ("avg", sizeof(float), 0);
        rciControlFxns (td, NULLF, taskFxn, NULLF, 0);
        rciScheduling (td, ON_CLOCK, NULL);
        rciRate (td, 10, 0);

        *avg = 0.0;
        rciStart (~0);
        for (t=0; t<10; t++)
         { printf ("average value is %g\n", *avg);
           sleep(1);
         }
        rciRelease (~0);
}
```

```
/* FILE control.c: */

#include <rci.h>

#define REG_ADDR        0xe000          /* bus address offset */

taskFxn(td)
{
        float *avg;                     /* pointer to shared mem */

        avg = (float*)getMemByName("avg");
        *avg = 0.9*(*avg) + 0.1*(*(short*)(rciBus+REG_ADDR));
}
```

Compilation is done with the **rcc** command, a front end to **cc** that contains RCI extensions. To allow RCI to keep track of what the control level needs, all control level modules should be placed to the right of the special keyword *CTRL* on the **rcc** command line. The example here could be compiled with the command

```
rcc plan.c CTRL control.c
```

which will produce a single executable named **plan**.

The program uses ***rciCreate()*** to set up a task on CPU 0. ***allocMem()*** is used to allocate a piece of shared memory to store the averaged data value. ***controlFxns()*** sets the control function to be ***taskFxn***. To arrange for the task to be run at 100 Hz., its scheduling discipline is set to *ON_CLOCK* using ***rciScheduling()***, and its execution interval is set to 10 msec. using ***rciRate()***.

After initialization, the RCI task is set running with a call to ***rciStart()*** (*~0* is a bitmask specifying all control tasks). The planning task then reads out the value computed by the control task for a few seconds before shutting it down with ***rciRelease()*** and exiting.

The control task consists of the function ***taskFxn()***. It uses ***getMemByName()*** to locate the shared memory. The quantity being averaged is read from location *REG_ADDR* on the system bus, which is available through mapped memory starting at the base address ***rciBus***.

3. Scheduling, Communication, and Utilities

An RCI task may be run according to one of four principal scheduling disciplines, established using ***rciScheduling()*** and a few auxiliary routines:

ON_CLOCK – The task is run off the system clock at a periodic rate.

ON_FUNCTION – The task is run whenever a polling function, driven by a device interrupt, returns true.

ON_TRIGGER – A combination of *ON_CLOCK* and *ON_FUNCTION*: a trigger function, executed off of the system clock, wakes up some external entity, and this in turn wakes up the task with an interrupt.

ON_SIGNAL – The task is run off of a UNIX signal, in user mode. This is used for simulation and debugging applications.

Inter-task communication is done primarily through shared memory. The functions

```
mem = allocMem (name, size, pool)
freeMem (mem)
```

may be used to allocate or free a piece of shared memory that is accessible to any of the program's tasks. The allocation routines can be called from the planning level only. Since RCI tasks may run on different CPUs, shared memory addresses may differ from one CPU to another. Because of this, shared memory addresses should be looked up using a reference function, such as `getMemByName(name)`, which uses the memory object's name as a reference, or `getMemByAddr(addr)`, which uses its planning level address. To avoid name space collisions, it is common to use `getMemByAddr()` and not name the memory objects at all.

Shared memory objects, or portions of them, may be atomically read or written using the primitives

```
readMem (mem, buf)
writeMem (mem, buf)
copyMem (mem, src, dst, size)
```

While one task is referencing an object through one of these primitives, all other references through these primitives to the same object are blocked.

The `allocMem()` facility is optimized for handling small chunks of memory (typically 1 Kbyte or less). To allocate larger amounts of memory, it may be better to use the primitives

```
rciGlobalMemory (name, size, pool)
rciSharedMemory (name, size, pool)
```

The memory allocated by these routines cannot be recycled. However, the memory allocated by `rciGlobalMemory()` can be accessed by other UNIX processes or RCI programs, which is useful for data logging or monitoring applications.

A simple message passing mechanism has also been built into RCI, but has not been used very much.

A primitive which has proven to be extremely useful is `rciSignal(num)`, which sends a UNIX signal to the planning task.

For communicating with the outside world, portions of the system bus are mapped into memory, with base locations at `rciBus` and `rciIOpage`. What part of the bus is mapped depends on the system: on MicroVAX-II systems, `rciBus` is the entire Q-bus and `rciIOpage` is the Q-bus I/O space. On Sun systems, `rciBus` is a portion of VME24d16 space, and `rciIOpage` is the VME16d16 space. Direct bus access removes the need to go through the system in order to interface with peripheral devices.

UNIX system calls, as well as most of the RCI allocation and task control primitives, are not available to the control level. However, the control level is provided with a simple set of utility functions, such as:

```
rciPrintf(fmt, ...)        -- diagnostic print routine
rciMalloc(size)            -- basic memory allocation
rciFree(mem)
```

```
rciTime()                  -- read system time (in msec.)
rciBadAddr(addr,size)      -- check memory on the bus
rciAbort(code,msg, ...)    -- abort the program
```

4. Implementation of RCI on the UNIX CPU

A good way to implement RCI tasks on the UNIX CPU would be to invoke the control function from interrupt level using UNIX signals. Unfortunately, this is too slow ([Holder 89]). Instead, the RCI task is executed directly at interrupt level, in kernel mode. This requires attending to the following issues:

1. *Context switching*. The memory context must be set to that of the RCI process (since the system may be inside any context when the interrupt arrives). The RCI interrupt code masks out all other system interrupts, switches to the RCI process context, executes the control function, restores the original context, and then unmasks interrupts. This switching can be done quickly since we are only concerned with the address space context. On the MicroVAX-II system, the page table base and length registers (*p0br*, *p0lr*, *p1br*, and *p1lr*) are switched. On Sun/SPARC systems, the memory management unit (MMU) context number is switched (it is also necessary to lock the context number to the process so that it won't be given to another process).

2. *Memory locking*. All memory referenced by an RCI task must be physically resident. Those portions of the program's text and data used by the control level must be locked down. To do this, a pair of locking primitives, *memlock()* and *memunlock()*, were added to the MicroVAX-II and Sun kernels.

3. *Exception Handling*. When an exception (such as a divide by zero or a memory access fault) occurs in kernel mode, the normal response of the system is to crash, since it indicates a kernel bug. Since this is unacceptable when running RCI applications, a hook was put in the system trap code to catch exceptions occurring within an RCI task and terminate the program more gracefully.

Shared memory between tasks running on the arbiter CPU is trivial to implement because the tasks and the UNIX process share the same address space. Creating shared memory is simply a matter of allocating it and locking it down. Shared memory between an RCI program and other UNIX processes is done by allocating memory out of a pool of kernel memory whose pages have been made user-accessible. The bus interface (*rciBus* and *rciIOpage*) is created by mapping the appropriate bus region into system virtual space and making it user accessible.

5. Implementing RCI on auxiliary CPUs

Implementing an RCI capability on auxiliary CPUs is much "cleaner" than piggy-backing RCI onto UNIX, since the auxiliary CPUs can be booted with a real-time kernel that provides the necessary performance characteristics. Communication between the auxiliary CPUs and the arbiter is done using shared memory over the system bus.

To simplify the implementation, each auxiliary CPU may be attached to only one RCI process at a time. When a call to ***rciCreate()*** first references a particular CPU, that CPU is checked for availability and booted with the real-time kernel if necessary. The CPU is then loaded with those portions of the RCI process text and data segments used by the control level*. A monitor is then started which accepts instructions from the UNIX CPU to set up RCI tasks and run them when the appropriate scheduling conditions arise.

Communication between auxiliary CPUs and the arbiter CPU is done through shared memory. Each CPU exports a region of memory onto the external bus for access by all the other CPUs. The other CPUs import this memory into a specific region of their virtual address space. It would also be possible to use separate memory modules, although this has not yet been done.

6. Discussion

RCI provides a connect-to-interrupt facility for UNIX, allowing processes to create high-priority tasks for doing data acquisition or process control. Its principal use has been to provide the trajectory generation task for RCCL. The system concept is very simple, but this has allowed it to be implemented on a variety of hardware systems.

The main reason the system has been useful is probably because of the well-defined scheduling structure of the tasks it runs: RCI control tasks are assumed (more or less) to execute periodically at a known sample rate. Such tasks are easy to synchronize with, and they are also conducive to shared memory communication (a certain frequency of data access can be assumed, double buffering is feasible, etc.). This is also the reason that the RCI message passing primitives have not been used much.

References

[Holder 89] Gaylord Holder, "A General Device Driver for ULTRIX or Leave the Driver to /dev/bus". Technical Report MS-CIS-89-10, Grasp Lab 174, Department of Computer and Information Science, University of Pennsylvania, Philadelphia, Pa.

[Hayward and Paul 86] Vincent Hayward and Richard Paul, "Robot Manipulator Control Under UNIX: RCCL, a Robot Control C Library". *International Journal of Robotics Research*, Winter 1986, pp. 94 – 111. (Vol. 5, No. 4)

[Lloyd et al. 88] John Lloyd, Mike Parker, and Rick McClain, "Extending the RCCL Programming Environment to Multiple Robots and Processors". *IEEE Conference on Robotics and Automation*, Philadelphia, Pa., April 24-29, 1988, pp. 465 – 469

[Lloyd 88] John Lloyd, "RCI User's Guide". Report CIM-88-22, McGill Research Center for Intelligent Machines, McGill University, Montreal, January 1988.

*This assumes that the auxiliary CPU can execute from the same image as the RCI process itself, which is true for implementations to date. In cases where this was not true, the **rcc** command could produce a separate image for the auxiliary CPU.

Toward a Reference Model Architecture for Real-Time Intelligent Control Systems (ARTICS)

James S. Albus and Richard Quintero
Robot Systems Division
National Institute of Standards and Technology

INTRODUCTION

This position paper is a condensed version of a paper titled "Concept for a Reference Model Architecture for Real-Time Intelligent Control systems (ARTICS)", [Al 90], which advocates the development of a reference model open-system architecture as a means to accelerate the pace of technological development in automation and robotics. We believe many of the major bottlenecks in the development of intelligent machine systems could be alleviated, if not eliminated, by the development of a set of ARTICS guidelines.

The pace of commercial and military technological advancement in the fields of robotics, intelligent machine systems and automation is falling short of expectations. Problem complexity is one of the major contributors to this problem. Intelligent robot systems projects typically require bringing together teams of technologists with a broad mix of engineering disciplines and a high level of expertise. Robotics and automation manufacturers must make large investments in both developing custom test-beds and in recruiting and training competent engineering teams in order to compete in this market area. A second problem is the lack of a widely accepted theory, or system architecture model that ties together the many disciplines involved in intelligent robot systems. This limits the dissemination of intelligent machine systems technology developed in different parts of the robotics community. This prevents new projects from building upon the foundations laid by previous efforts.

A set of ARTICS guidelines would reduce the impact of problem complexity and would provide an efficient means of transferring technology between projects. Manufacturers will adopt ARTICS guidelines if they believe that their potential profits would be enhanced by an expanded market. This must be driven by traditional market forces (user demand). We need a way to create automation building blocks so that more complex systems can be developed without making the technologies more difficult to understand and to apply and without "reinventing the wheel" each time a new project begins. We believe that a common hardware/software shell structure would facilitate the incremental improvement which would produce rapid advancement in automation and robotics technology.

To summarize the goals advocated by this paper, we suggest that an Architecture for Real-Time Intelligent Control Systems (ARTICS) is needed to:

* reduce the impact of problem complexity in the development of robotic applications

* expand the market for intelligent control system components through open-system interface guidelines and protocols.

* promote portability, inter-operability and modularity of intelligent control system software and hardware

* facilitate technology transfer between intelligent control system projects

* reduce the time, cost, risk, and initial investment required in bringing new, world class, intelligent machine systems and control system products into the market place

2. ARTICS VISION

ARTICS guidelines would specify a reference model infrastructure of hardware components, software components, interfaces, communications protocols, and application development tools. Such a set of guidelines would make it possible for industry to develop and market a diverse line of control system components which could be interchangeable and realizable on many different vendors' control systems platforms.

ARTICS would be designed to facilitate technology and component transfer among the various users and developers, taking advantage of commonalities among otherwise disparate applications such as manufacturing, construction, environmental restoration, mining, space exploration telerobotics, medicine, and military applications of air, land, space, sea-surface, and undersea robotics.

A commercially manufactured ARTICS implementation product would come with libraries of algorithms for planning, task execution, sensor processing and world modeling. These libraries would be user expandable and replaceable. An ARTICS implementation would be fully documented so that users could easily modify or replace any module with a minimum of effort. It would also be commercially maintained, so that users would be able to get help in fixing bugs and making system modifications. In addition, vendors would offer training services to help the user community apply ARTICS products to their applications.

Widely available ARTICS off-the-shelf products would include a target computer system with a backplane and bus configured as a card cage, a local area network to link distributed applications and interface workstations for human/computer interface and software development, a real-time multi-processor/multi-tasking operating system, compilers, debuggers, and CASE tools. ARTICS compliant products could be integrated into an extendible open-system architecture with complete documentation of all hardware and software components.

The ultimate goal would be for ARTICS to evolve into a set of standards for real-time intelligent control systems. Figure 1 illustrates a possible common system configuration for a Version 1 ARTICS system. It would be organized into three levels.

The top level would consist of a number of workstations on an Ethernet for off-line software development and testing. A number of Computer Aided Software Engineering (CASE) tools, shell programs, simulators, debugging and analysis tools, and compilers for at least C, Ada, and Common LISP would be available.

These workstations might include one or more SUNs, LISP machines, VAXes, Butterflys, Connection Machines, graphics engines and display and image processing machines.

One or more of the workstation machines might also be used for on-line real-time control of processes where response time can exceed 1 second, and in situations where weight, power, and other environmental requirements permit. A real-time, multi-computer, multi-tasking operating system such as real-time UNIX, or MACH would be provided to support this type of operation.

The top level would also support an interface to a gaming environment such as the DARPA SIMNET. This would provide a low cost means for testing and evaluating the performance of intelligent machines in a war gaming environment against manned systems, or other unmanned systems. It would also provide an environment for developing tactics and strategy for using large numbers of intelligent vehicles and weapons systems in large scale battle simulations.

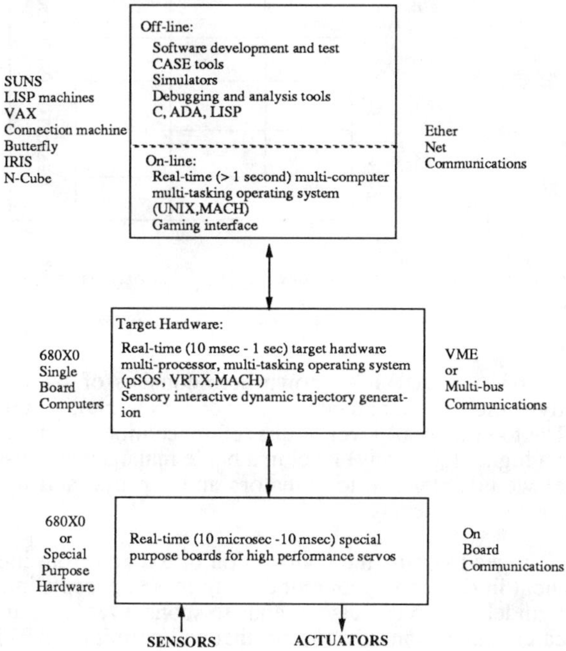

Figure 1: ARTICS Version 1 Common System Architecture

The middle level of the ARTICS system would consist of target hardware, such as single board computers and memory boards of the 680X0 variety, using VME or Multi-bus communications. More than one such bus might be connected via bus gateway cards. This middle level would have a real-time, multi-processor, multi-tasking operating system such as pSOS, VRTX, or MACH capable of supporting response times of ten milliseconds or greater.

The bottom level would consist of special purpose hardware which would interface to the VME or Multi-bus. This level would support high speed parallel processing for images, as well as servo controllers with response times between ten microseconds and ten milliseconds.

Figure 2 shows a possible reference model architecture based the Real-time Control System (RCS) concepts NIST has developed since 1980. These have been implemented in a number of applications including the Automated Manufacturing Research Facility (AMRF), the NASA/NBS Standard Reference Model for Telerobot Control System Architecture (NASREM) [Al 89], the Air Force Next Generation Controller for machine tools and robots, and the control system architecture research conducted for the NIST/DARPA Multiple Autonomous Undersea Vehicle (MAUV) project [Al 88].

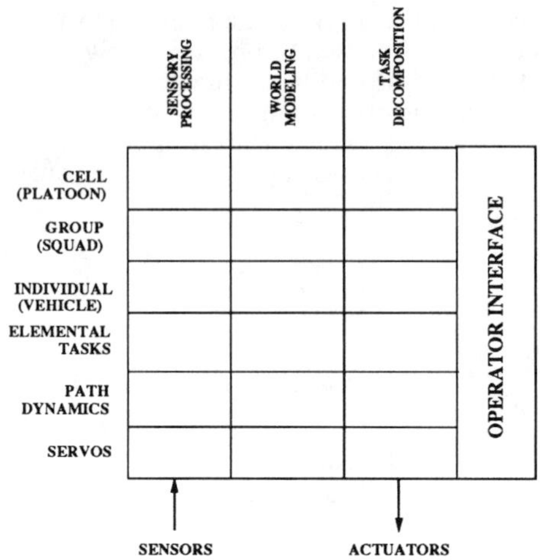

Figure 2: Reference Architecture for Real-Time Intelligent Control

The version of ARTICS shown here consists of six hierarchical levels: servo, path dynamics, elemental tasks, individual (vehicle), group (squad), and cell (platoon). The top (platoon) level of this reference model architecture would have interfaces to a higher (company) level in a battle management system. The bottom (servo) level would interface to actuators and sensors, and operator interfaces would be defined for all levels.

NIST hopes to enlist the cooperation of experts from industry, academia, and government in developing and modifying these concepts into an agreed upon initial set of guidelines. NIST also intends to sponsor research and enlist others to sponsor research, into advanced concepts that will permit the ARTICS guidelines to evolve as technology advances.

3. REQUIREMENTS

The following "strawman" list of requirements is intended to encourage the robotics community to begin the discourse.

Functional and temporal extensibility; human/computer interface flexibility, levels of automation (i.e. teleoperation, remote control, computer-aided advisory control, traded control shared control, human override, human supervised control, autonomous control, sensory interactive control, mixed mode control), real-time andtemporal reasoning, distributed systems, graceful degradation, software portability, ease of use, cost effectiveness, software development environment, simulation and animation.

A discussion of these requirements can be found in [Al 90].

4. APPROACH

To implement the ARTICS concept a consensus must be achieved in several areas of the common control system architecture. Such a conceptual framework would provide developers with a common design philosophy to guide the development of new robotic applications and control system products. A number of control architectures should be considered and evaluated against some set of agreed upon common control system requirements and finally a common conceptual architecture must be derived from the results of the process. More than likely such an architecture would include the ideas of a number of researchers as well as strong input from the user community. The following is a list of recent research in this area:

- Action Networks [Ni 89]
- Autonomous Land Vehicle (ALV) [Lo 86]
- Automated Manufacturing Research Facility (AMRF) [Si 83, Al 81]
- Control in Operational Space of a Manipulator-with-Obstacles System (COSMOS) [Kh 87]
- COmmunications Database with GEometric Reasoning (CODGER) [Sh 86]
- Field Materiel-Handling Robot (FMR) [Mc 86]
- Generic Vehicle Autonomy (GVA) [Gr 88]
- Hearsay II [Le 75]
- Hierarchical Control [Sk 89, Sk 87, Ko 88, Ko 88]
- Hierarchical Real-time Control System (RCS) [Ba 84, Al 81
- Intelligent Control [Sa 85]
- Intelligent Task Automation (ITA) [Bl 88]
- Manufacturing Automation System/Controller (MAS/C) [Ho 88]
- Multiple Autonomous Undersea Vehicles (MAUV) [Al 88]
- NASA/NBS Standard Reference Model for Telerobot Control System Architecture (NASREM) [Al 89]
- Pilot's Associate [Sm 87]
- Robot Control "C" Library (RCCL) [Ha 86]
- Robot Schemas [Ly 89]
- Soar: Architecture for General Intelligence [La 86]
- Subsumption Architecture [Br 86]
- Task Control Architecture (TCA) [Si 89]
- Tech-based Enhancement for Autonomous Machines (TEAM) [Sz 88]
- University of New Hampshire (UNH) Time Hierarchical Architecture [Ja 88]

There are a number of government efforts under way that should be factored into the process of defining an initial set of common architecture components. Some of these include:

- The NIST Federal Information Processing Standard (FIPS)
- The NIST Government Open Systems Interconnection Profile (GOSIP)
- The Navy's Next Generation Computer Resources (NGCR) program
- The Air Force's Next Generation Controller (NGC) program
- The Army's Standard Army Vetronic Architecture (SAVA) program

In addition there is a Department of Energy interest in establishing guidelines for robotic systems needed in their Environmental Restoration and Waste Management Program, the U.S. Bureau of Mines Pittsburgh Research Center is conducting research in automation systems for coal mining and there are a number of DARPA programs (past, present and on-going) which are producing relevant technologies.

5. REFERENCE MODEL DEVELOPMENT PLAN

ARTICS must be able to evolve as technological progress is made. It will be important to create an organizational structure that can coordinate the process of evaluating change and update proposals and a process for achieving consensus on the release of new versions of the ARTICS guidelines. Such an organization will need a steering committee made up of leading experts in the field of robotics, intelligent machines and automation from industry, academia and government.

An ARTICS development effort could take the form of a voluntary organization much like the Initial Graphics Exchange Specification (IGES)/Product DataExchange Specification (PDES) organization [Ig 89] chaired by NIST. Alternatively the reference model could be developed by a major user of the technology such as the Department of Defense in the form of a military specification (MILSPEC). In either case working groups will be needed to steer the ARTICS development and to document and distribute the results. Once an initial set of ARTICS guidelines has been agreed to it can be submitted to one or more national or international standards organizations as a proposed standard (e.g., ANSI, EIA, IEC, IEEE, ISO, RIA, etc.).

6. REFERENCES

[Al 90] J.S. Albus, R. Quintero, R. Lumia, M. Herman, R, Kilmer, and K. Goodwin, “Concept for a Reference Model Architecture for Real-Time Intelligent Control Systems (ARTICS),” NIST Technical Note 1277, April 1990.

[Al 89] J.S. Albus, H.G. McCain, and R. Lumia, "NASA/NBS Standard Reference Mode for Telerobot Control System Architecture (NASREM)," NIST (formerly NBS) Technical Note 1235, April 1989 Edition.

[Al 88] J.S. Albus, "System Description and Design Architecture for Multiple Autonomous Undersea Vehicles Project," NIST Technical Note 1251, September 1988.

[Al 81] J.S. Albus, "Brains, Behavior and Robotics," BYTE/McGraw-Hill, Petersborough, NH, 1981.

[Ba 84] A.J. Barbera, J.S. Albus, M.L. Fitzgerald, and L.S. Haynes, "RCS: The NBS Real-Time Control System," Robots 8 Conference and Exposition, Detroit, MI, June 1984.

[Bl 88] R. Blair, "Intelligent Task Automation Interim Technical Report," Report No. II-6, Honeywell Corporate Systems, Golden Valley, MN, August 1988.

[Br 86] R.A. Brooks, "A Robust Layered Control System For A Mobile Robot," IEEE Journal of Robotics and Automation, Vol. RA-2, No. 3, March 1986.

[Gr 88) M.D. Grover, et. al., "An Autonomous Undersea Vehicle Software Testbed,"Proceedings of the Fifteenth Annual Technical Symposium of the Association for Unmanned Vehicle Systems, San Diego, CA, June 6-8, 1988.

[Ha 86] V. Hayward, R.P. Hayward, "Robot Manipulator Control Under Unix RCCL: A Robot Control "C" Library," International Journal of Robotics Research, Vol. 5, No. 4, Winter 1986.

[Ho 88] Honeywell Inc., "MAS/C System Overview," MA70-100, Release C, Honeywell Inc., Phoenix, AZ, May 1988.

[Ig 89] "Welcome to IGES/PDES, Newcomer Material," July 1989, available through the Chairman, IGES/PDES, NIST, Bldg. 220, Room A150, Gaithersburg, MD, 20899.

[Ja 88] J. Jalbert, "EAVE III Untethered AUV Submersible," Proceedings of the Fifteenth Annual Technical Symposium of the Association for Unmanned Vehicle Systems, San Diego, CA, June 6-8, 1988.

[Kh 87] O. Khatib, "A Unified Approach for Motion and Force Control of Robot Manipulators: The Operational Space Formulation," IEEE Journal of Robotics and Automation, Vol. RA-3, No. 1, February 1987.

[Ko 88] W. Kohn, T. Skillman, "Hierarchical Control for Autonomous Space Robots,"Proceedings of the AIAA Guidance, Navigation, and Control Conference, Minneapolis, Minnesota, Aug 15-17, 1988.

[Ko 88] W. Kohn, "Declarative Theory for Relational Controllers," Proceedings of the IEEE Control Decision Conference, Austin, Texas, Dec 7-9, 1988.

[La 86] J. E. Laird, A. Newell, P. S. Rosenbloom, "Soar: An Architecture for General Intelligence," Carnegie-Mellon University paper CMU-CS-86-171, December 1986.

[Le 75] V.R. Lesser, et. al. "Organization of the Hearsay II Speech Understanding System," IEEE Trans. on ASSP, Vol. 7, No.1, 1975, p. 11.

[Ly 89] D. M. Lyons, M. A. Arbib, "A Formal Model of Computation for Sensory-Based Robotics," IEEE Transactions on Robotics and Automation, Vol. 5, No. 3, June 1989.

[Lo 86] J. Lowerie, et. al. "Autonomous Land Vehicle," ETL- 0413 Martin Marietta Denver Aerospace, Denver, Co, July 1986.

[Mc 86] H. G. McCain, et. al. "A Hierarchically Controlled Autonomous Robot for Heavy Payload Military Field Applications," Proceedings of the International Conference on Intelligent Autonomous Systems, Amsterdam, The Netherlands, December 8-11,1986.

[Ni 89] N. Nilsson, "Action Networks," Proceedings from the Rochester Planning Workshop: From Formal Systems to Practical Systems, J. Tenenberg, et al. (eds.), University of Rochester, New York, 1989.

[Ro 89] Robotic Technology Incorporated, "Standard Architecture for Real-Time Control Systems (SARTICS), Final Report," NIST Contract No. 43NANB909769, dated 21 Nov. 1988, Submitted to NIST Robot Systems Division, Gaithersburg, MD, 3 July 1989.

[Sa 85] G.N. Saridis, "Foundations of the Theory of Intelligent Controls," IEEE Workshop on Intelligent Control, Troy, August 1985, p. 23.

[Sh 86] S.A. Shafer, et. al. "An Architecture for Sensor Fusion in a Mobile Robot," Proceedings of the IEEE International Conference on Robotics and Automation, San Francisco, CA, April 7-10, 1986, pp. 2002-2010.

[Sk 89] T. L. Skillman, K. Hopping, "Dynamic Composition and Execution of Behaviors in a Hierarchical Control System," SPIE Mobile Robotics IV, Philadelphia, Nov 7- 8, 1989.

[Sk 87] T. L. Skillman, W. Kohn, et al."Blackboard Based Hierarchical Intelligent Control System," Proceedings of the AIAA Conference on Computers in Aerospace VI, Wakefield, MA, Oct 7-9, 1987.

[Si 89] R. Simmons, T. Mitchell, "A Task Control Architecture for Mobile Robots,"Computer Sciences Dept., Carnegie Mellon Univ., Pittsburgh, PA, AAAI '89 Spring Symposium, Robotic Navigation, 1989.

[Si 83] J.A. Simpson, R.J. Hocken, and J.S. Albus, "The Automated Manufacturing Research Facility of the National Bureau of Standards," Journal of Manufacturing Systems, Vol.1, No. 1, pg. 17, 1983.

[Sm 87] D. Smith, M. Broadwell, "Plan Coordination in Support of Expert Systems Integration", Proceedings of the Knowledge-Based Planning Workshop, Austin, TX, December 8-10, 1987.

[Sz 88] S. Szabo, H. A. Scott, R. D. Kilmer, "Control System Architecture for the TEAM Program," Proceedings of the Second International Symposium on Robotics and Manufacturing Research, Education and Applications, Albuquerque, NM, November 16-18, 1988.

SLIDING CONTROL OF ELECTROMAGNETIC SUSPENSION

DAN CHO
Dept. of Mechanical and Aerospace Engineering
Princeton University
Princeton, NJ 08544–5263

ABSTRACT

The application of sliding control to stabilize an electromagnetic suspension system is presented in this paper. Unlike conventional linear control techniques, the sliding control method can be applied to a class of nonlinear systems in the global sense. The performance and stability robustness as well as the practicality of the sliding controller are demonstrated using a single axis electromagnetic suspension system.

INTRODUCTION

Electromagnetic levitation and suspension systems have been successfully demonstrated for many applications, including frictionless bearings for inertial instruments (Downer,1980), vibration isolation tables (Dahlen, 1983), and high speed passenger trains (Limbert, et al., 1979; Yamamura and Yamaguchi, 1990). The above examples are typically implemented using feedback control techniques. In general, the governing equations of motion are linearized about some equilibrium point, and closed loop control algorithm is designed using an available linear control technique. Thus, the control force is composed of two parts: best open loop estimate for an equilibrium position (which is an unstable one) and closed loop perturbation control to locally stabilize the equilibrium position.

Note that totally open loop magnetic levitation is also possible. This can be achieved by the use of a dipole configuration (including the superconducting levitation by the Meissner effect), a radio frequency tuned–resonator configuration (Kaplan and Regev, 1976, and references contained therein), or a repulsive force field configuration. A recent survey on various levitation methods can be found in Brandt (1989).

Variable structure systems or sliding mode control (Utkin, 1977 and 1978) is relatively new but has gained much popularity in the recent years due to its applicability to nonlinear systems and its ability to consider robustness issues in the global sense. Application examples include servo systems (Young and Kwanty, 1982), robot manipulator tracking (Slotine and Sastry, 1983, Asada and Slotine, 1986), and automotive fuel injection systems (Cho and Hedrick, 1988).

The application of the sliding control method to globally stabilize an electromagnetic suspension system is investigated in this paper, and experimental results obtained for a single axis electromagnetic suspension are presented.

ELECTROMAGNETIC SUSPENSION HARDWARE

Figure 1 depicts the schematic of an electromagnetic suspension. The equation of motion is:

$$m\ddot{z} = -\,mg + F_{mag} \tag{1}$$

where m is the mass of the suspended object, and F_{mag} is the electromagnetic force. The electromagnetic force obeys the inverse distance square law, and calibration experiments are performed to empirically determine the force–distance relationship. Using a micro xyz positioner table capable of a 25μm accuracy and a computer, the voltage required to pick up objects at various levitation gaps is determined. The results are depicted in Figure 2 for three different objects, which are ping–pong balls with small permanent magnets bonded to them. The calibration results are curve–fitted using the least squares technique as $V_a = 0.02205z^2 - 1.180z + 19.12$, where z is measured from some datum point in mm, and V_a is the output of the amplifier in volts.

A photo emitter and detector pair is used to determine the position of suspended objects. Again, using the xyz positioner, the photo detector characteristics are calibrated as depicted in Figure 3. The results are least squares fitted as $V_d = -0.2378z + 10.92$, where z is measured from the identical datum point as before in mm, and V_d is the photo detector voltage in volts.

SLIDING CONTROLLER SYNTHESIS

Variable structure systems with sliding modes are based on the second method of Lyapunov (Utkin, 1977 and 1978). Consider a nonlinear system in canonical form:

$$\begin{aligned} \dot{x}_1 &= x_2 \\ \dot{x}_2 &= x_3 \\ &\vdots \\ \dot{x}_n &= f(\underline{x},t) + g(\underline{x},t)u + d(t) \end{aligned} \tag{2}$$

where $\underline{x} = [x_1, x_2, \ldots x_n] \in \mathbb{R}^n$ are the states, $f(x,t)$ and $g(x,t)$ represent nonlinear functional relationships, $u \in \mathbb{R}^1$ is the control, and $d(t)$ is the disturbance. The control objective is to track a desired trajectory x_{1d}. Define the tracking error as:

$$e \equiv x_1 - x_{1d} \tag{3}$$

Then a manifold defined by $S = 0$ where:

$$S \equiv (\frac{d}{dt} + \lambda)^{n-1} e \tag{4}$$

becomes "attractive", if the control is formulated to satisfy:

$$S = -\eta \operatorname{sgn}(S) \tag{5}$$

It is straightforward to show this by selecting a Lyapunov function as $S^2/2$. Substituting (2 – 4) into (5), the control that guarantees the attractiveness of $S = 0$ manifold, in the presence of disturbances and modeling uncertainties is (Slotine and Sastry, 1982; Cho and Hedrick, 1989):

$$u = -\hat{g}(\underline{x},t)^{-1}\left[n\lambda\frac{d^{n-1}e}{dt} + ... + n\lambda^{n-1}\dot{e} + \lambda^n - \frac{dt^n x_{1d}}{dt^n} + \hat{f}(\underline{x},t) + k\,\mathrm{sgn}(S)\right] \quad (6)$$

where the best model estimates of $f(\underline{x},t)$ and $g(\underline{x},t)$ are:

$$f(\underline{x},t) = \hat{f}(\underline{x},t) + \Delta f(t) \quad \text{and} \quad |\Delta f(t)| < \Delta F$$
$$g(\underline{x},t) = \hat{g}(\underline{x},t) + \Delta g(t) \quad \text{and} \quad |\Delta g(t)| < \Delta G \quad (7)$$

The robustness parameter k is:

$$k = \Delta F + \Delta D + \Delta G \cdot u_{max} \quad (8)$$

where:

$$|d(t)| < \Delta D \quad (9)$$

The induced chatter due to the use of sgn(S) in (6) can be ameliorated by the use of sat(S/ϕ) where:

$$\mathrm{sat}(S/\phi) = \begin{cases} \mathrm{sgn}(S) & |S| > \phi \\ S/\phi & |S| < \phi \end{cases} \quad (10)$$

and ϕ is the boundary layer thickness (Slotine and Sastry, 1982). The boundary layer thickness ϕ in (10) and the control bandwidth λ in (4) and (6) must be chosen such that $\phi \geq k/\lambda$. This is known as the static balancing condition (Slotine and Sastry, 1982). Then, a tracking accuracy of $e < \phi/\lambda^{n-1}$ is guaranteed. Note that the use of sat(S/ϕ) only guarantees the attractiveness of the boundary layer defined by $S = \phi$. Inside the boundary layer (i.e., $|S| < \phi$), the dynamics may "wander" about $S = 0$ manifold. The trade–off involved here is the chatter magnitude and tracking accuracy. However, by judiciously selecting the robustness parameter and boundary layer thickness, desired tracking performance can be achieved while eliminating or substantially reducing the chatter.

A straightforward application of sliding control to the electromagnetic suspension in (1) yields:

$$V_c = m(0.8342z^2 - 22.32z + 180.84)[g + \ddot{z}_d - \lambda(\dot{z} - \dot{z}_d) - k\,\mathrm{sat}(S/\phi)] \quad (11)$$

where V_c is the command voltage in integer volts of 0 to 2047 and:

$$S = \ddot{z} - \dot{z}_d + \lambda(z - z_d) \quad (12)$$

Figure 4 depicts experimental results for three different balls. The control parameters used are:

$$z_d = 10\text{mm}, \quad \lambda = 25, \quad k = 200, \quad \phi = 25$$

The desired control bandwidth $\lambda = 25$ is chosen to be sufficiently slower than the sampling frequency (100Hz) yet adequately high for good transient performance. The robustness parameter $k = 200$ corresponds to 1.46 volts at the amplifier output, which is approximately 10% of the maximum control voltage at the amplifier output. This particular value is selected because the least squares fit of the electromagnetic forces can have up to a ± 1 volt error in the vicinity of $z = 10$mm (cf. Figure 2). Finally, the boundary layer thickness $\phi = 25$ is selected in accordance with the static balancing condition. Then, a tracking accuracy of better than ± 1mm is guaranteed, and Figure 4 shows that the tracking accuracy as well as stability and robustness are easily satisfied. Note that the non–zero position error present for all three balls are largely attributed to the sensor inaccuracy (cf. Figure 3).

Further reduction of the output chatter is possible by increasing the boundary layer thickness. Figure 5 depicts the results for $\phi = 50$. It is interesting to note the slow frequency "wandering motion" near $z = 10$mm. This is inevitable in trading–off the high frequency chatter. The control voltage at the amplifier output is also depicted in Figure 6, which shows that neither control saturation nor excessive control authority is problematic.

CONCLUSIONS

A Sliding controller is synthesized to stabilize an electromagnetic suspension, which is an open loop unstable system or a non–minimum phase system. It is demonstrated experimentally that the designed sliding controller is robust to modeling errors by suspending three different objects, each with a different force–distance relationship. The sliding controller achieves desired levitation gaps within the prescribed error boundaries, with position sensors that are not particularly accurate. The chattering solution inherent in sliding control is significantly smoothed by the use of saturation function.

ACKNOWLEDGEMENT

The author would like to thank Messrs. Kevin Lynch, Christopher Marrison and Subrata Sircar for their assistance in the development of the hardware and in the calibration of the hardware.

REFERENCES

Asada, H., and Slotine, J.J.E., (1986), Robot Analysis and Control, John Wiley & Sons.

Brandt, E.H., (1989), "Levitation in physics", Science, 243, pp. 349–355.

Cho, D., and Hedrick, J.K., (1988), "Nonlinear control methods for fuel–injected automotive engines", ASME J. of Eng. for Gas Turbines and Power, 110(3), pp. 313–320.

Cho, D., and Hedrick, J.K., (1989), "Sliding mode control of systems with uncertain actuator models and its application in automotive powertrain systems", Proc. of American Control Conference.

Dahlen, N.J., (1983), "Magnetic active suspension and isolation", S.M. Thesis, Dept. of M.E., M.I.T.

Downer, J.R., (1980), "Analysis of a single axis magnetic suspension system", S.M. Thesis, Dept. of M.E., M.I.T.

Kaplan, B.Z., and Regev, D., (1976), "Dynamic stabilization of tuned–circuit levitators", IEEE Trans. on Magnetics, Mag–12(5), pp. 556–559.

Limbert, D.A., Richardson, H.H., and Wormley, D.N., (1979), "Controlled characteristics of ferromagnetic vehicle suspension providing simultaneous lift and guidance", ASME J. of Dynamic Systems, Measurement and Control, 101, pp. 217–222.

Slotine, J.J.E., and Sastry, S.S., (1984), "Tracking control of non–linear systems using sliding surfaces, with application to robot manipulators", Int. J. of Control, 38(2), pp. 465–492.
Utkin, V.I., (1977), "Variable structure systems with sliding modes", IEEE Trans. on Auto. Control, AC–22(2), pp. 212–222.
Utkin, V.I., (1978), Sliding Mode and Its Application in Variable Structure Systems, MIR Pub.
Yamamura, S., and Yamaguchi, H., (1990), "Electromagnetic levitation system by means of salient–pole type magnets coupled with laminated slotless rails", IEEE Trans. on Vehicular Tech., 39(1), pp. 83–87.
Young, K.K.D., and Kwanty, H.G., (1982), "Variable structure servo–mechanism design and applications to overspeed protection control", Automatica, 18(4), pp. 385–400.

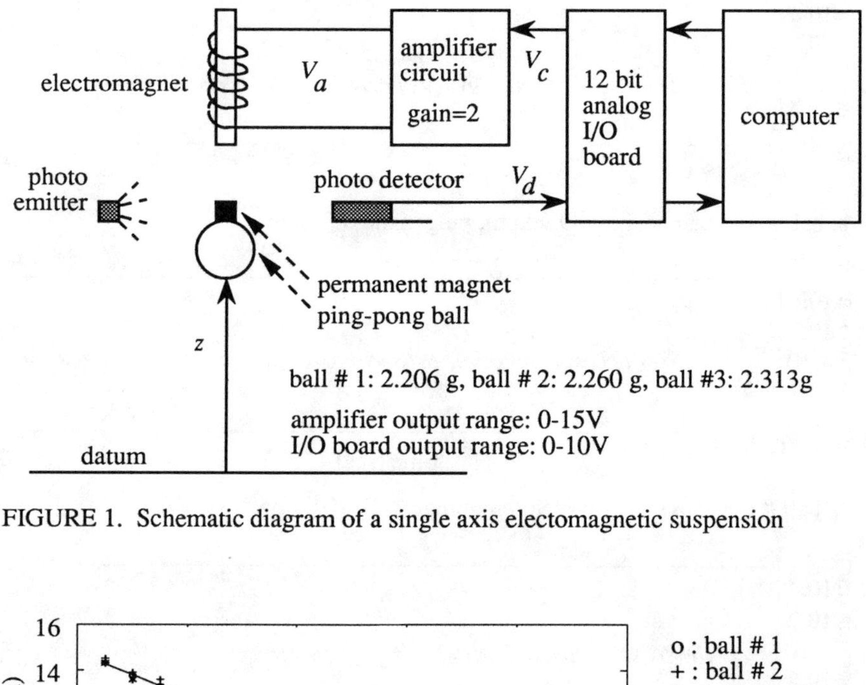

FIGURE 1. Schematic diagram of a single axis electomagnetic suspension

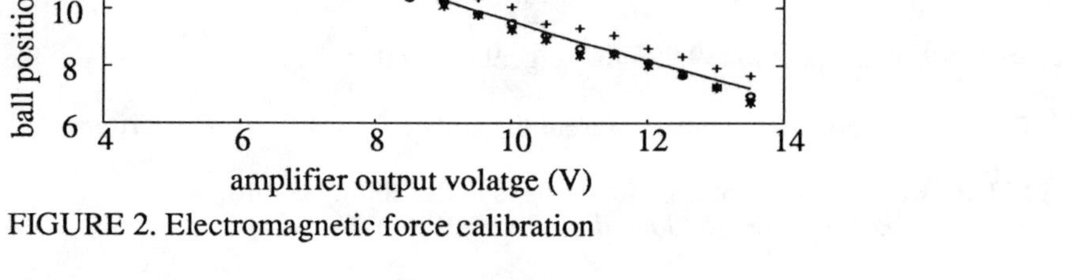

FIGURE 2. Electromagnetic force calibration

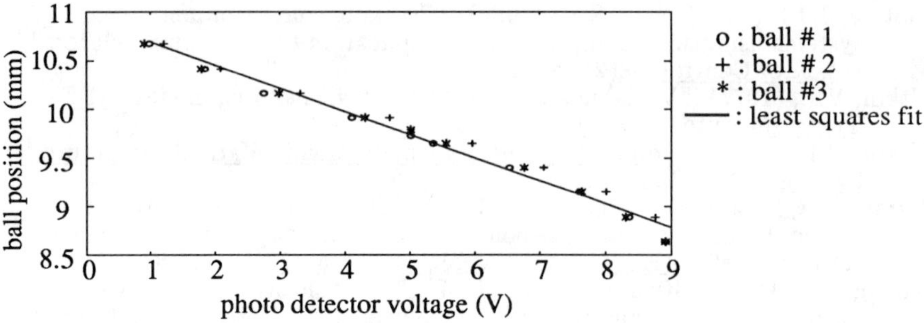

FIGURE 3. Photo detector voltage calibration

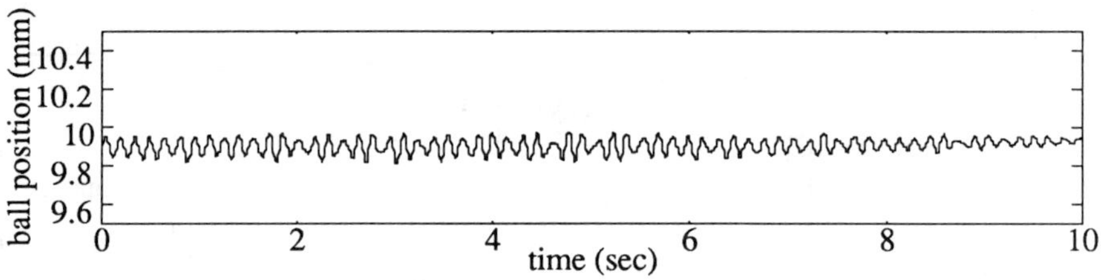

a. ball # 1: mean position=9.90mm, range:9.80-9.97mm

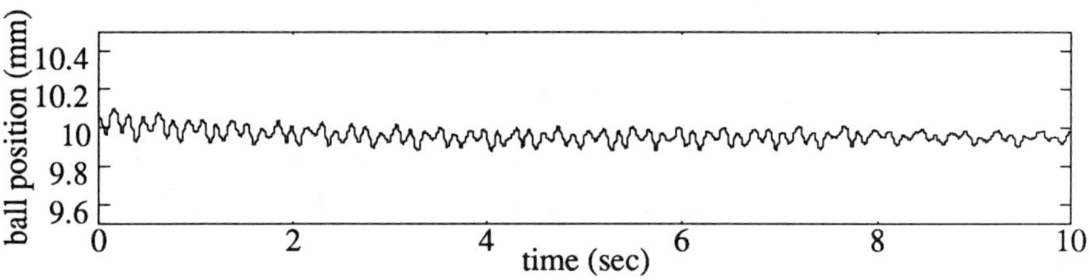

b. ball # 2: mean position=9.96mm, range:9.87-10.09mm

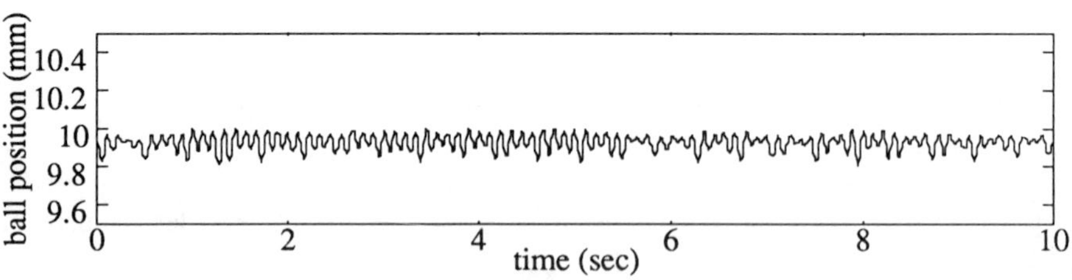

c. ball # 2: mean position=9.92mm, range:9.80-10.0mm

The depicted position ranges represent the allowable position error, ±1mm

FIGURE 4. Position of three different balls under sliding control
k=200, *phi*=25, *lambda*=25, *zd*=10mm

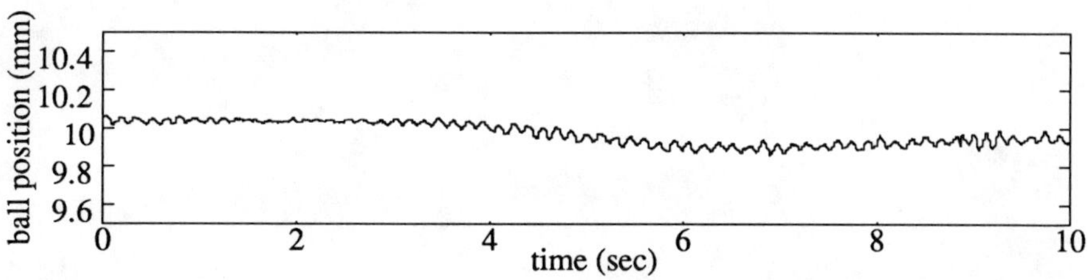

The depicted position range represents 1/2 of the allowable position error, ±2mm

FIGURE 5. Position of the ball #1 under sliding control
k=200, *phi*=50, *lambda*=25, *zd*=10mm

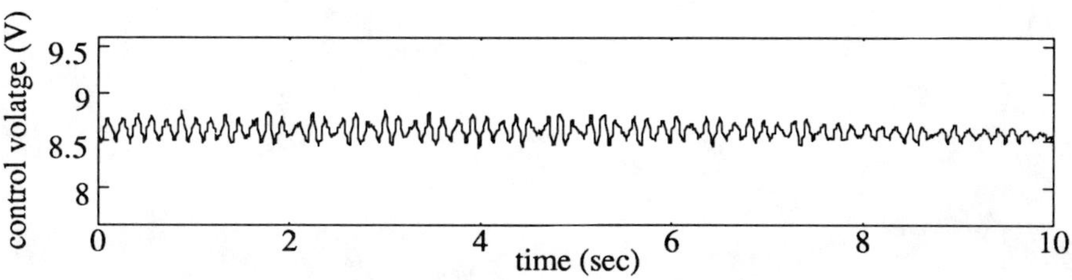

a. *k*=200, *phi*=25, *lambda*=25, *zd*=10mm

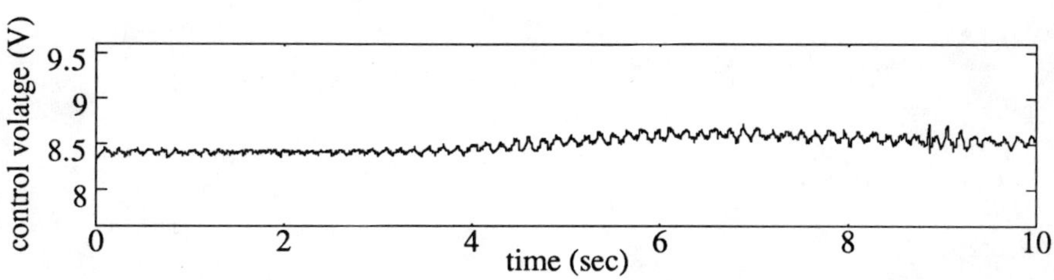

b. *k*=200, *phi*=50, *lambda*=25, *zd*=10mm

FIGURE 6. Control voltage at the output of the amplifier

On the Design and Control of Magnetically Levitated Robot Wrists

R.L. Hollis
Manufacturing Research
IBM Research Division
Thomas J. Watson Research Center
Yorktown Heights, NY 10598, USA

S.E. Salcudean
Department of Electrical Engineering
University of British Columbia
2356 Main Mall
Vancouver, BC V6T-1W5, Canada

Abstract

This paper describes a high performance six degree-of-freedom magnetically levitated fine motion robot wrist with programmable compliance. Aspects of design, modelling, and control required to build and operate such a device are discussed. The experimental results obtained with a prototype wrist include high bandwidth position control, compliant control and the emulation of several mechanisms through software gain setting. These establish the use of magnetically levitated robot wrists as a new and desirable option for manipulation tasks requiring high precision and fine compliant motions.

1 Introduction

It has long been recognized that the lack of accuracy, the poor dynamic response and the large apparent mass presented in contacts with the environment limit the use of conventional robots in manufacturing tasks requiring precise and fine compliant motions. It is also well known that these limitations (due to accumulated errors from joint position measurements, low actuator/encoder resolutions, bending, gear backlash, friction, etc.) are fundamental and cannot be overcome by sophisticated motion control because of inherent modelling errors [1]. Therefore, for high performance, the conventional position- or force-controlled robot must be augmented by a dextrous hand or fine motion device, dividing thus the robot manipulation task into coarse and fine domains, with the manipulator itself having redundant coarse and fine degrees of freedom [2]. The coarse manipulator acts merely as a position-controlled transportation system, while the fine manipulator attached to it can be position, compliance or force-controlled and actually carries out the work. For position control, some form of end-point sensing must be used to measure relevant task parameters directly in order to guide the manipulator system to achieve the desired goal.

The design and control of the fine manipulator device is crucial to success at the system level and motivated the work on a six degree-of-freedom magnetically levitated (maglev) wrist, the "Magic Wrist", presented for the first time at the 1987 ISRR [3]. This paper discusses several issues related to the design, modelling and control of such a device. Of the many approaches that have been taken to building fine manipulation devices (see [3] for references), the most relevant ones to the work presented in this paper are a levitated 5-DOF wrist using magnetic bearings which was used as a programmable RCC [4], and a magnetically levitated antenna pointing system using electromagnets [5].

2 Design and Modelling

The determining factor in the decision to design a 6-DOF wrist with active magnetic suspension was the desire to have a single, rigid, frictionless moving part. As an analogy to the term *rotor* of conventional motor terminology, the term *flotor* will be used for the wrist's moving part. Magnetic bearings (as in [4]) and/or toothed-structure actuators can be used for active magnetic levitation. However, such *electromagnetic* actuators are non-linear and are effective only with extremely narrow motion gaps, thus posing severe motion range limitations for flotors with more than 3-DOF. In addition, for electromagnetic actuators of the magnetic bearing type, the actuation forces are unidirectional and therefore each degree of freedom requires two antagonistic actuators, leading to fairly difficult design and control problems [6].

For a maglev wrist to be capable of "soft" compliant motions, a quite different *electrodynamic* moving coil approach should be used, with magnetic gaps wide enough to allow 6-DOF motion through a substantially larger workspace. The magic wrist prototype, which has a workspace of approximately ±5mm, ±4° from a nominal center, is structured as interleaved hexagonal rings and is shown schematically in Figure 1. A stator element comprised of an inner and outer ring of permanent magnet structures is fixed to the robot. Nested between these rings is a flotor element ring containing flat-wound coils. A hexagonal platform on top of the moving flotor (not shown in Figure 1) carries the gripper, tooling or end effector necessary to perform the task. A maglev wrist's basic electrodynamic actuator formed by the combination of an individual magnet structure and its associated coil will be referred to as a *forcer*. The magic wrist's forcer is shown in Figure 2. A commanded coil current passes through the polyimide-insulated flat copper wire and generates a "current sheet" which interacts with the flux (steered by permeable 1008 steel back-plates) of four powerful Nd 35 magnets to generate a Lorentz force of approximately 6 N/A.

The actuator shape and layout of an electrodynamic maglev device is dictated to a large extend by the application. As seen in Figure 1, in the case of the magic wrist there are six forcers alternately arranged horizontally and vertically around the hexagonal ring, with the ring shape providing easy access on one side for robot mounting, and on the other for end-effector mounting.

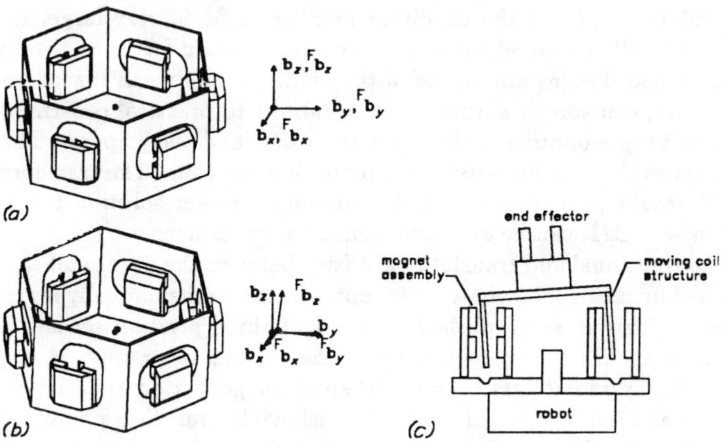

Figure 1. *Flotor Configuration: (a) at zero position and orientation, showing coincident stator and flotor frames, (b) displaced and rotated, (c) displaced and rotated, side-view.*

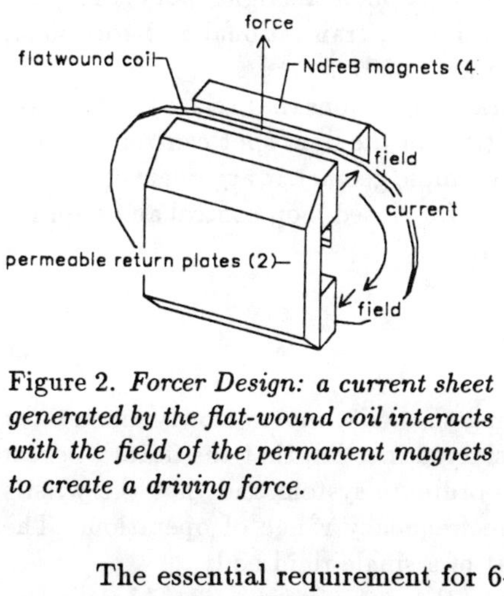

Figure 2. *Forcer Design: a current sheet generated by the flat-wound coil interacts with the field of the permanent magnets to create a driving force.*

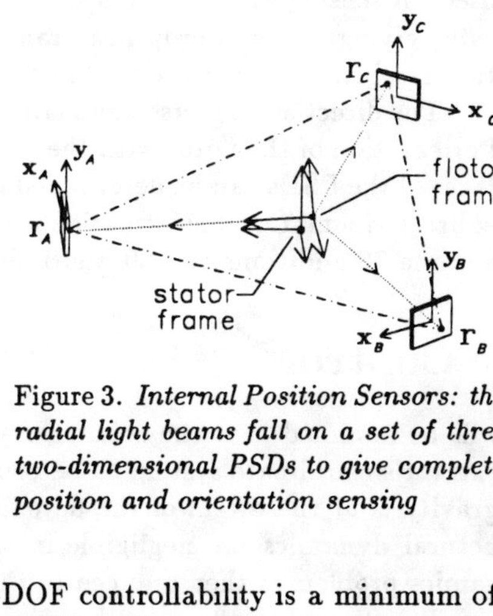

Figure 3. *Internal Position Sensors: three radial light beams fall on a set of three two-dimensional PSDs to give complete position and orientation sensing*

The essential requirement for 6-DOF controllability is a minimum of at least six electrodynamic actuators, arranged in such a manner that the resultant six-dimensional force-torque vector $\mathbf{w}$ acting between the stator and the flotor can have an arbitrary direction. Indeed, if we let $\mathbf{I} = [I_1...I_6]^T$ be the vector of coil currents, $\mathbf{I}$ and $\mathbf{w}$ are related by the transformation $\mathbf{w} = M\mathbf{I}$, where the 6 by 6 matrix M depends on the wrist geometry and the translational and orientational offsets between the flotor and the stator. The matrix M plays a key role in the design of electrodyanmic maglev wrists or other maglev motion devices. Indeed, it should be noted that, for the flotor to be controllable, M must be non-singular for any of the flotor's position and orientation within its workspace. In fact, for a given amount of power used to drive the wrist, the range of all possible force-torque vectors $\mathbf{w}$ generated between the wrist's stator and flotor is a six-dimensional ellipsoid with principal axis lengths given by the singular values of M. If the condition number of M is very large, it would take unrealistically high coil currents to generate certain forces and torques on the flotor. Good designs aim for M with condition number as low as possible, (after some type of torque normalization), subject to practical constraints, *e.g*, the need to have mounting surfaces for the robot and the gripper. The same analysis carries through for wrists with more than six coils. The transformation matrix M would be non-square and a minimum power solution for the coil currents in $\mathbf{w} = M\mathbf{I}$, can be obtained from its pseudo-inverse.

The rotational and translational offsets between the flotor and the stator are obtained by means of a non-contact optical sensor schematically represented in Figure 3. This is accomplished by a set of three position sensors located at 120° intervals around the periphery of the stator's inner ring. A triplet of narrow, coplanar, radial light beams 120° apart are generated by LEDs attached to the flotor and impinge on three two-dimensional lateral effect photodiodes (or PSDs, for Position Sensing Device). The centroids of the light spots projected onto the active areas of the PSDs are easily obtained by measuring the current through the PSD electrodes. It is clear that, depending on the geometry selected in the design of a maglev wrist, other geometries of LED beams and PSDs could be used for sensing. The position sensing resolution of the light spots is roughly 0.5 μm, giving, after appropriate transformation, translational and rotational motion resolutions of approximately 1 μm and 0.001°.

The direct and inverse kinematic transformations that relate the position and orientation of the flotor with the positions of the light spot centroids on the surfaces of the PSDs can be determined by simple geometric arguments (see [3]). The direct kinematic transformation is used for closed-loop control and requires more than 70 additions and 80 multiplications.

3 Control

It will be assumed that the robot carrying the wrist moves slowly enough for the stator's coordinate system to be considered inertial, that the flotor's center of gravity is at the origin of the flotor coordinate system and that the wrist's structural dynamics are negligible in the frequency range of operation. The dynamics problem is then reduced to that of a single rigid body.

Rotation of the flotor with respect to the stator is represented by the Euler parameter vector, defined by $\mathbf{p} = [\beta_0 \boldsymbol{\beta}^T]^T = [cos(\phi/2)\, sin(\phi/2)\mathbf{s}^T]^T$, where $\mathbf{s}$ and ϕ are the axis ($\|\mathbf{s}\| = 1$) and, respectively, the angle of rotation. This parametrization has several advantages, the most important one being the explicit representation of the axis of rotation.

It can be shown [7] that after linearization (approximate or by exact cancellation of the dynamics, as in computed torque control), the flotor's task space dyanmics is given by

$$\ddot{\boldsymbol{\beta}} = \frac{1}{2} J^{-1} \boldsymbol{\tau} = \mathbf{u}_1 \qquad (1)$$

$$\ddot{\mathbf{r}} = \frac{1}{m}\mathbf{f} + \mathbf{g} - \mathbf{r}_F \times J^{-1}\boldsymbol{\tau} = \mathbf{u}_2, \tag{2}$$

where $\mathbf{r}$ and $\mathbf{r}_F$ are the tool-point position with respect to the stator and flotor coordinate systems ($\mathbf{r}_F$ is constant), m and J are the flotor's mass and inertia matrix (in body coordinates and about principal axes, so J is diagonal and constant), $\mathbf{g}$ is the gravitational acceleration and $[\mathbf{f}^T\boldsymbol{\tau}^T]^T = \mathbf{w} = M\mathbf{I}$ is the force-torque vector acting between the flotor and the stator ($\boldsymbol{\tau}$ is in flotor coordinates) and is controlled by the vector $\mathbf{I}$ of coil currents.

Thus, linearizing the flotor dynamics leads to decoupled second order systems, which makes the flotor control very simple. For example, if we let $\mathbf{u}_1$ and $\mathbf{u}_2$ in (1) and (2) be specified by the PD control law

$$\mathbf{u}_1 = K_p(\boldsymbol{\beta}_d - \boldsymbol{\beta}) - K_v\dot{\boldsymbol{\beta}} \tag{3}$$

$$\mathbf{u}_2 = \tilde{K}_p(\mathbf{r}_d - \mathbf{r}) - \tilde{K}_v\dot{\mathbf{r}} \tag{4}$$

where $\mathbf{r_d}$ is the desired tool-point position, $\boldsymbol{\beta}_d$ describes the desired tool orientation and for rotation matrices R and $\tilde{R}$, the gain matrices $K_p, K_v, \tilde{K}_p, \tilde{K}_v$ are of the form $K_p = R^T diag(k_{p1}, k_{p2}, k_{p3})R$, $K_v = R^T diag(k_{v1}, k_{v2}, k_{v3})R$, $\tilde{K}_p = \tilde{R}^T diag(\tilde{k}_{p1}, \tilde{k}_{p2}, \tilde{k}_{p3})\tilde{R}$, $\tilde{K}_v = \tilde{R}^T diag(\tilde{k}_{v1}, \tilde{k}_{v2}, \tilde{k}_{v3})\tilde{R}$, the closed-loop system dynamics are given by decoupled second order systems in coordinate frames determined by R and $\tilde{R}$.

The gains $K_p, K_v, \tilde{K}_p, \tilde{K}_v$ in (3) and (4) can be adjusted in order to synthesize a number of useful mechanisms with different stiffneses. For example, if the stiffness is set to be high along all motion directions but one, then a unidirectional plunger or a single-axis rotator is synthesized. If all stiffnesses are large except two in translation, a slider mechanism results. By setting all the rotational stiffnesses high, only translational motion is allowed. Conversely, by setting the translational stiffnesses high, a ball and socket rotator is created. By selecting different tool-point displacement vectors $\mathbf{r}_F$, the center of rotation of the flotor can be programmed, allowing mechanisms with remote centers of compliance, as may be needed for part mating and peg-in-hole insertions.

4 Experimental Results

The direct kinematic transformations, a PID controller (with low-pass filtering), and the transformation determining the coil currents for a desired force-torque vector acting on the flotor were implemented on a VME-based real-time controller (a MC68020-based single-board computer with a Mercury Floating Point Digital Signal Processor used as a co-processor were used for the control calculations) and *the prototype flotor was flown successfully.* Both *position* and *compliant* control were demonstrated. Although interferometer measurements have not yet been performed, from the PSD readings and experience with such sensors, one can deduce that for low to medium stiffness, the translational repeatability for position control is better than 1 μm. Depending on the axis of measurement, position tracking control bandwidths (to - 3 dB) between 20 and 40 Hz were measured using a Hewlett Packard 3562A Dynamic Signal Analyzer.

Several mechanisms were synthesized -unidirectional plungers, rotators, gimbals, translators. Mechanism synthesis was achieved by executing small interpreted C functions that reside on a development workstation. These programs change the gains of the controller *while the wrist is in operation.* The "feel" provided by the wrist's flotor when emulating a mechanism is truly outstanding, due to the lack of friction, low mass and high acceleration.

5 Conclusion

It is expected that electrodynamic maglev wrists, with their frictionless magnetic suspensions, high accelerations, precise positioning capability, and programmable compliance will find many applications in automatic assembly and testing for microelectronics. Beyond immediate application, it would seem, in many respects, that maglev wrists like the magic wrist provide the basis for truly useful fine motion *universal mechanisms.*

References

[1] J.C. Doyle and G. Stein. Multivariable feedback design: concepts for a classical/modern synthesis. *IEEE Trans. Automat. Cont.*, AC-26:4–16, 1981.

[2] R.H. Taylor, R.L. Hollis, and M.A. Lavin. Precise manipulation with end-point sensing. *IBM Journal of Research and Development*, 4:363–376, July 1985.

[3] R. L. Hollis, A. P. Allan, and S. Salcudean. A six degree-of-freedom magnetically levitated variable compliance fine motion wrist. In R. Bolles and B. Roth, editors, *Robotics Research, Vol. 4*, pages 65–73, MIT Press, 1988. Presented at the 4th International Symposium on Robotics Research, Santa Cruz, CA., August, 1987.

[4] Toshiro Higuchi, Masahiro Tsuda, and Shigeki Fujiwara. Magnetic supported intelligent hand for automated precise assembly. In *Proceedings of the 13th Annual IEEE Industrial Electronics Conference (IECON'87)*, November 1987.

[5] K. Takahara, T. Ozawa, H. Takahashi, S. Shingu, T. Ohashi, and H. Sugiura. Development of a magentically-suspended, tetrahedron-shaped antenna pointing system. In *NASA CP-2506, 22nd Aerospace Mechanisms Symposium*, pages 133–147, Hampton, VA, May 1988.

[6] M. Tsuda, Y. Nakamura, and T. Higuchi. Design of magnetic servo levitation for robot mechanisms. In *Proc. of 20th ISIR*, 1989.

[7] S. Salcudean. On the control of magnetically levitated robot wrists. In *Proceedings of the 27th IEEE Conference on Decision and Control*, pages 186–191, Austin, Texas, December 1988.

APPLICATIONS OF MAGNETIC LEVITATION IN ROBOTICS AND MANUFACTURING

TOSHIRO HIGUCHI
Institute of Industrial Science
University of Tokyo

ABSTRACT

Magnetic levitation technique including magnetic bearings has the ability to support objects without mechanical contact. In addition it has the functions of controlling the stiffness and damping of levitation and of controlling the position and attitude of the supported object with an accuracy of a micro meter or submicrons. By taking advantages of these characteristics of magnetic levitation, robots are expected to improve their abilities and to expand the field of activity in clean rooms, vacuums, and space. Applying magnetic bearings to robots, they have good tools for the tasks which need precise manipulation in manufacturing. As the example, clean room robots with magnetic bearing joints, super clean actuators having both functions of stepping motors and magnetic bearings, and magnetically floated robot hand are introduced.

CLEAN ROOM ROBOT

Cleanliness of production environment is said to greatly affect the reliability and yield rate of semi-conductor circuits. Since the width of the circuit pattern reaching the order of sub-micron, prevention of dusts and oil vapor generation in clean rooms is becoming an even greater necessity. Human operators being the greatest source of contaminant, it is desirable to replace them by robots and automated machines. However, when even greater degree of cleanliness of environment is needed, dust and oil generation from the machines can not be overlooked. So, the machines to be used in super clean rooms should not contaminate the air.

There are two ways to avoid the dust generation from machines. The first and conventional one is applying sealing technique to separate the inside where dusts and oil vapor are generated from motors, gears, ball screws, and guides etc. from the outside of the robot where cleanliness is required. For example, most of the commercialized clean room robots are made

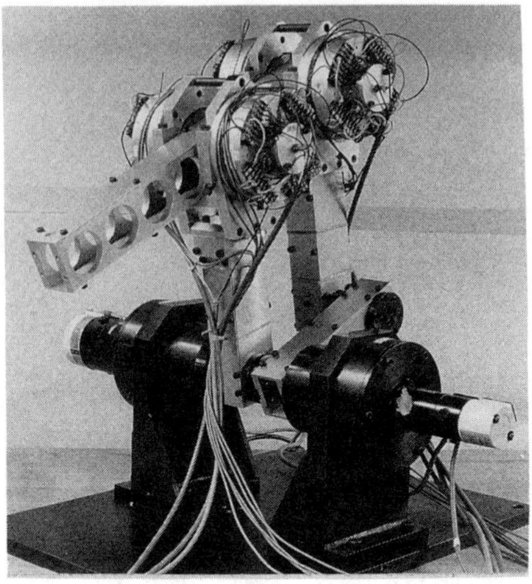

FIG. 1. Clean robot with MB joits.

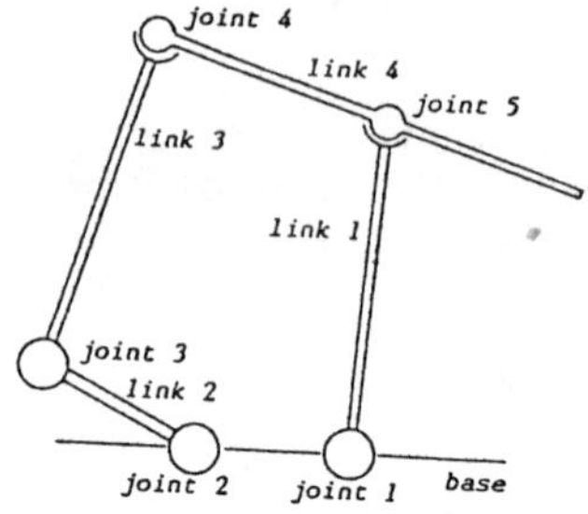

FIG. 2. Structure of robot.

based on the method and also keep the inside pressure of robots lower than the pressure of clean room to prevent escape of dusts. The second and ideal one is to develop inherently clean machines by using magnetic bearings and magnetic levitation to eliminate mechanical contacts which are the worst source of dusts.

The clean room robot installing magnetic bearings in its joints is more expansive than the conventional sealed robots. But, the increasing demand for perfection of cleanliness will choose the robot with magnetic bearings. And since the robots without mechanical contacts need no lubricant, they are also suitable for applications in a vacuum and in space.

In order to make the clean room robot, it is also necessary to use clean actuators, which will be described later. As the first step, we [1] designed and constructed the parallel link robot of which upper two joints are composed of magnetic bearings as shown in Fig. 1.

As shown in Fig. 2, links 1 and 2 are driven by DC motors at joints 1 and 2 respectively. Joint 3, 4, and 5 are passive joints without driving mechanism. Magnetic bearings shown in Fig. 3 are applied to joints 4 and 5. Joints 1, 2, and 3 are composed of usual ball bearings. As the air flow of clean room is made downward, joints 4 and 5 are the most important joints to have to be clean. Link 4 is completely levitated by the two magnetic bearings.

Beside the complete cleanliness, several advantages as follows are expected.

a) free from troublesome problems related to friction
b) precise manipulation and active vibration control
c) force sensing and stiffness control

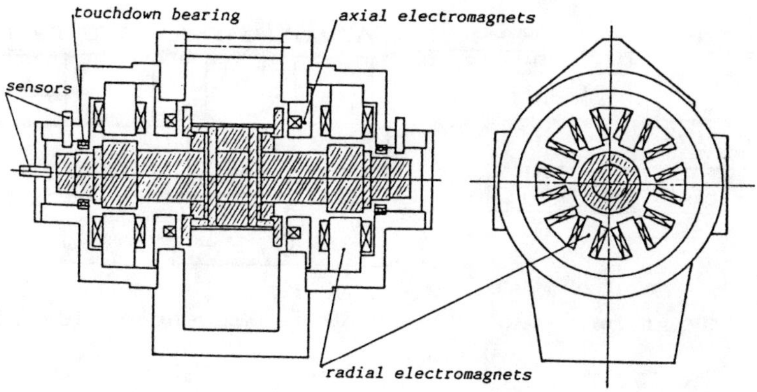

FIG. 3. Structure of the magnetic bearing joint.

SUPER CLEAN ACTUATORS

In order to complete cleanliness, inherent clean actuators are necessary to drive machines in clean rooms. For robots and conveying machines in clean rooms, vacuums, and space, we [2] developed innovated actuators which have the function of magnetic bearings and the function of stepping motors in one body. The developed actuators may be regarded as an amalgam of stepping motors and magnetic bearings.

Explanation of the idea and basic construction of the actuators is done in the case of a linear motor. A conventional linear PM (Permanent Magnet) stepping motor is shown in Fig. 4. Propulsive force is gained by superposition of the magnetic flux generated by the currents of coils onto the magnetic flux generated by the permanent magnets. Since a strong attractive force, approximately ten times of the maximum propulsive force, acts between the mover and stator, ball bearings, which are the source of dusts and oil vapor, are indispensable to maintain a constant gap.

The basic element of the super clean actuators is shown in Fig. 5. Its structure is similar to that of PM stepping motor except that the permanent magnets are replaced by electromagnets (main coils). By controlling the current of main coils based on gap sensor signal, mover is able to levitate keeping a constant gap in spite of the toothed pole structure inherent to stepping motors.

Six coils are wound to laminated core. A pair of lager coils (main coils) connected in series produce main magnetic flux. Coils A1 and A2 are connected in series and when one coil produces a magnetic flux in the same direction as the main flux, the other produces a magnetic flux against the main flux. The same applies for coils B1 and B2. The core has four

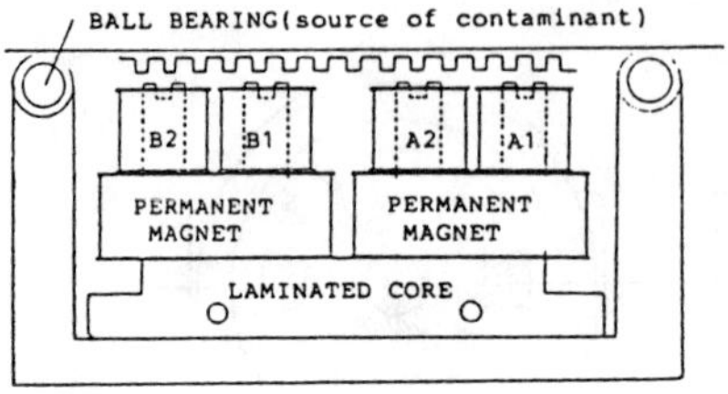

FIG. 4. PM stepping motor.

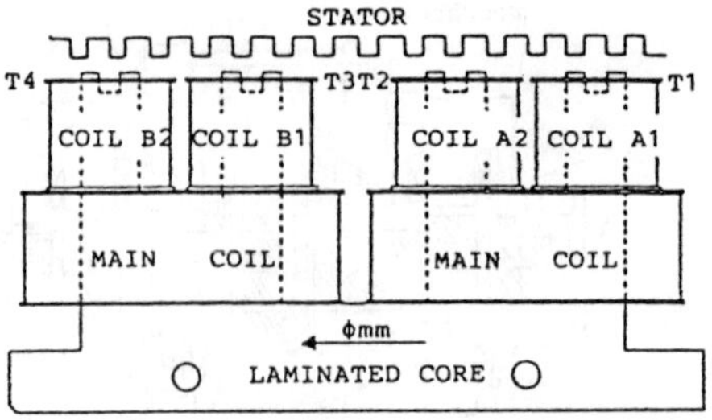

FIG. 5. Element of actuator.

toothed poles. Let one pitch of the teeth be expressed as 2pi radian, then the phase relation of the toothed poles are presented as T2-T1 = 1 pai rad. and T1-T3 = T4-T1 = $\frac{1}{2}$ pi rad.

Due to the structure, summation of the attractive forces of the four poles is rather constant and insensitive to the X position, whereas the attractive force on each toothed pole fluctuates severely along X direction. And the active control of attractive force of the basic element can be realized by only controlling the main coil current independently of the X position. Driving and positioning of the motor can be done in exactly the same manner as 2-phase PM stepping motors by using coil A and coil B. Micro-step drive method is preferable to this actuator for calm suspension, precise positioning, and smooth movement.

Since the basic element in itself has no function of attitude control, totally contactless levitation of a rigid body requires several number of the elements and/or co-use of controlled electromagnets. More than 10 types of the super clean actuators (linear motion type, rotary motion and composed type) have been made and tested [2],[3]. As a example of composed motion type, θ-Z actuator for wafer transferring is shown in Fig. 6. The actuator is outer rotor type. On the upper part of the inner surface of the rotor, teeth are machined so as to control the rotary motion. And on the lower part of the surface, circular grooves are machined to control the height position. Both the rotary and up-down motions are controlled by means of the function of stepping motors. The other freedoms are controlled actively by using four eddy current type gap sensors to achieve firm suspension like active magnetic bearings.

As the super clean actuators are completely free from friction, they can do precise force, torque, velocity, and position control. As to the actuator shown in Fig. 6, the accuracy of up-down positioning is ±1 um and ±2 sec for rotation.

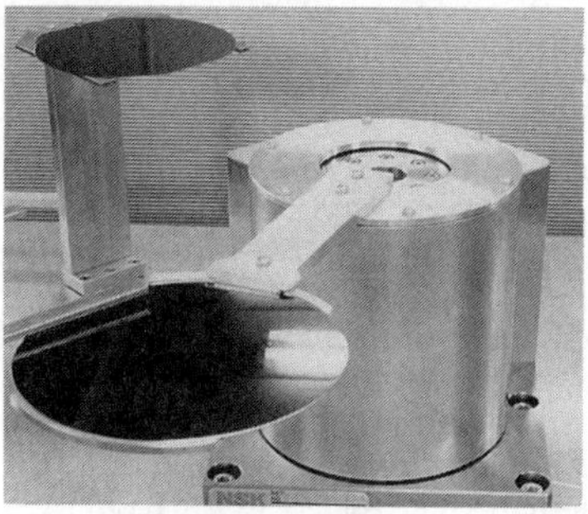

FIG. 6. θ-z super clean actuator.

MAGNETICALLY SUPPORTED INTELLIGENT HAND

Magnetically supported intelligent hand (MSIH) is a new end effector for robots to do precise assembling by using magnetic bearings.

In the assembling process, usually it is very difficult to align a part to another part. So it is necessary to correct the attitude and position of the part. The developed hand can achieve very precise assembling tasks by correcting the alignment errors automatically by means of both passive method and active method. For the passive method, the magnetic bearing supports one on the mates with appropriate compliances like RCC [4]. And for the active correction, the magnetic bearing manipulates the part according to information of forces acting between the mates. The force can also detected by the currents of coils and air gaps of the magnetic bearing without using additional force sensors.

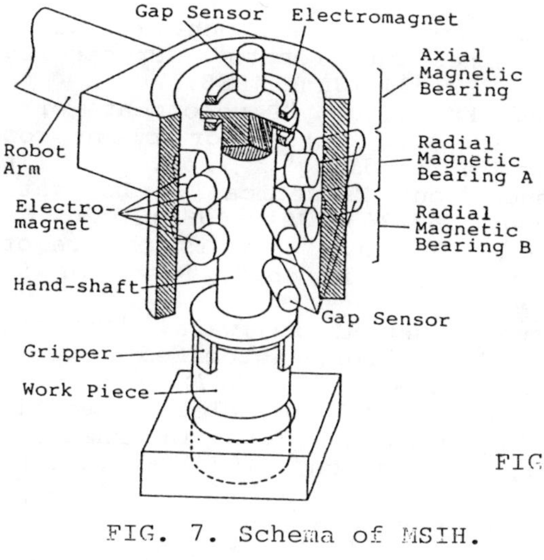

FIG. 7. Schema of MSIH.

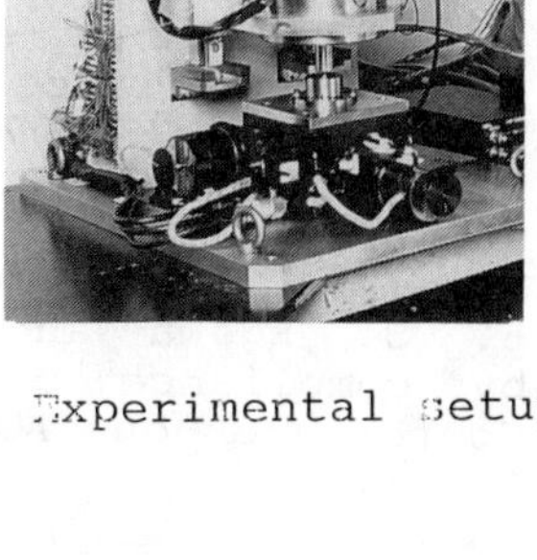

FIG. 8. Experimental setup.

The hand has the unique structure shown in Fig. 7 incorporating a magnetic bearing [5]. It is composed of a hand shaft, two radial magnetic bearing units and one axial magnetic bearing unit. The hand shaft which hold a work piece with its gripper is supported by the magnetic bearing. Five gap sensors are installed and five degrees of freedom of motions of the shaft are controlled actively by using DSP.

The MSIH can posses a function corresponding to that of the RCC. It also can easily and intentionally move the compliance center and change the support stiffness by altering the control parameters of the magnetic bearing. Whereas the mechanical RCC devices have a fixed compliance center and stiffnesses, the MSIH has the flexibility to be adapted to many kinds of parts.

The MSIH can realize more effective passive insertion by controlling the compliance during an insertion operation. And it is possible to support a work piece stiffly during transportation by the robot, whose arm is connected to the MSIH, to insertion spot, but flexibly during insertion in order to save the settling time of vibration of the hand shaft caused by the rapid robot motion.

We have succeeded in automated insertion such a task with 4um clearance by the MSIH experimental setup shown in Fig. 8. And chamferless insertions were also done by using MSIH and another hand based on the same concept [6].

REFERENCES

1. T. Higuchi and K. Oka, "Study of robot manipulators using magnetic bearing joints", Proceedings of the 20th International Symposium on Industrial Robots, 529-536 (October 1989).
2. T.Higuchi and H. Kawakatu, "Development of super-clean actuator for machines and robots", Proceedings of the 12th MOTOR-CON88, 322-334 (June 1988).
3. T. Higuchi and H. Kawakatu, "Development of a magnetically suspended stepping motor for clean room transportation and sample handling", Proc. of 11th International Conference on Magnetically Levitated Systems and Linear Drives, 363-368 (July 1989).
4. D.E. Whitney and J.L. Nevins, "What is the remote center compliance and what can it do?", Proc. of 9th ISIR, 135-152 (1979).
5. T. Higuchi, M. Tsuda, and S. Fujiwara, "Magnetic supported intelligent hand for automated precise assembly", Proc. of IECON'87, 926-933 (1987).
6. S. Egawa and T. Higuchi, "Chamferless insertion by compliance control", Proc. of 10th International Conference on Assembly Automation, 107-114 (October 1989).

SECTION 4 : ROBOT CONTROL

ISSUES IN FORCE CONTROL OF ROBOTS WITH NONLINEARITIES

Julio J. Gonzalez[†], Luciano Chirinos[†], Shaheen Ahmad[‡], and Glenn R. Widmann[†]

[†] Colorado State University
Institute of Robotic Studies
Department of Electrical Engineering
Fort Collins, CO 80523

[‡] Purdue University
School of Electrical Engineering
West Lafayette, IN 47907

ABSTRACT - In this paper, two important issues in the force control of real robot manipulators are investigated. First, the effect of control signal saturation on system stability, and secondly, the performance of force control in the presence of a moving environment. Concerning the former topic, the non-linearity of control signal saturation (present in all real systems) is found to be the most important reason for instability of discrete force control systems, as opposed to other non-linearities (like slip-stick friction), more commonly treated in literature. Regarding the latter topic, conception of environmental motion as a system disturbance leads to feedforward compensation with marked improvement of system performance. The theoretical analysis of both topics is followed by experimental verifications performed on real robots.

1. INTRODUCTION

Surprisingly, very little work has been performed on the effects of nonlinearities on the force controlled system performance. In [1],the authors examined the existence of limit cycles due to the discontinuous surface chattering. Also, a study of the instability caused by the non-linear, slip-stick friction on an integral-type feedback force control structure has been analyzed [2]. A central theme in the above investigations is that the analysis was performed assuming a continuous-time system. In a practical sense, the force controller must be implemented in discrete-time.

In this paper, the destabilizing effects that arise due to discrete-time control on the manipulator system which exhibit nonlinearities will be examined. Specifically, this paper will demonstrate that for the system using even a simple discrete proportional-type feedback force controller structure will lead to instabilities, even when the sampling theory is satisfied. This unstable behavior is actually not due to the non-linear slip-stick friction, but rather can be explained by another commonly occurring system non-linearity, namely the saturation of the control signal.

The second topic under study, force control performance in the presence of a moving environment is relatively new, and consequently has also received little attention, in spite of its potential importance in future industrial applications. Early works in this subject can be found in [3] and [4]. In this paper, the environmental motion is conceived as a system perturbation, and consequently feedforward compensation is attempted which successfully improves the system performance.

2. CONTROL SIGNAL SATURATION NON-LINEARITY

2.1. Model Representation

For the simplicity of experimental verification, only a 1 DOF case will be considered. Consider the system configuration, displayed in Figure 1, which uses only the shoulder joint of a PUMA 560 robot that touches a purely elastic static environment with a stiffness coefficient, K_E. Thereby, the robot configuration, $\underline{q}$, will remain approximately constant during a force experiment. A control voltage u (provided by a controller developed at Colorado State University [5]) is applied through a transadmittance amplifier to a D.C. motor that drives the joint, and the environment reacts with force f_{env}. As a result of the robot configuration, this reaction force is applied nearly along y_2.

In [6], it is shown that under all the above assumptions, the plant dynamics reduces to:

$$u = m\ddot{q} + v\dot{q} + g_u - \gamma l f_{env} + u_{SS} \qquad (1)$$

where

m	: mass	l	: length of link (.i.e., PUMA 560 Shoulder)
v	: viscous friction coefficient	γ	: conversion factor from torque to voltage
g_u	: voltage to overcome gravity	u_{SS}	: non-linear voltage representing slip-stick friction

Since the robot configuration remains approximately fixed during a force control experiment, then m, v and g_u can be considered constants. The characteristics of the environmental force becomes:

$$f_{env} = -K_E(y_2 - y_2^0) \quad \text{and} \quad \dot{f}_{env} = -K_E\dot{y}_2 . \tag{2}$$

Combining (1) and (2), using $\dot{q} = \dot{y}_2/l$, and setting $y_2^0 \equiv 0$, we have:

$$u = -\left\{\frac{m}{K_E l}\ddot{f}_{env} + \frac{v}{K_E l}\dot{f}_{env} + \gamma l f_{env}\right\} + \left\{g_u + u_{SS}\right\} \triangleq u_l + u_n . \tag{3}$$

The control input is comprised of a linear component, u_l (which contains the sign), and a non-linear component, u_n, as illustrated in Figure 2. The linear part of the dynamics is characterized by the type 0, second-order transfer function:

$$\frac{F_{env}(s)}{U_l(s)} = \frac{K_p\omega_n^2}{s^2 + 2\zeta\omega_n s + \omega_n^2} = \frac{[\gamma l]^{-1}\,\omega_n^2}{s^2 + 2\zeta\omega_n s + \omega_n^2} \tag{4}$$

where: $\omega_n = l\,[\gamma K_E m^{-1}]^{1/2}$ and $\zeta = \frac{1}{2} v\, l^{-1}\,[\gamma K_E m]^{-1/2}$.

The following are the parameter values experimentally identified for the PUMA 560 robot using the CSU Testbed facility.

$$K_p = 40\ (N/V); \quad \omega_n = 90\ (s^{-1}); \quad \zeta = 0.1; \text{ and } \quad g_u = -2.4\ (V). \tag{5}$$

For model discretization, the continuous-time system model representation, shown in Figure 2, is modified by adding a Zero Order Hold (ZOH), with a sampling period T, at the control output u(t). Consequently, we may denote this constant-held sampled signal as $u_h(k) \triangleq u(kT)$ for the sampled period $kT \leq t < (k+1)T$. Because of the actual physical presence of the D/A converter (i.e., ZOH), a ZOH mapping seems to be the most appropriate approach. However, this method will lead to an exact discretization of the linear portion of the plant (4) only if the input to the linear plant, namely u_l, remains constant during the sampling period $kT \leq t < (k+1)T$. In [6], we showed that this is indeed the case in spite of the existence of the internally generated nonlinear signal, u_n (i.e., a signal which can not be sampled) due to gravity and slip-stick friction.

2.2. Effects of Control Signal Saturation on Stability.

In this section, we present the underlying theoretical principles that will be used to demonstrate that a robotic system under the influence of a discrete-time proportional-type feedback force control structure can be driven to instability when the control input, u, saturates. This instability results from the interaction between sample and hold, low damping coefficient (inherent to force control) and saturation non-linearity.

Consider the closed-loop system of Figure 3, with a P-type controller of gain K_c. The limiting saturation value of the control input is denoted by, u_o. Furthermore, consider the case when the input to the system is a step function of amplitude F_d. Thereby, the system error, e, is defined as $e \triangleq F_d - f$. Lets initially consider the case when (1) non-linear slip-stick friction has been neglected, and (2) the system is continuous-time. (i.e., $\lim\{T\} \rightarrow 0$). These two idealized conditions will be eliminated, later in the experiment section, in order to observe their effects.

The piecewise linear characteristic curve of the saturation non-linearity defines three distinct linear regions, denoted by R_I, R_{II}, and R_{III}, for the control output, u.

$$u = \begin{cases} u_o & \text{if } eK_c > u_o \quad (R_I) \\ K_c e & \text{if } |e|\,K_c < u_o \quad (R_{II}) \\ -u_o & \text{if } eK_c < -u_o \quad (R_{III}) \end{cases} \tag{6}$$

Using standard phase plane analysis techniques, it is easy to determine that when the control output, u, is operating in the distinct region R_i (i = I,II,III), then the system is attracted to the respective stable convergence focus F_i (i = I,II,III). Defining the phase plane coordinate axes as the pair $(e,\dot{e}/\omega_n)$, then the three stable focuses corresponding to each region are shown in Table 1.

Table 1. Location of Stable Focus.

System Operation		Corresponding Stable Focus	
R_I	$u_o < eK_c < \infty$	F_I	$(e,\dot{e}) = ([F_d - K_p u_o],0)$
R_{II}	$-u_o \leq eK_c \leq u_o$	F_{II}	$(e,\dot{e}) = (F_d[1 + K_c K_p]^{-1},0)$
R_{III}	$-\infty < eK_c < -u_o$	F_{III}	$(e,\dot{e}) = ([F_d + K_p u_o],0)$

In [6], it is shown that for a practical system (i.e., $K_c K_p \gg 1$), focus F_I is located in region R_{III} (and vice-versa) and F_{II} is located in R_{II}. This means that the system is attracted to a point in R_{III} when operating in R_I, and to a point in R_I when operating in R_{III}. Consequently, if the system is stable, it will spiral in and eventually gets caught by the convergence focus F_{II}.

2.3. Experiments.

In order to illustrate the stable/unstable behavior of the system, consider the numerical example which uses the practical experimental values from (5) along with:

$$K_c = 0.25 \qquad u_o = 5\,V \qquad \text{and} \qquad F_d = 100\,N \ . \tag{7}$$

The phase-plane coordinates are $(e,\dot{e}/\omega_n)$ for the continuous case, and $(e(k),\{\Delta e(k)\}/\omega_n)$ for the discrete case, where Δ denotes the backward difference operator.

EXPERIMENT 1 (Figure 4). This case illustrates the phase-portrait response which corresponds to the continuous-time system (i.e., T = 0). The system is clearly stable, since after one cycle the distance from the phase-plane curve to focus F_{II} has decreased with respect to the initial distance.

EXPERIMENT 2 (Figure 5). It is a discrete-time case, with a sampling period T = 1.5 msec. Notice that this sampling period more than satisfies Shannon's theorem. That is, the sampling frequency is $f_s = 666.67\,Hz$, while the closed-loop resonant frequency is

$$f_N = [2\pi]^{-1}\omega_n \sqrt{1 + K_c K_p} = 47.5\,Hz \ . \tag{8}$$

This case illustrates that for even a extremely unpractical sampling period the system exhibits instability. The explanation of this phenomenon is that the ZOH process has destroyed the *harmonic sequence* (i.e., system in a specific region and convergence focus in the appropriate region). In fact, consider for example, the system at a point in R_I and just about to enter R_{II}. While in R_I, the control signal, u, is u_o which is driving the system to F_I. Now when the system enters R_{II}, before a sampling period elapses, the control signal is still remains at u_o (because of the hold action). That means that the system is still being driven to F_I, instead of F_{II}. The drastic effect of increasing T can be visualized from the fact that it takes a short time for the system to transit narrow region R_{II}.

EXPERIMENT 3 (Figure 6). Even at T = 3 msec is enough to produce a limit cycle. In practice such a cycle could never be observed, since e(k) growing beyond F_d means that the environmental force is negative. This physically means the manipulator has broken contact with environment.

EXPERIMENT 4 (Figure 7). The purpose of this experiment is to show that stable force control is possible when the control signal saturation non-linearity is avoided. The experiment shows the simulated close-loop step responses without (Curve a) and with (Curve b) the inclusion of non-linear friction. The sampling period used in the simulation is 15 msec (i.e., CSU testbed sampling period). A lead-lag algorithm whose low-pass filter action prevents the control signal from hitting the saturation bounds $\pm u_o$ is simulated. In curve (b), non-linear friction even adds a stabilizing effect, but this conclusion is, of course, only valid for a type-0 controller. Also, because of Coulomb friction, the steady-state error differs slightly from the theoretically predicted.

3. FORCE CONTROL IN THE PRESENCE OF A MOVING ENVIRONMENT

An important concern for a force controlled manipulator is the force tracking performance in response to variations of the environmental position in the direction of the applied force. The general expression for the closed-loop force control response developed in [3] demonstrates the environmental motion of the workpiece can be perceived as a disturbance to the system. Thereby, the transient force error resulting from this disturbance can be reduced through the use of feed-forward compensation. In [3], it is shown that for a single-link manipulator, the optimal feed-forward positional compensator is represented by a fifth order filter. For experimental verification however, servo compensation will only be shown for a reduced order approximation for this compensator filter. Thereby, the positional compensator will be implemented by a second order filter of the form: $Y_p(s) = Y_{pv}s + Y_{pp}$, where the coefficients of the compensator are chosen to minimize the ITAE criterion. In particular, simplification of this compensation filter yields $Y_{pp} \equiv 1$. On the other hand, the exact form of Y_{pv} is a complicated expression of the system parameters as shown in [3].

3.1. Experimental Set-Up.

The description of the surface motion trajectory is chosen to be uncomplicated, but still allowing valuable insight into the effects of dynamic compensation. Figure 8 presents the top view illustration of the experimental set-up. It shows a two bar linkage configuration which makes contact with the distal tip of a single link manipulator. A separate DC servo motor, under closed-loop speed control, is added to the overall system. The purpose of the second motor is to rotate the first linkage bar with a constant angular velocity ω. The second linkage is permitted to rotate freely about the first while the distal tip maintains contact with the manipulator. The motion of the linkage system resembles a piston action, with respect to the manipulator. The distal tip of the second linkage bar moves in a nearly perpendicular direction with respect to the link without permitting any relative lateral slippage. Thereby, the main advantage of this closed chain configuration is that lateral surface motion friction has been eliminated. The ensuing motion of the linkage at the point of contact (i.e., the contact surface motion trajectory) is theoretically determined [3] to be a waveform that is nearly sinusoidal in nature.

3.2. Experiments.

EXPERIMENT 1 (Figure 9). This figure displays the force tracking behavior of the uncompensated force controlled system for a rotating surface with an angular velocity of one Hz. Additionally, the force experiment is performed with a step input force demand of 1000 (unit-less). Using linear system analysis, the force response theoretically should be represented by an approximate sinusoidal waveform. However, the experimental force response is not sinusoidal in nature. The difference between the theoretical and experimental force can be attributed to static friction. This is because as the surface motion reaches its peak, the direction of link motion must also change. Thereby, the link velocity for an instant becomes zero. During this interval, the static friction prohibits further joint motion until a threshold torque is exceeded and joint motion ensues. Hence, in the neighborhood of motion peaks, the force error will also have a peak.

EXPERIMENT 2 (Figure 10). This experiment illustrates the improved force tracking behavior of the force control system which has been compensated for surface motion. Positional servo compensation is driven based on the theoretical motion trajectories. Thereby, this method assumes perfect knowledge of the surface motion. A second servo compensation (not shown) that assumes absolutely no apriori knowledge of the surface motion was also investigated. The motion trajectory of the surface was obtained using the motor encoder information and the third order Lagrangian estimate of its velocity. The results obtained were good (similar to the previous case) provided that the joint flexibility is negligible, in which case the actuator position is nearly the joint position.

4. CONCLUSIONS

Control signal saturation, a non-linearity present in all real robotic systems, has been identified as the main factor for instability in discrete force control systems. A control algorithm that prevents the control signal from hitting the saturation bounds leads to an stable system in spite of the presence of non-linear slip-stick friction.

Feedforward compensation of surface motion has proved successful as a way of keeping an acceptable performance when a manipulator acts on a moving workpiece.

REFERENCES

[1] S.D. Eppinger and W.P. Seering, "Understanding Bandwidth Limitations in Robot Force Control," *Proc. IEEE Int. Conf. on Robotics and Automation,* Raleigh, NC, pp. 904-909, April 1987.

[2] W.T. Townsend and J.K Salisbury, "The Effect of Coulomb Friction and Stiction on Force Control," *Proc. IEEE Int. Conf. on Robotics and Automation,* Raleigh, NC, pp. 883-889, Mar. 1987.

[3] G.R. Widmann, "Compliant Motion Control of Manipulators with Flexible Joints," Ph.D. Thesis, School of EE, Purdue University, W. Lafayette, IN, Dec. 1988.

[4] R.K. Roberts, "The Compliance of End Effector Force Sensors for Robot Manipulator control," Ph.D. Thesis, School of EE, Purdue University, W. Lafayette, IN, Dec. 1984.

[5] J.J. Gonzalez, L. Chirinos, G.K.F. Lee, and G.R. Widmann, "A Complete Experimental System to Analyze Compliant Motion Control Issues Using an Industrial Size Robotic Manipulator," *Proc. IASTED Int. Conf. on Adap. Knowledge-Based Cntrl Sig. Proc.,* Honolulu, HI, Aug. 1989.

[6] J.J. Gonzalez, L. Chirinos, G.K.F. Lee, and G.R. Widmann, " Experimental Issues in the verification of a Realistic Plant Model for Industrial Size Robots", *Proc. ISMM Int. Conf. on Computer Applications in Design, Simulation and Analysis,* New Orleans, Louisiana, Mar. 1990.

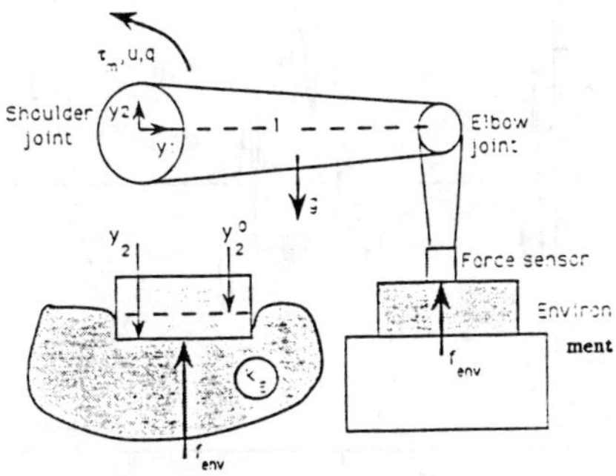

Fig 1. Configuration of Manipulator with Static Surface Contact.

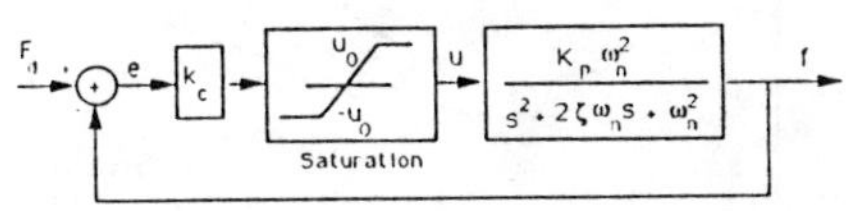

Fig 3. Closed-Loop System with Controller Saturation.

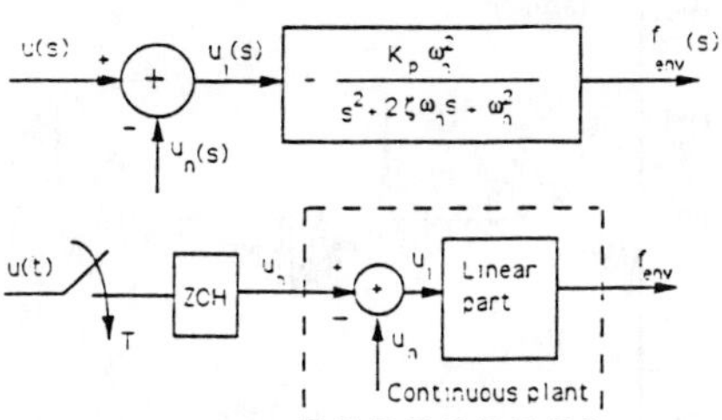

Fig 2. Open-Loop Force Model for a One DOF Robot.

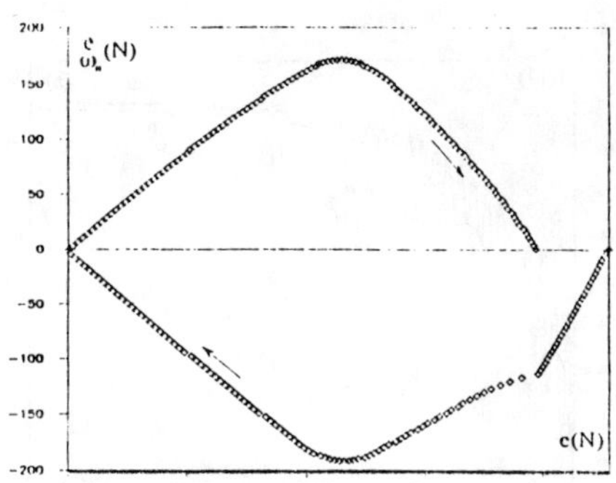

Fig 4. Phase Plane Portrait of Continuous-Time System.

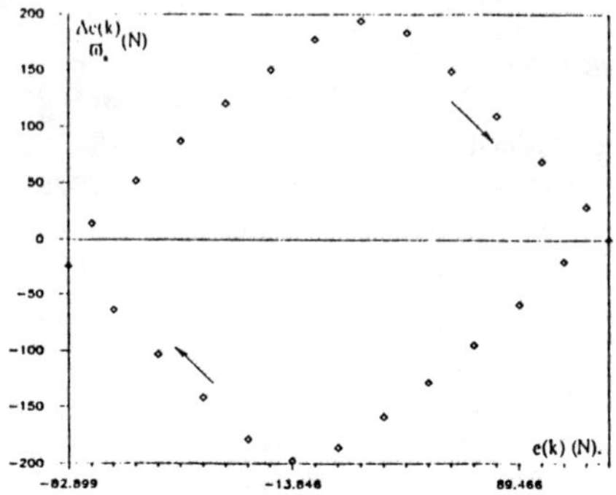

Fig 5. Phase Plane Portrait of Discrete-Time System with T = 1.5 msec.

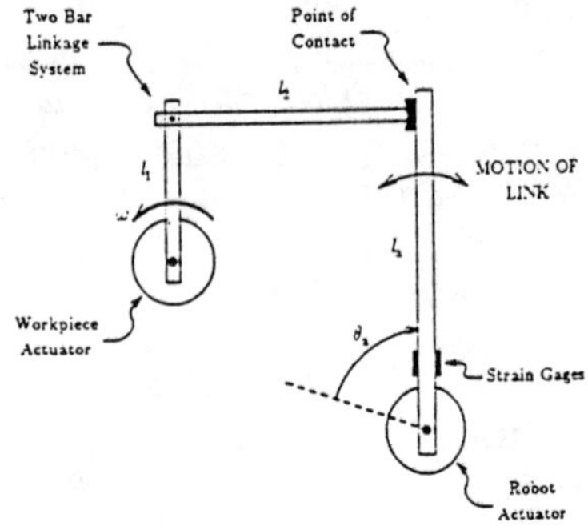

Fig 8. Top View Presentation of the Hardware Configuration for the Single Link Robot with a Non-Static Work Environment.

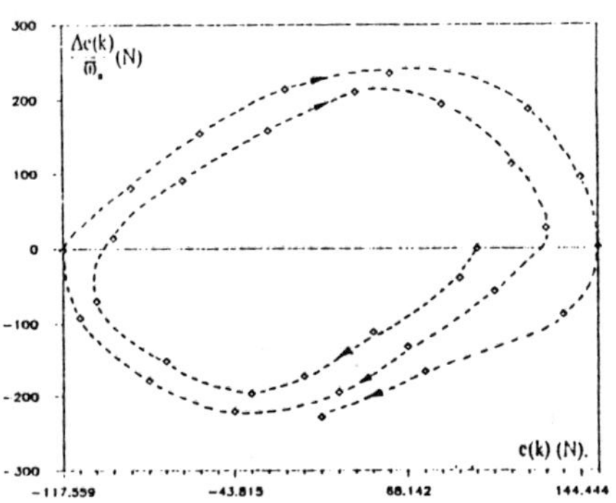

Fig 6. Phase Plane Portrait of Discrete-Time System with T = 3 msec.

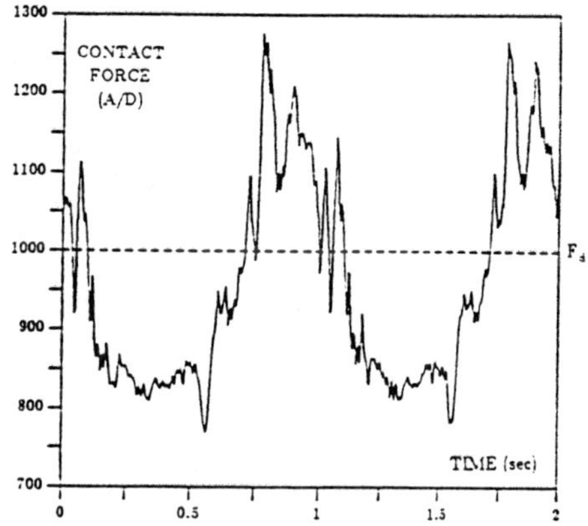

Fig 9. Force Response in a Moving Environment with no Compensation.

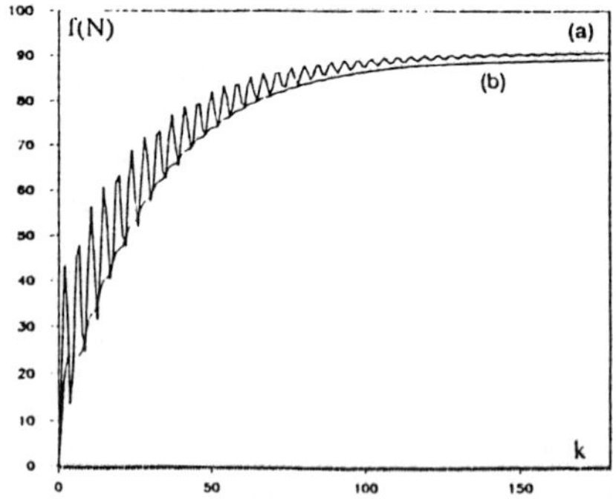

Fig 7. Step Response of System with Lead-Lag Compensation: (a) Without Slip-Stick Friction, (b) With Slip-Stick Friction.

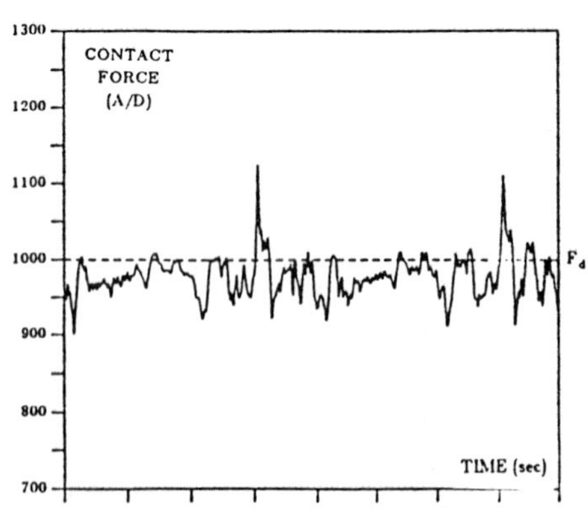

Fig 10. Force Response in a Known Moving Environment using Compensation.

Implementation of Robust Impedance and Force Control Using VSS Theory

Ziren Lu A.A. Goldenberg

Robotics and Automation Laboratory
Dept. of Mechanical Engineering
University of Toronto
Toronto, Ontario
Canada, M5T 1A4

ABSTRACT

This paper presents an approach to the problem of controlling the force for an impedance controlled robot and implementation of target impedance. The force control schemes for an impedance controlled robot in different cases are proposed. An off-line estimation method to the environment model is provided; it is needed by the control schemes in the presence of the environment model uncertainties. A sliding mode impedance controller is developed simultaneously and it provides the basis for the above force control schemes. A new continuous control law is introduced, which reduces the chattering in the control input while maintaining the system performance in the vicinity of the sliding surface.

1 INTRODUCTION

Compliant motion control is extensively studied in the recent years because of its importance to the application of robot in industry. For assuring compliant motion for manipulators in the constrained task, there are two approaches: force control, and impedance control.

One of the objectives in this paper is to address the problem of controlling the force for an impedance controlled manipulator. An off-line estimation method for an environment model including its position, and its mechanical properties like stiffness, damping, and mass, is developed. The method is needed by the above force control schemes when the environment model is not available in advance. A further and related objective in this paper is to develop a robust control scheme for the implementation of target impedance using sliding model control. The merits of this control scheme are: (1) the high-frequency components in the control input are reduced and the performance in the vicinity of the sliding surface is improved; (2) the robustness of stability of the sliding surface is ensured.

In the following Section 2, the schemes of the force control and estimation of the environment model are developed. The theoretical background of the design for the implementation of impedance control is presented in Section 3. The simulation results are detailed in Section 4.

2 FORCE CONTROL

In this section, we assume that the target impedance of a robot I_R has been specified and exactly implemented, the details on how to implement target impedance will be discussed in section 3. When an impedance controlled robot is in contact with the environment surface, see Fig. 1, its dynamic behavior can be represented by its target impedance I_R. The environment can be modeled as a mechanical impedance [Kazerooni, 1988]. In the frequency domain, the dynamic model of the system in Fig. 4 is represented by

$$I_R(X - X_c) = F_e\ ; \qquad (1)$$

and

$$I_E(X_e - X) = F_e\ , \qquad (2)$$

where I_R^* is the target impedance of the robot and I_E is the impedance of the environment. X_c is the command position of the robot, X is the actual position of the robot, and X_e is the position of interaction point before the deformation occurs. The block diagram for equations (1) and (2) is shown in Fig. 2.

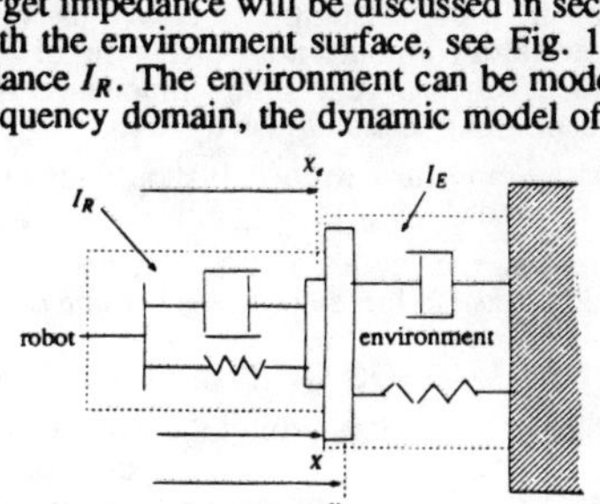

Fig. 1 Dynamic Model of Robot and Environment

2.1 CONTROL OF FORCE

From Fig. 2, we see that the interaction force F_e can be controlled by input command position X_c, but it is also affected by X_e and I_E. The force control problem is discussed in the following two different cases: the environment model is known and the model is not known.

Case 1: Both the position X_e and the impedance I_E of environment are known. Using them in the input X_c (see Fig. 3) will get

$$X_c = X_e - F_d\left(\frac{1}{I_R} + \frac{1}{I_E}\right) \quad (3)$$

where F_d is the desired force. Using equation (1), (2) and (3) leads to

$$F_e = F_d.$$

The interaction force F_e can be precisely controlled as long as an exact model of environment is available. However, in some circumstances, the environment model is difficult to know exactly; it is not always available used in X_c.

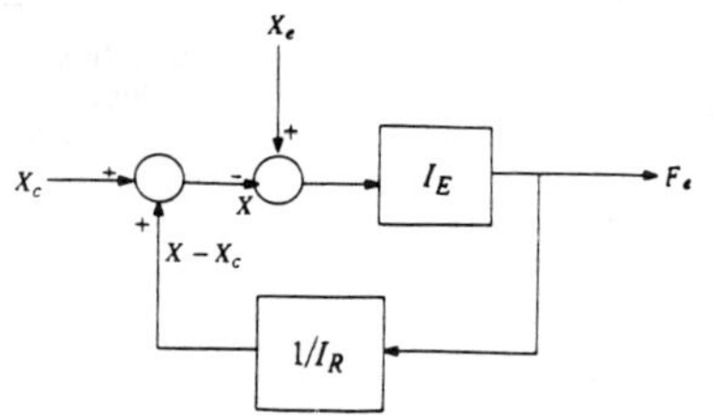

Fig. 2 Block Diagram of Robot and Environment

Case 2: Neither the position X_e nor the impedance I_E is known. We use $\hat{X}_e$, the estimation of X_e, in the control input X_c, which leads to

$$X_c = \hat{X}_e - \frac{F_d}{I_R}, \quad (4)$$

substituting X_c into (1) and using (2) and (4) leads to

$$F_e = \frac{1}{1 + \frac{I_R}{I_E}} F_d + \frac{I_R}{1 + \frac{I_R}{I_E}} \delta X_e,$$

where $\delta X_e = X_e - \hat{X}_e$. Again, if $\left|\frac{I_R}{I_E}\right| = |\alpha| \ll 1$, we will get

$$F_e \approx (1 - \alpha)(F_d + I_R \delta X_e) \approx F_d + I_R \delta X_e. \quad (5)$$

Equation (5) shows that the parameter α, a ratio of the target impedance of a robot to the environment impedance, is an important factor in the above force control. In the absence of the knowledge about the environment model, a smaller α; it implies a smaller I_R, will result in a smaller error in the force F_e in (5), and *vice versa.* However, when the environment is not very stiff, the assumption $\alpha \ll 1$ is no longer valid (I_R could not be very small in practice, since a very small impedance is corresponding to a sluggish dynamic response, and furthermore, it should also satisfy the stability requirement when the robot is contacting the environment).

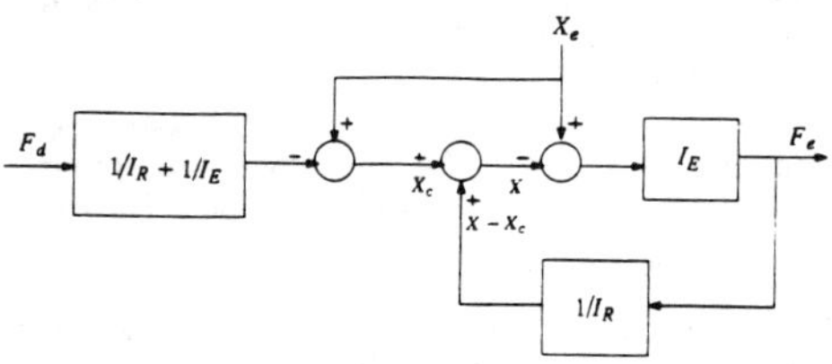

Fig. 3 Force Control

Therefore, the model of the environment have to be used in the above force control schemes to obtain a precise force control when the environment is not stiff.

2.2 ESTIMATION OF THE MODEL OF ENVIRONMENT

The basic idea behind the estimation method is that, consider the robot and environment as a whole system, the input and output variables are the desired force F_d and the interaction force F_e. Using the relationships between them, we obtain the environment model without measuring its deformation. The details are presented in the following.

Usually, the impedance $Z(s)$ is expressed as $\frac{F}{X} = sZ(s)$; hence, in this paper, the relation between the $I(s)$ and $Z(s)$ is $I(s) = sZ(s)$.

When the environment model is totally not known, we first use a preliminary estimation value $\hat{X}_e$ in the control input X_c(see Fig. 4);

$$X_c = \hat{X}_e - F_d \frac{1}{I_R}, \tag{6}$$

from Fig. 3, we get

$$F_e = \frac{1}{I_R} \frac{I_E}{1 + \frac{I_E}{I_R}} F_d + \frac{I_E}{1 + \frac{I_E}{I_R}} (X_e - \hat{X}_e), \tag{7}$$

and express F_e as

$$F_e = F_1 + F_2, \tag{8}$$

where $F_1 = \frac{1}{1 + \frac{I_R}{I_E}} F_d$; $F_2 = \frac{I_R}{1 + \frac{I_R}{I_E}} \delta X_e$. It is noted that, if F_1 and F_2 are both known, the values of I_E and δX_e can be obtained by solving them from F_1 and F_2. Actually, F_1 and F_2 can be achieved as follows. First, let $F_d = 0$; F_2 will equal to F_e as F_1 is zero, so, F_2 is obtained by measuring F_e.

Then, set F_d for a proper value, we'll get another value of F_e, since δX keeps constant, so does F_2; then F_1 is obtained by equation (8). At last, the environment model; I_E and δX_e is obtained by

$$I_E = \frac{1}{\frac{F_d}{F_1} - 1} I_R; \tag{9}$$

$$X_e = \hat{X}_e + \frac{F_d}{F_1} \times \frac{F_2}{I_R} \tag{10}$$

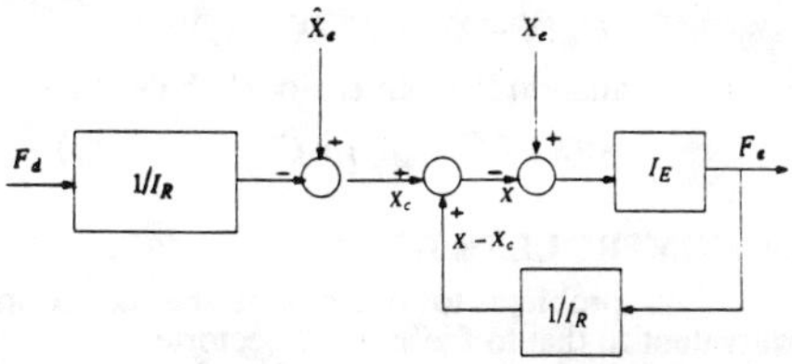

Fig. 4 Estimation of Environment Model

3 IMPLEMENTATION OF IMPEDANCE CONTROL

Last section is dedicated to the problem of controlling the force for an impedance controlled robot manipulator. We have discussed this problem in several different cases, and derived the estimation method for an environment model, under the assumption that the target impedance of a robot I_R has been specified and exactly implemented. Therefore, a robust implementation of the target impedance provides a strong basis to carry out the above force control schemes in the real situations.

3.1 DYNAMIC MODEL OF ROBOT

Consider a non-redundant rigid manipulator with n joints, we represent the position and orientation of its end-effector by n independent parameters x_i, $1 \leq i \leq n$ in a fixed Cartesian space. Then the dynamics of the manipulator, in the absence of friction and disturbances, can be expressed as

$$H(X)\ddot{X} + C(X,\dot{X})\dot{X} + G(X) = F + F_e \tag{11}$$

where $H(X)$ is a $n \times n$ inertia matrix with the property of symmetric positive definition, $C(X,\dot{X})$ is a $n \times n$ cariolis centripetal matrix and $G(X)$ is a $n \times 1$ vector of gravity forces, F is a $n \times 1$ vector of applied forces (or torques), F_e is a $n \times 1$ vector of external force acting on the end-effect of the robot.

3.2 TARGET IMPEDANCE OF MANIPULATOR

The general form of target impedance including mass, stiffness and damping terms, can be expressed as

$$M_m\ddot{X} + C_m(\dot{X} - \dot{X}_c) + K_m(X - X_c) = F_e \tag{12}$$

where M_m is a $n \times n$ positive definite inertial matrix, C_m and K_m are the $n \times n$ semi definite damping matrix and stiffness matrix respectively. X is the actual position of the robot , X_c is the command position of the robot and F_e is the reaction force acting on the end-effector of the robot.

3.3 SLIDING SURFACE

For a n-dimensional state-space system with m-dimensional control inputs, the sliding surface $\sigma(X) = 0$ is defined as [DeCarlo and Zak, 1988]:

$$\sigma(X) = [\sigma_1(X), \cdots, \sigma_m(X)]^T = 0$$

where $\sigma_i(X) = 0$ is the ith switching surface, X is the state vector. From this equation we know that the sliding surface $\sigma(X) = 0$ is a $(n - m)$-dimensional manifold in the system state space. The sliding surface is designed such that the system response restricted to $\sigma(X) = 0$ has a desired dynamic behavior. Since equation (12) expresses the desired dynamic response of robot, the sliding surface should be designed based on this equation. In order to implement the target impedance, we use the results in [Chan, 1989] to integrate equation (12), that leads to

$$M_m\dot{X} + C_m(X - X_c) + K_m \int (X - X_c)\, dt = \int F_e\, dt, \tag{13}$$

then we write sliding surface $\sigma(X)$ as

$$\sigma(X) = \dot{X} + M_m^{-1}C_m(X - X_c) + M_m^{-1}K_m \int (X - X_c)\, dt$$
$$- M_m^{-1} \int F_e\, dt = 0. \tag{14a}$$

When the state trajectories of system reach the surface $\sigma(X) = 0$, the dynamic behavior is determined by equation (14a), and also satisfies equation (12). It should be noted that when the target impedance only involves damping and stiffness;

$$C_m(\dot{X} - \dot{X}_c) + K_m(X - X_c) = F_e$$

since this equation is n-dimensional, the sliding surface $\sigma(X)$ can be directly written as

$$\sigma(X) = (\dot{X} - \dot{X}_c) + C_m^{-1}K_m(X - X_c) - C_m^{-1}F_e = 0. \tag{14b}$$

3.4 CONTROLLER DESIGN

The problem to implement the target impedance, as known from above discussion, is equivalent to that to force the trajectories of the system to reach to and remain on the sliding surface $\sigma(X) = 0$ (see (14a) and (12)). Therefore, the design task is to find a control scheme that allows the existence of a sliding mode (that means the trajectories of system are forced to come in contact with the sliding surface and maintained on the surface) in the system. The sufficient conditions for the existence of sliding mode can be briefly expressed as:

There exists a positive definite function $V(t,\sigma,X)$ with respect to σ, i.e., (i) $V(t,X,0) = 0$; for all $t \in R^+$ and $X \in R^n$; and (ii) $V(t,\sigma,X) > 0$; for all $\sigma \neq 0$ and $t \in R^+$, $X \in R^n$, and the total time derivative of V for the system (1) satisfies (iii): $\dot{V} < 0$; for all $\sigma \neq 0$. The more strict mathematical description and proof of above conditions are given in [Utkin, 1978].

From above arguments, The control problem can be formulated as: first to choose a proper function that satisfies the conditions (i) and (ii), and then to find a control law that makes the derivative of this function satisfy condition (iii).

1) *Sliding Mode Control* : Assume an estimated dynamic model of a robot is used with bounded uncertainties on its parameters. The sliding mode control law is

$$F = \hat{H}\ddot{X}_r + \hat{C}\dot{X}_r + \hat{G}(X) - F_e - K_v sgn(\sigma), \tag{15}$$

where $\dot{X}_r = -M_m^{-1}C_m(X - X_c) - M_m^{-1}K_m \int (X - X_c)\, dt + M_m^{-1} \int F_e\, dt$, and $\hat{H}(X)$, $\hat{C}(X,\dot{X})$ and $\hat{G}(X)$ refer to the modeled values of matrices $H(X)$, $C(X,\dot{X})$ and $G(X)$ respectively. F_e is the interaction force between the robot and environment and is measured by force sensor. K_v is a $n \times n$ diagonal matrix;

$$k_{vi} = \sum_{j=1}^{j=n} \delta h_{ij} |\ddot{x}_{r_j}| + \sum_{j=1}^{j=n} \delta c_{ij} |\dot{x}_{r_j}| + \delta g_i + d_i; \qquad 1 \le i \le n,$$

where δh_{ij}, δc_{ij} and δg_i are the bounds of uncertainty on the system parameters, these bounds satisfy $|\hat{h}_{ij} - h_{ij}| < \delta h_{ij}$; $|\hat{c}_{ij} - c_{ij}| < \delta c_{ij}$; $|\hat{g}_i - g_i| < \delta g_i$, and are assumed to be known; d_i is a positive constant. $sgn(\sigma)$ is a $n \times 1$ vector.

proof : we choose the function $V(\sigma,X)$ as

$$V(\sigma,X) = .5\sigma^T(X)H(X)\sigma(X), \tag{16}$$

since the inertial matrix $H(X)$ of a robot is positive definite; the function $V(\sigma, X)$ satisfies the conditions (i) and (ii). Now we prove that its derivative is negative. The differentiation of $V(\sigma,X)$ leads to

$$\dot{V}(X, \sigma) = \sigma^T[(\hat{H} - H)\ddot{X}_r + (\hat{C} - C)\dot{X}_r + (\hat{G} - G) - K_v\, sgn(\sigma)], \tag{17}$$

here the property of skew-symmetry of $(\dot{H} - 2C)$ is used. Using above assumption on the bounds of uncertainty and substituting K_v into above equation, we have

$$\dot{V}(X,\sigma) \le \sum_{i=1}^{i=n} |\sigma_i| \left(\sum_{j=1}^{j=n} (|\hat{h}_{ij} - h_{ij}| - \delta h_{ij}) |\ddot{x}_{r_j}| + \sum_{j=1}^{j=n} (|\hat{c}_{ij} - c_{ij}| - \delta c_{ij}) |\dot{x}_{r_j}| + |\hat{g}_i - g_i| - \delta g_i \right) - \sum_{i=1}^{i=n} d_i |\sigma_i| \le -\sum_{i=1}^{i=n} d_i |\sigma_i| < 0; \quad \sigma \ne 0. \tag{18}$$

The target impedance is realized without the requirement of an exact model in the control scheme; the derivative of $V(\sigma, X)$ is negative, and the conditions of existing a sliding mode are satisfied.

2) *Continuous Control Law*: Discontinuous control laws, such as those given in (15) will give rise to chattering in the control input along the sliding surface. Chattering is undesirable; it represents a high-frequency signal component in the control input, which may excite unmodeled high-frequency dynamics. Several continuous functions to replace the discontinuous sign function are proposed [DeCarlo, 1988] to remove the chattering. One of methods is to replace $sgn(\sigma)$ with a saturation function $sat(\frac{\sigma}{\varepsilon})$, suggested in the paper [Slotine, 1985]. As mentioned by the author, with the saturation function, the trajectories are directed toward the switching surface only when $|\sigma| > \varepsilon$, but when $|\sigma| < \varepsilon$, there is no such guarantee. Therefore the dynamics of the state trajectory inside $|\sigma| < \varepsilon$ is only an approximation to the desired dynamics on the sliding surface. Here, we introduce an integral and proportional function $F(\sigma)$ to replace the switching function $sgn(\sigma)$; the form of $F(\sigma)$ is

$$F(\sigma) = \begin{cases} sgn(\sigma); & |\sigma| > \varepsilon, \\ K_I \int_{t_i}^{t} \sigma dt + K_p \sigma; & |\sigma| \le \varepsilon. \end{cases} \tag{19}$$

where K_I and K_p are $n \times n$ diagonal constant matrices; $k_{pi} = \frac{1}{\varepsilon}$. When $|\sigma| > \varepsilon$; $F(\sigma) = sgn(\sigma)$; the system trajectories are restricted within a ε-neighborhood of the sliding surface. On the other hand, when the system trajectories ($t = t_i$) come into the ε neighborhood of the sliding surface, because of the term of integration in (19), the trajectories of the system will be confined to the sliding surface, $\sigma = 0$, so that the dynamics inside the $|\sigma| < \varepsilon$ is improved to close to the sliding surface no matter how large a ε we choose; usually a large ε is helpful to reduce the chattering in the control input. Simulation results, see Fig. 8 and 10, confirm the effectiveness of such continuous approximations; the steady trajectory measure errors σ_{x1} and σ_{x2} are only about 8.5×10^{-6} and -4.5×10^{-5} respectively, although a large $\varepsilon = 0.5$, is used.

4 SIMULATION

A simulation of two-degree freedom robot is considered to examine the validity of the force control scheme in section 2 and the impedance controller in section 3. The configurations of arm and environment surface are shown in Fig. 5. The target impedance I_R (damping and spring), see (20), is implemented in the direction of x_2, in the direction of x_1 is the position control; no force is measured in this direction.

$$C_m(\dot{X}_c - \dot{X}) + K_m(X_c - X) = F_e \tag{20}$$

where $C_m = \begin{bmatrix} 200 & 0 \\ 0 & 200 \end{bmatrix}$; and $K_m = \begin{bmatrix} 400 & 0 \\ 0 & 400 \end{bmatrix}$;

$X_c = \begin{bmatrix} x_{c1} \\ x_{c2} \end{bmatrix}$; $X = \begin{bmatrix} x_1 \\ x_2 \end{bmatrix}$; $F_e = \begin{bmatrix} 0 \\ f_{e2} \end{bmatrix}$.

The model of environment is

$$f_e = k_e(x_e - x_2)$$

where $k_e = 2 \times 10^4$ and $x_e = 0.95$ are the stiffness and the position of environment respectively.

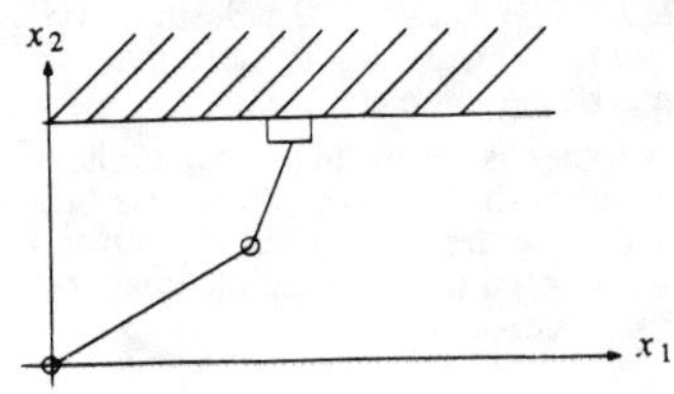

Fig. 5 Configurations of Robot and Environment

A task is designed for the purpose of simulation. It constitutes the robot moving freely, at first, toward the surface, and then, after collision with the surface, sliding along the surface in the direction of x_1 with a constant force in the direction of x_2. The actual position of the robot x, the command position x_c and surface position x_e are shown in Fig. 6.

As to the force control, the desired force f_d is defined as -20, and the position of the environment is supposed to be known. In the period of first 0.6 sec., the arm is moving freely toward the surface, when it reaches the surface, it undertakes a collision, and is forced to move along the surface. The change of velocity, resulting from the collision, in the direction of x_2, multiplied by the damping coefficient C_m causes a large reaction force between the end-effect and surface (Fig. 7). Then, after about 0.5 sec., the force f_e begins to follow the desired force f_d. The nominal steady value of f_e is calculated by (14b);

$$\bar{f}_e = \frac{1}{1+\frac{I_R}{I_E}} f_d = \frac{1}{1+\frac{400}{20000}}(-20) = -19.608.$$

The error between f_d and $\bar{f}_e$ is $-20 + 19.608 = -0.392$, which is from the absence of environment stiffness. Since the ratio $\alpha = \frac{I_R}{I_E}$ is only 0.02; the error in the force is small. The actual force f_e of robot in the simulation is -19.599. Using f_e and f_d, we obtain the estimated environment stiffness $k_e = 1.96\times10^4$. Then, it is incorporated in the control input; a more precise force $f_e = -20.002$ is obtained; it is very close to the desired force f_d.

By above example, we show that the interaction force of an impedance controlled robot can be controlled accurately using the methods in section 2.1 and 2.2. On the other hand, although the robot experiences two different kinds of motion; from the free motion to the constraint motion, the control input does not change, see Fig. 6, there is no switching of control modes, which is usually needed in the most of present force control schemes.

As to the controller, its robustness to the exogenous disturbances and parameter uncertainties in modeling is also shown through above process. Although there is a large exogenous disturbance from the collision between robot and environment, which causes a deviation in σ_{x_1} and σ_{x_2} (see Fig. 8 and 10), however, after about 0.5 sec., the trajectories of the system reach back to the sliding surface. The robustness to the uncertainty on parameters is also confirmed in the simulation by the fact that the values of the estimated robot model used in the controller are chosen to be only 50% of that of the robot model.

From Fig. 10 and 11, we see that the trajectories of the system, which are not initially on but within the ε-neighborhood of the sliding surface, are forced to come in contact with the sliding surface by the integral function $\int \sigma dt$. The steady trajectory measure errors, σ_{x_1} and σ_{x_2} are around 8.5×10^{-6} and -4.5×10^{-5}. The simulation results show that, by using the continuous function (10) to approximate the discontinuous function sign in the vicinity of sliding surface, the chattering in the control input is almost eliminated, meanwhile, the performance in the vicinity of sliding surface is improved.

5 CONCLUSION

When an impedance controlled robot manipulator is in contact with the environment, the interaction force is totally determined by its input position, target impedance and the environment mechanical properties. When environment is very stiff, force can also be controlled in a desired accuracy range without the environment model. When the stiffness of the environment is not much greater than that of the target impedance, it is necessary to estimate the model of the environment to control the interaction force. Considering the robot and environment as a whole system, and using the relationship between the desired force and the interaction force, we obtain the environment model without measuring its deformation. Then the estimated model is used and improves the performance of the force control.

Another issue in this paper is the implementation of the target impedance by using sliding model control theory. The robustness property is guaranteed by the sliding mode control. We also have dealt with the problem of chattering in the control input. A PI correction of trajectory measure error is proposed to eliminate the chattering while to maintain the performance in the vicinity of the sliding surface.

6 REFERENCES

1. Chan, S. P. and Gao, W. B., 1989, "Variable Structure Model-Reaching Control Strategy for Robot Manipulators", *Proc. IEEE Int. Conf. on Robotics Automat.*, pp. 1504-1508.

2. DeCarlo, R. A. and Zak, S. H., 1988, "Variable Structure Control of Nonlinear Multivariable Systems: A Tutorial", *Proc. IEEE* Vol. 76, NO. 3.

3. Kazerooni, H., Kim, S., Waibel, B.J., 1988, "Theory and Experiment on the Stability of Robot Compliance Control", *Proc. IEEE Int. Conf. on Robotics Automat.*, pp. 87-101.

4. Slotine, J. J. E. and Li, W., 1988, "Adaptive Manipulator Control: A Case Study", *IEEE Trans. on Automation Control,* Vol. 33, NO. 11, pp. 995-1003.

5. Utkin, V.I., 1978, "Sliding Modes and Their Application in Variable Structured System", Moscow, *Soviet Union: MIR Publishers*

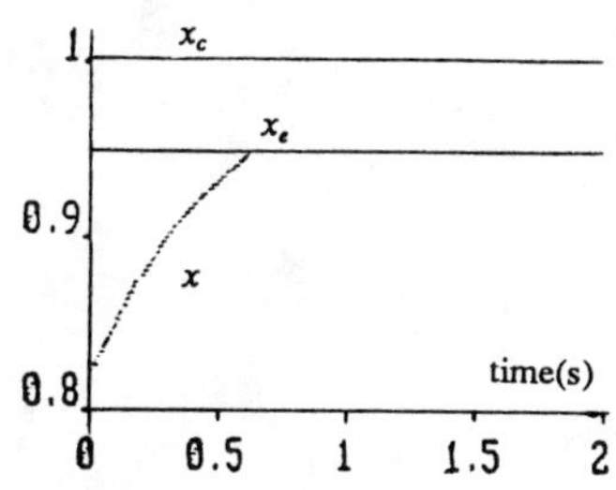

Fig. 6 Positions

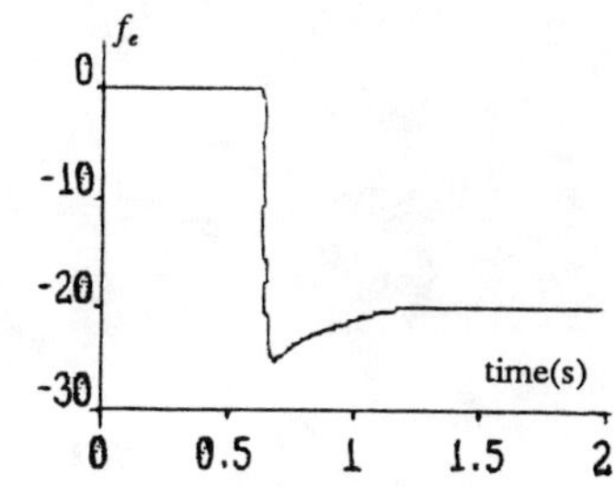

Fig. 7 Interaction Force

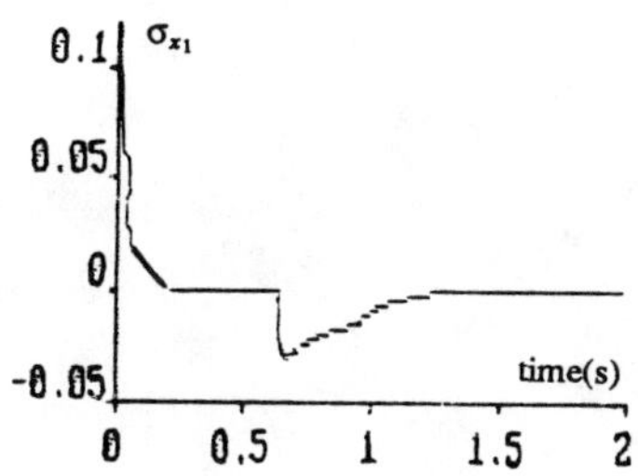

Fig. 8 Trajectory Measure Error

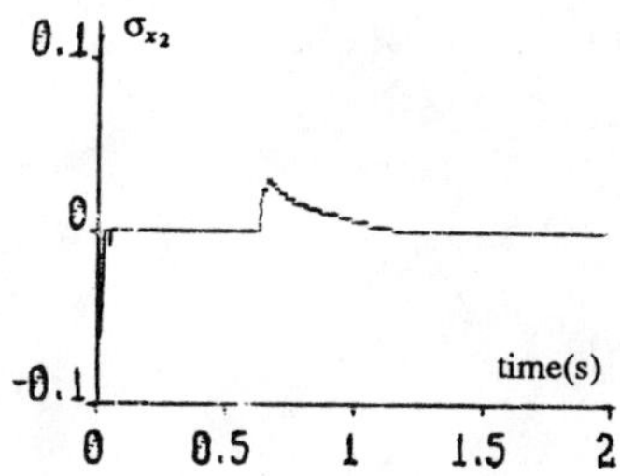

Fig. 9 Trajectory Measure Error

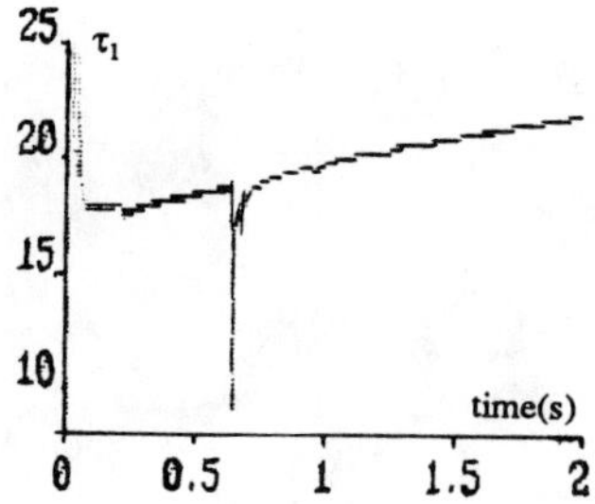

Fig. 10 Joint Torque

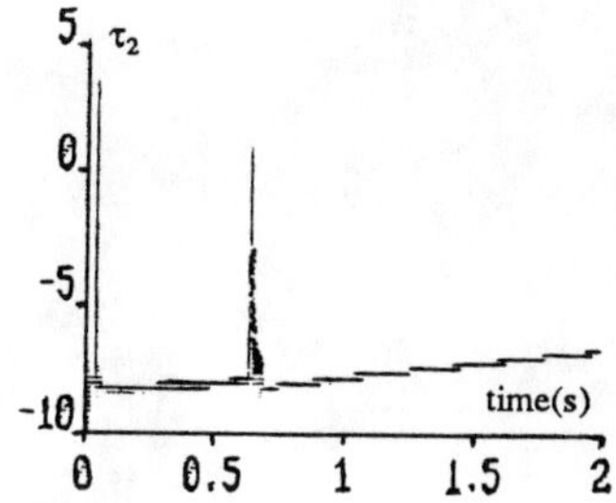

Fig. 11 Joint Torque

A Kinematically Stable Hybrid Control Formulation

Hong Zhang

Department of Computing Science
University of Alberta, Edmonton, Alberta, CANADA

ABSTRACT

In this paper, we address the problem of kinematic instability of hybrid control and present a kinematically stable hybrid control formulation. The formulation is based on a necessary and sufficient condition for the stability of force controlled robot manipulators under a linearized system model. It is obtained by eliminating forces along constrained Cartesian axes applied by the position controller of hybrid control. The validity of the theoretical development is demonstrated by simulation on a realistic two-jointed robot manipulator.

INTRODUCTION

Compliant motion is a prerequisite for robot manipulators to perform operations constrained by task geometry and in the presence of uncertainties and variations in their environments. Examples of these operations range from grinding, contour following, to mechanical assembly. Small uncertainty in relative position generates large and potentially damaging contact forces when rigid parts interact. A viable alternative is to control both force and position and modify motion according to contact conditions so as to reduce the mating forces. This approach, known as force control, has been widely studied in recent years [1].

Force control algorithms can be classified by the type of inner control loop employed by the joints [2]. Although algorithms based on a torque inner control loop [3][4] may ultimately yield better performance because of their explicit incorporation of dynamics in their formulation, their performance is often limited by the nonlinear, unmodeled disturbance torques [2] that exist in robot manipulators. The lack of effective joint torque sensing mechanisms also makes it difficult to close a torque loop in joints [5]. Algorithms based on a position inner control loop [6], on the other hand, provide a natural solution to implementing force control on industrial robots, which almost exclusively employ a position control inner loop. High bandwidth is achieved through the use of the high-gain position servo and the robustness of the resulting system to disturbances such as joint frictions are clearly demonstrated by the excellent performance of industrial robots in positioning tasks.

Hybrid control [6] is the most successful force control algorithm based on a position controlled inner loop. It is derived from the concept of partitioning the six Cartesian degrees of freedom (DOFs) of the hand frame into force controlled (constrained) axes and position controlled (unconstrained) axes [7]. Compliance is achieved by ignoring Cartesian position errors along the constrained axes. Upon removal of these errors, the Cartesian error vector is transformed to the equivalent joint errors, which the joint compensators use to compute corrective torques.

Unfortunately there is a major problem associated with using independent, constant joint position compensators to implement hybrid control. An and Hollerbach showed that the kinematic coupling among joints introduced by hybrid control may lead to what they referred to as *kinematic instability*, in certain manipulator configurations $\boldsymbol{\theta}$ [8]. [9] presented a necessary and sufficient condition in general under a linearized model for robot manipulators to be kinematically stable under force control. In this paper, present a solution to the problem by formulating a kinematically stable hybrid control (KSHC), and demonstrate the validity of the formulation using a two-link robot manipulator.

A KINEMATICALLY STABLE HYBRID CONTROL FORMULATION

In hybrid control compliance condition is specified by a n×n selection matrix **S**, which is diagonal with 1's or 0's on the diagonal corresponding to constrained or unconstrained Cartesian axes. n is the number of joints ≤ 6. In the joint space, compliance is achieved by computing command joint torque vector $\boldsymbol{\tau}$ with the following control law [10]:

$$\boldsymbol{\tau} = \mathbf{K}_d \mathbf{J}^{-1}\bar{\mathbf{S}}\mathbf{J}\delta\dot{\boldsymbol{\theta}} + \mathbf{K}_p \mathbf{J}^{-1}\bar{\mathbf{S}}\mathbf{J}\delta\boldsymbol{\theta} \tag{1}$$

where $\delta\boldsymbol{\theta}$ stands for joint position error vector, **J** for the Jacobian matrix, and $\mathbf{K}_p$ and $\mathbf{K}_d$ for the proportional and derivative gain matrices, respectively. All matrices are n×n. A state space model can be established with the state vector $\delta\mathbf{x} = [\delta\boldsymbol{\theta}, \delta\dot{\boldsymbol{\theta}}]$, and the corresponding system characteristic polynomial can be easily derived with the form [9]

$$\Delta_h(\lambda) = \Delta_h(\lambda) = |\lambda^2\mathbf{D} + \lambda\mathbf{B} + \mathbf{C}|, \tag{2}$$

where

$$\mathbf{B} = \mathbf{K}_d \mathbf{J}^{-1}\bar{\mathbf{S}}\mathbf{J}, \ \ \mathbf{C} = \mathbf{K}_p \mathbf{J}^{-1}\bar{\mathbf{S}}\mathbf{J}, \tag{3}$$

and **D** is the inertial matrix. Stability of the system depends on the locations of the roots of (2), which are manipulator configuration dependent. It has been shown [9] that the system is stable only if both **B** and **C** are positive semi-definite.

As previously mentioned, hybrid control approach to robot compliance is based on the idea that a manipulator is treated as a perfect Cartesian positioner in a certain Cartesian subspace. As such, the torque vector that is generated at the end-effector in reaction to position errors should contain zero components along the force controlled directions. A simple analysis of (1), however, indicates that this is not the case. Considering that (1) computes a joint torque vector whose components are non-zero in general, the static force transformation,

$$\mathbf{F} = (\mathbf{J}^T)^{-1}\boldsymbol{\tau} \tag{4}$$

allows one to establish easily that all components of **F** are also non-zero in general. Therefore, the joint-based position regulators (1) generate forces and torques along *all* Cartesian directions. In other words the control law (1) still results in force application along those directions in which the manipulator is supposedly constrained. This is a fundamental contradiction that causes the kinematic instability.

To eliminate forces along compliant directions that are applied by (1), one can use the selection matrix **S** once more. From (4) the Cartesian force that should be applied in order to control position only is obtained by

$$\mathbf{F}_r = \bar{\mathbf{S}}\,\mathbf{F} = \bar{\mathbf{S}}\,(\mathbf{J}^{-1})^T\boldsymbol{\tau}. \tag{5}$$

The final torque to be applied to joints is

$$\boldsymbol{\tau}_r = \mathbf{J}^T\mathbf{F}_r = \mathbf{J}^T\bar{\mathbf{S}}(\mathbf{J}^{-1})^{-1}\boldsymbol{\tau} = \mathbf{H}^T\boldsymbol{\tau}. \tag{6}$$

The position regulator of the hybrid control that is free of force application in compliant direction has the form

$$\boldsymbol{\tau}_r = \mathbf{J}^T\bar{\mathbf{S}}(\mathbf{J}^{-1})^T(\mathbf{K}_p\mathbf{J}^{-1}\bar{\mathbf{S}}\mathbf{J}d\boldsymbol{\theta} + \mathbf{K}_d\mathbf{J}^{-1}\bar{\mathbf{S}}\mathbf{J}d\dot{\boldsymbol{\theta}}) \tag{7}$$

or

$$\boldsymbol{\tau}_r = \mathbf{H}^T\mathbf{K}_p\mathbf{H}d\boldsymbol{\theta} + \mathbf{H}^T\mathbf{K}_d\mathbf{H}d\dot{\boldsymbol{\theta}}. \tag{8}$$

The characteristic polynomial of the kinematically stable hybrid control (KSHC) can be similarly derived with the form

$$\Delta_h(\lambda) = |\lambda^2\mathbf{D} + \lambda\mathbf{B} + \mathbf{C}| \tag{9}$$

where $\mathbf{B} = \mathbf{H}^T\mathbf{K}_d\mathbf{H}$ and $\mathbf{C} = \mathbf{H}^T\mathbf{K}_p\mathbf{H}$. When the gain matrices, $\mathbf{K}_p$ and $\mathbf{K}_d$, are positive definite (diagonal with positive gains, for example), it can be easily established that both **B** and **C** are positive semidefinite. Therefore the system under (8) is kinematically stable [9]. (8) is also computationally efficient, as the matrix **H** is already computed in hybrid control and the KSHC requires at most an additional $n \times n$ by $n \times 1$ matrix vector product.

In general for a robot manipulator of n joints, characteristic polynomial (2) or (9) is of order $2n$. It can be shown that of its $2n$ roots, there are always $2m$ roots at the origin, where m is the number of compliant degrees of freedom, while the rest $2(n - m)$ non-zero roots are confined to the left-half complex plane due to the stability constraint. This property is obtained by rewriting (2) or (9) in the following form:

$$\Delta_h(\lambda) = |\mathbf{D}|\,|\lambda^2\mathbf{I} + f(\mathbf{J}, \mathbf{D}, \mathbf{K}_d, \mathbf{S})\bar{\mathbf{S}}\lambda + g(\mathbf{J}, \mathbf{D}, \mathbf{K}_p, \mathbf{S})\bar{\mathbf{S}}\lambda| \qquad (10)$$

where *f( )* and *g( )* are $n \times n$ matrices. (10) can be evaluated in block matrix form according to directions that are constrained and those that are not, resulting in a $|\lambda^2\mathbf{I}_{m\times m}|$ term or $2m$ zeros at the origin.

SIMULATION RESULTS

To verify the results, we use MIT Serial Link Direct Drive Arm in [8] as an example. Its kinematics is shown in Fig. 1.

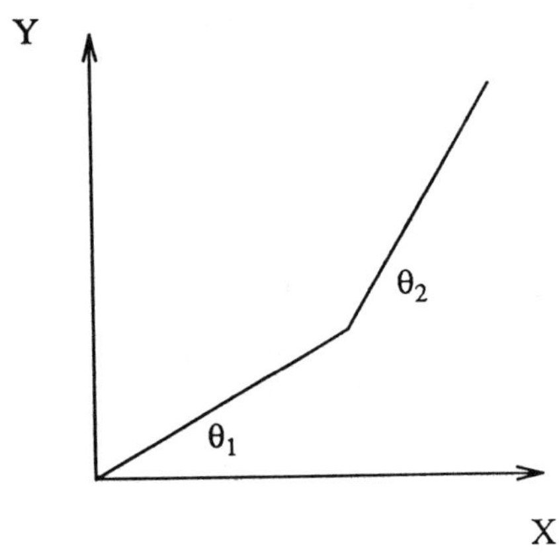

Fig. 1. Kinematics of A Two Link Manipulator

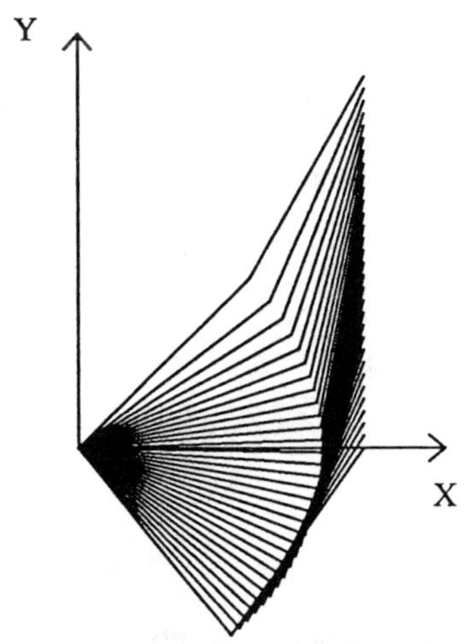

Fig. 2. Test Trajectory with $\mathbf{S} = diag[1, 0]$

Its inertial matrix, Jacobian matrix, and gain matrices can be found in [8]. The straight-line trajectory depicted in Fig. 2 is used to evaluate the performance of both the hybrid control and the kinematically stable hybrid control. In this case, $S = diag\,[1, 0]$, i.e., the robot moves upwards along the y direction and complies along the x direction of the base coordinate frame. (9) is a fourth order polynomial with two roots always at the origin. There are two non-zero roots, which change with the manipulator configuration and can be obtained by solving a second order polynomial. The root-loci of the characteristic polynomials corresponding to the test trajectory for hybrid control (1) and KSHC (8) are depicted in Figures 3 and 4, respectively.

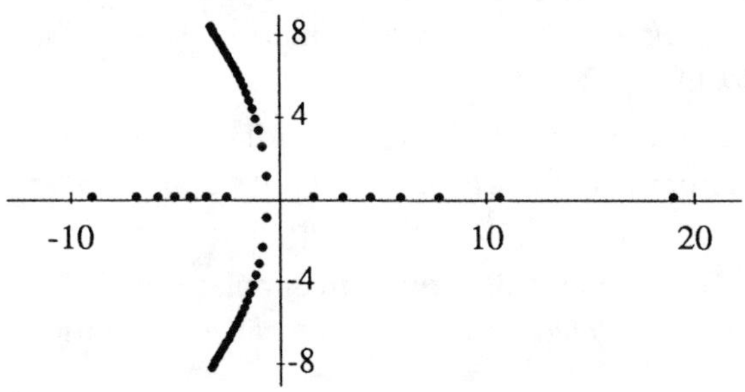

Fig. 3. Root-locus of the Test Trajectory in Fig. 1. under Hybrid Control

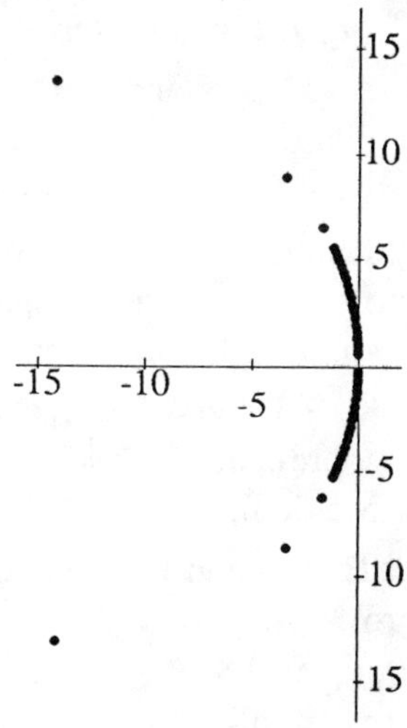

Fig. 4. Root-locus of the Test Trajectory in Fig. 1. under KSHC

As can be seen, the two non-zero roots of the characteristic polynomial depicted in Fig. 3 first move towards the origin and away along the real axis, and one of them becomes positive when the end-effector position is at (0.547, 0.547) or $\theta_1 \approx 17.6^\circ$ and $\theta_2 \approx 58.8^\circ$, implying the system becomes unstable from that point on. Roots of the characteristic polynomial of KSHC depicted in Fig. 4, however, are always confined to the left-half open plane, and is therefore stable, as we anticipated.

CONCLUSION

In order to take advantage of high disturbance rejection offered by hybrid control, a satisfactory solution must be found to the problem of kinematic instability. We offer but one possible solution and demonstrate its validity by both a theoretical analysis and simulation using real data taken from previous studies. We are currently conducting experiments on a six joint robot manipulator to further verify our results. In addition, we are investigating solutions based on the analysis of the system characteristic polynomial (2) whose coefficients are functions of joint positions and joint feedback gains. If non-diagonal $\mathbf{K}_d$ and $\mathbf{K}_p$ are employed, more flexibility is injected to the system and, at least in the low dimensional case, kinematic stability can be achieved by varying the joint feedback gains.

REFERENCES

1. D. Whitney, "Historical perspective and state of the art in robot force control," *Int. J. Robotics Research*, vol.6, no.1, 1987, pp.3-14.

2. J. A. Maples and J. J. Becker, "Experiments in force control of robotic manipulators," *Proc. 1986 IEEE Int. Conf. on Robotics and Automation*, San Francisco, CA, April, 1986, pp. 695-702.

3. O. Khatib, "A unified approach for motion and force control of robot manipulators: the operational space formulation," *IEEE J. Robotics and Automation*, vol. RA-3, no.1, February, 1987, pp.43-53.

4. J. K. Salisbury, "Active stiffness control of a robot manipulator in Cartesian coordinates," *Proc. IEEE Conf. Decision and Control* (Albuquerque, New Mexico), 1980.

5. C. H. Wu and R. P. Paul, "Manipulator compliance based on joint torque control," *Proc. 19th IEEE Conf. on Decision and Control,*, vol.1, December, 1980, pp.84-88.

6. M.H. Raibert and J.J. Craig, "Hybrid position/force control of manipulators," *J. Energy Resources Technology*, Vol.103, pp.126-133, 1981.

7. M.T. Mason, "Compliance and force control for computer controlled manipulators," Artificial Intelligence Laboratory, MIT, AIM-515, 1979.

8. C. H. An and J.M. Hollerbach, "Kinematic stability issues in force control of manipulators," *Proc. IEEE Conf. Robotics and Automation* (Raleigh, North Carolina), pp.897-903, 1987.

9. H. Zhang, "Kinematic stability of robot manipulators under force control," *Proc. 1989 IEEE Int. Conf. on Robotics and Automation*, Scottsdale, Arizona, May, 1989, pp.80-85.

10. H. Zhang and R. P. Paul, "Hybrid control of robot manipulators," *Proc. IEEE Int. Conf. on Robotics and Automation* (St. Louis, Missouri), vol.1, pp.602-607, 1985.

HYBRID CONTROL OF A SIX-DEGREE-OF-FREEDOM IN-PARALLEL ACTUATED MICRO-MANIPULATOR MOUNTED ON A SCARA ROBOT

C.REBOULET - R.PIGEYRE
Centre d'Etudes et de Recherches de Toulouse (CERT)
Département d'Etudes et de Recherches en Automatique (DERA)
BP 4025 - 31055 - TOULOUSE Cedex (France)

ABSTRACT
This paper describes a micro-manipulator system made up of a SCARA robot on which is mounted a micro-manipulator (a STEWART platform). A hybrid force-position control is implemented on the micro-manipulator whilst the SCARA is position controlled only. It is shown how the redundancies are used to obtain high performance force control in an extended working field. It will be seen that the stability of the whole system is assured by the decoupling of the SCARA and micro-manipulator servo-system bandwidths. In addition, it is shown how to take into consideration the main dynamic coupling in the hybrid control.

1 - INTRODUCTION
The carrying out of delicate assembly tasks with close tolerances is complex because of the interactions which exist between the robot's end effector and its environment.

There are two approaches to solving this problem :
• a quasi-static approach which exploits force information in a logical way in order to perform trajectory corrections
• a dynamic approach which exploits force information continuously in a servo-loop.

It is this latter approach which is used in this paper as it seems the best adapted to delicate assembly tasks. It involves the setting up of hybrid position-force control structures similar to those presented in [1].

As to the choice of a structural mechanical combination, two possibilities exist as we have shown in [2] :

• the most classic consists in equipping the manipulator with a force sensor and a compliant system. Active force control is obtained by controlling the deformation of the compliant system. This system, even though used frequently, does not allow a good level of performance to be obtained as the stability problems it creates do not allow very good bandwidths to be obtained.

• the other solution, which we propose in this paper, is based on the concept of a micro/macro-manipulator. It consists in associating a micro-manipulator to a main robot. This idea has already been presented in certain publications [3] [4] [5].

2 - ACTIVE COMPLIANCE BASED ON REDUNDANCY
2.1 Basic principle
A micro-manipulator or, more generally, a specialised device in the form of « fingers » [5] or «wrists» is generally well-adapted to active force control because of its low inertia. A certain number of these devices are organised around parallel structures [3] [7] [8]. These devices have an extremely restricted work

field which is why it is necessary to mount them on a conventional robot capable of greater movement.

In this paper, the only mode which require close cooperation is analysed. This problem is similar to the one found in two-arms manipulation systems [9].

By using redundancy it is possible to influence each degree of freedom (d.o.f.), either through the manipulator actuators or through those of the macro-manipulator. It will be shown how, thanks to a judicious coordination of the control of all the actuators, it is possible to extend the high performance of the micro-manipulator force control to the entire working field of the main robot.

2.2 - Description of the experimental test rig

In this paper, the new experimental test rig used (Figure 1) is made up of a new version of a Stewart platform type parallel micro-manipulator, constructed according to the same principles as that presented in [3] ; it is installed on a SCARA type robot (IBM robot 7576).

Briefly, this parallel robot is driven by six pneumatic actuators. These actuators are organised around a polyarticulated structure between two plates. Each actuator is linked to the upper plate by gimbals and to the lower by ball-and-socket joints. The unit is equipped with a six-component force sensor. Because of the compactness of the structure, dynamic effects and dry friction are reduced to a minimum. This structure, which can be classed in the category of lightweight robots, is perfectly adapted to assembly tasks calling for active force control. Moreover, it permits the creation of precisely controlled movements. Direct pneumatic drives have been chosen so as to reduce inertia.

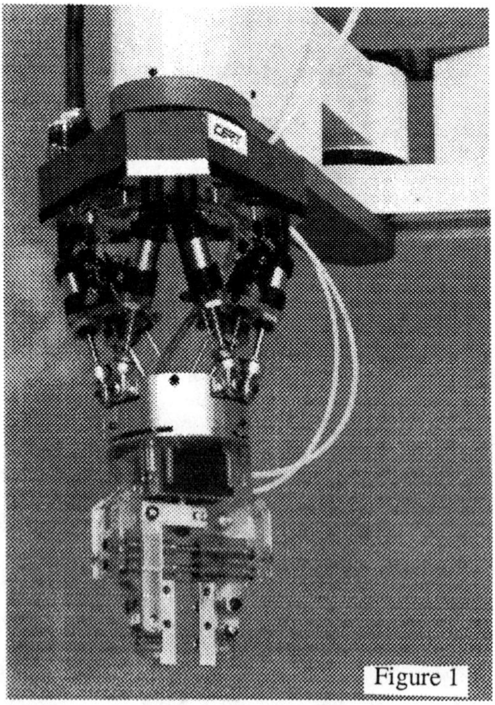

Figure 1

3 - HYBRID CONTROL OF THE WHOLE SYSTEM

3.1 Principle

In the case of simple mechanical structures, the outlines of hybrid position-force control, are now well-known [1]. It consists of two servo-loops, one force and the other position, working in parallel.

In our case, the problem is slightly different as we have a redundant structure with ten d.o.f. The position-controlled sub-space and the complementary force-controlled sub-space are considered separately.

There are no major difficulties as far as the position-controlled sub-space is concerned. It must simply be noted that not all the degrees of freedom are redundant. We know that a SCARA type robot has only four d.o.f., that is to say, space ξs (effectively, it is not possible to incline the end effector). These four d.o.f. are, therefore, redundant. The two complementary d.o.f. (that is, space $\xi_{s\perp}$) are not redundant and are obtained thanks to the micro-manipulator. Any trajectory can thus be broken down into two parts : the components relating to space ξ_s and those relating to space $\xi_{s\perp}$.

The problem of the force-controlled sub-space is a little more delicate. All the effort commands are naturally sent to the micro-manipulator which functions like a genuine force generator. However, any movement in the force-controlled sub-space must also be broken down into two parts : one relating to the non-redundant space $\xi_{s\perp}$ and the other to the redundant space ξ_s. This latter part represents the errors of the wrist in space ξ_s relative to its nominal working point. This error must be kept at zero by controlling the position of the main robot.

The coupling of two robots can, however, pose a problem of stability. We shall see that the mechanical architecture and the type of coupling chosen can solve this problem.

We shall first show the structure of the hybrid position-force control of the micro-manipulator and then how to solve the problem of coordination between the micro-manipulator and the SCARA robot.

3.2 Micro-manipulator hybrid control

Figure 2 shows the geometrical structure of a parallel micro-manipulator.

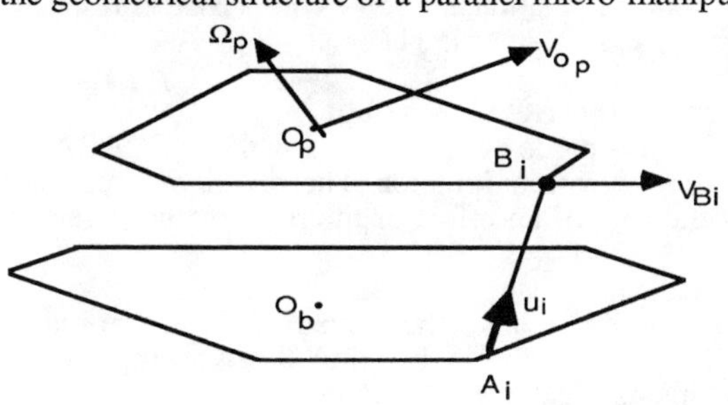

Figure 2

We have shown in [10] that the general dynamic model of a parallel manipulator, in its working space, takes the form :

$$\left[J_{O_p}^{-T}\right] f = [M]\,\dot{W}_{O_p} + \phi + \mu$$

$\left[J_{O_p}^{-T}\right]$ is the 6x6 jacobian matrix at point 0_p

[M] is the 6x6 inertia matrix

Φ is a 6 component vector related to W_{O_p}

μ is a 6 component vectorcontaining elements subjected to the flield of gravity

$W_{O_p} = [\dot{x}\ \dot{y}\ \dot{z}\ \Omega_x\ \Omega_y\ \Omega_z]^T$ represent the kinematic twist of the plateform in 0_p,which consists of vector V_{O_p} and Ω_p

f represents the drive force vector in joint space

Neglecting the gyroscopic torques and the gravitational effects, we can use the simplified dynamic model

$$\left[J_{O_p}^{-T}\right] f = [M]\,\dot{W}_{O_p}$$

The hybrid position-force control scheme (figure 3) which carries out the decoupling of all the d.o.f. can be inferred from the previous equation.

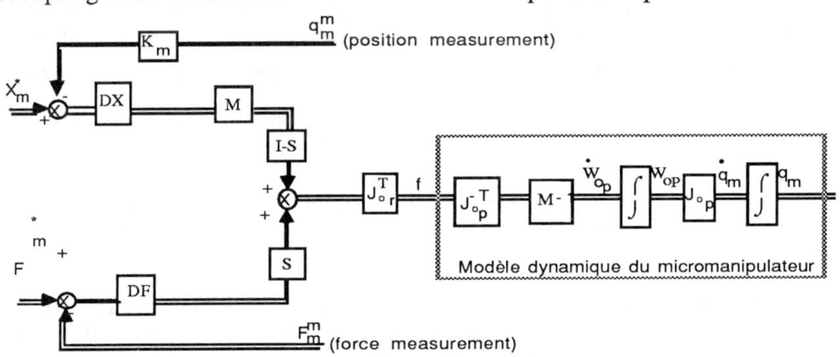

Figure 3

X_m^* , F_m^*	represent the hybrid commands
DX, DF	represent the position and force compensators respectively
S and I-S	have the same role as in [1]..

3.3 - Coupling with the SCARA robot

The problem is to define a mode of cooperation with the SCARA robot based on a hybrid control of the micro-manipulator. The position-force controlled micro-manipulator plays the role of an effort or micro-movement generator in a small space.

To enlarge its field of action, it has to be moved by the macro-manipulator. So the goal of this manipulator consist in keeping the working field of the micro-manipulator around his nominal operating point.

A command structure which fulfills these conditions is shown in Figure 4.

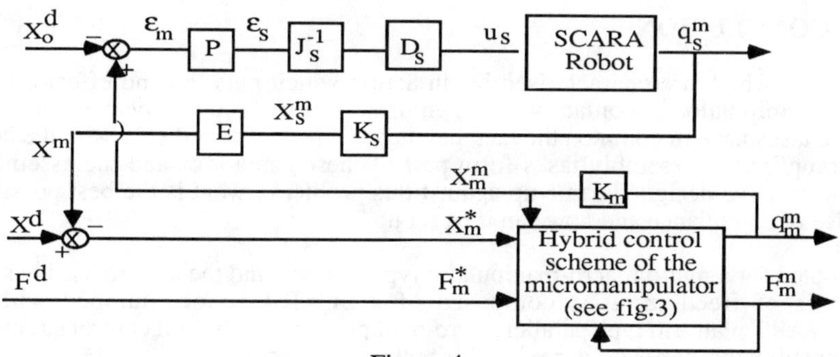

Figure 4

q_m, q_s	position in joint space of the micro-manipulator and the SCARA robot respectively
K_m, K_s	direct kinematic models of the micro-manipulateur and the SCARA robot respectively
J_s	Jacobian matrix of the SCARA robot
D_s	compensator of the position servo-loop
F	six components force vector
X_m, X_s, X	position/orientation vector of : - effector/micromanipulator basement - micromanipulator basement/reference coordinate frame - effector/reference coordinate frame
u_s	control variable of the SCARA robot
K	upper n, d indices mean measured and desired respectively
E	is a function which expands the vector X_s^m adding 0 in the non-redundant sub-space $\xi_{s\perp}$.
P	is a projection function in the redundant space ξ_s
X_o	represents the nominal working point of the micro-manipulator

The interest of this architecture lies in the fact that each manipulator is solicited only by a mode which is appropriate to it. The low inertia micro-manipulator gives short response times and good force control performance ; it is position-force controlled and has local control of interactions with its environment.

The SCARA robot, with greater inertia, does not have the same dispositions. On the other hand, it is capable of generating large movements : it is position controlled for micro-manipulator movements and does not act directly on the environment.

As far as the effects of dynamic interactions on stability are concerned, we have seen that the morphologies of the two manipulators give them dynamic properties of rather different orders. Moreover, the coupling installed (input-output relations only) is such that even if the micro-manipulator servo loops are very fast compared to those of the SCARA, the two robots are dynamically independent ; strictly, dynamic coupling exists but the bandwidths are sufficiently wide apart for stability problems to be avoided.

Thus, this architecture sets up a low level cooperation between the micro-manipulator and the SCARA which conserves the advantages of each, excluding their major disadvantages, and does not pose any particular stability problems.

4 - CONCLUSION

All tasks which are characterised by an action which puts the end effector of a robot manipulator in contact with its environment make up a special category. These tasks have in common the fact that they can be performed only with the help of compliance. Assembly tasks form part of these categories and the assembly robot we have designed came up against this problem : what is the best possible choice of compliance and how can it be set up ?

We opted for a micro/macro-manipulator type solution and the system used has 10 degrees of freedom. It is composed of a SCARA robot equipped with a STEWART platform type parallel micro-manipulator. The hybrid control structure presented in this paper is adapted to this redundant structure.

The functionalities offered by hybrid control and in particular the possibility of selecting any degree of freedom to perform either a force or position control, have been extended to the entire SCARA robot working space. These functionalities permit great flexibilty in programming an application thanks to a set of basic primitives described in [11].

This robot manipulator is at present being installed on Aerospatiale's gyroscope assembly industrial site at Bourges. The gyroscope is composed of 60 small mechanical parts, most of which have to be assembled with a clearance of just a few microns.

REFERENCES

[1] Raibert M.H., and Craig J.J. - Hybrid position/force control of manipulator ASME, Journal of Dynamic Systems, Measurement and Control, june 1981.

[2] Reboulet C. and Berthomieu T. - Hybrid position-force control : implementation on a parallel manipulator - Nato Advanced Research Workshop : Kinematic and Dynamic Issues in Sencor Based Control - Oct. 1987 - Italy.

[3] Reboulet C. and Robert A. - Hybrid control of a manipulator equipped with an active compliant wrist. 3rd ISRR - Oct. 1985 - Gouvieux (France).

[4] Waldron K.J., Raghavan M., Roth B. - Kinematics of a hybrid series-parallel manipulation system - Journal of Dynamic Systems, Measurement and Control - June 1989.

[5] Sharon A., Hogan N. and Hardt D. - High bandwidth force regulation and inertia reduction using a macro/micro manipulator system - IEEE Int. Conf. on Robotics and Automation - Philadelphie (USA), April 1988.

[6] Salisbury J.K. - Design and control of an articulated hand - Proc. Int. Symposium on Design and Synthesis - Tokyo (Japon), July 1984.

[7] Merlet J.P. - Force-feedback control of parallel manipulators - IEEE Int. Conf. on Robotics and Automation - Philadelphie (USA), April 1988

[8] Hayward V. - An analysis of redundant manipulators from several view points - Nato Advanced Research Workshop : Robot with redundancy, design, sensing and control - June 1988 - Italy.

[9] Dauchez P. and Uchiyma M. - Kinematic formulation for two force-controlled cooperating robots - Proc. 3rd Int. Conf. on Advanced Robotics Versailles (France), Oct. 1987.

[10] Reboulet C., Fuertes V. and Gasmi B. - Use of an active compliant device for industrial applications - Int. Workshop on Sensorial Integration for Industrial Robots - Zaragoza (Spain), Nov. 1989.

[11] Reboulet C., Berthomieu C. - Dynamic models of parallel manipulators - To appear in Int. Journal of Robotic Research.

Design of a Torque Controller for the Adept-2 Robot

Gregory P. Starr, Associate Professor

Christopher W. Wilson, Research Engineer

Department of Mechanical Engineering
The University of New Mexico
Albuquerque, NM 87131

Abstract

The ability to command actuator torque is necessary to perform research into robot control. Commercial manipulators using direct-drive actuators offer high performance, but do not allow specification of actuator torque. We experimentally characterize the motors in an Adept-2 manipulator, and develop a linearizing and decoupling controller which allows torque specification. Torque ripple is 13%.

INTRODUCTION

The work described herein is part of the development of an alternative controller for the Adept-2 robot. This controller will command motor torque, and will be used as part of a control testbed for research into force control and constrained motion.

Motivation

Stable, well-behaved robot force control is difficult to achieve. Force control typically has low bandwidth because of the inherent dynamics of the constrained mechanism. Direct-drive actuators promise greater torque/inertia ratio, and accompanying performance

One approach to force control is to control mechanical impedance, defined as the generalized relationship between force and motion, rather than force[1]. Impedance control may be realized by measuring motion error, then commanding corresponding actuator torque.

Controller Redesign

When designing an alternative controller for a commercial robot, there is the question of what to retain. There are several degrees of redesign that can be undertaken[2]. In designing the Adept-2 controller, we retained the motor power amplifiers, but nothing else. All other functions of the standard Adept controller, such as encoder quadrature decoding, motor drive waveform synthesis, etc., were replaced.

Previous Work

The Adept-2 robot is driven by direct-drive variable-reluctance motors (VRM) manufactured by NSK of Japan. These motors are capable of high torque and are electrically more efficient than other types of direct-drive motors[3]. They do, however, require a sophisticated approach to actuation. VRMs has been investigated by several researchers, see [4], [5], and [6]. The VRM is very nonlinear, and the effects of its (typically) three stator phases are coupled.

One of the primary measures of the quality of motor actuation is the variance of motor torque with motor rotation, commonly called torque ripple. Ideally, VRM linearization and decoupling eliminates torque ripple. Taylor *et al*[6]] presented experimental motor position step response data, but did not present torque ripple data.

EXPERIMENTAL APPARATUS AND PROCEDURE

Variable Reluctance Motor

A variable reluctance motor (VRM) consists of an inner and outer stator, with a rotor between. The stator has windings which are excited. The resulting magnetic field passes between the two stators and thus penetrates the ferrous rotor. The force developed on the rotor by this magnetic field generates the motor torque The stator windings are divided into groups of (usually) three phases, with sets of phases repeated around the stator circumference. The rotor teeth are shaped to focus the magnetic field and maximize torque. The motor used in joint 1 of the Adept-2 manipulator has 150 teeth on the rotor, and 144 sets of three stator phases.

The torque generated is a function of both rotor position θ and the three stator command intensities. Since the amplifiers used in the Adept-2 robot accept TTL-level PWM signals, it is appropriate to speak of a phase's pulse width rather than its current. Thus the torque may be described as

$$\text{Torque} = f(\theta) * g(pw) \tag{1}$$

where $f()$ expresses the torque dependence on rotor position, and $g()$ expresses the torque dependence on phase pulse width command intensity.

The dependence of torque on rotor position is periodic. The dependence of torque on phase pulse width was assumed (after Taylor *et al* [6]) to be linear, with an initial saturation. To realize a desired rotor torque which is independent of rotor position, one must actuate the three phases accordingly. The command of a particular phase will be dependent on (1) magnitude of desired torque, (2) sign of desired torque, and (3) instantaneous rotor position.

At a given position, there is a single "primary" phase which is most efficient at producing torque, hence only this phase should be driven. The pulse width of this primary phase must vary with rotor position to maintain constant torque, due to the changing stator/rotor geometry. In addition, the primary phase itself will change as the rotor moves past the stator poles, hence the drive must be switched from one phase to the next.

Motor Characterization Procedure

To synthesize the stator pulse width commands to generate a given torque, it is necessary to determine the torque dependence on both rotor position θ and pulse width pw, functions $f()$ and $g()$ in (1). The steps in motor characterization may be summarized as:

> (a) Obtain experimental motor torque *vs* θ and pw, (b) Perform multivariable fit of all data, (c) Construct table of torque vs θ and pw, (d) Invert the torque table to yield pw *vs* θ, torque, (e) Ramp pw down to zero at ends of table to ease transition.

These individual processes are described in the following sections. Unless otherwise noted, the discussion applies to motor 1. The method for motors 2, 3, and 4 is the same, but the numerical specifics may be different. Where the differences are significant, they will be stated.

Experimental Measurement of Motor Torque

To measure motor torque, the joints were constrained with an apparatus which allowed controlled rotation of the motor, with acquisition of position and torque. The specifications of the motors are:

motor 1: (VRM) 460,800 counts/rev (encoder 1), 150 poles, 3 phases, 3072 counts/cycle

motor 2: (VRM) 460,800 counts/rev (encoder 1)

230,400 counts/rev (encoder 2), 150 poles, 3 phases, 3072 counts/cycle when encoder 1 used to measure position.

motor 3: (Brushless) 7,680 counts/rev (encoder 3), 10 poles, 3 phases, 768 counts/cycle

motor 4: (Brushless) 7,680 counts/rev (encoder 4), 10 poles, 3 phases, 768 counts/cycle

The dependence of motor 2 on both encoders 1 and 2 illustrates the mapping between actuator space and joint space. The housing of both motors 1 and 2 are fixed to the base. Motor 1 drives link 1 directly, and its position is given directly by encoder 1. Motor 2 drives link 2 via a 4-bar linkage. Joint 3 (prismatic quill feed) is driven *via* a ball screw, while joint 4 is driven by motor 4 directly. Figure 1 shows the drive mechanisms of the Adept-2 links.

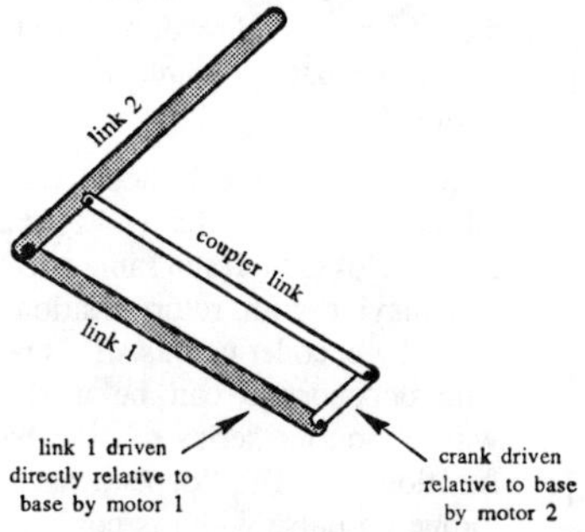

Figure 1. Joint Drive Mechanisms for the Adept-2 Manipulator.

Encoder 2 measures the angular displacement of joint 2, but the angular displacement of motor 2 is dependent on both encoders 1 and 2. For characterization of motor 2, joint 2 was fixed, so the angular position of motor 2 was given solely by encoder 1. Joint 3 is a prismatic joint, driven by motor 3 using a ball screw with a 1-inch pitch. Encoder 3 measures the rotation of motor 3 directly, thus the linear resolution of motor 3 is 7,680 counts/inch. Joint 4 is a revolute "twist" joint through which motor 4 drives link 4 directly. Encoder 4 provides a direct measurement of the angular position of motor 4. As link 4 is rotated, it will also "climb" up (or down) the ball screw nut unless motor 3 is driven in a compensatory manner. The mapping between actuator space (q_1, q_2, q_3, q_4) and joint space $(\theta_1, \theta_2, d_3, \theta_4)$ is given in equations (2)...(5):

$$\theta_1 = q_1 \quad (2)$$

$$\theta_2 = q_1 + q_2 \quad (3)$$

$$d_3 = q_3 + q_4 \quad (4)$$

$$\theta_4 = q_4 \quad (5)$$

Link Constraints.

The apparatus used to constrain joint 1 consisted of a piece of 0.5-inch aluminum plate which was driven by a lead screw. A torque sensor was interposed between the plate and link 1 so that the motor torque could be directly measured. A similar fixture was used for joint 2.

For all motors, we used pulse width duty cycles of 10, 20, ..., 60% when performing the characterizations. All motors were moved through one full commutation cycle to obtain a full period of torque data. The 60% duty cycle pulse width limit is somewhat conservative, but still yielded torques of over 1000 in.-lb. for motor 1, for example.

ANALYSIS OF MOTOR TORQUE DATA

To characterize a motor of the Adept-2 manipulator, we constrained the appropriate joint, applied a given phase pulse width command, and rotated the joint so at least one full motor commutation cycle was covered. Torque and motor position were continually recorded during this process. This process was repeated for both directions of motion, and at several pulse width commands. These data were then used to synthesize tables of pulse width *vs* desired torque and rotor angle that were then used to drive the motors directly.

Experimental Data

Torque data were taken at every 4 encoder counts for joints 1 & 2, and at each encoder count for joint 4. Each joint was moved in both directions to assess the effect of friction. Only joint 1 data will be shown.

A plot of experimental torque data for joint 1 is shown in Figure 2. Because of the method we used to align the rotor at the beginning of the run, the torque is always negative for the first half commutation cycle, as shown in Figure 2. The motion direction was positive in this plot. When the motion is positive, joint friction acts to oppose the motion, and thus the curves of Figure 2 are slightly biased downwards. When the motion is in the negative direction, friction biases the data upward, as seen in Figure 3.

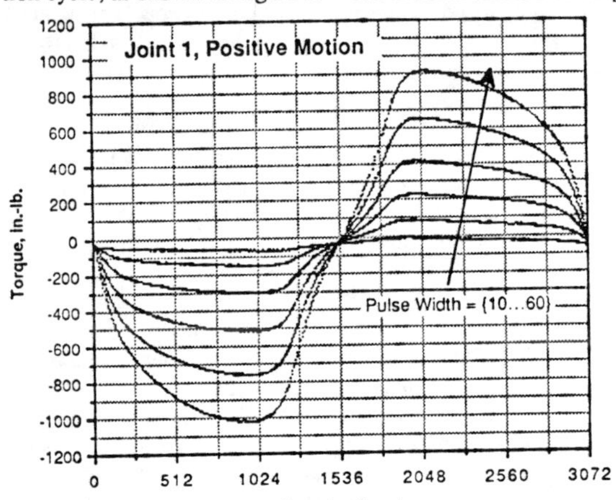

Figure 2. Experimental Torque vs Rotor Position, Pulse Width.

Referring to Figure 2, the curves for all six pulse widths (10%...60% duty cycle) exhibit the same periodic behavior with rotor position as given by encoder counts. This periodic dependence can be modeled with a Fourier series of sine basis functions. The dependence of torque on pulse width is not explicitly shown in Figure 2, since the data was not organized that way, but the dependence seems approximately linear, with a initial lag at low pulse widths. Note that this experimental data is quite smooth, reflecting constant joint velocity and rigid joint constraint while acquiring data.

Figure 3 on the next page shows the effect of friction in joint 1. In the positive motion direction, friction opposes joint torque and therefore subtracts from it, while in the negative motion direction friction adds to the joint torque. At the very slow joint speeds during data collection, the friction is probably a combination of static and stick-slip friction, rather than viscous friction.

Joint Data Analysis.

As mentioned earlier, the steps in motor analysis were:

1. Obtain experimental motor torque vs θ and *pw*
2. Perform multivariable fit of all data.
3. Construct table of torque vs θ and *pw*
4. Invert the torque table to yield *pw vs* θ, torque.
5. Ramp *pw* down to zero at ends of table to ease transition.

The multivariable fit of the data generates a single expression for rotor torque *vs* rotor position and individual stator pulse width. In Step 3, we use this equation to compute a table of rotor torques for all necessary values of rotor position and pulse width. In Step 4, this table is reorganized to yield a table of pulse widths for corresponding values of rotor position and desired torque. Finally, in Step 5 we identify commutation switching points and generate smooth transitions between phases with linear ramping functions.

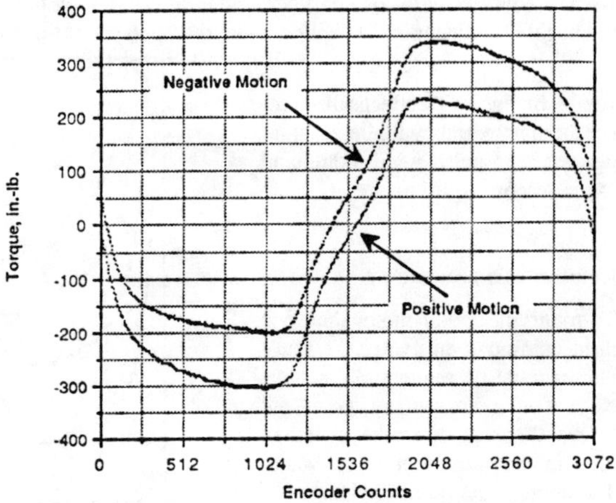

Figure 3. Torque vs Rotor Position, Positive and Negative Motion, Pulse Width 30.

Curve Fitting of Motor Data.

Fitting a set of multivariable data can be viewed as a single-variable fit, with the basis functions being all possible products of the individual basis functions, and with the data organized properly. Examination of the residual showed 10 sinusoidal terms and 4 linear/exponential terms to yield a good fit to the experimental motor data, thus giving a total of 40 multivariable basis functions.

To allow for the effect of friction, both the positive and negative motion data were fit at the same time, so that the fit would "average out" the effect of friction. The multivariable function is

$$\tau = \sum_{m=1}^{M} \sum_{n=1}^{N} c_{mn} \sin \frac{m\pi\theta}{\theta_{max}} \left[n \frac{pw}{pw_{max}} + \exp\left(-n \frac{pw}{pw_{max}}\right) - 1 \right] \qquad (6)$$

where $\theta_{max} = 3072$, and $pw_{max} = 60$. The upper limit of 60% on pulse width is arbitrary, and was selected conservatively.

Torque Table Generation

The multivariable function of equation (6) was used to construct a table of computed torques for a complete commutation cycle, and for pulse widths varying from 1 to 60% in steps of 1%. Thus the pulse width resolution here was much higher than for the experimental data. Encoder counts varied from 0 to 3072, with stepsize 4.

Torque Table Inversion

The torque table was then "inverted" (actually reorganized) to yield the pulse width necessary to achieve a desired torque at a given rotor position. Thus the table was the same size as the torque table, but expressed differently. Note that the pulse width tends toward infinity as the rotor approaches alignment with a stator pole, or approaches the halfway point between poles. This is because the torque generation capability of the motor goes to zero at these positions, regardless of commanded pulse width. For the torque table inversion, these singularities in pulse width were limited to the maximum pulse width of our experiments, 60%.

An example of these table waveforms is shown in Figure 4 for five representative torque values. Note the increase in pulse width near 0 counts (pole alignment) and 1536 counts (pole halfway point). This plot is for one half commutation cycle only; the plot for the second half cycle (producing positive torques) is a reflection about the 1536 count line.

The granularity of the pulse width values in Figure 4 reflects the integer hardware that will be used to generate the pulse width command to the Adept amplifiers.

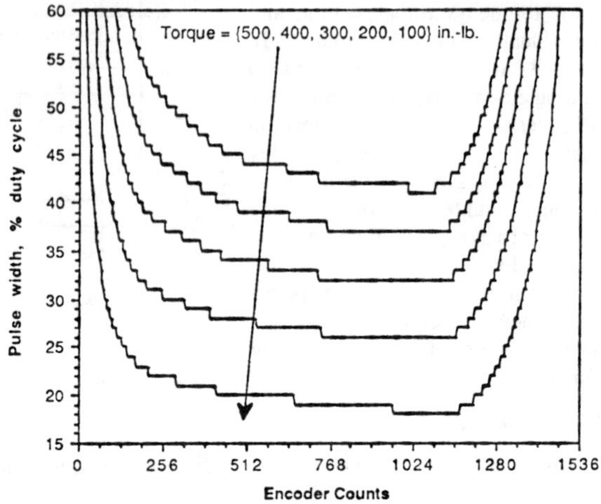

Figure 4. Motor drive waveforms for five representative torques.

Commutation Switching

To rotate the motor motor than one third commutation cycle, the stator phases must be sequentially actuated. At the appropriate point the "current" phase must be switched off and the "next" phase activated. The appropriate point to change active phases, *i.e.* the commutation switching point, may be determined by examining a plot of torque from all three phases. Such a plot is shown in Figure 5 for pulse with 40%.

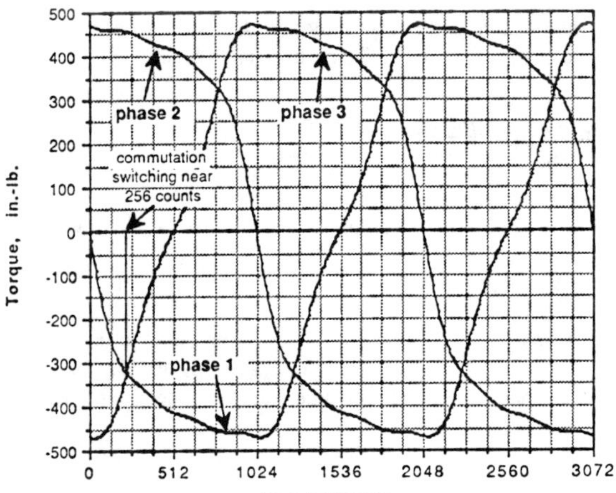

Figure 5. Torque from all three phases showing commutation switching point.

Referring to Figure 5, the point to switch phases is given by the intersection of the torque curves. When a phase becomes inefficient at generating torque of the desired sign (as shown when its decreasing torque curve intersects the increasing curve of the next phase) drive switches to the next phase. While Figure 5 represents the behavior for a pulse width of 40%, other pulse widths yielded similar results. Since the intersection point was near 256 encoder counts, this point (256) was used as the commutation switching point.

Phase Overlap.

Rather than switching from zero pulse width to a large pulse width instantly at commutation, one phase is ramped down and the oncoming phase ramped up. Thus during the transition regime two phases are activated.

The commutation switching point of 256 counts was used as the center of the overlap region, and the width of the overlap region was 256 counts. Thus from 128 to 384 counts the phase was being ramped up, and from 1152 to 1408 counts the phase is ramped down. The pulse width drive waveforms for the same five representative torque values are shown in Figure 6. Note that in the center portion of the plot the waveforms are the same as the non-overlapped versions of Figure 4. At the ends of the waveform, near the singularities, the linear ramping results in smooth transitions between phases.

The waveforms of Figure 6 were used to drive the three phases of the NSK variable reluctance motors. The experimental results of using these waveforms will be presented and discussed in the next section.

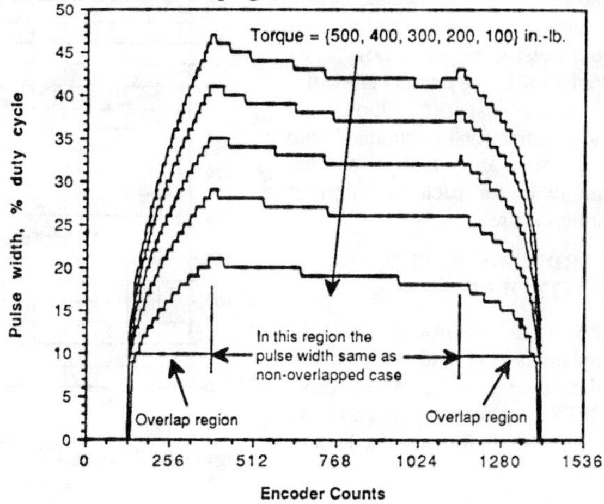

Figure 6. Overlapped motor drive waveforms for five representative torques.

EXPERIMENTAL RESULTS

The motors were driven with the same computational hardware as was used for data acquisition

As with the analytical procedure of the previous section, only joint 1 results will be presented here. Joints 2, 3, and 4 were driven, with similar results. A simulation of motor torque will first be presented, then the experimental data.

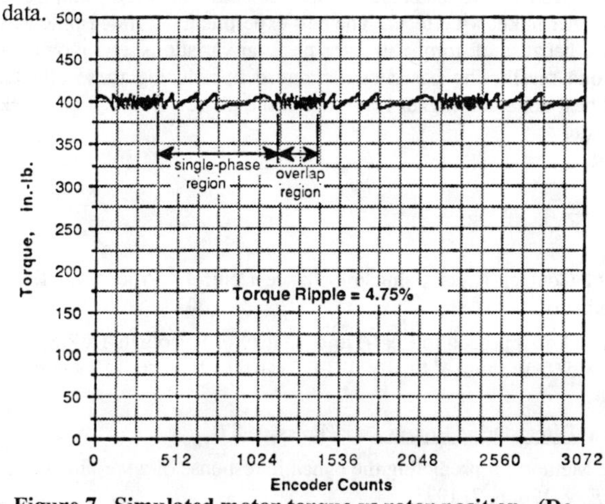

Figure 7. Simulated motor torque vs rotor position. (Desired torque = 400 in.-lb.)

Motor Simulation Results

To check our drive waveforms, we developed a motor simulation based on the multivariable curve fit. The simulation results are shown in Figure 7, with torque ripple of 4.75%. The sharp variations of torque are due to the granularity of the integer PWM command.

Motor Experimental Results

The initial torque *vs* rotor position measurement for joint 1 showed a torque ripple of 21%.

Based on the data, it appeared that the drive table could be adjusted to nearly eliminate torque ripple. However, due to variation of inductance with rotor motion, and the difference in the slope of the torque *vs* rotor position curve at either side of the commutation switching point (refer to Figure 5), this was not the case. Improvement in one direction of motion worsened the other.

Another approach is to use separate commutation tables for positive and negative motion, yet another approach is to compensate at the torque level—outside the drive table. The latter approach (torque prefiltering outside the drive table) was implemented, and the resulting torque plot appears in Figure 8.

The mean value of the torque in Figure 8 is still low, due to friction, but the ripple is now 13%. We feel that this method is worthy of further experimentation, and can be directly incorporated into the commutation table if a separate table for each direction of motion is used.

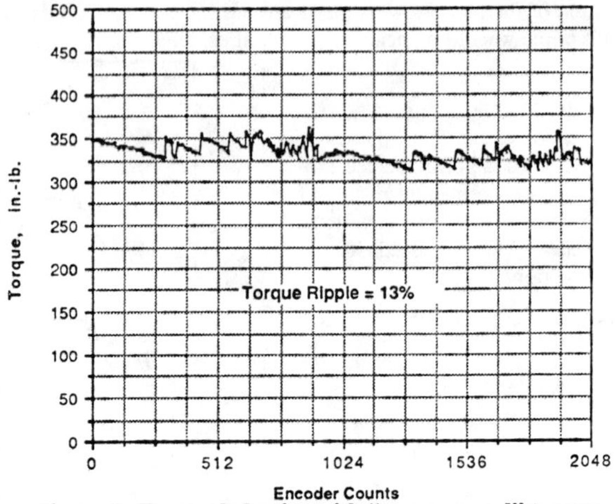

Figure 8. Torque behavior with "preprocessed" torque command, 400 in.-lb.

HARDWARE-BASED CONTROLLER

The final controller will be implemented in hardware to maximize speed, using a Motorola 68HC11 microcontroller and 128k × 8 EPROMs for drive table storage.

CONCLUDING REMARKS

This work has demonstrated capability to command torque of the NSK variable reluctance motor. We have shown techniques to compensate for modeling errors and other unknown effects. The best experimental torque ripple was 13%. Directional torque effects arise because of, torque *vs* rotor position variations, and motor inductance effects. Commanded torque prefiltering has shown promise in reducing these effects. Compensation for friction effects is best done at a higher level, not at the phase commutation level. This is typical in motor control practice.

REFERENCES

[1] Hogan, N., "Stable Execution of Contact Tasks Using Impedance Control," Proc. 1987 IEEE Int'l. Conf. on Robotics & Automation.

[2] Goldenberg, A. A., and L. Chan, "An Approach to Real-Time Control of Robots in Task Space. Application to Control of PUMA 560 Without VAL-II," IEEE Trans. of Industrial Electronics, v. 35, n. 2, May 1988.

[3] Asada, H., and K. Youcef-Toumi, Direct-Drive Robots, MIT Press, 1987.

[4] Pickup, I.E.D., and D. Tippins, "Method for predicting the dynamic response of a variable reluctance stepping motor," *Proc. IEE*, v. 120, n. 7, 1973.

[5] Taylor, D.G., M. Ilic-Spong, R. Marino, S. Peresada, "A feedback linearizing control for direct drive robots with switched reluctance motors," *Proc. 25th IEEE Conf. on Decision and Control*, December 1986.

[6] Taylor, D.G., M.J. Woolley, M. Ilic, "Design and implementation of a linearizing and decoupling feedback transformation for switched reluctance motors," *Proc. 17th Symposium on Incremental Motion Control Systems and Devices*, Champaign, IL, June, 1988.

LIAPUNOV DESIGN AVOIDANCE CONTROL FOR AN OPEN-END TWO-LINK ROBOT MANIPULATOR

George Bojadziev and Maria Bojadziev

Department of Mathematics and Statistics
Simon Fraser University

Mathematics Department
British Columbia Institute of Technology

ABSTRACT

A Liapunov control of motion program for a robot manipulator is established such that the manipulator links move in a desirable region and do not enter into a specified avoidance region.

1. INTRODUCTION

We study a mechanical system with two degree of freedom consisting of two open-end links in a vertical plane α. The links connected by a force, are subjected to gravity and damping. A controlled torque force is applied to each link at their joint point O. Using a Liapunov stability methodology (Leitmann and Skowronski [1]) we present a control program which guarantee that the manipulator links move in a desirable workspace in α and do not enter into a certain avoidance angle. Our study is also based on the book by Skowronski [2] on control of robot manipulators and the paper [3] by the authors on avoidance control for a manipulator with two links having equal length.

2. THE MODELLING EQUATIONS

In the vertical plane α we consider the double pendulum (Fig. 1) as a model for a robot manipulator with two open-end rigid links connected at O. The links rotate about a fixed axis perpendicular to α and have lengths ℓ_1 and ℓ_2. The masses m_1 and m_2 which also represent the grippers are centered at the points P_1 and P_2. The position of the links is determined by the angular coordinates φ_1 and φ_2.

We assume that the damping force $R_s = -\lambda_s \varphi_s'$, $s=1,2$, $\lambda_s > 0$, is acting on the s-th link and also that a damping force of the type $q = F(\varphi_2' - \varphi_1')$ is acting on the links at the joint point O. The force q reflects the existence of connection between the links. Each link at the joint point O is subjected to a controlled torque force $r_s = m_s\ell_s^2 u_s$, $s=1,2$, $u_s \in R$.

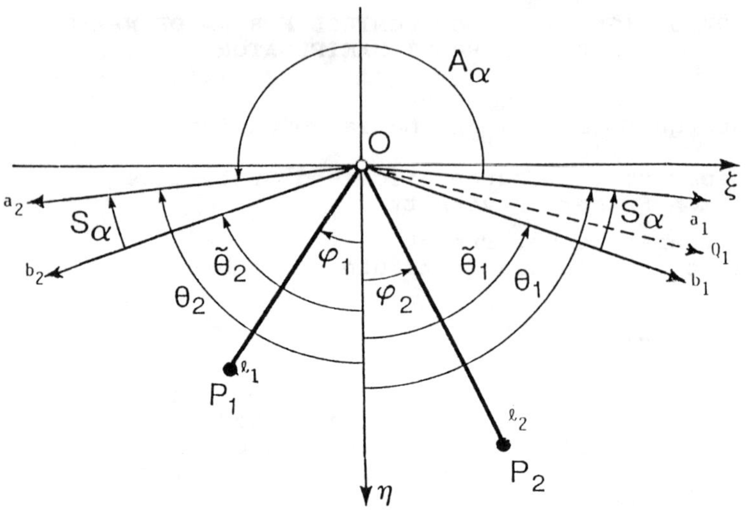

Fig. 1

Using the Lagrange equations with R_s, r_s, s=1,2, and q, we obtain the differential equations modelling the motion of the two-link robot manipulator

$$\begin{aligned} \varphi_1'' + \alpha_1\varphi_1' + \beta_1\sin\varphi_1 &= \gamma_1 F(\varphi_2'-\varphi_1') + u_1 , \\ \varphi_2'' + \alpha_2\varphi_2' + \beta_2\sin\varphi_2 &= -\gamma_2 F(\varphi_2'-\varphi_1') + u_2 , \end{aligned} \tag{1}$$

where the constants

$$\alpha_s = \frac{\lambda_s}{m_s\ell_s^2}, \quad \beta_s = \frac{g}{\ell_s}, \quad \gamma_s = \frac{1}{m_s\ell_s^2}, \qquad s=1,2, \tag{2}$$

are positive and g is the gravitational constant.

The control torque force $\mathbf{u}(t) = (u_1,u_2)^T \in U$ and the control objectives will be specified later. Setting $\varphi = \mathbf{y}$, $\varphi = (\varphi_1,\varphi_2)^T$, $\mathbf{y} = (y_1,y_2)^T$, we present (1) in normal form in the phase space R^4:

$$\begin{aligned} &\varphi_1' = y_1 , \qquad \varphi_2' = y_2 , \\ &y_1' = -\alpha_1 y_1 - \beta_1\sin\varphi_1 + \gamma_1 F(y_2-y_1) + u_1 , \\ &y_2' = -\alpha_2 y_2 - \beta_2\sin\varphi_2 - \gamma_2 F(y_2-y_1) + u_2 . \end{aligned} \tag{3}$$

For $u_1 = u_2 = 0$ in (3) we obtain the uncontrolled model (UCM). We adopt and modify some of the assumptions made in [4], where a more general but uncontrolled model has been investigated.

(a) The function $F(w)$, $w = y_2 - y_1$ is a C^1 function in the neighbourhood of the equilibria of the UCM (3).

(b) The function $F(w)$ is an odd function, i.e.

$$F(w) = -F(-w) \Rightarrow F(0) = 0 . \tag{4}$$

(c) $$(\gamma_2 y_2 - \gamma_1 y_1)\, F(y_2 - y_1) > 0 \ \forall\ |y_2 - y_1| > 0 . \tag{5}$$

3. EQUILIBRIA AND LIAPUNOV FUNCTION

The equilibria of the UCM (3) are located in the hyperplane (φ_1, φ_2) and are solutions of the system

$$\beta_1 \sin\varphi_1 = 0 , \qquad \beta_2 \sin\varphi_2 = 0 , \tag{6}$$

which is obtained from (3) by setting $\varphi'_s = y'_s = 0$, $s=1,2$, and taking into account (4). The system (6) has solutions of the type $(p\pi, p\pi, 0, 0) \in R^4$, where p is integer. Besides the trivial equilibrium $E_o(0,0,0,0)$ we deal only with the equilibria

$$E_{1,2}(\pm\pi,0,0,0), \quad E_{3,4}(0,\pm\pi,0,0). \tag{7}$$

There is a neighbourhood **B** of E_o, which does not contain inside any other equilibrium of the UCM.

Consider the function

$$V(\varphi,\mathbf{y}) = y_1^2 + y_2^2 + 2\beta_1(1-\cos\varphi_1) + 2\beta_2(1-\cos\varphi_2) , \tag{8}$$

which has in **B** the following properties:

(i) $V(\mathbf{0},\mathbf{0}) = 0 , \quad \mathbf{0} = (0,0)^T$;
(ii) $V(\varphi,\mathbf{y})$ is positive definite in $\mathbf{B}\backslash E_o$;
(iii) $\dfrac{dV}{dt} = \sum_{s=1}^{2} \left(\dfrac{\partial V}{\partial \varphi_s} \varphi'_s + \dfrac{\partial V}{\partial y_s} y'_s \right) < 0$;

φ'_s, y'_s are the right hand sides of the UCM (3). The inequality in property (iii) follows from assumption (5) since it can be represented in the form

$$-2 \sum_{s=1}^{2} \alpha_s y_s^2 - 2(\gamma_2 y_2 - \gamma_1 y_1)\, F(y_2 - y_1) < 0 ; \tag{9}$$

From (ii) and (iii) we conclude that (8) is a Liapunov function for the UCM in **B**. It describes in $V \times \mathbf{B} \subset R^5$ a "cup-shaped" hypersurface with minimum at E_o. The equation

$$V(\varphi,\mathbf{y}) = h , \quad h = \text{const} > 0 , \tag{10}$$

represents a family of level surfaces in $V \times \mathbf{B} \subset R^5$ whose orthogonal projection onto $\mathbf{B} \subset R^4$ generates hypersurfaces **H** in R^4 with the following properties: (a) They are closed; (b) Do not intersect; (c) Contain inside only the equilibrium E_o; (d) If $h_1 < h_2$, the hypersurface $\mathbf{H}_1$ is inside $\mathbf{H}_2$. The largest hypersurface $\mathbf{H}_*$ from the set **H** with the above properties (a)-(d) passes through one or more equilibria (7) of the UCM, whose coordinates minimize h. We find from (8) and (10)

$$h^* = \min h_s = \min V(E_s) = \min(4\beta_1, 4\beta_2), \qquad (11)$$

hence $\mathbf{H}_* = \{(\varphi, \mathbf{y}) \in R^4 : V(\varphi, \mathbf{y}) = h^*\}$. The function $V(\varphi, \mathbf{y})$ possesses the properties (i)-(iii) in larger regions than **B**. The largest region **D** with properties (i)-(iii) is $\mathbf{D} \triangleq \{(\varphi, y) \in R: V(\varphi, \mathbf{y}) \leq h^*\}$ with boundary $\partial \mathbf{D} = \mathbf{H}_*$.

4. BASIC DEFINITIONS

It is assumed that there is no danger of links collision when the links overlap ($\varphi_1 = \varphi_2$). In order to avoid collision with stationary or moving objects, the links may not be allowed to reach or cross a specified boundary defined by the segments Oa_s, s=1,2, which form the outer arms of the angles $\theta_s = \measuredangle \eta O a_s$ (Fig. 1). In a real-world process there are always small disturbances due to various errors which makes meaningful the introduction of safety segments Ob_s, s=1,2, whose role is to prevent the links OP_s of getting too close to the boundary segments Oa_s. The segments Ob_s form the angles $\tilde{\theta}_s = \measuredangle \eta O b_s$ (Fig. 1). For a particular study the difference $\theta_s - \tilde{\theta}_s$, s=1,2, can be selected as small as desired. For simplicity we assume that $|\theta_2| = \theta_1 = \theta$, $|\tilde{\theta}_2| = \tilde{\theta}_1 = \tilde{\theta}$, $\theta, \tilde{\theta} > 0$, $\theta > \tilde{\theta}$.

To formulate the control objectives more precisely we introduce, using [3], the following definitions concerning angles in the plane α.

Admissible angle $\Omega_\alpha \triangleq \measuredangle (a_2 O a_1)$: $O\eta \in \Omega_\alpha$, $Oa_s \notin \Omega_\alpha$;

Avoidance angle $A_\alpha \triangleq \measuredangle (a_1 O a_2)$: $O\eta \notin A_\alpha$, $Oa_s \in A_\alpha$;

Safety angle $S_\alpha \triangleq \measuredangle (b_1 O a_1) \cup \measuredangle (a_2 O b_2) \in \Omega_\alpha$, $Oa_s \notin S_\alpha$;

Desirable angle $\tilde{\Omega}_\alpha \triangleq \measuredangle (b_2 O b_1)$: $O\eta \in \tilde{\Omega}_\alpha$, $\tilde{\Omega}_\alpha = \Omega_\alpha - S_\alpha$.

We say that the two-link manipulator enters into the angle A_α if at least one of the connected links OP_1 and OP_2 enters into A_α.

The Control objective is to design a strategy according to which the control force **u**(t) secures that the two-link open-end robot does not enter into the avoidance angle A_α but preferably moves inside the desirable angle $\tilde{\Omega}_\alpha$ (Fig. 1).

The motion of the controlled two-link manipulator modelled by (1) in the plane α has dynamics described by the system (3) in R^4 whose solution for a given **u** is called response to **u**; we denote it by $\mathbf{z}[t] = \mathbf{z}(t,\varphi(t_o), \mathbf{y}(t_o),\mathbf{u})$, $\mathbf{z} \in R^4$, $z_s = \varphi_s$, $z_{2+s} = y_s$, s=1,2, $(\varphi(t_o), \mathbf{y}(t_o)) \in R^4$ is the initial state, $t_o \geq 0$.

Using (11) we introduce the values h_A and h_S:

$$h_A = h^* - \varepsilon \ , \qquad h_S = h^* - \delta \ , \tag{12}$$

where $0 \leq \varepsilon < \delta$, hence $h_A > h_S$; ε and δ are as small as desired for a particular study. Then with the help of (12) we define the following regions in R^4.

Admissible set $\Omega \triangleq \{(\varphi,\mathbf{y}) \in R^4: V(\varphi,\mathbf{y}) < h_A \leq h^*\}$;

Avoidance set $A \triangleq \mathbf{D}\backslash\Omega$ with boundary $\partial A = \mathbf{H}_A$;

Safety zone $S \triangleq \{(\varphi,\mathbf{y}) \in R^4: h_S \leq V(\varphi,\mathbf{y}) < h_A\}$ with boundary $\partial S = \mathbf{H}_S$;

Desirable region $\tilde{\Omega} \triangleq \Omega\backslash S$.

5. AVOIDANCE CONTROL

Definition. The region A is avoidable by the response $\mathbf{z}[t]$ of (3) if there is a control $\mathbf{u} \in U$ such that for all states $(\varphi(t_1), \mathbf{y}(t_1)) \in S$, $t_1 \geq t_o$, the response $\mathbf{z}[t] = \mathbf{z}(t, \varphi(t_1), \mathbf{y}(t_1), \mathbf{u})$ cannot enter A.

Theorem. The response $\mathbf{z}[t]$ of the dynamical system (3) is controllable in $\Omega \subset \mathbf{D}$ for avoidance of the region A if there is a control $\mathbf{u} \in U$ such that

$$\frac{dV}{dt} = \sum_{s=1}^{2} \left(\frac{\partial V}{\partial \varphi_s} \varphi'_s + \frac{\partial V}{\partial y_s} y'_s \right) \leq 0 \ , \tag{13}$$

where φ'_s and y'_s are determined by (3) and V by (8).

Proof. Assume that the region A is not avoidable. Then for some state $(\varphi(t_1), \mathbf{y}(t_1)) \in S$ and time $t > t_1$, the response $\mathbf{z}[t]$ enters A. This implies that there is a $t_2 > t_1$ such that $\mathbf{z}[t_2] \in \partial A$, the boundary of A. Hence $V(\varphi(t_1), \mathbf{y}(t_1)) < V(\varphi(t_2), \mathbf{y}(t_2))$, i.e. the Liapunov function $V(\varphi,\mathbf{y})$ in increasing thus contradicting (13) according to which $V(\varphi,\mathbf{y})$ is non-increasing along every response of (3).

The sufficient condition (13) for avoidance can be expressed in explicit form. Using (3) and (8), we get

$$\sum_{s=1}^{2} y_s u_s - \sum_{s=1}^{2} \alpha_s y_s^2 - (\gamma_2 y_2 - \gamma_1 y_1) F(y_2 - y_1) \le 0 . \qquad (14)$$

From (14) we see that the role of the control force is to balance the effect of the velocities. We can assume that $\mathbf{u}(t) \in U = \{u_s : u_s \in [-r,r]\}$, $r = \text{const}, s=1,2\}$ as long as (14) is satisfied.

6. CONTROL STRATEGY

We design an avoidance control strategy for the two-link robot manipulator modelled by (1) based on the results derived in Section 5 concerning the response of the dynamical system (3) in **D**.

Let us assume that the links OP_1, OP_2 subjected to fixed control $\mathbf{u}(t_0)$ oscillate in angle $\tilde{\Omega}_\alpha \subset R^2$. The corresponding response $\mathbf{z}[t]$ of (3) moves in R^4. To guarantee that the links will avoid entering into angle A_α, we monitor simultaneously: (i) The value $V(\mathbf{z}[t])$ which should be smaller than h_S given by (12), since we wish $\mathbf{z}[t] \in \tilde{\Omega}$; (ii) The angular coordinates φ_1 and φ_2 so that the links OP_1, OP_2 belong to $\tilde{\Omega}_\alpha$. The following two situations may occur.

A. The links OP_1, $OP_2 \in$ angle $\tilde{\Omega}_\alpha$ at time $t_1 > t_0$ while $h_S < V(z[t_1]) < h_A$, which means that the response $\mathbf{z}[t]$ has crossed the boundary ∂S and is into the safety zone S. Our strategy is at time t_1 to change the control $\mathbf{u}(t_0)$ to a new control $\mathbf{u}(t_1) \in U$ which satisfies the control law (14). Then the response $\mathbf{z}[t]$ would move "down" through decreasing levels of $V(\varphi(t), \mathbf{y}(t))$.

B. One of the links, say OP_2, enters the safety angle S_α, say angle(b_1Oa_1), and at time $t_1 > t_0$ coincides with the segment $OQ_1 \in$ angle(b_1Oa_1) (see Fig. 1), while the response $\mathbf{z}[t]$ reaches the state $\mathbf{z}[t_1] \equiv Q(\varphi(t_1), \mathbf{y}(t_1)) \in \tilde{\Omega}$, i.e. $V(\mathbf{z}[t_1]) < h_S$. To prevent OP_2 of entering into the avoidance angle A_α, a new control vector $\mathbf{u}(t_1) \in U$ which obeys (14) has to be selected at the switching point Q. The response $\mathbf{z}[t]$ to the control $\mathbf{u}(t_1)$ will be steered "down" through decreasing levels of $V(\varphi(t), \mathbf{y}(t))$ and the link OP_2 will move back into the desirable angle $\tilde{\Omega}_\alpha$.

Note. It is possible at time $t_2 > t_1$ one of the links either to enter again angle S_α, or the response to enter S. Then we apply the strategy in A or B.

7. CASE STUDY

Let us specify the model parameters in (2) and the function F in (3) as follows: $\ell_1 = 10$, $\ell_2 = 12$, $m_1 = 1$, $m_2 = 1.2$, $\lambda_1 = \lambda_2 = 10$, $F(y_2-y_1) = (y_2-y_1)^3$. Then (3) reduces to a system which we label by (RS) and (8) now is

$$V(\varphi,\mathbf{y}) = y_1^2 + y_2^2 + 2[(.98)(1-\cos\varphi_1) + (.818)(1-\cos\varphi_2)].$$

The condition for avoidance control (14) becomes

$$y_1u_1+y_2u_2-(.1)y_1^2-(.058)y_2^2-[(.0058)y_2-(.01)y_1](y_2-y_1)^3 \le 0.$$

From (14) with (11) and (7) we find that $h^* = 3.270$. We select $\theta = 1.571\ (90^\circ)$ and $\tilde{\theta} = 1.396\ (80^\circ)$, $\varepsilon = .01$, $\delta = .1$. Then from (12) we get $h_A = 3.260$, $h_S = 3.170$. Now we consider the system(RS) at time $t_0 = 0$ with initial state $(0, 0, -0.5, 0.5)^T$ and control $\mathbf{u} = (-0.5,1)^T$. A fourth order Runge-Kutta was used on time interval [0,5] with step .01 to integrate (RS), i.e. to find the response z[t]. At time $t = 1.22$ we get $V(\varphi,\mathbf{y}) = 3.175 > h_S$, $\varphi_1 = -0.745 < \tilde{\theta}$, $\varphi_2 = 1.131 < \tilde{\theta}$. Hence the response has entered S and according to case A in Section 6 we change the control at Q(-0.745, 1.131, -0.570, 1.180). The new control is $\mathbf{u} = (0.52, -1.15)^T$. The oscillations of the

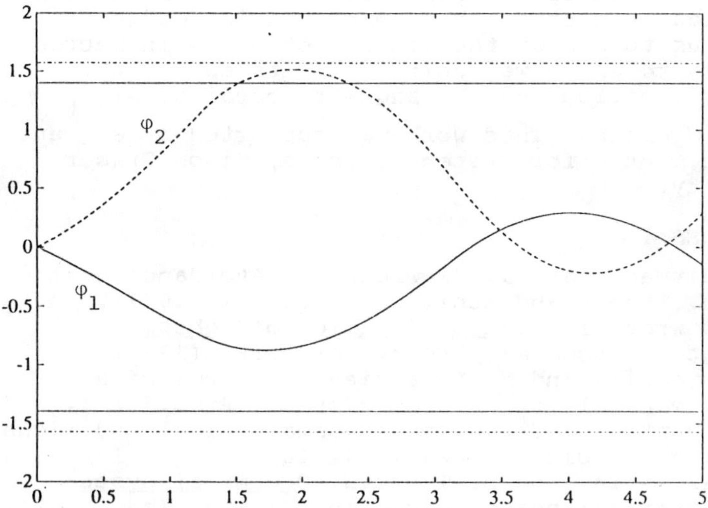

Fig. 2

angular coordinates versus time are shown in Fig. 2, and the level of V on Fig. 3. On $t \in [0, 5]$ there are three switching points, two are associated with case A which requires lowering the level of V (see Fig. 3) and one with case B (see Fig. 2).

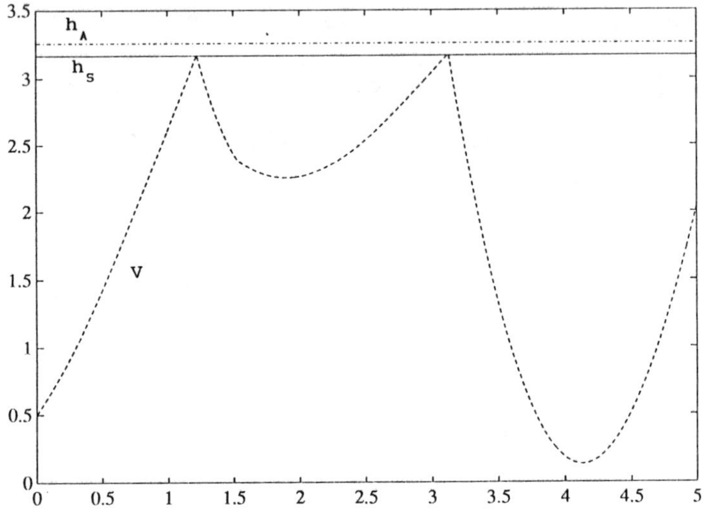

Fig. 3

8. CONCLUSION

A Liapunov design avoidance control technique for an open-end two-link manipulator has been presented. The avoidance control theorem proved in Section 5 enables us to design the control strategy in Section 6 according to which we monitor simultaneously the Liapunov function and the angular coordinates φ_1, φ_2.

Acknowledgements. This work was supported by a grant from the Centre for Systems Science, Simon Fraser University.

REFERENCES

1. G. Leitmann, and J. Skowronski, "Avoidance control", J. Optim. theory and Appl., 23, 581-591 (1977).
2. J. Skowronski, Control dynamics of robotic manipulators, Academic Press, New York, (1986).
3. G. Bojadziev and M. Bojadziev, "Control of a two-link manipulator with open ends", Proc. of IASTED Adaptive and Knowledge-Based Control and Signal Processing, Honolulu, Hawaii, 42-46 (Aug. 16-18, 1989).
4. E. Barbashin, and B. Tabueva, Dynamical systems with cylindrical phase space (in Russian, Izd. Nauka, Moskow, 1969).

REFLEX-ACTION POSITION CONTROL AS APPLIED TO CONSTRAINED MANEUVERS

J. N. Anderson **Y. L. Lee**
Center for Manufacturing Research and Technology Utilization
Tennessee Technological University
Cookeville, Tennessee

ABSTRACT

The reflex-action position (RAP) control method is effective when applied for unconstrained manipulator motion where the load is possibly time-varying and/or unknown. However, once the manipulator is constrained, it is not clear that the system will remain stable or have appropriate contact forces. In this paper, dynamic models appropriate for constrained motion are reviewed and extended. These models, which include static and Coulomb contact friction effects, are applied to the RAP control algorithm demonstrating the potential for unbounded control torques when setpoints are not reachable. Simulation provides verification, but also demonstrates compensation of tangential friction forces. Methods to compensate for the normal contact forces are discussed.

1. INTRODUCTION

For a broad class of closed loop manipulator control methods, performance can be severely degraded when maneuvers are constrained. The reflex-action position (RAP) control method is a promising solution to some of the control problems posed by unknown, time-varying loads [1]. It is a model-based control method that includes a load-decoupling force feedback loop. This has the effect of stiffening the manipulator considerably, even for low position loop gains. Simulation results for unconstrained motion of a RAP-controlled manipulator indicate excellent performance under various loading conditions. However, once the manipulator is constrained, it is not clear that the system will remain stable or have appropriate interaction (contact) forces. (The generic term forces used here includes forces and torques.)

In this paper, a dynamics model suitable for the simulation of a RAP-controlled manipulator performing constrained maneuvers is developed. The model includes effects due to static and sliding (Coulomb) contact friction. The closed-loop dynamics are examined to predict the RAP controller's response to motion constraints. Performance is also evaluated via simulation. Note that the problem of impact is not considered.

2. RAP CONTROL

The dynamics of a rigid n-link manipulator are given by [2]

$$(H(q) + J_m)\ddot{q} + (C(q,\dot{q}) + B)\dot{q} + G(q) + J^T(q)f + F(\dot{q}) = \tau\ , \qquad (1)$$

where τ = joint forces, q = joint displacements, $F(\dot{q})$ = forces due to Coulomb friction, $G(q)$ = joint forces due to gravity, $C(q,\dot{q})\dot{q}$ = joint forces due to Coriolis and centrifugal effects (all are $n \times 1$), f = external forces on the load (6×1), $J(q)$ = the manipulator Jacobian $(6 \times n)$, $H(q)$ = manipulator inertias ($n \times n$, positive-definite, symmetric), J_m = drive system inertias, and B = damping coefficients due to drive system viscous friction and back-EMF (both are $n \times n$, diagonal). All coordinate-dependent quantities, such as J and f, are relative to the base frame $\mathcal{F}_\circ$. With conventional model-based control (CMB) [2],

$$\tau = \tau_c + \hat{\beta}\ , \qquad (2)$$

where $\hat{\beta} = (\hat{C}(q,\dot{q}) + \hat{B})\dot{q} + \hat{G}(q)$ is intended to cancel the corresponding terms in (1). The control torques τ_c are often obtained by using PID control. Since load properties are unknown, a nominal load must be used when calculating $\hat{H}$, $\hat{C}$, and $\hat{G}$.

If f_s contains the forces at a wrist force sensor, then the dynamic effects of the tool/load and the external forces are included in (1) by $J_s^T(q)f_s$ so that

$$(H_\circ(q) + J_m)\ddot{q} + (C_\circ(q,\dot{q}) + B)\dot{q} + G_\circ(q) + J_s^T(q)f_s + F(\dot{q}) = \tau\ , \qquad (3)$$

where $H_\circ$, $C_\circ$, and $G_\circ$ are load-independent. In RAP control, the tool/load and external forces are compensated by using

$$\hat{\beta} = (\hat{C}_\circ(q,\dot{q}) + \hat{B})\dot{q} + \hat{G}_\circ(q) + \hat{J}_s^T(q)f_m \qquad (4)$$

to linearize *and* decouple the load [1]. (f_m contains the *measured* forces.) τ_c in (2) is defined as $\tau_c = (\hat{H}_\circ(q) + \hat{J}_m)\varepsilon_p + \hat{J}^T(q)\varepsilon_f$, where ε_p and ε_f are position and force feedback controls. ($\varepsilon_f = 0$ for single-arm control.) With PD control, $\varepsilon_p = \ddot{q}_d + K_v(\dot{q}_d - \dot{q}) + K_p(q_d - q)$. With an accurate model, steady-state position errors are zero while $\hat{J}^T f_m$ in (4) generates "reflex actions" to load variations [1]. A CMB/PD controller must have a perfect model *and nominal loading* for zero steady-state error.

3. THE EFFECTS OF MOTION CONSTRAINTS

When the manipulator's tip is constrained, kinematic (path) constraints and dynamic constraints (contact forces and torques) should be considered. Relative to $\mathcal{F}_\circ$, a *bilateral* (holonomic) constraint is $\phi(p) = 0$, where $\phi : R^6 \rightarrow R$ and $\phi(p)$ is twice differentiable, i.e., the constraint surface is smooth. A *unilateral* (nonholonomic) constraint is $\phi(p) \geq 0$. When the constraint is active ($\phi(p) = 0$), there exists a contact force and the constraint must be regarded as an intrinsic part of the dynamics. For the purpose of this study, the unilaterally constrained dynamic model is adopted.

The simplified dynamics of a unilaterally constrained manipulator that does not consider contact friction are modeled by [3]

$$(H(q) + J_m)\ddot{q} + C(q,\dot{q})\dot{q} + G(q) = \tau + J_c^T(q) f_c\,, \tag{5}$$

where $J_c(q)$ is the Jacobian from the point of contact and f_c is the vector of contact forces (normal to the constraint surface) given by

$$f_c = (\nabla\phi(p))^T \lambda\,. \tag{6}$$

In this equation, λ is a Lagrange multiplier and the gradient $\nabla\phi(p)$ is a 1×6 direction vector normal to the constraint surface in the position variables x, y, and z and the Euler angles φ, θ, and ψ. The value of λ required for the evaluation of f_c may be determined as shown in [3].

The complementary conditions for this model are

$$\phi(p) \geq 0\,,\ \lambda \geq 0\,,\ \text{and}\ \lambda\phi(p) = 0\,. \tag{7}$$

When $\phi(p) > 0$, (7) yields $\lambda = 0$. In this case (5) reduces to the unconstrained dynamic model, (1).

The unilaterally constrained dynamic model including static and sliding (Coulomb) contact friction effects, damping terms, and Coulomb joint friction for each can be formulated using d'Alembert's principle by interpreting these dissipative forces as active forces [4, 5]. The result is

$$(H(q) + J_m)\ddot{q} + (C(q,\dot{q}) + B)\dot{q} + G(q) + F(\dot{q}) = \tau + J_c^T(q) f_c\,, \tag{8}$$

where

$$f_c = (\nabla\phi(p))^T \lambda + f_\mu \tag{9}$$

and f_μ is either static friction ($\dot{p} = 0$) or sliding friction ($\dot{p} \neq 0$). Sliding friction is given by [5],

$$f_\mu(\dot{p} \neq 0) = -\mu \frac{\dot{p}}{\|\dot{p}\|} \|(\nabla\phi(p))^T \lambda\| = -\mu \mathrm{sign}(\lambda) \frac{\dot{p}}{\|\dot{p}\|} \|(\nabla\phi(p))^T\| \lambda\,, \tag{10}$$

where μ is the sliding friction coefficient. By (6) and (10), the sliding friction has magnitude proportional to that of the normal contact force and direction opposite to that of the tip velocity.

To find λ, an approach similar to that of [5] is used. First define

$$A \triangleq \nabla\phi(p) J_c(q) \quad \text{and} \quad \hat{A} \triangleq -\ \mu \mathrm{sign}(\lambda) \frac{\dot{p}^T}{\|\dot{p}\|} \|(\nabla\phi(p))^T\|\ J_c(q). \tag{11}$$

Thus (8) can be written as

$$(H(q)+J_m)\ddot{q}+(C(q,\dot{q})+B)\dot{q}+G(q)+F(\dot{q})=\tau+(A+\hat{A})^T\lambda\,. \tag{12}$$

Since $\phi(p)=0$, $\nabla\phi(p)\dot{p}=\nabla\phi(p)J_c(q)\dot{q}=A\dot{q}=0$. So

$$\dot{A}\dot{q}+A\ddot{q}=0\,. \tag{13}$$

From (12) and (13)

$$\begin{aligned}\lambda=&[A(H(q)+J_m)^{-1}(A+\hat{A})^T]^{-1}\\ &[A(H(q)+J_m)^{-1}((C(q,\dot{q})+B)\dot{q}+G(q)+F(\dot{q})-\tau)-\dot{A}\dot{q}]\end{aligned} \tag{14}$$

is the value of λ needed to find f_c.

Static friction is present when $\dot{p}=J_c(q)\dot{q}=0$. Thus $J_c(q)\ddot{q}+\dot{J}_c(q)\dot{q}=0$. From (8),

$$\begin{aligned}f_c(\dot{p}=0)=&[J_c(q)(H(q)+J_m)^{-1}J_c^T(q)]^{-1}[J_c(q)(H(q)+J_m)^{-1}\\ &((C(q,\dot{q})+B)\dot{q}+G(q)+F(\dot{q})-\tau)-\dot{J}_c(q)\dot{q}]\,,\end{aligned} \tag{15}$$

Since static friction is in the tangential direction,

$$f_\mu(\dot{p}=0)=f_c(\dot{p}=0)-\left(\frac{\nabla\phi(p)}{\|(\nabla\phi(p))\|}f_c(\dot{p}=0)\right)\frac{(\nabla\phi(p))^T}{\|(\nabla\phi(p))\|}\,. \tag{16}$$

For the purpose of simulation, if the static friction exceeds a preset threshold the simulation program should allow motion at the tip.

4. RAP CONTROL WITH MOTION CONSTRAINTS

To demonstrate the effects of motion constraints on the performance of RAP control, a different form of the dynamics equation is needed. If f_l contains the *inertial* forces and on the load and $J_l(q)$ is the Jacobian from the load's mass center, (8) may be written as

$$(H_\circ(q)+J_m)\ddot{q}+(C_\circ(q,\dot{q})+B)\dot{q}+G_\circ(q)+F(\dot{q})+J_l^T(q)f_l=\tau+J_c^T(q)f_c\,. \tag{17}$$

(Note the load-independent quantities.)

The measured forces (f_m) can be expressed as

$$f_m=f_s+\text{measurement error}\,. \tag{18}$$

Since f_s includes contributions from f_l and f_c,

$$J_s^T(q)f_s=J_l^T(q)f_l-J_c^T(q)f_c\,. \tag{19}$$

Thus f_m is transferred to joint space by

$$\hat{J}_s^T(q)f_m=\hat{J}_l^T(q)\hat{f}_l-\hat{J}_c^T(q)\hat{f}_c\,, \tag{20}$$

where it is assumed that the measurement errors may be apportioned betweeen f_l and f_c, as indicated by $\hat{f}_l$ and $\hat{f}_c$.

The form of the linearizing and load-decoupling feedback of RAP control may be expanded by substituting (20) into (4). The resulting control torque vector may be substituted into (17) to give

$$\begin{aligned}(H_\circ + J_m)\ddot{q} + (C_\circ + B)\dot{q} + G_\circ + F + J_l^T f_l - J_c^T f_c \\ = (\hat{H}_\circ + \hat{J}_m)\varepsilon_p + (\hat{C}_\circ + \hat{B})\dot{q} + \hat{G}_\circ + \hat{F} + \hat{J}_l^T \hat{f}_l - \hat{J}_c^T \hat{f}_c \,.\end{aligned} \tag{21}$$

By assuming; 1) that $(\hat{H}_\circ(q) + \hat{J}_m)\ddot{q}$, and $(\hat{C}_\circ(q, \dot{q}) + \hat{G}_\circ(q))$ do not change appreciably over one control interval, 2) that the manipulator's kinematic and dynamic parameters are computed accurately, and 3) that the PD control law is used for ε_p, (21) may be reduced to

$$\ddot{E} + K_v\dot{E} + K_pE = (H_\circ + J_m)^{-1}\left[J_l^T f_l - \hat{J}_l^T \hat{f}_l + \hat{J}_c^T \hat{f}_c - J_c^T f_c)\right] . \tag{22}$$

where $E = q_d - q$ are the joint errors. Thus with the proper choices of K_v and K_p, the system will be bounded-input, bounded-output stable. But, from (4), the control torques would be unbounded if the measurement forces are unbounded. This might occur if a position was commanded that could not be achieved due to the constraint.

Assume that a constant position trajectory is commanded whose setpoint is within a rigid surface. Hence, the robot cannot eliminate the position error. This results in nonzero contact forces. Assuming that; 1) the position remains constant (such that $\ddot{E} = \dot{E} = 0$), 2) $J_l^T(q)f_l - \hat{J}_l^T(q)\hat{f}_l = 0$, and 3) there is zero measurement error, then, after k sample periods, (22) can be approximated by

$$J_c^T(q)f_c^k \approx J_c^T(q)f_c^{k-1} - (H_\circ(q) + J_m)K_pE \,. \tag{23}$$

The full sample period delay is included to model the computational delay. After an additional n sample periods,

$$J_c^T(q)f_c^{k+n} \approx J_c^T(q)f_c^{k-1} - (n+1)(H_\circ(q^k) + J_m)K_pE \,. \tag{24}$$

Clearly, large contact (and measured) forces will be produced as time increases. This prediction is tested via simulation in the next section.

5. SIMULATION

Based on the models of the previous two sections, a constrained PUMA 562 manipulator operating under RAP control was simulated using ROBPACK [6] with modifications and additions to the subroutines as described in [7]. As indicated in [3], the condition $\lambda < 0$ indicates that the robot has lost contact with the constraint surface. For the purpose of simulation, if $\lambda < -\epsilon$ ($\epsilon = 10^{-7}$) the robot is considered to have lost contact and the simulation is terminated.

A problem in simulating constrained maneuvers is constraint violation due to round off and numerical errors. Methods have been proposed to correct for constraint violation (e.g., see [8]). The method used in this study is based on the fact that $\phi(p) = 0$, $\nabla\phi(p)\dot{p} = 0$, and $\ddot{\phi}(p) = 0$.

1. Position constraint violation ($\phi(p) \neq 0$) — At each integration step, the point (p_c) on the constraint surface where the surface normal intersects the tip position is used to replace p. The corrected joint positions (q_c) are found from the inverse kinematics.
2. Velocity constraint violation ($\nabla\phi(p)\dot{p} \neq 0$) — At each integration step, the nonzero values of $\nabla\phi(p)\dot{p}$ are used to correct $\dot{p}$ by

$$\dot{p}_c = \dot{p} - \left(\frac{\nabla\phi(p)}{\|\nabla\phi(p)\|}\dot{p}\right)\frac{(\nabla\phi(p))^T}{\|\nabla\phi(p)\|}\,. \tag{25}$$

Inverse kinematics are used to calculate the corrected joint velocities ($\dot{q}_c$).

The smooth, rigid constraint surface was modeled by $x = 0.567$ m with respect to $\mathcal{F}_\circ$. The load was a 0.1 Kg point mass located 0.1 m from the force sensor along the z-axis of the sensor frame. The limit of static friction was taken to be 3 Newtons (N) and μ was set to 0.01. The manipulator was commanded to follow a circular trajectory with radius 0.1 m that was 0.2 mm within the constraint surface. Initially the tip was located at (0.567, 0.3, 0.4, 0.0, $\pi/2$, 0.0). The commanded trajectory was $x = 0.5672$, $y = 0.3 + 0.1\cos(t)$, $z = 0.3 + 0.1\sin(t)$, $\varphi = 0.0$, $\theta = \pi/2$, and $\psi = 0.0$. The controller was initialized such that the velocity and acceleration errors were zero. The control update rate was 200 Hz, and the cutoff frequency of the force sensor filter was 80 Hz. The integration step size was 1 msec. Simulation output data were plotted at the control rate.

The plots in Figure 1 show that the position errors are stable, but the measured forces and actuator torques are increasing. Since the position error in the x direction cannot be eliminated, the positive force feedback ($\hat{J}_s^T(q)f_m$) in Equation (4) causes increasing control torques. These simulation results verify the special case discussed in the previous section. Their similarity to the case of a frictionless surface [7] (not included), indicate that RAP control is effective in compensating for the tangential contact forces, i.e., for those due to friction.

6. CONCLUSIONS

The RAP control as applied to constrained maneuvers was studied. The models developed here can easily be extended to multi-dimensional constraint functions. The simulation results show that the position tracking is stable and accurate, but the control torques, and hence the contact forces, can become unbounded if actutor torque limits are not modeled. Static and Coulomb friction are compensated by the RAP controller since they occur in unconstrained directions. A case including white noise and sinusoidal interference in the force sensor signals and parameter errors was presented in [7]. The results demonstrated good robustness and noise rejection and that parameter errors have a major effect on performance.

Using the ε_f input in calculating τ_c, a force controller could be included for constrained maneuvers to control the normal contact forces. However, an important advantage of RAP control is that knowledge of the load's mass properties and geometry are not required. Without this knowledge, the contact forces cannot be derived from the force measurements. This fact can be seen from (19). If the unknown load is time invariant, its characteristics could be estimated via unconstrained maneuvers. The contact forces and torques due to constrained maneuvers could then be estimated. Other approaches would be to mount a sensor on the constraint surface or to use a visual system to detect the constraint surface characteristics.

REFERENCES

1. Anderson, J. N. and M. E. Pittelkau, "Reflex-Action Position Control of Robots: Comparison with Impedance Control," *Proc. IEEE Southeastcon '88,* April 1988, pp. 408–412.
2. Craig, J. J., *Introduction to Robotics: Mechanics and Control*, Addison-Wesley, 1986.
3. Huang, Han-pang "Constrained Manipulators and Contact Force Control of Contour Following Problems," Ph.D. Dissertation, Robot Systems Division, The University of Michigan, 1986.
4. Sanfordon Groesberg, *Advanced Mechanics*, Wiley, 1968.
5. Kankaanranta, R. and H. N. Koivo, "A Model for Constrained Motion of A Serial Link Manipulator," *Proc. 1986 IEEE Int'l Conf. on Robotics and Automation*, Vol. 2, April 1986, pp. 1186-1191.
6. Anderson, J. N. and M. E. Pittelkau, "ROBPACK User's Manual and Report," Tech. Report MCTR-12-87-15, Ctr. for Manufacturing Research, Tennessee Technological University, December 1987.
7. Anderson, J. N. and Y. L. Lee, "Reflex-Action Position Control as Applied to Constrained Maneuvers," Tech. Report MCTR-10-89-01, Ctr. for Manufacturing Research, Tennessee Technological University, October 1989.
8. Baumgarte, J., "Stabilization of Constraints and Integrals of Motion in Dynamic Systems," *Computer Methods in Applied Mechanics and Engineering*, 1972, pp. 1–16.

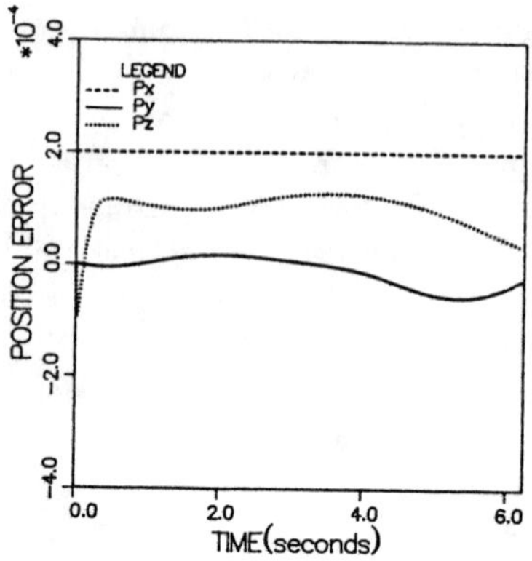

(a) Position Errors (m)

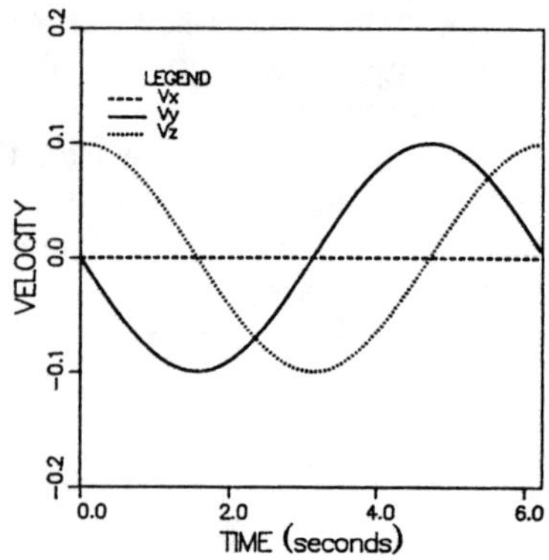

(b) Velocities (m/s)

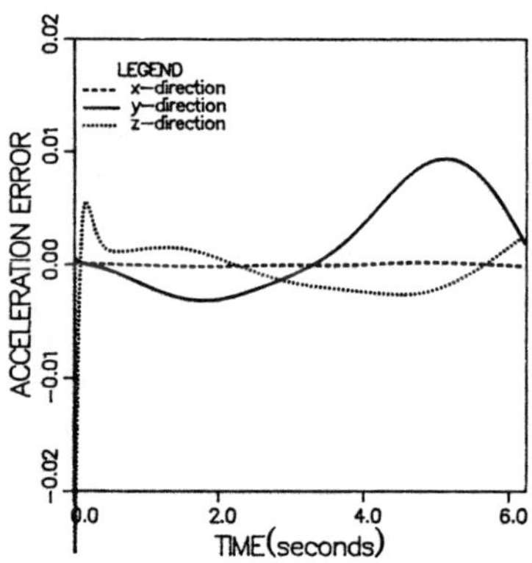

(c) Acceleration Errors (m/s^2)

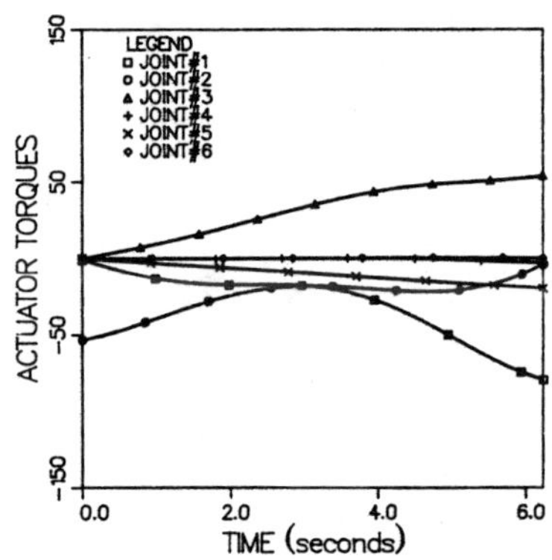

(d) Actuator Torques (N-m)

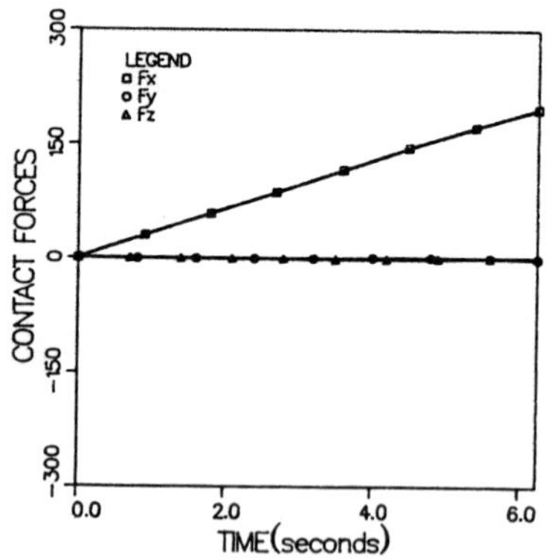

(e) Contact Forces (N)

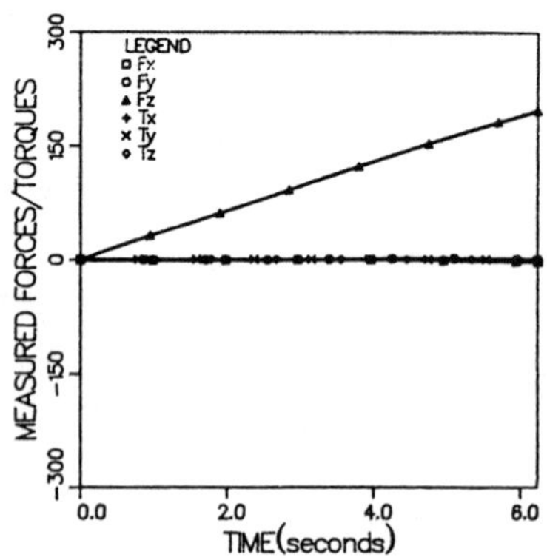

(f) F/T Measurements

Figure 1. Constrained RAP Control: Circular Trajectory and Contact Friction

A POSITIVE-REAL IMPEDANCE CONTROL DESIGN FOR ROBOTS

C. Abdallah, R. Jordan, & B. Horne
Department of Electrical Engineering and Computer Engineering
University of New Mexico
Albuquerque, NM, 87131 USA.

ABSTRACT

In this paper we study the robustness problem of impedance controllers for robotic manipulators. These controllers attempt to specify a desired relationship (impedance) between the applied forces and the corresponding velocities at the end-effector of a robotic manipulator.

I. INTRODUCTION

Given the following task-space description of a rigid robot [1]

$$D(x)\ddot{x} + h(x,\dot{x}) - F = f \tag{1.1}$$

this research attempts to guarantee the robust stability of impedance control of such robots despite large uncertainties in $D(x)$. Impedance controllers were introduced in [2], where the control objective was to specify an impedance relationship between the applied end-effector force and the resulting velocity. More recently [1], it was shown that impedance control is but one example of the inner/outer control startegy which, in turn, is a special case of the feedback-linearization approach of controlling nonlinear systems [3]. Unfortunately, inexact cancellations in the inner-loop may result in loss of stability which limits the applicability of these results as discussed in [1,7].

In this paper, we show that a combination of feedback-linearization and Strictly-Positive-Real (SPR) design of the impedance controller will guarantee the robust stability of the closed-loop system without the exact knowledge of the matrix $D(x)$. In section II we introduce the problem and our notation. The main results are given in section III, and our conclusions are presented in section IV.

II. PROBLEM STATEMENT

In this paper, we consider the class of rigid robotic systems described in task-space by the following equations

$$D(x)\ddot{x} + h(x,\dot{x}) - F = f \tag{2.1}$$

with $x(t) \epsilon R^n$, and the control $f(t) \epsilon R^n$. D and h contain the robot's parameters, some of which may be unknown. The vector F is an n vector representing the contact forces at the end-effector and is assumed to be available. The matrix D is a symmetric positive-definite inertia matrix, and h is the vector of centrifugal, Coriolis and gravity forces. A state-space description of equation (2.1) is given by

$$\begin{bmatrix} \dot{x} \\ \ddot{x} \end{bmatrix} = \begin{bmatrix} \dot{x} \\ D^{-1}(F-h) \end{bmatrix} + \begin{bmatrix} 0 \\ D^{-1} \end{bmatrix} f. \tag{2.2}$$

This nonlinear system is feedback-linearizable as described in [3]. Assuming D, h, and F are available, the controller design based on the linearizing transformation is given by

$$f = D(\ddot{x}_d - u) + h - F \tag{2.3}$$

where u is designed to obtain a desired closed-loop linear system. This will lead to

$$\begin{bmatrix} \dot{x} \\ \ddot{x} \end{bmatrix} = \begin{bmatrix} 0 & I \\ 0 & 0 \end{bmatrix} \begin{bmatrix} x \\ \dot{x} \end{bmatrix} + \begin{bmatrix} 0 \\ I \end{bmatrix} u \tag{2.4}$$

where

$$u = D^{-1}(f + F - h). \tag{2.5}$$

In fact, if one considers $e_1 = x_d - x$, $e_2 = \dot{e}_1$, where x_d and $\dot{x}_d$ are to be specified later, and the choice of u below

$$u = -Ke = -K_1 e_1 - K_2 e_2, \tag{2.6}$$

where

$$K_1 = M^{-1}R, \quad K_2 = M^{-1}N, \tag{2.7}$$

M, R and N are diagonal positive-definite matrices such that

$$M\ddot{x} + N\dot{x} + Rx = M\ddot{x}_d. \tag{2.8}$$

The error equation becomes

$$\ddot{e}_1 + K_2\dot{e}_1 + K_1 e_1 = 0, \tag{2.9}$$

which, by choosing $K_1 > 0$, and $K_2 > 0$ will guarantee that e_1 and e_2 go to zero asymptotically. Note that the above design is the impedance controller [2] which enforces a desired acceleration $\ddot{x}_d$, a desired velocity $\dot{x}_d$ and a desired position x_d based on the contact force F as follows

$$F = M \ddot{x}_d \tag{2.10}$$

and on (2.9). This controller is given by the following equation

$$f = D[\ddot{x}_d + K_2 e_2 + K_1 e_1] - (I + DM^{-1})F + h. \tag{2.11}$$

or, after substituting (2.10)

$$f = -M \ddot{x}_d + D[K_2 e_2 + K_1 e_1] + h. \tag{2.12}$$

III. ROBUST IMEPDANCE CONTROLLERS

Since D and h are usually unknown or too complex to be evaluated at every sampling instant, a computed version f_c of f is applied to the system (2.1) where

$$f_c = D_c(\ddot{x}_d - u) + h_c - F_c, \tag{3.1}$$

where D_c, h_c and F_c are computed versions of D, h, and F. One therefore obtains a calculated version u_c of the input u to the linear system (2.4). Let u_c be given by

$$u_c = \ddot{x}_d - D^{-1}(f_c + F - h) \tag{3.2}$$

Then, substituting (3.1) into (3.2), one gets

$$u_c = \Delta\ddot{x}_d + \delta + (\Delta - I)M^{-1}(Ne_2 + \mathrm{Re}_1 - F) + D^{-1}(F_c - F), \tag{3.3}$$

where

$$\Delta = I - D^{-1}D_c, \quad \delta = D^{-1}(h - h_c). \tag{3.4}$$

One is now concerned with the stability of a linear system with a nonlinear feedback described by

$$\dot{e} = (A - BK)e + B[\Delta(Ke + \ddot{x}_d - \delta + D^{-1}(F_c - F + D_c M^{-1} F)] \tag{3.5}$$

A more compact description of the error system is given by

$$\dot{e} = A_c e + Bv \tag{3.6}$$

The first step in our design is to choose the linear gain matrix $K = [K_1 \quad K_2]$, and an output matrix C, in order to guarantee that the closed-loop system (A_c, B, C) is SPR. Then, using the passivity results [5], one can show the asymptotic stability of (3.6) when h is known but D is not.

Theorem 1:

Let K_1, K_2 be two diagonal matrices with

i) $K_1 = diag(k_{1i})$; $k_{1i} > 0; i = 1, \ldots, n$

ii) $K_2 = diag(k_{2i})$; $k_{2i} > 0; i = 1, \ldots, n$

iii) $(K_2)^2 > K_1$; $i = 1, \ldots, n$

Then the system given by

$$\dot{e} = \begin{bmatrix} 0 & I \\ -K_1 & -K_2 \end{bmatrix} e + \begin{bmatrix} 0 \\ I \end{bmatrix} u = A_c e + Bu ,$$

$$y = [K_1 \quad K_2] = Ce$$

is Strictly-Positive-Real (SPR).

Proof:

Consider the Lyapunov equation

$$A_c^T P + PA_c = -Q$$

where

$$Q = \begin{bmatrix} 2K_1^2 & 0 \\ 0 & 2K_2^2 - 2K_1 \end{bmatrix}$$

Note that $Q > 0$ by the conditions of the theorem and solve for P as given below

$$P = \begin{bmatrix} 2K_1K_2 & K_1 \\ K_1 & K_2 \end{bmatrix}$$

It can be shown that $P > 0$. On the other hand, note that

$$C = B^T P.$$

Using the MKY lemma [6], the system is then SPR.

Note that the relationship $K_2^2 > K_1$ is easy to satisfy. Next, we show that when h is known and F is measurable, and using the passivity theorem [4,5], the asymptotic stability of the closed-loop system is guaranteed with a particular choice of M_c.

Theorem 2:

Suppose h is known. Then, the origin of (3.6) is an asymptotically-stable equilibrium point if

$$D_c = aI; \quad h_c = h; \quad F_c = (I - aM^{-1})F$$

where

$$aI > D;$$

Proof:

Given the choice of $h_c = h$ and $F_c = (I - aM^{-1})F$, one gets from (3.5,3.6)

$$\delta = 0$$

and

$$v(t) = -\Delta(Ke + \ddot{x}_d) = -\Delta w.$$

Consider then the block diagram of Figure 1. The output of the nonlinear block is given by

$$z(t) = -\Delta w(t).$$

To determine if the nonlinear block is passive, check that there exists some finite μ such that for all finite T

$$\int_0^T -w^T \Delta w dt \geq \mu,$$

If one chooses $\mu = 0$, one needs to show that

$$\int_0^T w^T [D^{-1} D_c - I] w dt \geq 0,$$

for all T finite. It is then sufficient to choose

$$D_c = aI > D$$

as specified in the theorem. Using Popov's hyperstability crieterion [5], one deduces that the the signals $z(t)$ and $e(t)$ in Figure 1 are bounded. Then noting that the linear block is SPR, one deduces that $e(t)$ goes to zero asymptotically.

Note that the choices made in the theorem result in

$$f_c = a(Ke + \ddot{x}_d) + h - F. \tag{3.7}$$

or, using (2.10)

$$f_c = aKe + (aI - M)\ddot{x}_d + h. \tag{3.8}$$

Note also that the above discussion shows that the nonlinear feedback due to the uncertainties in the inertia matrix D can be made passive when F is measured directly. This of course excludes the case when impulsive forces F are obtained because of contact of the robot with a rigid body. An interseting exstension would be to guarantee the robustness of the scheme when the contact forces F are not exactly known.

REFERENCES

[1] R.J. Anderson, M.W. Spong, "Hybrid Impedance Control of Robotic Manipulators," *IEEE J. Rob. & Auto.* , Vol. 4, No.5, pp.549-556, October 1988.

[2] N. Hogan, "Impedance Control: An Approach to Manipulation," *J. Dyn. Syst., Meas., Contr.* , Vol. 107, pp.1-7, March 1985.

[3] S. Sastry, M. Bodson, *Adaptive Control: Stability, Convergence, and Robustness,* Prentice Hall, Englewood Cliffs, New Jersey, 1989.

[4] C.A. Desoer, and M. Vidyasagar, *Feedback Systems: Input-Output Properties* , Electric Science Series, Academic Press, New York, 1975.

[5] M. Vidyasagar, *Nonlinear Systems Analysis* , Prentice-Hall Inc., Englewood Cliffs, New Jersey, 1978.

[6] R. Kelly, and R. Carelli, "Unified Approach to Adaptive Control of Robotic Manipulators," *Proc. 27th Conf. Dec. & Cont.* , Austin, Texas, December, 1988.

[7] R.J. Anderson, "Passive Computed Torque Algorithms for Robots," *Proc. 28th Conf. Dec. & Cont.* , Tampa, Florida, pp.1638-1644, 1989.

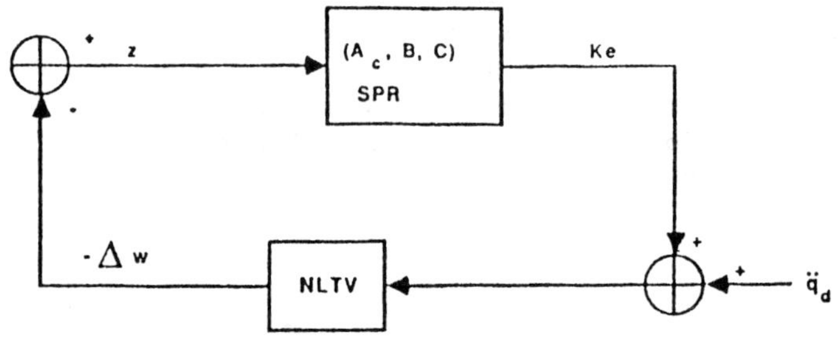

Figure 1: Hyperstability Block Diagram

ADAPTIVE IMPEDANCE CONTROL OF MECHANICAL MANIPULATORS*

W.-S. LU and **Q.-H. MENG**

Department of Electrical and Computer Engineering
University of Victoria, Victoria, B.C., CANADA V8W 2Y2

Abstract

Motivated by the adaptive control algorithm proposed by Craig et al for unconstrained motion control, an adaptive version of impedance control is presented. By introducing the concept of target-impedance reference trajectory which characterizes a desired dynamic relation of end-point with the environment, it is shown that an adaptive law can be derived in a manner similar to Craig's work, which works with the conventional impedance control of Hogan. Simulation results are also presented to illustrate the proposed algorithm.

I. INTRODUCTION

Stable execution of contact tasks by mechanical manipulators has been identified as one of the major challenges in robotics [1], and a great deal of researchers' attention has been attracted to the problem during the past decade. Among several proposed control schemes, it appears that the impedance control of Hogan [2][3] provides an unified approach to unconstrained motion control, obstacle avoidance, and constrained manipulation. In addition, it has theoretically been shown that the approach may also preserve motion stability during the contact between the end-effector and environment [1]. This is a property especially attractive to the field of contact-force control.

As in the computed-torque servo for unconstrained motion control [4], implementation of impedance control requires the use of accurate manipulator dynamics in its operational space [3]. On the other hand, however, some parameters in the manipulator's model may be uncertain due for example to the changing dynamic properties of different grasped load such as moments of inertia or position of the center of mass in the end-effector [5]. It is therefore of practical importance to clarify how performance as well as motion stability of a manipulator/controller system are affected by its model uncertainties. Even more importantly, compensation method should be sought so as to improve overall performance of impedance-controlled system and maintain motion stability in the presence of parameter uncertainties.

Motivated by the adaptive control algorithm suggested by Craig et al [6] for unconstrained motion control, an adaptive version of impedance control is presented in this paper. As mentioned in [6], the method proposed there "might also be extended to include an active force control servo", but such an extension has not been as yet available to date. The difference of an adaptive impedance control from an adaptive unconstrained motion control is that here one deals with a problem of regulating the mechanical impedance of the robot which involves external force/torque measurements rather than controlling end-effector's position/orientation which is independent of environment. As will be shown in section III, by introducing a concept of "target-impedance reference trajectory" (TIRT) which characterizes the desired dynamic relation of the end-point with environment, an adaptation mechanism can be established and an error analysis for the proposed algorithm can be carried out in a manner similar to [6].

II. PRELIMINARIES

2.1 The Dynamic Models

Dynamics of an n d.o.f. manipulator can be described by

$$H(q)\ddot{q} + C(q,\dot{q})\dot{q} + g(q) + f(\dot{q}) = \tau - J^T F_{ext} \tag{1}$$

When controlling the dynamic behavior of the end-effector/environment interaction comes to be a main concern, it is often desirable to describe the manipulator

* This work was supported by the Natural Sciences and Engineering Research Council of Canada.

dynamics in its operational (i.e. Cartesian) space as

$$H_x(x)\ddot{x} + C_x(x,\dot{x})\dot{x} + g_x(X) + f_x(\dot{x}) = J^{-T}\tau - F_{ext} \tag{2}$$

where

$$H_x = J^{-T}HJ^{-1}, \quad C_x(x,\dot{x}) = J^{-T}(C - HJ^{-1}\dot{J})J^{-1}, \quad g_x = J^{-T}g, \quad f_x(\dot{x}) = J^{-T}f$$

2.2 The Conventional Impedance Control

In impedance control the target impedance is usually specified by a higher-level supervisory system as a second-order dynamics [3]

$$M(\ddot{x} - \ddot{x_v}) + B(\dot{x} - \dot{x_v}) + K(x - x_v) = -F_{ext} \tag{3}$$

where x_v is the virtual trajectory that often coincides with the desired trajectory when no contact occurs. However, it will most likely correspond to positions beyond the robot's workspace during the contact in order to maintain a proper amount of contact force. We'll return to this point when discussing simulation results of the proposed control strategy in Section IV.

Given manipulator dynamics (2) and target impedance (3), it is quite natural to use a constrained motion control counterpart of the well-known resolved acceleration algorithm [7] to assign a control torque τ such that the overall system dynamics coincides with that given in (3). Obviously, such a torque is given by $\tau = J^T F$ with

$$F = F_{ext} + C_x\dot{x} + g_x + f_x + H_x\{\ddot{x_v} - M^{-1}[B(\dot{x} - \dot{x_v}) + K(x - x_v) + F_{ext}]\} \tag{4}$$

i.e., in the notation of [4],

$$\begin{aligned}\tau &= J^T\{W^{-1}[M^{-1}(K(x_v - L(\theta)) + B(\dot{x_v} - J\dot{\theta})) + \ddot{x_v} \\ &\quad + JH^{-1}(C\dot{\theta} + g + f) - \dot{J}\dot{\theta}] + [I - W^{-1}M^{-1}]F_{ext}\}\end{aligned} \tag{5}$$

where $L(\theta)$ represents the forward kinematics operator which maps a set of joint displacements into the corresponding end-point position/orientation, and $W = JH^{-1}J^T$ is the mobility tensor [2] whose inverse is the actual inertia of the robot end-effector [3]. The control law (5) has been known as impedance control for constrained motion of robots where the major control task is to regulate the dynamic relation of the end-effector with the environment to be contacted. It follows from above observation that the impedance control is indeed a duality of the resolved acceleration control in the domain of constrained motion control.

Note that control law (5) can also be used for redundant manipulators where the Jacobian J is a full row-rank rectangular matrix over its workspace. Indeed the introduction of mobility tensor W in this case avoids performing inverse operation for the rectangular Jacobian. On the other hand, however, the control torque expression in (5) is certainly not computationally efficient when the manipulator involved has six d.o.f. and its Jacobian are nonsingular. As a matter of fact, it can readily be shown that the control torque τ can be computed using the following formula

$$\tau = H\ddot{\theta}^* + C\dot{\theta} + g + f + J^T F_{ext} \tag{6}$$

where $\ddot{\theta}^*$ is given by

$$\ddot{\theta}^* = (MJ)^{-1}[K(x_v - L(\theta)) + B(\dot{x_v} - J\dot{\theta}) + M\ddot{x_v} - \dot{J}\dot{\theta} - F_{ext}] \tag{7}$$

One can therefore use the measurements from the joint position/velocity sensors as well as the wrist force sensor to compute $\ddot{\theta}^*$ and adopt in turn the recursive Newton-Euler computation scheme [8] to compute the sum of the first four terms on the right-hand side of (6).

2.3 Strictly Positive Real (SPR) Functions

The concept of SPR has been extensively studied for more than two decades and found numerous applications in network analysis, robustness analysis, nonlinear control and adaptive control. A proper rational function $f(s)$ is said to be SPR if the real part of $f(j\omega)$ is strictly greater than zero for all $\omega \in R$. Let us consider as an example the following second-order stable transfer function

$$f(s) = \frac{s+\psi}{s^2 + k_v s + k_p} \tag{8}$$

where $k_v > 0$, $k_p > 0$ are given and ψ is a system parameter. It is easy to verify that $f(s)$ is SPR if $0 < \psi < k_v$. When a stable SPR rational function is realized in a state-space, the passive nature of such a transfer function provides a system-theoretic property that is more valuable than a single Lyapunov equation does. Indeed, if function $f(s)$ given by (8) with $\psi \in (0, k_v)$ is realized as $\sum = (A, b, c)$ where

$$A = \begin{bmatrix} 0 & 1 \\ -k_p & -k_v \end{bmatrix}, \quad b = \begin{bmatrix} 0 \\ 1 \end{bmatrix}, \quad and \quad c = [\psi \quad 1]$$

It can then be shown that there exist two positive definite matrices P and Q such that

$$\begin{aligned} A^T P + PA &= -Q \\ Pb &= c^T \end{aligned} \tag{9}$$

The combination of the two equations in (9) plays a crucial role in deriving an adaptive control law for unconstrained motion control [6].

III. THE CONTROL ALGORITHM

3.1 Impedance Control with Estimated Dynamics

Let $p \in R^{r\times 1}$ denote the vector of all unknown parameters in the robot dynamic equation (2) and $\hat{p} \in R^{r\times 1}$ its estimate from an adaptation law to be specified later. The impedance control can be implemented if the estimated parameter values are used in (4) or (6). For the sake of simplicity, it is assumed that the Jacobian is a square and nonsingular matrix. The control torque can then be written as

$$\tau = \hat{H}\ddot{\theta}^* + \hat{C}\dot{\theta} + \hat{g} + \hat{f} + J^T F_{ext} \tag{10}$$

where $\hat{H}$, $\hat{C}$, $\hat{g}$, and $\hat{f}$ are evaluated using estimated parameters.

Similar to the discussion in section 2.2, the recursive Newton-Euler computation schemes [8] applies to (10), making it a numerically attractive formulation in implementing the proposed control algorithm.

In order to perform an error analysis, however, it is more convenient to write the control torque as $\tau = J^T F$ with

$$F = F_{ext} + \hat{C}_x\dot{x} + \hat{g_x} + \hat{f_x} + \hat{H}_x\{\ddot{x_v} - M^{-1}[B(\dot{x} - \dot{x_v}) + K(x - x_v) + F_{ext}]\} \tag{11}$$

3.2 The Target-Impedance Reference Trajectory

Given the desired end-point/environment relation (3), the TIRT is a differentiable vector function $x_r(t)$ of time, which solves the second-order differential equation (3) with a proper set of initial conditions, e.g. $x_r(0) = x_v(0)$ and $\dot{x_r}(0) = \dot{x_v}(0)$. Clearly, the TIRT as a function of time coincides with the desired tracking trajectory over the unconstrained motion duration simply because of $F_{ext} \equiv 0$. However, it may differ from the pre-assigned virtual trajectory considerably during the contact. This is particularly true when either the environment involved has a very high stiffness, or the target impedance are assigned such that its dynamics has a pole sufficiently close to the $j\omega$-axis. To be specific, the TIRT $x_r(t)$ is defined as the unique solution of the following initial-value problem

$$M\ddot{x_r} + B\dot{x_r} + Kx_r = -F_{ext} + (M\ddot{x_v} + B\dot{x_v} + Kx_v), \quad x_r(0) = x_v(0), \quad \dot{x_r}(0) = \dot{x_v}(0) \tag{12}$$

A couple of remarks on the concept introduced above are now in order. First, with measurements from the wrist force sensor and the given virtual trajectory, the reference trajectory can be formed very quickly during each sampling duration by numerically integrating equation (12). Further notice that such numerical integration is necessary only during the contact since one may otherwise simply take $x_r(t) = x_v(t)$. Second, the importance of this concept is due to the fact that analytically the essence of impedance control is to reform the dynamic relation for the end-point/environment interaction such that the end-point motion could follow the TIRT. The importance of equation (12) will become apparent when an attempt is made to derive an error equation for end-effector's position.

3.3 The Parameter Adaptation

Applying control law (10) to dynamic equation (2), simple algebra leads to

$$M(\ddot{x} - \ddot{x}_v) + B(\dot{x} - \dot{x}_v) + K(x - x_v) = F_{ext} - M\hat{H}_x^{-1}(\tilde{H}_x\ddot{x} + \tilde{C}_x\dot{x} + \tilde{g}_x + \tilde{f}_x) \quad (13)$$

where $\tilde{H}_x = H_x - \hat{H}_x$, $\tilde{C}_x = C_x - \hat{C}_x$, $\tilde{g}_x = g_x - \hat{g}_x$, and $\tilde{f}_x = f_x - \hat{f}_x$. As mentioned before, it is not the virtual trajectory but the TIRT that the end-point tries to follow. Consequently, the above equation is combined with (12) and an error equation is obtained as

$$\ddot{e} + B_n\dot{e} + K_n e = \hat{H}_x^{-1}W\phi \triangleq u \quad (14)$$

where $e = x - x_r$, $B_n = M^{-1}B$, $K_n = M^{-1}K$, and

$$W\phi = \tilde{H}_x\ddot{x} + \tilde{C}_x\dot{x} + \tilde{g}_x + \tilde{f}_x, \qquad \phi = \hat{p} - p \quad (15)$$

and W a $6 \times r$ matrix depending on $\ddot{x}$, $\dot{x}$, x, and the known manipulator parameters. On comparing equation (14) with the error equation (11) of [6], they look very much the same except the following distinctions: (i) equation (14) is set up in the operational space rather than the joint space; (ii) the left-hand side of (15) is a reflection of the desired target impedance rather than desired unconstrained motion; (iii) a friction term has been included in equation (15) with the assumption that only viscous friction and dynamic Coulomb friction effects are modeled.

It now follows from [6] and Section 2.3 that a positive linear combination of the position error e and velocity error $\dot{e}$ may be formed as

$$e^* = \dot{e} + \Psi e \quad (16)$$

with $\Psi = diag(\psi_1, \ldots, \psi_n)$, $\psi_i \in (0, b_{n_i})$, and b_{n_i} the ith diagonal element in B_n so that the input/output map $u \longmapsto e^*$ admits a state-space description given by

$$\begin{aligned} \dot{X} &= \mathcal{A}X + \mathcal{B}\hat{H}_x^{-1}W\phi \\ e^* &= \mathcal{C}X \end{aligned} \quad (17)$$

and the transfer function matrix of system (17) is diagonal with all diagonal elements SPR. Consequently, there exist $\mathcal{P} = diag(P_1, \ldots, P_n) > 0$ and $\mathcal{Q} = diag(Q_1, \ldots, Q_n) > 0$ such that

$$\begin{aligned} \mathcal{A}^T\mathcal{P} + \mathcal{P}\mathcal{A} &= -\mathcal{Q} \\ \mathcal{P}\mathcal{B} &= \mathcal{C}^T \end{aligned} \quad (18)$$

As in [6], using the Lyapunov function

$$V(X, \phi) = X^T\mathcal{P}X + \phi^T\Gamma\phi, \qquad \Gamma > 0$$

and property (18) lead to the adaptation law

$$\dot{\hat{p}} = -\Gamma W^T\hat{H}_x^{-1}e^* \quad (19)$$

Equations (10) or (11) and (19) provide now an adaptive version of Hogan's impedance control, which works for controlling the dynamic interactions between the environment and the end-point of the robot where certain parameter uncertainties are presented.

IV. SIMULATION RESULTS

A 2 d.o.f. planar robot was considered in the simulation to verify the proposed algorithm. As shown in Fig. 1, the robot motion begins at point (1.5, 0) moving along an elliptic curve until the end-point of the robot hits the wall at point a. The end-point is then required to be in contact with the wall while sliding down to point b along the environment surface. The end-point then leaves the wall moving back to point (1.5, 0) along the ellipse. It is assumed that the mass of each link is a point mass located at the distal end of the link. The true mass values are $m_1 = 2kg$, $m_2 = 1kg$, where m_2 represents $0.5kg$ link mass plus $0.5kg$ payload, and the link lengths are $l_1 = l_2 = 1m$. The major portion of the virtual trajectory $x_v(t) = [x_{1v}(t) \;\; x_{2v}(t)]^T$ is an ellipse described by $x_{1v}(t) = 1.5cos\pi t$, $x_{2v}(t) = sin\pi t$ for $0 \leq t \leq 2$ sec., but there is a small straight segment replacing the ellipse arc in order to maintain proper contact force.

Assuming that all parameters of the robot manipulator are precisely known, Fig. 2 shows the simulation results when the conventional impedance control (5) is employed, where the parameters in target impedance characterization (3) are chosen as $B = 2\sqrt{MK}$ with $M = 2$, $K = 4000$, and the environment is assumed to have a high stiffness $K_f = 10^5$. As the proposed adaptive control is concerned, our simulation is performed for two cases where the initial guess of the payload is $1kg$ (i.e. $m_2 = 1.5kg$) and $0kg$ (i.e. $m_2 = .5kg$), respectively. The target impedance in the adaptive control case is set as $B = 340$, $M = 2$, and $K = 4000$. Comparison of the actual trajectory with the pre-assigned virtual trajectory, the estimated values of m_2, the tracking error, and the contact force for both cases are shown in Fig. 3 and Fig. 4 respectively. It is observed that tracking error as well as contact forces of the manipulator with one unknown parameter have been kept within an acceptable level when the proposed adaptation law is employed in conjunction with the modified impedance control algorithm. Also notice that in both cases the estimated values of m_2 quickly approaches to its true value $1kg$ in the free motion duration but exhibit a 8% relative error in case I and 35% relative error in case II during the contact.

REFERENCES

1. N. Hogan, "On the Stability of Manipulators Performing Contact Tasks," IEEE J. Robotics & Automation, vol. RA-4, pp667-686, (Dec. 1988).
2. N. Hogan, "Impedance Control: An Approach to Manipulation: Part I - Theory, Part II - Implementation, Part III - Applications," J. Dyn. Syst., Meas., Cont., vol. 107, pp. 1-24, 1985.
3. N. Hogan, "Stable Execution of Contact Tasks Using Impedance Control," in Proc. IEEE Conf. Robotics & Automation, pp. 1047-1054, 1987.
4. R. P. Paul, "Modeling, Trajectory Calculation, and Servoing of a Computer Controlled Arm," Stanford Artificial Intelligence Lab., Memo AIM-177, 1972.
5. J. J. E. Slotine and W. Li, "Adaptive Manipulator Control: a case study," IEEE Trans. Automatic Control, Vol.33, pp.995-1003, 1988.
6. J. J. Craig, P. Hsu, and S. S. Sastry, "Adaptive Control of Mechanical Manipulators," in Proc. IEEE Conf. Robotics & Automation, pp. 190-195, 1986.
7. J. Y. S. Luh, M. W. Walker, and R. P. Paul, "Resolved Acceleration Control of Mechanical Manipulators," IEEE Trans. Automatic Control, Vol. AC-25, pp.468-574, (June 1980).
8. J. Y. S. Luh, M. W. Walker, and R. P. Paul, "On-Line Computational Scheme for Mechanical Manipulators," ASME J. Dyn. Syst., Meas., Contr., Vol.102, pp.69-76, (June,1980).

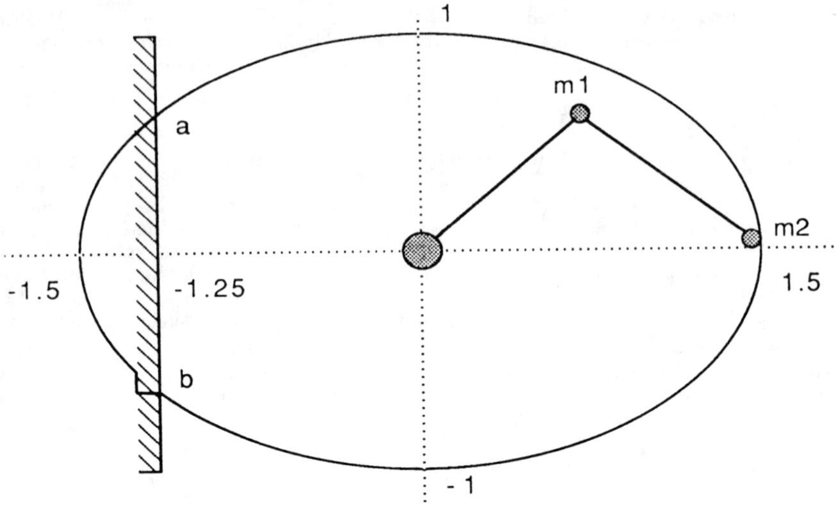

Fig. 1 The Manipulator and Assigned Virtual Trajectory

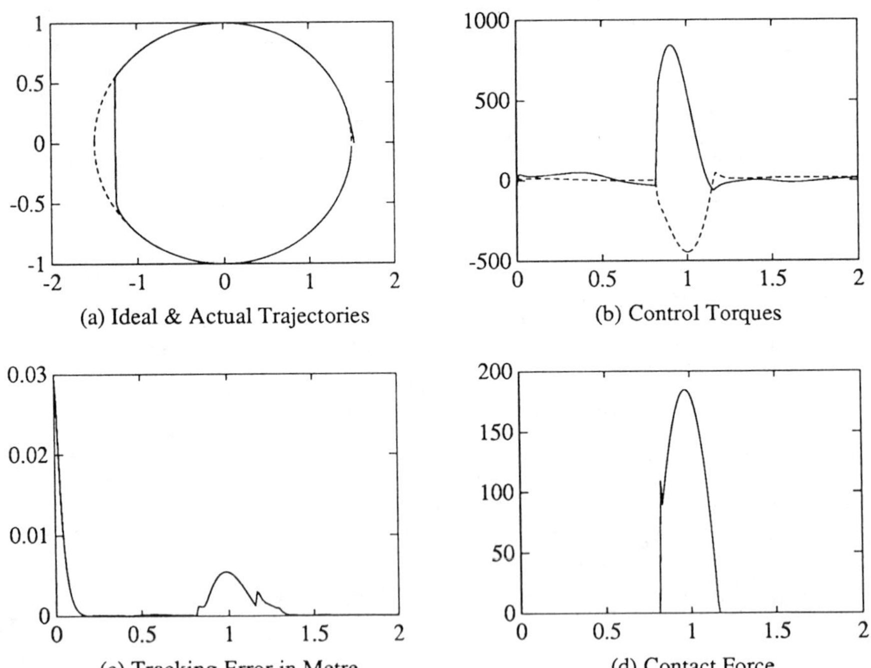

Fig.2 Conventional Impedance Control

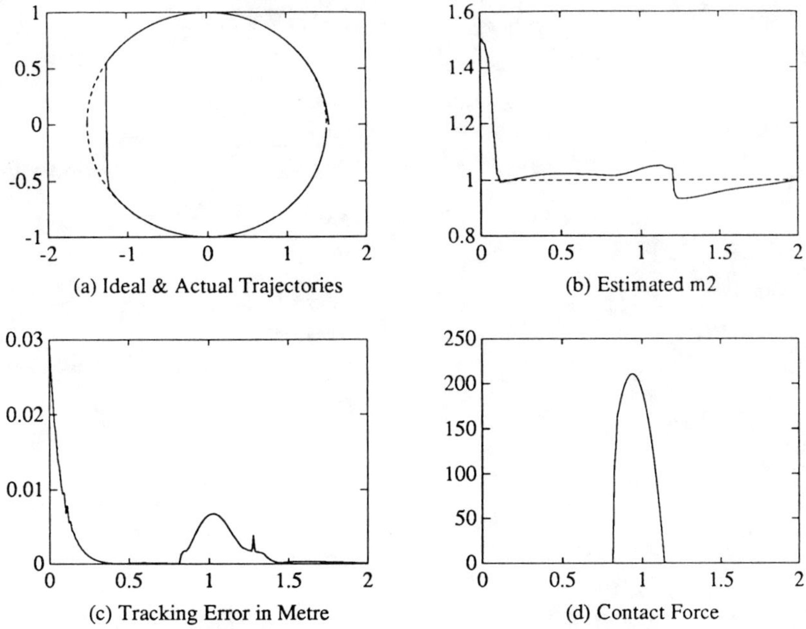

(a) Ideal & Actual Trajectories

(b) Estimated m2

(c) Tracking Error in Metre

(d) Contact Force

Fig.3 The Adaptive Impedance Control: Case I

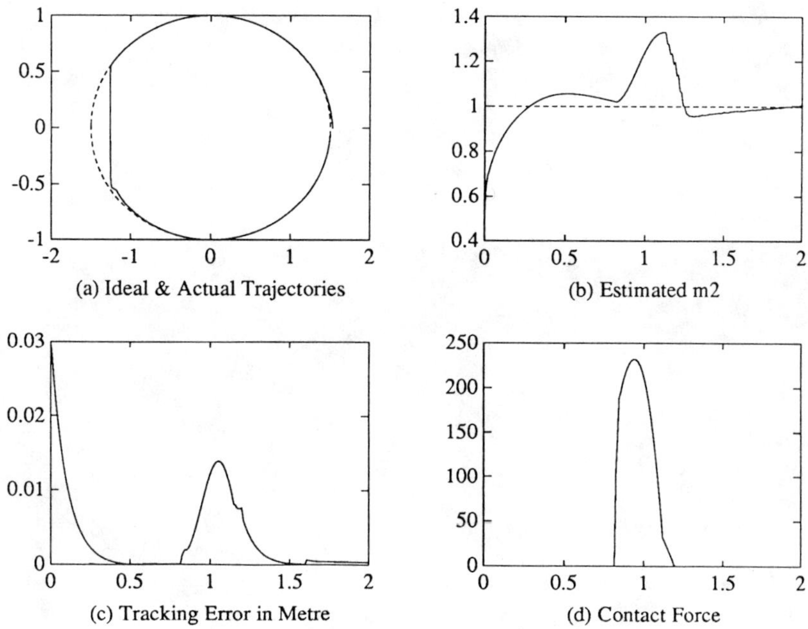

(a) Ideal & Actual Trajectories

(b) Estimated m2

(c) Tracking Error in Metre

(d) Contact Force

Fig.4 The Adaptive Impedance Control: Case II

Design and Application of Pole-Placement Adaptive Controller for Milling Process

F. Ordubadi

Research Engineer, Department of Electrical Engineering

Y.Altintas, F. Sassani

Assistant Professor, Department of Mechanical Engineering

The University of British Columbia, Vancouver, B.C., Canada

Abstract

Adaptive control of peak forces in milling with slender cutters is addressed. The dynamics of the milling process is modelled as a first order discrete system. Design of adaptive Pole-Placement controller is presented. The performance of the controller is compared using simulations and real time machining experiments. It is shown that when the time varying cutting process and time invariant feed drive transfer functions are accurately identified and decoupled, the adaptive controller can be successfully applied to machining processes.

Introduction

Development of Adaptive Control (AC) systems for machining process control has been on-going since the early stages of Computer Numerical Control (CNC) systems. This paper focuses on the design and analysis of parameter adaptive control algorithms to constrain the maximum cutting force on end milling cutters. The constraint or the reference is selected as a limiting force to avoid the shank breakage of the end mill. The early works treated the cutting process as a simple gain [1] or a gain and a delay [2]. Beadle and Bollinger [1] used AC system for a face milling process which was modelled as a gain with one sampling time delay. In early works the feedback controllers were tuned to be stable for extreme depth and width of cuts [1,2,4]. When the chip load exceeds the tuned value, the open loop gain of the controller becomes too large which leads to the instability of the feedback controller. The first true parameter adaptive control was explored by Tomizuka et al.[3] who applied model reference adaptive control (MRAC) scheme on milling process. They used a first order discrete transfer function model between the feed-rate and force changes.

In what follows modelling of the transfer function of the time invariant CNC feed drive and time varying cutting process is presented. The cutting process model is based on the influence of the end mill's static deflection on maximum chip thickness per tooth period. The description of experimental set-up is followed by the design of adaptive controller. The paper is concluded with the simulation and experimental proofs.

Discrete Transfer Function of the Milling Process System

The milling process to be controlled has two cascaded dynamics. The time-invariant feed drive servo control , and the time-variant cutting process dynamics. The continuous transfer function of the servo is identified as a second order dynamics.

$$G_s(s) = \frac{f(s)}{u(s)} = \frac{152591.6}{s^2 + 2000s + 152591.6} \tag{1}$$

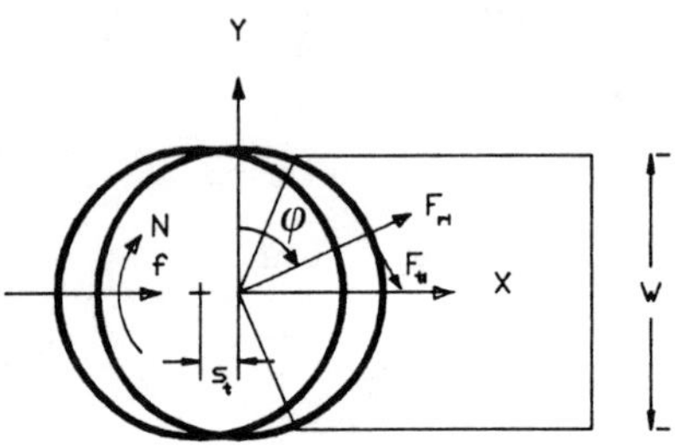

FIG. 1: Cutting forces in Milling

where $u(s)$ is the commanded feeding velocity (mm/sec) and $f(s)$ is the actual feeding velocity of the table (mm/sec) . In the present work, straight end milling cutter with two carbide teeth was used at a spindle speed of $n = 775rev/min$. The corresponding tooth passing period is $T = 38.7ms$. The feed-rate per tooth for a two teeth cutter is defined as;

$$s_t(s) = \frac{1}{2n} f(s) = 0.03871\, f(s) \tag{2}$$

The zero order hold equivalent of G_s can be derived as;

$$G_s(z) = \frac{s_t(z)}{u(z)} = \frac{0.0368445\,(z + 0.0020946)}{z(z - 0.0461992)} \tag{3}$$

For straight end milling cutters there are two fundamental cutting forces acting on the tool shank as shown in Figure 1. The tangential cutting force F_t is proportional to chip thickness (h), axial depth of cut (a) and the cutting pressure K_s.

$$F_t = K_s\, a\, h(\phi) \tag{4}$$

$$F_r = r_1\, F_t \tag{5}$$

r_1 is the radial to tangential force ratio. The resultant cutting force exerted on the cutter shank is ;

$$F(\phi) = \sqrt{F_t^2 + F_r^2} \tag{6}$$

The discrete dynamic model of the milling process is derived using geometric deformation model [5,3]. The Special case where one tooth is in 90 degrees immersion angle is dicussed here but the result is true for any cutter at any immersion angle[7]. The radial force acting on the cutter shank at tooth period i will deflect the cutter which has a static stiffness of K_x.

$$F_r(i) = K_x\, \delta(i) \tag{7}$$

where $\delta(i)$ is the tool deflection . The resulting chip thickness $h(i)$ is derived from the geometry shown in Fig. 2.

$$h(i) = s_t(i-1) + \delta(i-1) - \delta(i) \tag{8}$$

Substituting Eq. 8 into the radial cutting force yields;

$$F_r(i) = r_1 K_s a(i)\,(s_t(i-1) + \delta(i-1) - \delta(i)) \tag{9}$$

Equating and rearranging the Eqs. 7 and 9 gives;

$$\delta_i = \frac{r_1\,\mu}{1 + r_1\,\mu}\,(s_t(i-1) + \delta(i-1)) \tag{10}$$

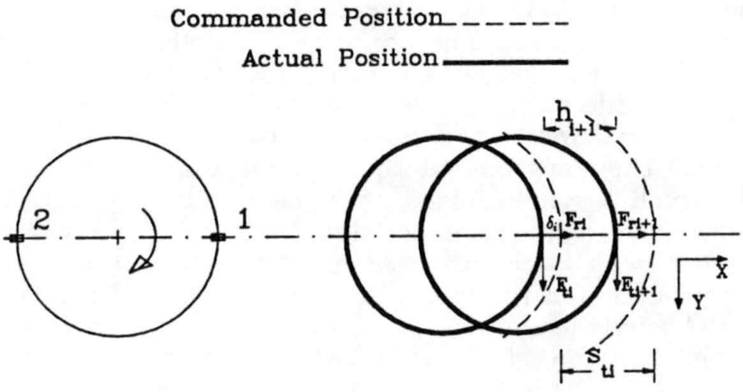

FIG. 2: Modelling of Milling Process

where $\mu = K_s\, a_i / K_x$. Substituting Eq. 10 into Eq. 7 gives;

$$F_{ri} = \frac{K_x\, r_1\, \mu}{1 + r_1\, \mu} s_t(i-1) + \frac{r_1\, \mu}{1 + r_1\, \mu} F_r(i-1) \tag{11}$$

$F_r(i)$ and $F_r(i-1)$ are radial forces at tooth periods (i) and $(i-1)$. Applying a forward discrete time shift operator (i.e. $F_{ri+1} = z\, F_{ri}$) to Eq. 11 and replacing by its equivalent in terms of the resultant force $F(i)$ results in;

$$\frac{F(i)}{s_t(i)} = \frac{[K_x\, \mu/1 + r_1\, \mu]\, \sqrt{1 + r_1^2}}{z - [\, r_1\, \mu/1 + r_1\, \mu]} \tag{12}$$

or

$$\frac{F(z)}{s_t(z)} = \frac{b_0}{z - a_1} \tag{13}$$

Equation 12 gives the transfer function of the milling process where the input is the feed-rate and the output is the maximum resultant cutting force at discrete tooth periods. The transfer functions of the time invariant feed drive servo, Eq. 3, and the time varying milling process dynamics, Eq. 13, can be combined to obtain plant's discrete transfer function.

$$G_p(z) = \frac{F(z)}{u(z)} = \left(\frac{0.0368445\,(z + 0.0020946)}{z(z - 0.0461992)}\right)\left(\frac{\hat{b_0}}{z - \hat{a_1}}\right) \tag{14}$$

where $u(z)$ is the control command velocity produced by the adaptive controller and $F(z)$ is the actual peak resultant cutting force at each tooth period. $\hat{b_0}$ and $\hat{a_1}$ are the estimates of their real values.

Experimental Set-up

The machining experiments to test the proposed adaptive control algorithm were carried out on a research CNC milling machine. The three axes of the milling machine were retrofitted with dc servo motors that can provide continuous 5 Nm torque during milling. The three feed drive dc servo motors are powered by Pulse Width Modulated amplifiers with 15Amp. continuous current supply. Each axis control unit has a velocity feedback sensor (tacho-generator) and a position feedback unit (encoder). Each encoder position feedback pulse has a resolution of $0.00127mm$ linear table travel. The position and velocity of the three axis are controlled by a MC68008 based motion controller which resides in a MULTIBUS card cage. Further detail of the CNC system is given in Reference [6]. The adaptive control and data acquisition system reside in a personal computer. The adaptive controller samples the feed-rate and peak force values at each tooth period. The tooth period is provided by a spindle shaft mounted encoder which has two slots for the two teeth cutter. The feeding and normal forces are measured by a piezo-electric Kistler force dynamometer, and they are passed through an analog peak resultant force detector circuit which is synchronized with the spindle encoder. The feed-rate was measured from the tacho-generator signals at each tooth period. The measured feed-rate and peak force values are used to estimate the unknown, time varying cutting process parameters. The adaptive control algorithm, which is resident in the PC, sends feeding velocity control commands to the axis motion control unit in the Multibus, through a bus adaptor at each tooth period.

Design of Adaptive Controller

The aim of the adaptive controller based on Pole-placement design is that the closed loop transfer function between the reference force $U_c(ref)$ and the actual force F obey the following desired model dynamics.

$$\frac{F(t)}{U_c(t)} = \frac{B_m(q^{-1})}{A_m(q^{-1})} \tag{15}$$

where q^{-1} is a backward shift time operator. For the stability of the system, the roots of the characteristic equation $A_m(z)$ is evidently selected within the unit circle. An ideal damping ratio of $\zeta = 0.7$ is selected to obtain minimum overshoot during the transients. A rise time of four tooth periods is selected, which corresponds to the natural frequency of 16 rad/sec. The desired closed loop transfer function of the system with a dc gain of unity is given as;

$$\frac{B_m(z)}{A_m(z)} = \frac{0.2512}{z^2 - 1.162\, z + 0.4132} \tag{16}$$

The plant's transfer function given in Eq. 14 is rewritten in a discrete form as;

$$\frac{F(t)}{u(t)} = \frac{B(q^{-1})}{A(q^{-1})} \tag{17}$$

where $F(t)$ and $u(t)$ are the force and control input signal respectively. It is assumed that A and B polynomials are prime and $\deg A \geq \deg B$. A general form of a self tuning regulator for milling process control is shown in Fig. 3. The polynomials $S(q^{-1})$, $T(q^{-1})$ and $R(q^{-1})$ represent feedback, feed-forward and error regulators respectively, which have to be determined adaptively at each tooth period. The plant to be controlled was given by Eq. 17. The closed loop transfer function of the controller is derived from Fig. 3 as;

$$\frac{F(k)}{U_c(k)} = \frac{B\,T}{AR + BS} \tag{18}$$

where k is the sampling counter. Note that for simplicity the q^{-1} operator is omitted as an argument of the polynomials. In order for the system to behave like the desired model, the following equivalence must hold.

$$\frac{B\,T}{AR + BS} = \frac{B_m}{A_m} \tag{19}$$

The design problem is defined as finding the polynomials R, S and T using the plant's and desired model's transfer functions. The relationship between the

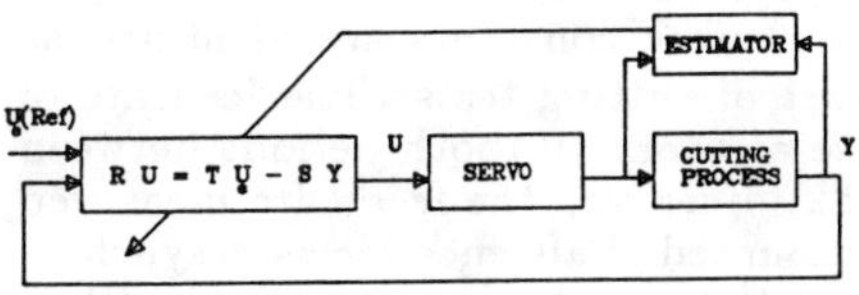

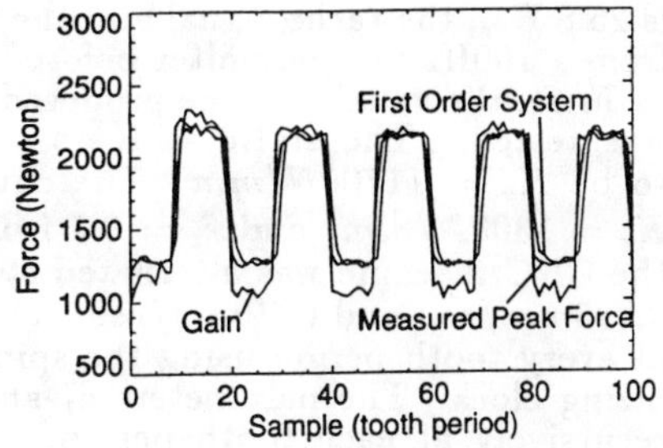

FIG. 3: Pole-Placement Control Loop   FIG. 4: Cutting Process Identification

desired reference force U_c and regulated feeding velocity u can be expressed as

$$R\,u(k) = T\;U_c(k) - S\;F(k) \tag{20}$$

The above eqution reduces to the *Diophantine* equation [8].

$$AR' + q^{-2}B^{-}S = A_m A_0 \tag{21}$$

The equation 21 has a unique solution when the following orders are assumed for the regulator and observer polynomials.

$$deg\ R \;=\; deg\ S \;=\; deg\ T \tag{22}$$

$$\begin{aligned} S &= s_0 z^2 + s_1 z + s_2 \\ R &= B^{+}\,(z + r_1) \\ A_0 &= z^2 \end{aligned}$$

Substitution of polynomials into the Eq. 21 gives the solution for the regulator parameters

$$\begin{aligned} r_1 &= -1.1158 + \hat{a}_1 \\ s_0 &= \frac{0.4132 + r_1\,\hat{a}_1 + 0.0461992\,(r_1 - a_1)}{K\hat{b}_0 0.9518} \\ s_1 &= \frac{-0.0461992\,(r_1\;\hat{a}_1)}{K\hat{b}_0 0.9518} \\ s_2 &= 0. \end{aligned} \tag{23}$$

The machining process parameters $\hat{b}_0$ and $\hat{a}_1$ are identified recursively, the parameters of the regulators R, S and T, and the control input signal are calculated at each tooth period according to Eqs. 23 and 20.

Simulation and Experimental Results

A two teeth straight carbide end mill with 25.4 mm diameter was used in milling Aluminum 7075T651 alloy at constant 775 RPM spindle speed. Since the bandwidth of the feed drive servo is $65Hz$ and the tooth passing frequency is 25.8 Hz., the tacho signal and the feed and normal cutting forces are passed from a 100Hz low pass filter before being processed by the peak detector.

First, the validity of the proposed discrete first order cutting process model was tested. The static stiffness of the tool on the spindle was measured to be $K_x = 11700 N/mm$. The cutter-workpiece constants are identified as $K_s = 1300 N/mm^2$ and $r_1 = 0.7$ from a set of cutting tests. The feedrate of the CNC machine was alternated stepwise at every 10 tooth periods between $0.295 mm/rev$ and $0.59 mm/rev$. The peak forces and the feeds are measured at every tooth period using the spindle mounted shaft encoder as a synchronizing clock. The parameters a_1 and b_0 of the cutting process are identified recursively at each tooth period. The actual measured and recursively reconstructed peak forces using the estimated parameters are shown in Fig. 4. When the cutting process is replaced just as a gain, there is a dc offset between the reconstructed and the actual forces. On the other hand the simulated peak force using the estimated first order cutting process is following the measured cutting forces reasonably well.

The proposed Pole-Placement controller has been tested using simulations and cutting experiments at identical cutting conditions. The test workpiece profile contains severe transients due to step changes in axial depth of cut as shown in Fig. 5. The changes in the axial depth of cut varies the parameters

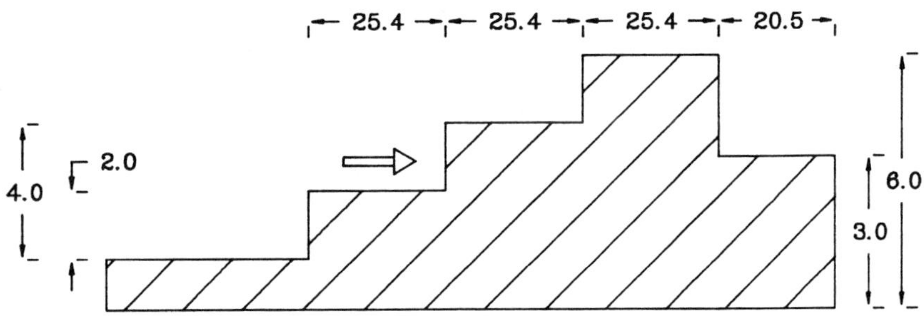

FIG. 5: Work-Piece Profile (Dimensions in mm)

a_1 and b_0 of the cutting process transfer function. The tests were carried out in full immersion slotting. In practice, the width of cut may also vary, however such transients are less severe than the sudden changes in the axial depth of cut. A constant reference force of $F_r = 1000N$ is selected for the tests.

The simulated response of the adaptive controller is shown in Fig.6. Note that one count is the resolution of the machine and equal to 0.000127 mm. The small oscillations on the cutting force is due to the run-out on the cutter which is simulated by imposing an alternating 10% depth of cut at each tooth period. It is noted that the larger the change in the axial depth of cut is, the longer time the adaptation takes. This is due to convergence time required by the Recursive Least Square (RLS) estimation algorithm when the measurements differ largely during transients. The Pole-Placement controller was experimentally tested using the same cutting conditions. The measured cutting force and the control input histories are shown in Fig. 7. The cutting force is always kept at 1000 N reference value. the transient periods are rather short. This is due to persistent excitation that exists during the milling process which prevents the parameter drifts during estimation with the RLS method. The fast response of the Pole-Placement controller comes with the penalty of large overshoots. Although the peak force was rapidly brought to 1000 N reference value, there is an overshoot of extra 600 N at transients.

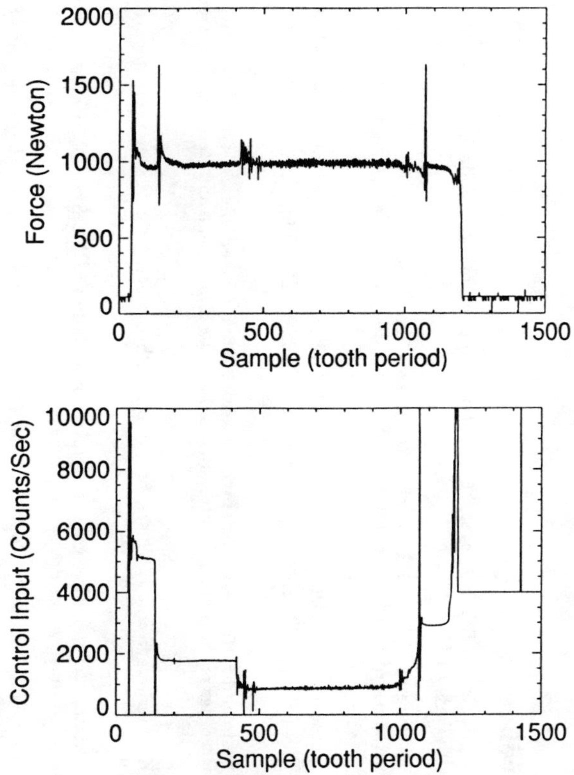

FIG. 7: Machining Results

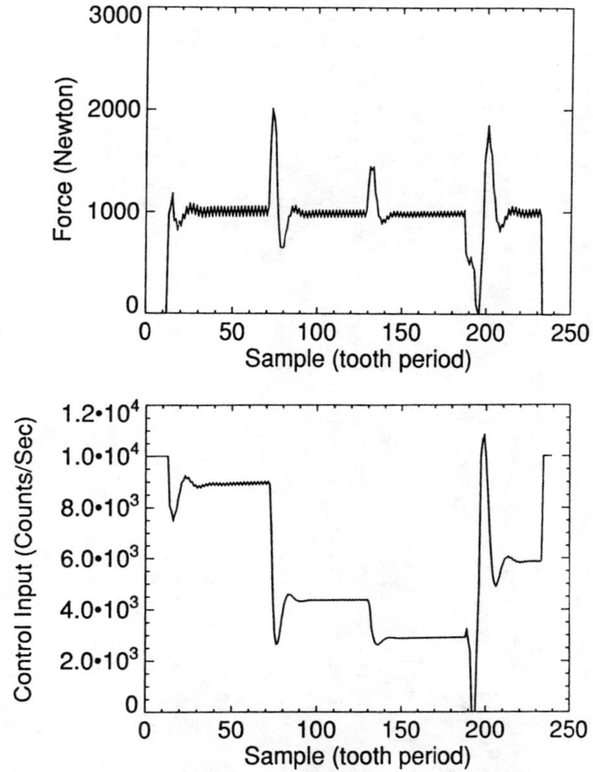

FIG. 6: Simulation Results

Conclusions

A systematic design and engineering application of an adaptive controller on the milling process was presented. It was shown that adaptive control mechanism can be successfully applied on metal removal processes provided that the CNC feed drive and the cutting process are accurately decoupled and identified. The cutting process is analytically and experimentally proven to have a first order dynamics, which is essential to know for the design of adaptive controller.

The proposed design can be extended to the control of geometric deformations (i.e. geometric adaptive control) in any machining system by cascading the stiffness of the tool to the proposed transfer function of the cutting process. The problem which remains to be solved is to remove excessive cutting force overshoots during transients, which can be eliminated only by providing priori information about the incoming transients to the controller.

Acknowledgement

This research was supported by the National Science and Engineering Research Council of Canada under the grants OGP-0006164 and OGP-0005542.

References

[1] B.R. Beadle and J.G. Bollinger, "Computer Adaptive Control of a Machine Tool", Annals of CIRP 16, 61-65 (1971).

[2] J. Tlusty and M.A.A. Elbestawi, "Analysis of Transients in an Adaptive Control Servomechanism for Milling with Constant Force", Journal of Engineering for Industry, 99, 766-772 (1977).

[3] M. Tomizuka, J.H. Oh and D.A. Dornfeld, "Model Reference Adaptive Control of the Milling Process", in Control of Manufacturing Processes and Robotics Systems, Edited by D.E.Hardt and W.J.Book (ASME Press, New York 1983).

[4] G. Stute and F.R. Goetz, "Adaptive Control System for Variable Gain in ACC Systems", Proceedings of the Sixteenth International Machine Tool Design and Research Conference, 117-121 (1975).

[5] F. Koenigsberger and J. Tlusty, Machine Tool Structure, Volume One (Pergamon Press 1970).

[6] Y. Altintas, J.Peng, "Design and Analysis of a Modular CNC System", Journal of Computers in Industry, (In Press, Dec. 1989).

[7] F. Ordubadi, "Adaptive Control of the Milling Process", CAMROL Report No. 89-3, Mechanical Engineering Department, UBC.

[8] K.J. Astrom and B. Wittenmark, Computer Controlled Systems, Theory and Design (Prentice-Hall 1984).

A Reduced-Order Discrete Time Adaptive Controller for Robotics Manipulators

Robert L. Carroll, and Hossein Bolandi
Department of Electrical Engineering & Computer Science
George Washington University
Washington, D.C. 20052, USA

ABSTRACT:

This paper presents an adaptive approach to control problem of robotic manipulators. A discrete-time adaptive control algorithm for a robotic manipulator is devised. Each joint of the robotic manipulator is controlled by an adaptive controller of the model-reference type. The order of the adaptive control system for each joint of the manipulator is reduced to unity for both a position feedback loop and a velocity feedback loop. The angular position of each link of the manipulator may be arbitrarily commanded, subject only to a limitation in available actuator torque. The adaptive algorithm is applied to the joint control of a two-link robot manipulator. A computer simulation of the performance of the discrete-time algorithm is presented for a continuous time nonlinear model of the manipulator under variable load conditions.

Introduction

The control problem of a robotic manipulator involves the determination of the suitable inputs for actuators such that a set of satisfactory values for the positions and velocities are achieved. Since the dynamic characteristics of robots are highly nonlinear functions of the positions and velocities of the manipulator elements and hence are trajectory dependent, there is a need for developing advanced control methods for robotic manipulators.

One of these approaches has been decentralized designs, in which the controller is dedicated for each joint. However, because of the nonlinear and time-varying dynamics of manipulators and also because the operating environment of the manipulators are poorly known, the control problem gets somewhat complicated.

To overcome these difficulties it has been shown that adaptive schemes are promising. Model-reference adaptive control of robotic manipulators was described in [1]. For the controller design therein, each link of the manipulator is modeled by a second-order linear ordinary differential equation having no input terms from other links. In [2], a continuous time adaptive control algorithm is developed for a 3-link robot, and simulation studies justify the decoupling of the links. In [3] and [4], a model-reference approach, using continuous linear system and PID controllers, is examined.

Discrete model-reference adaptive algorithms have also been employed to control robots. In [5] an adaptive model-following algorithm was proposed for a discrete linear time-invariant plant having known order but unknown parameters. In [6,7], these techniques are used to devise a discrete model-reference adaptive controller. A similar discrete self-tuning algorithm was proposed in [8].

In this paper a reduced-order discrete time decentralized adaptive controller is presented. The order of adaptive control for each joint is reduced to unity for both the position and velocity feedback loops. Then it is applied to the mathematical model of a continuse-time, nonlinear and coupled model of a two link manipulator under variable load conditions.

Model

The dynamic equations of an n-link robot can be written [9] as:

$$M(\theta)\,\ddot{\theta} + N(\theta,\dot{\theta}) + G(\theta) + H(\dot{\theta}) = T \qquad (1)$$

where:

$\theta(t)$ is n x 1 vector of joint angular positions.
$M(\theta)$ is a n x n symmetric positive definite inertia matrix.
$N(\theta,\dot{\theta})$ is n x 1 coriolis and configurational torque vector.
$G(\theta)$ is n x 1 gravity loading vector and $H(\dot{\theta})$ is n x 1 frictional torque vector and T is the n x 1 vector of applied joint torques.

The elements of M, N, G and H are highly complicated nonlinear functions of the manipulator configuration θ and the speed of motion $\dot{\theta}$ as well as the inertia parameters of the payload carried by the manipulators' end-effector. The following dynamic equation of a two link robot is considered:

$$\begin{bmatrix} T_1 \\ T_2 \end{bmatrix} = \begin{bmatrix} M_{11} & M_{12} \\ M_{21} & M_{22} \end{bmatrix} \begin{bmatrix} \theta_1 \\ \theta_2 \end{bmatrix} + \begin{bmatrix} 2\,M_{112}\,\dot{\theta}_1\,\dot{\theta}_2 + M_{112}\,\dot{\theta}_2^2 \\ M_{211}\,\dot{\theta}_1^2 \end{bmatrix} + \begin{bmatrix} M_1 \\ M_2 \end{bmatrix}$$

$$M_{11} = a_{11} + b_{11}\cos\theta_2; \qquad M_{12} = M_{21} = a_{22} + b_{12}\cos\theta_2 \qquad (2)$$

$$M_{22} = a_{22} \quad ; \qquad M_{112} = -\,b_{12}\sin\theta_2$$

$$M_1 = a_1\cos\theta_1 + b_1\cos(\theta_1 + \theta_2); \qquad (3)$$

$$M_{211} = b_{12}\sin\theta_2 \quad ; \qquad M_2 = b_1\cos(\theta_1 + \theta_2)$$

Note that the parameters(a's and b's) are function of mass, length and gravity. In order to write a state equation for the two link robot let :

$$\theta_1 = x_1 \quad ; \quad \theta_2 = x_2 \quad ; \quad \dot{\theta}_1 = x_3 \quad ; \quad \dot{\theta}_2 = x_4 \qquad (4)$$

where:

θ_1 is the position of link one, θ_2 is the position of link two.
$\dot{\theta}_1$ being the velocity of link one, $\dot{\theta}_2$ being the velocity of link 2

Then the nonlinear state equation for the two link robot manipulator is given by:

$$\dot{x}_1 = x_3 \quad ; \quad \dot{x}_2 = x_4 \quad ; \quad \dot{x}_3 = f_3 \quad ; \quad \dot{x}_4 = f_4 \qquad (5)$$

where :

$$f_3 = \frac{a_{22}\,R_1 - M_{12}\,R_2}{M_{11}\,a_{22} - (M_{12})^2}; \quad f_4 = \frac{M_{11}\,R_2 - M_{12}\,R_1}{M_{11}\,a_{22} - (M_{12})^2} \qquad (5.a)$$

$$R_1 = T_1 + (2\, b_{12} \sin x_2)\, x_3\, x_4 + (b_{12} \sin x_2)\, x_4^2 - a_1 \cos x_1 - b_1 \cos (x_1 + x_2) \quad (6)$$

$$R_2 = T_2 - b_{12} \sin (x_2)\, x_3^2 - b_1 \cos (x_1 + x_2) \quad (7)$$

As it follows from the equation, the manipulators' dynamic equation shows strong interactions between joint motions.

Controller Design

The manipulator control problem is to design a control scheme which will cause the robot joint angles to track,as closely as possible, state variables of a given reference model. In this paper, we have first design a discrete model reference adaptive controller for the velocity loop and then a model following adaptive controller for the position loop.

Instead of Equation (6), the controller is developed for the system: $d\, xp_i(t)/dt = xv_i(t)$

$$m_i(t)\, d\, xv_i(t)/dt = q_i(t) - d_i(t) \quad (8)$$

where xp_i is the joint angular position, xv_i the joint angular velocity of link i, $q_i(t)$ is the actuator torque input, $m_i(t)$ is the inertia load and $d_i(t)$ represents the corliolis centripetal accelerations due to the gravitational load.

Assuming that the changes in $m_i(t)$ and $d_i(t)$ in Eq. (8) is small over one sampling time, the discrete time equivalent of (8) is:

$$xp_i(k) = xp_i(k\text{-}1) + T\, x_{vi}(k\text{-}1) + .5\, (T^2/m_i)\left[q_i(k\text{-}1) - d_i(k\text{-}1)\right]$$

$$x_{vi}(k) = x_{vi}(k\text{-}1) + (T/m_i)\left[q_i(k\text{-}1) - d(k\text{-}1)\right] \quad (9)$$

where k denotes the kth sampling instant and T is the sampling time. To account for computational delay let: $q_i(k) = u_i(k\text{-}1)$

where $u_i(k)$ is the controlling input to link i.

The dynamics from the $u_i(t)$ to $x_{vi}(k)$ is described by:

$$(1\text{-}q^{-1})\, x_{vi}(k) = q^{-2}\left[bo_i\, u_i(k) - d_i'\right] \quad (10)$$

where $bo_i = T/m_i$ and $d_i' = bo\, d$ and q^{-1} is the delay operator.

If elements of the equation (10) are known then the controlling input can be determined such that the error between the $x_{vi}(k)$ and the desired arm angular velocity $x_{vdi}(k)$ tend to zero satisfying the equation:

$$Crv_i(q^{-1}) * \left[x_{vdi}(k) - x_{vi}(k)\right] = 0 \quad (11)$$

where $Crv_i(q^{-1}) = 1 + Crv_i q^{-1}$ and $-1 < Crv_i \leq 0$ (12)

Then the controlling input will be :

$$u_i(k) = \frac{C_{rvi}(q^{-1}) \quad X_{vdi}(k+2) - r_{oi} X_{vi}(k) + s_i(q^{-1}) d_i'}{bo_i s_i(q^{-1})} \tag{13}$$

where $S_i(q^{-1}) = 1 + s_{1i} q^{-1}$ (14)

roi is determined by solving: $C_{rvi}(q^{-1}) = (1 - q^{-1}) s_i(q^{-1}) + q^{-2} ro_i$

If the elements of equation (10) are unknown, we rewrite the equation (10) as:

$$(1-q^{-1}) x_{vi}(k) = \theta_i^T \phi_i(k-2) \tag{15}$$

where $\theta_i^T = \begin{bmatrix} bo_i & d_i' \end{bmatrix}$ (16)

$$\phi_i^T(k) = \begin{bmatrix} u_i(k) & o \end{bmatrix} \tag{17}$$

so the parameter vector θ_i can be estimated by the recursive parameter adaption algorithm expressed in the following form:

$$\hat{\theta}_i(k) = \hat{\theta}_i(k-1) + \frac{F_i(k-1) \phi_i(k-2) \varepsilon_i^{o}(k)}{1 + \phi_i^T(k-2) F_i(k-1) \phi_i(k-2)} \tag{18}$$

where

$$\varepsilon_i^{o}(k) = \left[x_{vi}(k) - x_{vi}(k-1)\right] - \hat{\theta}_i^T(k-1) \phi_i(k-2) \tag{19}$$

$$F_i(k) = \frac{1}{\lambda_{1i}(k)}\left[F_i(k-1) - \frac{F_i(k-1) \phi_i(k-2) \phi_i^T(k-2) F_i(k-1)}{\frac{\lambda_{1i}(k)}{\lambda_{2i}(k)} + \phi_i^T(k-2) F_i(k-1) \phi_i(k-2)}\right]$$

$o < \lambda_{1i}(k) \leq_1 \qquad o \leq \lambda_{2i}(k) < 2.$ (20) & (21)

so the resulting controller becomes:

$$u_i(k) = -s_{1i} u_i(k-1) + \frac{1}{\hat{bo}_i(k)}\left[Crv_i(q^{-1}) x_{vdi}(k+2) - r_{oi} x_v(k) + \hat{d}_i(k)\right] \tag{22}$$

satisfying $\lim_{k\to\infty} Crv_i(q^{-1}) \left[x_{vdi}(k) - x_{vi}(k)\right] = 0$ (23)

Considering that the velocity loop MRAC is achieving its role, the desired angular velocity $x_{vdi}(k+2)$ is:

$$x_{vd_i}(k) = x_{pdsi}(k-2) - x_{pi}(k-2) \tag{24}$$

$$x_{pds_i}(k) = \left[x_{pd_i}(k+3) + Z_2 x_{pd_i}(k+1) + Z_3 x_{pd_i}(k) + Z_1 x_{pd_i}(k-1)\right]/Z_4 \tag{25}$$

;and $Z_1 = .5;\ Z_2 = Z_1 - 1;\ Z_3 = 2Z_1;\ Z_4 = 2;$

$x_{pds_i}(k)$ is an estimate of the desired angular position of link i

To design a model following adaptive controller for the position loop, the transfer function from desired angular position $x_{pd_i}(k)$ to actual angular position $x_{p_i}(k)$ is:

$$\frac{x_{p_i}(k)}{x_{pd_i}(k)} = \frac{Z^2\left[Crp_i(q^{-1})\right]}{Z^2 + rp} = \frac{crp_i(q^{-1})}{1 + rp_i\, q^{-2}} \qquad (26)$$

where $crp_i(q^{-1}) = 1 + Crp_i(q^{-2});$ (27)

To achieve perfect tracking, the model following adaptive controller for position loop of link i becomes:

$up_i(k) = x_{pd_i}(k+2)\, Crp_i(q^{-1}) - rp_i\, x_{p_i}(k)$ & $crp_i = rp_i = c_{rvi}$

Simulation

After the development of the adaptive controllers , we will investigate the performance of the system under variable load to show the feasibility of the design. We will apply the proposed controllers to the continuous-time nonlinear model of a planar robot in Fig.1. To do so, the following numerical values was chosen for the link parameters of equation (3) : $a_1 = 118.90$; $a_{11} = 45.2065$; $a_{22} = 6.4141$; $b_{12} = 11.53$; $b_1 = 30.96$; $b_{11} = 23.0907$

The reference models for angular positions of link 1, and 2 was chosen as it is shown at Fig.(2). To test the adaptability at k=56, the load is suddenly changed . In Fig. (3, 4) both the velocity loop and feedback loop are in place.

Conclusions

This paper has proposed a reduced-order discrete time adaptive controller for robot manipulators. The order of the adaptive control system of manipulator is reduced to unity both for position and velocity loops. The response of the nonlinear model ofa two-link robot under the proposed adaptive control was shown to be satisfactory. For the extension of the purposed controller to n-link (n > 2) robot, we just need to add a required set of controllers.

REFERENCES

(1) S. Dubowsky and D.T. DesForges, "The Application of Model-Referenced Adaptive Control to Robotic Manipulators", *J. of Dynamic Systems Measurement, and Control.* Vol. 101, 193-200.

(2) S. Nicosia and P. Tokei, "Model Reference Adaptive Control Algorithms for Industrial Robots", *Automatica,* Vol. 20, No. 5, 635-644 (1984).

(3) B.J. Oh and M. Jamshidi. "Decentralized Adaptive Feedforward/Feedback Robot Manipulator Control", *Proc. IEEE Conference. Systems, Man, and Cyb.* Alexandria, VA, October 1987.

(4) M. Jamshidi, H. Seraji, and Y.T. Kim, "Decentralized Control of Nonlinear Robot Manipulators", *Robotics,* Vol.2, 1987.

(5) I.D. Landau and R. Lozano, "Unification of Discrete-Time Explicit Model-Reference Adaptive Control Designs", *Automatica,* Vol.

17, 593-611 (July 1981).

(6) R. Horowitz and M. Tomizuka, "Discrete-Time Model Reference Adaptive Controllers of Mechanical Manipulators", *Computers in Engineering*, Vol. 2, *Robots and Robotics*, ASME, 107-112, (1982).

(7) M. Tomizuka, R. Horowitz and G. Anwar, "Adaptive Techniques for Motion Controls of Robotic Manipulators", *Japan - USA Symposium of Flexible Automation*, 217-224.

(8) A.J. Koivo and Ten-Huei Guo, "Adaptive Linear Controller for Robotic Manipulators", *IEEE Transaction on Automatic Control,* Vol. AC-28, No. 2, 162-171 (February 1983).

(9) H. Seraji, "Decentralized Adaptive Control of Manipulators: Theory, Simulation, and Experimentation," IEEE Trans. on Robotics and Automation, 183-201 vol. 5, no. 2, April 1989.

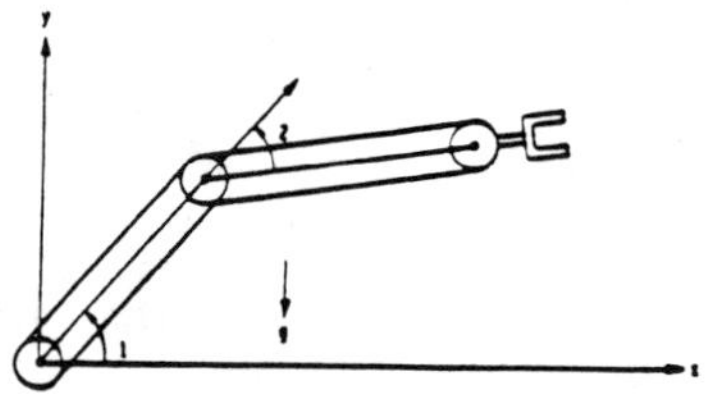

Fig. 1 Two link planar robot

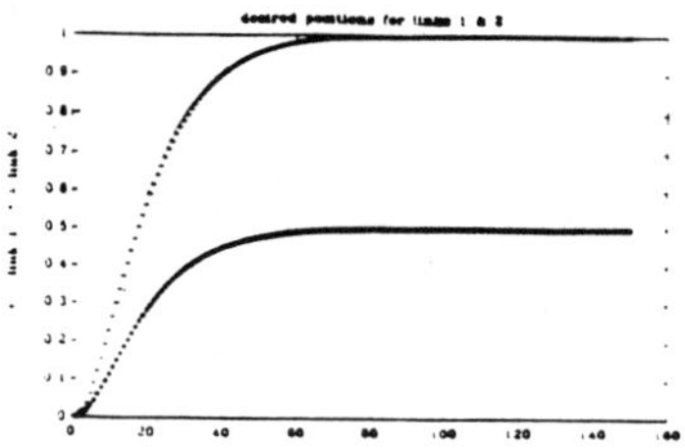

Fig. 2

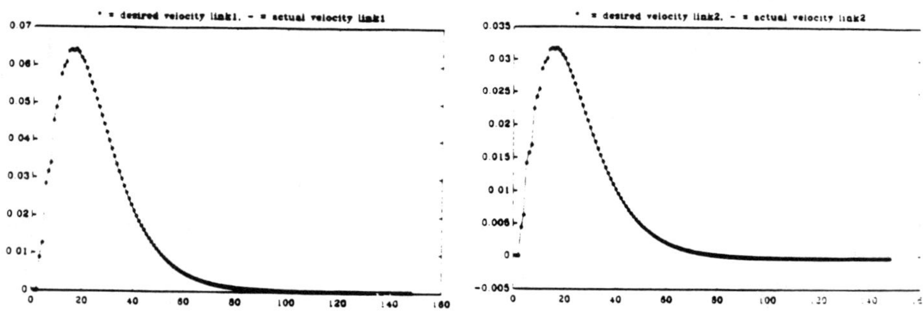

Fig. 3

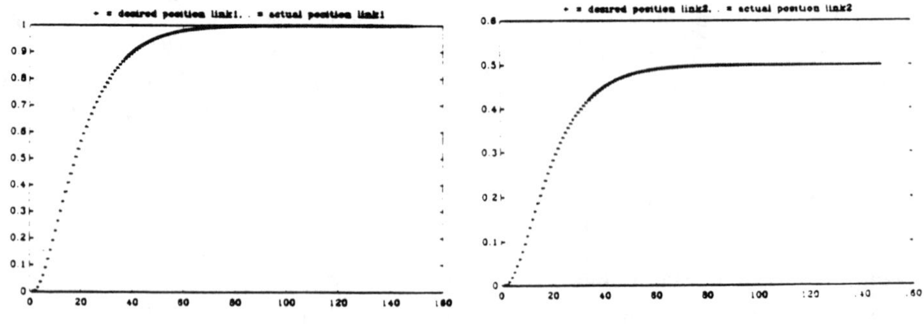

Fig. 4

Hybrid adaptive impedance control of robot manipulators

L. Mo and M.M. Bayoumi

Department of Electrical Engineering

Queen's University

Kingston, Ontario, Canada K7L 3N6

Abstract:

Adaptive impedance control of robot manipulators is discussed in this paper. The control law is continuous but the parameter estimation law is discrete. This reduces the on-line computational burden for the adaptive implementation of the impedance control. Unmodeled dynamics have also been discussed in this paper. Global stability of the adaptive system is ensured.

1) Introduction

Impedance control has been discussed in a number of papers (Kelly *et al* 1988, Kelly *et al* 1989, Niemeyer and Slotine 1989). In these papers, the control law and the parameter estimation law are both continuous. Because of the complexity of the adaptive control law, digital implementation may have to be employed. Hence a continuous time adaptive control law should be approximated through fast sampling. Because a reasonably large computation time may be required for those involved controllers, rapid sampling may not always be feasible. How the system will operate if a relatively slow sampling rate is used is still an open question. On the other hand, the skew property of the robot arm is used to ensure stability. It may be inappropriate to use the skew property without introducing the unmodeled dynamics since these unmodeled dynamics could destroy this property.

In this paper, a hybrid adaptive control scheme is presented. The word "hybrid" is used in the sense that the control law is continuous but the parameter estimation algorithm is discrete. The unmodeled dynamics is considered in the robot model in order to take account of the modelling error and measurement noise. The standard RLS algorithm is modified and this modified version is used to estimate the unknown manipulator parameters in the presence of the unmodeled dynamics. Those estimated parameters will update a continuous time control law very T seconds. Global stability of this hybrid controller can be ensured without any *a priori* information about the manipulator. The parameter estimation part is introduced to refine the performance of the system instead of stabilizing the system. The information of the joint acceleration is not necessary. If the true parameters of the robot manipulator are known, the goal of the impedance will be satisfied. In this control scheme, the measurement of force at the end point of the manipulator is required.

2) Robot model and assumptions

It is will known that the robot model can be expressed as

$$H_q(q)\ddot{q} + C_q(q, \dot{q})\dot{q} + G_q(q) + J^t(q)F + \Gamma_q = \tau \tag{1}$$

where q is the $n \times 1$ vector of the joint displacement. τ is the $n \times 1$ vector of the joint torques. The $n \times n$ matrix $H_q(q)$ is the inertial matrix of the robot. The $n \times 1$ vector $C_q(q, \dot{q})\dot{q}$ represents the centripetal and coriolis torques. $G_q(q)$ is the $n \times 1$ the vector of the gravitational torques. The Jacobian matrix of the manipulator is denoted as $J(q)$. The vector F is the $n \times 1$ vector of forces/moments at the end-effector. The $n \times 1$ vector Γ_q represents the unmodeled dynamics, which may include the friction and other factors. It is assumed that the manipulator is non-redundant and that the Jacobian matrix is always non-singular for all the vectors q considered.

If the arm is represented in the Cartesian space, we have the following dynamical equation (Kelly and Carelli 1988);

$$H(x)\ddot{x} + C(x, \dot{x})\dot{x} + G(x) + F + \Gamma = J^{-t}\tau \tag{2}$$

with

$$H(x) = J^{-t}H_q J^{-1} \tag{3}$$

$$C(x, \dot{x}) = -J^{-1}H_q J^{-1}\dot{J}J^{-1} + J^{-t}C_q J^{-1} \tag{4}$$

$$G(x) = J^{-t}G_q \tag{5}$$

$$\Gamma = J^{-t}\Gamma_q \tag{6}$$

It is obvious that the following property is true

$$y^t(\dot{H}(x) - 2C(x, \dot{x}))y = 0 \quad \forall\, y \in R^n \tag{7}$$

This is the so called skew property of the robot manipulator. In general, this property will be destroyed if non-ideal terms (e.q. friction) are present. Because of the introduction of the term Γ in eq(2) to represent the unmodeled dynamics, eq(7) can then be justified for the non-ideal case. Hence the use of the skew property without considering the unmodeled dynamics may not be appropriate in practice.

By restricting the robot under consideration to be such that is has revolute joints only, the following bounding conditions can be stated if J^{-1} is bounded for all the vectors x considered (Kelly *et al* 1989);

$$||H(x)|| \leq k_{1b} \tag{8}$$

$$||C(x, \dot{x})|| < k_{2b}||\dot{x}|| \tag{9}$$

$$||G(x)|| < k_{3b} \tag{10}$$

where k_{1b}, k_{2b} and k_{3b} are all finite positive constants and where $||.||$ denotes the L_2 norm. We further assume that the unmodeled dynamics satisfy the following condition

$$||\Gamma|| \leq m(t) = k_{4b}||\dot{x}||^2 + k_{5b}||\dot{x}|| + k_{6b} \tag{11}$$

where k_{4b}, k_{5b} and k_{6b} are finite positive constants. These coefficients can be obtained from physical reasoning and appropriate experiments.

For the adaptive control scheme under consideration, we assume that the dynamic vector parameter θ of the manipulator and payload is constant and unknown. The Jacobian matrix $J(q)$ is assumed to be known. The assumption that the matrix $J(q)$ is known is not restrictive because it does not depend on the dynamic parameters of the manipulator. Hence we have

$$Y(\ddot{x}, \dot{x}, x)\theta + \Gamma + F = J^{-t}\tau \tag{12}$$

If the above equation is sampled every T seconds, then eq(12) can be expressed as

$$Y(k)\theta + \Gamma(k) + F(k) = J^{-t}(k)\tau(k) \tag{13}$$

It should be noted that the above equation can not be used to estimate the dynamic parameters θ unless the accelerations of the manipulator are assumed to be measurable. To avoid the problem of measuring the acceleration, the filter which was introduced by Middleton and Goodwin (1988) can be used. In his case, eq(12) can be expressed as

$$Y_f(\dot{x}, x)\theta + \Gamma_f + F_f = (J^{-t}\tau)_f \tag{14}$$

where $(.)_f = W(s)(.)$ with $W(s) = \omega/(s + \omega)$. If eq(14) is sampled, then

$$Y_f(k)\theta + \Gamma_f(k) + F_f(k) = (J^{-t}\tau)_f(k) \tag{15}$$

Since the parameter estimation is preformed at discrete instants of time, then eq(13) or eq(15) will be used to update the parameters, which will be discussed later.

3) Controller design

The torque of the manipulator is assumed to be given by

$$\tau = J^t[\hat{H}(\ddot{x}_d + M_m^{-1}(B_m\dot{e} + K_m e + F)) + \hat{C}(\dot{x} - v) + \hat{G} + F - K_v v] \tag{16}$$

where the error e is given by $e = x_d - x$. This is a continuous time control law. But its parameters, $\hat{H}$, $\hat{C}$ and $\hat{G}$, are updated every T seconds. The $\hat{H}$, $\hat{C}$ and $\hat{G}$ are the estimated H, C and G based on the estimated dynamic parameter vector $\hat{\theta}$ at time kT, where $0 \le t - kT < T$. The vector v is given by

$$v = -\frac{1}{s + \lambda} M_m^{-1}(M_m\ddot{\xi} + B_m\dot{\xi} + K_m\xi); \ \lambda > 0 \tag{17}$$

$$\xi = e + (s^2M_m + sB_m + K_m)^{-1}F \tag{18}$$

For impedance control, the objective is to force ξ to be equal to zero $\xi = 0$. From eq(2), eq(16), eq(17) and eq(18), it is follows that

$$H(\dot{v} + \lambda v) + Cv = S(t) - K_v v \tag{19}$$

where

$$S(t) = (\hat{H} - H)(\ddot{x}_d + M_m^{-1}(B_m\dot{e} + K_m e + F)) + (\hat{C} - C)\dot{x} - (\hat{C} - C)v + \hat{G} - G - \Gamma \tag{20}$$

If we assume that

$$||\hat{\theta}(k) - \theta|| < \infty \qquad k = 1,2,... \tag{21}$$

then it is obvious that

$$\begin{aligned} ||S(t)|| \le K_{b1} &+ K_{b2}||\dot{e}|| + K_{b3}||e|| + K_{b4}||\dot{e}||^2 \\ &+ k_{b5}||v||\,||\dot{e}|| + K_{b6}||v|| + K_{b7}||F|| \end{aligned} \tag{22}$$

where k_{bi} for $i = 1,...,7$ are all positive and bounded.

Theorem 1

If

$$K_v = K_{v1} + K_{v2}||\dot{e}||^3 + K_{v3}||\dot{e}||^2||v||^2 + K_{v4}||v||^2 + K_{v5}||e||^2 + K_{v6}||F||^2 \tag{23}$$

where k_{vi} for $i = 1,...,6$ are all positive and bounded, then v is bounded.

Proof

Let $V = \frac{1}{2} v^t H v$, then

$$\dot{V} = v^t S(t) - \lambda v^t H v - K_v \| v \|^2 \tag{24}$$

Here we have used the condition that $y^t(\dot{H} - 2C)y = 0$ for any $y \in R^n$. If v is large, since K_v grows faster than $S(t)$, $\dot{V} < 0$. Hence v will be bounded.

Since v is bounded, it follows that ξ and $\dot{\xi}$ are bounded. If $\Gamma = 0$ and $\lim_{k\to\infty} \hat{\theta}(k) = \theta$, we have $\lim_{t\to\infty} S(t) = 0$ and hence $\lim_{t\to\infty} v = 0$. This implies that

$$\lim_{t\to\infty} \xi = 0 \tag{25}$$

This is the goal of the impedance control. For the non-ideal case where $\Gamma \neq 0$, this can only be approximately achieved. The advantage of this control law is that it will ensure stability regardless of the parameter estimation algorithm in case that the estimated parameters are bounded. The problem of how to ensure bounded estimates of the system parameters will be discussed in the next section. The problem of obtaining the true parameters in case $\Gamma = 0$ will also be handled in that chapter.

4) Parameter estimation

Here the parameter estimation will be based on equation (13). The results can be applied to the parameter estimation based on equation (15) directly. Dividing both sides of equation (13) by $m(k)$, which is defined in eq(11), we get

$$X^t(k)\theta_o + \epsilon(k) = \tau_n(k) \tag{26}$$

where

$$X(k) = Y(k)/m(k) \tag{27}$$

$$\epsilon(k) = \Gamma(k)/m(k) \tag{28}$$

$$\tau_n(k) = (J^{-t}\tau(k) - F(k))/m(k) \tag{29}$$

From (eq(13)), we have

$$\|\epsilon(k)\| \leq 1 \tag{30}$$

An RLS type algorithm will be used to estimate the unknown parameter vector θ_o, which is given by the following;

$$\hat{\theta}(k) = \hat{\theta}(k-1) + \beta(k)P(k)X(k)\xi(k) \tag{31}$$

where the *a priori* prediction error $\xi(k)$ is given by

$$\begin{aligned} \xi(k) &= \tau_n(k) - X^t(k)\hat{\theta}(k-1) \\ &= X^t(k)\tilde{\theta}(k-1) + \epsilon(k) \end{aligned} \tag{32}$$

and where the parameter error vector $\tilde{\theta}(k-1)$ is given by $\tilde{\theta}(k-1) = \theta_o - \hat{\theta}(k-1)$. The covariance matrix $P(k)$ will be updated using the equation

$$P^{-1}(k) = P^{-1}(k-1) + \beta(k)X(k)X^t(k) \tag{33}$$

The scalar $\beta(k)$ is given by

$$\beta(k) = \frac{\beta_1(k)}{1 + \gamma(k)} \tag{34}$$

with

$$\gamma(k) = \lambda_{\max}(X^t(k)P(k-1)X(k)) \tag{35}$$

where $\lambda_{max}(A)$ denotes the maximum eigenvalue of the matrix A. The scalar $\beta_1(k)$ will be given by

$$\beta_1(k) = \frac{\alpha f(\delta, ||\xi(k)||)}{||\xi(k)||} \qquad ||\xi(k)|| \geq \delta$$
$$= 0 \qquad \text{elsewhere} \tag{36}$$

with $\alpha \in (0,1)$ and $\delta = (1 + \alpha)^{1/2}$. The function $f(\delta, ||\xi(k)||)$ is defined as

$$f(\delta, ||\xi(k)||) = ||\xi(k)|| - \delta \qquad ||\xi(k)|| \geq \delta$$
$$= 0 \qquad \text{elsewhere} \tag{37}$$

Theorem 2

For the recursive estimation algorithm presented above, we have

$$1) \quad \lim_{k\to\infty} \frac{f(\delta, ||\xi(k)||)}{1 + \gamma(k)} = 0 \tag{38}$$

$$2) \quad ||\tilde{\theta}(k)|| \leq \kappa ||\tilde{\theta}(0)|| \tag{39}$$

where κ is the conditioning number of the matrix $P(-1)$.

Proof.

This is a multivariable version of the work by Leal and Ortega (1987). The proof is similar to that of Leal and Ortega (1987).

From eq(31) and eq(33), it follows;

$$\begin{aligned}\tilde{\theta}(k) &= \tilde{\theta}(k-1) - \beta(k)P(k)X(k)(X^t(k)\tilde{\theta}(k-1) + \epsilon(k)) \\ &= P(k)P^{-1}(k-1)\tilde{\theta}(k-1) - \beta(k)P(k)X(k)\epsilon(k)\end{aligned} \tag{40}$$

From eq(33), the matrix $P(k)$ can be expressed as

$$P(k) = P(k-1) - \beta(k)P(k-1)X(k)(I + \beta(k)X^t(k)P(k-1)X(k))^{-1}X^t(k)P(k-1) \tag{41}$$

Then $P(k)X(k)$ is given by

$$P(k)X(k) = P(k-1)X(k)(I + \beta(k)X^t(k)P(k-1)X(k))^{-1} \tag{42}$$

The *a posteriori* prediction error $X^t(k)\tilde{\theta}(k)$ can then be expressed as

$$X^t(k)\tilde{\theta}(k) = (I + \beta(k)X^t(k)P(k-1)X(k))^{-1}\xi(k) - \epsilon(k) \tag{43}$$

Define

$$V(k) = \tilde{\theta}^t(k)P^{-1}(k)\tilde{\theta}(k) \tag{44}$$

Then $(V(k) - V(k-1))$ can be expressed as

$$\begin{aligned}V(k) - V(k-1) &= \tilde{\theta}^t(k)P^{-1}(k)\tilde{\theta}(k) - \tilde{\theta}^t(k-1)P^{-1}(k-1)\tilde{\theta}(k-1) \\ &= \tilde{\theta}^t(k)P^{-1}(k-1)\tilde{\theta}(k-1) - \begin{matrix}\tilde{\theta}^t(k-1)P^{-1}(k-1)\tilde{\theta}(k-1) - \\ \beta(k)\tilde{\theta}^t(k)X(k)\epsilon(k)\end{matrix} \\ &= (\tilde{\theta}(k) - \tilde{\theta}(k-1))^tP^{-1}(k-1)\tilde{\theta}(k-1) - \beta(k)\tilde{\theta}^t(k)X(k)\epsilon(k) \\ &= -\beta(k)\xi^t(k)X^t(k)P(k)P^{-1}(k-1)\tilde{\theta}(k-1) - \beta(k)\tilde{\theta}^t(k)X(k)\epsilon(k) \\ &= -\beta(k)\xi^t(k)X^t(k)(\tilde{\theta}(k) + \beta(k)P(k)X(k)\epsilon(k)) - \beta(k)\tilde{\theta}^t(k)X(k)\epsilon(k)\end{aligned}$$

$$= -\beta(k)\xi^t(k)(I + \beta(k)X^t(k)P(k-1)X(k))^{-1}\xi(k) + \beta(k)\xi^t(k)(I - \beta(k)X^t(k)P(k)X(k))\epsilon(k) - \beta(k)\xi^t(k)(I + \beta(k)X^t(k)P(k-1)X(k))^{-1}\epsilon(k) + \beta(k)\epsilon^t(k)\epsilon(k)$$

It is then straight forward to show that

$$V(k) = V(k-1) + \beta(k)(||\epsilon(k)||^2 - \xi^t(k)(I + \beta(k)X^t(k)P(k-1)X(k))^{-1}\xi(k)) \tag{45}$$

Note that

$$\begin{aligned}\xi^t(k)(I + \beta(k)X^t(k)P(k-1)X(k))^{-1}\xi(k) &\geq ||\xi(k)||^2\lambda_{min}(I + \beta(k)X^t(k)P(k-1)X(k))^{-1}\\ &= ||\xi(k)||^2[\lambda_{max}(I + \beta(k)X^t(k)P(k-1)X(k))]^{-1}\\ &= ||\xi(k)||^2(1 + \beta(k)\gamma(k))^{-1}\\ &= ||\xi(k)||^2[1 + \beta_1(k)\gamma(k)/(1+\gamma(k))]^{-1}\\ &\geq ||\xi(k)||^2/(1 + \beta_1(k))\end{aligned}$$

From eq(30), eq(34) and eq(13), it follows that

$$V(k) \leq V(k-1) + \frac{\beta_1(k)}{1+\gamma(k)}(1 - \frac{||\xi(k)||^2}{1+\beta_1(k)}) \tag{46}$$

Note that (since $\alpha \geq \beta_1(k)$);

$$\beta_1(k)(1 - \frac{||\xi(k)||^2}{1+\beta_1(k)}) \leq \frac{\alpha f(\delta, ||\xi(k)||)}{||\xi(k)||}(1 - \frac{||\xi(k)||^2}{1+\alpha}) \quad ||\xi(k)|| \geq \delta \tag{47}$$

and

$$(||\xi(k)||^2 - \delta^2) \geq f(\delta, ||\xi(k)||)||\xi(k)|| \quad ||\xi(k)|| \geq \delta \tag{48}$$

Then

$$\frac{f(\delta, ||\xi(k)||)}{||\xi(k)||}(\delta^2 - ||\xi(k)||^2) \leq -f^2(\delta, ||\xi(k)||) \quad ||\xi(k)|| \geq \delta \tag{49}$$

Thus

$$\beta_1(k)(1 - \frac{||\xi(k)||^2}{1+\beta_1(k)}) \leq -\frac{\alpha f^2(\delta, ||\xi(k)||)}{\delta^2} \quad ||\xi(k)|| \geq \delta \tag{50}$$

Since $V(k) = V(k-1)$ if $||\xi(k)|| < \delta$, hence we have

$$V(k) \leq V(k-1) - \frac{\alpha}{1+\alpha}\frac{f^2(\delta, ||\xi(k)||)}{1+\gamma(k)} \tag{51}$$

Hence 1) and 2) in this theorem are true following the argument of Goodwin and Sin (1984).

It should be noted that this theorem indicates that the estimated parameters will be bounded and that the normalized prediction error will tend to unity. It should also be noted that the sequence $\gamma(k)$ in eq(13) can be replaced by its upperbound if it exists.

Since eq(39) does not depend on the stability of the system, then eq(21) can be ensured and hence the stability of the system under unmodeled dynamics can be guaranteed. If there are no unmodeled dynamics, then results similar to those obtained by Kelly *et al* (1988) can be obtained

5) Conclusion

A stable hybrid adaptive impedance control scheme for robot manipulators has been presented. The continuous time control law will stabilize the system in case the estimated parameters are bounded. The parameter estimation section is designed to get the estimates of the parameters of the robot manipulator in order to refine the performance of the system. Unmodeled dynamics have been introduced in the robot model in order to account for the model error and measurement noise.

References

G.C. Goodwin and K.S. Sin "Adaptive filtering, prediction and control" Prentice Hall, 1984.

R. Kelly and R. Carelli "Unified approach to adaptive control of robotic manipulators" Proc 27th CDC, pp.1598-1604, 1988.

R. Kelly, R. Carelli, M. Amestegui and R. Ortega "On adaptive impedance control of robot manipulators" IEEE International Conference on Robotics and Automation, pp.572-577, 1989.

H.R. Middleton and G.C. Goodwin "Adaptive computed torque control for rigid link manipulators", System & Control Letters, 10, pp.9-16, 1988.

G. Niemeyer and J.J.E. Slotine "Computational algorithms for adaptive compliant motion" IEEE International Conference on Robotics and Automation, pp.566-571 1989.

Stability Analysis and Adaptive Trajectory Control for Rigid Manipulators

K. Ha
Fairchild Space *
Germantown, MD

G. Mosier
NASA/GSFC
Greenbelt, MD

April 26, 1990

Abstract

This paper results from an effort at NASA/Goddard Space Flight Center (GSFC) toward the application of parameter adaptive control in manipulator trajectory tracking for the NASA Flight Telerobotic Services (FTS) Program. In this paper, a formulation of the joint error angle dynamics for the considered control scheme is established. From this error dynamics, discussions on a number of quantitative results related to the system stability are provided. A complete development of an adaptive trajectory control scheme with a general setting for the selection of the gain matrices for both the PD feeback and the adaptation are given.

1 INTRODUCTION

Adaptive control strategies have received considerable attention as robotics system have found more applications outside the structured environments of factories and laboratories. At NASA/GSFC, a telerobot system consisting of a 7-dof slave manipulator and a 6-DOF force-reflecting master hand-controller is used as a testbed for the NASA Space Station Robotics Program. An effort is currently underway to evaluate and select the most efficient and robust control scheme for autonomous operations. The present emphasis is on trajectory tracking control systems. The objective is to track a bounded trajectory specified by desired joint positions, rates and accelerations as functions of time. In autonomous operations, the uncertainties in the knowledge of mass and inertial of loads to be handled by the manipulators can be significant. Because control systems for autonomous operations must perform with greater tracking accuracy and possess the ability to learn and adjust to the changes in the environment (e.g.loads), they lend themself to the application of adaptive control techniques.

The adaptive tracking control algorithm considered in this paper has the same basic features as those given in [1]-[3]. However, a more general approach is taken here. The control law under consideration consists of a PD feedback plus a feedforward dynamics compensation. The gain matrices for the PD terms are not restricted to be constant. The adaptive nature of this control scheme is reflected in the time-varying PD gains and in the on-line updating of the dynamic parameters, which involves the computation of a symmetric positive definite matrix.

*Research sponsored by NASA Goddard Space Flight Center

2 JOINT ERROR DYNAMICS

In order to provide a framework in which the adaptive control scheme can be systematically developed and its performance can be investigated, the derivation of the dynamics of the joint error angle vector corresponding to the specified control scheme will be given first in this section.

Based on the equation of motion of a n-DOF rigid manipulator [4], the commanded torque vector generated by the control scheme is a vector function of the commanded joint angles, rates, accelerations (θ_c,$\dot{\theta}_c$,$\ddot{\theta}_c$) and the current known measured joint angles and rates (θ_m,$\dot{\theta}_m$). It has the following functional form:

$$\tau_c = \hat{H}(\theta_m)\ddot{\theta}_c + \hat{C}(\theta_m,\dot{\theta}_m)\dot{\theta}_c + \hat{Q} + K_d(\dot{\theta}_c - \dot{\theta}_m) + K_p(\theta_c - \theta_m) \tag{1}$$

Where $\hat{H}$ is the modelled inertial matrix, $\hat{C}$ represents the modelled effect of the coriolis and centripetal forces. $\hat{Q}$ represents torques induced by gravity, friction and other modellable disturbances. K_d and K_p are the chosen gain matrices.

Similarly, the actual response of the manipulator to this commanded torque can be written as:

$$\tau_c = H_t(\theta_m)\ddot{\theta} + C_t(\theta_m,\dot{\theta})\dot{\theta} + Q_t \tag{2}$$

Where H_t is the true inertial matrix, C_t is the coriolis and centripetal matrix, and Q_t represents the actual torque arising from all the remaining sources, including the unmodelled components. The vectors $\ddot{\theta}$ and $\dot{\theta}$ are the joint accelerations and rates experienced by the manipulator immediately after the commanded torque is applied to the manipulator. It should be noted that due to the finite time-delay in system response the joint rate vector $\dot{\theta}$ is not available to the control law at the current time step. For latter reference, we denote the difference between the current commanded joint angles θ_c and the current sensed data θ_m as $\tilde{\theta}$, i.e. $\tilde{\theta} \equiv \theta_c - \theta_m$, and between θ_c and θ to be the joint error angles ϵ, i.e. $\epsilon \equiv \theta_c - \theta$.

It should be noted that the vector $\tilde{\theta}$ denotes the differences between the current and the desired joint angles, and is used in the control law to produce the required torque to generate desired motion. Whereas, vector ϵ and its derivatives denote the errors of the actual motion from the desired motion of the manipulator, which is the quantity the control law nominally must maintain within some small specified magnitude at all time.

With the above definitions, by subtracting (2) from (1) the joint error angle dynamics is obtained:

$$\hat{H}\ddot{\epsilon} + \hat{C}\dot{\epsilon} + K_d\dot{\tilde{\theta}} + K_p\tilde{\theta} = (Q_t - \hat{Q}) + (H_t - \hat{H})\ddot{\theta}_m + (C_t - \hat{C})\dot{\theta}_m \tag{3}$$

From definitions of $\tilde{\theta}$ and ϵ, we obtain $\theta = \theta_m + (\tilde{\theta} - \epsilon)$, thus,

$$C_t(\theta_m,\dot{\theta}) = C_t(\theta_m,\dot{\theta}_m) + C_t(\theta_m,\dot{\tilde{\theta}} - \dot{\epsilon}) \tag{4}$$

Using the fact that:

$$C_t(\theta_m,\dot{\tilde{\theta}} - \dot{\epsilon})\dot{\theta} = C_t(\theta_m,\dot{\theta})(\dot{\tilde{\theta}} - \dot{\epsilon}) \tag{5}$$

Equation (5) can be expressed as:

$$\hat{H}\ddot{\epsilon} + \hat{C}\dot{\epsilon} + K_d\dot{\epsilon} + K_p\epsilon = (C_t(\theta_m,\dot{\theta}) - K_d)(\dot{\tilde{\theta}} - \dot{\epsilon}) - K_p(\tilde{\theta} - \epsilon) + Y\tilde{p} + \mathrm{d_n} \tag{6}$$

Where:

$$Y\tilde{p} \equiv (H_t(\theta_m) - \hat{H}(\theta_m))\ddot{\theta}_m + (C_t(\theta_m, \dot{\theta}) - \hat{C}(\theta_m, \dot{\theta})\dot{\theta}_m \qquad (7)$$

and:

$$\mathrm{d_n} \equiv \mathrm{Q_t} - \hat{\mathrm{Q}} \qquad (8)$$

Equation (7) follows directly from a well-known result [6] that the system dynamics can be expressed in a linear form in some suitably selected set of manipulator mass parameters. Assuming that both the modelled and the actual dynamics have the same functional form, the vector $\tilde{p}$ represents the uncertainty in the modelled robot inertial and mass parameters.

The torque $\mathrm{d_n}$ in Equation (8) basically contains all the unknown and unmodelled effects in the system. Unlike the term $Y\tilde{p}$ corresponding to the parameter uncertainty, there is no apriori knowledge about the composition of this torque vector, except that it is conceivably manipulator specific;i.e. the disturbance torque depends on the detailed characteristics of the manipulator mechanical and electrical components. Currently, NASA/GSFC is sponsoring several projects whose efforts are to develop more efficient techniques to accurately characterize robot systems.

3 UNMODELLED DISTURBANCES AND EFFECTS

Equation (6) shows that the joint error angle dynamics is essentially driven by the control torque as well as the three principal disturbance torques. In the case where one assumes that $(\epsilon,\dot{\epsilon})$ and $(\tilde{\theta},\dot{\tilde{\theta}})$ are identical (which implies zero-delay in the feedback loop) and that the disturbance $\mathrm{d_n}$ is zero, Equation (6) can be shown to yield results given in [1].

By multiplying both sides of Equation (6) by $\dot{\epsilon}^T$ and using the first order approximation of $(\tilde{\theta} - \epsilon)$ (i.e. $\tilde{\theta} - \epsilon \simeq \dot{\theta}_m T$, where T denotes the sampling period), and by using the skew-symmetric properties of the matrix $(\dot{\hat{H}} - 2\hat{C})$, Equation (6) becomes:

$$\frac{1}{2}\frac{d}{dt}(\dot{\epsilon}^T \hat{H}\dot{\epsilon} + \epsilon^T K_p \epsilon) = -\dot{\epsilon}^T K_d \dot{\epsilon} + \dot{\epsilon}^T((C_t - K_d)\ddot{\theta}_m - K_p \dot{\theta}_m)T + \dot{\epsilon}^T(\mathrm{d_n} + \mathrm{Y}\tilde{\mathrm{p}}) \qquad (9)$$

Equation (9) suggests that the stability of the system may be achieved if the dynamics gorverning the parameter uncertainty and disturbance are stable systems and the sampling rate is adequately high. These conditions however are not the sufficient conditions for system stability. Equation (9) more strongly indicates that global asymptotic stability of the system can never be established due to the fact that the unmodelled disturbances do not neccessarily go to zero, in any manipulator's configuration, as $\dot{\epsilon}$ approaches zero. In this section, we will provide some quantitative discussion of the disturbance torques to determine some guidelines with which limitations of the system can be identified and performance of a given control system can be more realistically evaluated.

3.1 TIME DELAY

The time-lag induced by the processing of the sensed data can play a significant role in the stability analysis of the control system, as indicated by Equation (9). Although its effect can be reduced, its presence can not be completely removed. In practice, it therefore prevents this type of control scheme from achieving an asymptotically stable solution. Moreover, it not only depends on the motion of the manipulator but also on the gain matrices K_p, K_d and on the coriolis and centripetal matrix $C_t(\theta_m, \dot{\theta})$. In short, the system design portion involving the sampling rate must take into account the maximum operating accelerations and speeds of the joint angles as well as the selection of the gain matrices. The general guidelines to the selection of the sampling rate can be inferred from Equation (13). For high joint rates and accelerations, the sampling period must be small to keep the disturbance torque induced by the discretization negligible. Also, to have large positional gain K_p for good disturbance rejection ratio [5], the sampling period must be adequately reduced.

Based on the above discussion, the disturbance due to the discretized nature of the sensed data will be included in the unmodelled disturbance term $\mathrm{d_n}$ in the subsequent sections.

3.2 Disturbances and Parameter Uncertainty

Information about $\mathrm{d_n}$ and $Y\tilde{p}$, which represent the unknown components of the system, can only be extracted from actual sensor data during the operation of the manipulator. The process by which additional information about these terms is derived in real-time can not by any means represent their true behavior. This is due to the fact that, as unknown components, there are no dynamics models nor any apriori knowledge of their dependencies on other known variables. Thus, any assumption about the relationship between these unknown components and measurable ones are very often not justifiable. One plausible approach is to recursively estimated $\mathrm{d_n}$ and $\tilde{p}$ using some linear estimation technique. This approach, however, is very computationally intensive, and in some instances, the system stability can not be formulated.

Although the parameter uncertainty vector $\tilde{p}$ is a constant quantity and has a known structure associated with it (i.e. Y), information about $\tilde{p}$ is not easily obtained. Any estimate of $\tilde{p}$ requires a sequence of proper data points to gather sufficient information about $\tilde{p}$ if an reliable estimation is to result and no unreasonable consequence when Y losses rank is to be encounted. This requirement is basically due to the fact that the matrix Y, although configuration dependent, has a non-trivial null space. The "rich" or "persistently exciting" trajectories memtioned in [1]-[3] and other works, which are desired for generating these proper data points, show the limitation of current existing approaches. In addition, the problem of separating effects due to the disturbances on the system will in practice prevent good estimation of any unknown disturbance parameters, particularly of the $\tilde{p}$ vector.

The approach taken in this paper essentially assumes that these two components,$\mathrm{d_n}$ and $Y\tilde{p}$, can be lumped into one and the effect of $\mathrm{d_n}$ is such that it can be modelled as the variation in the parameter uncertainty. In other words, because of the fact that the zero solution of the joint error angle equation is not attainable in the asymtoptically stable sense, as long as the effect due to the $\mathrm{d_n}$ term is small compared to the other terms, its exact compensation is not required. The consequence of the stated assumption will be elaborated below in the development of the parameter adaptation scheme.

4 ADAPTIVE CONTROL SCHEME

In this section, the proposed adaptive control scheme is given with detailed analysis. The results presented here are currently being tested through simulation and actual implemention on the NASA/GSFC robotics testbed.

4.1 PD Gain Matrices

Before proceeding with the discussion of the adaptive structure for the parameter uncertainty, the computation of the time-varying gain control matrices K_p and K_d in the feedback loop is first given. Since the joint error angle equation, Equation (9), does not implicitly depend on the joint error angles ϵ, stability results derived from this equation do not apply to the behavior of the joint error angles. In order to guarantee both the convergence of the joint error rates and angles, an approach similar to that in [1]-[2] is used here. By defining a new error state vector as $\mathrm{s} = \dot{\epsilon} + \Delta\epsilon$, where Δ is required to be some symmetric positive definite matrix, the modified error state differential equation directly derived from Equation (9) can be written as:

$$\hat{H}\dot{\mathrm{s}} + \hat{C}\mathrm{s} + (K_d - \hat{H}\Delta)\dot{\epsilon} + (K_p - \hat{C}\Delta)\epsilon = Y\tilde{p} + \mathrm{d_n} \tag{10}$$

By choosing Δ to be:

$$\Delta = (K_d - \hat{H}(\theta_m)\Delta)^{-1}(K_p - \hat{C}(\theta_m, \dot{\theta}_m)\Delta) \tag{11}$$

Equation (10) becomes:

$$\hat{H}(\theta_m)\dot{\mathrm{s}} + \hat{C}(\theta_m, \dot{\theta}_m)\mathrm{s} + (K_d - \hat{H}(\theta_m)\Delta)\mathrm{s} = Y\tilde{p} + \mathrm{d_n} \tag{12}$$

In the case where the disturbance $\mathrm{d_n} = 0$ and $\tilde{p} = 0$, the system is globally asymptotically stable provided that the matrix $(K_d - \hat{H}(\theta_m)\Delta)$ can be made to be a symmetric positive definite, i.e.

$$v^T(K_d - \hat{H}(\theta_m)\Delta)v > 0 \qquad \forall\, v \neq 0 \tag{13}$$

In this case, the gain matrix K_p is derived from Equation (11), which yields:

$$K_p = (K_d - \hat{H}(\theta_m)\Delta)\Delta + \hat{C}(\theta_m, \dot{\theta}_m)\Delta \tag{14}$$

The zero solution of the system specifying by Equation (10), if attained, ensures the convergence of the joint error angles ϵ and rates $\dot{\epsilon}$. This is because the convergence of the joint error rates is already ensured by the symmetric positive definiteness of the K_d matrix, thus, as s approachs zero ϵ must also approach zero.

One of the most obvious choices for K_d is a multiple of the matrix $\hat{H}\Delta$ (i.e. $= n\hat{H}\Delta$). With Δ chosen to be a constant multiple of the identity matrix (i.e.$=\alpha I$), K_p can be obtained from Equation (14). The control law in this case, from Equation (1), becomes:

$$\tau_c = \hat{H}(\ddot{\theta}_c + n\alpha\dot{\tilde{\theta}} + \alpha^2(n-1)\tilde{\theta}) + \hat{C}(\dot{\theta}_c + \alpha\tilde{\theta}) + \hat{Q} \tag{15}$$

Also, K_d can be chosen to be a constant matrix, H_{max}, representing an upperbound of the H matrix and conresponding to some known configuration of the manipulator. In this case, the control law becomes:

$$\tau_c = H_{max}(\dot{\tilde{\theta}} + \alpha\tilde{\theta}) + \hat{H}(\ddot{\theta}_c - \alpha\dot{\tilde{\theta}} - \alpha^2\tilde{\theta}) + \hat{C}(\dot{\theta}_c + \alpha\tilde{\theta}) + \hat{Q} \tag{16}$$

One point worth noting from the above development is the fact that when $\Delta = 0$, K_p is zero. This points out the fact that K_p does not contribute to the convergence of the joint error rates. The removal of the K_p in all the derivations in previous sections will not invalidate the results obtained there. Only when the convergence of the joint error angles is taken into account does the positional gain K_p comes into play.

4.2 PARAMETER ADAPTATION

The usual approach to the development of the parameter adaptation found in recent works chose to have the dynamics of the variation of the parameter uncertainty term $\tilde{p}$ follow the relation:

$$\frac{1}{2}\frac{d}{dt}(\tilde{p}^T \Phi \tilde{p}) = -\mathbf{s}^T Y \tilde{p} \tag{17}$$

where Φ is noted to be a constant symmetric positive definite matrix. The adaptation is then derived by expanding Equation (17):

$$\dot{\tilde{p}}^T \Phi \tilde{p} = -\mathbf{s}^T Y \tilde{p} \tag{18}$$

By assuming that the true parameter vector is a constant, the updating of the modelled parameter vector $\hat{p}$ is chosen to be:

$$\dot{\hat{p}} = \dot{\tilde{p}} = -\Phi^{-1} Y^T \mathbf{s} \tag{19}$$

In the discussions of unmodelled disturbances, it was pointed out that the relationship between $\tilde{p}$ and measurable quantities are not defined due to the nature of $\tilde{p}$. Therefore relations, such as the one specified by Equation (17), can not be verified.

The parameter adaptation given in this paper uses Equation (17) as a model, with variable parameters to be determined according to the actual sensed data, for the system variation associated with $\tilde{p}$. In this setting, the vector $\tilde{p}$ is assumed to be the constant error to be corrected by the adaptation at every time step. The matrix Φ is not a constant matrix here and will be computed based on the change of the total energy of the error system. With these definitions, the adaptation law becomes:

$$\tilde{p}^* = -\dot{\Phi}^{-1} Y^T \mathbf{s} \tag{20}$$

In order to show that the above parameter adaptation is neccessary for the stability of the system, consider the energy equation of the joint error dynamics:

$$\frac{1}{2}\frac{d}{dt}(\mathbf{s}^T \hat{H} \mathbf{s}) = -\mathbf{s}^T (K_d - \hat{H}\Delta)\mathbf{s} - \mathbf{s}^T Y \tilde{p} \tag{21}$$

As the effect of $\tilde{p}$ on the system varies, the change in the total energy of the system can become positive, which is an undesirable situation. This situation can only occurs when the effect due to the distubances overcomes the PD feedback control and generates a large disturbance torque that is essentially in the same direction as the vector s. In the case where the disturbance is bounded, the PD feedback control is sufficient to maintain the error state vector s within some bounded region about zero. However, this bounded region is not neccessarily small, thus, the required tracking accuracy may not be achieved. Alternatively, because global asymptotic stability is not guaranteed, there is a smaller region about zero in which s can not remain for all time.

The adaptation law provides the neccessary mechanism to keep these two regions as small as possible (constrained by sensor noise). By substituting the adaptation law given in Equation (20) in (21), the model of the error system's energy equation can be shown to be:

$$\frac{1}{2}\frac{d}{dt}(\mathbf{s}^T\hat{H}\mathbf{s}) = -\mathbf{s}^T(K_d - \hat{H}\Delta)\mathbf{s} \; - \; \mathbf{s}^T Y\dot{\Phi}^{-1}Y^T\mathbf{s} \tag{22}$$

It should be noted that the change in the total energy in the left hand side of the above equation is measurable. To obtain this quantity, we require either a numerical differentiation of the computable scalar quantity ($\mathbf{s}^T\hat{H}\mathbf{s}$) or measured joint accelerations. The term in the right hand side of the equation involving the derivative gain matrix K_d is also computable since it contains variables depend on sensed data. Thus, the matrix Φ is defined by this equation. Because of the complexity of and problems associated with the unmodelled disturbances described in previous sections, only the most simple form of the matrix Φ is considered. The selected form is $\Phi = f(t)I$, where $f(t)$ is a time scalar function whose derivative is derived from Equation (22) as:

$$\dot{f}(t)^{-1} = \frac{\frac{1}{2}\frac{d}{dt}(\mathbf{s}^T\hat{H}\mathbf{s}) + \mathbf{s}^T(K_d - \hat{H}\Delta)\mathbf{s}}{\mathbf{s}^T Y Y^T\mathbf{s}} \tag{23}$$

the function $f(t)$ can be derived from Equation (17) and (21) as:

$$f(t) = \dot{f}(t)^2\frac{\mathbf{s}^T\hat{H}\mathbf{s} + \int \mathbf{s}^T(K_d - \hat{H}\Delta)\mathbf{s}dt}{\mathbf{s}^T Y Y^T\mathbf{s}} \tag{24}$$

which is a positive function, hence, the matrix Φ is a symmetric positive definite matrix.

The stability induced by this adaptation can be seen by recognizing that in the absence of sensor noise, the generalized energy term (i.e. $\mathbf{s}^T Y\tilde{p}$) associated with the unmodelled disturbance is a measurable quantity. Although this additional piece of information, in a scalar form, does not provide any insight into the effect on the system due to each component of the unmodelled disturbances, it specifies the net effect of disturbances on the system. The model given by Equation (17) is used in determining the distribution of this net effect on the elements of the vector $\tilde{p}$.

The adaptation essentially attempts to nullify the computed disturbance energy through the updating of the manipulator parameters. By substituting Equation (23) into (20), we obtain:

$$\tilde{p}^* = -\frac{\frac{1}{2}\frac{d}{dt}(\mathbf{s}^T\Phi\mathbf{s}) + \mathbf{s}^T(K_d - \hat{H}\Delta)\mathbf{s}}{\mathbf{s}^T Y Y^T\mathbf{s}}Y^T\mathbf{s} \tag{25}$$

Equation (25) shows that the adaptation will generate a parameter correction vector at every time step to effectively counter the change in the total energy about some level determined by the control law. The function $\dot{f}(t)^{-1}$ in this case can be thought of as the differential change of the manipulator parameter vector with respect to the vector $Y^T\mathbf{s}$. It specifies the direction in which the energy changes and provides some "uniform" proportional value for all parameters with respect to $Y^T\mathbf{s}$.

By multiplying both sides of Equation (25) by the term $\mathbf{s}^t Y$, it is clear that the disturbance energy is compensated for by the proposed parameter adaptation, provided that measurements from sensors are not grossly corrupted by noise. Thus, the stability of the system is guaranteed.

It is important to note that the estimate of the $\tilde{p}$ by Equation (25) is not guaranteed to converge to any physical quantity. It only represents the disturbance components that are projectable to the set of functions of joint angles, rates and accelerations that constitute the matrix Y.

5 CONCLUSION

This paper has presented a detailed stability analysis for a rigid manipulator system with the specified control law. The main result indicates the fact that global asymptotic stability is essentially not attainable due to the inaccuracies in system models and the inherent time-delay associated with discrete systems. A new robust parameter adaptive control system is proposed and a general formulation for the selection of the feedback gains is given. The robustness of this control law is based on the derivation of the disturbance energy, which does not require any apriori knowledge of the disturbance structure. The weakness of this approach is in the assumption that sensor noise sources are negligible.

6 ACKNOWLEDGMENTS

The authors would like to express their appreciation to the NASA/GSFC for their support in this research. Thanks also go to J. Bjorge, R. Byrne, M. Ali for providing technical supports during the preparation of this paper.

References

[1] Slotine, J-J.E., Li, W., "Adaptive Manipulator Control: A Case Study," IEEE Trans. on Automatic Control, Vol.33, N0.11, 1988.

[2] Slotine, J-J.E., Li, W., "On the Adaptive Control of Robot Manipulators," Int.J.Robotics Research, Vol.6, N0.3, 1987.

[3] Craig, J.J., Hsu, P., Sastry, S.S, "Adaptive Control of Mechanical Manipulators," Proc. IEEE Int. Conf. on Robotics and Automation.

[4] Paul, R.P., "Robot Manipulators: Mathematics, Programming, and Control," MIT Press, Cambridge, Mass., 1981.

[5] Khosla,P., "Choosing Sampling Rates for Robot Control," Carnagie Mellon Univ., March 1987.

[6] Khosla,P., Kanade,T., "Parameter Identification of Robot Dynamics," IEEE Conf.Decision and Control, Fort Lauderdale.

CONTROL PARAMETER BOUNDS FOR ROBUST STABILITY OF AN ADAPTIVE COMPUTED TORQUE CONTROL SCHEME APPLICATION TO ROBOTICS

Aminul Karim[†], Mohammad S. Alfhaid[‡], and Glenn R. Widmann[†]

[†] Colorado State University
Institute of Robotic Studies
Department of Electrical Engineering
Fort Collins, CO 80523

[‡] King Abdulaziz University
Department of Electrical and Computer Engineering
Jeddah, Saudia Arabia

ABSTRACT - A manipulator control architecture based on a computed torque technique in cooperation with a model reference adaptive control scheme is examined. This control architecture is referred to as an Adaptive Computed Torque Control Scheme. It has been shown this scheme is unstable in the presence of unmodelled dynamics and bounded inputs. System stability is achieved by placing bounds on the adaptive control parameters. In this paper, a novel approach is discussed for obtaining the value of these bounds. Experimental results are presented on the Colorado State University 3 degree-of-freedom cylindrical robot.

1. INTRODUCTION

In recent years, the search for the development of a robust controller for robotic applications has been very active. A powerful theme in this development has been the adaptive control of a linear model of the robot. While a vast majority of such schemes depend on local linearization, a new trend in research has been to look towards global linearization of the complex, highly nonlinear robot dynamics by the method of computed torque control. A unique step in this approach is the Adaptive Computed Torque Scheme (ACTCS), suggested by the authors [1-4], which utilizes a Direct Model Reference Adaptive Control (MRAC) philosophy for adaptation. The objective of the ACTCS is to implement a control strategy which does not require as rich of an information base required for computed torque methods, but alternatively relies on a richer base than is necessary for most adaptive control strategies. The robustness issues of this scheme has been previously investigated in [1-4].

The adaptive control strategy of the ACTCS is based on an input/output type of Direct MRAC, as presented in [5], but with modifications. Theoretically, it has been shown [6] that global stability of the Direct MRAC algorithm can only be achieved in the absence of unmodelled plant dynamics (or are at least very weak) and when there is no constraints on the plant input. In practice, these restricted conditions are always violated since the plant dynamics are not strictly known and the saturation of the joint actuators results in bounded input joint torques. From the viewpoint of the ACTCS architecture, the accuracy of the global linearization scheme will determine the extent to which the unmodelled dynamics enters the problem. In any event, the system stability is always threatened. Consequently, in the presence of bounded input torques, the adaptive control parameters of the resultant MRAC controller will grow unbounded. In this paper, the authors will present experimental verification of such behavior.

It has been shown in [7], that a modified adaptive law that places limit bounds on the adaptive control parameters produces a stable system. In fact, the system discussed in [1-4] could only be stabilized when upper and lower bounds were placed on the MRAC control parameters. It is not discussed, however, how these bounds may be theoretically determined.

In this paper, a method for obtaining these bounded values experimentally is introduced. It will also be shown that the combination of parameters in the bounded set is unique. The impact of the proposed scheme on the robustness and stability of the ACTCS controller will be investigated. The performance of this technique will be evaluated by performing experiments on the Colorado State University (CSU) three degree-of-freedom (DOF) cylindrical manipulator, as shown in Figure 1.

2. ACTCS CONTROLLER ARCHITECTURE

The ACTCS control architecture is shown in Figure 2. It is composed of two major subsystems: (a) Global Linearization Scheme, and (b) Direct MRAC scheme.

2.1. Global Linearization

Under the assumption of rigid body dynamics, the dynamic equations of motion for a k degree-of-freedom robot is

$$\tau = D(q)\ddot{q} + c(q,\dot{q}) + g(q) \tag{1}$$

where:

τ : vector of applied joint forces
q : vector of joint coordinate positions
$D(q)$: symmetric positive definite inertia matrix
$c(q,\dot{q})$: vector of Coriolis and centripetal forces
$g(q)$: vector of gravitational loading forces.

It has been shown [8], that the robot model can be globally linearized by acceleration feedback control (Computed Torque Method). The global linearizing block converts the given coupled non-linear k-d.o.f robot plant (1) into k approximate globally linear decoupled plants whose parameters (e.g., poles, zeros, and gains) are determined by optimal feedback. Let's define the joint coordinate position error Δq as: $\Delta q \triangleq q - q_d$, where q_d is the desired joint coordinate position. The global linearization torque control law τ_{CT}, obtained by optimal linear feedback, can be expressed as

$$\tau_{CT} = -K\left[\Delta\ddot{q} + A_1\Delta\dot{q} + A_o\Delta q\right] \tag{2}$$

where K is a positive definite diagonal feedback gain matrix, and A_o and A_1 represents the LQ designed gain matrices. The optimally-designed, feedback-decoupled linearized robot can be obtained as described in [1].

A feedforward block may also be added to approximately compensate for the modeling mismatch in the non-linear dynamics of the Computed Torque scheme. Thereby, the positioning accuracy of the manipulator can be further improved if the acceleration feedback generated torque τ_{CT} is augmented with a feedforward compensation torque τ_{FF}, such that the total applied joint torque τ is defined by $\tau \triangleq \tau_{CT} + \tau_{FF}$, where

$$\tau_{FF} = \hat{D}(q)\ddot{q} + \hat{c}(q,\dot{q}) + \hat{g}(q) \ . \tag{3}$$

The parameters $\hat{D}(\cdot)$, $\hat{c}(\cdot,\cdot)$, and $\hat{g}(\cdot)$ represents the estimated model approximations of the actual robot parameters from (1). This knowledge of the robot dynamics is obtained by the designer from specification sheets and experiments on the robot. This description may vary from a very crude (or even non-existent) to a very detailed model. The description may be presented in a variety of forms, such as: logical description (including a rule-based decision system), compensation transfer function description, time-series description, or a mixture of the above.

2.2. Direct MRAC Algorithm

The main objective of the Direct MRAC controller is to attempt to force the k decoupled, approximately linear plants to behave asymptotically like their linear, time invariant reference models. The Direct MRAC scheme is described by the transfer function

$$W_p(s) = h_p^T\left(sI - A_p\right)^{-1} b_p \triangleq k_p \frac{Z_p(s)}{R_p(s)} \tag{4}$$

where "s" represents the Laplace operator, $W_p(s)$ being strictly proper, $R_p(s)$ is a monic polynomial of degree n, $Z_p(s)$ is an asymptotically stable polynomial of degree m ($\leq n-1$), and k_p is a constant gain parameter. It is also assumed that m, n, and $\mathrm{sgn}[k_p]$ are known.

The reference model which represents the desired robot behavior is assumed to have a uniformly bounded input $r(\cdot)$ and output $y(\cdot)$. The reference model is defined by the transfer function

$$W_M(s) \triangleq k_M \frac{Z_M(s)}{R_M(s)} \tag{5}$$

where $R_M(s)$ is a monic Hurwitz polynomial of degree n, $Z_M(s)$ is a monic polynomial of degree m ($\leq n$), and k_M is a constant.

According to the algorithms of Narendra et al [5], the controller structure will be of the form

$$\dot{v}^{(1)} = \Lambda v^{(1)} + bu \qquad \dot{v}^{(2)} = \Lambda v^{(2)} + by_p \tag{6a}$$

$$w^{(1)} = c^T v^{(1)} + c_o r \qquad w^{(2)} = d^T v^{(2)} + d_o y_p \tag{6b}$$

where $b^T \in \mathbb{R}^{n-1}$, $c^T \in \mathbb{R}^{n-1}$, $d^T \in \mathbb{R}^{n-1}$, $\Lambda \in \mathbb{R}^{(n-1)\times(n-1)}$, is a stable matrix, and $v^{(1)}$, $v^{(2)}$: $\mathbb{R}^+ \rightarrow \mathbb{R}^{n-1}$. The pair (Λ, b) form a controllable pair. The control output $u(t)$ becomes

$$u(t) = \overline{\theta}^T(t)\overline{w}(t) \tag{7}$$

where

$$\overline{w}^T \triangleq \left[r(t) \mid v^{(1)T}(t) \mid y_p(t) \mid v^{(2)T}(t)\right] \quad \text{and} \quad \overline{\theta}^T \triangleq \left[c_o(t) \mid c^T(t) \mid d_o(t) \mid d^T(t)\right] \tag{8}$$

Under the conditions given on $W_p(s)$, a unique constant parameter vector $\overline{\theta}^*$ exists such that the transfer function of the plant together with the controller matches exactly the transfer function of the model $W_M(s)$ when $\overline{\theta}(t) = \overline{\theta}^*$. The parameter error $\overline{\phi}$ is defined by: $\overline{\phi}(t) \triangleq \overline{\theta}(t) - \overline{\theta}^*$.

Let's define the error in the output to be

$$e(t) \triangleq y_p(t) - y_m(t) = \frac{k_p}{k_M} W_M(s)\, \overline{\phi}^T(t)\overline{w}(t) . \tag{9}$$

The adaptive design is robust if the error $e(t)$ satisfies the property: $\lim_{t\rightarrow\infty} |e(t)| \rightarrow 0$, in the presence of unknown disturbances and uncertain system dynamics. In order to obtain a differentiator-free formulation, the error $e(t)$ is augmented by a supplementary signal as described in [1]. Briefly, the augmented error $\overline{e}(t)$ is defined by using an additional adaptive control parameter $\psi(t)$, making a total of $(2n+1)$ adjustable control parameters. The adaptive control parameter ψ is used to identify the unknown plant gain k_p (Note: $\psi(t) \equiv 0$, if $k_p = k_M$). The adaptive law, at the kth time interval, can be expressed as

$$\Delta\overline{\overline{\theta}}_k = \Delta\overline{\overline{\phi}}_k = -\frac{\Gamma_k}{T_k} \int_{t_k}^{t_k+T_k} \frac{\overline{e}(\tau)\overline{\overline{\zeta}}(\tau)}{1 + \overline{\overline{\zeta}}^T(\tau)\overline{\overline{\zeta}}(\tau)} d\tau \tag{10}$$

with $\Delta\overline{\overline{\phi}}(t) = \Delta\overline{\overline{\phi}}_k$, for all $t \in [t_k, t_{k+1})$, and the signal vector $\overline{\overline{\zeta}}(t)$ as formulated in detail in [1]. Γ_k ($\Gamma_k > 0$) and T_k are the adaptation gain and the sampling period, respectively. The sampling period varies inversely with respect to the magnitude of the error as dictated by an adaptive sampling control law which uses the objective function as specified in [1].

3. DETERMINATION OF PARAMETER BOUNDS

In our previous experimentation with the ACTCS architecture, it has been shown that system stability was only possible if the adaptive control parameters were bounded. The instabilities are the result of bounded torque inputs and unmodelled dynamics. The ensuing discussion will outline a technique which can be implemented experimentally that can be used to determine how the bounded values (used either as a lower or upper bound) may be assigned.

The approach used to determine the control parameter bounds is based on a heuristic approach derived from certain arguments. The robot is initially placed at one of its limits and held physically stationary. The controller is now allowed to be run and the adaptive control parameters recorded. An examination of these results will show that some of the control parameters saturate while the others become unbounded. The same is done with the arm at its other extreme limit. By keeping the arm stationary at its extreme positions, the controller is being subjected to the "worst case" conditions, Thereby, we can determine our bounds on the parameters from the saturated adaptive controller parameter values obtained at the extreme positions. These saturated parameter values form the conservative bounded set on the control parameters. That is not to infer that some other set of values for the bounds will not result in a stable control behavior, but this technique will at least give the user some method to obtain the values.

4. EXPERIMENTS AND RESULTS

The experimental setup consists of the CSU three DOF cylindrical robot (See Figure 1) connected to an IBM-AT PC. A precision potentiometer is mounted at each joint of the robot to measure the joint position. The joint velocities are estimated using Kalman filters. The entire control scheme is implemented in software. Although the measurement sampling time was adaptively varying from 40 to 100 msec, the output of the adaptation algorithm was delivered at 400 msec intervals. For the sake of brevity, only the response of the radial prismatic joint (e.g., r-axis) will be shown.

In order to evaluate the stability, performance, and robustness of the ACTCS controller, the reference input signal r(t) must satisfy a persistency of excitation condition. It has been shown, by Mason et al [9], that the input signal must satisfy a persistency of excitation condition which requires ρ-independent exciting input variables in order to identify ρ-independent control parameters in the adaptive scheme (e.g., in our examination there are four parameters: ψ, c_1, d_0, and d_1). This has been shown to be a necessary condition, but not a sufficient condition in the presence of unmodelled dynamics. The signal space chosen for investigation was a sum of three sine-wave signals of the form

$$r(t) = A_1\sin(\omega_1 t) + A_2\sin(\omega_2 t) + A_3\sin(\omega_3 t) \tag{11}$$

where "A" represents the signal maximum magnitude and ω_0 is the fundamental frequency of oscillation. The description of the six independent sine-wave signal variables (A_i,ω_i) are:

$$(A_1,\omega_1) = (0.5,2\pi) \qquad (A_2,\omega_2) = (-0.4,2.4\pi) \qquad (A_3,\omega_3) = (0.3,2.8\pi) \tag{12}$$

EXPERIMENT ONE (Figure 3). This illustrates the r-axis response of y_p and y_m when the adaptive control parameters remain unbounded. A sustained biased oscillatory position tracking error exists which can not be overcome. As a result, the adaptive control parameters are constantly varying in an attempt to correct for the error. Eventually (e.g., $\simeq$ 9 sec) two of the parameters, ψ and d_o, makes a quantum leap in an attempt to diminish this error. This over-compensated corrective action ultimately resulted in the system going unstable and drives the link to its hard physical limit, and consequently loosing the tracking capability to follow the model reference signal. All experiments performed with unbounded control parameters have always shown that the system is unstable.

EXPERIMENT TWO (Figure 4). This experiment demonstrates selected results of the technique proposed for determining the control parameter bound. When the r-axis is held physically stationary at the lower physical limit (e.g., fully retracted), two of the parameters (e.g., d_0 and ψ) went unbounded (not shown). The other two remaining parameters had a saturated value of $c_1 \simeq 6.9$ and $d_1 \simeq -1.8$ (shown). On the other hand, when the r-axis is held stationary at the upper physical limit (e.g., fully extended), two of the parameters (e.g., c_1 and d_1) went unbounded (not shown). The other two parameters achieved a saturated value of $\psi \simeq 0$ and $d_0 \simeq 0$ (not shown). These saturated values form the conservative bounded set.

EXPERIMENT THREE (Figure 5). This experiment shows the system can be stabilized when the conservative bounded set limits the range of values the controller gains can attain. The position tracking error still exhibits oscillatory behavior but maintains an unbiased bounded error within an acceptable limit, and each control parameter eventually converges to a value. It is conjectured the error does not asymptotically go to zero because the parameters of the linearized system may be slowly time-varying. Extensive studies with other combinations of parameter bounds does not provide any better position error characteristics than if the conservative bound set is used. However, there are other combinations of parameter bounded sets which results in the system being unstable.

5. CONCLUSION

The stability of the ACTCS algorithm was demonstrates in response to the adaptive controller gain being bounded. System instabilities are the result of bounded torque inputs and unmodelled dynamics. Theoretically, the system can be stabilized if the adaptive control law is modified such that bounds are placed on the adaptive control parameters. It is unclear how these bounded limit values can be theoretically assigned. This paper investigated an experimental method on how the bounded values maybe assigned. That is not to infer that some other set of values for the bounds will not result in a stable control behavior, but this technique will at least give the user some method to obtain the values. Verification of this technique was demonstrated experimentally.

REFERENCES

[1] A. Karim and G.K.F. Lee, "Adaptive Computed Torque Control of a Cylindrical Robot using Linear Compensation," *Proc. 2nd Int. Symp. on Robotics & Manufacturing: Research, Education & Application,* Albuquerque, NM, Nov. 1988.

[2] A. Karim, G.K.F. Lee, G.R. Widmann, and E.M. Nebot, "Robustness Investigation of Adaptive Control Under Reference Model Switching for a Cylindrical Robot," *Proc. 1989 Amer. Control Conf.,* Pittsburgh, PA, pp. 31-36, June 1989.

[3] A. Karim, D. Shusterman, M. Thompson, and G.R. Widmann, "Input Signal Analysis for an Adaptive Computed Torque Control Scheme Applied to a Three DOF Cylindrical Robot," *Proc. 32nd IEEE Midwest Symp. on Circuit and Systems,* Urbana, IL, August 1989.

[4] A. Karim, G.K.F. Lee, and G.R. Widmann, "Inner-Loop PID Control within an Adaptive Computed Torque Control Scheme for Robot Manipulators," *Proc. 1989 IASTED Int. Conf. on Adaptive & Knowledge-Based Control and Sig. Proc.,* Honolulu, HI, August 1989.

[5] K.S. Narendra and I.H. Khalifa, Tech. Reports 8204 & 8206, Yale Univ., New Haven, CT, 1982.

[6] K.S. Narendra and A.M. Annaswamy, "A General Approach to the Stability Analysis of Adaptive Systems," *Proc. 23rd Conf. on Decision and Control,* Las Vegas, NV, Dec. 1984.

[7] K.S. Narendra and A.M. Annaswamy, *Stable Adaptive Systems,* Prentice Hall Information and Science Series, (pp. 306-308), 1989.

[8] G.L. Lou and G.N. Saridis, "L-Q Design of PID Controllers for Robot Arms," *IEEE J. of Robotics & Automation,* Vol. RA-1, No. 3, pp. 152-159, Sept. 1985.

[9] J.E. Mason, E.W. Bai, L.-C. Fu, M Bodson, and S.S. Sastry, "Analysis Identifiers in the Presence of Unmodelled Dynamics: Averaging and Tuned Parameters," Memo No. UCB/ERL M87/25 Electronics Res. Lab., Berkely, CA, 1987.

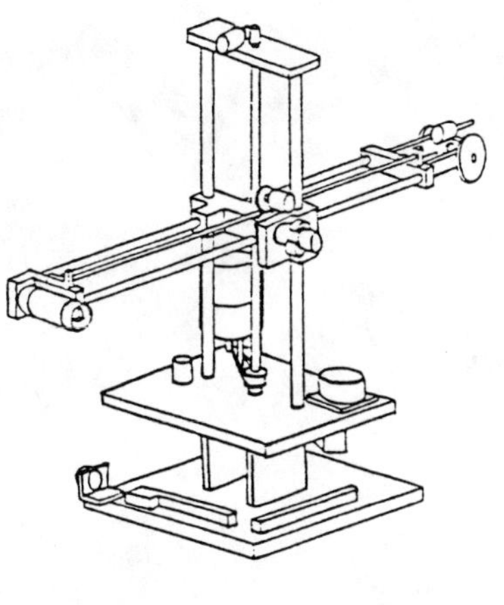

Fig 1. The CSU 3 DOF Cylindrical Manipulator.

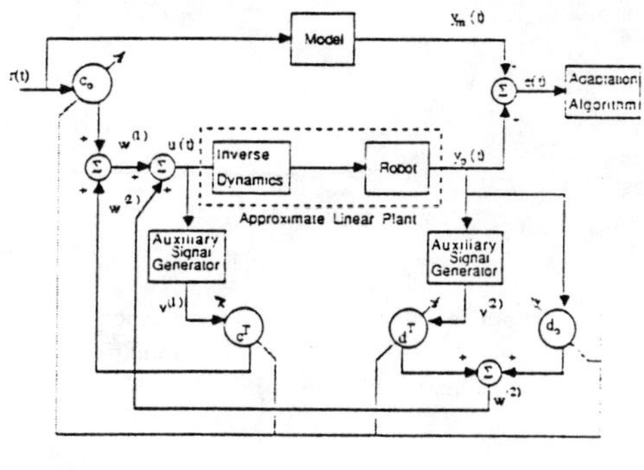

Fig 2. Block Diagram of the ACTCS Scheme.

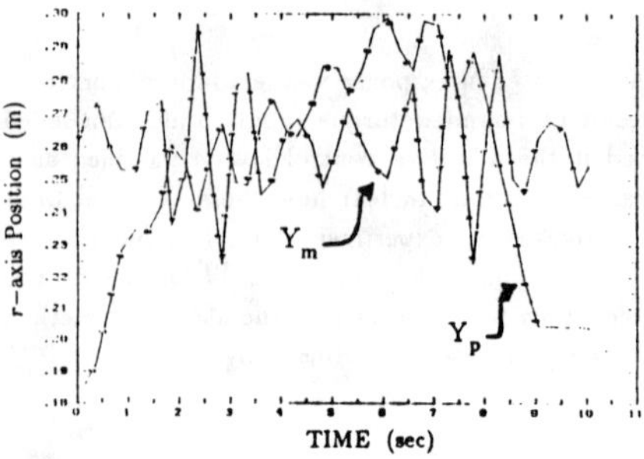

Fig 3. Unstable Position Response of r-axis with Unbounded Control Parameters.

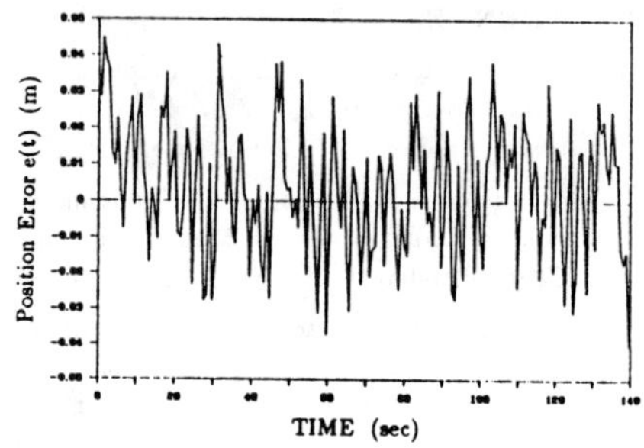

Fig 5. Stable Position Response of r-axis with Bounded Adaptive Control Parameters.

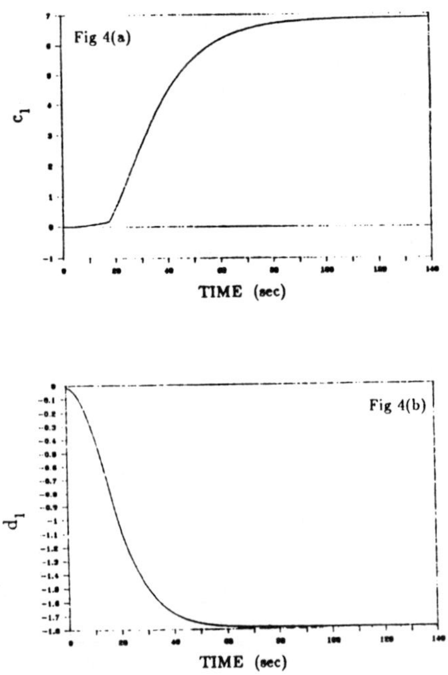

Fig 4. Adaptive Control Parameter Response using Proposed Bound Determination Technique: (a) c_1, and (b) d_1.

ON THE ADAPTIVE CONTROL OF A CONTOUR-FOLLOWING ROBOTIC SYSTEM

J.N. Liou and M. Jamshidi
CAD Laboratory for Systems/Robotics
Dept. of Electrical and Computer Engineering
University of New Mexico, Albuquerque, NM 87131 USA

ABSTRACT

To demonstrate the force sensing and control, Starr models a complete contour-following system as a fourth order linear continuous-time time-invariant system, in which stiffness k_t, robot natural frequency w_n, linear accommodation gain k_x, and angular accommodation gain k_θ are all constants obtained by measurement and experimentation. It works really well for following a short contour. To satisfactorily follow a longer contour, k_t, w_n, k_x and k_θ can be thought of as unknown constants or time-varying variables. When they are considered as unknown constants, a robust model reference adaptive controller can be used to achieve both stability and tracking purposes without having to know or find the true values of those constants, given bounded input or output disturbances and stable unmodeled dynamics. If w_n is assumed to be an unknown constant but k_t, k_x, and k_θ are assumed to be unknown variables, then one has a linear time-varying plant and other types of model reference adaptive controllers have to be used to achieve the same purpose. In this paper, the schemes from various robust model reference adaptive control design will be studied and comparison and suggestions will also be made based on the simulation results for the contour-following robotic system mentioned above.

1. INTRODUCTION

In the past ten years, extensive research in the robust model reference adaptive control has been active and several famous direct model reference adaptive control schemes have been appeared in the literature [1-5]. In sections 2 and 3 we briefly review them and then in section 4 apply these results to our continuous-time contour-following robotic system [6,7] to demonstrate the pros and cons of various adaptive control algorithms and hopefully stimulate new ideals in modification of existing approaches to improve the performance in both transient behavior of the output error and the convergence speed of the control parameters, while maintaining the stability in the overall adaptive control system.

2. MODEL REFERENCE ADAPTIVE CONTROL (MRAC)

2.1. Ideal Case Analysis (no disturbance and no unmodeled dynamics) [1]

A single-input single-output (SISO) plant to be controlled is represented by a linear time-invariant (LTI) state equation

$$\dot{\mathbf{x}}_p = \mathbf{A}_p \mathbf{x}_p + b_p u \tag{2.1}$$

$$y_p = h_p^T \mathbf{x}_p \tag{2.2}$$

with a transfer function $W_p(s) = h_p^T(s\mathbf{I} - \mathbf{A}_p)^{-1} b_p = k_p Z_p(s)/R_p(s)$, where $Z_p(s)$ and $R_p(s)$ are monic coprime polynomials of degree m and n ($> m$), respectively, and $Z_p(s)$ is Hurwitz. The sign of k_p and the values of m and n are specified but the coefficients of $Z_p(s)$ and $R_p(s)$ are unknown. A reference model is set up with a transfer function $W_m(s) = k_m / R_m(s)$, where $R_m(s)$ is a known Hurwitz polynomial of degree $n^* = n - m$. A piecewise continuous uniformly bounded reference input $r(t)$ to the model yields the desired output $y_m(t)$. The adaptive control problem is to determine a bounded control input $u_p(t)$ to the plant so that the output error $e_1(t) = y_p(t) - y_m(t)$ tends to zero asymptotically with all the signals in the closed-loop system remaining bounded at all time. Consider a plant with relative degree unity, i.e., $n^* = n - m = 1$, the adaptive controller is described by the following equations

$$\dot{w}_1(t) = \mathbf{\Lambda} w_1(t) + l u_p(t) \tag{2.3}$$

$$\dot{w}_2(t) = \mathbf{\Lambda} w_2(t) + l y_p(t) \tag{2.4}$$

$$w(t) \equiv [\, r(t)\; w_1^T(t)\; y_p(t)\; w_2^T(t)\,]^T \tag{2.5}$$

$$\theta(t) \equiv [\, k(t)\; \theta_1^T(t)\; \theta_0(t)\; \theta_2^T(t)\,]^T \tag{2.6}$$

$$u_p(t) = \theta^T(t) w(t) \tag{2.7}$$

where $k : R^+ \to \mathbf{R}^1$, $\theta_1, w_1, : R^+ \to \mathbf{R}^{n-1}$, $\theta_0 : R^+ \to \mathbf{R}^1$, $\theta_2, w_2, : R^+ \to \mathbf{R}^{n-1}$, $\mathbf{\Lambda}$ is an arbitrary asymptotically stable matrix, $(\mathbf{\Lambda}, l)$ is a controllable pair, and $\theta(t)$ is the control parameter vector. When $\theta(t) \equiv \theta^*$, where θ^* is a constant vector, the transfer function of the plant together with the controller matches that of the model exactly, the problem therefore can be rephrased as the determination of the adaptive law for adjusting $\theta(t)$ so that $\lim_{t \to \infty} e_1(t) = 0$ with all signals remaining bounded. If the plant is completely known, the constant parameter vector $\theta^* = [\, k^*\; \theta_1^{*T}\; \theta_0^*\, \theta_2^{*T}\,]^T$ can be calculated as follows: let

$\lambda(s) = Z_m(s) = det(s\mathbf{I} - \mathbf{\Lambda})$, $C^*(s) = \lambda(s) - Z_p(s)$, $k_p D^*(s) = R_p(s) - R_m(s)$, then scalars k^*, θ_0^* and vectors θ_1^* and θ_2^* exist such that $k^* = k_m / k_p$, $\theta_1^{*T}(s\mathbf{I} - \mathbf{\Lambda})^{-1} l = C^*(s)/\lambda(s)$, and $\theta_0^* + \theta_2^{*T}(s\mathbf{I} - \mathbf{\Lambda})^{-1} l = D^*(s)/\lambda(s)$.

2.2. Error Model And Standard Model Reference Adaptive Law

It has been shown that the stability of adaptive systems is most readily analyzed by considering error models relating an output error $e_1(t)$ to the parameter error vector $\Phi(t)$, where $\Phi(t)=\theta(t)-\theta^*$, they can be related by the equation $W_m(s)\Phi^T(t)w(t)=e_1(t)$. If $W_m(s)$ is a strictly positive real (SPR) function, i.e., (i) $W_m(s)$ is real for real s, (ii) $Re[W_m(s)]\geq 0$ for all $Re[s]>0$, and (iii) $W_m(s-\epsilon)$ also satisfies with (i) and (ii) for some $\epsilon>0$, a stable control law can be realized using the signals in the adaptive loop as

$$\dot{\theta} = \dot{\Phi} = -\, sgn(k_p)e_1 w \tag{2.8}$$

and the stability of the overall closed loop system has been analyzed and proved via Lyapunov direct method. If in additional to the output error $e_1(t)$ tending to zero, it is desired that the control parameters tend to their true values, i.e., $\lim_{t\to\infty}\Phi(t)=0$, then $w(t)$ needs to be persistently exciting (PE) [8]. The intuition behind the concept of PE is that the input should be *rich* enough to excite all the modes of the system that is being probed. However, it is clear that the PE of $w(t)$ cannot be demanded directly since it is a dependent variable and the output of a nonlinear system. This fact was recognized by many researchers [9] who later proved the convergence of $\theta(t)$ depends on the PE of the reference input $r(t)$. Based on these results it can be stated that $\lim_{t\to\infty}\theta(t)=\theta^*$ if $r(t)$ contains n distinct frequencies (or n spectral lines).

3. ROBUST MODEL REFERENCE ADAPTIVE CONTROL

3.1. e_1 Modification Scheme

The robustness of the adaptive system can be achieved either by modifying the adaptive law and/or by making the reference input sufficiently persistently exciting. Without loss of generality, k_p is assumed to be positive throughout this section. Narendra and Annaswamy [2] proposed the following adaptive law, known as e_1 modification scheme:

$$\dot{\theta} = \dot{\Phi} = -\, e_1 w \;-\gamma|e_1|\theta \tag{3.1}$$

where $\gamma>0$ is a design parameter, to assure the origin in the $[e_1,\phi]$ space is exponentially stable in the presence of a class of perturbation (bounded external disturbances) when the reference input is persistently exciting and has a sufficiently large amplitude.

3.2. σ Modification Scheme

Ioannou and Kokotovic [3] modified eq. (2.8) by adding a decay term $-\sigma\Gamma\theta$ and obtained the adaptive law, known as σ modification scheme:

$$\dot{\theta} = \dot{\Phi} = -\,\Gamma e_1 w \;-\sigma\Gamma\theta \tag{3.2}$$

where $\Gamma=\Gamma^T>0$, and $\sigma>0$ can be chosen by the designer. Unfortunately, in the absence of disturbances the output errer $e_1(t)$ does not tend to zero asymptotically. When the first derivative of the output disturbance is bounded the design parameter σ can be made arbitrarily small so that its effect on the ideal properties in the disturbance-free case is small. The σ modification scheme guarantees boundedness and convergence of the state/parameter error $\Phi(t)$ to a residual set. The size of the set depends on the magnitude of disturbances and the design parameter σ.

3.3. Switching σ Modification Scheme

Ioannou and Tsakalis [4] proposed a switching σ modification scheme which is robust with respect to stable unmodeled dynamics of the additive and multiplicative form. The controller is designed for the reduced-order plant with minimum phase and of known order and relative degree, but is applied to the overall plant, which, due to the presence of unmodeled dynamics, may be nonminimum phase and of unknown order and relative degree. The equation for the adaptive law to adjust the parameter vector θ is

$$\dot{\theta} = -\Gamma\frac{\epsilon_1\varsigma}{m^2} - \Gamma\sigma\theta\ ,\ \sigma = \begin{cases} 0 & \text{if } \|\theta\| < M_0 \\ \sigma_0(\frac{\|\theta\|}{M_0} - 1) & \text{if } M_0 \leq \|\theta\| \leq 2M_0 \\ \sigma_0 & \text{if } \|\theta\| > 2M_0 \end{cases} \quad (3.3)$$

where $\Gamma = \Gamma^T > 0$, $\epsilon_1 = y_p - y_m + \theta^T\varsigma - \nu$, $\varsigma = W_m(s)\mathbf{I}w$, $\nu = W_m(s)\theta^T w$, and $\dot{m} = -\delta_0 m + \delta_1(|u_p| + |y_p| + 1)$, $m(0) \geq \delta_1/\delta_0$. It has shown from their simulation that the normalizing signal $m(t)$ plays an important role during the transient by guaranteeing slow adjustment of the controller parameters where appropriate. The smooth adjustment of the controller parameters reduces the excitation of the unmodeled dynamics by the control input, and therefore improves the accuracy of adaptation. When the initial conditions, especially the initial parameter vector, are far away from the desired ones and the plant is initially unstable, the signals in the adaptive loop may have a bad transient and poor convergence.

3.4. Variable Structure Control (VSC)

The VSC approach is based on switching functions of the state variable. These switching functions can be used to create a "sliding surface" which is intended to constrain the dynamics of the system trajectories. When this occurs, the switching functions keep the trajectory on the surface, thus yielding the desired system dynamics, which becomes insensitive to parameters variations or nonlinearities. Hsu and Costa [5] proposed a variable structure model reference adaptive control (VR-MRAC) scheme to demonstrate its remarkable transient behavior and robustness with respect to bounded disturbances, non-linearities, plant parameter variations, and unmodeled dynamics. The control u_p to the plant is structured as

$$u_p = \theta_{eq}^T w \tag{3.4}$$

and the adaptive law is

$$\mu\dot{\theta}_{eq} = -\sigma\,\theta_{eq} + \sigma\, diag\,[\psi_i] \tag{3.5}$$

where $\psi_i = -\overline{\theta}_i\, sgn(w_i e_1)$, $\mu > 0$, $\sigma > 0$ are design parameters, and $\overline{\theta}_i > |\theta_i^*|$ for all i. Note that θ_{eq} is the low-pass filtered vector of ψ and the exponential stability of the origin of the error equations is verified independently of the reference input $r(t)$, i.e., no persistency of excitation is needed.

4. SIMULATION EXAMPLES AND RESULTS

The transfer function of the 4-th order contour-following robotic system in [6,7] is

$$\frac{k_t\, v_y (s + w_n)^2 s}{s^4 + 2w_n s^3 + w_n^2 s^2 + w_n^2 k_t\, k_x\, s + w_n^2 k_t\, v_y\, k_\theta} \tag{4.1}$$

Using the experimental values of $k_x = 0.00103\ m/N\ s$, $k_\theta = 0.123\ rad/N\ s$, $k_t = 13850\ N/m$, $v_y = 15\ mm/s$, and $w_n = 16\ rad/s$ to follow a short contour, one has the following LTI plant $\frac{207.75(s+16)^2 s}{s^4 + 32s^3 + 256s^2 + 3652s + 6541.6}$; a reference model is obtained via a least-squares system identification technique as $24.006/(s+24.006)$. Assume bounded input and output disturbances are $v_i(t) = v_o(t) = 0.2cost$, with e_1 modification scheme, all the signals in the adaptive loop are bounded and figure 1 shows the output error e_1 when the reference input $r(t)$ is persistently exciting (PE) and equal to $15cost + 20cos\,5t + 25cos\,25t + 40cos\,30t$; the design parameter γ in eq. (3.1) is chosen as 0.1.

Using the switching σ scheme and applying a non-PE reference input $r(t) = 10cos\,1.75t$, figures 2 and 3 display the output error e_1 with and without the output disturbance, respectively. In both cases, the modeled part of the plant or contour-following system is structured as $207.75(s+16)^2(s+10)/(s^4 + 32s^3 + 256s^2 + 3652s + 6541.6)$, and $\mu = 0.1$, the unmodeled multiplicative and additive dynamics are $-100/(s+10)$ and $1/(s+15)$, respectively. The design parameters in eq. (3.3) are chosen as $\Gamma = 10$, $\delta_0 = 0.7$, $\delta_1 = 1.0$, $M_0 = 20$, and $\sigma_0 = 0.1$.

To follow a long contour, the system in eq. (4.1) is treated as a linear time-varying plant with three variables $k(t) = 13850 + 100cos\,0.1t$, $k_x(t) = 0.00103(1 + 0.1cos\,0.1t)$, $k_\theta(t) = 0.123(1 + 0.1cos\,0.1t)$, v_y and w_n are the same constants as before. Also with the non-PE reference input $r(t) = 10cos\,1.75t$, figures 4-6 demonstrate the transient behavior of VS-MRAC scheme applied to the time-varying plant. The design parameters for the low-pass filtered vector in eq. (3.5) are chosen as $\mu = 0.05$ and $\sigma = 1.0$; in figures 4 and 5, no disturbance is present and $\overline{\theta}_i$ are chosen as 3 and 10, respectively. In figure 6 $\overline{\theta}_i = 5$ and both input and output disturbances $v_i(t)$ and $v_o(t)$ are assumed to exist.

5. CONCLUSION

Robust model reference adaptive control schemes are investigated and applications to continuous-time contour-following system has been presented. It is obvious from our simulation that there are rooms for further improvements in many areas. Research in combination and modification of robust adaptive control approaches [10,11] are currently under study and these theories and applications to our system will be reported in other occasions.

REFERENCES

1. K. S. Narendra, and L. S. Valavani, "Stable Adaptive Controller Design Direct Control", IEEE Trans. Automat. Contr., 23(4), 570-583 (August 1978).

2. K. S. Narendra, and A. M. Annaswamy, "A New Adaptive Law for Robust Adaptation without Persistent Excitation", IEEE Trans. Auto. Contr., 32(2), 134-145 (Feb. 1987).

3. P. A. Ioannou, and P. V. Kokotovic, Adaptive Systems with Reduced Models, (Springer-Verlag, N. Y. 1983).

4. P. A. Ioannou, and K. S. Tsakalis, "A Robust Direct Adaptive Controller", IEEE Trans. Automat. Contr., 31(11), 1033-1043 (November 1986).

5. L. Hsu, and R. R. Costa, "Variable Structure Model Reference Adaptive Control Using Only Input and Output Measurements: Part I", Int. J. of Control, 49(2), 399-416 (Feb. 1989).

6. G. P. Starr, "Control of Tool-Workpiece Interaction Forces: Contour-Following with a PUMA 560 Manipulator", Recent Trends in Robotics: Models, Control, and Education, M. Jamshidi, et al. (eds), 261-266 (Elsevier Science Publishing Co., N.Y. 1986).

7. J. N. Liou, M. Jamshidi, and G. P. Starr, "Adaptive Edge-Following Force Control of an Industrial Robot", Robotics and Computer-Integrated Manufacturing, 6(4), 331-337 (1989).

8. A. P. Morgan, and K. S. Narendra, "On the Uniform Asymptotic Stability of Certain Linear Nonautonomous Differential Equations", SIAM Journal on Control and Optimization, 15(1), 5-24 (January 1977).

9. S. Boyd, and S. S. Sastry, "On Parameter Convergence in Adaptive Control", Systems and Control Letters, 3(), 311-319 (December 1983).

10. K. S. Narendra, and J. D. Boskovic, "A Combined Direct, Indirect and Variable Structure Method for Robust Adaptive Control", 1990 American Control Conference, 543-548 (San Diego, CA 1990).

11. Li-Chen Fu, "A Robust Model Reference Adaptive Control Using Variable-Structure Adaptation for a Class of Plants", Proc. of the 1989 American Control Conference, 2291-2296 (Pittsburgh, PA 1989).

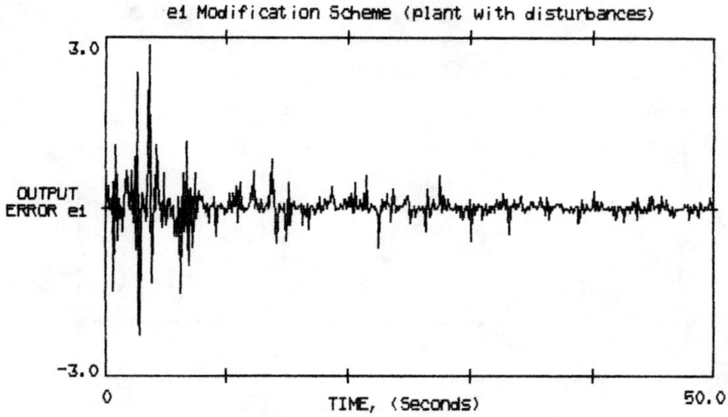

Fig. 1 ———

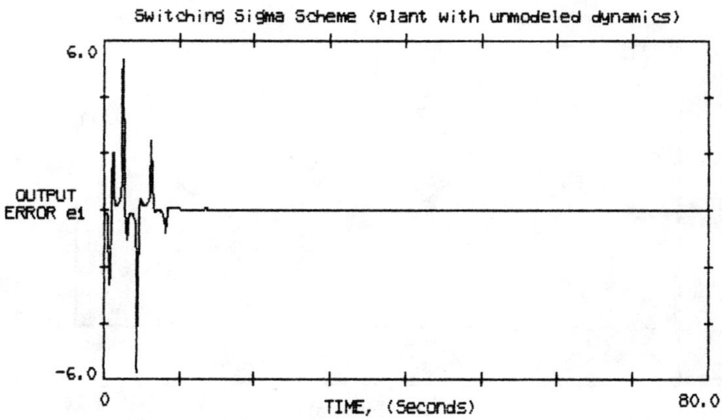

Fig. 2 ———

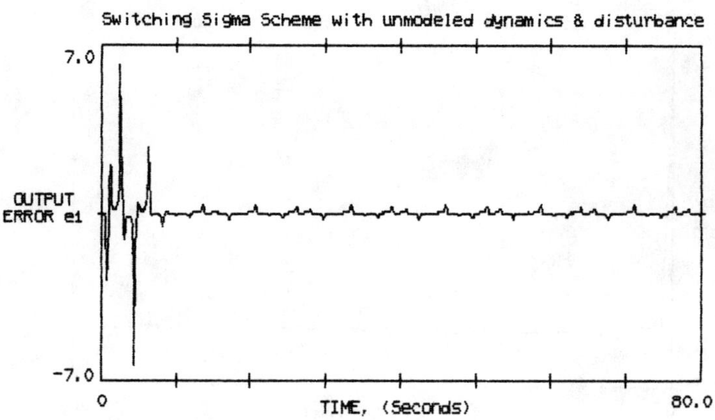

Fig. 3 ———

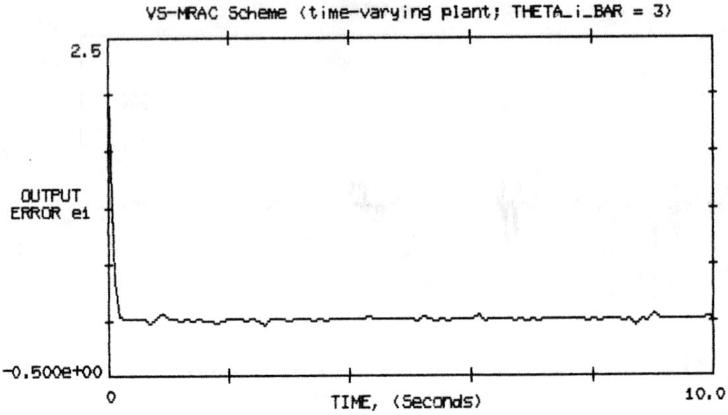

Fig. 4 ————

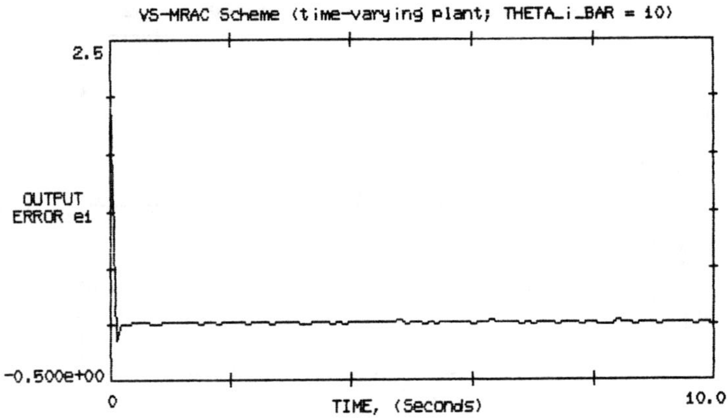

Fig. 5 ————

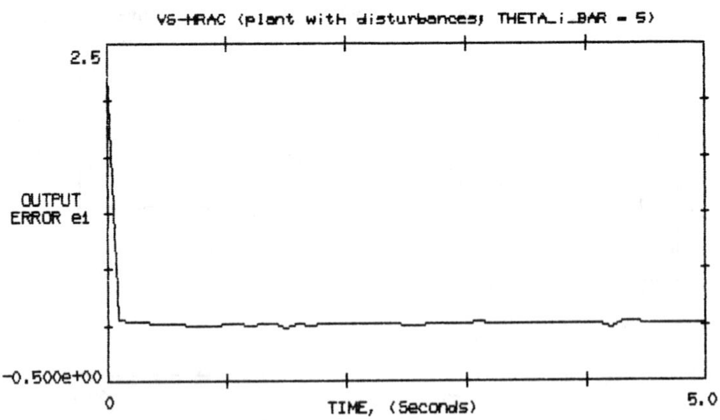

Fig. 6 ————

SECTION 5 : ROBOTS IN UNSTRUCTURED ENVIRONMENTS

Techniques for Robot Survivability[1]

Ronald C. Arkin* and George Vachtsevanos**
*School of Information and Computer Science
**School of Electrical Engineering
Georgia Institute of Technology
Atlanta, GA 30332

Abstract

In order to ensure the applicability of robots in hazardous and unstructured environments, certain measures must be taken to protect them. This paper describes four such techniques. The first incorporates reactive control to provide the ability for a robot to rapidly cope with changes in the environment. The second method is concerned with the management of a robot's available resources. The third technique involves intelligent fault-tolerant control techniques that deal with subsystem and component failure in real time. The last approach is to build reconfigurable and redundant robots that can replace hardware failures in a straightforward manner.

1. Introduction

From its inception, robotics has been concerned with placing machines into environments that are dangerous to people. The nature of these threats can range from operations in remote locations as found in extra-terrestrial exploration, to life-threatening situations (e.g., radiation in nuclear power plant accidents), dangerously high temperatures or pressures found in rescue operations, or highly unstructured and even random environments (e.g., cave-ins in mining operations). Although in principle actual hardware design can help to reduce the threat of some of these conditions to the robot itself, it must be recognized that these machines are vulnerable also, albeit in different ways than people.

In order to achieve the goal of robots supplanting people in hazardous situations, techniques must be developed to cope with the issues of machine survivability. Due to the nature of these environments and the desire to have relatively general purpose machines, novel approaches are required. Certainly traditional techniques involving redundant hardware and control have their place, but when viewed by themselves they are insufficient. For truly autonomous and intelligent machines, new methods for survivable systems are required in addition to the old.

What are these methods? We have been exploring multiple techniques in the area of robot survivability and this paper surveys them. They are complementary in nature; each dealing with a different aspect of the survivability issue. Reactive control (Sec. 2) enables a mobile robotic system to cope with unpredicted events in the world. In almost every situation, things are not exactly as expected. This is even more the case in a hazardous environment. Reaction to unexpected occurrences is crucial to both ensure the safety of a robotic vehicle and the successful completion of its mission.

In Section 3, homeostatic control as applied to robotic systems is described. With this approach, feedback regarding the robot's internal state and its available resources enables the robot to dynamically modify its behavior based upon its self-perception. Adaptive behavior that reflects the robot's own limitations is the product.

Specialized intelligent fault-tolerant control techniques, described in Section 4, cope with possible hardware failure in a robotic system. By using hybrid classical control and artificial intelligence techniques, real-time fault detection, isolation, propagation, and reconfiguration can be conducted. This assures that the robot will be

[1]This research is supported in part by the Office of Naval Research under Grant #N00014-89-J3113.

able, in most circumstances, to extricate itself from a threatening situation even after suffering limited hardware failure.

The last technique (Sec. 5) involves the use of multiple autonomous agents, each relatively simple, working together to achieve a task. This kind of agent redundancy, in contrast to subsystem or component redundancy that might be found in a complex robot, can treat some of the robots working together as expendable. A similar kind of redundancy is paralleled in nature and is found in colonies of insects such as bees or ants. This holistic or sociological approach enables a group of agents to work together, sacrificing the survival of any single agent to assist in the survival of the whole.

Space limitations preclude the use of figures for illustrating the experimental and simulation results available for these techniques. The reader is referred to the references within the sections following for these studies.

2. Reactive Control for Unstructured Worlds

A relatively recent development in robotic control involves the appearance of a new methodology for navigation. Reactive control is a technique intended largely to supplant traditional hierarchical planning mechanisms. Hierarchical planners (see [1] for a more complete exposition) have difficulty, in general, with partially modeled and unstructured domains. Reactive navigational techniques, on the other hand rely heavily on the immediacy of sensor data to continuously adapt the behavior of a robot to its surroundings. There is little or no reliance on abstract world representations and behaviors rather than plans are the means by which the robot interacts with the world. Representative reactive control schemes include Brooks' subsumption architecture [6], Kaelbling's Rex [10], and Payton's work[11]. Much of our research in mobile robot navigation[4] also fits into this category.

In the Autonomous Robot Architecture (AuRA) motor schemas provide the reactive component of navigation. Instead of planning by predetermining an exact route through the world and then trying to coerce the robot to follow it, motor schemas (behaviors) are selected and instantiated in a manner that will enable the robot to interact successfully with unexpected events while still striving to satisfy its higher level goals. Motor schemas are manifested as potential fields. Multiple active schemas are usually present each producing a velocity vector driving the robot in response to its perceptual stimulus. The robot only needs to compute the single vector at its current location. The individual schemas which are concurrently active, each post their contribution to the robot's motion at a centralized location. The individual schemas are each summed together and normalized to fit within the limits of the robot vehicle to yield a single combined velocity for the robot to move in. These vectors are continually updated as new perceptual information arrives with the result being immediate response to any new sensory data.

Potential fields are prone to problems with local maxima, local minima, and cyclic behavior. AuRA's navigational system takes this into account by detecting failure of schema-based navigation when it occurs (which is infrequent) and using a hierarchical planner to either reformulate the set of behaviors to be used, plan an alternate route, or perhaps even reformulate or abandon the mission.

This navigational methodology has been tested successfully in both simulation and experimentally on our mobile robot [4]. This methodology has also been tested in simulation for navigation over rough topography [2] and in three dimensional navigational situations as might be encountered in undersea or aerospace domains [5].

The advantages of this form of navigation are many. These include rapid computation and the ability to be mapped onto parallel architectures making real-time response easily attainable. Modular construction affords ease of integration of new motor behaviors simplifying both system maintenance and the ease of transfer to new problem domains. Motor schemas readily reflect uncertainty in perception, when such a measure is available, and also react immediately to environmental sensor data.

These factors all contribute to the needs of a navigational system that assist in the survival of a robot.

3. Dynamic Replanning Based on Internal Sensing

For survivable systems, environmental perception alone is inadequate. The robot must also be able to analyze its own internal state and use that information in a manner that is reflected in its outward behavior. Factors such as internal temperature, available fuel and rate of fuel consumption, pressure, and other considerations should all have a bearing on the actions of the robot. Additionally this analysis should be performed in a manner that allows for smooth, "unconscious" modification of these behaviors. AuRA's homeostatic control subsystem [3] provides the basis for the integration of dynamic behavioral modification based on internal sensing into the reactive control system described above.

Little work by the research community as a whole has been performed on dynamic planning based on the robot's available resources and current condition. Some research using worst case analysis based on available fuel reserves or other factors has been conducted, but dynamic replanning during plan execution is an area that is, unfortunately, largely unattended. The difficulty with this form of plan analysis is that worst case scenarios must be used to ensure the robot's survival. This precludes the execution of many missions that actually would be feasible. Our research on homeostatic control as a mechanism for coping with dynamic replanning in hazardous environments addresses this issue directly. Homeostatic ocntrol is concerned with the maintenance of a safe internal environment for the robot.

Our extension of the schema methodology described in Section 2 into homeostatic control permits dynamic replanning based on internal sensing to occur. Signal schemas have been defined in two forms: a transmitter schema associated with an internal sensor and broadcasting information throughout the system; and receptor schemas embedded within the motor schemas and modifying their internal parameters in response to messages received from the transmitter schemas.

This model is inspired by the control and communication mechanisms found in the mammalian endocrine system [3]. Summarizing the basic features: (1) A non-hierarchical broadcast communication mechanism is exploited; (2)Controller targetting occurs by the presence of receptor schemas embedded with in a motor schema; (3) Negative feedback is the basic control regime; (4) Only data regarding the maintenance of a safe and consistent internal environment are managed by the robot; and (5) Motor process rates are affected by the data generated and received by the signal schemas.

We have demonstrated in simulation [3] how this type of information management can affect not only the velocity of the robot but also the route which it takes. In these simulations, as the robot's fuel begins to deplete, it takes shorter but slower routes, eventually taking an entirely different path than it would otherwise have if given unlimited fuel capacity.

4. Hybrid Fault-tolerant Control

Intelligent robotic task control strategies must be demonstrated through autonomy and versatility. Attributes of autonomy and versatility for complex tasks imply survivability of the machines, sensors, and robots used to execute the tasks. Survivability, in this context, means effective control of the robot system which requires the continuous process of normally functioning sensors and actuators. It is important, therefore, to investigate the development of sensor algorithms and analytical tools that are capable of detecting errors or failure conditions, provide appropriate alarms, and possibly allow the system to operate in a degraded mode without endangering the task objectives.

The complexity and sophistication of autonomous robot systems require new robust and reliable automated techniques in order to detect failures, pinpoint the location and identity of the faulty component or actuator, and eventually restructure the system and reconfigure the controls so that the operational integrity of the system is maintained for the duration of the emergency.

We address this need by introducing a problem-focused system integration philosophy that capitalizes upon a blend of numerical and symbolic manipulations, thus combining concepts of control theory and artificial intelligence. AI techniques allow a marked reduction of the set of hypotheses, thus permitting traditional control-theoretic approaches to be applied efficiently.

The underlying factor of this methodology capitalizes upon structural features of the system. The fault-tolerant design procedure entails the following tasks: A method for fault diagnosis is introduced which is based upon a combination of signal redundancy and fuzzy logic. When a fault is detected, casual or qualitative reasoning is used to model its propagation to adjacent subsystems. The system is restructured next by isolating faulty components. Finally, a design method is suggested leading to a structural control law that reconfigures the original system controller in order to meet the primary objective of guaranteed stability while the system is operating in a degraded mode.

We focus attention in this paper on the fault diagnostics approach. It consists of: (1) a triggering mechanism to initiate the diagnostic algorithm; (2) a routine for determining a validate vector of direct and analytical measurements; (3) a fault detection algorithm based on parity space consideration; (4) a means of identifying the faulty component; (5) an error trending routine; (6) an algorithm for isolating the failed component; and (7) an estimate of the "safest" values of the system variables.

It is of interest that the fault diagnostics problem be initiated only when some evidence is present which implies the possible failure of a system sensor or component. Such triggering of the fault detection routines results in substantial computational savings and improves the reliability of the algorithm by reducing the frequency of false alarms. Sensor validation uses signal redundancy in a modified version of the parity space algorithm for signal validation developed by Draper Laboratories [8]. The validation procedure compares each reading to all other like readings. The "best" estimate from good measurements is defined as that value which gives the minimum of a particular function of the measurements.

Fault detection and identification uses validated sensor and analytic data to analyze the status of the system components. Fault diagnosis involves the incremental accumulation of evidence in the form of symptoms to determine the "health" of the system components. Parity space representation and analytic redundancy [7,8,12,13] in addition to limit checking, are applied to accumulate the required redundant measurements into a new array, called a parity vector, in such a way that the true value of the underlying variable is suppressed, leaving only components of the measurement errors as elements of the parity vector [8]. In parity space, the parity vector remains near zero when the redundant measurements are consistent, i.e., when no faults are present. When a fault occurs, the parity vector grows in magnitude, and its direction of growth is uniquely associated with the faulty measurement. The fault detection part of the algorithm capitalizes upon the behavior of the magnitude of the parity vector to detect the presence of a fault.

A rule base and inference scheme are intended to provide a quantitative measure of the severity of the fault depending upon prevailing system conditions. Uncertainty is accounted for in terms of fuzzy membership functions for the fault symptoms.

The fault identification part of the program is intended to identify which component is faulty. Following [8], we forward a technique based upon the concurrent checking of the relative consistency of smaller size subsets of projection of the parity vector (generated from all measurements) on a reduced dimensionality subspace spanned by the failure directions. The parity vector is directly generated from the measurements. The consistency of large subsets of measurements can be determined in terms of the consistency of all possible smaller subsets of two measurements. For m measurements, we form t subsets, $S_1, S_2, \ldots, S_t$, of (n+1) = 2 measurements each with all possible combinations of m_i, i=1,... For each subset S_i, we calculate a corresponding reduced parity vector P_i. For each subset S_i, a threshold value Θ_i is computed from the corresponding component normal error bounds. Finally, P_i is compared with Θ_i to determine whether the component associates with V_i measurement axis has failed or not. This procedure allows not only the identification of a faulty sensor or component but also the estimation of the "best" value for the corresponding variable from measurements which remain fairly consistent. Component fault evidence is accumulated by comparing each P_i with the corresponding Θ_i value. The rule base and inference scheme are constructed, for the fault identification algorithm, in a similar manner as for the fault detection case.

Further evidence for component or sensor failures can be obtained from stored error histograms. Suppose that after each measurement sampling period, the error between an individual sensor and the estimate from the collection of sensors is calculated, and that the last "k" of these errors is stored. A fuzzy variable may be defined as the average of these k values scaled by the error bound, providing an indication of bias. Drift may be indicated by another fuzzy variable. Although detailed statistical analysis could be used to calculate the error trending variable, good performance is achieved, in a simulated environment, by simply taking the difference between the averages of the last k/2 error values and the first k/2 error values, and again scaling by the error bound.

The fault isolation algorithm is provided with information about the fault type and the severity of the fault, and it decides as to the action to be taken to isolate the faulty component for repair and maintenance purposes. It issues a status report and appropriate commands to the maintenance module for further action. Corrective measures can be taken by transmitting the fault isolation report to the scheduling and maintenance modules.

The system is restructured next by isolating faulty or potentially faulty components. Finally, a design method is suggested leading to a structural control law that reconfigures the original controller in order to meet the primary objective of guaranteed stability while the system is operating in a degraded mode.

5. Redundant Robots

Our work in redundant robots is in the early stages of research. This approach involves the use of multiple autonomous agents, each of which configures itself in relation to the other robotic agents using schema-based techniques [4]. This coordination/cooperation allows complex robots to be constructed from simpler machines. Fukuda [9] has shown some interesting preliminary results where multiple robots combine into a chain.

The advantages of these techniques are many. These include the ability to replace failed robots from a stockpile of other units. Also, instead of building a single somewhat specialized machine to accomplish one type of task, multiple simple units could be combined to yield several such complex agents.

The problems associated with the development of such a system are formidable, requiring a significant basic research effort. Effective protocols must be developed to allow communication between these agents in an efficient manner. Peer-peer interaction is preferred over a centralized host issuing commands to slave units. This minimizes the risk of failure and maximizes survivability. The communication issue must be studied in two phases: during the link-up of the basic units and also during the operation of the more complex interacting multi-agent structure.

We are currently studying communication and cooperation of multiple agents in the domain of "space tugs": robots whose goal is to move objects that are considerably larger than themselves with the help of other similar robots. Studying the behavior and communication protocols of certain insects (e.g., ants) and humans (tug-boat captain - pilot interactions during the docking of ships) provides a basis for the communication schemes tested in this work. Additional reports on this research will appear in the near future.

6. Summary

In order to achieve the goal of placing robots in environments that are hazardous to humans, we must place them in situations that often threaten their own survival. Several different techniques have been developed in this paper describing how a robot can cope with the many problems it can face in such a world.

Reactive control using schema-based navigation provides rapid response to a changing world. This allows for reflex-like reaction to potentially dangerous events. An extension of this technique, utilizing signal schemas, provides for a link between the internal condition of the robot and its behavioral responses. This eliminates the need for worst-case analysis and enables the robot to complete missions and extricate itself from dangerous situations that it might not be able to handle otherwise.

Hardware failures are handled in two ways. The first involves a hybrid artificial intelligence - classical control technique that permits the real-time reconfiguration of a failed subsystem. Although not necessarily guaranteeing an optimal control configuration, a sub-optimal control mode developed rapidly is preferable by far to no control at all. Finally, by using multiple redundant cooperating agents, individual hardware failures can be swapped out in their entirety automatically, potentially with no human intervention. A major challenge involves developing effective non-hierarchical communication mechanisms that still afford intelligent cooperative behavior.

REFERENCES

[1] Arkin, R.C., "Towards the Unification of Navigational Planning and Reactive Control", *AAAI 1989 Spring Symposium on Robot Navigation*, Stanford Univ., CA., March 1989.

[2] Arkin, R.C. and Gardner, W., "Reactive Inclinometer-Based Mobile Robot Navigation", to appear in *Proc. 1990 IEEE Int. Conf. on Robotics and Automation*, Cincinnati, OH, 1990.

[3] Arkin, R.C., "Dynamic Replanning for a Mobile Robot Based on Internal Sensing", *Proc. 1989 IEEE Int. Conf. on Robotics and Automation*, Scottsdale, Arizona, pp. 1416-1421, May 1989.

[4] Arkin, R.C., "Motor Schema-Based Mobile Robot Navigation", *International Journal of Robotics Research*, Vol. 8, No. 4, August 1989, pp. 92-112.

[5] Arkin, R.C., "Three Dimensional Motor Schema Based Navigation", *Proceedings of the NASA Conference on Space Telerobotics*, Pasadena, CA, Jan. 1989.

[6] Brooks, R., "A Robust Layered Control System for a Mobile Robot", *IEEE Journal of Robotics and Automation*, Vol. RA-2, No. 1, pp. 14-23, 1986.

[7] Desai, M. and Ray, A., "A Fault Detection and Isolation Methodology", *Proc. of 20th IEEE Conf. on Decision and Control*, pp. 1363-1369, San Diego, CA, December 1981.

[8] Fraser, D.C., et al., "First Quarterly Progress Report", *Contract NAS9-4065, Task Order 41*, Charles Stark Draper Lab, MIT, Cambridge, MA, August 30, 1971.

[9] Fukuda, T., Nakagawa, S., Kawauchi, Y., and Buss, M., "Structure Decision Method for Self-Organising Robots Based on Cell Structures - CEBOT", *Proc. 1989 IEEE International Conf. on Robotics and Automation*, Scottsdale, AZ, May 1989, pp. 695-700.

[10] Kaelbling, L., "An Architecture for Intelligent Reactive Systems", *SRI Technical Note 400*, SRI International, Oct. 1986.

[11] Payton, D., "An Architecture for Reflexive Autonomous Vehicle Control", *IEEE Conf. on Robotics and Auto.*, pp. 1838-1845, 1986.

[12] Potter, J.E. and Suman, M. C., "Thresholdless Redundancy Management with Arrays of Skewed Instruments", *NATA AGARDODograph 224*, April 1977.

[13] Willsky, A.S., "A Survey of Failure Detection Design Methods in Dynamic Systems", *Automatica*, vol. 12, pp. 601-611, 1976.

PERFORMANCE OF MULTIPLE TASKS BY AN AUTONOMOUS ROBOT USING VISUAL AND ULTRASOUND SENSING

M. BECKERMAN, D. L. BARNETT, M. DICKENS and C. R. WEISBIN†
Oak Ridge National Laboratory, Center for Engineering Systems Advanced Research, Building 6025, MS-6364, P.O. Box 2008, Oak Ridge, TN 37831-6364

ABSTRACT

While there have been many successful mobile robot experiments, only a few papers have addressed issues pertaining to the range of applicability, or robustness, of robotic systems. The purpose of this paper is to report results of a series of benchmark experiments done to determine and quantify the robustness of an integrated hardware and software system of a mobile robot.

INTRODUCTION

HERMIES-IIB is one of a series of mobile robots developed at the ORNL Center for Engineering Systems Advanced Research (CESAR) [1]. The hardware for this robot includes a mobile platform, a sensor turret equipped with an array of twenty-four ultrasonic range sensors and three CCD cameras, and dual manipulator arms. It also includes a 16-node hypercube computer and an AT host computer.

The integrated system, to be described herein, was designed to perform a number of tasks sequentially. First, the robot had to use its ultrasound sensors to navigate through an obstacles strewn field to a preselected intermediate goal. Next, the robot had to use its CCD cameras to identify a control panel. It then had to navigate up to and dock in front of that control panel. In the last phase, the robot had to read analog meters and lamp conditions, and guided by the vision system, manipulate buttons and slides on the control panel.

By robustness we mean stability against sensor and mechanical errors, uncertainties and limitations for a specified range of environmental conditions. To determine the robustness of the integrated system we examined the robustness of each component subsystem. This was done in a series of benchmark experiments which delineated the environmental conditions under which successful operation could be achieved.

We begin by briefly describing the subsystems and connected tasks, noting the inputs and outputs from each component. We then turn to the experiments and results, summarizing the environmental conditions, problem areas and issues.

† Currently at Jet Propulsion Laboratory, Robotics and Automation Section, 4800 Oak Grove Drive, Pasadena, CA 91109-8099.

SUBSYSTEMS AND CONNECTED TASKS

Ultrasound Data Processing

The HERMIES-IIB ultrasound sensors provided 120 samples of range data covering 360 degrees in three degree steps. These data represented the distance to the nearest object intercepted by the beam. In processing the data, maps were not used. Instead, for each scan lists of paired and unpaired edges were constructed and then used to generate a list of corridors of free space.

Path Planning

The design of a path to a goal had to take into account the incompleteness of the information obtained from any single 360 degree ultrasound scan. Part of the space will lie beyond the effective horizon while other regions will be occluded by objects. Since the robot can only navigate through space known to be clear, the HERMIES-IIB paths were built in several stages using a local path planner. At each stage a new sonar scan was taken, and the next portion of the path was planned.

Panel Identification

The control panel identified by the robot contained a pair of analog meters, a danger light, four buttons and lamps, and two slides. At large distances the analog meters were the only features other than the outline of the panel itself which could be discerned. The function of the vision software during navigation was to identify the panel, and provide distance and angle information for navigation to the docking zone in front of the control panel (Fig. 1).

Fig. 1. HERMIES-IIB at the docking position, pushing a button.

The CCD cameras mounted on the HERMIES-IIB robot provided 256 x 256 pixels of 8-bit grey-level intensity values from a 60 degree field-of-view. Concurrent processing of the visual data was done on-board the robot using the 16-node hypercube computer. In the low-level parallel processing the grey-scale image was converted to a binary image. A list of the labelled, connected regions in the binary image was then constructed. Two lists were produced, one for black regions and one for white regions. The algorithms for doing this have been described in detail elsewhere [2]. Geometric properties and low-order moments which were then extracted for each region.

In order to identify the control panel with its two meters, the list of connected regions in the image was pruned until only one candidate remained. The entry in the list sought was a dark region (control panel) containing two light regions (meters), situated side-by-side in the upper part of the dark region. A typical pruning procedure is illustrated in Figs. 2-5.

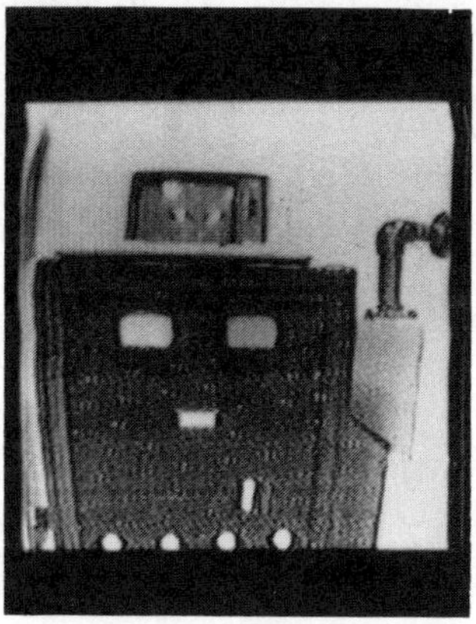

Fig. 2. Visual image of the control panel from a HERMIES-IIB CCD camera.

Fig. 3. Early intermediate processing stage: note that a shadow on the right and part of the object on the left are merged with panel.

Fig. 4. Later processing stage: the many features which have been pruned are shown as black areas.

Fig. 5. Final image: the only remaining features are the large connected region (panel) containing two smaller regions (meters) side-by-side.

Docking

Estimates of distance, angle and yaw were made by the robot using the geometric information about the two control panel meters, as extracted from the visual image. Pixel dimensions were converted to spatial dimensions using prior information on the physical dimensions of the meters and the optics of the detection system.

Below 10 degrees, yaw estimates were unreliable. As a result of the inaccuracies in determining yaw at small angles, estimates of this quantity were not used for docking. Navigation to the docking zone was done in several (3-5) steps. In each step, a picture was taken, and a new distance and angle were extracted. The last image was acquired from 3-4 ft., roughly in front of the panel, where highly accurate distance measures could be obtained by the robot.

Vision-Guided Manipulation

The manipulation tasks consisted of pushing one or more of the four buttons, and moving either one or both of the slides. There were two main problems associated with reliable operation of the Heathkit Hero arms used for manipulation, namely: (i) there were no absolute encoders, and (ii) there was considerable backlash in the shoulder and wrist joints. The magnitude of the backlash was reduced by mechanically loading the arms. However, the remaining backlash was still large enough to lead to failures in manipulation unless controlled in some manner. The vision system functioned within a closed-feedback loop to first locate the various control panel features and then guide the movement of the manipulator.

A three-step procedure was used to position the end-effector at the goal location. In each step two CCD cameras mounted on the robot chassis sensed the location of the buttons and slides, and (in the same scene) the position of a light-emitting-diode (LED) mounted near the tip of the two-jaw grippers of the manipulator. Two pictures were taken with each camera, one with the LED on and one with the LED off. The two images for each camera were subtracted from one another to produce images containing only the dot representing the location of the LED. A stereo algorithm [3] was then used to obtain the (x,y,z) coordinates of the end-effector and the features to be manipulated.

EXPERIMENTS AND RESULTS

Environment

1. Ultrasound: Four actions were taken to reduce the effects of ultrasound distortions and specular reflections. First, the boxes which served as obstacles in the ultrasound-guided navigation phase were covered with a material to reduce the specular reflections. Second, a 4.5 m sensing horizon was imposed. Third, there was always at least one corridor whose actual width was equal to or greater than 1.5 m. This was done to take into account both the finite size of the robot (0.9 m) and the distance-dependent distortion errors. Fourth, corridors near the hard, smooth (specular reflecting) walls of the lab were excluded from consideration.
2. Visual: Displayed in Fig. 6 is a photo of a typical setup for the vision-guided phases of the experiments. The control panel is the dark object placed against the wall, as seen in the center of the background of the photo. A monitor plus keyboard has been placed on top of the control

panel, and there is clutter to the left and the right. The photo was taken using floodlights. In the experiments only standard overhead fluorescent lighting was used. Some of the fluorescent lamps were situated near air conditioning ducts, and the average illumination varied considerably from place to place in the lab.

3. Manipulation: The length of each slide canal of the control panel was 193 mm. This length was partitioned into low, medium and high segments for the manipulation tasks. The depth accuracy required for the slide operations was given by the 5 cm interior opening of the gripper; the height of the slides was 5 cm. The button radius was 2 cm, and the allowed variation in depth was 1.5 cm about the mean value needed to activate the appropriate lamp. The buttons and levers were made of plastic, light enough to be manipulable, and no torque was applied to them by the manipulators.
4. Navigation: One to four boxes, tall enough to intercept the sonar viewing plane were used for testing obstacle avoidance. Navigational devices such as beacons were not added to the environment. Instead, the robot moved over the smooth, level floor of the laboratory using dead reckoning. The control panel yaw, however, was fixed for panel identification and docking.

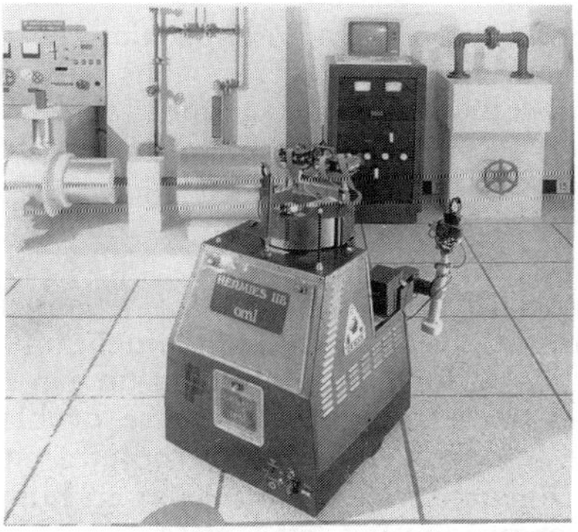

Fig. 6. Photo of the experimental setup used in the vision-guided navigation stages of the experiments.

Experimental Results

In order to perform the manipulation tasks the robot had to dock within a narrow zone centered in front of the control panel. This zone was determined experimentally to be three inches deep, 19 to 22 inches from the panel, and two inches wide. For both manipulators to operate the control panel the yaw had to be less than 2 degrees.

Distance errors, due to the intrinsic resolution of the CCD arrays and discretization errors of the digitizers, decreased from 10 to 20% at 4 m to 1% at 1 m. By using an incremental algorithm the robot was able to dock successfully from any transition point within the field-of-view of the control panel over distances from 4.5 m to 0.5 m.

At a distance of 60 cm the absolute accuracy of the stereo algorithm used for manipulation was approximately 50 mm with a repeatability of 5 mm. Although the absolute accuracy was marginal, good relative accuracy was achieved by locating the LED and panel feature in the same image. This relative accuracy was ultilized in the three-step positioning.

In the most recent series of experiments the robot was successful in carrying out all its tasks in 14 out of 20 trials. Three failures were attributable to one wheel sticking slightly to produce a 5 degree yaw at docking. Two of the three other failures were manipulation failures; in each case one of the buttons was not pushed in far enough to turn on the lamp. The remaining failure was due to an error in reading the condition of the control panel. These failures were typical of those encountered in other series of experiments done by us over a six month period. In the absence of mechanical problems the robot's success rate for the series of tasks was high, exceeding 90%.

The panel identification algorithm was able to recognize situations where the control panel was not in the field-of-view of the CCD camera. It was also able to handle lighting situations which produced panel shapes such as those illustrated in Figs. 2-5. However, panel identification problems were encountered when glare was produced from highly reflective metal surfaces. Other identification problems were produced by large shadows which were merged with the control panel. To achieve the above-mentioned high success rate some restrictions were imposed upon glare and shadows, and upon background clutter.

Similarly, ultrasound distortions and specular reflections led to failures in path planning for some obstacle configurations such as "picket fences" and "box canyons." In those instances the planner was unable to find either valid edges or free corridors.

CONCLUDING REMARKS

We have attempted to take into account sensor and mechanical errors, uncertainties and limitations in our development of the integrated system described above. The performance of tasks in several steps using sensor feedback in each step overcame many of these problems, and success rates in excess of 90% were achieved. It should be noted that, in large measure, the mechanical problems encountered pertain to an in-house-built prototype, and not to a commercial or production-level system.

Situations were encountered in interpreting the processed binary images and ultrasound data, where there was a lack of sufficient information for the robot to make intelligent decisions. In those situations we need to make better use of the available resources. In work being done in parallel to the present study [4,5] we are investigating ways to combine data from several scans, views, and sensor domains, and to extract more information from the data available.

We have described the factors in the experimental setup which could vary rather freely. With respect to these factors the performance of the robot was robust. We have also indicated those factors which were controlled to some extent and we have described the information about the environment built into the software. From this second viewpoint the performance was not as robust as we would wish. We are now attempting to develop the means to reduce these dependences.

ACKNOWLEDGMENTS

This research was supported by the Engineering Research Program at the Office of Basic Energy Sciences, of the U.S. Department of Energy, under

Contract No. DE-AC05-84OR21400 with Martin Marietta Energy Systems, Inc. The authors wish to thank S. M. Killough and J. P. Jones for helpful discussions.

REFERENCES

1. B. L. Burks, G. de Saussure, C. R. Weisbin, J. P. Jones and W. R. Hamel, "Autonomous Navigation, Exploration and Recognition Using the HERMIES-IIB Robot," IEEE Expert, Vol. 2, pp. 18-27, 1987.

2. J. P. Jones, "A Concurrent On-Board Vision System for a Mobile Robot," in Proceedings of the Third Conference on Hypercube Concurrent Computers and Applications, G. Fox (ed.), Vol. II, pp. 1022-1032, 1988.

3. S. M. Killough and W. R. Hamel, "Sensor Capabilities for the HERMIES Experimental Robot," Proceedings of the ANS Third Topical Meeting on Robotics and Remote Systems, Charleston, SC, Sec. 4.1, pp. 1-7, 1989.

4. M. Beckerman and E. M. Oblow, "Treatment of Systematic Errors in the Processing of Wide Angle Sonar Sensor Data for Robotic Navigation," IEEE Trans. on Robotics and Automation, Vol. RA-6, April, 1990.

5. M. Beckerman, L. A. Farkas and S. E. Johnston, "Treatment of Systematic Errors II: Fusion of Ultrasonic and Visual Sensor Data," Oak Ridge National Laboratory Technical Memo ORNL/TM-11349 (CESAR -89/38), (in preparation).

An Integrated FRHC-PUMA-EE Teleautonomous System

Edwin P. Kan
Jet Propulsion Laboratory
California Institute of Technology, Pasadena, Ca., USA

Jin Lee* and Michael Junod
Advanced Technology Laboratory
General Electric Aerospace Co., Moorestown, N.J., USA

ABSTRACT

The system discussed in this paper is an integrated stand-alone system with the full functional capabilities required of a telerobot system. It is complete with a force-reflecting 6-DOF hand controller, driving a PUMA 560 or 762 robot, with an integrated force-torque sensing wrist sensor and servo-driven parallel jaw gripper. A mix of custom and standard electronics, distributed computers and microprocessors, with embedded and downloadable software, have been integrated into the system, giving rise to a powerful and flexible teleautonomous control system.

INTRODUCTION

The telerobot system discussed in this paper was developed at the Jet Propulsion Laboratory (JPL), with a major integration and software subcontract support from the GE Advanced Technology Laboratory. It was developed for the Goddard Space Flight Center (GSFC) to respond to the needs for an integrated laboratory system that could be used as a baseline to compare with other telerobot systems available from the market or research community.

This system is a teleautonomous system, capable of force reflecting pure teleoperation at one end of the spectrum, and capable of fully autonomous force compliant control on the other end of the spectrum. A linear blend of the two ends of the spectrum can produce a powerful teleautonomous control system.

Being a research system in an ever evolving technological environment, and also bound by certain project constraints, this system comprises electronics and software technology elements that were developed independently. In some ways, this system is a hybrid between the JPL Advanced Teleoperation System [1] and the JPL Telerobot Testbed Manipulation Subsystem [2]. Its architecture can also be unified into a more homogeneous configuration such as in the Kali system [3]. In fact, the initial project intent was that this system development be followed by an upward compatible conversion to the JPL Telerobot System and/or the Kali system as they both mature in CY90.

The major components of this integrated system include: (1) FRHC - a six-axes force reflecting hand controller; (2) UC - a 'Universal Controller' for the low-level servo control of the FRHC and the hand trigger in the FRHC; (3) VME high level control electronics - for driving the UC and interfacing with the robot electronics; (4) RCCL - control software on a μVAX for the robot high level

* Jin Lee is now with the Pohang Institute of Science and Technology, Pohang, Kyungbuk, Korea.

controls; (5) Unimate Controller - the low level servo controller for the PUMA 560 or 762 robot; (6) PUMA 560 or 762 robot; (7) a 'GSEE' - a sensor-based end effector, which has wrist force-torque sensing and force-controlled finger gripping; (8) graphics generator - displaying real-time force-torque information; and (9) a user interface - a multi-window 3-tier menu system on a SUN 3/60 workstation.

Figure 1 shows the system configuration. In the foreground, from left to right are the Sun terminal (with 5 windows displayed), a color graphics display of force-torque and GSEE data, 2 video monitors showing camera views of the worksite, the UC driving the FRHC, the FRHC (right-handed), an IBM PC interfacing with the UC, and the VME electronics chassis. In the background from right to left (some items not visible) are the Unimate controllers for the PUMA 560 and 762, the μVAX driving the robots, a PUMA 762 robot with the GSEE, and a PUMA 560 robot with a pneumatic gripper.

The overall capabilites of this integrated system can be summarized as follows:

- o It is a telerobot system capable of force reflecting teleoperation and shared/teleautonomous control (i.e. teleop mixed with auto force compliant control).
- o The teleautonomous capability allows gain mixing per axis; gain (k) for teleop and gain (1-k) for auto control.
- o The man-machine interface is a 3-tier menu system, for modes and parameters selection.
- o It has a multi-window display on a SUN 3/60 workstation.
- o Teleoperation modes include: position (cartesian) mode in tool or mixed world-tool coordinates; rate (cartesian) mode; force reflection on/off mode; index on/off mode.
- o Control scaling includes: position, rate and force scaling; force-torque biasing.
- o Color graphics provide a display of force-torque data, the end-effector (GSEE) grip force, GSEE grip opening, GSEE internal temperature; graphics software also permits axes transformation and scaling.
- o GSEE gripper control is integrated with the FRHC hand trigger; a hold button on the hand trigger commands the GSEE to hold position.
- o GSEE has position control mode, rate mode, and force mode; a combined position/rate/force mode allows the closing of the gripper until object contact, when force control is effected.

For further details on the specifications of the GSEE and of the FRHC, refer to [4,5].

SYSTEM DESCRIPTION

Figure 2 shows the overview hardware block diagram of the system. For detailed board-level hardware configuration, data flow diagram and software flow diagrams, refer to [6].

The UC (Universal Controller, also known as UMC, Universal Motion Controller in [1]) provides the control and direct interface with the FRHC. It contains two NS 32016 processors, two joint interface cards, power amplifiers, and one XYCOM parallel interface card. One NS 32016 processor (also named UC processor) is in charge of executing servo algorithms and internal user interface, whereas the other NS 32016 processor (also known as the UC COMM processor) is in charge of communication with the VME chassis. In position mode, the UC receives six joint encoder postions and other analog readings from the FRHC. The UC processes and transmits this data to the VME processors via the XYCOM parallel interface. In the return loop, the UC receives the joint setpoint (encoder counts) from the VME processors, and drives the FRHC.

An IBM PC provides the development environment and downloading facilities to the UC processors. Software for the two UC processors are written in assembly language [1].

The VME chassis consists of three Motorola 68020 processor cards, one etheRnet card, one XYCOM parallel interface card, one BIT3 shared memory card, and a set of two Parallax graphics cards. The main 68020 processor card, known as the HACS processor card, is mainly responsible for computing the forward kinematics of the FRHC, hence the cartesian incremental positions/orientations for the PUMA robot. This same card also converts the robot force/torque sensor data to pseudo robot incremental setpoints, and computes the inverse kinematics of the FRHC to produce the desired joint positions for creating force reflection in the FRHC. The second 68020 processor card, known as the Interpolator I/O card (also called the UC/COMM card or FRHC/moper card), directs I/O with the UC via the XYCOM card, and computes the interpolated joint encoder setpoints for the FRHC. The third 68020 processor card, known as Graphics processor card, is in charge of the graphics display, driving the two Parallax graphics cards. The BIT3 shared memory card provides the communication medium among the VME cards and the μVAX on the robot side.

The data sharing and computation in the VME chassis are performed in the following manner. The Interpolator I/O processor receives the FRHC encoder values and analog readings (digitized in the UC) from the UC via the XYCOM interface and deposits them on the BIT3 shared memory. Then the HACS processor reads them from the shared memory, and computes the corresponding cartesian incremental transformations to derive the PUMA robot setpoints. In the return loop, the HACS processor reads the force-torque sensor readings on the shared memory, as deposited by the RCCL μVAX. A force gain matrix transformation provides the incremental psuedo PUMA setpoints, corresponding to the force/torque "position accomodation" errors. The HACS then computes the FRHC inverse kinematics to obtain the corresponding desired FRHC joint positions. The Interpolator I/O processor then processes and interpolates these joint encoder setpoints, and returns them to the UC, backdriving the FRHC.

A Sun 3/60 computer running on the VX Works commercial operating system provides the development environment for the VME processors. All programs were written in "C" and downloaded to the processors either via the ethernet or via serial ports.

For the robot control software execution and development, a μVAX running on the UNIX operating system and installed with RCCL [7] is used. This μVAX consists of one main CPU and one slave CPU. The main CPU provides the UNIX program development environment and the user interface to the operator/programmer. The slave CPU executes the real time portion of the RCCL code on a modified UNIX kernel. The interface with the VME chassis (and hence FRHC) is through the BIT3 shared memory. The interface with the PUMA Unimate controller, specifically the LSI-11 processor in the controller, is through a DRV-11 parallel interface. Every cycle (normally 20 milliseconds), the slave CPU interrupts the VME system and gets the cartesian setpoint for the PUMA robot. It also computes the inverse kinematics, converting the cartesian setpoint to the joint setpoints, then sending them over to the LSI-11 PUMA processor via the DRV-11 interface. In the return loop, the slave CPU receives the force/torque data from the LSI-11 processor and deposits them on the shared memory to be retrieved by the VME processors. Notice that no forward kinematics of the robot is performed on the μVAX, because the design does not specify force reflection due to robot position lag from the FRHC commanded position. That is, the present design derives the FRHC force reflection purely from the robot wrist sensor.

The LSI-11 processor in the PUMA controller performs I/O functions for the PUMA robot. It reads joint setpoints from RCCL/μVAX and sends them over to the lower level 6503 processors. Conversely, it reads the joint encoder positions of the PUMA's six joints and send them over to the μVAX. The LSI-11 also collects the force/torque data from either the GSEE (which has its own integrated force/torque sensor) on the PUMA 762, or from the Lord force/torque sensor on the PUMA 560. The data is shipped to the μVAX. Finally, the low level 6503 processors provide the direct interface with and control over the individual robot joints. They perform joint interpolation, run the servo algorithms, and drive the robot joints. They also process the data coming from the PUMA encoders, compute the encoder values and make them available for the LSI-11 processor to read.

SYSTEM OPERATION AND PERFORMANCE

This teleautonomous system is designed to be a turn-key system, to be used by application users who are not system programmers nor development programmers. Facilities are, however, very flexible and available to programmers for software modifications. All software for the VME processors can be edited, compiled and downloaded from the SUN workstation operating under VX Works™ development environment.

The user interface is a multi-window menu display on the SUN workstation (see Figure 1). These windows include the (a) FRHC_MAIN (also known as the HACS, HAnd Controller Software) user window, (b) RCCL_MAIN user window, (c) VME_UC_MAIN user window, and (d) Graphics window. Two other windows are the editing window and the console window (for system messages).

The user windows allow the user to change operation modes and specify parameters and configuration. Examples include the choice of position versus rate mode, manual non-force reflecting control versus shared control versus teleautonomous control. Force/position/loop gains, tool transformation parameters, end-effector grip force, rate, etc. are all selectable and changeable from the menus. Typical robot align, visit, and move commands are available at the RCCL window.

Figures 3, 4 and 5 show the 3-tier menu input system for the FRHC_MAIN, RCCL_MAIN, and VME_UC_MAIN processes. For further details, refer to [6].

Very satisfactory, but nonetheless limited, operational experience has been gained in the use of this teleautonomous system in the performance of generic telerobotic tasks, such as expected for the NASA's Flight Telerobotic Servicer. These tasks include peg-in-the-hole insertion, contour following, and door-opening. Certain combination of gains, teleautonomous scaling, and choice of shared control axes, have been determined to be efficient in the performance of these tasks (see reference [6]).

The performance of this system is mainly constrained by the speed of the μVAX RCCL hardware/software. The bottleneck there requires the whole system to be basically synchronized to 50 hertz. When the entire control goes around the loop, from the hand controller to the robot controller/end-effector and back to the hand controller (for force reflection), a time latency of some 100-125 milliseconds is estimated. For pure teleoperation control, such a time latency is not desirable. In fact, experiments on this system have shown potential control stability problems - not in the sense of control system stability when the system is on its own, or even affected by outside disturbances. Potential problems arise when the human operator is in the loop, depending on the operator's experience and "impedance" in his grip of the force-reflecting hand controller.

It is experienced, as also predicted by the theory, that the teleautonomous capability of the system can overcome such potential instability problems. A linear combination of the teleop control input with the autonomous force control has provided very satisfactory system response in the performance of the above tasks. Note that the autonomous mode of control is not performance limited by the same latency rate. Control there is local to the μVAX/RCCL, which has 50 hertz cycle rate. Low level servo rates are even higher, close to 1 Khertz. The teleautonomous performance has consistently ranked better than teleoperation or shared (hybrid force-position) control performance.

SUMMARY

A powerful and flexible telerobot control system has been integrated and tested, which has a teleautonomous capability that outperforms teleoperation systems and shared (hybrid force-position control) systems. It is complete with hand controllers, electronics, and integrated with the PUMA robots equipped with a custom servo driven end-effector. This system is also designed to be upward compatible with a higher-performance system, where the computation limitation of the μVAX/RCCL is replaced by a faster computer or cluster of distributed processors. As is, this system can serve as a flexible telerobot testbed whereby the human factors aspects and application experiments can be extensively investigated.

Acknowledgement

This work was performed at the Jet Propulsion Laboratory, California Insitute of Technology, under contract to the National Aeronautics and Space Adminstration. GE Aerospace / Advanced Technology Laboratory was subcontractor to JPL, responsible for the system integration and certain software development.

References

[1] A. Bejczy and Z. Szakaly, "Universal Computer Control System (UCCS) for Space Telerobots," Proc. 1987 IEEE Int. Conf. on Robotics and Automation, Raleigh, N. Carolina, Mar. 1987, pp-318-324.

[2] S. Hayati, T. Lee, E. Kan and J. Lloyd, "The JPL Telerobot Manipulator Control and Mechanization Subsystem," Proc. 2nd NASA Symposium on Space Telerobotics, Pasadena, Ca. Jan. 31- Feb. 2, 1989.

[3] P. Backes, S. Hayati, V. Hayward, and K. Tso, "The KALI Multi-Arm Robot Programming and Control Environment," Proc. 2nd NASA Symposium on Space Telerobotics, Pasadena, Ca. Jan. 31- Feb. 2, 1989.

[4] P. Blaire, F. Hawes, R. Killion, and L. Robinson, "Goddard Smart End Effector (GSEE): System Description and Operating Instructions," Jet Propulsion Laboratory, California Institute of Technology, Internal Document, Feb. 1988.

[5] D. McAffee, "Teleoperator Subsystem / Telerobot Demonstrator: Force Reflecting Hand Controller Equipment Manual," Jet Propulsion Laboratory, California Institute of Technology, Document #JPL D-5712, Jan. 1988.

[6] M. Junod, "Operation Manual for the GSFC Telerobot Control System 1," GE Aerospace / Advanced Technology Laboratory, developed under Contract #958149 with Jet Propulsion Laboratory, Pasadena, Ca., Sept. 1989.

[7] J. Llyod, M. Parker, and R. McClain, "Extending the RCCL Programming Environment to Multiple Robots and Processors," Proc. 1988 IEEE Int. Conf. on Robotics and Automation, Philadelphia, Pa., Apr. 24-29, 1988.

Figure 1. The Integrated FRHC-PUMA-EE Teleautonomous System

Figure 2. System Functional Block Diagram

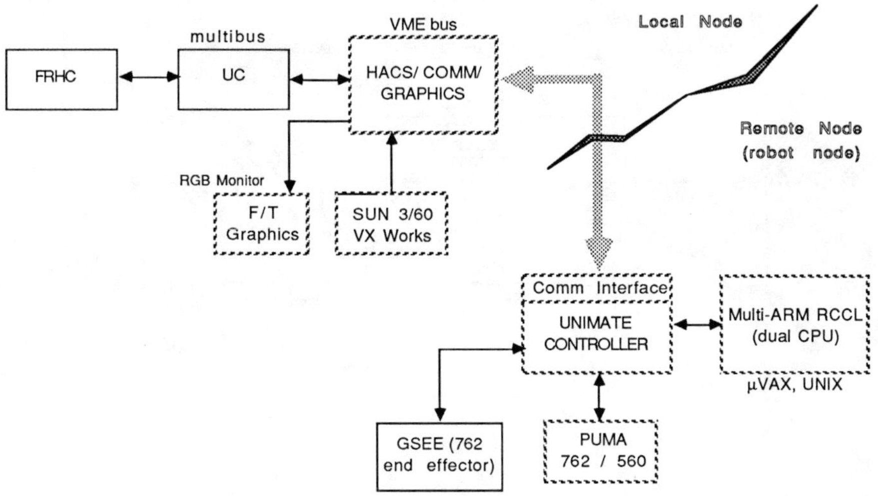

Figure 3. FRHC_MAIN Menu

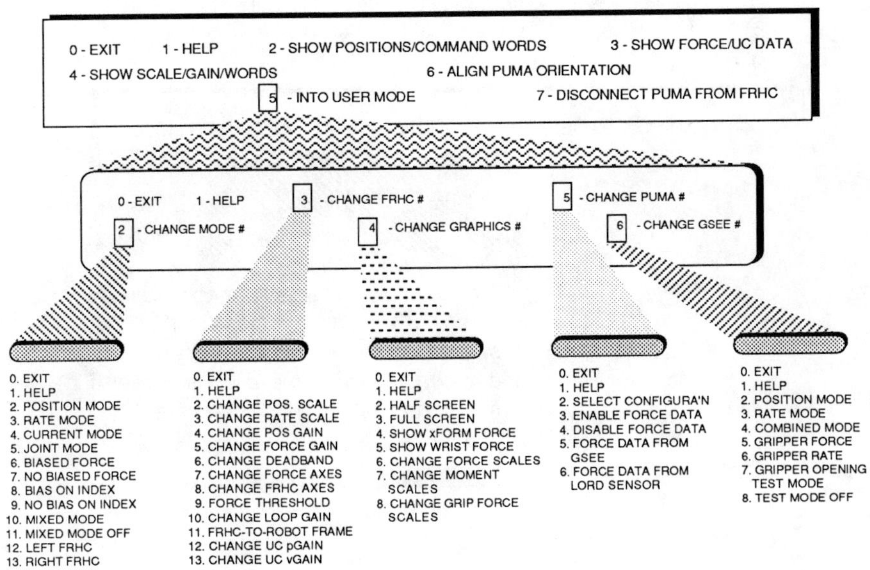

Figure 4. RCCL_MAIN Menu

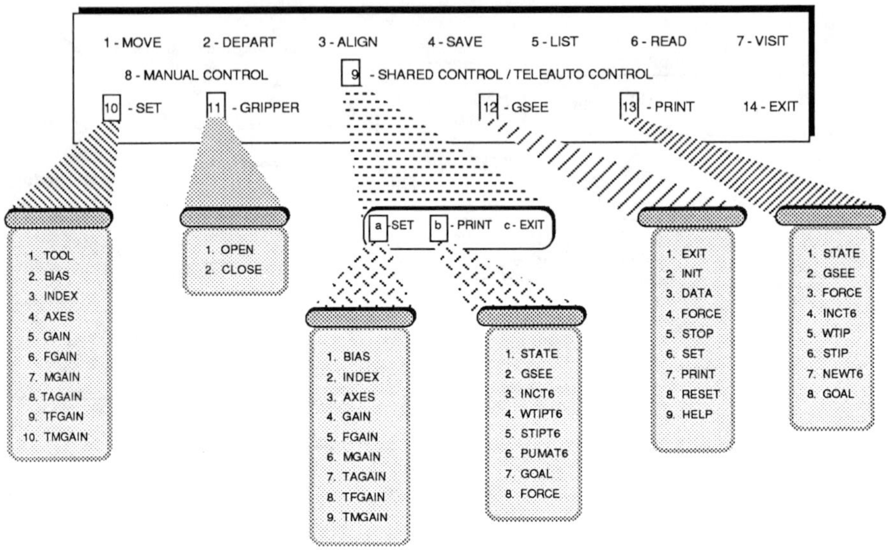

Figure 5. VME_UC_MAIN Menu

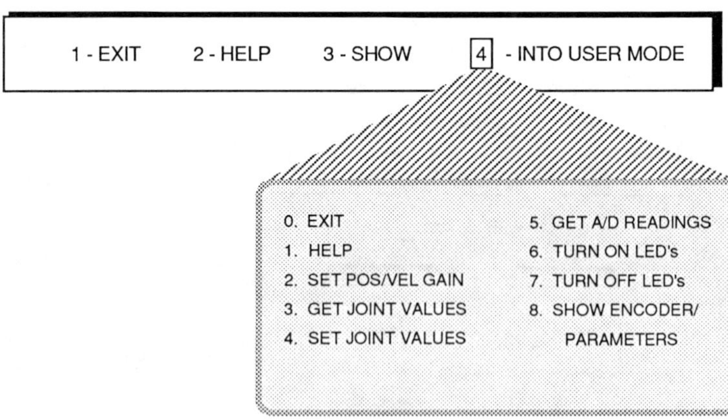

GEOMETRIC SPATIAL PLANNING FOR ROBOTIC MANIPULATORS WITH MULTIPLE LINKS

C. E. COLLINS and J. S. O'MEARA

Jet Propulsion Laboratory
California Institute of Technology
Pasadena, California

ABSTRACT

In robotics, the role of spatial planning is to determine collision-free paths. Automatic spatial planning is required when the endpoints of particular motions are unknown until the time of execution of tasks. This paper presents an automatic spatial planner, CENTER, which plans safe paths for all links of a robotic arm throughout each complete arm motion. Developed as one component of the automated planner for JPL's telerobotic research project, CENTER is applicable to manipulators with any number of links moving in three-dimensional task space.

This paper describes the general theory behind CENTER, including the representation of collision-free space, the construction and sparsification of the search graph, and the generation of requested paths. Also discussed are the performance results for two test scenarios which involve grasping and transporting objects with a six-link arm. These results show that CENTER can find a safe path quickly.

INTRODUCTION

CENTER stands for the Cluttered Environment Navigator for the TEleRobot. We begin by introducing some terminology. *Spatial planning* for a robot arm is the design of paths between given starting and ending robot configurations which avoid collisions with the objects in the workspace. *Gross motions* are those which affect the entire arm, whereas *fine motions* concentrate mainly on manipulating the end effector.

In the *configuration space*, or *joint space*, of the robot arm, each point corresponds to a specific arm configuration. A configuration of the arm is said to *collide* with an obstacle if that configuration is prohibited by the obstacle. The set of configurations which cause no collisions is called *free space*.

CENTER, the gross motion spatial planner for the JPL Space Robotics Project, efficiently generates collision-free geometric spatial paths for multi-link arms. Using a description of the workspace, it generates and uses a multi-dimensional mathematical graph of the object-free areas. This graph is a subset of free space consisting of a discrete number of nodes representing selected arm positions and edges representing arm movements between pairs of these positions.

The main question in spatial planning is how to represent the free space in a manner which will allow for fast path generation. The answer in CENTER is to ignore as much as possible and pay attention to only as much as is necessary. The concept of using a skeletal model of free space appears in other approaches to spatial planning, notably the Voronoi diagram approaches. The CENTER graph paradigm readily extends to six or more dimensions, whereas the Voronoi diagram approach does not.

Our main goal is to provide on-line spatial planning for a robotic manipulator with six or more links in a cluttered environment without sensor feedback. Furthermore, the system must be able to accommodate large held objects, such as instruments for the EOS platform. In order to achieve this goal, we make the assumption that all obstacles remain in fixed, predetermined positions during any arm movement. Also, the robot frame is considered to be fixed relative to the workspace, and this relative position is known in advance.

When we reviewed the other research being done in spatial planning, we found no existing system that can provide collision-free paths for a six-link arm carrying a large object. The system developed by Tomas Lozano-Perez [1] at MIT came closest to being useful in our testbed. The main reason we could not use the Lozano-Perez system directly is that when the arm is carrying a large object, the bounding region for the last three links would be so large that it would indicate collisions for most of the arm positions. CENTER and the Lozano-Perez system are similar in that both require a fixed workspace and that a database describing the collision-free portions of the workspace is preprocessed. The main difference lies in the detail included in the database. The Lozano-Perez system uses a complete description of the free space, whereas CENTER uses a sparse description and is therefore able to handle larger degrees of freedom efficiently.

By using a sparse description of the free space, the CENTER spatial planner quickly provides collision-free paths for multiple links, even when the arm is carrying a large object.

THEORY FOR AN N-LINK ARM

The theory is presented in a two-dimensional task space. The extension to three dimensions is straightforward, and our applications are performed in three-dimensional task space.

A simplifying assumption that we make for the discussion of the theory is that each link is a line segment of 0 width. The extension to reality is straightforward when we realize that a workspace with swollen obstacles and a thinned arm produces similar collision behavior to an original workspace. In our applications the actual arm and actual obstacle sizes are used.

Representation of Free Space

There are two ways to view the space in which the robot moves. One is *task space*; the three-dimensional space about us. The other view of the space is as *configuration space* in which each point is a vector of angular positions, one for each joint.

Suppose we have a manipulator arm with N rotational joints and with the base centered at the origin of the plane. For $n \in \{1,...,N\}$, let I(n) be the joint range for joint n. Let length(n) be the length of link n, and let lsum be the sum of the length(n):

$$\text{lsum} \equiv \sum_{n=1}^{N} \text{length}(n).$$

The configuration space is the subspace $C \equiv \bigtimes_{1}^{N} I(n)$ of N-dimensional Euclidean space $\mathbb{R}^N$; that is, C is the subspace of N-dimensional space in which dimension n has bounds defined by I(n), the angle range for joint n. A robot position, $\bar{\theta} \in C$, gives the value of the angle for each of the joints. The ith element of $\overline{\theta}$, $\theta(i)$, is the angle value for the ith joint relative to link i; for joint 0 the angle is relative to the x-axis.

In task space, the set of positions which the end effector may occupy is contained in the set $D = \{(x,y) : x^2 + y^2 \leq lsum^2\}$. For $\overline{\theta} \in C$, let $\overline{\phi} \in \mathbb{R}^N$ be defined by

$$\phi(i) \equiv \sum_{k=1}^{i} \theta(k)$$

the sum of the angle values for joints 1 through i. Then for $n \in \{1,...,N\}$, let $L(n,\overline{\theta})$ be the line segment for link n when the arm is in configuration $\overline{\theta}$:

$$L(n,\overline{\theta}) \equiv \{(\sum_{i=1}^{n-1} l(i)\cos\phi(i) + t\cos\phi(n), \sum_{i=1}^{n-1} l(i)\sin\phi(i) + t\sin\phi(n)) : 0 \leq t \leq \text{length}(n)\}$$

The portion of Cartesian space which contains the robot in the position $\overline{\theta} \in C$ is

$$R(\overline{\theta}) = \bigcup_{n=1}^{N} L(n,\overline{\theta}).$$

Suppose there are M obstacles. For $m \in \{1,...,M\}$, let $B(m)$ be the point set for obstacle m. Let $O \equiv \bigcup_{m=1}^{M} B(m)$. Then the portion of Cartesian space which is obstacle-free is $D \backslash O$. Correspondingly, the collision free part of configuration space is

$$F \equiv \{\overline{\theta} \in C : R(\overline{\theta}) \subseteq D \backslash O\}.$$

Construction of the Graph

To construct the graph, all possible configurations derived from specific angle sets for each joint form the vertices. All of the collision-free configurations are included in the final search graph. We build the graph as follows. First, fix a positive integer $A(n)$ for each joint n such that $A(n) > 2$. Then select $A(n)$ angles for joint n. For example, select the endpoints of $I(n)$ and $A(n) - 2$ evenly spaced interior angles. For $n \in \{1,...,N\}$, let $S(n)$ be the $A(n)$ selected angles.

The vertex set V and the edge set E for the graph are defined as follows. Let

$$V_0 \equiv \bigtimes_{1}^{N} S(n).$$

Then $V \equiv V_0 \cap F$. The set of possible edges E_0 is found by linearly ordering each $S(n)$ and taking all pairs (v,w) of vertices in V such that:

1. v and w are identical in every coordinate except one, call it j; and
2. v(j) and w(j) are adjacent in the linear order on S(n).

Then E is the set of possible edges which are collision-free:

$$E \equiv \{(v,w) \in E_0 : t\,v + (1 - t)\,w \in F \text{ for all } t \in [0,1]\}$$

Let $G = (V,E)$. This is our free-space graph which represents the collision-free portion of configuration space.

On-line Generation of the Paths

Given two collision-free arm positions s, e $\in$ F, find a collision-free path from s to e. There are three steps to finding the path:

Step 1. Find graph vertices v_s and v_e which are close to s and e, respectively.

Step 2. Find collision-free paths between s and v_s and between e and v_e.

Step 3. Find a path through the graph which connects v_s and v_e.

The desired path is then the concatenation of the three paths s to v_s, v_s to v_e, and v_e to e. The path is a function $\overline{\theta}(t)$ which is continuous for $0 \leq t \leq T$ such that $\overline{\theta}(0) = s$, $\overline{\theta}(T) = e$, and $\overline{\theta}([0,T]) \subseteq F$.

To accomplish Step 1, we use the linearly ordered angle sets S(i). For the ith coordinate of s, we find the nearest element in the set S(i), and let that be the ith coordinate of a new position w. After constructing w, we determine whether w is contained in the graph and thus is collision-free. If so, we let $v_s = w$. Otherwise, we vary the coordinates of w from the sets S(i), staying close to the values of the coordinates of s, until finding a collision-free vertex v_s.

In order to find the vertices in Step 1 such that Step 2 is possible, the graph must contain a sufficient number and distribution of graph nodes. For a given application, it may be necessary, prior to graph generation, to increase the values of the A(i) or alter the selection of the S(i).

To accomplish Step 2, it is useful to have a local motion planner such as the one demonstrated in the 1988 JPL telerobotic testbed. It was efficient for finding short paths but was unable to find paths across the workspace in a reasonable amount of time. Thus it was appropriate for finding the paths between s and v_s and between e and v_e.

To accomplish Step 3, we use an A* algorithm [2] to search the graph for a path. Two heuristics are used. One associates with each search node the actual length of the path from the start node to that node. The other estimates the remaining metric distance to the closest goal.

APPLICATIONS TO A SIX-LINK ARM

The CENTER system consists of two main parts, the off-line spatial graph generator and the on-line path finder. This system will work in any relatively static workspace containing one or more manipulator arms of any number of degrees of freedom. When more than one arm is used, an individual graph is generated for each arm with the other arms represented as part of the workspace in some known position.

The graph generator uses a database of the workspace to generate the graph (the database model of the free space for each arm), and the path finder uses the graph to generate paths as they are needed at run time. The graph generation code depends upon the number of degrees of freedom of the arm and the number of angles per joint being used in the graph. However, the rest of the system is independent of these variables, and all of the code is independent of the particular workspace under consideration.

In the JPL telerobotic testbed, we used Puma arms in three-dimensional task space. Since the Puma arm has six joints (also called six *degrees of freedom,* or *DOF*), its configuration space has six dimensions. CENTER built and searched six-dimensional graphs.

Graph Generation

The input to the graph generator consists of the number of joints in an arm, the selected angle values for each joint, and a calibrated constructive solid geometry

(CSG) model of the workspace. In a CSG model, objects are represented by unions, intersections and differences of geometric shapes. Their locations are given in terms of three rectangular dimensions and three orientation factors.

For each n in {1,...,N} we choose the A(n) numbers in S(n) to be numerically convenient ones. The set V_0 consists of all possible arm configurations based on the selected angles for each joint . Standard collision detection algorithms are used to generate V by testing each node in V_0 for collisions. The set E_0 of all edges between pairs of collision-free nodes can be listed given V. The collision detection algorithms are used to test each of these edges in order to generate E.

To build a graph for a manipulator that will be carrying objects a bounding box, which is as large as the largest object to be carried, is attached to the end effector before the collision detection was performed.

Path Finding

The other part of the CENTER spatial planner is the on-line path finder. The input to the path finder consists of three arguments: an indication of which arm is to be moved, the current configuration of that arm, and a set of acceptable goal configurations. The reason that the input contains a list of goals instead of a single goal is that for a desired position and orientation of the end effector, there are up to eight configurations of the arm which will achieve the objective. The subset of these configurations which are collision-free and kinematically possible for the arm forms the goal list.

In the TIPS system [3], the path request may come directly from the operator or from the automated task planning system. The output path is expressed as a sequence of configurations to be connected by linearly interpolated joint motion. This path is then passed to another telerobot subsystem which moves the manipulator along the path specified by the spatial planner. If there is no path to the desired goal, then a descriptive message is sent to the task planner or the operator.

There are three parts to each path: move to the graph; move along the graph; move off of the graph to a goal. On-line collision detection or fine-motion control is needed to assure the path between the starting configuration and nearest search graph node configuration is collision-free. To search the graph for an acceptable path, we use an A* search algorithm which has been modified to accept a set of possible goals.

Sparsification

It may be the case that in order to get the paths needed in a cluttered area of the workspace, we must increase the number of values per joint. However, when a larger number of angles per joint is used in generating the graph, there may be an abundance of nodes in relatively uncluttered areas where they are not needed. These unnecessary nodes can slow down the search time for finding paths. Thus an algorithm for making the graphs more sparse in areas away from obstacles was developed.

The sparsification module of CENTER takes as input a graph and generates as output an appropriately sparsified graph. This algorithm also allows the user to specify what relatively uncluttered means. In the graphs for a 6-degree-of-freedom arm, each node can have up to 12 neighbors - nodes which differ from the given node by one step in one joint. If a node is in an uncluttered area, it will have more collision-free neighbors than if it is near obstacles. The user can specify a number of neighbors, such as 9. Then the algorithm assumes that a node is in a relatively uncluttered area if it has at least 9 collision-free neighbors. If all of the neighbors of a given node are collision-free and each of those neighbors is in a relatively uncluttered area, that neighborhood is collapsed to the given point and appropriate edges out are determined.

RESULTS IN TWO SCENARIOS

The two scenarios in which performance evaluation was performed were the Solar Max satellite servicing testbed and the EOS (Earth Observing Satellite) instrument replacement testbed.

Solar Max Satellite Servicing

The servicer for the satellite repair testbed consisted of two Puma arms with a toolbox between them and a model of the Solar Max satellite in easy reach. The arms moved independently; when one was moving the other was placed in a known position.

In building the spatial graph for each of the arms,we found that, for this scenario, it worked best to select five angles for each of the six joints. This gave us 5^6, or 15,625, possible graph nodes, each with six to twelve possible neighbors. The nodes were checked for collisions and the graph edges were determined as described above. The time to check the graph nodes for collisions was about 12 hours using old collision detection software which takes 2 to 15 seconds to check each node.

Timing studies on test graphs showed that the time to find a path was less than 1 minute. If we select six angles per joint instead of five, the time was at most about 5 minutes. However, if we select seven angles for each joint, it took as long as 2 hours to find a single path.

In the integrated TIPS (Telerobot Interactive Planning System), removing the main electrical box from the satellite and storing it in the toolbox required 36 gross motion paths. Using the graph built on five angles per joint, the times for finding paths averaged 3.8 seconds, and ranged from 0.4 to 11.5 seconds. Thus for this telerobot testbed, the CENTER spatial planner generated paths much more quickly than they could be executed.

EOS Instrument Replacement

The goal of this spatial planning study was to determine the appropriate structure for a spatial graph for the CENTER spatial planner in the EOS platform servicing environment. The study concentrated on two aspects: the effects of varying the number of angles per joint in the generation of the graph and the effects of various degrees of sparsification on the efficiency and capabilities of the arm paths.

The workspace included two arms, a stowage rack, a work platform on the lathe bed, an instrument module, and an instrument subassembly. The stowage rack contained storage locations for the instrument module and the instrument subassembly; the work platform could contain either the module or the subassembly. The workspace was arranged so that the left arm could perform all of the motions. It was assumed that each object could be held in a single predetermined manner.

A motion scenario was developed to simulate instrument replacement on the EOS satellite. The scenario consisted of five motions, each performed with a different workspace configuration of the placement of the module and the subassembly either on the work platform, on the stowage rack, or in the end effector. Four sets of graphs were generated, one with 4 angles per joint, one with 5, one with 6, and one with 5 angles per joint which was then sparsified. Each set consisted of 5 graphs for the left arm, one for each workspace configuration.

The results varied widely. With 4 angles per joint, there were not enough nodes near the work areas to find the paths. With 6 angles per joint, no paths were found either. The statistics on the number of nodes which were collision-free in each graph indicated that the lack of paths was due to the particular angles which were chosen for each joint. Our later experience in other testbeds have shown that intelligent selection of the particular angles used is necessary.

With 5 angles per joint, the average time to find a path was 1.5 seconds, with a range from 0.2 to 3.4 seconds. Sparsifying this graph had no effect on the time for finding paths, but did decrease the ability to find paths.

This study suggests three areas of future investigation:

1. intelligent selection of the particular angles used;
2. building graphs with more angles per joint and sparsifying them to determine more accurately the effects of sparsification; and
3. addition to the graph of selected graph nodes in work areas.

CONCLUSION

In a manufacturing setting, the CENTER spatial planner would be a useful tool for finding paths which robot arms can follow through the workspace. Since these settings tend to be relatively static, the off-line graph generation and appropriate angle selection could be performed once, with the on-line path finding performed as the needs arise for the robot to reach different places in its workspace.

If the arm is carrying objects, or if the obstacles in the workspace vary in a predictable manner, then the list of arguments for the path finder may be expanded to include an indication of the bounding region for the object in the end effector and the current conditions of the variable workspace objects. These additional arguments would determine which graphs or graph parts to use.

In our applications, the spatial planner can be used by the operator either independently of the task planning system or in combination with task planning. Control between automatic spatial planning and teleoperation can be exchanged anywhere in the workspace. The teleoperator can use the spatial planner to perform motions between operator-specified configurations of the arm. For example, CENTER can be used to execute a potentially dangerous pose flip safely for the teleoperator since the CENTER paths are generated in configuration space rather than task space, where the teleoperator works.

Our testing has shown that CENTER can be a reliable, efficient spatial planner for a multiple-linked robot arm carrying various objects through a cluttered environment.

ACKNOWLEDGEMENTS

The original work from which this spatial planner design was developed was done by M. Starbird of the University of Texas and T. Starbird of the Mission Profile and Sequencing Section of JPL. The other members of the Sequence Automation Research Group who worked on the telerobot project are D. Mittman, M. Rokey, S. Peters, and E. Zamani. Other individuals participating in sequence automation research include R. Brill, J. Cruz, W. Dias, C. Eggemeyer, S. Grenander, M. Hollander, W. Lombard, J. Sisino, B. Streiffert and B. Zimmerman.

REFERENCES

1. T. Lozano-Perez, "A Simple Motion-Planning Algorithm for General Robot Manipulators", IEEE Journal of Robotics and Automation, RA-3(3), 224-238 (June 1987).

2. N. Nilsson, Principles of Artificial Intelligence (Morgan Kaufmann Publishers, Inc., Los Altos, California 1980).

3. C. Collins and M. Rokey, "Planning for the Jet Propulsion Laboratory Telerobotics Project", Proceedings of the Fourth Annual Artificial Intelligence and Advanced Computer Technology Conference, Long Beach, California, 429-437 (May 4-6, 1988).

ROBOTICS FOR SPACE APPLICATIONS

STANFORD OLLENDORF
Chief, Office of Telerobotic Engineering
NASA Goddard Space Flight Center

ABSTRACT

The NASA Goddard Space Flight Center (GSFC) robotics program is structured to increase the scope and efficiency of what man can accomplish in space through the use of robotics. The main focus is to assemble and service Space Station Freedom (SSF) and its payloads. The approach being taken to achieve this is by telerobotics, where manipulators can be controlled by human operators and later evolve to total computer control of the robot. This paper will describe, in an overview fashion, the current plan at GSFC to develop the Flight Telerobotic Servicer (FTS) for Space Station Freedom, in-orbit servicing of platforms, and ground demonstrations of robotics development within the GSFC Development, Integration and Test Facility (DITFAC). Robot evolution to support a Lunar/Mars mission will also be discussed.

INTRODUCTION

GSFC is one of the key NASA Centers developing state-of-the-art robotics for the Space Station era. The main focus of the robotics program is to assemble and service SSF and its payloads, as well as other free flyer missions. The approach being taken is to achieve this by robotics; where machine manipulators are controlled by human operators and/or computer programs.

Goddard is developing the Flight Telerobotic Servicer (FTS) in support of SSF. The FTS will be NASA's first operational robotic system. The system is a robotic device, initially controlled by humans via teleoperation, that will assist in Space Station assembly and maintenance operations while attached to the station or to the Space Shuttle. The FTS will also assemble spacecraft and instruments which have been flown on separate flights to SSF by the Shuttle.

In the near future, robot technology will be available to perform these on-orbit servicing missions autonomously without manned presence in space. The push towards robotics development in satellite servicing comes from several sources; a desire to perform tasks in a hazardous environment, reduce Extra Vehicular Activity (EVA) time for the astronauts, and the need to service satellites in polar and other orbits where manned presence is not available. The goal of the FTS and the Goddard robotics program is to move robotics concepts from the world of research into flight systems, providing spin-off benefits for the science and industrial communities.

GSFC FTS Servicer Program and System Description

FTS is a robotic device that is controlled by teleoperation or semi-autonomously as instructed by a computer but supervised by a human. Initially, FTS will be limited to minor SSF assembly and servicing tasks. Ultimately, the FTS mounted on an Orbiting Maneuvering Vehicle (OMV) will be able to reach, retrieve, and service unmanned satellites in both polar and equatorial orbits, and return these satellites to useful service. The FTS is shown on-board SSF in a typical operations scenario in Figure 1. Martin Marietta was recently selected for the FTS final design implementation contract. Scheduled to be launched in the 1994/1995 timeframe, FTS will improve reliability, productivity and safety, and decrease mission operation costs.

FIGURE 1. FTS Operations On-board Space Station Freedom

The FTS system, as proposed by Martin Marietta, consists of a human operated robot capable of working on Space Station, Shuttle or an OMV as well as the accommodation equipment for storing the telerobot and its spares. Included as part of the telerobot are two manipulator "arms"; an Attachment, Stabilizing and Positioning Subsystem (ASPS) or "leg"; cameras and lights; and all end-of-arm servicing tools. A detailed sketch of the FTS is shown in Figure 2. The stabilizing positioning leg can be seen at its lower end. The FTS will have a set of tools and end effectors that can be autonomously selected through the use of the end effector changeout mechanism located on the end of each manipulator. Tool holsters located on the front body will store extra end-of-arm tooling when not in use.

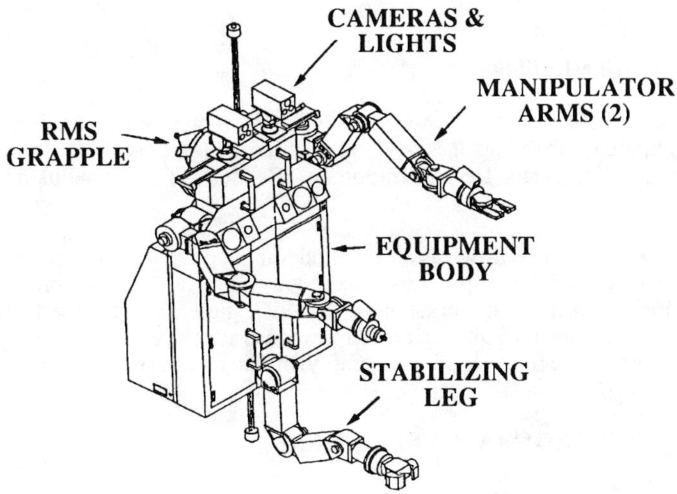

FIGURE 2. Flight Telerobotic Servicer (FTS)

The FTS is a system that is designed for growth and evolution over the years. Initially it will be a teleoperated device under Shuttle crew control with some autonomous activity. The vision system, which will start out as a closed circuit television system will later expand to stereo vision and include some autonomous image recognition and location capabilities. Path planning capabilities may be added to the robot control and execution software to enable the system to plan and execute simple tasks autonomously.

Workstations located on the Shuttle or the Space Station will be the principal point of human interaction for the control of the FTS. Each workstation will be equipped with two 6 degree-of-freedom (DOF) mini-master force reflecting hand controllers. Video displays in the workstation will be capable of displaying up to three video images simultaneously, or two video images with health and status information displayed on the third screen.

The total weight of the FTS and the Shuttle workstation is anticipated at 1500 lbs (681 kg). In the stowed configuration the FTS will fit in a volume that is 7 feet x 3.5 feet x 3 feet (2.1m x 1.1m x 0.91m). The power required by the FTS will typically not exceed 2000 watts peak power, 1000 watts average power and 350 watts standby power. [Ref. 1]

Space Station Freedom and Shuttle Operations

In order to help in the assembly and maintenance of the SSF, the FTS will operate from both the Shuttle orbiter and the SSF. Operation from the Shuttle will occur during early SSF assembly flights. FTS will make use of standard Shuttle interfaces, with perhaps the addition of a special umbilical for operations while attached to the existing Shuttle Remote Manipulator System (RMS). Operators will use an FTS workstation located in the Shuttle Aft Flight Deck (AFD) to interface with power, data and video systems located in the Shuttle payload bay. The RMS will be employed to move and position the FTS for operations on the Shuttle. This requires the FTS to be equipped with an RMS compatible grapple fixture for structural attachment to the

RMS.

Demonstration Test Flights

Prior to FTS Space Station assembly flights, there are two early Shuttle test flights anticipated. They are the Development Test Flight (DTF-1), scheduled for launch in mid-1991, and the Demonstration Test Flight (DTF-2), scheduled for launch in 1993.

The DTF-1 will be flown as an attached Shuttle payload to test the control and performance of the FTS manipulators in zero gravity as well as the human-machine interface through an in-orbit workstation. In 1993, the DTF-2 will be flown with a mature version of the FTS to verify and demonstrate the system's capabilities to perform representative Space Station assembly, maintenance, servicing and inspection tasks.

GROUND TEST PROGRAM AT GODDARD

Much of the technology planned for use in the FTS project and on the demonstration test flights is relatively new and untested. To provide the answers needed to design safe, reliable and fully functional robotics for flight, Goddard has developed a unique Development, Integration and Test Facility (DITFAC) shown in Figure 3. This highly sophisticated facility will be used by the Goddard robotics team to create, test, and evaluate new robotic technologies required to support Space Station.

FIGURE 3. Development, Integration and Test Facility

The DITFAC consists of two primary test beds; the FTS Functional Simulator and the FTS Engineering Test Bed.

FTS Functional Simulator

The FTS Functional Simulator (FTS/FS), located within the DITFAC, is being used to develop Space Station mock-ups, workstations and simulations to perform robot operational scenarios. This engineering effort will design Space Station mockups and perform FTS assembly tasks on simulated Space Station elements. [Ref. 2]

The FTS/FS consists of a 6-DOF gantry mounted robot that is capable of lifting up to 4000 pounds (1818 kg) of payload and applying 4000 foot-pounds (5460 N-m) of torque to lift and position large Space Station elements within 0.10 (0.28 mm) inch accuracy. Suspended from one mast of the gantry is a set of 6-DOF industrial arms, which will be used as the FTS/FS. The other mast has a grapple to emulate the Space Station RMS and is used primarily to transport payloads and structure to and from the worksite. An operator workstation, of the robot installed in a mockup of the Shuttle Aft Flight Deck, will permit teleoperation in the constrained environment of the Shuttle. This mockup is designed to be reconfigurable in order to determine the best positioning of hand controllers and displays.

The FTS/FS will be used to investigate operational techniques, camera configurations, operator interfaces, Orbital Replacement Unit (ORU) designs, end effectors and training techniques. Astronauts scheduled for DTF flights and FTS operational missions can use the facility to perform FTS assembly and servicing scenarios. Current operations being performed with the FTS/FS include ORU exchanges, alignment and servicing of SSF related mockups.

FTS Engineering Test Bed

Co-located with the FTS/FS is the FTS Engineering Test Bed (FTS/ETB). The FTS/ETB is being used to research issues such as robot control algorithms, dual-arm teleoperation, collision avoidance, dynamic simulation techniques, end effector design, and hierarchical control using high-level programming languages. [Ref 3]

The FTS/ETB is centered around the dual-arm teleoperation of a pair of 7-DOF manipulators. Each arm has an extension of almost 6 feet (1.9 m), mounted on one stand approximately 8 feet (2.46 m) from the ground, 1.5 feet (0.46 m) apart, simulating a right and left arm. Master-slave (teleoperation) control of the RRC arms is accomplished with a pair of 6-DOF mini-master hand controllers.

A safety system is being developed to ensure that operation of the robot does not damage or cause harm to an operator or bystander, to itself, or to any object in the robot's workspace.

Working closely with the National Institute of Standards and Technology (NIST), one of Goddard's goals has been to define and maintain a functional robot architectural standard. Such an architecture establishes the communications necessary for the incremental development and hierarchial evolution of a telerobotic system, eventually leading to greater autonomy. The NASA/NIST Standard Reference Model for Telerobot Control System Architecture or NASREM has been selected and developed by the FTS project to fill this need. Adopting this architecture standard will enhance our efforts to standardize robot development and link our development through NIST to U.S. industry, making technology transfer possible. It is planned to implement this architectural scheme into the FTS/ETB for evaluation.

OTHER RELATED PROGRAMS

Robot Friendly Instruments

Several projects are underway at Goddard to develop "robot-friendly" instrument designs for assembly and servicing operations in space. One study is defining design features and operational scenarios for the Astromagnetic Experiment (Astromag), a superconducting magnet cosmic ray facility. The Astromag instrument will be an attached payload on the Space Station. A representative mock-up of the instrument will be designed and tested in the Goddard DITFAC to develop the hardware designs, attachment methods, restraint systems, and equipment access that can support assembly and repair of Astromag by robot manipulation. Critical areas for robot handling include magnet to instrument interfaces, submodule interfaces, and latch handles for doors and actuators.

Space Based Materials Processing and Manufacturing

The microgravity space environment offers advantages in space-based materials processing. Thermal convection, buoyancy and sedimentation are eliminated in the absence of gravity and containerless processing is also possible, reducing the likelihood of contamination of a semiconductor. Other space processing advantages are high vacuum controlled atmosphere, cleanliness, and isolation of hazardous materials.

At GSFC, research is planned that could lead to a commercial space-based microelectronics manufacturing facility using robotics. Work in developing a telerobotic system is already under way to automate the processes. The next step will be to build a ground based test lab to demonstrate the capability to automate production tasks currently performed by skilled technicians. In the process of developing and demonstrating this capability, knowledge will be accrued that will reduce the risk of developing a space-based facility.

Robotic Assembly for Human Exploration Initiative

A manned mission to Mars or the construction of a manned lunar base is almost certain to require the assembly of large structures. EVA time is expensive and hazardous, and will probably be a very limited resource. Much of the work could be performed by teleroperated or autonomous robotic systems. The Mission Analysis/ Systems Engineering (MASE) of the Exploration Program has established the need for trade studies to develop technical and programmatic concepts for lunar/Mars human exploration. One purpose of the trade studies is to derive overall strategies, approaches, and techniques to conduct the exploration initiative using humans and robotics in the most efficient manner. Studies are currently being conducted to assess robotic assembly of a Lunar Transit Telescope (LTT). This lunar based telescope will allow an imaging survey of the universe with greater clarity than the operational Hubble Space Telescope. Emplacing the LTT on the lunar surface will require geological surveying, excavation, and the assembly of the LTT prior to human deployment. Robots will be essential in performing these tasks.

REFERENCES

1. J. Andary, K. Halterman, D. Hewitt, P. Sabelhaus, "The Flight Telerobotic Servicer (FTS) - NASA's First Operational Robotic System." Proceedings of the 1989 Space Operations, Automation and Robotics Conference, Johnson Space Center, July 25-27, 1989.

2. S. Ollendorf, "Research and Development Activities at the Goddard Space Flight Center for the Flight Telerobotic Servicer," Proceedings of the 1989 NASA Conference on Space Robotics, Pasadena, CA, January 31 - February 2, 1989.

3. R. Schnurr, M. O'Brien, S. Cofer, "The Goddard Space Flight Center (GSFC) Robotics Technology Testbed," Proceedings of the 1989 NASA Conference on Space Robotics, Pasadena, CA, January 31 - February 2, 1989.

DUAL-ARM VOICE-CONTROLLED FORCE-REFLECTING TELEOPERATION SYSTEM WITH A TIME DELAY FEATURE

Thomas S. Lee, Samad Hayati, Kam S. Tso, and Paul G. Backes
Jet Propulsion Laboratory, California Institute of Technology
Mail Station 198-219, Pasadena, California 91109

ABSTRACT

This paper describes the design and architecture for a dual-arm force-reflecting teleoperation system that exemplies a local and remote site paradigm (the operator is stationed at the local site and controls a manipulator which is located physically distant from the operator). It is a non-replica master/slave system which uses a differential-position-based (Jacobian) algorithm and has the feature of artificially introducing communication time delay between the local and remote sites. The operator utilizes three available teleoperation modes (tool, world, and camera) to perform the task and interacts with the system by using a voice input/output system and a graphics menu.

1. INTRODUCTION

We have been working on a general problem of controlling a remotely located manipulator, perhaps on the space station or in Mars, from a ground-earth station (local site). Our system was designed to investigate problems arising from such local/remote paradigm and to examine human interface issues such as effectively utilizing various teleoperation modes, force reflection, and the vision system to perform tasks in a reliable way. Another goal of the teleoperation system was to build a flexible hardware and software environment that can easily integrate with an autonomous system to be able to control the manipulator in a shared teleoperation/autonomous mode. Shared mode enables the capability of an autonomous planning system to be utilized to execute primitives that control robot motions and interaction forces without constant supervision from the operator. Emphasis was placed on making the system user-friendly yet providing the flexibility to easily alter parameters such as force reflection gains to perform tasks in different environments. In addition, to assess the feasibility of ground-remote operation, the system was designed to simulate transmission time delay which can be significant when the manipulator is controlled from a long distance in the order of many thousands of miles. In this paper, we will describe the design and implementation of this system.

The next section will describe the hardware and software environment at the Robotics and Teleoperators Research Laboratory in JPL. The operating environment and the available teleoperation modes are detailed in the third section. The fourth section will present timing results. Finally, the implementation of time delay is described.

2. SYSTEM DESCRIPTIONS

In this section, a brief description of the system will be given. The reader should refer to References [1,2] for more specific details about the hardware and software environment.

The hardware diagram is shown on Figure 1. Three Unimation PUMA 560 robot arms are located at the remote site in our laboratory. Two of them are used for dual-arm manipulation — each arm is equipped with a Lord Corporation wrist force/torque sensor and a TRI servo gripper. On the third PUMA arm, a pair of color stereo cameras is mounted for camera-view control. The operator observes the remote worksite via these cameras and can re-position the camera arm to view the task space from different angles. The VME chassis at the remote site collects sensor data such as force/torque and teleoperation information and communicates with the SUN-4 which generates robot trajectory in real-time.

At the local site, the operator moves two six degrees-of-freedom (DOF) universal Force-Reflecting Hand Controllers (FRHC) [4]. The FRHC that is integrated into the system is the third generation hand controller (version C) built at JPL which is capable of generating 8 lbs of force and 14 in-lbs of torque in each DOF. The VME chassis (either L-TELEOP or R-TELEOP) at the local site generates teleoperation information and controls the force/torque display.

All teleoperation software is written in the C language with the exception of the NSC32016 assembly language code that runs on the motor controller which servoes each of the robots and hand controllers. Code is developed on a SUN 3/60 UNIX computer utilizing SUN's C compiler and Wind River's VxWorks/Wind real-time library and is downloaded through Ethernet to the processor boards. The software at the servo level runs on MC68020-based processor boards in VME chassis.

At the remote site, the robot task planning and execution level software runs on a SUN 4/200 computer. The programs are written in the C language utilizing a multi-arm robot programming library (Robot Control "C" Library – RCCL) [7,8,3,9]. On the SUN-4, two level hierarchy programs run concurrently: the UNIX-level planning and the real-time control level. The planning level executes in the foreground in the sense that it interacts with the user, performs high-level computations, and communicates with the control level. At this level, the programmer specifies the task in Cartesian coordinate system in terms of homogeneous transformations and defines manipulator's desired end-effector trajectory to satisfy a kinematic constraint equation. Features such as task sequencing and motion planning/queueing are available in this level.

The RCCL control level program executes synchronously at every sampling interval (at the rate of 200 Hz for a single arm and 100 Hz for two arms) and performs trajectory generation (interpolating between specified via points) and inverse kinematics to solve for joint angle setpoints to be sent to the joint motor control processors. An important feature which is widely used at this level is the ability to update any of the matrices in the kinematic contraint equation in real-time. For example, a matrix which expresses a differential change between two frames in the kinematic equation can be modified at every sampling cycle by hand controller inputs to perform teleoperation or by force/torque readings to activate compliance control.

3. TELEOPERATION ENVIRONMENT

The operator is located at the local site in the Operator Control Station (OCS) [5,6] and moves two FRHC's to generate feed-forward commands consisting of differential positional and orientational setpoints for the manipulator. In the feedback path, the forces and torques that are sensed at the tool tip of each robot in the remote site are reflected back respectively to each of the hand controller handle. He wears a pair of polarized glasses to view the task space by looking at a stero display image of the cameras mounted on the third manipulator arm. He enters commands into the system by interacting with a graphics menu and is able to communicate with the system by speaking into a microphone which is attached to his headset.

3.1 TELEOPERATION MODES

The teleoperation system [1,2,3] accomodates the following modes of operation: joint, Cartesian, and force-feedback. Joint mode is implemented by having a one-to-one mapping of each of the hand controller's DOF to that of the robot. This mode is used to execute gross motions and to move through a manipulator singularity encountered in Cartesian space (i.e., to change the manipulator pose). In Cartesian mode, non-replica teleoperation is used, which has the advantage that the same hand controller can be used to teleoperate different arms independent of their size or geometry. Movements in Cartesian mode can be referenced with respect to the manipulator's end-effector frame (*tool* mode), an absolute world base frame (*world* mode), or the camera view frame (*camera* mode). *Tool* mode enables the the operator to move the joints of the robot in a coordinated manner such that he has the notion of moving the end-effector as if it is being held by his hand (e.g., holding onto a screwdriver and moving the handle to control the tip of the screwdriver). *World* mode is used when the operator wishes to move the robot relative to a fixed frame, for example, to move with respect to a defined world frame in the task space. A convenient feature is *camera* mode which is used to move the manipulator in a natural way with respect to the operator's camera view even if a mis-orientation exists among the hand-controller and robot control frames and the operator's viewing frame. A detail explanation on theory of operation under these three modes is presented in Reference [1].

3.2 TELEOPERATION USER INTERFACE

A user-friendly menu was implemented using the SUNVIEW library (see Figure 3). The top half portion of the menu is used to select the arm and teleoperation mode. The operator can select either single or dual arm operation and the desired teleoperation mode by clicking the appropriate buttons on the menu using a SUN mouse. If the system is placed under tool mode, he specifies the tool that he has picked up. The operator can also modify the desired positional scaling between his hand motion and the resulting motion of the robot arm. These scalings can be specified for each individual Cartesian or joint axis. The force reflection gains can also be modified by utilizing the menu.

The bottom portion of the menu displays status information. It shows observed joint angles, force/torque sensor readings, gripper opening, force reflection status, and a status which indicates whether or not the hand controller is engaged to move the manipulator. During Cartesian mode operation, force reflection is activated by toggling a button on the hand controller handle.

The system detects anomalous situations such as the robot reaching a singular comfiguration or a joint limit. At this point, the software will automatically place the system in joint mode and inform the operator to move the robot away from the anomaly region. Another user-friendly feature is the commercial voice recognition

(Verbex) and synthesizer (DECtalk) system which is part of the OCS. Vocal commands such as "switch the right arm to tool mode" are comprehended by the system, and messages such as "right arm reached singularity" are generated. A vocabulary of words and phrases associated with operating the teleoperation system is entered by the operator. The voice system recognizes the operator's command by matching it with a set of previously entered vocabulary and validating it in hardware. The voice input feature is convenient and timing-saving in the sense that the operator can generate an input to the system while he is teleoperating the manipulators without releasing the hand controller handles.

4. TIMING DATA

In this section, timing data of the implemented system are presented. In Figure 4, the closed-loop time can be referred to as the "around-the-loop" or critical path flow time, which is the time for data to proceed from hand controller to manipulator and then back to the hand controller (see also Figure 2). The shorter the closed-loop time, the greater allowable force reflection gains that can be set yet have a stable control system, which provides a better "crisp" kinesthetic feel to the operator as he performs dexterous operations. The closed-loop time of our present system is estimated to be 13.45 milliseconds.

To perform required computations, three processors (68020 Kinematics (KIN_H), 68020 Auxiliary (AUX_R), and 68020 Moper ($MOPER_R$) on Figure 1) coordinate to perform all necessary transfers of Cartesian and force/torque information. Each processor performs a set of designated computations (see adjacent description boxes in Figure 4). The feed-forward loop begins by having KIN_H communicate with the motor controller of the FRHC which takes 1.3 ms to read in the new joint encoder values. The hand controller Jacobian (J_H) is then calculated (in 1.2 ms) to compute the Cartesian information (ΔX), and KIN_H communicates with AUX_R to pass ΔX. AUX_R receives ΔX and deposits it (in 0.03 ms) into the shared memory. $MOPER_R$ reads in ΔX from the shared memory and passes it to the SUN4 to be processed by the RCCL teleoperation program (in 1.35 ms) and, later during the same cycle, receives the new desired encoder values for the manipulator which are then sent to the motor controller (1.7 ms). The feedback loop starts as the force/torque (raw hex) data are forwarded to AUX_R. Approximately 1.3 ms is used by AUX_R to receive the hex data and to compute resolved Cartesian force/torque values (F_R). Next, F_R is forwarded to KIN_H (in 0.9 ms) where it is transformed to the appropriate frame and pre-multiplied by the transpose of the hand controller Jacobian (J_H – computed previously in the feed-forward path). Finally, the desired motor torque values are forwarded to the FRHC controller to reflect sensed forces/torques back to the operator's hand.

5. TIME DELAY IMPLEMENTATION

At every sampling time, in the feed-forward path from the local to the remote site, data such as robot trajectory information and execution commands are sent, and in the feedback path, sensor information such as vision, forces and torques, and other measured data and command acknowledgements are received. Considering practical aspects of implementation, it is desirable to delay the relayed information only in the feed-forward and not in the feedback path. The reason for this is primarily to eliminate the need to store massive amounts of vision data which can be generated at the rate of a few megabytes per sampling time. The idea is to simulate a situation that has time delay of T along each forward and feedback path by delaying only the forward path by 2T. The assumption fails if the remote environment has objects which are not moving in constant speed. Since our laboratory environment contains only objects that are fixed and stationary or at least which move in a predictable fashion, this assumption is valid for our application.

To conduct realistic experiments, it is necessary to create both deterministic and variable time delays. With this in mind, we use the following form: $A + B(sin[wt] + 1) + C * random)$. Constant time delay A is associated with the length of time for the signal to travel in speed of light to reach its destination – this attributes to a large portion of time delay in the local/remote paradigm. In the case of Mars, even at its closest approach position to Earth, the signal would take about 3 minutes to reach us. Constant A also takes into account baseline time used by the receiving facilities to process the data. The sinusoidal behavior of time delay is introduced to consider the changing orbital relationship between the receiving ground-earth station and the transmitting remote site. For example, if the transmitter is located on Mars, then the distance from the earth will vary with a period of approximately 780 days. The random behavior of time delay is introduced because of an unpredictable amount of delay augmented to the baseline time to process data by software (e.g., buffers) and hardware (e.g., circuit delay elements) and to transfer data via networks to the final destination on Earth. For example, if the signal is routed to a NASA/TRDS satellite, it would first reach a set of receivers in Whitesands, New Mexico which would then proceed to another set of receivers at the station where the operator is located via various networks (e.g., land lines). In this example, even though data proceeds through three sets of different communication media, the magnitude of randomness (C) is still relatively small compared to constants A and B (e.g., C is in the order of milliseconds when communicating from a geo-stationary platform). On the other hand, randomness can be abrupt as a sudden infinite amount of time delay which simulates a situation when a “dropout” (an unplanned loss of communication) occurs typically due to a malfunction at the receiving site.

On the average, one way time delay of communicating from the space station (with FTS – Flight Telerobot Servicer) to a ground-earth station is around 2 seconds (communicating via the on-board space station bus which is slow, uplinking to a NASA/TRDS geo-stationary satellite since the space station is on low orbit, downlinking to White Sands TRDS tracking station in New Mexico, then finally using a satellite, e.g., WESTAR, to relay data from White Sands to a NASA center). When we consider communicating with a rover on Mars, one way communication time delay can range from 3 to 21 minutes (via a DSN station, DOMSAT, Goldstone, and then to a NASA center) – this poses a very challenging robot control problem.

We implemented the time delay feature in hardware by using a shared memory card (MM 6320D) which has 4 Megabytes DRAM and inserting it into each of the hand controller VME chassis (see Figures 1 and 2). Appropriate software changes were merged into the existing hand controller code – a circular ring buffer is used to adjust the length of time delay by storing ΔX data temporarily before being forwarded to the remote site. Simulated time delay can vary from 0 to 10 minutes in increments of a few milliseconds.

6. CONCLUSION

The design and implementation details of a dual-arm force-feedback teleoperation system with a time delay feature were presented in this paper. Future research plans include performing teleoperation experiments under various time delay situations and adding an impedance control feature into the system to overcome expected adverse effects of force-feedback teleoperation when the delay is in the order of a few seconds.

7. ACKNOWLEDGEMENT

The research described in this document was performed at the Jet Propulsion Laboratory, California Institute of Technology, under contract with the National Aeronautics and Space Administration. The authors would like to thank Jake Matijevic and JPL DSN operations personnel for providing information on time delay.

References

[1] Lee, T. S., Kim, W. S., "Control of a Dual-Arm Force-Reflecting Telemanipulator with Reference to the Camera View," *Proceedings of the First International Symposium on Measurement and Control in Robotics*, Houston, Texas, June 1990.

[2] Lee, T. S., "Implementation and Design of a Teleoperation System based on a VMEbus/68020 Pipelined Architecture," *Proceedings of NASA Conference on Space Telerobotics*, January 1989.

[3] Hayati, S., Lee, T., *et al.*, "A Testbed for a Unified Teleoperated-Autonomous Dual-Arm Robot System", *Proceedings of IEEE International Conference on Robotics and Automation*, Cincinnati, Ohio, May 1990.

[4] McAffee, D., Ohm, T., "Teleoperator Subsystem/Telerobot Demonstrator: Force Reflecting Hand Controller Equipment Manual", JPL D-5172 (internal document), Jet Propulsion Laboratory, Pasadena, California, January 1988.

[5] Kan, E., *et al.*, "The JPL Telerobot Operator Control Station: Part I - Hardware," *Proceedings of NASA Conference on Space Telerobotics*, Pasadena, California, Jan. 31 - Feb. 2, 1989.

[6] Kan, E., *et al.*, "The JPL Telerobot Operator Control Station: Part II - Software," *Proceedings of NASA Conference on Space Telerobotics*, Pasadena, California, Jan. 31 - Feb. 2, 1989.

[7] Hayward, V., Paul, R.P., "Robot Manipulator Control under Unix RCCL: A Robot Control C Library, " *The International Journal of Robotics Research*, Vol. 5, No. 4, MIT Press, Winter 1986, pp. 94-111.

[8] Lloyd, J., "Implementation of a Robot Control Development Environment," Master's Thesis, Department of Electrical Engineering, McGill University, Montreal, Canada, Dec., 1985.

[9] Seraji, H., Lee, T.S., Delpech, M., "Experimental Study on Direct Adaptive Control of a PUMA 560 Industrial Robot," *Journal of Robotic Systems*, Volume 7, No. 1, John Wiley & Sons, Inc., February 1990, pp. 81-105.

FIGURES

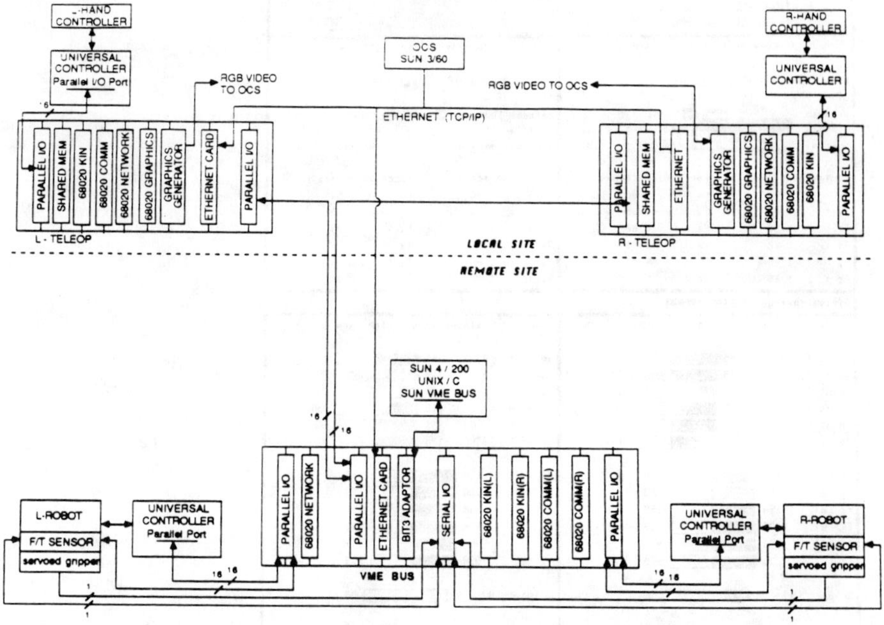

Figure 1: Hardware Diagram of the Teleoperation System.

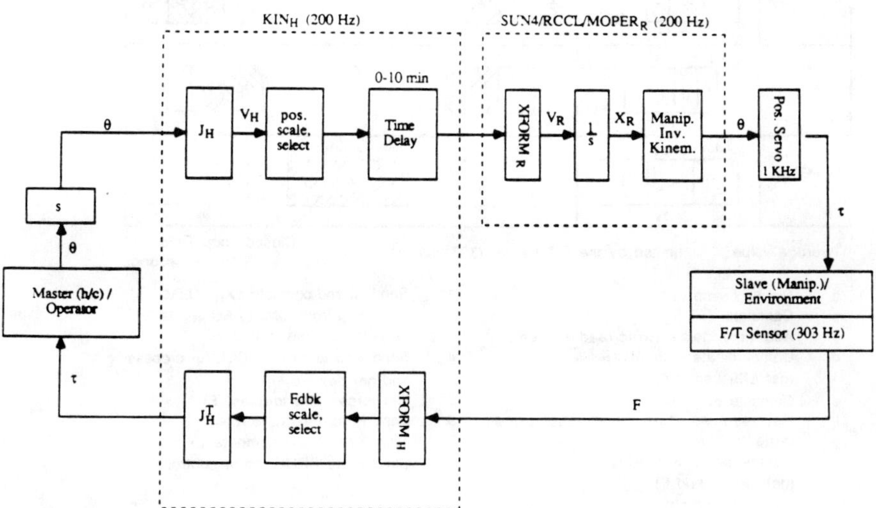

Figure 2: Teleperation System Control Block Diagram.

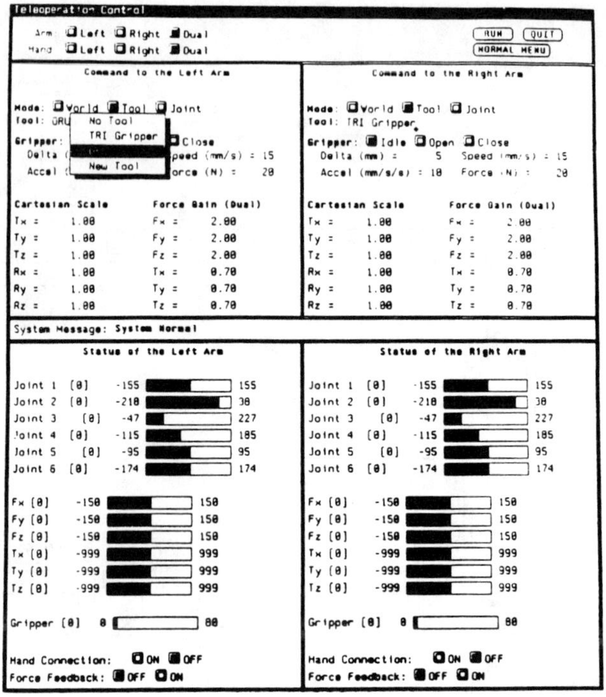

Figure 3: Teleoperation Menu with Voice-Commanding Capability

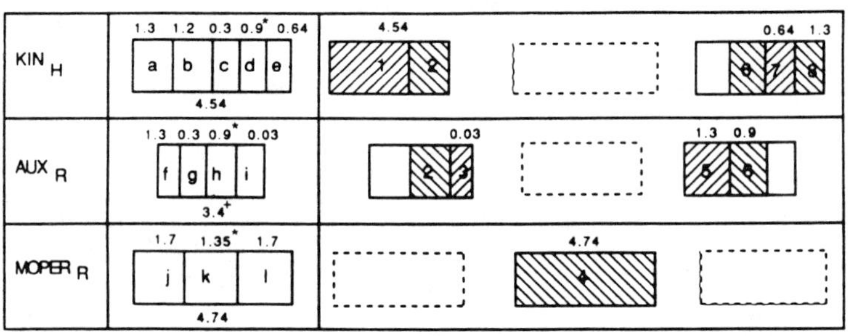

*Average value + limited by the F/T sensor (303 Hz)

Closed Loop Time = 13.45 milliseconds

a	Get θ_H / send τ_H	1	Send θ_H and compute $\Delta X_H = J_H \Delta\theta_H$
b	Calculate J_H	2	Send ΔX_H from KIN_H to AUX_R
c	Add time delay (write/read shmem)	3	Write to shared memory
d	Communicate with AUX_R (get ΔX_H / send F)	4	Send ΔX_H to SUN4/RCCL for processing and get desired θ_R
e	Compute τ_H	5	Get forces and torques (F)
f	Get resolved Cartesian forces/torques	6	Send F from AUX_R to KIN_H
g	Write to shmem	7	Transform F and compute τ_H
h	Communicate with KIN_H (get ΔX_H / send F)	8	Send τ_H to FRHC motor controller
i	Wrote to shmem		
j	Gather robot, f/t data, and ΔX_H (from shmem)		
k	Wait for SUN4 processing (transform, Inv. kinematics)		
l	Receive data from SUN4		

Figure 4: Timing Diagram

World Modeling for Sensory Interactive Trajectory Generation

L. Kelmar, R. Lumia

National Institute of Standards and Technology, Gaithersburg, MD 20899

1. Abstract

A major consideration in the design of sensory interactive trajectory generation software for a Flight Telerobotic Servicer (FTS) is the availability and maintenance of a current model of the manipulator's world. The NASA/NBS Standard Reference Model for Telerobot Control System Architecture (NASREM), which has been adopted by NASA for control of the FTS, provides a logical computing architecture for telerobotics. It defines a hierarchical control system in which complex tasks are decomposed into progressively simpler subtasks, or objectives. It contains a hierarchy of world modeling modules which maintain the system's internal model of the manipulator and its world by continuously updating the model based upon sensory data. Each world modeling module contains support processes or functions which simultaneously and asynchronously support sensory processing and task decomposition. This paper discusses the role of world modeling in support of trajectory generation and execution.

This paper discusses the world modeling modules of a hierarchical control system which facilitate sensory interactive trajectories by decoupling and supporting the sensory and manipulator planning processes. We define the types of information which should be included in the interfaces to the modules, as well as the modules' structure and function. Finally, we discuss the implementation and results of a particular sensory interactive algorithm performed in our laboratory.

2. Introduction

This paper describes the interfaces, structure, and function of the world modeling modules which support trajectory generation for a hierarchical control system for telerobots. As shown in Figure 1, the NASREM telerobot control architecture [ALB87] contains separate hierarchies of sensory processing, world modeling, and task decomposition modules. Sensory processing acquires and processes information about the state of the world, which may include data from position, force, tactile, vision, range, and other sensors. The task decomposition hierarchy decomposes high-level goals into simpler and simpler subgoals, and accomplishes them by planning and executing appropriate actions. The world model consists of modeling processes and a global data system; it serves as an intermediary between the sensory processing and task decomposition hierarchies. The world modeling processes perform calculations based on manipulator, object, and environment models to provide estimates of current and future world states. The operator control provides a means by which a human operator can access status information, supervise, and directly control the telerobot at any level of the system.

We are implementing the NASREM control architecture in our laboratory. Our telerobot system consists of a manipulator and a camera. The processing modules, for this subset of the NASREM architecture, can be recombined according to their function in the system, as shown in figure 2. The system consists of two main *branches*; the left branch contains the perception processes and the right branch contains the manipulation processes. The perception processes provide sensory

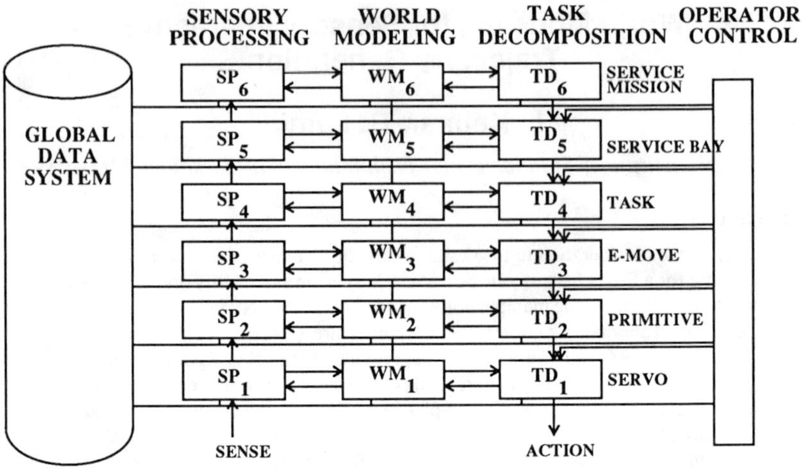

Figure 1. A Hierarchical Telerobot Control System Architecture.

feedback from the camera; the manipulation processes plan and execute manipulator trajectories. Note that while the two branches decompose tasks independently within each branch, communication between processes, both within a branch and across branches, occurs via the global data system. The perception and manipulation branches cooperate during trajectory generation and execution.

The second level of the task decomposition hierarchy is called the Primitive Level (Prim). It generates dynamic motion and force commands from a static description of the desired behavior of a device. Prim creates the time sequence of *attractor sets* needed to produce a *dynamic trajectory* and sends these as commands to the Servo level of the task decomposition hierarchy [WAV88].

Real-time trajectory generation based on sensory feedback, such as vision and

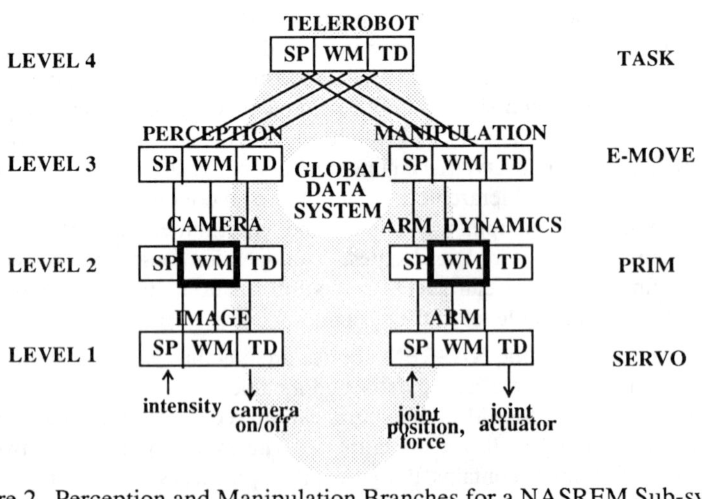

Figure 2. Perception and Manipulation Branches for a NASREM Sub-system.

proximity, allows performance of tasks based on sensed data rather than a priori knowledge. This provides for robustness of task execution amid incomplete or uncertain knowledge of the environment. Robot positioning systems base their corrective motions on the difference between the desired, or reference, joint positions and the actual ones. Because of possible errors in the manipulator kinematic model and/or the locations of object in the workspace, the relative position of the end-effector and the object may be in error. With the addition of perceptual sensory feedback, the control system can more reliably relate the relative pose of the manipulator's end-effector to objects in the world.

The discussion in this paper focuses on the world modeling modules highlighted in figure 2: one which supports the manipulator trajectory generation module (Prim) and one which supports visual perception. In the following sections we characterize the algorithm information passed to the support modules and the corresponding modeling information which is returned. Algorithm independent interfaces, which allow for both pre-planned and sensory interactive trajectories, will be defined. We also introduce the role of the operator interface for initializing perception processing. Section 6 discusses a particular sensory interactive algorithm executing on our system and shows how the modules defined accommodate the algorithm.

3. World Modeling to Manipulator Prim Task Decomposition Interface

To be able to perform tasks, FTS manipulators must be controlled by a motion generation system which enables the specification, planning, and execution of a wide range of dynamic trajectories. The task decomposition servo and trajectory generation software modules most affect these behaviors. Together, the modules define what types of trajectories, or large dynamic motions, the manipulators can perform (trajectory generation), and how well the manipulator will be able to perform these motions (servo) [LUM89].

The Prim task decomposition module generates and executes plans for dynamic trajectories for a manipulator [WAV88]. The plans may give explicit velocity, and acceleration profiles as a function of time. Alternatively, the plans may specify the shape or characteristics of the trajectory, without explicitly defining the path. For example, when using vision data in real time to perform a trajectory toward a moving object, the path that the manipulator follows in space is determined *as* the trajectory is being performed. In such cases, it may be more appropriate to simply command a goal state which defines *what* is to be achieved, along with an algorithm specification that defines *how* to achieve it. These types of algorithms perform what is referred to here as *sensory-interactive trajectory generation.* The world modeling modules provides the necessary connection between the manipulator and the perception sensory processing.

The goal of this section is to characterize the algorithm information computed and maintained by the Prim world modeling module. The information listed below is not specific to any one algorithm; it defines the module which supports a wide range of trajectory generation algorithms [KEL89]. We begin with support for the Prim in planning trajectories.

- **Forward and Inverse Kinematics** - The manipulator configuration specification may be the desired Cartesian end-effector pose or velocity, or it may include additional manipulator configuration parameters. For example, if the manipulator is redundant, it may be desirable to specify the configuration of the manipulator's elbow, as well as that of the end-effector.

- **Current (Actual) Arm Position and Velocity** - Prim bases its future movements on the current position and velocity of the manipulator.

- **Actuator Limits** - Prim requires the torque and acceleration limits, which are a function of the manipulator configuration (joint angles). Also, world modeling must provide the joint position limits for the manipulator.

- **Arm and Payload Dynamics** - The manipulator dynamics must be considered in order for the full capabilities of the manipulator to be available and to prevent it from exceeding actuator limitations. When the manipulator carries an object, the dynamic model should reflect the additional mass.

- **Gain Information** - Prim requires gain information for various motions. The information may be constant and stored in the global data system, or may be adjusted based upon experimental or run-time results. Often, the choice of gains is strictly task dependent and is maintained Prim.

- **Object Data** - The world modeling support module must be able to provide Prim with a variety of object data. Object data may be available in the database or may be the result of etrapolation or prediction by world modeling. For example, the position of an object may be constant or may be predictable based upon the velocity profile of the object. The exact nature of the data supplied by the world modeling support module depends on the algorithms implemented. The interface between Prim and its world modeling support module must be rich enough to support Prim in planning and executing trajectory generation algorithms. It should include the position and velocity of the object(s) of interest, as well as any tolerances and kinematic constraints for mating objects.

During execution of a trajectory, world modeling supplies Prim with various sensed values. The exact sensory information passed from world modeling to Prim depends on both the particular algorithm being executed and the sensors available in the system. Typically, sensory interactive algorithms require that world modeling provide continuously updated readings. For example, force and torque readings provide important feedback when the manipulator performs contact tasks, such as inserting a peg in a hole. The data used by Prim enable manipulation when a priori information is not sufficiently recent or precise to complete the task. Sensor data may include:

- **Input Device Information** - For teleoperated control, the manipulator trajectory is guided by an input device, such as a master or joystick. The values from the input sensor, such as desired manipulator joint velocities, are transformed (if necessary) and made available by World Modeling.

- **Manipulator Position and Velocity** - The Execution module requires the position and velocity of the manipulator as feedback. The values may be in joint or Cartesian space.

- **Object or Feature Position (relative displacement)** - Often the precise location of an object cannot be known with sufficient accuracy prior to execution of the task. For such situations, sensory feedback during task execution can enable successful completion of the task.

- **Force and Torque Values** - Feedback from manipulator force and torque sensors can be used to detect a manipulator's contact with the environment. The reading, together with knowledge of how to make corrective motions, also can improve the performance of manipulators in assembly operations.

Each World Modeling module also interfaces to Sensory Processing. The Manipulator Prim World Modeling module interfaces to the Manipulator Level 2 Sensory Processing module, as shown in figure 2. Manipulator sensor data includes readings from manipulator sensors, such as joint encoders, force/torque sensors, and proximity sensors. At this time, it appears that only one level of sensory processing is necessary for manipulator sensor data; it is filtered and stored at Level 1.

4. World Modeling to Level 2 Perception Sensory Processing Interface

The Level 2 sensory processing module for image data extracts features from iconic information received from Level 1 and converts the features to symbolic representations. The types of features include edges, corner points, centroids, and moments. The features are entered into the global data system to be used by a manipulator control module, by higher level sensory processing modules for scene interpretation, or by both. World modeling maintains a history of recent readings which it uses for filtering out spurious readings. The world modeling support module also provides predictions of expected features. The predictions may be based upon the histories of readings, the knowledge of the camera's location with respect to the object of interest, as well as knowledge of how the object and/or camera is moving.

For example, consider the task of tracking a ball moving on a planar surface. In one approach, Level 1 image sensory processing acquires two sequential images and subtracts them. The resulting iconic representation, of the difference in intensities between the images, represents the area of motion. Level 2 sensory processing computes the centroid of the difference image. If the ball is the only object which changes position between successive images, then the algorithm successfully locates the ball. However, if the integrity of the scene cannot be guaranteed, a model should be employed to isolate the motion of the ball from possible motion of other objects. World modeling supports sensory processing by providing a window of interest around the expected location of the ball in each image. The predictions are based upon previous reading(s) and knowledge about the movement of the objects between sampling instants.

5. World Modeling to Operator Interface

The operator interface provides a means by which human operators can observe, supervise, and directly control the system [ALB87]. The interface includes such input devices as a joystick, a mouse, and a keyboard. Operator interaction may be used in the absence of a complete world model. For example, the operator can indicate safe pathways or the location of an object to be manipulated. The transformation from the manipulator or world reference frame to the object of interest would be computed and stored in the world model. Similarly, the operator can assist the perception processing modules, especially during initialization of a task. Consider the task of tracking an object. The object, or features, of interest can be indicated via the operator interface. Thereafter, tracking the object can be accomplished autonomously by sensory processing (with windowing and predictive support from world modeling).

6. Current System Setup and Experimental Results

In this section we discuss an initial task performed in our laboratory involving the lowest two levels of the NASREM control hierarchy. The demonstration successfully integrates the sensory processing, world modeling and task decomposition modules, thus closing the sensory feedback loop. In the experiment, a small ball is released at the top of an inclined board which contains randomly placed pegs, as shown in figure 3. As the ball rolls down the board it is tracked by the single camera vision system and then caught by the manipulator at the bottom of the board. Our manipulator is a seven degree of freedom arm. The vision processing is performed by a real-time pipelined image processing machine, PIPE (Pipelined Image Processing Engine), which is commercially available through Aspex, Incorporated[1]. The software is written in Ada and runs on 68020 processors. This section details the role of the world modeling modules in closing the feedback loop between sensory processing and manipulation; figure 4 shows the world modeling computations for the task.

The Operator interface facilitates initialization of the system. The board is placed within the workspace of the manipulator. Its location is then determined by "teaching" the position of three corners of the board with the manipulator. From the three points, the position and orientation of the board, with respect to the manipulator, are computed. The camera is placed at the start of the task; the only constraint on its placement is that the entire face of the board must be in the field of view. The camera's location is determined using a four coplanar point algorithm [YEH88]. The transformations from camera to board and from manipulator to board are stored in the world model. After initialization, all computations are performed by cyclically executing processes which communicate via global read-write interfaces [FIA89].

The vision algorithm used for the experiment determines the location of the ball using the difference of successive images. The PIPE machine acquires successive images every 1/30 second. After smoothing the images, PIPE differences the entire images in one cycle. The resulting differenced image is thresholded and then passed to Level 2 for additional processing. At Level 2, the centroid of the area in the differenced image is computed for use by Prim world modeling. Currently, world modeling does not supply a window of interest around the expected location of the

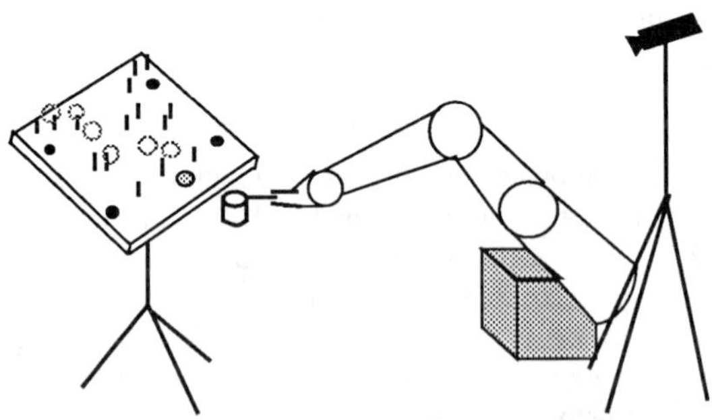

Figure 3. Ball Catching Task Setup.

----

[1]Products named in this paper are listed for purposes of information only. There is no implied endorsement of any products or implication that they are the best available for the purpose.

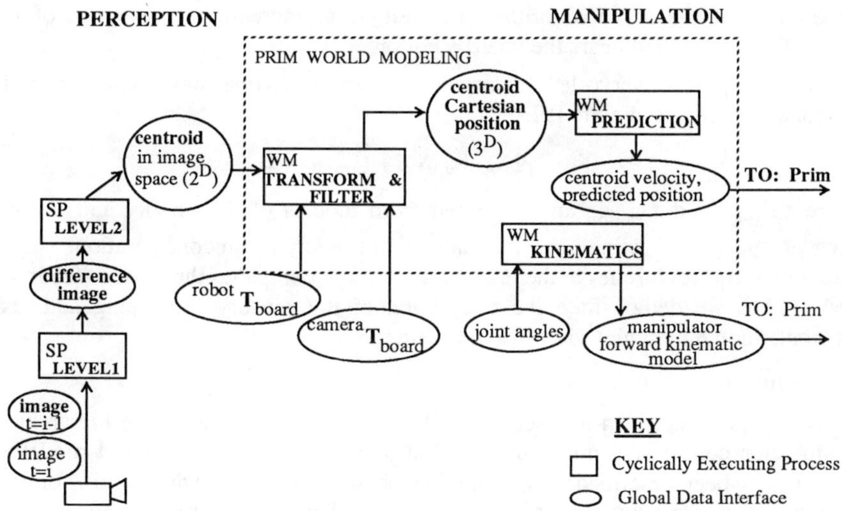

Figure 4. Level 2 World Modeling Support for Ball Catching Task.

ball.

The centroid value, as supplied by sensory processing, is a 2^D centroid in image space; Prim requires a 3^D position with respect to the manipulator. World Modeling performs several computations to transform the 2^D centroid data to a value usable by Prim. First the location of the ball's centroid in the image is represented as a 3^D point on the camera's image plane. The ball's position on the inclined board is computed by projecting a ray from the camera's optical center through the point on the image plane. The intersection of the ray with the inclined board represents the ball's location. World modeling then filters out spurious or invalid data points corresponding to locations beyond the boundaries of the inclined board. Finally, world modeling transforms the point to a convenient coordinate system for Prim, using the pre-stored transform.

The actual position of the ball, as computed from the image, is of little use to Prim. In order to catch the ball, the manipulator must move toward predicted locations of the ball. The algorithm world modeling uses to predict the location of the ball extrapolates linearly from the current and the last readings. No attempt is made to model the ball's collisions with the pegs on the board. Thus, the first order predicted ball position ($X_{t=i+1}$) sent to Prim is computed from the two most recent ball locations (X-ball) by:

$$\text{X-ball}_{t=i+1} = \text{X-ball}_{t=i} + (\text{X-ball}_{t=i} - \text{X-ball}_{t=i-1}),$$

where t represents the sampling instant. The Prim algorithm computes the projection of the (predicted) ball position on the base of the board. As the ball nears the bottom of the board, Prim also must determine how far out from the bottom of the board to catch the ball. The faster the ball is rolling, the further out from the board it must be caught. Prim sends the desired location (X_d), as well as the position and velocity control gains (K_p, K_v) for the move to Servo. The gains are adjusted to maintain the smoothness of the servo algorithm, as well as to increase the response of the manipulator as the ball nears the board's bottom.

The manipulator Servo level computes the control torques based upon errors in Cartesian space according to [FIA90]:

$$\tau = J^T [K_p (X_d - X) - K_v (\dot{X}] + \tau_{gravity} + \tau_{friction},$$

where $\tau_{gravity}$ and $\tau_{friction}$ are computed from models of the gravity and friction forces, respectively, acting upon the manipulator in any given configuration. Upon sending the control torques to the manipulator, Servo completes the sensory feedback loop. Thus we demonstrate the integration of the sensory processing and task decomposition hierarchies via world modeling.

7. Conclusions

This paper has given a description of Prim world modeling for a hierarchical manipulator control system. The function and interfaces of the world modeling module have been described. Algorithm independent interfaces, which allow for both pre-planned and sensory interactive trajectories, have been defined. Finally, we demonstrated how a sensory interactive trajectory generation algorithm is accommodated by the modules' structure and interfaces.

8. References

[ALB87] Albus, J.S., McCain, H.G., Lumia, R., *NASA/NBS Standard Reference Model Telerobot Control System Architecture (NASREM),* NASA Document SS-GSFC-0027, June 18, 1987.

[FIA89] Fiala, J., "Note on NASREM Implementation," NIST, Gaithersburg, MD, Dec. 1989.

[FIA90] Fiala, J. and Wavering, A., "Implementation of a Jacobian-Transpose Algorithm," ICG Note #23, NIST, Gaithersburg, MD, Jan. 1990.

[KEL89] Kelmar, L., "Manipulator Primitive Level World Modeling," NIST Technical Note 1273, NIST, Gaithersburg, MD, Oct. 1989.

[LUM89] Lumia, R., Wavering, A., "Trajectory Generation for Space Telerobots," Conference on Space Telerobotics, Pasadena, California, Jan. 30, 1989.

[WAV88] Wavering, A., "Manipulator Primitive Level Task Decomposition," NIST Technical Note 1256, NIST, Gaithersburg, MD, Oct. 1988.

[YEH88] Yeh, P.S., Barash, S., Wysocki, E., "A Vision System for Safe Robot Operation," *Proceedings of the 1988 IEEE Int'l Conference on Robotics and Automation*, pp. 1461-1465, April 24-29, 1988.

A Ground-Remote Telerobot Programming and Control System

Paul G. Backes, Kam S. Tso, Samad Hayati and Thomas S. Lee
Jet Propulsion Laboratory
California Institute of Technology
Pasadena, California

Abstract

A ground-remote telerobot programming and control system for space telerobotics has been conceptualized and an initial laboratory implementation tested. The system consists of a local site, where an operator creates and issues commands, and a remote site, where a system executive receives commands and controls task execution. The local site includes a user interface and a simulator where task execution is simulated before sending commands to the remote site. The remote site consists of three 6 degree of freedom manipulators, a task executive which receives commands, and task execution primitives which the task executive calls as commanded from the local site. Two of the remote site manipulators are used for task execution and the third is used to position cameras.

I. Introduction

Telerobotics will be increasingly utilized in space missions to execute tasks. A telerobot is a robot or group of coordinated robots that can be controlled via interaction with a human. Near-Earth applications include satellite servicing and servicing and assembly at Space Station Freedom [1]. More distant applications will occur in missions to the planets. Three types of control may be utilized in a telerobot: teleoperation where all robot motion is input by a human using hand controllers; autonomous control where all motion of the robots is generated by computer control; and shared control where inputs from an operator using hand controllers are merged with autonomous control inputs during task execution. Two further distinctions are traded control, where teleoperation and autonomous control can be used interchangeably during different parts of a task, and supervisory control where autonomous control is used for task execution but an operator specifies the task and task execution parameters and can intervene at any time.

Some space telerobotic tasks will be performed by an operator in space near the telerobot. But for many missions, it may not be feasible to colocate the operator control station and the telerobot in space. The costs in space qualifying, launching, and maintaining the operator control station hardware and software for a remote space site may be prohibitive. Also, there are many dangers inherent in space activities which can be eliminated by placing the operator at an Earth based station. The cost of supplying life support systems for long duration missions, e.g., missions to Mars, makes telerobotics with an Earth based control station a useful alternative, especially on unmanned precursor missions. Also, a space based operator control station will likely have less capability than an Earth based control station due to computation limitations at the space site.

With a colocated space based operator control station and telerobot, there will be little communication time delay between the issuing of commands at the operator control station and execution at the telerobot. In this case, teleoperation, shared, and autonomous control can all be used. As the distance between the operator and telerobot increases (for example, an Earth based operator control station and a space based telerobot), the communication time delay will increase with performance effects varying for the different modes of control. The most seriously affected mode of control is teleoperation. A delay in command (hand controller positions) and feedback information (e.g., video and force reflection) will reduce the operator's effectiveness. For near Earth tasks where round trip delays of a few seconds or less between the Earth based station and the remote telerobot may be expected, teleoperation may still be possible, but with greatly reduced capability. As the distances get even greater, such as a Mars mission, the time delays will prevent teleoperation as an option. Shared control may add some robustness to time delay to teleoperation, but delays of more than a few seconds will cause serious degradations of shared control also. Autonomous control is least affected by time delay. Autonomous commands from the operator control station can be in the form of task descriptions and task execution parameters. Closed loop task execution is then done completely at the remote site. A drawback with increasing time delay in any mode is the delay time for intervention from the operator control station.

Due to the limitations of teleoperation and shared control for long time delays and limited computation capability at the remote telerobot site, supervisory control telerobotic systems with a local-remote architecture may be the best near term solution for utilizing robotics in space. The local site is a ground (Earth) based operator control station and the remote site is the space telerobot. This paper describes such a ground-remote telerobot programming and control system and an initial implementation. The operator creates and issues commands at a local site which are sent for execution at the remote site telerobot. The remote site telerobot consists of two manipulators used for task execution and a third used to position cameras. The initial implementation has only single arm control. An extension to dual-arm cooperative control is expected in the coming year. Task execution can also be simulated at the local site.

This paper focuses on interactive task description and on task execution. Other important features such as task planning, modeling and vision are discussed only briefly. Section II describes the system architecture. Sections III and IV discuss task description and task execution. The implementation environment is described in section V and conclusions in section VI.

II. System Architecture

Depending on the capabilities of the system, different architectures may be utilized. A simple architecture is shown in figure 1 and a more complex architecture providing more features is shown in figure 2 [2]. The simple architecture of figure 1 would require the operator to perform most of the task planning and description. The more complex architecture of figure 2 is a proposed future architecture which utilizes more autonomous task planning, reasoning, and analysis to potentially relieve the operator of some of this burden. The system described in this paper fits into both architectures. The operator interface is the Synthesis subsystem of the simple architecture and a part of the Process-Level Operator Interface in the more complex architecture. The task execution system is the Execution subsystem of the simple architecture and is the Manipulation subsystem of the complex architecture. The architecture of figure 1 is the first

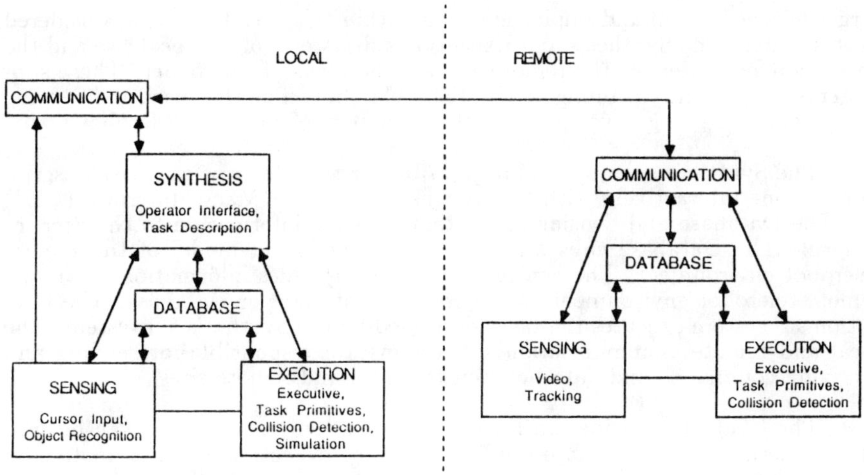

Figure 1: Simple local-remote telerobot architecture

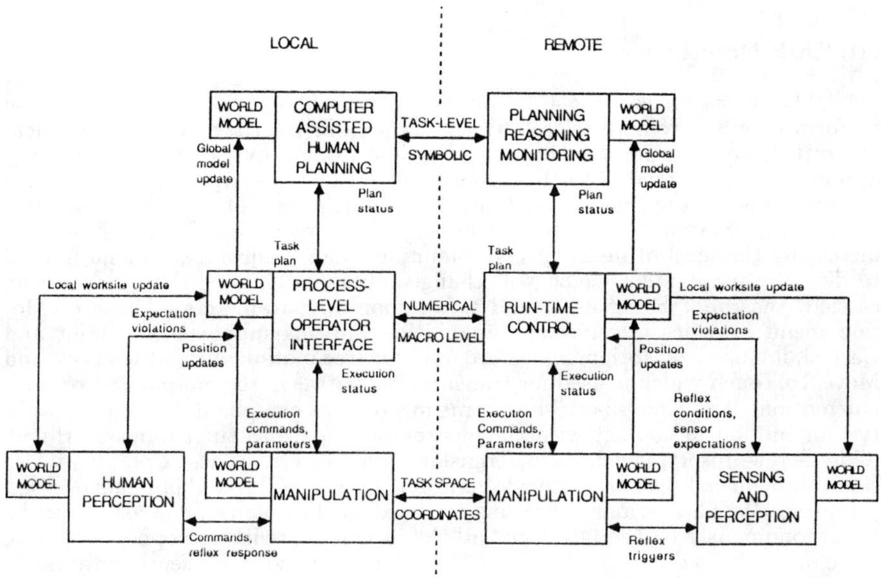

Figure 2: Alternative local-remote telerobot architecture

target for the system and implementation within this architecture is considered in this paper. The Synthesis and Execution subsystems of the local site and the Execution subsystem of thc remote site are the focus of this paper. These subsystems are presently running on the JPL Telerobot. The Database and Sensing subsystems are being developed by other members of the JPL Telerobot team.

The Synthesis subsystem is responsible for task description. Task description is done interactively with the user using the User Macro Interface (UMI) [3]. The Database and Sensing subsystems provide information to the user interface. The Database subsystem maintains the world model of the remote telerobot environment. The Sensing subsystem provides information about the remote telerobot environment to the operator interface or Database. The Execution subsystem executes the tasks described by the Synthesis subsystem. The local and remote Communication subsystems are responsible for sending and receiving commands and data between the local and remote sites.

The local site provides an Execution subsystem so that tasks can be simulated at the local site before sending them for execution at the remote site. The local and remote Execution subsystems are identical. The local Sensing subsystem provides video images of the remote site environment and interactive selection of objects or points in the environment to the operator interface. The remote Sensing subsystem gathers video and other sensor information and sends it back to the local site Sensing subsystem via the Communication subsystems. Some sensor information is also gathered and sent by the remote Execution subsystem. The local Database subsystem builds and maintains the world model of objects in the remote environment and the remote Database subsystem maintains an image of the world model developed at the local site.

III. Task Description

The operator uses the UMI operator interface to specify the task to be performed by the remote telerobot. The operator interface has a hierarchical design to lead the operator through task description, giving the operator only options which are related to the present level of task description. Seven menus are selectable at the highest level: Environment, Free Motion, Compliant Motion, Grasp, Status, Save, and Sequencing. Each menu has further submenus increasing the level of detail of task planning. The Environment menu is used to describe parameters which will change infrequently, e.g., the object being carried, the robot being used, and safety monitor parameters. The Free Motion menu provides free motion capabilities and has submenus for Joint and Cartesian Guarded Motion, which are used for free motion without contact, and Move To Touch which is used for transitioning between free motion and compliant motion. The Compliant Motion menu provides compliant motion capabilities for motion in contact with the environment and has the submenus Hinge, Slide, Screw, Insert, Level, Push, Translate, and General. These Compliant Motion submenus all set parameters for one general Generalized-Compliant-Motion primitive [4]. The various submenus are used so that many parameters can be set autonomously by the interface and the interaction with the operator can be more natural since the context of the task is more specific. Presently only single arm control is available. When dual-arm cooperative control is implemented (expected in the coming year), the menu hierarchy should be the same for single and dual-arm tasks since tasks are described by the action to be performed, independently of the number of arms attached to the object. Only specification of the specific single or dual-arm control parameters will be different depending on whether one or two arms is being used.

The Grasp menu provides free and compliant grasp capability. The Status menu provides current position and force information from the remote site. The Save menu allows the operator to save position information for later use. The Sequencing menu is used to save and execute commands which the operator has interactively built. Using sequencing, an operator can specify the parameters of a task command and then save the task command as a sequence. Later execution of the command is achieved by simply selecting the sequence name rather than specifying all of the parameters again. Also, sequences can be nested so that a new sequence can be made from a combination of previously built sequences.

Several aids are available to the operator. Default parameters are provided to the operator for each task type. The operator only needs to enter changes to the default values. Input values are checked for validity within expected bounds. In the planned future version of the system, the Sensing system will also provide the operator with 6 dimensional (position and orientation) destination information via a cursor overlayed on a stereo image of the remote scene and object selection using the cursor and the world model objects overlayed on the scene. Also, the result of task execution at the remote site is displayed to the operator.

There are various parts of task description. The operator describes the motion using the Free Motion, Compliant Motion, and Grasp menus. Along with the execution information, the operator also specifies the termination conditions and acceptable command results. The operator can also specify parameters for background monitoring which is done during task execution. The background monitors check for conditions such as force and joint limits.

The local site Execution subsystem is used for simulation of tasks on a graphics simulator. The local and remote Execution subsystems may also provide collision detection in future versions. Execution is directed to the local Execution subsystem rather than the remote site if the Environment menu simulation parameter is set.

IV. Task Execution

Task execution occurs within the local or remote Execution subsystem. The Execution executive receives task sequences which include task commands and monitor commands. Task commands come in the form of a primitive name, parameters, and acceptable command results. Task commands execute only once. Monitor commands have monitor parameters and are executed concurrently with succeeding task commands until a new monitor command is received. A direct task command from the local site which was not described as a sequence is packaged as a sequence before transmitting it to the remote site. The executive will successively execute the commands of a sequence until the result of a command which executed is invalid when compared to the acceptable command results given in the parameter list for that command. The executive will send command results to the user interface as well as periodical execution status.

V. Implementation Environment

The hardware and software implementation environment utilized for the system is described in [5]. The local site has a Sun3 for the operator interface and a Sun4 for task simulation. The remote site has several VME chassis and a

Sun4 for control and three PUMA 560 manipulators. Two of the manipulators are used for task execution and the third is used to position cameras. Task execution software is written utilizing the RCCL Robot Control C Library [6].

VI. Conclusions

Ground-remote telerobot control provides many benefits for space telerobot applications. Safety and cost savings as well as increased computing power are achieved by placing the operator control station at an Earth based site. The effects of communication time delay are minimized by operating in a supervisory control mode. The ground-remote telerobot programming and control system described in this paper is an initial step toward actual future space qualified ground-remote telerobot systems.

Acknowledgements

The research described in this paper was performed at the Jet Propulsion Laboratory, California Institute of Technology, under contract with the National Aeronautics and Space Administration. The authors would like to thank Mr. Brian Wilcox for his original ideas in ground-remote operations, and Mr. John Lloyd and Mr. Mike Parker for providing RCCL.

References

[1] Wayne Zimmerman. Telerobotic control issues related to real task applications in the space environment. In *13th Annual AAS Guidance and Control Conference*, Keystone, Colorado, 1990.

[2] Jake Matijevic. Update of the conceptual architecture for telerobot/rover. December 20 1989. Jet Propulsion Laboratory Interoffice Memorandum 3473-89-078 (internal document).

[3] Paul G. Backes and Kam S. Tso. Umi: an interactive supervisory and shared control system for telerobotics. In *Proceedings IEEE International Conference on Robotics and Automation*, 1990.

[4] Paul G. Backes. Generalized compliant motion task description and execution within a complete telerobotic system. In *Proceedings IEEE International Conference on Systems Engineering*, August 9–11 1990.

[5] Samad Hayati, Thomas Lee, Kam Tso, and Paul G. Backes. A testbed for a unified teleoperated-autonomous dual-arm robotic system. In *Proceedings IEEE International Conference on Robotics and Automation*, 1990.

[6] J. Lloyd, M. Parker, and R. McClain. Extending the rccl programming environment to multiple robots and processors. In *Proceedings IEEE International Conference on Robotics and Automation*, pages 465–474, 1988.

DESIGN OF A TELEROBOTIC CONTROLLER WITH JOINT TORQUE SENSORS USING 2-PORT NETWORK THEORY*

J. F. Jansen J. N. Herndon
Oak Ridge National Laboratory
Oak Ridge, Tennessee 37831-6005

ABSTRACT

The purpose of this paper is to show how to design a joint controller for a telerobotic system when joint torque sensors are available based on 2-port network theory. Mechanical manipulators with high levels of friction are assumed in this study. Very high levels of friction for the master have been reduced using high-gain feedback while avoiding limit cycles. Finally, the results are applied to a telerobotic system built for NASA and called the Laboratory Telerobotic Manipulator (LTM).

DEFINITIONS

The following definitions and symbols will be used in this paper:

n	$\triangleq$ gear ratio,	J_L	$\triangleq$ load or joint inertia,
J_m	$\triangleq$ motor inertia,	i	$\triangleq$ motor current,
KT	$\triangleq$ motor torque const.,	T_s	$\triangleq$ drive train torque,
F^s_{ro}	$\triangleq \dfrac{N^m_{co}\, KT^m\, G^m_{po}}{N^s_{co}\, KT^s\, G^s_{po}}$,	M_g	$\triangleq 1 + n\, KT\, G_{tor}$,
e	$\triangleq$ positional error,	T_L	$\triangleq$ load torque,
θ_m	$\triangleq$ motor pos.,	θ_L	$\triangleq$ joint pos.,
θ_{des}	$\triangleq$ desired joint pos.,	$N_c(s)$	$\triangleq \dfrac{K_s}{n\, Z_m\, s}$,
Z'_m	$\triangleq J_m s + K_{vm}$,	Z'_L	$\triangleq J_L s + K_{VL}$,
Z_m	$\triangleq Z'_m + KT\, K_{vin} + \dfrac{M_g K_s}{n^2 s}$,	M_{go}	$\triangleq 1 + n\, KTG_{tor}\vert_{s=0}$,
e^m	$\triangleq \theta^s_L - \theta^m_L$,	e^s	$\triangleq \theta^m_L - \theta^s_L$,
N_{co}	$\triangleq N_c(s=0) = \dfrac{n}{M_{go}}$,	Z_L	$\triangleq Z'_L + \dfrac{K_s}{s}\left(1 - \dfrac{N_c M_g}{n}\right)$,
Z_{12}	$\triangleq N^m_c\, KT^m \left(\dfrac{G^m_p}{s} + G^m_v\right)$,	Z_{21}	$\triangleq N^s_c\, KT^s \left(\dfrac{G^s_p}{s} + G^s_v\right)$,
K_{VL}	$\triangleq$ joint viscous frict.,	K_{vm}	$\triangleq$ motor viscous frict.,
τ^m_L	$\triangleq -(T^m_L + NLFL^m + N^m_c\, NLFm^m)$,	Z_{22}	$\triangleq Z^s_L + Z_{21}$,

---

*Research performed at Oak Ridge National Laboratory, operated by Martin Marietta Energy Systems, Inc., for the U.S. Department of Energy under Contract No. DE-AC05-840R21400, and sponsored by the National Aeronautics and Space Administration, Langley Research Center.

$\tau_L^S \triangleq -(T_L^s + NLFL^s + N_c^s\, NLFm^s)$, $\quad Z_{11} \triangleq Z_L^m + Z_{12}$.

INTRODUCTION

Merging the mechanical and control features necessary for a force-reflecting teleoperated manipulator and a robotic positioner into a single telerobotic system is a particularly difficult engineering task. A good force-reflecting servomanipulator designed for efficient human-in-the-loop control emphasizes end-effector speed for good master response to human control input, good slave tracking of the master, good joint backdrivability for force reflection, and low friction and inertia reflection to minimize operator fatigue. On the other hand, a good robotic positioner emphasizes end-effector accuracy, end-effector speeds, and mechanical and control stiffness. These design goals tend to conflict. One of the most significant differences is the issue of backdrivability for a teleoperated system. Reduction of apparent friction by means of a torque sensor was examined by Luh et al. [Luh,83]. While Luh examined only the robotic mode, teleoperation will be the main emphasis of this paper; however, the results can be applied to the robotic mode.

LINEAR ANALYSIS

The block diagram for a fairly generic single joint and its controller is shown in Figure 1. The blocks represented by G_p, G_v, and G_{tor} are the transfer functions of the joint position, velocity, and torque signals, respectively. The blocks represented by K_{vin} and K_{vm} are the "inner" velocity gain and the motor viscous friction gain, respectively. Unmodeled dynamics due to motor saturation, gear boxes, coulomb, and stiction friction effects are extremely significant in the design of a controller according to Good and Sweet [Good,85]. Six nonlinear terms are shown in the figure: (1) motor saturation (NLSat), (2) motor nonlinear friction (NLFm), (3) backlash of the gear box (NLB), (4) gear box nonlinearities (NLGB), (5) joint nonlinear friction (NLFl), and (6) nonlinear drive compliance (NKs). Linear analysis is helpful to physically understand how torque feedback can reduce the apparent friction levels.

For the linear analysis of the system shown in Figure 1, replace the nonlinear gear box (NLGB), backlash (NLB), and motor saturation function (NLSat) with a unit-gain block. Replace the nonlinear drive compliance with a linear one, K_s. The nonlinear friction terms (NLFm and NLFL) will be modeled as disturbance signals. Furthermore, set the calculated gravitational current signal i_{grav} to zero. The governing equations describing the joint dynamics as seen in the block diagram are:

$$e = \theta_{des} - \theta_L \ , \tag{1}$$

$$i = G_p\, e + G_v\, \dot{e} - G_{tor}\, T_s - K_{vin}\, \dot{\theta}_m \ , \tag{2}$$

$$\tau_e = KT\, i \ , \tag{3}$$

$$J_m\, \ddot{\theta}_m + K_{vm}\dot{\theta}_m = \tau_e - NLFm - \frac{T_s}{n} \ , \tag{4}$$

$$J_L\, \ddot{\theta}_L + K_{VL}\dot{\theta}_L = T_s - T_L - NLFL \ , \tag{5}$$

$$T_s = K_s \left(\frac{\theta_m}{n} - \theta_L\right) . \tag{6}$$

After taking the Laplace transform and after some tedious algebra, the governing Eqs.(1) to (6) can be written compactly as:

$$Z_L\,\theta_L = N_c\left[KT\,(G_p\,e + G_v\,\dot{e}) - NLFm\right] - (T_L + NLFL) \qquad (7)$$

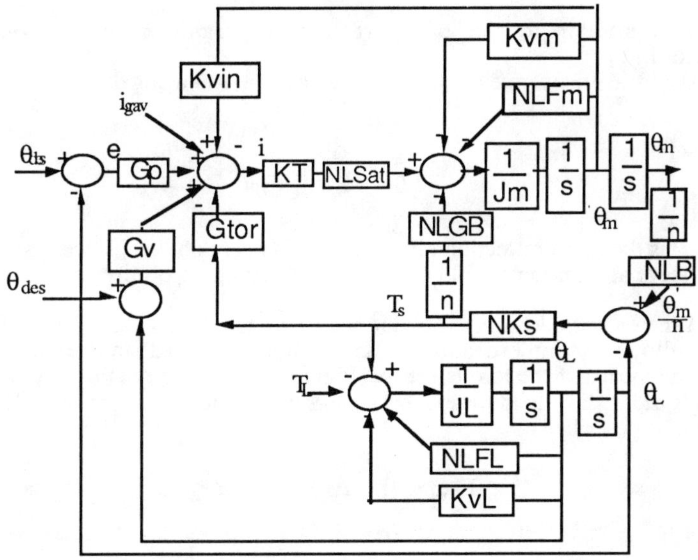

Figure 1 - Block Diagram Showing Joint Nonlinearities.

ANALYSIS OF TELEOPERATED SYSTEMS

The previous analysis can be extended to a teleoperated system that has a master/slave pair. A superscript in the upper right-hand corner of a variable is used to denote either a master (m) or a slave (s) term. The superscript i will denote either a master or slave term. A superscript ss in the upper left-hand corner will denote a steady-state term. The variable s will denote either a slave variable and/or the derivative operator in the s domain and will be clear from the context of the discussion. Also, the variable m will denote either a master variable and/or a motor term. Eq. (7) can now be modeled to denote the dynamics of both the master and the slave:

$$Z_L^m\dot{\theta}_L^m = N_c^m\left[KT^m(G_p^m\,e^m + G_v^m\dot{e}^m) - NLFm^m\right] - (T_L^m + NLFL^m), \qquad (8)$$

$$Z_L^s\,\dot{\theta}_L^s = N_c^s\left[KT^s\,(G_p^s\,e^s + G_v^s\dot{e}^s) - NLFm^s\right] - (T_L^s + NLFL^s). \qquad (9)$$

In matrix form these equations can be compactly written as

$$\begin{bmatrix} Z_{11} & -Z_{12} \\ -Z_{21} & Z_{22} \end{bmatrix}\begin{bmatrix} \dot{\theta}_L^m \\ \dot{\theta}_L^s \end{bmatrix} = \begin{bmatrix} \tau_L^m \\ \tau_L^s \end{bmatrix}, \qquad (10)$$

where the τ_L^i terms include the external torques and the nonlinear friction terms. Another convenient form is

$$\begin{bmatrix} Z_{22}/Z_{21} & -1/Z_{21} \\ \Delta / Z_{21} & -Z_{11}/Z_{21} \end{bmatrix} \begin{bmatrix} \dot{\theta}_L^s \\ \tau_L^s \end{bmatrix} = \begin{bmatrix} \dot{\theta}_L^m \\ \tau_L^m \end{bmatrix} . \qquad (11)$$

Eq.(11) is similar to the two-port hybrid parameter model found in [Hannaford,89].

An ideal teleoperated system should have the following dynamic form:

$$\begin{bmatrix} 1 & 0 \\ 0 & -F_r^m \end{bmatrix} \begin{bmatrix} \dot{\theta}_L^m \\ \tau_L^m \end{bmatrix} = \begin{bmatrix} \dot{\theta}_L^s \\ \tau_L^s \end{bmatrix} , \qquad (12)$$

where F_r^m is the force reflection ratio. Eq.(12) is fairly obvious since it says only that the slave and master should have perfect tracking capability and a set force-reflection ratio.

When a static load is applied to the slave, the steady-state load felt at the master follows a simple deadband curve where the deadband region can be reduced by means of the torque sensor. Figure 2 shows the relationship between the steady-state slave load torque, ${}^{ss}T_{Lo}^s$, and the master load torque, ${}^{ss}T_{Lo}^m$. The curve can be described as

$$-{}^{ss}T_{Lo}^m = \begin{cases} F_{ro}^s\,(\,{}^{ss}T_{Lo}^s - \Delta_B\,) & , \text{for } {}^{ss}T_{Lo}^s > \Delta_B \\ 0 & , \text{for } -\Delta_B \le {}^{ss}T_{Lo}^s \le \Delta_B \\ F_{ro}^s\,(\,{}^{ss}T_{Lo}^s + \Delta_B\,) & , \text{for } {}^{ss}T_{Lo}^s < -\Delta_B \end{cases} , \qquad (13)$$

where

$$\Delta_B = \frac{1}{F_{ro}^s}\left[\,|NLFL_o^m| + N_{co}^m\,|NLFm_o^m|\,\right] + |NLFL_o^s| + N_{co}^s\,|NLFm_o^s| \; . \qquad (14)$$

$$NLFL^i \rightarrow NLFL_o^i = \text{constant as } t \rightarrow \infty ,$$

$$NLFm^i \rightarrow NLFm_o^i = \text{constant as } t \rightarrow \infty .$$

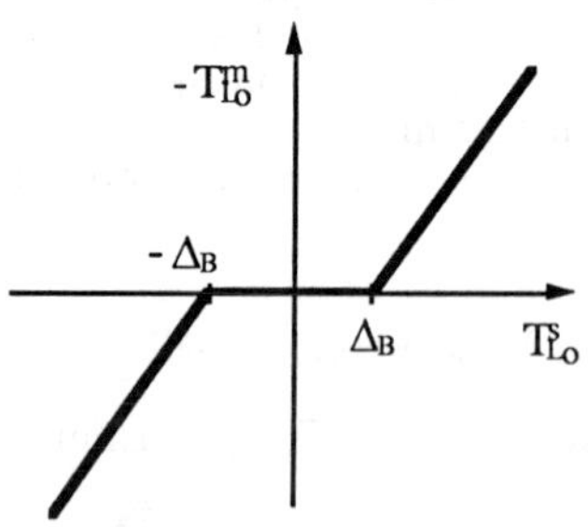

Figure 2 - Deadband Curve for Force Reflection.

Examining Eq.(14) reveals that the apparent motor nonlinear friction can be reduced by means of the torque signal being fed back to the motor because the

N^{i}_{co} terms are a function of the torque signal. High-gain torque feedback will reduce Δ_B, which means that force threshold sensitivity is increased.

DESIGN OF CONTROLLER

The controller design (i.e., determination of G_p, G_v, and G_{tor}) will now be formulated. From Eq.(11), the diagonal terms should dominant. Ignoring for now the nonlinear terms, let the slave move in free space, which means that $\tau^s_L = 0$. From Eq.(11),

$$\frac{Z_{22}}{Z_{21}} \dot{\theta}^s_L = \dot{\theta}^m_L \quad . \tag{15}$$

The poles of Z_{21} and the zeroes of Z_{22} must be in the LHP (left-hand plane). Eq.(15) can be rewritten as

$$\dot{\theta}^s_L = \frac{Z_{21}}{Z_{22}} \dot{\theta}^m_L = \frac{K_s\, KT\, (\, G_p + s\, G_v\,)}{Den} \dot{\theta}^m_L \quad . \tag{16}$$

where
$Den = K_s\, KT\, (\, G_p + s\, G_v\,) + n\, Z_m\, s^2\, (J_L\, s + K_{vL}) + n\, K_s\, s\, (J_m\, s + K_{vm} + KT\, K_{vin})$.
The s superscript has been dropped because only the slave manipulator is being discussed. The compensators, G_p and G_v, should be set as follows:

$$G_p = \frac{M_g}{n\,KT} K_p \quad \text{and} \quad G_v = \frac{M_g}{n\,KT} K_v \tag{17 and 18}$$

where K_p and K_v are positive constants. Eqs.(17) and (18) have the property that: (1) K_p will now be equal to the steady-state stiffness [Jansen,90], (2) G_p and G_v are model-dependent only on the gear box ratio n and the motor torque constant KT, and (3) the root locus plots are simplified considerably and will be shown next.

Using Eqs.(17) and (18), Eq.(16) can be recast into a root locus problem. If only M_g is allowed to vary, the root locus problem is:

$$1 + \frac{K_s \frac{M_g}{n} \left[J_L\, s^2 + \,(K_{vL} + K_v)\, s + K_p\right]}{n\, s\left(Z'_m + KT\, K_{vin}\right)\left(J_L\, s^2 + K_{vL}\, s + K_s\right)} = 0 \quad . \tag{19}$$

Let $M_g = M_{go}\, G_{Mg}(s)$ where M_{go} is the dc term of M_g, which can be varied, and $G_{Mg}(s)$ is the normalized transfer function of M_g. A detail discussion of the root locus analysis of Eq.(19) can be found in [Jansen,90]. The root locus plot of Eq.(19) when $G_{Mg}(s) = 1$ is insightful and is shown in Figure 3.

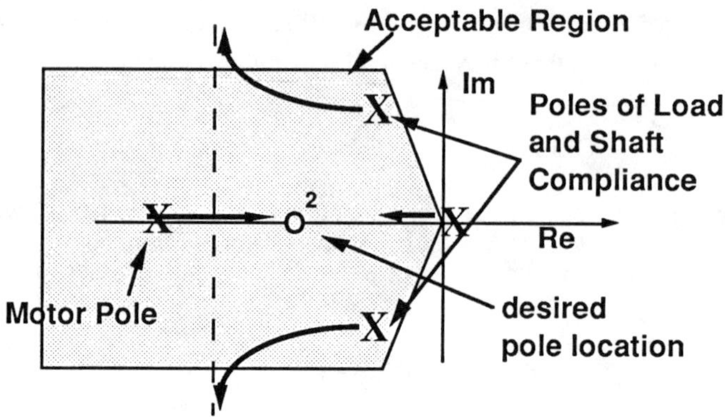

Figure 3 - Root Locus Plot without Torque Compensator.

According to the root locus plot shown in Figure 3, very large values of M_{go} will never make the system go unstable. Unfortunately, due to unmodeled dynamics (i.e., nonlinearities) the system, in practice, will go unstable for even moderate values of M_{go}. High positive values of M_{go} are needed in teleoperation to reduce the apparent friction levels around the motor. A lag-lead compensator for G_{tor} can be used to achieve high values of M_{go} [Jansen,90].

APPLICATION AND SUMMARY

The Laboratory Telerobotic Manipulator (LTM) was designed using a modular approach for joint construction, with common pitch-yaw differential joints implemented for the arm, shoulder, elbow, and wrist. This modularity is made possible through the use of distributed drive motors within the pitch-yaw modules with high local gear reductions (typically between 150:1 to 450:1). This approach is a significant departure from the normal practice for force-reflecting servomanipulators, where the reductions typically are maintained less than 50:1 for good backdriving efficiency.

The techniques described in this paper have been applied to the LTM. Due to the high-transmission ratio of each joint, the friction forces dominated the performance of the system. The most significant joints affected were the master/slave wrist pitch/yaw joints and are the only joints discussed. Frictional forces were measured by means of a portable force sensor, which gave the peak force required to overcome the retarding static-frictional forces. The frictional forces without any torque feedback are called the base friction level. Without any torque feedback, the LTM master wrist pitch/yaw joint was difficult to move. With feedback but without any compensators, the frictional forces were reduced to around 33% of the base friction level before limit cycles were observed; however, the master wrist pitch/yaw joint was still difficult to move. When the compensators were inserted, the frictional forces were reduced to around 5% of the base friction level with no limit cycles being observed.

Overall, the use of torque sensors to reduce the effective joint friction has been shown to be a very effective technique. Limit cycles were successfully avoided while using high-gain feedback.

REFERENCES

[Good,85] Good, M. C., Sweet, L. M. and Strobel, K. L., "Dynamic Models for Control System Design of Integrated Robot and Drive Systems," Dynamic Systems, Meas., Control, Vol. 107, pp. 53-59, March 1985.

[Hannaford,89] Hannaford, B., "A Design Framework for Teleoperators with Kinesthethic Feedback," IEEE Trans. Robotics and Automation, Vol. RA-4, No. 4, pp. 426-434, Feb. 1989.

[Jansen,90] Jansen, J. F. and Herndon, J. N. "Design of a Telerobotic Controller with Joint Torque Sensors," 1990 IEEE Internat. Conf. Robotics and Automation (to appear).

[Luh,83] Luh, J. Y. S., Fisher, W. D. and Paul, R. P. C., "Joint Torque Control by a Direct Feedback for Industrial Robots," IEEE Trans. Automatic Control, Vol. AC-28, No. 2, pp. 153-161, Feb. 1983.

An Expert System for Tuning Force-reflecting Servomanipulators

M. W. Noakes

Telerobotic System Section
Fuel Recycle Division
*Oak Ridge National Laboratory**
Oak Ridge, TN 37831-6304

ABSTRACT

This paper examines the use of an expert system to optimize the controls tuning of a gear and shaft force-reflecting servomanipulator designed to operate in hazardous environments without the need for hands-on maintenance. Early force-reflecting servomanipulators used cable- and pulley-driven actuation to produce low-backlash, low friction, high sensitivity operation. This configuration provides the operator with an excellent perception of feel; however, delicate mechanical maintenance procedures prevent the use of such designs in hazardous environments where contamination effects complicate or preclude human repair. These constraints led to the design of the Advanced Servomanipulator (ASM).

INTRODUCTION

The ASM project has produced the first remotely maintainable force-reflecting master/slave servomanipulator by using rugged gear and shaft drivetrains on the slave arms. This method of construction, while essential to the design for the sake of remote maintainability, creates serious controls problems relating to backlash, friction, and inertia. To provide the operator with a high-performance manipulator with adequate force-reflection, the controls engineer must make difficult decisions concerning the conflicting priorities of maximizing the operator's perception of force-reflection and minimizing operator fatigue while also dealing with complex nonlinear control algorithms and cross-coupled motions.

---

*Operated by Martin Marietta Energy Systems, Inc. for the U.S. Department of Energy under Contract No. DE-AC05-84OR21400.

Due to the heuristic nature of this tuning problem and the emphasis on human perception of performance, an expert system is considered as an alternative to the traditional trial and error methods or the more sophisticated algorithmic optimization of gains. Additionally, since the manipulator would be located in a hazardous environment, it may be called upon occasionally to operate in a degraded performance mode for an extended period while awaiting repair of damaged bearings, a failing motor, or corrosion. These conditions would call for retuning in order to maintain reasonable operation. Operators themselves may prefer to vary the tradeoffs between sensitivity and fatigue.

The primary focus of this development has been to produce a functional expert system that can be integrated with existing real-time control software. The development system is the Oak Ridge National Laboratory's Advanced Integrated Maintenance System (AIMS) which has the ASM as its principal functional component. The ASM, shown in Figure 1, is a prototypic, remotely maintainable, force-reflecting master/slave teleoperator developed for remote maintenance and handling studies within the Consolidated Fuel Reprocessing Program (CFRP) at the Oak Ridge National Laboratory's Fuel Recycle Division (FRD). One goal of the CFRP has been to advance the technology of in-cell systems planned for future nuclear fuel cycle facilities. The ASM, as part of the Advanced Integrated Maintenance System (AIMS), has provided a development tool to examine the needs and constraints necessary for hot-cell teleoperation. The expert system has been implemented using AIMS as the development platform[1,2].

The AIMS control hardware consists of fifteen single-board computers divided among seven Multibus backplanes[2,3]. Four out-of-cell control racks are interconnected with a fiber optic local area network. A Man-Machine Interface (MMI) rack handles video switching, color graphics touch screen interfacing, and controls the system hard disk and local area network. An Auxiliary Control System (ACS) rack provides positioning of the overhead manipulator transporter and 20-ton crane, as well as facility camera control. Calculations for each ASM master/slave pair are accomplished in each of the two master racks by five CPU boards acting in parallel. Three additional "slave" racks in the hot cell provide smart data acquisition systems for the ACS and master racks communicating

FIG. 1. The Advanced Servomanipulator.

through dedicated fiber optic links. The AIMS software is written entirely in multitasking polyFORTH® by FORTH, Inc., and encompasses the actual real-time controls, a graphics package, and a custom local area and control network (LAcN) in a multitasking environment.

MANIPULATOR CONTROLS

H. L. Martin, who developed the particular bilateral position-position control algorithms implemented on the ASM, established a set of goals for quality force-reflecting teleoperation, summarily to minimize operator effort and fatigue and to maintain good force-reflection sensitivity[4]. In order to meet these requirements with manipulators using a position-position control scheme, the drive gear trains should be highly back-drivable with low backlash, and manipulator weight, inertia, and drive friction must be minimized. However, even when given a good mechanical design, the various software gain parameters must still be adjusted properly before satisfactory operation is possible.

When tuning the ASM, master and slave position and velocity gains must be adjusted according to a set procedure. These values must not only be tuned to produce optimum control loop response, but they must also implement the various force-reflection ratios. The force-reflection ratios provided on the ASM include 1:1, 2:1, 4:1, 8:1, and 16:1. A ratio of 4:1 indicates that for every four pounds of force applied at the slave arm by the environment, only one pound of force is reflected back to the master arm. These force-reflection ratios are produced by the ratio of slave position gain to master position gain. This means that the controls engineer has the added restriction of maintaining the position gain ratio along with attempting to tune individual loop parameters. Nonlinear compensations that are software adjustable include master and slave deadband regions, which compensate for backlash, noise, and the potential for oscillation and limit cycling, and friction compensation, which provides a boost to overcome starting friction and to offset dynamic friction, reducing operator fatigue.

EXPERT SYSTEM IMPLEMENTATION

The controls tuning expert system developed for AIMS has been implemented using FORPS, a public domain expert system shell written by C. J. Matheus[1,5]. The largest advantage to using this software on AIMS as opposed to a conventional AI language is that FORPS is written in FORTH and may therefore call real-time control FORTH code directly as part of the action clause of a rule. FORPS uses a simple forward chaining, or data-driven inference engine where a rule table is used to organize the important information concerning the specific rules. Backward chaining, or goal-oriented search may be implemented by writing a set of rules which set intermediate goals that are then searched using forward chaining.

The actual controls tuning expert system is a composite of existing tuning procedure documentation and the author's personal experience. Knowledge required to represent and conduct the tuning process is extracted and compiled into a set of goals and rules. The tuning procedure is then broken down into the following simple components. Determine first the maximum allowable limit of velocity self feedback with no limit cycling; second, the maximum velocity feed forward that produces adequate master to slave tracking even with position gains set low; and third, determine the maximum position gains that do not cause oscillation. Position and velocity gains should be set as high as possible without instability to increase control loop stiffness and reduce master to slave phase margin. The secondary gains are then used to reduce operator fatigue due to friction, but every gain made in reducing fatigue also degrades force-reflection sensitivity.

The whole tuning process, then, is a trade-off that must be balanced based on the operator's needs for each force-reflection ratio. The first pass prototype has been narrowed down to one simple upper arm joint to simplify the scope and to make the problem achievable in a reasonable amount of time without large numbers of rules. The shoulder roll joint has been selected since it requires no consideration of counterbalancing or cross-coupling with other joints.

RESULTS AND CONCLUSION

The expert system prototype consisted of 59 rules governing the tuning process for only one of six joints. Much more work would have been necessary to implement a full tuning system for the ASM. Many expert systems require multiple iterations with years of investment. This simple expert system could not hope to have the results that an investment of years of effort would produce. However, the implementation of this prototype did provide a significant improvement in force-reflection sensitivity. In a test run to document expert system performance, the force-reflection sensitivity threshold was improved from a prior value of approximately ten pounds to less than two pounds, which is very close to the theoretical mechanical limits indicated by initial ASM design studies. The exercise of knowledge acquisition, knowledge representation, and rule generation has also improved understanding of the overall tuning process. This is apparently a typical side effect of the development of expert systems since the problem must be thoroughly analyzed, understood, and broken down into a logical format to a high degree before the expert system may be created. The goal of this project has been to produce an expert system with the capability of assisting an operator in the tuning of force-reflecting servomanipulators. Even in the preliminary stages of development, drastic improvements in manipulator performance have been realized and the principles of expert systems and where they may best be applied have been understood in the process.

REFERENCES

1. Noakes, M. W., "An Expert System Applied to the Controls Tuning of Digital Force-reflecting Servomanipulators", a Thesis presented for the Master of Science Degree, The University of Tennessee, Knoxville, December, 1989.

2. Noakes, M. W., G. K. Corbett, and C. T. Kring, "A Flexible Controls Architecture That Supports Servomanipulator Development", *Proceedings of the ANS Third Topical Meeting on Robotics and Remote Systems,* Charleston, SC, 1989.

3. Rowe, J. C., R. F. Spille, and S. D. Zimmermann, "Integrated Digital Control and Man-Machine Interface for Complex Remote Handling Systems", *ANS Remote Systems Topical,* 1987.

4. Martin, H. L., "Reduction of Friction and Inertia Effects in Digitally Controlled Force-Reflecting Servomanipulators", a Dissertation presented for the Doctor of Philosophy Degree, The University of Tennessee, Knoxville, August, 1986.

5. Matheus, Christopher J., and H. Lee Martin, "FORPS: A FORTH-Based Production System and Its Application to a Real-Time Robot Control Problem", Office of Facilities, Fuel Cycle, and Test Programs, U. S. Department of Energy, under Contract No. DE-AC05-84OR21400.

A Human Factors Research Environment for Ground-Vehicle Telerobotics Research

G. K. Corbett, S. M. Killough, and B. S. Richardson
*Oak Ridge National Laboratory**
Oak Ridge, Tennessee 37831-6005

M. Herman
National Institute of Standards and Technology
Gaithersburg, Maryland 20899

ABSTRACT

Preliminary experimentation in remote control of wheeled vehicles has been performed by the U.S. Army Human Engineering Laboratory. An early human factors testbed was demonstrated by remotely driving the soldier robot interface project vehicle and a high-mobility/ maneuverability wheeled vehicle. Experimentation with this testbed provided valuable information for developing a new human factors evaluation platform. These testbed designs and their research applications are described.

INTRODUCTION

The modernization of U. S. Army weapons and vehicles has increased the need for human factors research as applied to combat control stations. The Technology-Based Enhancement for Autonomous Machines (TEAM) Project, sponsored by the U.S. Army Human Engineering Laboratory (HEL), is concerned with fielding autonomous and telerobotic defense vehicles and studying the human-machine interface (HMI) requirements for control of these vehicles. A human factors testbed prototype was designed to perform initial human factors studies aimed toward solving the problems associated with teleoperation of wheeled combat vehicles. The testbed's capabilities were demonstrated by remotely driving the high-mobility/maneuverability wheeled vehicle (HMMWV) and soldier robot interface project (SRIP) telerobot (Fig. 1). The SRIP vehicle is skid-steered and can be driven from only the HMI; it is designed to support a 3-degree-of-freedom (DOF), 300-lb capacity platform manipulator that mates with a separate 3-DOF wrist. [1] SRIP research has focused on driving dissimilar vehicles and integrating manipulator and vehicle control from a common HMI.

---

*Operated by Martin Marietta Energy Systems, Inc., for the U.S. Department of Energy under Contract No. DE-AC05-OR21400.

FIG. 1. Soldier robot interface project vehicle (*left*) and high-mobility/maneuverability wheeled vehicle (*right*).

The HMMWV is a standard Army vehicle that is remotely or manually driven. Experiments with the HMMWV have concentrated on improving operator performance while driving under various degrees of sensory deprivation. Sensory deprivation can take two forms: one that is imposed upon the operator by the constraints of battlefield requirements and another by the limitations of remote-sensor technology. The latter form of sensory deprivation can be compensated for by improving the HMI through better sensor placement and coordinating sensor positioning with advanced operator interfaces (e.g., cameras moving in coordination with teleoperator head position). Previous research with the HMMWV did not implement such techniques, but it indicated several additional types of sensors and interfaces that would help the operator in particular situations. An example of the former type of sensory deprivation is that resulting from the use of low-detection (spread-spectrum or frequency-hopping) radios for communication in the battlefield. Standard video information requires a wide bandwidth and direct line of sight between the transmitter and receiver.[2] Video information for battlefield telerobotics must be compressed into a baud rate standard for data-radio channels so that low-detection radios can be employed for all remote-vehicle communication; however, temporal and spatial compression of video data greatly distorts the scene being viewed. Various video-compression techniques can be implemented for telerobotic operation, and the testbeds described herein have (and will be) used to determine the optimum compression technique for remotely driving wheeled vehicles. The sensory deprivation resulting from video compression can also be compensated for by improving the HMI as indicated above.

A human factors evaluation platform (HFEP) is being developed to provide a research environment flexible enough to accommodate a wide variety of interface scenarios. Both the HMMWV and SRIP vehicle will be remotely driven from the HFEP. Major areas of research will focus on

(1) determining optimum physical layouts and functional specifications of HMI components for specific combat tasks and environments, (2) optimizing operator performance while driving telerobotic combat vehicles with various degrees of sensory deprivation, and (3) integrating manipulator and vehicle control within one HMI.

TESTBED DESIGN AND RESEARCH

The HMI for remote control of the HMMWV and SRIP vehicle is shown in Fig. 2. Detailed design information is described in ref. [1] for the SRIP vehicle and ref. [3] for the HMMWV. The vehicles use a 19.2-kilobaud, half-duplex data radio and two standard broadcast television channels to communicate with the HMI. A VME-based computer is installed in the vehicles to interface sensors, actuators, and the data radio. The VME computers in both of the vehicles and the HMI use a 32-bit bus with Motorola 68020 CPUs and the OS-9™ real-time operating system by Microware Systems Corporation. The graphics interface for the HMI provided by a Macintosh II communicates with the VME computer over an RS232 serial interface. The vehicles are remotely controlled through an interface similar to that in a standard vehicle. Speed and direction are derived from brake, accelerator, and steering wheel positions. Starting or stopping the engine, gear selection, and camera controls are provided through the graphics interface, which also displays vehicle sensor information and status messages. The VME computer at the HMI processes the graphics and analog operator inputs and transmits vehicle commands through the data radio; after the commands are sent, the computer receives the vehicle status packet and communicates this information to the graphics interface. The Macintosh and data-radio interface drivers are run as parallel tasks to minimize serial-communication delays. The Macintosh and VME computer programs are written in C; thus program and database constructs can be preserved across different operating systems.

FIG. 2. Testbed human-machine interface.

Color video information is displayed on two 13-in. monitors (one for each camera). One of the two monitors can be switched to view a compressed (black and white) version of the real-time video signal. Use of compressed video data to remotely drive a vehicle is termed low-data-rate driving (LDRD). True LDRD digitizes and compresses the video information at the remote vehicle and decodes and displays the images at the HMI. This testbed simulates LDRD in that the video is digitized, compressed, and displayed–all within the HMI. Given that the communication delays are programmed into the simulation, no discernable difference exists between teleoperation of a vehicle with either simulated or actual LDRD.

The HMMWV was remotely driven on a 0.125-mile gravel track 8 h/day for seven days. Obstacles and potholes were placed on the track to test the maneuverability of the vehicle. PIPE (a real-time, pipelined, image-processing machine) was used for the initial LDRD experiments with this testbed.[4] Image compression and reconstruction is achieved on PIPE by decimation and Poisson interpolation.[5] The video refresh rate during LDRD is approximately seven frames per second.

Teleoperation was impaired by the unrealistic appearance of the scene being viewed. The camera placement and field of view tended to give the impression at the HMI that the vehicle was traveling much faster than its actual speed. In addition, depth perception was obscured so that discerning between potholes and ground discoloration was difficult. This problem was enhanced in the LDRD mode because the scene was viewed without color, at fewer frames per second, and with some distortion due to the compression algorithm. Several methods can be used to improve the video quality at the HMI. For example, more cameras with greater fields of view can be used in conjunction with larger monitors; camera position adjustment can be coordinated with operator eye or head position to simulate seeing directly where the operator is driving; helmet-mounted display can also be used, but it must be coordinated with specific tasks because the device restricts the operator's view to just the video (and associated overlay graphics). To improve the LDRD, the video can be displayed in color; other compression techniques such as the discrete cosine transform can be employed; simulation of the imagery between frames at the HMI can enhance the quality of the video as well as allow images to be transmitted at a slower rate; and graphic overlay can be used to simulate the front of the vehicle, the horizon, and parallel lines extending forward from the sides of the vehicle. Additional recommendations for improving the HMI include adding an audio sensor to give engine-state information, embedding force-feedback in the steering mechanism, and employing additional sensors for detecting obstacles outside the normal viewing range.

The SRIP vehicle will be used in an early field demonstration and characterization of a buried waste site survey.[6] An ultrasonic ranging and data system is being added to the SRIP vehicle to determine xy-position so that autonomous path following can be applied to the surveying task. The demonstration will provide a means for evaluating the applicability of sensor systems to remote operation and for improving productivity through future automation.

HUMAN FACTORS EVALUATION PLATFORM

The HFEP is being designed for experimentation in the area of combat robotics. The HFEP will operate telerobotic and supervised autonomous combat vehicles (RCVs). RCVs will have the ability to perform basic combat frontal-area operations such as surveillance, reconnaissance, and enemy attack functions. The HFEP will provide an environment whereby a command officer and operational staff can deploy and operate the functions in the RCV. The HMMWV and the SRIP vehicle will be the first RCVs to be demonstrated with the HFEP. The SRIP vehicle will provide a research platform for studying the role of robotic manipulators in combat environments. The HMMWV will be used to demonstrate LDRD, automatic target acquisition (ATA), and weapons systems operation.

The HFEP will be a tractor/trailer enclosure with two operator control stations, data communications equipment, software development equipment, and service equipment necessary for operation of the RCVs. The operator stations will be modular and reconfigurable so that critical human-machine interface issues can be evaluated. A schematic of a proposed operator control station is shown in Fig. 3. Three vertical equipment racks with adjustable-angle side racks comprise the control station's frame. An angled shelf will provide an area to mount interface equipment such as joysticks, buttons, switches, and a keyboard. The center shelf will be reserved for primary driving equipment such as a steering wheel, yoke, or joystick; gas and brake pedals can be mounted under the center shelf. The interface equipment will be reconfigurable so that different methods for RCV control can be tested. The video monitors will be mounted at eye level; and the graphics monitors, tilted up toward the operator, will be mounted below the video shelf. The initial configuration will have six 13-in. monitors: three for display of standard video information and three for the graphics interface. During the LDRD mode, the center video monitor will display the LDRD images, and the two side monitors will display graphics or digitized video. The lower three graphics monitors will have touchscreen and mouse (or trackball) inputs for the two side monitors and only mouse inputs for the monitor behind the primary driving equipment. The center graphics monitor will be reserved for critical driving information display and input.

FIG. 3. Schematic of proposed operator control station.

The display-design software will be interactive so that graphics interfaces for human factors research can be created and modified at the control station.

The RCVs and the HFEP will be controlled by VME multi-CPU computers running the VxWorks™ real-time operating system by Wind River Systems, Inc. The software is written in the C programming language, compiled on a Sun workstation, and downloaded over Ethernet to the target for debugging (also on the Sun). A Sun workstation will be located in the HFEP to facilitate software modifications. Other computer subsystems that are not crucial for vehicle or manipulator control will interface the primary HFEP and RCV computer over a high-speed data link. The initial subsystems foreseen for the HFEP are communications, navigation, weapons, LDRD, and ATA. The architecture for all software modules will adhere to the National Institute of Standards and Technology (NIST) real-time control system (RCS) methodology. [7] A supervisory control module will coordinate the interfaces between the HFEP and the RCVs and also between the subsystem interfaces.

SUMMARY

A human factors testbed prototype was demonstrated by remotely driving the SRIP vehicle and the HMMWV with standard and compressed video techniques. This initial experimentation provided insight into some of the key problems associated with remote driving, and several applications to improve the HMI for teleoperation of RCVs have been suggested. A human factors evaluation platform has been designed to facilitate studies of basic human-machine interface issues involved in the operation of teleoperated vehicles using both the HMMWV and SRIP vehicle. The modular and reconfigurable characteristics of this platform will provide an ideal environment for the integration of HMI improvements resulting from initial experimentation with the testbed prototype.

REFERENCES

1. B. S. Richardson, G. A. Armstrong, and W. W. Manges, "Telerobotic Vehicle for Evaluating Remote Battlefield Support Operations," pp. 607–12 in *Proc. 2d Int. Symp. Robotics Manuf.: Res., Educ., and Applica.*, Albuquerque, N.M., (November 1988).
2. M. Herman et al., "Low Data Rate Remote Vehicle Driving," pp. 376–81 in *Proc. 3d IEEE Int. Symp. Intelligent Control*, Arlington, Va., (August 1988).
3. G. K. Corbett, S. M. Killough, and B. S. Richardson, "A Human Factors Testbed for Ground-Vehicle Telerobotics Research," pp. 618–20 in *Proc. IEEE Southeastcon*, New Orleans, La., Vol. **2**, April 1990.
4. E. W. Kent, M. O. Shneier, and R. Lumia, "PIPE (Pipelined Image Processing Engine)," *J. Parallel Dist. Comput.* **2**, pp. 50–78 (1985).
5. M. Herman, K. Chaconas, M. Nashman, and T-H. Hong, "Video Compression for Remote Vehicle Driving," pp. 136–43 in *Proceedings of SPIE Advances in Intelligent Robotics Systems: Mobile Robots III*, Vol. 1007, Cambridge, Mass., November 1988.

6. B. S. Richardson et al., "Remote Site Survey and Characterization for the National ER&WM Program Using the SRIP Vehicle," Proceedings of the Annual Meeting of the American Nuclear Society, Nashville, Tenn., to be presented, June 1990.
7. S. A. Leake and R. D. Kilmer, "The NBS Real-Time Control System User's Reference Manual," *NBS Technical Note* 1250, August 1988.

SECTION 6 : REDUNDANT AND FLEXIBLE ROBOTS

Modeling Static Behavior of Redundant Manipulators

Jin-Oh Kim and Pradeep K. Khosla
The Robotics Institute, Carnegie Mellon University

Wan-Kyun Chung
Mechanical Engineering, POSTECH, Korea

ABSTRACT

This paper provides a theoretical background for the control of a redundant manipulator when the end-effector of a redundant manipulator is in contact with the environment and the motion of a redundant manipulator is small enough to be considered as a static displacement. We introduce the *Premultiplier Diagram* to describe the static behavior of a redundant manipulator. This diagram provides the optimal decomposition of redundancy into particular and homogeneous solutions. Redundancy expressions for joint displacement, joint torque and joint stiffness matrix are presented. The orthogonality properties between particular and homogeneous solutions are discussed. We use the insight provided by the *Premultiplier Diagram* to propose a technique to stabilize stiffness control of a redundant manipulator.

1 INTRODUCTION

In this paper, we study static behaviors of redundant manipulators, where the joint motion is slow enough to be approximated as a static displacement. This assumption is quite reasonable for small displacements of joint angles and especially when the manipulator is in contact with the environment.

The static motion of redundant manipulator can be decomposed into two parts - particular and homogeneous. The homogeneous motion corresponds to the joint motions which do not create any end-effector motion and the particular motion corresponds to the joint motions that creates end-effector motion (total joint motion = particular + homogeneous). This decomposition of redundant motion into a particular and a homogeneous solution includes two steps. First, we have to decide which generalized inverse (g-inverse) will be used. There are infinite number of possible g-inverses, depending on the weight used in the computation. In general and for simplicity, Moore-Penrose g-inverse has been used [4, 5, 6, 10]. This is obtained by setting the weight to the identity matrix [8]. The space and magnitude of the particular solution, and the space of the homogeneous solution are decided at this step. Next, the magnitude of homogeneous solution is determined by applying a secondary criterion such as singularity avoidance and obstacle avoidance [6, 10]. This approach with Moore-Penrose g-inverse is termed as a kinematic control of a redundant manipulator.

In this paper, we show this kinematic control of Moore-Penrose g-inverse is not desirable and propose the stiffness control of a redundant manipulator. We address the problem of choosing the best weight for the g-inverse, which decouples the redundant motion into independent particular and homogeneous motions. This decoupling is essential to optimize and stabilize the control of a redundant manipulator. We introduce the *Premultiplier Diagram* to derive this best weight of a g-inverse. We also derive the redundancy expression for the joint displacement, joint torque and joint stiffness matrix K_θ in Section 2. We propose a technique to stabilize active stiffness control of a redundant manipulator by making the joint stiffness matrix nonsingular (positive definite). The orthogonality properties between the particular and the homogeneous solutions are examined in detail in Section 3. In Section 4, we derive the totally decoupled and decomposed static motions (particular and homogeneous) based on the g-inverse obtained from the *Premultiplier Diagram*, and we summarize this paper in Section 5.

2 PREMULTIPLIER DIAGRAM

2.1 Basic Operations

The instantaneous kinematics are used to relate the joint velocities and the end-effector velocities in operational space through the Jacobian matrix J, of the system. When the displacements of all the joints are sufficiently small, then the corresponding infinitesimal operational displacement δx, can be related to the joint displacement $\delta\theta$, by the Jacobian J. The infinitesimal operational and joint displacements are related to the operational force F, and the joint torque τ, via the operational stiffness K, and joint stiffness K_θ, respectively. These relations are summarized as follows:

$$\delta x = J\delta\theta, \tag{1}$$
$$\tau = J^T F, \tag{2}$$
$$\tau = K_\theta \delta\theta, \tag{3}$$
$$F = K\delta x, \tag{4}$$

where Jacobian matrix J, is mxn for a n degree of freedom (dof) manipulator working in m dof operational space.

The above four equations are substituted into the *Premultiplier Diagram* and two new matrices are introduced to give inverse relationships of J and J^T. This diagram is used to provide an elegant method of deriving one quantity from some other given quantities. For instance, a direct way to get $\delta\theta$, from a specified δx, can be replaced by the circuitous way following the path $\delta x \rightarrow F \rightarrow \tau \rightarrow \delta\theta$. The operation used in *Premultiplier Diagram* is basically the same as that of the transformation graph of [7].

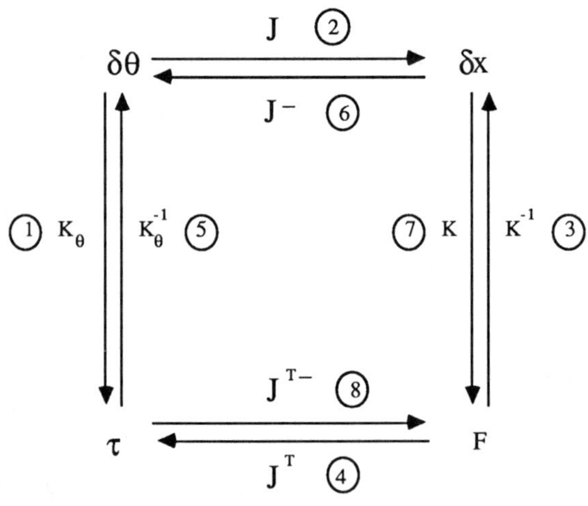

FIG.1. Premultiplier Diagram

The following example describes how the *Premultiplier Diagram* can be used to obtain the desired quantity from the given quantities. By using a circuitous path, the relations among matrices in FIG.1. can be derived as follows;

Example 1 *When we want the active stiffness controller of a redundant manipulator, the operational stiffness K is given and the joint stiffness K_θ needs to be derived [9].*
Path 2 → 7 → 4 = Path 1

$$\tau \overset{4}{=} J^T F \overset{7}{=} J^T K\delta x \overset{2}{=} J^T K J\delta\theta \overset{1}{=} K_\theta \delta\theta, \tag{5}$$

$$\Rightarrow K_\theta = J^T K J. \tag{6}$$

As in Example 1, we can derive K, J^- and J^{T-} using their equivalent circuitous paths. The results are summarized;

$$K = (JK_\theta^{-1}J^T)^{-1}, \tag{7}$$
$$J^- = K_\theta^{-1}J^T K \quad \text{and} \tag{8}$$
$$J^{T-} = KJK_\theta^{-1}. \tag{9}$$

The *Premultiplier* is so named because the matrix of the current path is premultiplied to the matrix of path taken just before, as we follow the circuitous path. It can be shown that J^- in (8) is a Joint Stiffness (K_θ) Weighted Minimum Norm G-inverse of J, and J^{T-} in (9) is a Joint Compliance (K_θ^{-1}) Weighted Least Square G-inverse of J^T [3]. This *Premultiplier Diagram* provides an optimal weighted g-inverse of J, which decouples the particular and homogeneous static motions of a redundant manipulator as shown in Section 5. From now on, we use J_m^+ and J_l^{+T} instead of J^- and J^{T-}, respectively, to say explicitly what the meaning is.

In Equations (7) - (9), K_θ is assumed to be nonsingular. But if K_θ is derived from Equation (6) for the active stiffness control of a redundant manipulator, the K_θ is always singular (positive semidefinite). This implies that we cannot use Equations (8) - (9). The transformation graph of Morasso and Mussa-Ivaldi [7] does not give any answer for this singular joint stiffness K_θ. That is, their transformation graph could be used only when the nonsingular K_θ is given. But in this paper the K_θ is made nonsingular (positive definite) artificially while its equivalent operation stiffness is still the same as the given operational stiffness K for active stiffness control. The method is discussed in detail in Section 2.2.

2.2 Redundancy Expressions

The redundancy of a redundant manipulator can be expressed by the use of two weighted g-inverses derived from the *Premultiplier Diagram*. The redundancy expression for $\delta\theta$ is

$$\delta\theta = J_m^+ \delta x + (I - J_m^+ J)k, \ k \text{ arbitrary} \tag{10}$$

where J_m^+ is a Stiffness Weighted Minimum Norm G-inverse, which minimizes instantaneous potential energy (PE) that is given by

$$PE = \frac{1}{2}\delta\theta^T K_\theta \delta\theta. \tag{11}$$

Similarly, we can derive the redundancy expression for τ as

$$\tau = J^T F + (I - J^T J_l^{+T})k, \ k \text{ arbitrary} \tag{12}$$

where J^T has meaning of a Compliance Weighted Minimum Norm G-inverse that minimizes

$$\frac{1}{2}\| \tau \|_{K_\theta^{-1}}^2 = \frac{1}{2}\tau^T K_\theta^{-1}\tau. \tag{13}$$

Finally, let's derive the redundancy expression for the joint stiffness matrix when the operational stiffness matrix K is given for active stiffness control. There are infinite number of joint stiffness matrices K_θ satisfying the given stiffness operational stiffness matrix K. The K_θ in (6) corresponds to the particular solution of a joint stiffness matrix. The derivation is different from the ones for τ and $\delta\theta$ because K_θ is a matrix. Before deriving an expression for K_θ, it would be useful to look at the properties of K_θ obtained from Example 1. For a positive definite stiffness K, K_θ is always singular whenever there is redundancy. That is, K_θ is positive semidefinite (p.s.d.) and has the same number of zero eigenvalues as the degree of redundancy ($n - m$).

For simplicity, the 3 dof planar manipulator is used to derive a positive definite K_θ from K and we assume that both stiffness matrices are symmetric. Hereafter the singular K_θ in (6) will be termed as $K_{\theta p}$.[1] By the use of similarity transform we can get a diagonal matrix $D_{\theta p}$,

$$D_{\theta p} = E^T K_{\theta p} E = \begin{bmatrix} \lambda_1 & 0 & 0 \\ 0 & \lambda_2 & 0 \\ 0 & 0 & \lambda_3 \end{bmatrix}, \tag{14}$$

[1] Subscript p indicates a matrix or a vector related to the particular solution of linear system. And the subscript h indicates the homogeneous part of general solution. The general solution will be denoted without specifying any subscript.

where the columns of the orthogonal matrix E (i.e. $E^{-1} = E^T$) are the eigenvectors corresponding to eigenvalues λ_1, λ_2 and λ_3, respectively, and

$$\lambda_1 \geq \lambda_2 \geq \lambda_3 = 0. \tag{15}$$

When λ_3 is shifted to an arbitrary positive value h, the orthonormal matrix E remains unchanged. This property enables us to construct a new nonsingular matrix from K. The new matrix D_θ is defined as

$$\begin{aligned} D_\theta &= D_{\theta p} + D_{\theta h} && (16)\\ &= E^T K_{\theta p} E + hI_3 && (17) \end{aligned}$$

where

$$D_{\theta h} = hI_3 \qquad \text{and} \qquad I_3 = \begin{bmatrix} 0 & 0 & 0 \\ 0 & 0 & 0 \\ 0 & 0 & 1 \end{bmatrix}.$$

The matrix $D_{\theta p}$, is always orthogonal to the matrix $D_{\theta h}$ and the new nonsingular matrix K_θ becomes

$$\begin{aligned} K_\theta &= ED_\theta E^T = ED_{\theta p}E^T + ED_{\theta h}E^T \\ &= J^T K J + hS = K_{\theta p} + K_{\theta h} && (18) \end{aligned}$$

where

$$S = EI_3E^T. \tag{19}$$

The matrix S can be obtained by using the eigenvector corresponding to $\lambda_3 = 0$. That is, if $\vec{e} = [e_1\ e_2\ e_3]^T$ is the eigenvector corresponding to the eigenvalue λ_3, then

$$S = \vec{e}^T\vec{e} = [e_1\ e_2\ e_3]^T[e_1\ e_2\ e_3]. \tag{20}$$

This matrix S enables us to assign a new nonzero eigenvalue h in the direction perpendicular to the hyperplane in which the three columns of $K_{\theta p}$ lie. The scalar h should be positive to ensure the positive definiteness of K_θ.

In this section, we derived an expression for nonsingular K_θ from a given K by assigning an arbitrary nonzero eigenvalue. This makes the *Premultiplier Diagram* complete and allows us to define J_m^+ and J_l^{+T} in (8) and (9), which could not be defined earlier because of a singular K_θ. Note that the method used to derive nonsingular K_θ in this section, can be applied to a redundant manipulator with more than one degree of redundancy. In the following section, the properties of these weighted G-inverses (J_m^+ and J_l^{+T}) are discussed.

2.3 Properties of Weighted G-inverse

When Jacobian J, is a mxn matrix with an inequality ($m < n$) for a redundant manipulator, J_m^+ is a nxm matrix, J_l^{+T} is a mxn matrix, and J_l^+, which is the transpose of J_l^{+T}, is a nxm matrix. A new weighted G-inverse can be defined from the system ($\delta x = J\delta\theta$). $J_{KK_\theta}^+$ is a minimum K_θ-norm K-least square G-inverse which provides $\delta\theta$ such that $\delta\theta^T K_\theta \delta\theta$ is a minimum subject to $\tau = J^T F$ [8]. The following are the properties of the above three G-inverses (J_m^+, J_l^{+T} and $J_{KK_\theta}^+$) assuming that K_θ and K are symmetric and positive definite:

1. $J_m^+ = J_l^+$.
2. $J_m^+ = J_l^+ = J_{KK_\theta}^+$ if rank k of Jacobian is equal to m(the number of rows of Jacobian).
3. $J_m^+ = J^+$ (Moore-Penrose) if $K_\theta = cI$ where c is a constant and I is a identity matrix.
4. $J_m^+ = J^+$ if K is given and K_θ is derived by using the method in Section 2.2.
5. If the redundant manipulator is in singular position(i.e. $k < m$) and the system equation is inconsistent, only $J_{KK_\theta}^+$ can be used to give a unique solution.

6. If $G = J^+$, then it satisfies

$$\begin{array}{rclcrcl} AGA & = & A & \quad & GAG & = & G \\ (AG)^T & = & AG & & (GA)^T & = & GA. \end{array}$$

7. If $G = J^+_{KK_\theta}$, then it satisfies

$$\begin{array}{rclcrcl} AGA & = & A & \quad & GAG & = & G \\ (AG)^T K & = & KAG & & (GA)^T K_\theta & = & K_\theta GA. \end{array}$$

From now on, we use $J^+_{K_\theta}$ instead of J^+_m in order to express explicitly the weight used. The particular part of the redundancy expressions will be termed as 'net' and the homogeneous part termed as 'null'. In the following section, we discuss about the orthogonality between two solutions using the properties discussed here.

3 ORTHOGONALITY BETWEEN TWO SOLUTIONS

In this section, we discuss about the orthogonality between the net and the null solutions of joint displacement, torque and stiffness matrix. We consider two cases - first, when K is given and, second, when K_θ is given.

3.1 When K is given.

When an arbitrary positive definite and symmetric operational stiffness K is given for the active stiffness control, we assume that a nonsingular K_θ is obtained as shown in Section 2.2. This nonsingular K_θ has a property that the particular part $K_{\theta p}$ is orthogonal to the homogeneous part $K_{\theta h}$. This is because we assign nonzero eigenvalue of the homogeneous part in the orthogonal direction of nonzero eigenvalues of the particular part.

When K is given, we can derive the redundancy expressions for joint displacement, joint torque and joint stiffness for 3 dof planar manipulator as follows;

$$\delta\theta = \delta\theta_p + \delta\theta_h = J^+_{K_\theta} J \delta\theta + (I - J^+_{K_\theta} J)\delta\theta, \tag{21}$$

$$\tau = \tau_p + \tau_h = J^T J^{+T}_{K_\theta} \tau + (I - J^T J^{+T}_{K_\theta})\tau, \tag{22}$$

$$K_\theta = K_{\theta p} + K_{\theta h} = J^T K J + hS. \tag{23}$$

Now we list without proofs three theorems about the orthogonal properties between net and null parts of joint dispacement, torque and stiffness matrix without proof. (See [3] for the proof.)

Theorem 1 *$\delta\theta_p$ and $\delta\theta_h$ are always orthogonal to each other, and in addition they are orthogonal to each other with respect to the joint stiffness matrix K_θ.*

Theorem 2 *τ_p and τ_h are always orthogonal to each other, and additionally they are orthogonal to each other with respect to the inverse of stiffness matrix K_θ^{-1}.*

Theorem 3 *$K_{\theta p}$ and $K_{\theta h}$ are always orthogonal to each other.*

These three theorems are valid only when K_θ is made nonsingular by the use of the method in Section 2.2. Note that the given K determines only $K_{\theta p}$, and $K_{\theta h}$ is made orthogonal to $K_{\theta p}$ artificially. The orthogonal properties in this section are used to decouple the static behavior into net and null motion which are independent of each other.

3.2 When K_θ is given.

When a joint stiffness K_θ is given, we can still decompose a given joint displacement and a joint torque as in (21) and (22). But this given joint stiffness does not satisfy Theorem 3 except for one special case, in which K_θ is equal to a scalar times *I*dentity matrix. For this special case, we can get the same expression as (23) and there exists an orthogonality between $K_{\theta p}$ and $K_{\theta h}$. But in general, K_θ is symmetric and positive definite, and is not equal to a scalar times *I*dentity matrix.

The general positive definite and symmetric joint stiffness K_θ can be decomposed as;

$$K_\theta = K_{\theta p} + K_{\theta h} = J^T K J + (K_\theta - J^T K J) \tag{24}$$

where K is an equivalent operational stiffness matrix determined by (7). However, $K_{\theta p}$ is not orthogonal to $K_{\theta h}$. Property 4 in Section 2.3 is not applicable for this general case, that is, $J^+_{K_\theta} \neq J^+$. The orthogonal properties of this case are stated in the following theorems.

Theorem 4 *$\delta\theta_p$ is always orthogonal to $\delta\theta_h$ with respect to the joint stiffness matrix K_θ.*

Theorem 5 *τ_p is always orthogonal to τ_h with respect to the inverse of the joint stiffness matrix K_θ^{-1}.*

In the following section, we discuss the decoupling of static motion of a redundant manipulator into two motions - net and null. This decoupling excludes the interaction between net and null motion.

4 STATICS OF A REDUNDANT MANIPULATOR

Decoupling of redundant static motion into net and null motion implies that the net torque affects only net motion and null torque affects only null motion. This independent behavior is essential to stabilize the control and to eliminate unnecessary control efforts.

4.1 When K is given

When a symmetric positive definite operational stiffness K is given, we have shown, in Section 3.1, that the corresponding joint stiffness K_θ can be decomposed into a net part and a null part which are orthogonal to each other. The null part is artificially made orthogonal to the net part by assigning a new nonzero eigenvalue in the direction of zero eigenvalue of the net part in Section 2.2.

By the use of Equations (21) - (23), we get the following result;

$$\begin{aligned} \tau &= K_\theta \delta\theta && (25)\\ &= (K_{\theta p} + K_{\theta h})(\delta\theta_p + \delta\theta_h) && (26)\\ &= K_{\theta p}\delta\theta_p + K_{\theta h}\delta\theta_h + K_{\theta p}\delta\theta_h + K_{\theta h}\delta\theta_p && (27)\\ &= \tau_p + \tau_h. && (28) \end{aligned}$$

It can be proved that the two cross multiplication terms $K_{\theta p}\delta\theta_h$ and $K_{\theta h}\delta\theta_p$ in (27), are always zero. This property enables us to describe the net part and the null part separately. Comparing (27) and (28), we can write

$$\tau_p = K_{\theta p}\delta\theta_p \tag{29}$$

and

$$\tau_h = K_{\theta h}\delta\theta_h. \tag{30}$$

Equations (29) and (30) describe the net motion and the null motion of the redundant manipulator, respectively. That is, the net subsystem and the null subsystem do not affect each other in static control and act as two independent systems. The net torque creates only net displacement, so it creates a minimal total *PE* in a sense that it does not create any null *PE*. And the null torque creates only null displacement with a minimal total *PE* and no net *PE*.

4.2 When K_θ is given

When the joint space stiffness K_θ is given, there exists a similar decoupling between net and null motions. Again, this decoupling can be obtained only with the stiffness weighted g-inverse of Jacobian J. Morasso and Mussa-Ivaldi [7] intuitively proposed to use stiffness weighted g-inverse as a natural solution to control redundancy. This is true because the natural static system has an orthogonal property with respect to joint stiffness matrix.

For this case, the kinematic control with Moore-Penrose g-inverse can not be applied except a special case when K_θ is equal to a scalar times *I*dentity matrix. For this special case, we can derive the same decomposed and decoupled subsystems as if K is given. But for a general case, if Moore-Penrose g-inverse is used, the net and the null motion are coupled and affect each other, because the net and null displacements are orthogonal to each other, but the net and null torque are not orthogonal to each other. The net and null torque lose their orthogonality when an arbitrary K_θ is multiplied by desired and orthogonal net and null joint displacements. This dependence on each other implies unnecessary control effort to eliminate the interaction and can cause instability due to excessive interaction.

In this section, we have shown that we can always derive two decoupled net and null motions when stiffness weighted g-inverse of Jacobian matrix is used instead of Moore-Penrose g-inverse.

5 SUMMARY

In this paper, the static behavior of redundant manipulator is expressed in the *Premultiplier Diagram*, which provides one optimal (joint stiffness) weighted g-inverse. The importance of this optimal weighted g-inverse is explained in terms of decoupling of the redundant motion. The decoupling of redundant motion into net and null motions is essential to optimize and stabilize the control of a redundant manipulator. By the use of joint stiffness weighted g-inverse, we have derived redundancy expression for joint stiffness matrix, which makes the *Premultiplier Diagram* complete and stabilizes active stiffness control of a redundant manipulator.

The *Premultiplier Diagram* is based on statics. The results can be extended to control robot fingers which are redundant and whose motion is small enough to be considered as static. That is, for the control of redundant fingers, we propose to use stiffness weighted g-inverse instead of Moore-Penrose g-inverse. In addition, the methodology used with our diagram can be easily extended to obtain the inertia weighted g-inverse in free space to include dynamics. Even more, an extension to the impact problem can be made. We can extend the result of this paper to propose to use the inertia weighted g-inverse of Jacobian matrix for the dynamic control of redundant manipulator.

ACKNOWLEDGEMENT

This research was funded in part by NASA under grant NAG-1-1075, the Department of Electrical and Computer Engineering, and The Robotics Institute, Carnegie Mellon University.

REFERENCES

1. Asada H. and Slotine J-J.E. Robot Analysis and Control, John Wiley and Sons, 1986
2. Khatib O. "Reduced Effective Inertia in Macro-/Mini-Manipulator System" Robotics Research: the fifth int. symp., pp.329-334, 1989
3. Kim J.-O., Khosla P.K. and Chung W.K. On the Static Premultiplier Model of Redundant Manipulators, Technical Report, Carnegie Mellon University, CMU-RI-AML-89-01, September, 1989
4. Klein C.A. and Huang C. "Review of Pseudoinverse Control for Use with Kinematically Redundant Manipulator", IEEE trans. on Systems, Man, and Cybernetics, SMC-13(3), pp.245-250, 1983
5. Liégeois A. "Automatic Supervisory control of the Configuration and Behavior of Multibody Mechanisms", IEEE trans. on Systems, Man, and Cybernetics, SMC-7(12), pp.868-871, 1977
6. Maciejewski A.A. and Klein C.A. "Obstacle Avoidance for Kinematically Redundant Manipulators in Dynamically Varying Environments", The International Journal of Robotics Research, 4(3), pp.109-117, Fall, 1985
7. Morasso P. and Mussa-Ivaldi F.A. "The Role of Physical Constraints in Natural and Artificial Manipulation", IEEE conf. on Robotics and Automation, pp.1901-1905, 1986.
8. Rao C.R. and Mitra S.K. Generalized Inverse of Matrices and its Application, John Wiley and Sons, 1971
9. Salisbury J.K. "Active Stiffness Control of a Manipulator in Cartesian Coordinate", IEEE conf. on Decision and Control, pp.95-100, 1980
10. Yoshikawa T. "Analysis and Control of Robot Manipulators with Redundancy", Robotics Research: The first int. symp., ed. Brady and Paul, MIT press, pp. 735-747, 1984

A COMPUTATIONALLY EFFICIENT ERROR-BASED ADAPTIVE CONTROL SCHEME FOR KINEMATICALLY REDUNDANT MANIPULATORS

Charles C. Nguyen* **Zhen-Lei Zhou**†
Catholic University of America
Washington DC, 20064

Gary E. Mosier
NASA/Goddard Space Flight Center
Greenbelt, Maryland 20771

Abstract

A Cartesian-space control scheme is developed to control the motion of kinematically redundant manipulators with 7 degrees of freedom (**DOF**). The control scheme consists mainly of proportional-derivative (**PD**) controllers whose gains are adjusted by an adaptation law driven by the errors between the desired and actual trajectories. The adaptation law is derived using the concept of model reference adaptive control (**MRAC**) and Lyapunov direct method under the assumption that the manipulator performs non-compliant and slowly-varying motions. The developed control scheme is computationally efficient because its implementation does not require the computation of the manipulator dynamics. Computer simulation performed to evaluate the control scheme performance is presented and discussed.

INTRODUCTION

Considerable research effort has been spent to study kinematics and control of kinematically redundant[1] manipulators [1]-[8] during the last decade. In [1], the problem of programming and control of redundant manipulators was investigated and the method of extended Jacobian was applied to optimize an object function. The concept of *self motion* was introduced by the authors of [2], who proposed a method to resolve the redundancy by directly controlling a set of self-motion parameters. The global optimization problem of joint rates and kinetic energy in redundant manipulators was considered in [5], and the results of the global optimization results were compared with those of the local one. The method of pseudo-inverse was reviewed in [6] for application in redundant robots. Recognizing the challenging difficulties in the control of position and force because of redundancy, numerous researchers have focused their effort on developing control schemes for redundant robot manipulators. Cartesian-space control schemes were developed in [3] to control position/force of redundant manipulators, and in [4] to control a hydraulic redundant robot. Joint-space and Cartesian-space adaptive control schemes were developed in [7] and in [10], respectively for a redundant telerobot manipulator. Recently the so-called *configuration control* which utilizes the manipulator redundancy was introduced in [8] to control the manipulator configuration directly in task space.

In this paper, we first present the development of a Cartesian-space adaptive control scheme for redundant manipulators. We then present and discuss the computer simulation performed to study the performance of the developed control scheme implemented to control the non-compliant motion of a 7 DOF redundant manipulator. Finally a conclusion will describe some current research activities and outline the future research direction. Matrix

---

*Associate Professor of Electrical Engineering

†Graduate Research Assistant

[1] The term "redundant" is often used instead of "kinematically redundant".

notations used in this paper are: $\mathbf{M}^T$=transpose of the matrix $\mathbf{M}$; $\mathbf{0}_n$=(nxn) matrix whose elements are all zero; $\mathbf{I}_n$=(nxn) identity matrix; and $tr[\mathbf{M}]$=trace of matrix $\mathbf{M}$.

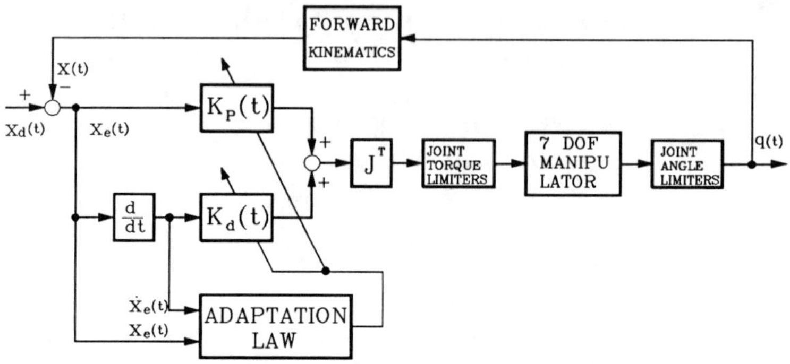

Figure 1 The Cartesian-space adaptive control scheme.

THE ADAPTIVE CONTROL SCHEME

Figure 1 presents the control scheme proposed for controlling the non-compliant motion of a 7 DOF redundant manipulator. As the figure shows, the actual joint variables measured by joint position sensors, are transformed into the corresponding Cartesian variables via the forward kinematic transformation. The errors between the actual and desired Cartesian variables serve as inputs to a PD controller whose gains are adjusted by an adaptation law to be derived. The Cartesian forces available at the output of the PD controllers are transformed into the corresponding joint forces using the Jacobian transpose.

The dynamics of the 7 DOF redundant manipulator performing non-compliant motion can be expressed in Cartesian space as [10]

$$\mathbf{F}(t) = \boldsymbol{\Lambda}(\mathbf{x},\dot{\mathbf{x}})\,\ddot{\mathbf{x}}(t) + \boldsymbol{\Phi}(\mathbf{x},\dot{\mathbf{x}})\,\dot{\mathbf{x}}(t) + \boldsymbol{\Gamma}(\mathbf{x},\dot{\mathbf{x}})\,\mathbf{x}(t) \tag{1}$$

where $\mathbf{x}(t)$ denotes the (6x1) Cartesian position[2] vector of the slave arm end-effector, and $\mathbf{F}(t)$, the (6x1) Cartesian force applied to the end-effector. $\boldsymbol{\Lambda}(\mathbf{x},\dot{\mathbf{x}})$, a symmetric positive-definite matrix, $\boldsymbol{\Phi}(\mathbf{x},\dot{\mathbf{x}})$ and $\boldsymbol{\Gamma}(\mathbf{x},\dot{\mathbf{x}})$ are of order (6x6) with elements that are highly complex nonlinear functions of $\mathbf{x}$ and $\dot{\mathbf{x}}$.

The outputs of the PD controller constitute the Cartesian force vector $\mathbf{F}(t)$ computed by

$$\mathbf{F}(t) = \mathbf{K}_p(t)\,\mathbf{x}_e(t) + \mathbf{K}_d(t)\,\dot{\mathbf{x}}_e(t) \tag{2}$$

where $\mathbf{x}_e(t)$ given by

$$\mathbf{x}_e(t) = \mathbf{x}_d(t) - \mathbf{x}(t) \tag{3}$$

denotes the Cartesian error vector between the actual Cartesian position vector $\mathbf{x}(t)$ and the desired Cartesian position vector $\mathbf{x}_d(t)$, $\mathbf{K}_p(t)$ and $\mathbf{K}_d(t)$, are gain matrices of the proportional and derivative controllers, respectively.

Substituting (2) into (1) yields

$$\boldsymbol{\Lambda}\,\ddot{\mathbf{x}}_e + (\boldsymbol{\Phi} + \mathbf{K}_d)\,\dot{\mathbf{x}}_e + (\boldsymbol{\Gamma} + \mathbf{K}_p)\,\mathbf{x}_e = \boldsymbol{\Lambda}\,\ddot{\mathbf{x}}_d + \boldsymbol{\Phi}\,\dot{\mathbf{x}}_d + \boldsymbol{\Gamma}\,\mathbf{x}_d \tag{4}$$

where the dependent variables of the matrices and vectors were dropped for simplicity.

[2] In this paper, position implies both position and orientation and force implies both force and torque.

The state equation of (4) can be obtained by defining the state variable vector $\mathbf{z}(t)$

$$\mathbf{z}(t) = \left[\mathbf{x}_e^T(t) \quad \dot{\mathbf{x}}_e^T(t)\right]^T . \tag{5}$$

so that from (4), we obtain

$$\dot{\mathbf{z}}(t) = \begin{bmatrix} \mathbf{0}_6 & \mathbf{0}_6 \\ -\mathbf{B}_1 & -\mathbf{B}_2 \end{bmatrix} \mathbf{z}(t) + \begin{bmatrix} \mathbf{0}_6 & \mathbf{0}_6 & \mathbf{0}_6 \\ \mathbf{B}_3 & \mathbf{B}_4 & \mathbf{I}_6 \end{bmatrix} \mathbf{u}(t) \tag{6}$$

where

$$\mathbf{B}_1 = \boldsymbol{\Lambda}^{-1}(\boldsymbol{\Gamma} + \mathbf{K}_p), \quad \mathbf{B}_2 = \boldsymbol{\Lambda}^{-1}(\boldsymbol{\Phi} + \mathbf{K}_d), \tag{7}$$

and

$$\mathbf{B}_3 = \boldsymbol{\Lambda}^{-1}\, \boldsymbol{\Gamma}, \quad \mathbf{B}_4 = \boldsymbol{\Lambda}^{-1}\, \boldsymbol{\Phi}, \tag{8}$$

and

$$\mathbf{u}(t) = \left[\mathbf{x}_d^T(t) \quad \dot{\mathbf{x}}_d^T(t) \quad \ddot{\mathbf{x}}_d^T(t)\right]^T . \tag{9}$$

In order to achieve a minimal computational burden in real-time control, a *reference model* which characterizes the desired manipulator performance, should be selected as follows:

$$\ddot{x}_{ei}(t) + 2\, \xi_i\, \omega_i\, \dot{x}_{ei}(t) + \omega_i^2\, x_{ei}(t) = 0 \tag{10}$$

where ξ_i and ω_i denote the damping ratio and the natural frequency of x_{ei}, and $x_{ei}(t)$ for i=1,2,...,6 are the elements of the tracking error vector $\mathbf{x}_e(t) = [x_{e1}(t)\, x_{e2}(t) \ldots x_{e6}(t)]^T$. As seen from Equation (10), the reference model is a linear time-invariant system consisting of 6 uncoupled systems, each of which represents the desired error behavior in the corresponding Cartesian position.

From (10), the state equation of the reference model can be obtained as

$$\dot{\mathbf{z}}_m(t) = \boldsymbol{\Delta}\, \mathbf{z}_m(t) = \begin{bmatrix} \mathbf{0}_6 & \mathbf{I}_6 \\ -\boldsymbol{\Delta}_1 & -\boldsymbol{\Delta}_2 \end{bmatrix} \mathbf{z}_m(t), \tag{11}$$

where $\boldsymbol{\Delta}_1$=diag(ω_i^2) and $\boldsymbol{\Delta}_2$=diag($2\xi_i\omega_i$) are constant (6x6) diagonal matrices, and

$$\mathbf{z}_m(t) = \left[\mathbf{x}_m^T(t) \quad \dot{\mathbf{x}}_m^T(t)\right]^T \tag{12}$$

with

$$\mathbf{x}_m = (x_{e1}\; x_{e2}\; \ldots x_{e6})^T . \tag{13}$$

Now solving (11), we obtain

$$\mathbf{z}_m(t) = e^{\boldsymbol{\Delta} t}\, \mathbf{z}_m(0) \tag{14}$$

where the initial value of $\mathbf{z}_m(t)$ is denoted by $\mathbf{z}_m(0)$. Here if we assume that the initial values of the actual and desired Cartesian position and velocity vectors are identical, i.e. $\mathbf{z}_m(0) = 0$, then $\mathbf{z}_m(t) = 0$. Otherwise we can properly select ξ_i and ω_i so that all eigenvalues of $\boldsymbol{\Delta}$ are stable to make $\mathbf{z}_m(t) \to 0$ as $t \to \infty$.

Next the adaptation error vector $\mathbf{E}$(t) is defined as

$$\mathbf{E}(t) = \mathbf{z}_m(t) - \mathbf{z}(t). \tag{15}$$

Then from (6) and (11), an *error system* is obtained by

$$\begin{aligned} \dot{\mathbf{E}}(t) \;&=\; \begin{bmatrix} \mathbf{0}_6 & \mathbf{I}_6 \\ -\boldsymbol{\Delta}_1 & -\boldsymbol{\Delta}_2 \end{bmatrix} \mathbf{E}(t) + \begin{bmatrix} \mathbf{0}_6 & \mathbf{0}_6 \\ \mathbf{B}_1 - \boldsymbol{\Delta}_1 & \mathbf{B}_2 - \boldsymbol{\Delta}_2 \end{bmatrix} \mathbf{z}(t) \\ &\quad + \begin{bmatrix} \mathbf{0}_6 & \mathbf{0}_6 & \mathbf{0}_6 \\ -\mathbf{B}_3 & -\mathbf{B}_4 & -\mathbf{I}_6 \end{bmatrix} \mathbf{u}(t). \end{aligned} \tag{16}$$

Now a Lyapunov function $v(t)$ is selected such that

$$\begin{aligned} v(t) \;&=\; \mathbf{E}^T\mathbf{P}\mathbf{E} + tr\left[(\mathbf{B}_1 - \boldsymbol{\Delta}_1)^T \boldsymbol{\Psi}_1 (\mathbf{B}_1 - \boldsymbol{\Delta}_1)\right] \\ &\quad + tr\left[(\mathbf{B}_2 - \boldsymbol{\Delta}_2)^T \boldsymbol{\Psi}_2 (\mathbf{B}_2 - \boldsymbol{\Delta}_2)\right] + tr[\mathbf{B}_3^T \boldsymbol{\Psi}_3 \mathbf{B}_3] + tr[\mathbf{B}_4^T \boldsymbol{\Psi}_4 \mathbf{B}_4], \end{aligned} \tag{17}$$

where $\mathbf{P}$ and $\boldsymbol{\Psi}_i$ for i=1,2,...,4, are positive definite matrices which will be determined.

After differentiating (17) with respect to time and simplifying the resulting expression, we arrive at

$$\begin{aligned}\dot{v}(t) &= \mathbf{E}^T(\mathbf{P}\boldsymbol{\Delta}+\boldsymbol{\Delta}^T\mathbf{P})\mathbf{E}+2tr\left[(\mathbf{B}_1-\boldsymbol{\Delta}_1)^T(\boldsymbol{\Upsilon}\mathbf{x}_e^T+\boldsymbol{\Psi}_1\dot{\mathbf{B}}_1)\right]-2tr\left[\mathbf{B}_3^T(\boldsymbol{\Upsilon}\mathbf{x}_d^T-\boldsymbol{\Psi}_3\dot{\mathbf{B}}_3)\right]\\ &\quad +2tr\left[(\mathbf{B}_2-\boldsymbol{\Delta}_2)^T(\boldsymbol{\Upsilon}\dot{\mathbf{x}}_e^T+\boldsymbol{\Psi}_2\dot{\mathbf{B}}_2)\right]-2tr\left[\mathbf{B}_4^T(\boldsymbol{\Upsilon}\dot{\mathbf{x}}_d^T-\boldsymbol{\Psi}_4\dot{\mathbf{B}}_4)\right] \qquad (18)\end{aligned}$$

where

$$\boldsymbol{\Upsilon}=[\mathbf{P}_2\ \ \mathbf{P}_3]\mathbf{z}(t)=-\mathbf{P}_2\mathbf{x}_e-\mathbf{P}_3\dot{\mathbf{x}}_e \qquad (19)$$

and $\mathbf{P}$ is given by

$$\mathbf{P}=\begin{bmatrix}\mathbf{P}_1 & \mathbf{P}_2\\ \mathbf{P}_2 & \mathbf{P}_3\end{bmatrix} \qquad (20)$$

and we note that $\mathbf{z}_m(t)=0$ results in $\mathbf{E}(t)=-z(t)$.

Now if ξ_i and ω_i of (10) are selected so that $\boldsymbol{\Delta}$ is a *Hurwitz* matrix [9], i.e. all eigenvalues of $\boldsymbol{\Delta}$ have negative real parts, then according to *Lyapunov theorem*, there exists a positive definite symmetric matrix $\mathbf{P}$ satisfying the Lyapunov equation given by

$$\mathbf{P}\boldsymbol{\Delta}+\boldsymbol{\Delta}^T\mathbf{P}=-\mathbf{Q} \qquad (21)$$

for any given positive-definite symmetric matrix $\mathbf{Q}$.

Now in (18) letting

$$\boldsymbol{\Upsilon}\mathbf{x}_e^T+\boldsymbol{\Psi}_1\dot{\mathbf{B}}_1=\boldsymbol{\Upsilon}\dot{\mathbf{x}}_e^T+\boldsymbol{\Psi}_2\dot{\mathbf{B}}_2=0;\ \ \boldsymbol{\Upsilon}\mathbf{x}_d^T-\boldsymbol{\Psi}_3\dot{\mathbf{B}}_3=\boldsymbol{\Upsilon}\dot{\mathbf{x}}_d^T-\boldsymbol{\Psi}_4\dot{\mathbf{B}}_4=0, \qquad (22)$$

changes (18) to

$$\dot{v}(t)=-\mathbf{E}^T\mathbf{Q}\mathbf{E} \qquad (23)$$

which is a negative definite function of $\mathbf{E}(t)$. Also from (22), we obtain

$$\dot{\mathbf{B}}_1=-\boldsymbol{\Psi}_1^{-1}\boldsymbol{\Upsilon}\mathbf{x}_e^T;\ \ \dot{\mathbf{B}}_2=-\boldsymbol{\Psi}_2^{-1}\boldsymbol{\Upsilon}\dot{\mathbf{x}}_e^T;\ \ \dot{\mathbf{B}}_3=\boldsymbol{\Psi}_3^{-1}\boldsymbol{\Upsilon}\mathbf{x}_d^T;\ \ \dot{\mathbf{B}}_4=\boldsymbol{\Psi}_4^{-1}\boldsymbol{\Upsilon}\dot{\mathbf{x}}_d^T. \qquad (24)$$

According to (17), since $\mathbf{P}$ is a positive definite matrix by properly selecting $\boldsymbol{\Delta}$ and $\mathbf{Q}$, the error system described in (16) is asymptotically stable if we can show that $\boldsymbol{\Psi}_i$ for i=1,2,...,4, are also positive definite matrices. In other words, the *adjustable system* (6) will follow the reference model very closely, or $\mathbf{z}(t)$ approaches $\mathbf{z}_m$ asymptotically as $t\to\infty$.

Before proceeding, we assume that the end-effector of the manipulator performs slowly varying motion so that the elements of $\boldsymbol{\Lambda}$, $\boldsymbol{\Phi}$ and $\boldsymbol{\Gamma}$ can be considered *almost constant.* As a result, from (7) and (8) we have

$$\dot{\mathbf{B}}_1\simeq\boldsymbol{\Lambda}^{-1}\dot{\mathbf{K}}_p;\ \ \dot{\mathbf{B}}_2\simeq\boldsymbol{\Lambda}^{-1}\dot{\mathbf{K}}_d;\ \ \dot{\mathbf{B}}_3\simeq 0;\ \ \dot{\mathbf{B}}_4\simeq 0. \qquad (25)$$

Substituting (25) into (24) yields

$$\boldsymbol{\Lambda}^{-1}\dot{\mathbf{K}}_p=-\boldsymbol{\Psi}_1^{-1}\boldsymbol{\Upsilon}\mathbf{x}_e^T;\ \ \boldsymbol{\Lambda}^{-1}\dot{\mathbf{K}}_d=-\boldsymbol{\Psi}_2^{-1}\boldsymbol{\Upsilon}\dot{\mathbf{x}}_e^T, \qquad (26)$$

and

$$0\simeq\boldsymbol{\Psi}_3^{-1}\boldsymbol{\Upsilon}\mathbf{x}_d^T;\ \ 0\simeq\boldsymbol{\Psi}_4^{-1}\boldsymbol{\Upsilon}\dot{\mathbf{x}}_d^T. \qquad (27)$$

Now in (26), if we let

$$\boldsymbol{\Psi}_1=-\frac{1}{\beta_1}\boldsymbol{\Lambda};\ \ \boldsymbol{\Psi}_2=-\frac{1}{\beta_2}\boldsymbol{\Lambda}, \qquad (28)$$

where β_1 and β_2 are arbitrary positive scalars, and solving for $\dot{\mathbf{K}}_p$ and $\dot{\mathbf{K}}_d$, we arrive at

$$\dot{\mathbf{K}}_p=\beta_1\boldsymbol{\Upsilon}\mathbf{x}_e^T;\ \ \dot{\mathbf{K}}_d=\beta_2\boldsymbol{\Upsilon}\dot{\mathbf{x}}_e^T. \qquad (29)$$

In (28), obviously $\boldsymbol{\Psi}_1$ and $\boldsymbol{\Psi}_2$ are positive definite matrices that can be considered as nearly constant because $\boldsymbol{\Lambda}$ is positive definite and slowly time-varying. In addition, $\boldsymbol{\Psi}_3$ and $\boldsymbol{\Psi}_4$ should be chosen to be positive definite matrices whose determinants approach ∞ in order to satisfy (27). To achieve this, we can select Ψ_3 and Ψ_4 to be diagonal matrices whose main diagonal elements assume very large and positive values.

Now integrating both sides of the equations given in (29) results in

$$\mathbf{K}_p(t) = \mathbf{K}_p(0) + \beta_1 \int_0^t (\mathbf{P}_2\mathbf{x}_e + \mathbf{P}_3\dot{\mathbf{x}}_e)\mathbf{x}_e^T\, dt \tag{30}$$

and

$$\mathbf{K}_d(t) = \mathbf{K}_d(0) + \beta_2 \int_0^t (\mathbf{P}_2\mathbf{x}_e + \mathbf{P}_3\dot{\mathbf{x}}_e)\dot{\mathbf{x}}_e^T\, dt \tag{31}$$

where $\mathbf{K}_p(0)$ and $\mathbf{K}_d(0)$ are initial conditions of $\mathbf{K}_p(t)$ and $\mathbf{K}_d(t)$, respectively and can be set arbitrarily.

The development of the adaptive controller is now completed. As shown in (30) and (31), the adaptation law is designed based on the errors of the Cartesian variables of the slave arm end-effector and the submatrices of $\mathbf{P}$. The adaptation law is very computationally efficient because $\mathbf{P}$ is a constant matrix and $\mathbf{x}_e$ can be easily computed from the desired and actual Cartesian variables which are specified by the user and given from position sensors, respectively. Consequently, high sampling rates up to 1 KHz can be utilized in the implementation of the control scheme while the manipulator dynamics can be considered *almost constant* during each sampling interval of typically about 1 ms. Another attractive feature is that the implementation of the control scheme does not require the computation of the manipulator dynamics, which is very difficult, if not possible to model accurately.

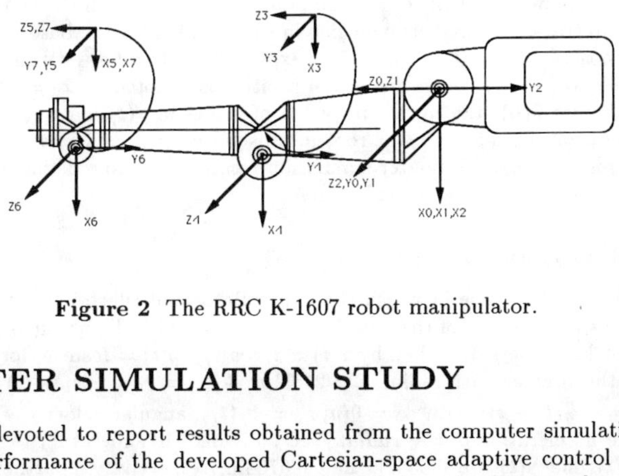

Figure 2 The RRC K-1607 robot manipulator.

COMPUTER SIMULATION STUDY

This section is devoted to report results obtained from the computer simulation conducted to study the performance of the developed Cartesian-space adaptive control scheme which was applied to control the motion of a RRC K-1607 robot manipulator as shown in Figure 2. The RRC K-1607, manufactured by Robotics Research Corporation is a 7 DOF kinematically redundant manipulator. The manipulator serves as a slave arm of a dual-arm telerobot system recently developed by Goddard Space Flight Center to investigate research issues in telerobotics such as advanced robot control algorithms, dual-arm teleoperation, end-effector design, and hierarchical control using high-level programming languages, etc. The slave arm motion can be controlled in a teleoperation mode by a human operator using the master arm system or autonomously controlled by a set of computer programs. A complete analysis of forward kinematics and differential motion of the RRC K-1607 can be found in [11]. Following the convention in [12], 8 coordinate frames are assigned to the manipulator as illustrated in Figure 2 showing the manipulator in its home configuration with all joint angles being zero. Each ith frame {i} is characterized by its coordinate axes $\mathbf{x}_i$, $\mathbf{y}_i$, $\mathbf{z}_i$ and its origin $\mathbf{0}_i$ for i = 0,1,2,...,7. The Denavit-Hartenberg parameters are given in the following table:

i	α_{i-1}	a_{i-1}	d_i	θ_i
1	0°	0.000in	0.0in	θ_1
2	-90°	0.000in	0.0in	θ_2
3	90°	5.625in	27.0in	θ_3
4	-90°	4.250in	0.0in	θ_4
5	90°	-4.250in	27.0in	θ_5
6	-90°	3.125in	0.0in	θ_6
7	90°	-3.125in	0.0in	θ_7

Table 1 D-H parameters of the RRC K-1607 manipulator.

For the simulation, ξ_i and ω_i for i=1,2 were selected so that 2 characteristic roots of (10) are -1 and -2. Thus we have $\mathbf{D}_1=2\mathbf{I}_7$ and $\mathbf{D}_2=3\mathbf{I}_7$. Selecting $\mathbf{Q}_i = \mathbf{I}_{14}$ for i=1,2,...,6 and solving (21) gives $\mathbf{P}_2 = \mathbf{P}_3 = 0.25\mathbf{I}_7$. The adaptive controller gains can be now computed by substituting the derived values of $\mathbf{P}_2$ and $\mathbf{P}_3$ into (30)-(31) where $\mathbf{K}_p(0)$ and $\mathbf{K}_d(0)$ can be arbitrarily set. The scalars β_1 and β_2 can be adjusted to improve the tracking quality provided that they are positive values. The payload was set to zero at the beginning of the simulation and suddenly changed to full payload of 10 lb at 1/3 of the simulation time and suddenly dropped to zero again at 2/3 of the simulation time. The sudden change in payload was modeled by delayed step functions. We considered two cases: tracking a straight line and tracking a circular path during a step change in payload. A modified version of Manipulator Simulation Program (**MSP**) [13] was employed for the computer simulation.

Tracking A Straight Line

Figure 3(a)-(b) present the results of the computer simulation in which the robot end-effector was controlled to track a desired straight line in the x-y plane of the base frame. The desired straight line is described by $x(t) = x_0 + 9[1 + 3\exp(-\frac{3.5}{3}t) - 4\exp(-\frac{3.5}{4}t)]$ and $y(t) = y_0 + 9[2 + 6\exp(-\frac{3.5}{3}t) - 8\exp(-\frac{3.5}{4}t)]$ where the initial position is denoted by $x_0 = 33.996in, y_0 = 0in$. According to Figure 3(b), the maximum value of errors in x(t) and y(t) are 0.3473in and 0.1599in, respectively. In addition, the root-mean-square errors of x(t) and y(t) are 0.1536in. and 0.1047in, respectively. Computer simulation results also showed that position errors in z(t) and orientation errors were negligible.

Tracking A Circular Path

Simulation results of the case in which the manipulator end-effector was to track a desired circular path in the x-y plane of the base frame are reported in Figure 4(a)-(b). The circular path consists of 3 segments described by $x(t) = R\cos\Phi_i$; $y(t) = R\sin\Phi_i$ for $t_{i-1} \leq t \leq t_i$ for i=1,2,3 where the circular path radius $R = 24.32in$, $\Phi_1(t) = \phi_0 + \frac{\beta}{2}t^2$, $\Phi_2(t) = \phi_1 + \omega(t - t_1)$, and $\Phi_3(t) = \phi_0 - \frac{\beta}{2}(t_3 - t)^2$ with $\phi_0 = 0in; \phi_1 = \Phi_1(t_1)$, angular velocity $\omega = \frac{2\pi}{9}radian/sec$ and the angular acceleration $\beta = \frac{2\pi}{9}radian/sec^2$. Figure 4(b) shows that the maximum value of errors in x(t) and y(t) are 0.6551in and 0.6764in, respectively. Furthermore, the root-mean-square errors of x(t) and y(t) are 0.2730in and 0.3657in, respectively. According to the computer simulation results, position errors in z(t) and orientation errors were negligible.

CONCLUDING REMARKS

In this paper, an adaptive control scheme in Cartesian space was developed to control non-compliant motion of kinematically redundant robot manipulators having 7 DOFs or more. The development of the control scheme employed the concept of model reference adaptive control and the Lyapunov theorem under the assumption that the robot end-effector performs slowly-varying motion. The control scheme was mainly composed of PD controllers whose gains are adjusted by an adaptation law to effectively react to dynamic and static coupling between joints and sudden change in payloads. Computer simulation conducted for two study cases: tracking a straight line and tracking a circular path showed that the developed control provided good tracking quality with root-mean-square errors of about 0.16in for the straight line and about 0.36in for the circular path, despite step changes in payloads. Current research activities focus on implementing the developed control scheme to control the real-time motion of the RRC K-1607 manipulator. Experimental results on the implementation will be reported in future papers. A hybrid control scheme is currently under development for simultaneous control of position and force of redundant robots. In addition, utilization of the manipulator redundancy for avoidance of obstacles and joint limits is currently an active research area.

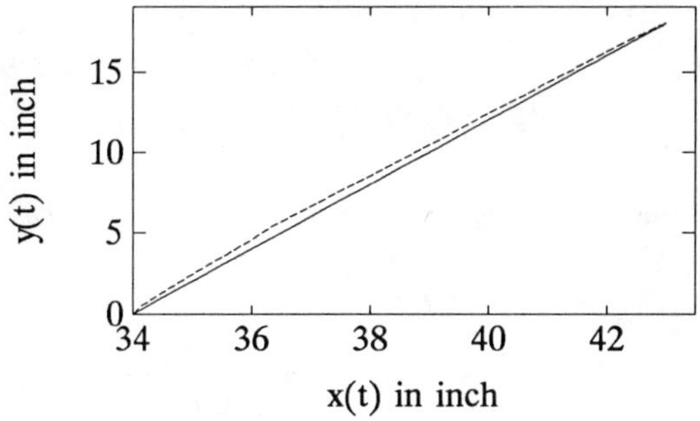

Figure 3(a) Tracking a straight line *(solid=desired; dashed=actual).*

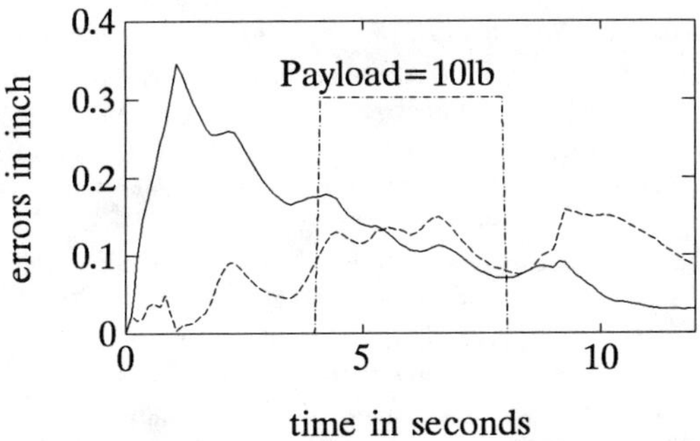

Figure 3(b) Errors in tracking a straight line *(solid=x-error; dashed=y-error).*

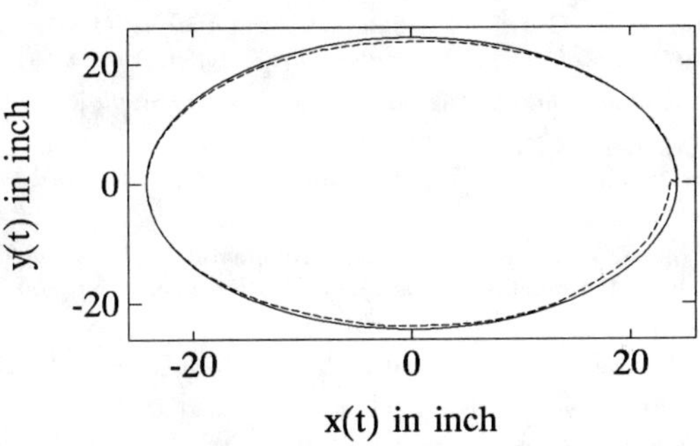

Figure 4(a) Tracking a circular path *(solid=desired; dashed=actual).*

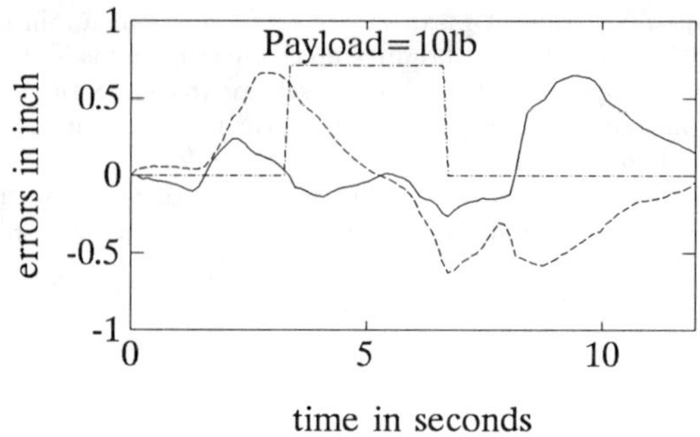

Figure 4(b) Errors in tracking a circular path *(solid=x-error; dashed=y-error).*

ACKNOWLEDGEMENT

The research presented in this paper was conducted at the Catholic University of America under a research grant, Grant Number NAG 5-1124, funded by NASA/GSFC.

References

[1] **Baillieul, J., Hollerbach, J., Brockett, R.**, "Programming and Control of Kinematically Redundant Manipulators," *Proc. 23rd IEEE Conf. on Decision and Control*, pp. 768-774, 1984.

[2] **Burdick, J., Seraji, H.**, "Characterization and Control of Self-Motions in Redundant Manipulators," *Proc. NASA Conference on Space Telerobotics*, Pasadena, January 1989.

[3] **Colbaugh, R.D.**, "Adaptive Position and Force Control of Redundant Robot Manipulators," in *Robotics and Manufacturing: Recent Trends in Research, Education, and Application*, edited by M. Jamshidi et al, ASME Press, New York, pp. 319-328, 1988.

[4] **Egeland, O.**, "Cartesian Control of a Hydraulic Redundant Manipulator," *Proc. IEEE Intern. Conf. on Robotics and Automation*, Raleigh, pp. 1081-1087, April 1987.

[5] **Kazerounian, K., Wang, S.**, "Global versus Local Optimization in Redundancy Resolution of Robotic Manipulators," *Int. J. Robotics Research* Vol. 7, No. 5, pp. 3-12, 1988.

[6] **Klein, C.A., Huang, C.H.**, "Review of Pseudoinverse Control for Use with Kinematically Redundant Manipulators," *IEEE Trans. Sys., Man, and Cyber.*, pp. 245-250, March 1983.

[7] **Nguyen, C.C., Zhou, Z.L., Mosier, G.E.**, "Joint-Space Adaptive Control of a Redundant Telerobot Manipulator," *Proc. Fourth IEEE International Symposium on Intelligent Control*, Albany, New York, pp. 59-65, September 1989.

[8] **Seraji, H.**, "Configuration Control of Redundant Manipulators: Theory and Implementation," *IEEE Transactions on Robotics and Automation*, pp. 472-490, August 1989.

[9] **Landau, Y.D.**, *Adaptive Control: The Model Reference Approach*, Marcell Dekker, 1979.

[10] **Nguyen, C.C., Mosier, G.E.**, "Model Reference Adaptive Control of a Telerobot System," *Proc. ISMM Intern. Symp. Computer Applications in Design, Simulation and Analysis*, Nevada, pp. 282-285, Feb. 1989.

[11] **Nguyen, C.C., Zhou, Z.L., Mosier, G.E.**, "Kinematic Analysis and Control of A 7-DOF Redundant Telerobot Manipulator," *Proc. 22nd Southeastern Symposium of System Theory*, Cookeville, Tennessee, pp. 71-77, March 1990.

[12] **Craig, J.J.**, Introduction to Robotics, 2nd Edition, *Addison-Wesley, Inc.*, Reading, Ma. 1989.

[13] **Chen, R., Ou, Y.J.**, "Dynamic Formulation for Efficient Digital Simulation of Telerobotic Manipulation," *Final Report, Grant No. NAG 5-1019*, NASA/GSFC, December 1988.

ADAPTIVE IMPEDANCE CONTROL OF REDUNDANT MANIPULATORS

R. Colbaugh† K. Glass† H. Seraji‡ K. Perryman†

†New Mexico State University, Las Cruces, NM 88001

‡Jet Propulsion Laboratory, Pasadena, CA 91109

Abstract
This paper presents an adaptive scheme for controlling the mechanical impedance of redundant manipulators. The proposed control system consists of two subsystems: an adaptive impedance controller which generates the Cartesian-space control input $\mathbf{F} \in \Re^m$ required to provide the desired end-effector impedance characteristics, and an algorithm that maps this control input to a robot joint torque vector $\mathbf{T} \in \Re^n$. The $\mathbf{F} \rightarrow \mathbf{T}$ map is constructed so that the robot redundancy is utilized to improve either the kinematic or dynamic performance of the robot. The impedance controller does not require knowledge of the complex robot dynamic model or parameter values for the robot, the payload, or the environment, and is implemented without calculation of the robot inverse kinematic transformation. As a result, the scheme is very general and is computationally efficient for on-line implementation.

1. Introduction

Increasing the applicability and utility of robot manipulators requires the development of strategies for controlling the robot when it is in contact with the environment. Robots are subject to interaction forces whenever they perform tasks involving motion which is constrained by the environment, such as assembling parts or deburring edges. In such motion, the interaction forces must be accommodated rather than resisted, so as to comply with the environmental constraints.

Recognizing the importance of compliant motion for robots, many researchers have investigated this problem. As a result of these studies, two broad approaches to achieving compliant motion have emerged. The first approach, usually called *hybrid position/force control*, is based on the observation that when the robot end-effector is in contact with the environment, the Cartesian-space of the end-effector coordinates may be naturally decomposed into a "position subspace" and a "force subspace" [1]. The position subspace and force subspace are orthogonal, and correspond to the Cartesian directions in which the end-effector is, respectively, free to move and constrained by the environment. The hybrid position/force control approach to compliant motion control is to track a position (and orientation) trajectory in the position subspace and a force (and moment) trajectory in the force subspace.

The second approach to compliant motion control for robots is called *impedance control*, and proposes that the control objective should not be the tracking of position/force trajectories, but rather should involve the regulation of the mechanical impedance of the robot end-effector [2]. Thus the objective of the impedance controller is to maintain a desired dynamic relationship between the end-effector position and the end-effector/environment contact force. In this paper, the impedance control approach to compliant motion control for robots is considered.

Following the pioneering work of Hogan [2-5], several researchers have studied impedance control of robots and have demonstrated the potential of this approach to compliant motion control [e.g., 6-12]. However, much of this research is either conceptual in nature, or else proposes control schemes designed using "classical" robot control techniques, such as the Computed Torque method [2-10]. Thus the controllers developed in these studies require exact knowledge and real-time computation of the robot dynamic model and inverse kinematic transformation, as well as precise information concerning the parameter values for the robot, the payload, and the environment. It has only been very recently that researchers have applied adaptive control methods to the impedance control problem [11,12]. This work has removed the need for knowledge of the parameter values for the robot and payload. However, the adaptive control approach taken in [11,12] yields control algorithms which still require knowledge and calculation of the robot dynamics and robot inverse kinematics. Additionally, in all of the previous studies [2-12], very little is done to extend the application of impedance control methods to kinematically redundant robots.

This paper presents an adaptive impedance control scheme for redundant manipulators. The controller is developed by considering the end-effector impedance control problem and the redundancy resolution problem separately. The end-effector impedance control problem is solved by using an adaptive scheme to generate the Cartesian-space control input $\mathbf{F} \in \Re^m$ required to realize the desired end-effector impedance characteristics; observe that this control problem is nonredundant even for redundant robots. The Cartesian-space impedance controller is derived using a Model Reference Adaptive Control (MRAC) approach, and does not require knowledge of the complex robot dynamic model or parameter values for the robot, the payload, or the environment. As a result, the scheme is very general and is computationally efficient for on-line implementation. The redundancy resolution problem is solved by appropriately mapping the control input $\mathbf{F}$ to a robot joint torque vector $\mathbf{T} \in \Re^n$. Note that for redundant robots the $\mathbf{F} \rightarrow \mathbf{T}$ mapping problem is underdetermined, so that it is at this stage of the control problem that the redundancy can be utilized to improve the kinematic or dynamic performance of the robot. Resolving the redundancy through the construction of an appropriate $\mathbf{F} \rightarrow \mathbf{T}$ map represents a departure from conventional approaches, and provides a general framework within which the redundancy can be effectively utilized.

This paper is organized as follows. In Sections 2 and 3 the adaptive impedance control algorithm for redundant robots is developed in the partitioned form indicated above, with the Cartesian-space impedance controller being derived in Section 2 and the redundancy resolution scheme being described in Section 3. Finally, Section 4 summarizes the paper and draws some conclusions.

2. Adaptive Cartesian-space Impedance Control Algorithm

The adaptive Cartesian-space impedance controller is developed within the MRAC framework in two steps. First, the impedance control objective is used to construct the "reference model" utilized in the MRAC approach, and then a Lyapunov-based design method is employed to derive an adaptive controller which ensures that the robot closely tracks this reference model. Consider an n DOF robot manipulator performing tasks in an m-dimensional Cartesian-space, with $m \leq n$. These tasks will, in general, involve motion of the robot end-effector in certain directions and accommodation of end-effector/environment interaction forces in the remaining directions. Let $\mathbf{x} \in \Re^m$ define the position (and orientation) of the end-effector in Cartesian-space and $\theta \in \Re^n$ be the vector of joint coordinates. The relationship between the end-effector coordinates $\mathbf{x}$ and joint coordinates θ is

$$\mathbf{x} = \mathbf{f}(\theta) \tag{1}$$

or

$$\dot{\mathbf{x}} = J(\theta)\dot{\theta} \tag{2}$$

where $\mathbf{f} : \Re^n \rightarrow \Re^m$ represents the forward kinematics of the robot and $J = \partial\mathbf{f}/\partial\theta \in \Re^{m\times n}$ is the end-effector Jacobian matrix. The dynamic model of the robot with its end-effector in contact with the environment may be written in joint-space as [e.g., 14]

$$\mathbf{T} = H(\theta)\ddot{\theta} + \mathbf{V}_{cc}(\theta,\dot{\theta}) + \mathbf{V}_f(\theta,\dot{\theta}) + \mathbf{G}(\theta) + J^T\mathbf{P} \tag{3}$$

where $\mathbf{T} \in \Re^n$ is the joint torque vector, $\mathbf{P} \in \Re^m$ is the end-effector/environment contact force and/or moment, $H(\theta) \in \Re^{n\times n}$ is the symmetric positive-definite inertia matrix, and $\mathbf{V}_{cc}, \mathbf{V}_f, \mathbf{G} \in \Re^n$ represent the torque vectors due to Coriolis and centripedal acceleration, friction, and gravity, respectively. Alternatively, the robot dynamic model can be expressed in Cartesian-space as [e.g., 15]

$$\mathbf{F} = H_x\ddot{\mathbf{x}} + \mathbf{V}_x + \mathbf{P} \tag{4a}$$

where $\mathbf{F} \in \Re^m$ is the generalized force vector corresponding to the Cartesian coordinate vector $\mathbf{x}$, and $H_x \in \Re^{m \times m}$ and $\mathbf{V}_x \in \Re^m$ are defined as follows:

$$H_x = (JH^{-1}J^T)^{-1} \tag{4b}$$

$$\mathbf{V}_x = (JH^{-1}J^T)^{-1}JH^{-1}[\mathbf{V}_{cc} + \mathbf{V}_f + \mathbf{G}] - (JH^{-1}J^T)^{-1}\dot{J}\dot{\theta} \tag{4c}$$

The Cartesian-space impedance control scheme can be developed with either a centralized or a decentralized control structure. In this section, the decentralized structure is adopted; however, the centralized controller can be obtained in a similar manner. For the development of a decentralized control algorithm, it is convenient to view each Cartesian component of the robot dynamic model (4) as a subsystem of the complete robotic system; these subsystems are interconnected by the dynamic cross-couplings between the Cartesian degrees-of-freedom. The dynamic models for these subsystems can be written as

$$F_i = h_{ii}\ddot{x}_i + \sum_{\substack{j=1 \\ j \neq i}}^{m} h_{ij}\ddot{x}_j + v_i + P_i \qquad i = 1, 2, \ldots, m \tag{5}$$

where F_i, x_i, v_i, P_i, and h_{ij} are the elements of $\mathbf{F}, \mathbf{x}, \mathbf{V}_x, \mathbf{P}$, and H_x in (4). Note from (4b) that H_x is symmetric positive-definite, so that $h_{ii} > 0$ for $i = 1, 2, \ldots, m$ for any robot configuration. Defining

$$d_i = \sum_{\substack{j=1 \\ j \neq i}}^{m} h_{ij}\ddot{x}_j + v_i$$

allows (5) to be written as

$$F_i = h_{ii}\ddot{x}_i + d_i + P_i \qquad i = 1, 2, \ldots, m \tag{6}$$

Equations (6) represent the dynamic models for the m subsystems corresponding to the m Cartesian components of the complete robot dynamics (4).

Let $\mathbf{x}_r(t) = [x_{ri}(t)] \in \Re^m$ define the twice-differentiable, bounded reference position trajectory for the robot end-effector and $\mathbf{e} = [e_i] = \mathbf{x}_r - \mathbf{x}$ denote the Cartesian tracking error. The impedance control approach to compliant motion control proposes that the control objective should not be asymptotic trajectory tracking, in which the tracking error $\mathbf{e}$ satisfies $\mathbf{e} \to \mathbf{0}$ as $t \to \infty$ [e.g., 3]. Rather, this approach to robot control asserts that the control objective should involve the achievement of a desired dynamic relationship between the tracking error and the end-effector/environment contact force. Thus, given a reference position trajectory $\mathbf{x}_r(t)$ and a contact force $\mathbf{P}$, the goal is to ensure that the end-effector response mimics the dynamics of a Cartesian mass-spring-damper system. In other words, the desired end-effector response $\mathbf{x}_m(t) = [x_{mi}(t)] \in \Re^m$ to the reference trajectory $\mathbf{x}_r$ and contact force $\mathbf{P}$ is defined by the *desired impedance dynamics* [e.g., 3]:

$$m_i\ddot{x}_{mi} + n_i(\dot{x}_{mi} - \dot{x}_{ri}) + k_i(x_{mi} - x_{ri}) = -P_i \qquad i = 1, 2, \ldots, m \tag{7}$$

where m_i, n_i, and k_i are positive scalar constants chosen to reflect the desired mass, damping, and stiffness in the ith Cartesian direction. Defining $\mathbf{e}_m = [e_{mi}] = \mathbf{x}_r - \mathbf{x}_m$ and $\mathbf{E}_{mi} = [\ e_{mi} \quad \dot{e}_{mi}\ ]^T$ allows the ith component of the desired impedance dynamics (7) to be written as

$$\dot{\mathbf{E}}_{mi} = \begin{bmatrix} 0 & 1 \\ -\omega_i^2 & -2\zeta_i\omega_i \end{bmatrix} \mathbf{E}_{mi} + \begin{bmatrix} 0 \\ \ddot{x}_{ri} \end{bmatrix} + \begin{bmatrix} 0 \\ P_i/m_i \end{bmatrix} \tag{8}$$

where $\omega_i^2 = k_i/m_i$ and $2\zeta_i\omega_i = n_i/m_i$.

The impedance control objective is to devise a control scheme which, when combined with the robot dynamics (6), will exhibit a closed-loop response with dynamics identical to the desired impedance dynamics (7). Toward this end, consider the ith component of a simple decentralized adaptive control law [13]:

$$F_i(t) = f_i(t) + k_{pi}(t)e_i + k_{vi}(t)\dot{e}_i \tag{9}$$

where F_i is the ith Cartesian control input, f_i is a scalar adaptive auxilliary signal, and k_{pi} and k_{vi} are scalar adaptive position and velocity feedback gains. Applying the control law (9) to the robot dynamics (6) yields the error dynamics for the ith subsystem:

$$\dot{\mathbf{E}}_i = \begin{bmatrix} 0 & 1 \\ -k_{pi}/h_{ii} & -k_{vi}/h_{ii} \end{bmatrix} \mathbf{E}_i + \begin{bmatrix} 0 \\ \ddot{x}_{ri} \end{bmatrix} + \begin{bmatrix} 0 \\ (P_i + d_i - f_i)/h_{ii} \end{bmatrix} \tag{10}$$

where $\mathbf{E}_i = [\ e_i \quad \dot{e}_i\ ]^T$. The impedance control objective is therefore realized if $\mathbf{E}_i = \mathbf{E}_{mi}$ $\forall t$ for each component i, since this implies that $\mathbf{x}(t) = \mathbf{x}_m(t)$ and thus that the dynamics of the closed loop system (6),(9) is identical to the desired impedance dynamics (7). In practice, ensuring that $\mathbf{E}_i = \mathbf{E}_{mi}$ for all time is difficult since this implies precise knowledge and computation of the system dynamic model (6) at the control servo-rate. The practical solution to the impedance control problem is to require the output of the error dynamics (10) to track the output of the desired impedance dynamics (8) *asymptotically*, so that $\mathbf{E}_i \to \mathbf{E}_{mi}$ as $t \to \infty$. Observe that this control objective can be formulated within the MRAC framework by defining (8) to be the "reference model" and (10) to be the "adjustable system". The control objective can then be stated as follows: determine adaptation laws for updating the gains f_i, k_{pi}, and k_{vi} in such a way that the adjustable system (10) tracks the reference model (8) asymptotically. These adaptation laws are derived in [13] using a standard Lyapunov-based MRAC design method, and have the following form:

$$\begin{aligned} f_i(t) &= f_i(0) + \alpha_{1i} \int_0^t q_i dt + \beta_{1i} q_i \\ k_{pi}(t) &= k_{pi}(0) + \alpha_{2i} \int_0^t q_i e_i dt + \beta_{2i} q_i e_i \\ k_{vi}(t) &= k_{vi}(0) + \alpha_{3i} \int_0^t q_i \dot{e}_i dt + \beta_{3i} q_i \dot{e}_i \\ q_i(t) &= w_{pi}(x_{mi} - x_i) + w_{vi}(\dot{x}_{mi} - \dot{x}_i) \\ m_i \ddot{x}_{mi} + n_i \dot{x}_{mi} + k_i x_{mi} &= n_i \dot{x}_{ri} + k_i x_{ri} - P_i \end{aligned} \tag{11}$$

where the α_{ji} and β_{ji} are positive and nonnegative scalar adaptation gains, respectively, which are chosen to provide the desired adaptation rates for the controller terms, and the w_{pi} and w_{vi} are positive scalar constants selected to reflect the relative significance of the errors $x_{mi} - x_i$ and $\dot{x}_{mi} - \dot{x}_i$. Note that since the adaptation laws depend on the desired impedance dynamics (7), this equation is included for completeness.

The decentralized adaptive control law (9),(11) ensures that each component of the robot error dynamics (10) tracks the desired impedance error dynamics (8) globally and asymptotically [13]. Thus $\mathbf{x} \to \mathbf{x}_m$, where $\mathbf{x}_m$ is the desired impedance trajectory of the robot end-effector defined in (7). The controller (9),(11) has a simple structure and can be implemented on-line with a high sampling rate using modest computing power. The control input $\mathbf{F}$ is computed entirely based on the observed performance of the robot, so that knowledge of the robot dynamics is not required.

Observe that the control force vector $\mathbf{F} \in \Re^m$ computed in (9),(11) cannot physically be applied to the robot end-effector; therefore this desired control input must be mapped to a joint torque vector $\mathbf{T} \in \Re^n$. For nonredundant robots ($m = n$), this $\mathbf{F} \to \mathbf{T}$ map is uniquely defined and is given by [e.g., 15]:

$$\mathbf{T} = J^T(\theta)\mathbf{F} \tag{12}$$

For redundant robots ($m < n$), however, the $\mathbf{F} \to \mathbf{T}$ mapping problem is underdetermined, so that it is at this stage of the control problem that the redundancy can be utilized to improve robot performance. This redundancy resolution problem is considered in the following section.

3. Resolution of Redundancy

In order for the Cartesian-space impedance control scheme (9),(11) to be implemented with a robot, the control input $\mathbf{F}$ must be mapped to a joint torque vector $\mathbf{T}$. The $\mathbf{F} \to \mathbf{T}$ mapping problem is underdetermined for redundant robots, so that there exists an infinite number of joint torque vectors corresponding to a given control input $\mathbf{F}$. By properly selecting a particular joint torque vector from the possible candidates, the robot redundancy can be utilized to improve the robot's performance. Resolving the redundancy through the construction of an appropriate $\mathbf{F} \rightarrow \mathbf{T}$ map represents a departure from the conventional approach in which the redundancy is resolved at the inverse kinematics level; a general and comprehensive discussion of this approach to redundant robot control may be found in [16-19]. In this section, the problem of constructing an appropriate $\mathbf{F} \rightarrow \mathbf{T}$ map is considered within the context of impedance control.

Recall that for a redundant robot, $\mathbf{F} \in \Re^m$ and $\mathbf{T} \in \Re^n$ with $m < n$. The problem of constructing an appropriate $\mathbf{F} \rightarrow \mathbf{T}$ map may be formulated in terms of inverting the known $\mathbf{T} \rightarrow \mathbf{F}$ map, which is unique even for redundant robots. The $\mathbf{T} \rightarrow \mathbf{F}$ may be shown to be [15,16]

$$\mathbf{F} = (JH^{-1}J^T)^{-1} JH^{-1}\mathbf{T} \equiv M(\theta)\mathbf{T} \tag{13}$$

where it is easily verified that $M \in \Re^{m \times n}$. Inversion of the $\mathbf{T} \rightarrow \mathbf{F}$ map (13) may be achieved in two ways: "direct" inversion using the theory of generalized inverses [20], or "indirect" inversion in which both M and $\mathbf{F}$ are augmented with $r = n - m$ additional rows and the resulting fully-determined system is inverted utilizing standard methods. These approaches are now outlined briefly. Additional details concerning this problem are given in [16-19].

The direct approach to inverting (13) is useful for realizing dynamic performance objectives, primarily because inverting (13) using generalized inverse theory readily permits optimization of objective functions involving joint torque $\mathbf{T}$ and joint acceleration $\ddot{\theta}$. For example, consider the problem of globally minimizing robot kinetic energy subject to the constraint that the end-effector motion $\mathbf{x}(t)$ closely tracks the desired impedance trajectory $\mathbf{x}_m(t)$. This problem may be formulated as follows:

$$\text{minimize} \qquad \int_0^{t_f} \frac{1}{2}\dot{\theta}^T H(\theta)\dot{\theta}\, dt$$

$$\text{subject to the constraint} \qquad \mathbf{x}_m - \mathbf{f}(\theta) = \mathbf{0} \tag{14}$$

where t_f is the trajectory completion time. The solution to the constrained optimization problem (14) can be obtained by computing the control input $\mathbf{F}$ using the algorithm (9),(11), and then mapping this input to the joint torque vector $\mathbf{T}$ with the following $\mathbf{F} \rightarrow \mathbf{T}$ map [16]:

$$\mathbf{T} = J^T\mathbf{F} + [I_n - J^T(JH^{-1}J^T)^{-1}JH^{-1}](\mathbf{V}_f + \mathbf{G}) \tag{15}$$

where I_n is the $n \times n$ identity matrix. Now it can be shown that the vector $[I_n - J^T(JH^{-1}J^T)^{-1} JH^{-1}](\mathbf{V}_f + \mathbf{G})$ is typically much smaller than the vector $J^T\mathbf{F}$, so that the $\mathbf{F} \rightarrow \mathbf{T}$ map

$$\mathbf{T} = J^T\mathbf{F} \tag{16}$$

provides a close approximation to the global minimum kinetic energy map (15) [16,17]. Observe that the map (16) is computationally efficient and requires no knowledge of the robot dynamic model. The $\mathbf{F} \rightarrow \mathbf{T}$ map (16) has been shown to significantly (and stably) reduce the joint torque requirements compared with other redundancy resolution schemes [16,17].

The indirect approach to inverting the $\mathbf{T} \rightarrow \mathbf{F}$ map (13) is useful primarily for realizing kinematic performance objectives. The idea of augmenting M and $\mathbf{F}$ with r additional rows to obtain a fully-determined system is conceptually simple, and can be accomplished by augmenting the vector of end-effector generalized coordinates $\mathbf{x}$ with an additional generalized coordinate vector $\psi = [\psi_i] \in \Re^r$. The vector ψ is chosen so that the robot redundancy is resolved in some useful manner. For example, many kinematic performance objectives can be quantified by specifying that, in addition to performing the desired m-dimensional end-effector task, the robot must perform an r-dimensional additional task defined by the following kinematic relationship [18]:

$$\psi = \mathbf{g}(\theta) = \psi_r(t) \tag{17}$$

where $\mathbf{g} : \Re^n \to \Re^r$ is a kinematic function whose desired evolution is defined by $\psi_r(t) = [\psi_{ri}(t)] \in \Re^r$. Let $\mathbf{F}_c \in \Re^r$ denote the generalized force vector corresponding to the generalized coordinate vector ψ. Then the desired trajectory $\psi_r(t)$ for the additional task (17) can be tracked by computing $\mathbf{F}_c$ using the algorithm (9),(11). Note that for the additional task, $P_i = 0$ in (7) so that the impedance model (7) reduces to

$$m_i \ddot{\psi}_{mi} + n_i(\dot{\psi}_{mi} - \dot{\psi}_{ri}) + k_i(\psi_{mi} - \psi_{ri}) = 0 \qquad i = 1, 2, ..., r$$

which is a standard MRAC reference model for trajectory tracking control.

Augmenting the end-effector coordinates $\mathbf{x}$ with the additional task coordinates ψ allows the $\mathbf{T} \to \mathbf{F}$ map (13) to be simplified considerably. Letting $\mathbf{y} = [\ \mathbf{x}^T \ \vdots\ \psi^T\ ]^T \in \Re^n$ define the augmented task-space coordinate vector and $J_a = \partial \mathbf{y}/\partial \theta \in \Re^{n \times n}$ denote the augmented Jacobian matrix, it can be seen that in terms of this new augmented task-space the $\mathbf{T} \to \mathbf{F}$ map (13) becomes

$$\begin{bmatrix} \mathbf{F} \\ -- \\ \mathbf{F}_c \end{bmatrix} = (J_a^T)^{-1}\mathbf{T} = [\,J^T \ \vdots\ J_c^T\,]^{-1}\mathbf{T} \tag{18}$$

where $J_c = \partial \mathbf{g}/\partial \theta \in \Re^{r \times m}$ is the additional task Jacobian matrix, and it is assumed that J_a is nonsingular. Inverting (18) yields

$$\mathbf{T} = [\,J^T \ \vdots\ J_c^T\,] \begin{bmatrix} \mathbf{F} \\ -- \\ \mathbf{F}_c \end{bmatrix}$$

or

$$\mathbf{T} = J^T\mathbf{F} + J_c^T\mathbf{F}_c \tag{19}$$

As an example of using the $\mathbf{F} \to \mathbf{T}$ map (17)-(19) to realize a kinematic performance objective, consider the problem of optimizing the general kinematic objective function $G(\theta) : \Re^n \to \Re$. This goal can be achieved with the map (17)-(19) by setting

$$\mathbf{g} = N_e \partial G/\partial \theta \tag{20}$$

and $\psi_r(t) = 0$, where $N_e \in \Re^{r \times n}$ is any matrix whose rows form a basis for the null-space of J [16-18]. One formulation for N_e is [16]:

$$N_e = [\ -J_2^T(J_1^{-1})^T \ \vdots\ I_r\ ] \tag{21}$$

where $J_1 \in \Re^{m \times m}$ and $J_2 \in \Re^{m \times r}$ are the partitions of J defined by $J = [\ J_1 \ \vdots\ J_2\ ]$. The $\mathbf{F} \to \mathbf{T}$ map (17)-(21) has been applied to a variety of kinematic objective functions with excellent results [16-18].

In summary, the proposed impedance control algorithm for redundant robots consists of the control law (9) and adaptation laws (11), together with the appropriate $\mathbf{F} \to \mathbf{T}$ map. The decentralized adaptive impedance controller (9),(11) ensures that the desired end-effector impedance is achieved, and the $\mathbf{F} \to \mathbf{T}$ map can be chosen (as described above) to effectively utilize the redundancy to improve robot performance.

4. Conclusions

This paper presents a scheme for controlling the mechanical impedance of kinematically redundant manipulators. The proposed control system consists of two subsystems: an adaptive impedance controller which generates the Cartesian-space control input $\mathbf{F} \in \Re^m$ required to provide the desired end-effector performance, and an algorithm that maps this control input to a robot joint torque vector $\mathbf{T} \in \Re^n$. The $\mathbf{F} \to \mathbf{T}$ map is constructed so that the robot redundancy is utilized in an appropriate manner. The resulting control scheme is computationally fast and does not require knowledge of the complex dynamic model or the inverse kinematic transformation for the robot. Future research will address the implementation of the proposed control scheme on more complex redundant robotic systems, including dual-arm robots.

5. Acknowledgements

The research described in this paper was supported in part through grants from the Waste Management Education and Research Consortium and the General Motors Corporation, and in part through contracts with Sandia National Laboratories, the Jet Propulsion Laboratory, and the National Aeronautics and Space Administration. The authors gratefully acknowledge many helpful interactions with Patrick Eicker of Sandia National Laboratories.

6. References

1. Raibert, M. and J. Craig, "Hybrid Position/Force Control of Manipulators", *ASME Journal of Dynamic Systems, Measurement, and Control*, Vol. 102, No. 2, 1981, pp. 126-133
2. Hogan, N., "Mechanical Impedance Control in Assistive Devices and Manipulators", *Proc. Joint Automatic Controls Conference*, San Francisco, CA, Aug. 1980
3. Hogan, N., "Impedance Control: An Approach to Manipulation, Parts I-III", *ASME Journal of Dynamic Systems, Measurement, and Control*, Vol. 107, No. 1, 1985, pp. 1-24
4. Hogan, N., "On the Stability of Manipulators Performing Contact Tasks", *IEEE Journal of Robotics and Automation*, Vol. 4, No. 6, 1988, pp. 677-686
5. Sharon, A., N. Hogan, and D. Hardt, "Controller Design in the Physical Domain (Application to Robot Impedance Control)", *Proc. IEEE Intern. Conference on Robotics and Automation*, Scottsdale, AZ, May 1989
6. Kazerooni, H., T. Sheridan, and P. Houpt, "Robust Compliant Motion for Manipulators, Parts I and II", *IEEE Journal of Robotics and Automation*, Vol. 2, No. 2, 1986, pp. 83-105
7. Kazerooni, H., "On the Robot Compliant Motion Control", *ASME Journal of Dynamic Systems, Measurement, and Control*, Vol. 111, No. 3, 1989, pp. 416-425
8. Kazerooni, H., "Compliant Motion Control for Robot Manipulators", *International Journal of Control*, Vol. 49, No. 3, 1989, pp. 745-760
9. Anderson, R. and M. Spong, "Hybrid Impedance Control of Robotic Manipulators", *IEEE Journal of Robotics and Automation*, Vol. 4, No. 5, 1988, pp. 549-555
10. Goldenberg, A., "Implementation of Force and Impedance Control in Robot Manipulators", *Proc. IEEE Intern. Conference on Robotics and Automation*, Philadelphia, PA, May 1988
11. Niemeyer, G. and J-J. Slotine, "Computational Algorithms for Adaptive Compliant Motion", *Proc. IEEE Intern. Conference on Robotics and Automation*, Scottsdale, AZ, May 1989
12. Kelly, R., R. Carelli, M. Amestregus, and R. Ortega, "On Adaptive Impedance Control of Robot Manipulators", *Proc. IEEE Intern. Conference on Robotics and Automation*, Scottsdale, AZ, May 1989
13. Colbaugh, R., K. Glass, and H. Seraji, "Adaptive Impedance Control of Robot Manipulators", to appear
14. Tarn, T., A. Bejczy, and X. Yun, "Robot Arm Force Control Through System Linearization by Nonlinear Feedback", *Proc. IEEE Intern. Conference on Robotics and Automation*, Philadelphia, PA, May 1988
15. Khatib, O., "A Unified Approach for Motion and Force Control of Robot Manipulators: The Operational Space Formulation", *IEEE Journal of Robotics and Automation*, Vol. 3, No. 1, 1987, pp.43-53
16. Colbaugh, R., "Dynamic Performance Optimization of Redundant Robot Manipulators", *Preprints 842nd Meeting of the American Mathematical Society*, Las Cruces, NM, April 1988
17. Colbaugh, R. and K. Glass, "Cartesian Control of Redundant Manipulators", *Journal of Robotic Systems*, Vol. 6, No. 4, 1989, pp. 427-459
18. Seraji, H., "Configuration Control of Redundant Manipulators: Theory and Implementation", *IEEE Trans. Robotics and Automation*, Vol. 5, No. 4, 1989, pp. 472-490
19. Colbaugh, R., H. Seraji, and K. Glass, "Obstacle Avoidance for Redundant Robots Using Configuration Control", *Journal of Robotic Systems*, Vol. 6, No. 6, 1989, pp. 721-744
20. Lawson, C. and R. Hanson, *Solving Least Squares Problems*, Prentice-Hall, Englewood Cliffs, NJ, 1974

OPTIMAL POSITIONING OF REDUNDANT MANIPULATOR-PLATFORM SYSTEMS FOR MAXIMUM TASK EFFICIENCY

François G. Pin and Jean-Christophe Culioli
Center for Engineering Systems Advanced Research, Oak Ridge National Laboratory, P.O. Box 2008, Building 6025, MS-6364, Oak Ridge, TN 37831-6364

ABSTRACT

Mobile manipulators are attracting significant interest in the industrial, military and public service communities because of the potential they provide for increased efficiency in material handling and manipulation tasks. Corresponding interest has arisen in the robotic research community since the combination and coordination of the mobility of an autonomous platform with the robotic motion of a manipulator introduce complex problems. This paper focusses on a subset of these problems, namely the utilization of the redundancy occurring in most mobile manipulators to optimize their initial positioning and configuration in the accomplishment of given tasks. Optimization schemes are developed for cases when force and position constraints are applied at the end-effector. Various optimization criteria are investigated for optimizing the position of the platform and the configuration of the manipulator with respect to obstacle avoidance, maneuverability and several torque functions. Sample results are presented for a system involving a three-link manipulator on a mobile platform. The various optimization schemes are discussed and compared, and several directions including multi-criteria optimization, tunneling algorithms, and minimax optimization are outlined for further extensions of the methods to the general problem of optimal positioning and configuring of redundant robotic systems with combined mobility and manipulation capabilities.

INTRODUCTION

Autonomous mobile manipulators, i.e. manipulators mounted on mobile platforms, are attracting significant interest in the industrial, military and public service communities because of the potential they provide for increased efficiency in material handling and manipulation tasks (e.g. warehouse management, plant surveillance, maintenance or repair, material transfer and delivery, etc.) and their capability to perform tasks in environments unaccessible to or very hazardous for humans (e.g. nuclear plants, battlefields, chemical, fire fighting, etc.). A requirement of such mobile manipulators is to incorporate both an autonomous large-scale mobility capability and autonomous dexterous manipulation abilities. Although significant amounts of research and development have been performed in the area of autonomous navigation for mobile platforms and in the area of autonomous control of manipulator systems very little has been reported on the topic of coupling these two functions for the purpose of developing an autonomous robot with combined large-scale mobility/dexterous manipulation [1-5]. The Center for

Engineering Systems Advanced Research (CESAR) at the Oak Ridge National Laboratory (ORNL) is working toward this goal, focussing on methodology development and proof of principle experiments using the HERMIES-III autonomous mobile robot [3].

An important characteristic of practical mobile manipulators is their particular kinematic redundancy, created by the addition of the degrees of freedom of the platform to those of the manipulator. This redundancy, quite desirable for dexterous manipulation and transport functions in cluttered environments, allows the systems to be optimally positioned and configured for maximum performance when given task requirements and/or system, environment and task constraints are imposed. The coupling of the mobility and manipulation functions, however, introduces a new order of complexity in the motion planning and control problem, which arises mainly from the highly redundant kinematics of the system (e.g. a 3 d.o.f. platform and a 6 d.o.f. manipulator produce a 9 d.o.f. system) and from the differences in the motion control of non-serial link platform systems (typically including non-holonomic constraints) and that of serial link (possibly redundant) manipulators.

In the accomplishment of sequences of tasks, the resolution of the redundancy generally involves two steps: the global optimization of the motion and configuration of the system during execution of a task with given requirements and constraints, and the local optimization of the position and configuration of the system during task commutation when changes in both task requirements and constraints occur. This paper is concerned with the latter topic of optimizing the initial position and configuration of both the platform and manipulator for performance of tasks where load and position constraints are applied at the end-effector and obstacle avoidance, maximum manipulability, and torque requirements are imposed on the tasks.

OPTIMIZATION APPROACH

The kinematic relations for the mobile manipulator system are given by

$$\overline{X}_e = \overline{X}_p + \overline{X}_{m/p} = \overline{X}_p + \overline{F(\overline{\Theta})} \tag{1}$$

where $\overline{X}_e$ is the "task vector" representing the position and orientation of the end-effector in the absolute reference frame, $\overline{X}_p$ is the "platform vector" representing the position and orientation of the platform reference frame in the absolute reference frame, $\overline{X}_{m/p}$ represents the vector of end-effector position and orientation with respect to the platform reference frame and $\overline{\Theta}$ is the vector of articular positions of the manipulator. The components $Xp_i, i = 1, n$ and $\Theta_j,\ j = 1, m$, of the vectors $\overline{X}_p$ and $\overline{\Theta}$, thus represent the system variables generating a space, S, of dimension typically greater than 6.

Assuming that a point load $\overline{f}$ is applied at the end effector, the contributed manipulator actuator torque $\overline{\tau}_m$ on the static system can be calculated as:

$$\overline{\tau}_m = J^{\tau}(\overline{\Theta})\overline{f} \tag{2}$$

where $J^{\tau}(\overline{\Theta})$ denotes the transpose of the manipulator jacobian matrix. The friction-induced torques $\overline{\tau}_p$ on the platform wheel actuators can be calculated from the specific platform kinematics.

The problem of finding an optimum commutation position and configuration for the mobile manipulator can be formulated as the constrained optimization in S of a given criterion E subject to a set of constraints C_i. The criterion E corresponds to some requirement for performance of the task which involves avoiding obstacles, not exceeding actuator torque limits and/or maintaining some maneuverability in all directions (e.g. by staying away from singular configurations). Redundancy resolution schemes generally involve minimization of the norm of some system variable or of the actuator torques. A similar approach can be utilized here by minimizing the load-induced actuator torque norm, $E_1 = \|\overline{\tau}_t\|^2$, where the first components of $\overline{\tau}_t$ are those of $\overline{\tau}_m$ and the last components are those of $\alpha\overline{\tau}_p$ with α being a given parameter weighing the relative importance of the platform actuator torques with respect to those of the manipulator.

Minimization of E_1 in cases where the load forces are very high, however, can lead to very uneven distributions of the joint actuator torques, sometimes exceeding the limit of an actuator while the others remain relatively low. We thus considered a criterion, $E_2 = Max_i|B_i\tau_{ti}|$, where the coefficients B_i weigh the individual torques by the inverse of their relative limit value. Utilization of this criterion leads to much more evenly distributed optimal actuator loads; however, at the expense of a much more complex mathematical treatment: minimization of E_2 essentially consists in solving a "minimax" problem [6]. The Lagrangian in this type of problem is not differentiable and numerical approaches such as those described in [6], [7], or [8] need to be utilized. For the results described in this paper, a projected subgradient algorithm has been utilized [8,9].

For obstacle avoidance schemes, a criterion involving a repulsive potential [10], $E_{obs} = \frac{\lambda}{\|\overline{X}-\overline{X}o\|^2}$ is used, where λ is a scalar and $\overline{X}$ and $\overline{X}o$ represent the cartesian position vector of any given point on the mobile manipulator and on an obstacle, respectively.

To provide commutation configurations avoiding as much as possible the singular configurations of the manipulator, a manipulability index inspired from that proposed in several redundant manipulator studies, e.g. see [11], $E_{man} = \frac{1}{Det(JJ^T)}$ can also be utilized. Here, it has been assumed that the platform is omnidirectional such as that of HERMIES-III [3] and that, consequently, the maneuverability requirement involves only the manipulator (thus the use of a manipulability index).

The constraints applied in the optimization can be of various types and need to be expressed in terms of the system variables. For example, the most common constraint for material handling tasks imposes a given position and orientation of the end-effector at task commutation: $\overline{X}e = \overline{X}c$. With this constraint, the Lagrangian for an optimization criterion E becomes:

$$L(Xp_i, \Theta_j, \mu_k, \nu_l) = E + \sum_k \mu_k(Xe_k - Xc_k) + \sum_l \nu_l C_l \tag{3}$$

were the C_l $l = 1, p$ represent p additional constraints on the system. The optimality conditions (except when E_2 is involved as discussed previously) can be written as:

$$\frac{\partial L}{\partial Xp_i} = \frac{\partial L}{\partial \Theta_j} = \frac{\partial L}{\partial \mu_k} = \frac{\partial L}{\partial \nu_l} = 0$$

$$\forall i = 1, n \quad \forall j = 1, m \quad \forall k = 1, 6 \quad \forall l = 1, p \tag{4}$$

and this system of $n + m + p + 6$ equations can be solved with a Newton algorithm.

It should be noted that Eq. (4) represents only necessary conditions for local optimality. One must check that a global minimum has in fact been found through the Newton algorithm and, as necessary, use an iteration or Tunneling-type algorithm [12] to avoid poor local minima. Also note that the criterion E can be modified to consider cases when several requirements exist on the upcoming task [4,5]. For this, a "global criterion" [5] can be formulated as:

$$E_g = \sum_j \sum_i \gamma_{ij} E_{obs}(\overline{X}o_i) + \delta E_{man} + \epsilon E_1 + \ldots \tag{5}$$

where the first term on the right-hand side of Eq. (5) represents the requirements for avoidance of i obstacles located at $\overline{X}o_i$ by j chosen points on the mobile manipulator external surface. Each criterion contributing to the global criterion is weighted by a coefficient $(\gamma_{ij}, \delta, \epsilon, \ldots)$ representing the relative importance given to the corresponding requirement (e.g. risk of collision versus manipulability) in the upcoming task.

SAMPLE RESULTS

Results of sample problems involving a four d.o.f. planar mobile manipulator are shown in the following figures. These results were obtained on a Macintosh II using *Mathematica*, a general symbolic manipulation package produced by Wolfram Research, Inc., which contains built-in functions for most of the low-level algorithms needed for our optimization problem. In each of the following examples, the system includes three links of length $L_1 = L_2 = 1, L_3 = \frac{1}{2}$, with revolute joints, and a platform moving along the horizontal axis. In the figures, the platform is represented by the bold rectangle on the x axis. Two degrees of redundancy are provided to the system by imposing constraints on the end-effector position only as $(Xc, Yc) = (5, 1.4)$. A force $\overline{f} = (0, -5)$ is imposed at the end-effector.

Figure 1 shows the results of the optimization scheme for a least torque norm criterion. In this configuration, the values of the manipulator torques are $\overline{\tau}_m = \{-2.6, -1.2, 0.8\}$ with a torque norm value of $\|\overline{\tau}_m\|^2 = 9.1$. The maximum torque value is at the first joint while the ratio of maximum to minimum actuator torque is 3.25. If a limit is placed on the torque of the first joint, say $|\tau_1| \leq 2.0$, the solution space is correspondingly reduced, and the new commutation configuration shown in Fig. 2, quite different from that of Fig. 1, is obtained. The torque values are found as $\overline{\tau}_m = \{-2.0, 2.8, -1.1\}$ with a least norm $\|\overline{\tau}_m\|^2 = 12.76$ and a maximum to minimum torque ratio of about 2.5. It is interesting to note that the solution shown in Fig. 2 was found as a local minima during the optimization iterations leading to the solution shown in Fig. 1, confirming that with the least norm criteria, the solutions corresponding to highly constrained problems with limits on individual torque values, are generally found faster than the global optimum for the unconstrained (by individual torque limits) problem. This feature may prove significant in problems where stringent time constraints exist, requiring that sub-optimal solutions be rapidly obtained and examined for viability.

Figure 3 shows the commutation configuration when the "minimax" criterion, i.e. minimization of the maximum actuator torque, is utilized with the same limit applied to all individual torques (i.e., all coefficients $Bi = 1$). The resulting torque values are $\overline{\tau}_m = \{-2.45, 2.45, 2.45\}$. This represent a much more uniform torque distribution than those of the two previous examples, confirming that the minimax criterion produces optimal commutation configurations that allow best utilization of all available individual torques. The run time to convergence in this example, however, was about an order of magnitude greater than that of the two cases previously discussed.

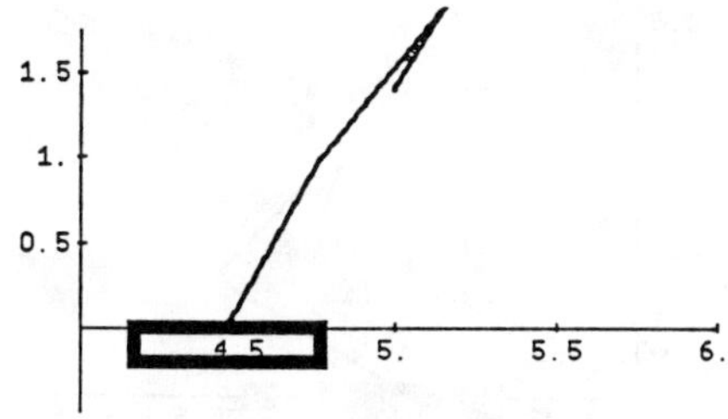

Fig. 1. Commutation configuration for a least torque norm requirement.

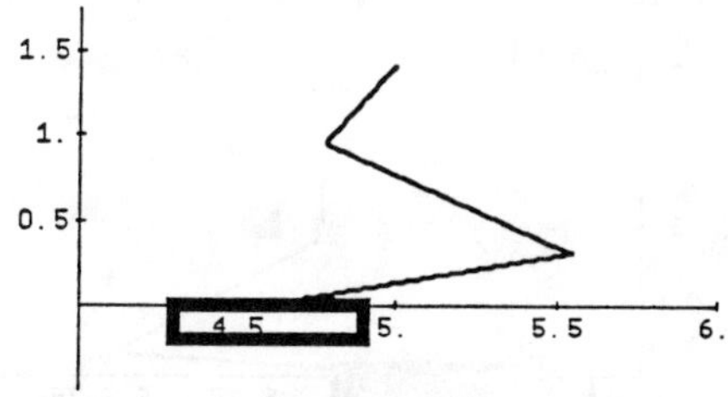

Fig. 2. Commutation configuration for a least torque norm requirement and a limit of 2.0 imposed on the torque of the first joint.

Figure 4 shows the commutation configuration when the manipulability index is utilized as the optimization criterion. The resulting torque norm reaches $\|\overline{\tau}_m\|^2 = 128$, which indicates the high price generally paid (in terms of torque demand) to maximize manipulability.

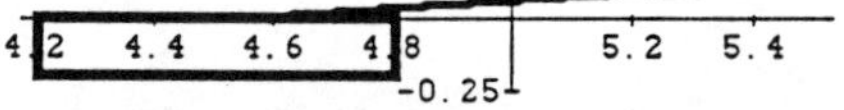

Fig. 3. Commutation configuration for the minimax problem.

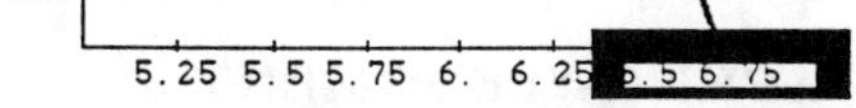

Fig. 4. Commutation configuration for a maximum manipulability requirement.

Figures 5 and 6 show the commutation configurations in cases when the optimization criterion considers avoidance of an obstacle by the endpoint of the second link (Fig. 5) and by a point on the platform (Fig. 6). Surprisingly, these configurations avoiding obstacles are not very penalizing in terms of torque demand, with torque norms equal to 14 and 18, respectively. These results suggest that in general, obstacle avoidance and least torque norm requirements may be quite compatible with each other in a multi-criteria optimization scheme, whereas maximum manipulability and least torque norm would be rather conflicting, typically "working against each other" in most configurations [5].

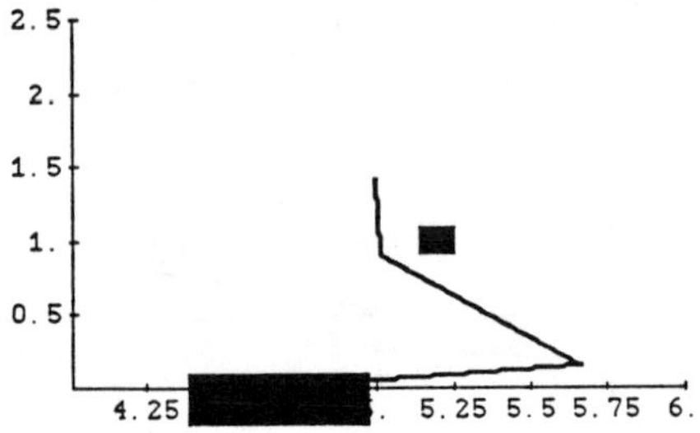

Fig. 5. Commutation configuration for avoidance of an obstacle located at (5.2, 1.0).

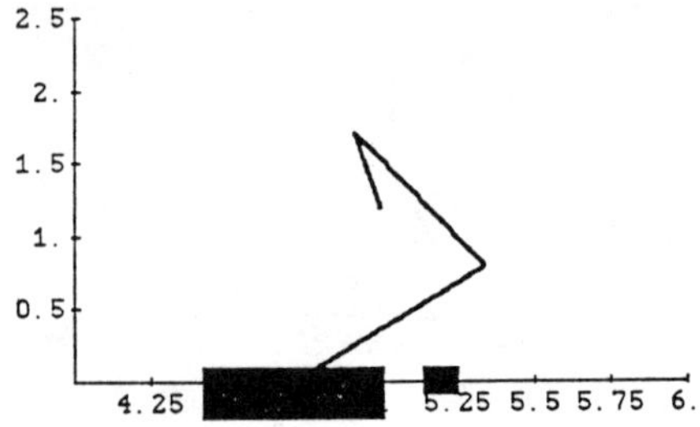

Fig. 6. Commutation configuration for avoidance by the platform of an obstacle located at (5.2, 0.0).

CONCLUSION

An approach has been presented for the optimization of the task commutation configuration of mobile manipulators in the accomplishment of given tasks. Optimization criteria and solution schemes have been introduced to optimize the position of the platform and the configuration of the manipulator with respect to obstacle avoidance, optimum manipulability, least torque norm, and maximum actuator torque (minimax) requirements. The general approach has also been shown to be extendable to cases involving multi-criteria requirements. Sample results of the methods for a system including a three-link planar manipulator mounted on a mobile platform have been presented. The results indicate that for torque-limit related requirements, the minimax approach represents an adequate optimization technique, although the run time of the algorithms is typically an order of magnitude greater than that for the least norm approach. Further work is pursuing in this area to address improved algorithms for rapidly reaching the global optimum in any situation.

ACKNOWLEDGMENT

This research was sponsored by the Engineering Research Program of the Office of Basic Energy Sciences, of the U.S. Department of Energy, under Contract No. DE-AC05-84OR21400 with Martin Marietta Energy Systems, Inc., by the U.S. Army Advanced Concepts and Technology Program under LABCOM Interagency Agreement No. 1495-CO62-A1 with the U.S. DOE, and by the Fort Belvoir Research Development and Engineering Center under Interagency Agreement No. 0883-CO11-A1 with the U.S. Department of Energy. The second author was supported in part by an appointment to the U.S. Department of Energy Postgraduate Research Program administered by Oak Ridge Associated Universities.

REFERENCES

1. S. F. Sousk, F. G. Pin, and C. D. Crane, "Motion Planning for the Universal Self Deployable Cargo Handler (USDCH)," to appear in Proceedings of the 17th U.S. Army Science Conference, Durham, North Carolina, June 12–15, 1990.

2. H. G. McCain et al., "A Hierarchically Controlled Autonomous Robot for Heavy Payload Military Field Applications" Proceedings of the International Conference on Intelligent Autonomous Systems, The Netherlands, December 8–11, 1986.

3. C. R. Weisbin et al., "HERMIES-III: A Step Toward Autonomous Mobility, Manipulation, and Perception," *Robotica*, Vol. 8, pp. 7–12 (1990).

4. W. F. Carriker, P. K. Khosla, and B. H. Krogh "An Approach for Coordinating Mobility and Manipulation," in Proceedings of the IEEE International Conference on Systems Engineering, Dayton, Ohio, pp. 40–44, August 1989.

5. F. G. Pin and J.-C. Culioli, "Multi-Criteria Position and Configuration Optimization for Redundant Platform/Manipulator Systems," to appear in Proceedings of IROS '90, the Third IEEE International Workshop on Intelligent Robots and Systems, Tsuchiura, Japan, July 3–6, 1990.

6. V. F. Dem'yanov and V. N. Malozemov, *Introduction to Minimax*, Wiley, New York (1974).

7. S. R. K. Dutta and M. Vidyasagar, "New Algorithms for Constrained Minimax Optimization," *Mathematical Programming*, pp. 140–155 (1977).

8. G. Cohen, "An Algorithm for Convex Constrained Minimax Optimization Based on Duality," *Appl. Math. Optim.* **7**, pp. 347–372 (1981).

9. G. Cohen, "Minimization of Convex Functionals Involving Nested Minima: Nonconcave Duality and Algorithms," *J. of Optimization Theory and Applications* **36**(3), pp. 203-233 (1982).

10. O. Khatib, "Real-Time Obstacle Avoidance for Manipulators and Mobile Robots," 1985 IEEE International Conference on Robotics and Automation, St. Louis, Missouri, pp. 500–505 (March 1985).

11. T. Yoshikawa, in *Analysis and Control of Robot Manipulators with Redundancy,* MIT Press, Cambridge, Massachusetts, pp. 735–748 (1984).

12. A. V. Levy and S. Gomez, "The Tunneling Method Applied to Global Optimization," in *Numerical Optimization 1984,* SIAM Press, Philadelphia, Pennsylvania, pp. 213-244 (1984).

CONSTRAINED TRAJECTORY OPTIMIZATION FOR KINEMATICALLY REDUNDANT ARMS

Craig R. Carignan and Janice M. Tarrant
ST Systems Corp.
Lanham, MD 20706

ABSTRACT

Two velocity optimization schemes for resolving redundant joint configurations are compared. The Extended Moore-Penrose Technique minimizes the joint velocities and avoids obstacles indirectly by adjoining a cost gradient to the solution. A new method can incorporate inequality constraints directly to avoid obstacles and singularities in the workspace. A four-link arm example is used to illustrate singularity avoidance while tracking desired end-effector paths.

1 INTRODUCTION

Kinematic redundancy can be defined as having more degrees of freedom than necessary to perform a particular task. For a three-dimensional task, the maximum number of end-effector positions which can be specified is six, three positions and three orientations. Thus a manipulator with more than six degrees of freedom is generally considered "redundant." Though redundancy imposes greater burden during trajectory planning, it is often a desirable property from the viewpoint of avoiding obstacles and gaining workspace accessibility.

Approaches to resolving the excess degrees of freedom in a kinematically redundant arm typically fall into one of two categories: (a) specify an additional number of equality constraints equal to the degree of arm redundancy, or (b) optimize a cost which reflects some desirable local property of the manipulator. The former case is called the "Extended Jacobian" technique [1] because the Jacobian is extended until it becomes a square 1-to-1 mapping between the extended end-effector velocity space and joint velocity space. The latter case is often called the "generalized inverse" technique because the resulting solution involves a weighted pseudoinverse of the Jacobian.

This paper focuses on local velocity optimization approaches for solving the redundancy problem. First, the unconstrained problem is solved resulting in the well known Moore-Penrose solution. The constrained case is then examined for both a direct and an indirect solution technique: the direct method is a modified version of Hildreth's quadratic programming procedure, and the indirect method is the Extended Moore-Penrose Technique. A four-link arm example serves to illustrate these approaches for avoiding singularities while minimizing joint velocities along a straight line end-effector path.

2 REDUNDANCY PROBLEM

The redundancy problem can be formulated as follows. Let $\underline{x}$ be a vector of dimension m representing the desired positions of the end-effector in workspace coordinates, and let $\underline{\theta}$ be a vector of dimension n representing the joint positions of the manipulator. Typically, the θ_i represent joint angles, but they many also represent distances as in the case of a prismatic joint. If n>m, the manipulator is "kinematically redundant".

The constraint that the end-effector must follow a particular trajectory through the workspace can be expressed by the set of m equations:

$$\underline{x} = \underline{f}(\underline{\theta}) \qquad (1)$$

This constraint can also be expressed as a velocity constraint by differentiating (1):

$$\dot{\underline{x}} = J\,\dot{\underline{\theta}} \qquad (2)$$

where J, the partial of $\underline{f}$ with respect to $\underline{\theta}$, is called the "Jacobian".

A trivial way of resolving the redundancy is to simply choose a set of n-m constraints $\underline{\alpha}(\underline{\theta})$. This approach is often referred to as the "Extended Jacobian Technique" because the additional constraints can be treated in a manner exactly analogous to $\underline{x}$ in equation (1). Thus the derivative of

$\underline{\alpha}$ can be rewritten as

$$\dot{\underline{\alpha}} = A\,\dot{\underline{\theta}} \qquad (3)$$

where A is the partial of $\underline{\alpha}$ with respect to $\underline{\theta}$. Combining (2) and (3) yields

$$\begin{bmatrix} \dot{\underline{x}} \\ \dot{\underline{\alpha}} \end{bmatrix} = \begin{bmatrix} J \\ A \end{bmatrix} \dot{\underline{\theta}} \qquad (4)$$

The coefficient of $\dot{\underline{\theta}}$ in (4) is the called the"extended Jacobian" because it is obtained by adding rows to the original Jacobian to account for the $\underline{\alpha}$ constraints. Since the extended Jacobian, J_{ext}, is square, $\dot{\underline{\theta}}$ can be determined directly from (4) by inverting J_{ext} and multiplying by the end-effector/constraint velocity vector.

Though simple in concept, the extended Jacobian approach is fraught with problems unless the additional constraints are chosen carefully. If the additional constraint has a simple geometric interpretation, then the approach may be viable. Otherwise, it is very difficult to show analytically that the extra constraint can always be achieved given the end-effector path to be followed. Walker and Marcus [2] present successful results for composite constraints which realize singularity avoidance and obstacle evasion, but it is not clear whether their constraint is reachable throughout the workspace. Another possible side-effect of introducing nonphysical constraints is the inducement of large joint velocities due to complex joint interactions. In addition, the constraint generally does not realize a performance objective.

One nice feature about this approach is that cyclic motion is attained through the invertible mapping between the extended end-effector coordinates and joint coordinates. If the subset of the workspace in which paths can be tracked is simply connected, Baker and Wampler [3] have shown that closed paths in the workspace produce closed paths in the joint space. Thus when the end effector returns to its starting position, the arm linkages are also in their initial configuration. This will prevent the manipulator from drifting when the end-effector is performing repetitive tasks involving closed paths. This feature is not unique, however, as position-based optimizations also exhibit this behavior.

3 TRAJECTORY OPTIMIZATION

Instead of resolving the n-m excess degrees of freedom by adding constraints, the optimization approach takes advantage of the excess degrees of freedom by maximizing a performance objective. The computational complexity will depend upon the cost used and how constraints are incorporated in the optimization. Since the effect of the contraint is only strong near a constraint boundary, undesirable side effects in the arm's behavior are often averted. In addition, extra singularities in the generalized inverse are introduced only when the inequality constraints are active.

The presence of inequality constraints and computational considerations often lead researchers to choose local over global optimization procedures. Sometimes minimizing a local cost may yield better performance than minimizing a global cost. For example, it may be better to minimize the instantaneous velocity rather than the integral if one is interested in limiting the kinetic energy present in the system.

A local optimization involves minimizing an instantaneous or "local" cost. One cost function which has received a lot of attention is the velocity norm

$$C(\dot{\underline{\theta}}) = \frac{1}{2}\dot{\underline{\theta}}^T S\,\dot{\underline{\theta}} \qquad (5)$$

Often the weighting matrix S is chosen to be the n x n identity matrix so that one is seeking the minimum norm solution for $\dot{\underline{\theta}}$. Adjoining the velocity constraint (2) to the cost (5) yields the Hamiltonian

$$H(\dot{\underline{\theta}}) = C(\dot{\underline{\theta}}) + \underline{\lambda}^T(\dot{\underline{x}} - J\dot{\underline{\theta}}) \qquad (6)$$

Minimizing H with respect to $\dot{\underline{\theta}}$ and $\underline{\lambda}$, and solving the resulting equations for $\dot{\underline{\theta}}$ yields

$$\dot{\underline{\theta}} = S^{-1}J^T(J\,S^{-1}J^T)^{-1}\,\dot{\underline{x}} \qquad (7)$$

The coefficient matrix of $\dot{\underline{x}}$ is the weighted right pseudoinverse, J_s^{-1}, of the Jacobian since $JJ_s^{-1}=I$. The solution in (7) is usually referred to as the "Moore-Penrose Technique" but is sometimes referred to as "resolved motion rate control" [4].

The advantages of the velocity-based algorithm include minimization of joint velocities or kinetic energy (if S is the inertia tensor), and zero joint rates when the end-effector velocity is zero. A natural byproduct of the first property is that the arm is repelled by most singularities because of their associated high joint rates. The latter property is a consequence of the minimum norm solution for the joint rate which only contains components that contribute to the end-effector velocity. A disadvantage of the Moore-Penrose solution is that it does not have the cyclic property of position-based solutions. The "Extended Moore-Penrose Technique" (to be discussed later) is a means of limiting arm drift in performing repetitive tasks but also does not result in cyclic motion.

The requirement that the manipulator not get too close to a singularity is graphically depicted in Fig. 1 and can be expressed as the union of two inequality constraints

$$\alpha \leq \alpha_{min} \cup \alpha \geq \alpha_{max} \tag{8}$$

where α_{min} and α_{max} represent upper and lower bounds on α, respectively. Since the minimization is to be performed over the joint velocity not position, the singularity position constraint in (8) is converted to a velocity constraint

$$\begin{aligned} \dot{\alpha} &\geq 0 \qquad \text{for} \qquad \alpha=\alpha_{max} \\ \dot{\alpha} &\leq 0 \qquad \text{for} \qquad \alpha=\alpha_{min} \end{aligned} \tag{9}$$

which is enforced at every time step. Unfortunately, (9) produces a harsh effect when α hits the constraint boundary, and a smoother transition is achieved through the constraint

$$\begin{aligned} \dot{\alpha} &\geq b\,(\alpha_{max}-\alpha) \qquad \text{for} \qquad \alpha > (\alpha_{max}+\alpha_{min})/2 \\ \dot{\alpha} &\leq b\,(\alpha_{min}-\alpha) \qquad \text{for} \qquad \alpha < (\alpha_{max}+\alpha_{min})/2 \end{aligned} \tag{10}$$

where $b>0$ is a design parameter which specifies the maximum rate of change in α for a given change in α and is chosen sufficiently large so that (10) only becomes active near the border.

The local minimization problem with equality constraints for the end-effector and inequality constraints for the singularity avoidance can now be formulated from (5), (2) and (10):

$$\min_{\dot{\underline{\theta}}} \quad C = \frac{1}{2}\dot{\underline{\theta}}^T S(\underline{\theta})\,\dot{\underline{\theta}} \tag{11}$$

$$\text{subject to} \quad J(\underline{\theta})\,\dot{\underline{\theta}} = \dot{\underline{x}}$$

$$A(\underline{\theta})\,\dot{\underline{\theta}} \leq \underline{c}$$

where A is the gradient of α with respect to $\underline{\theta}$ and $\underline{c}$ is the right side of the inequality constraint in (10). (A negative sign precedes these quantities in A and $\underline{c}$ for the "≥" constraints.) Hildreth's solution [5] to the above problem modified to account for the equality constraints [6] is given by

$$\dot{\underline{\theta}} = -\,S^{-1}\,Q^T\,\underline{\nu} \tag{12}$$

$$\nu_i^{k+1} = \begin{cases} w_i^{k+1} & i=1\ldots m \\ \max\,(0\,,\,w_i^{k+1}) & i=m+1\ldots m+p \end{cases}$$

where

$$w_i^{k+1} = -\frac{1}{P_{ii}}\left[d_i + \sum_{j=1}^{i-1} P_{ij}\nu_j^{k+1} + \sum_{j=i+1}^{m+p} P_{ij}\nu_j^{k}\right]$$

and

$$P \equiv Q\,S^{-1}Q^T\,,\quad \underline{\nu} = \begin{bmatrix} \underline{\nu}_E \\ \underline{\nu}_I \end{bmatrix},\quad \underline{d} = \begin{bmatrix} \dot{\underline{x}} \\ \underline{c} \end{bmatrix},\quad Q = \begin{bmatrix} J \\ A \end{bmatrix}$$

where $\underline{\nu}_E$ are the Lagrange multipliers for the end-effector equality constraints in (11), $\underline{\nu}_I$ are the multipliers for the p inequality constraints, and k is the iteration number.

An alternative way of including position inequality constraints is the Extended Moore-Penrose Technique. Unlike the direct Lagrangian approach just developed, the position constraint is realized indirectly by including a position term in the unconstrained Moore-Penrose solution which forces a gradient descent away from the obstacle

$$\dot{\underline{\theta}} = J_s^{-1}\dot{\underline{x}} + \rho\,(I - J_s^{-1}J)\,\nabla P(\underline{\theta}) \tag{13}$$

where ρ is a positive scalar constant, and $P(\underline{\theta})$ is an objective function to be maximized. The first term in (13) is the minimum norm solution for $\dot{\underline{\theta}}$ which is the unconstrained solution to (5). The second term is in the nullspace of J and thus contributes nothing to the end-effector motion. Thus ∇P can have any value and still not perturb the constraint in (2).

Liegeois [7] experiments with objective functions for a six joint revolute manipulator to minimize joint deviations while tracking a circle with the end-effector. Klein and Huang [8] report results for a similar cost using a three-link manipulator being commanded by a joystick. Baillieul et al [9] consider a manipulability criterion for a planar four-link mechanism and demonstrate the ability of the extended method to keep the manipulator away from the singular initial configuration. Only Klein and Huang discuss the effects of varying the size of ρ, emphasizing the need to scale it for different end-effector positions.

4 CONSTRAINED VELOCITY OPTIMIZATION EXAMPLE

A schematic of the manipulator used in our study is shown in Fig. 2. The K1607 Robotics Research® Arm has seven revolute joints in an alternating roll/pitch sequence beginning with the shoulder roll at the base. The Denavit-Hartenberg (D-H) parameters for the first five joints of this arm are given in Table I. In this study, we will only be concerned with controlling the cartesian position of the wrist, the origin of frame 5 in Fig. 2 which is only a function of the first four joint angles.

Table I: D-H Parameters for Robotics Research K1607 Manipulator.

joint # i	a_i (cm)	α_i (deg)	d_i (cm)
1	0.00	-90	0.00
2	14.29	+90	0.00
3	10.80	-90	68.58
4	-10.80	+90	0.00
5	-	-	68.58

In the simulation examples, the trajectory planner will be attempting to avoid the singularities formed by the intersection of the two planes [6]

$$\frac{\sin(\theta_2)}{\cos(\theta_2)} = \frac{-a_2 f_2(\theta_4)}{b_3 f_2(\theta_4) - a_3 f_1(\theta_4)} \quad \text{and} \quad \cos(\theta_3) = \frac{a_2 f_1(\theta_4)}{b_3 f_2(\theta_4) - a_3 f_1(\theta_4)} \tag{14}$$

where $f_1(\theta_4) \equiv d_5 c_4 + a_3 s_4$ $\quad c_i \equiv \cos(\theta_i)$

$f_2(\theta_4) \equiv d_5 s_4 - a_3 c_4$ $\quad s_i \equiv \sin(\theta_i)$

These conditions can be used to form a manipulability measure which goes to zero at the line singularity:

$$\alpha(\underline{\theta}) = |f_a| + |f_b| \tag{15}$$

where $f_a \equiv c_3(d_3 f_2 - a_3 f_1) - a_2 f_1$

$f_b \equiv s_2(d_3 f_2 - a_3 f_1) + c_2 a_2 f_2$

In the direct Lagrangian method, α was used for the inequality constraint in the problem formulation (11). The upper and lower bounds on α were 0.20 m and -0.20 m, respectively, and the slope parameter, b, was 10.0 s^{-1}. Because the Lagrange multiplier solution is numerical, a sufficiently close approximation may take several iterations. In the extended Moore-Penrose method, α was used for P, the position objective. The scalar weighting for the nullspace component, ρ, was chosen to be 0.15 which yielded a minimum manipulability of about 0.2 m (the same as the direct Lagrangian approach). In both cases, the unweighted velocity norm was used for the objective function.

The end-effector was commanded to follow a straight-line path in the workspace from an initial joint configuration of $\underline{\theta}^T$ = [-0.1 -1.0 0.8 1.3] rad which corresponds to $\underline{x}^T$ = [-0.40 0.68 1.08] m to a final position of $\underline{x}^T$= [0.6 0.7 0.5] m. The path acceleration was constant to the midpoint followed by a constant deceleration of the same magnitude to the end point.

Ten or more iterations on the Lagrange multipliers were often needed before the Lagrange algorithm converged. The primary driving force behind the convergence rate was the size of the Lagrange multiplier for the inequality constraint relative to the end-effector multipliers. The larger the constraint violation in the unconstrained case, the larger the multiplier and the slower the convergence. Thus the drastic change in the multiplier for the first few iterations propagated through the other multipliers and caused errors in the end-effector velocities as well.

The degree to which the unconstrained (Moore-Penrose) solution was affected by the manipulability constraint in the Lagrangian approach is seen in Fig. 3. The joint velocities were unconstrained until about 1.3 s when the manipulator got close to the line singularity. At this point, the manipulability constraint took effect preventing further acceleration toward the preset threshold at 0.2 m. Before the manipulability reached this boundary, the unconstrained solution no longer violated the constraint, and the solution reverted back to Moore-Penrose. The curves do not coincide exactly from this point because the joint configurations were altered by the constraint.

The shoulder and elbow joint velocities (joints 2 and 4) in the Lagrangian-constrained (converged) and unconstrained cases are shown in Fig. 4. These curves more dramatically illustrate the effect of the manipulability constraint on the arm. All of the unconstrained velocities peaked when the manipulability reached the minimum in Fig. 3. When the constraint went active, joint 2 underwent a discontinuous jump in acceleration resulting in a decrease in velocity. The constraint caused a significant but less abrupt decrease in speed for joint 4. Joints 1 and 3 (not shown) underwent behavior paralleling that of joints 2 and 4, respectively.

The Extended Moore-Penrose approach displayed smoother behavior than the Lagrangian method. Since the nullspace component was being continuously applied rather than only when an inequality constraint was being violated, the manipulability measure for the Extended Moore-Penrose solution in Fig. 5 always differed from that for the unconstrained solution (ρ=0). At the beginning of the trajectory, the manipulabilities did not differ greatly, but as the arm drew closer to the singular configuration, the nullspace component became dominant, driving the manipulator away from the singularity. When tracking the end-effector trajectory placed the singularity out of reach, the Extended Moore-Penrose solution once again approached the Moore-Penrose solution.

The pitch joint velocities in the Extended Moore-Penrose case are plotted along with the unconstrained results in Fig. 6. The velocity of joint 4 was remarkably similar to that for the Lagrangian approach, but the velocity for joint 2 did not make the same sharp transition near the singularity. This is because the activity of the inequality constraint in the Lagrangian solution is binary, and thus does not possess the continuous properties of the Extended Moore-Penrose approach. Thus, whilst the constant application of the nullspace component is less efficient, it does yield the advantage of continuous joint velocities.

The cycle computation times for the Extended Moore-Penrose and Lagrangian Techniques were 2.8 msec and 2.9+0.3k msec, respectively, where k is the number of Lagrange multiplier iterations. By comparison, the unconstrained solution took 2.2 msec, and the extended Jacobian method took 2.0 msec. These times are for Microsoft® C 5.1 running on an Intel® 80386/7 and do not include the 1.4 msec it takes to compute trigonometric functions of the joint angles.

Although the Extended Jacobian Method would provide the fastest solution, the measure in (15) would not be a good choice for the constraint because of its limited range. For example, attempting to hold α constant for the same initial configurations used above fails at 1.3 sec because it cannot satisfy the specified end-effector trajectory. Kreutz-Delgado et al [10] suggest using an elbow roll angle for α in their work with a similar arm, but this further restricts the workspace and does not avoid singularities without knowing a priori which α's cause singularities.

5 CONCLUSIONS

It was discovered that position constraints can be included in the optimization either directly or indirectly. For a quadratic cost and linear constraints, the problem can be solved via Hildreth's quadratic programming algorithm. This approach incorporates the inequality constraints directly by iterating on the Lagrange multipliers. A second approach, the Extended Moore-Penrose method, realizes the constraint by adding a gradient component to the solution which steers the arm away from the constrained region.

These two approaches were examined for the case of a four-link arm operating in 3-D space. The goal for this example was to prevent the arm from entering a region surrounding a line singularity while following a straight line path with the end-effector. The Lagrangian solution avoided the singularities but had sharp velocity transitions. The Extended Moore-Penrose solution smoothly avoided the singularity but at the expense of optimality in the velocity norm. The Extended Moore-Penrose method was more than twice as fast as a ten-iteration Lagrangian when a single constraint was being applied.

The extension of optimization-based algorithms to include inequality constraints makes them a more attractive option for redundancy resolution. Instead of simply adding more constraints to reduce the excess degrees of freedom to zero, the designer can now utilize the redundancy to his advantage by maximizing some performance objective critical to the task while avoiding undesirable regions of the workspace. In addition, the penalty for including these constraints may be less than 50% of the total trajectory computation time for the extended Jacobian solution. Coupled with new results on their global characteristics, local optimization methods have now become a viable contender for real-time trajectory planning.

ACKNOWLEDGEMENTS

The authors would like to express their appreciation to the Space Station Project Office (Code 400.6), NASA Goddard Space Flight Center for their support of this research.

REFERENCES

1. Baillieul, J., "Kinematic Programming Alternatives for Redundant Manipulators," Proc. IEEE Int. Conf. on Robotics and Automation, St. Louis, March 25-28, 1985, pp. 722-728
2. Walker, I., and Marcus, S., "Subtask Performance by Redundancy Resolution for Redundant Robot Manipulators," IEEE J. Robotics and Automation, Vol. 4, No. 3, June 1988
3. Baker, D., and Wampler, C., "On the Inverse Kinematics of Redundant Manipulators," Int. J. Robotics Research, Vol.7, No. 2, Apr. 1988
4. Whitney, D., "Resolved Motion Rate Control of Manipulators and Human Prostheses," IEEE Trans. on Man-Machine Systems, Vol. MMS-10, No. 2, June 1969, pp. 47-53
5. Luenberger, D., Optimization by Vector Space Methods, Wiley & Sons, New York, ©1969, pp. 299-300
6. Carignan, C., "Trajectory Optimization for Kinematically Redundant Arms," Tech. Rep. No. STX/ROB/90-01, STX Corp., Lanham, Md., January 1990
7. Liegeois, A., "Automatic Supervisory Control of the Configuration and Behavior of Multibody Mechanisms," IEEE Trans. on Systems, Man, and Cybernetics, Vol. 7, No. 12, 1977
8. Klein, C., and Huang, C.-H., "Review of Pseudoinverse Control for Use with Kinematically Redundant Manipulators," IEEE Trans. on Systems, Man, and Cybernetics, Vol. SMC-13, No. 3, March 1983, pp.245-250
9. Baillieul, J., Hollerbach, J., and Brockett, R., "Programming and Control of Kinematically Redundant Manipulators," Proc. of 23rd Conf. on Decision and Control, Las Vegas, Dec. 1984, pp. 768-774
10. Kreutz, K., Long, M., and Seraji, H., "Kinematic Analysis of 7 DOF Manipulators," JPL, Pasadena, (in progress)

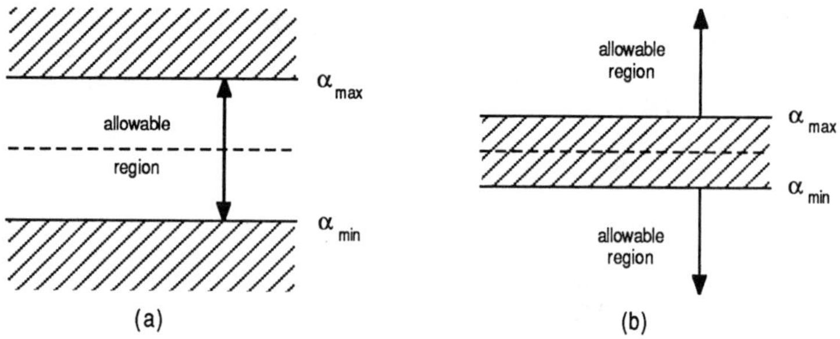

Figure 1: Allowable regions for (a) joint limit and (b) singularity constraint types.

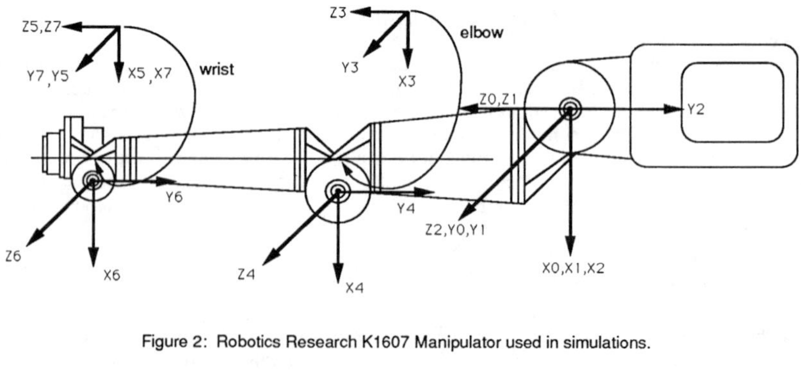

Figure 2: Robotics Research K1607 Manipulator used in simulations.

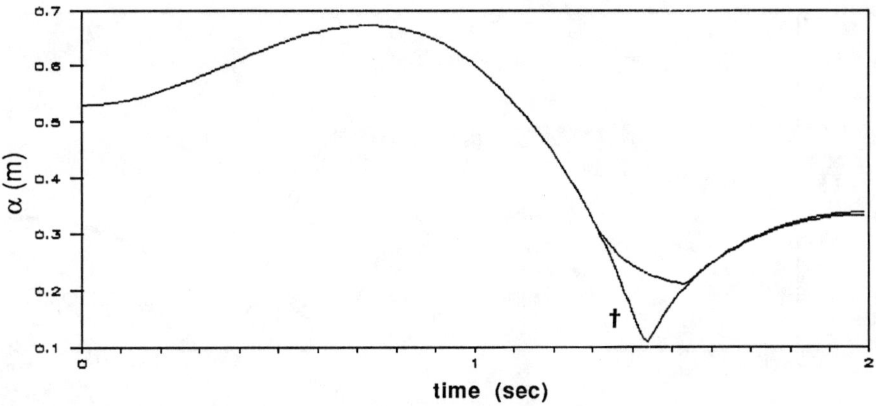

Figure 3: Manipulability for the constrained Lagrangian case is plotted versus the unconstrained case (†).

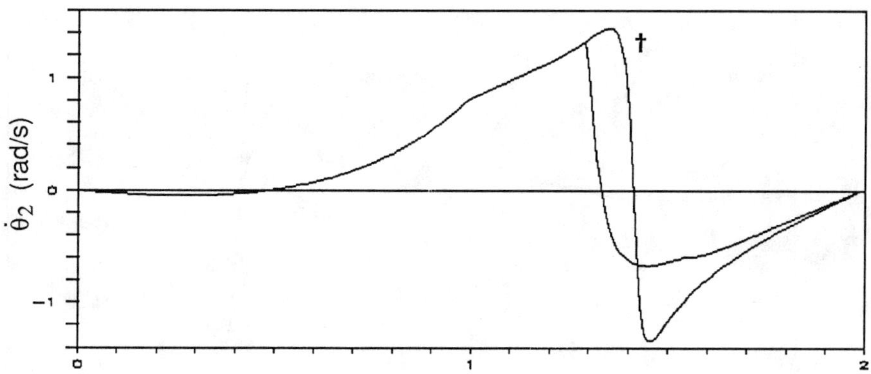

(a)

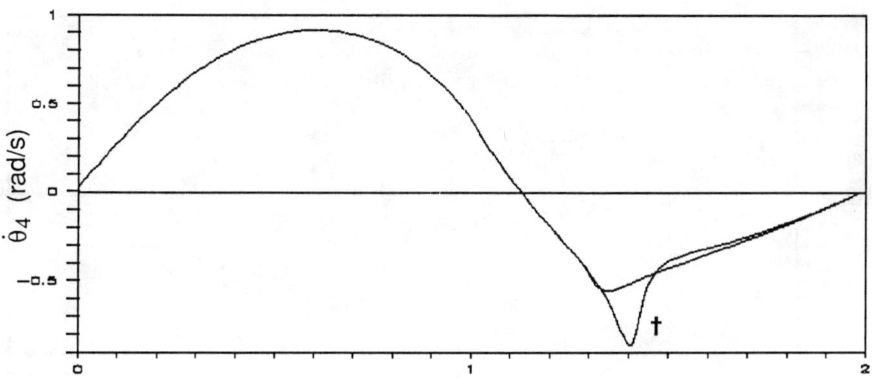

(b)

Figure 4: (a) Shoulder and (b) elbow pitch velocities for constrained Lagrangian case versus unconstrained (†).

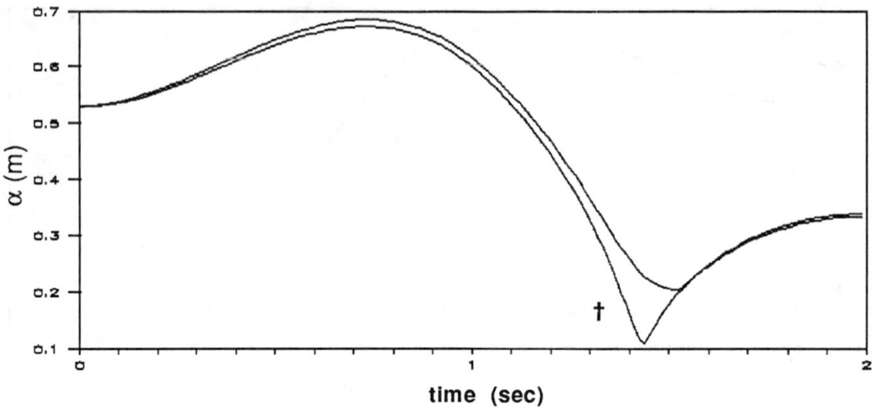

Figure 5: Manipulability for the constrained Moore-Penrose case is plotted versus the unconstrained case (†).

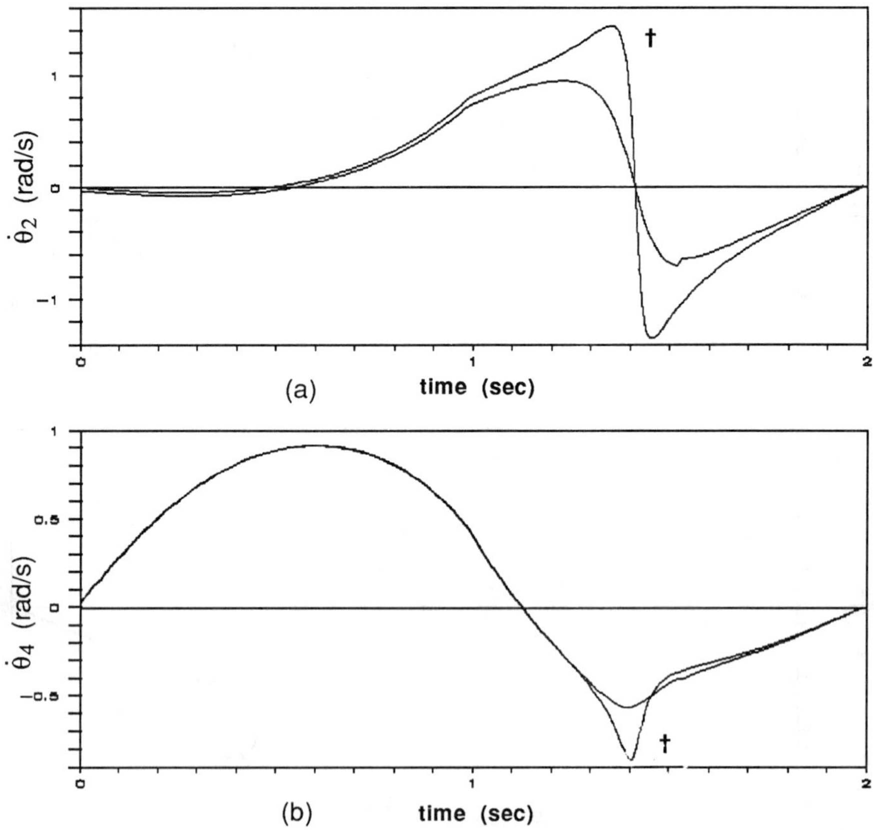

Figure 6: (a) Shoulder and (b) elbow pitch velocities for constrained Moore-Penrose case versus unconstrained (†).

COMPLIANT CONTROL OF A REDUNDANT TELEROBOT MANIPULATOR VIA A POSITION/FORCE CONTROL SCHEME

C.C. Nguyen* Z-L. Zhou* G.E. Mosier†
Catholic University of America
NASA/Goddard Space Flight Center

Abstract

This paper presents the development of a position/force scheme proposed to control the compliant motion of kinematically redundant manipulators with 7 degrees of freedom (DOF). The control scheme consists of a position controller and a force controller whose controller gains are adjusted by the corresponding adaptation laws which are derived based on the concept of model reference adaptive control and the Lyapunov theorem. Case study using computer simulation conducted on the slave arm of the NASA telerobotic testbed is presented.

1 Introduction

Kinematically redundant manipulators have attracted tremendous researchers' attention because the extra degrees of freedom (**DOF**) can help the manipulator to avoid singularities, joint limits and obstacles [1]-[7]. However, the manipulator redundancy has complicated the problem of control of force and position since the inverse of the non-square Jacobian matrix must be approximated by the pseudo-inverse method in Cartesian-space control schemes [1]. Adaptive Cartesian-space control schemes were proposed in [2] to deal with the problem of position/force control of redundant manipulators. The global optimization of joint rates and kinetic energy in redundant manipulators was treated in [3]. Adaptive control schemes were designed in [4] and in [5] in joint space and Cartesian space, respectively for a redundant telerobot manipulator having 7 DOFs. The concept of *configuration control* which utilizes the manipulator redundancy was introduced in [6] to control the manipulator configuration directly in task space.

This paper is organized as follows. Section 2 describes the hybrid control scheme proposed to control the compliant motion of redundant manipulators. Then Section 3 presents the development of the adaptation laws which adjust the gains of the force and position controllers. After that, Section 4 discusses the computer simulation conducted to study the performance of the hybrid control scheme implemented to control the compliant motion of the slave arm of the NASA telerobotic testbed. Finally Section 5 gives a summary of the paper and outlines some future research work. In this paper, the following matrix notations will be used: $\mathbf{M}^T$=transpose of the matrix $\mathbf{M}$; $\mathbf{0}_n$=(nxn) matrix whose elements are all zero; $\mathbf{I}_n$=(nxn) identity matrix; and $tr[\mathbf{M}]$=trace of matrix $\mathbf{M}$.

2 The Adaptive Hybrid Control Scheme

Compliant motion occurs when the robot end-effector is in contact with the working environment during the motion. For each task to be performed in compliant motion described in Cartesian space, the user should define a *compliance frame* which is an orthogonal coordinate system usually fixed with respect to the robot end-effector. The compliance frame assignment

*Catholic University of America
†NASA/Goddard Space Flight Center

depends mostly on the particular contact situation between the end-effector and the environment. The task freedoms are defined to be translation and rotation along and about each of the compliance frame axes. Depending on the nature of the task to be performed, the user may then specify which axes are to be position-controlled[1] and which to be force-controlled provided that the coordinate in which position is to be controlled is not force-controlled and vice versa. Thus in the compliant frame, the n-dimensional Cartesian space **X** can be decomposed into an m-dimensional position subspace **Y** and an k-dimensional force subspace **Q** where m + k = n.

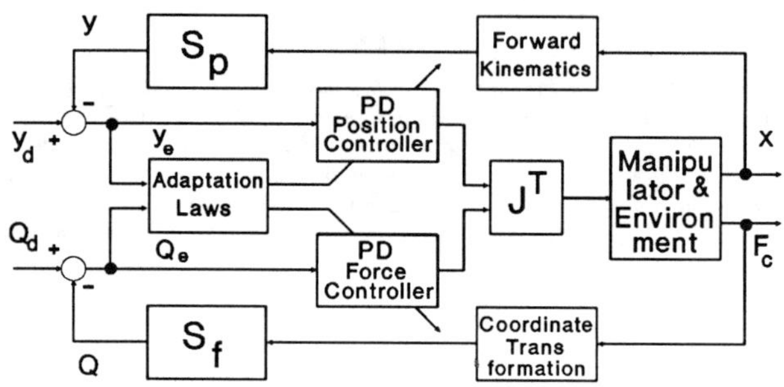

Figure 1 The adaptive hybrid position/force control scheme.

The implementation of the above concept is presented in Figure 1 illustrating a control scheme which is the modified hybrid position/force control scheme introduced in [8]. As the figure shows, the control scheme consists of 2 control loops which allow position to be controlled in the upper loop and force to be controlled in the lower loop. The selection of which positions and forces to be controlled is done by the *selection matrices* $\mathbf{S}_p$ of order (mxn) and $\mathbf{S}_f$ of order (kxn) in the position and force control loops, respectively. The gains of the position and force controllers are adjusted by adaptation laws which are mainly based on the errors in positions and forces. The problem of this paper is to develop the adaptation laws to effectively react to the highly nonlinear coupling between manipulator joints and uncertainties in payloads and environment stiffness so that the manipulator end-effector can track a desired motion while applying desired contact forces on the environment with minimum errors.

3 Development of the Adaptation Laws

This section presents the development of the adaptation laws for the position and force controllers.

3.1 The Force Control Scheme

For the ease of analysis, let us consider the case in which force is controlled in all Cartesian coordinates and consequently the hybrid control scheme in Figure 1 becomes a pure force control scheme. In this case, the upper position control loop in inactive. The force dynamics of a manipulator performing compliant motion can be described by [9]:

$$\mathbf{F}_f(t) = \mathbf{M}(\mathbf{x},\dot{\mathbf{x}})\ \ddot{\mathbf{Q}}(t) + \mathbf{D}(\mathbf{x},\dot{\mathbf{x}})\ \dot{\mathbf{Q}}(t) + \mathbf{S}(\mathbf{x},\dot{\mathbf{x}})\ \mathbf{Q}(t) \qquad (1)$$

where $\mathbf{x}(t)$ denotes the (6x1) Cartesian position vector of the manipulator end-effector, $\mathbf{F}_f(t)$, the (6x1) Cartesian force vector applied to the end-effector and $\mathbf{Q}(t)$, the (6x1) Cartesian force

[1] In this paper, "position" implies both "position and orientation" and "force" both "force and torque".

vector applied by the end-effector on the environment. In addition, $\mathbf{M}(\mathbf{x},\dot{\mathbf{x}})$, a symmetric positive-definite matrix, $\mathbf{D}(\mathbf{x},\dot{\mathbf{x}})$ and $\mathbf{S}(\mathbf{x},\dot{\mathbf{x}})$ are of order (6x6) with elements that are highly complex nonlinear functions of $\mathbf{x}$ and $\dot{\mathbf{x}}$. Consider a PD control law defined by

$$\mathbf{F}_f(t) = \mathbf{K}_{fp}(t)\,\mathbf{Q}_e(t) + \mathbf{K}_{fd}(t)\,\dot{\mathbf{Q}}_e(t) \tag{2}$$

where $\mathbf{Q}_e(t)$ given by $\mathbf{Q}_e(t) = \mathbf{Q}_d(t) - \mathbf{Q}(t)$ is the error vector between the actual Cartesian contact force vector $\mathbf{Q}(t)$ and the desired Cartesian force vector $\mathbf{Q}_d(t)$ and the gain matrices of the proportional and derivative controllers are denoted by $\mathbf{K}_{fp}(t)$ and $\mathbf{K}_{fd}(t)$, respectively.

Substituting (2) into (1) and defining a state vector $\mathbf{v}(t) = [\mathbf{Q}_e^T(t)\ \ \dot{\mathbf{Q}}_e^T(t)]^T$, we obtain the following state-space representation

$$\dot{\mathbf{v}}(t) = \begin{bmatrix} \mathbf{0}_6 & \mathbf{0}_6 \\ -\mathbf{B}_1 & -\mathbf{B}_2 \end{bmatrix} \mathbf{v}(t) + \begin{bmatrix} \mathbf{0}_6 & \mathbf{0}_6 & \mathbf{0}_6 \\ \mathbf{B}_3 & \mathbf{B}_4 & \mathbf{I}_6 \end{bmatrix} \mathbf{u}(t) \tag{3}$$

where

$$\mathbf{B}_1 = \mathbf{M}^{-1}(\mathbf{S}+\mathbf{K}_{fp});\ \ \mathbf{B}_2 = \mathbf{M}^{-1}(\mathbf{D}+\mathbf{K}_{fd});\ \ \mathbf{B}_3 = \mathbf{M}^{-1}\,\mathbf{S};\ \ \mathbf{B}_4 = \mathbf{M}^{-1}\,\mathbf{D}, \tag{4}$$

and

$$\mathbf{u}(t) = \left[\mathbf{Q}_d^T(t)\ \ \dot{\mathbf{Q}}_d^T(t)\ \ \ddot{\mathbf{Q}}_d^T(t)\right]^T. \tag{5}$$

We proceed to choose a *reference model* described by

$$\ddot{Q}_{ei}(t) + 2\,\xi_i\,\omega_i\,\dot{Q}_{ei}(t) + \omega_i^2\,Q_{ei}(t) = 0 \tag{6}$$

where ξ_i and ω_i denote the damping ratio and the natural frequency of Q_{ei}, for i=1,2,...,6 which are the elements of the force error vector $\mathbf{Q}_e(t) = [Q_{e1}(t)\ Q_{e2}(t)\ldots Q_{e6}(t)]^T$.

From (6), the reference model can be converted into the state-space form as

$$\dot{\mathbf{v}}_m(t) = \boldsymbol{\Delta}\,\mathbf{v}_m(t) = \begin{bmatrix} \mathbf{0}_6 & \mathbf{I}_6 \\ -\boldsymbol{\Delta}_1 & -\boldsymbol{\Delta}_2 \end{bmatrix} \mathbf{v}_m(t), \tag{7}$$

where $\boldsymbol{\Delta}_1$=diag(ω_i^2) and $\boldsymbol{\Delta}_2$=diag($2\xi_i\omega_i$) are constant (6x6) diagonal matrices, and $\mathbf{v}_m(t) = [\mathbf{Q}_m^T(t)\ \dot{\mathbf{Q}}_m^T(t)]^T$ with $\mathbf{Q}_m(t) = [Q_{e1}\ Q_{e2}\ \ldots Q_{e6}]^T$. Now the solution of (7) is obtained as $\mathbf{v}_m(t) = \exp(\boldsymbol{\Delta}t)\mathbf{v}_m(0)$ where $\mathbf{v}_m(0)$ denotes the initial value of $\mathbf{v}_m(t)$. Furthermore, under the assumption that $\mathbf{v}_m(0) = 0$, we obtain $\mathbf{v}_m(t) = 0$. Otherwise, ξ_i and ω_i can be properly selected so that all eigenvalues of $\boldsymbol{\Delta}$ are stable to make $\mathbf{v}_m(t) \to 0$ as $t \to \infty$.

Next defining the adaptation error vector $\mathbf{E}(t) = \mathbf{v}_m(t) - \mathbf{v}(t)$, we obtain from (3) and (7), an *error system* given by

$$\begin{aligned} \dot{\mathbf{E}}(t) &= \begin{bmatrix} \mathbf{0}_6 & \mathbf{I}_6 \\ -\boldsymbol{\Delta}_1 & -\boldsymbol{\Delta}_2 \end{bmatrix} \mathbf{E}(t) + \begin{bmatrix} \mathbf{0}_6 & \mathbf{0}_6 \\ \mathbf{B}_1 - \boldsymbol{\Delta}_1 & \mathbf{B}_2 - \boldsymbol{\Delta}_2 \end{bmatrix} \mathbf{v}(t) \\ &+ \begin{bmatrix} \mathbf{0}_6 & \mathbf{0}_6 & \mathbf{0}_6 \\ -\mathbf{B}_3 & -\mathbf{B}_4 & -\mathbf{I}_6 \end{bmatrix} \mathbf{u}(t). \end{aligned} \tag{8}$$

We proceed to select a Lyapunov function $L(t)$ such that

$$\begin{aligned} L(t) &= \mathbf{E}^T\mathbf{P}\mathbf{E} + tr\left[(\mathbf{B}_1 - \boldsymbol{\Delta}_1)^T\boldsymbol{\Phi}_1(\mathbf{B}_1 - \boldsymbol{\Delta}_1)\right] \\ &+ tr\left[(\mathbf{B}_2 - \boldsymbol{\Delta}_2)^T\boldsymbol{\Phi}_2(\mathbf{B}_2 - \boldsymbol{\Delta}_2)\right] + tr[\mathbf{B}_3^T\boldsymbol{\Phi}_3\mathbf{B}_3] + tr[\mathbf{B}_4^T\boldsymbol{\Phi}_4\mathbf{B}_4], \end{aligned} \tag{9}$$

where $\mathbf{P}$ and $\boldsymbol{\Phi}_i$ for i=1,2,...,4, are positive definite matrices to be determined. Differentiating (9) with respect to time and simplifying the resulting expression result in

$$\begin{aligned} \dot{L}(t) &= \mathbf{E}^T(\mathbf{P}\boldsymbol{\Delta} + \boldsymbol{\Delta}^T\mathbf{P})\mathbf{E} + 2tr\left[(\mathbf{B}_1 - \boldsymbol{\Delta}_1)^T(\boldsymbol{\Lambda}\mathbf{x}_e^T + \boldsymbol{\Phi}_1\dot{\mathbf{B}}_1)\right] - 2tr\left[\mathbf{B}_3^T(\boldsymbol{\Lambda}\mathbf{Q}_d^T - \boldsymbol{\Phi}_3\dot{\mathbf{B}}_3)\right] \\ &+ 2tr\left[(\mathbf{B}_2 - \boldsymbol{\Delta}_2)^T(\boldsymbol{\Lambda}\dot{\mathbf{x}}_e^T + \boldsymbol{\Phi}_2\dot{\mathbf{B}}_2)\right] - 2tr\left[\mathbf{B}_4^T(\boldsymbol{\Lambda}\dot{\mathbf{Q}}_d^T - \boldsymbol{\Phi}_4\dot{\mathbf{B}}_4)\right] \end{aligned} \tag{10}$$

where $\Lambda = [\mathbf{P}_2 \ \ \mathbf{P}_3]\mathbf{v}(t) = -\mathbf{P}_2\mathbf{Q}_e - \mathbf{P}_3\dot{\mathbf{Q}}_e$ and $\mathbf{P}$ is given by

$$\mathbf{P} = \begin{bmatrix} \mathbf{P}_1 & \mathbf{P}_2 \\ \mathbf{P}_2 & \mathbf{P}_3 \end{bmatrix} \tag{11}$$

and $\mathbf{E}(t) = -v(t)$ from the fact that $\mathbf{v}_m(t) = 0$.

Now if ξ_i and ω_i of (6) are selected so that all eigenvalues of $\boldsymbol{\Delta}$ have negative real parts, then according to the *Lyapunov Theorem* [10], for any given positive-definite symmetric matrix $\mathbf{N}$ there exists a positive definite symmetric matrix $\mathbf{P}$ satisfying the Lyapunov Equation given by $\mathbf{P}\boldsymbol{\Delta} + \boldsymbol{\Delta}^T\mathbf{P} = -\mathbf{N}$.

Now in (10) we proceed to let

$$\Lambda\mathbf{Q}_e^T + \Phi_1\dot{\mathbf{B}}_1 = \Lambda\dot{\mathbf{Q}}_e^T + \Phi_2\dot{\mathbf{B}}_2 = 0; \ \ \Lambda\mathbf{Q}_d^T - \Phi_3\dot{\mathbf{B}}_3 = \Lambda\dot{\mathbf{Q}}_d^T - \Phi_4\dot{\mathbf{B}}_4 = 0, \tag{12}$$

so that (10) becomes $\dot{L}(t) = -\mathbf{E}^T\mathbf{N}\mathbf{E}$ which is a negative definite function of $\mathbf{E}(t)$. Also from (12), we get

$$\dot{\mathbf{B}}_1 = -\Phi_1^{-1}\Lambda\mathbf{Q}_e^T; \ \ \dot{\mathbf{B}}_2 = -\Phi_2^{-1}\Lambda\dot{\mathbf{Q}}_e^T; \ \ \dot{\mathbf{B}}_3 = \Phi_3^{-1}\Lambda\mathbf{Q}_d^T; \ \ \dot{\mathbf{B}}_4 = \Phi_4^{-1}\Lambda\dot{\mathbf{Q}}_d^T. \tag{13}$$

According to (9), since $\mathbf{P}$ is a positive definite matrix by properly selecting $\boldsymbol{\Delta}$ and $\mathbf{N}$, the error system described in (8) is asymptotically stable if we can show that Φ_i for i=1,2,...,4, are also positive definite matrices. In other words, the *adjustable system* (3) will follow the reference model very closely, or $\mathbf{v}(t)$ approaches $\mathbf{v}_m$ asymptotically as $t \to \infty$.

We proceed to assume that the end-effector of the manipulator performs slowly varying motion so that elements of $\mathbf{M}$, $\mathbf{D}$ and $\mathbf{S}$ can be considered *almost constant*. Consequently, from (4) we obtain

$$\dot{\mathbf{B}}_1 \simeq \mathbf{M}^{-1}\dot{\mathbf{K}}_{fp}; \ \ \dot{\mathbf{B}}_2 \simeq \mathbf{M}^{-1}\dot{\mathbf{K}}_{fd}; \ \ \dot{\mathbf{B}}_3 \simeq 0; \ \ \dot{\mathbf{B}}_4 \simeq 0. \tag{14}$$

Substituting (14) into (13) yields

$$\mathbf{M}^{-1}\dot{\mathbf{K}}_{fp} = -\Phi_1^{-1}\Lambda\mathbf{Q}_e^T; \ \ \mathbf{M}^{-1}\dot{\mathbf{K}}_{fd} = -\Phi_2^{-1}\Lambda\dot{\mathbf{Q}}_e^T; \ \ 0 \simeq \Phi_3^{-1}\Lambda\mathbf{Q}_d^T; \ \ 0 \simeq \Phi_4^{-1}\Lambda\dot{\mathbf{Q}}_d^T. \tag{15}$$

Now in (15), letting $\Phi_1 = -\frac{1}{\beta_1}\mathbf{M}$; $\Phi_2 = -\frac{1}{\beta_2}\mathbf{M}$, where β_1 and β_2 are arbitrary positive scalars, and solving for $\dot{\mathbf{K}}_{fp}$ and $\dot{\mathbf{K}}_{fd}$, result in

$$\dot{\mathbf{K}}_{fp} = \beta_1\Lambda\mathbf{Q}_e^T; \ \ \dot{\mathbf{K}}_{fd} = \beta_2\Lambda\dot{\mathbf{Q}}_e^T. \tag{16}$$

We observe that Φ_1 and Φ_2 are positive definite matrices that can be considered as nearly constant because $\mathbf{M}$ is positive definite and slowly time-varying. In addition, Φ_3 and Φ_4 should be chosen to be positive definite matrices whose determinants approach ∞ in order to satisfy (15). To achieve this, we choose Φ_3 and Φ_4 to be diagonal matrices whose main diagonal elements are very large and positive.

Finally integrating both sides of the equations given in (16) results in

$$\mathbf{K}_{fp}(t) = \mathbf{K}_{fp}(0) + \beta_1\int_0^t(\mathbf{P}_2\mathbf{Q}_e + \mathbf{P}_3\dot{\mathbf{Q}}_e)\mathbf{Q}_e^T dt \tag{17}$$

$$\mathbf{K}_{fd}(t) = \mathbf{K}_{fd}(0) + \beta_2\int_0^t(\mathbf{P}_2\mathbf{Q}_e + \mathbf{P}_3\dot{\mathbf{Q}}_e)\dot{\mathbf{Q}}_e^T dt \tag{18}$$

where $\mathbf{K}_{fp}(0)$ and $\mathbf{K}_{fd}(0)$ are arbitrary initial conditions of $\mathbf{K}_{fp}(t)$ and $\mathbf{K}_{fd}(t)$, respectively.

Equations (17) and (18) represent the adaptation law for the force control scheme, whose implementation does not require intensive computation because $\mathbf{P}$ is a constant matrix and $\mathbf{Q}_e$ can be easily computed from the desired and actual Cartesian forces specified by the user and given from the force sensors, respectively.

3.2 The Position Control Scheme

Similar to development of the force control scheme, the adaptation laws for the position control scheme are developed by considering the case in which all Cartesian positions are controlled. In this case, the lower force control loop in Figure 1 is inactive and the hybrid control scheme becomes a pure position control scheme. The dynamics of a redundant manipulator performing non-compliant motion is described by [5]

$$\mathbf{F}_p(t) = \boldsymbol{\Upsilon}(\mathbf{x},\dot{\mathbf{x}})\ \ddot{\mathbf{y}}(t) + \boldsymbol{\Psi}(\mathbf{x},\dot{\mathbf{x}})\ \dot{\mathbf{y}}(t) + \boldsymbol{\Omega}(\mathbf{x},\dot{\mathbf{x}})\ \mathbf{y}(t) \tag{19}$$

where $\mathbf{y}(t) = \mathbf{x}(t)$ denotes the (6x1) vector containing Cartesian positions to be controlled and $\mathbf{F}_p(t)$, denotes the (6x1) Cartesian force applied to the end-effector. $\boldsymbol{\Upsilon}(\mathbf{x},\dot{\mathbf{x}})$, a symmetric positive-definite matrix, $\boldsymbol{\Psi}(\mathbf{x},\dot{\mathbf{x}})$ and $\boldsymbol{\Omega}(\mathbf{x},\dot{\mathbf{x}})$ are (6x6) matrices whose elements are highly complex nonlinear functions of $\mathbf{x}$ and $\dot{\mathbf{x}}$.

Consider a PD control law defined by

$$\mathbf{F}_p(t) = \mathbf{K}_{pp}(t)\ \mathbf{y}_e(t) + \mathbf{K}_{pd}(t)\ \dot{\mathbf{y}}_e(t) \tag{20}$$

where $\mathbf{y}_e(t)$ given by $\mathbf{y}_e(t) = \mathbf{y}_d(t) - \mathbf{y}(t)$ denotes the Cartesian error vector between the actual Cartesian position vector $\mathbf{y}(t)$ and the desired Cartesian position vector $\mathbf{y}_d(t)$, $\mathbf{K}_{pp}(t)$ and $\mathbf{K}_{pd}(t)$, are the gain matrices of the proportional and derivative controllers, respectively.

We observe that (19) and (20) assume the same form as that of (1) and (2), respectively. Therefore without repeating the same development as in previous section, we present the following adaptation law for the position control scheme:

$$\mathbf{K}_{pp}(t) = \mathbf{K}_{pp}(0) + \alpha_1 \int_0^t (\bar{\mathbf{P}}_2\mathbf{y}_e + \bar{\mathbf{P}}_3\dot{\mathbf{y}}_e)\mathbf{y}_e^T dt \tag{21}$$

$$\mathbf{K}_{pd}(t) = \mathbf{K}_{pd}(0) + \alpha_2 \int_0^t (\bar{\mathbf{P}}_2\mathbf{y}_e + \bar{\mathbf{P}}_3\dot{\mathbf{y}}_e)\dot{\mathbf{y}}_e^T dt \tag{22}$$

where $\mathbf{K}_{pp}(0)$ and $\mathbf{K}_{pd}(0)$ are arbitrary initial conditions of $\mathbf{K}_{pp}(t)$ and $\mathbf{K}_{pd}(t)$, respectively and α_1 and α_2 are arbitrary positive scalars. In addition, $\bar{\mathbf{P}}$ assuming the same form as that of (11) is the solution of the Lyapunov Equation $\bar{\mathbf{P}}\bar{\boldsymbol{\Lambda}} + \bar{\boldsymbol{\Lambda}}^T\bar{\mathbf{P}} = -\bar{\mathbf{N}}$ where $\bar{\boldsymbol{\Lambda}}$ represents the reference model for the position errors under the same form as given by (7).

3.3 The Position/Force Control Scheme

There exists internal cross coupling of the manipulator dynamics when the hybrid position/force control scheme is employed to control both position and force of the manipulator end-effector. Consequently the dynamical equation given in (1) and (19) must be modified to include the corresponding disturbances so that

$$\mathbf{F}_f(t) = \mathbf{M}(\mathbf{x},\dot{\mathbf{x}})\ \ddot{\mathbf{Q}}(t) + \mathbf{D}(\mathbf{x},\dot{\mathbf{x}})\ \dot{\mathbf{Q}}(t) + \mathbf{S}(\mathbf{x},\dot{\mathbf{x}})\ \mathbf{Q}(t) + \mathbf{H}_p(\mathbf{y}) \tag{23}$$

$$\mathbf{F}_p(t) = \boldsymbol{\Upsilon}(\mathbf{x},\dot{\mathbf{x}})\ \ddot{\mathbf{y}}(t) + \boldsymbol{\Psi}(\mathbf{x},\dot{\mathbf{x}})\ \dot{\mathbf{y}}(t) + \boldsymbol{\Omega}(\mathbf{x},\dot{\mathbf{x}})\ \mathbf{y}(t) + \mathbf{H}_f(\mathbf{Q}) \tag{24}$$

where $\mathbf{H}_p(\mathbf{y})$ and $\mathbf{H}_f(\mathbf{Q})$ represent the disturbances introduced by the position loop into the force loop and by the force loop into the position loop, respectively. In this paper we assume that the above disturbances are negligible so that the adaptation laws given in (17)-(18) and (21)-(22) can be employed.

4 Case Study Using Computer Simulation

The developed adaptive hybrid control is now applied to control the compliant motion of the slave arm of the NASA/GSFC telerobotic testbed, as shown in Figure 2. The slave arm, a RRC K-1607 robot manufactured by Robotics Research Corporation, is a 7 DOF kinematically redundant manipulator whose kinematics and differential motion analysis can be found in [11]. A modified version of Manipulator Simulation Program (MSP) [12] was employed to simulate the manipulator dynamics. Using the hybrid control scheme in Figure 1, the manipulator end-effector is controlled to apply a constant force of 50 lbs on a vertical frictionless reaction surface located at y=y_0 = 50.31in, with respect to the base frame, while tracking a desired vertical path specified by $z(t) = 39.447 - 10[1 - \exp(-0.025t^2)]$. To evaluate the effectiveness of the adaptive control scheme, we also consider the case in which the reaction table stiffness undergoes step changes. For the computer simulation, the contact force is obtained by $Q_y(t) = 0$ if $y(t) < y_0$ (no contact) and $Q_y(t) = K_e[y(t) - y_0]$ if $y(t) \geq y_0$ (during contact) where K_e denotes the table stiffness. For the simulation, ξ_i and ω_i for i=1,2 were selected so that 2 characteristic roots are -1 and -2 for both force and control loops.

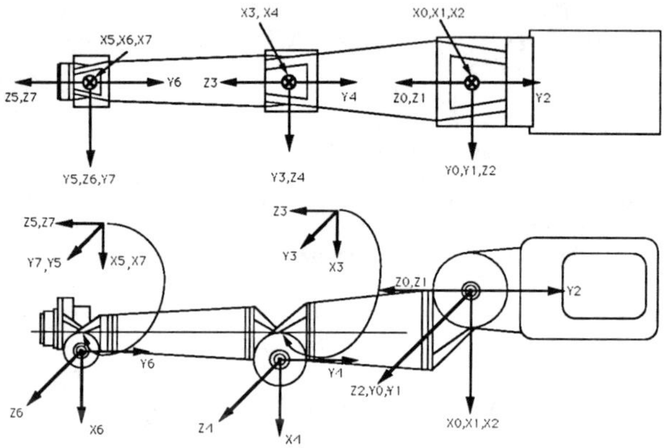

Figure 2 The Robotics Research Corporation K-1607 manipulator.

Figure 3(a)-(b) present the computer simulation results for the case of constant stiffness of 1000lb/in. As the figures show, the maximum errors of vertical position and contact force are 0.315in and 0.8556lb, respectively. The root-mean-square (**RMS**) errors are 0.133in and 0.1495lb for vertical position and contact force, respectively. Figure 4(a)-(b) report the results for case of sudden change in stiffness which was set to 1000lb/in at the beginning of the simulation and suddenly reduced to 714lb/in at 1/3 of the simulation time and suddenly increased to 800lb/in at 2/3 of the simulation time. The figures show that the force controller provides remarkable performance despite the sudden changes in stiffness. We observe that the force errors are relatively big at the time of the stiffness changes and are significantly reduced as the hybrid controller quickly adapted to the new stiffness. For this case, the maximum errors for position and force are 0.315in and -4.4908lb, respectively. In addition, the rms errors are 0.1379in for position and 0.1916lb for force.

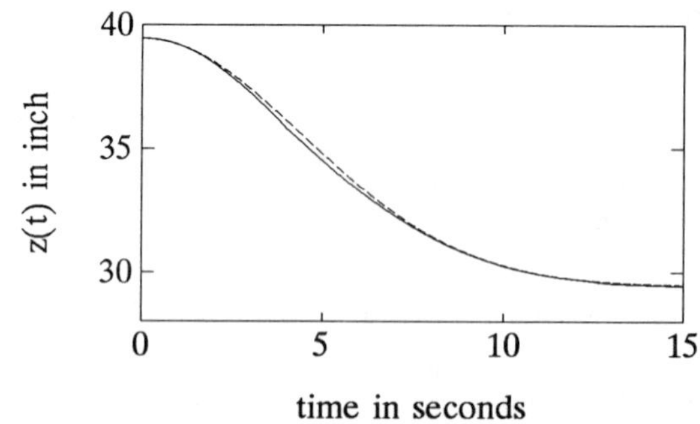

Figure 3(a) Vertical trajectories for constant stiffness (*solid=desired; dashed=actual*).

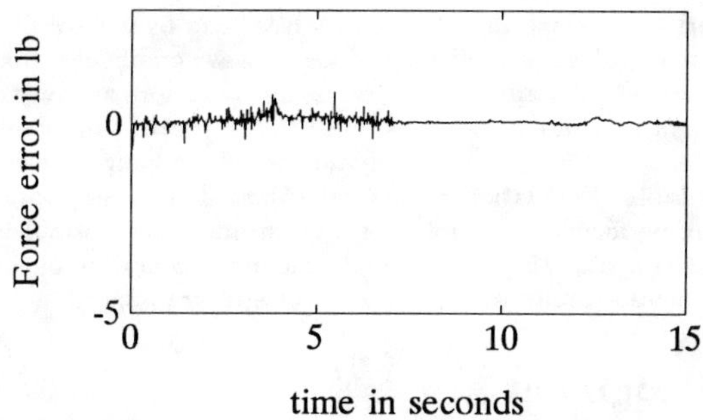

Figure 3(b) Horizontal force error for constant stiffness.

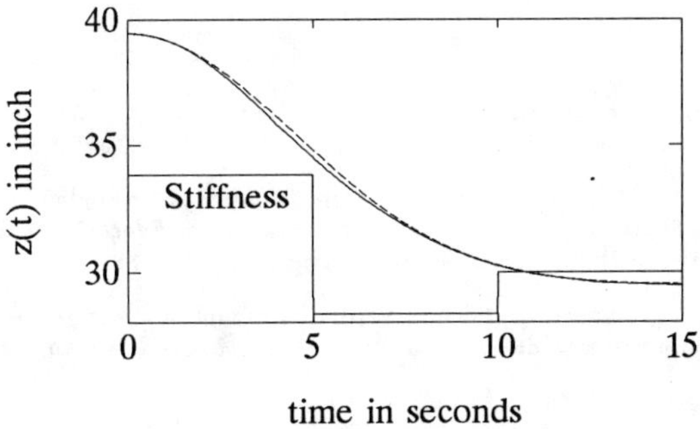

Figure 4(a) Vertical trajectories for variable stiffness (*solid=desired; dashed=actual*).

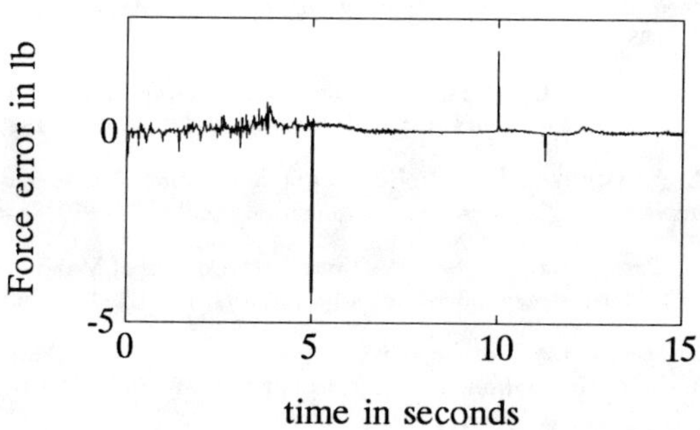

Figure 4(b) Horizontal force error for variable stiffness.

5 Conclusion

This paper has presented a hybrid position/force control scheme proposed to control compliant motion of kinematically redundant manipulators. The control scheme is composed mainly of a position control loop and a force control loop whose PD controller gains are adjusted by adaptation laws derived by using the concept of MRAC and Lyapunov Theorem under the assumption that the manipulator end-effector performs slow compliant motion. The adaptation laws are mostly based on force and position errors and therefore are very computationally efficient. The developed control scheme was applied to control the compliant motion of a 7DOF redundant manipulator to track a vertical motion while applying a desired contact force on a vertical reaction table. Simulation results have showed that the control scheme provided remarkable tracking performance in both constant stiffness and variable stiffness cases. Future activities will be devoted to study the real-time implementation of the developed control scheme and experimental results will be reported in a future paper.

6 Acknowledgment

The research presented in this paper was conducted at the Catholic University of America and has been supported by NASA/Goddard Space Flight Center under a research grant, Grant Number NAG 5-1124.

References

[1] **Klein, C.A., Huang, C.H.**, "Review of Pseudoinverse Control for Use with Kinematically Redundant Manipulators," *IEEE Trans. Sys., Man, and Cyber.*, pp. 245-250, March 1983.

[2] **Colbaugh, R.D.**, "Adaptive Position and Force Control of Redundant Robot Manipulators," in *Robotics and Manufacturing: Recent Trends in Research, Education, and Application*, edited by M. Jamshidi et al, ASME Press, New York, pp. 319-328, 1988.

[3] **Kazerounian, K., Wang, S.**, "Global versus Local Optimization in Redundancy Resolution of Robotic Manipulators," *International Journal of Robotics Research* Vol. 7, No. 5, pp. 3-12, 1988.

[4] **Nguyen, C.C., Zhou, Z.L., Mosier, G.E.**, "Joint-Space Adaptive Control of a Redundant Telerobot Manipulator," *Proc. Fourth IEEE International Symposium on Intelligent Control*, Albany, New York, pp. 59-65, September 1989.

[5] **Nguyen, C.C., Mosier, G.E.**, "Model Reference Adaptive Control of a Telerobot System," *Proc. ISMM Intern. Symp. Computer Applications in Design, Simulation and Analysis*, Nevada, pp. 282-285, Feb. 1989.

[6] **Seraji, H.**, "Configuration Control of Redundant Manipulators: Theory and Implementation," *IEEE Transactions on Robotics and Automation*, pp. 472-490, August 1989.

[7] **Yoshikawa, T.**, "Analysis and Control of Robot Manipulators with redundancy," *Proc. 1st Intern. Symp. on Robotics Research*, New Hampshire, pp. 735-747, 1983.

[8] **Raibert, M.H., Craig, J.J.**, "Hybrid Position/Force Control of Manipulators," *ASME, Journal of Dynamic Systems, Measurement and Control*, 102, pp. 126-133, June 1981.

[9] **Khatib, O., Burdick, J.**, "Motion and Force Control of Robot Manipulator," *Proc. IEEE Conf. on Robotics and Automation*, San Francisco, pp. 1381-1386, 1986.

[10] **Landau, Y.D.**, *Adaptive Control: The Model Reference Approach*, Marcell Dekker, New York, 1979.

[11] **Nguyen, C.C., Zhou, Z.L., Mosier, G.E.**, "Kinematic Analysis and Control of A 7-DOF Redundant Telerobot Manipulator," *Proc. 22nd Southeastern Symposium of System Theory*, Cookeville, Tennessee, March 1990.

[12] **Chen, R., Ou, Y.J.**, "Dynamic Formulation for Efficient Digital Simulation of Telerobotic Manipulation," *Final Report, Grant No. NAG 5-1019*, NASA/GSFC, December 1988.

FORCE-REFLECTING CONTROL OF A TELEOPERATED SYSTEM COUPLING A NONREDUNDANT MASTER WITH A REDUNDANT SLAVE*

R. Kress, J. Jansen, F. DePiero, and S. Babcock
Oak Ridge National Laboratory
Oak Ridge, Tennessee 37831-6285

ABSTRACT

A bilateral, force-reflecting controller was formulated for a 6-Degree-of-Freedom (6-DOF) master operating a 7-DOF slave. Six degrees of freedom are required to arbitrarily position and orient a manipulator end effector. A 7-DOF slave manipulator is desirable because the additional (or redundant) degree of freedom potentially allows for an infinite number of manipulator configurations for a given end effector position and orientation, thus permitting the manipulator to reach around obstacles to perform a task, to avoid joint limits, or to optimize its confinguration in some other manner. A master/slave control system was formulated using a stiffness controller for the master and a gradient-projection algorithm for the slave. The complete control system provides scaled position tracking and indexing between slave and master and force reflection between master and slave.

INTRODUCTION

In the late 1940s, Goertz and his colleagues at Argonne National Laboratory developed the mechanical master/slave manipulator with force-reflecting capabilities. In the early 1950s, Goertz and his colleagues developed an electric master/slave manipulator in which each slave joint servo was tied directly to a corresponding master joint servo because both the master and slave were kinematically similar. The control structure for these manipulators was the classical position-position controller. The positional difference between the master and the slave drives the slave. This difference is also reflected back as a drive signal to the master to provide the human operator with some feel of the environment. Goertz's work is summarized in Goertz and Thompson [1]. The position-position control scheme has been the basic controller for almost all master/slave servomanipulators used by industry. When a redundant slave is controlled by a nonredundant master, the kinematic dissimilarity makes the design of the controller particularly difficult. Three major issues are associated with this control problem: (1) redundancy resolution, (2) accurate force-refection, and (3) orientation representation. Because of space limitations, only the first two objectives are elaborated upon in this paper; however, proper direction for the third objective is provided.

Salisbury [2] developed a stiffness controller that allows an operator to specify the three translational and three rotational stiffnesses about a frame located arbitrarily in an end-effector coordinate frame. This type of controller is a good

* Operated by Martin Marietta Energy Systems, Inc., for the U.S. Department of Energy under Contract No. DE-AC05-840R21400.

choice for the master side of a master/slave teleoperated system because an operator could increase the desired stiffness in directions relevant to a task (e.g., along the axis of a hole during pin insertion) to enhance the feel in that direction, and an operator could reduce the desired stiffness in directions not important to the task (i.e., parallel to a work surface) to lessen the influence of contacts in these directions.

To traverse a desired path in three-dimensional space including both desired end-effector location and orientation, at least six joints of a robot must move simultaneously in a coordinated fashion. In the case of a 6-Degree-of-Freedom (6-DOF) manipulator, the joint velocities required to produce the desired end-effector path are unique. Six DOFs are required to produce three independent linear velocities and three independent angular velocities. In the case of a redundant manipulator with 7 or more DOFs for three-dimensional operation, the solution of the kinematic equations relating unknown joint velocities to desired linear and angular velocities is not unique. An infinite number of solutions exist for most positions. This feature allows one or more joint velocities to be selected to satisfy an external criterion, for example, minimizing the joint velocities, minimizing the joint torques, avoiding joint limits, or maximizing the distance of the manipulator from an obstacle.

Many control schemes have been proposed to determine the joint velocities for a redundant robot [3]. Some schemes take a global approach to optimize criteria over the entire trajectory while others take a local approach optimizing criteria for an "instantaneous" set of joint motions. Local schemes, because of their decreased computational requirements, seem best suited for real-time control. A local redundancy resolution scheme will be used for the slave side of the controller.

CONTROLLER FORMULATION

Master Stiffness Controller

To formulate the master controller as a stiffness controller, use the following control laws:

$$\tau_m = J_m^T (\sigma F_s) + \tau_{grav} + \tau_m^c \quad , \tag{1}$$

$$\tau_m^c = J_m^T [K_{pm} (x_s - x_m) + K_{vm} (\dot{x}_s - \dot{x}_m)] \quad , \tag{2}$$

where

τ_m = (6×1) vector of complete master controller signals,

J_m^T = transpose of the master Jacobian (6×6),

σ = positive force sensitivity constant,

F_s = (6×1) vector of measured slave forces,

τ_{grav} = (6×1) vector of gravity compensation signals,

τ_m^c = (6×1) vector of master controller signals defined in Eq. (2),

K_{pm} and K_{vm} = (6×6) position and velocity gain matrices respectively,

x_s and $\dot{x}_s$ = (6×1) slave position and velocity vectors respectively,

x_m and $\dot{x}_m$ = (6×1) master position and velocity vectors respectively.

Effective force feedback is achieved by using a six-axis force/torque sensor mounted at the wrist of the slave manipulator to produce F_S. To avoid operator fatigue, gravity forces can be essentially eliminated with proper mechanical design employing counterbalance weights, lightweight links, and joints oriented parallel to the gravity vector. If velocity is not measured at the master joints, the velocity gain matrix K_{vm} must be set to the zero matrix. Damping would then be achieved by means of joint and actuator friction, and human operator damping. The gain selection should be such that the operator's dynamics dominate the master arm's dynamics.

The preceding simplifications make the master controller a stiffness controller of the following form:

$$\tau_m = {}^0J_m^T [K_{pm} (x_s - x_m)] + {}^0J_m^T (\sigma F_s) \quad , \tag{3}$$

where ${}^0J_m^T$ is the (6×6) transpose of the master Jacobian and where the 0 superscript has been added to denote clearly that we are working in a fixed base frame; all other terms have been previously defined.

The master position can be determined by solving the master kinematics. For the master Jacobian, it is convenient to convert from Cartesian angles to Euler angles because of the natural physical associations of pitch, yaw, and roll. The slave position, x_s, and master position, x_m, are (6×1) vectors where the last three entries are the Cartesian angles. To convert to Euler angles for the last three entries, the Jacobian, ${}^0J_m$, needs to be multiplied by

$${}^0J_e = \begin{bmatrix} I^{(3\times3)} & 0^{(3\times3)} \\ 0^{(3\times3)} & {}^{\dot{\alpha}}_{\omega}R \end{bmatrix} {}^0J_m \quad , \tag{4}$$

where ${}^0J_e$ is the (6×6) master Jacobian using Euler angles in the zero or base frame and ${}^{\dot{\alpha}}_{\omega}R$ is the (3×3) transform that converts Cartesian angular rates into Euler angular rates [4, p.235] given by

$${}^{\dot{\alpha}}_{\omega}R = \sec\beta \begin{bmatrix} C\gamma & S\gamma & 0 \\ -S\gamma C\beta & C\gamma C\beta & 0 \\ C\gamma S\beta & S\gamma S\beta & C\beta \end{bmatrix} . \tag{5}$$

Note that Eq. (3) still represents the master controller torque but ${}^0J_m$ in the first term on the right-hand side is replaced by ${}^0J_e$. Note also the weakness when Euler angles represent orientation in the stiffness controller. Equation (5) shows that an artificial singularity is introduced at $\beta = \pm 90^o$. For simplicity, we chose to avoid the singularity by modifying the secant function so that it did not become infinity at $\pm 90^o$. This approach is not an optimal solution and a better way to circumvent this artificial singularity is by means of quaternions [5].

Assuming a rigid-bodied master with a mechanical design providing completely balanced gravity forces, the dynamics of the master can be described as

$$M(\theta)\ddot{\theta}_m + C(\theta,\dot{\theta})\dot{\theta}_m + ({}^0J_m)^T F_{hand} = \tau_m \quad , \tag{6}$$

where,

$M(\theta)$ = (6×6) inertia tensor of the master,

$C(\theta,\dot{\theta})$ = (6×6) Coriolis and centrifugal force matrix,

θ = (6×1) master joint angle vector,

τ_m = (6×1) vector of controller signals defined in Eq. (3),

F_{hand} = (6×1) hand force vector exerted by operator.

When the master is at rest or steady state, Eq. (6) combined with the definition of τ_m [Eq. (3)], simplifies to

$$R^T K_{pm} (x_s - x_m) + \sigma F_s - F_{hand} = 0 \quad , \tag{7}$$

where ${}^0J_e = R\ {}^0J_m$ from Eq. (4). Equation (7) indicates that the steady state force at the master gripper is related to the difference in the slave and master positions by a stiffness matrix K_{ps}.

Slave Gradient-Projection Algorithm

The null space of a manipulator is defined as those joint motions that produce no change in the manipulator end-effector position and orientation. The slave control algorithm uses a gradient-projection method to add a component of motion in the null space of the redundant, 7-DOF manipulator. This component of motion can take advantage of the manipulator's redundant DOFs to optimize a predetermined criterion. A control algorithm to determine the null space motion is described in Dubey and Luh [6] for a general manipulator and in Dubey, et al. [7] for a manipulator with a spherical wrist. A brief summary of this algorithm follows.

A manipulator with n joints that controls m independent variables is governed by the following kinematic equation:

$$\dot{x} = J\dot{\theta} \quad , \tag{8}$$

where $\dot{x}$ is the (m×1) velocity vector, J is the (m×n) Jacobian matrix, and $\dot{\theta}$ is the (n×1) joint velocity vector. If J is square (m = n) and it is nonsingular, it can be inverted and Eq. (8) can be solved for a unique set of joint velocities. If J is singular or rectangular with m < n, then the joint velocities may be determined by:

$$\dot{\theta} = J^+\dot{x} + (I - J^+J)\dot{\phi} \quad , \tag{9}$$

where J^+ is (n×m) the Moore-Penrose generalized inverse of the Jacobian [8]. The matrix I is the (n×n) identity matrix and $(I - J^+J)$ is the (n×n) null space projection matrix. The vector $\dot{\phi}$ is an (n×1) joint velocity vector with arbitrary magnitude. Its magnitude is arbitrary because it is directed in the null space of the manipulator and, as noted earlier, joint motions in the null space produce no change in the end-effector position. The first term on the right-hand side of Eq. (9), $J^+ \dot{x}$, is the least-norm solution, that is, the solution that minimizes the Euclidean norm of $\left(J\,\dot{\theta} = \dot{x}\right)$ for a given $\dot{x}$. The second term on the right hand side of Eq. (9), $(I - J^+J)\,\dot{\phi}$, represents the null space motion. It is orthogonal to $J^+\dot{x}$ and is called the homogeneous solution or the manipulator self-motion.

The homogeneous solution should be selected to improve the performance of the robot. To use a gradient projection method, define a cost function, H(θ). The homogeneous solution is then given by $k\nabla H(\theta)$, where the coefficient k is a real scalar constant bounded by joint velocity limits and $\nabla H(\theta)$ is the gradient of the cost function given by:

$$\nabla H(\theta) = \left\{\partial H/\partial\theta_1\,,\ \partial H/\partial\theta_2\,,\ \ldots\partial H/\partial\theta_n\right\}^T \quad . \tag{10}$$

The scalar k is taken as positive if H(θ) is to be maximized and negative if H(θ) is to be minimized.

IMPLEMENTATION

The stiffness controller is implemented at ORNL on a Kraft master controller. The Kraft KMC 9100-MC is a lightweight 6-DOF master arm designed, manufactured, and sold by Kraft Telerobotics, Inc., of Overland Park, Kansas. Its joints are (from base to gripper): shoulder azimuth, shoulder pitch, elbow, wrist yaw, wrist pitch, and wrist roll. Position is measured at each joint by potentiometers. The first five joints are actuated by ac servomotors for force feedback. (Wrist roll is not actuated.) The Kraft arm is supplied with the KMC 9100S electronics interface unit which includes amplifiers and communications interfaces.

The signal required to produce impending motion was measured for each joint of the Kraft master. These experimental results show that each joint requires ~10 to 20% of full signal in order to move. To optimize between good backdrivability and force sensitivity when using a master with this much deadband some compensation is required. A simple, but effective, form of compensation is an offset function called the preload function (PLF) which is the inverse of a deadband function [9]. An analysis was performed using describing function methods to determine whether limit cycles would be present when applying the preload function. A limit cycle is defined to be an initial condition-independent periodic oscillation occurring in dissipative systems [9]. The results of the analysis indicate that a limit cycle does not occur.

The PLF was implemented on the master controller. For the shoulder azimuth, shoulder pitch, and elbow joints, the preload was set to ~10%. For the wrist pitch and yaw joints, the preload was reduced to ~5% to avoid a chattering at the switching line between plus and minus preload. The PLF did not introduce any instabilities, including limit cycles, and it improved the force reflection.

The gradient-projection algorithm is implemented on the Center for Engineering Systems Advanced Research Manipulator (CESARm) [7]. CESARm is a custom-designed 7-DOF arm. Its joints are (from base to gripper): waist, shoulder pitch, shoulder roll, elbow, wrist pitch, wrist yaw, and wrist roll. The waist, shoulder pitch, and shoulder roll are actuated with dc servomotors coupled to gear reducers. The elbow is actuated by a dc servomotor coupled through a 4-bar linkage. Wrist pitch and yaw are actuated by dc servomotors connected through cables to a 2-DOF differential. Wrist roll is actuated by a dc servomotor connected via cables. Position is measured at each motor by optical encoders and velocity is measured for analog feedback by tachometers mounted on the motors. Three components of force and three components of torque are measured at the CESARm wrist by a JR^3 universal force sensor communicating through a parallel interface.

The master and slave control algorithms were programmed in C and implemented on three Motorola 68020s having 68881 floating point coprocessors. The algorithms were partitioned as follows. CPU 1 had all low-level routines. This included low-level control routines (e.g. sending motor signals, enabling motors and brakes, reading encoders, etc.), PID control routines, simple joint and cable limit checks, and input/output routines. CPU 2 had the gradient-projection algorithm and an obstacle avoidance routine. CPU 3 had the master stiffness routine, the JR^3 force/torque sensor routine, and master and JR^3 communications.

RESULTS AND CONCLUSIONS

Initial experiments indicate that the master/slave controller operates stably in a bilateral, force-reflecting mode; however, without the JR^3 force/torque sensor, force reflection was not very sensitive. Algorithms on CPU 1 and CPU 2 run at 100 Hz, whereas, algorithms on CPU 3 run at ~30 Hz and ~50 Hz with and without the JR^3 routines respectively. The slower speed of the routines on CPU 3 was due to communications overhead. The resulting master/slave system is configured to provide a smooth transition between the teleoperated mode and the robotic mode to allow an operator (or an intelligent supervisory algorithm) to choose the most efficient mode for a given task.

The master controller formulation allows for adjustable stiffness along Cartesian axes. This is shown for two directions in Fig. 1.

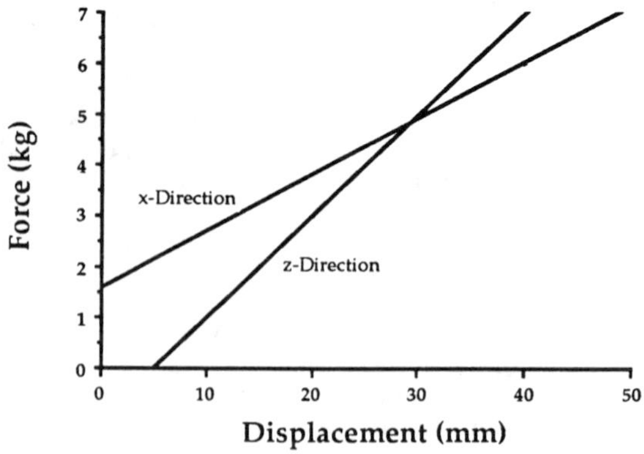

Fig. 1. Spring rates measured at the master arm for fixed slave (k=1).

Each of the gains in Eq. (3) are variable such that the springs are effectively decoupled. For example, if a (6×6), diagonal K_{pm} was input and the slave is held in place, then the master has the feel of being "attached" to a home position with a set of linear and torsional springs. Different gains yield similar results.

The slave controller has been previously demonstrated for a robotic mode by Dubey et al. [7]. It has also been demonstrated as a nonforce-reflecting teleoperated mode by Babcock et al. [10]. It worked in a force-reflecting teleoperated mode with the master stiffness controller.

This system is by no means an optimal master/slave controller. Improvements are part of an ongoing effort at ORNL. These include using quaternions and formulating the slave controller as a stiffness or an impedance type controller [11].

ACKNOWLEDGMENTS

This work was partially funded by the Biological Acoustics Branch, Biodynamics and Bioengineering Division, Armstrong Aerospace Medical Research Laboratory, Wright-Patterson Air Force Base.

REFERENCES

1. R. C. Goertz and W M. Thompson, "Electronically Controlled Manipulator," *Nucleonics*, Nov. 1954, pp. 46-47.
2. J. K. Salisbury, "Active Stiffness Control of a Manipulator in Cartesian Coordinates," *Proc. of the 19th IEEE Conf. Decision and Control,* Albuquerque, N. M., Nov. 1980, pp. 95-100.
3. D. N. Nenchev, "Redundancy Resolution Through Local Optimization: A Review," J. of Robotic Systems, 6, 6, Dec. 1989, pp. 769 - 798.
4. K. S. Fu, R. C. Gonzalez, and C. S. G. Lee, *Robotics: Control, Sensing, Vision, and Intelligence*, McGraw-Hill, Inc., New York, N. Y., 1987.
5. J. S. C. Yuan, " Closed-Loop Manipulator Control Using Quaternion Feedback," *IEEE Transactions on Robotics and Autom.*, Vol. 4, No. 4, August, 1988.
6. R. V. Dubey and J. Y. S. Luh, "An Efficient Gradient Projection Optimization Scheme for a Seven-Degree-Of-Freedom Redundant Robot with Spherical Wrist," *Proc. of the 1987 IEEE Int. Conf. on Robotics and Autom.*, Raleigh, N. C., **2**, 1066 - 1072, March 31 - April 3, 1987.
7. R. V. Dubey, J. A. Euler, and S. M. Babcock, "An Efficient Gradient Projection Optimization Scheme for a Seven-Degree-Of-Freedom Redundant Robot with Spherical Wrist," *Proc. of the 1988 IEEE Int. Conf. on Robotics and Autom.*, Philadelphia, Pa., **1**, 28 -36, April 24-29, 1988.
8. A. Albert, *Regression and the Moore-Penrose Pseudo-Inverse*, Academic Press, 1972.
9. A. Gelb and W. E. V. Velde, *Multiple-Input Describing Functions and Nonlinear System Design*, McGraw-Hill, Inc., New York, N. Y., 1968.
10. S. M. Babcock, R. V. Dubey, J. A. Euler, R. L. Glassell, R. L., Kress, and W. R. Hamel, "Telerobotic Control of the Seven-Degree-of-Freedom CESAR Manipulator," *NATO Advanced Research Workshop on Robotics with Redundancy: Design, Sensing, and Control*, Salo, Lago di Garda, Italy, June 27-July 1, Proc. published by Springer-Verlag, 1989.
11. J. F. Jansen and S. M. Babcock, "Bilateral Force-Reflection for Teleoperators with Masters and Slaves with Dissimilar and Possibly Redundant Kinematics," ORNL/TM-11326, Oak Ridge National Laboratory, 1989.

Learning of Visuomotor-Coordination of a Robot Arm with Redundant Degrees of Freedom

Thomas Martinetz, Helge Ritter and Klaus Schulten

Beckman Institute and Department of Physics
University of Illinois at Urbana-Champaign
405 North Mathews Ave., Urbana, IL 61801, USA

Abstract

An improved version of an earlier introduced algorithm for learning of visuomotor-coordination has been successfully applied to a simulated robot arm system with five degrees of freedom, two of which are redundant. The learning algorithm was not affected by this redundancy, and with the improved version the robot arm system is able to reduce its positioning error to about 5% of its linear dimensions after only 200 learning steps, and finally to about 0.3% after 6000 learning steps. Learning proceeds without the need of an external teacher by a sequence of trial movements using input signals from a pair of cameras. The topology conserving map used for the representation of the input-output transformation leads to an automatic resolution of the redundancy of the arm as it tries to minimize the variation of the joint angles over the work space. The improved learning algorithm incorporates some immediate feedback, so that now the robot arm is able not only to adapt to slowly occurring miscalibrations, but also can compensate for sudden changes in its geometry, such as picking up e.g. a tool.

Introduction

Compared to present-day robots, biological motor control systems excel with an enormous degree of flexibility. This flexibility is to a large extent due to two major factors. The first is the use of highly developed sensors, most notably vision, capable to monitor a multitude of different aspects of ongoing movements simultaneously. The second factor is the presence of a high amount of redundancy in the musculo-skeletal system, offering for most movement goals a wide range of alternative realizations. Both strengths are enhanced further by the high degree of adaptability that neural systems exhibit for both sensory perception and selection and control of movements.

There have been many attempts to capture these remarkable features in biologically inspired neural network models, see e.g. [1-8]. Our own work has previously considered the problem of adaptive visuo-motor coordination for a robot, using a

An earlier version of this paper was presented at the International Conference on Parallel Processing in Neural Systems and Computers (ICNC), Düsseldorf, FRG, March 19-21, 1990.

non-redundant (simulated) three-link robot arm [4,5,8]. In this contribution, we want to focus on the additional issue of controlling a redundant arm and to resolve its redundancy by the use of a topology conserving map imposing a "smoothness constraint" on the arm configurations. In addition, we present an improved version of the learning rule used in [4,8], which incorporates a more accurate visual feedback.

For a redundant robot arm, specification of a target location does not yet uniquely determine the configuration of the arm, but instead allows a wide range of possibilities still compatible with the given end effector location. The general problem of redundancy resolution is then to choose one of the infinity of possible configurations such that some overall cost-function or some performance measure is extremalized. Frequently, however, the precise form of the performance measure is of secondary interest, and serves merely as a tool to impose a smoothness constraint upon the system, i.e. to ensure that the selected configurations for two nearby end effector locations differ from each other as little as possible.

However, this is precisely the kind of task solved "naturally" by topology conserving maps. If such map is used to represent the mapping from task space to joint coordinates, neighboring task coordinates will activate neighboring nodes in the network for output. Due to the natural tendency of the map to smooth out any unnecessary variations in the outputs of neighboring nodes, any redundant degrees of freedom will be used by the map to make the variation between joint configurations for neighboring target points as small as possible. This method differs from previous approaches, like the pseudo-inverse technique, in that the optimization of smoothness comes "for free" by the natural learning dynamics of the network and does not require any auxiliary computations beyond the usual adaptation steps.

The Model

In Fig.1 we see the robot system, consisting of a robot arm with five degrees of freedom and a pair of cameras providing the spatial information about the location of the object the robot arm shall reach for. For each trial movement, the target location is chosen randomly within the work space. Each target within the three-dimensional work space corresponds to a pair of two-dimensional vectors, namely the locations of the images of the object on the two camera "retinas". Both locations are combined to a four-dimensional vector $\mathbf{u}$ which then carries the visual information necessary for the network to extract the spatial position of the object. These four-dimensional vectors form the input signals for a vector-quantization network of Kohonen-type [9,10]. This network is able to adaptively discretize the relevant three-dimensional submanifold W within the four-dimensional input space U (see Fig.2). The highly nonlinear transformation $\vec{\theta}(\mathbf{u})$ from camera input $\mathbf{u}$ to joint angles $\vec{\theta}$ is then linearized in the vicinity of each discretization point [4,5,8]. Below we describe how the local linear transformations are learnt from trial movements simultaneously with the adaptive discretization.

As depicted in Fig.2, to each formal neuron $\mathbf{r}$ of a Kohonen-network with a three-dimensional topology we assign a discretization point $\mathbf{w_r}$ within the four-dimensional input space U. If the target location $\mathbf{u}$ is closer to the discretization point $\mathbf{w_s}$ than to any other $\mathbf{w_r}$, $\mathbf{r} \neq \mathbf{s}$, the formal neuron $\mathbf{s}$ becomes responsible for providing the output, i.e. it then determines the appropriate joint angles to reach the target object. To perform this task, to each formal neuron additionally a five-

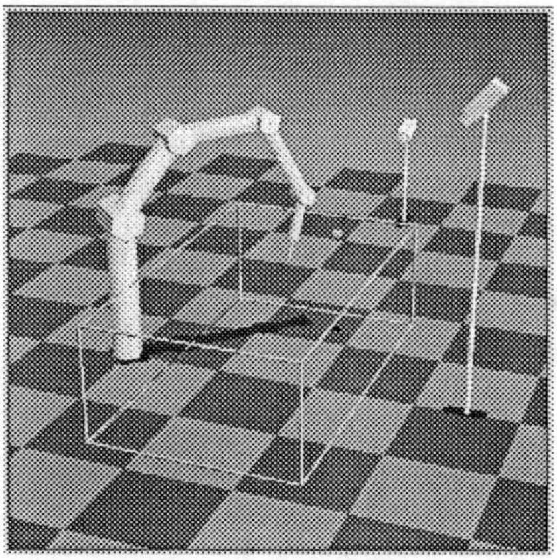

Fig.1: The robot arm with the two cameras, the workspace, and the target object.

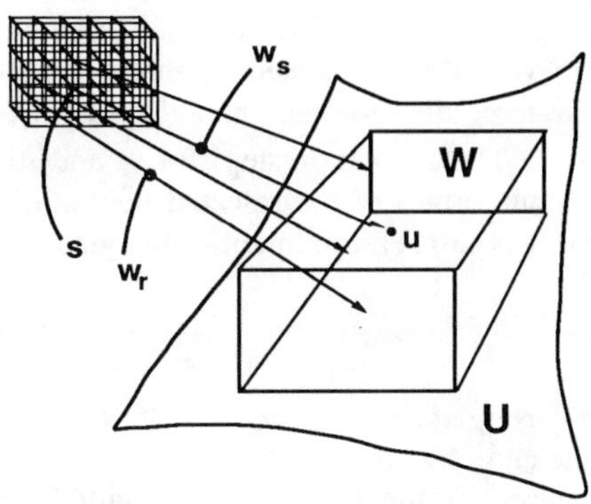

Fig.2: The three-dimensional "neural grid" for discretizing the workspace.

dimensional vector $\vec{\theta}_s$ of joint angles and a 5×4 matrix $\mathbf{A}_s$ is assigned. $\vec{\theta}_s$ and $\mathbf{A}_s$ approximate the zero order term and the Jacobian of the Taylor-expansion of $\vec{\theta}(\mathbf{u})$ around $\mathbf{w}_s$ and are iteratively refined after each trial movement.

After the formal neuron s responsible for the momentary target location has been selected, the positioning of the end effector is carried out in two phases: (i) a gross-positioning movement, and (ii) a subsequent, usually only small correction. For the gross-positioning movement the robot arm system uses the five-dimensional vector $\vec{\theta}_s$, each element of which determines one joint angle. After this gross-movement the position of the end effector on both camera "retinas" will be denoted by a four-dimensional vector $\mathbf{v}_i$, which is usually already close to the four-dimensional retinal target location $\mathbf{u}$. Departing from the ansatz in [4,8], we now use the linear correction term

$$\Delta\vec{\theta} = \mathbf{A}_s(\mathbf{u} - \mathbf{v}_i) \tag{1}$$

for the final corrective joint movement. $\Delta\vec{\theta}$ is usually sufficient to compensate the error of the gross-movement very precisely. A further decrement of the residual positioning error can be achieved by repeating the corrective fine movement (1) several times.

The presentation of target objects at locations randomly chosen within the three-dimensional workspace, followed by trial movements of the robot arm, drives an adaptation process for the locations of the discretization points $\mathbf{w}_r$ as well as for the output values $\vec{\theta}_r$ and $\mathbf{A}_r$. This adaptation process is described by the adaptation steps

$$\begin{aligned} \mathbf{w}_r^{new} &= \mathbf{w}_r^{old} + \epsilon h_{rs}(\mathbf{u} - \mathbf{w}_r^{old}) \\ \vec{\theta}_r^{new} &= \vec{\theta}_r^{old} + \epsilon' h'_{rs}(\vec{\theta}^* - \vec{\theta}_r^{old}) \\ \mathbf{A}_r^{new} &= \mathbf{A}_r^{old} + \epsilon' h'_{rs}(\mathbf{A}^* - \mathbf{A}_r^{old}) \end{aligned} \tag{2}$$

with h_{rs} and h'_{rs} as Gaussians (see eq.(6) below) determining the spatial decrease of the learning step within the "neural grid", and ϵ, ϵ' as scaling factors for the overall size of the adaptation steps. The first of these equations is the learning step used in Kohonen's algorithm for the formation of topologically correct feature maps [9,10]. This learning step results in a topologically correct mapping from the relevant three-dimensional submanifold of the four-dimensional input space U to the three-dimensional "neural grid" [4].

The adaptation steps for $\vec{\theta}_{\mathbf{r}}$ and $\mathbf{A}_{\mathbf{r}}$ use improved estimates $\vec{\theta}^*$ and $\mathbf{A}^*$ for the output values of the selected neuron via a linear error correction rule of Widrow-Hoff-type [11], minimizing the quadratic error

$$E = \frac{1}{2}(\Delta\vec{\theta} - \mathbf{A}_{\mathbf{s}}\Delta\mathbf{v})^2 \tag{3}$$

by using steepest descent. In Eq.(3), $\Delta\mathbf{v} = \mathbf{v}_f - \mathbf{v}_i$ with $\mathbf{v}_f$ as the final position of the end effector seen by camera 1 and camera 2 after completion of both movement phases (i) and (ii). Together with (1) and by choosing the optimal step size we obtain as an improved estimate $\mathbf{A}^*$

$$\mathbf{A}^* = \mathbf{A}_{\mathbf{s}} + ||\Delta\mathbf{v}||^{-2} \cdot \mathbf{A}_{\mathbf{s}}(\mathbf{u} - \mathbf{v}_f)\Delta\mathbf{v}^T. \tag{4}$$

If we denote by $\mathbf{A}(\mathbf{w}_{\mathbf{s}})$ the correct Jacobian associated with the discretization point $\mathbf{w}_{\mathbf{s}}$, and by $\vec{\theta}(\mathbf{w}_{\mathbf{s}})$ the correct joint angles, we have (to linear order) the relation $\vec{\theta}(\mathbf{w}_{\mathbf{s}}) - \vec{\theta}_{\mathbf{s}} = \mathbf{A}(\mathbf{w}_{\mathbf{s}})(\mathbf{w}_{\mathbf{s}} - \mathbf{v}_i)$. Since $\mathbf{A}(\mathbf{w}_{\mathbf{s}}) \approx \mathbf{A}_{\mathbf{s}}$, we can obtain an improved estimate $\vec{\theta}^*$ for the correct joint angles $\vec{\theta}(\mathbf{w}_{\mathbf{s}})$ by

$$\vec{\theta}^* = \vec{\theta}_{\mathbf{s}} + \mathbf{A}_{\mathbf{s}}(\mathbf{w}_{\mathbf{s}} - \mathbf{v}_i). \tag{5}$$

The learning steps for $\vec{\theta}_{\mathbf{r}}$ and $\mathbf{A}_{\mathbf{r}}$ in (2) have a similar form as the adaptation steps for $\mathbf{w}_{\mathbf{r}}$ with the result, that $\vec{\theta}_{\mathbf{r}}$ and $\mathbf{A}_{\mathbf{r}}$ during the learning phase adapt to values which are ordered topologically, too. The output $\vec{\theta}_{\mathbf{r}}$ and $\mathbf{A}_{\mathbf{r}}$ of formal neurons neighbored in the three-dimensional grid (which also means neighbored within the input space) adapt to values which are neighbored in the space of all five-dimensional vectors $\vec{\theta}$ and all 5×4 matrices $\mathbf{A}$, respectively, assuming an euclidian metric within the space of these values. This mapping, topologically correct also with respect to the output space, now enforces a smoothness constraint. Out of an infinite number of possible arm configurations the neural network adapts to arm postures which vary smoothly over the "neural grid". Additionally, by "spreading" the adjustments to all neighbored neurons of s the learning steps for $\vec{\theta}_{\mathbf{r}}$ and $\mathbf{A}_{\mathbf{r}}$ provide an enormous increase in speed and stability of the convergence of the learning algorithm [8].

Simulation Results

In the following, we describe the results of a simulation using a $7 \times 12 \times 4$ lattice of 336 neurons. Function $h_{\mathbf{rs}}$ was taken to be the Gaussian

$$h_{\mathbf{rs}} = \exp(-||\mathbf{r} - \mathbf{s}||^2/2\sigma^2(t)) \tag{6}$$

and $h'_{\mathbf{rs}}$ likewise. Parameters ϵ, ϵ' and the widths σ, σ' all had the same time dependence $p(t) = p_i(p_f/p_i)^{t/t_{max}}$ with t as the number of the already performed learning steps and $t_{max} = 10\,000$ [4]. In contrast to our previous work [4,8], we

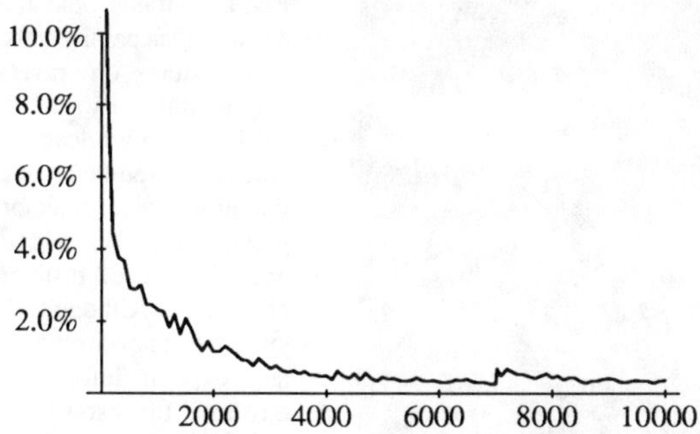

Fig.3: Positioning error versus number of trials. After 7000 trial movements, the length of the last arm segment was increased by 5%. Due to the corrective fine-movement, the actual error caused by this perturbation is less than 1%. With further trial movements, the neural net adapts to the new situation and the positioning error returns to its previous value.

now employ the improved adaptation rules (1), (4), (5) and consider the robot arm with five degrees of freedom, shown in Fig.1. The initial output values for each neuron were chosen at random, but subject to the constraint that the resulting shape of the robot arm be convex in all cases.

Figure 3 shows the development of the remaining positioning error after each trial. The error decays very rapidly to about 0.3 percent of the robot's dimensions. Although the number of links is now larger (5 vs. 3), the decay of the error is faster than in [4,8], which clearly shows the superiority of the new learning rule over that employed in [4,8]. This stems mainly from basing the corrective fine-movement on the actual distance $\mathbf{u} - \mathbf{v}_i$ remaining after the preceding gross-movement, instead of using the distance $\mathbf{u} - \mathbf{w_s}$ between the target and the closest discretization point. This provides the system with an accurate feedback about its residual error, and enables the fine movement to compensate even for sudden and unexpected changes in the geometry of the arm. To demonstrate this, we suddenly increased after 7000 trials the length of the last arm segment by about 5% of its linear dimension, a situation which could result from picking up e.g. a wrench. As we see from Fig.3, even this perturbation causes only a small increase in the error, which rapidly returns to its baseline value.

Due to the redundancy of the arm, each target location is compatible with a continuous subset of different arm postures. Initially, as a result of the random initialization of the network, mutually close target locations may give rise to very different arm postures selected for reaching. As learning proceeds, those posture differences not contributing to differences in end effector locations become more and more wiped out and only postures which vary smoothly over the target space emerge. An example is provided by Fig.4, which shows successive postures traversed if the end effector follows a path along a diagonal of the workspace.

Conclusion

We have presented a learning algorithm for the visuomotor-coordination of a redundant manipulator with five degrees of freedom. The method is an improved version of an earlier algorithm [4,8], based on Kohonen's self-organizing feature maps [9,10]. As a novel feature, the resolution of the arm's redundancy is based on the topology-preserving property of the mapping algorithm. We explored the method in computer simulations and obtained good results, in regard to both learning speed and robustness against external perturbations.

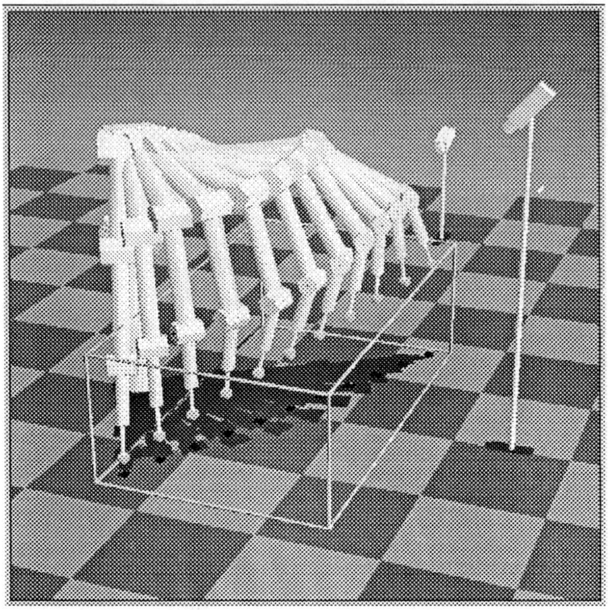

Fig.4: Stroboscopic rendering of successive robot arm postures on traversing a trajectory in the workspace after learning. Due to the redundancy of the arm, each trajectory point can be reached by an infinitude of different postures. Although the system has never obtained any explicit information on how to resolve this redundancy, the topology conserving property of the map used for the learning algorithm has enforced a solution, leading to a particularly smooth movement.

Acknowledgement

This work has been supported by the University of Illinois, the German Ministry of Science and Technology (ITR-8800-G9), and by the Volkswagen Foundation.

References

[1] Graf D.H., LaLonde W.R., (1988) A Neural Controller for Collision-Free Movement of General Robot Manipulators. Proceedings of the IEEE Int. Conf. on Neural Networks '88, San Diego 1988, Vol.I:77-84.

[2] Kuperstein M., Rubinstein J. (1989) Implementation of an Adaptive Neural Controller for Sensory-Motor Coordination. IEEE Control Systems Magazine, Vol.9, No.3, pp. 25-30.

[3] Lane, Handelman, Gelfand (1988) A Neural Network Computational map approach to reflexive motor control. Proceedings of the 1988 IEEE conference on intelligent control.

[4] Martinetz T., Ritter H., Schulten K. (1990) Three-Dimensional Neural Net for Learning Visuomotor Coordination of a Robot arm. IEEE Transactions on Neural Networks, Vol.1, No.1, pp:131-136.

[5] Martinetz T., Schulten K. (1990) Hierarchical Neural Net for Learning Control of a Robot's Arm and Gripper. Proceedings of the Int. Joint Conf. on Neural Networks '90, IJCNN-90, San Diego 1990, (in press).

[6] Mel B.W. (1987) A Robot that Learns by Doing. AIP Proceedings 1987, Neural Information Processing System Conf., Denver, CO, 1987

[7] Miller W. T. (1989) Real-Time Application of Neural Networks for Sensor-Based Control of Robots with Vision. IEEE Transactions on Systems, Man, and Cybernetics, Vol.19, No.4, pp 825-831.

[8] Ritter H., Martinetz T., Schulten K. (1988) Topology-Conserving Maps for Learning Visuomotor-Coordination. Neural Networks 2, pp. 159-168.

[9] Kohonen T. (1982) Self-organized Formation of Topologically Correct Feature Maps. Biological Cybernetics 43:59-69.

[10] Kohonen T. (1984) Self-Organization and Associative Memory. Spinger Series in Information Sciences 8, 3rd. ed., Springer, Heidelberg 1984.

[11] Widrow B., Hoff M.E. (1960) Adaptive switching circuits, WESCON Convention Record, part IV pp. 96-104.

A Dynamically Redundant Parallel Robot Arm

Theodore Kokkinis and Pierre Millies
Department of Mechanical and Environmental Engineering
and Center for Robotic Systems
University of California, Santa Barbara, CA 93106, U.S.A.

Abstract

A dynamically redundant in-parallel-actuated robot arm is considered for fine motion and high payload applications. The arm has a fully parallel kinematic structure with exclusively revolute joints and two closed loops. Redundant ground-based actuation is provided with compact two-actuator drives located at the roots of the three legs of the linkage. Closed-form expressions for the inverse and differential kinematics of the arm are shown. The kinetostatic performance of the arm is evaluated and compared to that of a similar, but not dynamically redundant arm.

INTRODUCTION

In 1942 Willard Pollard was awarded a patent for what he called a "position controlling apparatus," proposing to use it for spray painting of automobile bodies. The apparatus was perhaps the first ever robot arm; based on a closed-chain kinematic structure, it had fully parallel actuation of its regional structure. We have no evidence that the arm was constructed or operated; it appears today that its control system, based on the technology of the era, would have been its most serious problem. The first working robot arms that were introduced in the 60's and 70's [1] had serial kinematic structures. Interest in closed-chain kinematic structures for robot arms was revived only in the 80's. The planar 5-bar linkage [2] was the structure that initially received most of the attention, along with the Stewart platform mechanism [3]. Hunt [4] studied the structural kinematics of fully parallel multi-DOF closed-chain linkages.

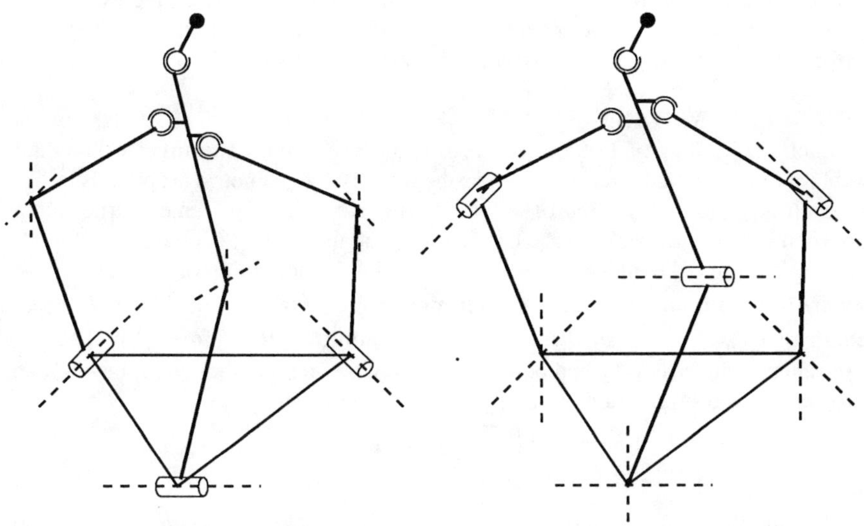

Figure 1: 3DOF Robot Arm Regional Structures

Earl and Rooney [5] considered the synthesis of closed-chain linkages from actuation and distribution components. Several closed-chain arms have been considered in the last 5 years ([6],[7],[8],[9]). In an earlier work [10] we studied the kinematics of the linkage of Pollard's arm, shown in Fig. 1 (left). Its regional structure consists of three legs, each of which is an RU chain (R means revolute joint, U universal), connected with 2 spherical (S) joints. This linkage is a member of a family of 7 fully parallel 3DOF linkages with revolute joints. Its kinematic structure, shown in Fig. 2 using interchange graph symbolism, is obtained from the general 2-loop 15R linkage by coalescing joint freedoms to form U and S joints. This coalescence can occur in a number of ways; another possibility is also shown in Fig. 2. This is the structure of the linkage shown in Fig. 1 (right), which is the subject of the present work.

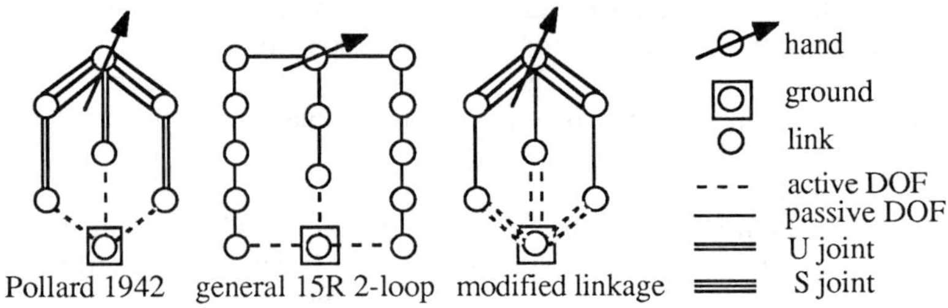

Figure 2: Kinematic structure of 3DOF linkages

Pollard's arm had three ground-mounted motors. To make it actuationally redundant, we would considerably increase the inertia of the arm as the additional actuators would have to drive joints remote from the ground. The modified linkage can be actuated redundantly without significant inertia increase. The U joints at the roots of the legs are actuated either by using a compact arrangement of two motors with intersecting axes or a differential drive. We consider this modified linkage as a parallel redundantly actuated regional structure for a robot arm. We evaluate its potential for fine motion and severe payload applications by studying its kinetostatic performance, and comparing it to a similar nonredundant arm.

KINEMATIC MODEL

Each of the legs of Pollard's linkage is an RU chain. The inverse kinematic problem can be solved in closed form as the underlying geometric problem is to find, for each leg, the intersection of a sphere with a circle. The same is true for the modified linkage, whose legs are UR chains. The underlying geometric problem in this case is to find the intersection of a plane with the intersection of two spheres, one centered at the root, the other at the terminal point. For any leg i, let ℓ_1^i, ℓ_2^i be the length of its two links, $\theta_1^i, \theta_2^i, \theta_3^i$ be the three joint angles, and ${}^{i0}x, {}^{i0}y, {}^{i0}z$ be the cartesian coordinates of the end point in a frame with origin at the root of the leg. The zero-reference configuration of the leg is fully extended along the x axis. Then:

$$\begin{aligned} \cos(\theta_1^i)\cos(\theta_2^i+\theta_3^i)\ell_2^i + \cos(\theta_1^i)\cos(\theta_2^i)\ell_1^i &= {}^{i0}x \\ \sin(\theta_1^i)\cos(\theta_2^i+\theta_3^i)\ell_2^i + \sin(\theta_1^i)\cos(\theta_2^i)\ell_1^i &= {}^{i0}y \\ \sin(\theta_2^i+\theta_3^i)\ell_2^i + \sin(\theta_2^i)\ell_1^i &= {}^{i0}z \end{aligned} \tag{1}$$

It is easy to solve (1) for θ_3^i:

$$\cos(\theta_3^i) = {}^{i0}x^2 + {}^{i0}y^2 + {}^{i0}z^2 - (\ell_1^i)^2 - (\ell_2^i)^2 / 2\ell_1^i\ell_2^i,\ \theta_3^i = \tan^{-1}\left(\pm\sqrt{1-c^2(\theta_3^i)} / c(\theta_3^i)\right) \quad (2)$$

Then θ_1^i (defining the plane) is easily found from (1):

$$\theta_1^i = \arctan\left({}^{i0}y / {}^{i0}x\right) \quad (3)$$

Finally θ_2^i (defining the intersection of the two spheres) is found from the first two equations in (1), observing that with θ_1^i, θ_3^i known they constitute a linear system of equations in $\sin(\theta_2^i), \cos(\theta_2^i)$. There are one or two solutions if:

$$\left|({}^{i0}x^2 + {}^{i0}y^2 + {}^{i0}z^2 - (\ell_1^i)^2 - (\ell_2^i)^2) / 2\ell_1^i\ell_2^i\right| \geq 0 \quad (4)$$

(the spheres intersect or are tangent). An infinite number of solutions is found for θ_2^i if $\ell_1^i = \ell_2^i$ and $\theta_3^i = \pi$, and θ_1^i is undetermined. This means that when the terminal point of a leg with two equal links is at its root, the universal joint is free to move without displacing the terminal point.

Given the position of the terminal point the joint angles are found first for the center leg (the one carrying the hand); from those the cartesian coordinates of the centers of the two S joints are found and are used for finding the joint angles of the side legs. There may be up to $2^3 = 8$ solutions, corresponding to different modes of the linkage. In contrast the inverse kinematics for Pollard's linkage has up to $4^3 = 64$ solutions. The direct kinematic problem has no meaning for a redundantly actuated arm; for a nonredundant arm the problem reduces to finding the intersection of three tori, and therefore has no closed form solution.

DIFFERENTIAL MOTIONS

For simplicity we take a symmetric linkage with identical legs stemming out from the vertices of an equilateral base triangle, and meeting at the terminal point of the arm. A world frame is put at the center of the base triangle, with the y axis pointing in the direction of one of the vertices, and the z axis normal to the plane of the base. The differential kinematics for leg i (i=1,2,3) are given by $\dot{\mathbf{x}}_i = \mathbf{J}_i\dot{\Theta}_i$ with jacobian:

$$\mathbf{J}_i = \quad (5)$$

$$\begin{bmatrix} -s(\theta_1^i+\phi_i)[c(\theta_2^i+\theta_3^i)\ell_2^i + c(\theta_2^i)\ell_1^i] & -c(\theta_1^i+\phi_i)[s(\theta_2^i+\theta_3^i)\ell_2^i + s(\theta_2^i)\ell_1^i] & -c(\theta_1^i+\phi_i)s(\theta_2^i+\theta_3^i)\ell_2^i \\ c(\theta_1^i+\phi_i)[c(\theta_2^i+\theta_3^i)\ell_2^i + c(\theta_2^i)\ell_1^i] & -s(\theta_1^i+\phi_i)[s(\theta_2^i+\theta_3^i)\ell_2^i + s(\theta_2^i)\ell_1^i] & -s(\theta_1^i+\phi_i)s(\theta_2^i+\theta_3^i)\ell_2^i \\ 0 & c(\theta_2^i+\theta_3^i)\ell_2^i + c(\theta_2^i)\ell_1^i & c(\theta_2^i+\theta_3^i)\ell_2^i \end{bmatrix}$$

The angles ϕ *are* $-90°, 30°, 150°$ for the three legs. The differential constraints imposed by the connection of the legs with S joints centered at the terminal point are:

$$\begin{bmatrix} \dot{\mathbf{x}}_1 \\ \dot{\mathbf{x}}_2 \\ \dot{\mathbf{x}}_3 \end{bmatrix} = \mathbf{J}\begin{bmatrix} \dot{\Theta}_1 \\ \dot{\Theta}_2 \\ \dot{\Theta}_3 \end{bmatrix} = \begin{bmatrix} \mathbf{I} \\ \mathbf{I} \\ \mathbf{I} \end{bmatrix}\dot{\mathbf{x}}_T,\ \mathbf{J} = \begin{bmatrix} \mathbf{J}_1 & 0 & 0 \\ 0 & \mathbf{J}_2 & 0 \\ 0 & 0 & \mathbf{J}_3 \end{bmatrix} \quad (6)$$

where $\dot{\mathbf{x}}_T$ is the velocity of the terminal point. Notice that this is an implicit expression of the constraints; $[\mathbf{I} \;\; \mathbf{I} \;\; \mathbf{I}]^T$ is a matrix whose columns are basis vectors of the null space of the constraint mapping. If none of the legs is in a singular configuration, this equation may be solved for $\dot{\Theta} = \left[\dot{\Theta}_1^T \;\; \dot{\Theta}_2^T \;\; \dot{\Theta}_3^T\right]^T$, the joint velocity vector of the arm:

$$\dot{\Theta} = \mathbf{K}\dot{\mathbf{x}}_T, \qquad \mathbf{K} = \mathbf{J}^{-1}[\mathbf{I} \;\; \mathbf{I} \;\; \mathbf{I}]^T \quad (7)$$

These are the velocities induced to the joints of the arm by $\dot{\mathbf{x}}_T$, and are consistent with the constraints. **K**, the jacobian of the inverse kinematic map has rank ≤ 3. If any of the legs is singular, the arm is at a stationary singularity at the boundary of the workspace. $\mathbf{K}_a$, the jacobian of the inverse kinematic map to actuated joint velocities consists of the rows of **K** corresponding to actuated joints:

$$\dot{\Theta}_a = \mathbf{K}_a \dot{\mathbf{x}}_T \quad (8)$$

Rank deficiency (<3) of $\mathbf{K}_a$, signifies uncertainty singularities of the arm.

KINEMATIC PERFORMANCE

The kinematic performance of the redundantly actuated arm is evaluated on the basis of the ratio γ of the minimum to the maximum singular value of $\mathbf{K}_a$. Both axes of the U joints at the root of the legs are actuated, resulting in a total of 6 actuators for 3 DOF. The accuracy of positioning the terminal point in fine motion is expected to be best when γ is one, degrading as γ approaches zero. The value zero is taken at stationary singularities (max singular value goes to ∞) and at uncertainty singularities (min singular value goes to 0). The variation of γ over a typical vertical cross section of the workspace is shown in Fig. 3; because of symmetry only one quarter of the cross section is shown; the size of the dot indicates the magnitude of γ at that point, with the smallest dots corresponding to values of γ under 0.01.

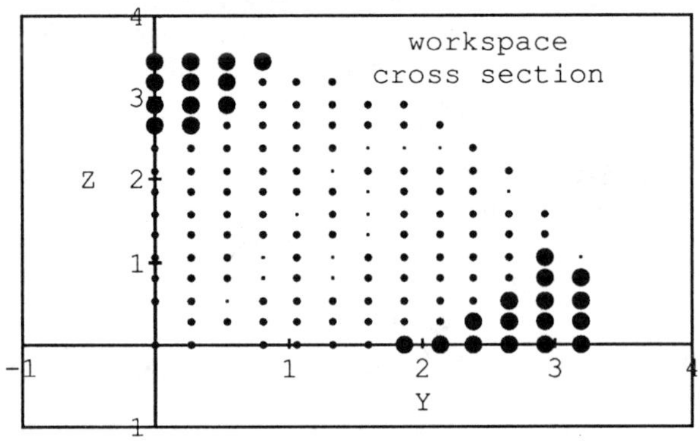

Figure 3: Kinematic Performance of Redundant Arm

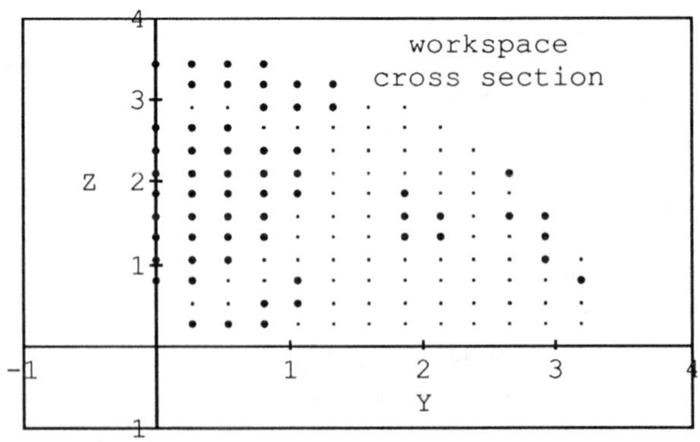

Figure 4: Kinematic Performance of Nonredundant Arm

γ is also shown for a similar nonredundant arm in Fig.4, moved by 3 actuators, each actuating the horizontal axis of the U joints at the root of the legs. If the actuators were made to actuate the vertical axis of the U joints, $\mathbf{K}_a$ (in this case a 3x3 matrix) is everywhere singular, and the vertical position of the terminal point is unobservable. Comparison of these results for the two arms shows that the redundant arm is superior, even if not outstanding, especially near the boundary of the workspace.

STATIC PERFORMANCE

Let $\mathbf{T}_a, \mathbf{F}_T$ respectively be the actuated joint torque vector and the external force applied to the terminal point of the arm. Using duality of velocity and force and the fact that the torques of the unactuated joints are zero we can show that:

$$\mathbf{F}_T = K_a^T \mathbf{T}_a \tag{9}$$

For a nonredundant arm in a nonsingular configuration the vector of joint torques may be uniquely determined from (9) as:

$$\mathbf{T}_a = K_a^{-T} \mathbf{F}_T \tag{10}$$

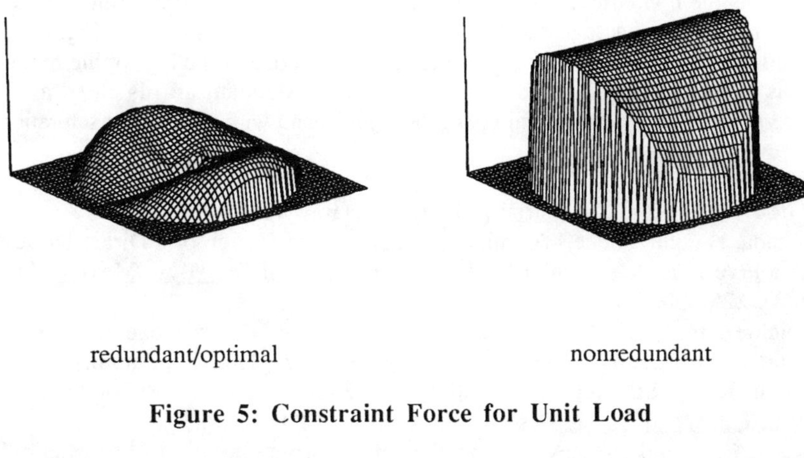

redundant/optimal nonredundant

Figure 5: Constraint Force for Unit Load

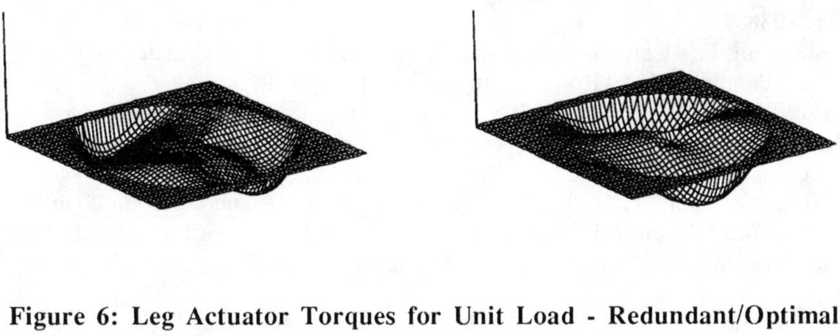

Figure 6: Leg Actuator Torques for Unit Load - Redundant/Optimal

Figure 7: Leg Actuator Torque for Unit Load - Nonredundant

In the redundant case there is no unique solution for $\mathbf{T}_a$ for given $\mathbf{F}_T$. This, however, is an advantage of redundant actuation: we may select the solution that optimizes a given criterion. It is sometimes of interest to minimize the constraint forces, to avoid having the legs fighting each other instead of resisting the load. Fig. 5 shows the magnitude of the vector of constraint forces in the S joints for a horizontal unit load applied at the terminal point, for the redundant arm (after optimal load distribution), as well as the nonredundant arm, over a horizontal cross section of the workspace. The corresponding actuator torques for one of the legs are shown in Fig. 6 for the redundant arm. The torque for the single actuator at one of the legs of the redundant arm is shown in Fig. 7. The superiority of the redundant arm is clear, as it can support higher loads without failure of the S joints and with no actuator saturation.

REFERENCES

1. Groover, M., et al., Industrial Robotics, McGraw-Hill, 1986.
2. Asada, H., and Youcef-Toumi, K., "Analysis and Design of a Direct-Drive Arm with a Five-Bar-Link Parallel Drive Mechanism," J. of Dyn. Sys., Meas. and Cont., 106(3): 225-230, Sept. 1984.
3. Fichter, E.F., "A Stewart-Platform-Based Manipulator: General Theory and Practical Construction," The Int. J. of Rob. Res., 5(2): 157-182, Summ. 1986.
4. Hunt, K.H., "Structural Kinematics of In-Parallel-Actuated Robot Arms," J. of Mech., Trans. and Aut. in Design, 105(4): 705-712, Dec. 1983.
5. Earl, C.F., and Rooney, "Some Kinematic Structures for Robot Manipulator Designs," J. of Mech., Trans. and Aut. in Design, 105(1): 15-22, Mar. 1983.
6. Gosselin, C., and Angeles, J., "The Optimum Kinematic Design of a Planar 3-DOF Parallel Manipulator," J. of Mech., Trans. and Aut. in Design, 110(1): 35-41, Mar. 1988.
7. Sterneim, F., "Computation of the Direct and Inverse Kinematic Models of the Delta-4 Parallel Robot," Robotersysteme, 3:199-203, 1987.
8. Burkhardt, C.W., and Ming, L.S., "Scaling Rules in Robotics and the Design of a Mini-Robot," Int. Symp. on Aut. and Rob., Xi'an, China, May 1988.
9. Kokkinis, T., and Stoughton, R.S., "Optimal Parallel Actuation Linkage for 3-DOF Elbow Manipulators," 20th ASME Mech. Conf., Orlando, Florida, Sept. 1988.
10. Kokkinis, T., and Millies, P., "Kinematics of Linkages for Fully Parallel Robot Arms," 1st Nat. Conf. on Applied Mech. and Rob., Cincinatti, Ohio, Nov. 1989.

EFFICIENT KINEMATIC COMPUTATIONS FOR 7 DOF MANIPULATORS

M. Long*, K. Kreutz-Delgado†, H. Seraji*

* Jet Propulsion Laboratory
California Institute of Technology
4800 Oak Grove Drive
Pasadena, CA 91109

† University of California at San Diego
AMES Department, R-011
La Jolla, CA 92093

ABSTRACT

This paper presents kinematic algorithms for anthropomorphic seven degree-of-freedom serial link spatial manipulators with revolute joints. To uniquely determine joint angles for a given end-effector position and orientation, the redundancy is parameterized by a scalar variable which corresponds to the angle between the arm plane and a reference plane. Efficient algorithms are given for the forward kinematic mappings from joint-space to end-effector coordinates and arm angle, and the augmented Jacobian matrix which gives end-effector and arm angle rates as functions of joint rates. These algorithms are discussed from the computational viewpoint.

1. INTRODUCTION

Robot manipulators which have more joint degrees-of-freedom (DOF) than the minimum number needed to perform some end-effector tasks of interest are referred to as "redundant" to indicate the existence of the excess joint degrees-of-freedom. Although redundancy is obviously a task-dependent concept, manipulators with greater than six DOF are usually called redundant since the classical problem of end-effector position and orientation control for a "spatial" manipulator can be handled by a six DOF robot arm [1,2]. Redundancy can be exploited for a variety of applications — singularity avoidance, collision avoidance, enhancement of mechanical advantage, manipulability enhancement, subtask performance, and so on — which greatly increase the flexibility and use of redundant robot arms.

The Robotics Research arms form a family of commercially available 7 DOF revolute joint serial link manipulators which offer one degree of joint-space redundancy over that needed for the fundamental task of end-effector placement and orientation. In this paper, a scalar parameterization, ψ, is given of the redundancy. In particular, for the Robotics Research arm, ψ is defined as the "arm angle" (a natural redundancy parameter for anthropomorphic arms [1]), which is the angle between the plane passing through the arm and a reference plane. The forward kinematic mappings from joint-space to end-effector coordinates and ψ are then given. We also present the augmented Jacobian, J^A, which gives end-effector rates and $\dot{\psi}$ as a function of joint rates. Although in this paper a particular emphasis is placed on the Robotics Research arms (in particular, the model K-1207 arm) and the use of the arm angle for resolving redundancy, many of the results developed here apply equally to any arbitrary 7 DOF manipulator with the use of *any* appropriate redundancy resolving scalar parameter, and not necessarily the arm angle.

Due to the displacement of joint axes ("non-zero joint offsets"), the Robotics Research arms have no known analytic closed-form inverse kinematic solutions for specified end-effector coordinates and redundancy parameter ψ. Consequently, the importance of differential kinematics, i.e. of the Jacobian, is increased since at this time most approaches to solving the inverse kinematics, and controlling the end-effector motions, of nonsolvable arms are based on differential methods. For example, a "resolved-rate" (or "inverse Jacobian") kinematic control approach used for non-redundant arms [3], can be extended for general task-space control as well as to find joint angles for given end-effector coordinates and ψ. Similarly, an augmented "Jacobian-transpose" approach can be used for control [4]. Alternatively, end-effector motions can be controlled using pseudoinverse techniques [5] which are based on the use of the pseudoinverse of the end-effector Jacobian rather than on using the inverse (or transpose) of the augmented Jacobian.

Since kinematic calculations are often performed at high rates (typically 40-200Hz), it is worthwhile to investigate efficient kinematic algorithms. These algorithms are even more important in controlling redundant manipulators because of additional calculations required to resolve the redundancy. Due to the large amount of overlap in the various kinematic quantities mentioned above, it is generally not efficient to derive very complex symbolic expressions for each quantity we wish to compute and calculate them separately. Also, complex symbolic expressions do not generally exploit the special structure associated with many kinematic calculations.

Because of space limitations, the discussions in this paper are necessarily brief and derivations are omitted. A complete description containing all the derivations may be found in [6].

2. FORWARD KINEMATICS

2.1. Mapping from Joint-Space to End-Effector Coordinates

The Robotics Research model K-1207 arm is a 7 DOF manipulator with nonzero offsets (denoted by the nonzero link lengths a_i, $i = 1, \cdots, 6$) at each of the joints, as shown in Figure 1. Denavit-Hartenberg (D-H) link frame assignments are given in accordance with the convention described in [7]. The D-H parameters for the K-1207 arm are given in Table 1. If the link length parameters a_i, $i = 1, \cdots, 6$ are set to zero, the 7 DOF all-revolute anthropomorphic arm described in [1] is retrieved and we call this arm the "zero-offset" arm. Rather than construct and implement the symbolic expression, it is more efficient to numerically compute the forward kinematic function ${}^0T_7$ via a link-by-link iteration of the form

$$ {}^0T_i = {}^0T_{i-1} \cdots {}^{i-1}T_i\,, \quad i = 1, \cdots, 7 \tag{2.1} $$

exploiting special structural properties of the homogeneous transformation matrices during each link update [8,9]. The link $i-1$ to i homogeneous transformation ${}^{i-1}T_i$ is given by [7]:

$$ {}^{i-1}T_i = \begin{bmatrix} {}^{i-1}\mathbf{X}_i & {}^{i-1}\mathbf{Y}_i & {}^{i-1}\mathbf{Z}_i & {}^{i-1}\mathbf{P}_{i-1,i} \\ 0 & 0 & 0 & 1 \end{bmatrix} \tag{2.2} $$

$$ = \begin{bmatrix} C_i & -S_i & 0 & a_{i-1} \\ S_iC\alpha_{i-1} & C_iC\alpha_{i-1} & -S\alpha_{i-1} & -S\alpha_{i-1}d_i \\ S_iS\alpha_{i-1} & C_iS\alpha_{i-1} & C\alpha_{i-1} & C\alpha_{i-1}d_i \\ 0 & 0 & 0 & 1 \end{bmatrix} \tag{2.3} $$

where C_i, $S_i = \cos(\theta_i)$, $\sin(\theta_i)$ and $C\alpha_{i-1}$, $S\alpha_{i-1} = \cos(\alpha_{i-1})$, $\sin(\alpha_{i-1})$ respectively. Important quantities (such as the vectors $\mathbf{w}$, $\mathbf{e}$, and $\mathbf{p}$ defined later) are often direct by-products of the intermediate steps of the iteration (2.1). With respect to the non-trivial elements of 2.1 and 2.3, the iteration 2.1 can be restated as:

$$ \begin{aligned} {}^0\mathbf{X}_i &= {}^0\mathbf{X}_{i-1} \cdot C_i + S_i({}^0\mathbf{Y}_{i-1}C\alpha_{i-1} + {}^0\mathbf{Z}_{i-1}S\alpha_{i-1} \\ {}^0\mathbf{Z}_i &= -{}^0\mathbf{Y}_{i-1}S\alpha_{i-1} + {}^0\mathbf{Z}_{i-1}C\alpha_{i-1} \\ {}^0\mathbf{Y}_i &= {}^0\mathbf{Z}_i \times {}^0\mathbf{X}_i \\ {}^0\mathbf{P}_{i-1,i} &= a_{i-1}{}^0\mathbf{X}_{i-1} + d_i\,{}^0\mathbf{Z}_i \end{aligned} \tag{2.4} $$

where the vector ${}^c\mathbf{P}_{a,b}$ is the position vector from the origin of frame a to the origin of frame b written in the c coordinate system.

Most manipulators today are designed to have link twists, α, that are 0° or ±90°. This increases the efficiency of algorithm (2.4) which simplifies significantly. For example, when

$\alpha = 90°$, (2.4) becomes:

$$\begin{aligned} {}^0\mathbf{X}_i &= {}^0\mathbf{X}_{i-1} \cdot C_i - {}^0\mathbf{Z}_{i-1} S_i \\ {}^0\mathbf{Z}_i &= {}^0\mathbf{Y}_{i-1} \\ {}^0\mathbf{Y}_i &= {}^0\mathbf{Z}_i \times {}^0\mathbf{X}_i \\ {}^0\mathbf{P}_{i-1,i} &= a_{i-1}{}^0\mathbf{X}_{i-1} + d_i\, {}^0\mathbf{Z}_i \end{aligned} \tag{2.5}$$

The iteration in (2.4) or (2.5) is most efficiently implemented not in a loop, but explicitly so that multiplications by zero (due to sparse vectors in the first few links, zero link lengths, or zero offsets) are avoided. The intermediate results of the forward kinematics iteration are of direct use in subsequent computations such as calculation of the kinematic redundancy resolution parameter (to be discussed shortly), obstacle avoidance calculations, gravity compensation calculations (link position vectors given from forward kinematics), etc.

2.2. Mapping from Joint-Space to Arm Angle

When the arm is in a kinematically nonsingular configuration, there will generally exist one excess joint degree-of-freedom for the task of end-effector control since there are seven joint angles available to position and orient the end-effector — a task which requires only six degrees-of-freedom. As a result, for a fixed end-effector frame, there is generally a one-dimensional subset of joint space (a "self-motion") which maps to this configuration. The extra degree-of-freedom represented by a self-motion can be used to attain some additional task requirement, provided that this task can be performed independently of end-effector placement [4,10]. Furthermore, the imposition of an auxillary task constraint can provide sufficient additional information to uniquely determine all the joint angles (to within the multiplicity of distinct solutions represented by the pose [3,4]). We say that the "basic task" of end-effector placement has been *augmented* by the "additional task," which will be here represented by ψ. In essence, the concept of the forward kinematic map is generalized to be the mapping from $\theta \in R^7$ to $({}^0T_7, \psi)$.

Although ψ can be any additional scalar parameter which is independent of end-effector frame, we define and use the "arm angle" to resolve the manipulator redundancy. Refer to Figures 1 and 2 where $S = O_1$, $E = O_4$, and $W = O_7$ denote the origins of link frames 1, 4, and 7 attached to the shoulder, elbow and wrist respectively. The arm angle ψ is defined by the angle from the reference plane containing any unit vector $\hat{\mathbf{V}}$ and the shoulder-wrist line SW to the shoulder-elbow-wrist plane SEW in the right-hand sense about the vector $\mathbf{w} = W - S$. With the arm angle ψ as a parameterization of manipulator redundancy, a self-motion is described by a rotation of the plane SEW about the line SW. Note that the arm angle ψ is undefined when the wrist point W is anywhere on the line through the shoulder point S and containing $\hat{\mathbf{V}}$ — even though this is generally not a singular configuration — since in this case the reference plane is not uniquely defined. The angle ψ is also undefined when $\mathbf{e}$ and $\mathbf{w}$ are colinear since then the plane SEW is not uniquely defined. In the latter case, the arm is either nearly fully outstretched, or folded, and is therefore near or at an "elbow singular" configuration [2,11]. In such cases, ψ ceases to parameterize the redundancy and an alternative must be chosen.

The forward kinematic function which maps the joint angles θ to the arm angle ψ is [6]:

$$\psi = \operatorname{atan2}(\widehat{\mathbf{w}} \cdot (\widehat{\mathbf{V}} \times \mathbf{p}), \widehat{\mathbf{V}} \cdot \mathbf{p}) \tag{2.6}$$

where $\mathbf{w} = W - S = {}^0\mathbf{P}_{0,7}$, $\mathbf{e} = E - S = {}^0\mathbf{P}_{0,4}$, and $\widehat{\mathbf{V}}$ denotes the fixed reference unit vector. In this paper, $\widehat{\mathbf{V}}$ is chosen in the shoulder frame along the z axis and when represented in the shoulder frame $\widehat{\mathbf{V}} = \hat{\mathbf{u}}_3 = (0,0,1)^T$. The projection of $\mathbf{e}$ onto $\mathbf{w}$ is given by $\mathbf{d} = \widehat{\mathbf{w}}(\widehat{\mathbf{w}}^T\mathbf{e})$, $\widehat{\mathbf{w}} = \mathbf{w}/||\mathbf{w}||$. The minimum distance from the line SW to the point E is along the vector $\mathbf{p} = \mathbf{e} - \mathbf{d}$. The reference plane is the plane which contains both $\mathbf{w}$ and the unit vector $\widehat{\mathbf{V}}$. The unit vector in the reference plane which is orthogonal to $\mathbf{w}$ is given by $\widehat{\mathbf{h}} = \mathbf{h}/||\mathbf{h}||$, with $\mathbf{h} = (\mathbf{w} \times \widehat{\mathbf{V}}) \times \mathbf{w}$. We also define the unit vector $\widehat{\mathbf{p}} = \mathbf{p}/||\mathbf{p}||$. The augmented forward kinematics mapping, $\theta \to ({}^0T_7, \psi)$, is then given by (2.1) and (2.6).

The calculation of ψ from (2.6) is straightforward. For example, if ${}^0J^\psi$ is to be calculated subsequently, use:

$$\hat{\mathbf{w}} \cdot (\hat{\mathbf{V}} \times \mathbf{p}) = -(\hat{\mathbf{w}} \times \mathbf{p}) \cdot \hat{\mathbf{V}} = -({}^0\hat{\mathbf{w}} \times {}^0\mathbf{p})^T \hat{\mathbf{u}}_3 = -({}^0\hat{\mathbf{w}} \times {}^0\mathbf{p})_3 \tag{2.7}$$

in computing ψ (since the entire vector $\hat{\mathbf{w}} \times \mathbf{p}$ is needed in the ${}^0J^\psi$ calculation as shown below), where $\mathbf{a}_3 = \mathbf{a} \cdot \hat{\mathbf{u}}_3$ and $\hat{\mathbf{u}}_3 = (0,0,1)^T$. Otherwise, it is more efficient to directly compute the third component of $\hat{\mathbf{w}} \times \mathbf{p}$.

$$\hat{\mathbf{w}} \cdot (\hat{\mathbf{V}} \times \mathbf{p}) = ({}^0\hat{\mathbf{w}} \times {}^0p)_3 = {}^0\hat{\mathbf{w}}_2\, {}^0p_1 - {}^0\mathbf{w}_1\, {}^0p_2 \tag{2.8}$$

Depending on the task application, a variety of other kinematic functions may be defined for the additional task. Again, many of these kinematic functions can be computed easily as by-products of iteration (2.5).

For the special case of the zero-offset arm discussed in [1], corresponding to $a_i = 0, i = 1, \ldots, 6$, the position vector for the elbow is found to be

$${}^0\mathbf{e} = {}^0\mathbf{P}_{0,4} = d_3 \begin{bmatrix} C_1 S_2 \\ S_1 S_2 \\ C_2 \end{bmatrix} \tag{2.9}$$

and the arm angle (2.6) reduces to

$$\psi = \operatorname{atan2}\left(S_2 S_3 S_4,\ \frac{d_5 S_4}{||\mathbf{w}||} [C_2 S_4 + S_2 C_3 (1 + C_4)] \right) \tag{2.10}$$

where $||\mathbf{w}|| = (d_3^2 + d_5^2 + 2 d_3 d_5 C_4)^{1/2}$.

3. DIFFERENTIAL KINEMATICS

3.1. Manipulator End-Effector Jacobian, J^{ee}

To present actual values for the end-effector Jacobian, J^{ee}, it is first necessary to choose a velocity reference point, as well as a frame in which to represent the vectorial quantities which define the columns of the Jacobian. In this section, to simplify notation, we will suppress the trailing superscript and write the end-effector Jacobian simply as $J = J^{ee}$. When a velocity reference point, a, and a representation frame, $\mathcal{F}_r$, have been chosen, we write ${}^rJ_a = {}^rJ_a^{ee}$.

Let ω_a and v_a be the angular and linear velocities of a coordinate frame, $\mathcal{F}_a$, located at a point a and fixed with respect to the manipulator end-effector. The point a is known as a "velocity reference point" of the end-effector. The Jacobian, $J_a(\theta) \in R^{6\times 7}$, relates joint rates to the rate of change of frame $\mathcal{F}_a$ via the linear relationship $(\omega_a^T, v_a^T)^T = J_a(\theta)\dot{\theta}$ and is given by [12]

$$J_a = \begin{pmatrix} \hat{\mathbf{z}}_1 & \cdots & \hat{\mathbf{z}}_7 \\ \hat{\mathbf{z}}_1 \times \mathbf{P}_{1,a} & \cdots & \hat{\mathbf{z}}_7 \times \mathbf{P}_{7,a} \end{pmatrix} \tag{3.1}$$

In (3.1), $\hat{\mathbf{z}}_i$ denotes the unit vector corresponding to the z-axis of link frame i (i.e. of $\mathcal{F}_i$). Note that $\mathbf{P}_{i,i} = 0$.

Let $\mathcal{F}_b$ denote an alternative frame fixed with respect to the end-effector and located at the velocity reference point b. Let $\mathcal{F}_r$ and $\mathcal{F}_s$ be frames which are *not* necessarily fixed with respect to the end-effector. Note that we have defined a and b to be end-effector reference points, i.e. to be *fixed* with respect to the end–effector, while we have placed no constraints on r and s. The Jacobian, ${}^rJ_a$, giving the rate of change of $\mathcal{F}_a$ represented in $\mathcal{F}_r$, is related to ${}^sJ_b$, the Jacobian giving the rate of change of frame $\mathcal{F}_b$ represented in frame $\mathcal{F}_s$, by [8]

$${}^rJ_a = \begin{pmatrix} {}^rR_s & 0 \\ 0 & {}^rR_s \end{pmatrix} \cdot \begin{pmatrix} I & 0 \\ {}^s\tilde{P}_{a,b} & I \end{pmatrix} \cdot {}^sJ_b \tag{3.2}$$

where, for a 3-vector $\mathbf{x}$, $\tilde{x}$ denotes the 3×3 skew symmetric matrix defined by $\tilde{x}\mathbf{y} = \mathbf{x} \times \mathbf{y}$ for every $\mathbf{y} \in R^3$.

An important aspect of the decomposition (3.2) is that s and b can sometimes be chosen to make the Jacobian matrix have a particularly simple structure for the purposes of singularity analysis and efficient inversion [8]. The computation of

$$ {}^0J_7 = \begin{pmatrix} {}^0\hat{\mathbf{z}}_1 & {}^0\hat{\mathbf{z}}_2 & \cdots & {}^0\hat{\mathbf{z}}_7 \\ {}^0\hat{\mathbf{z}}_1 \times {}^0\mathbf{P}_{1,7} & {}^0\hat{\mathbf{z}}_2 \times {}^0\mathbf{P}_{2,7} & \cdots & {}^0\hat{\mathbf{z}}_7 \times {}^0\mathbf{P}_{7,7} \end{pmatrix} \tag{3.3} $$

naturally fits in with the forward kinematics iteration (2.5), since ${}^0\hat{\mathbf{z}}_i$ and ${}^0\mathbf{P}_{i-1,i}$ are directly available.

The position vectors required for any Jacobian computation (3.3) can be easily found from the vectors ${}^0\mathbf{P}_{i-1,i}$ $(i = 2 \ldots 7)$ found in iteration (2.5). For example, if we wish to compute ${}^0J_7$, the position vectors ${}^0\mathbf{P}_{i,7}$ $(i = 1 \ldots 7)$ are needed which can easily and efficiently be computed by implementing the following recursive assignment statement, where computation begins at the innermost assignment.

$$ \begin{aligned} {}^0\mathbf{P}_{1,7} := \Bigg[{}^0\mathbf{P}_{2,7} := \bigg({}^0\mathbf{P}_{3,7} := \Big\{ {}^0\mathbf{P}_{4,7} := [{}^0\mathbf{P}_{5,7} := {}^0\mathbf{P}_{6,7} + {}^0\mathbf{P}_{5,6}] \\ + {}^0\mathbf{P}_{4,5} \Big\} + {}^0\mathbf{P}_{3,4} \bigg) + {}^0\mathbf{P}_{2,3} \Bigg] + {}^0\mathbf{P}_{1,2} \end{aligned} \tag{3.4} $$

where ${}^0\mathbf{P}_{7,7} = (0,0,0)^T$. Similarly, for the computation of ${}^0J_4$

$$ \begin{aligned} {}^0\mathbf{P}_{1,4} &:= ({}^0\mathbf{P}_{2,4} := {}^0\mathbf{P}_{2,3} + {}^0\mathbf{P}_{3,4}) + {}^0\mathbf{P}_{1,2} \\ {}^0\mathbf{P}_{7,4} &:= ({}^0\mathbf{P}_{6,4} := -{}^0\mathbf{P}_{4,5} + {}^0\mathbf{P}_{5,6}) - {}^0\mathbf{P}_{6,7} \end{aligned} \tag{3.5} $$

The results of the Jacobian calculations themselves may also be useful in subsequent computations. If computing Inertia Decoupling Matrices or other dynamic or energy quantities the columns of the Jacobian are easily seen to be equivalent to the partial linear and angular velocities found in [13] (for the serial manipulator with generalized coordinates defined as the joint angle positions).

3.2. Arm Angle Jacobian, J^ψ, and Augmented Jacobian, J^A

Let the relationship between the rate of change of *any* scalar additional task variable, ψ, and the joint rates be given by $\dot{\psi} = J^\psi \dot{\theta}$. The "augmented" Jacobian is given by [4]

$$ J^A = \begin{pmatrix} J^{ee} \\ J^\psi \end{pmatrix} \tag{3.6} $$

where J^{ee} is the end-effector Jacobian discussed in Section 3.1. For the task of positioning and orienting the end-effector *augmented* by an additional task represented by ψ, the augmented Jacobian relates joint rates to the simultaneous rates of change of the end-effector coordinates and ψ. Having the end-effector Jacobian, J^{ee}, the augmented Jacobian J^A is obtained once J^ψ has been determined for a given task variable ψ. In this section, J^ψ is constructed for the case where ψ describes the arm angle between a reference plane and the elbow plane SEW as defined in Section 2.

Before proceeding further, it is necessary to define the Jacobians $\mathbf{E}$ and $\mathbf{W}$ which relate joint rates to $\dot{\mathbf{e}}$ and $\dot{\mathbf{w}}$ respectively via $\dot{\mathbf{e}} = \mathbf{E}\dot{\theta}$ and $\dot{\mathbf{w}} = \mathbf{W}\dot{\theta}$, where $\mathbf{e}$ and $\mathbf{w}$ are defined in Section 2.2. Note that $\dot{\mathbf{e}}$ is the linear velocity of the manipulator elbow point $E = O_4$ and $\dot{\mathbf{w}}$ is the linear velocity of the wrist point $W = O_7$. We have

$$ \mathbf{E} = (\hat{\mathbf{z}}_1 \times \mathbf{P}_{1,4},\ \hat{\mathbf{z}}_2 \times \mathbf{P}_{2,4},\ \hat{\mathbf{z}}_3 \times \mathbf{P}_{3,4},\ 0,\ \cdots,\ 0) \tag{3.7} $$

$$\mathbf{W} = (\,\widehat{\mathbf{z}}_1 \times \mathbf{P}_{1,7}\,, \quad \cdots, \quad \widehat{\mathbf{z}}_6 \times \mathbf{P}_{6,7}\,, \quad 0\,) \tag{3.8}$$

Note that (3.7) and (3.8) are given in coordinate-free form and that to provide values for $\mathbf{E}$, or $\mathbf{W}$, a choice of reference frame for representing $\widehat{\mathbf{z}}_j$ and $\mathbf{P}_{i,j}$ must be made. Also note that in many cases (3.8) is identical to the last three rows of (3.3).

The relationship between $\dot{\theta}$ and $\dot{\psi}$, where ψ is the arm angle as defined in Section 2.2, is given by $\dot{\psi} = J^{\psi}\dot{\theta}$ where from [6]

$$J^{\psi} = \frac{(\widehat{\mathbf{w}} \times \mathbf{p})^T}{\|\mathbf{p}\|^2}\mathbf{E} + \left\{ \frac{\widehat{\mathbf{V}}^T\mathbf{w}}{\|\mathbf{h}\|^2}(\widehat{\mathbf{w}} \times \mathbf{h})^T - \frac{\widehat{\mathbf{w}}^T\mathbf{e}}{\|\mathbf{w}\|\,\|\mathbf{p}\|^2}(\widehat{\mathbf{w}} \times \mathbf{p})^T \right\}\mathbf{W} \tag{3.9}$$

Note that J^{ψ} is independent of the reference frame chosen to represent the quantities in the right-hand side of (3.9).

From the forward kinematic mapping algorithms given in Section 2, the quantities $\mathbf{w}$, $\mathbf{p}$, $\|\mathbf{w}\|$, $\hat{\mathbf{w}}$, $\hat{\mathbf{w}}^T\mathbf{e}$, and $(\hat{\mathbf{w}} \times \mathbf{p})$ are directly available making evaluation of J^{ψ} very efficient.

4. COMPUTATIONS

For the special case of the Robotics Research Model K-1207 manipulator, the implementation of these forward and differential kinematic algorithms has resulted in the following approximate computation time on a Sun Microsystems Model 3/60 and a Silicon Graphics IRIS 4D70 GT.

RRC-K1207 Kinematics Algorithms

	Operation Count	SUN 3/60	IRIS
Sine & Cosine of Joint Angles	14t	2.74ms	.13ms
Forward Kinematics	86m 47a	1.15ms	.10ms
Arm Angle ψ	18m 10a 1s 1t	.64ms	.034ms
End Effector Jacobian, ${}^0J_7$	36m 18a	.58ms	.064ms
Arm Angle Jacobian, J^{ψ}	72m 44a	.99ms	.097ms
	212m 119a 1s 15t	6.1	.42

where a, m, s, t refer to additions, multiplications, square root, and trigonometric operations respectively.

5. CONCLUSIONS

In this paper, for the 7 DOF Robotics Research arms and the related zero-offset manipulator of [1], we have given efficient algorithms for the forward kinematic functions which map from joint-space to end-effector coordinates and arm angle and their corresponding Jacobians. Future research will focus on computational issues associated with kinematic functions for other subtasks along with numerical detection and avoidance of singularities. Future analytical effort will be concerned with the relationship between singularities and pose (in the sense of [2]) and with the categorization of singularities into avoidable and non-avoidable singularities. Global solvability of the K-1207 arm is also an important research issue.

It must be pointed out that the results of this paper apply to robot arms used as spatial manipulators [2], i.e., for the case where the basic task is the spatial placement (position and orientation) of the end-effector in Cartesian space. The statements given here about the transformability of the end-effector Jacobian do not, in general, apply to common special cases such as a planar manipulator used only for positioning in the plane.

6. ACKNOWLEDGMENTS

This research was performed at the Jet Propulsion Laboratory, California Institute of Technology, under contract with the National Aeronautics and Space Administration. The suggested use of intermediate Jacobians and comments on the forward kinematics algorithm by Dr. Amir Fijany of JPL is gratefully acknowledged. Thanks are also due to Dr. Joel Burdick of Caltech for many fruitful and clarifying discussions.

7. REFERENCES

[1] J.M. Hollerbach, "Optimum Kinematic Design for a Seven Degree of Freedom Manipulator", 2nd *Int. Symp. Robotics Research*, Kyoto, Japan, Aug. 20–23, 1984, pp. 215–222.

[2] J.W. Burdick, "Kinematic Analysis and Design of Redundant Robot Mechanisms", Ph.D. Thesis, Dept. of ME, Stanford, March 1988.

[3] D.E. Whitney, "Resolved motion rate control of manipulators and human prostheses," IEEE Trans. Man-Machine Syst., Vol. MMS-10, No. 2, pp. 47-53, 1969.

[4] H. Seraji, "Configuration Control of Redundant Manipulators: Theory and Implementation", *IEEE Trans. Robotics and Automation*, Vol. 5, No. 4, pp. 472-490, 1989.

[5] A. Liegeois, "Automatic Supervisory Control of the Configuration and Behavior of Multibody Mechanisms," *IEEE Trans. Systems, Man, and Cybernetics*, Vol. SMC-7, pp. 868–871, 1977.

[6] K. Kreutz-Delgado, M. Long and H. Seraji, "Kinematic Analysis of 7 DOF Manipulators," JPL Engineering Memorandum 347-90-275, January 1990.

[7] J.J. Craig, *Introduction to Robotics*, Addison–Wesley, 1986.

[8] A. Fijany and A.K. Bejczy, "Efficient Jacobian Inversion for the Control of Simple Robot Manipulators", *1988 IEEE Int. Conf. Robotics and Automation*, Philadelphia, 1988, pp. 999–1007.

[9] D.E. Orin and W.W. Schrader, "Efficient Computation of the Jacobian for Robot Manipulators", *Int. J. Robotics Research*, 1984.

[10] O. Egeland, "Task–Space Tracking with Redundant Manipulators", *IEEE J. Robotics and Automation*, Vol. RA–3, 1987, pp. 471–475.

[11] C.W. Wampler, II, "Inverse Kinematics of a 7 DOF Manipulator", *NATO Advanced Research Workshop on Robots with Redundancy: Design, Sensing, and Control*, June 27–July 1, 1988, Salo, Salo di Garda, Italy.

[12] H. Asada and J.-J.E. Slotine, *Robot Analysis and Control*, Wiley–Interscience, 1986.

[13] T.R. Kane and D.A. Levinson, "Dynamics: Theory and Applications," *McGraw-Hill Book Company*, 1985

i	α_{i-1}	a_{i-1}	d_i	θ_i	
1	0°	0	0	θ_1	$a_1 = -4.850$ in $= -12.319$ cm
2	-90°	a_1	0	θ_2	$a_2 = +4.250$ in $= +10.795$ cm
3	+90°	a_2	d_3	θ_3	$a_3 = -3.125$ in $= -\ 7.938$ cm
4	-90°	a_3	0	θ_4	$a_4 = +3.125$ in $= +\ 7.938$ cm
5	+90°	a_4	d_5	θ_5	$a_5 = -1.937$ in $= -\ 4.920$ cm
6	-90°	a_5	0	θ_6	$a_6 = +1.937$ in $= +\ 4.920$ cm
7	+90°	a_6	0	θ_7	$d_3 = d_5 = 21.5$ in $= 54.61$ cm

TABLE 1: Denavit-Hartenberg Parameters of the K-1207 Arm

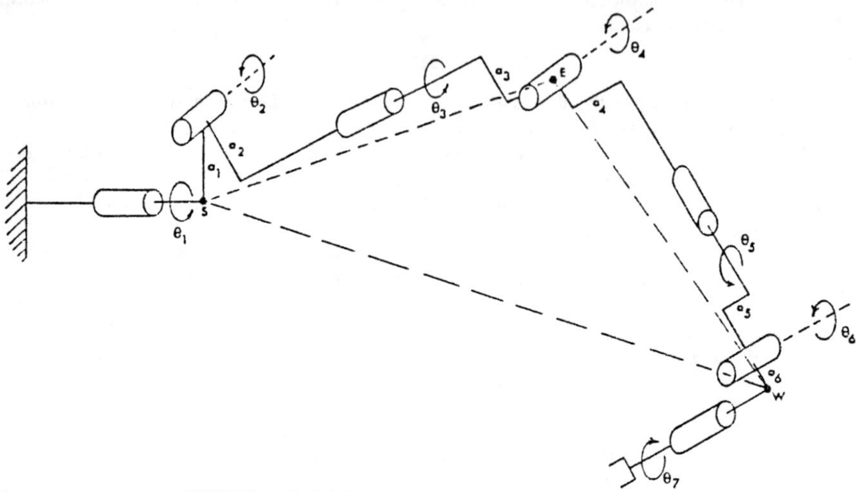

FIGURE 1: Definition of the Elbow Plane SEW

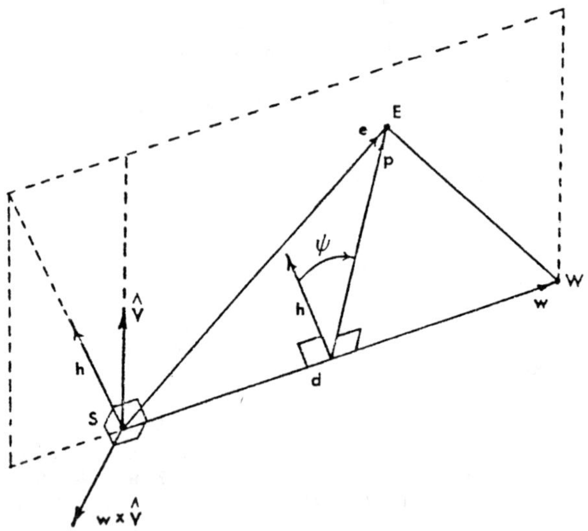

FIGURE 2: Definition of the Elbow Angle ψ

Software Package ROBSIM for the Dynamics and Control of Elastic Manipulators

Klaus Richter   Friedrich Pfeiffer
TU München, Lehrstuhl B für Mechanik
Arcisstr. 21, D-8000 München 2, FRG

Abstract

The paper introduces a software package for the dynamics and control of an arbitrary tree-structured manipulator with elastic joints and links. A multistage concept is applied. The first step is the generation of a reference path while neglecting elasticity. Secondly, the equations of motion linearized around that path are calculated for the system including elastic deformations. A controller is optimized to counterbalance elastic vibrations. Simulation results illustrate the diversity of the concept.

1 Introduction

For special applications, a manipulator might be lighter or less bulky than a conventional industrial robot. In that case, elasticity of joints and links cannot be neglected and the dynamics and control become much more complicated. If an elastic robot follows a path that was planned by taking into account only the properties of a rigid robot, deviations from that path will occur. These deviations are compensated in several steps.

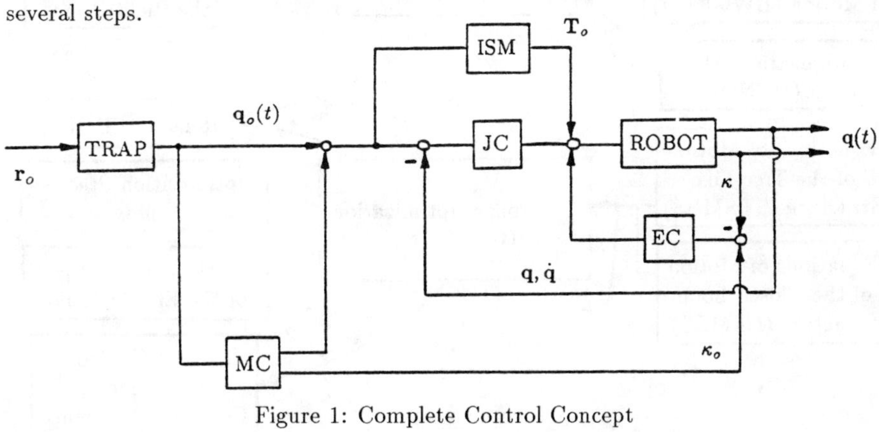

Figure 1: Complete Control Concept

Fig. 1 shows the structure of the multistage control concept, which is described in detail in [1,2,3,4]. In a first stage the reference trajectory is generated from a given path $\mathbf{r}_o$ in a world coordinate frame (TRAP). The path is then represented by a set of curves of the joint coordinates $\mathbf{q}_o$ with respect to time. The actuator torques $\mathbf{T}_o$

that are necessary to realize that motion are computed by the inverse system model (ISM) of the rigid robot.

The next step is the precalculation and correction of the influence of elasticity. The reference path $\mathbf{q}_o$ is adapted by a model of correction in order to compensate for the elastic deformations. All steps so far can be performed off-line. The on-line controller consists of two parts: the joint control (JC) and the elasticity control (EC). In the elasticity control block the curvatures κ of the elastic links are measured by strain gauges. The differences to the reference deformations κ_o (calculated by MC) are then fed back. The torques generated by the controller are added to the precalculated actuator torques $\mathbf{T}_o$.

If the robot makes contact with the environment it will be necessary to calculate the equations of motion of the structure with a closed kinematical loop. They can be derived from the equations of the tree-like structure by additionally considering a contact force and an equation that describes the geometrical constraint. The controller has to be laid out to influence the behaviour of the contact force in a predefined way. The on-line control concept remains the same [2].

2 Software Structure

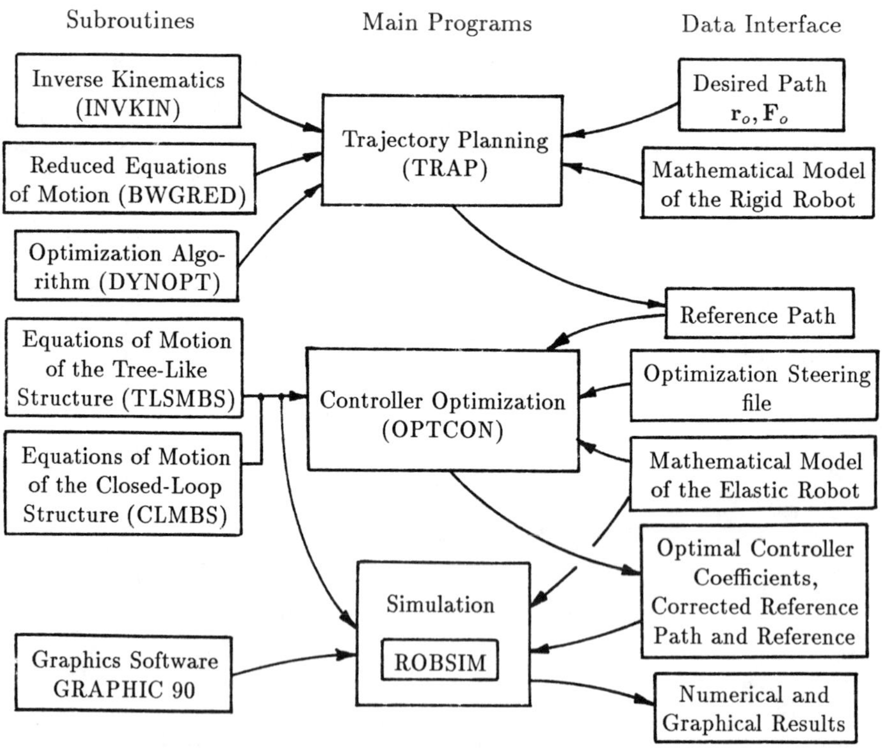

Figure 2: Global Structure of the Software Package ROBSIM

The software-package ROBSIM is a set of modular FORTRAN-programs that work independently and communicate via several interface files. Fig. 2 shows the global software structure that reflects the indiviual steps of the theoretical multistage approach. Main emphasis was laid on high flexibility concerning the structure of the elastic multibody system.

If a new situation is considered, the user will have to perform the following steps. The parameters of the rigid robot (masses, inertias, ratios of the joint transmissions and actuator capacities, etc.) have to be stored on a data-file. The desired path that consists of a trajectory described in world coordinates with hermitian splines and forces at the gripper must be defined. When starting the trajectory planning module TRAP, the user has to additionally supply a subroutine that provides information about the kinematical structure of the chosen robot. TRAP produces an output file that contains the reference path in terms of the joint coordinates with respect to time.

Before the next main program OPTCON is entered, the mathematical model of the robot has to be completed with respect to elasticity in joints and links. After having chosen an optimality criterion and several constraint values concerning the properties of the controller, OPTCON delivers optimal controller coefficients and a corrected reference path including reference deformations of the links. The results of OPTCON are the input for the main program ROBSIM that simulates the behaviour of the manipulator following the corrected reference trajectory. OPTCON and ROBSIM are supported by two subroutine libraries TLSMBS and CLMBS that contain the generation of the equations of motion and represent a substantial part of the entire structure.

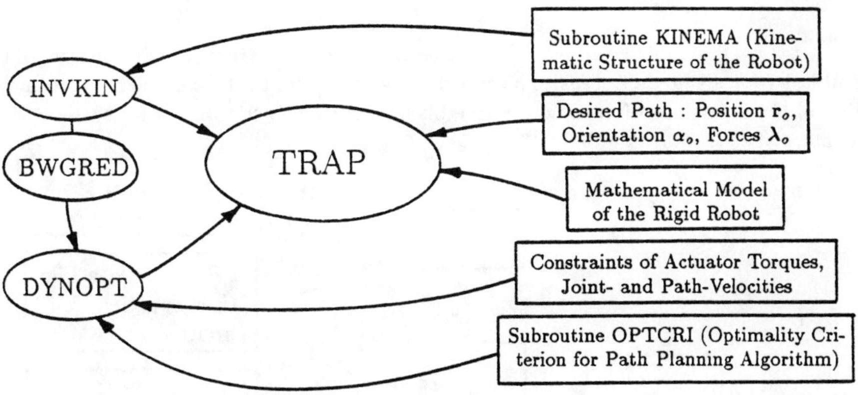

Figure 3: Detailed Structure of TRAP

2.1 Trajectory Planning

The reference trajectory for the rigidly assumed robot may be prescribed by any of the existing path planning methods [5,6]. In the simplest case it can be chosen by applying some physical considerations with regard to smooth path velocities and acceleration [3,4]. The algorithm that is used in TRAP is based on the theory of 'dynamic programming' [7] and a topological method that is described in [5]. Fig. 3 illustrates the detailed structure of TRAP.

In defining a specific manipulation task, the position and orientation of the gripper is given in an inertial cartesian coordinate frame. Hermitian cubic splines are used to define a continuous path with respect to a path parameter s. This trajectory has to be transformed into joint coordinates, which is done by an inverse kinematics algorithm that is briefly introduced in [8]. The forward kinematics have to be defined by the user, who supplies a subroutine that establishes the kinematics of the manipulator in a recursive way by using standard routines of the library INVKIN (e.g. rotation around an axis of the local coordinate frame; translation in the direction of a given vector).

Considering that the whole reference path is given now in terms of the joint angles $\mathbf{q}_o$ with respect to s, the equations of motion can be reduced to, [5]

$$\mathbf{A}_i(s)(\dot{s}^2)' + \mathbf{B}_i(s)\dot{s}^2 + \mathbf{C}_i(s) = \mathbf{T}_{Ai}. \quad (1)$$

Eq. 1 is a first order differential equation in $\dot{s}^2$ and s with only one degree of freedom, the path parameter s. The algorithm that generates these reduced equations of motion is the subroutine BWGRED that uses again the kinematic-subroutines of the library INVKIN. Based on a description of the robot dynamics as in eq. 1, the subroutine DYNOPT computes the reference path $\mathbf{q}_o(t)$ taking into account limitations of actuator torques and path- and joint-velocities. DYNOPT requires a user supplied subroutine that calculates any reasonable optimality criterion, e.g. time optimality or time-energy-optimality [5].

2.2 Elastic Multibody Systems with Tree-Like Structure

After path planning the robot has to be regarded as a structure with elastic joints and links for all further steps. The subroutine-library TLSMBS supports the generation of the equations of motion of a tree-like structured elastic mutibody system. Fig. 4 shows the structure of TLSMBS.

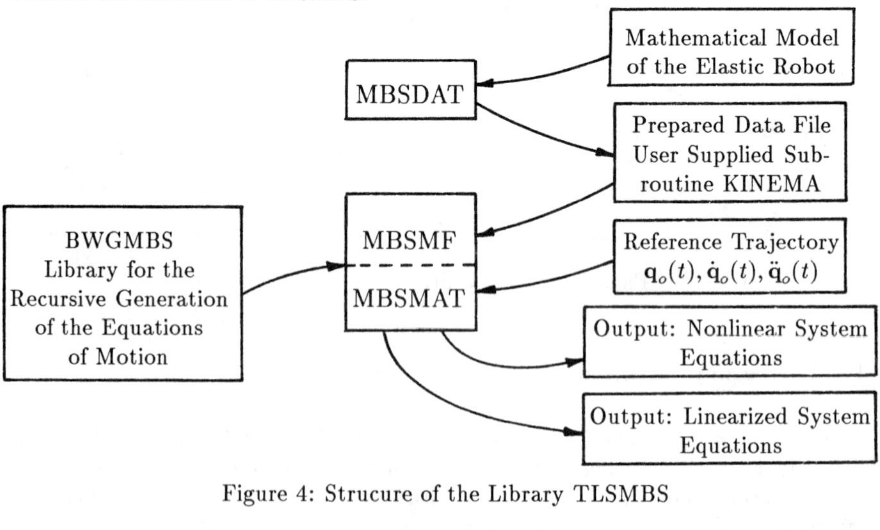

Figure 4: Strucure of the Library TLSMBS

Before the subroutines of TLSMBS can be used, the user has to prepare the input data that describe the model of the elastic robot by calling the program MBSDAT.

Calculations that have to be done only once are performed by MBSDAT and stored on a data-file. Thereafter, the user can obtain the equations of motion by calling the subroutines MBSMAT or MBSMF. MBSMAT supplies the equations of motion linearized around the reference trajectory and MBSMF gives them in the completely nonlinear form. In either case the user has to provide an external that defines the kinematics of the robot as described before.

The generation of the equations of motion is based on the well-known principle of Jourdain. A Ritz-approach is used for describing the elastic deformations of the links. According to [1,3,9], where the generation of the equations of motion of an elastic multibody system is explained in much detail, we arrive at a standard set of nonlinear dynamic equations

$$\mathbf{M}(\mathbf{q},t)\ddot{\mathbf{q}} + \mathbf{f}(\dot{\mathbf{q}},\mathbf{q},t) = \mathbf{B}\,\mathbf{T}_A(t). \tag{2}$$

The vector of the generalized coordinates $\mathbf{q}^T = (\mathbf{q}_r^T, \mathbf{q}_e^T)$ consists of two parts. $\mathbf{q}_r$ represents the degrees of freedom that describe the rigid-body-structure and contains f actuator coordinates and f joint coordinates. Actuator- and joint-coordinates have of course to be distinguished because of the elasticity in the joints. $\mathbf{q}_e$ stands for all elastic or modal coordinates describing the elastic deformation of the links which are approximated by a Ritz-series [3]. $\mathbf{M}$ is the inertia matrix and $\mathbf{f}$ is a vector that contains Coriolis and centrifugal forces as well as the influence of gravity. The input matrix $\mathbf{B}$ is not quadratic and the vector $\mathbf{T}_A$ represents all actuator forces and torques.

For the development of a control system, a linearization of eq. 2 around the nominal path is required, i.e. $\mathbf{q} = \mathbf{q}_o + \mathbf{y}$. The rather lengthy linearization process results in a set of linear, time-variant differential equations in the form [2,9]

$$\mathbf{M}(\mathbf{q}_o)\,\ddot{\mathbf{y}} + \mathbf{P}_o(\mathbf{q}_o,\dot{\mathbf{q}}_o)\,\dot{\mathbf{y}} + \mathbf{Q}_o(\mathbf{q}_o,\dot{\mathbf{q}}_o,\ddot{\mathbf{q}}_o)\,\mathbf{y} = \mathbf{h}_o(\mathbf{q}_o,\dot{\mathbf{q}}_o,\ddot{\mathbf{q}}_o) + \mathbf{B}_o\,\mathbf{T}_A \tag{3}$$

that can be obtained by calling the subroutine MBSMAT. The main program OPTCON calls MBSMAT, because for example the calculation of eigenvalues at the endpoint of a trajectory can only be done while regarding the linearized system.

2.3 Control of an Elastic Manipulator with a Closed Loop

If the robot makes contact with the environment, the entire structure of the mathematical model changes and becomes more complicated. A geometrical closed loop condition has to be added and constraint forces act on the gripper. An additional call of the subroutine CLMBS after a call of MBSMAT or MBSMF transforms the equations of the tree-like structure into a set of equations that describe the dynamics of the closed-loop system. Controller optimization and simulation can then be done without any additional modifications.

Fig. 5 shows the situation of a robot moving on a contact surface. The closed loop condition can be written in linearized form according to eq. 3 , [2]

$$(\mathbf{n}^T\mathbf{J}_T)\,\mathbf{y} = 0 \tag{4}$$

where $\mathbf{n}$ is the surface normal and $\mathbf{J}_T$ is the Jacobian matrix of translation of the endeffector. In order to consider the closed-loop condition in the equations of motion, it is necessary to differentiate eq. 4 twice. The geometrical constraint can then be

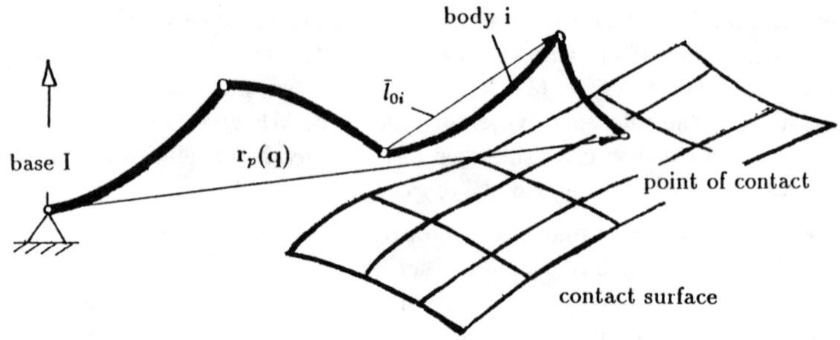

Figure 5: Closed Kinematical Loop

combined with eq. 2 or 3 by applying the idea of a Lagrangian multiplier λ as described in detail in [2]. λ represents exactly the constraint force between endeffector and contact surface. The state space matrix of the equations of motion of the closed-loop structure (for a time-invariant case, e.g. the endpoint of a trajectory) will in this case always have a double zero-eigenvalue because one degree of freedom gets lost when contacting a surface.

After having generated the equations of motion, which is done automatically by the main program OPTCON, the controller optimization can be performed. Satisfactory results could be achieved with an optimality criterion concerning the eigenvalues of the controlled system [4]

$$\sum_i \frac{1}{(Re\lambda_i)^2} \longrightarrow min\ . \tag{5}$$

When the user initiates the program OPTCON, he has to choose which coefficients of the controller shall be involved in the optimization process. The controller-structure applied here is a PID-controller that feeds back the actuator angles and the curvatures of the links.

When contacting a surface, the controller can be given a certain stiffness perpendicular to the contact surface. An optimization according to eq. 5 with the additional mathematical constraint

$$\frac{\partial\lambda}{\partial(\mathbf{n}^T\Delta\mathbf{r})} = c \quad (\text{cart. stiffness}) \tag{6}$$

yields the desired behaviour of the controller. $(\mathbf{n}^T\Delta\mathbf{r})$ indicates the position error of the constraint surface in the direction of the surface normal $\mathbf{n}$. As in the case of the inverse kinematics, the optimization routines of OPTCON are based on standard algorithms of the International Mathematical and Statistical Library (IMSL).

3 Results

Table 1 shows some characteristic magnitudes concerning the size of the software package ROBSIM. The required processor time for the individual modules depends heavily on the complexity of the problem, i.e. the number of degrees of freedom of the manipulator and the number of controller coefficients to be optimized. Two examples may give the reader a rough picture of the capabilities of the software package. Calculations were done on a Perkin Elmer 3212 computer (Concurrent Computer Corporation).

Module	No. of included Subroutines	Total No. of FORTRAN-Statements	Required Storage Capacity of the Main Programs
TRAP	50	3000	573 kByte
TLSMBS	46	5000	-
CLMBS	7	400	-
OPTCON	26	2000	413 kByte
ROBSIM	25	2000	262 kByte

Table 1: Characteristic Magnitudes of ROBSIM

3.1 Trajectory Planning

The first step in the control concept is the path planning. Based on a rigid robot model and a given path including forces at the gripper, a reference trajectory is precalculated as described in section 2.1.

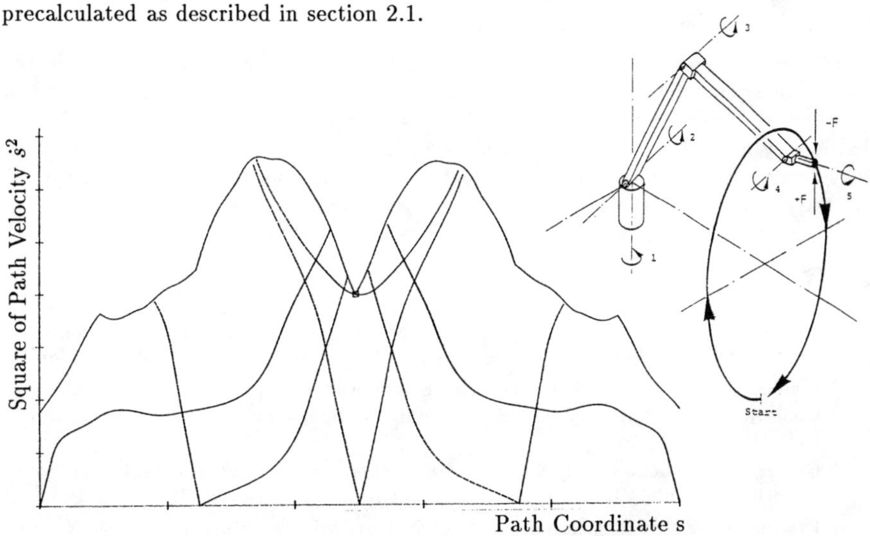

Figure 6: Time-Optimal Topology for a Circular Path with Vertical Forces

Fig. 6 shows a robot with five degrees of freedom following a circular path and the time-optimal topology according to that path. These results are discussed more thoroughly in [8]. The module TRAP requires approximately 120 sec processor time for the calculation of the reference trajectory and the nominal actuator torques.

3.2 Simulation of an Elastic Manipulator in Contact with a Surface

In the simulation presented here, force control is performed in a reference point on a plane, rigid surface. A controller was optimized that feeds back the position

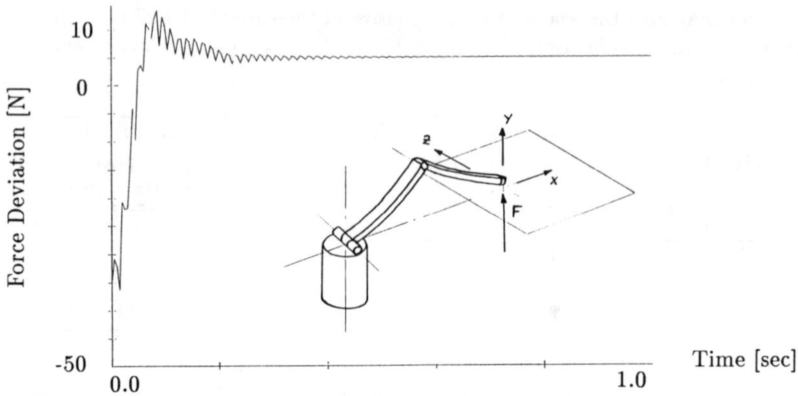

Figure 7: System Response to a Step in the Desired Contact Force

of the actuator angles and the elastic deformations of the links. Simulation results are shown in Fig. 7. The robot model has 15 degrees of freedom (including the elastic degrees of freedom). The endeffector was given the cartesian stiffness of $c = 0.5$ N/mm according to eq. 6. The optimization module OPTCON requires 30 min for the calculation of 8 controller coefficients.

For the simulation, an error of 10mm was assumed for the contact surface. In the simulation a remaining force error of 5 N can be observed according to the prescribed cartesian stiffness of the endeffector. In the force diagram, a high-frequency oscillation that is very underdamped can be seen. Nevertheless we do not expect problems from this effect in our future experiments mainly because unmodelled friction in the joints and the contact surface will tend to increase the stability of the system.

4 References

[1] Pfeiffer, F., Gebler, B., Kleemann, U., A Multistage Approach to the Dynamics an Control of Elastic Robots, to appear in IEEE Journal of Robotics and Automation.

[2] Pfeiffer, F., Richter, K., Wapenhans, H., Elastic Robot Trajectory Planning with Force Control, to appear in Proc. of IFIP International Symp. on Theory of Robots, Rome, Italy (1990)

[3] Gebler, B., Modellbildung, Steuerung und Regelung für elastische Industrieroboter, Fortschrittberichte VDI, Reihe 11, Nr. 98, VDI-Verlag, Düsseldorf 1987

[4] Kleemann, U., Regelung elastischer Roboter, Fortschrittberichte VDI, Reihe 8, Nr. 191, VDI-Verlag, Düsseldorf 1989

[5] Pfeiffer, F., Johanni, R., A Concept for Manipulator Trajectory Planning, IEEE Journal of Rob. Autom., Vol. RA3, No. 2, April 1987

[6] Shin, K., McKay, N., Minimum-Time Control of Robotic Manipulators with Geometric Path Constraints, IEEE Trans. Automat. Contr. 30 (1985), pp. 531-541

[7] Bellmann, R. E., Dynamic Programming. Princeton: Princeton University Press 1957

[8] Pfeiffer, F., Richter, K., Optimal Path Planning including Forces at the Gripper, to appear in Journal of Intelligent and Robotic Systems, Kluwer Academic Publishers Netherlands

[9] Johanni, R., On the Automatic Generation of the Equations of Motion for Robots with Elastically Deformable Arms, Proc. of the IFAC/ IFIP/ IMACS Internat. Symp. on Theory of Robots, Vienna, Austria (1986)

ROBUST ADAPTIVE CONTROL FOR FLEXIBLE JOINT MANIPULATORS*

K. Khorasani
Department of Electrical and Computer Engineering
Concordia University
Montreal, Quebec H3G 1M8

Abstract

This paper is concerned with the problem of designing an adaptive control strategy for flexible joint robot manipulators. By utilizing the concept of integral manifolds, reduced order models of the flexible system are obtained. This enables us to develop adaptive control schemes for the flexible full order system at the reduced order level which would have been difficult to obtain due to at least ill-conditioning and curse of dimensionality. Two common adaptive control strategies are considered: i) the adaptive inverse dynamics; and ii) the Slotine and Li algorithm (although any other algorithm can be treated similarly). It is shown how these standard rigid adaptive control techniques can be generalized and improved upon for successful application to flexible joint manipulators. The result is robust adaptive control laws which take both parametric and dynamic uncertainties into account. Numerical simulations for a two link flexible joint manipulator are carried out to illustrate the potential and advantages of the new adaptive schemes as compared to the two standard algorithms in the literature.

I. INTRODUCTION

Different forms of adaptive control strategies for rigid manipulators have been proposed in the literature [5]-[7] to circumvent the difficulties arising from parametric uncertainty. We can essentially classify these works into two different categories. In the first category, the adaptive control scheme is based upon the generalization of the computed torque or the inverse dynamics. The work of [5] , [6] falls into this category, which is commonly referred to as adaptive inverse dynamics scheme. The objective of the adaptive controller is to render the closed loop system linear and decoupled even though the parameters are unknown. The second category is conceptually different from the adaptive inverse dynamics in the sense that the controller will not feedback linearize the closed loop system but rather preserve the passivity properties of the rigid robot. The work of [7] , [8] falls into this category. Since the above mentioned adaptive control schemes are based upon rigid models, they are not directly applicable to flexible joint manipulators. This is basically due to the presence of unmodeled high frequency dynamics which can lead to severe stability and robustness problems. In addition, the lack of an independent control input for each degree of freedom and loss of passivity requirements [7] prevent the use of adaptive control schemes of the second type that are developed based upon these requirements. The use of high-gain feedback commonly used to suppress the effects of the nonlinearities and fast adaptation could eventually lead to the excitation of unmodeled dynamics and cause stability complications. The deficiencies alluded to leave us with practically no adaptive control scheme to satisfactorily control flexible joint manipulators for high performance manipulations subject to extensive parameter variations, uncertainties and disturbances.

In this paper we present two adaptive control schemes for flexible joint manipulators. The first scheme is based on the generalization of adaptive inverse dynamics and the second scheme is based on the generalization of Slotine and Li scheme. A design procedure is outlined at the reduced order level to improve the performance of the adaptive control schemes up to any arbitrary degree of accuracy. The simulation results for a two link flexible joint manipulator are included which illustrate the potential of the new control schemes.

* This work was supported in part by the Natural Sciences and Engineering Research Council of Canada and by the Fonds Pour la Formation de Chercheurs et l'Aide la Recherche, Programme Etablissement de Nouveaux Chercheurs.

II. FLEXIBLE JOINT MANIPULATOR MODEL: SINGULAR PERTURBATION APPROACH

The flexible model briefly described here has already appeared in [1-3]. The manipulator has $n+1$ rigid links interconnected by n flexible joints. We let q_i, $i = 1, \dots, n$ denote the position of the ith link, and let q_i, $i = n+1, \dots, 2n$ denote the position of the ith actuator (divided by the gear ratio). It is assumed that the elasticity at the ith joint can be modelled as a linear torsional spring with spring constant k_i. We introduce an additional variable $z_i := k_i (q_i - q_{i+n})$, to represent the elastic force at the ith joint. The Euler-Lagrange equations of motion are

$$M(\tilde{q})\,\ddot{\tilde{q}} + h\left(\tilde{q},\dot{\tilde{q}}\right) + e(\tilde{q}) = Pu \tag{1}$$

where $u := (u_1, \dots, u_n)^T$, (T denotes transposition), u_i is the generalized force delivered by the ith actuator, $\tilde{q} := (q_1, \dots, q_{2n})^T$, $e(\tilde{q}) = (e_1(\tilde{q}) \dots, e_{2n}(\tilde{q}))^T$, $e_i(\tilde{q}) := k_i\, z_i$, $1 \le i \le n$; and $e_i(\tilde{q}) := -k_i\, z_{i-n}$, $n+1 \le i \le 2n$, $M(\tilde{q})$ is a symmetric positive definite $2n \times 2n$ inertia matrix, $h(\tilde{q}, \dot{\tilde{q}})$ is the vector of Coriolis, Centripetal, and gravitational forces and $P := [\,0_{nxn}{}^T \;\; I_{nxn}{}^T\,]^T$. A two-time scale arises with k_i's large. Without loss of generality we take $\tilde{k}_i \equiv 1$ and $\mu := 1/k$. Moreover we partition matrix M and vector h as follows

$$M(q) := \begin{bmatrix} M_1(\tilde{q}) & M_2(\tilde{q}) \\ M_2(\tilde{q}) & M_3(\tilde{q}) \end{bmatrix} \qquad \text{and} \qquad h\left(\tilde{q},\dot{\tilde{q}}\right) := \begin{bmatrix} h_1\left(\tilde{q},\dot{\tilde{q}}\right) \\ h_2\left(\tilde{q},\dot{\tilde{q}}\right) \end{bmatrix},$$

where $M_i \in R^{nxn}$, $i = 1, \dots, 3$, and $h_j \in R^n$, $j = 1, 2$. In view of the above definitions the equations of motion (1) can be rewritten as

$$(M_1(q,\mu z) + M_2(q,\mu z))\,\ddot{q} - \mu M_2(q,\mu z)\ddot{z} + h_1(q,\dot{q},\mu z,\mu\dot{z}) + z = 0 \tag{2}$$

$$(M_2(q,\mu z) + M_3(q,\mu z))\,\ddot{q} - \mu M_3(q,\mu z)\ddot{z} + h_2(q,\dot{q},\mu z,\mu\dot{z}) - z = u \tag{3}$$

In the next section we address the problem of constructing reduced order models from the flexible joint full order model (2)-(3).

III. REDUCED ORDER FLEXIBLE MODELS

The dynamical system (2)-(3) is extremely complex to control because of high dimensionality, ill conditioning, severe nonlinearities and strong coupling effects. In lieu of these, a need for order reduction and simplification is inevitable. A reduced order model of the manipulator is obtained by formally setting μ to zero (neglecting elasticity) in (2)-(3) and solving for z from (2) and substituting it in (3) to get the *rigid model*

$$(M_1(q,0) + 2M_2(q,0) + M_3(q,0))\,\ddot{q} + h_1(q,\dot{q},0,0) + h_2(q,\dot{q},0,0) = u\,. \tag{4}$$

Improvements on this approximate model of the flexible manipulator is indeed possible by utilizing the concept of integral manifolds [3]-[4]. We define the integral manifold for system (2)-(3) by a surface M_μ in $(z, \dot{z}, t)$ space as $M_\mu := \{(z(t), \dot{z}(t)) : z(t) = \Phi(q(t), \dot{q}(t), u(t), \varepsilon), \dot{z}(t) = \dot{\Phi}(q(t), \dot{q}(t), u(t), \varepsilon)\}$, and $\varepsilon = \sqrt{\mu}$ with the property that if $(z(t), \dot{z}(t)) \in M_\mu$ at $t = t_0$ then $(z(t), \dot{z}(t)) \in M_\mu$ $\forall t \ge t_0$. Dynamics of the full order system restricted to this manifold defines the *exact reduced flexible model*

$$\begin{aligned}&\left(M_1(q,\mu\Phi) + 2M_2(q,\mu\Phi) + M_3(q,\mu\Phi)\right)\ddot{q} \\ &+\left(h_1\left(q,\dot{q},\mu\Phi,\mu\dot{\Phi}\right) + h_2\left(q,\dot{q},\mu\Phi,\mu\dot{\Phi}\right)\right) - \mu\left(M_2(q,\mu\Phi) + M_3(q,\mu\Phi)\right)\ddot{\Phi} = u\end{aligned} \tag{5}$$

with Φ satisfying the so-called *manifold condition* given by

$$(M_1 + M_2)\,\ddot{q} - \mu M_2\left[\frac{\partial\dot{\Phi}}{\partial q}\dot{q} + \frac{\partial\dot{\Phi}}{\partial\dot{q}}\ddot{q} + \frac{\partial\dot{\Phi}}{\partial u}\dot{u}\right] + h_1 + \Phi = 0 \tag{6}$$

$$(M_2 + M_3)\,\ddot{q} - \mu M_3\left[\frac{\partial\dot{\Phi}}{\partial q}\dot{q} + \frac{\partial\dot{\Phi}}{\partial\dot{q}}\ddot{q} + \frac{\partial\dot{\Phi}}{\partial u}\dot{u}\right] + h_2 - \Phi = u\,, \tag{7}$$

where $\Phi = \Phi(q,\dot{q},u,\varepsilon)$, $\dot{\Phi} = \dot{\Phi}(q,\dot{q},u,\varepsilon) = \dfrac{\partial\Phi}{\partial q}\dot{q} + \dfrac{\partial\Phi}{\partial\dot{q}}\ddot{q} + \dfrac{\partial\Phi}{\partial u}\dot{u}$, and

$\ddot{q} = \left(M_1 - M_2 M_3^{-1} M_2\right)^{-1}\left[\left(I + M_2 M_3^{-1}\right)\Phi + M_2 M_3^{-1} u + h_1 - M_2 M_3^{-1} h_2\right]$ (obtained by eliminating $\ddot{\Phi}$ from (6)-(7)). It can be shown easily that the exact reduced order flexible system (5) reduces to the rigid model (4) as $\mu \to 0$. The significance of this fact is that the exact reduced order flexible model is well-posed with respect to the perturbation parameter μ. Due to the complexity involved in solving for the exact manifold Φ from (6)-(7), it is then necessary to develop a procedure for obtaining an approximate solution. To this end, first expand Φ and u in power series about $\mu = 0$ as $\Phi = \Phi_0 + \mu \Phi_1 + O(\mu^2)$ and $u = u_0 + \mu u_1 + O(\mu^2)$. From the manifold condition (6)-(7) at $\mu \neq 0$ and $\mu^2 = 0$ we get

$$\left(M_1(q, \mu\Phi_0) + M_2(q, \mu\Phi_0)\right)\ddot{q} - \mu M_2(q, 0)\ddot{\Phi}_0 + h_1\left(q, \dot{q}, \mu\Phi_0, \mu\dot{\Phi}_0\right) + \Phi_0 + \mu\Phi_1 = 0 \tag{8}$$

$$\left(M_2(q, \mu\Phi_0) + M_3(q, \mu\Phi_0)\right)\ddot{q} - \mu M_3(q, 0)\ddot{\Phi}_0 + h_2\left(q, \dot{q}, \mu\Phi_0, \mu\dot{\Phi}_0\right) - \Phi_0 - \mu\Phi_1 = u_0 + \mu u_1 \tag{9}$$

Now by combining (8)-(9) we get

$$\begin{aligned}\left(M_1(q, \mu\Phi_0) + 2M_2(q, \mu\Phi_0) + M_3(q, \mu\Phi_0)\right)\ddot{q} + h_1\left(\left(q, \dot{q}, \mu\Phi_0, \mu\dot{\Phi}_0\right)\right. \\ \left. + h_2\left(q, \dot{q}, \mu\Phi_0, \mu\dot{\Phi}_0\right)\right) - \mu\left(M_2(q, 0) + M_3(q, 0)\right)\ddot{\Phi}_0 = u_0 + \mu u_1\ ,\end{aligned} \tag{10}$$

where $\dot{\Phi}_0$ and $\ddot{\Phi}_0$ are to be computed along the rigid model dynamics (4). We denote model (10) as the *corrected reduced order model*. This model is an approximation to the exact reduced order flexible model (5) up to order μ^2. It turns out that it is possible to "make" the dynamics of the reduced order flexible model identical to that of the rigid model so that the flexible manipulator behaves identical to the rigid model. The key to the solution of this problem is to use the extra control terms u_1, u_2, etc., to "cancel" order μ and higher order terms in the reduced order flexible model.

The problem we consider can be stated as follows: Given the corrected reduced order model (10), design a corrected control u_1 such that the dynamics of the flexible joint manipulator restricted to the integral manifold M_μ is identical to the dynamics of the rigid manipulator up to $O(\mu^2)$. From equations (4) and (8)-(9) we get

$$\left(\partial_2 M_1 + \partial_2 M_2\right)\ddot{q} - M_2(q, 0)\ddot{\Phi}_0 + \left(\partial_3 h_1 + \partial_4 h_1\right) + \Phi_1 = 0 \tag{11}$$

$$\left(\partial_2 M_2 + \partial_2 M_3\right)\ddot{q} - M_3(q, 0)\ddot{\Phi}_0 + \left(\partial_3 h_2 + \partial_4 h_2\right) - \Phi_1 = u_1\ , \tag{12}$$

where $\partial_2 M(x, \mu y) := \partial M / \partial(\mu y)\big|_{\mu=0}$, $\partial_3 h(x, y, \mu z, \mu w) := \partial h / \partial(\mu z)\big|_{\mu=0}$ and $\partial_4 h(x, y, \mu z, \mu w) := \partial h / \partial(\mu w)\big|_{\mu=0}$. Now by combining equations (11) and (12) we have

$$\Phi_1 = -\left(\partial_2 M_1 + \partial_2 M_2\right)\ddot{q} + M_2(q, 0)\ddot{\Phi}_0 - \left(\partial_3 h_1 + \partial_4 h_1\right), \tag{13}$$

and the corrected control is defined as

$$u_1 = \partial_2(M_1 + 2M_2 + M_3)\ddot{q} - (M_2(q, 0) + M_3(q, 0))\ddot{\Phi}_0 + \partial_3(h_1 + h_2) + \partial_4(h_1 + h_2). \tag{14}$$

It now follows that application of (14) to (10) renders the exact reduced order flexible system feedback equivalent to the rigid model (4) up to $O(\mu^2)$. The significance of this result is that by adding an extra control term to the rigid control we have been able to force the dynamics of the flexible system to behave very close to that of the rigid model dynamics. Higher order feedback equivalences can be carried out in a similar way. Implicit in the above result is the condition that integral manifold is an attractive set so that the restriction of the full order system dynamics on the manifold is meaningful. If this is not satisfied, a fast control has to be added to the slow control to guarantee asymptotic stability of the fast subsystem (details are omitted due to space limitations).

In the next section we introduce uncertainty in the parameters of the flexible system (2)-(3) and describe how robust adaptive control schemes can be developed.

IV. ROBUST ADAPTIVE CONTROL STRATEGY FOR FLEXIBLE SYSTEM

The uncertain flexible joint manipulator model considered in this section is described by system (2)-(3) with M_i, i=1, ... , 3, and h_j, j=1, 2, containing either uncertain constant parameters or slowly time varying parameters. To obtain an adaptive scheme for controlling the flexible joint manipulator we assume that the estimate of the matrix and vector functions M_i and h_j, i = 1, ... , 3; j =1, 2 are given by $\widehat{M}_i$ and $\widehat{h}_j$ respectively. Since all the parameters of interest such as mass, inertia, etc. appear as coefficients of well defined functions of the generalized coordinates we can express the difference between M_i and $\widehat{M}_i$ and h_i and $\widehat{h}_i$ as follows:

$$\left(M_i(q, \mu z) - \widehat{M}_i(q, \mu z)\right)\ddot{q} = \psi_i(q, \mu z)\left[\ddot{q}\right]\left(P - P^*\right) \qquad , i=1, \cdots, 3 \tag{15}$$

$$h_j(q,\dot q,\mu z,\mu\dot z)-\hat h_j(q,\dot q,\mu z,\mu\dot z)=\phi_j(q,\dot q,\mu z,\mu\dot z)(P-P^*),\ j=1,2\ ,\tag{16}$$

where $P^*\in R^\rho$ is a constant vector of uncertain (unknown) parameters, $P\in R^\rho$ is an estimate of the unknown parameter vector p^*, and ψ_i and ϕ_j are $n\times\rho$ matrices. Also we have used the notation $\psi_i(q,z)[\ddot q]$ to emphasize the dependence of ψ_i on $\ddot q$. By adding and subtracting $\hat M_i$ and $\hat h_i$ to and from (2)-(3) and using the expressions in (15)-(16) we get the uncertain flexible joint manipulator model

$$(\hat M_1+\hat M_2)\ddot q-\mu\hat M_2\ddot z+\hat h_1+z+[(\psi_1+\psi_2)[\ddot q]-\mu\psi_2[\ddot z]+\phi_1](P-P^*)=0\tag{17}$$

$$(\hat M_2+\hat M_3)\ddot q-\mu\hat M_3\ddot z+\hat h_2-z+[(\psi_2+\psi_3)[\ddot q]-\mu\psi_3[\ddot z]+\phi_2](P-P^*)=u.\tag{18}$$

The *uncertain rigid model* can be obtained by formally setting $\mu=0$ to get

$$(\hat M_1+2\hat M_2+\hat M_3)\ddot q+\hat h_1+\hat h_2+[(\psi_1+2\psi_2+\psi_3)[\ddot q]+\phi_1+\phi_2](P-P^*)=u_0\tag{19}$$

where u_0 denotes rigid control. To design the adaptive control strategy we rewrite $\hat h_i$ in (19) as $\hat h_i(q,\dot q,0,0)=\hat C_i(q,\dot q)\dot q+\hat g_i(q)$, $i=1,2$, where $\hat C_i(q,\dot q)\dot q$ is the $n\times 1$ vector of Centripetal and Coriolis forces, and $\hat g_i(q)$ is the $n\times 1$ vector of gravitational forces.

A. <u>Adaptive Inverse Dynamics: Rigid Model</u>

If we momentarily neglect the uncertainties in the rigid model (19), the rigid control u_0 can be designed as a state feedback that renders the certain system feedback equivalent to a linear controllable system. The *certain rigid model* (with $P=P^*$) is

$$(\hat M_1+2\hat M_2+\hat M_3)\ddot q+(\hat C_1+\hat C_2)\dot q+\hat g_1+\hat g_2=u_0.\tag{20}$$

If we choose the state feedback control as an inverse dynamics scheme

$$u_0=(\hat M_1+2\hat M_2+\hat M_3)\upsilon+(\hat C_1+\hat C_2)\dot q+\hat g_1+\hat g_2\ ,\tag{21}$$

with υ a new input designed as

$$\upsilon(t)=\ddot q^d(t)-K_1(q(t)-q^d(t))-K_2(\dot q(t)-\dot q^d(t)),\tag{22}$$

where $q^d(t)$ and $\dot q^d(t)$ are desired joint position and velocity trajectories and K_1 and K_2 are gain vectors properly selected for desirable transient characteristics, the closed loop system $\ddot q=\upsilon$ is linear, decoupled and controllable with $q(t)\to q^d(t), \dot q(t)\to\dot q^d(t)$ as $t\to\infty$.

To derive the adaptive inverse dynamics scheme we apply the control (21)-(22) to the uncertain rigid model (19) which yields

$$(\hat M_1+2\hat M_2+\hat M_3)[\ddot q+K_2\dot q+K_1q-K_1q^d(t)-K_2\dot q^d(t)-\ddot q^d(t)]$$
$$+[(\psi_1+2\psi_2+\psi_3)[\ddot q]+\phi_1+\phi_2](P-P^*)=0.\tag{23}$$

Defining $e(t):=q(t)-q^d(t)$ and $\dot e(t):=\dot q(t)-\dot q^d(t)$, we can rewrite (23) as

$$\ddot e+K_2\dot e+K_1e=-(\hat M_1+2\hat M_2+\hat M_3)^{-1}[(\psi_1+2\psi_2+\psi_3)[\ddot q]+\phi_1+\phi_2](P-P^*),\tag{24}$$

where it is assumed that the estimated inertia matrix $(\hat M_1+2\hat M_2+\hat M_3)$ remains nonsingular during the adaptation process. Now consider the Lyapunov function candidate

$$V(x,P)=x^TRx+(P-P^*)^T\Gamma^{-1}(P-P^*)\tag{25}$$

where $x:=[e^T\dot e^T]^T$, $A:=\begin{bmatrix}O & I\\ -K_1 & -K_2\end{bmatrix}$, and $R:=\begin{bmatrix}R_1 & R_2\\ R_2^T & R_3\end{bmatrix}$ is the unique positive definite solution of $A^TR+RA=-S$, S and Γ^{-1} are symmetric positive definite matrices. From the time derivative of (25) along the trajectories of (24) it follows that if the update law is

$$\dot P=-\Gamma\Psi^T(R_2e+R_3\dot e),\tag{26}$$

where $\Psi:=-(\hat M_1+2\hat M_2+\hat M_3)\Big|_{\mu=0}^{-1}[(\psi_1+2\psi_2+\psi_3)[\ddot q]+\phi_1+\phi_2]\Big|_{\mu=0}$, then

$$\dot V(x,P)=-x^TSx\le 0.\tag{27}$$

Since $\dot V$ is negative semi-definite and V is lower bounded by zero, V tends to a constant as $t\to\infty$ and remains bounded. This implies that x and P are bounded, therefore q and $\dot q$ are bounded. This in turn implies, using (21) and (23), that $\ddot e$ (or equivalently $\dot x$) is bounded. Hence, x is uniformly continuous and since $\dot{V}$ is bounded, $\dot{V} \to 0$ as $t \to \infty$. Consequently the equilibrium of (24), namely $e = 0$, $\dot{e} = 0$, and $P = P^*$ is stable for all the initial states (e (0), $\dot{e}$ (0), P (0)) starting in the set $T := \{(e, \dot{e}, P) : \dot{V}(x, P) \leq 0, V(x, P) \leq c > 0\}$, for some positive constant c. Moreover, it follows that $e(t) \to 0$ and $\dot{e}(t) \to 0$ as $t \to \infty$.

B. Slotine and Li Scheme: Rigid Model

The scheme recently proposed by Slotine and Li [7] and [8] for rigid manipulators is briefly summarized below. Consider an uncertain rigid model (4) rewritten as

$$(M_1 + 2M_2 + M_3)\ddot{q} + (C_1 + C_2)\dot{q} + (g_1 + g_2) = u_0, \tag{28}$$

subject to the control law (at $\mu = 0$)

$$u_0 = (\hat{M}_1 + 2\hat{M}_2 + \hat{M}_3)\left(\ddot{q}^d - \Lambda(\dot{q} - \dot{q}^d)\right) + (\hat{C}_1 + \hat{C}_2)\left(\dot{q}^d - \Lambda(q - q^d)\right) + (\hat{g}_1 + \hat{g}_2) - K_D\left(\dot{q} - \dot{q}^d + \Lambda(q - q^d)\right), \tag{29}$$

and parameter update law

$$\dot{P} = -\Gamma Y^T\left(\dot{q} - \dot{q}^d + \Lambda(q - q^d)\right), \tag{30}$$

where Λ is a diagonal positive definite matrix, K_D and Γ are symmetric positive definite matrices, and $Y = Y(q, \dot{q}, q^d, \dot{q}^d, \ddot{q}^d) = -(\psi_1 + 2\psi_2 + \psi_3)\left[\ddot{q}^d - \Lambda(\dot{q} - \dot{q}^d)\right] - (\hat{\phi}_1 + \hat{\phi}_2)\left(\dot{q}^d - \Lambda(q - q^d)\right) - (\bar{\phi}_1 + \bar{\phi}_2)$ where $C_i(q, \dot{q}, \mu z, \mu\dot{z}) - \hat{C}_i(q, \dot{q}, \mu z, \mu\dot{z}) := \phi_i(q, \dot{q}, \mu z, \mu\dot{z})(P - P^*)$ and $g_i(q, \mu z) - \hat{g}_i(q, \mu z) := \bar{\phi}_i(q, \mu z)(P - P^*)$, $i = 1, 2$. By choosing the Lyapunov function candidate

$$V(q, \dot{q}, P) = \frac{1}{2}\left(\dot{q} - \dot{q}^d + \Lambda(q - q^d)\right)^T (M_1 + 2M_2 + M_3)\left(\dot{q} - \dot{q}^d + \Lambda(q - q^d)\right) + \frac{1}{2}(P - P^*)^T \Gamma^{-1}(P - P^*), \tag{31}$$

then it can be shown that $\dot{V}(q, \dot{q}, P) = -\left(\dot{q} - \dot{q}^d + \Lambda(q - q^d)\right)^T K_D\left(\dot{q} - \dot{q}^d + \Lambda(q - q^d)\right) \leq 0$, which guarantees stability of (28)-(30) following arguments similar to those used for (27). Moreover, $q(t) \to q^d(t)$ and $\dot{q}(t) \to \dot{q}^d(t)$ as $t \to \infty$.

Now if one attempts to apply the rigid adaptive control schemes (29)-(30) or (21)-(22), (26), to the flexible joint manipulator one would be confronted with severe stability and convergence difficulties (see section V). The remainder of this section is devoted to the development of adaptive control schemes for flexible joint manipulators based on generalization of the two adaptive laws developed for rigid manipulators. To this end we consider the uncertain flexible model (17)-(18) with its restriction to the integral manifold up to order μ^2 by

$$(\hat{M}_1 + \hat{M}_2)\ddot{q} - \mu\hat{M}_2\ddot{\Phi}_0 + \hat{h}_1 + \Phi_0 + \mu\Phi_1 + \left[(\psi_1 + \psi_2)[\ddot{q}] - \mu\psi_2[\ddot{\Phi}_0] + \phi_1\right](P - P^*) = 0 \tag{32}$$

$$(\hat{M}_2 + \hat{M}_3)\ddot{q} - \mu\hat{M}_3\ddot{\Phi}_0 + \hat{h}_2 - \Phi_0 - \mu\Phi_1 + \left[(\psi_2 + \psi_3)[\ddot{q}] - \mu\psi_3[\ddot{\Phi}_0] + \phi_2\right](P - P^*) = u_0 + \mu u_1. \tag{33}$$

The corrected control u_1 is designed to cancel estimated $O(\mu)$ terms

$$u_1 = \partial_2(\hat{M}_1 + 2\hat{M}_2 + \hat{M}_3)\ddot{q} - (\hat{M}_2(q, 0) + \hat{M}_3(q, 0))\ddot{\Phi}_0 + \partial_3(\hat{h}_1 + \hat{h}_2) + \partial_4(\hat{h}_1 + \hat{h}_2). \tag{34}$$

Note that in the above control $u_1 = u_1(q, \dot{q}, \ddot{q}, P, q^d, \dot{q}^d, \ddot{q}^d, \dddot{q}^d)$.

C. Adaptive Inverse Dynamics: Flexible Model

By using u_1 in (32)-(33) the uncertain flexible model is approximated up to $O(\mu^2)$

$$(\hat{M}_1 + 2\hat{M}_2 + \hat{M}_3)\ddot{q} + (\hat{C}_1 + \hat{C}_1)\dot{q} + \hat{g}_1 + \hat{g}_2 + \left[(\psi_1 + 2\psi_2 + \psi_3)[\ddot{q}] - \mu(\psi_2 + \psi_3)[\ddot{\Phi}_0] + \phi_1 + \phi_2\right](P - P^*) = u_0, \tag{35}$$

where $\hat{h}_1 + \hat{h}_2 = (\hat{C}_1 + \hat{C}_2)\dot{q} + \hat{g}_1 + \hat{g}_2$. Model (35) is exactly the same as the rigid model up to $O(\mu^2)$ except for the second term in the bracket. It is now shown that the rigid control (21)-(22) can indeed be used for controlling (35) with a modification to the update law (26). Let us choose the adaptation law

$$\dot{P} = -\Gamma\psi_c^T(R_2 e + R_3\dot{e}) \tag{36}$$

where $\Psi_c := -(\hat{M}_1 + 2\hat{M}_2 + \hat{M}_3)^{-1}\Big|_{\substack{\mu \neq 0 \\ \mu^2 = 0}}\left[(\psi_1 + 2\psi_2 + \psi_3)[\ddot{q}] - \mu(\psi_2 + \psi_3)[\ddot{\Phi}_0] + \phi_1 + \phi_2\right]\Big|_{\substack{\mu \neq 0 \\ \mu^2 = 0}}$

and R_2, R_3, Γ, e and $\dot{e}$ are as defined earlier for (26). The controls u_0, u_1 and the above adaptation law guarantee the convergence of the position and velocity tracking errors to zero. To see this consider the Lyapunov function candidate (25). Arguments similar to those used for (27) can be applied to show asymptotic stability of the error signals, namely $e(t) \to 0$ and $\dot{e}(t) \to 0$ as $t \to \infty$. What we have accomplished is the following.

Lemma 1. The flexible joint manipulator (2)-(3) with parameter uncertainty characterized by (15)-(16) can achieve robust tracking of a reference trajectory $q^d(t)$ and $\dot{q}^d(t)$, provided that a fast control, a corrected slow control, and a slow corrected adaptive control, given by (34) and (21)-(22), (36), respectively, are utilized. More specifically, for all $\mu \in [0, \mu^*]$, $\mu^* > 0$, and $(q(0), \dot{q}(0), z(0), \dot{z}(0), P(0)) \in \mathscr{S} \subset R^{4n+p}$, we have $q(t) \to q^d(t) + O(\mu^2)$, $\dot{q}(t) \to \dot{q}^d(t) + O(\mu^2)$, $\eta(t) \to 0 + O(\mu)$, $\dot{\eta}(t) \to 0 + O(\mu)$, and $P(t) \to \overline{P}$ (a constant vector) as $t \to \infty$.

Proof: The result is immediate by construction as outlined in the text and from composite Lyapunov stability analysis of singularly perturbed systems .

D. <u>Slotine and Li Scheme: Flexible Model</u>

Adaptive algorithm based on Slotine and Li scheme for the flexible joint manipulator is now developed. To derive the update law for the flexible model (32)-(33) let us set the Lyapunov function candidate the same as in (31). By setting the update law as

$$\dot{P} = -\Gamma\left(Y^T - \mu Z^T\right)\left(\dot{q} - \dot{q}^d + \Lambda\left(q - q^d\right)\right) \tag{37}$$

where $Z = Z\left(q, \dot{q}, \ddot{q}, q^d, \dot{q}^d, \ddot{q}^d, \dddot{q}^d, \ddddot{q}^d\right) = (\psi_2 + \psi_3)\left[\ddot{\Phi}_0\right] - \partial_2(\psi_1 + 2\psi_2 + \psi_3)[\ddot{q}] - \partial_3(\phi_1 + \phi_2) - \partial_4(\phi_1 + \phi_2)$, and Y and $\ddot{\Phi}_0$ as defined previously, it follows that $\dot{V} \le 0$ and consequently based on arguments similar to those used earlier it can be shown that $q(t) \to q^d(t)$ and $\dot{q}(t) \to \dot{q}^d(t)$ as $t \to \infty$. As can be seen from the adaptive scheme (37), the direct measurement of the joint acceleration is now required. This distinctive feature for adaptive control of flexible joint manipulator is not present in the case of rigid manipulator.

The above result can now be formally stated as follows:

Lemma 2. The flexible joint manipulator (2)-(3) with parameter uncertainty characterized by (15)-(16) can achieve robust tracking of a reference trajectory $q^d(t)$ and $\dot{q}^d(t)$ provided that a fast control and the rigid control (29) and the corrected control u_1 given by (34) with corrected adaptation law (37) are utilized. In other words, there exists a real number $\mu^* > 0$ and a set $\mathscr{S} \subset R^{4n+p}$ such that for all $\mu \in [0, \mu^*]$ and $(q(0), \dot{q}(0), z(0), \dot{z}(0), P(0)) \in \mathscr{S}$ we have $q(t) \to q^d(t) + O(\mu^2)$, $\dot{q}(t) \to \dot{q}^d(t) + O(\mu^2)$, $\eta(t) \to 0 + O(\mu)$, $\dot{\eta}(t) \to 0 + O(\mu)$, and $P(t) \to \overline{P}$ (a constant vector) as $t \to \infty$.

Proof: Follows along the same lines as the proof of Lemma 1.

V. A TWO-LINK EXAMPLE

Consider the two-link manipulator with remotely driven links as shown in Figure 1, which has appeared in [9]. The desired position trajectory for links 1 and 2 are specified as $q_1^d(t) = 1 + \sin t$ and $q_3^d(t) = 2 + \cos t$. Figure 2 shows the response of the flexible system using only the rigid control. The flexible system is unstable. This is due to the fact that the integral manifold is not an attractive set. Therefore a fast control has to be added to the rigid control to ensure asymptotic stability of the fast dynamics. Figure 3 shows the response of the flexible system using the composite control consisting of the rigid control and the fast control. The flexible system is now stable, however, there is a considerable amount of tracking error. The error is compensated by adding the corrective control. The result is shown in Figure 4. Note that after an initial transient during which the fast variables converge to the manifold, tracking is achieved.

Next it is assumed that all the parameters of the flexible system are gone through 50% change from their nominal values. Figure 5 shows the response of the flexible system using the rigid control, the corrected control, and the fast control. Figure 6 shows the response of the flexible system using the fast control and the rigid adaptive inverse control. The tracking error is considerably reduced as a result of the rigid adaptive control. The error is reduced even further by using the corrected adaptive control and the corrected slow control. Figure 7 shows the response of the flexible system using the fast control, the rigid control, the corrected slow control, and the corrected adaptive inverse control.

Figure 8 shows the response of the flexible system using the fast control, the rigid control and the rigid Slotine and Li update law. As can be seen the flexible system is unstable. It is interesting to observe that the instability has occurred long after the fast transients have settled down to their quasi-steady state values. The instability quite resembles those that occur in standard adaptive control techniques which are designed by ignoring the effects of unmodelled dynamics. There is a correlation between the amount of joint elasticity and the instability. As discussed in Lemma 2, there is an upper bound on μ for which tracking is achieved. To see this, μ is reduced to .0025. The response of the flexible system is now shown in Figure 9. The instability is not present anymore and stable tracking is achieved. To illustrate the advantage of the corrected adaptive control law, we repeated the simulation with $\mu = .01$, using fast control, corrected control, and the corrected adaptive control. The response is shown in Figure 10. In contrast to the response by the rigid adaptive control law which is unstable (Figure 8), the response by the corrected adaptive control law is stable and tracking is achieved. This shows that the modified controller is more robust to uncertainty than the standard rigid controller of Slotine and Li.

REFERENCES

[1] Nicolo, F., R. Marino, and S. Nicosia, "A Singular Perturbation Approach to the Control of Elastic Robots", Proceedings of the 21st Allerton Conference on Communication, Control, and Computing, University of Illinois, 1983.

[2] Khorasani, K., and M.W. Spong, "Invariant Manifolds and Their Application to Robot Manipulators with Flexible Joints", Proceedings of the IEEE International Conference on Robotics and Automation, St. Louis, March 1985.

[3] Spong, M.W., K. Khorasani, and P.V. Kokotovic, "An Integral Manifold Approach to the Feedback Control of Flexible Joint Robots", IEEE Journal of Robotics and Automation, Vol. RA-3, No. 4, pp. 291-300, August 1987.

[4] Khorasani, K., and P.V. Kokotovic, "Feedback Linearization of a Flexible Manipulator Near its Rigid Body Manifold", Systems and Control Letters, Vol, 6, pp. 187-192, August 1985.

[5] Craig, J.J., P. Hsu, and S. Sastry, "Adaptive Control of Mechanical Manipulators", Proceedings of the IEEE International Conference of Robotics and Automation, San Francisco, March 1986.

[6] Middleton, R.H., and G.C. Goodwin, "Adaptive Computed Torque Control for Rigid Link Manipulators", Systems and Control Letters, Vol. 10, pp. 9-16, 1988.

[7] Slotine, J.J.E., and W. Li, "Adaptive Manipulator Control: A Case Study", IEEE Transactions on Automatic Control, Vol. AC-33, No. 11, pp. 995-1003, 1988.

[8] Slotine, J.J.E., and W. Li, "On the Adaptive Control of Robot Manipulators", International Journal of Robotics Research, Vol. 6, No. 3, pp. 49-59, 1987.

[9] Sira-Ramirez, H., and M.W. Spong, "Variable Structure Control of Flexible Joint Manipulators", International Journal of Robotics and Automation, Vol. 3, No. 2, 1988.

Legend: ——— is q_1^d, - - - - is q_3^d, · · · · is q_1 and – · – · is q_3. Each horizontal point in graphs is equal to .05 seconds. Fig.1: Two-link flexible joint manipulator; Fig.2: Response using rigid control, μ=.04; Fig.3: Response using rigid control and fast control; Fig.4: Response using rigid control, corrected control and fast control, μ=.04; Fig.5: Response using rigid control, fast control and corrected control with 50% variation in parameters μ= .01; Fig.6: Response using fast control and rigid adaptive inverse dynamics control, μ=.01; Fig.7: Response using fast control, corrected slow control and corrected adaptive inverse dynamics control, μ=.01; Fig.8: Response using fast control and rigid adaptive control of Slotine and Li, μ=.01. Simulation was halted at point 620 (=31 sec.) due to instability;

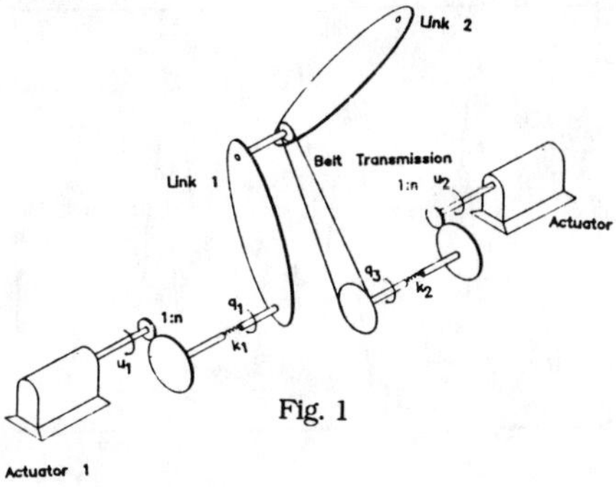

Fig. 1

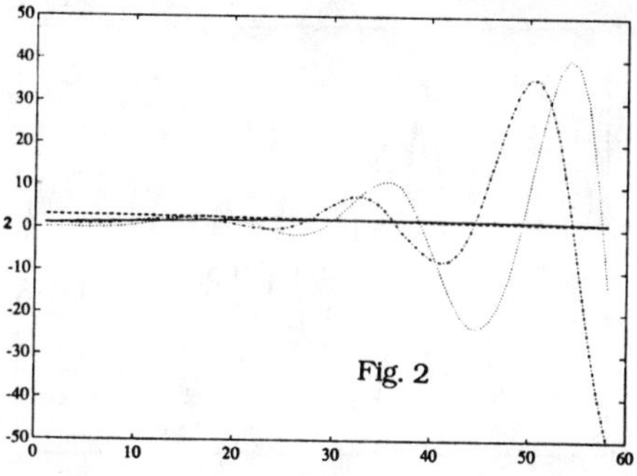

Fig. 2

Fig.9: Response using the same control as in Fig.8 except μ is reduced to .0025 (system is more rigid now); Fig.10: Response using fast control, slow corrected control, and corrected adaptive control of Slotine and Li. μ is increased to .01 from .0025.

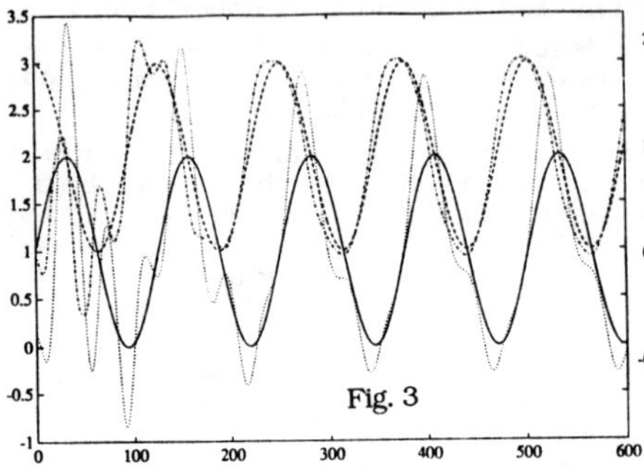

Fig. 3

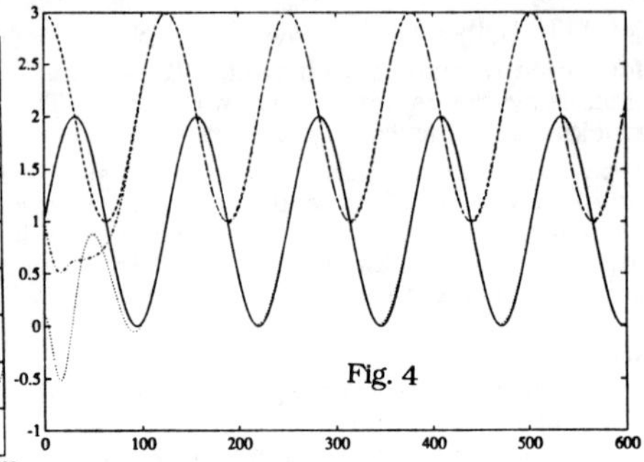

Fig. 4

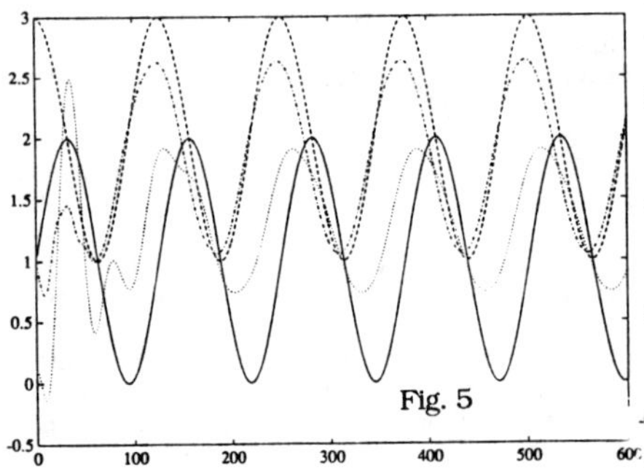

Fig. 5

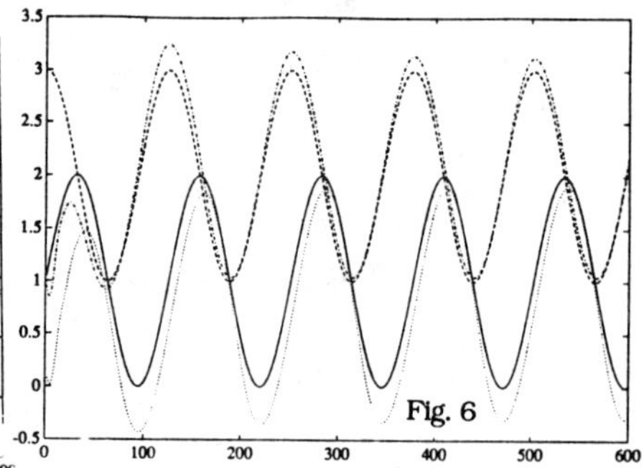

Fig. 6

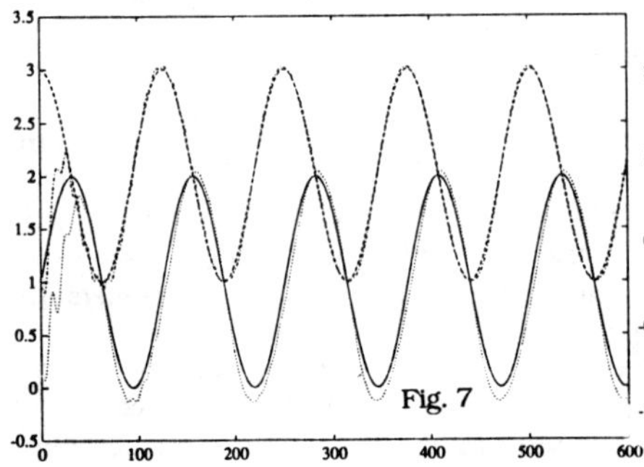

Fig. 7

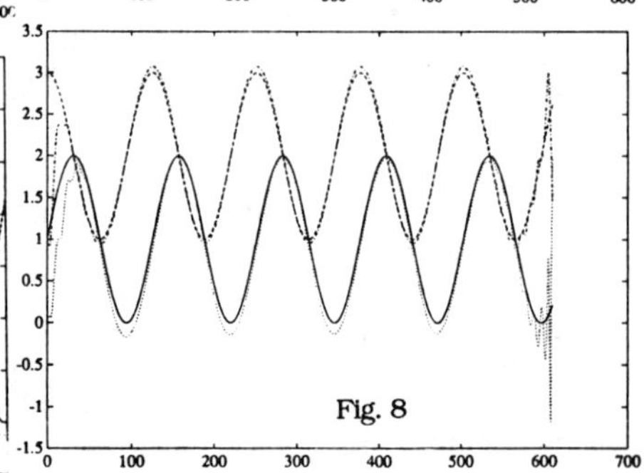

Fig. 8

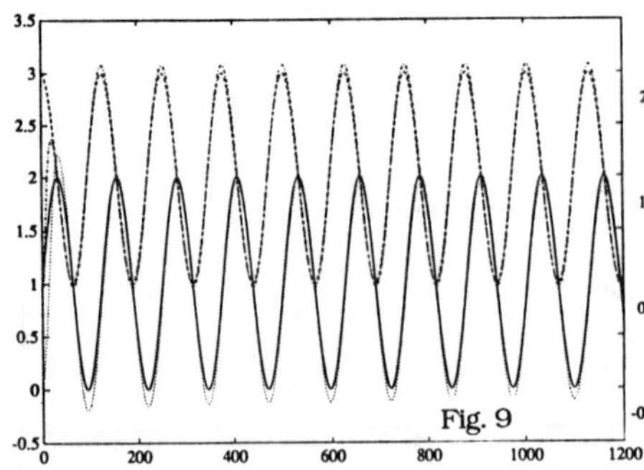

Fig. 9

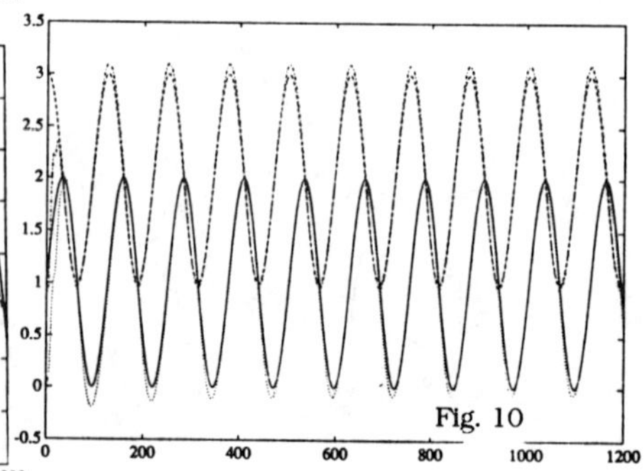

Fig. 10

Improper Controllers for the Single Flexible Link[1]

D. Vinke and D. Wang

Department of Electrical and Computer Engineering,
University of Waterloo,
Waterloo, Ontario, Canada N2L 3G1

Abstract

It is demonstrated in this paper that it is possible to improve the control of the single flexible link by doing two things. First, the constraint that the stabilizing controller must be proper is removed and, second, a transfer function which is different from the conventional one is used. It is shown via simulations that improved step responses and lower torques are achievable. These are only preliminary studies but indicate that improved control strategies may still be possible for the single flexible link.

Introduction

The modelling and control of single flexible links has been studied in detail in the past few years. In [1], conventional LQG controllers were used to successfully control a single flexible link. In [2], the modelling aspects of the single flexible link were more thoroughly studied. An initial attempt to use H_2-optimal controllers was presented in [3] with good simulation results. An H_2-optimal controller chooses, from the set of all stabilizing controllers, the one which minimizes the mean square of the tracking error. The design algorithm is also simple. However, the attempts to implement H_2-optimal controllers were unsuccessful due to the unrealistically high inputs demanded by the controller. Using a bounded-input H_2-optimal controller, it was shown in [4] that excellent experimental results are achievable albeit at the expense of a very complicated design strategy. This process was simplified somewhat in [5].

Initial research has concentrated on using a transfer function which has the motor torque as its input and the net tip movement as its output. However, in [6], several problems with the conventional transfer function were pointed out and suggestions were made for alternate transfer functions. A result of that research was the observation that it is sometimes possible to choose a model of this system which is passive [7]. A passive system is essentially one that dissipates or stores energy. The resulting experimental work on the passive model indicated that surprisingly good results were possible from a controller as simple as just a derivative feedback. Since derivative feedback is an improper controller, this has raised the question of whether superior performance may be available from the use of improper controllers.

In Section 2 of this paper, we present the alternate transfer function of [6] and, in Section 3, one method of obtaining an improper controller is shown. A comparison of the simulation results with those using the conventional techniques of [3] is given in Section 4 and, in Section 5, some interesting observations and suggestions for future work are discussed. It should be emphasized that the work

[1] This research is supported by the Manufacturing Research Corporation of Ontario, Canada

presented here is only in its preliminary stages but we will continue to explore the interesting possibilities raised in this research.

2. Model of the Single Flexible Link

In this section, the model to be used is briefly described. For more details, refer to [6]. The single flexible link of length L is assumed to rotate horizontally. The damping in the beam is ignored. The height of the link cross-section is chosen to be greater than the width, thus constraining the vibrations to also be horizontal. The angular rotation of the link is denoted as $\theta(t)$ while the elastic deflection at the tip of the link is indicated by $w(t)$ (see Figure 1).

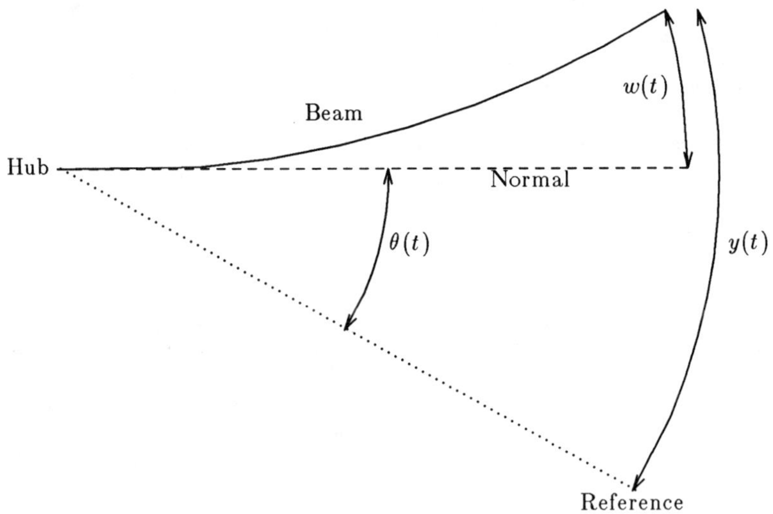

Figure 1 *Beam Deflection*

The state space model for the single flexible link is formed using the assumed modes approach [6]. To obtain a better approximation for the elastic deflection, one simply increases the number of modes in the expansion for $w(t)$. This method is widely used and can be easily used to form a transfer function for this system.

The input $u(t)$ to the system is assumed to be the motor torque $T(t)$ while the output $y(t)$ is assumed to be the rigid body deflection *minus* the elastic deflection of the tip. That is, $y(t)=\theta(t)L-w(t)$. This is different from the conventional output used which is the *sum* of the rigid body deflection and the elastic deflection. It is straightforward to show that the transfer function is of the following form

$$\frac{y(s)}{u(s)}=\frac{a_n s^{2n}+a_{n-1}s^{2n-2}+\cdots+a_0}{s^2(s^{2n}+b_{n-1}s^{2n-2}+\cdots+b_0)}. \tag{1}$$

If the conventional transfer function were used, then one encounters problems as n is increased to improve the modelling accuracy. For example, as n increases, then a_n decreases to zero while a_{n-1} increases to infinity. When the alternate output is used, this problem disappears and the a_n approach a constant value.

3. Controller Design Strategy

In this section, the design of an improper controller is described in detail. The alternate transfer function of the previous section will be used. The design methodology of this section is admittedly not very general. However, future work will examine how the design of improper controllers might be extended into a more general framework.

To obtain an improper controller, the plant transfer function is first differentiated. A controller $C'(s)$ is then designed for the differentiated plant $sP(s)$. The resulting controller for the plant $P(s)$ is then $sC'(s)$ which can be improper.

There are several reasons for taking this simple method to design an improper controller. First, it is shown in [8] that it is possible to use a controller of the form ks, where k is a gain, to control the single flexible link. Experimental results were promising. What is done here is to replace the gain k with an H_2-optimal controller which is described below. Another advantage of multiplying $P(s)$ by s is that the order of the plant is reduced due to the cancellation of one of the plant poles at the origin. In other words, although $P(s)$ is of order $2n+2$, $sP(s)$ will only be of order $2n+1$.

The method used to design a controller for the system $sP(s)$ is now examined. This method is called H_2-optimization. The feedback configuration is displayed in Figure 2.

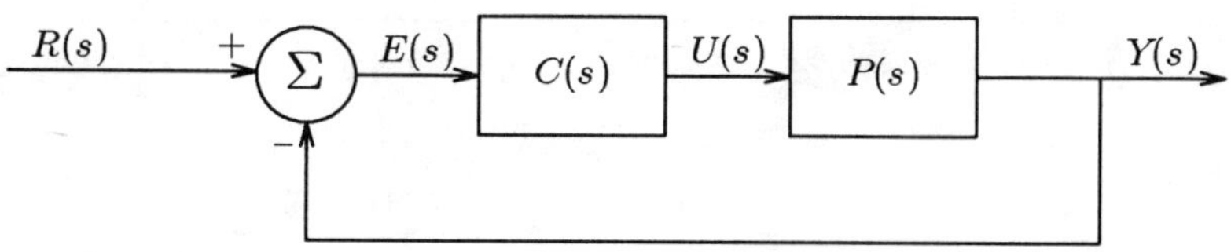

Figure 2 *Feedback Configuration*

It is possible to parametrize the entire set of stabilizing controllers of a plant. This is known as the Youla parametrization. The objective is to find the controller from this set which minimizes the mean square of the tracking error $e(t)$. This optimization process is relatively complicated and is described in detail in [8]. It turns out that, for continuous time systems, one cannot find the optimal controller but must approximate it arbitrarily closely with a suboptimal controller. As well, one can redefine the region of stability to increase the speed of the response. For example, by defining the region of stability to be to the left of -1, then the convergence to the steady state must be as fast as $\exp(-t)$.

Once this suboptimal controller C' is found, then it is obvious that the suboptimal controller $C(s)$ of the plant $P(s)$ is of the form sC'. This will be improper in general for the single flexible link since C' is nonstrictly proper. There should be more general methods of finding an improper optimal controller but this will be examined in future research.

4. Simulation Results

In this section, we compare the H_2-optimal controller using the conventional transfer function to the design strategy suggested above. To make the comparison valid, all the design parameters of the controllers were made equal and, in both cases, the region of stability was chosen to be to the left of -1. The plant was chosen to be of order 6.

The step and torque responses using the conventional methods are shown in Figures 3 and 4. The results using the methods suggested in this paper are shown in Figures 5 and 6. Note the tremendous improvement in the torques applied. The torques in the first case are unrealizable but are definitely realizable in the second case. As well, the response is much smoother in the second case.

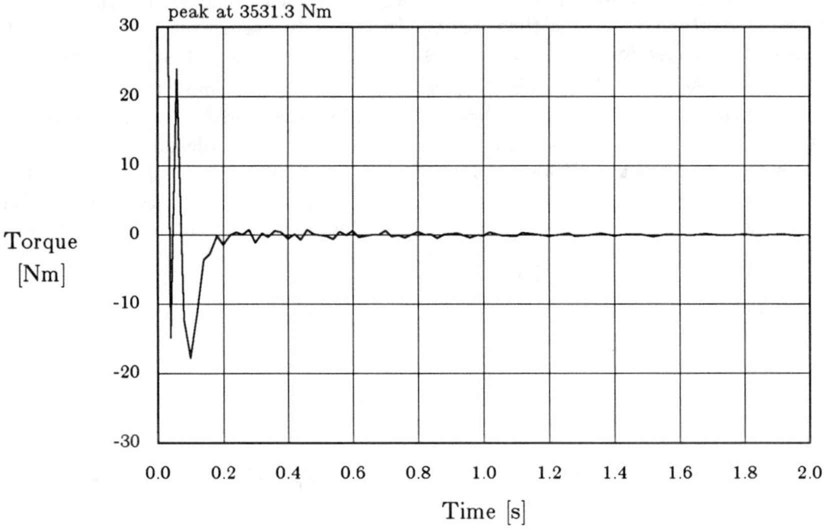

Figure 3 *Compensator using* True *Model* (*Torque*)

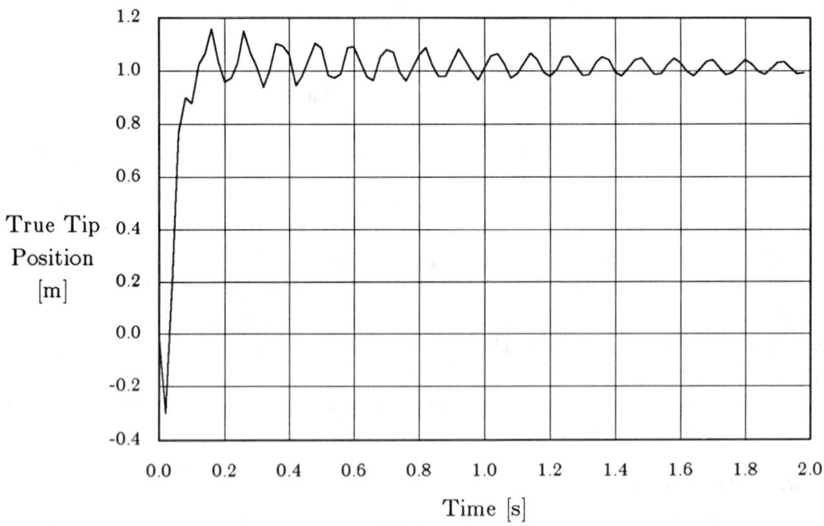

Figure 4 *Compensator using* True *Model* (*Position*)

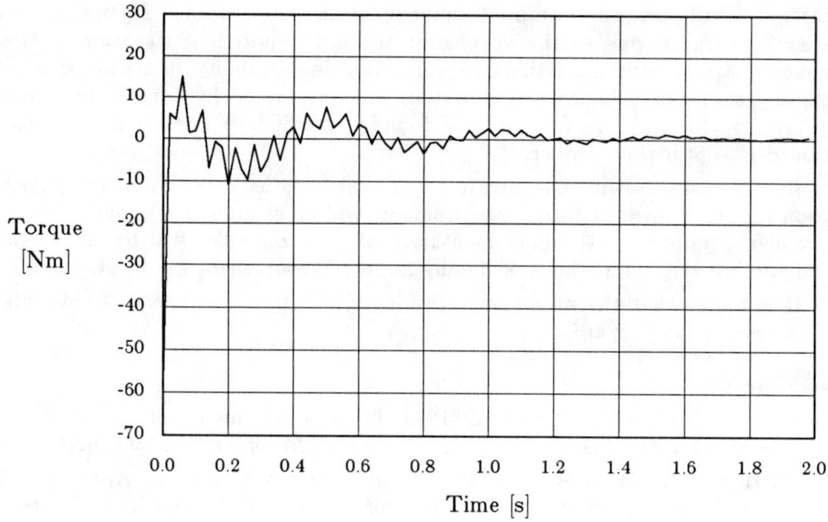

Figure 5 *Compensator using Differentiated Model (Torque)*

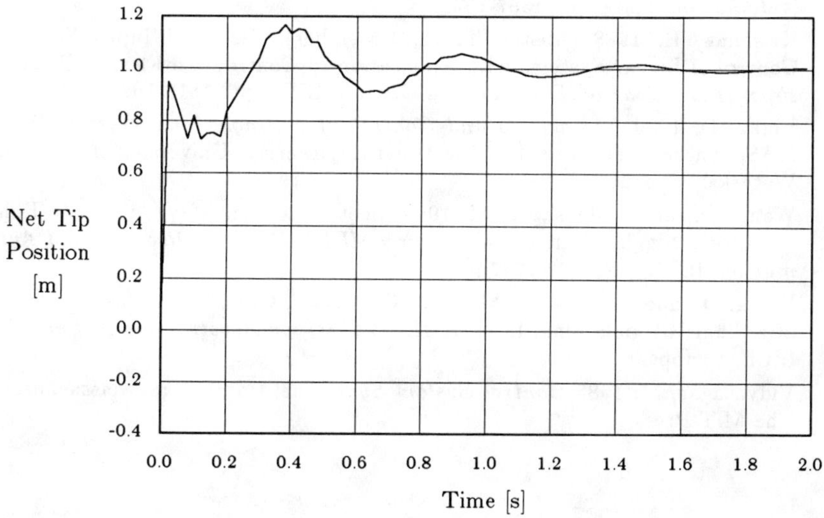

Figure 6 *Compensator using Differentiated Model (Position)*

5. Conclusions and Future Research

There are several interesting findings in this research. First, it is possible to design controllers with reduced torque requirements by removing the constraint that the controller be proper. The smoothness of the resulting responses are also desirable. This is reflected in the simulation results of Section 4. Second, it may now be possible to design an unconstrained H_2-optimal controller without having to resort to using the very complicated bounded-input H_2-optimal control strategy. Until now, all attempts to implement an unconstrained H_2-optimal controller have failed due to the very large torques required. Saturation of these torques due to motor limitations have resulted in instability in simulations. As well, any small disturbance in the system is counteracted by very large torques. We are hopeful that we have at last found a method to implement an unconstrained H_2-optimal controller.

Future research will concentrate on the following aspects. First, the physical reason for the improved torque requirements will be examined. Second, since the controller is improper, the implementation of this controller will be challenging. Finally, more general methods of finding improper controllers will be studied.

It is hoped that the preliminary results of this paper will serve to stimulate future research in this area.

References

1. Cannon, R.H., and Schmitz, E. 1984. Initial Experiments on the End-Point Control of a Flexible One-Link Robot. *Int. J. Robotics Res.* **3**(3):62-75.
2. Hastings, G., and Book, W.J. 1986 (San Francisco, California, April). Verification of a Linear Dynamic Model for Flexible Robotic Manipulators. In *Proc. IEEE Conf on Robotics and Automation,* IEEE, p. 1866-1870.
3. Wang, D., and Vidyasagar, M. 1987 (Rayleigh, NC, April). Control of a Flexible Beam for Optimum Step Response. In *Proc. IEEE Conf on Robotics and Automation,* IEEE, pp. 1567-1572.
4. Krishnan, H. 1988 (Austin, Texas, December). Bounded Input Feedback Control of Linear Systems with Application to Control of a Flexible Beam. In *Proc. IEEE Conf on Decision and Control,* IEEE, pp. 1619-1626.
5. Vinke, D. 1989. *Modelling and Control of a Single-Link Flexible Beam.* MASc Thesis, Department of Electrical Engineering, University of Waterloo, Waterloo, Canada
6. Wang, D., and Vidyasagar, M. 1989 (Scottsdale, Az, May). Transfer Functions for a Single Flexible Link. In *Proc. IEEE Conf. on Robotics and Automation,* IEEE, pp. 1042-1047.
7. Wang, D., and Vidyasagar, M. 1990 (Cincinnati, Ohio, May). Passive Control of a Single Flexible Link. In *Proc. IEEE Conf. on Robotics and Automation,* IEEE, to appear
8. Vidyasagar, M. 1985 *Control System Synthesis,* Cambridge, Massachusetts, The MIT Press.

CONTROL OF FLEXIBLE JOINT ROBOTS

FOUAD MRAD and SHAHEEN AHMAD
Real-time Robot Control Laboratory
School of Electrical Engineering
Purdue University
West Lafayette, IN 47907-0501
U.S.A.

ABSTRACT

Almost all industrial robots exhibit joint flexibility due to mechanical compliance of their gear boxes. In this paper we outline a design of an adaptive controller for flexible joint robots based on the arms energy deviating from the desired trajectory. The first stage drives the actuator and the joints to a desired manifold, the second controller then seeks to drive the joints to their desired trajectory. Our control scheme does not require link acceleration or jerk measurements, numerical differentiation of the velocity signal, or inversion of the inertial matrices. Simulations are presented to verify the validity of the control scheme, and superiority over existing rigid robot adaptive schemes.

INTRODUCTION

Recent trend is towards high-technology direct-drive robots, where the actuators are directly connected to links. In addition, the increasing demand for high-speed operation, requires the control system to compensate for the dominant nonlinear link dynamics. Experimental evidence indicates that joint flexibility should be accounted for in both modeling and control of manipulators [1, 4, 15]. Numerous techniques to control Flexible Joint Robots have been suggested [2, 3, 4, 6, 12, 15]. One approach is based on the idea of feedback linearization, which requires the measurement of joint acceleration and jerk to be used in the feedback loop [2, 12]. Another method is based on the concept of reduced order system and requires the restriction of the system to a suitable integral manifold in the state space [6].

Similar to the work on rigidly jointed robots [8, 11, 12], our controller design starts by selecting a Lyapunov function which is similar to the energy of the FJR. At this time, the only adaptive control scheme for flexible joint robots that we are aware of that uses position and velocity feedback is the one derived from singular perturbation arguments [4].

MANIPULATOR AND TRAJECTORY MODELS

The dynamic equations of the flexible joint robots are given as:

$$\tau = D_m \ddot{q}_m + B_m \dot{q}_m + K_s(q_m - q) \tag{1}$$

$$0 = D(q)\ddot{q} + C(q,\dot{q})\dot{q} + g(q) + K_s(q - q_m) \tag{2}$$

where, an n-link manipulator becomes a 2n-degrees of freedom system, and where D_m: Diagonal motor inertia matrix $\in \mathbb{R}^{n\times n}$; B_m: Diagonal motor damping matrix $\in \mathbb{R}^{n\times n}$; K_s: Diagonal drive shaft stiffness matrix $\in \mathbb{R}^{n\times n}$; q_m: Vector of sensed motor angles $\in \mathbb{R}^{n\times 1}$; $D(q)$: Link inertia matrix $\in \mathbb{R}^{n\times n}$; $C(q,\dot{q})$: Centrifugal and Coriolis terms matrix $\in \mathbb{R}^{n\times n}$; $g(q)$: Gravitational vector term $\in \mathbb{R}^{n\times 1}$; q: Vector of link joint angles $\in \mathbb{R}^{n\times 1}$.

Matrices D_m, B_m, K_s, $D(q)$, are positive definite matrices. Further, $D(q)$ is symmetric, and both $D(q)$ and $D^{-1}(q)$ are both bounded as function of q [4, 14]. The first most important property shows that $D(q)$ and $C(q,\dot{q})$ are not independent, but the matrix $(\dot{D}-2C)$ is skew symmetric, this can be easily derived from the Lagrangian formulation of the manipulator dynamics. The second property confirms that the individual terms of the right hand side of equation (2), excluding the $K_s(q-q_m)$ term, could be represented by a linear relationship between a suitably selected set of unknown manipulator and load parameters [4, 11, 12, 14], in other words equation (2) could be rewritten as:

$$0 = Y(\ddot{q},\dot{q},q)\,P + K_s\,(q - q_m) \tag{4}$$

where $Y(\ddot{q},\dot{q},q) \in \mathbb{R}^{n\times r}$, is called the regressor matrix of known functions, and $P \in \mathbb{R}^{r\times 1}$ is a vector of unknown parameters.

Let $q_d(t) \in C^4$ denote a desired link trajectory in which case $q_d(t), \dot{q}_d(t), \ddot{q}_d(t), \dddot{q}(t)$ are all bounded and continuously differentiable. The set of desired motor trajectory can be derived using equation (4). We assume the link parameters and the load handled by the end effector are time invariant, i.e.

$$P = \text{Constant vector, thus, } \dot{P} = \ddot{P} = 0 \tag{5}$$

The above assumption is valid in a large class of applications. The desired motor trajectory may now be computed as follows:

$$q_{md}(t) = K_s^{-1}Y(\ddot{q}_d,\dot{q}_d,q_d)P + q_d(t) \tag{6}$$

$$\dot{q}_{md}\,(t) = K_s^{-1}\dot{Y}(\ddot{q}_d,\dot{q}_d,q_d)P + \dot{q}_d(t) \tag{7}$$

$$\ddot{q}_{md}\,(t) = K_s^{-1}\ddot{Y}(\ddot{q}_d,\dot{q}_d,q_d)P + \ddot{q}_d(t) \tag{8}$$

The subscript "d" is used to denote the desired trajectory.

Notice that the desired motor trajectory $q_{md}(t)$, $\dot{q}_{md}(t)$ and $\ddot{q}_{md}(t)$ are dependent on the desired link trajectory $q_d(t)$, $\dot{q}_d(t)$ and $\ddot{q}_d(t)$, on the unknown parameters P, and on the link dynamic structure represented by Y_d, $\dot{Y}_d$ and $\ddot{Y}_d$.

SELECTION OF AN ENERGY BASED LYAPUNOV FUNCTION

Excluding the potential energy of the FJR, the energy of the robot arm along a prespecified trajectory is :

$$E(t) = \tfrac{1}{2}\dot{q}_d^t D\dot{q}_d + \tfrac{1}{2}(q_d - q_{md})^t K_s (q_d - q_{md}) + \tfrac{1}{2}\dot{q}_{md}^t D_m \dot{q}_{md} \tag{9}$$

Likewise, the energy in the system which causes the FJR to deviate from the desired trajectory is given as :

$$V(t) = \tfrac{1}{2}\dot{e}^t D\dot{e} + \tfrac{1}{2}(e - e_m)^t K_s (e - e_m) + \tfrac{1}{2}\dot{e}_m^t D_m \dot{e}_m \tag{10}$$

where, we define the error terms as: $e = (\,q_d - q\,)$ and $e_m = (\,q_{md} - q_m)$.

Throughout the trajectory it is desired to have $\frac{dV}{dt} \leq 0$, furthermore $\dot{V}(t)$ and $V(t)$ should be dependent on e and e_m as well as $\dot{e}_m$ and $\dot{e}$. We can make V(t) dependent on $\dot{e}$, e, $\dot{e}_m$ and e_m by selecting:

$$\begin{aligned} V(t) = {} & \tfrac{1}{2}\dot{e}_m^t D_m \dot{e}_m + \tfrac{1}{2}\dot{e}^t D\dot{e} + \tfrac{1}{2}(e - e_m)^t K_s (e - e_m) \\ & + \tfrac{1}{2}e^t K_p e + \tfrac{1}{2}e_m^t K_{pm} e_m \end{aligned} \tag{11}$$

where, $K_p \in \mathbb{R}^{n\times n}$, $K_{pm} \in \mathbb{R}^{n\times n}$ are some positive gain matrices.

CONTROL AND ADAPTATION LAW DESIGN

As the dynamic parameters of the arm and load are unknown, we can define the parameter error vector as $e_p = \tilde{P} - P$, where $\tilde{P}$ is the estimated parameter vector. Notice $\dot{e}_p = \dot{\tilde{P}}$, as $\dot{P} = 0$. Based on the estimated value of the parameter vector $\tilde{P}$, we obtain an estimate of the desired motor position as $\tilde{q}_{md}$ using equation (6). Similarly, we can compute the estimated motor velocity and acceleration. Define the following motor error as $\tilde{e}_m=(\tilde{q}_{md}-q_m)$. Similar terms for $\dot{\tilde{e}}_m$ and $\ddot{\tilde{e}}_m$ can be defined. Based on the above Lyapunov function (11), we can find the energy of the trajectory deviating from the desired trajectory as:

$$V(t) = \tfrac{1}{2}\dot{\tilde{e}}_m^t D_m \dot{\tilde{e}}_m + \tfrac{1}{2}\dot{e}^t D\dot{e} + \tfrac{1}{2}(e-\tilde{e}_m)^t K_s (e-\tilde{e}_m)$$
$$+ \tfrac{1}{2}e^t K_p e + \tfrac{1}{2}\tilde{e}_m^t K_{pm}\tilde{e}_m + \tfrac{1}{2}e_p^t M e_p \quad (12)$$

The last term in (12) is added to account for parameter adaptation, where $K_p, K_{pm} \in \mathbb{R}^{n\times n}$, and $M \in \mathbb{R}^{r\times r}$ are some positive gain matrices.
For convenience let us define:

$$D(q)\ddot{q}_d + C(q,\dot{q})\dot{q}_d + g(q) = \Psi(\ddot{q}_d, \dot{q}_d, \dot{q}, q)P \quad (13)$$

where, $\Psi \in \mathbb{R}^{n\times r}$, and $Y_d = Y(\ddot{q}_d, \dot{q}_d, q_d)$ where, $Y_d \in \mathbb{R}^{n\times r}$, and let Λ be some positive diagonal matrix $\in \mathbb{R}^{n\times n}$, then we let

$$\Gamma(\ddot{q}_d, \dot{q}_d, \dot{q}, q)P = D(q)[\ddot{q}_d + \Lambda\dot{e}] + C(q,\dot{q})[\dot{q}_d + \Lambda e] + g(q) \quad (18)$$

where, $\Gamma \in \mathbb{R}^{n\times r}$. Furthermore let us define the following variables:
$s \in \mathbb{R}^n$, $s=(\dot{e}+\Lambda e)$. Let us also define a region where $\ddot{\tilde{e}}_{mi} = \ddot{\tilde{q}}_{mdi} - \ddot{q}_{mi}$ as:

$$\mu_{min}(i) \leq \ddot{\tilde{e}}_{mi} \leq \mu_{max}(i) \quad \text{for } i=1,2,\ldots,n \quad (15)$$

where, $\mu_{min}(i)$, and $\mu_{max}(i)$ are real scalars. Let us also set vector $\lambda = (\lambda_1, \ldots \lambda_n)^t \in \mathbb{R}^n$ be defined such that:

$$\lambda_i = \tfrac{1}{2}D_{mi}\{Sgn(s_i)[\mu_{min}(i) - \mu_{max}(i)] + \mu_{min}(i) + \mu_{max}(i)\} \quad (16)$$

for $i=1,2,\ldots,n$. where, $Sgn(s_i) = \left\{ +1 \text{ if } s_i > 0;\ -1 \text{ if } s_i < 0;\ 0 \text{ if } s_i = 0 \right\}$

Theorem 1:

The system given by the dynamical model (1) and (2), subjected to the following two stage control and adaptation laws, achieves desired trajectory tracking.

$$\tau = D_m\ddot{\tilde{q}}_{md} + B_m\dot{q}_m + K_s(\tilde{q}_{md} - q_d) + K_{pm}\tilde{e}_m + K_{dm}\dot{\tilde{e}}_m$$
$$+ \frac{\dot{\tilde{e}}_m}{\|\dot{\tilde{e}}_m\|^2}[\dot{e}^t((\Psi - Y_d)\tilde{P} + K_p e) + \dot{e}^t K_d \dot{e} + e^t K_p e + \dot{\tilde{e}}_m^t K_{pm}\tilde{e}_m] \quad (17)$$

if $\|\dot{\tilde{e}}_m\|^2 > \epsilon > 0$, for a scalar ϵ, and

$$\text{otherwise,} \quad \tau = \Gamma(\ddot{q}_d, \dot{q}_d, \dot{q}, q)\tilde{P} + D_m\ddot{\tilde{q}}_{md} + B_m\dot{q}_m + K_d s - \lambda \quad (18)$$

where, K_{dm}, $K_d \in \mathbb{R}^{n\times n}$ are some positive gain matrices. Corresponding to the two stage control laws we have the following adaptation laws.

$$\text{if } \|\dot{\tilde{e}}_m\|^2 > \epsilon, \quad \dot{\tilde{P}}(t) = M^{-1}\Psi^t(\ddot{q}_d, \dot{q}_d, \dot{q}, q)\,\dot{e} \quad (19)$$

and otherwise,

$$\dot{\tilde{P}}(t) = M^{-1}\Gamma^t(\ddot{q}_d, \dot{q}_d, \dot{q}, q)[\dot{e} + \Lambda e] \quad (20)$$

Proof of Theorem 1:

Differentiation of the candidate Lyapunov function V(t) in equation (12) yields the following:

$$\dot{V}(t) = \dot{\tilde{e}}_m^t[D_m\ddot{\tilde{e}}_m + K_s(\tilde{e}_m - e) + K_{pm}\tilde{e}_m]$$
$$+\dot{e}^t[D\ddot{e} + \tfrac{1}{2}\dot{D}\dot{e} + K_p e + K_s(e - \tilde{e}_m)] + e_p^t M\dot{e}_p - \dot{e}^t(\tfrac{1}{2}\dot{D} - C)\dot{e} \tag{21}$$

In order to simplify equation (21), we have subtracted the term $\dot{e}^t(\frac{1}{2}\dot{D}-C)\dot{e} = 0$. Simplifying equation (21) and substituting the dynamic equation of the FJR given by (1) and (2), we have

$$\dot{V}(t) = \dot{\tilde{e}}_m^t[D_m\ddot{\tilde{q}}_{md} + K_s(\tilde{q}_{md} - q_d) + K_{pm}\tilde{e}_m + B_m\dot{q}_m - \tau]$$
$$+ \dot{e}^t[D\ddot{q}_d + C\dot{q}_d + K_s(q_d - \tilde{q}_{md}) + K_p e + g(q)] + e_p^t M\dot{e}_p \tag{22}$$

Let us now assume that $||\dot{\tilde{e}}_m||^2 > \epsilon > 0$, then by substituting the controller (17) into (22), using the definition of Ψ given by equation (13), and by using the fact that $K_s(q_d - \tilde{q}_{md}) = -Y_d P$ derived from (6), we get:

$$\dot{V}(t) = \dot{\tilde{e}}_m^t\{-K_{dm}\dot{\tilde{e}}_m - \frac{\dot{\tilde{e}}_m}{||\dot{\tilde{e}}_m||^2}[\dot{e}^t((\Psi - Y_d)\tilde{P} + K_p e) + \dot{e}^t K_d\dot{e} + e^t K_p e + \tilde{e}_m^t K_{pm}\tilde{e}_m]\}$$
$$+ \dot{e}^t[\Psi P - Y_d\tilde{P} + K_p e] + e_p^t M\dot{e}_p$$
$$= -\dot{\tilde{e}}_m^t K_{dm}\dot{\tilde{e}}_m - \dot{e}^t K_d\dot{e} - e^t K_p e - \tilde{e}_m^t K_{pm}\tilde{e}_m + e_p^t[M\dot{e}_p - \Psi^t\dot{e}] \tag{23}$$

Since, $\dot{e}_p = \dot{\tilde{P}} - \dot{P}$, and as, $\dot{P} = 0$ we can substitute the adaptation law (19) into (23) and the final expression for the derivative of the Lyapunov function is given as:

$$\dot{V}(t) = -\dot{\tilde{e}}_m^t K_{dm}\dot{\tilde{e}}_m - \dot{e}^t K_d\dot{e} - e^t K_p e - \tilde{e}_m^t K_{pm}\tilde{e}_m \le 0 \tag{24}$$

Which guarantees the convergence of $\dot{\tilde{e}}_m$, $\dot{e}$, $\tilde{e}_m$, and e as time goes to infinity.

Let us now consider the case when $||\dot{\tilde{e}}_m||^2 \le \epsilon$. We cannot use τ defined in equation (17) as $||\dot{\tilde{e}}_m||^2 \to 0$, then $\tau \to$ large value. Notice now as $||\dot{\tilde{e}}_m||^2 \to 0$, **a structural reduction in the system is** apparent as the "Lyapunov" function V(t) in (12) resembles that of a rigid robot, as the first term is zero. We will exploit this property in the second stage control. The dynamic equations (1) and (2) can be combined by simple addition to obtain a single system equation:

$$\tau = D(q)\ddot{q} + C(q,\dot{q})\dot{q} + g(q) + D_m\ddot{q}_m + B_m\dot{q}_m \tag{25}$$

Let us define a surface, $s = (\dot{e} + \Lambda e) \in \mathbb{R}^n$, along which we desire the link trajectory to track. Let us consider the Lyapunov function candidate W(t):

$$W(t) = \tfrac{1}{2}(\dot{e} + \Lambda e)^t D(\dot{e} + \Lambda e) + \tfrac{1}{2}e_p^t M e_p \tag{26}$$

Differentiating W(t) with respect to time, substituting the defined value of Γ given by (14), and using the dynamic equation (25) leads us to:

$$\dot{W}(t) = s^t[D(\ddot{e} + \Lambda\dot{e}) + \tfrac{1}{2}\dot{D}(\dot{e} + \Lambda e)] + e_p^t M\dot{e}_p$$
$$= s^t[\Gamma\, P + D_m\ddot{\tilde{q}}_{md} - D_m\ddot{\tilde{e}}_m + B_m\dot{q}_m - \tau] + e_p^t M\dot{e}_p \tag{27}$$

Substituting for τ from equation (18) into (27) yields,

$$\dot{W}(t) = s^t[-\Gamma\, e_p - K_d s - D_m\ddot{\tilde{e}}_m + \lambda] + e_p^t M\dot{e}_p$$
$$= -s^t K_d s - s^t[D_m\ddot{\tilde{e}}_m - \lambda] + e_p^t[M\dot{e}_p - \Gamma^t s] \tag{28}$$

Now substitute $\dot{\tilde{P}}(t)$ given by equation (20) into (28),

$$\dot{W}(t) = -s^t K_d s - s^t[D_m \ddot{\tilde{e}}_m - \lambda] + e_p^t[MM^{-1}\Gamma^t s - \Gamma^t s]$$
$$= -s^t K_d s - \sum_{i=1}^{n} s_i (D_{mi} \ddot{\tilde{e}}_{mi} - \lambda_i) < 0 \qquad (29)$$

SIMULATION RESULTS FOR A TWO-LINK PLANAR FJR

We now describe the computer simulation for a two-link planar manipulator with revolute joints (see Figure 1). We selected a sampling period of 10×10^{-3} seconds corresponding to a servo rate of 100 Hz. We selected, $\epsilon = 1$, and second controller was activated when, $\|\dot{\tilde{e}}_m\|^2 < \epsilon$. The following bounds were used in the definition of λ_i $-2 \leq \|\ddot{\tilde{e}}_{mi}\| \leq 2$, for i =1,2. Table S-1 shows the numerical values of the parameter vector P, and the known motors parameters. As seen from table (S-1), the robot considered

TABLE S-1. Actual parameters values

Parameters	Vector P
$K_{s1}=k_{s2}=50$ N m rad^{-1}	$P_1 = 1.66$
$D_{m1}=d_{m2}=.05$ kg m^2	$P_2 = 0.42$
$D_{m1}=b_{m2}=.05$ N m sec rad^{-1}	$P_3 = 0.63$
$l_1=l_2=.25$ m	$P_4 = 3.75$
$m_1=m_2=5$ kg	$P_5 = 1.25$

here has extremely flexible joints. In the below simulations, we assume the manipulator is initially at rest with $q_1 = -90°$, and $q_2 = 0°$.

Case #1: In order to show that current rigid robot adaptive schemes are ineffective when applied to FJR. In this case, we applied the elegant adaptive control schemes suggested by [12] to the FJR described in table (S-1). The controller gains were found to be $\Lambda=K_p=K_{pm}=0.25I_{2\times2}$, $M=K_d=K_{dm}=I_{2\times2}$. The response of the manipulator to Slotine and Li's adaptive control law was unstable. We expect that all other rigid robot adaptive control schemes would also produce unstable responses when applied to control FJR's with such low joint stiffness.

Case #2: In order to show the effectiveness of the results derived in this paper given by equations (17) through (20), we applied our control scheme to the FJR. The response is given in Figures (2a, 2b, and 2c). We can see that the motor and the link tracking errors go to zero. The parameters also do not diverge, although they do not converge to their exact actual values, they oscillate about their true values. The controller gain matrices were $K_p=K_{pm}=10I_{2\times2}$, $K_d=K_{dm}=3I_{2\times2}$, $\Lambda=M=2I_{2\times2}$.

CONCLUSION

In this paper we have presented an adaptive control scheme for the FJR without employing linearization techniques such as [3]. Acceleration and jerk measurements, as well as inertia matrix inversion were not needed. Adaptive controller for the FJR was derived using Lyapunov's second method. From the simulation results, it is clear that the improvement in the tracking and parameter estimation is significant over rigid robot adaptive schemes, and therefore it is necessary to account for joint flexibility effects when deriving control schemes for industrial robots with such compliance. We compared our adaptive scheme to the one developed in [4], their scheme was derived from a singular perturbation argument and gave similar results to our energy based adaptive controller.

It is obvious that some correction scheme could be added to the derived adaptation law to improve the robustness of our controller in the presence of bounded disturbances or unmodeled dynamics [5].

REFERENCES

1. S. Ahmad, "Analysis of Robot Drive Train Errors, their Static Effects and their Compensations," *IEEE Journal of Robotics and Automation,* Vol. 4, No. 2, April 1988.
2. A. De-Luca, "Dynamic Control of Robots with Joint Elasticity," *Proceedings of 1988 IEEE International Conference on Robotics and Automation,* Philadelphia, PA, pp. 152-158, April 1988.
3. L. Fu and K. Chen, "Nonlinear Adaptive Motion Control for a Manipulator with Flexible Joints," *Proceedings of 1989 IEEE International Conference on Robotics and Automation,* Phoenix, Az, pp. 1201-1206, May 1989.
4. F. Ghorbel, W. Spong and J. Hung, "Adaptive Control of Flexible Joint Manipulators," *Proceedings of 1989 IEEE International Conference on Robotics and Automation,* Phoenix, Az, pp. 1188-1193, May 1989.
5. P. Ioannou, " Robust Adaptive Controller with Zero Residual Tracking Errors," *IEEE Trans. Automatic Control,* Vol. 31, No. 8, pp. 773-776, Aug 1986.
6. K. Khorasani, "Robust Adaptive Stabilization of Flexible Joint Manipulators," *Proceedings of 1989 IEEE International Conference on Robotics and Automation,* Phoenix, Az, pp. 1194-1199, May 1989.
7. P. Khosla, "Categorization of Parameters in the Dynamic Robot Model," *IEEE Trans. Robotics and Automat.,* Vol. 5, No. 3, pp. 261-268, June 1989.
8. D. Koditschek, "Lyapunov Analysis of Robot Motion," *Tutorial Workshop of 1987 IEEE International Conference on Robotics and Automation,* Raleigh, NC, pp. 1B-3-1 to 1B-3-56, March 1987.
9. R. Paul, *Robot Manipulators: Mathematics, Programming and Control,* MIT Press, 1981.
10. S. Sastry, P. Hsu, M. Bodson and B. Paden, "Adaptive Identification and Control for Manipulators without Using Joint Accelerations," *Proceedings of 1987 IEEE International Conference on Robotics and Automation,* Raleigh, NC, pp. 1210-1215, March 1987.
11. H. Seraji, "Decentralized Adaptive Control of Manipulators: Theory, Simulation, and Experimentation," *IEEE Journal of Robotics and Automation,* Vol. 5, No. 2, pp. 183-201, 1989.
12. J. Slotine and W. Li, "On the Adaptive Control of Robot Manipulators," *Int. J. Robotics Research,* Vol. 6, No. 3, pp. 49-59, Fall 1987.
13. M. Spong, K. Khorasani and P. Kokotovic, "An Integral Manifold Approach to the Feedback Control of Flexible Joint Robots," *IEEE J. of Robotics and Automat.,* Vol. RA-3, No. 4, pp. 291-300, Aug 1987.
14. M. Spong, "Modeling and Control of Elastic Joint Robots," *Transactions of the ASME,* Vol 109, pp. 310-319, Dec 1987.
15. G. Widmann and S. Ahmad, "Control of Robots with Flexible Joints," *Proceedings of 1987 IEEE International Conference on Robotics and Automation,* Raleigh, NC, pp. 1561-1566, March 1987.

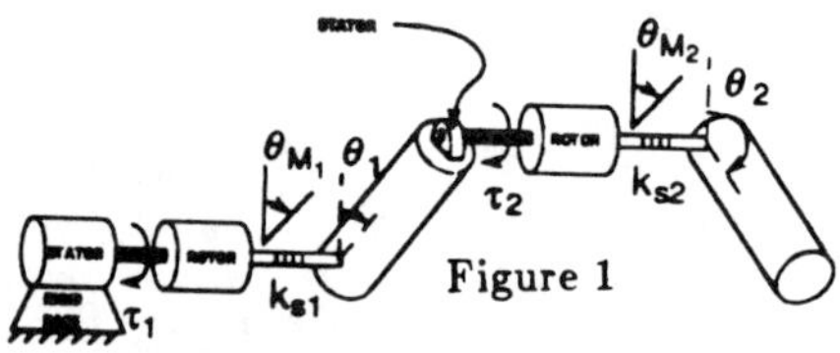

Figure 1

Figure 2b: Motors tracking

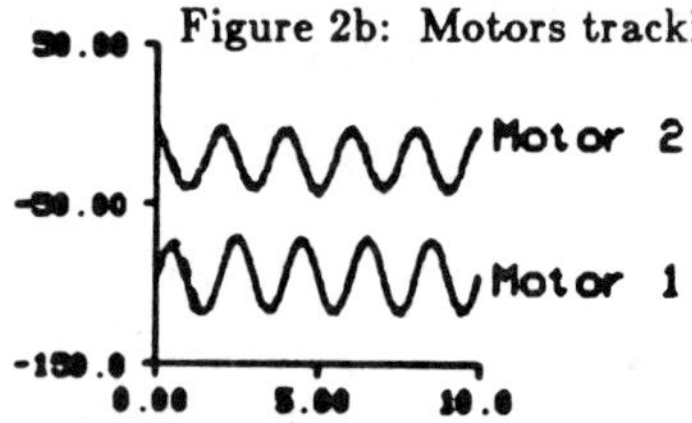

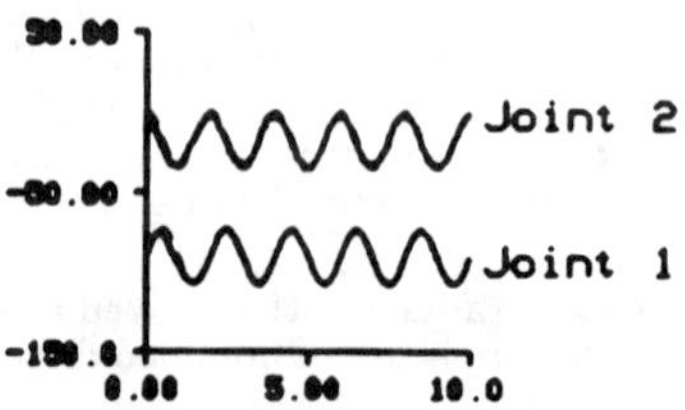

Figure 2a: Links tracking

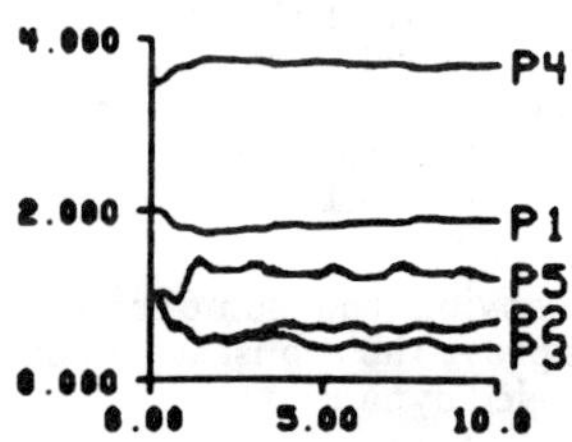

Figure 2c: Parameters

Force Control for a Flexible Link Manipulator Using Generalized Predictive Control

D.J. Latornell[1] and D.B. Cherchas[2]
CAM/Robotics Laboratory (CAMROL)
Department of Mechanical Engineering
University of British Columbia
Vancouver, B.C., Canada

Abstract

The problem of controlling the force exerted by a structurally flexible manipulator link on its environment is addressed. An adaptive implementation of the Generalized Predictive Control (GPC) algorithm is used to design control laws which give rapid response, good setpoint tracking and zero steady–state error. A simulation of a single flexible manipulator link contacting a deformable surface has been developed to test the algorithm. The simulation includes a beam–type continuum model of the manipulator link and a lumped–parameter model of the surface dynamics in the region of contact. Simulation results are presented which demonstrate that the adaptive GPC algorithm has the ability to produce complex specified trajectories of contact force and provide smooth operation during making and breaking of contact.

Introduction

The ability to sense and control the forces and torques exerted by a manipulator on its environment has, for some time, been recognised as a necessary and beneficial extension of robotic system capabilities. Force sensing and control offers improved performance of tasks which involve contact of the end–effector with the environment by providing robustness to variations in location and shape of the objects being manipulated. In addition, force control makes possible tasks not possible under purely positional control.

Manipulators with structurally flexible links such as the Space Shuttle Remote Manipulator System (or Canadarm) will, in the near future, be required to operate under conditions where force control is essential. Among the the autonomous tasks which have been proposed for the space station

---

[1] Research Assistant, supported by Post-graduate Scholarships from The Natural Sciences and Engineering Research Council of Canada (NSERC) and the British Columbia Advanced Systems Institute.

[2] Professor and Founding Fellow of the British Columbia Advanced Systems Institute.

manipulators are maintenance activities on the exterior of the station such as cleaning of the view ports, clearly requiring control of contact forces in the presence of significant structural flexibility of the manipulator. Light weight earth–based manipulators designed for high speed motion can also be expected to exhibit structural flexibility which will need to the accounted for in the design of both force and motion controllers.

The importance of compliance, associated with either the manipulator or the environment, in achieving stable force control with existing robots and controllers is well established [7]. Manipulators with structurally flexible links naturally provide compliance which is distributed over the entire manipulator structure. This distributed compliance is helpful in achieving high performance force control by absorbing the impulse which occurs when contact is made. Large initial contact forces are currently avoided in most robotic force control operations by approaching the environment at very low speed. Clearly, to take advantage of this shock absorption ability, the compliant nature of the links must be included in the controller design.

GPC Algorithm

The Generalized Predictive Control (GPC) algorithm [2] is a single–input, single–output discrete time algorithm which yields control laws with such desirable characteristics as inherent integral action and robustness to inaccuracies in both model order and delay time. The algorithm is based on prediction of the plant output for a series of future time steps. The predictions are expressed in a form which allows solution for an optimal (in the sense of minimisation of a quadratic cost function) set of future inputs that will produce the desired response. This procedure is conducted in a receding horizon context, that is, a new set of inputs are calculated at each time step and only the first member of the set is used. Receding horizon predictive control predates GPC in the process control literature but, as the name suggests, the GPC algorithm is a generalized form of the method. Indeed, it has been shown [2] that, depending on the choice of design parameters, the GPC algorithm results in control laws identical to a variety of well known algorithms [1, 5, 8].

The GPC algorithm is based on the input–output representation of the of the controlled plant by a Controlled Auto–Regressive and Integrated Moving–Average [6] model in the discrete time domain:

$$A(q^{-1})y(t) = B(q^{-1})u(t-1) + \frac{\xi(t)}{\Delta} \tag{1}$$

where A and B are polynomials, of degree na and nb respectively, in the backward shift operator q^{-1} (i.e. $q^{-1}x(t) = x(t-1)$), $\xi(t)$ is an uncorrelated random sequence and $\Delta = 1 - q^{-1}$ is the differencing operator. This model assumes that the disturbances to which the plant is subject can be modelled as steps of random magnitude occurring at random times.

Using the well known Diophantine identity (1) can be manipulated to obtain a j–step ahead predictor of the plant output:

$$y(t+j) = F_j(q^{-1})y(t) + E_j(q^{-1})B(q^{-1})\Delta u(t+j-1) + E_j(q^{-1})\xi(t+j). \quad (2)$$

The predicted output $y(t+j)$ is the sum of filtered sequences of output, input and disturbances. Given the assumed random nature of $\xi(t)$ and the fact that the degree of E_j is $j-1$ the contribution to $y(t+j)$ of the disturbance sequence is not only unknown but unknowable. Hence, the optimal predictor of $y(t+j)$ is:

$$\hat{y}(t+j) = F_j y(t) + G_j \Delta u(t+j-1), \quad (3)$$

where $G_j = E_j B$. This prediction of $\hat{y}(t+j)$ can be written as a term, $f(t+j)$, which consists entirely of past inputs and outputs (in effect the free response of the system) and a term due to the future input increments which are to be determined.

Considering predictions N steps into the future the following vectors are defined:

$$\begin{aligned} \hat{\mathbf{y}} &= [y(t+1) \quad y(t+2) \quad \ldots \quad y(t+N)]^T \\ \tilde{\mathbf{u}} &= [\Delta u(t) \quad \Delta u(t+1) \quad \ldots \quad \Delta u(t+N-1)]^T \\ \mathbf{f} &= [f(t+1) \quad f(t+2) \quad \ldots \quad f(t+N)]^T \end{aligned}$$

and a set of N prediction equations of the form of equation (3) are written as:

$$\hat{\mathbf{y}} = \mathbf{G}\tilde{\mathbf{u}} + \mathbf{f}, \quad (4)$$

where $\mathbf{G}$ is an $N \times N$ lower triangular matrix whose elements are the coefficients of the G_j polynomials corresponding to the yet to be determined future inputs.

The quadratic cost function $J = (\hat{\mathbf{y}} - \mathbf{w})^T(\hat{\mathbf{y}} - \mathbf{w}) + \lambda\tilde{\mathbf{u}}^T\tilde{\mathbf{u}}$ is minimised to obtain an optimal set of future inputs which will produce the desired plant response. The vector $\mathbf{w} = [w(t+1)\ w(t+2)\ \ldots\ w(t+N)]^T$ is a sequence of future setpoints, and λ is a cost weighting associated with changes in the future inputs. Substituting from (4) and minimising J with no constraints on future inputs yields the control law:

$$\tilde{\mathbf{u}} = (\mathbf{G}^T\mathbf{G} + \lambda\mathbf{I})^{-1}\mathbf{G}^T(\mathbf{w} - \mathbf{f}) \quad (5)$$

Note that the first element of $\tilde{\mathbf{u}}$ is $\Delta u(t)$ which, given the receding horizon method of applying the control law, is the only input increment which will be used.

The control law represented by (5) yields an offset free response but is subject to certain restrictions and deficiencies [2], not the least of which is the computational load of inverting the $N \times N$ matrix $(\mathbf{G}^T\mathbf{G} + \lambda\mathbf{I})$ at each timestep (in an adaptive implementation). To improve the robustness of the algorithm, and at the same time reduce its computational load, the

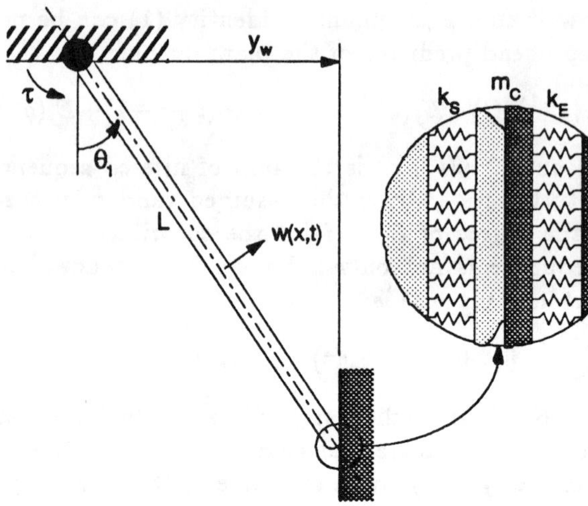

Figure 1: Geometry of a single flexible link in contact with a surface

concept of a control horizon [3] is introduced [2]; that is, after some number of time steps, $N_U < N$, the input level is held constant. Imposition of this condition results in the vector $\tilde{\mathbf{u}}$ being truncated from N to N_U elements and thus the matrix $\mathbf{G}$ in (4) being replaced by the $N \times N_U$ matrix $\mathbf{G}_1$ which is the matrix $\mathbf{G}$ with the left most $N - N_U$ columns truncated. The resulting control law has the same form as (5) but with $\mathbf{G}_1$ substituted for $\mathbf{G}$ and only requires the inversion of an $N_U \times N_U$ matrix. Furthermore, the non-triangular structure of the matrix $\mathbf{G}_1$ guarantees that $\mathbf{G}_1^T\,\mathbf{G}_1$ will be nonsingular. It is the inclusion in the GPC algorithm of the control horizon concept that is at the heart of its ability to control difficult plants such as those with unknown or variable plant delay, non-minimum phase characteristics or lightly damped, poorly modelled modes.

Implementation of the GPC algorithm as an explicit adaptive controller is straight forward. At each time step the parameters of the plant transfer function are estimated using a Recursive Least Squares (RLS) estimation algorithm with U/D factorisation of the covariance matrix. The estimated parameters are assumed to be the true plant parameters and the input increment is calculated using (5). At the next time step a new set of plant parameters is estimated and used to calculate a new input increment.

Dynamic Model

Figure 1 shows a schematic of a single flexible link in contact with a surface. The orientation of the link is given by the angle θ_1. The shape of the structurally deformed link is given by $w(x_1, t)$, the displacement of actual centreline from the nominal centreline of the undeformed link. The location of the contact point relative to the joint is y_w.

The link is modelled as an elastic beam clamped to a rotary actuator at the joint end. Rotary inertia and shear deformation of the beam are neglected as is axial foreshortening due to deflection. The motion of the beam is assumed to be such that the positional and temporal variation of its shape can be separated, that is, $w(x_1, t)$ is modelled as a linear combination of a set of orthogonal mode shape functions, $\phi_i(x_1)$, and a set of modal amplifiers, $\psi_i(t)$. The contacting surfaces of the link and the environment are modelled as simple masses on damped elastic foundations. The deformation of the surfaces during contact is represented by the deflection, ε_C of the contact mass m_C (the sum of the surface masses).

The kinematics of the system are described by a set of homogeneous transformation matrices and Lagrange's equations are employed to derive the equation of motion of the system. Selecting ε_C, θ_1 and ψ_i, $i = 1, 2, \ldots$ as generalized coordinates the equations of motion are [4]:

$$m_C\ddot{\varepsilon}_C = -b_E\dot{\varepsilon}_C - (k_S + k_E)\varepsilon_C + k_S(Ls_1 + w_o c_1 - y_w) \tag{6}$$

$$\begin{aligned}(I_0 + \sum_{i=1}^{\infty} I_{2i}\psi_i^2)\ddot{\theta}_1 + \sum_{i=1}^{\infty} I_{1i}\ddot{\psi}_i + m_C\ddot{\varepsilon}_C &= \tau_1 - b_{\theta_1}\dot{\theta}_1 \\ &-2\sum_{i=1}^{\infty} I_{2i}\psi_i\dot{\psi}_i\dot{\theta}_1 - k_S(Lc_1 - w_o s_1)(Ls_1 + w_o c_1 - y_w - \varepsilon_C) \\ &-\frac{1}{2}ML(g_x s_1 - g_y c_1) - (g_x c_1 + g_y s_1)\sum_{i=1}^{\infty} I_{4i}\psi_i\end{aligned} \tag{7}$$

$$\begin{aligned}I_{1i}\ddot{\theta}_1 + I_{2i}\ddot{\psi}_i = -b_i\dot{\psi}_i + I_{2i}\dot{\theta}_1^2\psi_i - I_{3i}\psi_i - (g_x s_1 - g_y c_1)I_{4i} \\ -k_S\phi_{io}c_1(Ls_1 + w_o c_1 - y_w - \varepsilon_C), \quad i = 1, 2, \ldots\end{aligned} \tag{8}$$

where $s_1 = \sin\theta_1$, $c_1 = \cos\theta_1$, $w_o = w(L, t)$, L is the link length, M is its mass and g_x and g_y are components of the gravitational field.

The Is in these equations represent integrals whose values depend on the geometry, mass distribution and structural characteristics of the link. They are defined as follows:

$I_0 = \int_M (L - x_1)^2\, dm$; the mass moment of inertia of the undeformed link;

$I_{1i} = \int_M (L - x_1)\phi_i\, dm$; the coupling inertia between mode i and the rigid body motion of the link;

$I_{2i} = \int_M \phi_i^2\, dm$; the second moment of modal mass of mode i;

$I_{3i} = \int_0^L (EI)(\frac{\partial^2}{\partial x_1^2}\phi_i)^2\, dx_1$; the modal stiffness of mode i; and

$I_{4i} = \int_M \phi_i\, dm$; the first moment of modal mass of mode i.

Note that in the absence of contact equation (6) disappears as do the terms in (7) and (8) which involve k_S and the equations of motion reduce to those for a flexible link in free motion.

Simulation Results

The equations of motion and the GPC algorithm have been coded in a simulation called FLEXLINK. In most configurations of practical importance the natural frequencies of the system span several orders of magnitude and the equation of motion are stiff. To deal with this situation efficiently FLEXLINK uses Gear's algorithm to numerically integrate the dynamic model.

The GPC algorithm is used to design a control law for a torque servoed joint actuator. This approach allows control of both the contact force and, by controlling the joint angle, the gross motion of the link. The transfer function models used in these two modes of control are necessarily distinct and of different order. The two models and the corresponding output history data are identified and maintained in parallel so that switching between control modes can be accomplished smoothly. Switching of the controller from one mode to the other occurs at specified times which are selected as part of the setpoint program generation. This form of control mode switching is a simplified model of a trajectory generator that uses information from an ancillary world model to calculate the desired timing of mode switches. As will be seen in the simulation results presented below it is not necessary to switch from motion to force control at the exact moment of contact.

Figure 2 shows results for a test case in which the link was modelled as an aluminium rod $2.0cm$ in diameter and $1.0m$ long. The first two structural modes of the link were included in the calculation. The natural frequencies of these modes when the link is considered as a simple cantilever beam are $14.6Hz$ and $91.4Hz$. The surface stiffnesses k_E and k_S in the contact model were both set at $5000N/m$. This results in the dynamic model of the link in contact with the environment having open loop natural frequencies of $14.4Hz$, $66.4Hz$, $306.5Hz$ and $357.5Hz$. The gravitational field was assumed to be perpendicular to the plane of the link motion.

The test involves control of both gross motion of the link and the contact force exerted on the environment. The calculated response of the joint angle and contact force are shown as solid lines while the dashed lines represent the setpoint levels. Table 1 summarises the series of events specified in the joint angle and contact force setpoint programs. During the initial joint angle ramp command the controller provides excellent tracking. As the joint angle approaches $15°$ the link tip makes contact with the environment at an approach velocity of approximately $0.5m/s$. A contact force spike of approximately $6N$ occurs followed by the tip briefly leaving the surface and recontacting it with a lesser impact force. Note that the link is still under joint angle control during these contact events. As the response settles to the $15°$ setpoint level the link deflects to accommodate the presence of the environment and a steady contact force of approximately $2N$ results.

Following the switching of the controller from joint angle mode to con-

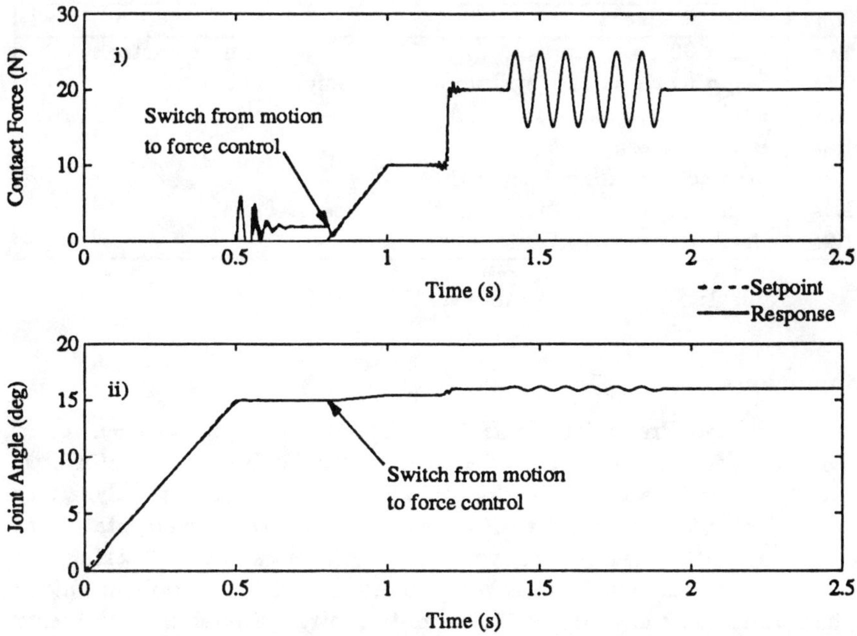

Figure 2: Setpoint and response plots for FLEXLINK test case; i) Contact force; ii) Joint angle

tact force mode at $0.8s$ tracking of a force ramp command is demonstrated. Next, the very quick step response of the force controller is exhibited. This quick response comes at the price of an initial overshoot of approximately 5%. Finally, the ability of the controller to track a sinusoidally varying setpoint is shown. Note that the $12Hz$ frequency of the setpoint signal is very close to the first natural frequency of the open loop system.

An interesting observation regarding this and other tests of the GPC algorithm in this context is that it is not necessary to change the set of horizon values and the control weighting factor when control is switched from one mode to the other. The horizons and weighting factor are effectively design parameters for the GPC algorithm since their choice determines the structure of the feedforward and feedback terms in the control law. The fact that the appropriate choices of values for these parameters, and hence for the controller structure, is apparently independent of whether or not contact exists is somewhat surprising and not yet fully understood.

Time	*Setpoint Event*
0.0s	Begin $30°/s$ ramp from 0° to 15° with joint angle control.
0.5s	Begin dwell at 15° (still under joint angle control).
0.8s	Begin $50N/s$ ramp from $0N$ to $10N$ with contact force control.
1.0s	Begin dwell at $10N$.
1.2s	Step contact force from $10N$ to $20N$.
1.4s	Begin $\pm 5N$ oscillation of contact force at $12Hz$.
1.9s	Begin dwell at $20N$.

Table 1: Setpoint program events for FLEXLINK test case

Conclusions

The Generalized Predictive Control (GPC) algorithm [2] has been found to be well suited to the control of a structurally flexible manipulator link. Using FLEXLINK, a newly developed code for simulation of the dynamics of a flexible link interacting with an environment, an implementation of the GPC algorithm has been shown to give excellent setpoint tracking and very quick step response. Moreover, exploitation of the compliant nature of the link in the controller design permits control through impact events at relatively high tip velocity.

Further work aimed at developing an understanding of the relationship between the structural characteristics of the link and selection of appropriate design parameters (horizons) for the control algorithm is underway. Also, extension of the control algorithm and the FLEXLINK simulation to permit study of multiple link manipulators is being undertaken.

References

1. D.W. Clarke and P.J. Gawthrop, "Self-tuning controller", IEE Proc. 122(9), 929–934 (1975).

2. D.W. Clarke, C. Mohtadi, and P.S. Tuffs, "Generalized predictive control — Parts I and II.", Automatica 23(2), 137–160 (1987).

3. R.C. Cutler and B.L. Ramaker, "Dynamic matrix control — A computer control algorithm", 1980 JACC, San Francisco (1980).

4. D.J. Latornell, "Force and motion control of a single flexible manipulator link", UBC CAMROL Report 90-17 (March 1990).

5. V. Peterka, "Predictor-based self-tuning control", Automatica 20(1), 39–50 (1984).

6. P.S. Tuffs and D.W. Clarke, "Self-tuning control of offset: A unified approach", IEE Proc. 132(3), 100–110 (1985).

7. D.E. Whitney, "Historical perspective and state of the art in robot force control", Int. J. Robotics Res. 6(1), 3–14 (1987).

8. B.E. Ydstie, "Extended horizon adaptive control", 9th IFAC World Congress., Budapest (1984).

DEVELOPMENT OF POSTECH 7-DOF DIRECT DRIVE ROBOT

K.W. Jeong*
* Research Scientist, Division of Mechatronics,
Research Institute of Industrial Science and Technology

W.K.Chung** and Y.Youm**
** Professor, Department of Mechanical Engineering
Pohang Institute of Science and Technology and
Research Scientist, Division of Mechatronics, RIST
Pohang, Korea

ABSTRACT

A 7-degrees-of-freedom redundant robot was developed using a direct drive mechanism for all 7 joints. In this paper we discuss the design of the POSTECH arm. A design criterion to determine the lengths of each link was proposed to give isotropic motion and the selected link ratio was found to follow the well known Fibonacci sequence for the first planar 3 links. To control the robot a real time computing environment based on the SUN 3/260 workstation was used and the performance of the robot system is discussed through experiments.

INTRODUCTION

The first prototype of the direct drive robot was developed in 1981 at the Robotics Institute of Carnegie Mellon University[1] as a possible alternative to overcome the mechanical uncertainties in the robot drive mechanism. As manufacturing technology of direct drive motors and rare-earth magnets improves, robots with a direct drive mechanism [2] have become widely used in industrial applications. The direct drive robot is defined to be an arm which eliminates transmission between actuators and linkgages. Due to the advantages of direct drive mechanism, the robot can move faster than the conventional robot and can be easily used for compliant motions such as assembly jobs where the robot contacts with environments. In addition to the research of this new drive mechanism, research on a redundant manipulator is also actively going on to handle the obstacle avoidance problem and to give more flexibility to the conventional robot [3,4]. The POSTECH direct drive robot which is a redundant robot was designed to have 7 dof (4 dof for positioning and 3 dof for wrist motion) using the direct drive mechanism for all the joints. In the following, the robot's design specifications, controller architecture and performance of the robot system will be explained.

Design of the Robot

The overall configuration of the robot is shown schematically in Fig.1 when the robot is fully stretched. This robot can cover a very large cylindrical workspace with an outer radius of 1330 mm. Due to the use of redundant link in design the flexibility of motion is increased. The robot is a SCARA type one considering the maximum power of the state-of-the-art of the direct drive motors. The hardware specifications of the POSTECH direct drive robot are listed as follows:

1) Number of degrees of freedom : 7 DOF
2) Configuration : SCARA type
3) Drive Mechanism : Direct drive
4) Actuator : AC torque ring motor
5) Sensor : 16 bit absolute encoder
6) Payload : 3 kg at high speed
7) Control : real time MC68020 based computing environment

In designing the robot, our first consideration was to determine the lengths of each link. For the non-redundant manipulator, the manipulability proposed by Yoshikawa[3] which was defined by

$$W = \sqrt{\det (JJ^T)} \tag{1}$$

is a generally accepted criterion for kinematic performance measure, where J is the manipulator Jacobian. It is maximum when the lengths of the forearm and upper arm are the same for a planar 2 link manipulator. For a redundant manipulator, however, there is no generally accepted criterion to select the link lengths of the manipulator. For a planar 3 link manipulator as shown in Fig.2, if we select a point (x,y) and θ_4 as parameters, then we can obtain the inverse kinematics solution as follows:

$$x' = x - l_3 \cos\theta_4$$

$$y' = l_3 \sin\theta_4 - y$$

$$\theta_2 = -\cos^{-1}\left(\frac{x^2 + y^2 - l_1^2 - l_2^2}{2 l_1 l_2}\right) \tag{2}$$

$$\theta_1 = \operatorname{atan}\left(\frac{y'}{x'}\right) - \operatorname{atan}\left(\frac{l_2 \sin\theta_2}{l_1 + l_2 \cos\theta_2}\right)$$

$$\theta_3 = -\theta_4 - \theta_2 - \theta_1$$

If we fix l_1 as 0.5 meter and give a constraint as $l_2 + l_3 = 0.5$, then the maximum of the manipulability index is shown in Fig.3. This figure shows that the manipulability index increases as l_2 goes to zero. If we consider a large manipulability index as an indicator of good kinematic performance, then this implies that the selection of link lengths based on this index is not acceptable because a manipulator with a very small l_2 will move ackwardly. This point can be illustrated in Fig.4 where the possible range of configuration that the robot can take for different l_2 is shown. As can be seen in this figure, the flexibility in robot motion will decrease as l_2 goes to zero. Based on the above results, we can conclude that the manipulability measure is not a meaningful determinant in selecting the lengths of the links for a redundant manipulator, although it can give quantitative information on the kinematic performance for a given robot. Instead, another criterion, i.e., condition number is proposed in this paper as

$$C = \mathrm{cond}(JJ^T) = \frac{\lambda_{max}}{\lambda_{min}} \tag{3}$$

whose physical meaning is the isotropy of manipulator ellipsoid. The result using this criterion is shown in Fig.5. To minimize the maximum condition number that the robot can take for a given point, the second link length was chosen as 0.3 meter under the constraint of $l_2 + l_3 = 0.5$. Based on the above results, the link ratio l_1: l_2: l_3 = 5: 3: 2 was determined in our robot design for isotropic motion of the robot and interestingly this ratio is the well known Fibonacci sequence.

After selecting the lengths of the links, the motors were selected considering weight and power. The selected motors are shown in Table 1. The magnets of the motors are neodymium-dysprosium-iron-boron, which can offer the values of energy product equal to, and in some cases, higher than that of samarium cobalt magnets. This magnet has a high coercivity even at high temperatures and has excellent mechanical properties. The robot is driven by PWM power amplifiers which generate sine wave commutation currents using the position data from the encoder. The motor and power amplifiers were calibrated to produce maximum power for the given command input (+/- 10 VDC). The links and motor housings of the robot was fabricated using aluminum alloy 2024 and 7075 to reduce the weights and inertias of the links. In the design stage, a hole was made in the housing 1) to define the zero position of the robot and 2) to fix selected joints by inserting a steel pin through the hole in the housing into a matching slot in the shaft. A 16 bit absolute encoder was used as a position sensor for every joint. These encoders are electrically isolated for the rest of the manipulator with insulator spacers. The design of the control system including the main controller, single board computer and hardware interface circuits will be explained in the following section.

Control System Design

The main purpose of the control system design is to develop a computing environment, which enables research on advanced control schemes for the redundant arm and the direct drive robot. The schematic of the overall robot control system is shown in Fig.6. The SUN 3/260 is a file server of the SUN 3/160 which is configured as a client to the SUN3/260 through ethernet. The Ironics board, which is a single board computer using 68020/68882 processors and a parallel I/O board which is interfaced to joint-interface boards, are placed in the VME bus backplane of the SUN 3/160. The control software developed in the SUN is downloaded to the Ironics board through the VME bus under the CHIMERA[5] environment. The CHIMERA is a real time multi-processing operating system which enables a real time control and programming environment. Through the parallel I/O board, the control software reads the joint angles which are decoded from each joint absolute encoder, and writes torque commands to each corresponding servo motor. Under this environment it is possible for a user to interact with the control software via the SUN system, while the real time control software is running on the Ironics board[6]. Using the control software, the robot can be controlled with a user selected servo rate ranging from 10Hz to 1000Hz.

Experiments

The developed manipulator is shown in Fig.7 whose total weight is 81 kg. The performance of the robot was tested using the simple PID control scheme

$$\tau = K_p(\theta_d - \theta) + K_v(\dot{\theta}_d - \dot{\theta}) + K_i\int(\theta_d - \theta)dt \tag{4}$$

controlled at a 333Hz servo rate. For the given sinusoidal trajectory, the output response for the first 3 links are shown in the Fig.8. The tracking error is shown to be within 0.03 rad during the entire trajectory. The proportional and derivative gains used are : K_p = (600,300,150) and K_v = (100, 30, 20).

CONCLUSION

A 7-DOF direct drive robot was developed for the research on the direct drive mechanism and redundant robot. The criterion for selecting the link ratio for the redundant manipulator was proposed based on the isotropic motion of the robot. The resulting link ratio was found to be 5:3:2 for the first planar 3 links. This link ratio is a well known Fibonacci sequence which can be frequently observed in nature. The robot control system using the SUN workstation as a real time computing environment was used to control the robot.

Acknowledgement : This project was supported by the research grants from POSTECH and RIST.

REFERENCES

[1] T.Kanade and D.Schmitz, " Development of CMU Direct Drive Arm II", Technical Report CMU-RI-TR-85-5, CMU, Robotics Institute, 1985.

[2] H.Asada and K.Youcef-Toumi, Direct Drive Robots, The MIT Press, 1987.

[3] T. Yoshikawa," Analysis and Control of Robot manipulators with Redundancy," Robotics Research: The First International Symposium, MIT Press, pp.735-747, 1984.
[4] P.Hsu, J.Hauser and S.Sastry," Dynamic Control of Redundant Manipulators", 1988 IEEE Conf. on Robotics and Automation, pp.183-187, 1988.
[5] D.Stewart, D.Schmitz and P.K.Khosla, " CHIMERA II, Real Time Programming Environment", Robotics Institute, CMU, 1990.
[6] Y.Youm, "POSTECH 7-DOF Direct Drive Robot", Technical Report, POSTECH, 1990.

Table 1 Motor specifications

Description	cont. torque $(N \cdot m)$	K_t $(N \cdot m/A)$	O. D. (mm)	length (mm)	mass (kg)	torque/weight $(N \cdot m/kg)$
Jonit 1	25.00	5.69	177.8	102.6	12.1	2.1
Jonit 2	16.40	4.28	177.8	51.0	6.9	1.7
Jonit 3	11.40	2.85	177.8	51.0	6.9	1.7
Jonit 4	16.40	4.28	177.8	76.3	9.6	1.7
Jonit 5	2.00	1.14	40.6	51.0	1.5	1.3
Jonit 6	2.00	1.14	40.6	51.0	1.5	1.3
Jonit 7	0.81	0.22	60.4	48.5	0.6	1.4

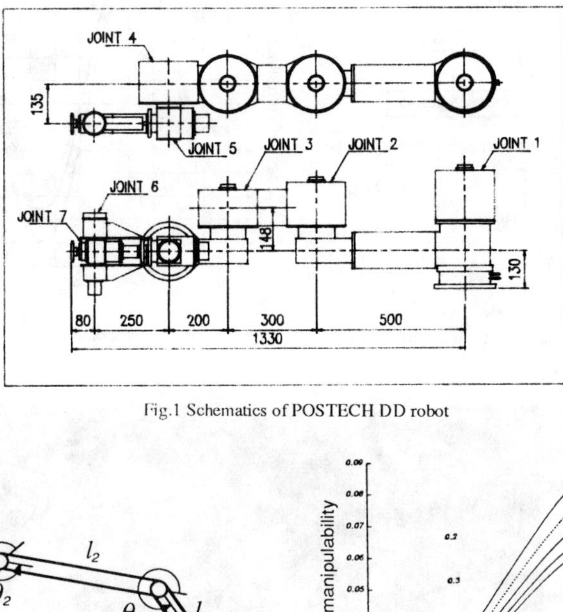

Fig.1 Schematics of POSTECH DD robot

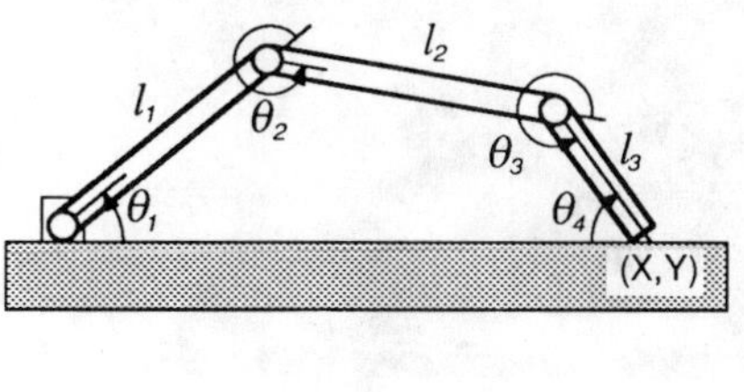

Fig.2. Planar three link manipulator

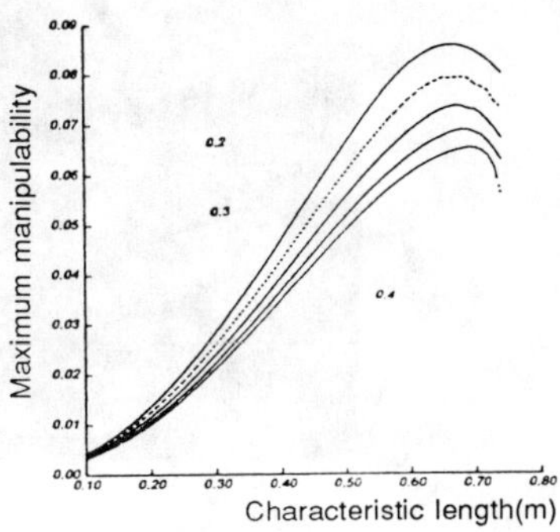

Fig.3 Maximum manipulability for various link length l_2

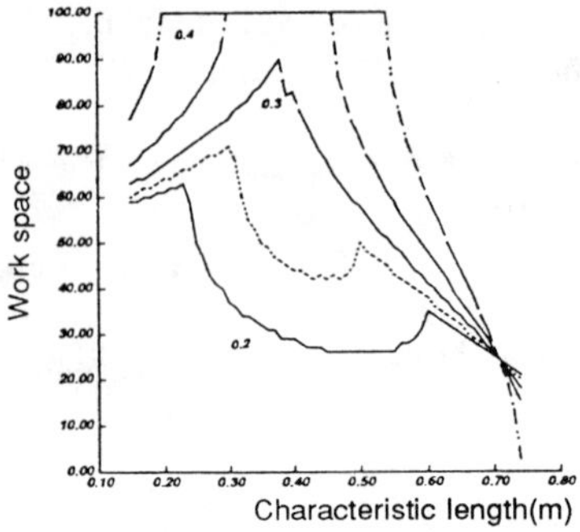

Fig.4 Workspace for various link length l_2

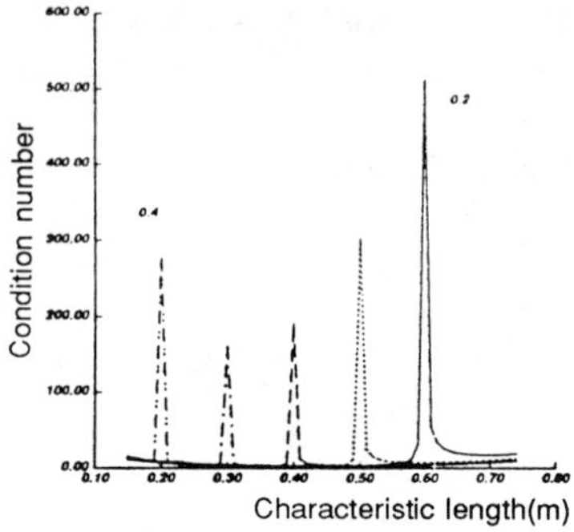

Fig.5 Maximum condition number for various link length l_2

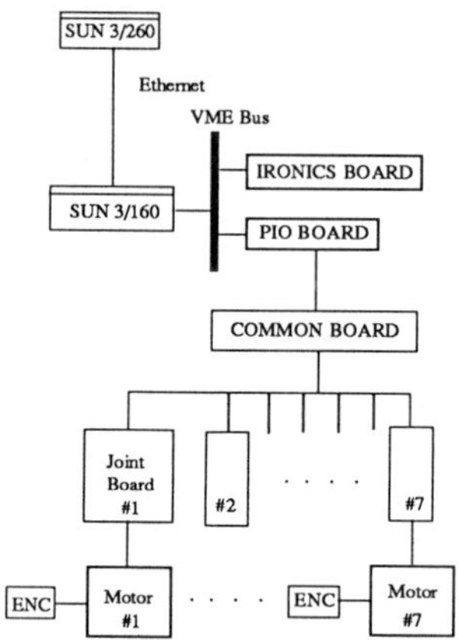

Fig.6 Schematic diagram of robot control system

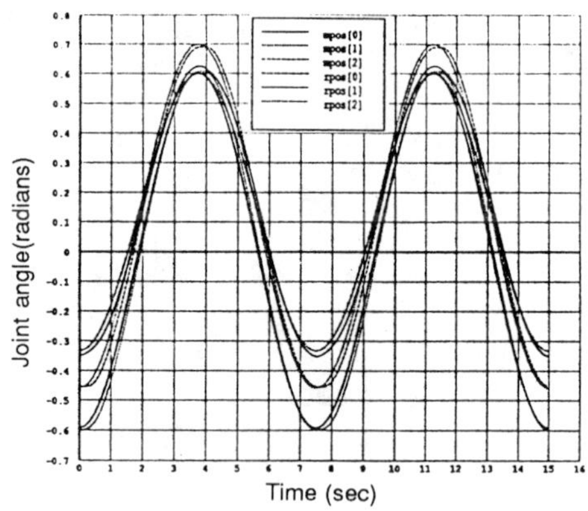

Fig.8 Experimental responses for sinusoidal input trajectory

Fig.7 Overall view of POSTECH DD robot

NONLINEAR OBSERVERS FOR ROBOTS WITH RIGID AND ELASTIC JOINTS*

S. NICOSIA and P. TOMEI

Dipartimento di Ingegneria Elettronica, Seconda Università di Roma, *Tor Vergata*, Via O. Raimondo, 00173 Roma, Italy.

ABSTRACT

The problem of the state estimation is considered for robots having either rigid or elastic joints. Nonlinear observers are designed by exploiting some structural properties of the robot dynamics. An estimate of the region of asymptotic stability is also given.

INTRODUCTION

Usually, feedback control of robot manipulators is based on the complete knowledge of the state variables. Even in the case of simple independent PD controllers, all the state variables need to be measured [1]. Such requirements become also more stringent when we deal with robots having elastic joints, since four state variables characterize each joint [6].

The design of suitable observers for robot manipulators is a complex task, due to the strong nonlinearity of the corresponding model. Indeed, a general theory of observers for nonlinear systems is not available, even though some works have been already done [4,5,9,14]. As far as robot manipulators with flexible joints and/or links is concerned, approximate observers have been derived whose validity is restricted to the neighborhood of the operating points [7,8].

In this paper, by exploiting some structural properties of the robot dynamics, various observation algorithms are proposed for rigid and elastic joint robots. The more interesting one allows us to estimate the state of an elastic joint robot by using only one position sensor for each joint.

ROBOT MODEL

The dynamics of robot manipulators having n elastic joints can be written as [6,12]

$$B(q_1)\ddot{q} + C(q_1,\dot{q})\dot{q} + Kq + F\dot{q} + e(q_1) = m \tag{1}$$

where $q = [q_1^T, q_2^T]^T$ is the $2n \times 1$ vector grouping the link and actuator relative displacements, and $m = [0, \ldots, 0, u_1, \ldots, u_n]^T$, with u_i being the torque applied to joint i. The inertia matrix $B(q_1)$, that is symmetric positive definite

*This work was supported by CNR under contract Nr. 89.00531.67.

and bounded for any q_1, owns the form

$$B(q_1) = \begin{bmatrix} B_1(q_1) & B_2(q_1) \\ B_2^T(q_1) & B_3 \end{bmatrix} \tag{2}$$

where B_3 is a constant diagonal matrix depending on the actuator inertias and gear ratios. Matrix B enjoys the following properties.

*P*1. Matrix $B_2 + B_3$ is always nonsingular [13].

*P*2. Matrix $G_1 = (B_1 - B_2 B_3^{-1} B_2^T)$ is symmetric positive definite. Indeed, since B is symmetric positive definite, G_1 exists and is symmetric. Besides,

$$\begin{bmatrix} x_1 \\ x_2 \end{bmatrix}^T B \begin{bmatrix} x_1 \\ x_2 \end{bmatrix} > 0 \qquad \forall \begin{bmatrix} x_1 \\ x_2 \end{bmatrix} \neq 0$$

Taking $x_2 = -B_3^{-1} B_2^T x_1$, the property follows.

Matrix $C(q_1, \dot{q})$ takes into account the Coriolis and centrifugal terms. It is characterized by the following properties.

*P*3. If $C(q_1, \dot{q})$ is suitably defined, then matrix $\dot{B} - 2C$ is skew-symmetric [3,11]. A possible definition for the elements of C is [11]

$$C_{ij}(q_1, \dot{q}) = \frac{1}{2}\left[\dot{q}^T \frac{\partial B_{ij}}{\partial q} + \sum_{k=1}^{2n}\left(\frac{\partial B_{ik}}{\partial q_j} - \frac{\partial B_{jk}}{\partial q_i}\right)\dot{q}_k\right] \tag{3}$$

*P*4. If C is defined according to (3), then $C(q, x)y = C(q, y)x$ [15].

*P*5. From (3) we can see that matrix C is linear with respect to $\dot{q}$ and bounded with respect to q_1. Therefore, a suitable constant k_c exists such that

$$\|C(q_1, \dot{q})\| \leq k_c \|\dot{q}\|, \qquad \forall q_1 \in R^n$$

Matrix K takes the joint elasticity into account and is structured as follows

$$K = \begin{bmatrix} K_e & -K_e \\ -K_e & K_e \end{bmatrix}$$

where $K_e = \text{diag}[k_1, \ldots, k_n]$, with k_i being the elastic constant of joint i.

Finally, $e(q_1)$ represents the gravitational torques and F is the symmetric positive definite matrix of the viscous friction coefficients. If the joints are rigid (i.e. $k_i = \infty$, i=1, ...,n), the dynamic model can be recovered from (1) by imposing the constraint that the elastic displacements $q_1(t) - q_2(t)$ vanish for any t [6]. This way, the order of the dynamic model reduces to n.

NONLINEAR OBSERVERS

When we deal with elastic joint robots various choices are possible for the output variables, related to the available sensors. In the sequel, we will show how an observer can be derived for three different sets of output variables. The case in which the link positions and velocities are available from measuraments is not considered, since a stronger result ensuring the global asymptotic stability can be derived [13]. The following notations are used: $\tilde{x}_1 = q_1 - \hat{x}_1$, $\tilde{x}_2 = q_2 - \hat{x}_2$, $\tilde{x} = [\tilde{x}_1^T, \tilde{x}_2^T]^T$, $\hat{x} = [\hat{x}_1^T, \hat{x}_2^T]^T$, $x = [\tilde{x}^T, \dot{\tilde{x}}^T]^T$. Moreover, we use A_m and A_M, respectively, to indicate the minimum and the maximum eigenvalue of a symmetric positive definite bounded matrix $A(x)$, for any x.

Measuraments of q_1 and q_2

The structure of the observer is

$$\begin{aligned} \begin{bmatrix} \dot{\hat{x}}_1 \\ \dot{\hat{x}}_2 \end{bmatrix} &= \begin{bmatrix} \hat{x}_3 + k_D \tilde{x}_1 \\ \hat{x}_4 + k_D \tilde{x}_2 \end{bmatrix} \\ \begin{bmatrix} \dot{\hat{x}}_3 \\ \dot{\hat{x}}_4 \end{bmatrix} &= B^{-1}(q_1)(-C(q_1, \dot{\hat{x}})\dot{\hat{x}} - K\hat{x} - e(q_1) + K_P \tilde{x} - F\dot{\hat{x}} + m) \end{aligned} \tag{5}$$

where k_D is a positive constant and K_P is such that $K_P + K$ is symmetric positive definite. From (5) and (1), we obtain the error equation

$$B(q_1)\ddot{\tilde{x}} + C(q_1, \dot{q})\dot{\tilde{x}} + C(q_1, \dot{\hat{x}})\dot{\tilde{x}} + F\dot{\tilde{x}} = -(K + K_P)\tilde{x} - k_D B(q_1)\dot{\tilde{x}} \tag{6}$$

Defining $H_1(q_1) = \text{block diag}[K_P + K, B(q_1)]$, we can give the following theorem.

Theorem 1. Assume that $\|\dot{q}(t)\| \leq M$, for any $t \geq 0$. If $k_D > k_c M / B_m$, then the equilibrium point $x = 0$ of (6) is asymptotically stable, and a region of attraction is given by

$$S = \left\{ x \in R^{4n} : \quad \|x\| < \sqrt{\frac{H_{1_m}}{H_{1_M}}} \left(\frac{B_m k_D}{k_c} - M \right) \right\}$$

Proof. Consider the candidate Lyapunov function

$$v(x, t) = \frac{1}{2} x^T H_1(q_1) x \tag{7}$$

whose time derivative, recalling Properties P3 and P4, is

$$\dot{v}(x, t) = -\dot{\tilde{x}}^T (k_D B(q_1) + F)\dot{\tilde{x}} - \dot{\tilde{x}} C(q_1, \dot{\tilde{x}})\dot{\tilde{x}} \tag{8}$$

Owing to Properties P4 and P5, we can write

$$|\dot{\tilde{x}}^T C(q_1, \dot{\tilde{x}})\dot{\tilde{x}}| \leq \|\dot{\tilde{x}}\|^2 k_c (\|\dot{\tilde{x}}\| + M)$$

that substituted in (8) implies

$$\dot{v}(x, t) \leq -\|\dot{\tilde{x}}\|^2 [k_D B_m - k_c(\|\dot{\tilde{x}}\| + M)]$$

Therefore, if

$$\|\dot{\tilde{x}}\| < \frac{B_m k_D}{k_c} - M \tag{9}$$

then

$$\dot{v}(x,t) \leq -\beta\|\dot{\tilde{x}}\|^2 \tag{10}$$

with β being a positive constant. The right side of (9) is positive, by hypothesis. In addition, v is a positive definite decrescent function. Indeed,

$$\frac{1}{2}H_{1_m}\|x\|^2 \leq v(x,t) \leq \frac{1}{2}H_{1_M}\|x\|^2 \tag{11}$$

Since $\dot{v}$ is negative semidefinite for all x satisfying (9), the uniform stability of $x = 0$ follows [2, p. 199]. In view of (9), (10) and (11), if

$$\|x(0)\| < \sqrt{\frac{H_{1_m}}{H_{1_M}}}\left(\frac{B_m k_D}{k_c} - M\right) \tag{12}$$

then

$$v(x,t) \leq v(x(0),0), \qquad \forall t \geq 0 \tag{13}$$

$$\dot{v}(x,t) \leq -\beta\|\dot{\tilde{x}}\|^2, \qquad \forall t \geq 0 \tag{14}$$

The inequality (13) implies the boundedness of $\tilde{x}(t)$ and $\dot{\tilde{x}}(t)$. Since $\dot{q}(t)$ is bounded by hypothesis, it follows that $\ddot{\tilde{x}}$ is also bounded (see (6)), and consequently $\dot{\tilde{x}}(t)$ is uniformly continuous. The inequalities (13) and (14) enable us to write

$$\beta\int_0^\infty \|\dot{\tilde{x}}(t)\|\, dt \leq -\int_0^\infty \dot{v}(t)\, dt = v(0) - v(\infty) < \infty$$

and, by Barbalat lemma [10, p. 210],

$$\lim_{t\to\infty} \|\dot{\tilde{x}}(t)\| = 0$$

Remark. Obviously, with proper modifications, the observer (5) can be also used to estimate the joint velocities of rigid joint robots, on the basis of joint position measuraments.

<u>Measurements of q_1 and $\dot{q}_1$</u>

The observer equations are modified as follows

$$\begin{bmatrix}\dot{\hat{x}}_1\\ \dot{\hat{x}}_2\end{bmatrix} = \begin{bmatrix}\hat{x}_3 + k_D\tilde{x}_1\\ \hat{x}_4\end{bmatrix}$$

$$\begin{bmatrix}\dot{\hat{x}}_3\\ \dot{\hat{x}}_4\end{bmatrix} = B^{-1}(q_1)(-C(q_1,\dot{\hat{x}})\dot{\hat{x}} - K\hat{x} - e(q_1) + K_A\tilde{x} - F\dot{\hat{x}} + m) + k_D\begin{bmatrix}0\\ \dot{\tilde{x}}_2\end{bmatrix} \tag{16}$$

in which k_D is a positive constant and K_A = block diag$[K_a, 0]$, with K_a being a $n \times n$ symmetric positive definite matrix. The corresponding error dynamics is

$$B(q_1)\ddot{\tilde{x}} + C(q_1,\dot{q})\dot{\tilde{x}} + C(q_1,\dot{\hat{x}})\dot{\tilde{x}} + F\dot{\tilde{x}} = -(K + K_A)\tilde{x} - k_D B(q_1)\dot{\tilde{x}} \tag{17}$$

Let $H_2(q_1) = \text{block diag}[K_A + K, B(q_1)]$. We can state the following theorem.

Theorem 2. Assume that $\|\dot{q}(t)\| \leq M$, for any $t \geq 0$. If $k_D > k_c M / B_m$, then the equilibrium point $x = 0$ of (17) is asymptotically stable, and a region of attraction is given by

$$S = \left\{ x \in R^{4n} : \quad \|x\| < \sqrt{\frac{H_{2_m}}{H_{2_M}}} \left(\frac{B_m k_D}{k_c} - M \right) \right\}$$

Proof. Adopting as candidate Lyapunov function

$$v(x,t) = \frac{1}{2} x^T H_2(q_1) x$$

the proof is similar to that of Theorem 1.

Measurement of q_1

The observer described in the sequel needs only one position sensor for each joint to reconstruct all the robot state variables. We assume that matrix B_2, that is defined by (2), is constant. This assumption is surely true for a large class of robots. For instance, it is true for robots having shoulder, upper arm and forearm structure, whose motor rotors have rotation axes perpendicular to the axes of the corresponding actuated links [13]. The observer dynamics is given by

$$\begin{aligned} \begin{bmatrix} \dot{\hat{x}}_1 \\ \dot{\hat{x}}_2 \end{bmatrix} &= \begin{bmatrix} \hat{x}_3 + k_D \tilde{x}_1 \\ \hat{x}_4 - k_D B_3^{-1} B_2^T \tilde{x}_1 \end{bmatrix} \\ \begin{bmatrix} \dot{\hat{x}}_3 \\ \dot{\hat{x}}_4 \end{bmatrix} &= B^{-1}(q_1)(-C(q_1, \dot{\hat{x}})\dot{\hat{x}} - K\hat{x} - e(q_1) + K_A \tilde{x} - F\dot{\hat{x}} + m) \end{aligned} \tag{18}$$

Subtracting (18) from (1), we obtain

$$B(q_1)\ddot{\tilde{x}} + C(q_1, \dot{q})\dot{\tilde{x}} + C(q_1, \dot{\hat{x}})\dot{\tilde{x}} + F\dot{\tilde{x}} = -(K + K_A)\tilde{x} - k_D G(q_1)\dot{\tilde{x}} \tag{19}$$

where $G(q_1) = \text{block diag}[G_1(q_1), 0]$, with G_1 being defined in Property P2.

Theorem 3. Assume that $\|\dot{q}_1(t)\| \leq M$, for any $t \geq 0$. If $k_D > k_c M / G_{1_m}$, then the equilibrium point $x = 0$ of (19) is uniformly asymptotically stable, and a region of attraction is given by

$$S = \left\{ x \in R^{4n} : \quad \|x\| < \sqrt{\frac{H_{2_m}}{H_{2_M}}} \left(\frac{G_{1_m} k_D}{k_c} - M \right) \right\} \tag{20}$$

Proof. Choose as candidate Lyapunov function

$$v(x,t) = \frac{1}{2} x^T H_2(q_1) x \tag{21}$$

Differentiating (21), we get

$$\dot{v}(x,t) = -k_d \dot{\tilde{x}}_1^T G_1(q_1) \dot{\tilde{x}}_1 - \dot{\tilde{x}}^T F \dot{\tilde{x}} - \dot{\tilde{x}}_1^T C_1(q_1, \dot{\hat{x}}_1) \dot{\tilde{x}}_1 \tag{22}$$

Since B_2 is constant, from (3) we obtain $C(q_1, \dot{q}) =$ block diag$[C_1(q_1, \dot{q}_1), 0]$, with C_1 a $n \times n$ matrix. This fact and Properties P4 and P5 imply

$$|\dot{\tilde{x}}_1^T C_1(q_1, \dot{\tilde{x}}_1)\dot{\tilde{x}}_1| \leq \|\dot{\tilde{x}}_1\|^2 k_c(\|\dot{\tilde{x}}_1\| + M) \tag{23}$$

Owing to (23),

$$\dot{v}(x,t) \leq -\|\dot{\tilde{x}}_1\|^2[k_d G_{1_m} - k_c(\|\dot{\tilde{x}}_1\| + M)] - \dot{\tilde{x}}^T F \dot{\tilde{x}}$$

Hence, if $\|\dot{\tilde{x}}_1\| < G_{1_m} k_d/k_c - M$ then $\dot{v} \leq -\beta\|\dot{\tilde{x}}_1\|^2 - \dot{\tilde{x}}^T F \dot{\tilde{x}}$. The uniform stability of $x = 0$ is proved [2, p. 199]. As is shown in proof of Theorem 1, if $x(0)$ belongs to the region (20), then

$$v(x,t) \leq v(x(0), 0), \qquad \forall t \geq 0 \tag{24}$$

$$\dot{v}(x,t) \leq -\beta\|\dot{\tilde{x}}_1\|^2 - \dot{\tilde{x}}^T F \dot{\tilde{x}}, \qquad \forall t \geq 0 \tag{25}$$

which, recalling that $\dot{q}_1(t)$ is assumed bounded, imply

$$\lim_{t\to\infty} \|\dot{\tilde{x}}_1(t)\| = 0 \tag{26}$$

$$\lim_{t\to\infty} \|F\dot{\tilde{x}}(t)\| = 0 \tag{27}$$

Substituting (26) and (27) into (19), we have

$$\lim_{t\to\infty} [B_2\ddot{\tilde{x}}_2 + (k_e + K_a)\tilde{x}_1 - K_e\tilde{x}_2] = 0$$

$$\lim_{t\to\infty} [B_3\ddot{\tilde{x}}_2 - K_e\tilde{x}_1 + K_e\tilde{x}_2] = 0$$

By virtue of Property P1, we can solve the previous equations with respect to $\ddot{\tilde{x}}_2$ and $\tilde{x}_2$, yielding

$$\lim_{t\to\infty} \ddot{\tilde{x}}_2 = -(B_2 + B_3)^{-1} K_a \lim_{t\to\infty} \tilde{x}_1$$

$$\lim_{t\to\infty} \tilde{x}_2 = K_e^{-1}[B_3(B_2 + B_3)^{-1} + K_e] \lim_{t\to\infty} \tilde{x}_1$$

that imply

$$\lim_{t\to\infty} \|\tilde{x}\| = 0, \qquad \lim_{t\to\infty} \|\dot{\tilde{x}}_2\| = 0$$

CONCLUSIONS

The problem of observing the state variables of robot manipulators has been investigated. Robots with and without elasticity in the joints have been considered. Nonlinear observers have been presented and their convergence has been proved. We have shown that the region of convergence can be arbitrarily enlarged by increasing the gain constants of the proposed observers. One of the major advantages of our approach is that there is no need of coordinate change and the observer implementation is, therefore, straightforward.

Our future work will regard the extension of the proposed approach to robots having flexible links.

REFERENCES

1. S. Arimoto and F. Miyazaki, Stability and robustness of PID feedback control for robot manipulators of sensory capability, Robotics Research, eds. M. Brady and R.P. Paul, (MIT Press, Cambridge 1984).
2. W. Hahn, Stability of motion, (Springer-Verlag, Berlin 1967).
3. D.E. Koditschek, Natural motion of robot arms, 23rd IEEE Int. Conf. Decision and Control (Las Vegas, 1984).
4. A.J. Krener and A. Isidori, Linearization by output injection and nonlinear observers, Systems Control Lett., 3 47-52 (1983).
5. A.J. Krener and W. Respondek, Nonlinear observers with linearizable error dynamics, SIAM J. Control Optim., 23, 197-216 (1985).
6. R. Marino and S. Nicosia, Singular perturbation techniques in the adaptive control of elastic robots, IFAC Symp. Robot Control, (Barcelona 1985).
7. S. Nicosia, P. Tomei and A. Tornambè, A nonlinear observer for elastic robots, IEEE J. Robotics Automation, RA-4, 47-52 (1988).
8. S. Nicosia, P. Tomei and A. Tornambè, Non-linear control and observation algorithms for single-link flexible robot arm, Internat. J. Control, 49, 827-840 (1989).
9. S. Nicosia, P. Tomei and A. Tornambè, An approximate observer for a class of nonlinear systems, Systems Control Lett., 12 43-51 (1989).
10. V. M. Popov, Hyperstability of control systems, (Springer-Verlag, New York 1973).
11. J.E. Slotine and W. Li, On the adaptive control of robot manipulators, The Int. J. of Robotics Research, 6, 49-59 (1987).
12. M.W. Spong, Modeling and control of elastic joint robots, ASME J. Dynamic Syst., Meas., Contr., 109 310-319 (1987).
13. P. Tomei, An observer for flexible joint robots, IEEE Trans. Automatic Control, to appear.
14. B.L. Walcott, M.J. Corless and S.H. Zak, Comparative study of non-linear state observation techniques, Internat. J. Control, 45, 2109-2132 (1987).
15. S. Nicosia and P. Tomei, Robot control by using only joint position measuraments, IEEE Trans. Automatic Control, to appear.

ROBUST CONTROL STRATEGIES FOR A SINGLE LINK FLEXIBLE MANIPULATOR USING VARIABLE STRUCTURE CONTROL

Dmitry Shusterman, Luciano Chirinos, and Glenn R. Widmann

Colorado State University
Department of Electrical Engineering
Institute of Robotic Studies
Fort Collins, CO 80523

ABSTRACT: In this paper a Variable Structure Controller is developed with an application to a flexible link manipulator. Simulation results show that this controller is robust in the presence of plant parameter uncertainties where pole placement algorithms fail. Although the tracking performance of this controller is limited, it can be used as a back-up controller to the primary high performance controller.

1. INTRODUCTION

In our previous experiments with the control of a flexible link manipulator [1], we have determined that pole placement control algorithms are very sensitive to the uncertainties in the plant parameters. One way to provide high performance under such conditions is to implement a self tuning controller. Although such controllers remain robust, their performance is greatly reduced in the transitional adaptation stage due to the sluggishness in the speed of adaptation where the order of the controller is kept high in order to provide good performance.

As suggested in [2], this shortcoming can be overcome if a very robust low performance controller is used as a back-up controller controller during this adaptation stage. In [2] a simple PD-type compensator is used for this purpose. Here we present a Variable Structure Controller to serve the same purpose. The advantages are that a VSC controller provides higher robustness and has the potential of providing higher performance in comparison to such traditional control algorithms. Moreover, it can be naturally blended with the adaptation process such that it provides higher performance when the uncertainties are reduced.

2. CONTROLLER DESIGN

The true model for a flexible arm is inherently of infinite order. However, the system can be approximated by a finite order model with unmodeled dynamics. In this section, a controller based on this reduced order model will be developed. The mathematical description of the SISO system can be previewed by a nth order Controllable Canonical form given by:

$$x_p(k+1) = A_p x_p(k) + B u_p(k) \quad (1)$$

$$y_p(k) = C x_p(k+1) \quad (2)$$

$$A_p = \begin{bmatrix} 0 & 1 & \cdots & 0 \\ \vdots & \ddots & \ddots & \vdots \\ 0 & 0 & \ddots & 1 \\ -a_{p_n} & -a_{p_{n-1}} & \cdots & -a_{p_1} \end{bmatrix} \qquad \begin{array}{l} B = \begin{bmatrix} 0 & 0 & \cdots & 1 \end{bmatrix}^t \\ \\ C = \begin{bmatrix} c_{p_n} & c_{p_{n-1}} & \cdots & c_{p_1} \end{bmatrix} \end{array}$$

where $x_p(k) \varepsilon \Re^n$ is the plant state vector, $u_p(k)$ is the plant input, $y_p(k)$ is the plant output, and A_p, B and C are the system matrices.

In this section, we initially consider the case when the SISO system can only have uncertainties in the a_{p_i} parameters. The effect of the uncertainties in the C vector due to the uncertainties in the zeroes of the plant will be considered in a later section. Thereby, the A_p matrix can be decomposed into the known and unknown parts such that $A_p = \hat{A}_p + \tilde{A}_p$, where $\hat{A}_p$ is the best estimate of the plant and $\tilde{A}_p$ is the uncertainty in the plant. The generalized format of $\hat{A}_p$ and $\tilde{A}_p$ is the same as A_p but with elements $\{\hat{a}_{p_i}\}$ and $\{\tilde{a}_{p_i}\}$, respectively.

The objective of the controller design is to find a plant input $u_p(k)$ such that the states of the plant $x_p(k)$ will closely follow the states of the desired model $x_m(k)$ given by:

$$x_m(k+1) = A_m x_m(k) + B u_m(k) \quad (3)$$

where $u_m(k)$ is the reference input, and A_m has the same form as A_p, but with elements $\{a_{m_i}\}$.

Let us define the error in the states as $\sigma(k) \triangleq x_m(k) - x_p(k)$, and the difference in the inputs as $\Delta u(k) \triangleq u_m(k) - u_p(k)$. Then, by using (1) and (3), the error dynamics becomes:

$$\sigma(k+1) = A_m\sigma(k) + [A_m - \hat{A}_p]x_p(k) - \tilde{A}_p x_p(k) + B\Delta u(k) \tag{4}$$

Using a similar approach as used in state feedback control, the objective is to define that part of the input signal $\Delta u(k)$ needed to compensate for the known dynamics of the plant as follows:

$$\Delta u_1(k) \triangleq \sum_{i=1}^{n}[(\hat{a}_{p_i} - a_{m_i})x_{p_i}(k)] \tag{5}$$

Upon defining $\Delta\overline{u}(k) \triangleq u_m(k) - \Delta u_1(k)$, then the error dynamics (5) simplifies to:

$$\sigma(k+1) = A_m\sigma(k) + \tilde{A}_p x_p(k) + B\Delta\overline{u}(k) \tag{6}$$

In order to speed up the error dynamics, another component needs to be added to the input signal

$$\Delta u_2(k) \triangleq \sum_{i=1}^{n}[(a_{c_i} - a_{m_i})\sigma_i(k)] \tag{7}$$

where a_{c_i} are the elements of the desired error dynamics matrix A_c, which has the same format as A_p. By defining $\Delta\overline{\overline{u}}(k) \triangleq \Delta\overline{u}(k) - \Delta u_2(k)$, then the error dynamics equation (6) becomes:

$$\sigma(k+1) = A_c\sigma(k) - \tilde{A}_p x_p(k) + B\Delta\overline{\overline{u}}(k) \tag{8}$$

Due to the uncertainty in the error term $\tilde{A}_p x_p(k)$, a variable structure input will be developed in order to guarantee the stability and convergence of (8) over the uncertainty range. The error term can be rewritten as:

$$\tilde{A}_p x_p(k) = E\rho(k) \tag{9}$$

where $E = \begin{bmatrix} 0 & 0 & \cdots & 1 \end{bmatrix}^t$ and $\rho(k) = \sum_{i=1}^{n} \tilde{a}_{p_i} x_{p_i}(k)$. A necessary assumption in the design of the VSC controller and the observer (which is discussed in the next section), is that all the elements $\tilde{a}_{p_i}$ are bounded such that $|\tilde{a}_{p_i}| < \Delta a$ for all i, where Δa is a design parameter representing an upper bound for the plant parameter uncertainty.

Defining the Liapunov function $V(k) = \sigma^t(k)P\sigma(k)$, where P must satisfy the following 3 conditions:

$A1$: P is a positive definite matrix.
$A2$: $P - A_c^t P A_c = Q$, where Q is positive definite.
$A3$: $B^t P = C$.

According to Kalman-Yacubovich lemma [3], these conditions can always be satisfied if the triple $\{A_c, B, C\}$ is both controllable and observable, and A_c has stable eigenvalues. Then, in order for the error $\sigma(k)$ to be monotonically decreasing, we need to find a $\Delta\overline{\overline{u}}(k)$ such that $V(k)$ is decreasing. The incremental change in $V(k)$ can be written as:

$$V(k) - V(k-1) = -\sigma^t(k)[P - A_c^t P A_c]\sigma(k) - 2\rho(k)E^t P\sigma(k) + 2\Delta\overline{\overline{u}}(k)B^t P\sigma(k) \tag{10}$$

When condition $A2$ is satisfied, we need to only guarantee the following inequality:

$$-2\rho(k)E^t P\sigma(k) + 2\Delta\overline{\overline{u}}(k)B^t P\sigma(k) < 0 \tag{11}$$

Since $E = B$, then (11) becomes:

$$[-\rho(k) + \Delta\overline{\overline{u}}(k)]B^t P\sigma(k) < 0 \tag{12}$$

By using condition $A3$, one can arrive at:

$$[-\rho(k) + \Delta\overline{\overline{u}}(k)][Cx_m(k) - y(k-1)] < 0 \tag{13}$$

One way to guarantee (13) is to choose $\Delta\overline{\overline{u}}(k)$ as:

$$\Delta\overline{\overline{u}}(k) = \sum_{i=1}^{n} \xi_i x_{p_i}(k), \text{ where } \xi_i = \begin{cases} -\Delta a & if\ [Cx_m(k) - y(k-1)]x_{p_i}(k) > 0 \\ +\Delta a & if\ [Cx_m(k) - y(k-1)]x_{p_i}(k) < 0 \end{cases} \tag{14}$$

3. OBSERVER DESIGN

The implementation of the VSC controller described in Section 2 requires the knowledge of the states of the reduced order plant $x_p(k)$. Here we will develop a variable structure observer by using an approach similar to the one developed in [4]. Though a continuous-time observer has been developed in [4], in this paper, however, the theory will be extended to the discrete-case.

Defining $\hat{x}_p(k)$ as the estimate of the state vector, then the observer will have the following structure:

$$\hat{x}_p(k+1) = \hat{A}_p\hat{x}_p(k) - K[C\hat{x}_p(k) - y(k-1)] + B[u(k) + \Delta u_0(k)] \tag{15}$$

where $\Delta u_0(k)$ is a discontinuous control signal to be determined, K is such that it satisfies the relationship $\hat{A}_p - KC = A_0$, and A_0 is the desired observer matrix. The error dynamics will be given by:

$$q(k+1) = \hat{A}_0 q(k) + \tilde{A}\hat{x}_p(k) + B\Delta u_0(k) \tag{16}$$

where $q(k) \triangleq x_p(k) - \hat{x}_p(k)$, $\hat{A}_0 \triangleq \hat{A}_p - KC$, and $\hat{A}_0$ is the actual observer dynamics matrix. Due to the uncertainties in the A_p matrix, $\hat{A}_0$ is different from the desired matrix A_0. However, if this uncertainty is small, then $\hat{A}_0$ will not change significantly from A_0.

The design objective is to find a signal $\Delta u_0(k)$ such that $lim_{t\to\infty}\|q(k)\| = 0$. As before, we introduce the Liapunov function $V(k) = q^t(k)Pq(k)$, where P must satisfy the following 3 conditions:

$B1 : P$ is a positive definite matrix.
$B2 : P - \hat{A}_0^t P\hat{A}_0 = Q$, where Q is positive definite.
$B3 : PB = C^t$

The incremental change in $V(k)$ can be written as:

$$V(k) - V(k-1) = -q^t(k)[P - \hat{A}_0^t P\hat{A}_0]q(k) + 2\rho(k)E^t Pq(k) + 2\Delta u_0(k)B^t Pq(k) \tag{17}$$

Thus, in order to guarantee that $V(k)$ is monotonically decreasing, we must satisfy:

$$2\rho(k)E^t Pq(k) + 2\Delta u_0(k)B^t Pq(k) < 0 \tag{18}$$

Knowing that $E = B$ and using condition $3B$, (18) becomes:

$$[\rho(k) + \Delta u_0(k)]Cq(k) < 0 \tag{19}$$

Since $y(k) = Cx_p(k+1)$, we get:

$$[\rho(k) + \Delta u_0(k)][y(k-1) - C\hat{x}_p(k)] < 0 \tag{20}$$

Consequently, the following control law can be implemented to satisfy (20)

$$\Delta u_0(k) = \sum_{i=1}^{n} \xi_i \hat{x}_{p_i(k)}, \text{ where } \xi_i = \begin{cases} -\Delta a & if\ [y(k-1) - C\hat{x}_p(k)]\hat{x}_{p_i}(k) > 0 \\ +\Delta a & if\ [y(k-1) - C\hat{x}_p(k)]\hat{x}_{p_i}(k) < 0 \end{cases} \tag{21}$$

4. EFFECT OF UNCERTAINTIES IN THE PLANT ZEROES

In the previous section the robustness of the VSC controller was demonstrated when plant poles are unknown. This section will briefly discuss the effects of uncertainties in zeroes and show that only a bounded error is introduced in plant output while stability is always maintained.

The actual error in the state vector estimate $\hat{x}_p(k)$ and the plant output $y_p(k)$ in the presence of uncertainties in the plant zeroes arises from the fact that the sign of the switching functions $[Cx_m(k) - y(k-1)]\hat{x}_{p_i}(k)$ and $[y(k-1) - C\hat{x}_p(k)]\hat{x}_{p_i}(k)$ in (14) and (21) can not be known in some regions of the state space. The resulting error from this effect can be estimated if we introduce a similarity transformation that transforms the couple $\{\hat{A}_p, C\}$ into couple $\{\hat{A}_p, \hat{C}\}$, where $\hat{C}$ is the estimate of the vector C. We define a transformed plant state vector as $\overline{x}_p(k) \triangleq Tx_p(k)$ and its estimate as $\hat{\overline{x}}_p(k) \triangleq T\hat{x}_p(k)$, where T is the corresponding transformation matrix.

If C is replaced by $\hat{C}$ in the observer control law (21), then it is possible to show that the actual estimate is $\hat{\overline{x}}_p(k)$. This result is obtained by using $V(k) = [\overline{x}_p(k) - \hat{\overline{x}}_p(k)]^t P[\overline{x}_p(k) - \hat{\overline{x}}_p(k)]$ as a Lyapunov function in the design of the observer. Here, P satisfies the condition $B^t P = \hat{C}$, which is similar to the conditions $A3$ and $B3$.

When the estimate $\hat{\overline{x}}_p(k)$ is used to construct the control signal $u_p(k)$, an additional error term is added to the error dynamics equation (8). The effect of this term would be to change the closed loop response of the system. However, this can be easily compensated for in the framework of the VSC control law (14) by increasing the uncertainty bound Δa.

Using the Liapunov function $V(k) = \overline{\sigma}^t(k)P\overline{\sigma}(k)$, where $\overline{\sigma}(k) = x_m(k) - \overline{x}_p(k)$, it is easy to show that only $\overline{\sigma}(k)$ will be driven to zero by the control law (14) rather than the actual plant state error $\sigma(k)$. This will result in the output $y_p(k)$ deviating from the ideal response. The amount of deviation is always bounded and, thereby, verify that the system remains stable in the presence of uncertainties in the plant zeroes.

5. BOUNDARY LAYER CONTROL

Due to a finite sampling time, the control signals are expected to chatter in a vicinity of the switching surface. One way to reduce that effect is to slow the error dynamics of both the observer and the controller in order to slow down the approach of the states to the switching surface.

The alternate approach, proposed in [4], modifies the Variable Structure Control law by introducing a boundary layer. With this modification, the control laws (14) and (21) become:
For controller:

$$\Delta\overline{\overline{u}}(k) = \sum_{i=1}^{n} \xi_i^c \hat{x}_{p_i}(k) \tag{22}$$

$$\text{where } \xi_i^c = \begin{cases} +\Delta a & if\ [\hat{C}x_m(k) - y(k-1)]\hat{x}_{p_i}(k) > \epsilon \\ +\Delta a\frac{[\hat{C}x_m(k) - y(k-1)]\hat{x}_{p_i}}{\epsilon} & if\ |[\hat{C}x_m(k) - y(k-1)]\hat{x}_{p_i}(k)| < \epsilon \\ -\Delta a & if\ [\hat{C}x_m(k) - y(k-1)]\hat{x}_{p_i}(k) < -\epsilon \end{cases}$$

For the observer:

$$\Delta u_0(k) = \sum_{i=1}^{n} \xi_i^0 \hat{x}_{p_i}(k) \tag{23}$$

$$\text{where } \xi_i^o = \begin{cases} +\Delta a & if\ [y(k-1) - \hat{C}\hat{x}_p(k)]\hat{x}_{p_i}(k) > \epsilon \\ +\Delta a\frac{[y(k-1) - \hat{C}\hat{x}_p(k)]\hat{x}_{p_i}(k)}{\epsilon} & if\ |[y(k-1) - \hat{C}x_p(k)]\hat{x}_{p_i}(k)| < \epsilon \\ -\Delta a & if\ [y(k-1) - \hat{C}\hat{x}_p(k)]\hat{x}_{p_i}(k) < -\epsilon \end{cases}$$

where ϵ is a small positive number, which is a design parameter that determines the width of the boundary layer.

It has already been shown in [4] that the states of the system will stay within a bounded region around the switching surface and the overall system will remain stable, although, the tracking performance will be somewhat reduced.

6. SIMULATION RESULTS

Simulations were performed on the models obtained from the Colorado State University Flexible Link Manipulator Testbed (*Fig.*1) described in detail in [5]. It was constructed from two thin cantilevered beams interconnected along their length by eight bridges that attach to the side plates through flexible hinges. This structure provides flexibility in the horizontal plane but maintains rigidity in the vertical plane.

An off-line identification procedure developed in [5] was used to obtain the hub angle/motor torque transfer functions of the plant for different loading conditions. The digitized transfer functions were then obtained with a sampling rate of 50 ms. In previous experiments, it was shown that one flexible and a rigid mode are adequate for control of the plant. Thus, a reduced fourth order model of the plant was obtained for payloads of $150g$ and $200g$ and digitized with a sampling rate of $50ms$. The poles for two transfer functions are located at $0, -.2, (-.029 \pm j9.5)$ for $150g$, and at $0, -.2, (-.03 \pm j9.05)$ for $200g$ with the corresponding zeroes at $(-.023 \pm j2.5)$ and $(-.021 \pm j2.25)$, respectively.

The simulations were performed by using the $150g$ model as the actual plant and the $200g$ model as the best estimate of the plant. The poles of the desired closed loop response (reference model) were placed at $-2, -2, (-2, \pm j9.5)$. All the poles of the observer matrix A_0 were placed at -15 and all the poles of the controller error dynamics matrix A_c at -10. The vector K in the observer was computed by using Ackerman formula. From the transfer functions, the uncertainty Δa was found to be .008

Fig.2 shows the output of the controller without the boundary layer vs. desired response. As expected, there is some steady state error due to mismatch in plant zeroes and considerable amount of chatter. Same chatter can also be noticed in the controller input signal in *Fig*.3.

Fig.4 and *Fig*.5 show plant output and input when the boundary layer is present with $\epsilon = .05$. It is clear that the amount of chatter is considerably reduced.

The mismatch between actual and model output in *Fig*.2 and *Fig*.4 is due to the uncertainties in the plant zeroes. By looking at these figures, one can note that there is some degradation of tracking performance. However, the corresponding pole placement controller diverges when is confronted to this amount of uncertainty.

7. CONCLUSIONS AND FUTURE RESEARCH

The presented results show that a variable structure controller can be used as a robust controller for a flexible arm manipulator. Though its performance is not very high, the results show that it might be adequate to use it as a back-up controller for higher performance, but more sensitive controllers, such as self tuning controllers.

The investigation of the best application of this control algorithm is needed in the future. Also, there has to be found a way to minimize the effect of unmodeled dynamics (such as the spillover effect present in the flexible systems) on the performance of the variable structure controller. This is particularly important since the high frequency chattering might excite enough high order dynamics to significantly reduce the performance or make it unstable.

REFERENCES

[1] L.Chirinos, D.Shusterman, J.J.Gonzales, G.R.Widmann, "Practical Aspects on Robust Control Strategies of a Single Link Flexible Manipulator", International Symposium on Circuits and Systems, New Orleans, Louisiana ,May 1990.

[2] M.D.Sidman,"Adaptive Control of a Flexible Structure", Ph.D Thesis, Stanford University, June 1986.

[3] S.Lefschitz, "Stability of Nonlinear Control Systems", New York: Academic ,1965 (*p*.115).

[4] B.L.Walcott, "Observation and Control of Dynamical System in the Presence of Nonlinearities/Uncertainties", Ph.D. Thesis, Purdue University, December, 1986.

[5] E.M.Nebot, "Modeling and Control of a Single Link Flexible Manipulator", Ph.D. Thesis, Colorado State University, Summer 1988.

[6] E.P.Ryan, "A Variable Structure Approach to Feedback Regulation of Uncertain Dynamical Systems", Int. J. Control, Vol. 38, No 6, *pp*.1121 – 1134.

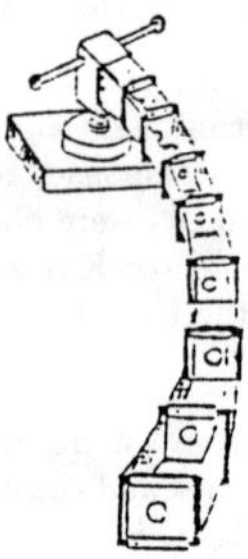

Fig 1. CSU Flexible Link Manipulator.

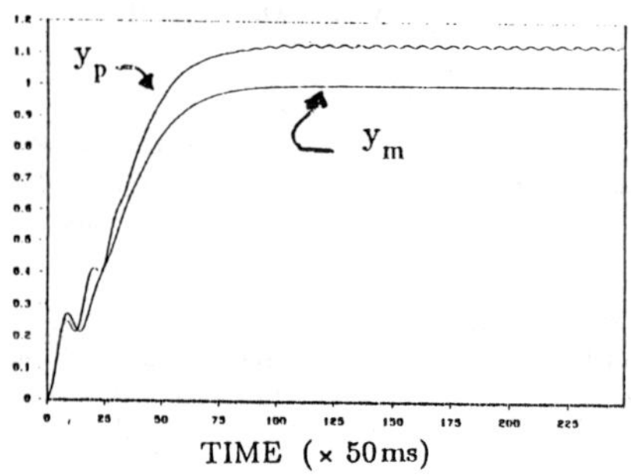

Fig 2. Comparison of Model and Plant Output Response for a Controller Without a Boundary Layer.

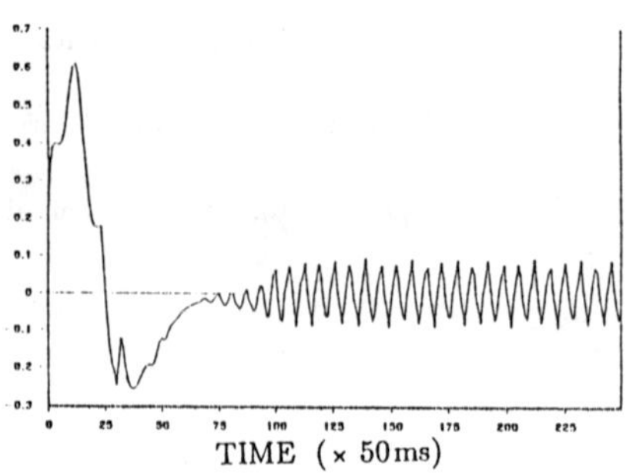

Fig 3. Plant Input Response for a Controller Without a Boundary Layer.

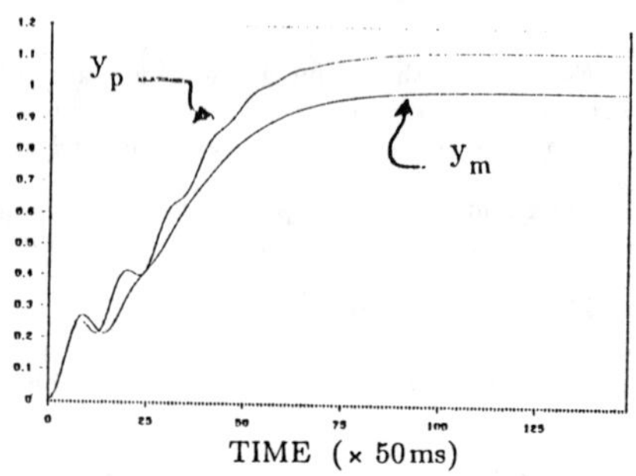

Fig 4. Comparison of Model and Plant Output Response for a Controller With a Boundary Layer.

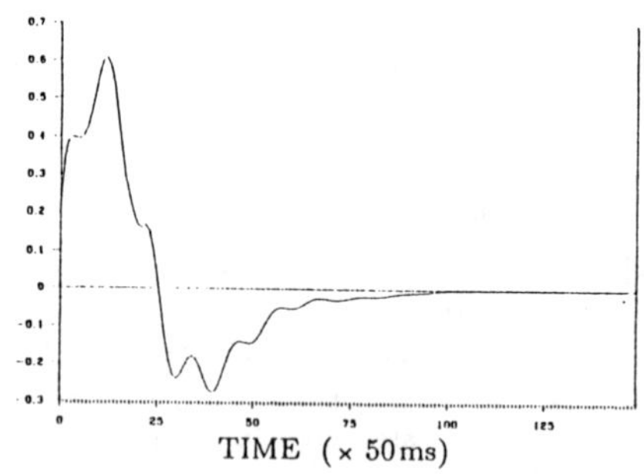

Fig 5. Plant Input Response for a Controller With a Boundary Layer.

SECTION 7 : AUTOMATED AND FLEXIBLE MANUFACTURING

DISTRIBUTED MOTION CONTROL

TAPIO VIRVALO and PERTTI PUUSAARI
Tampere University of Technology, Institute of Hydraulics and Automation, P.O.B. 527, 33101 Tampere, FINLAND

ABSTRACT

Today it is possible to buy compact ac and dc servo motion control packets with sophisticated programming aids. They are easy to use and they make overall machine construction easier. But still we are quite far from idea of the independent actuator. The independent actuator is a unit that includes the basic hardware and software needed: an actuator, a transducer, a control part and a high level communication interface. In this paper the idea of the hydraulic and pneumatic independent actuator is discussed.

INTRODUCTION

Distribution is a word of today. In the near future the machine designer has some standardized components for machine construction. These components are called the independent actuators. Their communication and controlling capabilities are optimized for distributed systems. They are for example hydraulic cylinders with integrated valves, control and communication units as depicted in Figure 1.

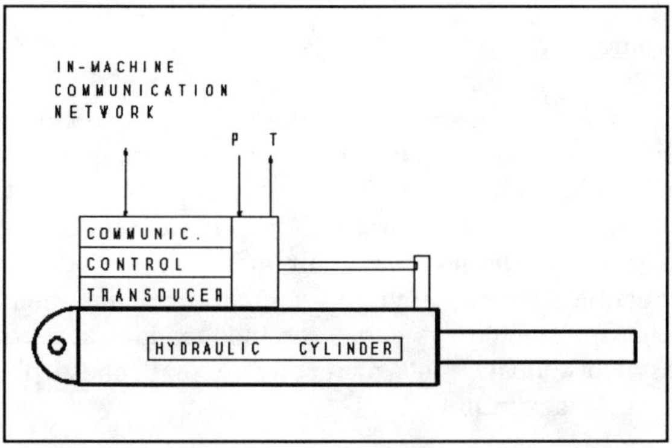

FIG. 1. The independent actuator.

THE INDEPENDENT ACTUATOR

The control of an automatic manufacturing plant is a hierarchical system. At the lowest level there are the independent actuator. It is a task programmable device. Because it has sophisticated control programs resident within, it is simple to use the unit for economical and fast machine design.

Programming

The actuator programming language is of high level. It includes commands for position control, parameter changing, synchronization, on/off control and sensing, sensor handling and error control. The programming language takes care of the several independent actuators simultaneously, which makes it applicable also for large systems.

Control

"Intelligence" is needed to operate independent actuators reliably in varying conditions with few instructions. Main problem is that increasing adaptivity needs a lot more computing power. In embedded systems this is many times a fatal problem because of limited computing power of microcontrollers. Our solution is to use robust controller, parameter tables, parameter changing and simple compensation functions. These can be premeditated in most applications. The system programmer can tell to an independent actuator about variations in operation conditions. We use a term "premeditated adaptivity" to describe the method.

The system software of the independent actuator includes such functions as error management, sensing of operation phase, supervisory functions, communication and command interpreting. This part should be as tiny as possible to reduce system overhead. However, special care is needed to achieve correct operation of the system.

Communication

Communication has a special role. It connects independent actuators to the system controller. Distances can be from meters up to hundreds of meters for example from manipulators to complete production lines respectively. Most of the traffic is between a system controller computer and the independent actuators. The information traffic is mainly bursty and consists of very short messages (some bytes). Synchronization demands are hard. Different transmission delays cause the independent actuators to start their movements asynchronously. This phenomenon shall cause position error in movement of several actuators.

Hardware

With technology of today it is possible to minimize the space needed for electronics part including controlling microcomputer and communication part. Use of surface mounting, flexible pc boards, programmable logic arrays and asic technology guarantee integration level needed. Problems arise when we also need reliable operation; electromagnetic interference, varying operating conditions, dust, oil and pollution makes conditions very hard. The small space and short wires used are fortunately minimizing problems of electromagnetic disturbances.

When sophisticated control algorithms instead of intelligent ones are used it is possible to realize our system on commercial microcontrollers. Today there are two possibilities; to choose 8-bit or 16-bit microcontroller. When a hydraulic actuator is used the time for control calculations is in optimum only 1 - 2 ms. It is proved that computing power of 8-bit microcontrollers is too small and also with 16-bit microcontrollers we have to do something to floating point calculations. One solution instead of microcontrollers is signalprocessors, but they are not applicable yet.

The sensor technology is still in the beginning. Sensors that operate reliably, are easy to integrate and still cheap are seldom met. Today the potentiometer type sensor as an integrated part of a hydraulic cylinder is available. It seems that applicable digital sensors are coming into the market.

THE DEVELOPMENT SYSTEM

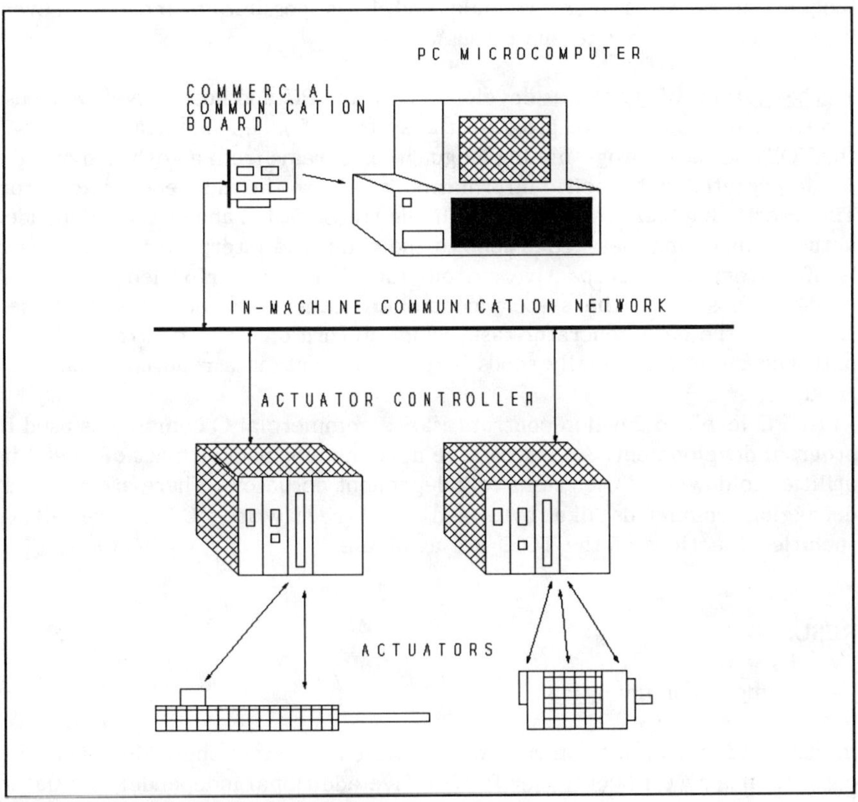

FIG. 2. The development system.

To develop systems based on the idea of the independent actuators a development system was made. At the moment it is made up of three boards, namely a microprocessor, input/output and a communication board. As a user interface an PC/XT/AT microcomputer with some software is used. The development system hardware is depicted in Figure 2.

The microprocessor board is based on 16-bit microcontroller. The board has some external memory; 32 kbytes eprom for system programs, a monitor program and user control programs and 32 kbytes ram for data and user loadable programs. Microcontroller's internal serial interface is used for the RS-232 communications.

The I/O board was made for basic interfacing purposes like connecting incremental encoder, potentiometers and switches.

The communication board is used for in-machine communications. Intel's Bitbus specification is used as a communication standard. This board replaces the RS-232 interface of the processor board. The Bitbus communication interface is of better performance; transmission speed is higher, distances are longer, interface is more reliable and it is possible to connect several independent actuators to one Bitbus cable.

The system software was developed both for PC and board level. With user interface it is possible to program the system of independent actuators with the TOKI actuator programming language. Programs from a keyboard or a disk are interpreted and specific information is send to the independent actuators. The interfacing tasks take care of reliable transmission and at the independent actuator incoming message is sent to the command interpreter. If a message is of system level, respective system functions are performed and a reply message is sent to the system controller. Otherwise the message is send further to the path generator task, which in turn operates control tasks. The path generator task finally sends a reply message as a report of completed task.

At PC level, to develop control tasks, a commercial C compiler is used for program development. An interactive development packet (part of TOKI) has utilities to download user tasks to independent actuators. There are also some debugging commands like byte and word read/write. Of course all the specified functions of the TOKI are available.

RESULTS

In-Machine Communication

In our tests and simulation runs we have used a system shown in Figure 3. It consists of a system controller PC and five additional independent actuators. In simulations it is assumed that PC is the only sending station apart from the

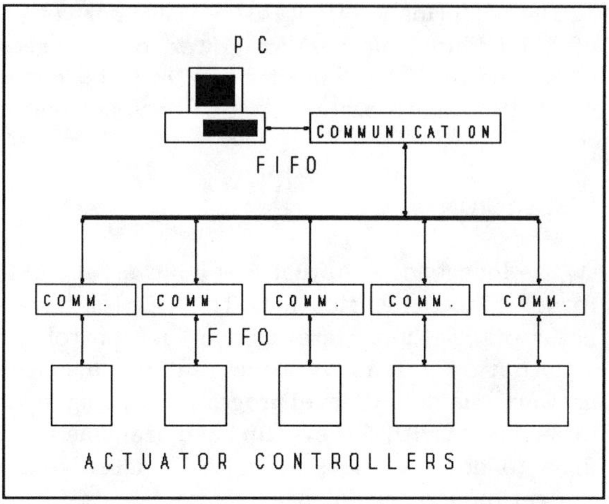

FIG. 3. The simulated system.

replies. This matches reality when the system controller continuously transmits position commands to the independent actuators, while the independent actuators have very seldom need for communication with the system controller. In simulation it is supposed that PC can generate messages at a desirable rate regardless of the PC's processing capability. The Bitbus network is used in the synchronous operation mode with a rate of 2.4 Mbps. The parallel FIFO type interconnections between controllers and communication part operate at a rate of about 80 kBps (both independent actuators and PC).

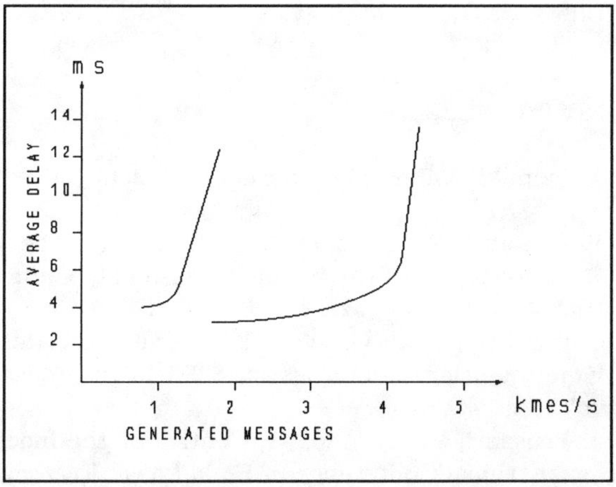

FIG. 4. Average delay versus generated messages (Bitbus 2.4Mbps, FIFO 80kBps, message 12B, 5 actuators).

The Figure 4 shows the average delay versus generated messages in two cases. The average delay is the time from offering the message to the FIFO queue in the PC to the incoming reply from the independent actuator. In these cases we have used the average message length of 12 bytes. In Case 1 we have assumed that about 10 per cent of messages need immediate processing such as read of parameters before a reply is sent. This generates an additional delay to that constant delay of normal operation of the system programs. In Case 2 no immediate processing is needed. This matches with robotic like path control of a manipulator where new position information is transmitted every 20 ms to 40 ms [1]. The figures show that in Case 1 it is possible to transmit messages without blocking the system at the rate of 1000 messages/s with average delay about 4 ms. In Case 2 the same value is about 3000 messages/s with delay 3.8 ms.

The performance figures shown are not yet reached. This is mainly because of the limited processing power of PC type microcomputers and time consumed to TOKI, but with new installations with workstation level system controllers those performance figures are easier to fulfil.

DISCUSSION

The Independent Actuator is a powerful building block when realized in the form of the specification. The results have been encouraging also with actuators without intelligence in control programs when premeditated adaptivity is used. We have realized also the in-machine communication network and the PC-level programming language. Simulation runs and practical experiments with this system have been encouraging. More work should be still done to develop the system to be intelligent one and to fulfil the synchronization demands of path control.

ACKNOWLEDGEMENT

The project is supported in part by the Academy of Finland and Tampere University of Technology.

REFERENCES

1. Puma 500MK2 robot system technical 506.9055, Unimation (Europe) pp 3-7
2. Puusaari, P., Sintonen, L., Virvalo, T., The Hierarchy of Communication Networks in the Programmable Assembly Cell: an Experimental Framework. IEEE Network Magazine (1988) May.
3. Puusaari, P., Virvalo, T., Developing Intelligent Actuator Systems: Experimenting with Bitbus and 8096. Microprocessors and Microsystems (1989), vol 13 no 4 May.
4. Puusaari, P., Virvalo, T., Control of the Independent Intelligent Actuator. International Conference of Fluid Power Transmission and Control, Hangzhou, China (1989).
5. Puusaari, P., Virvalo, T., Control of the Independent Intelligent Actuator. First JHPS International Symposium on Fluid Power Tokyo, Tokyo, Japan (1989).
6. Puusaari, P., Sintonen, L.,Virvalo, T., In-Machine Communication - A General Solution for Interconnecting Independent Intelligent Actuators. The International Conference on Communications, Boston, USA (1989).

ARCHIMEDES: AN EXPERIMENT IN AUTOMATING MECHANICAL ASSEMBLY

DAVID STRIP
Sandia National Laboratories
Albuquerque, N. M.

ANTHONY A. MACIEJEWSKI
Department of Electrical Engineering
Purdue University
West Lafayette, IN.

Abstract

Archimedes is a prototype mechanical assembly system which generates and executes robot assembly programs from a CAD model input. The system addresses the unrealized potential for flexibility in robotic mechanical assembly applications by automating the programming task. Input is a solid model of the finished assembly. Using this model, ***Archimedes*** deduces geometric assembly constraints and then produces an assembly plan that satisfies the geometric constraints, as well as other constraints such as stability and accessibility. A retargetable plan compiler converts the generic plan into robot and cell specific code, including recognition routines for a vision system. In the prototype system the code is executed in a workcell containing an Adept Two robot, a vision system, and other parts handling equipment.

1 Introduction

Archimedes is a prototype system for automating mechanical assembly. It accepts a solid model of an assembly as input and generates a robot program for carrying out the assembly.

Although robots offer the potential for truly flexible manufacturing systems, their application has usually been limited to situations in which the robot can operate in a relatively fixed manner for a reasonably long production run. The complexity of programming robots, designing fixtures, arranging workcells, and interacting with the output of the design process are the principle reasons for this limitation. The ***Archimedes*** system is motivated by the desire to significantly simplify the task of programming and setting up a mechanical assembly task.

Our goal is a system which accepts a set of plans and parts as inputs and produces an assembled product. In more practical terms, this translates into providing a solid model as input to a computer program and producing as output the necessary manipulation and sensing programs for carrying out the assembly on a robot and ancillary systems. As we are concerned with manufacturing systems, we assume that the environment can be be structured in order to simplify the task, although we must obviously restrict the degree of structure in order to maintain our goal of flexibility.

This work was performed at Sandia National Laboratories and supported by the U. S. Department of Energy under contract DE-AC04-76DP00789.

Figure 1: Pattern Wheel Assembly

Early efforts in automating assembly include Fahlman's BUILD System [1], which focused on planning, but contains a very interesting approach to stability analysis. Several task level programming systems have been proposed, such as Lama [4] and AML/X [8], both of which use the idea of skeleton strategies. These are both textually oriented and procedural in nature. Rapt [7] moves away from the procedural format, and uses a textual input describing the geometric relationships among the parts. These task-level systems are primarily conceptual designs; little experience is reported on their application to assembly tasks using physical robots. Somewhat more closely related to our work are approaches that use a geometric representation as part of their input. Autopass [2] is an early example in this direction. Again, little is reported on the actual implementation or application of these concepts. To our knowledge, Lozano-Perez's Handey system [5] is the only other planning system that we are familiar with that emphasizes implementation. Related work in a different vein includes de Mello's [3] ideas on AND-OR graphs for finding feasible subassembly and related planning issues. In our work we assume that important information on the breakdown into subassemblies is part of the design information, although provision is made for identifying additional subassemblies that are important for assembly purposes.

2 Overview of the *Archimedes I* Experiment

Figure 1 show the pattern wheel assembly from the dual stronglink, a safety component designed at Sandia National Laboratories. This assembly contains 13 parts (mostly pins and cogs) and is representative of a large fraction of the over 250 parts in this clockwork-like mechanism. For assembly planning, the parts can be modeled as stepped cylinders with holes in them. This simple model is sufficient for all the parts in the pattern wheel assembly, as well as a substantial fraction of the other parts in the dual stronglink. The motions needed for this particular assembly are uni-directional, as are a great many of the motions needed for assembling other parts of the larger mechanism. This is consistent with the frequently cited study of Nevins and Whitney [6] on assembly. The goal of the ***Archimedes I*** experiment is to accept a CAD model of the assembled pattern wheel and the parts as input and to produce an assembled pattern wheel as output

The workcell for assembling this component consists of an Adept Two manipulator with a tool changer, a two-finger pneumatic gripper, a vacuum gripper, a remote center compliance, a force sensor, an arm mounted video camera, and a

VMEbus-based computing system to supplement the commercial controller.

The modeling and planning software, written in Common Lisp, are executed on a Symbolics Lisp machine. The plan compiler, which takes the output of the planner and produces code for the robot controller, is written in C++ and executed on a Sun 3. The VMEbus system contains several processors; some are specialized for image processing, while others are single-board 68020 systems programmed in C++ and operating under VxWorks. The Adept controller is connected to the VMEbus system via a serial interface. Our concept is to develop a robot program offline and then download the program to a commercial controller for execution, using the robot controller to handle the (anticipated) error conditions and take the appropriate branching action. In line with this concept, the Adept program is generated offline on a Sun and downloaded to the controller via a serial link.

3 The modeler

The role of the solid modeler in ***Archimedes*** is to represent parts in the assembly and to answer the types of queries necessary to generate an assembly plan. Most available solid modelers are unable to respond to our queries in a reasonable amount of time. In addition, most are unable to represent assemblies and variational information such as tolerances. Fortunately, the relative simplicity of the part domain makes it possible to write a very fast, domain-specific modeler to address these issues.

The modeler we developed is able to model stepped cylinders with arbitrarily located stepped holes, although all cylinder and hole axes must be parallel. This makes queries such as detection of interference (null object detection of the intersection of two components) computationally straightforward. Collisions of two objects on certain simple paths are also very easy to detect in this domain. While this initially sounds like a very restrictive parts domain, examination of the service manual for a car of one of the authors reveals that the bulk of the drivetrain, as well as many other sub-assemblies of the car could be modeled in this domain.

In order to generate the set of allowable assembly sequences, we need to first find which parts interfere with others when moved in given directions. Here the uni-directional nature of the assembly speeds the process by reducing the number of interference checks that we need to make. The 2 1/2 dimensional nature of this domain allows the interference test to be reduced to a series of simple planar tests, which can be carried out quickly. Applying this operation to the pattern wheel assembly we can find the constraints which parts place on one another. One way to represent these constraints (with a loss of some of the information, however) is as an exploded diagram. (Exploded drawings are currently made by artists or by the operators of CAD systems, not automatically.) Figure 2 is an automatically generated exploded view. Using the solid modeler, the constraints among the parts are determined. The drawing is made by placing each part at the lowest possible level in the drawing such that it is above all parts that constrain it from below.

Moving from this simple domain to a more general domain will occur in several stages. We are now working on a generalized 2 1/2D modeler. This will retain much of the speed of the existing system while enriching the domain to include

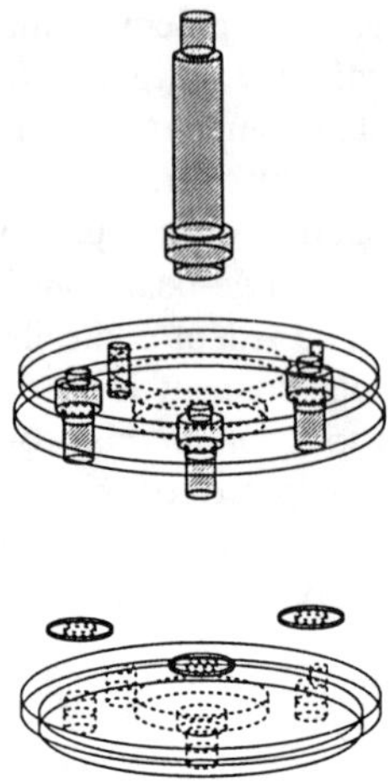

Figure 2: Automatically generated exploded diagram of pattern wheel assembly.

many currently manufactured designs. A general 3D domain requires the use of a true 3D solid modeler. Many such modelers exist, and their design is a research topic in itself. Many of the tests for part interference that we now use can be couched as queries to a general solid modeler, although they may take longer to evaluate than in the specialized modeler. The uni-directional assumption of this example cannot be expected to hold in general. For the 3D case, we are looking at a number of approaches for finding candidate directions for motion, both algorithmic, based on the part geometry, and knowledge-based, using information such as part categories or rules relating to previously discovered assembly directions.

4 The planner

There are many possible orders in which the parts may be assembled based solely on the geometric constraints determined by the solid modeler. Some of these orders may be inefficient; for example, many tool changes are required. Others may be infeasible due to instability of intermediate stages. The role of the planner is to evaluate candidate orders, determine their quality, and try to find a good order. Since we view the context of our problem as the batch manufacturing environment, planning is really a one-time offline task, so there is a strong motivation to spend the time to optimize the plan, which is to be executed many times.

The planner currently recognizes 4 types of infeasibilities: instability of intermediate stages, inaccessibility of weld surfaces, placing objects through the support surface, and out-of-order assembly. The planner takes a candidate sequence and evaluates it for feasibility against these four criteria. (In fact, the system is arranged to guarantee that only valid orderings are generated, so the last criterion does not have to be checked.)

For each type of infeasibility, there are one or more fixes that the planner can try. For instability, the alternatives are a different sequence or a fixture. Inaccessible surfaces may be made accessible by inverting parts or subassemblies. Fixtures or resequencing are used when parts penetrate the support surface. The search among the various possible assembly orders is controlled by assigning a "score" to a sequence based on the type and number of infeasibilities, and backtracking when the an operation fails to improve the score. Currently, the search stops when a feasible solution is found; no attempt is made at further optimization.

It is possible that no feasible sequence can be found using these operators. This is because no allowance is made for inverting a partially completed sub-assembly. The system allows the definition of new sub-assemblies as a solution to this problem.

Some assemblies, including the pattern wheel assembly in our example, cannot be assembled by a single manipulator without the use of fixtures. Our planner is capable of designing certain simple fixtures when necessary to allow development of feasible plans. The fixture design is output as standard dimensioned 3-view drawings, familiar to any machinist.

5 The compiler

The compiler is the part of the system that translates a generic, high-level plan into detailed instructions that can be directly executed in a robot workcell. The central principle in the development of ***Archimedes*** is modular design to maintain flexibility. Fundamental to this concept is the separation of the compiler front-end, which deals with a workcell-independent specification of allowable assembly plans, from the back-end which is responsible for generating the machine-specific control commands for the robot(s) and ancillary systems. As with traditional computer language compilers, information is passed between the modules through the use of a symbol table.

Input to the compiler is generated by the planner, and is in the form of an assembly plan which consists of a sequence of calls to assembly primitives. This assembly plan, together with the solid model of the completed assembly, specifies all of the *relative* motions between individual parts required to complete the assembly. One of the primary responsibilities of the compiler is to bind these relative part positions to absolute workcell locations. This mapping is performed by a workspace manager module which is a three-dimensional analog of the run-time memory management found in computer systems.

The workspace manager is responsible for maintaining the status of the reachable workspace of the robot for which the plan is being compiled. Parts are assigned absolute positions within the workspace based on their bounding spheres and order of use. The strategy seeks a compromise between minimizing workspace fragmentation (which would dictate a global best fit policy) and minimizing manipulator motion based on the proximity of related parts. The workspace manager is also able to reclaim portions of the workspace to be reused later in the assembly. This space is most commonly used to store subassemblies that have been created from the individual parts which had occupied this area.

While the workspace manager is responsible for assigning the position of a part, the nominal orientation of a part is specified by the grasp planning module. The philosophy of the grasp planner is to choose the initial orientation at which a part is presented to the workcell in such a way that regrasping is unnecessary. The grasp planner generates an initial set of grasps based on the no regrasp criterion. The resulting set of grasps is then mapped to the part's starting position and pared on the basis of stability. The set of orientation trajectories resulting from pairing each valid starting configuration with the set of valid ending grasps is then compared to the kinematic capabilities of the robot being used in the assembly. This procedure identifies all of the initial part orientations which result in a physically achievable motion trajectory that does not require regrasping. A unique solution from this set can be obtained by using the stability ranking, a dexterity measure, or a combination of the two.

The outputs of the workspace manager together with the grasp planner specify a nominal workcell design. However, due to the positional uncertainty inherent in material handling systems, the actual position and orientation of the parts presented to the workcell will vary. To accurately identify the locations of these parts at assembly time the compiler generates code with calls to the vision system. Since the nominal position of a part is already known, the system does not need to deal with part identification except to flag errors. The majority of this calculation is performed off-line at compile time. A single matrix multiply is all that is required at assembly time in order to perform a least-squares fit of the data to the part model. The advantages of this approach are faster execution time and flexibility in the target hardware.

After establishing absolute workcell locations for all parts in the assembly, the compiler translates the planner language commands into the control language for a specific system controller. The compiler is currently capable of outputting V code to run the Adept manipulator and GSL code for the IGRIP computer graphic simulation system, which was used in debugging the compiler code. The ability to easily modify the compiler for alternate manipulator controllers is a direct result of the separation between the assembly-dependent responsibilities of the compiler from the robot-dependent responsibilities.

6 Experimental Results

The planner and compiler described here were initially applied to the pattern wheel assembly presented earlier. We first compiled the plan into GSL code for the IGRIP simulation system and executed the assembly as a simulation. As noted, this was of great assistance in debugging the code generator. When the workcell was completed, we began compiling into Adept's V programming language. The primary difficulties that arose in this stage were similar to those encountered when manually programming a robotic task: accurately defining fixed locations in the workcell, getting various pieces of equipment to talk to one another, etc. Instead of using the compiler generated locations, we used a manually designed part tray that would allow part kitting. The vision system was used to locate the large cog-like parts, while the small pins were precisely located in holes in the part kit delivery tray. This corresponds to a situation in which the cogs are provided by an outside manufacturer and are manually added to the kits, while the pins are fabricated in house and are precisely kitted as they leave the turning station. The force sensor in the initial system is only used for guarded moves (and searches) rather than force-directed actions. A passive RCC on the arm provided the required compliance for the tight insertions. In spite of the limited capabilities of the workcell, the compiler generated code allows consistently successful assembly of the pattern wheel assembly. The plan creation and compilation each take less than one second to exectute. The assembly requires a few minutes in the workcell.

The ***Archimedes I*** system is capable of dealing with a large variety of interesting real-world assemblies, including a parts of the automobile drivetrain such as the transmission and differential; many subassemblies in machine tools; many household appliances like blenders, mixers, drills, etc; consumer electronics like VCRs and personal stereos ("Walkman"); and so on. We are taking advantage of the modular implementation of the system to expand its capabilities. We are extending the modeler to more general 2 1/2 D capabilities, and working on a true 3D approach. We are also adding sophistication to the planner, and adding capabilities for motion planning and error handling.

References

[1] S. E. Fahlman, "A Planning System for Robot Construction Tasks," Ph.D. Thesis, M. I. T., Cambridge, MA., 1973.

[2] L. I. Lieberman and M. A. Wesley, "AUTOPASS: An Automatic Programming System for Computer Controlled Mechanical Assembly," *IBM J. Research and Development*, Vol. 21, no. 4, July, 1977.

[3] L. S. Homem de Mello and A. C. Sanderson, "And/Or Graph Representation of Assembly Plans," *Proc. AAAI 86,* Philadelphia, PA, 1986.

[4] T. Lozano-Perez and P. Winston, "LAMA: A language for automatic mechanical assembly," *Proc. 5th Int'l Joint Conf. Artificial Intelligence,* Cambridge, MA., 1977.

[5] T. Lozano-Perez, *et. al.,* "Handey: A Robot System that Recognizes, Plans, and Manipulates," Proc. IEEE Int'l Conf. on Robotics and Automation, Raleigh, NC., 1987.

[6] J. L. Nevins and D. E. Whitney, "Assembly Research," *Automatica,* Vol. 16, 1980, pp. 595–613.

[7] R. J. Popplestone, A. P. Ambler, and I. Bellos, "An Interpreter for a Language Describing Assemblies," *Artificial Intelligence,* Vol. 14, No. 1, pp 79–107.

[8] R. H. Taylor, P. D. Summers, and J. M. Meyer, "AML: A Manufacturing Language," *Int'l. J. Robotics Research*, Vol. 1, No. 3, Fall, 1982.

Acknowledgements

The authors would like to thank Cliff Loucks and Colin Selleck for their assistance in making the robotic systems work.

High Velocity Manufacturing of Complex Electronic Systems

by

Larry Kinney AT&T, Denver, Co.

George Foo AT&T, Little Rock, Ar.

ABSTRACT

The Private Branch Exchanges (PBX) business, according to industry observers, is the most competitive, non-consumer electronics business in the world. To be profitable and survive in this business environment, it is a requirement to have "World Class" operations. The AT&T Denver Works manufactures several products including the Definity™ Communications System line of PBXs. The PBXs are complex, high option systems which are build-to-order. There are almost a limitless number of end-item configurations. Furthermore, there are also week-to-week volume fluctuations (as much as +/- 50%). Several years ago, the Denver Works embarked on an ambitious program focused at reducing cost and intervals but still provide flexibility to the customer. The initiative had three phases:

1. Implement a Material Requirements Planning (MRP) system for planning
2. Simplify processes by implementing Just-In-Time/Total Quality Control (JIT/TQC) systems for execution
3. Integrate the MRP (Push) planning system with the JIT (Pull) execution system

1. Phase I - Material Requirements Planning (MRP)

Material Requirements Planning (MRP) is a formal, computer based production and inventory management system. The principles of MRP are well documented in the literature[1] [2] and will not be reviewed here. What will be discussed will be a fundamental change in the product structure/bill-of-material (BOM) which made the successful implementation of MRP feasible.

The master production schedule (MPS) is the driver of the MRP system and the proper implementation of the MPS is required for successful material management. Therefore, the decision on what to master schedule is crucial. For build-to-order systems such as PBXs, there is almost a limitless number of end-item possibilities made from combinations of the basic components and sub-assemblies. The recommended[3] approach for master production scheduling of high option, build-to-order products, is to establish the MPS at the intermediate product or buildable option level where the buildable option will be defined fully in the bill of material files as a single-level bill of material. Thus, the modular bill of material structure has an architecture that couples component parts to buildable options, but it does not link either buildable options or components to end-item configurations. If the buildable options are simply sub-assemblies, then all that is required for the new architecture is to treat the sub-assemblies as end-items.

Fortunately, the AT&T PBXs were designed in a modular fashion and a re-design of the product was not required. Rather, what was required was a simple but fundamental change in the documentation to make more visible the sub-assemblies or buildable options which will result in the minimum number of master production scheduled items. That type of buildable option has no optionality within it but may be an option in itself.

The basic objective of this change was to minimize the number of master production scheduled items but still offer to the customer an almost limitless number of configurations. Additional details of this work are discussed by Kinney et. al[4].

2. Phase II - Shop Floor Improvements

With the successful implementation of an MRP system, the next step was to improve shop floor performance. Production control of shop floor operations can either be "push" or "pull". In a "push" mode, the shop is building in anticipation of a future demand while in a "pull" mode, the shop is building in reaction to a present demand.

2.1 Push vs. Pull Systems

2.1.1 Push Systems

MRP and MRP II (Manufacturing Resources Planning) systems are usually identified as "push" systems. Although, both have many planning features which are essential for efficient manufacturing, they are associated with "push" because they contain features such as shop order generation and shop order launch. MRP II systems also contain the shop floor control (SFC) module which is also associated with "push" systems.

Classical MRP II can be viewed as a "push system", both in its planning (MRP) and execution (SFC) phases. The description "push" is derived from the use of quantity scheduling. A vendor or work center is expected to deliver a specified quantity of a given item at a particular time, regardless of the schedule positions or actual needs of downstream operations. A great deal of effort is required to constantly monitor and change order due dates in a "push" environment to maintain balance between supply and actual demands (especially required for high-velocity manufacturing) since these systems are not always responsive to changes. In addition, routings data changes can be devastating to a system which operates on shop order due dates.

2.1.2 Pull Systems

In the broadest sense, Just-In-Time (JIT) is defined as a process which strives to eliminate waste and one of the JIT techniques used to identify waste is the kanban system. As mentioned previously, "pull" production control is when one is building in reaction to a present demand and kanban systems represent one form of "pull" systems. The operational specifics of kanban systems are well documented[5] [6] and will not be discussed here except to say that with respect to production control, kanban systems are inventory replenishment systems. This implies that in a kanban system, inventories of the buildable item must always be in stock since production of the buildable item is triggered only when the inventory stock falls below a certain threshold. When applied to also expose quality problems, kanban

LOW ← LEAD-TIME VARIABILITY → HIGH

	MATERIAL PLANNING	CONTROL SHOP ORDER RELEASE	SHOP FLOOR
PULL: CONTINUOUS FLOW	1 JIT	2 RATE BASED	3 JIT-PULL
HYBRID PUSH-PULL: BATCH REPETITIVE	4 JIT-MRP	5 PULL OR MRP	6 PULL
HYBRID PUSH-PULL: BATCH DYNAMIC	7 MRP	8 MRP	9 PULL OR ORDER SCHEDULING
PUSH: CUSTOMER ENGINEERING	10 MRP	11 ORDER SCHEDULING	12 OPERATION SCHEDULING

SOURCE: U. KARMAKAR, UNIV. OF ROCHESTER

FIGURE 1: TAILORED PRODUCTION CONTROLS

systems become powerful tools in improving manufacturing performance. Kanban systems are simple, cost effective and usually manual and should be used whenever possible.

Kanban systems, however, have their limitations. Karmakar [7] maintains that production management systems must be selected to match the characteristics of the product and product demand. Karmakar's concepts are illustrated in Figure 1. According to Karmakar, production processes which are dedicated to one or a few similar products and where the production is continuous, level and the production lead times are predictable and uniform, kanban is the system of choice for material planning, order release and shop floor management. At the other extreme where the product is low-volume, complex engineered or custom manufactured with no regularity in production patterns, MRP II for materials planning and order scheduling (push) for order release and shop floor management is the system of choice. However, most products and production systems fall between the two extremes and generally the production control system of choice is a hybrid of "push-pull" that uses the strengths of both approaches.

Reflecting the criteria established by Karmaker, kanban systems were implemented in the manufacture of the PBX cabinets, power units, fans etc. In addition, kanban was also used as the production control system for inexpensive items such as cables. On the other hand, the application of kanban systems to circuit pack manufacturing was uneconomical due to the high demand volatility and high cost of the circuit packs. However, even before a production control system could be put in place for the circuit pack manufacturing line, it was necessary to first characterize the manufacturing process and re-engineer the flow for maximum velocity. A major effort was undertaken to address the quality problems and reduce variation. Computer based simulation tools were also used to conduct analysis of line performance. Physical re-layouts of the facilities were also implemented to improve flow and reduce the need for shop floor scheduling. The end result of these improvements was a predictable, high velocity shop for the manufacture of circuit packs. The remaining issue was the selection and design of a production control system. The final design was a "hybrid" pull system known as Integrated Pull Manufacturing (IPM).

3. Phase III - Integrated Pull Manufacturing (IPM)

At the start of phase III, although the MRP system and the shop floor JIT/TQC operations were successfully implemented, the various factory operations were still not fully integrated. A missing link was an execution system that not only initiated circuit pack starts but more importantly synchronized all operations in the factory. That execution system was called Integrated Pull Manufacturing (IPM).

The IPM execution operations architecture is shown in Figure 2. IPM synchronizes and integrates MRP planning, the manual kanban systems in the equipment feeder shops, the storeroom and circuit pack manufacturing operations with the final assembly schedule. Critical concepts considered during the design of IPM were:

1. MRP is for planning while Pull is for execution.

   That is, all shop floor operations are pulled by a final assembly sequence which is driven by **actual** *customer orders as opposed to an MPS forecast* There are no shop orders!

2. Smooth and efficient manufacturing requires a level load which is driven by final assembly.

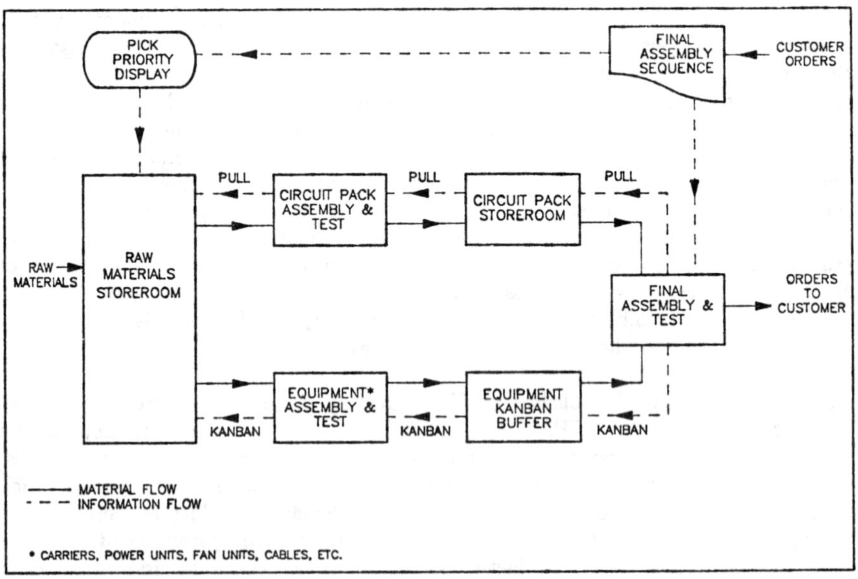

FIGURE 2: IPM OPERATIONS ARCHITECTURE

3.1 MRP for Planning, Pull for Execution

At the Denver Works, the belief was that MRP should be for planning while a hybrid, pull system (IPM) should be used for execution. The key to this separation, was the development of a "bank account" feature in the MRP system.

The "bank account" feature is a capability which allows the Master Scheduler to deposit or authorize a maximum level of material which can be withdrawn from for each planning unit or master scheduled item. The "bank account" feature has several positive attributes. First, it takes the Master Scheduler out of the execution process and allows him/her to focus in on planning and maintaining the balance between demand and supply of sub-assemblies. Second, the "bank account" also insures that the shop will not withdraw more than the amount authorized by the Master Scheduler's material plan. Third, it supports the pull environment since only material that is required will be picked and delivered based on a downstream pull signal (ultimately the customer order) rather than being pushed on to the factory floor by the MPS. Storeroom picking activities are prioritized by IPM and pulled by the shop.

3.2 Level Loading

The need to level the manufacturing load is driven by the desire to optimize resource (e.g. work force, facilities, etc.) utilization and to minimize manufacturing intervals. A close to level or smooth load was achieved through the use of a specially developed algorithm embodied in a program called the Final Assembly Sequencer (FAS)[8]. PBX orders to customers are scheduled for shipment by the end of the week. Therefore, the factory has the flexibility to build a week's order of cabinets/systems in any sequence it desires. Figure 3 shows a typical output of the FAS for eighty cabinets orders.

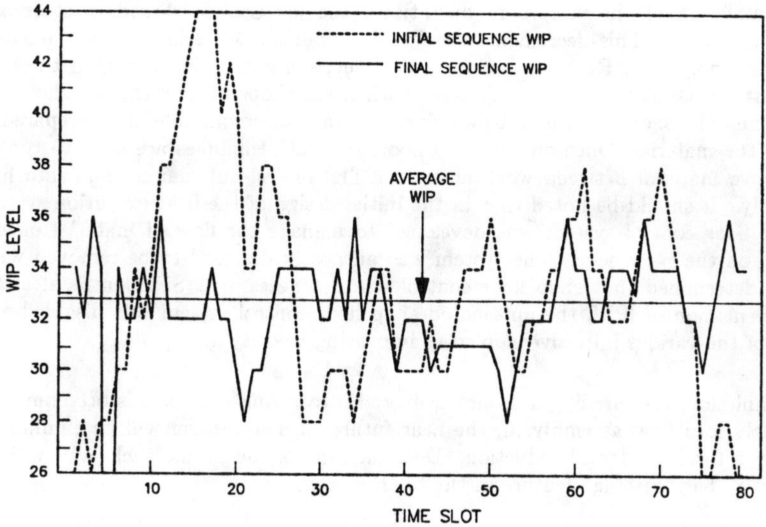

FIGURE 3: FINAL ASSEMBLY SEQUENCER OUTPUT

3.3 IPM Process

The IPM process begins when the actual PBX system and spare parts orders are received by the Denver Works' order processing system, where customer requirements are converted into manufacturable buildable items (planning units). The list of buildable items are then, on a nightly basis, translated into planning unit demands which are then transferred to the MRP system where the demands are displayed on the master production scheduling grid and consume the forecast based on customer order scheduled ship dates.

Under the IPM scenario there is a "demand" generator which consists of the final assembly sequencer and a net circuit pack requirements module. The final assembly sequencer, based on the final assembly schedule, first processes the weekly cabinet demand into daily requirements for the final assembly shop with the objective of generating the smoothest load on the feeder shops. Inputs come from the customer order database. Based on the final assembly sequence, "equipment" shop sub-assemblies (e.g. carriers, power distribution sub-assemblies etc.) are pulled into final assembly via a manual kanban system. For the circuit packs, the final assembly sequencer explodes the end-item requirements into circuit pack requirements. Based on the work-in-process (WIP) information contained in the master production schedules for circuit packs, net circuit pack requirements are determined. Manufacturing lead times for the circuit packs are then used to backschedule when the manufacturing of the item should begin. From this calculation, a priority list of "starts" is generated for the circuit packs. The manufacturing process begins in the storeroom. A priority display, which receives input from the IPM demand generator, defines a storeroom pick order. However, before an order can be picked, there must be material deposited in the "bank account" to withdraw against. The master production scheduler is responsible for establishing the bank account level by logically "depositing" picklists into the storeroom "bank account".

On completion of the storeroom operations, the material is staged in storeroom outbound buffer. This decrements the "bank account". Material is never "pushed" onto the shop floor. Rather, maximum WIP levels are established within the shop and material is delivered to the shop only when the shop is below the threshold and only when the shop transmits a pull signal to the storeroom that it is prepared to accept the material. Once on the shop floor, manual techniques are used to control and move material between work areas in a first-in-first-out manner and into final assembly. It should be noted that in the initial design of the IPM execution system, a shop floor control system was developed to manage the flow of material once it arrived on the shop floor. The system was purposely designed to be removed when it was determined that shop floor control was not necessary. Six months after the implementation of IPM, the automated shop floor control system was disabled as a result of the various initiatives aimed at improving velocity and quality.

On completion, the circuit packs are delivered into a finished goods storeroom and ultimately into final assembly. In the near future, this storeroom will be eliminated to allow a further interval reduction. Decrementing of the circuit pack WIP will be done by a "backflushing" feature in the MRP system.

4. Summary

At the AT&T Denver Works, a three step initiative was implemented to increase it's manufacturing competitiveness. During the course of the project, several key principles and concepts were demonstrated:

a. Designing products and their documentation for modular planning is crucial for successful material management.

b. Production control systems should be tailored to match the characteristics of the product design, cost and demand volatility.

c. A diverse set of production control systems can be complementary and can be integrated into a hybrid, unified, flexible manufacturing control system.

d. Process simplification and streamlining can eliminate the need for complex and expensive information systems such as shop floor control, tracking, etc.

The Denver experience confirms the evolutionary nature of information systems for manufacturing operations. In order to manage and synchronize complex, lengthy manufacturing operations, complex systems are required. As the processes are simplified, the information systems must follow suit to take full advantage of the process improvements.

e. MRP and JIT/Pull are complementary provided that MRP is used only for planning and JIT/Pull is for execution

The proof of these conclusions can be demonstrated by the results achieved over the three years of the project. Total inventory in the factory was reduced by 90%, inventory turnover ratio was increased by a factor of 4, the system (end-item) manufacturing interval was reduced by a factor of 4, the circuit pack manufacturing interval was reduced by a factor of 7 and circuit pack quality levels, as measured by first pass yields, increased from 55% to greater than 90%.

REFERENCES

1. J. Orlicky, "Material Requirements Planning", McGraw-Hill Book Company, 1975

2. O.W. Wright, "Production and Inventroy Management in the Computer Age" Van Nostrand Reinhold Company, New York, 1984

3. T. Vollmann, W.L. Berry, D.C. Whybark, "Manufacturing Planning and Control Systems", 2nd Edition, Irwin, 1988

4. L. Kinney, J.P. Clancy, G. Foo, C. Lindemulder, "Simplified Documentation Structure - Design for Material Logistics", AT&T Technical Journal, May/June 1990.

5. R.J. Schonberger, "Japanese Manufacturing Techniques - 9 Hidden Lessons in Simplicity", Free Press, New York, 1982.

6. Y. Monden, "Toyota Production System - Practical Approach to Production Management", Industrial Engineering and Management Press, Institute of Industrial Engineers, 1983

7. U. Karmakar, University of Rochester, "Getting Control of Just-In-Time", Harvard Business Review, September-October 1989, pp. 122 - 131.

8. H. Groeflin, H. Luss, M.B. Rosenwein and E.T. Wahls, "Final Assembly Sequencing for Just-In-Time Manufacturing", International Journal of Productivity Research, Vol 27, No.2, pp 199-213, 1989.

ON THE PROCESS PLAN SELECTION IN AUTOMATED MANUFACTURING SYSTEMS

Wu Naiqi and Xue Jinsong

Shenyang Institute of Automation
Academia Sinica
Shenyang 110003, P.R.China

ABSTRACT

In [1] Kusiak and Finke discussed the process plan selection problem in automated manufacturing systems. In this paper a modified formulation is proposed for this problem and efficient algrithm is presented, it is done by finding the shortest pathes in an acyclic digraph. So the computational complexity of solving the problem reduces very much.

INTRODUCTION

In automated manufacturing systems, generally, for each part a number of different process plans is given, and the process plan selection problem rises in the planning and scheduling. A prefered objective of doing that is to mininize the cost and the types of tools and auxiliary devices to be used. The model in [1] is based on this objective. But because of the high computational complexity in solving the problem so formulated they presented two heuristic algorithms.

In this paper a modified formulation is proposed, and it is easy to solve.

PROBLEM DESCRIPTION

Process plan problem can be seen in [1], [2] and [3]. In order to describe the process plan selection problem define for each process plan P_i the following incidence column vector:

$$x_i = (x_{1i}, \ldots, x_{ai}, x_{bi}, \ldots, x_{ci})'$$

where

$$x_{ti} = \begin{cases} 1, & \text{if a tool } t \text{ is used in } p_i,\ t \in \{1, \ldots, a\} \\ 0, & \text{otherwise} \end{cases}$$

$$x \ = \begin{cases} 1, & \text{if a fixture } f \text{ is used in } p_i,\ f \in \{b, \ldots c\} \\ 0, & \text{otherwise} \end{cases}$$

For any two process plans P_i and P_j define the weighted Hamming distance

$$d_{ij} = \sum_{q=1}^{c} w_q \delta(x_{qi}, x_{qj}), \forall i, j \quad (1)$$

where

$$\delta(x_{qi}, x_{qj}) = \begin{cases} 1, & \text{if } x_{qi} = x_{qj} \\ 0, & \text{otherwise} \end{cases}$$

and w_q is the weight coefficient of tool (or auxiliary device)q. The weighted Hamming distance d_{ij} measures dissimilarity between process plans P_i and P_j .

The following notation is needed for the graph-theoretic formulation of the process plan selection problem.

K = {1,2,...,m} is the set of parts to be manufactured.
N_k is the set of process plans for a part k, $k \in K$.
N is the set of all process plans (vertices), where

$$\bigcup_{k \in K} N_k = N$$

A is the set of arcs connecting process plans (vertices) from set N_k to set N_l, $(k,l) \in K \times K$ and $k \neq l$.
d_{ij} is the weighted Hamming distance, $\forall (i,j) \in A$.
c_i is the cost of process plan P_i, $\forall i \in N$.

From the definitions and notation above a complete m-partile graph G=(N,A) is obtained to describe the process plan selection problem. If S_k denotes an node from N_k then the process plan selection problem is to select one set from all the sets $S=\{S_1, S_2, \ldots, S_m\}$ such that the following performance function

$$H = \sum_{(i,j) \in A^S} d_{ij} + \sum_{i \in S} c_i \quad (2)$$

is minimized. Where $A^S \in A$ is the set of all (nondirected) arcs between the points of S. This problem so formulated cannot be solved exactly.

THE MODIFIED FORMULATION

One way to simplify the an optimization problem is to simplfy the model such that the problem can be solved efficiently. We do so here and attempt to modify the formulations given in [1].

Considering that the first term in the right hand of (2) does not describe the different types of tools and auxiliary devices used exactly, so to modify the formulation we may redefine the weighted Hamming distance d_{ij} in (1) as follows.

$$d_{ij} = \begin{cases} \sum_{q=1}^{c} w_q \delta(x_{qi}, x_{qj}), & i \in N_k, j \in N_{k+1} (k \in K), \text{or } i \in N_m, j \in N_1 \\ 0, & \text{otherwise} \end{cases} \quad (3)$$

And set A has a corresponding change.

A is the set of arcs connecting process plans (vertices) from set N_k to N_l, $(k,l)\in K\times K$, $k\in N_k$, $l\in N_{k+1}$, or $k\in N_m$, $l\in N_1$.

With this definition of distance d_{ij} in (3) we keep the performance function in (2) unchanged in our formulation. It is clear that by doing so all the different types of tools and auxiliary devices used for the selected process plans still have a contribution to the first term in (2), the only difference between it and that in [1] is that the contribution made is not same. So in the new formulation the first term in (2) is served to minimize the number of types of tools and auxiliary devices. This justifies our modification.

But according to the definition of the distance d_{ij} in (3) we obtain a noncomplete m-partile graph description instead of a complete m-partile graph. Fig.1 shows a noncomplete 4-partile graph. In the figure a_i $(i\in N)$ denotes the vertex of process plan P_i, c_i in the parenthesis is the cost of plan P_i. In our formulation, except that the graph describing the problem is a noncomplete m-partile graph, the process plan selection problem can be stated in the same way as the formulation given in [1].

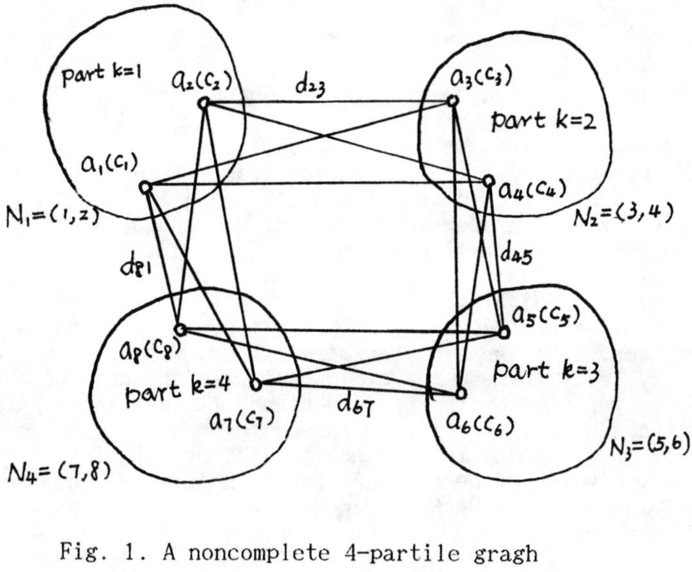

Fig. 1. A noncomplete 4-partile gragh

THE PROCEDURE FOR SOLVING THE PROBLEM

Because of the special structure of the noncomplete m-partile graph obtained in our model an efficient graph-theoretic approach can be developed to solve the problem.

For the noncomplete m-partile graph we make a transformation as follows. Replace vertex a_i in the graph by two vetices a_i and

b_i, let c_i be the distance from a_i to b_i and the direction of arc (a_i, b_i) be incident from a_i to b_i, the direction of arc $(b_i, a_j) \in A$

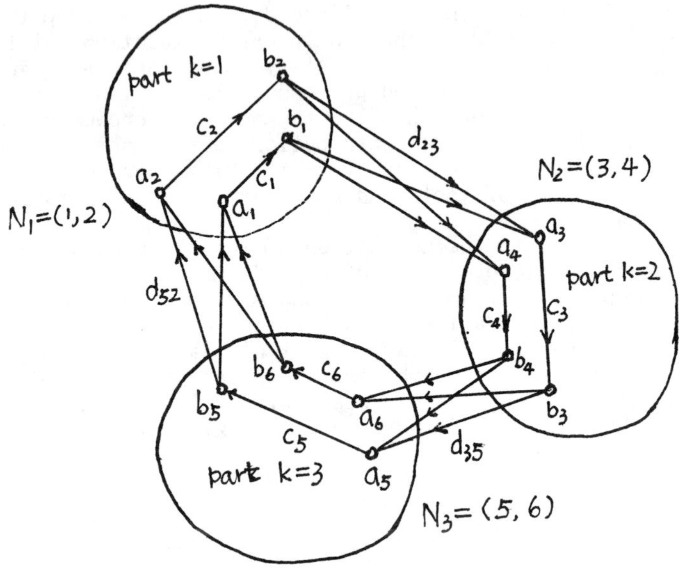

Fig.2. The equivalent digraph of a noncomplete 3-partile graph

($i \in N_k$, $j \in N_{k+1}$, or $i \in N_m, j \in N_1$) be incident from b_i to a_j. So doing the noncomplete m-partile graph can be transformed into an equivalent digraph (directed graph) G'(N,A), and it is a cyclic digraph. Fig. 2 shows such an equivalent digraph G'(N,A) obtained from a noncomplete 3-partile graph.

According to the definition of the process plan selection problem and the equivalent digraph description it is clear that in graph G'(N,A) start from any vertex a_i ($i \in K$) and go along with the directed arcs, through a number of vertices, finally come back to a_i and a cycle is formed, the cycle must pass through one and only one of the vertices of the process plans of each part $k \in K$. This implies that the set of vertices passed through by such a cycle is a set of process plans in S, and the distance H_i of the cycle is equal to the value of performance function (2) for the selected process plan set, or $H_i = H_S$. Hence if we know process plan P_i is one of the elements of the optimal process plan set then the process plan selection problem is to find the shortest cycle passing through vertex a_i of process plan P_i in the equivalent digraph G'(N,A).

Since one and only one process plan must be selected from the process plan set of any part $k \in K$ $F_k = \{P_i\}$, $i \in N_k$, there must exist a plan P_i ($i \in N_k$) in F_k such that $P_i \in S^*$, where S^* denotes the optimal process plan set in S. It follows from this observation that if for any plan $P_i \in F_k$ we find the shortest cycle passing through vertex a_i of plan $P_i \in F_k$ and let its distance be H_i, we obtain a set $H = \{H_i\}$, $i \in N_k$, and take the minimal one from set H, then the process plan selection problem is solved.

For the digraph G'(N,A) we can make further transformation as follows. Select a $k \in K$ arbitrarily and for any vertex a_i of process plan $P_i \in F_k$ replace it by two vertices a_i and a_i', let the distance

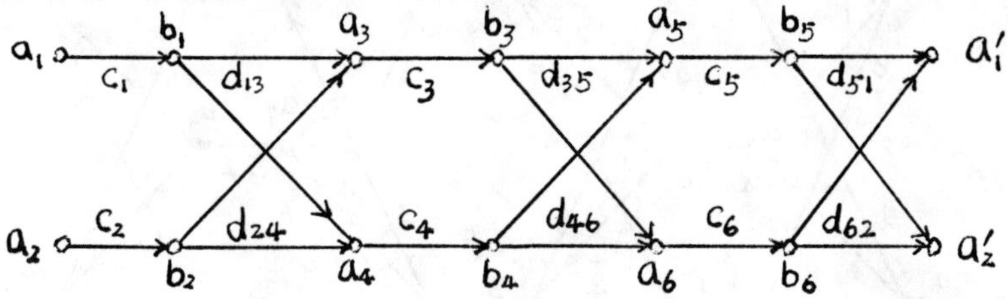

Fig. 3. An acyclic graph for the process plan selection

from a_i to a_i' be infinite, or there is no arc connecting a_i and a_i' in the new graph, and all the arcs which are incident from a_i in the original graph are incident from a_i and the arcs which are incident to a_i in the original graph are incident to a_i'. Doing so we transform the cyclic digraph G'(N,A) into an acyclic digraph G(N,A). Fig. 3 shows such an acyclic digraph. It is clear that the shortest path from a_i to a_i' in the acyclic graph G(N,A) is equivalent to the shortest cycle passing through a_i in the original graph G'(N,A), and the distance of the path is equal to the distance of the cycle. Without confusing, we use the same notation to denote them. In this way we can obtain the shortest cycle by finding the shortest path in an acyclic digraph.

Now we can find the solution as follws. For each vertex a_i of process plan $P_i \in F_k$ find the shortest path from a_i to a_i' in G=(N,A), a shortest path set $H=\{H_1,\ldots,H_n\}$ and a corresponding solution set of the problem $D=\{D_1,\ldots,D_n\}, D_i \in S$, are obtained, where $n=|N_k|$ is the number of the elements in N_k. Then let $LH = \min\{H_i\}$, and the corresponding solution LD of process plan selection problem is obtained.

Note that the main computation involved in the procedure is to find the shortest paths. Because of the acyclic property of graph G=(N,A) there are efficient and standard techniques to do that, for example, the methods given in [4]. Furthermore we have the following complexity theorem.

THEOREM The method given for solving the process plan selection problem modeled in this paper is $O(IM^2m)$ in computational complexity, where $n_k=|N_k|$ denotes the number of elements in set N_k, $I=\min\{n_1,\ldots,n_m\}$, $M=\max\{n_1,\ldots,n_m\}$.

The proof is omitted because of limited space.

AN ILLUSTRATIVE EXAMPLE

It comes from [1] and the data can be found in [1].

In this example I=2 and n =2, we select k=1 and obtain an acyclic digraph for the problem, it is shown in Fig. 4.

In Fig. 4 we find the shortest paths from a_1 to a_1' and from a_2 to a_2' respectively, and obtain $H_1=27.8$, $H_2=39.1$, $D_1=\{1,4,7,9\}$ and

$D_2=\{2,5,7,9\}$. Because $H_1<H_2$, $LH=H_1=27.8$ and the optimal process plan set is $LD=D_1=\{1,4,7,9\}$.

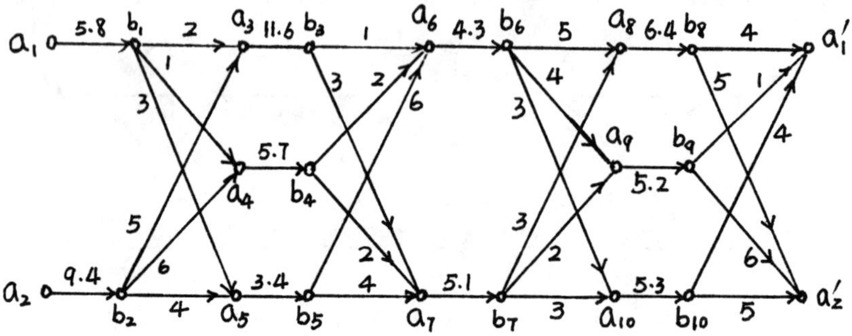

Fig. 4. The acyclic digragh of the example

From the result it is seen that for this example the solution obtained here is also optimal to the formulation given in [1], this justifies our modification made to some extent. By this example we also see that it is very simple to solve the problem modeled by our formulation, for this example it even can be solved manually without any difficulty.

CONCLUSIONS

In the practical manufacturing systems the process plan selection problem proposed in [1] is important and efficient method is needed. In view of the high computational complexity in finding the optimal solution of the problem modeled in [1] we made a resonable modification and a new formulation has been presented. For the new formulation a very efficient algorithm was developed.

It is necessary to point out that the solution of the process plan selection problem obtained from our formulation is dependent on the order of the number of the part, but considering the great reduction of the amount of computation and the applicablity in the practice we think it is worth doing so.

REFERENCES

[1] A. Kusiak and G. Finke, "Selection of process plans in automated manufacturing systems", IEEE J. of Robotics and Automation, Vol.4, 397-402 (1988).

[2] T. C. Chang and R. Wysk, An introduction to automated process planning systems, (Englewood Cliffs, NJ: Prentice-Hall, 1985).

[3] G. Halevi, The role of computers in manufacturing process, (New York, NY: Wiley, 1980).

[4] E. L. Lawler, Combinatoral optimization: Networks and matroids, (Holt, Rinehart and Winston, New York, 1976).

Hybrid Synthesis of Petri Nets for Manufacturing Systems: A Case Study

MengChu Zhou and Frank DiCesare
Department of Electrical, Computer, and Systems Engineering
Rensselaer Polytechnic Institute, Troy, NY 12180-3590, USA

ABSTRACT

This paper presents a hybrid methodology for systematic synthesis of bounded, live, and reversible Petri nets for automated manufacturing systems. Starting with a Petri net with the desirable system properties, we synthesize Petri nets by refining operation places in top-down modular ways, then adding non-shared resource places stepwise, and finally adding shared resource places in a bottom-up manner. A system consisting of three machines, a robot, and an AGV system is used to illustrate the method.

INTRODUCTION

Petri nets have been used to model discrete event dynamic systems such as automated manufacturing systems [2,6,9,13]. They can also lead to Petri net controllers for manufacturing systems [3,5,8]. In addition, there exist validation, simulation, and performance analysis tools for Petri nets.

Since in a moderately sized manufacturing system the complexity of design and analysis at the implementation level of detail may be unreasonable, the challenge is to develop methods to systematically construct a Petri net model such that it has the desirable qualitative properties for a manufacturing system.

Previous work on the synthesis of Petri net models can be summarized into two approaches: top-down, and bottom-up. Top-down synthesis is characterized by the stepwise refinement of an aggregate Petri net model. Each successive refinement includes more detail until the implementation level is reached. This method is very attractive because detail is introduced in an incremental manner, thereby reducing the complexity to be coped with in the process. Bottom-up design is characterized by the merging of Petri nets of the component processes into a complete net model for the entire process that captures the interactions between them. The merging of subnets is accomplished by sharing their common places [1,9], transitions, or paths [6]. The invariants of the resulting net can be determined from those of the subnets so that the analysis is conducted using these invariants.

We developed a top-down modular approach to systematic Petri net synthesis of manufacturing systems [14]. However, the method fails to generate the desirable properties when some shared resources are included in systems. It was recognized that the major difficulty of the problem lies in the method of dealing with the shared resources [15]. The goal of this research is to solve this problem. This paper presents a first step in this direction. Two objectives of this paper are: (1) To introduce a hybrid approach to design of Petri nets; and (2) To illustrate the approach by modeling a manufacturing system containing three machines, a shared robot, and a shared AGV system.

A HYBRID METHOD

A good tutorial and review on Petri nets and their applications was provided by Murata [7], Silva [10], Al-Jaar and Desrochers [2], and Jeng and DiCesare [4].

Definitions for Synthesis

In the manufacturing environment, the *specification* for a system can be considered as a list of resources, including machines, a list of operations, and their precedence relationships. Operations and resources are distinguished as two elementary components [6,9,13,14]. Each operation needs some resources, such as robots, machines, and/or parts. In the proposed method, operations and resources are modeled as places in Petri net models. Every transition represents start and/or termination of some operation. There are basically three kinds of places which depend on the initial marking of the Petri net and they are defined as follows.

Suppose that a marked Petri net Z=(P, T, I, O, m), where

1. P is a finite set of places;
2. T is a finite set of transitions with $P \cup T \neq \emptyset$, $P \cap T = \emptyset$.
3. I: $P \times T \rightarrow \{0, 1\}$ is an input function defining directed arcs from P to T;
4. O: $P \times T \rightarrow \{0, 1\}$ is an output function defining directed arcs from T to P;
5. m: $P \rightarrow \{0, 1, 2, ...\}$ is a marking whose i^{th} component represents the number of tokens in the ith place.

Also, suppose that m_0 is the initial marking in Z and p is a place in Z.

Definition 1. A place p is called an *A-place* iff $m_0(p)=0$, a *B-place* iff $m_0(p)$ is a constant, or a *C-place* iff $m_0(p)$ is a variable.

An A-place represents an operation place. A B-place models the availability of a fixed number of resources. A C-place represents the availability of raw materials, fixtures, pallets, etc. The number of the initial tokens in C-places should be determined so that the system is neither blocked nor starved.

In our method, instead of general Petri net design modules [11,12], *basic design modules* are introduced to aid Petri net synthesis for manufacturing systems [14]. Here we show only two of them.

Definition 2. *A sequence PN* module is defined as a Petri net (Fig. 2.1) which consists of n+1 places and n transitions (n>0) with an initial marking of zero. Its place PN block $S_P=(\{p_1, p_2, ..., p_{n+1}\}, \{t_1, ..., t_n\}, I, O, 0_{n+1})$ where 0_{n+1} is an n+1 dimension zero vector and with $p_{in} = p_1$, $p_{out} = p_{n+1}$,

$$I(p_i,t_j) = \begin{cases} 1 & \text{if } i=j \text{ and } 1 \le i \le n; \\ 0 & \text{otherwise.} \end{cases} \qquad O(p_i,t_j) = \begin{cases} 1 & \text{if } i=j+1 \text{ and } 1 \le j \le n; \\ 0 & \text{otherwise.} \end{cases}$$

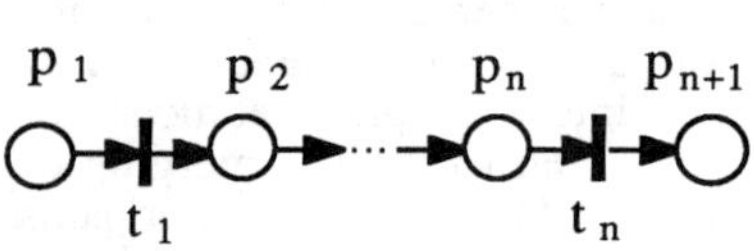

Fig. 2.1. A sequence PN (n>0)

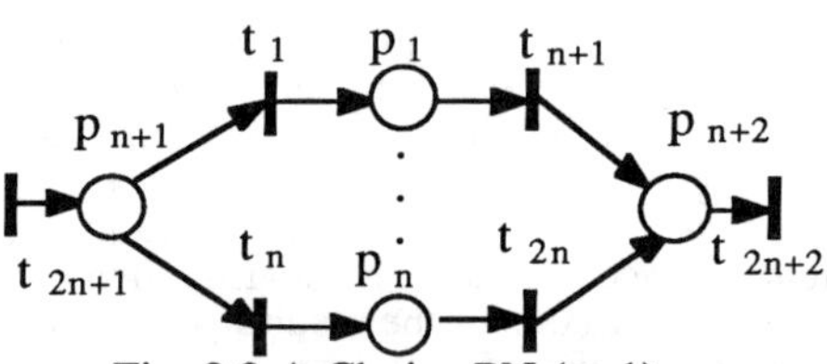

Fig. 2.2 A Choice PN (n>1)

Definition 3. *A choice PN* module is defined as a Petri net (Fig. 2.2) which consists of 2n+2 transitions and n+2 places with n parallel paths (n>1), and with an initial marking of zero. Its place PN block $S_P=(\{p_1, p_2, ..., p_{n+2}\}, \{t_1, ..., t_{2n}\}, I, O, 0_{n+2})$ and with $p_{in} = p_{n+1}$, $p_{out} = p_{n+2}$,

$$I(p_i,t_j) = \begin{cases} 1 & \text{if } i=n+1 \text{and } 1 \le j \le n \text{ or } 1 \le i \le n \text{ and } j=n+i; \\ 0 & \text{otherwise.} \end{cases}$$

$$O(p_i,t_j) = \begin{cases} 1 & \text{if } i=n+2 \text{ and } n+1 \le j \le 2n \text{ or } 1 \le i \le n \text{ and } j=i; \\ 0 & \text{otherwise.} \end{cases}$$

A sequence PN means a series of successive operations. A choice PN represents n choices for a successive operation.

In manufacturing, there may exist shared resources. For example, two or more machines share a robot for loading and unloading. Modeling the shared resources in the system is a major challenge for synthesis methods. The following gives the formal definitions and their explanations are referred to in [15].

Definition 4 Given a marked Petri net $Z=(P,T,I,O,m_0)$, $P=P_A \cup P_B \cup P_C$, a *k-generalized parallel mutual exclusion* is defined as k-GPME=(p_E, D) such that

1) D is a set of pairs of the transition sets or $D=\{(T_{a1}, T_{b1}), (T_{a2}, T_{b2}), ..., (T_{ak}, T_{bk})\}$, and $k \geq 1$ with $T_{ai}, T_{bi} \subset T$, $T_{ai} \cap T_{bj} = \emptyset$, $T_{ai} \cap T_{aj} = \emptyset$, and $T_{bi} \cap T_{bj} = \emptyset$ when $i \neq j$, $\forall$ i, j $\in N_k$, and there is an A-path between t and t', $\forall$ $t \in T_{aj}$, $t' \in T_{bj}$. $p_E \in P_B$ with $m_0(p_E)=1$, $I(p_E, t)=O(p_E, t')=1$, $I(p_E, t')=O(p_E, t)=0$, $\forall$ $t \in T_{aj}$, $t' \in T_{bj}$, $j \in N_k$, and $I(p_E, t_u)=O(p_E, t_u)=0$ when $t_u \notin T_a \cup T_b$ where $T_a=T_{a1} \cup T_{a2} \cup ... \cup T_{ak}$ and $T_b=T_{b1} \cup T_{b2} \cup ... \cup T_{bk}$; and
2) P_B and P_C satisfy that if $m_0(p)=0$, the output transitions are never enabled for any p in $P_B \cup P_C$ and there is no A-path between t and t' $\forall$ $t \in T_b$, $t' \in T_a$.
3) $\exists m_0$, and a sequence of transitions g that contains none in T_a, $\ni$ $g(m_0)$ enables t, $\forall$ $t \in T_a$. If $t \in T_{aj}$ fires, then $\exists$ h, $t' \in T_{bj}$, $\ni$ ht(m) enables t', and p_E is marked after one transition in T_{bj} fires, $\forall$ $j \in N_k$. If $m(p_E)=1$ and t is enabled, then $t \notin T_b$.

Note that an A-path refers to a simple elementary path containing only A-places and transitions.

Definition 5 Given a marked Petri net $Z=(P,T,I,O,m_0)$, $P=P_A \cup P_B \cup P_C$, a *generalized sequential mutual exclusion* (GSME) exists iff

1) $\exists$ (p_E, D) such that $D=D^1 \cup D^2 \cup ... \cup D^L$ where $L \geq 2$, $D^i=\{(T^i_{a1}\ T^i_{b1}), (T^i_{a2}, T^i_{b2}), ..., (T^i_{ak_i}, T^i_{bk_i})\}$, and (p_E, D^i) is a k_i-GPME in $Z_i=(P,T,I,O,m_0)$ that results from Z with $I(p_E,t)=O(p_E,t)=0$ when $t \in D-D^i$, $\forall$ $i \in N_L$;
2) $\forall$ m_0, a sequence of transitions g_i, and $i \in N_{L-1}$, if $g_i(m_0)$ enables t, $\forall$ $t \in T^{i+1}_a$, then $\exists$ $t' \in T^i_a$, $\ni$ g_i contains t' where $T^i_a=T^i_{a1} \cup T^i_{a2} \cup ... \cup T^i_{ak_i}$
3) $\exists$ m_0 and a sequence of transitions f_i, $\ni$ $f_i(m_0)$ enables $t \in T^i_a$ and $t' \in T^{i+1}_a$, $\forall$ $i \in N_{L-1}$;

Summary of Theoretical Results

Some previous theoretical results are summarized. For the complete proofs, a reader is referred to [11,14,15].

Theorem 1 (Refinement) If a place or transition in a bounded (safe), live and reversible Petri net Z is replaced by a basic design module, the refined Petri net is also bounded (safe), live and reversible.

Theorem 2 (Generalized Parallel Mutual Exclusion) $Z=(P_A \cup P_B \cup P_C, T,I,O,m_0)$ is a subnet of $Z'=(P',T',I',O',m_0')$, $D=\{(T_{ai},T_{bi}), i \in N_k\}$, $P'=P_A \cup (P_B \cup \{p_E\}) \cup P_C$, and T'=T. If (p_E, D) is a k-GPME, then (1) Z' is bounded (safe) if Z is so; (2) Z' is live iff Z is so; (3) Z' is reversible iff Z is so.

Theorem 3 (Generalized sequential mutual exclusion) $Z=(P_A \cup P_B \cup P_C, T,I,O,m_0)$ is a subnet of $Z'=(P',T',I',O',m_0')$, $D=D^1 \cup D^2 \cup ... \cup D^L$, $k_i=|D^i| \geq 1$, $L \geq 2$, $D^i=\{(T^i_{a1},T^i_{b1}), (T^i_{a2},T^i_{b2}), ..., (T^i_{ak_i},T^i_{bk_i})\}$, $i \in N_L$, $P'=P_A \cup (P_B \cup \{p_E\}) \cup P_C$, and T'=T. If (a) (p_E, D) is a GSME, and (b) $m_0(p) \geq \beta(p)$, $\alpha(p)m_0(p) \leq C$, $\forall$ $p \in P_C$, then (1) if Z is bounded (safe) and live, so is Z'; and (2) if Z is reversible, so is Z'.

Note that $C=\underset{1 \le i < L}{\mathrm{Min}}$ {achievable $C^{i\ i+1}$} where $C^{i\ i+1}$ is the token capacity between D^i and D^{i+1}. The definitions of $C^{i\ i+1}$, α, and β are given in [15]. Also, Theorem 2 is applicable for multiple GPME coping with shared resources, while Theorem 3 needs some minor modification for multiple GSME cases.

Procedure for Petri Net Synthesis of Manufacturing Systems

Step 1: Choose a bounded, live, and reversible Petri net as an aggregate model of a system, and determine the C-place set, P_C. **Step 2**: Decompose this system into several subsystems expressed as operation places using the defined basic design modules. Replace these operation places by more detailed basic design models until one reaches a point where the further division is unnecessary. **Step 3**: Appropriately add the non-shared resource places at each stage when an operation requires the resources. **Step 4**: Add the shared resource places which form GPMEs. **Step 5**: Add the shared resource places which form GSMEs and calculate the token capacities between its GPMEs; determine $m_0(p)$, $\forall p \in P_C$ by using Theorem 3.

PETRI NET MODELING OF A MANUFACTURING SYSTEM

System Description

This system comprises three machines M_1, M_2, and M_3; a robot for unloading among them; and an AGV system which delivers the final parts to assembly (see Fig. 3.1). There are several pallets in the system. The raw materials fixed with pallets enter the system at M_1. After they are fed to and processed by M_1, they may go to M_2 or M_3 for further processing. The system produces two types (G and F) of final parts which are sent to assembly and the pallets are released by the AGV system. An unlimited source of raw materials is assumed. The system is required to be neither blocked nor starved.

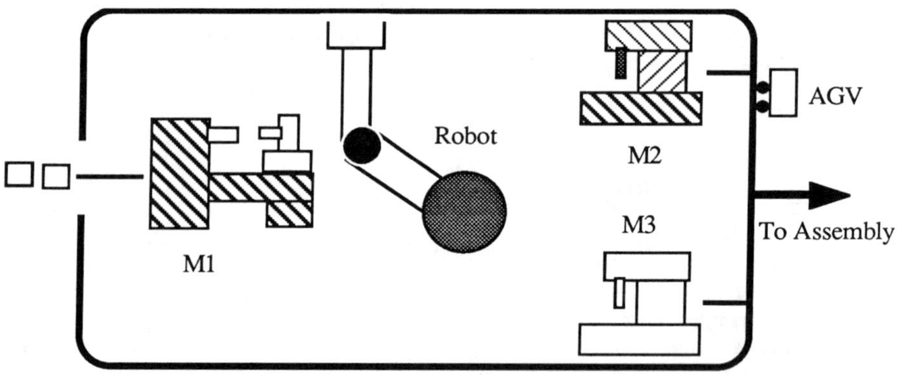

Fig. 3.1 A Manufacturing System

Modeling Process

Step 1: Choose a Petri net with two places and two transitions as shown in Fig. 3.1. $P_C=\{p_1\}$. The net is bounded, live, and reversible whenever $m_0(p_1)>0$.

Step 2 (Refinement): As shown in Fig. 3.2-3.3, we obtain a series of bounded, live, and reversible Petri nets using refinement of places. None of the replacements changes the desirable properties of the Petri nets by Theorem 1.

Step 3 (Addition of non-shared resources): Adding all machine resource places, we get the net in Fig. 3.3.

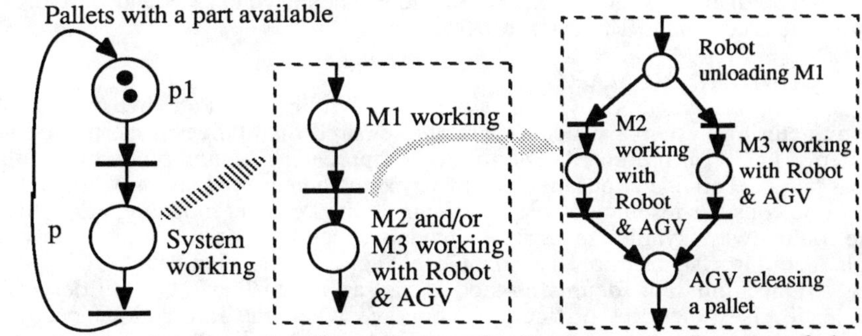

Fig. 3.2 First-level bounded, live, and reversible Petri net and refinement of places

Step 4 (Addition of GPMEs): Considering the AGV in the system, 1-GPME, i.e., (p_{14}, {({t_7, t_8}, t_{11})}) is formed in Fig. 3.4.

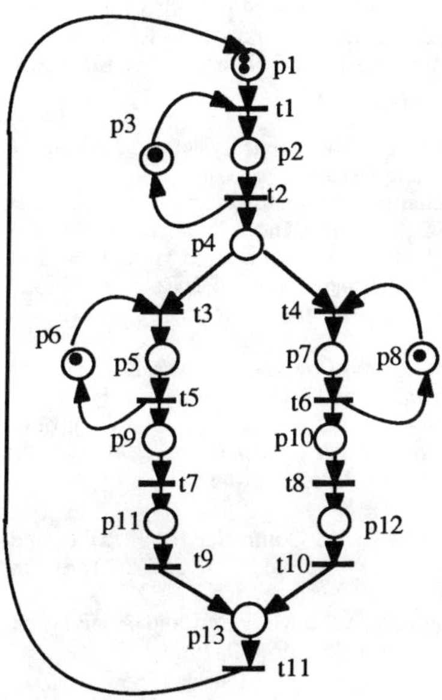

Fig. 3.3 After Steps 3 & 4.

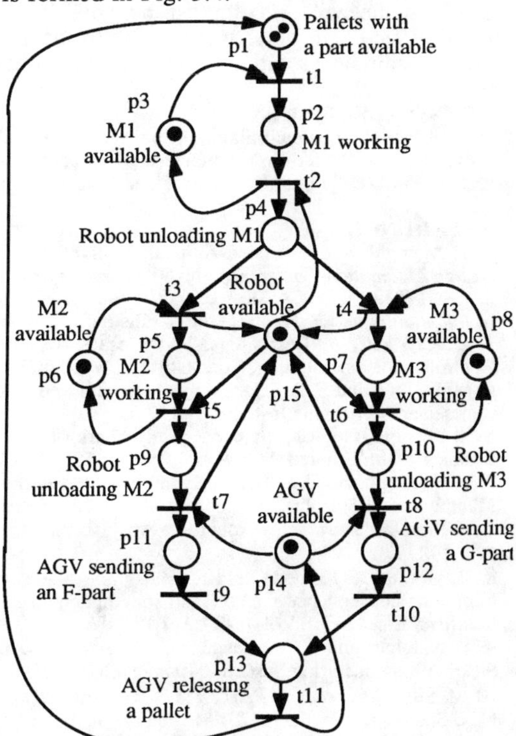

Fig. 3.4 A final Petri net.

Step 5 (Addition of GSMEs): First, consider part of the robot's task, i.e., when it unloads between M_1, and M_2 or M_3. (p_{15},{(t_2, {t_3, t_4})}) denoted by $GPME_1$ is an 1-GPME. Then, add arcs (p_{15}, t_5), (t_7, p_{15}), (p_{15}, t_6), and (t_8, p_{15}) to get the net in Fig. 3.4. An GSME (p_{15}, {(t_2, {t_3, t_4})}∪{(t_5, t_7), (t_6, t_8)}) is formed and can be easily verified. The token capacity can be calculated between $GPME_1$ and $GPME_2$ or (p_{15},{(t_5, t_7), (t_6, t_8)}). $C = C(t_3,t_5) + C(t_4, t_6) = 2$ and is achievable, and $\alpha(p_1)=1$, $\beta(p_1)=1$. Using Theorem 3, we have: $1 \le m_0(p_1) \le 2$.

The final Petri net is synthesized and is guaranteed to be bounded, live, and reversible according to Theorems 1 to 3.

CONCLUDING REMARKS

A hybrid methodology for systematic synthesis of Petri net models for manufacturing systems, which ensures the desired qualitative properties of the systems, has been presented. The proposed approach adopts top-down modular ideas to model all operations involved in a system, then uses a bottom-up method to stepwise consider resource places. These make the Petri net modeling easy. Also, one could always synthesize Petri nets models which have desirable properties for manufacturing systems from an abstract level to a desired level of detail.

One limitation for this method is that an initial Petri net with desirable properties is sometimes difficult to conceive when the interactions between subsystems become very complex for a practical manufacturing system.

Also, when we have several types of jobs which need processing by different machines in an FMS, all machines will become shared resources plus some shared robots and material movement systems among different machines. For such a system, the synthesis of a Petri net with desirable properties will become more and more difficult. In other words, how to incorporate planning or scheduling to the current approaches will become a new thrust in the synthesis of Petri nets.

ACKNOWLEDGEMENTS

The authors would like to acknowledge the facilities and assistance provided by the CIM Program in the Center for Manufacturing Productivity and Technology Transfer at RPI. They are grateful to M.-D. Jeng, I. Koh, J. Joshi, and Dr. A. A. Desrochers for their beneficial comments.

REFERENCES

1. T. Agerwala and Y. Choed-Amphai, "A Synthesis Rule for Concurrent Systems," *Proc. 15th Design Automation Conference,* pp.305-311, Las Vegas, June 1978.
2. R. Y. Al-Jaar and A.A. Desrochers, "Petri Nets in Automation and Manufacturing," In *Advances in Automation and Robotics*, G. N. Saridis (ed.), Volume 2., JAI Press Inc., Connecticut, 1990.
3. D. Crockett, A. Desrochers, F. DiCesare, and T. Ward, "Implementation of a Petri Net Controller for a Machining Workstation," *Proc. of IEEE R&A Conf.*, 1861-1867, Raleigh, 1987.
4. M. D. Jeng and F. DiCesare, "Synthesis and Reduction of Petri Nets". Working Paper, ECSE, Rensselaer Polytechnic Institute, 1989.
5. E. Kasturia, F. DiCesare, and A. Desrochers, "Real-time Control of Multilevel Manufacturing Systems Using Colored Petri Nets," *Proc. of IEEE R&A Conf.*, 1114-1119, Phildephia, 1988.
6. B. H. Krogh and C. L. Beck, "Synthesis of Place/Transition Nets for Simulation and Control of Manufacturing Systems," *Proc. 4th IFAC/IFORS Symp. Large Scale Systems*, Aug. 1986.
7. T. Murata, "Petri Nets: Properties, Analysis and Application," Proc. of the IEEE, 77, 4, 541-579, 1989.
8. T. Murata, N. Komoda, and K. Matsumoto, "A Petri-Net Based Controller for Flexible and Maintainable Sequence Control and its Applications in Factory Automation," *IEEE Trans. on Industrial Electronics,* Vol. IE-33, 1, 1-8, 1986.
9. Y. Narahari and N. Viswanadham, "A Petri Net Approach to the Modeling and Analysis of Flexible Manufacturing Systems," *Annals of Operations Research*, 3, 449-472, 1985.
10. M. Silva, *Las redes de Petri en la automatica y la informatica*, Ed. AC, Madrid, 1985.
11. I. Suzuki and T. Murata, "A Method for Stepwise Refinements and Abstractions of Petri nets," *J. of Com. and Sys. Sci.*, 27, 1, 51-76, 1983.
12. R. Valette, "Analysis of Petri Nets by Stepwise Refinements," *J. of Com. & Sys. Sci.*, 18, 35-46, 1979.
13. M. C. Zhou and F. DiCesare, "Adaptive Design of Petri Net Controllers for Error Recovery in Automated Manufacturing Systems," *IEEE Trans. on Sys, Man, and Cyb*, 19, 5, 963-973, 1989.
14. M. C. Zhou, F. DiCesare, and A. A. Desrochers, "A Top-down Modular Approach to Synthesis of Petri Net Models for Manufacturing Systems," *Proc. of IEEE Robotics and Automation Conference*, pp. 534-539, Scottsdale, AZ, May 1989.
15. M. C. Zhou and F. DiCesare, "Parallel and sequential mutual exclusions for Petri net modeling of manufacturing systems with shared resources," Working Paper, ECSE, Rensselaer Polytechnic Institute, Troy, NY, January 1990.

AN ADVANCED DEBURRING AND CHAMFERING SYSTEM

KARL N. MURPHY and FREDERICK M. PROCTOR
Robot Systems Division
National Institute of Standards and Technology

ABSTRACT

A second-generation automated deburring cell is currently under development at the National Institute of Standards and Technology. This cell will be capable of automatically generating, from part geometry and machining information, a series of robot paths for deburring aerospace parts made from superalloys. In order to compensate for the inaccuracy of robots when following computed coordinates, an actively-controlled end effector will be used to precisely position a rotary file in response to forces sensed at the part edge.

INTRODUCTION

Research has been underway at the National Institute of Standards and Technology since 1983 on methods to automate the deburring of machined metal parts. The goal of this research is a fully automated deburring cell capable of processing parts based only on their geometry and machining history, with no human intervention.

Such a deburring cell must contain several key elements: a process planner, which determines the sequence of events to take place; a trajectory planner, which computes actual deburring trajectories; a workcell controller, which schedules activities in the cell; and one or more robots to position the parts and perform the deburring [1]. Comprehensive computer-aided design (CAD) models are assumed to exist which define the part geometry and the machining operations completed prior to deburring.

The process planner can be effectively implemented as an expert system which encapsulates the knowledge of a human. The role of the process planner is twofold. First, it must determine the likely nature and location of burrs based on the part geometry and machining history. Second, it must consult models of burr removal to determine which deburring processes will effectively remove the burrs.

The trajectory planner uses the sequence of part features and the processes to be applied to these features to develop a set of robot trajectories that will accomplish the deburring. These trajectories consist of coordinates corresponding to the part features to be deburred, augmented with housekeeping data such as tool commands, approach points, and depart points.

This work is partially funded by the U. S. Navy Manufacturing Technology Program. Equipment listings are given for clarity and do not imply a recommendation by NIST.

Unfortunately, the use of computed coordinates places stringent requirements on the robot's accuracy (the ability of the robot to move to a specified point in space) rather than its repeatability (the ability of the robot to return to a previously recorded position). This is a major problem, since the accuracy of commercial robots is often several orders of magnitude worse than the repeatability. Many programmers do not know the poor accuracy of their robots, mainly because most robots are taught by showing, a method that does not require high robot accuracy.

Several methods can be used to accommodate for inaccuracy: calibration, sensory feedback, and the selection of a forgiving process. Calibration methods include kinematic model calibration where parameters that define the link lengths and joint offsets are precisely measured for a particular robot. Another calibration methods is error mapping where the robot's errors are tabulated in various regions of interest. Unfortunately, calibration does not account for dynamic loading, deformation, or external uncertainties such as part misplacement, errors in machining, or part tolerances. The use of sensory feedback can reduce these external uncertainties, but often the robot controller's large time delays and joint friction render closed-loop control of the robot end-effector unsatisfactory. Therefore, it is important to select a deburring process which is sufficiently forgiving to accommodate the remaining inaccuracies. This can be accomplished with passive compliance, or an actively-controlled end effector.

A typical scenario of a fully-automated deburring cell would be as follows: a CAD description of a part is developed, which at deburr time contains a complete machining history as well as any specific deburring requirements. Next, the expert system process planner consults models of burr formation and the CAD model to determine probable burr types and locations. The process planner then associates tools and tool parameters with these burrs by consulting models of burr removal. The trajectory planner computes the sequence of robot paths required to acquire, fixture, and deburr the part based on the edges, tools, and parameters selected. To ensure satisfactory deburring, real-time sensory feedback and end effector compliance account for robot inaccuracy. The workcell controller supervises the overall process, scheduling cell tasks and communicating with the higher levels of the factory.

Researchers at NIST have been developing workcells which approach the ideal of full automation. The Cleaning and Deburring Workstation (CDWS), begun in 1985, demonstrates the feasibility of deburring soft metals based on a graphically developed process plan and robot trajectories computed automatically from CAD data. A next-generation workcell capable of processing high-strength aerospace superalloys using an actively-compliant end effector is currently under development. This workcell, the Advanced Deburring and Chamfering System, will demonstrate the same off-line programming benefits shown by the CDWS, while furthering its capabilities.

THE CLEANING AND DEBURRING WORKSTATION

A first-generation automated deburring cell, the Cleaning and Deburring Workstation, began as part of the Automated Manufacturing Research Facility. Research in the CDWS encompasses hierarchical real-time robot control, task scheduling, and off-line programming. Information on various aspects of the CDWS can be found in References [2, 3, 4].

The CDWS consists of a workstation controller, two robots, various quick-change deburring tools, a rotary vise for part fixturing, a buffing wheel system, and a part washer and dryer. The workstation controller is a set of software modules which perform process planning and task scheduling. A Unimate 2000 six-axis hydraulic robot performs part handling and buffing. A PUMA 760 six-axis electric robot, fitted with a quick-change wrist, selects end-effectors from a tool rack containing a rotary end brush, a countersink tool, a hole brush, and a gripper. The vise can be rotated, reducing robot motion required for part refixturing and deburring.

A human expert develops the process plan at a graphics terminal displaying CAD images of a part. Using a mouse pointing device, the user selects features to be deburred, and associates tools and tool parameters to each feature. Part handling information is similarly generated. Once this process plan is developed, a trajectory planner computes robot coordinates necessary to handle the part and to deburr the specified features.

Various methods are used to accommodate for robot inaccuracy. Error mapping reduces part handling error to an extent that can be tolerated by the fixturing devices. For deburring, a two-pass technique known as self-teaching is used. On the first pass, the robot approaches the computed points, moves toward the part edge until a desired force is met, and updates each computed point with the current position. On the second pass, the robot deburrs the part, using this new set of points.

While this system demonstrates a substantial savings in time when compared with teach programming, it is not fully automated. A human expert is still required to select the edges, tools, tool parameters, and part handling data. Furthermore, only soft metals such as aluminum and brass can be deburred since deburring harder metals require non-compliant cutters. When used with the CDWS robot, hard cutters generate chatter, a scalloped edge. The use of hard cutters in an improved system is being jointly developed by NIST and industry under the sponsorship of the U.S. Navy's MANTECH program, as part of a solution for the automated deburring of aerospace parts.

THE ADVANCED DEBURRING AND CHAMFERING SYSTEM

The Advanced Deburring and Chamfering System (ADACS) incorporates CAD-based programming automation, with the addition of active hard-tooling to compensate for robot errors. Robot programs are developed automatically from CAD part data, and downloaded to a Cincinnati Milacron T3 robot at run time. The end effector, a TriKinetics Adaptive Deburring Tool (ADT), consists of a high-speed spindle and rotary file mounted in an active housing, which is gimbaled to provide two degrees of freedom. Sensors monitor the deburring forces and the position of the tool tip relative to its center location. Two independent force control laws allow for compliance to be varied in the normal and tangential directions. In addition, the orientation of the tool's normal vector can be specified continuously, allowing it to be changed quickly while following complicated edges and sharp corners. The addition of this active end effector results in two distinct yet coupled control problems: a macromanipulator which executes gross motion at relatively low frequency, and a micromanipulator which executes fine motion at a much higher frequency. A block diagram of the ADACS is shown in Figure 1.

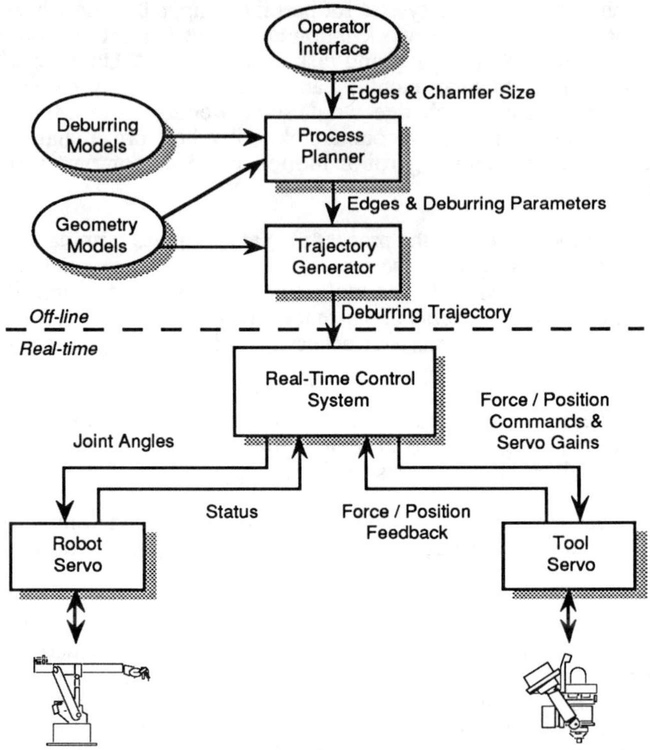

Figure 1. The Advanced Deburring and Chamfering System. A user selects edges and chamfer sizes. A process planner consults deburring models to determine how the part should be deburred. A trajectory generator computes deburring paths, which are sent to the real-time control system for division into robot and tool tasks.

Tooling

Hard cutters require compliant tool holders, either passive or active, which perform two main functions: reducing chatter and keeping the tool on the part despite robot and system inaccuracies. In contrast to machine tools, whose structural stiffness is high in all directions, the relatively low stiffness of robot arms allow large-amplitude resonances which cause chatter. Asada and Goldfine [5] have analyzed the mechanics of robotic grinding, and shown that chatter is reduced when the normal and tangential stiffnesses differ by a factor of ten, as shown in Figure 2. This difference in relative magnitudes and direction of stiffness reduces the otherwise strong coupling between the normal and tangential directions. Since the overall stiffness depends on both the tool and the robot, a tool must have an axis with stiffness less than a tenth of the stiffness of the robot.

Robot accuracy is not sufficient for edge following when deburring machined edges with hard cutters. In this case, the normal direction must be made compliant, so that the cutter will remain in contact with the edge and sustain the small normal forces needed for edge breaking. Often, the burr variation is small enough so that the depth of cut is unaffected.

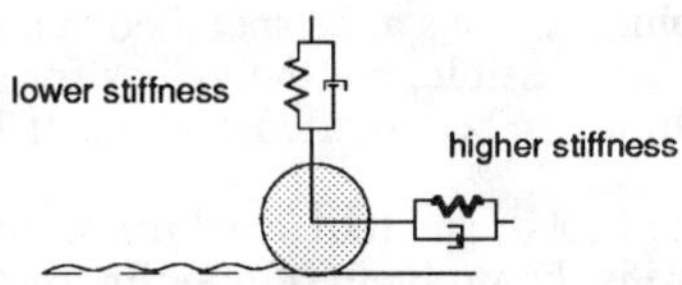

Figure 2. The 10:1 Rule. To reduce chatter, normal and tangential stiffness of the tool holder should differ by an order of magnitude or more. When deburring machined edges, the tangential stiffness is often chosen higher, as shown.

Compliance can be implemented either passively or actively. The ADT, shown in Figure 3, is an actively compliant tool which has several advantages over a passive tool holder. The compliance and desired force in both the normal and tangential directions are inputs to the tool controller and, therefore, can be easily changed for different materials, burr conditions, or desired chamfer size. Furthermore, a robot would have to physically rotate a passive tool holder to change the normal axis orientation. In contrast, the axis orientation of the ADT is determined electronically and can be rotated at high speeds. This allows for following complex edges without slowing the feed rate during tool rotation. The control of the ADT is described by Hollowell [6].

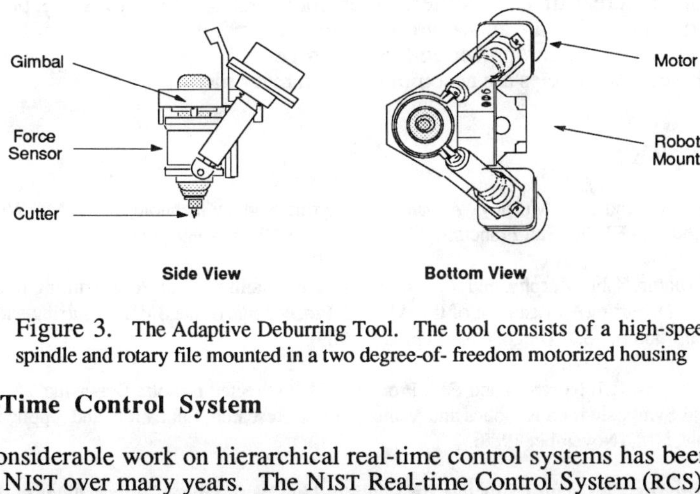

Figure 3. The Adaptive Deburring Tool. The tool consists of a high-speed spindle and rotary file mounted in a two degree-of- freedom motorized housing

Real-Time Control System

Considerable work on hierarchical real-time control systems has been carried out at NIST over many years. The NIST Real-time Control System (RCS) is being demonstrated on many platforms including the CDWS robot controller, the robot controller in an automated horizontal machining workstation, the mobility controller for teleoperated military vehicles, and the control system for the NASA Flight Telerobotic Servicer [3, 7, 8].

RCS is a hierarchy of three sets of computing modules: task decomposition, world modeling, and sensor processing. The task decomposition modules perform real-time planning, control, and monitoring functions. They decompose task goals both spatially and temporally. The sensory processing modules detect, filter, and correlate sensory information. The world modeling modules estimate the state of the external world and makes predictions and evaluations based on these estimates.

The ADACS RCS (See Figure 1.) decomposes deburring trajectories into commands for the robot and tool controllers. Although the robot servo control law is supplied by the vendor and is inaccessible at the lowest levels, provision has been made to allow for joint positions to be specified at a rate of 65 Hz. Similarly, the tool control laws are inaccessible, but the values for stiffness, damping, force setpoints, and axis orientation can be modified at a rate of 120 Hz.

There are two classes of robot and tool synchronization problems encountered in the ADACS. First, there is the problem of coordinating commands to the robot and tool controllers during the traversing of complicated geometries. RCS must rotate the tool normal vector as the tool traverses corners. Second, there is the problem of accommodating tool-tip straying. Since the tool controller maintains cutter contact with the edge, the tool-tip will stray from the center of its actuation range when the error in the robot's trajectory increases. RCS must correct the computed trajectory in the robot controller when the tool strays too far.

SUMMARY

A second-generation automated deburring system capable of processing aerospace materials is currently being developed at the National Institute of Standards and Technology. This system consists of an off-line programming interface which allows complex deburring trajectories to be developed from CAD data without the need for robot teach programming. These trajectories are downloaded to a real-time controller, which coordinates the motions of a position-controlled robot with an actively-compliant deburring tool. The tool can be commanded to maintain contact with a part edge with a prescribed force, so that robot and part inaccuracies are accommodated in real time.

REFERENCES

1. F.M. Proctor and K.N. Murphy, "Advanced Deburring System Technology," ASME Winter Annual Meeting, PED 38, San Francisco, CA, (December 10-15, 1989).

2. F.M. Proctor, K.N. Murphy, and R.J. Norcross, "Automating Robot Programming in the Cleaning and Deburring Workstation of the AMRF," Proceedings of the SME Deburring and Surface Conditioning, San Diego, CA, (February 1989).

3. K.N. Murphy, R.J. Norcross and F.M. Proctor, "CAD Directed Robotic Deburring," Second International Symposium on Robotics and Manufacturing Research, Education, and Applications, Albuquerque, NM, (November 1988).

4 R.J. Norcross, "A Control Structure for Multi-Tasking Workstations," Proceedings of the IEEE Conference on Robotics and Automation, Philadelphia, PA, (April 1988).

5. H. Asada and N. Goldfine, "Optimal Compliance Design for Grinding Robot Tool Holders," IEEE Conference on Robotics and Automation, St. Louis, MO, (March 1985).

6. R. Hollowell, "Hybrid Force/Position Control for Robotic Light Machining," Robotics and Remote Systems Conference, Charleston, SC, (March 1989).

7. S. Szabo, H.A. Scott, and R.D. Kilmer, "Control System Architecture for the TEAM Program," Second International Symposium on Robotics and Manufacturing Research, Education, and Applications, Albuquerque, NM, (November 1988).

8. R. Lumia and J. Albus, "Teleoperation and Autonomy for Space Robotics," Robotics 4(1), 27-33 (March 1988).

A MODELING AND OPTIMIZATION ENVIRONMENT FOR A COMPONENT PLACEMENT MACHINE

Javad AHMADI[1]***, Anna SCIOMACHEN***[2]

1. Karl Hermann Manufacturing Fellowship Project, IBM, Austin, TX

2. Dept. Information Sciences, University of Milan, Italy

ABSTRACT

Accurate performance prediction and optimization of robotic assembly tools are essential for planning and execution of production in high volume environments. Here we present a Petri Net model of a Surface Mount Technology placement machine for performance evaluation, and utilize this characterization for development of an optimization heuristic that is based on modification of the Traveling Salesman Problem.

1. INTRODUCTION

The Panasert MK-1 [3] is a high speed pick-and-place machine for surface mount technology (SMT) assembly of electronic boards. The burst rate performance of the machine is nearly at 12,000 placements per hour. Making it an attractive tool for high volume production of electronic boards which employ the higher density packaging feature of SMT devices. However, this burst rate performance represents a highly idealized setting in which the many concurrent functional operations of the machine are nicely synchronized. Two user controllable elements, setup of the components feeder carrier and the sequence of the pick-and-place operations (tool staging and process synchronization), significantly affect the machine cycle time in completion of a given assembly. These are complex decisions which unaided require extensive engineering overhead. The burst rate estimates of the cycle time or other average or nominal estimates are inadequate for assessment of the machine performance in the production environment for purposes of line balancing, production planning or other capacity sizing evaluations. As such, for the high volume production environment both optimization of the machine thruput and correct estimation of the machine cycle time are essential [1,2].

The effective specification of the machine staging and synchronization requires a through understanding of the machine's functional operations and its control logic. The MK-1 machine relies heavily on cams to synchronize and partially overlap its many-fold operations. We select and assess a subset of the operations which affect the thruput rate of the machine. We use the formalism of Petri Nets (PNs) to characterize the concurrency, synchronization, and control aspects of this set of operations, and provide a tool emulation system for accurate thruput estimation of the machine for a given setup and sequence of operation. For determination of optimized machine staging and pick-and-place sequencing we present a heuristic procedure that is based on modeling of this problem as a variant of the well known Traveling Salesman sequencing Problem (TSP) [5].

We envision that in the future, such assembly tools will feature the cycle time estimation, setup, and operation optimization capability automatically, in addition to standardized procedures for interface to product design, and bill of materials data bases. This capability will eliminate the need for such off-line data acquisition, optimization and formatting.

2. MACHINE DESCRIPTION

Figure 1 shows schematically the MK-1 machine featuring ten pick-and-place heads mounted on the circumference of a revolvable turret. The turret pivots or indexes forward in the clockwise direction through ten fixed stations [3]. Station 1 is the pickup position. Stations 2-5 are for components' centering, orientation, and defect detection. Station 6 is the placement position. Station 7 returns the head to its original orientation. Station 8 is where improper components are ejected. Station 9 affects the large/small nozzle selection. Station 10 is not currently used.

Tape reels containing the components required to populate the printed circuit board are mounted on the component delivery carrier. Each tape reel contains only components of a single type. There are 100 feeder stations on the component delivery carrier to accommodate these tape reels. The carrier moves so as to align the desired feeder station under the nozzle head at Station 1 for the pickup operation (Z-axis motion) [3]. The workboard also moves so as to align the desired placement location on the printed circuit board under the nozzle head at Station 6. This movement is affected by an X-Y table.

A myriad of operations are activated/deactivated and/or controlled by a single, variable speed, drive shaft via thirteen cams (Θ-axis). The operations governed by the cams are: pivot the turret forward one station, open/close the shutter plate, index the tape reel forward for the next component, affect the downward/upward pickup stroke of the nozzle head, activate/deactivate the pneumatic valve for pickup and retention of the component, make the selection of the large/small nozzle, activate/deactivate the pneumatic valve for the placement of the component, affect the downward/upward stroke of the nozzle head to the large centering plate, open/close tweezing action of the large centering plate, rotate the nozzle head to achieve the placement orientation and the corresponding reverse rotation, affect the downward/upward stroke of the nozzle head to the small centering plate, open/close tweezing action of the small centering plate. Figure 2 summarizes the cam profiles and timing charts for what we believe to be the most pertinent operations for accurately assessing the thruput rate.

The top four profiles describe the angles through which these operations are activated and controlled by the corresponding cams for: indexing the turret, the shutter opening, the pickup stroke, and the placement stroke. The bottom two profiles describe timing signals, which are transmitted to the

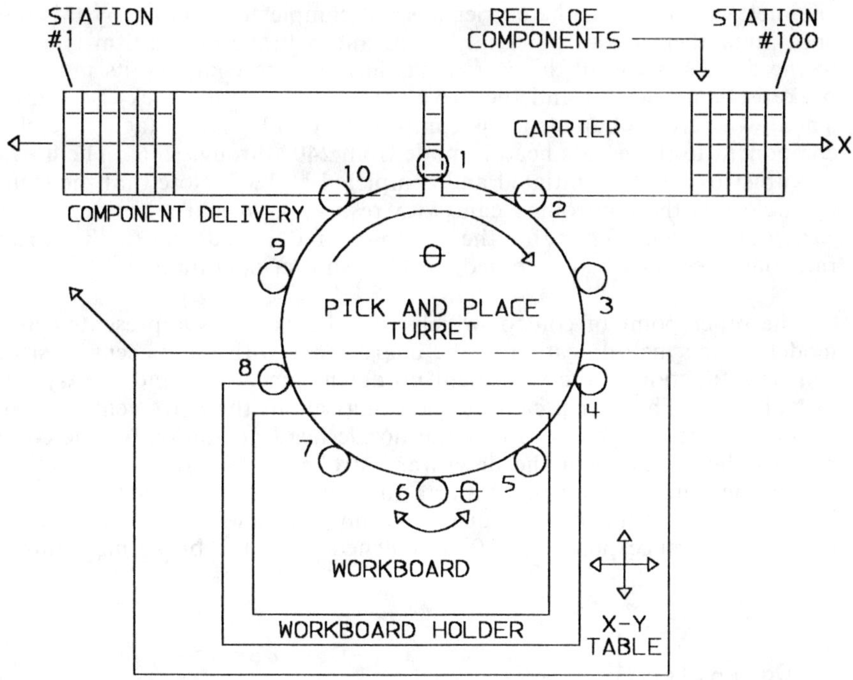

Figure 1. **Schematic of high speed placement machine**

controller. They operate on negative logic. That is, the corresponding axes are moveable during the on period.

The turret indexing advances the turret one station and is brought about while the drive shaft rotates from 10°through 130°, and all other operations are subsequent to this pivoting. The components on the tape reels are seated in recessed cups; a shutter mechanism closes over the cup to stabilize the component within it. The second cam profile opens this shutter mechanism to expose the component for the pickup operation. This profile is synchronized with the z interlock signal. That is, movement of the component delivery carrier is prohibited while this shutter mechanism is even partially open. The shutter begins to open at 170°. It is completely open at 232° and remains open through 300° and then begins to close. Closure is brought about by 346°. At this reference point of 346°, the control signal is sent releasing the carrier to align the next feeder location under Station 1. That is, the Z-axis start transmission is sent to the controller when this cycle timer is turned on. Closure of the feed shutter is maintained through the rotation back to 170°. The third cam profile affects the pickup of a component by the nozzle head at Station 1. The downward stroke of this nozzle head is initiated at 140° and continues through 170°. From the z interlock channel, we see that if the carrier has not completed its move by this reference point then a control signal is sent and a brake mechanism is applied to the drive shaft until the carrier has had time to complete its move. The brake is then released and the drive shaft continues its clockwise rotation. The downward pickup stroke is completed at 240°. The attachment of the component to the nozzle head is made from 240° through 266°. The upward pickup stroke is then initiated and completed at 320°. Note that the shutter open/close is the controlling cam with respect to synchronization with the carrier movement. Thus, for the purposes of PN modeling of the thruput rate, one need only be concerned with the shutter operation.

The other point of control which must be explicitly represented in the model is the synchronization of the board move with the placement stroke. This is a function of the placement up/down cam profile and the x-y interlock channel. The fourth cam profile brings about the placement of a component on the workboard under the nozzle head at Station 6. The control permits the initiation of the downward stroke of this nozzle head at 130° and it continues through 190°. From the x-y interlock channel, we see that if the workboard has not completed its move to new place location by the time the reference point of 190° is reached, then the brake mechanism is

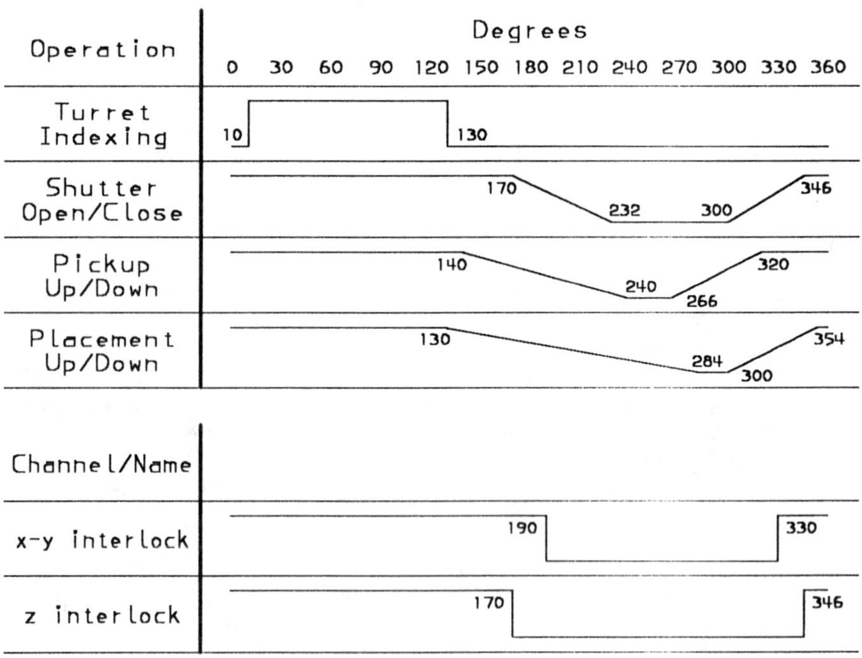

Figure 2. **Cam profiles and timing channels**

again applied to the drive shaft. The brake remains on until the workboard has completed its move. The brake is then released and the drive shaft continues its rotation from 190° through 300° completing the placement. The upward placement stroke is then initiated at 300° and proceeds through 330°. At this reference point, a control signal is again sent releasing the workboard to align the next place location. The rotation from 330° through 354° completes the upward placement stroke.

3. PN CHARACTERIZATION

Figure 3 shows the PN model of a pick-and-place cycle [4]. In the net there are 11 places representing logical conditions of the machine and 9 transitions representing the main activities of the machine, namely, turret indexing, shutter open/close, pickup up/down, placement up/down, board and carrier movement, and their synchronization aspects.

In the initial marking M_1 of the net there is a token in p_1 (pick-and-place cycle complete / new pick-and-place cycle). Starting from M_1, the net can be executed, i.e., the machine can be emulated, by moving tokens around, obtaining the following 16 states (only the marked places are reported):

$M_1 = (p_1)$, $M_2 = (p_2, p_3)$, $M_3 = (p_3, p_4, p_5)$, $M_4 = (p_2, p_6)$,
$M_5 = (p_4, p_5, p_6)$, $M_6 = (p_3, p_5, p_7)$, $M_7 = (p_3, p_4, p_8)$, $M_8 = (p_5, p_6, p_7)$,
$M_9 = (p_4, p_6, p_8)$, $M_{10} = (p_3, p_7, p_8)$, $M_{11} = (p_6, p_7, p_8)$, $M_{12} = (p_3, p_9)$,
$M_{13} = (p_6, p_9)$, $M_{14} = (p_3, p_{10})$, $M_{15} = (p_6, p_{10})$, $M_{16} = (p_{11})$.

The meaning of each of the above states can be easily derived from the tokens distribution. For instance, in M_1, t_1 fires in zero time placing tokens in p_2 (close feed shutter) and p_3 (initiate board move). The new marking M_2 enables the two timed transitions T_3 (board movement) and T_2 (the drive shaft rotates 16 degrees). The corresponding operations are being performed concurrently and are synchronized prior to the placement (T_9) by the immediate transition t_8, when there is a token both in p_6 (board move complete) and in p_{10} (proceed with downward placement stroke), and resulting in a token in p_9 (continue downward pickup stroke). After the firing of transition T_2 tokens in p_4 (initiate carrier move) and in p_5 (maintain feed shutter closure) enable the pickup operation, consisting of the two concurrent operations T_4 (carrier movement) and T_5 (closure of the shutter). These concurrent operations are synchronized by the immediate transition t_6 prior to transition T_7 (continuation of the downward placement stroke and the downward pickup / feed shutter continues to open), when there is a token both in p_7 (carrier move complete) and in p_8 (initiate opening of feed shutter), and resulting in a token in p_{11} (complete pickup and placements of components). Note that carrier movement is concurrent with the board movement and the drive shaft rotation from 346° to 170°; board movement is concurrent with the drive shaft rotation from 330° to 190°.

All the timed transitions, emphasized with capital letters, are data dependent [7]; that is, data streams have to be provided to determine the time required for each firing of the corresponding transition. In particular,

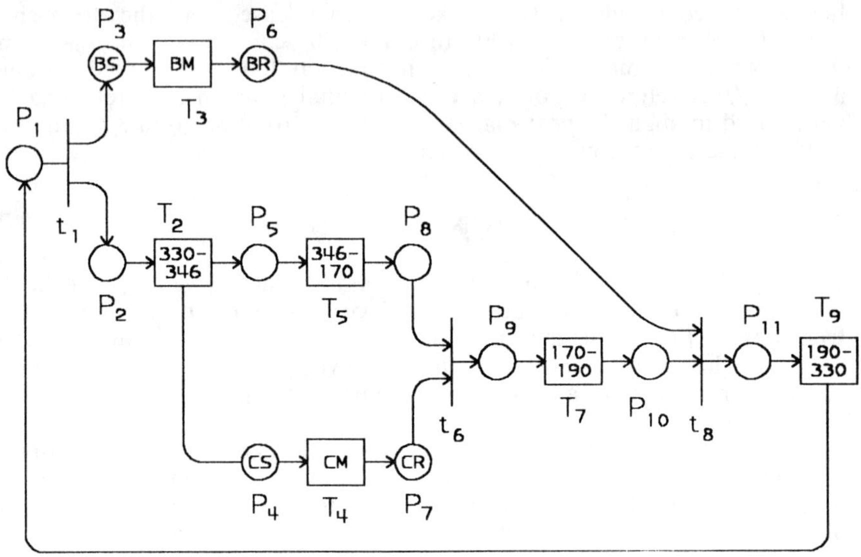

Figure 3. **PN model of a pick-and-place cycle**

T_2, T_5, T_7, T_9 depend on which of the three speed modes of the drive shaft is being used; T_3 depends on the previous place location on the workboard relative to the new place location; T_4 depends on the feeder station from which the previous component was acquired relative to the feeder station from which the current component is to be acquired.

The time required to complete a pick-and-place cycle is given by

$$S = \text{Max}\,(S_3, S_2 + \text{Max}\,(S_4, S_5) + S_7) + S_9$$

where S_i refers to the time required to complete transition T_i. The PN based emulator also allows identification of the operations which cause potentially avoidable delays [7]. Therefore, it is possible to interactively analyze the sequence of firings for a given product. Note that in the case of the MK-1 machine, both board movement and carrier movement delays can occur. In particular, a board move delay occurs when $S_2 + \text{Max}\,(S_4, S_5) + S_7 \leq S_3$. If $S_4 \leq S_5$ then no delay is incurred from the carrier movement.

4. THRUPUT OPTIMIZATION

Optimization of the machine requires specification of the arrangement of the feeder reels and the pick-and-place sequence. These are interrelated problems that require simultaneous solutions. If we assume zero access delay from the rack, the sequencing problem becomes a pure TSP, in which the placement locations represent the "cities" and the inter-city delays are determined by the X-Y table movement delays. For the size of the applications considered (over 400 placements) we confine our approach to heuristic procedures.

Although, complete freedom in selection of the pick-and-place sequence exists, there is a preferred class of solutions that minimizes the feeder rack movements. The heavily loaded feeder rack can experience high velocity movements with quick starts and stops which generate a great deal of inertia and can affect machine wear and calibration. This implies sequencing solutions in which all placements from each feeder reel are completed consequently. In this manner only a minimum number of rack movements occur. This Engineering concern in conjunction with the observation that the feeder rack movement between adjacent stations is masked by the turret indexing delay (i.e., in general $S_4 \leq S_5$ as verified by PN model characterization) results in a simplified method for solution of the staging and synchronization problem. In other words, once the placement sequence has been determined, the feeder reels are staged in the same order that the placements are organized. In this case, in the TSP analogy, each feeder reel represents a "state," and the tour of the "cities" must be done one state at a time.

4.1 Solution Procedure

Two heuristic procedures are offered for solution:

1. First the inner-state sequence is determined and then the inter-state ordering follows.

2. First the inter-state sequence is established and then the inner-state sequence follows.

Since we cannot establish a preference between the two methods, we offer both solution methods to the user and leave the final arbitration to the machine PN emulator. The optimized solutions provide the data values (T_3 and T_4) for the PN of Figure 3. The PC based solution package is thus an effective system that also provides a final format program file (NC data) to be down loaded to the machine RAM via a RS-232 port.

For solution of the TSP sub-problems we offer both the 2-opt and the 3-opt procedures. The user can select the limit on problem sizes for 3-opt applications for run time considerations.

4.1.1 The Inner State First Procedure

1. Determine the optimal Hamiltonian path for each state independently.

2. Reduce each Hamiltonian path to a single arc.

3. Construct a new graph involving the nodes of the arcs defined in the above step.

4. With the X-Y table datum position (home position) as the origin, determine the optimal tour of the above graph. This supplies the ordering of the states.

4.1.2 The Inter-State First Procedure

1. For each pair of states I and J (inclusive of the datum position), find the shortest distance d_{ij}, such that $i \in$ I and $j \in$ J.

2. With each state represented as a node and the inter-state distances d_{ij} defined as above, find the optimal tour of the states with the board datum position as the origin.

3. The optimum sequence of the states will supply at most two nodes for each state as entering and departing nodes.

4. With this restriction determine the inner-state's optimal Hamiltonian path.

References

1. J. Ahmadi, S. Grotzinger, D. Johnson, "Component Allocation and Partitioning for a Dual Delivery Placement Machine," *Operations Research*, Vol.36, (1988).
2. J. Ahmadi, S. Grotzinger, D. Johnson, "Emulation of Concurrency in Circuit Card Assembly Machines," *Research Report RC 12161*, IBM T. J. Watson Research Center, Yorktown Heights, NY, (1986).
3. Anonym, "Operating Instructions: Panasert MKI-LL Chip Component Placement Machine Model NM-2562B," Panasonic Industrial Company, Industrial Sales Division, One Panasonic Way, Secaucus, NJ 07094, (1985).
4. S. Grotzinger, A. Sciomachen, "A Petri Net Characterization of a High Speed Pick and Place Machine," *Proceedings of the 38th Electronics Components Conference,* Los Angeles, (1988).
5. E. L. Lawler, et al. (Eds.), "The Traveling Salesman Problem," (John Wiley & Sons, 1986).
6. J. L. Peterson, "Petri Net Theory and the Modelling of Systems," (Prentice Hall Inc., 1981).
7. A. Sciomachen, S. Grotzinger, F. Archetti, "Petri Net Based Emulation for a Highly Concurrent Pick and Place Machine," to appear in *IEEE Transactions on Robotics and Automation*, (1990).

MULTIDISCIPLINARY PROBLEMS IN MECHATRONICS AND SOME SOLUTIONS

Mr. Vesa Salminen
Federation of Finnish Metal, Engineering and Electrotechnical Industries, Helsinki, Finland

Mr. Timo Ropponen
Helsinki University of Technology
Espoo, Finland

Prof. Arto Verho
Lappeenranta University of Technology
Lappeenranta, Finland

ABSTRACT

The paper describes experiences in designing intelligent machines and manufacturing systems. Multidisciplinary character of mechatronic system creates new difficulties during design process. A model for carrying out design process of automatic production system is represented. Some guidelines in solving multidisciplinary problems in mechatronics are suggested.

INTRODUCTION

Mechatronics is a new smart way of combining electronics, mechanics and data processing. Increased flexibility, adaptability, safety and reliability are the gains which can be achieved by applying the mechatronics design concept to products and manufacturing systems, i.e. mechatronic systems. This goal is reached by optimizing all the possibilities of the available technologies.

The multidisciplinary concept of mechatronics causes new design problems. In solving these problems we need a specific method which takes into account all the available technologies. In this paper some important problems are presented and their solution models are given.

The experiences have been gathered during the Finnish mechatronics research program in the years 1987-1990 /1/. The general research topics were

- Integration of intelligence into machine part
- Internal data transfer with fibre optics
- Development of reliable, distributed control
- Systematic design of a mechatronic products.

The industrial applications included

- Intelligent robot grippers

- Fibre optical sensors and data transfer in electrolytic copper sheet handling crane
- Electropneumatic control of motion and force in an automatic manipulation system
- Distributed control and synchronization of hydraulic (rotating and linear) actuators in paper handling machinery
- High-density packaging of electronics into an elevator door
- Hydraulic stabilization of high inertia loads.

Fig.1 indicates the status of the mechatronics program within the hierarchical division of automation and limits the area of topics included in the program.

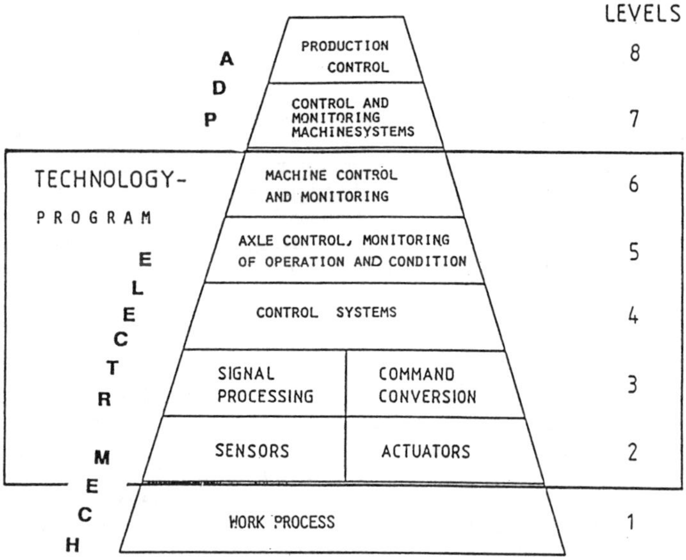

Fig. 1 Scope of the mechatronics program within the hierarchy of automation /1/

MULTIDISCIPLINARY PROBLEMS AND SOLUTIONS

Mechatronic products and manufacturing systems consist of several different elements which complicate design procedures. Multidisciplinary problems emerge as soon as the products has functions requiring a combination of mechanics, control and electronics. Such problems can be categorized as, e.g.

- technological problems (knowledge)
- conceptual design problems (method)
- safety and reliability (function).

Technological problems are easily recognizable, low level elements. To solve special technical tasks one must apply the technology available in that field. More complicated problems arise from conceptual design, when the objective is to define the tasks and search for technical solutions and combine them successfully. Safety and reliability factors also proved to be significant problems in automatic machines and systems.

In the following chapters two different design problems typical of mechatronic systems are discussed. First, we state the common problems in the task definition phase. Subsequently, we suggest solutions and tools for improving the task definition. The other example is related to imbedded control systems in mechatronic systems. In the same manner, some design guidelines are given.

Task definition

The design of a new flexible and adaptive production system demands a multidisciplinary approach. Discussion between the various field experts is a must; they must define and describe the functions in a generally applicable way already at the task definition phase, ensuring that everybody involved understand it in the same way /2/.

The importance of cooperation becomes even more obvious at the demand list and specification phase. No alterations during the design process are necessary in a jointly compiled demand list.

When analyzing the manufacturing process and defining the tasks to be performed, the designer can make use of the problem hierarchy method. In this problem hierarchy we descend from the top level basic problem down to detail problems. When all the potential problems are listed, the final specification list can be refined for the preliminary design.

Fig. 2 illustrates the hierarchical problem definition used during the process of designing an intelligent gripper /3/. The basic problem of the gripper task is divided to subproblems. This systematic approach helps a design engineer to write down the specification list for preliminary design.

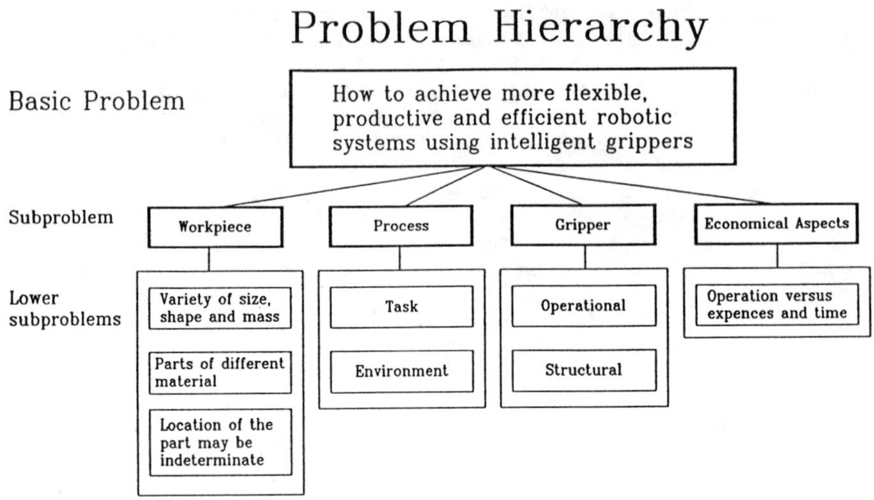

Fig. 2 Problem hierarchy of an intelligent gripper

Design of imbedded distributed control systems

It is the increase in adaptivity and flexibility in mechatronic systems that makes the distribution of intelligence necessary. The processing of data increases to such an extent that it is advantageous to treat information in a distributed manner because of the speed, reliability and safety required.

The way in which integrated systems control their operating environment, the mechatronic system, is becoming increasingly distributed and reactive.

In the multidisciplinary design of a real-time system, consisting of independent actuators, it is difficult to optimize all the existing technologies and solution models. The control of a mechatronic system is based on distributed data processing. The designer must be able to divide the construction into independent, separate, easy-to-control entities. Software has become more important and the increase in control functions has made the control systems more complex. This problem would not exist if ready-made package solutions consisting of an actuator, control, sensor functions and control electronics were available on the actuator level.

The technical specifications of an integrated control system can be defined e.g. by the QFD (Quality Function Deployment) method /4/. This method may also be used in software subcontracting and in the definition of the degree of automation required in a machine.

The SA/SD method /5/ can be used at the conceptual design phase for modelling the system. A context diagram illustrates communication between the system and its environment. In its support, an event list in verbal form can be compiled, especially to clarify technical safety details. The functions in a mechatronic machine (transformations, events, processes) and the flows between them (material, energy, information) can be described with hierarchical flow diagrams. By means of the flow diagrams, the functions can be divided into their elements, providing a model of the processes, the connections between the processes and their control.

Each function in the flow diagram can be described by textual specification, minispecs. The lowest level of minispecs normally describes the control algorithm and can also be written with ADA-language (SOKRATES-software)/6/.

The logical operation and especially the future behaviour of the machine can further be described with a state diagram, illustrating the changes in the operation as a function of the time elapsed and the impulses received by the system. This diagram is used as a model for the function, the sequence of the various functions to be performed, the actions between state shifts and the conditions under which state shifts occur. The state diagram model must be brought down to a level where the structure of the system is distinctly recognizable. Fig. 3 shows the state transition diagram of the function EXECUTE GRIPPING /7/.

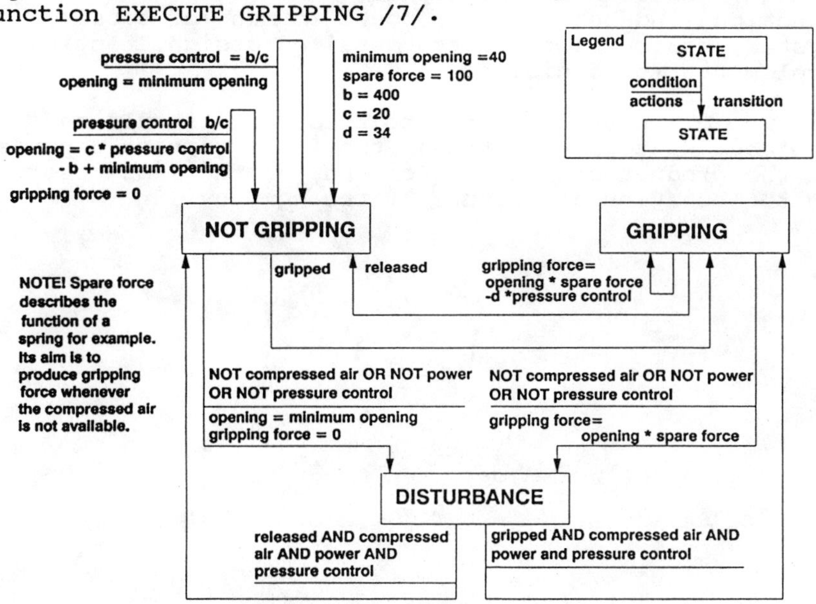

Fig. 3 State transition diagram of the function EXECUTE GRIPPING

Software designers may begin preparing a preliminary program code (e.g. a pseudo code) on the basis of flow and state diagrams. The state diagram is important as an illustration of the logical behaviour in complex systems and a prestage of a program code. For electronics designers it serves as a basis for the logic diagram. Generally speaking, the state diagram helps us internalize and understand the function of the machine.

Building up and checking a logic model essentially diminishes the number of faults and errors found at the test phase. This helps us avoid redesign and unnecessary costs.

The logic model serves as basis for the rest of the design process. The created models and minispecs will be fed to code generator, which creates C-language code.

Design model for a mechatronic production system

Above we already noted that the control of a custom-made mechatronic system is distributed. The more complex and flexible a system is, the more difficult it is to design in a harmonious way.

Fig. 4 illustrates the general process of designing a machine or system, used and tested in the R&D projects and the production redesign project of the mechatronics program.It consists of process analysis, preliminary design, decision of specification, description of functions and behaviour, concept forming of total system, choice of concepts, detailed design, testing and implementation decision.

Too close an adherence to this model should be avoided. What is accepted as the final version depends on the product and the system and, as the last resort, on the behavioural culture of the company.

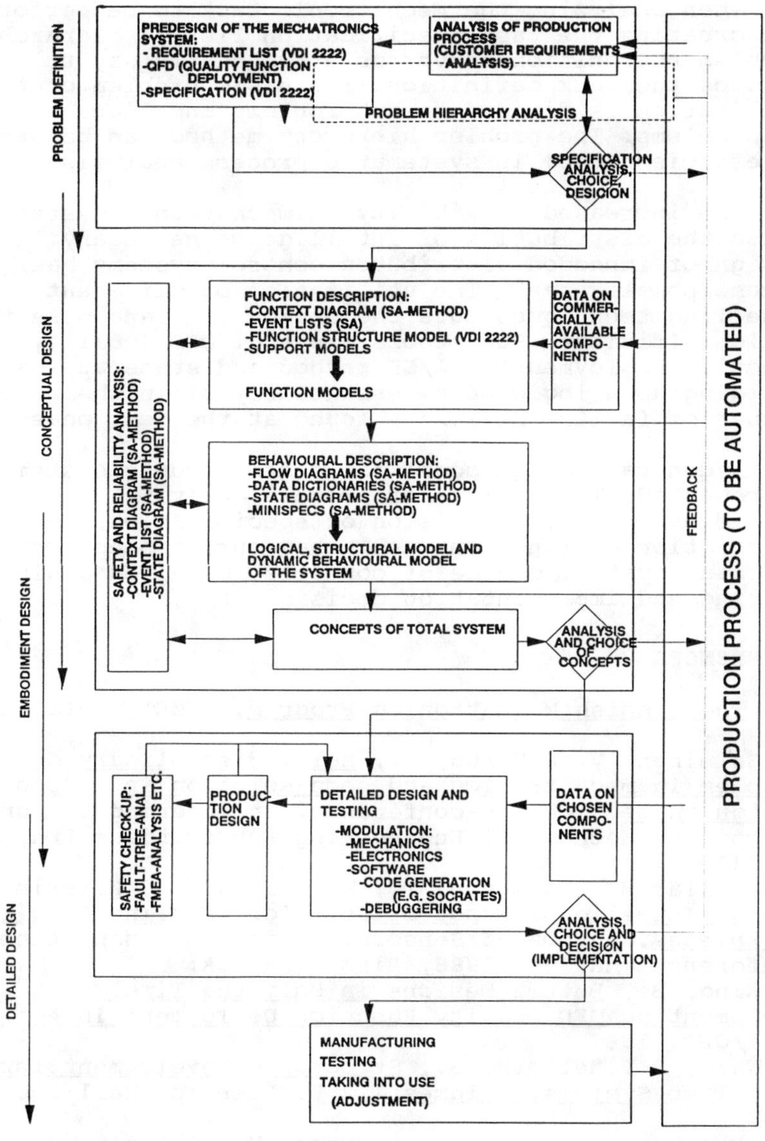

Fig. 4 General process of designing a machine or a system

CONCLUSION

The design of flexible and adaptive mechatronic systems, i.e. mechatronic products and manufacturing systems, requires a multidisciplinary approach. Special attention should be paid to certain design problems, which are characteristic to mechatronic systems.

When analyzing the mechatronic task to be performed and creating the task specification list, a hierarchical problem recognition serves as a systematic way to produce the task definition. The basic problem of the application is divided to subproblems and lower subproblems. The problem hierarchy method can be used as an efficient tool in systematic process analysis.

The increased flexibility in mechatronic systems makes the distribution of intelligence necessary. The design of imbedded distributed control systems has become complex task. The utilization of different modelling techniques cuts the design time and make the design efficient. Such tools are e.g. QFD (Quality Function Deployment), SA/SD method and state approach. Building up a logic model essentially diminishes the number of faults and errors found at the test phase.

A general design model for a mechatronic system is represented. It consists of process analysis, preliminary design, decision of specification, description of functions and behaviour, concept forming of total system, choice of concepts, detailed design, testing and implementation decision.

REFERENCES

1. The Finnish Mechatronics Program, FIMET, Helsinki 1989

2. Salminen, V. & Verho, A., Multi-disciplinary design problem in mechatronics and some solution in conceptual design phase. ICED 89-conference. Engineering Design, Harrogate. Mechanical Engineering Publications Ltd, 1989 p. 533-544

3. Airila, M., Ropponen, T., Merilinna, J., Kiiskinen, M., An Intelligent Robot Gripper for Handling Fragile Workpieces. Automan-5/Sensors, Vision and Inspection-conference. May 9th 1988, Birmingham UK.

4. King, B., Better Designs in Half the Time. Implementing QFD Quality Function Deployment in America. Goal/QPC, 1987

5. Ward, B & Mellors, S., Structured Development for Real-Time Systems, Volumes 1,2,3. Prentice-Hall Inc., 1985

6. Okkonen, A., Auer, A., Levanto, M., Okkonen, J., Kalaoja, J., Socrates SA -A Formal Method for Specifying Real Time Systems. Microprocessing and Microprogramming. The Euromicro Journal 27, North Holland, Amsterdam, 1989

7. Reunanen, M., Heikkilä, J., Hänninen, S., On Safety and Reliability Engineering during the Conceptual Design Phase of Mechatronic Products. Proceedings of the 10th Annual Symposium of the Society of Reliability Engineers, Stavanger, Norway, Oct 9-11, 1989. Elsevier, London.

AMRF COMPOSITE FABRICATION WORKSTATION [1]

Richard J. Norcross
W. Robert Bunch
Roger D. Kilmer
National Institute of Standards and Technology
Robot Systems Division
Gaithersburg, MD 20899

ABSTRACT

The Robot Systems Division of the National Institute of Standards and Technology (NIST) has initiated development of an advanced manufacturing workstation for the fabrication of composite parts. The focus of this project will be the application of sensor-based hierarchically controlled robot systems to automated fiber placement of thermoplastic composite materials. In such a system, continuous fiber reinforced thermoplastic polymer is fed from a spool and applied onto a mandrel or part form. The orientation of the fiber is based on the part's geometry and structural design. As the composite material is applied, it is heated to melt the polymer matrix, and then consolidates in place as it cools. The intention of this project is to demonstrate that the use of fiber placement via two cooperating robot manipulators and in-situ consolidation, can provide the capability to efficiently produce complex shaped composite parts.

INTRODUCTION

The aircraft and defense industry's demand for light-weight, high performance components has lead to rapid increases in both the development and use of fiber reinforced polymer composite materials. Unfortunately, the manufacture of high performance composite parts remains very time and labor intensive. New manufacturing methods are needed for the efficient production of parts using these materials.

The Automated Manufacturing Research Facility (AMRF) is a testbed at the National Institute of Standards and Technology (NIST) in which many factory automation issues are being addressed. The AMRF consists of several workstations which are used to study machining, deburing, inspection and handling of metal parts, as well as the associated control, data processing, machine programming, and scheduling issues. A new composites workstation is being developed and incorporated into the AMRF in order to apply AMRF principles to the fabrication of high-performance, complex-shaped thermoplastic composite parts. This paper gives a background on the manufacturing of high-performance composite parts, and outlines the research goals and status of this workstation.

---

[1] This work is sponsored by the U.S. Navy MANTECH Program.

BACKGROUND

The following three sections present a brief overview of composite materials and the associated manufacturing processes. This material is discussed in order to provide the necessary background information which led to the development of the research goals for the AMRF Composites Workstation.

Composite Materials

High-performance composite parts are generally laminated structures. Each ply is normally reinforced with continuous, unidirectional fibers. The designer specifies the fiber orientation in each ply in order to optimize the structural properties of the part. This results in a part that is lightweight and strong, but expensive to produce. Conversely, consumer goods do not have the stringent performance requirements, and they must be produced at lower cost. Typically, consumer products use discontinuous, randomly oriented fiber reinforcement and are not laminated. The Composites Workstation project will focus on automation techniques for producing high-performance composite components.

Fiber reinforced polymer composite materials, or simply composites, consist of a polymer matrix material and a fiber reinforcing material. The most commonly used fiber materials are glass, carbon, and aramid. In general, carbon fibers have greater specific strength and stiffness than other fiber materials and hence, high performance components most often use carbon fibers. There are many polymer matrix materials. These materials are often classified as either thermoplastic or thermoset. As the terms imply, thermoplastic polymers can be formed and reformed with the application of heat, while thermoset polymers undergo a cure reaction which forms a rigid and permanent polymer network.

In order to facilitate the laminated construction of high performance components, material manufacturers have combined the fiber and matrix materials to form what are referred to as prepregs. Prepregs are available in various fiber/matrix material combinations and in various flat roll forms for use by part manufacturers. Prepregs are typically about 0.005 inch thick and range in width from one-eighth inch tow forms, through three and six-inch tapes, up to 60-inch wide broadgoods.

Until recently, the demands of high-performance applications could not be met by using thermoplastic-based composites. As a result, virtually all high-performance composite components are currently made from thermoset polymers. Recently, thermoplastic prepregs have been developed which meet the demands of many aircraft and defense components. Thermoplastic-based prepregs have several potential advantages over the corresponding thermoset prepregs. The thermoset prepregs must be cured in an autoclave (pressurized oven), which typically requires several hours. Prior to cure, the thermoset prepregs are unstable and have a limited shelf life even when refrigerated. On the other hand, thermoplastics are stable at room temperature and have an unlimited shelf life. Thus, if high performance composite parts can be produced using on-line consolidation of thermoplastic prepregs, considerable savings can be realized in terms of both equipment costs and overall processing times.

Fiber Placement

Success with composites has led to demands for increasingly complex parts. These demands often exceed the capabilities of existing manufacturing technology. Until new methods are developed, these complex parts must be produced by hand layup. In hand layup, consecutive plies are manually applied onto a part form in specified locations and orientations. Periodically, a plastic sheet (called a "bag") is placed over the laminate and a vacuum is applied to compact the laminate. The bag is then removed and the hand layup continues. After all plies have been laminated, the part is again bagged and readied for the autoclave process. Hand layup, while very flexible, is very time and labor intensive, exposes workers to potential health hazards, and is less repeatable then an automated process.

Filament winding and tape laying are two popular mechanized processes used to fabricate high-performance parts. In filament winding, a fiber bundle or tow is guided under tension onto a rotating mandrel. The tension and the part's convex surface combine to produce a compaction force. Thus, filament winding is usually limited to axisymmetric shapes having convex surfaces. Tape laying generates the compaction force directly through a lay-down head. In tape laying, consecutive bands of 3 to 6-inch wide prepreg tape are rolled onto a part form. Tape laying is limited to surfaces having flat or mild contours.

To address these deficiencies, some organizations in the composites industry have directed research efforts toward developing a fiber placement process. Fiber placement combines aspects of both tape laying and filament winding. Like filament winding, fiber placement uses the narrow bands of prepreg tow. Like tape laying, the tow is applied onto the part form under direct pressure. With contact placement, the tow's path is not limited to convex geodesic paths. Thus, very complex shapes and fiber orientations can be produced.

In-situ Consolidation

In-situ consolidation refers to the welding or consolidating of the thermoplastic prepreg during lamination. As the band of prepreg approaches the bond zone or "nip", the band and the substrate, or previously applied laminae, are heated above the melt temperature. At the nip, pressure is applied to insure intimate contact between the two surfaces. As the material cools, the two surfaces consolidate into a single structure. Several problems arise in this scenario. For example, the desire for an economic, and therefore, rapid lay-down rate conflicts with the requirement to maintain an effective processing temperature and pressure long enough to achieve successful consolidation.

Since virtually all high-performance composite parts have been made of thermoset prepregs, the processing of thermoplastic prepregs is not well understood. Satisfactory consolidation has yet to be universally defined. Furthermore standard methods for the direct measurement of consolidation in thermoplastic prepregs do not exist. Despite problems, the pursuit of in-situ consolidation promises significant rewards. When in-situ thermoplastic consolidation is achieved, expensive autoclaves and associated cure cycles, as well as the refrigeration and special handling which are required for thermoset prepregs may no longer be necessary.

COMPOSITES WORKSTATION PROJECT GOALS

The purpose of this project is to enhance the automated production of high-performance parts made of composites. The project's goal is to use the control techniques of the AMRF to demonstrate the feasibility of a robotic thermoplastic fiber placement process. In this process, the prepreg tow or tows are consolidated in place as they are applied to complex shaped part forms. Much of this work will concentrate on developing the techniques required to fabricate parts whose geometry currently makes automated production difficult. These parts are defined here as being non-axisymmetric with concave surfaces and/or multiple axes of rotation.

Fiber placement and in-situ thermoplastic consolidation are both emerging manufacturing technologies [1-4]. While several organizations are exploring in-situ consolidation techniques, only a few have reported success, and they are reluctant to disclose the details of their discoveries. Before a fully robotic system for fabricating complex geometries can be implemented, a successful in-situ consolidation process must be developed.

Because of these uncertainties, a testbed system is being implemented which will be used to investigate and develop suitable consolidation and fiber placement techniques. Based upon the testbed results, a prototype robotic fiber placement system will be designed and installed in the AMRF. This fiber placement system will consist of two cooperating manipulators, each likely to have four to six degrees of freedom. While one manipulator maneuvers the part form, the other manipulator will apply the prepreg. This arrangement is expected to permit far greater access to difficult part geometries and greater flexibility than is possible with current equipment.

The manipulators' link dimensions and joint ranges have yet to be determined. Graphics simulation will be used to evaluate various alternatives for effective work volume and dynamic limitations. Since the system will be evolutionary in the sense that untested mechanical and thermal systems are being developed, graphics simulation and modeling will be used extensively throughout the project. This will assist the understanding of the effect of the process variables and to predict the performance of the system.

The control system will be based on the NIST-developed Real-time Control System (RCS) [5-7]. The result of over a decade of research on advanced manipulator and systems control, RCS is a hierarchially structured control system which utilizes sensory feedback to make real-time modifications of the robot motion. RCS provides logical separation and organized integration of the various functions of the controller. Adhering to the hierarchy also allows for alterations and enhancements to be easily made.

The fundamental paradigm of RCS is a three-legged hierarchy of computing modules (Figure 1), serviced by a communications system and a global memory. The three legs consist of Task Decomposition (TD) modules, World Modeling (WM) modules, and Sensory Processing (SP) modules. The task decomposition modules perform real-time planning and task monitoring functions; they decompose task goals both spatially and temporally. The sensory processing modules filter, correlate, detect, and integrate sensory information over both space and time in order to recognize and measure patterns, features, objects, events, and relationships in the external world. The world modeling modules answer queries, make predictions, and compute evaluation functions on the state space defined by the information stored in global memory. Global memory is a database which contains the system's best estimate of the state of the external world.

In this application the control system must simultaneously coordinate the two manipulators and the consolidation equipment. It is anticipated that material temperatures at various locations, will be used to make adjustments to the fiber placement process.

PROJECT STATUS

Using a review of current research programs and literature as a guideline, current efforts have been focused on the design and development of the research testbed. An informal survey of composite part manufacturers indicated that there is significant interest in the automated processing of carbon/PEEK (Poly-ether-ether-ketone) prepreg. Therefore, carbon/PEEK was selected as the thermoplastic prepreg to be used for initial

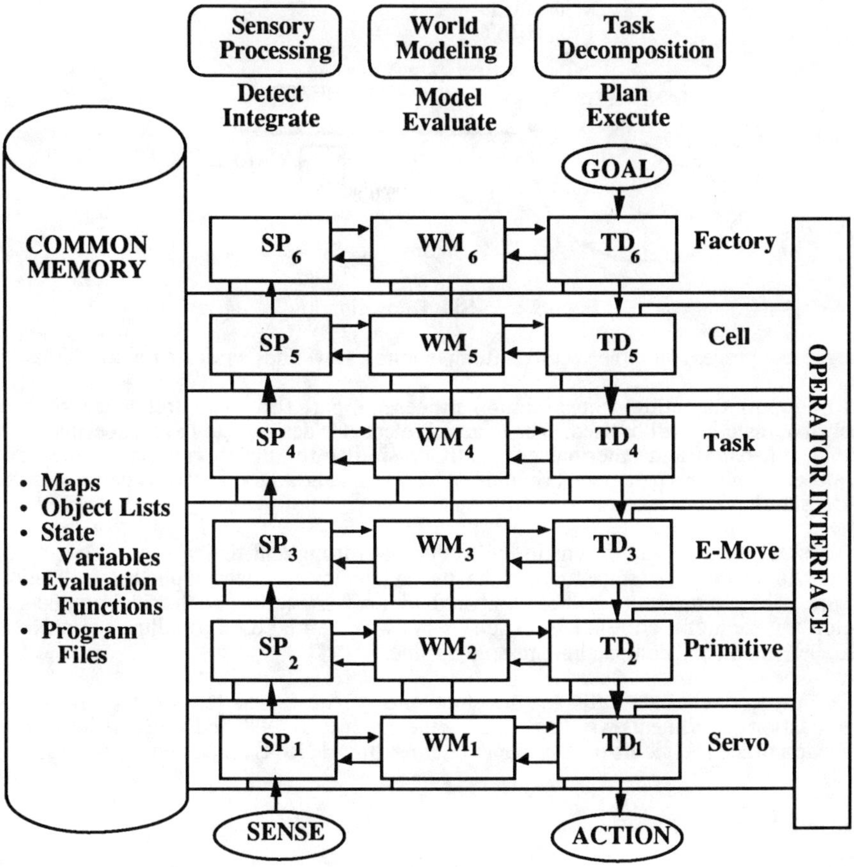

Figure 1. RCS Control Hierarchy.

consolidation experiments. A simple "contact placement winder" is now being developed as the testbed hardware so that the various process parameters involved in in–situ consolidation can be studied. This process is illustrated in Figure 2. The intention is to determine various combinations of temperature, pressure, and tow feed rate which result in successful consolidation of the tow prepreg. In order to determine an initial set of parameters, a thermal model of the process is being developed using a finite element analysis package.

Various organizations have experimented with a number of heating methods. Among the more popular are: lasers, hot gas, resistive, and infrared heaters. One method which appears both flexible and cost effective is a hot gas heater. The selected heater discharges up to 1400°F air through many small orifices which produce small but high velocity air streams (260 mph by 40 mil diameter). The high speed air cuts through the laminar air layer adjacent to the tow and quickly and efficiently transfers heat into the composite. The hot gas heater concept has been successfully used with thermoplastics by others,

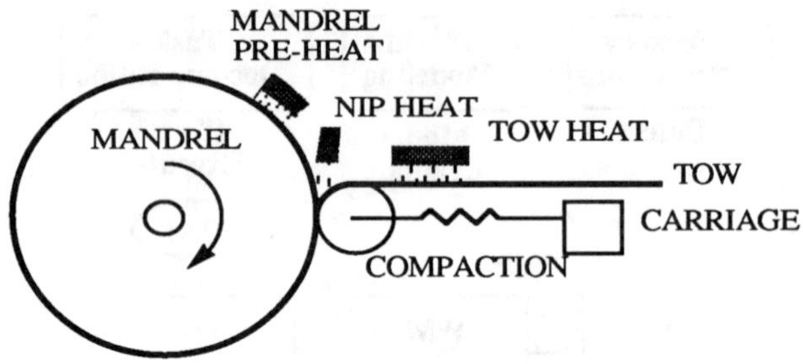

Figure 2. In-Situ Consolidation.

however, the process variable specifics (temperatures, flow rates, etc.) are not available.

To support the initial consolidation process, a part form mandrel made from a dissolvable material will be used. The material selected is designed to have a coefficient of thermal expansion and thermal conductivity similar to the thermoplastic prepreg. Therefore, the process parameters should not have to change significantly as part thickness increases during lamination .

Investigation of various sensors will continue throughout the project. The initial sensors will measure the temperature of the tow and laminate, and the applied consolidation pressure. These parameters will be compared with off-line testing of the final products to develop a meaningful correlation. The sensors will then be used to adjust the process speed, heat input, and compaction pressure on-line.

The project schedule calls for development of the initial testbed by fall 1990. Consolidation experiments will begin during the summer of 1990 and continue into 1991. Installation of the robotic fiber placement system in the AMRF is scheduled for late 1991.

REFERENCES

1. R.H. Stone, et al., "Manufacturing Science of Complex Shape Thermoplastics", Fourth Quarterly Interim Report AFWAL/MLBC Contract F33615-86-C-5008, (Wright-Paterson AFB, OH).
2. D. Hauber, "Filament Winding with Thermoplastics", SME Technical Paper EM87-552, (September 1987).
3. B.R. Latz, and J.O. Stine, "Automated Thermoplastic Filament Winding", Proceedings of the SME Clinic Productive Filament Winding Technology for Manufacturing Engineers and Managers, Los Angeles, CA, (August 1989).
4. R.J. Gardiner, "Filament Winding's Succession To Fiber Placement", Proceedings of the SME Clinic Productive Filament Winding Technology for Manufacturing Engineers and Managers, Los Angeles, CA, (August 1989).
5. J.S. Albus, A.J., Barbera, and R.N. Nagel, "Theory and Practice of Hierarchical Control", Proceedings of the Twenty-third IEEE Computer Society International Conference, Washington, DC, (September 1981).
6. A.J. Barbera, M.L., Fitzgerald, J.S., Albus, and L. S., and Haynes, "RCS: The NBS Real-Time Control System", Proceedings of the Robots 8 Conference and Exposition, Volume 2 - Future Considerations, Detroit, MI, (June 1984).
7. J.S. Albus, H.G. McCain, and R. Lumia, "NASA/NBS Standard Reference Model for Telerobot Control System Architecture (NASREM)", NIST Technical Note 1235, 1989 Edition, (National Institute of Standards and Technology, Gaithersburg, MD, April 1989) (supersedes NBS Technical Note 1235, July 87).

COMPUTER-CONTROLLED ASPHALT PLANT

AN Poo, KB Lim, GS Hong and CL Teo
Department of Mechanical and Production Engineering
National University of Singapore

ABSTRACT

This paper describes a computer-controlled asphalt plant. An IBM PC-AT compatible based system is used to control the composition of the pre-mix to ensure consistent quality of the mixture. The PC controls the speeds of the belt feeders to ensure the correct mix proportion. The system is able to compensate for the presence of moisture in the mix and temperature variations in the bitumen. Fully graphic displays are used to provide a friendly user interface. Field tests show that accuracies lie within 2 RPM at speeds of 1000 RPM.

INTRODUCTION

Continuous pre-mix plant is one of the technique to produce road-mix products. In such a plant, a continuous belt feeder unit is used to move the aggregate from the aggregate storage bins to a central belt conveyor. The conveyor transports the aggregate to a mixing drum where the aggregate is mixed with bitumen to form the bitumen mix used for roads and pavement laying. The proportions of each aggregate has to be controlled accurately to obtain various desired blends of different aggregate gradations.

This paper describes a computer-controlled component mixing system developed in the Department of Mechanical and Production Engineering at National University of Singapore.

The system has been successfully applied to such a road-mix plant. The heart of the system is an IBM PC-AT compatible equipped with analog input and output capabilities. It is used to control up to 8 different components and monitor up to 16 variables. The computer controlled system automatically control the continuous belt feeder unit to ensure the exact proportions of aggregate for different desired blends.

PROCESS DESCRIPTION

Bituminous surfaces for roads and pavements consist of various compacted aggregates combined with bitumen to form a flexible resilient, waterproof medium for traffic. The aggregates usually consist of crushed stones (from 10mm sizes to 38mm), gravel, sand, quarry dust or other similar material and cement filler. These aggregates are blended in different proportions with bitumen (and sometimes latex to increase flexibility) to produce various gradations. The gradation of the particles largely determines the mechanical stability of the bituminous mix.

Figure 1 shows a schematic of a continuous-mix bitumen plant. The various aggregates are stored in cold feed storage bins (on the left of the figure). They are then deposited onto a cold feed conveyor and moved to a rotating dryer-drum mixer.

There the aggregates are dried by heating and combined with bitumen (maintained at 150° C) to form the hot bitumen mix The dryer-drum mixer is typically big enough so that the aggregates have sufficient time to dry and be coated evenly with bitumen. The bitumen mix product is then moved by a hot-mix conveyor to a mix silo where it will be loaded onto trucks to be transported to buyers.

In such a continuous-mix bitumen plant, the aggregate and bitumen are introduced into the mixer and the mix discharges continuously. The devices feeding the aggregate and bitumen must maintain automatically the correct proportions to ensure good quality mix. In the following, we will discuss the devices used to feed the aggregate and the bitumen.

The cold aggregate feed is one of the critical control points in the production of bitumen mixes. Wide variations in quantity of a specific aggregate at the cold feed may cause a considerable change in the temperature of the aggregate leaving the dryer. This may cause uneven coating of the aggregate resulting in a loss in quality of the mix. Feeder units should therefore have controls that can be set and secured to produce a uniform flow of aggregate to the cold feed conveyor.

A common type of cold aggregate feeder unit is shown in Fig. 2. The amount of material discharged onto the cold feed conveyor depends both on the gate opening and the speed of the belt feeder. A bigger gate opening or higher belt feeder speed will increase the amount of material discharged and vice versa.

Material flow rate can be controlled by the use of a variable speed motor to drive the feeder belt with a fixed gate opening. The mass flow rate for the fixed gate opening has the relation with the drivers speed as such,

$$\mathrm{TPH} = F_{\mathrm{TPH}} * \frac{\text{motor Speed}}{1000 \text{ rpm}} .$$

where the F_{TPH} is factor which is a function of the material's density and gate opening at the motor speed of 1000 rpm.

For the bitumen, a positive fixed displacement bitumen pump is used to control the amount of bitumen introduced into the dryer-drum mixer. The bitumen is normally heated to around 150° C before it is able to be pumped into the drum mixer. The fixed displacement pump should be calibrated at the normal operating temperature of the bitumen.

Bitumen is a thermoplastic material that decreases in viscosity with increasing temperature. To maintain constant bitumen proportioning, the temperature of the bitumen going through the pump must be known at all times so that any necessary adjustments can be made to compensate for weight changes in the bitumen.

The conventional dryer-drum mixer is a revolving cylinder usually ranging from 3 to 10 feet (0.9 to 3 m) in diameter and from 15 to 40 feet (4.5 to 12 m long), in which the aggregate is dried and heated by an oil or gas burner. The dryer vaporizes and removes moisture from the aggregate and heats the aggregate to the mixing temperature. Most problems in dryer operation are caused by crowding more material through the dryer than it can properly handle. Hence control of the cold aggregate feeder to ensure the proper quantity of material coming into the dryer is very important.

CONVENTIONAL MANUAL PLANT

In the conventional manual plant, the production rate (tonnage) in Ton per Hour (TPH) cannot be easily changed. Calibration is normally done at approximately the designed production rate of the plant. The belt feeder speed is then manually set to achieve that production rate. Changing the desired TPH entails recalculating the set point speed of the cold aggregate belt feeders. There are a lot of waste during startup and shutdown as the composition of the mix is not of the desired proportions since calibration is only done for the normal operating point of the plant. During startup, it is desirable to start the plant at a smaller production output in order not to overload the dryer immediately. The output is then slowly increased to the normal operating point. This is not easily done using manual setting of the belt feeder speed. An extensive series of settings need to be provided to bring the plant on line.

The presence of surface moisture in the cold aggregate must be taken into account. The weight of the surface moisture on the aggregate must be deducted from the total weight of the aggregate to get the actual quantity of aggregate. This is especially important for fine aggregates since they tend to absorb more moisture. Also the aggregate tends to settle more and becomes more compact with moisture. This cannot be easily handled in the manual plant.

COMPUTER-CONTROLLED PLANT

The inclusion of a computer to control the plant gives it a great deal more flexibility and versatility. Once a blend has been selected and the plant properly calibrated, the plant will continue to produce the correct asphalt mix at the desired tonnage until the operator stops it.

At the heart of the system is an IBM PC-AT compatible computer equipped with a 12-bit 8 channel digital-to-analog (D/A) converter card and a 14-bit 16 channel analog-to-digital card. It is desired that the plant can still be operated manually should the computer breaks down. As such, external analog PI controllers are used for belt feeder motors. The plant can then be switched over to manual operate if so desired or when the computer fails. The computer is supervisory in nature (Fig. 3) where the inputs to the computer are the tacho-generators' outputs and the outputs of the load cells attached to the cold aggregate belt feeders. Based on the load cells readings, the selected blend and the desired production rate, the computer calculates the desired speed of each feeder belt motor and output the set point through the D/A converter. The analog PI controller ensures that there is no steady-state error between the desired and actual motor speeds.

The system can control up to 8 different stations with a virtually unlimited number of blends. In the actual plant, 6 different aggregates (mining sand, granite dust, 10mm chips, 19mm aggregate, 38mm aggregate and cement filler). latex and bitumen are automatically mixed in the correct proportions. The first 5 stations are the cold feed aggregates of mining sand (station 1), granite dust, 10mm chipping, 19mm aggregate and 38mm aggregate. Station 6 contains the cement filler while station 7 and 8 are the latex and bitumen stations respectively.

SPEED COMPENSATION

Using the computer, we can also compensate for the surface moisture that may be present in the aggregate. Load cells are placed at the feeder belts (see Figure 2) for this purpose and the readings relayed to the computer. A calibration curve of the weight of the aggregate versus the moisture content is stored in the computer. During TPH calibration, the material collected is weighed and then dried and weighed again to determine the moisture content. While the feeder belt is running during the TPH calibration, the computer also reads the load cell readings. Using this reference load cell reading, the moisture content of the collected material, and the calibration curve stored previously, the computer can adjust the speed of the feeder belt during actual run time to compensate for the moisture by

$$\text{Desired Speed} = \frac{1000\ \text{rpm}}{F_{TPH}} * \frac{\text{Desired TPH}}{(1 - \text{moisture content})} * \frac{\text{Weight}_{Cal}}{\text{Weight}},$$

where Weight_{Cal} is the load cell reading during calibration.

The temperature of the bitumen is not directly controlled by the computer. But the temperature is monitored by the computer so that the speed of the bitumen pump can be automatically adjusted if the temperature changes to compensate for the change in viscosity of the bitumen with temperature. With a computer, it is easy to store the calibration curve of density against temperature for bitumen. The speed of the bitumen pump can then be changed according to the calibration curve for any change in temperature as such.

$$\text{Desired Speed} = \frac{1000\ \text{rpm}}{F_{TPH}} * \frac{\rho(150°C)}{\rho(T)} * \text{Desired TPH}$$

where $\rho(T)$ is the density of bitumen at the temperature T.

In addition, an alarm will sound to warn the operator if the temperature of the bitumen should fall outside a certain range.

COMPUTER ASSISTED CALIBRATION

Computer assisted calibration of the plant parameters greatly improves the accuracy and reduces the time required. The speed of the feeder belt is calibrated by using the computer to generate the 2 desired speed set points and compared with the actual motor speeds. Using this information and the voltage read in from the A/D card, the computer will automatically adjust its calibration factors accordingly. Field tests indicate that once calibrated, the system is able to accurately control the speed of the belt feeders to within 2 RPM at a speed of 1000 RPM.

Calibration of the amount of aggregate discharged at a particular station is also relatively straightforward and more accurate. In the manual plant, the feeder belt motor has to be running normally (ie. at steady-state) before the material can be collected and weighed to get the TPH (ton per hour) factor for a particular station. With a computer, it is no longer necessary to wait until the feeder belt run normally (hence no waste of material). The effect of transient behaviour of the motor during starting and shutdown is eliminated via the integration of velocity-time curve to achieve the total tonnage.

OTHER FEATURES

The computer also acts as a data acquisition and data logging unit. The tacho-generator converts the speed of the belt feeder into an equivalent voltage and this is then fed to the computer through the A/D converter. The computer then calculates the actual production total based on the speed readings that it sampled. This value is then stored and displayed at the end of the run in the form of easy-to- read bar charts for record purposes. Using the computer, it is easy to manipulate the data. For example, today's tonnage data can be displayed by blend (ie the total tonnage produced for the different blends selected) or by aggregate (ie. the total amount of aggregate used up in a day) in the form of bar charts. This enables tighter control on the amount of mix produced and accurate accounting of the cold feed aggregate used.

A unique feature of the computer-controlled plant is the programmed starting and stopping sequences of the plant. From the physical plant (Figure 1), it can be seen that the distances of the aggregate storage bins to the dryer drum mixer are not equal.

During startup, material from the station closest to the dryer drum mixer will reach the drum mixer first while material from the station furthest from the mixer will be last. Hence the proportion of aggregate in the mix during startup will not be correct and have to be discarded. To avoid this wastage, a starting sequence is implemented into the system to start each station appropriately. For the same reason, a shutdown sequence is also implemented.

A user-friendly menu-driven software has been written to make the system easy and simple to operate. Selecting an item is easily done by either using the up/down arrow keys to move the highlighted bar to the particular entry simply striking the number corresponding to that item. An ENTER key confirms the selection of a particular command.

Graphics display are used throughout to provide an easy visual interface between the plant and the operator. During the run, tracking of the motor speed for each station can be easily visualized via the display of a horizontal bar (actual speed) following a triangle marker (desired speed). If the actual motor speed is not within a specified range of the desired motor speed, an alarm will sound to alert the operator. The speeds and the corresponding amount of material produced (TPH) are also displayed numerically on the right for easy readability. Other useful information like the date and time of day, and the time that the plant started running are also displayed to enable to user to monitor the plant closely.

A separate data page is also provided that contains the numerical data in tabular form rather than the graphic display. This shows the material at the stations, the blend chosen, the percentage of aggregate for the stations, and other useful information. This page has most of the relevant information on the plant to enable troubleshooting. User can switch between these graphics or data display easily.

The available commands are indicated below the speed display. IBM soft keys (F1 to F10) are used for each command. User can call up a help page to get a more detailed explanation of the commands associated with the soft keys

The user is able to change the production rate (TPH), change the blend, increase or decrease the production rate slowly by 1 or 10 tons, shut down or stop the plant in an emergency.

CONCLUSION

Two such automated plants have already been built, installed and operational.

To date, they have been running satisfactorily, producing up to 250 tons per hour. Further enhancements to the software such as including the capability to display and print the tonnage and blend data for the past 2 years are now under development. This will aid in quality control of the bitumen mix as any drift in the system can be picked up from these data.

REFERENCES

1. Paving and surface operations, Department of the Army Technical Manual TM5-337, Feb 1966.
2. Asphalt Plant Manual, The Asphalt Institute, March 1979.

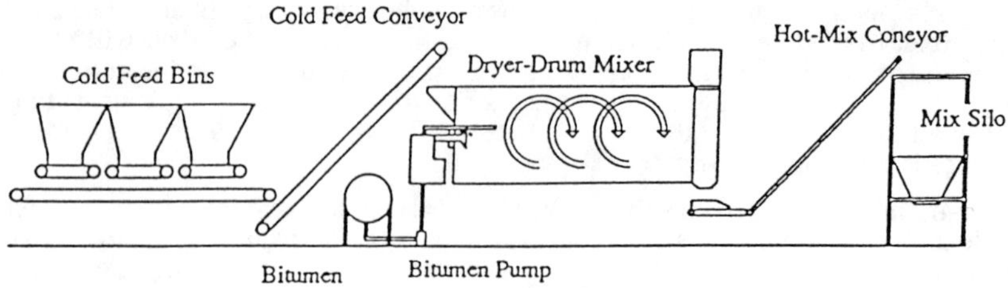

Figure 1. Schematics of a continuous-type asphalt plant.

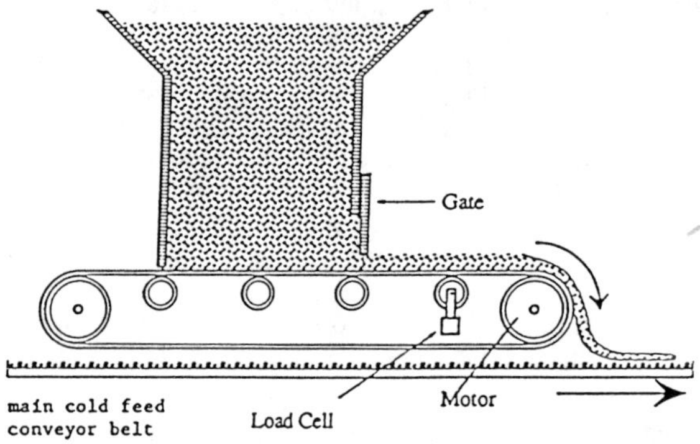

Figure 2. Continuous belt feeder unit for cold feed material.

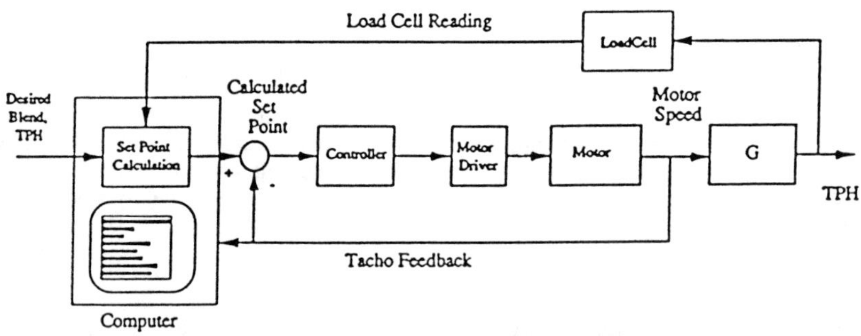

Figure 3. Schematics of a single speed control loop of the computer-controlled system.

CONFIGURABLE CONTROLLER FOR HYDRAULICS

Pertti Peussa
Machine Automation Laboratory, Technical Research Centre Of Finland
P.O.Box 192, SF-33101 Tampere, Finland *Tel +358-31-163606*
Internet: pertti.peussa@kau.vtt.fi *Fax +358-31-163494*

ABSTRACT

The time available for the design and assembly phase of machines is decreasing; therefore, the designer should be able to concentrate on essential things more, like how the drive must function, instead of striving with the details of different components.

This is also true in demanding fluid power applications, where sophisticated valves are used to control the force or motion of different hydraulic drives. The basic design of the electronics used to control these valves is more than ten years old, causing a lot of adjustment work, especially, when components of several manufacturers are used together.

A prototype of the next generation device, which can control most proportional and servo valves plus some typical hydraulic servo systems, is under development in Technical Research Centre of Finland. It will consist of a 16-bit microcontroller, flash memory, various analog components and of a so called 'SmartPower' chip, all on a single Europe 1 -size printed circuit board. Its configuration and properties can be changed by software, and it has a bi-directional serial bus to give commands and to ask for e.g. diagnosis information.

INTRODUCTION

When companies make offers, the delivery time is made so short as possible as it is a notable competition benefit. Thus the designer is facing increasing expectations with decreasing time margins for working out solutions. As new machines are developed, or when old machines are modernized, the designer is usually going to be dealing with more and more advanced control systems and their interfaces [1]. It is frustrating to use a lot of valuable design time to check details of different components, trying to figure out, how they could be used together.

One possibility is to use more-ready subsystems, which easily can be configured to various tasks, and to have a built-in serial interface. The communication with different subsystems is obviously going towards serial, well-standardized field buses, which can tolerate e.g. hostile industrial environments. If the subsystem has a general-purpose nature, the same component can be used elsewhere in the machine, too, to implement other required functions. This kind of strategy eases the system level planning, as well as spare part service and troubleshooting. A general-purpose device may also be very economical, as it can be manufactured in larger scale.

Valves in fluid power applications

Typical electrically controlled on/off valves are inadequate in those fluid power applications where a high performance operation is required. Proportional and servo valves are more sophisticated electrically controlled valves having faster and more accurate response than on/off ones. Their output is directly proportional to input guidance, enabling e.g. smooth acceleration and decelerat-ion of hydraulic cylinders and motors.

The difference in their names comes from the applications to which they were developed: the servo valve was obviously made for closed loop servo systems, while the proportional valve was intended to replace manually operated hydraulic valves in applications where a remote control was desirable. The properties of most of today's proportional valves are good enough for cost effective servosystems, so the difference between proportional and servo valves is fading all the time. Nowadays, the essential differences between these valves concern only their structural details, so there is no longer any actual need for two separate names, but they have been established in use.

All proportional valves and most servo valves are commanded by so called valve amplifiers which very often are implemented by analog electronics, and usually made by the valve manufacturer. In some cases the amplifier is built inside the valve, but typically it is a separate device, see Fig. 1. Amplifiers of different manufacturers are not compatible, although they are basically quite similar, differing only in details.

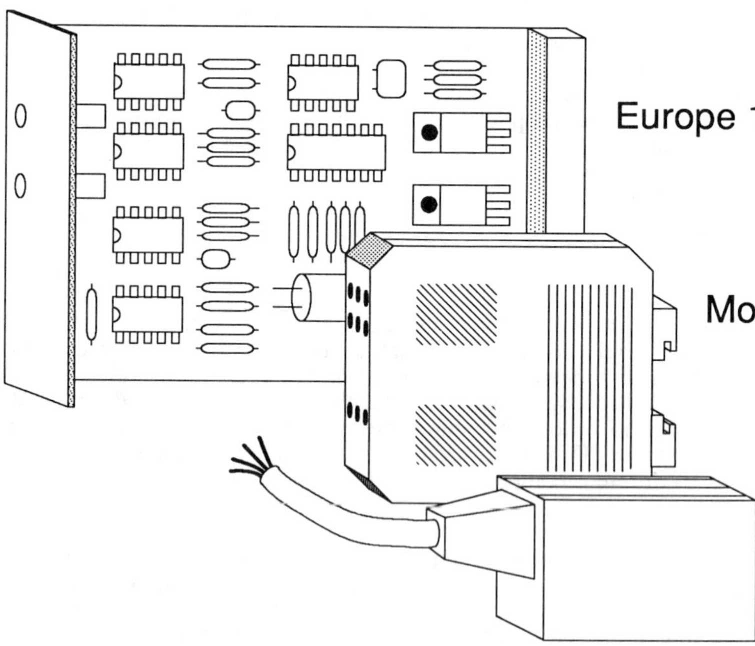

Fig. 1. Appearance of present valve amplifiers. Europe-1 is the most common whereas the module alternative is becoming more popular. Plugs are simpler amplifiers, suitable for low cost solutions.

Present valve amplifiers are typical current amplifiers, added with functions like ramps, dead zone compensation etc. to improve their applicability to proportional and servo valves. The basic design of today's amplifiers was developed over ten years ago, and this proven structure has been used in new amplifiers ever since, because it works well in simpler applications. This is still true, however, these amplifiers are becoming more and more limited when compared with modern control systems.

When more properties are required from a valve, a drive or a servosystem, several additional cards are needed. The amplifier inputs are practically always transferred as an analog signal, usually voltage. When a control system (e.g. a programmable logic controller) commands many valves and servosystems located in different places, the wiring scheme becomes complex, expensive and sensitive to noise.

The biggest drawbacks of today's amplifiers are:

1. A certain controller can handle only a few valve types.
2. The commanding signal is voltage, which means that:
   a) you can only give commands. You can not ask for information.
   b) voltage is susceptible to noise.
3. To get a typical additional function you need additional card(s).
4. Connectors are not compatible.
5. Parameters and set points are given with trimmers.
6. The tolerance of different ambient conditions are poorly specified.

ABOUT THE NEXT GENERATION OF VALVE ELECTRONICS

Traditional valve amplifiers are still selling well, so there may seem no need to develop more advanced electronics. But in order to keep hydraulics competitive, its amplifiers should meet the requirements of the future, and probably their next generation will be based on microprocessors. The whole concept of 'amplifier' needs updating, and because the next generation devices will be so much more advanced than today's amplifiers, the term '**controller**' would perhaps be more appropriate.

Processor-based control systems can be made very reliable but not without careful design and testing. One prerequisite for a thorough planning is that the volumes of production are large enough, which in hydraulics means that **the controller should be a general-purpose device - not a valve-specific amplifier.** A general-purpose device can also be produced in larger scale, which gives several advantages, as described earlier. It may be economical to implement parts of the controller as ASIC (Application Specific Integrated Circuit) circuits.

The controller should have the following features:

1. Its configuration can be changed by software (no jumpers, switches or additional parts) to meet most hydraulic proportional and servo valves.
2. It has a bi-directional, well specified serial field bus to give commands and ask for e.g. diagnosis information.
3. The most typical additional functions can be accomplished without extra components.
4. Connectors are compatible.
5. Parameters and set points can be given in numerical form.
6. The controller withstands hostile environments as specified.

The one who enters the market first can lay the foundations of a possible industrial standard. The most important thing, which may not be solved without market forces, is the communication standard. Usually the communication standard is ruled by the (factory, etc.) control system which the (valve) controller interfaces. That is why there should be several controller models, one for each standard, which is not an inordinate demand if the amount of notable field buses is reasonable.

CONFIGURABLE CONTROLLER

A prototype of the controller outlined above is under development in the Technical Research Centre of Finland, Machine Automation Laboratory. One important aspect in the project is to find a good compromise between the reliable simplicity and all the possibilities modern electronics offers.

The project began with a survey of the expectations which designers, users, salesmen and research engineers had about the progress of valve amplifiers. Generally, people agreed with the opinions above. But with regard to the particulars, the points of view are so different in mobile hydraulics and in industrial hydraulics that they can not be satisfied with only one kind of controller. However, two different kind of devices might satisfy most requirements adequately, namely a *valve controller* for mobile applications, and for industrial use a *configurable controller*, prototype of which will be realized according to Fig. 2.

In mobile sector the capability to control one stage proportional valves with no electrical feedback was considered significant; other valve types were not so important. Module size (see Fig. 1), low cost, simplicity, and of course reliability were preferred. Knowing the progress in microelectronics it may be possible to implement the whole valve controller with one or two ASIC chip(s) within a few years which promotes its integration in valve body.

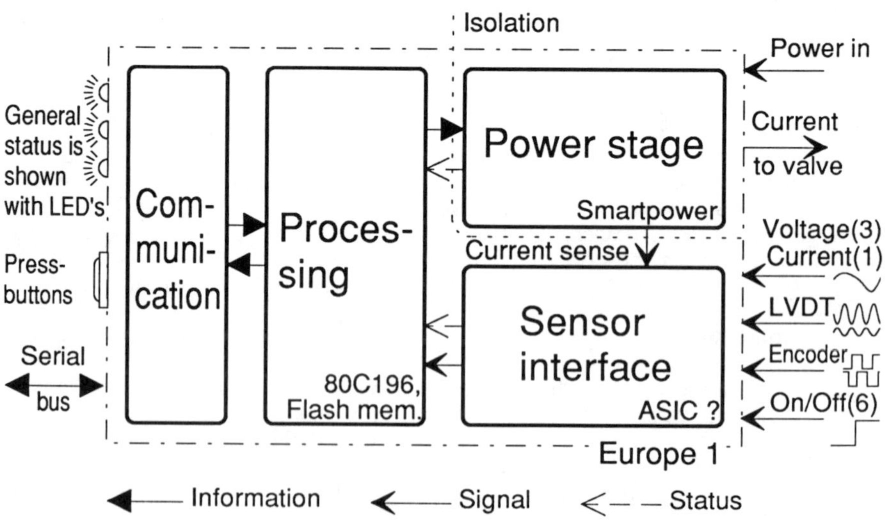

Fig. 2. The prototype of the configurable controller.

In industrial hydraulics the most important properties were good serviceability, and that it can be put to use easily. These were favored even if it means a slight increase in price. The device should be capable of controlling almost any one or two stage servo valves, and all one or two coil proportional valves. Since the controller is going to be programmable, it can be programmed to use its input signals as required, for instance to take valve set points from voltage input. In this configuration it can be used to replace existing valve amplifier applications. A servo system can be implemented by programming the controller to take the set point via its serial bus interface, to read e.g. encoder interface, to run the control algorithm, and to give the command to the valve.

The power stage of the controller must be adjustable so that it can supply electrical current in a way the valves require it. New 'SmartPower' components are used in the power stage which is capable of providing continuous currents up to 3 Amperes for fast proportional valves, but can also supply 20 mA accurately enough for two stage servovalves. The power stage is totally isolated, so if it fails, it will not damage other components of the controller.

The sensor interface consists of 3 voltage inputs, 1 current signal input, 1 LVDT (Linear Variable Differential Transformer) interface, 1 encoder interface, 6 digital on/off signal inputs, and one for emergency stop. The encoder interface supplies +5 V for the encoder, and the LVDT interface provides an oscillation signal for proper operation of the LVDT sensor. Current sense -signal is also handled here. At present, the sensor interface must be realized by many simple standard components, so it is an obvious subsystem for ASIC implementation, if a large scale production is considered.

The processor must be reasonably fast so that it can manage the controller efficiently in real time. A fast 16*16 multiplication, PWM (Pulse Width Modulation) output, facilities for serial communication, watchdog, extended temperature range, A/D converter, and capability for stand-alone operation were the characteristics examined from several processors, e.g. i80C196, NS46104, NEC78310 and 78320, MC68332 and TI320E14. I80C196 was chosen, because it covers well the features above, and has a good development support. Variables will of course be kept in static RAM, while flash memory will be used for program memory, and it can be reprogrammed via the serial bus. A log memory is useful in complicated systems, because it helps in e.g. maintenance and troubleshooting. A new memory technology called FRAM [2] seems a good choice for log memory, but since the availability of FRAM was questionable when hardware decisions were made, we decided not to implement any log at all.

Communication interface is dictated by the system, to which it is connected, as mentioned earlier, but it also depends on the commands we apply to the controller. Sophisticated continuous path guidance, which is used e.g. in robotics, requires high data transfer rates. A typical hydraulic application can typically be characterized as point-to-point control strategy, requesting only reasonable data transfer rates.

There are several field buses which can be applied to the point-to-point concept. They are typically industrial, not official standards, often based on RS-485 or RS-232, so the microcontroller can handle the interface with software and its own UART (Universal Asynchronous Receiver Transmitter). One example is Modbus [3] which was developed about 20 years ago to be used in Gould's programmable logic controllers. By using 19.2 kbits/s it can transfer 6 data bytes in 8.6 ms. Modbus and RS-485 are potential candidates for the serial bus interface implementation of the prototype.

DISCUSSION

Integrating valve electronics into the valve body is an increasing tendency. Of course, careful planning is required when developing reliable electronics for field environment, but the valve body - protective, only moderately vibrating housing cooled by hydraulic oil - is not an infeasible site for electronics. The wiring of valves becomes easier, especially those with electrical feedback. Also the quality of valves can be made more regular by adjusting valve electronics in production line to compensate some imperfections in valves hydraulic characteristics.

However, a separate device is a better solution when controlling hydraulic cylinders or motors, because the needed sensor interfaces, connectors, etc., would increase the size of the integrated valve and make it very bulky. Therefore it is sensible to have a separate controller capable of handling systems, which e.g. consist of a valve, a cylinder (or a motor), and of some sensors.

A valve and its amplifier are typically manufactured by the same company. When something has failed, the responsibility questions have been clear. A general-purpose controller, which is not manufactured by any valve company, will easily become a 'scapegoat' in malfunction situations. The development in computer networks is towards open systems, and similar trends are rising also in field bus networks. An analogous progress would be more than welcome in the field of advanced hydraulics, too. If there would be a precisely specified interface between network and controller (the situation here is improving due to standardization), as well as between controller and valve, the responsibility questions would be quite clear, no matter who the manufacture is.

CONCLUSIONS

In order to keep hydraulics competitive, especially with electrical drives, the old valve amplifier concept must be updated. The present amplifiers do those tasks well, which they are designed for, but the problem is that those tasks are becoming more and more limited when compared with the capabilities of modern control systems. There is a need for a general-purpose controller for hydraulics: a subsystem accessible via serial field buses, and configurable for various tasks.

It seems that the development in car electronics, smartpower components, ASIC technologies, and in field buses will direct the development of the next generation of valve amplifiers. It would be sensible that all the next generation devices would be more compatible than the present valve amplifiers are. It will be interesting to see how the market forces and the technical development gradually gives form to the next generation controller; our prototype is one attempt to find out what kind of a device this next generation controller might be. Other microprocessor based controllers for hydraulics are also under development [4].

REFERENCES

1. H. Kirrmann, "Europe's industrial bus scene", IEEE Micro, Dec. 1988, pages 6 - 7.
2. Nonvolatile Ferroelectronic Technology & Products, Technical Report, 1988, Ramtron Corporation, 12 pages.
3. Gould Modicon Modbus Protocol Reference guide, 1983, Gould Electronics Inc.
4. Karash, M., "Neue Schritte in der Entwiclung von mobiltauglicher, digitaler Elektronik", O+P 'Ölhydraulik und Pneumatik', Nr. 11 (1989), pages 854 - 859.

MAGNETIC LEVITATION-BASED MICRO-AUTOMATION OF MECHANICAL PROCESSES IN SEMICONDUCTOR FABRICATION

I. J. BUSCH-VISHNIAC, S.-J. CHEN, I.-Y. A. WANG, S.-H. LI, and J. LEIFER
Mechanical Engineering Dept., The University of Texas, Austin, TX 78712

ABSTRACT

Some operations in semiconductor manufacturing require moving objects with submicron precision and high speed. Such mechanical processes have traditionally resisted automation. We discuss magnetic levitation approaches to automation of micron scale mechanical operations concentrating on the applications of magnetic levitation-based micro-automation in semiconductor fabrication and designs appropriate to specific applications.

INTRODUCTION

Ideally, automation of small scale processes in semiconductor fabrication will involve small scale equipment with high speed and precision. The required accuracy and repeatability will be dictated by the size of features on the semiconductor. Currently, feature sizes are about 1μm in size and the industry expects features of about 0.5μm or 0.35μm to become the standard within about five years. We refer to machines capable of micron or submicron precision as micro-machines, and to the entire automation process as micro-automation. At a minimum, micro-automation systems must be able to deal with material transport and mechanical and electrical probing. They must be compatible with clean room operation and depending on the process, might be required to withstand high temperatures and low pressures.

Micro-automation poses significantly different problems from automation on a large scale. In particular, large scale automation systems typically deal with sizable friction whereas micro-automation systems must minimize friction in order to obtain submicron precision and to avoid particulate production. Further, the typical goals in large scale automation are an ability to handle high payloads and to apply large forces, whereas in micro-automation one is more likely to be interested in the ability to deliver well controlled, delicate forces and to transport small payloads.

In recent years there have been attempts to examine various micro-machine technologies, including piezoelectric, hydraulic, magnetic, shape memory alloy, and electrostatic approaches. We are examining a technology capable of robust three-dimensional motion using magnetically levitated manipulators. The major advantages of levitation are that the manipulator can operate as a rigid body rather than using jointed parts, and that friction can be dramatically reduced. This means that position errors don't compound, the dynamic behavior is simple to model, fabrication is straightforward, and harsh environments need not pose a problem since the device can be coated. The arguments in favor of magnetic levitation rather than electrostatic levitation are that greater field energy densities may be achieved, that forces tend to be greater on scales of 10μm or larger, that it is more compatible with harsh environments (such as water) and with clean room operation since magnetic fields don't attract dust, and that it is less susceptible to wear problems. The serious potential drawbacks are that magnetic levitation-based micro-machines are nonlinear so control can be computationally intensive, require fast feedback for both stable levitation and motion control, and easily overheat due to the substantial power requirements.

APPLICATIONS

In order to effectively design a magnetic levitation-based micro-automation system it is necessary first to determine the task to be performed since the task will influence the system design by determining the scale lengths, numbers of degrees of freedom, speed, and accuracy required. We generated a list of potential applications which are described in detail in Busch-Vishniac[1]. Among the applications areas are three topics on which this paper focuses: positioning, analytical probing, and bond quality control.

Positioning refers to applications which involve high speed and accurate motions but little or no mechanical contact between the moving object and a fixed workpiece. Such tasks span several scales and include mask alignment, in which one desires total position errors of less than 100nm over an 8in range (accuracy of roughly 5×10^{-7}, where accuracy is defined as the needed resolution normalized by the range), and wafer transport, in which one moves wafers smoothly from one processing location to another without regard to motion accuracy except at the path endpoints.

Analytical probing of semiconductor devices permits electrical probing of the interior nodes. Such probing operations are used increasingly for design and fabrication process debugging. The speed of motion is not an issue, but the ability to locate multiple probes in close proximity (roughly 10μm separations), each with a resolution of 0.1μm is desired (accuracy of about 10^{-5}).

Bond quality control testing refers to the automated mechanical testing of the quality of the bonds made to semiconductor devices in the packaging process. Ideally one desires the ability to test nondestructively all bonds on all devices without slowing down production. Hence speed and delicacy are crucial issues. However, since bond pads are the largest features of semiconductor devices, only low accuracy (roughly 10^{-3}) is required for this operation.

The attributes of these three types of applications are summarized in Table I. Designs to address these applications are discussed in detail below.

Table I: Comparison of various applications of magnetic levitation micro-automation in semiconductor fabrication

Application	Degrees of Freedom	Accuracy	Speed?	Scale Length
Positioning - Mask Alignment	5-6 (3 trans., 3 rots.)	5×10^{-7}	Yes	20 cm
Positioning - Wafer Transport	1 (lineal trans.)	N/A	Yes	meters
Bond Testing	3 min. (x,y,z)	10^{-3}	Yes	2-3 cm
Analytical Probing	3 min. (x,y,z)	10^{-5}	No	2-3 cm

WAFER TRANSPORT

Magnetic levitation-based wafer transport is attractive when compared with standard belts, linear motors, and robot handling techniques because friction can be significantly reduced while maintaining compactness, ease of reconfiguration, and wide range operation. We are examining the use of one-dimensional transport paths which could connect work stations. The bulk of the path is comprised of track along which wafers move quickly and smoothly. The track endpoints are designed to permit an automated seamless handoff from a process station to the path and vice versa.

The transport path we have studied is shown in Fig. 1 and described in further detail in Wang et. al.[2]. This path uses six wires wrapped repeatedly in a single layer around an aluminum block. By controlling the currents in each of the six wires, we control the forces and moments felt by the four permanent magnets on the flat rectangular manipulator.

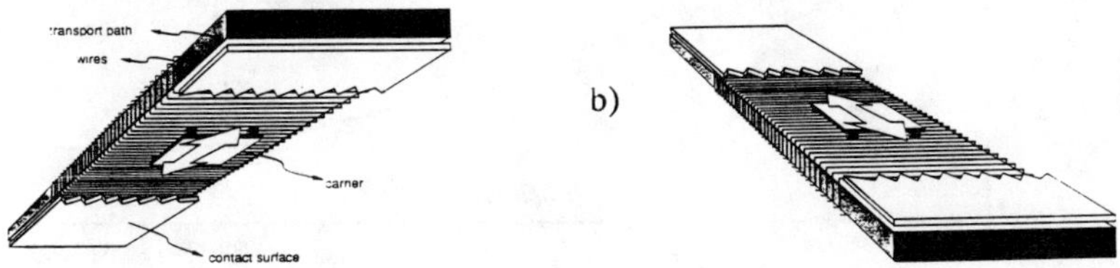

Figure 1: Magnetic levitation transport path with a) manipulator operating below and b) manipulator operating above the path.

One may determine the magnetic field **dB** established by a wire segment dS carrying current i using the Biot-Savart law:

$$\mathbf{dB} = \frac{\mu_0 i}{4\pi} \frac{\mathbf{dS} \times \mathbf{R}}{R^3} \quad , \tag{1}$$

where **R** is the location at which **dB** is determined, as measured from the center line of the wire segment, and μ_0 is the magnetic permeability. By summing the magnetic field contributions from each wire, we determine the total magnetic field **B**. We then determine the force **F** and moment τ felt by a point dipole moment **u** using

$$\mathbf{F} = (\nabla\mathbf{B}) \bullet \mathbf{u} \quad , \tag{2}$$

$$\tau = \mathbf{u} \times \mathbf{B} \quad . \tag{3}$$

As the manipulator is a rigid body, **F** and τ lead immediately to knowledge of the manipulator trajectory.

We operate the transport path with the manipulator moving either above or below the path. For below the path operation, we permit the manipulator to drop for an instant, and then activate the path to pull the manipulator forward and back into contact with the path. This step is quickly repeated to obtain smooth, rapid motion. For above the path operation, the path first is activated momentarily in such a manner as to produce a levitating force normal to the path. Then the manipulator is pulled forward as it returns to the path. Iteration of this sequence produces a rapid, smooth motion.

We have tested many varieties of this path for speed and smoothness as a function of payload. Our test path limits the current in each wire to no more than 3A, and uses a 0.176 gm manipulator. A typical result is shown in Fig. 2, which shows speed versus added weight of the payload (in terms of manipulator weight) for a configuration with the manipulator moving above the path, and the magnetic dipole moments of the manipulator magnets oriented along the path. Results for two current drive configurations are shown: two negative and four positive currents in the six wires (2:4), and three positive and three negative currents (3:3). Note that it is possible to carry payloads up to three times the carrier unloaded weight, which is roughly a factor of five greater than the maximum load capacity of U.S. made trucks.

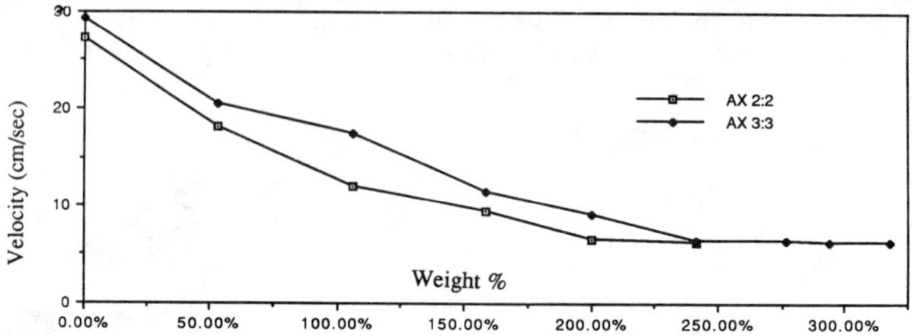

Figure 2: Speed versus payload for a typical transport path

In the path described above we stabilize the motion through frictional contact between the manipulator and the path. We are currently investigating a linear transport path which will yield stable, free-floating levitation, and will allow us to control the vertical location of the manipulator as well as its position along the path.

ANALYTICAL PROBE STATION

Analytical probing of semiconductors is very useful in microelectronics device research and development, where probe stations are used to determine the physical parameters of the device, and in failure analysis, where probing can isolate and determine the nature of a problem. Two general classes of analytical probing stations are currently available: manual or semiautomatic contact probers, and automatic noncontact probers. Manually controlled contact probe stations are difficult to use and can easily destroy the device being tested because the probes tend to skate as they contact the device. Noncontacting probe stations can pinpoint the location of a failure with great speed and accuracy, but cannot determine the nature of a fault since they cannot inject a signal. They are also typically an order of magnitude more expensive than contact probe stations. Magnetically driven analytical probers can rapidly position probes and then make contact with great delicacy (2-3gmf is needed to penetrate the oxide[3,4]).

Figure 3 shows a prototype design for a magnetic levitation analytical prober. The manipulator is about 35mm x 35mm x 45mm and holds 8 permanent magnets. The coils which drive the manipulator are set into a fixed frame which surrounds the manipulator, and are constructed in a manner similar to those used in the wafer transport path described above. The lower solenoids in Fig. 3 control the motion of the manipulator in the horizontal plane, and the upper solenoids control the vertical motion.

At the top of the manipulator there are infrared LEDs shining light onto lateral effect two-dimensional position sensitive devices fixed on the outer frame. They are employed to determine the precise position of the manipulator. The sensor signal processing circuit (described in Wang and Busch-Vishniac[5]) yields a resolution of better than $0.1\mu m$ with a working range of about 1mm in each direction.

The frame which surrounds the micro-manipulator is mounted on a coarse x-y positioner so that the total range of the system may be extended to a wafer-sized area. Operation of this system thus requires coordinated motion of the coarse and fine positioners. Such coordinated movement has been discussed by Taylor et. al.[6] and by Jeong[7], who showed that when the fine positioner mass is much lower than the coarse positioner mass it is possible to control each independently.

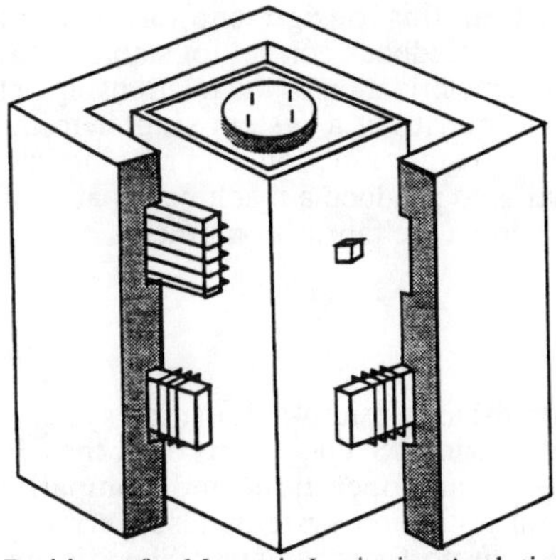

Figure 3: Fine Positioner for Magnetic Levitation Analytical Probe Station

Work is currently underway to construct and test a magnetic levitation-based analytical prober similar to that described here.

BOND TESTING

A significant number of the semiconductor devices which pass all current in-plant inspections fail shortly after being put into operation. It is suspected that substandard inner-lead bonds are a large factor in these early failures. Attempts to characterize bonds have found a correlation between mechanical pull-tests and bond-quality (see Harman[8] for example); however, the bond testers currently used by the semiconductor industry work off-line to test all of the bonds in only a very small sample of fabricated devices. These testers usually provide a known force to the bond, and record in a binary fashion whether the bond survives the applied stress. More useful bond quality testers, incorporating magnetically levitated actuators and position sensors to simultaneously pull on the bond and monitor its deflection, could be used to produce more quantitative information regarding the quality of the bond.

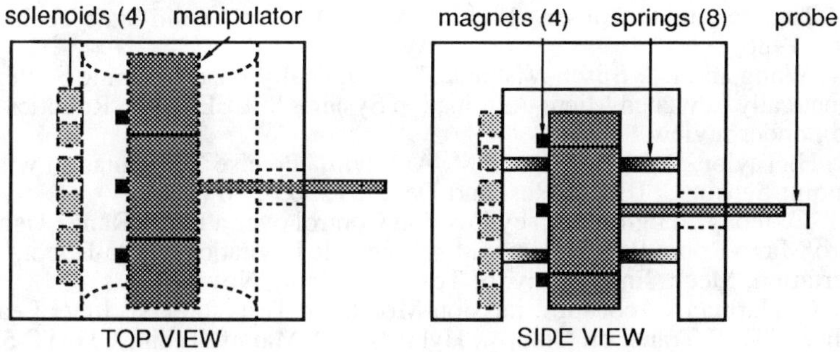

Figure 4: Test structure for determination of bond stress/strain ratio

Our work in this area has been divided into two stages. First we are testing nondestructively a number of bonds to determine how well one can correlate stress/strain ratios with "known" bond quality (bonds from a batch which have been graded statistically). This work is being accomplished using the system shown in Fig. 4. In this system conventional air-core solenoids are used to move a manipulator along a straight path. A probe on the manipulator supplies a known shear force to the bond while its deflection is continuously monitored. The cantilever springs used in this design support the manipulator, making it unnecessary to use real-time feedback control for stable frictionless operation.

Upon complete characterization of the relationship between bond stress/strain ratio and quality, we will construct a system similar to that shown in Fig. 4, but capable of three degree-of-freedom motion, and ranging in operation over die-sized areas. Our ultimate goal is to produce a machine capable of testing all bonds in all devices without slowing down the fabrication process.

CONCLUSIONS

Magnetic levitation-based micro-automation systems are well suited for use in semiconductor manufacturing. Such systems offer the advantages of reduction of friction, ease of analysis and operation, and compatibility with clean room environments. Magnetic levitation systems also permit integration of precision processing stations with transport mechanisms. Specific applications include wafer transport, analytical probing, and bond quality testing.

ACKNOWLEDGMENTS

The authors wish to acknowledge support from NSF and SRC (the Semiconductor Research Corp.). They also wish to thank all of their colleagues in the Micro-Automation Group of the University of Texas.

REFERENCES

1. I. J. Busch-Vishniac, "Applications of Magnetic Levitation-Based Micro-Automation in Semiconductor Manufacturing," IEEE Trans. on Semiconductor Manuf., accepted Nov. 1989.
2. I.-Y. A. Wang, S.-H. Li, and I. J. Busch-Vishniac, "A Magnetic Levitation Transport Path", IEEE Trans. on Semiconductor Manuf., under review.
3. "Special Report: Wafer Probing Equipment," Microel. Manuf. and Testing, **13**(4) (March, 1990).
4. M. Lautner and G. Bieser, "Probing ASIC Devices," Solid State Tech., **29**(8) (Aug., 1986).
5. W. Wang and I. J. Busch-Vishniac, "An Optically Based Sensing System for Magnetically Levitated Micro-Automation Systems", IEEE Trans. Robotics and Auto., under review.
6. R. H. Taylor, R. L. Hollis, and M. A. Lavin, "Precise Manipulation with Endpoint Sensing," IBM J. Res. and Dev., **29**, 363-376 (1989).
7. M.-C. Jeong, "High Accuracy Position Control over a Large Range Using a Macro/Micro Robot Pair System with a Magnetic Levitation Micro-Robot," Ph.D. dissertation, Mech. Engr., Univ. of Texas at Austin, Nov. 1989.
8. G. G. Harman, "Acoustic-Emission-Monitored Tests for TAB Inner Lead Bond Quality," IEEE Trans. on Comps., Hybrids, and Manuf. Tech., **CHMT-5**, 445-453 (1982).

APPLICATIONS OF PETRI NET FOR PERFORMANCE EVALUATION OF FMS

Shen Meie, Peng Wei, Wu Naiqi and Xue Jinsong

Shenyang Institute of Automation
Academia Sinica
Shenyang 110003, P. R. China

ABSTRACT

In this paper the Timed Event Graph is used to medel the FMS and evaluate the performance. The conditions under which a part sequencing is feasible and the relationship between the number of pallets in the system and the productivity of the system are presented. And the quantitative analysis of production period, flow time and work in process is made.

INTRODUCTION

Being suitable to describe the concurrent events, Petri Net is widely used to model the FMS and evaluate the performance [6][7]. Up to now a great attention is paid to the system description by usinvg Petri Nets, so various extended Petri Nets, such as temed Petri Net, colored Petri Net and stochastic Petri Net, are developed so as to increase the description functions.

In [4][6] the authors did other way, they used the timed Petri Net, whose description functions are relatively weak, to do the quantitative analysis of FMS, some meaniful results has been obtained.

Based on the results obtained in [4], in this paper, We used the timed Petri Net to model the FMS, and evaluate the performance, some new results are obtained.

THE TIMED EVENT GRAPH

For the definitions, properties and notations of the Petri Nets, Event Graph and Timed Event Graph can be found in [4]. Here we will give the basic theorems which will be used in this paper.

It is known that in a strongly connected and live Timed Event Graph the firing of transitions is periodic. Let Tr denote the period of directed elementary circuit r and t denote the time needed when the transition which take the longest time in r is fired, or

$$t_{m,r} = \max_{t_i \in r} \mu(t_i)$$

thereafter we call this transition as bottleneck transition, then we have the following result.

LEMMA In the Timed Event Graph for any directed elementary circuit r the following inequality holds.

$T_r \geqslant t_{m,r}$

Because of the space limit we omit the proof for all conclusions in this paper.

THEOREM 2.1 In a strongly connected and live Timed Event Graph there exists finite positive integer n_0, λ and k such that for all transitions t

$$s_t(n+k)=s_t(n)+k\lambda, \qquad n \geqslant n_0 \qquad (1)$$

and $$\lambda=\max_r \max(\mu(r)/M(r),\ t_{m,r})$$

where s (n) denotes the instant of the nth firing initiation of transition t, k is the periodicity factor , and

$$\mu(r)=\sum_{t_i \in r} \mu(t_i), \qquad M(r)=\sum_{p_i \in r} M(p_i)$$

In theorem 2.1 if all circuits contain only one taken , then k=1, $\lambda=\max_r(\mu(r))$, λ is a simple period. If k>1 is the average period. Clearly, the number of tokens in the circuits affects the period of a Timed Event Graph.

THEOREM 2.2 In the Timed Event Graph for any directed elementary circuit the following inequalities hold.

1) If $M(r) >= \mu(r)/t_{m,r}$, $T_r = t_{m,r}$

2) If $M(r) < \mu(r)/t_{m,r}$, $\lambda_r = \mu(r)/M(r)$ and the possible value of T_r is in $[t_{m,r}, \mu(r)-[m(r)-1]t_{m,r}]$.

Theorem 2.2 is for single circuit, and theorem 2.1 gives the period of overall net. It is clear intuitively that in a net the circuit which has longest period is dominant.

MODELING FMS BY TIMED EVENT GRAPH

Let R : $\{M_1^i, \ldots, M_{k_i}^i\}$ denote the process path of part R_i then it can be described by an event graph shown in Fig.1, where $\mu(M_{k_i}^i)$ for transition $t_{k_i}^i$ is the processing time for part R_i at machine

Fig.1

$M_{k_i}^i$, the places denote the buffers before machines and the number of tokens in a a place denotes the number of parts to be machined.

This event graph describes the processing process effectively.

If the process is a repeated process then a resource place r can be used to connect the input and output places, and the processing circuits are formed. It means that when a part is finished, it leaves the system, and the pallet which held the part before will be taken by anther part of the same type for processing. In this case the pallet is the critical resource provided the law matenal is infinite. If there are n parts to be machined then there are n processing circuits in the graph. Connecting all the transitions corresponding to the operations processed by the same machine according to the processing sequence and puting only one token in the circuit, this contructs the commad circuits. This Timed Event Graph obtained describes production system exactly and we the proformance of the system can be evaluated by analysing the Timed Event Graph.

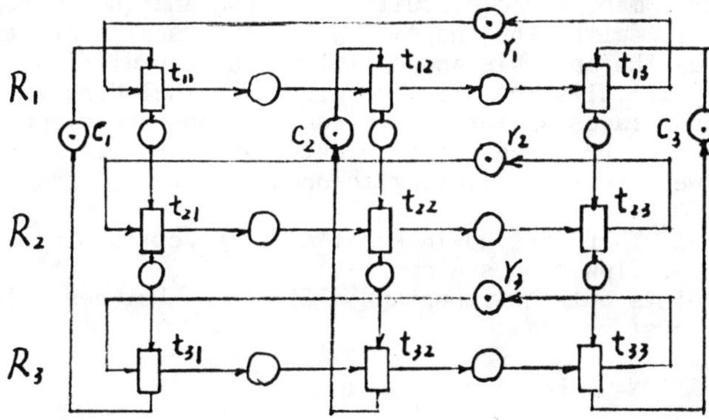

Fig. 2

Example 1 There are three types of parts R_1, R_2, R_3 and three machines M_1, M_2, M_3, and all of types of parts are to be processed by M_1, M_2, M_3 in series, or R_1, R_2, R_3: $\{M_1, M_2, M_3\}$, t_{ij} is the processing time for R_i at machine M_j. The Timed Event Graph for this system is shown in Fig.2.

Example 2 This is a job shop system and there are two types of parts R_1, R_2, and two machines M_1, M_2, and

R_1: (M_1, M_2), R_2: (M_2, M_1)

M_1: (R_1, R_2), M_2: (R_1, R_2)

The timed event graph for this system is shown in Fig.3.

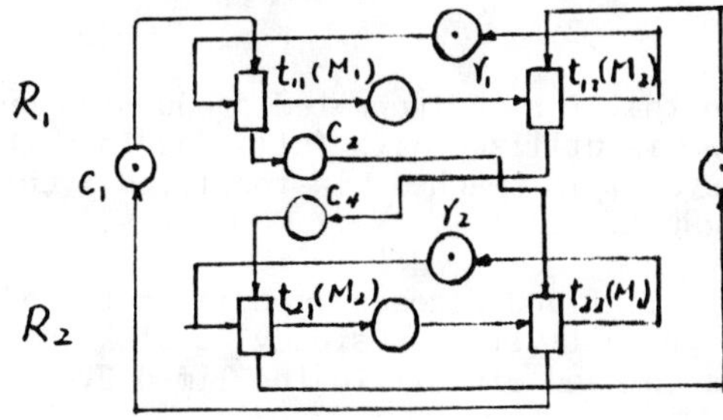

Fig.3

For the modeling we make several remarks as follows.

REMARK 1 The processing circuits give the sequence of the parts and the number of tokens in the resource place r_i denotes the initial number of pallets for R_i.

REMARK 2 In each command circuit there is only one token which is in one of the places in the circuit. This means that a machine can only process one part at a time. The command circuit and the place of token determine the sequence of the parts before the machine.

REMARK 3 The number of processing citcuits equals the number of types of parts and the number of command circuits is as same as the number of machines.

REMARK 4 Besides processing circuits and command circuits there is another class of directed elemetary circuits---the mixed circuit, it reflects the complex relation in the manufacturing system. There are many mixed circuits in an FMS and the number increases extremely with the increasing of the scale of the problem. This makes the analysis and optimization very difficult.

In the mixed circuits we call the circuits including only buffer and command places as internal circuits, and the circuits including at least one resource place as external circuits. From these properties we have the following theorem.

THEOREM 3.1 There are no internal mixed circuits in the Timed Event Graph of flow shop systems.

Theorem 3.1 points out the important difference between flow shop and job shop systems.

PERFORMANCE EVALUATION OF FMS

Sequence feasibility

Sequence feasibility is important for a manufacturing system. We have the following condition of feasibility.

THEOREM 4.1 A part sequence is infeasibile if and only if in the Timed Event Graph there exists at least one circuit which has no token.

But in flow shop infeasible sequence never occurs.

THEOREM 4.2 If a flow shop system has initial resources then any sequence in the process is feasible.

Productivity

It is well known that for the repeated production process if the bottleneck machine is utilized with full load then the maximal productivity of the system is reached. In the Timed Event Graph we have the following result.

RESULT The sufficient and necesary condition for a system to reach the maximal productivity in steady state is that the critical circuit is a command circuit in the Timed Event Graph of the system.

If the condition above is satisfied then the period of this circuit is the period of the system. For the FMS of flow shop, we have the following theorem.

THEOREM 4.3 In an FMS of flow shop, for any part sequence there exists a distribution of finite pallets such that the maximal productivity of the system is reached.

If for some reason, for example, the buffers are limited then the productivity will be affected.

COROLLARY In the flow shop any sequence is optimal in the sence of steady state productivity, provided there are no other limits.

But for job shop system this is not true.

THEOREM 4.4 For job shop systems it is not always possible to reach the maximal productivity by increasing the number of pallets.

Other performance indeces

Machine Utilization Machine utilization is defined as the ratio of real processing time to the total production time. The steady state utilization of machine j can be given by

$$u_j = \sum_{i=1}^{n} t_{ij} / \lambda \qquad i=1,\ldots,m$$

where m is the number of machines in the system, and n is the number of parts.

Clearly, when the bottleneck machine b is in full load then

$$\lambda = c_0 = \sum_{i=1}^{n} t_{ib}, \text{ and } u_b = \sum_{i=1}^{n} t_{ib} / \lambda = 100\%$$

When the bottleneck machine is not in full load, the loss factor is $(1-c_0/\lambda)$. In this case $\lambda > c_0$, so $c_0/\lambda < 1$.

Productivity Productivity of the system μ is the number of parts machined by the system in a time unit, we have

$$\mu = n/\lambda$$

When the bottleneck machine is in full load then $\mu = n/c_0$.

Flow Time From the processing circuits we know that the flow time of a part T is dependent on the number of pallets in the processing circuits and system period . So the flow time of the parts in processing circuit r is given by

$$T_{f_i} = M(r_i)\lambda$$

Work in Process When a part is finished it is unloaded from the pallet, then a new part is loaded into it, or at any time all pallets are not free. So the work in process is given by

$$n_{w_i,p} = \sum_{r=1}^{n} M(r)$$

THE NUMBER OF PALLETS IN THE SYSTEM

From the analysis of the previous section we see that the maximal próductivity, machine utilization and work in process are all dependent on the number of the pallets in the system. It is desired that the maximal productivity is reached and mean while the pallets used is as few as possible. In [4] and [6] some algorithms have been proposed to solve this problem. But they are complex and cannot meet the demand of application in the practice. In this section we will give the upper and lower bounds of the number of pallets used.

THEOREM 5.1 For the flow shop system of repeated production with 2 machines and n parts the upper and lower bounds of the number of pallets needed are n+1 and n repectively.

Furthermore we have

THEOREM 5.2 For the flow shop system of repeated production with m machines and n parts the upper bound of number of pallets needed is m+(n-1)(m-1).

CONCLUSIONS

In this paper we use the Timed Event Graph to model the FMS and analyse the system. Based on this model we evaluate the main performances of the system such as productivity, production period, work in process, flow time and so on, and give the upper and lower bounds of the number of pallets need for reaching the maximal productivety.

But using this model we can only consider the common pallet in analysing the number of pallet and bounds given in this paper are conservative. Basides further research of how to consider the machine breakdowns and the transportation time between machines by this model is needed.

REFERENCES

1. J.L.Peterson, "petri Net theory and the modeling of system", (Prentice--Hall, Inc., 1981).
2. W. Reising, "Petri Net, An Introduction", (GMD, 1985).
3. C.V.Ramamoorthy et al., "Performance evaluation of asynchronous concurrent systems using Petri Nets", IEEE Trans. on Software Ingineering, vol.5E-6(5) (1980).
4. H.P.Hillion, J,M.Proth, "Performance evaluation of job-shop systems using timed event-graphs".
5. M. Shen and K. Liu, "Performance evaluation of job shop systems based on timed event graph", Proceedings of 1988 DEDS meeting (1988).
6. M. Shen, "Applications of Petri Net theory in FMS", Master Thesis, Shenyang Institute of Automation (1989).
7. W.G. Zhai and Y.P. Zheng, "The derelopment of applications of Petri Net theory in DEDS", Proceedings of 1988 DEDS meeting (1988).

Optimal Equipment Selection for Flexible Manufacturing Systems Using Closed Queueing Network Theory

Ulrich A. W. Tetzlaff
Technische Hochschule Darmstadt, Federal Republic of Germany

ABSTRACT

A mathematical programming model and its solution procedure to perform optimal equipment selection for a flexible manufacturing system is shown, i.e. the selection of the types and numbers of CNC-machines, load/unload stations, transportation vehicles etc. and the number of pallets. The objective is to minimize equipment costs whilst fulfilling the production requirements for an average period.

INTRODUCTION

In the past, several models for the optimization of flexible manufacturing systems were developed. They can be divided into models which consider static and those which consider dynamic system behavior. Models which refer to static system behavior neglect the queueing processes in front of machines as well as part flow interactions. These models roughly estimate the system's throughput by the bottleneck machine(s), i.e. the machine(s) with the highest workload. Models which can be assigned to this category are for example that of Graves and Lamar [3,4], which optimizes the equipment selection and operation assignment for flexible assembly systems, or that of Whitney and Suri [12], which performs equipment and part selection.

Alternatively, if dynamic system behavior, i.e. queueing processes and part flow interactions are included, models can determine the system's throughput more precisely by using queueing network theory. Models of this kind are that of Vinod and Solberg [11] or of Dallery and Frein [1], which both perform capacity optimization to achieve minimal server costs (costs for machines, load/unload stations, vehicles etc.) whilst fulfilling production requirements. Furthermore Shanthikumar and Yao developed models which perform the optimal assignment of a given amount of servers, either to maximize throughput [7] or to maximize profits [8]. So far none of these models which consider dynamic system behavior have allowed the selection of equipment, i.e. the determination of the capacities for given types of equipment was possible, but not the inclusion or exclusion of certain types of equipment. However, in practice this is of major importance, because the system designer can often choose between different types of equipment with different costs but very similar capabilities (for a complete overview of design models for flexible manufacturing systems see [10]).

MODEL FORMULATION

This new model for optimal equipment selection incorporates dynamic system behavior. Different pieces of equipment are selected. The objective of the model is to minimize in-process inventory costs and equipment costs, i.e. costs for the cells, the transportation system and pallets. Each cell consists of one or several servers, i.e. identical CNC-machines, load/unload machines etc..

Operating costs are neglected (see [10] for an extended model which also includes operating costs). The flexible manufacturing system is configured so as to allow a given minimum production rate R_{min} of a given part family to be attained. Here the part family consists of a set of parts produced over a longer time horizon, i.e. an average period. To allow the inclusion or exclusion of certain types of equipment, the part types can be produced on one or several routes. A part route consists of a set of workloads requiring certain types of equipment, i.e. CNC-machines, load/unload stations etc. for production. To reduce data and thus the size of our model, part types which are produced on the same (alternative) route(s) and with almost identical workloads at each cell (for each route) are aggregated to a part group. Furthermore, the product rate (the fraction $frac_g$ each part group g has to have of total production T) is given and the number of pallets is restricted.

Indices:

b	:	index for number of servers (machines, set-up tables, carriers etc.)
g	:	part group index
m	:	index for cells or transportation system
r	:	index for routes

Decision variables:

N	:	number of pallets in the system
q_{gr}	:	flow fraction of part group g on route r
z_{bm}	:	binary integer variable which is one if cell (transportation system) m with the amount of b machines (vehicles) is included in the configuration of the flexible manufacturing system.

Parameters:

$D_N(\underline{q})$	:	delay, i.e. throughput time, when N pallets are in the system
CC_g	:	capital costs for a part of part group g
$frac_g$	:	product rate, i.e. the fraction part group g has of total production
I_{bm}	:	equipment costs for cells or transportation systems having b machines or respectively b vehicles
IP	:	pallet costs
IFI_g	:	fixture costs for parts of part group g
N	:	number of pallets
N^b	:	lower bound for the number of pallets
N_{max}	:	maximal number of pallets
N_+	:	ABA-bound estimate of N
R_{min}	:	required production rate for the system
SW_{min}	:	minimum workload of the system
$T(N,\underline{z},\underline{q})$	:	throughput rate of the system
W_{gmr}	:	workload of part group g at cell m on route r

Objective function:

$$\min \sum_{m=1}^{M} \underbrace{\sum_{b=1}^{B_m} I_{bm} \cdot z_{bm}}_{\text{cell or transportation system costs}} + \underbrace{IP \cdot N + N \cdot \sum_{g=1}^{G} (IFI_g + CC_g) \cdot frac_g}_{\text{pallet costs, fixture costs and in-process inventory costs}} \quad (1)$$

Constraints:

$$\sum_{r=1}^{R_g} q_{gr} = frac_g \qquad \forall g, \quad \text{with} \quad \sum_{g=1}^{G} frac_g = 1 \qquad (2)$$

└ product rate of part group g
fraction of part group g which is produced on route r

$$T(N,\underline{z},\underline{q}) \geq R_{min} \qquad (3)$$

└ minimum required production rate
└ throughput rate of the system

$$N \leq N_{max} \qquad (4)$$

└ maximal number of pallets
└ number of pallets in the system

$$0 \leq q_{gr} \leq 1 \qquad (5)$$

$$z_{bm} \in \{0,1\}, \; N \in Z^{+} \qquad (6)$$

In the objective function the annual equipment costs for cells and the transportation system are considered first. The binary decision variable z_{bm} decides whether b machines, set-up tables or vehicles of type m causing I_{bm} costs are included in the system. Here it is assumed, that cell or transportation system costs I_{bm} increase with higher b.

Next the annual pallet costs IP for one pallet are multiplied by the number of pallets N in the system to yield total pallet costs.

To obtain annual in-process inventory costs, first the capital costs CC_g for a single part of part group g is multiplied by the fraction $frac_g$ each part group has of total production. The summation over all part groups can be considered as the average annual inventory costs for one part. Multiplying it by the average number of pallets N which are constantly in the system yields the annual in-process inventory costs. Fixture costs IFI_g are obtained in an analogous way .

In the first constraint set (2) the summation of all fraction q_{gr} over all routes of part group g must be equal to the fraction $frac_g$ part group g has of total production.

Next the throughput rate $T(N,\underline{z},\underline{q})$, which is evaluated by closed queueing network theory [2], is equal to or greater than a required production rate R_{min} (3), and in (4) the number of pallets is less than or equal to the maximal number of pallets N_{max} allowed in the system.

Finally the decision variables are defined in the remaining constraint sets, i.e. q_{gr} is a real number between one and zero, z_{bm} is a binary variable and N is a positive integer number.

PROCEDURE FOR SOLUTION

The suggested procedure for solution is to use implicit enumeration. First a lower bound for the necessary equipment is evaluated. Based on this lower bound, a feasible solution is generated to provide an upper bound. Then, for an improvement of the feasible solution is searched. Finally the procedure enumerates over all possible configurations between the two bounds to improve the best solution found so far, which is at the beginning the improved feasible solution and at the end the optimal solution.

STEP 1: The lower bound vector can be generated by ABA-bounds (on ABA-bounds see [13]). However, depending on the part flow selected, different workloads can be assigned to the machines, load/unload stations, transportation vehicles etc. and thus different ABA-bounds are possible. Of these bounds, the one with the lowest server costs must be chosen. The following simple linear programming model is designed to find the lowest cost bound (for a proof see [10]).

Objective function:

$$\min \sum_{m=1}^{M} \sum_{b=1}^{B_m} l_{bm} \cdot z_{bm} \tag{7}$$

Constraints:

$$\sum_{g=1}^{G} \sum_{r=1}^{R_g} x_{gr} \cdot W_{gmr} \leq \sum_{b=1}^{B_m} b \cdot z_{bm} \qquad \forall\, m \tag{8}$$

- $\sum_{b=1}^{B_m} b \cdot z_{bm}$: number of machines/transport vehicles of type m
- W_{gmr}: workload of part group g at cell/transportation system of type m on route r
- x_{gr}: production rate of part group g on route r

$$\sum_{r=1}^{R_g} x_{gr} = frac_g \cdot R_{min} \qquad \forall\, g \tag{9}$$

and (5),(6)

Thus the model can be interpreted as a generator to find the lowest cost ABA-bound.

For the number of pallets the lower bound is given by:

$$N_+ = SW_{min} \cdot R_{min}$$

STEP 2: To obtain a feasible solution the configuration vector $\underline{z}_+$ of the lower cost bound vector is taken. Based on the following Theorem 1, the number of pallets is increased in steps, starting from N_+, until the required production rate is met.

Theorem 1: The ∞-monotonicity property holds for the throughput function of closed queueing networks with flow-optimal workloads W^* for each number of pallets N:

$$T(N, W^*) \leq T(N+1, W^*) \qquad \forall\, N$$

Proof: For constant workloads this was already proved by Suri [9]. Hence, if with each increase in the number of pallets, the workload is changed in such a way that the throughput is improved, Theorem 1 must hold.∎

For each new N the delay, i.e. the throughput time of the system is minimized. By Little's law this is equivalent to a throughput maximization:

Objective function:

$$\min_{\underline{q}} D_N(\underline{q}) \tag{10}$$

Constraints:

(2) and (5)

A reformulated version of the FDC-Algorithm of Kobayashi and Gerla [5] is taken as a solution procedure for this model (a complete description can be found in [10]).

To reduce the number of times this model must be applied until the required production rate R_{min} is reached, a lower bound N^b for the number of pallets to achieve R_{min} can be provided:

$$N^b = \frac{R_{min} - T(W^*,N)}{T(W^*,N)-T(W^*,N-1)} + N \leq N^*$$

It is derived by dividing the remaining gap between the actual and the required production rate (R_{min} - $T(W^*,N)$) by the increase in the production rate caused by adding the last pallet. However, it is assumed, that the following conjecture holds.

Conjecture 2: The throughput function of closed queueing networks with multiple-server stations and optimal workloads W^* is concave in the number of pallets N (see fig.1).

This was proved by Shanthikumar and Yao [6] for constant workloads W.

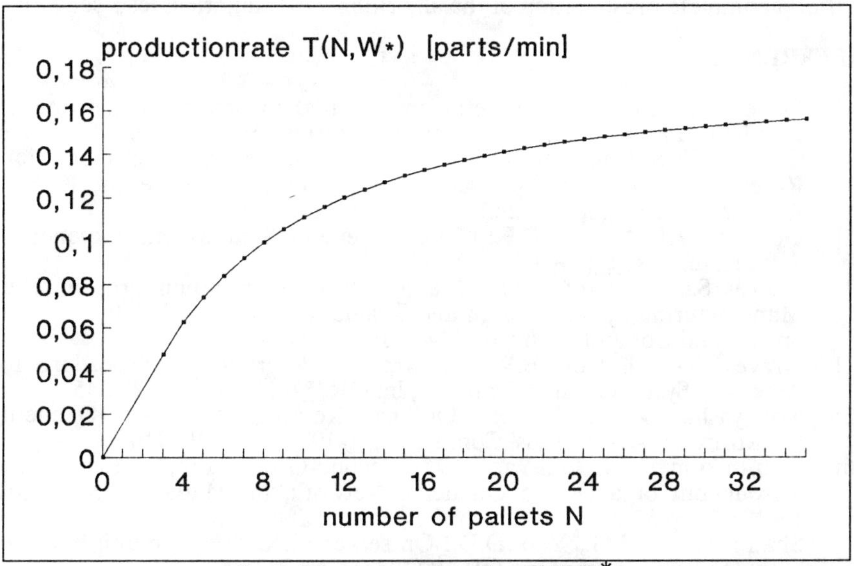

fig. 1: Throughput function T(N,W)*

When the required production rate is reached, it is known from Theorem 1 that this must be the cost optimal number of pallets N^* for the given configuration vector $\underline{z}$. However, it is possible that the maximal number of pallets is exceeded before the required throughput is obtained, or that the lower bound of the number of pallets N^b is larger than N_{max}. In that case the bottleneck cell is increased by another server, the number of pallets reset to N_+, and throughput is again maximized while the number of pallets is increased in steps. This is repeated until the first feasible solution is found.

STEP 3: Next an improvement of the feasible solution is tried, when the configuration vector $\underline{z}$ of the feasible solution differs from the static solution $\underline{z}_+$. This is done by a simple exchange procedure, i.e. from each type of machine, vehicle etc. one server is subtracted (if possible) and added successively to each of the other types. Each time, the optimal flow $\underline{q}$ and number of pallets N is evaluated in the same way as before for the feasible solution, i.e. the number of pallets is increased in steps and the throughput maximized. Moreover at each step it is checked whether system costs with the lower bound number of pallets ($\underline{z}$,N^b) are larger then system costs for the best solution found so far, or whether the lower bound number of pallets N^b is larger than N_{max}. If one (or both) of these conditions is (are) fulfilled, the configuration is discarded. Otherwise the number of pallets is increased and throughput maximized until either one (or both) of these conditions is (are) fulfilled, or the minimal production rate R_{min} is achieved and thus an improved feasible solution is obtained.

STEP 4: In the last step an implicit enumeration is performed. Starting from the $\underline{0}$-vector, all possible combinations of server configurations $\underline{z}$ are checked. A lower cost bound on the servers is provided with the equipment costs of $\underline{z}_+$ from step 1, an upper cost bound on system costs by the improved feasible solution of step 2+3. In addition a check is made to see whether all parts can be produced on the current configuration. When all these checks have been completed, the procedure continues in the same manner as in step 3 and evaluates the optimal flow vector $\underline{q}$ and the number of pallets N. Provided Conjecture 2 holds, in the end the best solution found is the optimal solution.

For a complete presentation of the algorithm with numerical tests see [10].

REFERENCES

[1] Dallery, Y., Frein, Y.: An efficient Method to determine the optimal Configuration of a Flexible Manufacturing System, in: Proc. 2nd ORSA/TIMS Conf. on Flexible Manufacturing Systems: Operations Research Models and Applications, edited by K.E. Stecke and R. Suri, Amsterdam 1986, pp.269-282

[2] Gordon, W.J., Newell, G.F.: Closed Queueing Systems with Exponential Servers, in: OR, 15(1967)2,pp.254-265

[3] Graves, S.C., Lamar, B.W.: A Mathematical Programming Procedure for Manufacturing System Design and Evaluation, in: Proc. IEEE Int. Conf. on Cir. and Comput., 1980, pp.1146-1149;

[4] Graves, S.C., Lamar, B.W.: An Integer Programming Procedure for Assembly System Design Problems, in: OR, 31(1983)3, pp.522-545

[5] Kobayashi, H., Gerla, M.: Optimal Routing in Closed Queueing Networks, in: ACM Trans. Comp. Syst., 1(1983)4, pp.294-310

[6] Shanthikumar, J.G., Yao, D.: Second-Order Properties of the Throughput of a Closed Queueing Network, in: Mathematics of OR, 13(1988)3, p.525

[7] Shanthikumar, J.G., Yao, D.D.: On server allocation in multiple center manufacturing systems, in: OR, 36(1988)2, pp.333-342

[8] Shanthikumar, J.G., Yao, D.D.: Optimal server allocation in a system of multi-server stations, in: MS, 33(1987)9, pp.1173-1180

[9] Suri, R.: A Concept of Monotonicity and Its Characterization for Closed Queueing Networks, in: OR, 33(1985)3, pp.606-624

[10] Tetzlaff, U.: Optimal Design of Flexible Manufacturing Systems, Dissertation der Technische Hochschule Darmstadt, Institut für Betriebswirtschaftslehre, Darmstadt 1990

[11] Vinod, B., Solberg, J.J.: The optimal design of flexible manufacturing systems, in: IJPR, 23(1985)6, pp.1141-1151

[12] Whitney, C.K., Suri, R.: Algorithms for Part and Machine Selection in Flexible Manufacturing Systems, in: Annals of OR, 3(1985), pp.239-261

[13] Zahorjan, J., Sevcik, K.C., Eager, D.L., Galler, B.: Balanced Job Bound Analysis of Queueing Networks, in: Comm. ACM, 25(1982)2, pp.134-141

Exception Handling and Task Level Programming of Flexible Manufacturing Systems

G.R. Meijer* **and U. Negretto****

* FWI, University of Amsterdam, Netherlands

**IPR, University of Karlsruhe, FRG

ABSTRACT

The objective of the work described in this paper is to integrate exception handling mechanisms into production cell control. A production cell is a Flexible Manufacturing System (FMS) with several active and passive components. A hybrid architecture model for an autonomous cell control is presented. In this architecture production planning and dispatching are embedded in a hierarchical structure and exception handling is integrated in a non-hierarchical structure. From the outside the cell controller behaves as a level in the hierarchical factory control model.

1. INTRODUCTION

The application of robot systems in a production cell is hampered by the sensitivity of the robot for exceptions such as calibration and operational errors. Because the robot is part of the production process, an exception occurring at the robot level can have direct consequences for the operation flow in the production cell. The cell control system will have to evaluate these consequences and take appropriate actions. The goal of these actions is to circumvent the use of the faulty component and to continue production with the available components. Similarly, if the exception has been dealt with, the cell control system has to take up the component in the production process again. For these reasons, it is our view that exception handling mechanisms have to form an integral part with the cell control system. The first part of the paper deals with architecture models for FMS control. Existing control architectures are reviewed and a new model is proposed. The second part addresses the exception handling functions and their integration in the proposed model.

2 CONTROL ARCHITECTURE AND EXCEPTION HANDLING (OVERVIEW)

There are three possible mechanisms to react to failures occurring in cell process execution. We argue that the order in which these mechanisms are applied is application dependent. To apply the proper mechanism at the right moment, an exception handling control function is forseen.

2.1 Models for Manufacturing Planning and Control

Basically models for manufacturing planning and control can be devided in hierarchical and non-hierarchical models.
A global reference model for manufacturing planning and control is proposed by Biemans[1]. Meystel [2] describes a functional model for sensor driven robotics which decomposes the control problem in three areas: sensor data processing, sensor data interpretation/representation and planning and control. A similar model is the NASA Reference Model (NASREM). It is based on a strict hierarchical structure and used as reference for developing software for space applications. In the Esprit project 623 a control architecture is developed for single components and assembly cells [3,4]. In this model three levels are identified. At the highest level a global cell task is decomposed into tasklists for all active components (robots, agv´s) in the cell. On the intermediate level, the active components are executing the tasks given by the cell controller. The tasks are expanded into a sequence of elementary operations (EO). At the lowest level of the hierarchy, the EO's are executed. An alternative approach to the architectural problem is coming from the mobile robots community. This approach is not based on a hierarchical decomposition of control functions, but on a decomposition of task achieving behaviours. For an overview see [5]. Brooks [6] proposes such a decomposition for the control of a mobile robot. Although this poses an interesting alternative to the classical hierarchical control models, some questions have to be clarified. It has to be studied how a global planning (which is needed in for example assembly tasks or inspection) can be achieved. Currently these models are developed and applied at low level control problems.

2.2 Exception handling

Earlier work on exception handling for a single robot system show that two mechanisms for exception handling can be identified [7]:
- (Sub-) Task rescheduling. In this case the exception was due to an error in the parts involved in the production task. A selection of the possible operations is made by inspecting the precedence relation between the parts involved in the manufacturing task.
- Recovery planning. Exception handling by recovery planning is based on the assumption that the environment can be manipulated in such a way that the pre-planned flow of actions can be continued. As a result a recovery plan is build consisting of EO's. After the recovery plan is executed, the pre-planned flow of operations is continued.

2.3 Discussion

A hierarchical control model imposes a master slave relation between two adjacent levels. From the point of view of the data-flow, this is correct. The initial task data is decomposed into suitable low level control

commands, which in turn represent the behaviour of the servo controllers. On the perception side, the higher levels of the hierarchies represent a compact, possibly fused version of the raw sensor data processed at the lower levels. However, from a control point of view the imposed master-slave relation is questionable.
Hierarchical control models provide good possibilities for planning and optimization. Each level is shielded from the lower level details. Under the assumption that tasks will be performed according to the plan, production scheduling and task synchronization can be carried out. Because hierarchical models are often the result of a top-down analysis of a control problem, they are easily extended to the lowest levels of control. However, at those levels the execution of pre-planned flow of actions is much less important then the proper reaction to the actual environment state. Therefore non-hierarchical models which apply a collection of specific control functions are much more adapted to perform control in a partly unstructured environment, where each pre-planned action is likely to be interrupted by an unforseen event.
In a hierarchical model the control functions are decomposed and assigned to different levels of the model. In particular with respect to exception handling this introduces a source of problems. Starting with the control architecture developed in the 623 project, an attempt was made to extend the exception handling capabilities to the cell level. The exception handling mechanisms were decomposed and assigned to the most appropiate level in the control hierarchy. To perform the decomposition succesfully, the assumption was made that there is a fixed order in the application of the exception handling mechanisms. This assumption does not hold. The control system has to react adequately to unexpected events. Such a system needs to apply the most suitable handling mechanism as efficient as possible, without bothering if it is a cell or single robot level task.

3 AUTONOMOUS ROBOT CELL ARCHITECTURE

3.1 Approach

The problem of a suitable control architecture is a question to what extend hierarchical and non-hierarchical models can be combined. The proposed solution is based on this observation. Identifying the planning requirements on the one hand and the required autonomous behaviour on the other, it is possible to build a suitable control architecture of hierarchical respectively non-hierarchical modules. The approach followed is based on an analysis of possible exception handling techniques in a manufacturing cell. To accomodate the exception handling functions, a suitable cell control architecure is needed. The architecture is based on a hybrid hierarchical - nonhierarchical model. The proposed hybrid model for the cell controller (ARCCON) is shown in figure 1.

3.2 The Model

The exception handling functions are not assigned to one component in particular. Instead they are shared by all components. This implies that all components can be taken into account when exception handling is needed. The cell controller receives production requests, communicates with transportation controllers and provides status information for the factory controller.

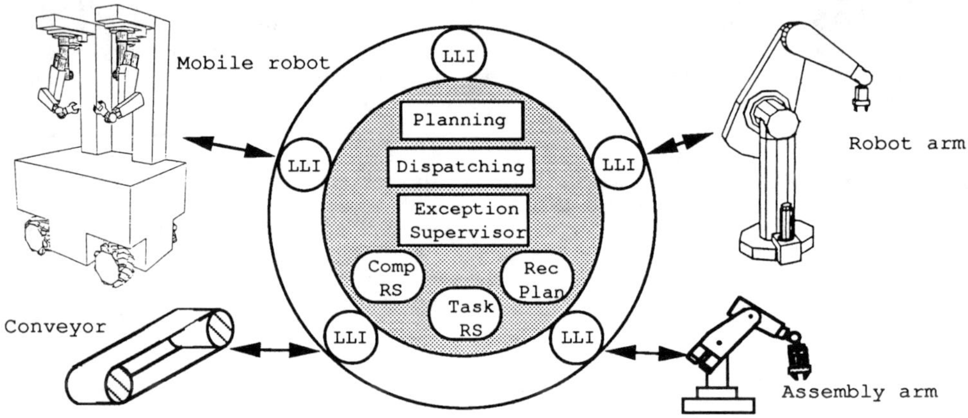

Figure 1. Control architecure for the Autonomous Robot Cell CONtroller (ARCCON)

Each of the components of the manufacturing cell is controlled by a Low Level Interpreter (LLI). The LLI takes care of the execution of single elementary instructions to the component in question. Within certain boundary conditions local handling of geometric deviations is performed. The nucleus of the controller consists of a group of modules, each performing a specific function. Among these functions are production planning and dispatching. The exception handling functions are also located here and are not assigned to a particular component. Both the LLI's and the nucleus are embedded in the information system. Through the information system, each control component can directly access the necessary information for its function.

4. EXCEPTION HANDLING FUNCTIONS

4.1 Recovery planning

Knowledge of the task structure of a robot program enables two approaches to exception handling: recovery planning and task re-scheduling [7]. In the first approach, an attempt is made to recover from the exception by re-planning the robot actions. The goal of the re-planning activity is to restore the operating conditions in such a way that the off-line generated or normal program can continue. Based on the formalized description of possible exceptions, strategies are used to specify the required steps to handle the exception at hand. Each strategy lists a number of motions or grasp planning activities which are executed one by one. In this way a detailed action plan is built, which is then executed by the robot.

4.2 Task rescheduling

If the exception was not identified, or it was impossible to generate a recovery plan from the selected handling strategy, the cell can not perform it's production task successfully. The remaining elementary operations of the task are skipped and an alternative exception handling scheme is applied. By using a description of the dependencies between subtasks of a production task as given by a precedence graph, the exception handler attempts to find a subtask which is not affected by the exception. This scheme ensures the continuation of the production process as far as possible, while the restoring of the failed subtask can be performed by a human operator.

4.3 Component rescheduling

As was stated in [4] the scheduling consists in generation of sequences of (sub)tasks and sysnchronization conditions for each cell component. Rescheduling at this level addresses the exception handling due to component failure. The component has to be taken out of production and the task has to be scheduled over the remaining components. Obviously the precondition is to have redundant functionalities in at least one of the remaining components. The rescheduling will then assume the cell status at the end of the last running subtask without interrupting the current process. Starting from this state production planning will reorder the remaining subtasks to the available components. This solution is not optimal due to transitory idle times of components, but it ensures satisfactory process continuation and termination.

4.4 The Exception Handling Controller

The three types of exception handling are not exclusively restricted to one type of exception or to one order in which they can be applied. On the contrary, the different ways to handle exceptions are inherently redundant. Different orders in which the mechanisms are applied and also the scope of their operation can influence the autonomous behaviour of the cell considerably. For instance the decison to start recovery planning before a task rescheduling is attempted can greatly reduce the productivity of the cell if an other component can take over the job without loss of time. In the case the cell is handling special or custom made parts, recovery planning can be more appropriate before the part is considered lost and rescheduling is started. For these reasons the exception handling behaviour has to be controlled. The Exception Handling Supervisor takes care of this function. This module is an agent which is capable to reschedule a task over the operational cell components and meanwhile plan corrective actions for an other robot to fix the failed component.

5. CONCLUSIONS

Further work is directed towards the development of the Exception Handling Supervisor and the integration of existing functions for exception handling. The work described in this paper is based on the two systems CECOSI (CEll COntrol SImulator) developed at the University of Karlsruhe and REXSIM (Robot EXception SIMulator) developed at the University of Amsterdam. The former environment allows the specification of control software for FAS defining the concurrency and the synchronization between the system components. The latter provides exception handling strategies for robot control systems. The previously generated robot programs are expanded automatically with sensor dependent monitoring conditions. Both systems are under development.

6. ACKNOWLEDGEMENT

This work is based on two CEC sponsored ESPRIT projects (Project No. 623 : Operational Control of Robot System Integration into CIM, Project No. 2202 PLATO : CIM Planning Toolbox). The work was performed at the Institute of Real-Time Computer Systems and Robotics (Prof.Dr.-Ing. U. Rembold and Prof.Dr.-Ing. R. Dillmann) at the University of Karlsruhe, Germany and the Computer Systems group of the University of Amsterdam (Prof.Dr.L.O. Hertzberger).

7 REFERENCES

1 Biemans, F. P. M., Vissers, C. A., Reference Model for Manufacturing Planning and Control Systems, Journal of manufacturing systems, Vol8 No 1, 1989.
2 Isik, C., Meystel, A.M., Pilot level of a hierarchical controller for an unmanned mobile robot, IEEE journal of robotics and automation, Vol 4 No 3, June 1988, pp 241
3 Meijer, G.R., Negretto, U., Structure and functionality of the cell programming system and the cell controller, Working document EP-UVA/UKA-03.88/1, Esprit project nr. 623, Amsterdam/Karlsruhe, March 1988.
4 Negretto, U., Process Control for Flexible Assembly Systems, Proceedings of ISRAM '90, Vancouver, July 1990
5 Malcolm, C., Smithers, T., Hallam, J., An emerging Paradigm in robot architecture, Proceedings of IAS2 conference, Amsterdam, December 1989, pp 545, ISBN 90-800410-1-7
6 Brooks, R.A., A robust layered control system for a mobile robot, IEEE journal of robotics and automation, VOL RA-2, No1, March 1986, pp14
7 Meijer G. R. and Hertzberger L.O. : "Exception Handling for Robot Manufacturing Process Control", CIM Europe Conference, Madrid, IFS Publications, 1988.

EXECUTION PLANNING OF ROBOTIZED MANUFACTURING PROCESSES

R. Bernhardt, H. Münch, G. Schreck
Fraunhofer-Institute for Production Systems and Design Technology, Pascalstr. 8-9, 1000 Berlin 10, Germany

ABSTRACT

This paper gives a survey of the developments in the off-line programming area for industrial robots. In this context off-line programming is to be understood as the complete planning of the execution of manufacturing processes, starting from a description of the production task and resulting in tested and optimized application programs. For this whole procedure the term process execution planning is used. The global structure of a realized system, the programming and optimization procedures as well as the spectrum of applications are briefly described.

INTRODUCTION

Robots are important components for flexible automation. The enlargement of their application area as well as their integration into CIM require the availability of computer-aided planning and off-line programming means. For their realization a research project with the title ESPRIT 623 in which European companies and research institutes co-operate was launched. The project was conducted in the frame of the European Strategy Program for Research and Development in Information Technology (ESPRIT) and partly funded by the European Communities [1].

This paper deals with the second part of the above mentioned CIM requirements: off-line programming. This part includes a model-based derivation of all necessary data for application programs, their test via simulation models, their transfer to the real work cell, and their execution on the shop floor level with a minimum of operator interaction. The last item can be fulfilled by applying calibration methods. With the realized system it has been proved for different manufacturing tasks that application programs can be generated off-line and executed at the shoop floor without any interaction of an operator.

Furthermore automatic trajectory planning functions have been realized allowing an optimization of application programs with regard to cycle times or minimal mechanical stress of the robot. The integration of these automatic planning features is a step towards implicit programming.

SYSTEM STRUCTURE

The off-line programming system is a part of an integrated planning and programming system. This overall system allows the planning of manufacturing processes on a logical level, the selection of components and their arrangement, the generation of application programs and the simulation and test of the manufacturing task execution [2]. The integration of components and tools is achieved primarily by information exchange and management, performed via a relational data base system. This enables to handle data safely and leads to a most flexible system structure with regard to the integration of functional system units and a quick adaption to user needs.

As a part of the overall system the programming system strucure is described in more detail. As can be seen in figure 1 it consists of the main components programming, code generation and transfer and simulation, which are accessed via a dialogue-oriented user interface offering graphical I/O functions, too.

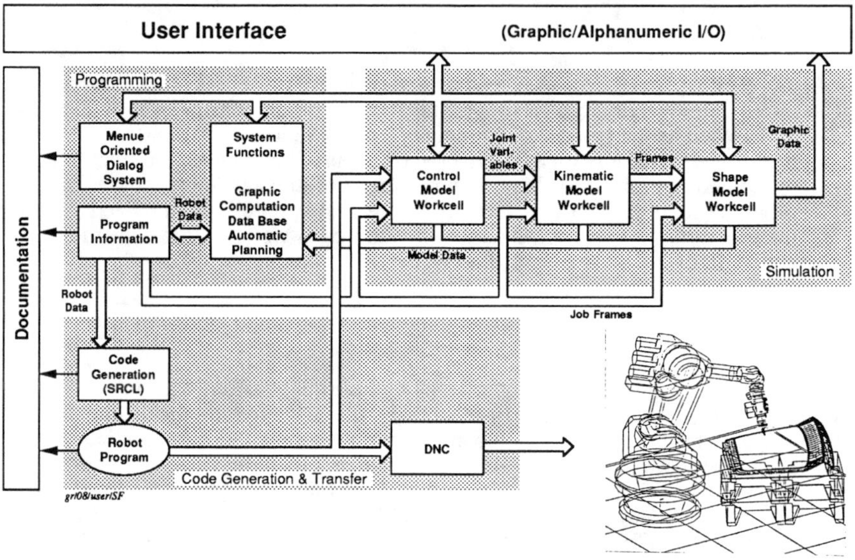

FIG. 1. Strucure of the realized off-line programming system

The system is characterized by a task-oriented execution of programming made possible by a user interface adapted to manufacturing tasks for which the necessary system functions are provided. In a dialogue with the system the user defines all geometrical and technological data describing the execution of the manufacturing task.

Geometrical data are derived from the shape models of work pieces, tools etc. by CAD functions. Technological data are supplied by the user with the help of pre-defined screen forms.

All information accumulated up to this point are integrated within the data structure "program information" which is still independent of the used robot controller. In the subsequent code generation step a robot program in the target language is created. The robot program can be transferred to the actual controller via an IRDATA interface.

For the support of the programming procedure as well as for testing the robot programs, simulation functions are available. They on the one hand allow a direct check of the user´s inputs and on the other hand guarantee a high quality of the robot programs with regard to their executability by the real manufacturing system. For the simulation system computer internal models are necessary which reproduce the components involved in the manufacturing task with regard to their motion behaviour (control model), kinematic (kinematic model) and their shape (shape model).

SYSTEM OPERATION

As shown in figure 2, the programming process is to be executed in four phases. At the workpiece-oriented programming level, the task geometry is defined on the basis of shape models provided by a CAD system using graphical-interactive or calculation functions.

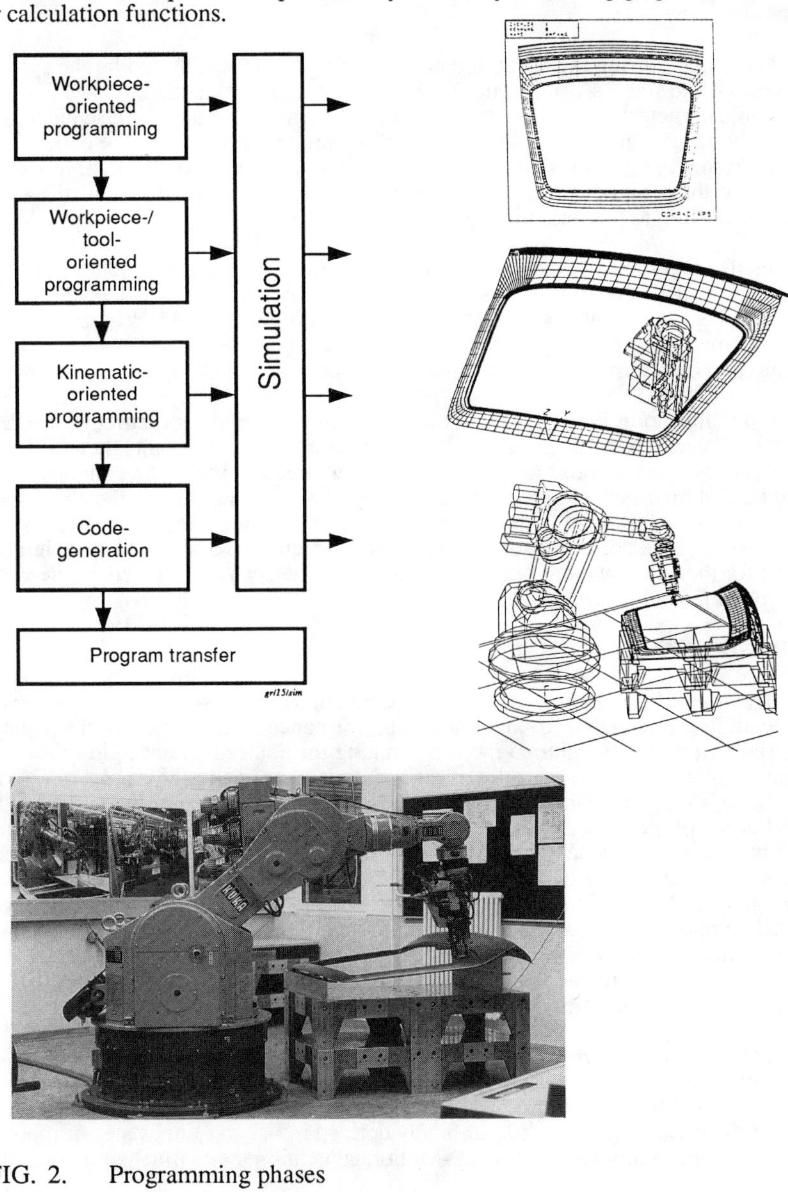

FIG. 2. Programming phases

In the phase of workpiece-/tool-oriented programming the tool configuration for correct task execution is done in relation to the workpiece. Furthermore all technological data essential for the process have to be input at this stage via screen forms.

During the kinematic oriented programming phase, the kinematic and in case of optimal motion planning the dynamic behaviour of the robot itself as well as the linkage with the work cell is taken into account. Due to kinematic frame conditions or in order to avoid collisions, additional auxiliary motions can be defined in this programming phase.

Data input during the first three programming phases are stored by the system in so-called "job frames". In a dialogue this information is extended by the execution parameters (motion parameters, peripherical instructions and sequence) and finally integrated in the data structure "program information". In the code generation step the program information is converted into a robot program in the language of the actual controller. This conversion is done automatically by the system according to predefined mapping instructions.

In all programming phases graphically supported simulation functions are available which also enables the handling of time variant kinematics. They allow the verification of user inputs in the different programming phases. Thereby the final test of the manufacturing task execution is only based on the robot application program and on the models of the simulation system.

Up to this point it was assumed that absolutely accurate geometrical data are available. Tolerances of the robot itself and its environment were not taken into account. These ideal conditions are of course never really given. The conclusion would be that off-line generated programs which have been tested by the simulation system cannot be executed in the real work cell and that their execution can only be made possible by a correction of the program by teach-in. To solve this problem, a method for the automatic calibration of industrial robots was developed, realized and applied [3].

OPTIMAL MOTION PLANNING

The use of intelligent motion planning functions is a basic precondition for a significant improvement of robots´ motion performance. With respect to the actual industrial requirements, optimal motion planning for non-redundant rigid robot systems is of prime interest as a lot of the industrially applied robots are working at the limits of their admissible mechanical stress as a consequence of the poor insight of the human programmer into the system´s dynamics. To overcome this problem - which means to reach a nearly optimal motion behaviour, characterized by reduced reaction forces and torques in combination with a decrease of energy consumption or by minimized execution time - all kinematic features of the robot behaviour have to be taken into account for motion planning. An efficient solution of the mentioned task can only be generated by intelligent motion planning functions including optimization procedures which lead to a decrease of mechanical stress, energy consumption or cycle times by an appropriate choice of task related cost criteria.

For the development of optimal motion planning functions it must be distinguished between two basic types of motion known as Continuous Path (CP) and Point to Point (PTP) motion. In case of PTP-motion generation all admissible trajectories connecting the task-specifically defined initial and final states of motion are available for optimization. For CP-Motion generation the optimal velocity behaviour along the task-specifically prescribed geometric path is determined taking into account the set of allowed velocity profiles.

The model-based planning functions for CP and PTP motion are based on parameter and trajectory optimization procedures. The optimal determination of the motion path described by six position parameters is solved primarily by constraint parameter optimization procedures. The known methods for the solution of the trajectory optimization problem are Calculus of Variation, Maximum Principle, and Dynamic Programming. Comparing the mentioned methods for trajectory optimization, Dynamic Programming is the most appropriate when taking into account the reduction of the problem complexity proposed by [4]. It allows the optimization with respect to all constraints of the real system and the direct exchange of cost functions or kinematic configurations.

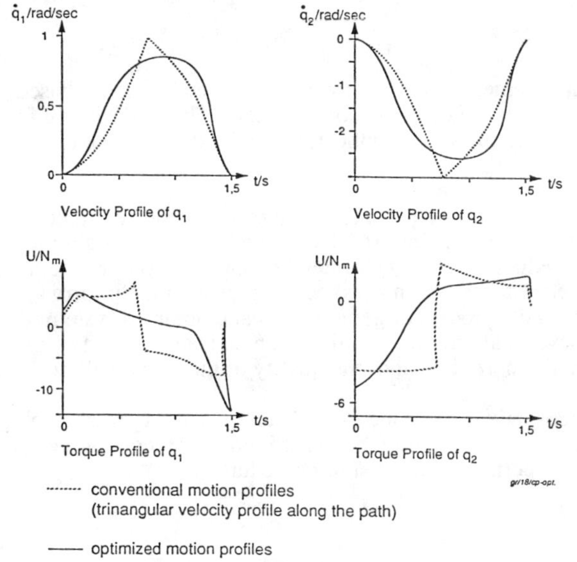

FIG. 3. Comparison of conventional and stress minimized motion profiles of CP-motion

Simulation results of optimal PTP-motion generation for a typical scara type industrial robot shows that the energy consumption can be reduced by 40 percent. In case of execution time optimization a 30 percent decrease of cycle time was effected [5]. But while considering these figures it should be taken into account that they are dependent of the defined task and the used robot.

In figure 3, velocity and torque profiles are shown for a stress minimized CP-motion task of the previously mentioned scara robot. The reduction of stress is accompanied by a reduction of energy consumption of about 25 percent.

As a summary it can be stated that the simulation results emphasize the importance of intelligent motion planning functions for improving the attainable motion performance of industrial robot systems which perform defined tasks with high rates of repetition over long time periods.

APPLICATIONS

The overall system and parts of it have been applied to a broad spectrum of applications in different industries. Some of them are listed below:

- assembly of dot matrix printers
- spot welding
- seam welding of ship sections
- inspection of underwater structures
- robot technology experiment in the German spacelab mission

As the system has been realized in cooperation of European companies and research institutions, there are a lot of applications within the involved companies from the automotive industries.

CONCLUSION

As a result of the improvement of the motion performance by model-based execution planning, applying robots becomes more efficient. This is desirable for the manufacturing industries and off-shore applications and indispensable for automation in space.

For the manufacturing industry whose requirements are mainly reflected in the ESPRIT 623 project, the developed and industrially applied system brought the advantages of a significant reduction of costs for design, optimization programming and workshop tests. At a very early stage in a project, highly accurate information is available which accelerates the process of getting the manufacturing system ready for operation. Moreover, several alternative solutions can be analysed and compared which results in an improvement of the quality of the found solution.

In the R&D area further steps have to be done towards task oriented programming including e.g. technological process planning by using innovative technologies to integrate the knowledge of experts into the overall procedure.

REFERENCES

[1] G. Spur et al., "Planning and Programming of Robot Integrated Production Cells" in Proceedings of the 4th ESPRIT Conference and Exhibition (Brussels, 28-30 Sept. 1987)

[2] R. Bernhardt, "Integrated Planning and Off-Line Programming System for Robotized Work Cells" in Proceedings of the 6th ESPRIT Conference (Brussels, 27 Nov. 1989)

[3] U. Kirchhoff, J. Held, K. Schröer, "Automatisierte Kalibirierung von Industrierobotern" in Komponenten für fortgeschrittene Roboter- und Handhabungssysteme, KfK-PFT 142 (Karlsruhe 1988)

[4] S. Dubrowsky, Z. Shiller, "On the Optimal Control of Robotic Manipulators with Actuator and End-Effector Constraints" in IEEE Int. Conf. on Robotics and Automation, pp. 614-620 (1985)

[5] M. Dlabka, J. Held, Y. Zhang, "A Practical Approach for Planning and Realization of Optimal Trajectory for Industrial Robots" in Proceedings SYROCO (Karlsruhe, Oct. 1988)

Process control for Flexible Assembly Systems

U.Negretto
Institute for Realtime Computer Systems and Robotics
Prof.Dr.-Ing. U. Rembold, Prof. Dr.-Ing. R. Dillmann
IPR, University of Karlsruhe

ABSTRACT

Control design and programming of flexible assembly systems (FAS) becomes more and more important with the increase of variety of product types and of small batch production. To cope with this requirement the control of FAS has to be adaptable to different tasks. To ensure independency of system control from the task to be executed is therefore of key importance. The presented method allows a well defined and consistant control specification for such systems. Process planning and simulation based on the specified control is shown on an assembly example to demonstrate the proposed method.

1. INTRODUCTION

The design of an FAS involves choosing suitable machines to manufacture or assemble certain types of products, define the needed operations and elaborate an appropiate control system. For discrete event dynamic systems like FAS, a well defined theory for the control design is not yet available. This in contrary is the case for linear, continuous or synchronous systems.
The problem is to integrate the cell resources into an efficently working contol structure taking into account concurrent and asynchronous processes [1,2,4]. Essential for the control design is the consideration of important FAS properties like absence of overflows and deadlocks and the capability of reconfiguration of the system. The latter property addresses the reacheability of the initial system state or an other working state from any current (i.e.also exceptional) state. The reconfigurability plays an important role in the context of exception handling and has to be supported by the control model [5,6].
This paper proposes a method for the design of flexible assembly systems control based on the representation of the system in augmented Predicate Transition nets. The focus is put on the representation of the static and dynamic functionality of the real system and in the verification of the control algorithm through simulation of the planned process.

1.1 Predicate Transition nets and their relation to FAS

Petri nets and Predicate Transition nets (Pr/T) have often been used to model discrete event dynamic systems like distributed systems, distributed data bases, FMS and others [1,3,7,10]. The advantage of these formal methods is the well defined theory wich offers methods for analysis, simulation and evaluation of the modeled systems.
Predicate Transition nets are directed graphs with two different types of nodes, predicates and transitions. In our method the predicates represent states of resources and the transitions represent events and cause the state changes of the real system.
The nodes are connected by arcs representing the functional relations between the assembly system states. An arc always connects a predicate with a transition or vice versa.
Pr/T have a structure to define predicates, operators and a set of individual tokens. To each transition and each arc in the net, an inscription is associated specifying either variables (to arcs) or logic equations (to transitions). The arc labels define type and quantity of assembly parts which can be handled. The transitions inscriptions specify conditions which have to be true in order to enable a change in the state and firing rules defining how the status changes.
A marking, i.e. the system state, is defined through the individual-token tuples in each predicate at a definite time, also called partial states. The tokens specify the parts, the tasks and the operations related to the part. A capacity relation is associated to each predicate defining the maximal amount of tokens for each individual type. With this criteria the problem of overflows in the FAS is reduced to the modeling problem.
For a more detailed introduction of Pr/T nets the reader is referred to [1,8].

2. PROBLEM STATEMENT

Given the specification of an assembly system or cell, to model the system such that it can be controled, programmed and simulated realistically. The specification refers to

- the layout of the FAS (list of resources)
- resources descriptions (functionality, capacity)
- task description (assembly graphs, material data)
- operation capabilities of active devices
- system start and goal state

2.1 Objectives

The correct and complete representation of the system and its functionality requires:

- A task independent model of the topology of the Flexible Assembly System

- The static representation of: layout, component properties and relations between components
- The dynamic representation of concurrency and synchronization of resources
- A workplan description of the task coherent to the functional FMS model

The objectives of the work are to support the design offering standard resource types (modules) for the configuration of new systems and enabling instanciation of the standard types through parameter specification.

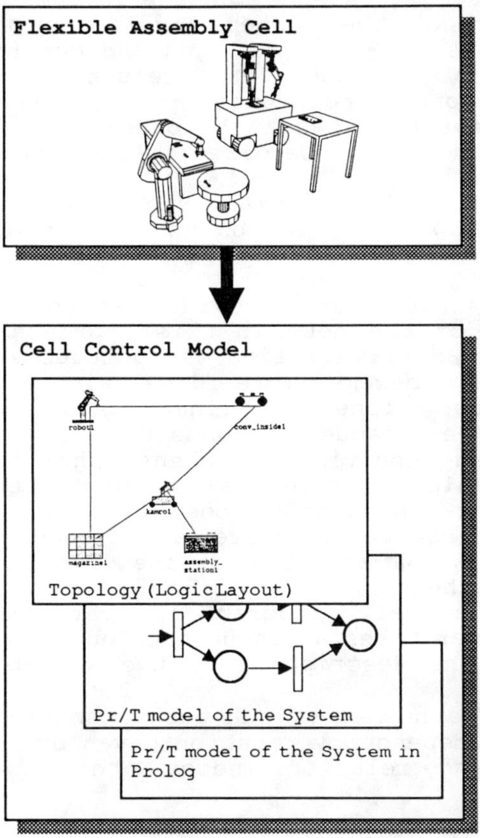

Fig. 1: Model levels of the FAS

The design of a flexible assembly cell consists in choosing, changing and extending the standard modules and relate those which can interact dependent to the required behaviour. The designer should concentrate on the design problem without being confronted with Pr/T net implementation details. The aim is therefore to provide a good user interface with interactive system modeling functions (i.e. standard modules, automatic consistancy check, storing and loading of modeled systems) and means for random interrupt generation during simulation. The interrupt is necessary to check the system state, change parameters and resume of execution with the new values.

3. PROPOSED METHODOLOGY

3.1 Approach

The first step of the method consists in the task independent representation of the flexible assembly system (fig.1). This discrete event system is modeled through a formalism embedded in augmented Predicate

Transition nets. Using this formal method the functionality of each component is modeled in a predicate transition net, forming a local subnet. The components (i.e. subnets) are related to each other through synchronization transitions. The connection of all subnets is the overall net model. The overall net model represents the functional model of the cell describing the allowed concurrency, the synchronization between components, the material flow and the relations between the devices. In a second internal step the Predicate Transition net model is transformed in a Prolog notation. The prolog inference machine is used as trigger mechanism to check the system conditions and start appropriate actions [7,9]. Simulation and control of processes in FAS with such nets requires the fulfillment of two additional conditions:

1. The informal representation has to be detailed to a formal level, where the model can be directly associated to processes and signals.
2. The execution of cell tasks requires the representation of the task work plan on the cell model (cell programming).

To 1.: Predicate transition nets provide methods to refine or coarse nodes of the net. In this approach a top-down method to gradually refine structures is followed. The inscriptions becomes more formal with each step. The nodes can be refined to linear predicate transition sequences or be expanded to subnets.
To 2.: The introduced individual tokens and the inscriptions with variable parameters allow to use the cell model for the task-dependent process simulation. The variables in the inscriptions are associated to task-specific constants, which direct the flow of individual token within the net.
The generation of the cell program is performed automatically. The planner takes as input the functional model of the cell and the description of the assembly task.
An assembly consists of a number of sub-tasks (abstract operators) which have a dependency relation given by the application. The dependency relations indicate the

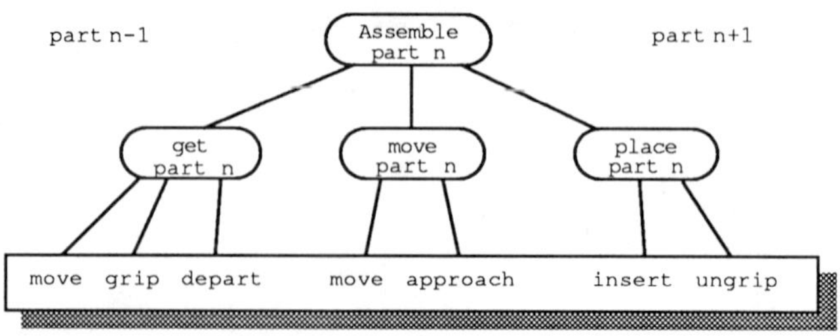

Fig.2:Elementary operations and task knowledge for an assembly operation.

necessary order in which two tasks have to be carried out. A cell program is generated by transforming a global task description into a sequence of subtasks. This process is repeated until the subtasks are detailed enough to be directly expressed in the operating primitives of the cell components. These operating primitives are called elementary operations. These elementary operations serve as the standard projection level of all planning and programming modules. The task decomposition is done analogously for all active components of the cell. As output of this phase for each active component a sequence of subtasks is planned which is expanded to elementary operation sequences during runtime.

4. RESULTS

A cell control simulator, CECOSI, was implemented in prolog, based on the presented method and offering functions for modeling the control and for analyzing and validating the planned cell program. The interaction is performed through an user interface in a mouse based window environment on a SUN (Sunviews) under Unix.
For the emulation of the specified control algorithm the system is connected with a graphic simulation system developed at the University of Karlsruhe. The results improve the quality of the former generated program in respect of performance and time.

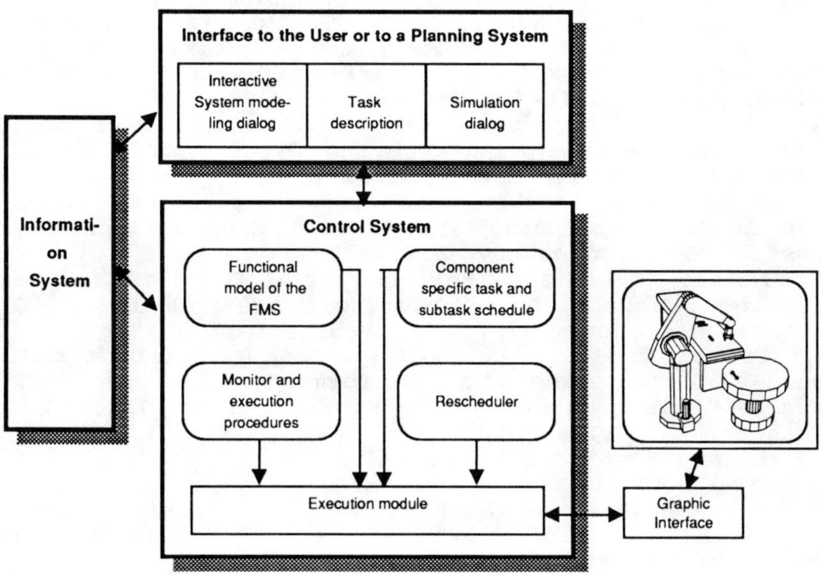

Fig. 3: Structure of the cell control simulator (CECOSI)

5. CONCLUSION

Concluding this method allows to pass directly from a functional specification of FAS to the control and monitoring of such systems. A consistent simple model representation of synchronization and concurrency of processes is given
The future work will emphasize the generalization of strategies dedicated to rescheduling, the verification of new generated models and the integration of sensor components with dedicated elementary operations.

6. ACKNOWLEDGEMENT

This work is based on two CEC sponsored ESPRIT projects (Project No. 623 : Operational Control of Robot System Integration into CIM, Project No. 2202 PLATO : CIM Planning Toolbox). The work was performed at the Institute of Real-Time Computer Systems and Robotics (Prof.Dr.-Ing. U. Rembold and Prof.Dr.-Ing. R. Dillmann) at the University of Karlsruhe, Germany.

REFERENCES

[1] Peterson J.L. :"Petri Nets Theory and modelling of Systems", Prentice Hall Inc., 1981.
[2] Zhou M., DiCesare F. and Desrochers A. : "Top Down Approach to Systematic Synthesis of Petri N. Models for Manufacturing Systems",IEEE Robotics and Autom., 1989.
[3] Dungern O.v., Schmidt G.K.:"A Comprehensive Architecture for Flexible Assembly Cell Control with Rule-Based Error Recovery", Int. Workshop on Intelligent Robots and Systems, Tokio, Nov.1988.
[4] Stienen H., van der Weerd P.R.:"Object-Oriented Design of a Flexible Manufacturing System" IFIP working Conference, Galway, Ireland, 1988.
[5] Meijer G.R. and Negretto U.:"Exception Handling and Task Level Programming of Flexible Manufacturing Systems", ISRAM, Vancouver, July 1990.
[6] Meijer G.R. and Hertzberger L.O. : "Exception Handling for Robot Manufacturing Process Control", CIM Europe Conference, Madrid, IFS Publications, 1988
[7] Camarinha-Matos L.M., Negretto U., Meijer G.R., Moura-Pires J., Rabelo R.: "Information Integration for Assembly Cell Programming and Monitoring in CIM", ISATA, Wiesbaden, Nov. 1989
[8] Reisig W. :"Petrinetze - Eine Einführung", Springer Verlag, 1986.
[9] Sahraoui A., Atabakhche M., Courvoisier M. and Valette R. : "Joining Petri Nets and Knowledge Based Systems for Monitoring Purposes",
Proceedings of 1987 IEEE International Conference on Robotics and Automation, Raleigh, USA, 1987.
[10] Valette R. :"Nets in Production System", GMD Advanced Course on Petri Nets, Bad Honnef, 1986.

INTEGRATION OF AN ARTICULATED DEXTEROUS HAND IN A ROBOT ARM

G. Vassura
DIEM - Università di Bologna - Viale Risorgimento 2, 40136 BOLOGNA, ITALIA

ABSTRACT

The goals and main problems in mechanical design concerning the installation of a remotely-actuated multi-fingered hand in a robot arm are examined and guidelines are suggested as to the evaluation of alternative configurations of transmission lines from the finger joints to the actuators. The design of a robotic forearm, including a roll-bend self-compensating wrist and an articulated hand (version II U.B. Hand) is then described in its main features.

INTRODUCTION

The state of art concerning the installation of dexterous articulated hands on robot arms shows that in most cases such devices have been put together after separate design and development. On the one hand, the commercial robot arms have been designed keeping the end-effector interchangeability concept in mind, not the fitting to a particular hand, on the other, efforts in dexterous hand development have aimed mainly at solving problems as to their configuration and control rather than at improving their fitness for installation on robot arms.

The integration of hand, wrist and arm means simultaneous and compatible design of these sub-systems right from their development phase. Whenever a really dexterous hand is available, end-effector interchangeability may be abandoned and a special-purpose arm design justified. Besides all the advantages of mechanical integration a very important consequence may be that the hand control, too, will be integrable in the robot control system, thus bringing great benefits in architecture simplification, system performance improvement and cost reduction.

As for mechanical design, the main goals of integration are: to reduce the weight and size of both the hand and the arm, thus avoiding structural redundancy; to improve performance by adopting built-in transmission lines; to reduce overall complexity and cost.

As relocation of actuators far from the arm's end becomes necessary in order to get a favourable payload ratio for the arm (that is to avoid the use of a huge arm to carry a heavy hand that can manipulate a very small object), the design of transmission lines and compatible wrist articulations becomes a critical point in the development of integrated systems.

TENDENCIES ON HAND-ARM INTEGRATION

Three main tendencies can be identified in recent research work on robotic hands and their installation on arms.

- Development of hands with self-contained actuation, intended as autonomous modules that can be attached to any arm with the appropriate payload; their bulk is reduced by limiting the number of actuated joints, at the expense of a reduction in dexterity, so that they can operate mainly in the field of adaptable static grasp. Significant examples of this design concept are the Belgrade-USC Hand [1] and Ulrich's Hand.[2] Some other designs, such as the Karlsruhe Hand [3] exhibit good suitability for controlled manipulation, but their bulk is increased by local actuation.

-Installation of multi-d.o.f.s remote-actuated hands on commercial robotic arms, superposing the transmission and the actuator box on to existing arm structures. The most frequently used configuration presents the motor box placed near the forearm joint of RRR-type arms in order to achieve static balancing and the transmission line made with sheath-guided tendons (an example is Salisbury's Hand [4]).In the case of installation on a gantry robot, as in the U.B.(University of Bologna) Hand case [5],the need to locate the motor box on the x-y carriage greatly emphasizes performance limits.

A remarkable case is that of the Utah-MIT Hand [6], with tendons routed only by pulleys along a passive remotizer arm from a stand-alone actuation box: an external robot arm is needed to move the hand, the integration with the carrier arm being not complete.

-Installation of complex, remote-actuated hands in purposely designed arms, based on design concepts that look at the robot hand as an integral part of the whole manipulation system, [7], [8].

An intermediate approach between the last two tendencies, that is, partial re-design of a commercial arm for the better housing of dexterous hands, can be of great interest and it can allow for short-term application. The main problems to be solved in mechanical design are:

-to properly house the remote actuation box within the arm;

-to define a transmission scheme with acceptable performance,

-to design wrist articulations that can be passed through by the transmission lines, without increasing the computational charge and overall complexity.

ON THE CHOICE OF A TENDON TRANSMISSION CONFIGURATION

Whenever a finger joint is driven by a couple of antagonistic tendons operated by a single actuator, it is mandatory for the routing length to be kept constant, that means adopting a transmission scheme with self-compensating capability, so that changes in the configuration of the wrist and arm do not generate tension variation in tendons.

As to the means for tendon routing, three basic configurations can be adopted:

-the tendons are routed only by pulleys or equivalent devices all along the arm from the finger joints to the actuators; the articulations can be passed through and changes in configuration of the kinematic chain usually result in joint angle variation, so that mechanical coupling does exist [9]; tendon length compensation can be obtained by proper routing over pulleys: design complexity may be remarkable, but good performance can be obtained in force and position control [10];

-the tendons are routed only by flexible, axially unextensible sheaths; the articulations are passed outside and changes in configuration of the kinematic chain are allowed by sheath bending, without mechanical coupling; no length compensation is needed, the design complexity is low, but the performance can be greatly affected by non-linear friction, high compliance and hysteresis [11];

-the tendons are routed by a mixed, intermediate configuration, alternating stretches with pulleys and others with sheaths; both coupling and need of length compensation exist and an acceptable compromise between design complexity and transmission efficiency may be sought in a number of alternative solutions that may be of some interest in practical application.

Aiming to aid the choice between alternative configurations, a merit criterion is being tested by the author, based on the comparison of a global functionality index defined for each of them:

$$I_F = W_C I_C + W_D I_D + W_M I_M$$

where W_C, W_D, W_M are evaluation weights ($W_C+W_D+W_M=1$) by which design priorities can be imposed, $I_C \leq 1$ is a quality index referred to the computational charge introduced by the adopted solution in transforming joint space into actuator space, $I_D \leq 1$ is an index of the suitability to force control accuracy, $I_M \leq 1$ is a parameter that allows evaluating design complexity and cost. Each of them is defined with respect to a reference value, that is, for I_M the computational charge of the decoupled full-sheath configuration, for I_D the mechanical parameters of the transmission model in full pulley routing and for I_M the complexity index of the full-sheath transmission. In this way, the expression of I_F can work well in the limit cases of $W_D=1$ or $W_M=1$ (absolute priority of the best performing or of the lowest-complexity solution).

As any method including arbitrary choice of influential parameters, it can be a tool for pointing out macroscopic differences and conflicting factors rather than giving an exact solution to the evaluation problem. The method is mainly intended to ascertain whether, when W_D is no longer the predominant factor and W_M assumes a consistent weight (in most cases W_C can be considered negligible), mixed transmission configurations with sheath- and pulley- tendon routing can be acceptable compromise solutions. This evaluation criterion was used to analize the following proposal for installation of the tendon-actuated U.B.hand on a Puma-like forearm: assumed W_C=0.1, W_D=0.5, W_M=0.4, the proposed configuration was found to be competitive with the equivalent full-sheath routing and full pulley-routing.

INTEGRATION OF THE U.B.HAND INTO A PUMA 560 ARM

The version II U.B.Hand is an eleven D.O.F., three-fingered hand now at the sub-system prototype stage that aims to enhance the performance of the previous version I U.B. hand (fig.1) in whole-hand manipulation procedures [12]. It presents a new mechanical design, with distributed intrinsic-type force-torque sensors.

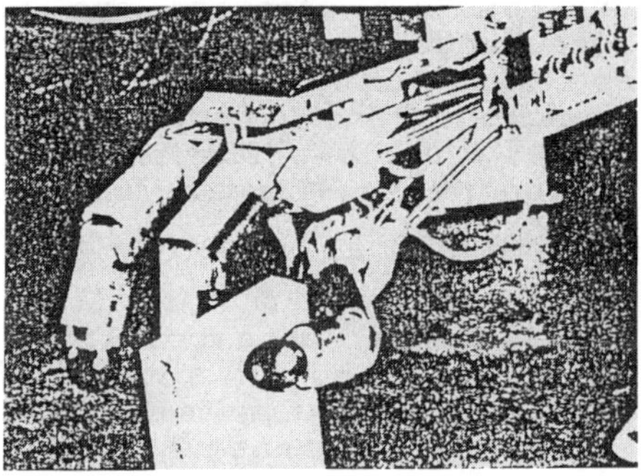

Fig.1 the U.B.Hand

Experience in designing and testing the previous version suggested overcoming some performance limits by re-locating the actuators in the forearm and connecting the hand to the arm through a purposely designed wrist.

A sketch of the simple architecture now under development is shown in fig.2.

A bend d.o.f. in the wrist is realized by an articulation between the hand and the forearm, while a roll d.o.f.is obtained by a revolute pair at the end of the intermediate link: the actuator box is located on this link so that only the bend articulation must be crossed by the tendons. Couples of pre-loaded antagonistic tendons, each driven by a single actuator, move the joints: the transmission is mixed as the tendons are routed with pulleys inside each finger, and with sheaths in their path from each finger to the hand-frame, to allow adduction-abduction displacement of the upper fingers and proximal rotation of the thumb, avoiding the more complex pulley-based design.This solution gives acceptable results in terms of mechanical performance, because the bending angles and their variation are small and the length of the sheathed stretch is short, thus reducing the effects of compliance and friction.

From the hand to the actuator the antagonistic tendons are guided parallel to each other and are routed,

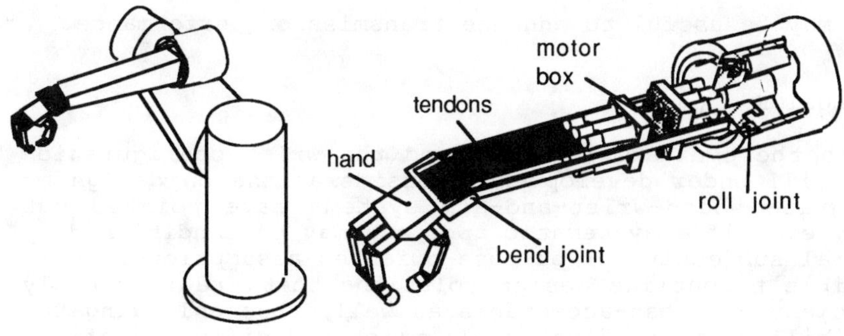

Fig.2 integration of the U.B. Hand in a Puma arm

across the wrist articulation, by the same side of a floating-shaft pulley train: this makes the joint motion independent of the wrist motion, as the two antagonistic tendons are subject to the same path variation in winding or unwinding over the pulleys. A controlled variation of the pulley-shaft position results in compensation of tendon tension at any rotation angle of the wrist.A simple wrist-driven cam mechanism (fig.3) can provide for constant tendon tension: after optimization of geometrical parameters, a cam was designed and tested and the preload variation in Kevlar tendons due to cam cutting errors proved to be negligible.

The proposed solution, though very simple, presents some actractive features, as the hand is mechanically decoupled from the wrist, the negative influence of the sheaths is limited, the total number of pulleys is low and the design complexity is reduced.

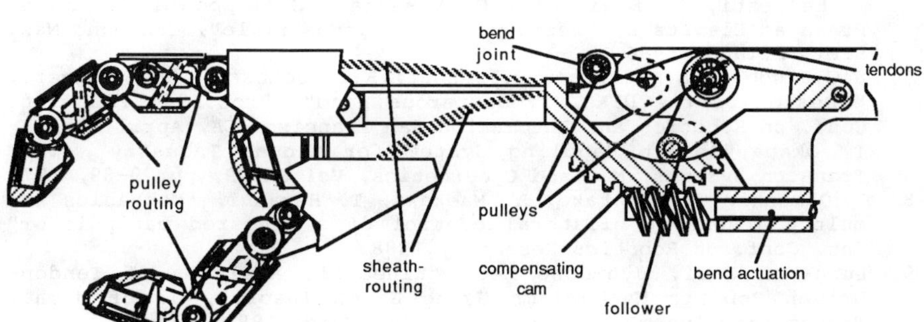

Fig.3 wrist scheme

Besides, an interesting feature is that the pre-loading of all the tendons may be regulated by means of active variation of the floating shaft position with respect to the position determined by the cam (e.g. by means of a simple S.M.A. actuator that can vary the length between the axes of the floating shaft and the follower): a tuning of tendon pre-load according to external load variation may be useful to enhance transmission performance.

CONCLUSION

Both the evaluation method and the wrist configuration are still under development: considerations on design of integrated hand-wrist-and-arm systems have pointed out that, even if a systematic approach may be conditioned by several subjective or purpose-oriented assumptions, it is possible to outline design solutions that are reasonably efficient and cost-acceptable as well, thus enlarging the possibility of practical application and diffusion of articulated hands.

ACKNOWLEDGEMENTS

Appreciated contributions to this work came from A. Avallone and D. Roncuzzi. Financial support was provided by CNR Progetto Finalizzato Robotica grant N. 89.00561.PF.67.3.

REFERENCES

1. B&L Engineering "The Belgrade/USC Robot Hand", Brochure, Preliminary Release, May 1989.
2. N. Ulrich, R. Paul, R. Bajcsy, "A Medium-Complexity Compliant End Effector", Proc. IEEE Int. Conf. on Robotics and Automation, Philadelphia, PA, 1988.
3. T. J. Doll, H. J. Schneebeli," The Karlsruhe Hand", Proc. IFAC SYROCO '88, Karlsruhe, West Germany, 1988.
4. J.K.Salisbury,"Design and Control of an Articulated Hand", Int. Symp. on Design and Synthesis, Tokyo, Japan, 1984.
5. M. Belletti, C. Bonivento, G. Vassura, "Sviluppo di Organo di Presa ad Elevata Destrezza per Robot Industriale", Convegno Naz. SIRI, Milano, 1986.
6. S.C.Jacobsen,E.K.Iversen,D.F.Knutti,R.T. Johnson, K.B. Biggers, " Design of the UTAH/MIT Dexterous Hand", Proc.1986 IEEE Int. Conf. on Robotics and Automation,San Francisco,CA, April 1986.
7. T. Okada,"Object Handling System for Manual Industry", IEEE Trans.on Systems, Man and Cybernetics, Vol SMC 9, pp.79-89, 1979
8. T. Oomichi, T. Miyatake, A. Maekawa, T. Hayashi, "Mechanics and multiple sensory bilateral control of a fingered manipulator" Int. Conf. on Robotics Research, 1988
9. Lung-Wen Tsai, Jyh-Jone Lee, "Kinematic Analysis of Tendon-Driven Robotic Mechanisms Using Graph Theory", J. of Mech., Transm. and Autom. in Des., Vol.111, (March 1989)
10. R. Vossoughi, M. Donath, "Robot Finger Stiffness Control in the Presence of Mechanical Nonlinearities", J.of Dyn.Syst., Meas. and Cont.,Vol. 110, 236-245,(Sept.1988)
11. W.T.Townsend, "The Effect of Transmission Design on Force-Controlled Manipulator performance", MIT AI Lab, Tech.Report 1054, April 1988.
12. G. Vassura, A.Bicchi, "Whole Hand Manipulation: design of an articulated hand exploiting all its parts to increase dexterity", Proc. Nato Advanced Research Workshop on Robots and Biological Systems, June 26-30, 1989, Il Ciocco, Tuscany, Italy.

SECTION 8 : AI AND SIMULATION

PARADIGMS FOR CONTROL IN COMPUTER INTEGRATED MANUFACTURING: SYNTHESIZING ORGANIZATIONAL AND ARTIFICIAL INTELLIGENCE PERSPECTIVES

William I. Bullers, Jr., Ph.D.
Richard A. Reid, Ph.D.
Howard L. Smith, Ph.D.

Anderson Schools of Management
University of New Mexico
Albuquerque, N.M. 87131

ABSTRACT

Both organizational and artificial intelligence (AI) paradigms provide useful conceptual frameworks that can serve to guide the development of information systems (IS) for organizational planning and control in Computer Integrated Manufacturing (CIM). Consideration of operational, tactical, and strategic perspectives provide an appropriate foundation from which CIM IS requirements can be analyzed and integration support provided. Basic paradigms from AI can then be invoked in order to provide intelligent reasoning in support of automated control and management decision support in CIM.

INTRODUCTION

Computer Integrated Manufacturing (CIM) requires the integration of Computerized Manufacturing Systems (CMS) on the factory-floor with higher-level Management Information Systems (MIS) in manufacturing organization. The CMS provide for real-time operations and automated control of flexible manufacturing systems, assembly lines, material storage and handling devices, and other industrial robots or processes. The MIS provide planning and control information about manufacturing operations in order to support organizational decision making by operational, middle, and upper management.

ORGANIZATIONAL PARADIGMS

Operational Perspectives

At the production processing level in an organization, the CMS constitute complex, distributed networks that include process control computers, numerically controlled machines, and robots. These CMS networks are characterized by the execution of concurrent production processes, perhaps requiring synchronous control under certain conditions, while employing asynchronous control at other times. Likewise, the nature of a given process operation may or may not be deterministic. A CMS invariably requires real-time control software to maximize utilization of limited manufacturing system resources which are subject to rapidly changing work loads and the availability of production processing capabilities [12].

Strategic/Tactical Perspectives

At higher organizational levels, the MIS must integrate production data from a variety of perspectives in order to support organizational planning and control. Vertical integration is required to control multi-level operations encompassing CMS facilities, shops, cells, work stations, and production processing equipment. Horizontal integration of the manufacturing functions is necessary to permit coordination among product design and engineering, inventory control, physical distribution, cost accounting, purchasing, personnel, and marketing. Each of these functional information systems may be characterized by multi-level operations which typically includes operational control over material and process flow, tactical planning and control over manufacturing operations, and strategic planning over production facilities.

An Integrative Perspective

Integration of information processing activities is also required to permit the capabilities of different types of information systems for transaction processing, routine and repetitive reporting, decision support, or providing expert opinions to be accessed by managers throughout the organization [2,3]. In addition, temporal integration of past, present, and future production data, along with physical integration across geographically dispersed production facilities must be considered. The multi-faceted nature of planning and control in this type of CIM environment requires a complex, coordinated, control system to be applied [5,9,10].

Research in the areas of organization structure and strategic planning has produced several concepts that are appropriate for high-level organizational planning and control in CIM. The identification of Strategic Business Units (SBUs) and their Critical Success Factors (CSFs) are important factors in determining essential activities that must be performed by the individual manufacturing units if the overall organization is to be successful [13]. For example, a specific product line and the market share associated with each product represents a relevant manufacturing-oriented SBU and its CSFs. Thus, consideration must be given to the pertinent decisions associated with the CSF, and the information requirements necessary to support those decisions.

Three paradigms associated with decision making are particularly useful in analyzing IS requirements: Simon's structured versus unstructured decision type [14]; Anthony's hierarchical level of managerial activity encompassing strategic planning, management control, and operational control [1]; and Simon's four-stage decision making process depicting the sequence of decision activities required between recognition of a problem and implementation of a solution [14]. These paradigms provide conceptual understanding and insight needed in developing the kinds of control mechanisms required in CIM.

ARTIFICIAL INTELLIGENCE PARADIGMS

Research in Artificial Intelligence (AI) has contributed a variety of knowledge representation schemes and models of control for inference processing in expert systems. Experience with the development of expert systems has suggested that simple inference methods such as generate and test, backward chaining, and forward chaining are very effective for implementing control in a wide variety of problem domains **if** they are coupled with appropriate knowledge representation schemes [6].

Three basic paradigms for knowledge representation - frames, rules, and logic - have emerged to support the complex task of storing knowledge in expert systems in order to provide intelligent reasoning [6]. Frame-based representation utilizes an object-centered view of knowledge specification, whereby knowledge is partitioned into discrete structures (frames) having individual properties called slots [4]. Frames can be used to represent broad concepts, classes of objects, or individual instances or components of objects. Frames can be related in an inheritance hierarchy that provides for the sharing of common properties among the frames without multiple specification of those properties.

Rule-based modeling relies on the implementation of production rules of the form IF conditions THEN consequences and IF conditions THEN actions to construct a knowledge base [8]. Logic programming relies on the principle of programming by description in the predicate calculus. A key feature of a logic program is the explicit separation of the problem's logical description (**what** is to be done) from the control components of an algorithm suitable for solving the problem (**how** it is to be done) [7,11]. All three paradigms for knowledge representation can contribute to a CIM system's ability to reason.

SOME ILLUSTRATIVE PARADIGM RELATIONSHIPS

The AI perspective that simple inference mechanisms are very effective when coupled with appropriate knowledge representation may have important application to organizational control issues in CIM. This view suggests that appropriate specification of, for instance, a manufacturing organization's CSF could simplify the planning and analyses associated with strategic decision making. Furthermore, proper partitioning of the CSF knowledge base could serve as the basis for implementing the CSF in the form of lower-level tactical and operational control activities.

Frame-based representation may be an appropriate knowledge specification scheme to utilize when determining the decision characteristics associated with multilevel operations encompassing CMS facilities, shops, cells, work stations, and production equipment. Clearly, in this hierarchical view of a manufacturing organization, the lower-level manufacturing units in a CMS inherit the policies, goals, and objectives of upper-level units. Frames, with their natural inheritance of properties, insure that operational-level CMS control routines adhere more closely to policies established in the strategic, higher-level organizational units.

A rule-based approach with its If conditions Then implications rules may also prove suitable, either by themselves or in conjunction with frames for the expression of high-level policies and their potential implications on lower-level manufacturing units and operations. Likewise, logic programming's separation of the what of a decision from the how of its implementation with may help manufacturing management to more clearly focus first on the overall policies and plans designed to address the organizational unit CSF, then to develop the operational procedures and activities necessary to achieve success in those areas.

NORMATIVE APPROACH IN CIM DEVELOPMENT

Real-time, automated control of operations on the factory-floor, and information processing support for higher-level organizational planning and control in CIM requires significant integration along the traditional dimensions of vertical, functional, information processing, temporal, and geographical units, along with the organization structure and AI perspectives on control. Low-level perspectives on knowledge representation and control from the field of AI must be synthesized with ideas of organizational planning and control in order to realize the objectives of CIM. A top-down, decision-oriented approach to CIM control will focus first on the manufacturing unit CSF, then on the identification of pertinent decisions, and finally on the derivation of information and low-level control requirements.

Initially, paradigms of organization structures must be considered to assure support for management planning and operational control functions that are required in CIM. Subsequently, the developers of the component CMS and MIS software must consider models of control from AI in order to implement intelligent information processing programs in support of automated control and management decision support in CIM.

REFERENCES

1. R.N. Anthony, Planning and Control Systems: A Framework for Analysis, (Harvard Graduate School of Business Administration Studies in Management Control, Boston, 1965).
2. W.I. Bullers, Jr., and R.A. Reid, "An Information System Architecture for Computer Integrated Manufacturing,", in Robotics and Manufacturing: Recent Trends in Research, Education, and Applications, edited by M. Jamshidi, J.Y.S. Luh, H. Seraji, and G.P. Starr, 983-992, (ASME Press, New York 1988).
3. W.I. Bullers, Jr., and R.A. Reid, "Toward a Comprehensive Framework for Computer Integrated Manufacturing," Information & Management, 18(2), 57-67, (Feb. 1990).
4. R. Fikes and T. Kehler, "The Role of Frame-Based Representation in Reasoning," Communications of the ACM, 28(9), 904-920, (Sept. 1985).
5. M.S. Fox, "The Intelligent Management System: An Overview," in Processes and Tools for Decision Support, edited by H.G. Sol, 105-130, (North-Holland, Amsterdam 1983).
6. P. Friedland, "Special Section on Architectures for Knowledge-Based Systems," Communications of the ACM, 28(9), 902-903, (Sept. 1985).
7. M.R. Genesereth and M.L. Ginsberg, "Logic Programming," Comm. of the ACM, 28(9), 933-941, (Sept. 1985).
8. F. Hayes-Roth, "Rule-Based Systems," Comm. of the ACM, 28(9), 921-932, (Sept. 1985).
9. T.L. Johnson, "Design of Hierarchical Control Systems For Automated Job Shops," Proceedings of the 1984 American Control Conference, 1062-1064, (June 1964).
10. A.W.T. Jones, and C. McLean, "A Proposed Hierarchical Control Model for Automated Manufacturing Systems," Journal of Manufacturing Systems, 5(1), 15-25.
11. R. Kowalski, "Algorithm = logic + control," Comm. of the ACM, 22(7), 424-436, (July 1979).
12. R.E. Larson, P.L. McEntire, and J.G. O'Reilly, "The Theory of Real-Time Control," in Handbook of Software Engineering, edited by C.R. Vick, 92-118, (Van Nostrand Reinhold, New York 1984).
13. J.F. Rockart, "Chief Executives Define Their Own Data Needs," Harvard Business Review, 81-93, (March-April 1979).
14. H.A. Simon, The New Science of Management Decision, (Harper & Brothers, New York 1960).

Automatic Inspection Planning Within a Feature-Based CAD System

Francis L. Merat*#
Gerald M. Radack+#
Jeremy F. Jacobsohn*#
*Department of Electrical Engineering and Applied Physics
+Department of Computer Engineering and Science
#Center for Automation and Intelligent Systems Research
Case Western Reserve University
Cleveland, Ohio 44106

ABSTRACT

This paper describes a CIM system which incorporates automatic generation of inspection plans for coordinate measuring machines (CMMs). Design is done in terms of form features. Associated with each feature are inspection code fragments which are instaces of inspection procedures. There are three basic types of code fragments : (1) fragments for checking the internal dimensions of a feature, (2) fragments for checking the relationships between features, and (3) modified fragments for use when the feature to be inspected intersects another feature. Using this approach, a list of suitable inspection points and approach vectors is generated. The system is consistent with the accepted ANSI Y14.5M standard for dimensioning and tolerancing as well as with observed industrial practice.

INTRODUCTION

The goal of this work is to develop an intelligent inspection planning subsystem for Computer Integrated Manufacturing (CIM) system for mechanical parts. An inspection planner takes the part model—consisting of a specification of the part geometry as well as dimensioning and tolerancing information—and produces an inspection plan, which gives detailed instructions as to how to inspect a manufactured part to determine whether it is within tolerance. An inspection plan might consist of printed instructions to a Quality Assurance person, or it might consist of codes that can be executed by automated inspection equipment such as a coordinate measuring machine (CMM) or a vision system. Ideally, it would consist of a combination of both manual and automatic steps, so that each aspect of a part is inspected in the optimal way in order to achieve the minimum time inspection plan.

Our system is intended to support "feature-based design." In a feature-based design system, a part is described in terms of "features." These features represent higher-level concepts than the geometric primitives used in traditional systems. One class of features includes slots, ribs, bosses, through holes, blind holes and pockets. These are called "form features" and correspond to specific geometric configurations on the surface of the part. Another class of features includes chamfers and fillets. These serve to modify the geometry specified by the form features. (For example, chamfers cause a square edge to be rounded.) The term "feature" is also sometimes used for attributes such as surface roughness; however, we avoid that usage here.

One of the novel aspects of our system is the ability to represent dimensioning and tolerancing information within the part model. Our representation for dimensions and tolerances conforms to system called "Geometric Dimensioning and Tolerancing" or GD&T. This system is a U.S. standard (ANSI Y14.5M) and is essentially the same as that specified in ISO standard 1101. Each GD&T datum or callout is represented by a feature within the part model.

The specific problem that we address here is how to use feature information to guide the inspection planning process. Features represent a higher-level description of the part than surfaces or primitive volumes that are traditionally used in CAD systems. In our system, each feature has a set of prepackaged GD&T specifications, as well as a set of "inspection code fragments" which can be used to check whether those specifications are met. However, a mechanism is also provided to allow the designer to override these prepackaged specifications, and work directly at the level of individual GD&T callouts. The inspection planner must integrate this information to produce a complete inspection plan. In this paper, we will assume that all inspection is to be done on a CMM. In a separate paper, we described an approach to producing plans for combined manual and automatic inspection [1].

This work is part of a larger project, sponsored by the U.S. Air Force, to develop a "rapid design system" (RDS). RDS is intended to support the fast and economical design of mechanical parts. RDS is being developed with the cooperation of the 4950th Test Wing, a design and manufacturing organization which specializes in custom modification of aircraft, e.g., to add new instrumentation and producing replacement parts which are not available from the manufacturer. For this type of application, turnaround time is more important than minimizing total machine time for manufacture and inspection.

The objective of the RDS project is to speed the design process by providing (1) an intelligent CAD interface which enables the designer to get his or her design into the computer faster than current systems allow; (2) integrated tools to check a design for manufacturability and inspectability.

The component of the RDS which comprises the intelligent CAD interface mentioned above is called the "Feature-Based Design Environment" or FBDE. This will be described in a later section.

GEOMETRIC DIMENSIONING AND TOLERANCING

Geometric dimensioning and tolerancing (GD&T) is a term applied to all dimensioning and tolerancing and inspection which adheres to the ANSI Y14.5M-1982 standard [2]. This standard applies to both engineering drawings (how the dimensions and tolerances are to be shown) and to the actual inspection (how is the part to be gaged and evaluated for acceptability). The fundamental concept of geometric dimensioning and tolerancing is that the specified tolerance and measurement procedure should relate to the part's function. For example, GD&T allows the specification of a positional tolerance of a hole such as might be typical of a requirement for parts to mate. In general, GD&T will specify a tolerance for a form (examples of this are the cylindricity of a round hole or the flatness of a surface) or a geometric relationship between features (an example of this is the specification of the position of a hole or the perpendicularity of a surface with respect to another surface). The dimensioning and tolerancing of a simple block with a hole in it is illustrated in FIG. 1.

GD&T is very well adapted to feature based design and inspection plan generation for automated equipment. First, it is inherently compatible with a feature-based part description. A basic dimension (used to position a tolerance zone) is readily generated from the part geometry. Tolerances as defined by GD&T are readily applied to design features such as holes, e.g. the position of a hole, and can be easily added to a feature-based CAD system as separate features. Since GD&T also provides a language for defining how a tolerance is to be measured (inspected) such information can be readily incorporated into the definition of a tolerance feature. The actual instance of an inspection procedure, i.e. how a particular feature tolerance will be measured and evaluated, is separately represented as an inspection feature in our system.

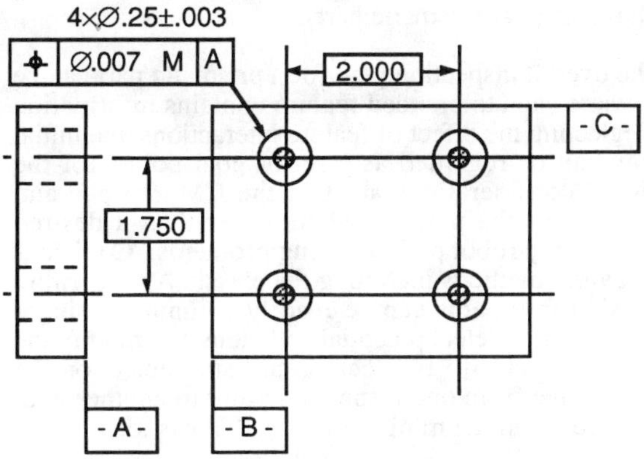

FIG.1 Plate with holes dimensioned and toleranced according to GD&T

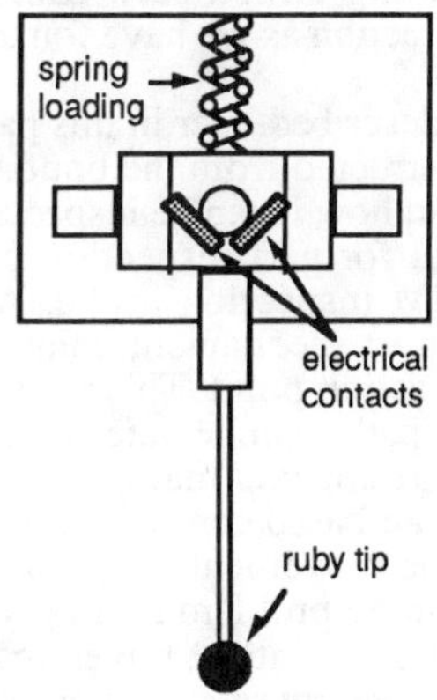

FIG.2. CMM touch probe.

COORDINATE MEASURING MACHINES

A major goal of the RDS is to enable the fast, turnaround of parts in small lot sizes. This implies that the majority of time allocated for inspection would be spent manually inspecting a produced part or writing inspection programs for automated equipment. We have rejected both approaches in favor of automatically generating inspection programs from the geometric and tolerance description of the part. The actual inspection will be performed with a computer-controlled coordinate measuring machine (CMM). Such machines are becoming very popular in industry and essentially are a gantry robot with a (optional) two degree-of-freedom wrist. This mechanical arm can position a touch probe to determine the location of a point on the inspected part's surface. The touch probe closes a pair of electrical contacts whenever the probe tip contacts a surface as shown in FIG.2. Because coordinate measuring machines are precision devices, the probe is constrained to approach a surface to be measured along an "approach direction" corresponding to a calibrated orientation. Typically such calibrations are performed for the probe either parallel or perpendicular to the measurement surface.

The result of the use of a calibrated touch probe produces several constraints upon how such a probe can be used for automated inspection. The first constraint is that the probe can only touch a surface to be inspected at a finite number of inspection points. The ANSI Y14.5M standard was originally defined for manual inspection using hard gages. The reconciliation of point sampling with tolerances specified according to ANSI Y14.5M is not straightforward. As a result we have used accepted industrial practices for the inspection and evaluation of tolerances in designs produced with the RDS.

PREVIOUS WORK

Two aspects of this work have been previously described in the literature. Elmargathy [3] has described a feature-based system for the automated inspection of turned parts. Her work used the ANSI Y14.5 tolerances as applicable to cylindrical parts and represents one of the first published uses of the ANSI Y14.5 tolerance specification as other than a drafting tool. The system was a rule-based system implemented in Prolog and concentrated upon how to access for inspection a free-standing turned part. Because of the nature of turned parts there was no feature interaction as we have found with rectangular prismatic parts.

As described later in this paper the overall inspection plan for a prismatic part can be constructed from the bottom up where each toleranced feature contains information about how it can be inspected. Neglecting the effect of feature interactions, the initial input for part inspection planning can be regarded as a set of goal points for the CMM inspection probe. Additional considerations are that the CMM probe and support mechanism cannot collide with the part anywhere other than a desired inspection point. This is identical to many robot path planning problems. Collision-free paths can be determined by several methods including the classic A* algorithm [4]. It appears that efficient collision-free paths can be readily computed using a feature-based configuration space planner to detect potential collisions and modify the probe movements to avoid such collisions [5]. Neglecting collision detection the planning problem is simply how to move from one inspection point to another with the constraints of traversing the entire set in the minimum time. This is identical to the traveling salesman problem.

THE FEATURE-BASED DESIGN ENVIRONMENT

The Feature-Based Design Environment (FBDE) is used by a designer to create or modify the part model. The part model consists of a collection of programming objects representing the design-with features, as well as others that describe manufacturing and inspection information, and still others that are concerned with constraint checking and constraint satisfaction. Associated with each design feature is a set of alternate schemes for dimensioning and tolerancing the feature, both internally and with respect to other features. The designer picks one of these schemes when he creates an instance of that feature. Each scheme translates to a set of GD&T callouts which can be displayed. If he is unhappy with the choices presented, he may specify different GD&T callouts for the feature by creating instances of "GD&T features."

PLANNING

The basic strategy of the inspection planner is to generate inspection code fragments (ICFs), which represent the instructions required to inspect individual features, then piece them together to form a complete inspection plan. Inspection planning is done at several levels. With each GD&T scheme within a feature class, a precomputed inspection code fragment generator (ICFG) is stored. The ICFG is a macro which, when called with the parameters of the feature, expands to an ICF which can be used to inspect a particular feature. The ICF is valid when the feature does not interact with other features. We say that a feature interacts with other features if they intersect geometrically in such a way that not all feature surfaces are present in their entirety. When a feature does interact with other features, planning is done on a surface-by-surface basis as follows: Each GD&T scheme has associated with it a set of sample requests and evaluation functions to be applied to the sampled points. A sample request is a request to sample one or more points from a given surface, along with constraints on the points (e.g., that two points must be at least a given distance away from one another). Since there may be several callouts referring to the same surface, it may be possible to use the same physical samples to satisfy several sample requests.Stated formally, the problem is:

> Given a surface S, let $P = \{P_i;\ 1 \le i \le m\}$, and $Q = \{Q_j;\ 1 \le j \le n\}$ be sets of variables representing points on S. Let C be a set of constraints on the points in P and let D be a set of constraints on the points in Q. We want to find values for all P_i and Q_j such that (1) all constraints in C and D are satisfied, (2) the set $P \cup Q$ is as small as possible—i.e., as many points of Q are identified with points of P, and vice versa.

This is a very difficult problem computationally. At present it is solved using the following heuristic:

```
select values of Pi that satisfy C.
for j from 1 to n do
        Find a value for Qj that satisfies C and D. Use Qjt= Pi for
        some i if it satisfies the constraints.
```

The algorithm backtracks as necessary when it gets "stuck." In the case of individually-specified GD&T callouts, sample requests are also generated and handled in an analogous manner. Finally, these sample requests and evaluation functions are turned into ICFs that can be merged into the overall inspection plan.

Inspection code fragments for checking the internal dimensions of features

Any of several possible inspection code fragments may be generated by the ICFG with the selection being determined by the feature's geometry or manufacturing process. Examples of inspection code fragments for inspection of basic form tolerances are discussed below.

Manufacturing processes also can change how a feature is inspected. Consider the flat surface produced by milling in Figure 4(a). Such a surface has a larger z-axis surface variation in the y-direction than in the x-direction. The surface variation in the x-direction depends upon the exact size of the cutting tool, how many flutes it has, the tool feed rate, the material being cut, etc. The surface variation in the y-direction will also depend upon the feed mechanism and the tool path overlap. The geometrically identical surface shown in (b) is produced by a grinding tool moving in the x-direction. The number of inspection points in the x-direction will be smaller than that for a milled surface due to the greater surface precision of a ground surface; however, the number of inspection points in the y-direction will not change from (a) because of the surface variation produced by the feed mechanism and tool path overlap.

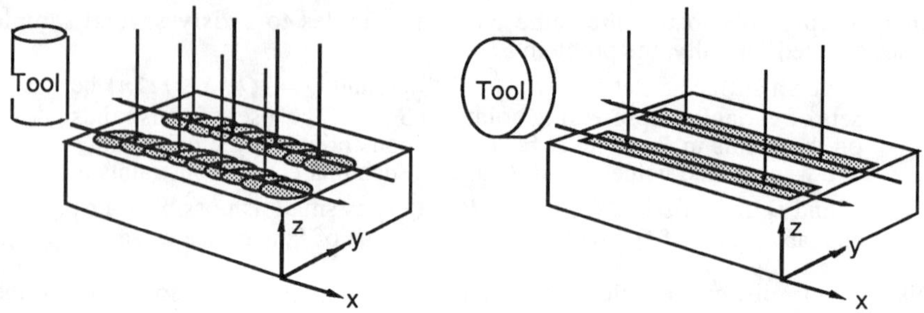

FIG.4 (a) Machining of a flat surface using a vertical mill (b) Machining of a flat surface using a grinder

Inspection code fragments for checking relationships between features

An inspection code fragment for a positional tolerance on a cylinder is shown in FIG.3(a). In practice the number and location of inspection points is a function of the hole "thinness," i.e., a tall and narrow hole might be inspected using two measurement planes with three inspection points each as shown in FIG.3(b).

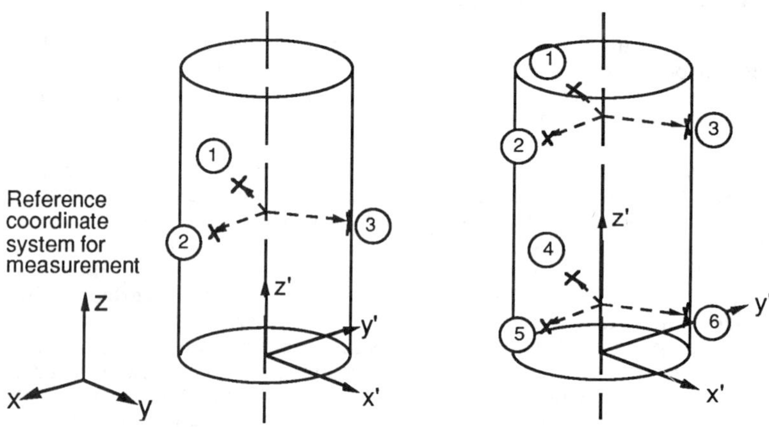

FIG.3.(a) Positional inspection of a hole (b) Positional inspection of a tall, thin hole

Inspection code fragment generation when features intersect

The previous two examples considered only individual features. Consider the pin bracket shown in FIG.5. The details of the part design are not necessary to this example; the hole is specified as a single hole passing through both supports. If the the position of the hole was toleranced, the nominal inspection plan would be similar to that shown in FIG.3(a). However, the hole feature is intersected by a pocket feature (which removed the material between the supports) requiring the inspection code fragment for the hole to be modified from that shown in FIG.3(a). The relevant constraint in this example is that the sample points must be on a physical surface. For the evaluation of a position tolerance it is not necessary that the sample points be on both supports. However, if we had specified a cylindricity tolerance in addition to the position tolerance a second sample request would have been generated. The cylindricity sample request would have required sample points on both supports. The resultant sample set might be like that shown in FIG.5 which can be used to satisfy both sample requests.

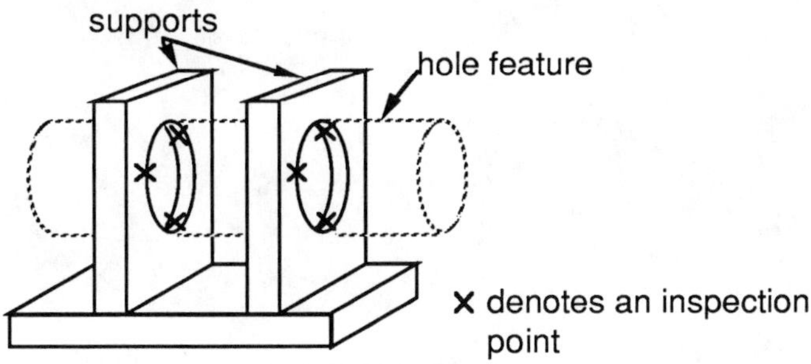

FIG.5 Pin bracket

SUMMARY AND CONCLUSIONS

We have described an approach to intelligent generation of inspection plans for CMM inspection of mechanical parts. Tolerances are specified according to the ANSI Y14.5M format. For features that do not interact with other features, inspection code fragments are generated based on encapsulated knowledge about the features including how they were manufactured. For features that do interact, sample requests are generated for each specified tolerance on the surface to be inspected. The resultant composite sample request will satisfy all requested tolerance evaluations as well as minimize the total number of sample points for that surface. The inspection plan is then generated on a surface-by-surface basis. This approach shows promise for fast generation of reasonably efficient inspection plans suitable for small lot-size manufacturing applications. We are presently pursuing the development of better heuristics for optimization as well as for generation of the overall part inspection plan.

REFERENCES

1. Radack, Gerald and F.L. Merat, "The Integration of Inspection into the CIM Environment", Hawaiian International Conference on System Science, January 1990.

2. Dimensioning and Tolerancing ANSI Y14.5M - 1982, ASME, 345 East 47th Street, New York, NY 10017.

3. H.A. ElMaraghy and H.P.Gu, Expert System for Inspection Planning, Annals of the CIRP 1988, p.85-89.

4. Pete E. Hart, Neil J. Nillson and Bertram Rafael, "A Formal Basis for the Heuristic Determination of Minimum Cost Paths," IEEE Transactions on Systems, Man and Cybernetics, MSSC-4, No.2, July 1968, p.100-107.

5. Michael Branicky and Wyatt Newmann, Rapid Computation of Configuration-Space Obstacles, to be presented at the 1990 International Conference on Robotics and Automation

THE APPLICATION OF DISTRIBUTED PROBLEM SOLVING IN CIM

S.J.Zhang G.N.Song Y.Zhang L.Li W.Wang Z.Gao H.Peng
Shenyang Institute of Automation
Chinese Academy of Sciences
Shenyang, 110003, P.R.China

ABSTRACT

This paper presents a distributed artificial intelligence system (DAIS) which is applied to computer integrated manufacture (CIM). It integrates CAD, CAPP and CAM into one system. Because of adopting distributed problem solving and communication technique, the computer resources located in different sites are fully used. The distributed artificial intelligence system which consists of several processing nodes can operate in parallel. Architecture of system running on three nodes are identical based on blackboard. At present,the system has been working on SUN computers connected by ETHERNET.

INTRODUCTION

Distributed problem solving is the cooperative solution of problems by a decentralized and loosely coupled collection of knowledge sources(KS's)(procedure, sets of rules, etc.) located in a number of distinct processing nodes. The KS's cooperate in the sense that no one of them has sufficient information to solve the entire problem; mutual sharing of information is necessary to allow the group as a whole to produce an answer. By decentralized we mean that both control and data are logically and geographically distributed; there is neither global control nor global data storage. Loosely coupled means that individual KS's spend the great persentage of their time in computation rather than communication. Distributed problem solvers offer advantages of speed, reliability,extensibility, the ability to handle applications with a natural spatial distribution. Because such systems are highly modular they also offer conceptual clarity and simplicity of design [4].

Using the idea of distributed problem solving, we developed a system into which CAD, CAPP and CAM are integrated.

This application is looks like a serial distributed processing. In order to improve processing efficiency, make an optimal use of processor resourses, reduce the work of user, we have developed them into distributed problem solving system based on blackboard. In this way, the subroutines of CAD,CAPP and CAM can work together serially or parallelly. Any processing node can be taken as a main processing node. Using communication technique, KS's in different processing node can be triggered in data-driven manner to execute special task. So, the flexibility of system is increased. In our system internode communications are carried out by means of a blackboard. A blackboard, an active data structure located at each processing node, allows information sharing among the nodes by storing tasks, plans, and partial results and transmitting them at the appropriate time in the problem-solving process. In addition, the blackboard provides a means for sharing information about the different problem-solving processes within the node.

The architectures of system in different processing nodes are identical. It consists of knowledge acquisition module, front-end processor which is in charge of receiving and transmitting data, task decomposition, evaluater and post-end processor and interpreter. The system on each processing node is possessed of the following capabilities for problem solving:

1) A dynamic planning ability which adjusts the problem-solving plan and guides it;

2) An internode communication facility that permits the different processing nodes to share tasks and results;

3) A problem-deduction ability that solves tasks by invoking required knowledge sources.

The system architecture in one processing node is shown in Fig.1.

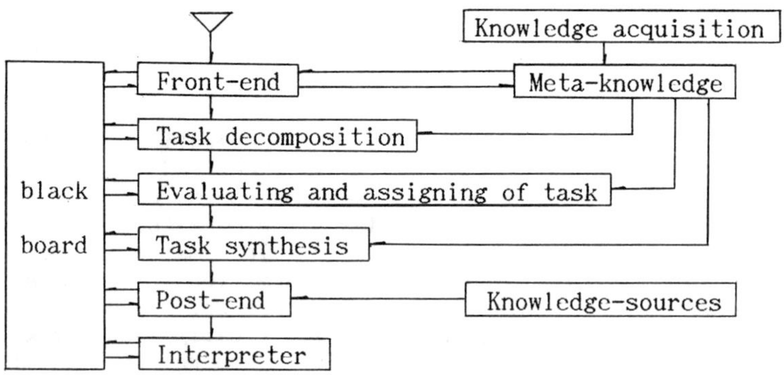

Fig. 1. The architecture of system

KNOWLEDGE ACQUISITION

The meta-knowledge is about how to decomposed the task and the resources of each computer. Meta-knowledge and knowledge about the resource of individual node are represented by frame. For example, the knowledge about one task is represented as follows:

```
TASKNAME     : identifier
FUNCTION     : sequence
SUBTASK1     : identifier
SUBTASK2     : identifier
SUBTASK3     : identifier
REQUIREMENT  : yes
DATA         : file-name
```

MEMORY : digit
EXECUTE-TIME : digit
LOAD-BORDEN : digit

In order to make computer processing conveniently, the system provides a knowledge acquisition subsystem. It translates the meta knowledge into internal representation of computer and build up knowledge base for the system. The menu provides an interface between user and computer. The main framework of menu is shown in Fig. 2.

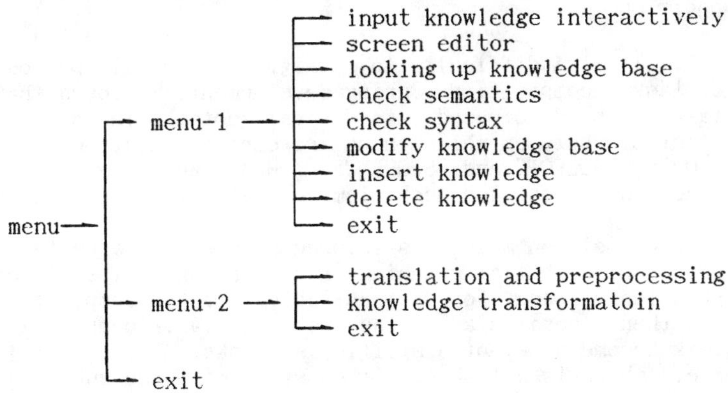

Fig. 2. Framework of Menu

During knowledge acquisition period, the prompt will be given on screen. Then user may choose corresponding items, do some required processing. It's very convenient for the user to acquire knowledge interactively. Subsystem provides also some modules of semantics and syntax checking, adding, deleting and modifying. These functional modules guarantee the correctness of meta-knowledge input by user, and the easy of meta-knowledge base extention.

FRONT-END PROCESSOR

The main function of front-end processor is to allow processing nodes to communicate between each other, share tasks and results. Front-end processor consists of transmitter and receiver. The transmited and received messages are stored in the two smaller blackboards before processing. Namely they are called transmitting blackboard and receiving blackboard.

A common language is required for all processing nodes to communicate their intentions and share information with each other [1].

Broadcast transmissions and direct transmissions are used in the systems. Five types of message are used during communication.

(1) evaluation: Asking for the information about resources and load of other processing nodes.

(2) response: Answering the information about resources and load

of other processor nodes.

(3) task: Transmitting or receiving task.

(4) answer: Giving task executing status and results.

(5) data: Indicating datafile or information needed by the task.

When processing nodes receive message from another node, the front-end processor will do some necessary processing depending on the type of messages, then store it into blackboard according to appropriate format. When a processing node transmits a message, front-end processor puts the message in the transmiting blackboard, then transmits the message to another nodes.

TASK DECOMPOSITION

In general, it is difficult to design a general purpose algorithm of task decomposition. During development, we found that the major task can be decomposed into several subtasks in the CIM and other fields, between which there are certain relations such as "AND", "OR", "SELECT" and "SEQUENCE". Hence we designed the function-guided decomposition mechanism to deal with this kind of problems.

The task decomposition module is invoked when there are tasks whose solutions need initial planning or replanning. It searches the meta-knowledge base for a suitable task-decomposition mechanism and then generates a new task-solving plan. In this way, the major task become a set of simplified subtasks.

For example, there is a task A. It consists of three subtasks, B,C and D, and there is a "AND" relation among them. That is, we can solve task A by breaking it down into three smaller subtasks, each of which can be solved using a smaller collection of specific knowledge. Task B consists of E,F and G. there is a "SEQUENCE" relation. It means that system executes them sequentially. Task C consists of H,I and J. Either I or J will become subtask of them, and it is depending on the value of H. So there is "SELECT" relation between I and J.

Figure 3 shows the task tree that will be generated by the process of task decomposition.

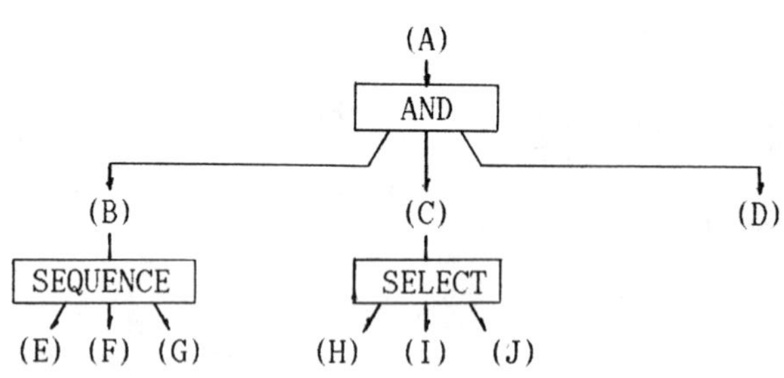

Fig. 3. Task Tree

When leaf node is generated during task decomposition, the

corresponding program which will perform special task is invoked. The whole decomposition process is executed in a global database (blackboard). Any result, information which is about whether a subtask is solvable,solved, or unsolved, decomposition status, evaluating information are also contained in the blackboard. Thus, the blackboard contains all the relevan t task-level problem-solving information. Under the control of the planner, a set of knowledge sourses are triggered, the task is decomposed, task is solved finally.

TASK EVALUATION AND ASSIGNATION

In order to make full use of computer resources, and distribute task onto processing nodes reasonably, it is necessary to evaluate the task according to requirements about resourses, status and load of each processing node. Pertinent to the evaluating score, the evaluator may decide where the task will be assigned in this case.

The memory, disk, load and availability of software are all considered by the evaluator. These information are stored in the blackboard. During evaluation, the evaluator checks whether the task itself has evaluating requirment or not. If the value of "REQUIREMENT" is 'yes', the evaluating algorithm is invoked, then the task is assgined to the appropriate processing node. Meanwhile user can directly assign the task to the processing node he wants.

TASK SYNTHESIS

Task synthesis is the opposite of task decomposition. It synthesizes the results from knowledge sources on different processing nodes (or the one processing node) and judges whether the task is solved or not. This process is also function-guided. The direction of synthesis is from bottom to top.

We illustrate synthesis process by use of the above-mentioned example. When synthesis module is working, it checks whether task E, F and G are solved. If E, F and G are solved task B is solved. If value of H shows that task I will be excuted, C is solved after task I is solved. If the B, C and D are all solved, task A is solved and the system is stopped.

When tasks are executing, they may be executed on the different processing node. In this way, results come from several processing nodes, so it requires that the functions possess with the ability of the distributed problem solving.

POST-END PROCESSOR

Post-end processor is in charge of scheduling and communicating within a node. It has two main functions:

A) Putting tasks into executing task list;

B) Performing task and result processing;

According to a set of constraints post-end processor put tasks with priority into an executing list. Before executing task, post-end processor prepares the datafile the task needs, executes the task and then put the tasks into result file.

Another job of the post-end processor is result processing.

Because of concurrent programming, post-end processor put the content of result file into blackboard, set executed mark. On the other hand, it judges whether the result of task will be sent to another processing node. If it does, corresponding processing is done. Meanwhile post-end processor needs to check whether the task has output datafile. If it has, post-end processor broadcasts data-typed message to all nodes, record the position of datafile. These information will be used by system lately.

INTERPRETER

Interpreter is embedded in the system of each processing nodes. By means of interpreter user can well know the status of task and working environment of each processing node.

Interpreter based on blackboard structure adopts the intuitive way perceived to display entire tree structure of task. It can obtain the information from other processing nodes, and display the contents dynamically, such as the status of task decomposition, task assignation, task execution and valid cycle. Because of providing the system with graph display, transparency of the system is highly improved.

CONCLUSION

A prototype system of distributed problem solving is running on SUN computers located in AI lab of Shenyang Institute of Automation. It intergrated CAD, CAPP, and CAM into one system. The record of working shows that the structure of control and communication is reasonable. The result is satisfying. Menu-based interface provides very convenient tools for user to acquire knowledge. Subtasks are written by C and FORTRAN. The communication between different languages is achieved. Flexible representation and data structure make the system extent easily and apply it to other fields.

REFERENCES

1. J.D. Yang, M.N. Hunns and L.M. Stephens, " An architecture for control and communications in distributed artificial intelligence system," IEEE Trans. Syst., Man, and Cybernet., Vol. SMC_15 No.3 MAY/JUNE 1985.

2. V.R.Lesser and D.D.Corkill,"Functionally accurate,cooperative distributed systems," IEEE Trans. Syst., Man, Cybernet., vol. SMC_11,no.1, pp.81_96, Jan. 1981.

3. D.D.Corkill and V.R.Lesser, "The use of meta-level control for coordination in a distributed problem solving network," in Proc. Eighth Int. Joint Conf. on Artificial Intelligence, Karlsruhe, West Germany, vol.2, Aug. 1983, pp.748_756.

4. R.G.Smith and R.Davis, "Frameworks for cooperation in distributed problem solving," IEEE Trans. Syst., Man, Cybernet., vol.SMC_11, no.1, pp.61_70, Jan. 1981.

5. R.G.Smith and R.Davis, "Negotiation as a metaphor for distributed problem solving," Artificial Intelligence, vol.20, no.1, pp.63_109, Jan. 1983.

Mission planning and execution for a mobile robot using artificial intelligence techniques

Jean-Marc Martinez *
Jean-Philippe Nominé **
Commissariat à l'Energie Atomique
** DRN/SERMA/LETR*
*** DTA/LMRS*
CEN Saclay - 91191 Gif-sur-Yvette Cedex - France

ABSTRACT

Intervention mission planning in nuclear power plants using mobile robots can be improved by artificial intelligence techniques. This work emphasizes the interest of designing the robot control system as a multi-agent software system. The purpose is to better specify and plan missions, and to perform navigation management, while allowing an operator to oversee the mission and to possibly intervene during its execution. An example of mission simulation is presented.

INTRODUCTION

The aim of this study is to develop a demonstration simulation prototype to qualify and test appropriate methods and techniques for solving some problems in the field of mobile robots. A model mission for the robot is specified within the scope of a research and development program at CEA : SYROCO (CEA stands for French Atomic Energy Commission). This mission consists in an intervention in the Nuclear Service Building (NSB) of a 1300 MW Pressurized Water Reactor (PWR) after a light incident. The diagnosis of the incident is considered done : a preliminary analysis of fluid circuitry showed for instance that a valve has to be closed, or it is necessary to go and test the current state of a valve. State of the art solutions involve teleoperation in some way, for manipulation as well as for motion. Constraints of different kinds (cluttered and possibly hostile environment, moreover not prepared to help robot locomotion and action in general) oblige to keep some human presence in control loops at different levels. So we only deal here with the sub-problem of improving the robot capacities with respect to motion and general mission management. Roughly, a typical mission will be for us : "go near valve Nr N", regardless of the effective destination activity with respect to the valve.

Such a mission involves some of the problems encountered in mobile robotics, such as navigation, piloting, perception and task planning. The purpose of the study was triple:

- take into account both off-line an on-line knowledge about the robot environment, and plan a mission with respect to this knowledge

- set a framework, that is a software architecture, for having the different needed skills, implemented through specialized agents, cooperate towards mission execution

- perform simulations for preliminary test and validation of the developments achieved from the previous ideas

The studied case uses a simple two wheel robot equipped with a range finder. Though not realistic with respect to locomotion constraints in a NSB (which would certainly require tracks instead of wheels due to the non flat nature of the floor), this example highlights some of the problems that have to be taken into account in such an approach.

Developments in progress use some existing software components developed at CEA. ARES (Robotics Simulation Tools) is a software environment dedicated to developing and testing of robotics task programs. It mainly consists in a simulator taking into account moving and sensory polyarticulated systems [1]. SPIRAL is a toolbox for building knowledge based systems [2]. In addition to a PROLOG-like formalism, it provides mechanisms for object representation (classes, simple inheritance, daemons), concurrent multitasking, integrated fast graph search procedures, and inclusion of external precompiled procedures.

IDEAS FROM THE FIELD OF AUTONOMOUS MOBILE ROBOTICS

A quick overview of approaches of some problems in autonomous mobile robotics will help choosing a number of concepts that we will include in our own approach.

A mobile robot operating totally or partially autonomously during a mission needs a number of skills that have each to bring their contribution to the mission. The way they will work together is a question of functional architecture, whereas the skills in themselves, once identified, rely on various techniques from geometry, graph search, plan generation, for instance.

Given the nature of our mission, we will focus on the following skills:

- navigation and path planning
- management of a model of the world for decisional purposes
- mission planning

before raising the problem of organizing these skills together.

Navigation and path planning

Navigation implies the ability to locate oneself and to find a path to reach a given aim. The localization problem for mobile robots is a difficult one. Various approaches have been proposed and tried, using odometers, inertial systems, external beacons or landmarks, natural or artificial. In an environment such as NSB, where pre-equipment with beacons is excluded, the robot has to rely upon its own internal capacities. In this first step, we only assumed odometric localization. This approach is very limited and error bound, but we only wish to bring the robot in the vicinity of a point of interest, where the control has to be switched to another mode for the destination activity (e.g. valve recognition and inspection or manipulation). Using natural landmarks (recognized pieces of the environment itself) would be a richer and complementary approach [3].

Path planning consists in finding a safe path in the cluttered environment. This environment has to be known some way. According to the nature of this knowledge and the way it is acquired, considered and used, one can distinguish between local and global approaches. A global approach uses a given static knowledge about the environment to compute a safe path for the mobile object among obstacles. Different kinds of reasoning are possible [4]: in the configuration space, by structuring free space. A global planning may be more or less approximating or heuristic, in order to accelerate computations, which are long if they are to be systematic, and is correct as long as the model used for planning corresponds to the current state of the world.

Local planning can generate finer motions from a model issued from suitable a priory knowledge of the immediate environment, possibly mixed with sensor data. Local planning allows optimization, refinement or on-line adjustment of motions.

Eventually, planned motions must be executed, knowing that kinematic constraints may forbid the mobile robot to perform arbitrary motions. Global planning may be performed more independently from such constraints than local planning.

Assuming our a priory knowledge about the environment, a global approach has been chosen for path planning. It uses a network of decision points described as a graph, and an A* algorithm to find optimal paths. A NAVIGATOR task performs the graph processing. A combination of object representation with SPIRAL and of precompiled graph search procedures integrated besides PROLOG basic mechanisms makes this operation very fast. So replanning can be performed often (if the global model is updated) without much loss of efficiency.

At the present day, motion execution is directly derived from global planning: the robot moves between decision points on straight lines. It can change its orientation on these points. These motions are compatible with the robot kinematics (two independent driving wheels), although far from optimal. Nevertheless, stop points are well suited to range finding, to decide whether to replan a path or to go on moving along the previously planned path. A PILOT task executes global motions piecewise. Though no local planning has been implemented yet, this PILOT task could be an actor of it, together with a perception task. For instance, ultrasonic range finders could be used for local obstacle avoidance, in order to adjust locally a piece of the global path between two decision points.

Management of a model of the world for decisional purposes [5]

Initializing and maintaining a model of the robot environment is a key capacity, around which a number of skills will articulate. We chose a global revisable model, built from a priory knowledge and range data read at a relatively low frequency. This approach is correct for global planning, and should be completed by a lower and faster acquisition loop, in order to improve security, local planning and possibly localization.

The robot internal representation of the world has three levels :

- a geometrical model is made of segments (polygonal 2D image of the obstacles). The initial set of segments is obtained by translating available maps of the NSB. This model is generally inaccurate and incomplete. So on-line range data (read at the decision points of the robot motion) may be used to update this model. The navigation process is initialized from the set of a priory data, which can anyway be empty. The raw data are filtered, projected and gathered into plausible segments added to the initial set.

- a topological model is derived from this set of segments. This model consists in a graph of decision points, obtained heuristically by combining informations from the set of segments and from a grid. It ensures that at a given time, assuming a static world between two decision steps, the robot can travel safely between the decision points. Any modification of the geometrical model causes this graph to be updated. This level of modelization organizes the free space for direct use in path planning (A* algorithm).

- some additional data are gathered in a more semantic level. These data are mainly the position of objects of interest (e.g. valves) with respect to the geometrical model. They are used in the mission general planning.

These models are all described in an object centered way, on which SPIRAL general inference and specific graph search mechanisms can act directly. Some parts of processing (sensor data, from raw to segments) are however coded in procedural external precompiled language, properly interfaced.

A PERCEPTION task manages range data acquisition and processing and is able to update the geometrical model directly, which induces topological model modifications.

Mission planning and software architecture

Planning at a task level in robotics has first been approached by plan generation techniques, likely to produce totally or partially sets of actions to be performed in order to achieve a given goal. More or less hierarchical schemes have been proposed. It must be pointed that a planner in itself only provides a canvas of actions to be performed, whose execution must be controlled and monitored in order to detect any problem that would arise, and to repair or replan if necessary [6]. Then another concept of planner was proposed, after the introduction of the blackboard structure [7] [8]. A blackboard is a common database through which specialized modules coordinate. Which such a control structure, the elaboration of a plan is incremental and the planning decisions are selected according to their opportuneness.

More generally, it appears that decision autonomy for a robotics system requires the organization and cooperation of numerous and different kinds of knowledge sources. A suitable architecture must be defined to support and coordinate these different knowledge sources. Present trends are based on distributed artificial intelligence techniques, with which problems are handled by a decentralized group of agents [9] [10]. An agent is an entity which may range from a simple processing element to a complex entity exhibiting rational behavior of its own. Information is shared some way to contribute to the solution of the problem. Information sharing may range from a full sharing through a common database (blackboard) to a minimal sharing by message passing.

We chose a solution inspired by the latter case. It is a more simple and modular approach, but it requires to define a communication protocol. Some agents have already been naturally introduced by the identification of needed skills and the natural sharing of actions and processing to perform. In addition, a SUPERVISOR was defined, which knows 'who can do what', plus an OPERATOR agent, playing the part of the human interface.

The OPERATOR defines the mission and can intervene during execution to change the operating mode (between 'autonomous' or 'step by step', that is by slices of agent actions), stop the execution, and even manually provide additional information during the mission. The SUPERVISOR first ensures the semantic interpretation of the mission defined by the operator, then organizes the message exchanges between the other agents.

Classes of objects Agent and Message have been defined. Each agent has a mailbox which it monitors periodically. The Agents are tasks managed concurrently by SPIRAL, each task possessing its own inference cycles. The general planning activity results from the interaction of the five agents : OPERATOR, SUPERVISOR, NAVIGATOR agent (path planning), PILOT agent (motion control), and PERCEPTION agent (verifications before PILOT motions, and world models updating).

Some facts and data are useful to more than one specialist. In order to avoid full duplications of information, a partial sharing of world model(s) has been allowed. For instance, the PERCEPTION and NAVIGATOR agents have both access to the models of the world, but not in the same mode nor at the same level (PERCEPTION reads/writes the geometrical level, NAVIGATOR reads the topological derived level).

SIMULATION

Before trying to implement any real case benchmark, a first version of this software has been developed around SPIRAL in a simulation context. Simulation consists in running a specific module concurrently with the above described software. This module built within ARES environment acts as a virtual robot in a CAD world. To avoid binding the prototype software with the simulation kernel, and confusing internal decisional models of the robot used for mission synthesis with external CAD models used for mission analysis, there is no sharing of mod-

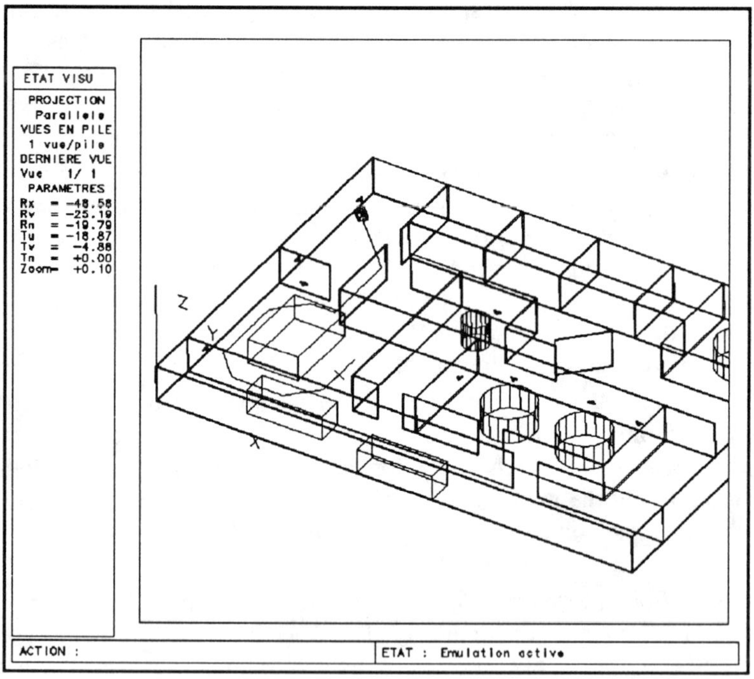

Figure 1 : the simulator module showing an achieved mission in a part of NSB els and data between the simulator and the robotics software, but through emulation channels representing actuation and perception of the robot.

More precisely, the simulator offers 3D modeling, animation and visualization functions, and accepts requests of two kinds :

- motion requests : increments of positions of the robot wheels computed by the PILOT agent are integrated to follow the trajectory

- perception requests : emulated raw range data are returned to the PERCEPTION agent for further processing

The simulation can be run on two workstations communicating by networking mechanisms. One runs the simulator, as an external witness of the robot behavior (Figure 1). The other uses SPIRAL graphics interfaces to highlight the robot behavior and internal models evolution in a didactic way (Figure 2).

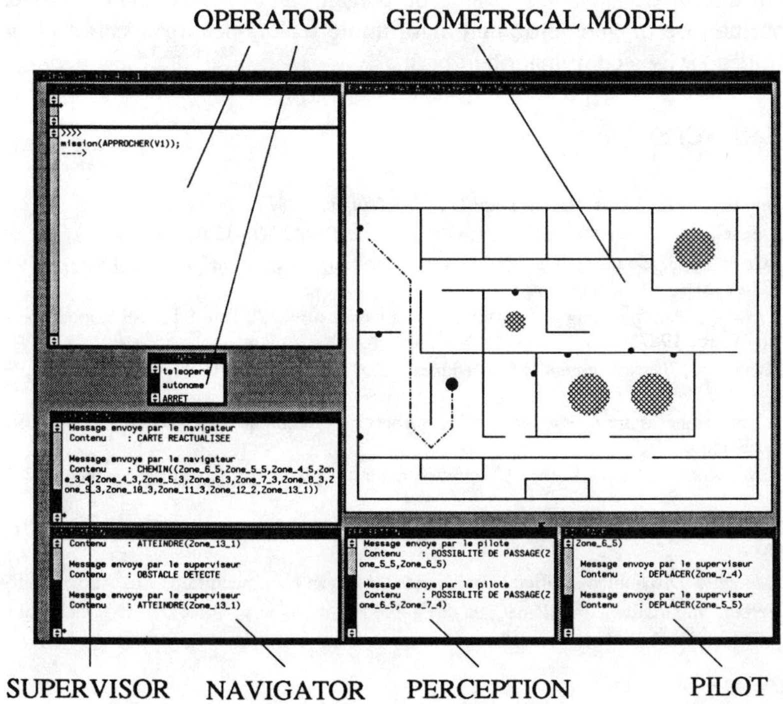

Figure 2 : intermediate state of the mission execution, from the internal point of view of the robot - a piece of initially unknown obstacle has already been added to the geometrical model and the path replanned - the path will still be replanned after turning the corner and discovering the rest of the obstacle, before producing the effective path of Figure 1.

In a real case, the simulation workstation could still play an interesting part, as a remote operator interface where various informations could be displayed or prompted for (as an auxiliary or extension of the OPERATOR agent).

CONCLUSION

This paper presented a mobile robotics problem handled by a set of decentralized agents, among which a human operator, in order to improve an intervention mission, both globally at the general level of mission management, and locally with respect to some basic skills, such as path planning. It stressed the importance of functional architectures and proved the flexibility of such an approach.

Once the general canvas defined, each agent can be individually improved by appropriate use of geometric and algorithmic techniques, of specific planning techniques, or by a combination of both.

REFERENCES

1. D. Chaigne, CEA Internal Report DEMT/87/318 (August 1987)

2. Y. Souchet, Spiral User Manual, CEA Internal Report DEMT/87/412 (October 1987)

3. J.L. Crowley, J. Couttaz, "Navigation and modelling for a mobile robot", TSI 5-5, 391-402 (AFCET, 1986)

4. J.L Crowley, "Path planning and obstacle avoidance (A survey)", Tutorial IEEE Robotics and Automation (March 1987)

5. J.P. Laumond, "Raisonnements géométriques", PRC-GRECO, 43-108 (Toulouse, France, March 1988)

6. L.M. Camarinha-Matos, "Plan generation in robotics", Techniques of Robotic Control, 291-328 (Elsevier Science, 1987)

7. B. Hayes-Roth, F. Hayes-Roth, "A cognitive model of planning", Cognitive Science 3, 275-310 (1979).

8. E. Crochon, A. Koenig, "Temporal reasoning in TRAM", AIE/IEA 590-597 (Tullahoma Tennessee, 1988)

9. M.N. Hunhs, "Distributed Artificial Intelligence" (Morgan Kaufmann Pub., Los Altos, California)

10. J. Ferber, M. Ghallab, "Problématique des univers multi-agents", PRC-GRECO, 295-320 (Toulouse, France, March 1988)

A PROGRAMMING ENVIRONEMENT FOR DECLARATIVE AND/OR PROCEDURAL PROGRAMS

D.DUHAUT, P.BIDAUD, D.FONTAINE

Laboratoire de Robotique de Paris
CNRS - Université Paris 6 - ENSAM
Tour 66, 4 place Jussieu, 75252 Paris Cedex 05

ABSTRACT

We present a programming environment for robots which we have implemented, based on the Ada language. With a logic programming language adapted to real-time constraints and an interpreted language (IAda) to solve problems of dynamic configuration of software, we have, with Ada, a complete system for robot programming. We show how we use it to increase the reliability of robot software.

INTRODUCTION

Because a robot must face unexpected failure, we believe that a law to be respected for robotic software systems is:

Reliability of robotic software
=
Efficient concurrent programs
+
A quality system of control

Efficient concurrent programs execution requires a powerful programming language with real-time facilities.
A quality system of control needs, to supervise the execution of low level tasks, declarative programs to perform lesser-known algorithms and interpreted programs to adapt to an environment dynamically (the software system to the real world).
Our aim is to construct a software environment which takes into account these constraints. Of the few languages which can be used to construct this environment [CO89] we have chosen the Ada language. This specification constitutes, for us, a good base because Ada is the only language which includes the following properties :

- it supports concurrency and allows static or dynamic task scheduling,
- the 'Rendez-vous' mechanism on tasks is very powerful and the Pragma Shared utility can be used to protect access to shared data on which one can easily assign addresses,
- the interrupt 'entry' for Ada tasks can be performed to solve hardware dependency,
- the language is standardized [DOD83] this means that there are 'good chances' on portability,
- the exception treatment is useful to keep software control in unexpected situations,
- the cross compiler of Ada gives independence from the operating system.

But Ada is only a procedural compiled language [BA82], and must be enhanced in order to take into account declarative programs and interpreted computation.

To accomplish this, we first add a Prolog inference engine written in Ada, which we developed, and adapted to real-time computation. This was inspired from that of Van Canaghem[CA 86]. Our second task was the construction of the IAda (Interpretor on Ada) language which contains a set of instructions to activate or deactivate Ada compiled primitives and control schemes to build programs. In the first part of this paper we present the IAda language.

In order to program a flexible manufacturing cell, we would like to construct programs which include both the aspects of "acting/sensing" and "control". By the "acting/sensing" aspect we mean all software tasks with an operative action in the cell organisation be it a low level control task, a trajectory generation program, a picture extraction, etc ... The "control" aspect is all software systems verifying the validity of "acting/sensing" tasks.

We distinguish two ways to combine acting and control:

- the "control" is included in the "acting/sensing" system:

Each program supervises it's own action. There are multiple inconveniences to this approch:

° the algorithm must be completely deterministic to known where and what kind of control must be performed.

° where do we stop verification (do we supervise the actual control?.)

° the control is reduced to one particular task and can not be easily distributed on a set of tasks.

- the "control" is disjunct with the "acting/sensing" system:

To each "acting/sensing" system we associate a "control" which runs independently. Here control is logically at a meta-level of the acting system. We can imagine that "control" needs also to be supervised. In this case we have a "meta-control" of the preceeding "control".

In the second part of this paper we will define the notion of "agent" and "meta-agent" to introduce this kind of programming.

In a third part, we show how the preceding programming environnement is useful to develop this kind of programming.

1. INTERPRETED COMPUTATION

To solve the problem of interpreted computation, we construct a language IAda which manipulates low-level primitives, like a debugger. These primitives are Ada compiled elements:

-Ada tasks and their entries,

-Ada procedures and their parameters, (an Ada procedure can be a Prolog inference engine on a prolog program input)

-global variables.

Let S be the Ada source program which contains all the primitives. By rewriting S in an Ada source program D we construct an IAda interpretor taking account of the primitives of S. The computation of the D program interprets IAda programs in which key words are IAda instructions and the Ada primitives defined in S named by their interface.

For instance:
if we define in S a task:

```
  task type ENGINE is
        Entry MOVE ( VALUE : in VECTOR );
                          ....
```

```
   End ENGINE;                  -- Define the type ENGINE ...
GRASPER_ENGINE : ENGINE;        -- Define a task variable GRASPER_ENGINE
```

We construct the D program and then we can write an IAda program with the instruction:

RENDEZ_VOUS (GRASPER_ENGINE.MOVE (POSITION))

RENDEZ_VOUS is an IAda instruction which demand an Ada rendez-vous with the task GRASPER_ENGINE on the entry MOVE and gives in data the value of the variable POSITION.

1.1. ADA PRIMITIVES MANIPULATIONS

Manipulation of procedure primitives:

```
EXECUTE ( subroutine );
        -- which compute the subroutine using passed arguments
```

Manipulation of task primitives:

```
DETRUIT ( task )
        -- abort a task
DESACTIVE ( task )
        -- suspend the task
ACTIVE ( task )
        -- setting a task
RENDEZ_VOUS ( task_entry )
        -- demand an Ada rendez-vous on the entry of an Ada task.
```

Manipulation of global variables:

Global variables are defined in the S program. To use them in a multi-processor environment the following informations must be defined in IAda:

-variable type to specifie the size of the variable
-variable utilisation (read, write)
-the updating period(if it is located on another CPU)
-variable access (RS 232, Memory address, network ...) to know how to fetch the values which are not directly accessible by Ada primitives.

The IAda interpretor will update all the global variables.

1.2. CONTROL SCHEMES

We keep in IAda the classical:

Loop:
```
LOOP
...
EXIT WHEN <test>
...
END LOOP
```

Alternative:
```
IF <test> THEN
....
ELSE
....
END IF
```

```
<test> is         BOOLEAN              -- to test expressions
                  or
                  TERMINEE( task )     -- to test if a task is over
                  or
                  <Exception_Name>     -- for treatement of Ada exceptions in IAda
                                       -- programs. This is important to react to
                                       -- unexpected problems in computation
```

Carriage return is used to separate instructions.

We did not define assignments on variables. However, if it is needed in an IAda program, it is possible to add it by using the EXECUTE() instruction on an Ada assignment primitive.

1.3. IADA AND DYNAMIC CONFIGURATION

Because IAda is an interpreted language, it is possible to dynamicaly rewrite the IAda program. To make this feature easy to use, we have introduced in IAda the following four instructions :

```
-EXECUTE_IADA ( reference )
            --which introduces the notion of subroutine in IAda

-RETOUR
            -- return from subroutine

-RESET
            -- restart the IAda program at the beginning

-STOP
            -- suspend the IAda interpretation
```

A more complete description of IAda laguage and its implementation is in [DU 89]

2. THE AGENT / META-AGENT STRUCTURE

In the following we will mean, by "control-power", actions like: DETRUIT, DESACTIVE, ACTIVE described earlier.

The software architecture of a robot is a set of cooperating tasks in which we can distinguish two classes of tasks:

-the tasks having no control-power on the other tasks

-the others which have control-power on the other tasks.

Let an agent be a task. A meta-agent is an agent able to change the action of, at least, one other agent. A meta-agent must have the control-power over other agents and its duty is to decide if their activity is valid. This means that a meta-agent must look at an agents activity.

In program proof [MA 74] we have similar levels when we assign a property to an instruction. This assignement has the form: "at this position in the program, a property $P(x_1,.. ..,x_n)$ is true". The parallel between the agent/meta-agent notion is given, in programs, by language/meta-language definition.

In a robotic system the program may be proved statically, but external failures may introduce invalid data that a classical proof cannot take into account. So we have to verify dynamically the properties of the programs. This dynamic control is done by the meta-agent who will run a dynamic proof on an agent or a set of agent.

Since the meta-agent needs to read the agents behaviour, the agent's actions must be visible to others agents. This leads to the notion of visibility. The visibility is a set of information which can always be accessed.

We introduce, for a agent A_i, in our programs two kinds of variables to define visibility:

- D_i a set of data of A_i.
- P_i a set of program pointer of A_i.

The meta-agent reading P_i, for an agent A_i, knows what state is reached in the computation and then what properties must be verified on the data D_i. In general, a meta-agent can act on an agent A_j by the knowledge of $<D_k,P_k>$ ... $<D_{k+n},P_{k+n}>$ for agents $<A_k,\ldots\ldots,A_{k+n}>$. In this case the set $<A_k,\ldots\ldots,A_{k+n}>$ of agents may include an agent which is already a meta-agent. So we can develop a hierarchy of control in software with a ladder of meta-control.

3. EXPERIMENTATION

With the IAda language, Prolog inference engine (which can be considered like an IAda primitive) and Ada, we have a complete environement for agent/meta-agent programming.

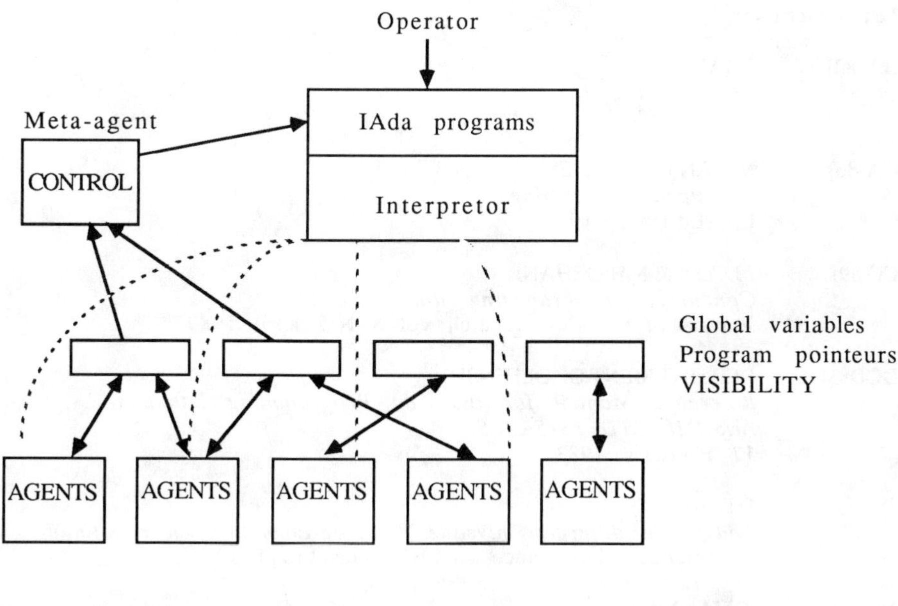

In our development:

- agents A_i are Ada procedural tasks "acting/sensing" in the cell.
- meta-agents Ma_j are Prolog declarative tasks which supervises the activity of A_i for i in [1..N].

The visibility of A_i is defined in the Ada interface of the tasks. In this desciption, we define the names and the memory addresses of the sets D_i and P_i associated to A_i. The expert system Ma_j will have a complete vision of the all cell using the $<D_i , P_i>$ for i in [1..N]. It is then easy to program the Ma_j if we know what properties must be verified. Ma_j will act on the A_i by the IAda interpretor by creating instructions DETRUIT(A_i), DESACTIVE(A_i), ACTIVE(A_i) which will by included in the IAda program and executed by the interpretor.

4. CONCLUSION

We have developed a complete environement for robot programming, actually implemented on a cell controller in our laboratory. All abstract levels of robot programming can be integrated:

- Ada for low-level loops,
- IAda for task programs,
- Prolog in IAda programs to perform planning and control at a low or high level by a ladder of control.

In the future we will extend IAda language to interrupt treatements to be able to dynamicaly configure the interrupt device on one of the task entries.

In addition, we will study the communication of IAda global variables between distributed computers to give to this language the power to be executed on a multi-computer architecture.

Finally, we are presently looking at an algorithm to construct an IAda program from a Petri Net input.

References:

[BA 82] J.BARNES
Programming in Ada
Addison-Wesley 1982

[CA 86] M.VAN CANAGHEM
L'anatomie de Prolog
InterEditions 1986

[CO 89] I.J.COX & N.H.GEHANI
Concurrent programming and robotics
Journal of robotics research vol 8 N°2 april 1989

[DOD83] DEPARTEMENT OF DEFENSE
Reference Manual for the Ada Programming Language
ANSI/MIL-STD 1815A,U.S.
17 February 1983

[DU 89] D.DUHAUT
Ada et le différents niveaux de programmation en robotique
Conférence ADA-France AFCET décembre 1989.

[MA 74] Z.MANNA
Mathematical theory of computation
MacGraw-Hill Computer Science Serie 1974.

A VISION OF PLANNING IN ROBOTICS

J.-C. MOUREAU
Equipe Logiciels pour la Productique
Laboratoire d'Automatique de Besançon
15, impasse des Saint-Martin
25000 BESANCON - FRANCE
Tel: 81.88.53.44 - Fax: 81.88.65.02

A. HAURAT, Y. MESLANS
Laboratoire Logiciels pour la Productique
Université de Savoie
41, avenue de la Plaine
BP 806
74016 ANNECY CEDEX - FRANCE
Tel: 50.57.40.70 - Fax: 50.57.48.85

ABSTRACT

Within the framework of task level robot programming, we have studied plan generation in robotics. This Artificial Intelligence technology is maladapted to the real worlds. So, the aim of this paper is to define an approach in order to generate actions plans tacking into account required robotics knowledge.
This approach lies on three main parts: a robotics actions model which is proposed in the form of an elementary robotics tasks typology, an model of the universe established in a functional shape and relating to each universe element, and a mechanism, composed of heuristics, which allows, thanks to universe elements, to choose the various elementary tasks and to build from them a task achieving the objective. It integrates the concept of object oriented programming.

INTRODUCTION

Because of the difficulty in developing complex robotized tasks, we must design new software tools in order to make the operator's work easier. He would benefit of a software environment to automatize its programming task. This kind of programming is often called task level programming. This programming is characterized by the fact that the operator specifies its robotized task with a goal to achieve and independently of the robot controller.

At this step, we have to manipulate a certain knowledge to take into account robots task level programming and therefore three fundamental questions are put:
(1) what is the required knowledge for developing a given robotized task?
(2) How to represent such a knowledge?
(3) How to use this knowledge?

The aim of this paper is to present an approach permitting to take into account robots task level programming in harmony with the three points evoked above.

1- PROBLEM POSITION.

Usually, robots tasks level programming is taken into account by AI technics called "Plan Generation" (Planning). Plan generator is a software tool which permits to produce a set of actions from a model of universe and actions in order to make evolute the world from an initial state to a final one [1].

A review of these technics has been carried out in [2] and we've been able to establish the fact that academic plans generation concepts are not really suitable to the robotics specificities:

- Indeed it is difficult to design a domain independent system, and thus we are tempted to integrate expert systems concepts to academic plan generation concepts.

- These plans generators are not well suited to manipulate robotics knowledge which is composed by ability (heuristics) and deterministic algorithms.

We're going to define, in this paper, an approach allowing to identify, to represent and to manipulate the robotics knowledge on the basis of the above established facts. This approach lies on four points:

- The formulation of an objective in a triplet form: **OBJECTIVE (action, element, place)** which defines the global action to be achieved on a robotics universe element and in a given place (which indicate the local nature of the task),

- The decomposition of the objective into macro-actions in order to identify the robotics universe elements setting in action during the task,

- The representation of the robotics universe elements by objects (object oriented concept),

- The representation of the robot by an elementary robotics tasks typology.

From these four points, a mechanism, called "unificator", containing a certain ability, will allow to link among themselves the actions that the diverse robotics universe elements called into the play during the task achieving the objective have to submit.

Section 2 of this paper presents the elementary robotics tasks typology, the representation of the robotics universe elements and also the unificator.

Section 3 states precisely the use of our approach.

2- FORMALIZATION OF THE ROBOTICS KNOWLEDGE TO TAKE INTO ACCOUNT ROBOTS TASK LEVEL PROGRAMMING.

2-1- An elementary robotics tasks typology.

The elementary robotics tasks typology aim is to represent, in a structured shape, robot abilities. It is founded on a hierarchical taxonomy of heterogenous robotics functionalities, which is organized in three main branches: action, perception and communication. This taxonomy is enriched of attributes allowing to identify the proper characteristics of the studied task.

We call elementary robotics tasks typology a set of tasks classes coming from the taxonomy and respecting the following properties:

- The contextualization of the action with regard to its environment,

- The links - not only hierarchical - between functionalities,

- The separation between the off-line functionalities and the on-line ones,

- The homogeneity and the sufficiently high level of the functionalities (object level of robot programming [3]),

- The respect of the basic concepts of AI: perception, reasoning (allowing the action), language (communication here), and common sense (quasi impossible to represent).

Figure 1 is an extract of this typology.

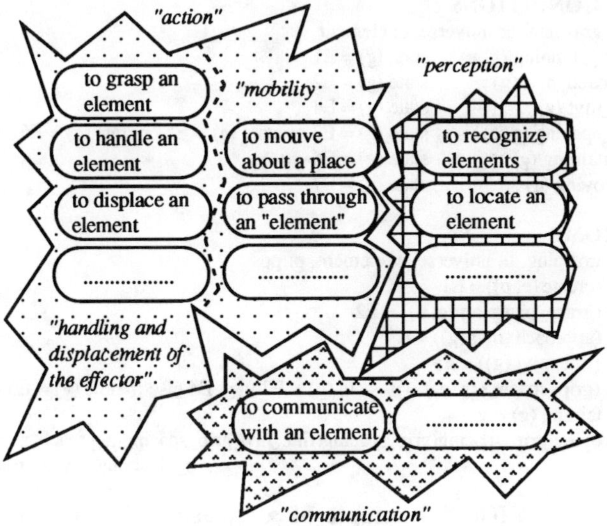

FIG. 1. An extract of the elementary robotics tasks typology.

To take into account these various properties and to specify correctly each elementary task, we have studied Data Abstract Types (DAT). An elementary robotics task DAT specification allows to describe task change during its execution.

A DAT specification comprises generally four parts: Types, Functions, Preconditions, and Axioms. Beyond the main type which is described in the specifications, Types part comprises others related types. These types are either the basic types (which are not described, like for instance Reals, Booleans, etc...) or types which are described elsewhere. This part displays the links - others than hierarchical ones - existing between various types. The Functions part comprises the various functions allowing to create, to modify or to access to the type elements. The functions are represented by their signatures, in the formalism: "name_function: domain → co-domain". The preconditions are useful to limit the applicability of certain functions. Finally, the axiomatic allows to specify the described type element behavior, which is the result of the functions application. The notion of inheritance appearing here naturally in the hierarchical taxonomy is available too because a "son" type is supposed to have at its disposal the functions of its "father". This essentially appears in the axioms part of the specifications. Figure 2 gives an example of DAT corresponding to a "grasping" elementary task.

TYPES

Grasping, effector, grip, **element**, **universe**, *pos*, *spacing*, *effort*

FUNCTIONS

create:	element X pos → grasping
grasp_planning:	grasping X universe → grasping
approach_mvt:	grasping → grasping
grip_spacing:	grasping X element → grasping
fine_mvt:	grasping → grasping
maintaining:	grasping → grasping
destroying:	grasping → ()
state:	grasping → {E1, E2, E3, E4, E5, E6, E7}

PRECONDITIONS
∀ g: grasping, u: universe, e: element
grasp_planning (g, u) : state (g) = E1
approach_mvt (g) : state (g) = E2
fine_mvt (g) : state (g) = E4
grip_spacing (g, e) : state (g) = E3 or state (g) = E5 or state (g) = E6
maintaining (g) : state (g) = E6
destroying (g) : state (g) = E7

AXIOMS
∀ g: grasping, u: universe, e: element, p: pos
state (create (e, p)) = E1
state (grasp_planning (g, u)) = E2
state (approach_mvt (g)) = E3
state (fine_mvt (g)) = E5
state (grip_spacing (g, e)) = if state (g) = E3 then E4 else if state (g) = E5 then E6 else E7
maintaining (g) = g
destroying(grip_spacing(grip_spacing(fine_mvt(grip_spacing(approach_mvt(
grasp_planning(create(e,p),u)),e)),e),e))= ()

FIG. 2. Example of a "grasping" task DAT.

Basic types correspond to real values. **Element** and **universe** are external types. Effector and grip are related types. With this type, we describe the states succession corresponding to a grasping task. So grasping an element requires: to create a type instance (E1), to plan grasping (E2), to move the arm to an approach configuration (E3), to space the grip to grasp the element (E4), to move finely the arm to reach the element (E5), and then to grasp the element with a certain effort (E6). After the handling of the element, which is maintained by the maintaining function, it is necessary to space the grip in order to put it down (E7). Then the type instance can be destroyed. The instance of the type comprises all the required knowledge during its life (E1 → E7): spacing value, effort value, approach configuration, which are given by the **element**, and the planned trajectory that exists from E2 state.

Robotics is a too broad domain for this typology to claim exhaustibility. So, several points have not been treated in detail. Nevertheless, the processing of these points reveals itself useful and efficient during the DAT implementation step. Consistently, it seems for us more advisable to maintain a sufficiently high and general level in order to preserve credibility and perenniality of the typology. In the form of proposed specification, it is however not so difficult to create new ones depending on operator's needs: extensibility can be preserved like that.

2-2- Representation of the robotics universe elements.

Robotics universe representation consists in describing each physical element of the robot environment **by the actions it is able to submit**. Because all these elements neither have the same importance nor the same role during the task, we propose to reason in the form of knowledge points of view. So, the robot will see a robotics universe element as an obstacle, an object to handle (to operate, to displace, etc...), or an absolute guide mark. The robotics universe representation will be compound, consistently, by the strict required knowledge for a given task.

The representation of a robotics universe element consists in the representation, perhaps partial, of the physical element with some attributes describing its characteristics and with some methods which are actions that this element is able to submit.

In consequence of these exigencies, the choice of an object oriented representation [5] seems natural. For each object of the real world corresponds a logical object of the system. This logical object is completely defined and can evolute in the system by being in relation or in communication with the others objects or in a completely autonomous way. The essential for us, is to equip ourselves of a representation mode in order to make the possibility of expression easier.

2-3- An expert mechanism: the unificator.

From an objective, from an object oriented representation of the universe elements and from an elementary robotics tasks typology (representing the robot), the approach target is to generate robotics actions plans. Thus it consists in finding, in the typology, the various elementary tasks to set in play and to connect them in order to obtain an actions plan: it is the mission of the unificator.

The unificator is equipped of a certain ability, which is represented in the form of heuristics. It links universe elements with the robot by:

- Choosing the elementary tasks with the help of the operations that must submit the universe element during the task,
- Parametrizing the elementary tasks with the help of the knowledge that contains universe elements (attributes, methods parameters of the object),
- Building the task that achieves the given objective.

3- THE USE OF OUR APPROACH.

The finality of our work is to generate automatically a robotics task from an objective. For instance, a robotics task would be able to be expressed by the objective: "close the N°X00A water-gate in the building of nuclear auxiliaries (BNA)".

It is important to detect in this objective the universe elements called into play. In that way, it must be required to decompose the triplet (action, element, place) into a totally ordered sequence of the macro-actions that may be taken into account by the unificator. For instance the previous cited objective decomposes itself into "move the robot to the water-gate" and "close the water-gate".

At this step, the following outstanding problematic must be solved: **"How, from the robotics macro-actions, to identify and to organize the required knowledge?"**. In order to answer, we feel the necessity of an object oriented development step [6].

Objects identification.
From a macro-action, we identify the object(s) corresponding to the universe elements which participate to the task achieving the objective.

Operations identification for each object.
We'll define the functional aspects of the elements during the task. For instance, *to close* and *to open* are operations of a water-gate element.

Establishment of the relations between the objects.
First of all, the DAT specifications of the typology must be transformed into objects so that we can have at our disposal a set of "elementary robotics tasks" objects.
The unificator chooses, at this step, the "elementary robotics tasks" objects. Then it parametrizes these "elementary robotics tasks" objects in organizing communication with the "universe elements" objects. By this it has defined the objects organization which allows a plan generation.

Plan generation concepts in our approach.

The proposed approach allows us to define a plan generator. The **universe model** is composed by a set of universe elements that can be used to achieve a given robotics task. The **actions model** is given by the elementary robotics tasks typology, set in the form of DAT specifications. The mechanism (**motor**) of our system, which can choose and order the actions, is the unificator; in fact it permits to link the universe and actions model of an objective (macro-actions).

The fundamental difference between our unificator and the academic plan generators motors resides in reasoning in a more shallow way with the help of heuristics, rather than reasoning in a deep way on the actions (to choose the actions by unification of the goal and the current state). These heuristics must allow to build the task which achieves the initial objective from, elementary tasks giving, a representation of the robot potentialities and from a partial knowledge of the universe elements concerning state variables and actions they can submit.

The contribution of our approach is the integration of the robotics ability in the plan generation:

- In the universe model, the even incomplete elements representation contains the required informations for their use in order to achieve a given robotics task,
- In the actions model, elementary robotics tasks specification describes their behavior in regard to any universe elements,
- In the unificator, heuristics represent the interaction between the robot and the universe elements.

From an oriented robotics functional universe model integrating the oriented object concept and a robotics actions model given by an elementary robotics tasks typology, the so defined universe evolutes from an initial state to a final one specified by an objective, with the help of an objects organization.

Our plan generator is then linear (or totally ordered), hierarchical, separatist (the plan generator liberates, off-line, a sequence of actions).

Finally, let us underline that each "elementary task" object supports a specific spatial planner allowing to generate elementary sequence of actions admissible by the robot controller.

CONCLUSION

The just established approach lies on the use of object oriented concept in order to take into account robot task level programming. Even if our approach is original in regard to the others various plan generation technics, it is interesting because it can manipulate a real universe knowledge: Robotics.

The objective analysis has allowed the identification of robot/"universe elements" interactions: this defines, with the help of a robotics ability, an objects organization representing a robotics plan generator.

REFERENCES

1. M. Dumas, "Génération de plans d'actions : principes et méthodes", DEMT report 88/043 SERMA/LETR/88/977, CEN Saclay, February 1988.
2. J.C. Moureau, "Génération de plans d'actions en robotique : situation, complexité et définition d'une démarche basée sur la notion d'objet fonctionnel", PhD thesis, Université of Besançon, To be published.
3. T. Lozano-Perez, "Robot programming", Proceedings of the IEEE, Vol. 71, July 1983.
4. G. B. Dunn, J. Segen, "Automatic Discovery of Robotics Grasp Configurations", Acts IEEE 1988, pp. 396-401, 1988.
5. P. De Vettor, "Bilan de l'étude des systèmes à objets pour la supervision des équipements automobiles", INAROB/RNUR report-X04 project, September 1989.
6. B. Meyer, "Object Oriented Construction", Ed. Prentice Hall, 1988.

RULE-BASED SYNTHESIS IN CONCEPTUAL DESIGN

ANDREW KUSIAK and
EDWARD SZCZERBICKI
Department of Industrial and Management Engineering, University of Iowa, Iowa City, Iowa 52242, USA

ABSTRACT

In this paper a model synthesis algorithm for conceptual design of mechanical systems is presented. The algorithm is integrated with production rules that generate connections for models and components. The connections are used to match the elements in the model base. Hierarchically structured conceptual design models are developed at various levels of abstraction that parallel the designer's expertise. The algorithm introduced in this paper is applied to synthesize the conceptual model of a holding device.

INTRODUCTION

The modeling approach to design includes model synthesis (e.g., [1, 2, 3, 4, 5, 6, 7]) that is based on the general systems theory [8] and uses hierarchical structures introduced in [9] as well as model base concepts presented in [10, 11, 12].

In this paper an algorithm for model synthesis process is presented. The algorithm is integrated with production rules that are used to generate connections between elements of a model base. The algorithm builds an overall model of the designed system beginning at the lowest level of abstraction (component level). It is based on a working hypothesis that it is possible to synthesize hierarchically structured mechanical systems by reasoning with components, models, and their connections.

INPUT TO THE SYNTHESIS PROCESS

Specification and representation stage of conceptual design provide an input to the design synthesis process [13]. For example, if the overall design requirement is formulated as R1: develop conceptual design of a holding device, the set of functions specified using the approach presented in [14] is given by {F1, F2, F3, F4, F5, F6, F7, F8, F9, F10, F11, F12, F13}, where:

F1 - apply human force,
F2 - apply human force by hand,
F3 - apply force rotationally,
F4 - hold material,
F5 - hold material by fixed part of the device,
F6 - hold material by movable part of the device,
F7 - transform radial force,
F8 - transform forcc by one screw,
F9 - change radial force into horizontal force,
F10 - change radial velocity into horizontal velocity,
F11 - fix the unmovable part of the device,
F12 - guide the motion of movable part,

F13 - fix the guide of movable part.
The above functions are assigned components as presented in Figure 1.

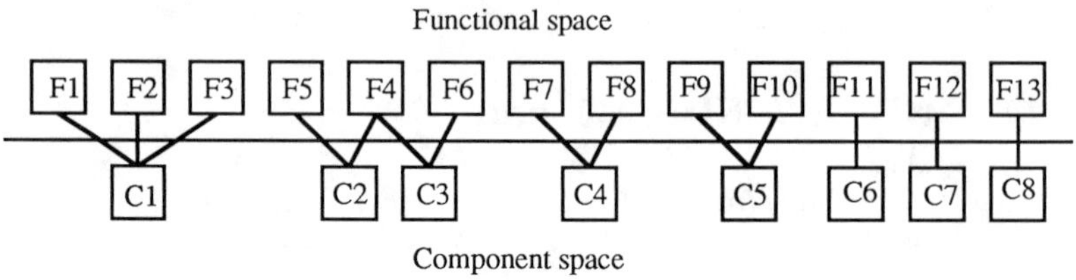

Figure 1. Components and the corresponding functions

For each component input and output variables are defined by the corresponding function(s). Figure 2 presents eight components together with their inputs and outputs.

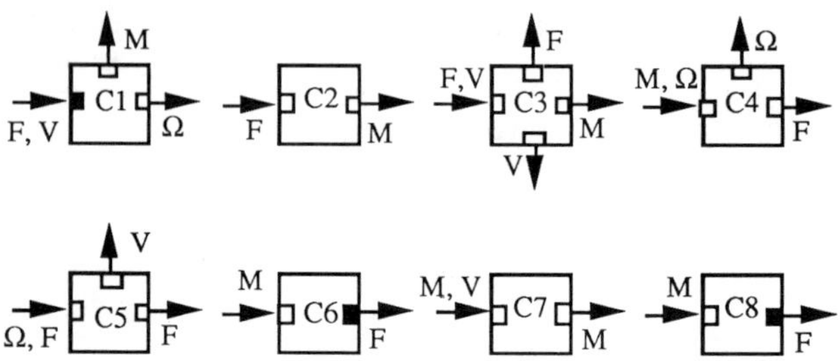

Figure 2. Components

Note that component C1 accepts input from an outside of the system boundary and components C6 and C8 send the output to the outside of the system boundary. Components with specified input and output variables represent the initial state of the model base that initiates the design synthesis process. Since the synthesis algorithm requires that component connections be known, in the next section production rules supporting generation of these connections are presented.

GENERATION OF CONNECTIONS

A set of production rules is used to generate connections between the elements stored in the model base. Since model base may include components and/or models, the rules specify connections of the following types:

(i) component -----> component,
(ii) model -----> model,
(iii) component -----> model,
(iv) model -----> component.

The components and models in the production rules are referred to as elements.

Rule 1

IF there is only one element in the model base
THEN do not generate connections

Rule 2

IF the model base includes more than one element
THEN select a connection for an input boundary element

Rule 3
IF in the model base there are elements other than the boundary elements
THEN do not specify any connections that involve boundary elements only
Rule 4
IF an element is an input boundary element
THEN it can not accept an input from any other element
Rule 5
IF an element is an output boundary element
THEN it can not provide an input to any other element
Rule 6
IF two elements in the model base have identical output and input variables
AND there are no production rules that prevent from connecting them
THEN specify the connections for these elements
Rule 7
IF a connection for an input boundary element has been specified
THEN continue with selecting connections for elements that have not been listed in the specifications
Rule 8
IF the model base contains boundary elements only
THEN specify connections between them

In the above rules, an input boundary element is the one that accepts an input from an environment of the designed object, and an output boundary element is the one that provides output to the environment.

SYNTHESIS ALGORITHM

Synthesis of components and models into an overall model of the designed object is performed according to the following algorithm:

Step 1. Open a set MODEL_BASE consisting of all components generated at the specification and representation stage.
Set level = 1.
Step 2. Apply production rules to generate connections between elements in MODEL_BASE.
Step 3. If no connections are generated, stop;
Otherwise, match elements in MODEL_BASE into pairs using the existing connections.
Step 4. Define input and output variables for models generated by the matching process.
Step 5. Remove from MODEL_BASE all elements that have taken part in the matching process.
Step 6. Add to MODEL_BASE all models generated by the matching process.
Step 7. Set level = level + 1 and go to Step 2.

The algorithm terminates at the level at which it is not possible to match components and/or models into pairs (no connections are generated). Two cases are possible. First, matching is no longer possible because all components and models have been exhausted and an overall model has been built. Second, although there are models and/or components available, it is not possible to match them. In this case, the synthesis path explored has not led to the overall model and none of the models at the highest level includes all functions specified in the functional space.

ILLUSTRATIVE EXAMPLE

The synthesis algorithm is applied to build an overall model of a holding device.

Step 1. MODEL_BASE = {C1, C2, C3, C4, C5, C6, C7, C8},

level = 1.

Step 2. C1 -----> C4 (rules 2, 3, 4, and 6),
C2 -----> C6 (rules 7, 6, and 5),
C5 -----> C3 (rules 7, and 6),
C7 -----> C8 (rules 7, and 6).

Step 3. Connections specified in step 2 allow to develop the following models
M1 = C1.C4, M2 = C2.C6, M3 = C5.C3,
M4 = C7.C8.

Step 4. Input and output variables for models generated in step 3 are presented in Figure 3.

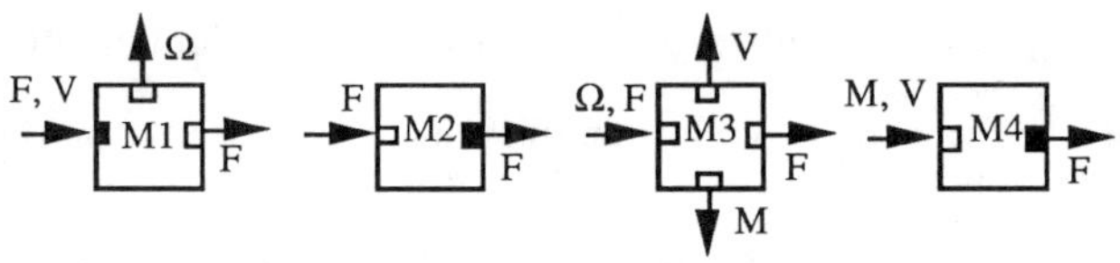

Figure 3. Models with input and output variables

Step 5. MODEL_BASE = {M1, M2, M3, M4}.
Step 6. Level = 2.
Step 2. M1 -----> M3 (rules 2 and 6).
Step 3. M5 = M1.M3.
Step 4. Figure 4 presents model M5 with its input and output variables.

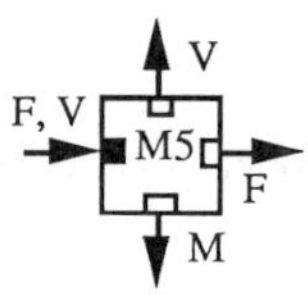

Figure 4. Model M5 with input and output variables

Step 5. MODEL_BASE = {M2, M4, M5}.
Step 6. Level = 3.
Step 2. M5 -----> M4 (rules 2 and 8).
Step 3. M6 = M5.M4.
Step 4. Figure 5 presents model M6 with its input and output variables.

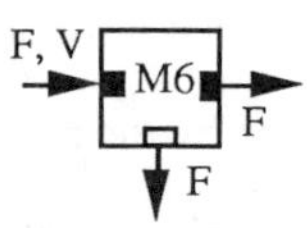

Figure 5. Model M6 with input and output variables

Step 5. MODEL_BASE = {M2, M6}.
Step 6. Level = 4.
Step 2. M6 -----> M2 (rules 2 and 8).
Step 3. M7 = M6.M2.
Step 4. Figure 6 presents model M7 with its input and output variables.

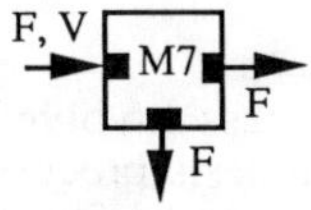

Figure 6. Model M7 with input and output variables

Step 5. MODEL_BASE = {M7}.
Step 6. Level = 5.
Step 2. No connections are generated (rule 1).
Step 3. Stop.

Note that the overall model M7 is a valid synthesis solution as all its input and output variables represent the connections with the system environment. The five level hierarchical tree of model M7 is presented in Figure 7.

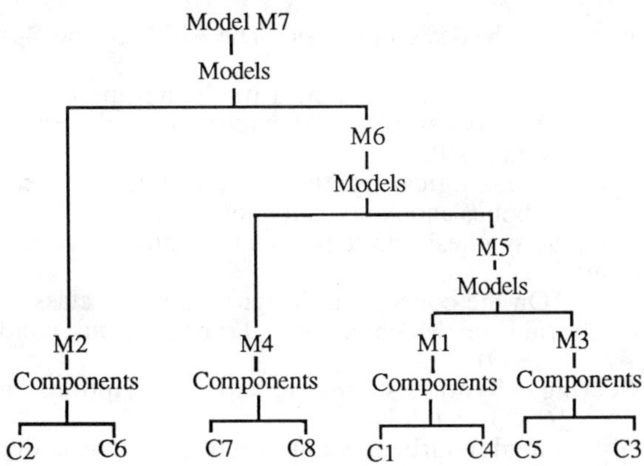

Figure 7. Five level hierarchical tree of the overall model M7

Physical Representation

One may consider the physical representation of the overall synthesized model at various levels of abstraction. This is probably one of the most significant advantages of the presented synthesis scheme, as it allows the designer to adjust his/her abstraction preferences to the levels that are available in the hierarchy of the overall model. Figure 8 presents a schematic representation of the synthesized solution .

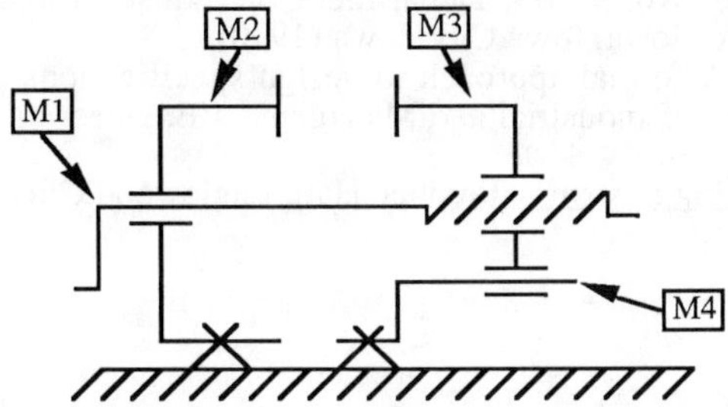

Figure 8. Schematic representation of the synthesized solution

CONCLUSION

The major activities of the conceptual design process involve specification, representation, and synthesis. The outcome of the synthesis process is the design solution. In this paper, the synthesis algorithm was presented for synthesis at the conceptual level. Elements of a model base are synthesized using the algorithm into an overall model of the designed object. The overall model has hierarchical structure and represents the designed system at various levels of abstraction. General production rules that support generation of component and model connections are an important element of the synthesis scheme. The algorithm and production rules were applied for building of an overall conceptual model of the holding device.

REFERENCES

1. T. Tomiyama, et al., "Metamodel: a key to intelligent CAD systems", Res. Eng. Des. 1(1), 19-34 (1989).
2. J. Mostow, "Toward better models of the design process", The AI Magazine, 44-57 (Spring 1985).
3. A. Kusiak and E. Szczerbicki, "Conceptual design: a modeling approach", Working Paper No. 90-02, Department of Industrial and Management Engineering, The University of Iowa, Iowa City, Iowa (1990).
4. S.L. Chang and L.W. Tsai, "Topological synthesis of articulated gear mechanisms", IEEE Transactions on Robotics and Automation, 6(1), 97-103 (1990).
5. S.P. Hoover and J.R. Rinderle, "A synthesis strategy for mechanical devices", Res. Eng. Des. 1(1), 87-103 (1989).
6. L.W. Tsai and F. Freudenstein, "On the conceptual design of a novel class of robot configuration", ASME Journal of Mechanisms, Transmissions, and Automation in Design, 111(1), 47-53 (1989).
7. K.T. Ulrich and W.P. Seering, "Synthesis of schematic descriptions in mechanical design", Res. Eng. Des. 1(1), 3-18 (1989).
8. B. Schmidt, "Systemanalyse, modellaufbau, simullation grundlagen des simulationssystems Simplex-II", IMMD IV, Universitat Erlangen-Nurnberg, Erlangen, F.R. of Germany (1987).
9. M.S. Mesarovic, Theory of Hierarchic Multilevel Systems (Academic Press, New York 1970).
10. T.I. Oren, Simulation and Model Based Methodologies: An Integrative View, edited by Oren, Zeigler, and Elzas (North-Holland, Amsterdam 1984).
11. B.P. Zeigler, "Hierarchical modular discret-event modelling in an object oriented environment", Simulation, 49(5), 219-230 (1987).
12. B.P. Zeigler, "Multifacetted Modelling and Discret Event Simulation (Academic Press, New York 1984).
13. A. Kusiak and E. Szczerbicki, "A methodology for conceptual design of mechanical systems", Working Paper No. 90-05, Department of Industrial and Management Engineering, University of Iowa, Iowa City, Iowa (1990).
14. A. Kusiak and E. Szczerbicki, "A formal approach to design specifications", Working Paper No. 90-07, Department of Industrial and Management Engineering, University of Iowa, Iowa City, Iowa (1990).
15. A. Kusiak, Intelligent Manufacturing Systems, (Prentice Hall, Englewood Cliffs 1990).

Strategies for Problem Decomposition in Multiagent Planning

M. Kamel and A. Syed
University of Waterloo
Waterloo, Canada

ABSTRACT

Automated planning systems, that are capable of generating synchronized concurrent actions for multiple agents, have an important role in robotics and automated manufacturing. This paper proposes different strategies to decompose the problem among the agents with similar and disparate operational capabilities. The problem of agent assignment to optimize the completion time is discussed and an approach for solving it is proposed

INTRODUCTION

Most previous work in planning has assumed a single agent in a relatively static environment. To address the planning problems in real applications involving robotics and automated manufacturing, we need to consider multiple agent environment. In multiagent planning since agents' actions can affect one another, it is important to represent these interactions in addition to the usual knowledge about the agents and the domain. Synchronization of multiagent plans involves coordination of concurrent actions of agents by taking advantage of helpful interactions and avoiding the harmful ones. In helpful interaction, an agent helps to satisfy the requirements (preconditions) of another agent's action. Whereas, in harmful interaction, an action of an agent may result in negating the effect of an earlier action of another agent. A critical problem in multiagent planning is to detect, resolve and synchronize interactions between agents' actions. Some of these issues have been addressed in [1], [2], and [7]. In [4], the authors have proposed knowledge representation schemes and plan generation approach which represent interactions as interagent constraints.

Solving multiagent planning problems also requires recognizing the different constraints and characteristics that may be associated with the multiagent environment. Among these, the constraint due to the number of agents available in the environment and their capabilities is germane to the way the problem is decomposed among the agents. There has been very little effort devoted to the area of problem decomposition in multiagent planning. Lansky and Fogelsong [6] proposed a strategy that partitions the planning domain based on the notion of locality. In this strategy, the domain properties and constraints are localized into independent regions of activity, in order to achieve fewer, manageable, and explicitly stated interactions between these regions.

This paper proposes strategies to decompose the planning problem among the agents. Two specific problem decomposition strategies are discussed. One is based on object oriented decomposition, which is directed to multiagents with equal operation capabilities, and the other is based on functional oriented decomposition which is more suitable for disparate capability agents. These strategies allow arbitrary number of implicit interactions between agents, and they support different configurations of multiagent environment. Based on the object oriented strategy, an approach to determine agents' assignments that allow the generation of plans with minimum completion time is also proposed.

THE APPROACH

For the purposes of this paper the domain is viewed as consisting of objects, relations, and agents. At any given instant, the domain is assumed to be in a particular state. Each state is represented by the objects, agents, and relationships each object holds with other objects and agents. The relationships in a state are defined in terms of relations. For example, a relation *inside* may be used to define the relationship *(A inside B)* which states that object A is inside object B. There exists a set of operators that agents can perform to change the relationships between objects. The planning problem is to generate a multiagent plan consisting of operations for agents to achieve the goal (final) state from the given initial state.

Agents in the domain are viewed as subplanners which generate plans to solve subproblems. The subproblems for agents (subplanners) are produced by partitioning the domain into subdomains, and the initial and goal states into initial and goal substates, for each subplanner. This partitioning is based on specific assignment of objects and relations to agents. The particular choice of assignment of objects or relations to agents is made by a *problem decomposition strategy* which takes into account the constraints and characteristics (such as the capability of agents) associated with the multiagent environment.

The initial and goal substates are represented by the objects and relationships in the initial and goal states that involve the objects and the relations assigned to the agent. Thus, the subplan of an agent consists of a sequence of actions (operations) on its subdomain to achieve the goal substate starting from the initial substate.

The approach allows each subplanner to generate a subplan involving only those actions that affect its assigned objects or relations, and pose actions that affect objects or relations assigned to others as constraints to their corresponding subplanners. While, the subplanner needs to satisfy the constraints for its objects or relations, from other subplanners, it considers the constraints it generates to be the responsibility of other subplanners. Thus, each subplan has to accommodate the constraints required by other subplanners as part of its plan to solve the assigned subproblem. The overall plan is a sequence of concurrent actions of all the subplans, with constraints representing the coordination points between the agents. The agents solve the planning problem by *concurrently* executing the operations in their subplans.

PROBLEM DECOMPOSITION STRATEGIES

The approach described above allows for the use of different types of problem decomposition strategies to deal with different environment constraints. The problem decomposition strategies specify the assignment of objects and relations to agents. In the following, we consider two strategies for multiagent problem decomposition. The first strategy, called *object oriented problem decomposition*, is designed for environments where agents have equal capabilities. The second strategy, referred to as *functional oriented problem decomposition*, is intended for environments where agents have disparate capabilities.

Object Oriented Problem Decomposition Strategy

The main objective of this decomposition strategy is to exploit the concurrency between agents with equal capabilities. The object oriented strategy is similar to the data parallel approach in parallel processing [3]. In data parallel approach, the aim is to extract parallelism that exists in a program at the data level; the computations (operations) are performed simultaneously across sets of data. Similarly, the object oriented approach treats the objects in the domain as data and generates the parallel operations to be performed by the agents in the domain.

The problem decomposition is based on assigning each agent specific objects to act on. Because, agents are assumed to have equal capabilities, each agent is capable of performing all of the required operations on their assigned object. The subdomain of an agent consists of objects assigned to the agent, relations from the domain, and agent's operators. The subproblems are defined in terms of the initial and goal substates involving the objects assigned to subplanners. Thus, the subplan of an agent consists of a sequence of actions (operations) on objects, assigned to that agent, in order to achieve the final state of these objects.

Using this strategy, each subplanner generates a subplan for its agents involving only those actions that manipulate its objects and poses actions that affect other objects as constraints to their corresponding subplanners. Each subplanner satisfies the constraints for its objects, from other subplanners.

By viewing the domain as a set of objects and agents to manipulate the objects, the planning problem becomes the problem of finding the sequence of concurrent actions by the agents to transform the initial state (an instantiation of objects, agents and their relationship) of the domain into the desired goal state.

Functional Oriented Problem Decomposition Strategy

In many applications, such as automated manufacturing and assembly, it is often more economically feasible to have agents with different capabilities than equal capabilities. For instance, a Flexible Manufacturing Cell (FMC) is typically realized by one or more Computer Numerical Control machines, each responsible for different functions (such as, drilling, grinding), and a robot for transporting material within the cell. For such an environment, the *object oriented strategy* approach would not be appropriate, as an agent may not have the capability to perform certain required operations on the assigned objects. A different problem decomposition approach, called the *Functional Problem Decomposition strategy* is more suitable for generating multiagent plans when the agents have disparate capabilities.

While, the object oriented strategy is similar to the data parallel approach, this strategy is based on the functional parallel approach used in parallel processing, where concurrent functions in a program are assigned to different processors for parallel execution.

The capability of each agent in the domain is represented by a set of operators which the agent can apply to change the relationships between objects in the domain. The planning problem is decomposed into subproblems by assigning specific relations in the domain to agents based on their capabilities. A particular relation is assigned to the agent which has an operator that can realize the relationship described by the relation.

The subdomain of an agent is comprised of objects from the domain, relations assigned to the agent, and agent's operators (corresponding to its capabilities). The subproblems are defined in terms of the initial and goal substates involving the relations assigned to subplanners. Thus, the subplan consists of a sequence of actions (operations) that manipulate relationships between objects, defined by the assigned relations, to achieve the final substate. Operations in each agent's subplan which affect relations assigned to other agents are posed as *Interagents Constraints* to the corresponding subplanners. Each subplanner must satisfy the constraints, on its assigned relations, produced by other subplanners.

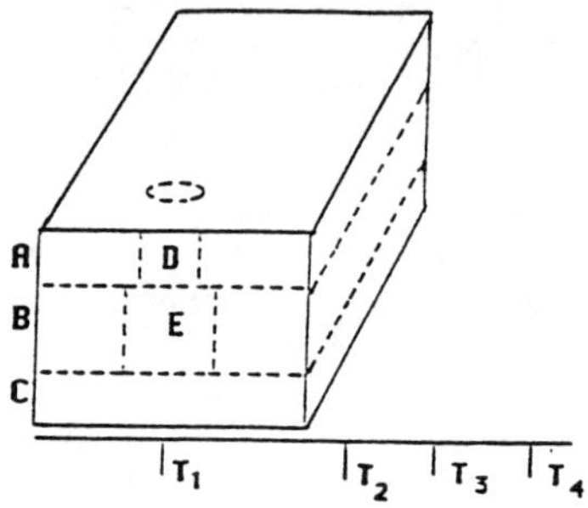

Figure 1: Initial state

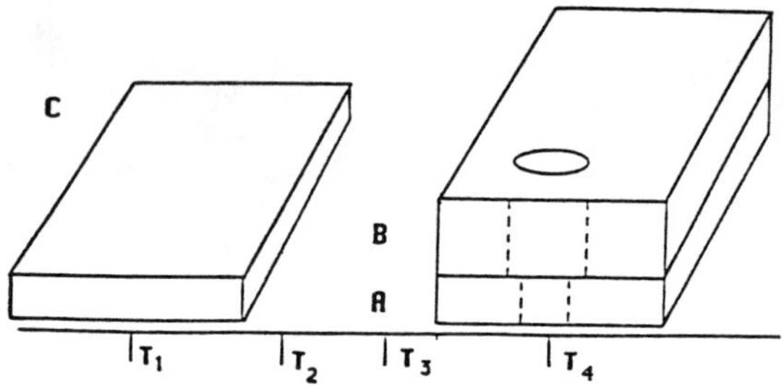

Figure 2: Final state

AN EXAMPLE

To illustrate the object and functional oriented problem decomposition strategies, let us consider the metal processing problem of producing parts from raw material. The raw material which is a rectangular prism stock, rested on table location T1, is to be processed to produce three rectangular objects A, B and C with A and B having holes D and E, as shown in figure 1. The final goal is to produce these parts and to stack B on top of A at position T4 on the table, as shown in figure 2. Thus, the initial state is described as (A attachedto B) (A on B) (D inside A) (B attachedto C) (B on C) (E inside B) (C on T1), and the final state is given as (A on T4) (B on A) (C on T1). T2 and T3 are table locations. There are four different types of relations (on, inside, heldby, and attachedto) used in this example to describe the relationships between objects. The relation *on* is used to define the fact that object A is on top of object B. The relation *heldby* is used when an object is heldby a robot. The relation *inside* is used to describe an object being inside another object and the relation *attachedto* may be used to indicate that an object is attached to another object. The operators, *Pickup(Robot X)*, *Remove(X Y)*, *Detach(X Y)*, and *Puton(X Y)*, are used in the example to pickup object X, remove object X from the inside of object Y, detach object X from object Y, and put object X on top of object Y. These operators are high level specifications of detailed machine operations on objects. For example, the high level operation *Remove(X Y)* may be translated into two specific sequences of robot operations such as *DRILL* and *BORE*.

Object Oriented Decomposition

For this configuration, we will assume the availability of two robots with equal capabilities: R1, and R2. Consider the assignment of objects A, D, T3, and T4 to robot, R1, and objects B, E, C, T1, and T2 to robot, R2. Based on this assignment, the appropriate initial and final substates are specified to the subplanners associated with the three robots. The initial substates are (A attachedto B) (A on B) (D inside A), for R1, and (A attachedto B) (A on B) (B attachedto C) (B on C) (E inside B) (C on T1), for R2. The final substates are (A on T4) for R1, and (B on A) (C on T1) for R2. Table 1 shows the final plan generated using the object-oriented decomposition strategy (execution of

Table 1: Object-oriented Plan

T	Robot 1 (R1)	Robot 2 (R2)
t1	Detach(A B)	Remove(D A)
t2	Pickup(R1 A)	waiting for R1
t3	Puton(A T4)	Detach(B C)
t4	Remove(E B)	waiting for R1
t5	idle	Pickup(R2 B)
t6	idle	Puton(B A)

T = time

Table 2: Functional oriented plan

T	Robot 1 (R1)	Robot 2 (R2)	Robot 3 (R3)
t1	Detach(A B)	waiting for R3	waiting for R1
t2	waiting for R3	waiting for R3	Pickup(R3 A)
t3	Detach(B C)	waiting for R3	Puton(A T2)
t4	idle	Remove(D A)	Pickup(R3 B)
t5	idle	waiting for R3	Puton(B T3)
t6	idle	Remove(E B)	Pickup(R3 A)
t7	idle	idle	Puton(A T4)
t8	idle	idle	Pickup(R3 B)
t9	idle	idle	Puton(B A)

T = time

each operation is assumed to consume one unit of time).

Functional Oriented Decomposition.

We will now consider three robots (R1, R2, and R3) with disparate capabilities, instead of two robots with equal capabilities. R1 is capable of cutting the raw metal, R2 is capable of drilling but its mobility is restricted to the table locations T2 and T3, and R3 is used for transporting objects from one location to another.

Based on the Functional Problem Decomposition concept, we include in the robots' subdomains operators (based on their capabilities) and relations that these operators can realize or negate. Thus, R1 is assigned the function *Detach* and the relation *attachedto*. R2 is given the function *Remove* and the relation *inside*. To restrict the mobility of R2, the precondition list of the function *Remove* is augmented to ensure that objects must be on T2 or T3 before applying the function. R3 uses the functions *Pickup* and *Puton* to manipulate relations, *on* and *heldby*. Based on this assignment, the appropriate initial and final substates are specified to the subplanners associated with the three robots. The initial substates are (A attachedto B) (B attachedto C) for R1, (D inside A) (E inside B) for R2, and (A on B) (B on C) (C on T1) for R3. The final substate for R3 consists of (A on T4) (B on A) (C on T1). The final substates for R1 and R2 only include the objects without relationships since the final state does not involve their assigned relations. The final plan using this strategy is shown in Table 2.

MULTIAGENT ENVIRONMENT CONFIGURATION

The decomposition strategies described above are capable of supporting various multiagent environment configurations. A configuration for multiagent environment can be defined in terms of number of agents, their capabilities, and the assignment of objects or relations to each agent. Thus far in our discussion, we have assumed that the specific assignment of objects or relations to agents in a configuration are known prior to plan generation. In the case of functional oriented decomposition the assignment of relations to agents with non-overlapping capabilities can be obtained from the specified operations of individual agents. However, in the case of object-oriented decomposition one may wish to find an assignment of objects that satisfies certain planning objectives.

Let us consider the objective of finding an (optimal) assignment of objects that generates a multiagent plan requiring minimum overall completion time for K agents and N objects ($K \leq N$). The optimal assignment of objects can be found exhaustively by generating plans for each possible assignment. However, the exhaustive approach, which requires $K^{N-1} - 1$ plans, can be prohibitive as the cost of generating a plan for a given assignment of objects is often computationally expensive.

One alternative to the exhaustive approach is to identify the basic inherent operations (and their sequences), by generating a plan for a given assignment of objects, and to use them to estimate the completion time of the plans corresponding to all possible assignments. There are two problems associated with this approach. One problem is how to select an appropriate configuration and assignment of objects which generate the initial plan. Another problem is how to estimate, based on the initial plan, the operations (and their sequences) which are particular to the plans corresponding to other assignments of objects. Below, we discuss an approach which addresses these problems.

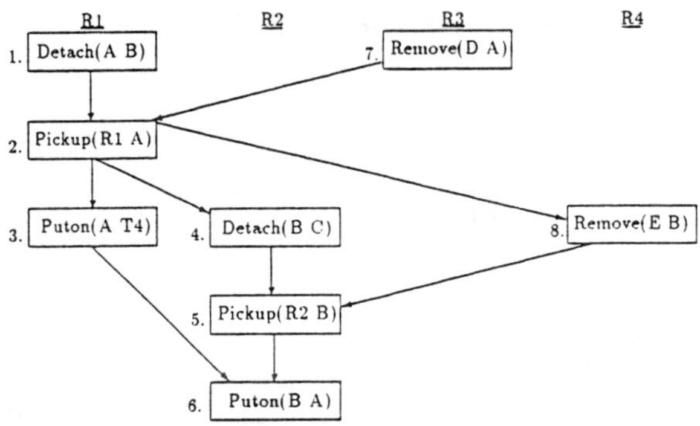

Figure 3: Task precedence graph

In the proposed approach, the initial configuration consists of as many agents as the given number (N) of objects. In this configuration each object is assigned to a unique agent in order to generate a plan (referred to as N-N plan). One advantage of this plan is that it only contains the basic operations and their sequences; operations (and their sequences) which are generated due to restrictions on number of agents and their assignments do not appear in this plan. Another advantage is that one can easily identify, from the analysis of the N-N plan, the number of objects (L) that need to be manipulated; which may be less than or equal to N. In cases where L is less than N, the number of possible assignments of objects are reduced from $K^{N-1}-1$ to $K^{L-1}-1$. Assignment of objects which are used (but not manipulated) in the N-N plan do not affect the completion time of the plan. These objects can be arbitrarily assigned to agents since they may serve to detect and resolve interactions in generating a plan. Based on the analysis of the N-N plan, one can construct a task precedence graph containing operations (and their sequences) of agents whose objects are manipulated.

Let us again consider the example problem. We can generate the N-N plan by assigning objects A, B, D, E, C, T1, T2, T3, and T4 to robots R1, R2, R3, R4, R5, R6, R7, R8, R9, respectively. The task precedence graph for the example problem is shown in figure 3. The nodes in the graph are numbered and they represent the operations (tasks), while the directed edges indicate the precedence relations (sequences) between operations. For instance, the edge from node 2 to node 4 indicates that the operation, *Pickup(R1 A)*, of R1 must be done before the operation, *Detach(B C)*, of R2. In other words, node 2 is a predecessor of node 4. The analysis of the N-N plan reveals that only four objects from the given nine objects need to be manipulated. That is, objects A,

B, D, and E are manipulated by the operations of robots R1, R2, R3, and R4. Objects C and T4 are used (node 3 and 4) but not manipulated by their robots, and therefore they are not included in figure 3. Objects T1, T2, and T3 are not involved in any operations and they are also not included in figure 3.

The task precedence graph is then used to estimate the completion time for each of the $K^{L-1} - 1$ assignments. From the comparison of these estimates, the optimal assignments of objects to agents are identified. Finally, with the estimated optimal assignment of objects to agents, the actual plan is generated in order to determine its feasibility. If the selected assignment of objects to agents is found to be infeasible, the next best assignment is selected until a feasible plan can be generated.

A detailed algorithm to estimate the completion time for a given assignment of objects and agents is discussed in [5]. Herein, we describe the steps involved in the algorithm for a configuration of two agents (K = 2) for the example problem, with the assumption that each operation consumes one unit of execution time:

Step 1. Consider a possible assignment of four objects to the two agents. For instance, we may assign objects A and E to robot, R1, and objects D and B to robot, R2.

Step 2. Distribute the operations from the task precedence graph to both agents according to their assigned objects. According to the assignment above, operations 1, 2, 3, and 8 should be assigned to R1 and operations 7, 4, 5, and 6 must be given to R2.

Step 3. For each agent, find the smallest distance of each of its operation to the operations of the other agent. Distance between two nodes is measured as the number of edges leading from one node to the other. For example, based on the distribution of the operations above, operation 2 of R1 has a distance of 1 to operation 4 and distance of 2 to operation 5 of R2. The smallest distance of operation 2 is, therefore, 1.

Step 4. Find the operation sequence for agents to execute in each time step: Starting from the initial step, an agent either executes an operation or remains idle if there are no ready operations. An operation becomes ready if all its predecessors have been executed at earlier time steps. If there are more than one ready operations available for an agent at a current time step, it selects the operation which avoids additional operations (discussed below) in order to minimize the completion time; otherwise, it chooses the operation which has the smallest distance value to the operation of the other agent. The smallest distance value is selected in an attempt to reduce the idle time of the other agent as a result of waiting for completion of operations which are assigned to this agent. Initially, for the example task precedence graph, operation 1 and 7 are considered to be ready for execution, and they are allocated the first time step to their agents for execution. But, in the next time step R1 is allocated operation 2 while R2 remains idle as none of its operations are ready for execution.

Step 5. Compute the completion time: The completion time corresponding to the given assignment is obtained when all the operations in the task precedence graph have been assigned a time step to execute by their agents. The completion time is the total number of time steps allocated to execute all operations.

Steps 1 to 5 are repeated for each assignment of objects. For the example problem, these steps are repeated for 7, that is (2^3-1), assignments because there are two agents (K = 2) and four manipulated objects (L = 4). The completion time estimated for all assignments of objects range from six to eight units. The plan of table 1 represents a feasible and optimal assignment of objects to robots R1 and R2 as it requires a completion time of six time units.

The restriction of two agents and their assignments in the example problem does not result in any more operations than those appearing in the task precedence graph. In other cases, however, additional operations may be needed for a certain assignment of objects to agents. In order to allocate the additional operations to the agents (in step 4 of the algorithm above) for a given assignment of objects, we need to know when the additional operations are generated. In general, the additional operations may occur when an agent's operations involve a transition from one object to another. That is, the object manipulated by the operation in the current time step is different from the object which the next operation will manipulate. If the current operation does not release the agent, additional operations are needed on the current object to release the agent for the next operation by temporarily discarding the current object. Similarly, additional operations are necessary to acquire the temporarily discarded object to complete the remaining operations on the object.

Let us, for instance, introduce in the example problem a new object F which is initially placed on location T4 and it is expected to be finally moved to the top of object A. We assign object F to R1 in addition to objects A and E. In order to place object A on top of object T4 (operation 3) and object F on object A, R1 may first pick up object F and perform an additional operation to temporarily place it on another location, followed by picking up and placing object A on object T4, and then performing another additional operation to pick up object F from its temporary location before placing it on top of object A. Both of these additional operations need to be included in estimating the completion time of agent R1 plan.

CONCLUSION

Two decomposition strategies were proposed in this paper for generating plans for agents with similar and disparate capabilities. An approach was also proposed to determine the assignment of agents (having similar capabilities) that generates a plan with minimum completion time. A system to support these strategies has been implemented on Xerox A.I. workstation, and it has been used to experiment with a number of assembly and process planning examples. Our future efforts will be directed towards methods for finding assignment of agents with overlapping capabilities that generate plans with minimum completion time.

REFERENCES

1. Georgeff, M. P., "The Representation of Events in Multiagent Domains," Proc. of AAAI, 70-75 (1986).
2. Georgeff, M. P., "Communication and Interaction in Multiagent Planning," Proc. of AAAI, 115-119 (1983).
3. Hillis, W. D. and Steele Jr., G. L., "Data Parallel Algorithms," Communications of the ACM 29, 1170-1183 (Dec. 1986).
4. Kamel, M. and Syed, A., "An Object-oriented Multiple Agent Planning System," Distributed Artificial Intelligence, Vol. II, 259-290, edited by Michael N. Huhns (Pitman Publishing Limited and Morgan Kaufman 1989).
5. Syed, A., "Strategies for Multiagent Planning," PhD. Thesis, Dept. of Systems Design, University of Waterloo, expected (1990).
6. Lansky, A. L., and Fogelsong, D. S., "Localized Representation and Planning Methods for Parallel Domains," Proc. AAAI, 240-245 (1987).
7. Stuart, C., "An Implementation of a Multiagent Plan Synchronizer," Proc. IJCAI, 1031-1033 (1985).

AN EXPERT SYSTEM FOR THE DIAGNOSIS OF FAULTS OF INDUSTRIAL ROBOTS

K. Hörmann, T. Hübner, U. Rembold

Universität Karlsruhe, IPR, Postfach 6980, D–7500 Karlsruhe 1

ABSTRACT

This contribution presents an Expert System which has been developed within a joint project of the University of Karlsruhe and the Nuclear Research Center Karlsruhe.The long term goal of this cooperation is to develop a generic tool for the diagnosis of faults of technical systems. In order to make such a tool generic enough to be easily adaptable to new applications, a deep reasoning approach has been developed. A prototype of the system has been implemented using the Knowledge Craft^{TM} environment. The prototype has been tested with industrial robots of UNIMATION, Inc. and is currently being applied to the much more complex system COMETOS for the cleaning of large castings consisting of several robots, end–effectors, sensors and different controllers. Both applications are described.

1. INTRODUCTION

Many industrial applications require an extremly high reliability of the equipment used. This is the case, for instance, with telemanipulators in nuclear power plants or with industrial robots in production lines. In such areas faults can cause emergencies or at least high cost. Hence it is desirable to detect faults as early as possible and to plan recovery procedures. For this purpose an Expert System has been developed. Based on input by human operators, it is able to support the operator in finding the reason for the fault. This is a non–trivial problem. One fault can produce several symptoms and one symptom can be caused by several faults. Faults of one component can cause malfunctions of other components, so it becomes difficult to distinguish between original faults and their effects.

The system developed uses the model–based paradigm. This requires a so–called *deep knowledge* or model of the target system. Advantages of this approach are that these models can be easily adapted to new applications. When building a new system from standard components, the corresponding diagnostic expert system can be constructed from a diagnostic library of these standard components. This is in contrast to the traditional heuristic expert systems, which require expert knowledge based on long-term experiences with the operation of the system.

The system has been implemented using the Knowledge Craft (^{TM} Carnegie Group, Inc.) environment. It has been tested with two different applications which are also described in this paper (see also [1], [2], [3]).

2. PRINCIPLES OF MODEL-BASED DIAGNOSIS

Model–based diagnosis is based on *deep knowledge* of the structure and the functions of a system. One can distinguish mainly between *pathological* models and *functional* models. Pathological models describe malfunctions of the system. They contain an explicit representation of the causal dependencies between diagnoses and symptoms. They describe how different malfunctions are related to each other. Pathological models are mainly used in medicine, where functional models are not known or would be too complex.

In contrast to this, a functional model describes the correct behaviour of the system. It comprises a description of structure and behaviour of the system. The structural description contains the system components and their relations to each other, their characteristics, system parameters and so on. Behaviour means the way in which the system functions and how the system parameters are related to each other. In technical systems the prerequisites for functional models are much better than in medicine, because the models can be built during the construction of the system.

The basic idea of diagnosis with functional models is to observe the behaviour of the real system, to simulate its behaviour using the model and to detect deviations between the two types of behaviour. Assuming a correct model and correct observations one tries to localize the malfunctions by adjusting the model to the real system. This means that one tries to model the malfunctioning of the system. If this succeeds, one possible explanation for the wrong behaviour has been found. This can be done by hypothesizing faults of particular system components in the model. If the simulated behaviour then equals the observed behaviour, the candidate components in the model can be the defective components in reality.

3. KNOWLEDGE REPRESENTATION AND INFERENCE MECHANISM

The basic components of our system are the model which contains the diagnostic knowledge and the inference mechanism which comprises the diagnostic algorithm. Both are based on the theory of Reiter [5]. Reiter uses *conflict sets* containing assumptions about the system which can not be true all at the same time. A typical assumption is, for instance, that a particular system component is defective or not. In contrast to Reiter a different modelling technique has been used. This technique allows a priori computation of the conflict sets which makes the system much more efficient. In addition, the system has been enhanced by features for incremental diagnosis based on [4]. This allows a step-wise refinement of the diagnosis through successive input of observations. The overall diagnostic algorithm is as follows.

```
CANDIDATES := {}
while still too many candidates do begin
   accept next observation
   CONFLICTS := all new minimal conflicts
   for each Q∈CONFLICTS and K∈CANDIDATES do
       if K∩Q = {} then begin
       delete K from CANDIDATES
       extend K with each assumption from Q and add them to CANDIDATES
       delete all non–minimal candidates from CANDIDATES
       end {if}
   report CANDIDATES
   propose observation
   end {while}
```

When a new observation is accepted, the old conflicts are updated. Then the old candidates are investigated to see whether they explain the new observation (i.e., the new conflicts) or not. A candidate is a particular hypothesis for how the system differs from the model. More precisely, a candidate is a set of components which are suspected of being defective and would, in this case, explain the system behaviour. A candidate explains a conflict if their intersection is non-empty. If a candidate does not explain a new conflict it is deleted. This candidate is extended by one assumption from q (which renders a non–empty intersection) and is thus able to explain the conflict. Afterwards non–minimal candidates are deleted. The reason for this is that minimal candidates are of particular interest. The assumption, for instance, that all components of a system are defective would explain every system behaviour but, of course , would have no practical value.

The model representation uses a graph structure with nodes representing entities of the system and arcs representing the diagnostic knowledge. An arc from A to B is interpreted as logical implication : if A is true then B must also be true. Vice versa, if B is false then A must also be false. This structure is used to formulate the causal dependencies in the form of complex networks. Naturally, circular structures are not allowed in the model. In spite of this we will see later on an example of how circular structures can be treated. These causal structures are only the mechanisms offered by the system. It is up to the user to define the interpretations of these structures with respect to reality. In our applications we have used two different interpretations: A causal structure model and a causal fault model.

The **causal structure model** is used to model the propagation of faults within the component structure of the system. If there is an arc from A to B, for instance, then A→B is true. In this case the truth values of the components a and b are interpreted in the opposite way: A or B being true denotes a malfunctioning of the components a or b. If A or B are false, the components a or b. must show correct behaviour. A→B means that faulty behaviour of a implies a faulty behaviour of b. Vice versa, a correct behaviour of b implies a correct behaviour of a.

Another interpretation is the **causal fault model** (i.e., a kind of pathological model). The model entities are representations of faults. Faults can be symptoms (i.e., faults which are caused by other faults) or diagnoses (i.e., faults which cause other faults). D→S now has the following interpretation. If diagnosis d is true then symptom s must occur. If symptom s does not occur, then d cannot be the cause. Due to our modelling technique the generation of the conflicts is very efficient. Consider the causal structure model depicted in figure 1.

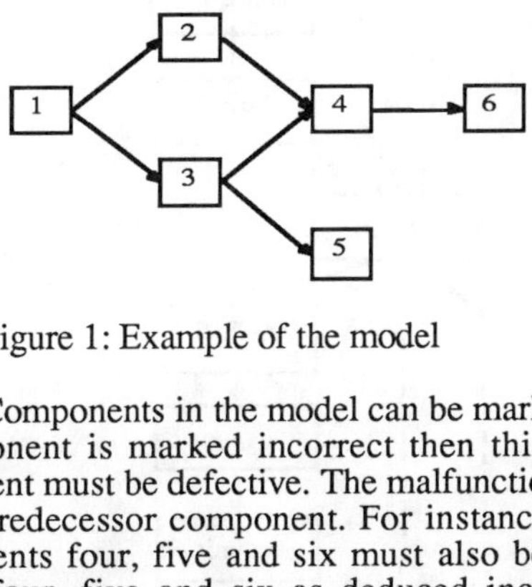

Figure 1: Example of the model

Components in the model can be marked by the user as correct or incorrect. If a component is marked incorrect then this does not necessarily imply that this component must be defective. The malfunctioning can also be caused by a defect of a causal predecessor component. For instance, if component three is incorrect then components four, five and six must also behave incorrectly. In this case we will denote four, five and six as deduced incorrect. On the other hand, all causal predecessors of a correct component are deduced correct. For instance, if five behaves correctly, then one and three must be deduced correct.

Let us assume that component four is incorrect. This must be due to a defect of four or of one of its predecessors. Hence the conflict for the observation "behaviour of four incorrect" is simply the set consisting of four and its predecessors.

If we observe that a component behaves in a correct way, this does not result in a new conflict, but an already existing conflict might not be minimal any more. If we observe in our example that component two behaves correctly, then the conflict {1, 2, 3, 4} is not minimal because {1, 3, 4} is also a conflict now. But this is still not minimal because of the deduced correct component one. Thus the new minimal conflict is {3, 4}. We obtain all new minmal conflicts by deleting all correct and deduced correct components from the conflicts computed so far.

4. IMPLEMENTED APPLICATIONS

Up to now the system has been tested for two different industrial applications. The pilot application was the fault diagnosis of a puma 200 type robot. The second application was a complex system (named COMETOS) for the cleaning of large castings.The system consists of three stations, one for measuring and two for cleaning the castings. For each of the two applications two different diagnostic models have been built: a causal structure model and a causal fault model. We will illustrate this in several examples.

A causal structure model for a puma 200 type robot

Figure 2 shows the cross structure of the robot system.

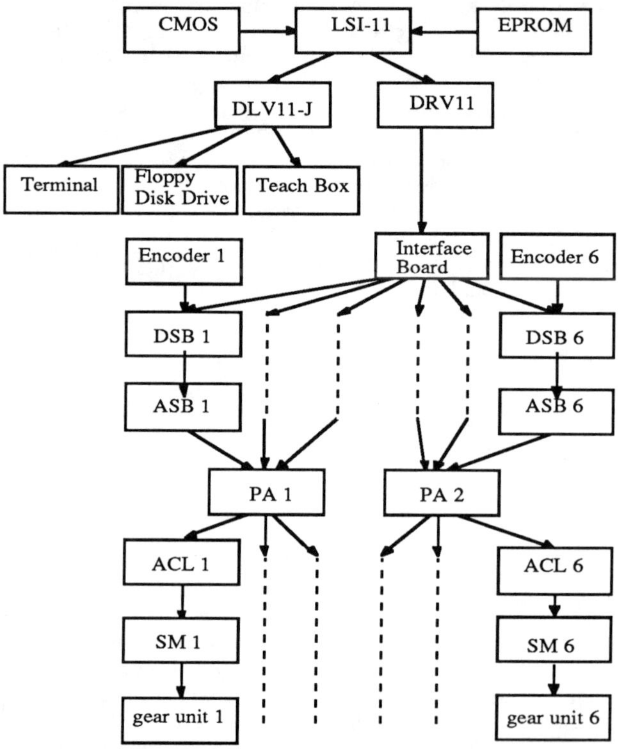

Figure 2: Causal structure model for a puma Robot

Suppose that joint two moves incorrectly. We will feed this observation into the system by marking gear unit two as incorrect. The conflict for this observation is the set {gear unit 2, SM2, ACL2, PA1, ASB2, DSB2, Encoder2, Interface–Board, DRV11, LSI11, CMOS, EPROM}. To focus on the cause for this incorrect behaviour we will have to gather additional information concerning other components in the model. If, for instance, joint three is also incorrect we can concentrate on the common predecessors of joints two and three. Assuming a single fault only, this will render the following conflict: {PA1, Interface-Board, DRV11, LSI11, CMOS, EPROM}. Suppose now that motor sm6 works properly. Since this implies that its predecessors also work correctly, we can remove these components from our conflict. This results in the single fault diagnosis {pa1}.

A causal structure model for the COMETOS–System

The structure model for the part of the system which has been selected for diagnosis contains about 470 entities and even many more arcs. Normally, it is not possible to describe circular structures using our representation form. The following example will show how this can be done inspite of this. Figure 3 showes the DNC–Interface between the controller and the superior personal computer.

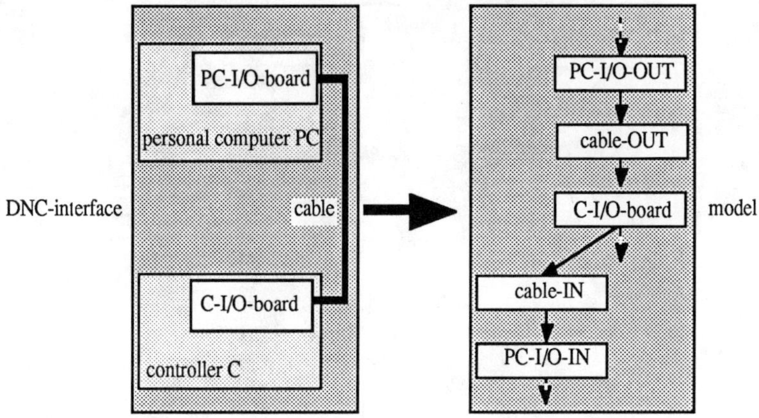

Figure 3: Modelling of a bidirectional data transfer

If an I/O–board in the personal computer is defective it is usually completely replaced by the user. A model representing the two I/O boards and the link between them seems sufficient. But since we have bidirectional data transfer (programs are sent from the PC to the controller and error messages are sent in the opposite direction) using only one model entity for the I/O–board would render a circular structure. Hence we split the I/O–board into two parts: an input part (I/O–in) and an output part (I/O–out). Also, the cable is modeled by two entities: cable–in and cable–out. The right hand side of figure 3 shows the resulting loop-less structure.

Causal fault model of the COMETOS–System

Model entities now represent faults of the system. Faults can be symptoms (faults which are caused by other faults) or diagnoses (faults which cause other faults). Figure 4 shows the top level structure of this fault model.

The complete representation contains about 300 symptoms and diagnoses. Actually it is not a tree structure as the figure implies, but a graph structure. This is due to connections between the controller and the robot which are omitted for clarity. The model is organized in several subsequent levels. Two different types of knowledge have been used: knowledge of the component structure of the system and knowledge contained in the self-diagnosis of the component systems. These two knowledge sources have been combined in a way which simulates the problem-solving process of human experts. Humans usually first try to roughly localize the fault inside the component structure and then step by step descend from component to subcomponent to find out the cause.

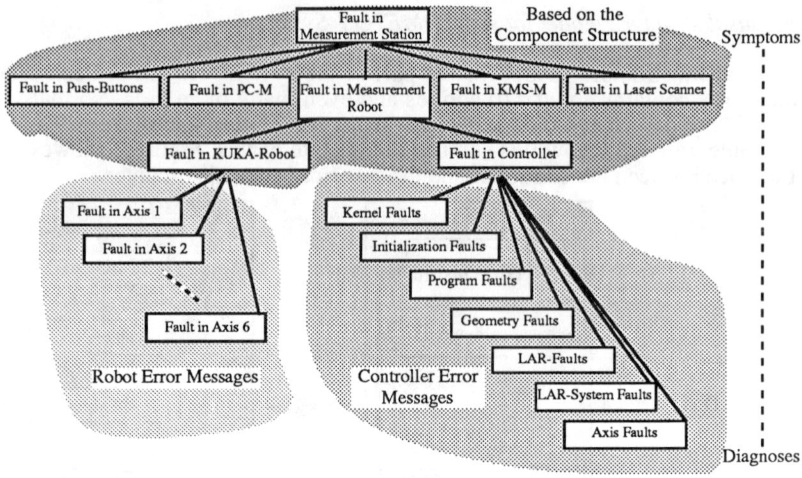

Figure 4 : Top-level layer of the pathological model

5. DISCUSSION

Compared with the classical heuristical approach ("shallow apprach") it is worth highlighting the following items. Heuristical systems can only solve cases which have already occurred in the past. They are incomplete with respect to all possible cases and can also contain contradictory knowledge. Theoretically, assuming a perfect model, this limitation is not true for the model-based approach, because every difference between the behaviour of the real system and the model can be detected. Heuristical systems require that experiences have been gathered during operation of the system. Often it is a troublesome process to inquire this knowledge from the human operators. In contrast to this, structure knowledge for the diagnosis can be gathered by the designers of the system. Thus a diagnostic Expert System can be put into operation together with a new machine. Causal models are easier to maintain and to update than the often weird structures of heuristical knowledge bases.

On the other hand, model-based diagnosis techniques are still in the early stage. In the future development of our system we will try to combine the advantages of both deep and shallow reasoning. We want to investigate how these can be integrated and how switching between the corresponding different inference techniques can be realized. With respect to the knowledge representation formalism we will start to build up a library of descriptions of standard components of our application domain. Another important topic will be the generation of observation proposals which has not yet been implemented.

REFERENCES

1. Hammes, D., Hörmann, K.: "Expertensystem zur Diagnose technischer Anlagen und Werkzeuge zu ihrer Entwicklung", Bericht zum Stand der Forschung, Universität Karlsruhe, 1988.
2. D.Hammes, K.Hörmann,"Entwicklung eines Expertensystems zur Diagnose von Fehlfunktionen an Handhabungssystemen", Abschlußbericht für das Projekt zwischen der Kernforschungszentrum Karlsruhe GmbH und der Universität Karlsruhe
3. Hübner, T., Hörmann, K., Cord, T.: "Entwicklung eines Expertensystems zur Diagnose von Fehlfunktionen eines Handhabungssystems zum Putzen großer Gußstücke", Abschlußbericht, Universität Karlsruhe, 1989.
4. de Kleer, J./Williams, B.C.: *Diagnosing Multiple Faults*, Artificial Intelligence 32 (1987) 97-130.
5. Reiter R.: A Theory of Diagnosis from First Principles, Artificial Intelligence 32 (1987) 57-95.

Graphical Programming and Simulation of Parallel Processes in Complex Robot Manufacturing Cells

Bernhard Hornung
Institute for Real-Time Computer Control Systems and Robotics
Faculty of Computer Science
University of Karlsruhe

ABSTRACT

This paper describes a system for programming of parallel actions in robot manufacturing cells. The system is integrated as a programming module into the robot simulation system ROSI.
The system is characterized by the use of a tree-structured topological model of the robot workcell and the classification of active components with the help of object types such as robot, sensor, peripheral, vehicle. The assignment of actions to a global system reference time which continues with equidistant time steps allows the synchronization and coordination of the processes. The parallel execution of commands on the interpolation level and the ability to reset the program pointer to any earlier moment of program execution facilitates programming and fine tuning of movements and actions.
The functionality of the system is presented and discussed with the help of a benchmark example, which illustrates the special capabilities of this programming system for complex robot workcells.

Introduction: Robot Simulation

Robot simulation systems have become a well-known tool for offline-programming of manufacturing workcells. By the programming of robots with the aid of a computer graphic system much time and effort can be economized which is necessary for the testing of a new application in the real workcell. Damages of machines and workpieces are avoided in the software environment, and the time consuming process of an optimal arrangement of the components in the cell can be realized very fast by using models and simulation techniques. At last, the use of a simulation system for robot planning tasks is one more step towards a factory with computer integrated manufacturing (CIM).
For geometric modeling of objects a CAD-system is used. It is an integral part of the simulation system. Robots, effectors, peripheral devices are stored in system specific libraries and can be built into a robot workcell with the help of modeling functions. Communication between the application programmer and the system is performed by a dialogue interface often realized in menu technique to support a comfortable interaction. Programming of manufacturing processes is performed either in a high-level language or directly in the control language of the particular robots. Generated workcell layouts and programs are stored in the simulation database. The visualization and animation of cells and the program flow is performed by use of a graphic workstation with a high solution screen, sometimes even in real time. The download of verified robot programs can be achieved by specific postprocessors, which generate control code for the real robots and load it to the workcell in the factory.

The Simulation System ROSI and the Programming Module SP^3R

At the Institute for Real-Time Computer Control Systems and Robotics, a robot simulation system called ROSI has been developed [1]. ROSI disposes of modeling functions to generate any desired robot workcell layout. Graphical input techniques help to facilitate the specification of objects and coordinate parameters. The application programmer can select from a library of multiple robots the best-suited one for the special manufacturing process. Other libraries contain peripheral devices like a conveyor, an automatic guided vehicle, a lift for workpieces and statical objects like a working table, a platform and other components of robot workcells.
For geometric design, the CAD-systems ROMULUS and EUCLID can be used. The relational database Rdb on a VAX host computer is utilized to represent the data schemes for ROSI. An Evans&Sutherland PS 390 graphic station is employed for the graphical visualization .

In this paper, the programming module of ROSI, SP^3R, is described [2]. SP^3R means *S*imulation and ***P***rogramming of ***P***arallel ***P***rocesses in ***R***obot Manufacturing Cells. The aim of this programming module is the description and execution of parallelism in a robot workcell on a very detailed level. Hereby, the robot workcell may consist of different kinds of actor types, and it may be structured according to a complex topological model.

Robot Workcell Model and Classification of Components in SP^3R

The initial point for programming of a robot application is a tree-structured topological model of the cell. Each node of the tree corresponds to an autonomous component of the cell which may be active (programmable like a vehicle) or passive (like a table). The vertices represent geometric affix-relations of the components to each other in frame notation. Such a frame denotes in terms of a (4x4)-matrix the position of the local coordinate system of the lower component relative to the higher component. The values of the frame may be fixed or time-dependent. In case of a robot mounted on a pedestal, the relation is fixed; in case of a sensor mounted to the last joint of a robot, the relation is time-dependent. The root of the robot workcell tree is the reference coordinate system of the cell.
The components of the workcell are classified according to their object type. Active components may be robots with their endeffectors, vehicles, peripherals like a conveyor, and sensors. Passive components are static objects and workpieces, which may be handled by active components. Each node in the tree-structured model fits into exactly one component class. For each active class, there exists a set of functions for programming of motion respectively measurement, which is typical for this type of class.
Robots can be programmed either in cartesian or in joint coordinates. The resulting three-dimensional trajectory of the tool center point of the robot is of interest, because the robot interacts with this trajectory with his environment. With the help of effector statements, workpieces can be grasped, ungrasped or treated. Vehicles perform two-dimensional motions on the floor of the workcell; they may be a platform for robots or a transport medium for workpieces. Peripherals with a low degree of freedom accomplish individual or continuous joint motions like a turning table or a conveyor. Sensors, at last, control the program flow of other components with the help of measuring functions. They generate information about the actual state in the workcell.
Fig. 1 shows an example of a robot workcell, Fig. 2 illustrates the corresponding topological model and Fig. 3 presents the active component classes of SP^3R and their functionality.

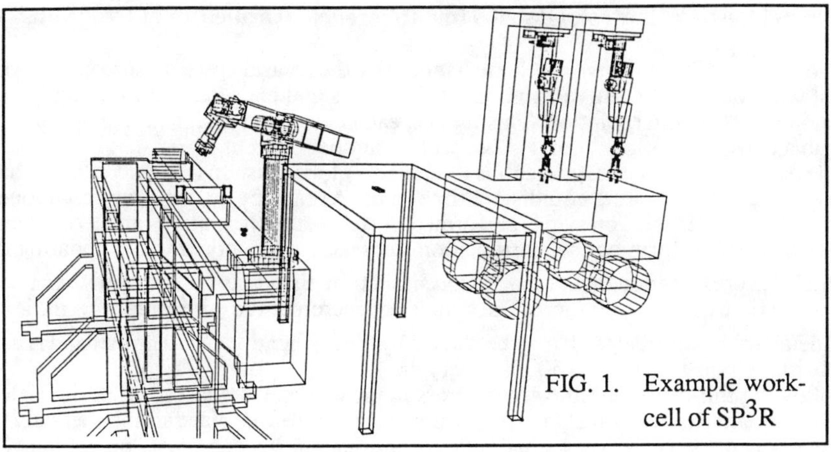

FIG. 1. Example workcell of SP^{3}R

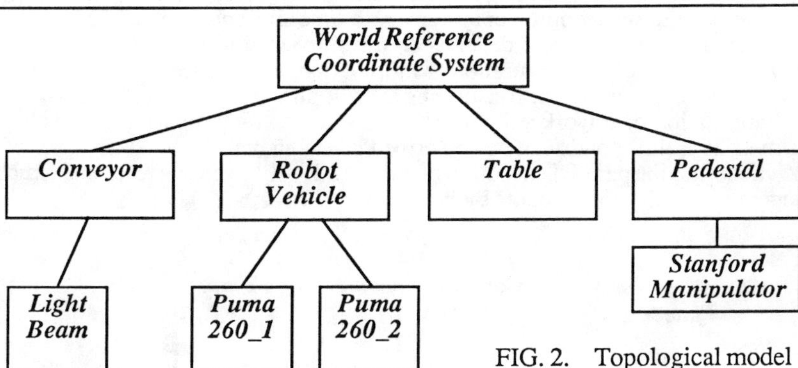

FIG. 2. Topological model of the example workcell

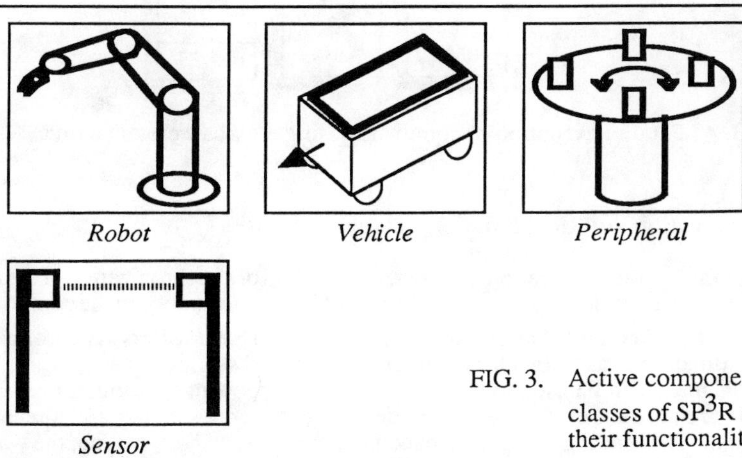

FIG. 3. Active component classes of SP^{3}R and their functionality

Central System Control, Global Time Reference, Parallelism of Programs

The components of the workcell are handled by the central system control. This central unit controls the active components in cycles and executes the functions of the components by interpolation. This means, each command assigned to a component with a certain duration is carried out with a number of steps.
The program flow in the workcell is related to a global system reference time, which keeps pace with discrete equidistant time steps. At each cycle, for every component exactly one interpolation step is performed. Assume, the simulation run has reached a time t_n; after performing the interpolation steps for the components and updating the graphical presentation on the screen, the simulation time may be switched to the next point of time $t_{n+1} = t_n + \Delta t\_sim$. The simulation increment Δt_sim represents the level of detail of the simulation. It may be chosen by the application programmer and is usually in the range of 5 up to 50 milliseconds.
Following this procedure the given actors in the workcell a quasi parallel simulated. The sequential execution of interpolation steps seems to take place simultaneously for the user because of the very small increment Δt_sim. This degree of "parallelism" in the simulation achieved on a computer with only one processor can be compared with an interactive multiuser computer system and a time slice mechanism. The time slice is for every user the same and depends on the number of users.
It is evident that the effective duration of a simulation run increases with the number of participated components. But considering the computing power of today`s workstations, a simulation of a workcell with about 10 active components and a cycle time Δt_sim of 25 milliseconds can be performed nearly in real-time.
Fig. 4 illustrates the parallel execution of commands using a global reference time and the operation of the components by interpolation.

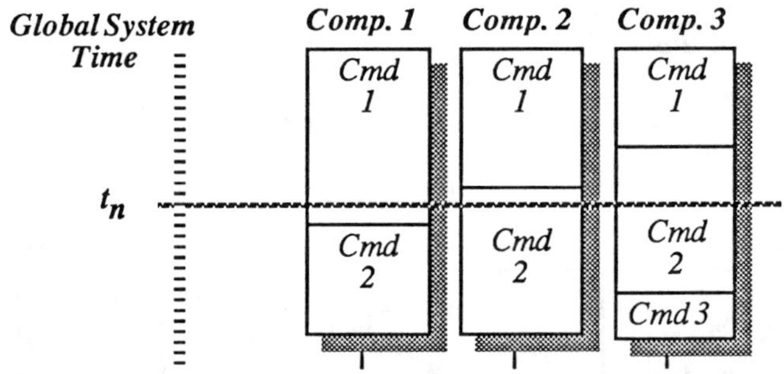

FIG. 4. Parallel execution of commands using a global reference time

Synchronization and Coordination Features

Related to any parallel operation of actors, the problem of synchronization of the components arises. For this purpose, the definition of interaction between the components of the robot workcell must be emphasized. The system SP^3R offers synchronization and coordination capabilities for complex interactive tasks.
Signal channels can be used to trigger a waiting component by another one. These channels are abstract objects. They are defined between a sending and a receiving component. The sending component executes a "set channel"-command; the waiting one executes a "receive channel"-command to continue program execution after

being triggered by the sender. This mechanism relates to the semaphor principle already well-known from operating systems.
An event object allows the synchronization of two or more components at a specific point of time. All the participating components continue simultaneously with their individual program after the occurence of the event. A "set and wait for event"-command realizes the operation.
For a coordinated motion, e.g. the transport of a workpiece by two robots, components may be driven in a so called "coordination mode". The definition of a coordinated motion orientates at the local coordinate system of a workpiece, the so-called coordination frame. A "master"-motion of the coordination frame is defined, and the subordinated robots follow this motion as "slaves" to perform the transport of the workpiece.

Program flow control; Offline-Programming and Simulation

For the programming of a workcell and the simulation of the defined actions, the programming system uses two modes of operation.
In the program definition mode of the system, the application programmer defines individual command sequences for each active component. Hereby, a single command is specified with its parameters using the menu driven dialogue interface of ROSI. The commands are stored in lists affixed to the component description in the topological model of the robot workcell.
In simulation mode, the commands of the components are executed in parallel by the central system control as described above. The interpolation of the functions is realized by software emulators which imitate the behaviour of the real component by software routines. These routines include trajectory and motion planning for robots, vehicles and peripherals, coordinate transformation procedures, the preparation of a coordinated motion or the measuring function of a sensor.
For an efficient development of complex programs the system is able to reset the program pointer. For that purpose the system reference time is reset to an earlier time instant of execution by restoring the valid cell information. This includes the individual program pointers of all the components as well as the dynamic state information of the cell (actual joint angles of robots, reference position of vehicles, state of a synchronization channel and so forth). Resetting is possible because of the storage of program and interpolation information in lists referring to the components. After resetting the program pointer, the user can specify new commands in the program definition mode for the desired components to correct the program.

Run Off of a Short Program for the Example Workcell

The following benchmark demonstrates the functionality and the highlights of the programming system.
In the robot workcell the following actions should happen:
A workpiece "shaft" is transported on the conveyor and passes the light beam. The sensor detects the object and triggers the grasping of the "shaft" by the Stanford manipulator, while the conveyor is in motion (this is realized using the coordination mode of SP^3R with an implicit given motion of the coordination frame, namely the "shaft"). The Stanford manipulator assembles the "shaft" with the workpiece "plate" on the table. The termination of this operation is signaled to the mobile robot CPV which is waiting in front of the table. The two Puma 260 robots mounted on the CPV grasp together the assembled workpieces, perform a coordinated transport of the workpieces, and start the departure of the CPV using a synchronization channel.
This is a very short program fragment, but it demonstrates the difficulties in handling actors operating in parallel and synchronizing at specific time points. With this method very complex multirobot interactive tasks can easily be programmed and fine tuned to reach better performance and to enlarge the robot skills.

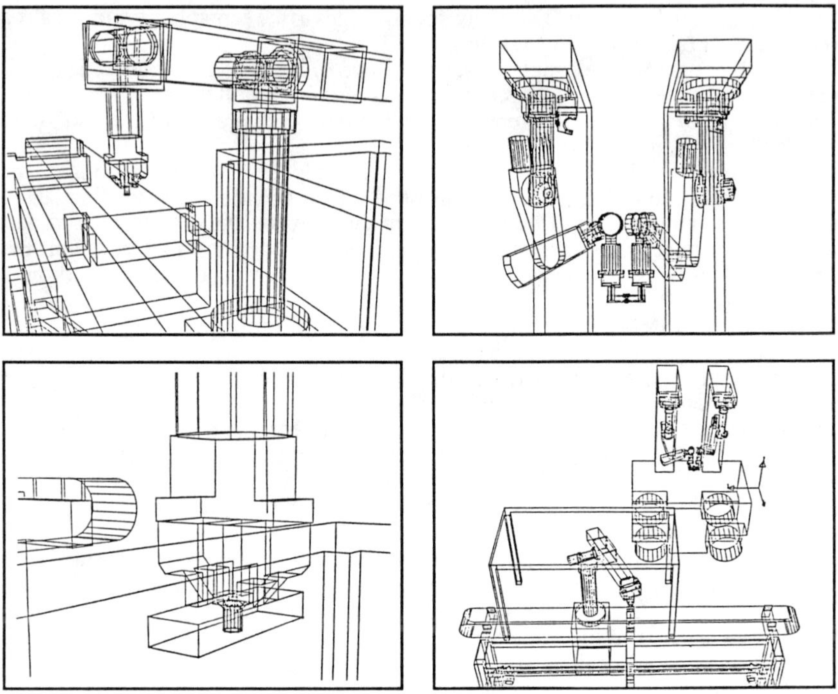

FIG. 5: Snapshots from the simulation run in the example robot workcell

Conclusion

In this paper, the programming module SP^3R of the robot simulation system ROSI has been presented. The special purpose of this programming tool is the parallel control of actions in a robot workcell on the very detailed level of interpolation. The system allows different kinds of interaction between the components and an efficient development of component programs using a reset mechanism. The graphical presentation supports the application programmer to verify and to tune the interrelated parallel programs of components in complex robot manufacturing cells.

Acknowledgement

This work was performed at the Institute for Real-Time Computer Control Systems and Robotics, Prof.Dr.-Ing. U.Rembold and Prof.Dr.-Ing. R.Dillmann, Faculty of Computer Science, University of Karlsruhe, West Germany.
The work was being funded by the state of Baden-Württemberg.

Literature:

1. Dillmann,R.; Hornung,B.; Huck,M.; "Interactive Programming of Robots Using Textual Programming and Simulation Techniques", Proc. 16th ISIR, Brussels 1986

2. Hornung,B.; "SP^3R: A System for the Graphical Programming and Simulation of Parallel Processes in Complex Robot Manufacturing Cells" (in German) , Ph.D. Thesis, Faculty of Computer Science, University of Karlsruhe, July 1990

World Model Registration for Effective Off- Line Programming of Robots

Nicholas Tarnoff, Adam Jacoff and Ronald Lumia
Robot Systems Division, National Institute of Standards and Technology
Gaithersburg, Maryland 20899 U.S.A.
January 1990

ABSTRACT

This paper discusses progress in world model calibration and environmental modeling of sensory information in the context of the Off-Line Programming (OLP) project at the National Institute of Standards and Technology's (NIST) Automated Manufacturing Research Facility (AMRF). The project began in 1988 to demonstrate an integrated OLP implementation. This baseline OLP system accepts IGES CAD data, enables graphical simulation of objects and devices and produces VAL II robot trajectories. The robot trajectories, however, are based on a world model that is inherently inaccurate. A basic OLP system and most existing commercial OLP systems, therefore, are most effective only as a prototyping tool. Consequently, we are now developing our OLP system into a production tool for generating robust robot control programs. We recently completed work on world model calibration and environmental modeling of force and torque in an attempt to improve world model registration of the real world. Environmental modeling of other sensory input, such as vision, is continuing. The eventual integration of an advanced sensory interactive controller will enable off-line programming of sensor servoing to compensate for inherent world model inaccuracies and real world changes.

I. INTRODUCTION

An Off-Line Programming (OLP) system provides a simulation of the real world for programming robotic workcells without the use of the real workcell. OLP of industrial robots potentially is very beneficial in improving the safety and cost-effectiveness of robotic workcells. Using OLP improves human safety by minimizing physical interaction between the operator and a powered robotic workcell. OLP also protects equipment by using simulations to check for robot programming errors that might otherwise have catastrophic results. In addition, off-line generated programs reduce robot down time which ordinarily occurs when programs are generated using the teach-pendant method. OLP, therefore, minimizes robot downtime both during and after robot programming, thus achieving significant economic benefits.

OLP systems are recognized as very useful for fast prototyping and debugging of complex robotic workcells and several commercial systems are available for that purpose [1]. However, OLP applications rarely produce the final robot control programs. Despite the significant benefits of using OLP as a production tool, OLP largely remains a prototyping tool. Therefore, our research is focused on typical OLP deficiencies. We have implemented a generic integrated OLP system with an emphasis on well known integration issues [2]. The baseline OLP world model, however, is not adequate for robust robot programming. The second and current phase of the project involves enhancing the effectiveness and functionality of the basic OLP system by implementing calibration and force/torque simulations.

Funding support for this project was provided, in part, by the Navy Manufacturing Technology Program. Equipment listings in this paper do not imply a recommendation by NIST nor that the equipment is best for the purpose.

An OLP system was implemented at the Cleaning and Deburring Workstation (CDWS) in the AMRF [3]. The CDWS, shown in figure 1, consists of two robots for deburring, buffing and cleaning of machined parts [4]. The OLP project involves programming the Unimate 2000 robot for buffing parts. Both the Unimate and the Puma 760 robots perform part handling to and from the supply trays, deburring vise and washer/dryer. In addition the Puma 760 performs deburring operations. The workcell control scheme allows for concurrent tasks providing that they do not lead to a collision. An Automated Guided Vehicle (AGV) delivers and retrieves parts between workcells and the material buffering system. The goal is to program off-line the gross and, more importantly, fine motions of the Unimate 2000 buffing process. While the application area is very specific, the intent for the project is to develop generic capabilities of OLP which are transferable to any robot programming task [5]. In fact, this generic OLP technology is being applied to a new composites manufacturing workstation now under development in the AMRF.

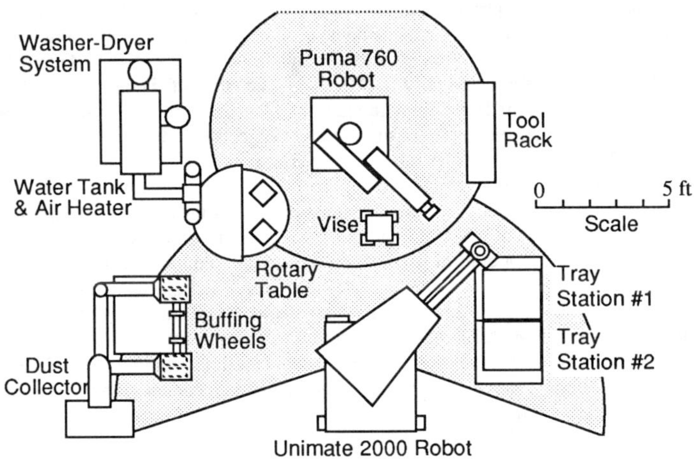

Figure 1 - The Cleaning and Deburring Workstation at the AMRF.

II. BASIC OLP IMPLEMENTATION

The lack of standard interfaces to and from OLP systems is a well known integration issue which exists in all components of Computer Integrated Manufacturing [6]. The flow of information from the CAD system, through the OLP system, down to the controller and back to the OLP system is interrupted by the lack of interfaces as illustrated in figure 2. The intent for the initial implementation was not to tackle an entire interface "barrier" but rather to chip away at it. Implementation of a basic OLP system focused on the problems of a CAD interface. The left hand barrier illustrates the need for a CAD interface to transfer accurate part design information to the OLP system.

Using the OLP system first involves transferring a CAD wireframe model of the part from a commercial CAD system (CADDS from ComputerVision) to a commercial OLP system (CimStation from Silma, Inc.) via the Initial Graphics Exchange Specification (IGES) [7]. IGES 3.0 is a standard file format for passing CAD data between systems. The imported IGES part model is added to the OLP world model. The operator then models the remaining workstation components within CimStation. Robot trajectories are created by defining frames on the graphics screen's rendering of the simulated workcell using a mouse and an on-screen pointer. The operator programs robots (for simulation) by the teach method or by entering absolute or relative tool coordinates and orientations. The user then generates programs textually and graphically to operate and synchronize all workstation devices such as robots, material handlers, fixtures and tooling. The resulting programs are tested and revised based on collision detection and other criteria. The final simulation is saved and the cartesian or joint space robot trajectories are post-processed for downloading to the VAL II robot controller. CimStation provides the simulation environment upon which we are developing additional modules such as environmental modeling. This seamless OLP implementation enables CAD and control information to travel quickly and reliably from design, through simulated testing iterations, to the robot controller.

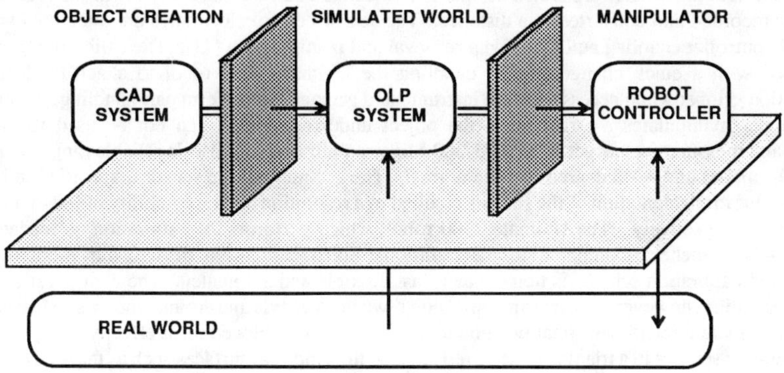

Figure 2 - Off-Line Programming Issues.

III. OLP ENHANCEMENTS TO THE WORLD MODEL

Integrating OLP to other systems such as CAD and robot controllers is both a necessity and a challenge. The major impediment to effective OLP, however, is the inconsistency between an OLP system's world model and the real world. The world model acts as an link to the real world for both the OLP system and control system [8]. The first part of the solution lies in improving the world model's geometric fidelity using calibration. The second part of the solution lies in having a world model that facilitates the off-line programming of sensors for execution by the control system. An adequate world model for generating robust off-line programs must include environmental models such as range (calibration), force, and vision. The first phase OLP implementation confirmed the importance of the world model and our work in implementing calibration and force simulation is illustrated in figure 3.

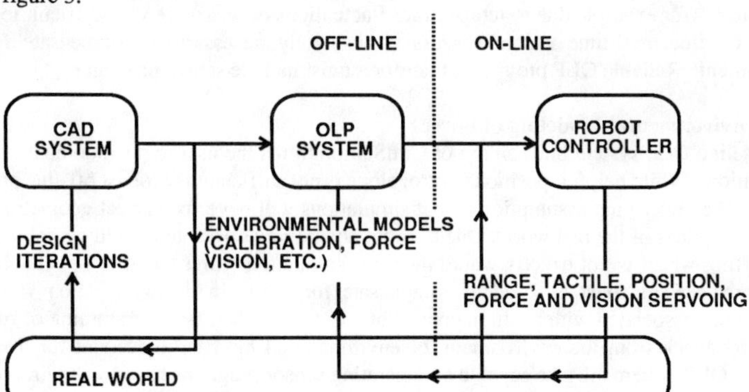

Figure 3 - OLP World Model Enhancements.

III.1 Environmental Modeling of Relative Positions - Calibration

The first step in improving the world model is to calibrate relative object positions based on measurements taken of the real world. An array of ranging sensors mounted on three normal surfaces of a fixed device is traditionally used to perform calibration[9]. Robot mounted non-contact calibration, however, is more flexible. This calibration method is made possible by compact ranging sensors such as ultrasound or recently developed laser triangulation sensors [10]. An instrumented gripper was built that consists of two ultrasound ranging sensors on each gripper finger. The ultrasound ranging sensors are aimed normal to a fixed hard surface and the delay of the return sound pulse is recorded and converted to a distance. The sensor control electronics are integrated with the CDWS controller enabling automatic data retrieval and manipulation [11]. The calibration gripper is equipped with a quick change adapter enabling the Unimate 2000 robot to attach or detach the calibration gripper when necessary. The instrumented gripper can perform part handling. In addition, the gripper manipulates a target when the object under calibration can not be used itself as an ultrasound measurement target. The RPS-300 Migatron ultrasound sensor's reliable range is between 3 and 14 inches and +/- 6 degrees off the normal. The accuracy of the sensor is specified to be 0.03 inches. Our implementation of the sensors resulted in a resolution and repeatability of 0.045 inche and 0.08 inche respectively. The Unimate 2000 robot, which performs the calibration, is repeatable to within +/-0.1 inche. Compared to the traditional fixed calibration device, our robot mounted ultrasound calibration scheme is non-contact, fast, flexible and automated. The configuration of our implementation, however, can measure only along two normal axis and around one axis. Measurement of the complete three dimensional position and orientation is achieved most effectively by mounting three ranging sensors in a triangle pattern, repeated on three normal surfaces such as that of a cube [12]. This configuration is also implemented easily as a non-contact robot mounted device.

World model calibration using ultrasound sensors consists of two steps; measuring the pose (location and orientation) of objects relative to the robot; and generating a three dimensional scalar map of robot end effector positioning errors in a volume surrounding a work area of interest. Both of these calibration procedures transform primarily joint space errors into cartesian space errors and are assumed valid for small distances and angles. Calibration results, therefore, are valid only in the vicinity of the calibrated object or volume. Measurements of object poses are uploaded to the OLP system for off-line calibration of object locations and robot trajectories, as illustrated in Figure 3. An error map, on the other hand, consists of several pose measurements within a volume of interest such as around a fixturing device. VAL II uses the error map on-line to compensate for positioning errors. Positioning errors for the Unimate 2000 in the vicinity of the CDWS vise, for example, vary between 10 and 20 percent.

Static calibration reduces the static errors of the world model only to within the limit of the calibration sensor's accuracy. Not only do calibration sensors have a finite accuracy, but also dynamic changes in geometry (for example due to temperature fluctuations or wear) are very difficult to model and/or track. On-line, real-time use of sensors is absolutely necessary to compensate for a dynamic environment. Reliable OLP programs, therefore, must include sensor programming.

III.2 Environmental Modeling of Forces

The baseline OLP system built on top of CimStation limits the user to position control of robots in a workstation. While helpful, position control alone is not sufficient for robust off-line programming of robots. The underlying assumption is that simulations will never be perfect geometric, kinematic or dynamic replicas of the real world. Due to robot positioning errors and world model inaccuracies, the real buffing wheel is not exactly where the OLP simulation represents it to be relative to the robot. Therefore, sensors must be used to compensate for errors in the workstation world model and inaccuracies associated with position control of robots[13]. Sensor programming of robot tasks uses sensor feedback from the environment, or environmental models, to account for these errors. An effective OLP system must be capable of generating sensor programs based on environmental models. The goal of environmental modeling is to map certain features in a subset volume of the workstation world model. Environmental models are developed using sensors such as proximity, vision and force/ torque sensors. These environmental models are then incorporated into the OLP process to enhance the world model by providing detailed information about regions of interest within the world model.

An example of environmental modeling is world model calibration via ultrasound proximity sensing introduced above. Development of a vision sensing model, to compensate for unknown part placement, is a future goal of this project. The following sections discuss work recently completed on environmental modeling of forces at the AMRF's Cleaning and Deburring Workstation.

Force/Position Model

The buffing of machined parts in our CDWS was chosen as the application on which to develop an environmental model[4]. The buffing application simply served as a vehicle for our research into environmental modeling. Robot errors in part placement into the rotating buffing wheel can result in inferior buffing or dangerously high forces on the part and the robot. The environmental model which was developed and implemented was a force/position(F/P) model of the buffing wheel. The F/P model relates buffing forces on a part to position of the part within the buffing wheel. The buffing forces were gathered experimentally using the force sampling end-effector shown in figure 4(a). The force sampling end-effector was designed and built at NIST specifically to obtain the relationship between buffing forces and relative position within the rotating buffing wheel. The force sampling end-effector is a box-like tool which contains a force/torque sensor. One of three different size probes, attached to the force sensor, peeks out through the faceplate. The probe tip is flush with the faceplate to avoid vertical (y axis) forces on the probe while the faceplate is large enough to look like an infinite plate to the spinning wheel. The force-sampling end-effector is also equipped with a quick-change adapter enabling the Unimate 2000 to attach/detach the end-effector when a new force/position model becomes necessary. Force readings were taken at 12.7 mm intervals both vertically (y axis) and horizontally (x axis), relative to the buffing wheel, and at 5 mm intervals into the buffing wheel (z axis). Surface plots generated at each y level of entry into the buffing wheel are illustrated in figure 4(b). These surface plots show the wave-like representation of the forces incurred during buffing as

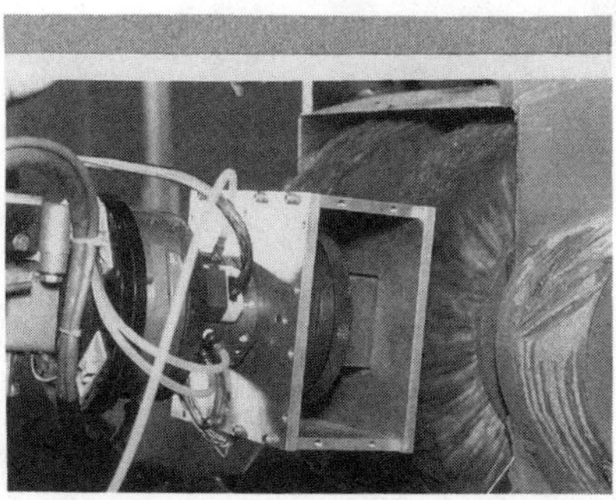

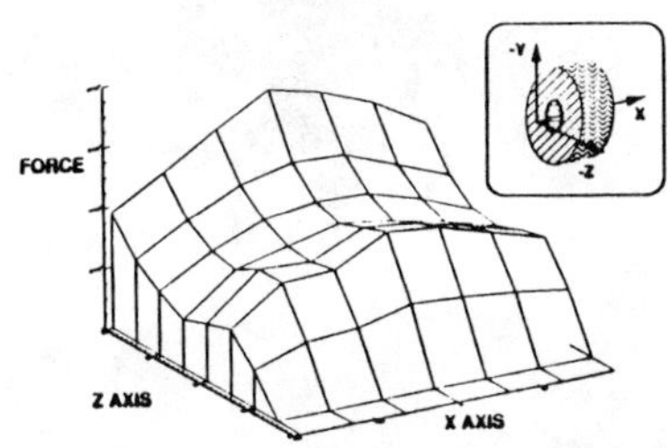

Figure 4(a)- Force sampling end-effector **(b)- Surface Plot of Forces (y=0)**

the probe penetrates the wheel. The buffing wheel F/P model consists of a three dimensional compilation of these surface plots. Linear interpolation was used to determine forces at locations between the experimentally obtained nodes in the F/P model. The buffing wheel F/P model is applied off-line during simulation of the buffing trajectory to monitor the expected forces on the part. Given a part position within the buffing wheel, the force/position model outputs the expected force that the part will feel based on the experimental data of that buffing wheel. It should be noted that the force/ position model is intended as an intermediate step toward on-line, real-time, sensor based robot programming to compensate for inherent world model inaccuracies and real world changes. Real-time force feedback, to a sensory interactive controller, would allow the robot to be programmed off-line according to the expected force/position model, and to actively look for that force on-line.

Force Display

The most visible feature of a graphical OLP system is the operator interface. The enhanced OLP system was built on top of the simulation capabilities of CimStation. Figure 5 shows CimStation's representation of the CDWS. With the simulation, the user can visually follow the movement of the robot as it executes its programmed trajectory. However, programs which make use of environmental models must report data from that model to the user. The environmental model gives the user more insight into how the robot interacts with its environment. In this case, the environment is the buffing wheel itself and the model represents the buffing forces acting on the part. The bar graph force display, as seen in the CDWS simulation, enables the user to adjust the programed buffing trajectory, off-line, according to the allowable forces of the task. The force/position model of the buffing wheel is applied during the robot's simulated buffing trajectory. The position of the part within the simulated buffing wheel is known and is monitored continuously using CimStation's object monitor feature. The part face is divided into smaller facets, in simulation, which are each tracked for collisions with the buffing wheel. CimStation senses a collision between any facet on the face of the part and the buffing wheel, and brightly colors the part and the wheel in simulation. The position of each facet within the

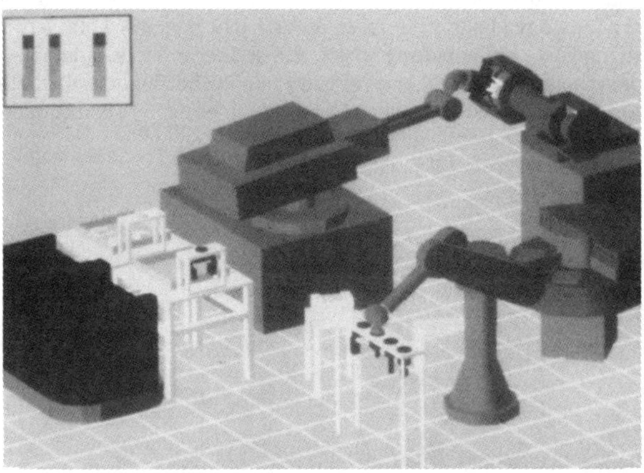

Figure 5 - CimStation Simulation of CDWS with Force Display

simulated buffing wheel corresponds to a force within the buffing wheel's force/position model. According to the F/P model, the expected forces on each facet colliding with the buffing wheel are summed over the whole part. The total force on the part is represented in the simulated force display shown in figure 5. The expected forces in the x, y, and z directions are monitored using color coded scales. The color code is as follows: a green or yellow force is acceptable, a red force is unacceptable or dangerous. The expected forces on the part also can be monitored numerically in CimStation's message window. The force display graphically represents the force/position model and advises the user of excessive or inadequate forces incurred during a particular buffing trajectory. Therefore, the force display allows the user to visually examine the expected forces associated with a programmed trajectory and adjust the trajectory as necessary. By applying the force/position model off-line, we utilize environmental modeling to help develop robot programs and evaluate them for their safety and effectiveness.

Mosaic

Another trajectory evaluation tool, which we have developed for off-line use, is the mosaic pattern shown in figure 6. Each facet on the face of the simulated part is colored according to the length of time it is exposed to the buffing forces within the wheel. The mosaic pattern represents the time exposure of each facet over the entire buffing trajectory. The mosaic color code is similar to the force display. Facets which are red in color indicate excessive buffing time. Facets which are yellow and green indicate acceptable buffing time. Other colors indicate inferior buffing and white facets are untouched by the buffing wheel. Therefor, the mosaic pattern distinguishes the areas of most concentrated buffing for a given trajectory, allowing the user to adjust a particular trajectory according to the shape of the part or the desired buffing pattern. The user can develop programed trajectories which isolate sections of the part for safety reasons. For example, if the leading (top) edge of the actual part were to contact the rotating buffing wheel, excessive y forces would immediately slap the part from the gripper. This dangerous situation can be avoided using the mosaic pattern, off-line, to identify and change trajectories which include contact between the simulated part's leading edge and the buffing wheel. The mosaic pattern is a useful evaluation tool for analysis of off-line programmed buffing trajectories based on buffing time.

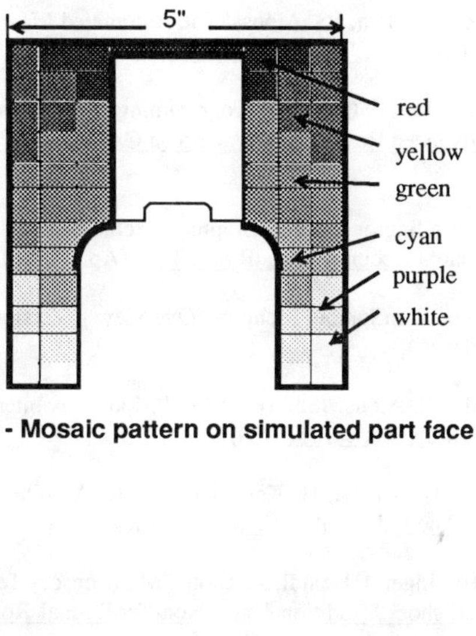

Figure 6 - Mosaic pattern on simulated part face

IV. FUTURE OF OLP

Most OLP research consists of automatically generating robot trajectories and part handling sequences. In comparison, we at NIST are concerned with improving the effectiveness of currently available OLP technology. An OLP system can function as an effective man-machine interface to a sophisticated robot controller provided that the OLP system's world model is relatively accurate and flexible. Experience in implementing a basic OLP system has shown that environmental models of relative positions (calibration) and of sensory input are essential for generating robot programs off-line. Geometric calibration provides a basic level of world model accuracy in specific regions of interest. The flexibility needed to operate in a dynamic environment requires the use of sensors. Likewise, a robust OLP program must include sensor programs based on environmental modeis of position, force and vision. Environmental models of sensory input are used to generate flexible and robust programs off-line for execution by a sensory-interactive controller.

REFERENCES

Important Note:
The National Bureau of Standards (NBS) became the National Institute of Standards and Technology (NIST) on August 23, 1988.

1. A.A. Tseng, "Assesment of Robotic Simulation Software", Proceeding of the 1988 ASME International Computers in Engineering Conference, (San Francisco, California, July 1988).

2. N. Tarnoff and R. Lumia, "The Role of Off-Line Robot Programming in Hierarchical Control", Second International Symposium on Robotics and Manufacturing Research, Education and Applications, ISRAM, (Albuquerque, NM, November 1988).

3. K.M. Murphy, R.J. Norcross and F.M. Proctor, "CAD Directed Robotic Deburring", Second International Symposium on Robotics and Manufacturing Research, Education and Applications (ISRAM), (Albuquerque, NM, November 1988).

4. J.A. Simpson, R.J. Hocken, J.S. Albus, "The Automated Manufacturing Research Facility of the National Bureau of Standards", Journal of Manufacturing Systems, 1, (1), 17 (1982).

5. R. Lumia, "CAD-Based Off-Line Programming Applied to a Cleaning and Deburring Workstation", NATO Advanced Research Workshop on CAD Based Programming for Sensor Based Robots, (Il Ciocco, July 1988).

6. B. M. Smith, J. Wellington, "Initial Graphics Exchange Specification (IGES), Version 3.0", National Bureau of Standards Document no. IR 86-3359, (April 1986).

7. J. J. Craig, "Silma CimStation Technical Overview - Version 3.0", Silma Inc. (Los Altos, California, August 1987).

8. E. W. Kent and J. S. Albus, "Servoed World Models as Interfaces Between Robot Control Systems and Sensory Data", Robotica, 1, (1984).

9. A. Nowrouzi, Y. B. Kavina, H. Kochekali and R. A. Whitaker, "An Overview of Robot Calibration Techniques", The Industrial Robot, 12 (1988).

10. J. Dietrich, G. Hirzinger, J. Heindl, J. Schott, "Multisensory Telerobotic Techniques", NATO Advanced Research Workshop "Traditional and Non-Traditional Robot Sensors", (Maratea, Italy, August 1989).

11. R. Bostelman, "Electronic Design of the Infrared/Ultrasound Sensing for a Robot Gripper", National Institute of Standards and Technology, NISTIR 89-4223 (November 1989).

12. D. L. Riley, "Robot Calibration and Performance Specification Determination", Proceedings of the 17th Symposium on Industrial Robots, (April 1987).

13. C. Crane, J. Duffy, "An Interactive Animated Display of Man-Controlled and Autonomous Robots", Proceeding of the ASME Computers in Engineering Conference, (Chicago, IL, July 1986).

Methods for Automating Tolerance Analysis and Synthesis in CAD Environments

Z. Dong
Department of Mechanical Engineering
University of Victoria
Victoria, B.C. V8W2Y2 Canada

A. Soom
Department of Mechanical and Aerospace Engineering
State University of New York at Buffalo
Amherst, N.Y. 14260 U.S.A.

Abstract

Analysis and synthesis of mechanical tolerances are tools supporting better mechanical designs with lower production costs. A method for automating the problem formulation of tolerance analysis and synthesis in CAD environments is presented. This method applies the graph representation and search technique to automatically identify related tolerances and to analysis tolerance relations. It can significantly simplify the tolerance calculation task and combine tolerance analysis and synthesis with Computer-Aided Design.

1 Introduction

Size and geometric tolerances are used in mechanical design to specify allowable deviations from a nominal or ideal part shape. The tolerance values that are assigned directly affect the manufacturing cost and the functionality of a component or an assembly. Methods of tolerance analysis (or tolerance stack-up) and tolerance synthesis (or optimal tolerance design) have been continuously developed as tools to produce better mechanical designs with lower product costs [1]-[9]. This research has significantly contributed to the field of mechanical tolerancing.

Contemporary tolerance analysis and synthesis techniques rely largely on a designer to formulate a tolerancing problem and to provide equations to the developed software tools. However, this formulation requires identification of a group of related tolerances from a large number of design tolerances, and analysis of their relations. It is quite time-consuming and error-prone, especially for an assembly design with multiple component drawings. The tolerance analysis and synthesis techniques have not been widely adopted in design partially due to this problem. With the increasing demand of computer-automated design and manufacturing systems, the importance of making tolerance analysis and synthesis as automated functions in the Computer-Aided Design (CAD) environments has been recognized [10].

This paper presents a method for automating the related tolerance identification, which plays a crucial part in the problem formulation of tolerance analysis and synthesis. Methods for the automatic retrieval and representation of dimension and tolerance information from CAD databases, using an IGES interface and a dimension and tolerance graph, called Three Branch Graph (TBG), and for the automatic identification of related tolerances using graph search, are discussed. A relevant issue, addressing the automatic formulation of an optimal tolerance design problem using an expert system, is discussed in the author's other publications [11, 12, 13].

2 Graph Representation and Search for Related Tolerance Identification

A graph search (or tree search) technique [14] is a very convenient tool to solve exploring type problems. It normally consists of a graph (or tree) representation of the problem, combined with a search scheme appropriate for the representation. Identification of related tolerances is readily structured as an exploring problem. The mechanical part dimension/tolerance and assembly hierarchy naturally form a graph representation. The availability of efficient search methods associated with the graph search technique makes it attractive for tolerance relation analysis.

A graph consists of a set of nodes (n_i, $n_i \epsilon N$) and arcs (a_i, $a_i \epsilon A$) connecting these nodes. A tree is a particular form of graph with only successors of its nodes. The tree representation of part dimension hierarchy, called dimensional tree, has been introduced earlier by Requicha to represent composition relations of a solid model [2]. In this work, a graph representation of design dimension and tolerance hierarchy, called Three Branch Graph (TBG), has been introduced. It is an extension and a generalized representation of Requicha's dimensional tree. for tolerance relation analysis use.

2.1 A Part TBG Branch

The part TBG has three branches. Each branch represents the design dimension and tolerance hierarchy for one of the part's three orthogonal Cartesian coordinates. An arc (a_i) represents a size dimension and its tolerance. The two ending nodes of the arc, (n_{2i-1}) and (n_{2i}) represent a dimensioned feature and a reference feature. The coordinates of the intersection of the dimension witness or leader line with the profile of a part feature (surface, surface edge, or centerline), called nodal coordinates, are node values. Nodes which share identical nodal coordinates are treated as identical nodes. By overlapping identical nodes of different arcs, related dimension and tolerance pairs are connected. Relations among related dimensions and tolerances are represented. An example part TBG branch is illustrated in Figure 1, in which the part (front cap) dimensions and tolerances along X direction and the corresponding X branch part TBG are shown.

2.2 An Assembly TBG Branch

Individual mechanical parts are eventually used in assemblies. The same graph form that has been suggested for individual components is also used for assemblies. Each component TBG becomes a part of the assembly TBG.

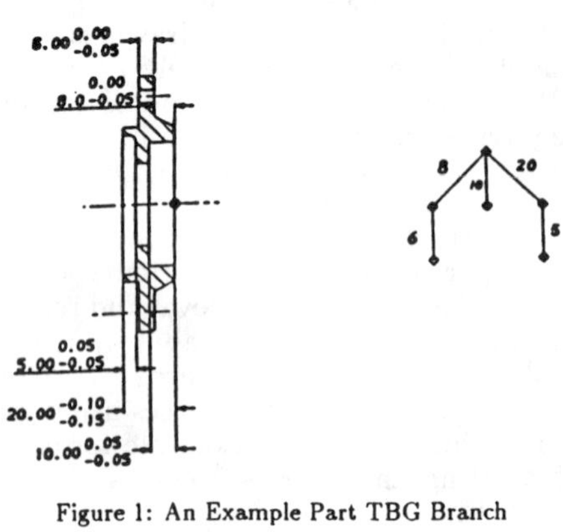

Figure 1: An Example Part TBG Branch

The root of an assembly TBG is the center of a global coordinate system of the assembly drawing. Nodes and arcs of the part TBG retain their original meaning and values. Part TBG's are connected together by additional arcs which represent assembly relations. Two kinds of assembly relations are included: i) physical contact of the mating surfaces of components, represented by arcs of zero dimension and finite tolerance; ii) two separated surfaces of parts, represented by arcs with the dimensions and tolerances specifying the separation. Since the nodal coordinates defined by the nodes of part graph are specified in the part local coordinate system, it becomes necessary to transform these values into the global (assembly) coordinate system in assembly TBG generation.

In generating an assembly TBG, the component list and the locations of the component part local coordinate centers in the assembly coordinate system are required. Instead of interpreting these kinds of information from the assembly drawing, an interactive program has been developed to support the assembly design and to obtain them during the assembly design. The program actually constrains the procedures of assembly drawing generation. Complicated geometric information extraction is avoided.

2.3 Generalized TBG's for 3D Mechanical Parts and Assembles

Methods have also been developed for decomposing and combining oblige size dimensions and tolerances, in order to represent the dimension and tolerance hierarchy of an ordinary three-dimensional tolerance design, using three independent graphs of the TBG. Geometric tolerance of location, which contributes to the tolerance accumulation, is converted into equivalent dimensional tolerance for tolerance calculation purpose. All of these transformations are based on the functional equivalent and conservative criteria. Variation sacrifices introduced in these transformation are minor. Readers may refer to [13] for details.

3 TBG Construction and Interface to CAD Systems

Dimension and tolerance information of a mechanical design lie on an assembly drawing and all its component part drawings. Obtaining all dimension and tolerance information from these drawings and representing their connections using the proposed TBG are the pre-requisite for tolerance relation analysis.

An interface to ordinary CAD systems has been developed in this work. This interface retrieves and updates dimension and tolerance information of CAD databases through the IGES (Initial Graphics Exchange Standard) neutral data format[15], which are available from most CAD systems for exchanging graphics data. In an IGES file, the CAD data are organized in subsections with all graphics and text data on the drawing represented as annotated entities. Certain entities are used for dimension and tolerance information. Nominal dimension, its upper and lower tolerances, and coordinates of its leader lines and witness lines, are organized in one data block. A linear dimension and tolerance pair $90^{+0.0}_{-0.1}$ specifying the surface vertexes at (70, 170) and (160, 170) has the following IGES format:

```
212. 1. 4. 14. 39999771. 4 0. 1. 1. 57079625. 0. 0. 0. 0. 99. 93519592. 148 0.   00000003P0000003
0. 0. 4H90. 0. 0. 0.                                                            00000003P0000004
212. 1. 2. 7. 19999886. 4 0. 1. 1 57079625. 0 0. 0. 0. 115 79917908.            00000005P0000005
151. 33197021. 0 0. 2H0. . 0. 0.                                                00000005P0000006
212. 1. 4. 14. 39999771. 4 0. 1. 1. 57079625. 0 0. 0. 0. 115. 79917908.         00000007P0000007
144. 66799927. 0 0. 4H-0. 1. 0. 0.                                              00000007P0000008
214. 1. 6. 0. 3. 0. 0 0. 70. 0. 150 0. 99. 13598633. 150 0. 0. 0;               00000009P0000009
214. 1. 6 0. 3 0. 0 0. 160 0. 150 0. 130 86398315. 150. 0. 0. 0;                00000011P0000010
106. 1. 3. 0. 0. 70. 0. 170. 0. 70. 0. 169 0. 70. 0. 149. 0. 0. 0;              00000013P0000011
106. 1. 3. 0 0. 160 0. 170 0. 160 0. 169 0. 160 0. 149 0. 0. 0;                 00000015P0000012
216. 3. 9. 11. 13. 15. 0. 0;                                                    00000017P0000013
                                                                                00000019P0000014
```

nominal dimension; upper tolerance; lower tolerance; lead lines; witness lines; linear dimension cluster; cluster data entry number

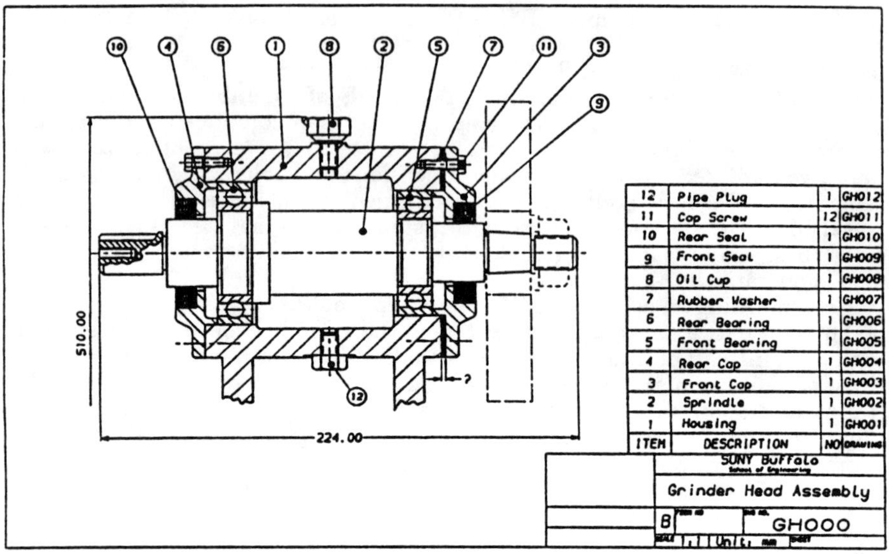

Figure 2: Grinder Head Assembly

All dimensions and tolerances of a design are retrieved and stored in the TBG form. The part TBG's are generated based on the relations of the obtained nodal coordinates. This interface also modifies the IGES files by inserting the calculated tolerances. When the drawing is regenerated from a modified IGES file, the calculated tolerances are fed back to a designer.

4 Related Tolerance Identification Using Graph Search

The TBG graph representation permits related dimensions to be identified and isolated from other irrelevant design dimensions and tolerances. An interrogated dimension in tolerance analysis or a specified clearance between two surfaces in tolerance synthesis is treated as any other dimension and is represented by two nodes on the TBG. One surface is the starting node (n_s), and another surface is the ending node (n_e). Performing a node-oriented, path-remembering, depth-first search through an assembly TBG, yields the detected paths $p(n_s, n_e)$. This path represents a related tolerance group (or a dimension chain). Related dimensions and tolerances are the values of the arcs on that path. Tolerance calculations for analysis or synthesis with these identified dimensions and tolerances then become relatively straightforward.

Another application of the graph representation is design dimension redundancy verification. By specifying two surfaces in a design, using a starting node n_s and an ending node n_e, and then performing all possible depth-first searches through the graph, one can identify a proper dimensional specification by the presence of only one path. Multiple paths imply over-dimensioning and the absence of a path represents under-dimensioning.

5 An Example

A sample tolerance analysis problem of a grinder head design, as shown in Figure 2, is presented to illustrate related tolerance identification. The design

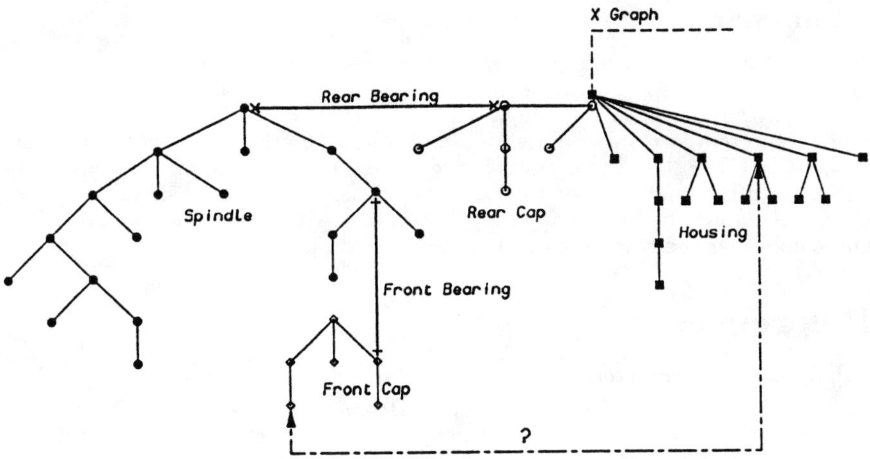

Figure 3: X Branch Assembly TBG (of Associated Parts)

details with dimensions and tolerances of each component are specified in part drawings, such as the front cap design illustrated in Figure 1.

A tolerance analysis problem can be stated as: Given the designs of the housing, spindle, front and rear caps, and knowing the dimensions and tolerances of purchased front and rear bearings, (both $20^{+0.01}_{-0.01}mm$), what is the possible range of variation of the gap between the front cap and the housing of the assembly, and will the rubber washer (No.7) properly seal the grinder head assembly ?

A tolerance analysis of the clearance between the housing and the front cap has been conducted. A dimension line is placed across the two surfaces with the question mark. Dimension and tolerance data are retrieved from the drawings through the developed interface to IGES files of the CAD system. TBG graphs of the parts, of which names are shown on the component list of the assembly drawing, are constructed. An assembly TBG is formed, based on its component's TBG's and the assembly relations shown on the assembly drawing. The X branch assembly TBG, which is automatically generated, is shown in Figure 3.

Through the graph search, only nine dimension and tolerance pairs of six parts, out of approximately two hundred dimensions and tolerances of twenty-three parts, have been found to be related to the gap in question. The tolerance stack-up calculation is performed only with these nine related dimensions and tolerances. Using a worst-case approach, in this instance, the result shows that the possible variation of the gap is $2^{-0.1}_{-0.52}$ (or $1.48 - 1.9mm$). The rubber washer is $2mm$ thick, with a $0.1 - 0.6mm$ depression required to seal properly. Since the actual depression will be $0.1 - 0.52mm$, the tolerance design is verified. Otherwise, modifications of the nine identified design tolerances would be required. With the IGES database post-processor developed, this calculated value $2^{-0.1}_{-0.52}$ replaces the "?" on the grinder head assembly drawing, providing feedback to the designer.

6 Summary

A method for automating the formulation of a tolerance analysis or synthesis problem is presented. The dimension and tolerance information is retrieved from ordinary CAD databases through an IGES interface. The retrieved information is represented using a dimension and tolerance graph, called Three Branch Graph (TBG). Related tolerances are identified using search on the constructed TBG. With this method, the tolerance calculation task can be significantly simplified. A major obstacle in applying tolerance analysis and synthesis techniques and combining them with a CAD system is eliminated.

References

[1] Bjorke, O., *Computer-aided Tolerance*, Tapir Publishers, 1978.

[2] Requicha, A.A.G., "Part and Assembly Description Language: I - Dimensioning and Tolerancing," *Technical Memorandum 19*, Production Automation Project, University of Rochester, 1977.

[3] Parkinson, D.B., "Tolerancing of Component Dimensions in CAD," *Computer-Aided Design*, Volumn 16, Number 1, January, 1984, pp.25-32.

[4] Dong, Z. and Soom, A., "Automatic Tolerance Analysis from a CAD Database," paper presented at the 12th Design Automation Conference in Columbus, Ohio, ASME 86-DET-36, October 5-8, 1986.

[5] *An Introduction to VSA^{TM} Variation Simulation Analysis Software*, Applied Computer Solutions, Inc., Clair Shores, MI.

[6] Requicha, A.A.G., "Toward a Theory of Geometric Tolerancing," *International Journal of Robotics Research*, Vol.2, No.4, 1983, pp.45-60.

[7] Srinivasan, V. and Jayaraman, R., "Issues in Conditional Tolerance for CAD Systems," *IEEE International Conference on Robotics and Automation*, 1985, pp.373-375.

[8] Chase, K.W. and Greenwood, W.H., "Design Issues in Mechanical Tolerance Analysis," *Manufacturing Review*, Vol.1, No.1, March 1988, pp.50-59.

[9] Wu, Z., Elmaraghy, W.H. and Elmaraghy, H.A., "Evaluation of Cost-Tolerance Algorithms for Design Tolerance Analysis and Synthesis," *Manufacturing Review*, Vol.1, No.3, October 1988, pp.168-179.

[10] Turner, J.U. and Wozny, M.J., "Solid Modeling for Automated Tolerance Analysis," paper presented on the SIAM Conference on Applied Geometry, July 20-24, 1987.

[11] Dong, Z. and Soom, A., "Optimal Tolerance Design with Automatic Incorporation of Manufacturing Knowledge," Proceeding of *1989 IIE Integrated Systems Conference and Society for Integrated Manufacturing Conference*, Nov. 12-15, 1989. pp. 409-414.

[12] Dong, Z. and Soom, A., "Some Applications of Artificial Intelligence Techniques to Automatic Tolerance Analysis and Synthesis," *Artificial Intelligence in Design*, D.T. Pham ed., IFS/Springer-Verlag AI in Industry Series, A. Kusiak Series Editor, 1989.

[13] Dong, Z., *Automatic Tolerance Analysis and Synthesis in CAD Environments*, Ph.D Dissertation, State University of New York at Buffalo, 1989.

[14] Charniak, E. and McDermott, D., *Introduction to Artificial Intelligence*, Addison-Wesley, 1985.

[15] Smith, B. and Wellington, J., *Initial Graphics Exchange Specification (IGES)*, Version 3.0, NBSIR 86-3359, U.S. Department of Commerce, April, 1986.

MANIPULATOR FREE WORKSPACE : A NEW APPROACH FOR ROBOTIC CAD SYSTEMS

Ph. Wenger and P. Chedmail
L.A.N. (C.N.R.S. u. n°4/823) and L.M.S.
Ecole Nationale Supérieure de Mécanique
1, rue la Noë 44072 Nantes-FRANCE

ABSTRACT

This paper presents a new connectivity analysis of the free workspace , which permits to characterize the ability of the robot to move within it. Stringent formalizations are given which lead to all the connected subspaces. This makes it possible to ensure the feasability of tasks such as point to point motions or continuous trajectories tracking. The connectivity analysis has been implemented in the robotized cells design package SITROB. The first results supplied by SITROB represent important improvement in comparison with classical robotics CAD systems.

1- INTRODUCTION

Computer Aided Design of robotic cells involves several geometrical problems[1] such as :

1- the choice of the robot : the difficulty is to find the morphology that is best suited to the task to be achieved (type of joints, lenght of the links, ...)

2- the geometrical layout of the cell : the relative position of the robot and the other components of the site must ensure the feasability of the tasks (accessibility of the specified workareas as well as mobility within them)

3- collision-free path planning : the problem is to find a continuous path among obstacles between one location to another.

Our study is a contribution to the solution of the first two problems. In a previous work[2], a method has been described to perform the automatic design and positioning of a robot among obstacles. The basic idea was, thanks to Free-workspace characterization, to analyse the ability of the robot to reach a given area describing the task to be achieved. This paper presents new connectivity analyses of manipulator free-workspace. We mean by free-workspace the set of positions and orientations that the end effector can reach according to its joint limits and the various obstacles. Stringent formalizations are given which lead to the characterization of all the connected subspaces. This makes it possible to ensure the feasability of tasks decribed as sets of points to be joined or continuous trajectories to be tracked. The connectivity analyses have been implemented in the robotized cells design package SITROB.

2- FREE-WORKSPACE OF A MANIPULATOR

2.1- Definition

Let Qf be the configuration free space. It describes the set of all the configurations that the robot can reach without collision[3],[4].

The free-workspace Wf can be defined as the image under the geometrical operator of the configuration free space : Wf = **f**(Qf).

The free-workspace of a robot is defined in the operational space. It describes the set of positions and orientations that the end effector can reach without collision between any part of the robot and the obstacles of the environment.

2.2- Description of the robot's performances in the operational space

Obstacles effects can be studied in the space of joint coordinates or in the space of operational coordinates. The first one is abstract whereas the second one is the "physical" space where the tasks are commonly described. When only the position coordinates are required, the operational space coincides with the cartesian space. The free-workspace is of great interest for the designer of a robotic cell.

A lot of work have been carried out on workspaces analysis, but few of them really takes into account the effects of obstacles. Most often, determination of robots workspaces do not take into account the obstacles[5],[6]. Some authors simply remove the space occupied by obstacles[7],[8], which can lead to important errors. Shiller describes the "obstacles shadows" (regions of obstacles obstruction in the workspace[9]), which do not characterize the topology of the free-workspace.

2.3 - Ability of a robot to perform point to point motions

In figure 1, the task to be achieved is the assembling in $\mathbf{P}_1$, $\mathbf{P}_2$, and $\mathbf{P}_3$ of pieces to be grasped from the table (which lies under the robot's workspace and consequently has no effect on the free-workspace). The free-workspace of the robot is path-connected, and the table has been placed in such a way that a part of it (A) be reachable. However, this layout does not enable the feasability of the tasks: as will be proved in part 3.2, no path exists between A and point $\mathbf{P}_1$.

This example shows that the topology of the free-workspace cannot be characterized by the classical path-connectivity property.

2.4 - Ability of a robot to track continuous trajectories

Tasks such as continuous welding or cutting need to track continuous trajectories. Then, it is necessary to verify that the robot does not change its configuration during the operation. This problem has already been studied in the case of environments without obstacles[10]. The author has defined the notion of "aspect". The aspects are related to the different classes of configurations of a manipulator. For example, a two-revolute joints planar manipulator may reach a position with aspects $\mathbf{q}_2>0$ or $\mathbf{q}_2<0$. When tracking a trajectory, the joint limits as well as an obstacle may oblige the robot to change aspect, as shown in figure 2.

The aspects realize a partition of the configuration space. Similarly, we have defined the free aspects in the configuration free space.

2.5 - Definition of the n-connected subspaces

The regions of the free-workspace where the robot can move between any n points are called the n-connected subspaces. We have proved[11] that the maximal n-connected subspaces are the images under the geometrical operator of the connected components of the configuration free space.

2.6 - Definition of the t-connected subspaces

The regions of the free-workspace where the robot can track any continuous trajectory are called the t-connected subspaces. In the case of non-redundant robots, we have proved[11] that the maximal t-connected subspaces are the images under the geometrical operator of the connected components of the free-aspects.

3- APPLICATION TO ROBOTICS CAD SYSTEMS

3.1- Determination of the connected subspaces[11]

The free-workspace is obtained via an optimized sweeping of the configuration free space, as in [12]. Collision detection is achieved with a given accuracy, based on a C.S.G. (Constuctive Solid Geometry) C.A.D. description of the environment and of the robot. The connected subspaces are obtained using octree model[13].

3.2- Application to off-line tasks programming

Softwares of current robotics CAD systems do not provide a priori information about global capabilities of the robot in its environment[2]. Programming a trajectory consists in simulating the movements of the robot between each work points that the human operator has placed (collisions are often automatically detected). However, the operator has to modify the layout of the cell if the trajectory is unfeasible, until a solution exists. The connected subspaces give straightforward information : according to the task to be achieved, trajectories will have to be defined in a n-connected subspace or in a t-connected subspace.

3.2.1- Using the n-connected subspaces

Let us recall the example of figure 1 : no trajectory exists between the table and point $\mathbf{P}_1$. Figure 3 shows that the table has to be put on the left side, in order that an n-connected subspace exists which recovers both point $\mathbf{P}_1$ and a part of the table.

3.2.2- Using the t-connected subspaces

When the robot's end-effector has to track a continuous trajectory, the problem is more difficult as no configuration changing must occur during the motion. The entire trajectory has to be enclosed in a t-connected subspace. This is very important for tasks such as continuous cutting or welding. Figure 4 depicts the free-workspace of a Scara robot which has to perform the continuous cutting of an L-shaped piece. Figure 5 shows how to position the piece in the workspace so that a t-connected subspace exists which recovers the outline. This choice permits to carry out the task.

3.3- Application to robotized cells design

The geometric layout of a robotized cell should ensure the accessibility of the various workareas as well as the ability to move within them. This is not easy to carry out, as complex geometric reasonning in 3D is required. Some robotics CAD systems as ROBCAD involve an automatic positioning function, but obstacles are not taken into account : the operator has to evaluate the solution. We present in this part the first results obtained using the robotized cells design package SITROB, in which the calculation of the connected subspaces has been implemented.

3.3.1- Description of SITROB[1]

A global description of the task to be achieved is used. The task, called "target space", is an area to be reached and travelled through among obstacles by the robot's end effector. The software aims at finding a position and an orientation of the robot's base that ensure the feasability of the task defined as previously. An optimization method is employed, in which one of the following criterion is minimized :

- the distance d between the free-workspace Wf of the robot and the target space $\mathcal{T}$ if $Wf \cap \mathcal{T} = \varnothing$,

- the volume v of $T-C_m$ as soon as overlapping appears, where C_m means the connected subspace with maximal overlapping.

The optimization method minimizes the above criterion using Hooke and Jeeves algorithm [15]. The algorithm stops as soon as v=0 (in which case the task is achievable), or when a given exploring accuracy is reached.

The target space, as well as the free-workspace and the connected subspaces, is modelled with octree structures.

3.3.2- Optimal positioning using t-connected subspaces

The following example depicts an industrial case. The site is composed of a car body and a robot whose first three joints are respectively two vertical revolute joints and one prismatic joint (see figures 6-7). We wish to find a positioning of the robot which enables the tracking of continuous trajectories in an area located under the top of the car. The aim is to achieve various tasks such as glue spreading and continuous welding. The target space is defined as a rectangular planar area to be enclosed in a t-connected subspace (see figure 6). The tool to be used must remain perpendicular to the car top. Thus, the analysis can be achieved by considering only the three first joints of the robot (the wrist is locked in such a way that the tool remains vertical).

Figure 8 shows the optimal positioning supplied by SITROB. The recovering t-connected subspace is represented in the target space section. The running time was of 30' on a VAX-8700 computer (accuracy : 2% of the total lenght of the robot).

4- CONCLUSIONS

This paper has presented new connectivity analyses of the free-workspace for a robot among obstacles, which supplies usefull help for C.A.D. in robotics. The maximal connected subspaces have been characterized. They are computed with octree structure. Application to robots automatic positioning has been described using the robotized cells design package SITROB. The solution ensures the feasability of tasks described as areas to be reached and to be travelled through. According to the task to be achieved, n-connected subspaces or t-connected subspaces are considered. N-connected subspaces are defined as areas of the free-workspace where the robot is able to move between any n points ; t-connected subspaces describe the regions where any continuous trajectory can be tracked by the end-effector. An industrial case has been studied, showing that this new approach leads to real improvements in comparison with classical robotics CAD systems.

REFERENCES

1. E. Dombre et al. : "Trends in CAD / CAM systems for robotics", I.E.E.E. Conf. on Rob. and Aut., 1986, pp 1913-1918.
2. P. Chedmail and Ph. Wenger : "Design and positioning of a robot in an environment with obstacles using optimal research", I.E.E.E. Conf. Rob. and Aut., May 14-19 1989, Scottdale, Arizona, U.S.A.
3. T. Lozano-Perez :"Spatial planning : a configuration space approach", I.E.E.E. Transaction on Computers, 32(2), p : 108-120, Feb. 1983.
4. B. Faverjon : "Object level programming of industrial robots", I.E.E.E. Int. Conf. on Robotics and Automation, Vol. 3, April 1986, p : 1406-1412.
5. A. Kumar, K. Waldron : "The workspaces of mechanical manipulators", ASME J. Mechanical Design, 103,1981, p : 665-672.
6. K.C. Gupta: "On the nature of Robot workspace", The Int. Jour. of Robotics Research, Vol. 5, N° 2, p : 112-121, 1986.
7. V. Hayward, : "Fast collision detection scheme by recursive decomposition of a manipulator workspace", IEEE Conf. on Rob. and Aut., pp 1044-1049, 1986.
8. R.A. Brooks : "Solving the path-finding problem by good representation of free space", I.E.E.E. Transactions on System, Man. and Cybernetics, Vol. S.M.C-13, p : 190-197.

9. Z. Shiller: "Interactive time optimal robot motion planning and work-cell layout design", I.E.E.E. Conf. on Robotics and Automation, p964-969, May 1989, Scottdale, Arizona, U.S.A.
10. P. Borrel : "A study of multiple manipulator inverse kinematic solutions with applications to trajectory planning and workspaces determination", I.E.E.E. Conf. on Rob. and Aut., 1986, p : 1181-1185.
11. Ph. Wenger and P. Chedmail: "Ability of a robot to travel through its free workspace", Int. Journ. of Rob. Research, to appear.
12. Z.Q. Yu and W. Khalil : "Table look up for collision detection and safe operation of robots", Symposium on Theory of Robots. Vienna, Dec. 3-5, 1986.
13. H. Samet : "Connected component labeling using quadtrees", Computer Science Department, University of Maryland, College Park, April 1979.
14. T. Lozano-Perez et al., :"Handley : a robot systems that recognises, planes and manipulates", I.E.E.E. Int. Conf. Rob. and Aut., pp : 843-849, March 1987.
15. R. Hookes and T. Jeeves : "Direct search solution of numerical and stastitical problems". J.A.C.M. 8, 1961; pp. 212-229.

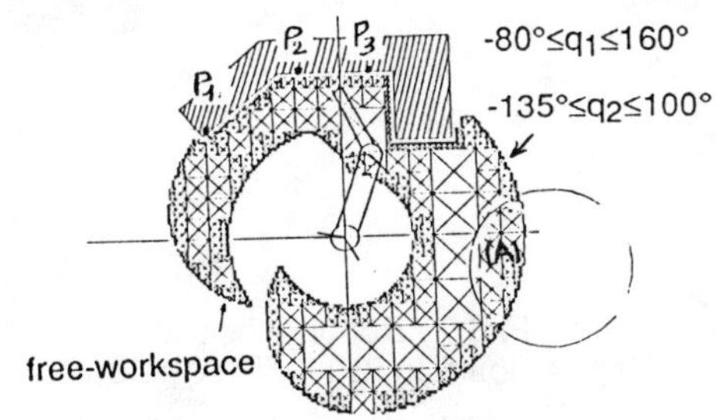

FIG 1 : Point P_1 cannot be reached from the table

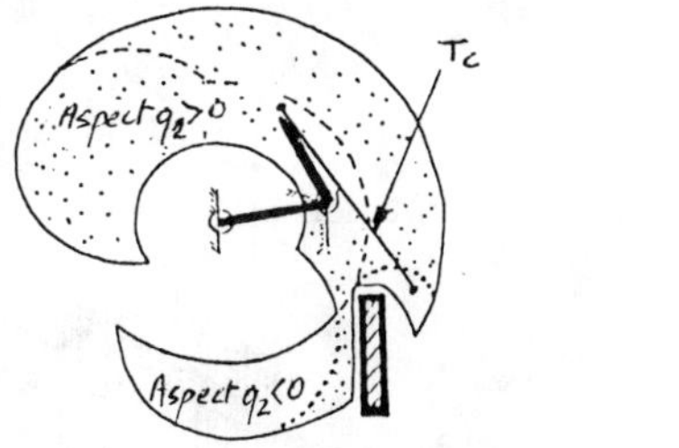

FIG 2: Unfeasible trajectory

FIG 3 : Position of the table for which the task is achievable

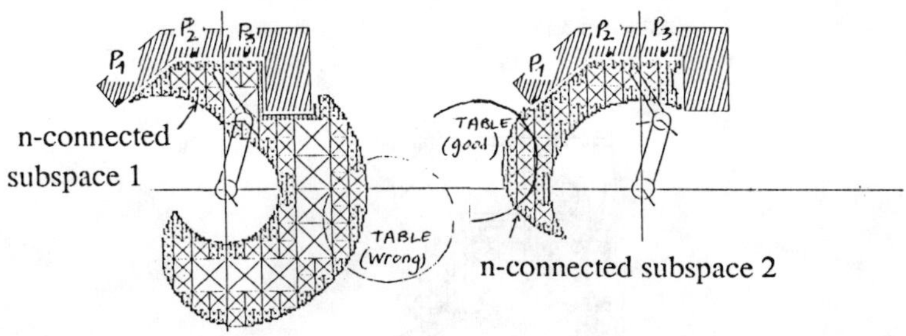

FIG 4 : cutting along an L-shaped oultine

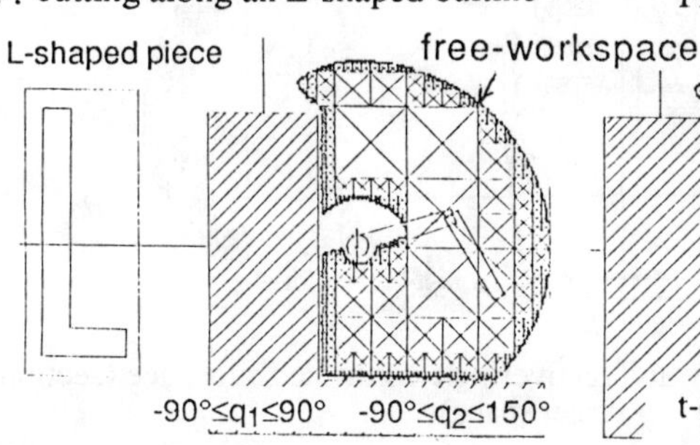

FIG 5 : A possible positioning of the piece

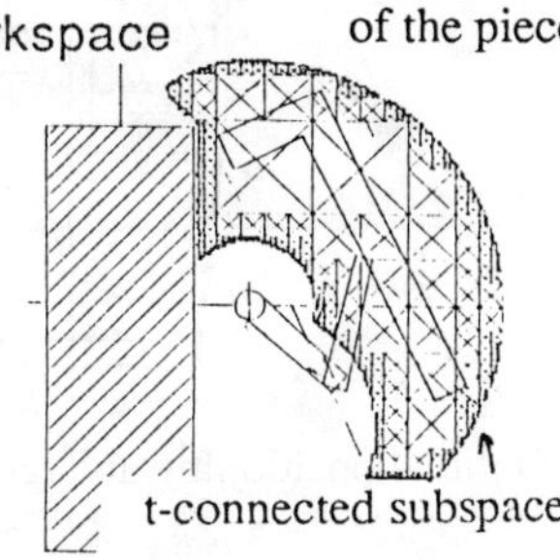

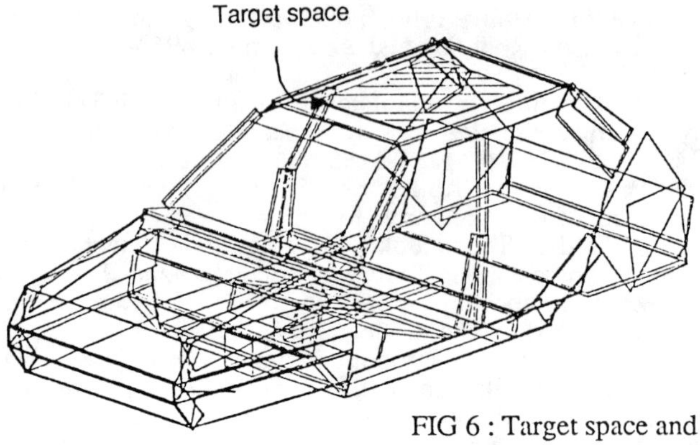

FIG 6 : Target space and environment

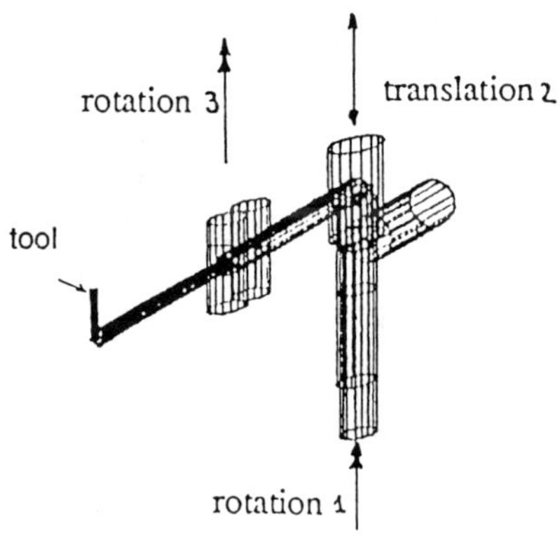

FIG 7 : Robot to be used (joint limits : $-130° \leq q_1 \leq 130°$, $0 \leq q_2 \leq 1m$, $-150° \leq q_3 \leq 150°$)

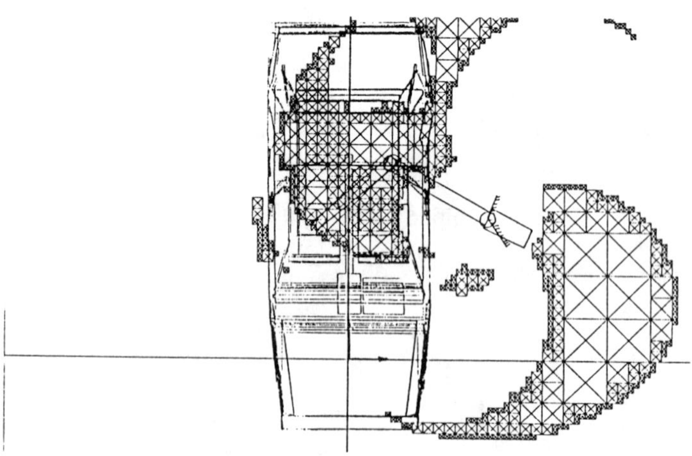

FIG 8 : Optimal positioning and recovering t-connected subspace (section in the target space)

COMPUTER-AIDED ELABORATION OF ALL ASSEMBLY SEQUENCES

J.M. Henrioud, A. Bourjault

Laboratoire d'Automatique de Besançon (URA CNRS 822)
Institut de Productique -
15, Impasse des Saint-Martin - 25000 BESANCON - France

ABSTRACT

In this paper, the authors present the first stage of a complete strategy to design a flexible assembly system for any product. The basic idea is to divide the assembly actions in two classes; the dedicated ones (specific of the product) and the positional ones (specific of the assembly process). The proposed method includes an algorithm which generates all the valid combinations of the dedicated actions, in the form of assembly trees ; the awkward ones are then rejected by the introduction of some set of constraints. The remaining trees are then transformed by the assignment of equipments to its different nodes, which introduces the positional actions. A comparison of the different solutions is then achieved by the way of a simulation.

INTRODUCTION

Before the eighties, the automatisation of the assembly tasks were most of the time the substitution of a man by a robot ; that is an approach called "work-station" ; the economic results were not always satisfying and today, it is preferable to have a global vision, i.e. a system-like approach. In this way, the most difficult is certainly to obtain and apply a methodology which permits to design an efficient flexible assembly system.

For several years we have been working on this problem, more precisely, on the elaboration of a computer-aided work-station, to design flexible systems ; whose elements could also be used to control the system on line.The resulting method is divided in two principal stages : the determination of a set of assembly plans, then, simulations.

The aim of this paper is to presents our last results about the assembly plans determination. Up to this days different works have been dedicated to this problem ; some of them propose an automatic method [7],[8],[10], while the others propose a computer-aided method [3],[4],[6] . Whatever the approach, the main problem lies in the number of feasible assembly plans, ranging from some 10^3 to some 10^5 for products having about fifteen components. Obviously this is prohibitive without some efficient way to select the best ones, since for each assembly plans there are several possible assembly systems (some tens) which have to be simulated before the final choice. It is mainly this selection aspect which requires the help of an human expert.

PRODUCT AND ASSEMBLY PLAN MODELISATION

Any assembly process for a given product is a set of actions, realized in serie and/or in parallel which can be parted in two great classes

- The dedicated actions which create some feature of the end product
- The positional actions which transport the different parts of the product inside the assembly system.

The first ones depend only upon the products, while the others depend upon the assembly process. That's why we have chosen to define the assembly plans as structured sets of dedicated actions, postponing the introduction of the positional ones to the stage of the assembly process determination. In order, to find out the valid orders for the dedicated actions, it is necessary to define a model for the product, including the set of features which are to be set up. Previously, the authors have proposed a simple model (the liaison graph model) [1]. In this paper we present a complete model describing all the features of the product (the operative model).

To any product P, we list the set of all its characters and we define a 5-uple:

$$<C, \Gamma, \Sigma, \Delta, f>$$

-C is the set of the elementary components

-Γ is the set of the liaisons,there is a liaison $\{c_i, c_j\}$ between two components c_i and c_j iff there is, at least, one mechanical liaison between them.

-Σ is the set of the joinings, a joining is the set of all the elements which contibute to secure, at least, one liaison.

-Δ is the set of the complementary characters : check, label, machining...

-f is a function mapping $\Sigma \cup \Delta$ into $\mathcal{P}(C) \times \mathcal{P}(\Sigma) \times \mathcal{P}(\Delta)$; it defines,for each joining or each complementary character,the set of elementary parts and characters concerned.

The couple $[C, \Gamma]$ defines a graph called the liaisons graph.

The function f has two purposes ; the first one is to define a necessary condition for the joinings and complementary characters to be set up, the second one being to define some precedence conditions between them. Thus if :

$$d \in \Delta \quad \text{and} \quad f(d) = (X, \sigma, \delta)$$

every character in σ or in δ has to be set up before the character d.

Figure 1 represents a quartz with its liaison graph ; it has:

-three elementary components : a crystal (a), a base (b) , a cover (c).

-two joinings : the welding of a and b : (w1), the welding of b and c : (w2)

- two complementary characters :

-an electric control of the welding w1 : (el)

-a labelling of the cover c : (la) ; this labelling assigns the quartz to a certain class of quality, depending upon the result of the electrical control, so we consider in the model that (la) is a labelling of the whole product.

The function f is thus defined by : f(w1) = ({a, b}, f, f)
f (w2) = ({b, c}, f, f)
f(el) = ({a, b}, {w1}, f)
f(la) = ({a, b, c}, {w1, w2}, {el})

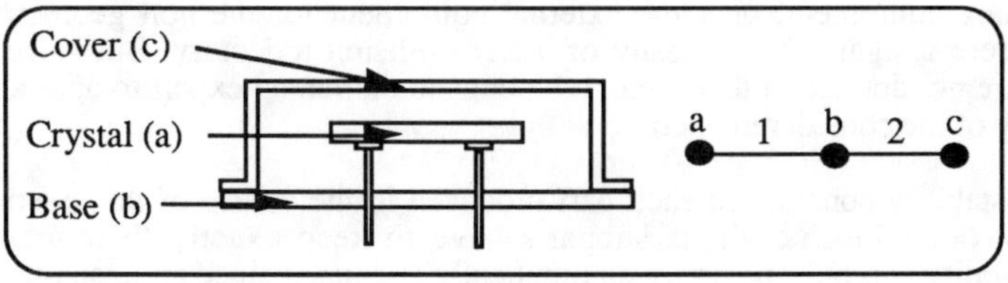

Figure 1 : Quartz and its liaison graph .

The assembly plans associated to this modelisation of the products are represented by assembly trees, which are rooted trees that we assume to be binary, such that:

- the root represents the product
- the nodes represent the sub-assemblies or parts
- the leaves represent the elementary parts

and such that each part associated to a node Ni can be produced from the parts associated to the successors of Ni by theexecution of one or several dedicated actions.

Figure 2 shows two assembly trees available for the quartz presented above. The nodes are labeled by the character created by the associated operation.

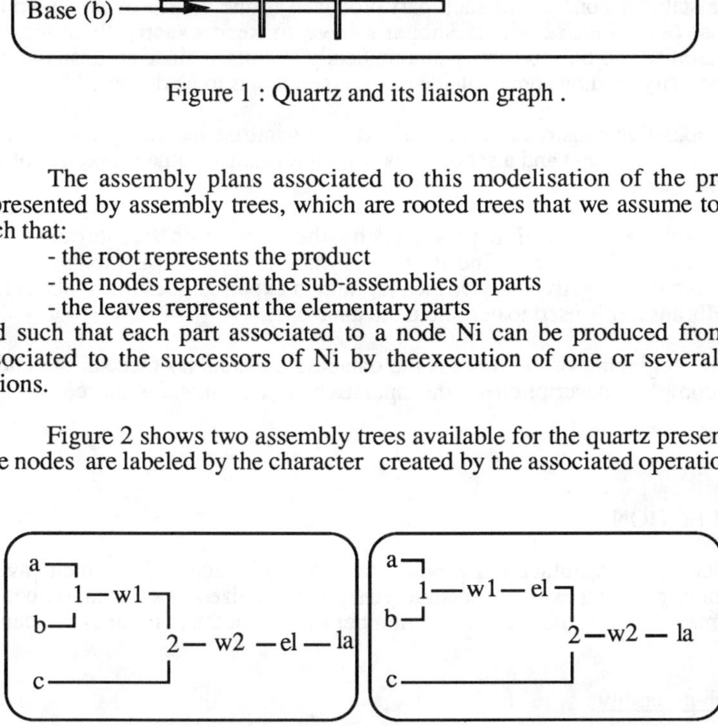

Figure 2 : Assembly trees for the quartz depicted in figure 1 .

In order to evaluate the assembly trees available for a given product we have elaborated a software LEGA based on a top-down algorithm. Starting from the product, we define all the valid operations having the product for result, which determines one or two parts ; then we reiterate for each of these parts till we obtain all the elementary components. An operation being valid when it agree with the set of the operative constraints.

<u>The operative constraints</u> : define whether any given operation is feasible ; we distinguish :

- the geometrical constraints : when mating two parts, there must exist at least one collision-free trajectory to bring the two parts in contact. This constraint depends only upon the product and can be automatically evaluated

- the material constraints : the effective realization of any operation involves some equipments, there is at least one part to hold (two for the geometrical operations), which requires some suitable parts surfaces for the holding ; moreover, since every operation involves some spatial relation between two objects : each part with its holding equipment, or some external equipment for the non geometrical operations, there is again the necessity of a free-collision trajectory. But since the equipments are not defined at this stage, it belongs to an human expert to appreciate the feasability of the considered operation

- the stability constraint : each part produced in the course of the assembly process must be stable, i.e all its subparts have to keep exactly their relative positions. Stability can only be partly automatically evaluated since an unstable part can be temporarily hold by some outside device, which has to be designed.

The operative constraints are described by a database including a set of facts (the unfeasible operations) and a set of rules which formalize some properties of the operative constraints.

For each new operation produced by the generation tree process, the database is searched in order to find if this operation is feasible ; if the datas don't allow a positive or a negative answer then an human expert is questioned about this operation. His answer is used to complete the database.

At the beginning of the process the database contains no fact ; at the end it contains a complete description of the operative constraints for the considered product.

TREES SELECTION

Generally, a manufactured product has some characteristics which favour some assembling strategies. If these strategies are formalized as constraints before the determination of assembly trees, the number of obtained trees will significantly decrease.

Imposed sub-assembly

Very often it is possible to define, for the product to assemble, some sub-assemblies which are to be produced in the course of the assembly process (for stockage purpose, because they are to be used as maintenance spare parts, for functional or stability reason) . This leads to a strategic constraint such that only the assembly trees including these sub-assemblies are produced.

This constraint can be reinforced by the interdiction that any sub-assembly, other that those which have been chosen, are produced ; this has for effect to reduce the parallelism of the resulting trees. This is of peculiar interest when a transfer system is to be used. In such a case it is generally possible to define an evident base component, which reduces still more the number of resulting assembly-trees.

Group of components

In a manufactured product, some components have similar shape or morphology, (they can be handled with the same gripper), an identical securing or the same assembly direction, so that it may be advisable to group the operations in which they are involved. These leads to a strategic constraint which imposes that in each assembly tree the leaves standing for the components belonging to a same group are consecutive . This concept of group of components is equivalent to the one of cluster presented in [2] .

If we want to consider several incompatible or even contradictory strategies, we must run the method several times with each strategic constraint, in order to obtain different sets of assembly trees corresponding to the different strategies.

Two basic structures for some groups of components, studied in [8] and concurrently in [2], are presented here, they lead to more selective constraints.

Stacks

A stack is a group of k ($k \geq 3$) constituents Ci (which are elementary components or sub-assemblies) such that a liaison exists beetween Ci and Ci+1 ($1 \leq i < k$), and such that the assembly of Ci ($1 < i < k$) on a part of the stack which contains C1 (respectively Ck) requires the presence of Ci-1 (respectively Ci+1) in this part. Each time a stack is detected in a product it may be declared by the expert, which creates a strategic constraint such that the group of constituents Ci is assembled either in the order C1,...Ck or in the order Ck,...,C1 and wihout any sub-assembly.

Layers

Sometimes the combinational on the assembly of a group components is not interessant because there is none precedence relationship beetween the assembly order of the components and all the combinations are equivalent. By example, on a printed circuit, the idendical components can be regrouped, but the combinational on this group is not primordial. So we define a particular structure of components, the layer, in which the assembly order is not specified at this level of description.

A layer is a group of k constituents Ci which all have a liaison with a same constituent B, the base (which can be a sub-assembly) and such that the liaisons (Ci , B) are independant . Let's notice that each part of a layer is a layer. The liaisons graph of a layer and its base is generally a star shaped graph.

There are k! assembly trees for a layer with k constituents. And the difficulty to realize each tree is obviously identical. So, in order to avoid this combinational, we propose a strategic constraint wich regroups all the assembly trees of a layer in a n-uple tree where the primary constituent is the base B and the secondary constituents are the elements of the layer without precedence choice.

CONCLUSION

This study describes the first stage of a complete method allowing the computer-aided design of flexible assembly systems. This stage is concerned by the determination of a set of valid and coherent assembly plans. The validity is guaranted by their agreements with the operative assembly constraints identified by an human expert, who will soon be supplied with an automatic system identifying the geometric constraints from a 3D representation of the products. The coherence is given by a set of strategic constraints, also introduced by the expert. The proposed method has been tested with different industrial products, having from ten to twenty components. Without any strategic constraint the number of valid assembly trees was always prohibitive. Nevertheless it has always proven to be possible to define some strategic constraints, deduced from the product structure, allowing to reduce the number of resulting assembly trees to a manageable size (from some tens to about one hundred). It's worth to emphasize that very few strategic constraints are necessary to reduce greatly the number of available solutions.

REFERENCES

1. A.Bourjault,"Contribution à une approche méthodologique de l'Assemblage Automatisé : Elaboration Automatique des Séquences Opératoires" ; Thèse d'Etat, Besançon, 1984.
2. N.Boneschanscher,C.J.M.Heermskerk,"Gouping Parts to Reduce the Complexity of Assembly Sequences Planning", INCOM'89, MADRID, pp.287-292.
3. A.Delchambre,"Data structures for Computer-Aided Assembly Planning: a Survey", INCOM'89,Madrid,pp.-523-528.
4. T.L.De Fazio,D.E.Whitney,"Simplified Generation of all mechanical assembly sequences" IEEE Robotics and Automation, Vol. RA.3, pp 640-658.
5. J.M.Henrioud, A. Bourjault,D. Chappe, "Elaboration des gammes d'assemblage : Approche composants" GAMI - Les Journées de la Productique,Paris,1987..
6. J.M.Henrioud, "Contribution à la conceptualisation de l'assemblage automatisé: nouvelle approche en vue de la détermination des processus d'assemblage", thèse d'état, Besançon 1989.
7. W.Jentsch, F.Kaden, "Automatic Generation of Assembly Sequences", Third Int.Conf.on A.I. and Information Control Systems of Robots,North-Holland, Amsterdam, pp.197-200,1983.
8. H.Ko, K.Lee, "Automatic Assembling Procedure Generation from Mating Conditions",CAD Journal, Vol 19,n°1,pp.3-10,1987.
8. M.Meunier,"Eléments méthodologiques pour la conception des systèmes flexibles d'assemblage", thèse de l'Université de Franche-Comté,1989.
9. H.Sekigushi , K.Kojima,K.Inoue, "Study on Automatic Determination of Assembly Sequences", Annals of CIRP, Vol.32/1,pp.371-374,1983.

THE OPTIMUM SELECTION OF THE MECHANISMS OF 7R ROBOT ARM

Yuan peizang, Gui tong, He yigang, Xiu daoxi

Shenyang Institute of Automation
Academia Sinica
Shenyang 110003, P.R.China

ABSTRACT

In this paper,the mechanisms of articulated robot arm are designed using workspace and the measure of manipulability of robotic mechanisms as performance evaluation index. In order to provide a designer a reference, a data library for eighty four mechanisms of 7R robot arm is built and the relative simulation graphs are drown.

INTRODUCTION

The optimum selection of the mechanisms of robot arm is an important problem that We must first resovle in the process of designing robot manipulators. The more perfect mechanisms for multiple rotary joint robot arm are 6R manipulators. Roth argued that the 6R manipulator having a superior workspace is the one shown in FIG.1(1980). Y.C.Tsai[1] found out the optimum geometry for robot arm (ROS) and the optimum geometries for robot wrist (OOP) shown in FIG.2 (1984). Y.Hasegawa [2] designed six robotic mechanisms according to the geometric characteristic of their workspaces shown in FIG.3 (1988). For the research on the optimum selection of 7R manipulator mechanisms, John M.Hollerbach has done a resultful work. He proposed a point of view that We can do the optimum kinematic design for a 7R manipulator. According to the point of view, the optimum geometry for 7R robot manipulator can be yielded by

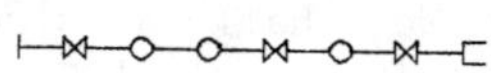

FIG.1. Roth'6R manipulator

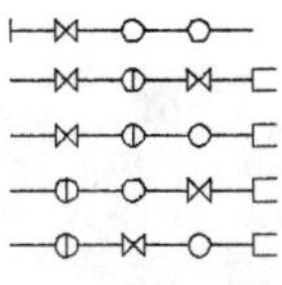

FIG.2. The arm (ROS) and the wrists (OOP)

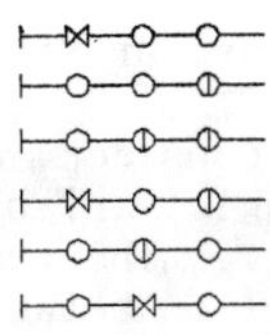

FIG.3. Robot arms designed by Y.Hasegawa

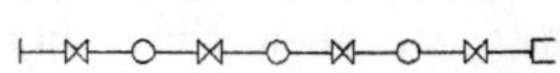

FIG.4 The optimum 7R robot manipulator

adding a rotary joint to the shoulder, the elbow or the wrist of the optimum 6R manipulator and the best is shown in FIG.4. Nakano [4] suggested that a optimum 7R manipulator can be formed by using the geometry that We add a rotary joint to one of the optimum robot arm found out by Y.C.Tsai as the manipulator arm and the optimum wrist (OOP) as the manipulator wrist. Its geometry is the same as the one shown in FIG.4.

Based on the analysis of the point of view described above, We provided two suggestions for discussion bellow.

(1) A 7R manipulator yielded by adding a rotary joint to the optimum 6R manipulator are not certainly the geometry having the optimum mechanism performance. Hence, a 7R manipulator shoud be formed and evaluated in another way.

(2) It is important to evaluate the mechanism performance using workspace. But there are variant singular states in the same volume workspace. Therefore a new evaluation index should be suggested.

THE EVALUATION INDEX FOR THE MECHANISMS OF ROBOT MANIPULATOR

In order to yield a optimum mechanisms of robot arm, We must use a practical and numerical performance evaluation index. The mechanism performance relates to the kinematics ,dynamics and the property of control etc. The kinematics property is the most basic and important among them. The workspace and the ability of manipulation of robot arm are the main characteristic of robot kinematics. The research on workspace is dicussed in [1]-[6]. In this paper ,We shall lay stress on the description of the ability of manipulation. The study of the numerical evaluation of manipulability is stated in [7]-[11] and the measure of manipulability is proposed. Here We use the measure of manipulability [9]-[12] as the numerical performance evaluation index for the optimum selection of 7R manipulator mechanisms.

The Measure of Manipulability

We consider a manipulator with n degrees of freedom which joint angle variables are denoted by q ($q \in R^n$), joint velocity is $\dot{q}$ ($\dot{q} \in R^n$). The velocities on the point of the wrist is $\dot{r} \in (\dot{r}\ R^m)$. The relation between $\dot{q}$ and $\dot{r}$ are given by

$$\dot{r}=J(q)\dot{q} \tag{1}$$

Where $J(q) \in R^{m \times n}$ is called the Jacobian.

If $\|\dot{q}\| =(\dot{q}^T\dot{q})^{1/2} \leqslant 1$, then $\dot{r}^T(J^+)^T J^+ \dot{r} \leqslant 1$. Where J^+ is called the general inverse of the Jacobian.

We can show that the subset of the realiziable velocities r in the space R is an ellipsoid. This ellipsoid is called the manipulability ellipsoid and could be a good means for the analysis, design and control of robot manipulators. The volume of ellipsoid is directly proportinal to W. Where W is called the measure of manipulability given by

$$W=\sqrt{\det(JJ^T)} \tag{2}$$

In general, $W \geq 0$. If $W=0$, $\mathrm{rank}(J(q))<m$ and the manipulator is at singular states. Therefore W can be regarded as a distance from the singular states.

Manipulability

The paper [3] presents the concept of manipulability. It can be described as follow. The equation (1) can be written by

$$\dot{r} = \begin{bmatrix} \dot{P} \\ \dot{R} \end{bmatrix}, \quad J = \begin{bmatrix} J^{a}_{x1} & J^{n}_{xa+1} \\ J^{a}_{z1} & J^{n}_{za+1} \end{bmatrix} \quad \dot{q} = \begin{bmatrix} \dot{q}_a \\ \dot{q}_n \end{bmatrix} \tag{3}$$

where $\dot{q}_a = (\dot{q}_1, \ldots, \dot{q}_a)$ and $\dot{q}_n = (\dot{q}_{a+1}, \ldots, \dot{q}_n)$.

The sufficient and necessary condition that ensures the manipulators having manipulability is given by

$$\mathrm{rank}(J^{a}_{x1})=3 \quad \text{and} \quad \mathrm{rank}(J^{n}_{za+1})=3 \tag{4}$$

Obstacle Avoidance

The purpose of the research on 7R redundant manipulators is to improve the ability of the manipulators to avoid singularity and obstacles. Hanafusa [12], Gui [13] have deeply discussed the problem of obstacle avoidance for a redundant manipulator. They proposed the concept of subtasks with order of priority and ultilized this concept and the general inverse of the Jacobian to obtain the solution of obstacles avoidance for a manipulator. The concept can be described by the formula given bellow.

$$q = J^{+}\dot{r}1 + (I - J^{+}J)\dot{r}2 \tag{5}$$

Where r1 represents the subtask with the first priority.
r2 represents the subtask with the second priority

A DATA LIBRARY AND SIMULATION GRAPHICS FOR ROBOT MECHANISMS

A multiple joint manipulator can be formaed with three typical joints -Φ-, -O-, -⋈- which rotate three axises X, Y, Z in the joint coordinates. 2187 types of 7R manipulator can be constructed using the three typical joints. The evaluation and optimum selection of them could be done with two approaches stated bellow.

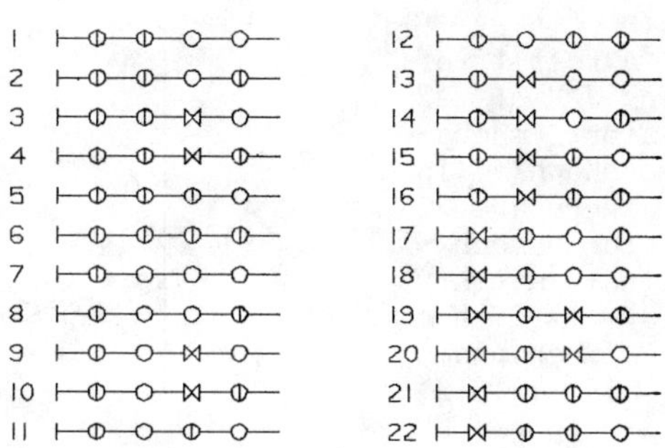

FIG.5. Four d.o.f. robot arms

The Use of Workspace

In order to improve the ability of manipulators to avoid obstacles. We choose four degrees of freedom to locate the position reached by the manipulator arm and three degrees of freedom to obtain the posture of the manipulator wrist. 81 available types of manipulator arms can be formed. Except for the repeat and degeneracy among them, the remained 22 mechanisms are shown in FIG.5. A data library and simulation graphs for their work spaces are built up at CAD workstation. The photograph (a) in FIG.6 shows the workspace of the robot arm mechanism proposed by M.Hollerbach and the photograph (b) shows the workspace of the robot arm mechanism signed NO.9 in FIG.5. With reference the photographs in FIG.6, it is recognized that the volume of the workspace of the mechanism proposed by M.Hollerbach is not the largest of them because there are some holes in the workspace.

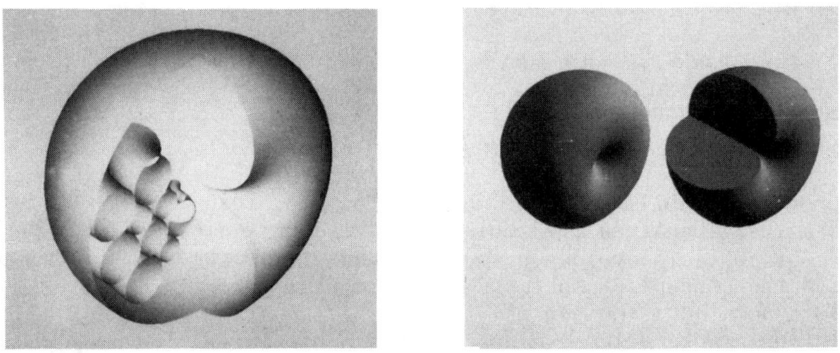

(a) The workspace of the robot arm designed by M.Hollerbach

(b) The workspace of the robot arm signed No.9 in FIG.5

FIG.6. The photographs of the workspaces of two typical robot arms

The Use of the Measure of Manipulability

As a result of computing $J_{x_1}^{a}$ of 22 manipulator mechanisms shown in FIG.5, there are 21 manipulator mechanisms that satisfy rank $(J_{x_1}^{a})=3$. For the OOP mechanisms shown in FIG.2, there are four OOP mechanisms that satisfy rank$(J_{z_{a+1}}^{n})=3$. 84 types of 7R manipulators shown in FIG.8 can be formed by linking up them.

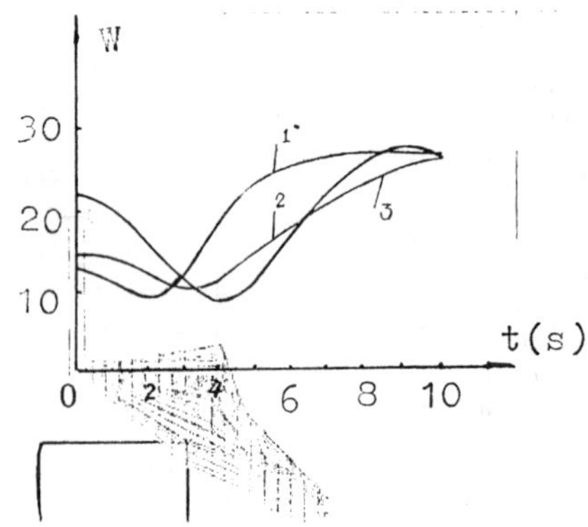

FIG.7. The variance of W of 7R manipulators for obstacle avoidance

At CAD workstation, a data library and simulation graphs are built up to deal with obstacle avoidance for 7R manipulators using W as the performance evaluation index. Three curves in FIG.7 show

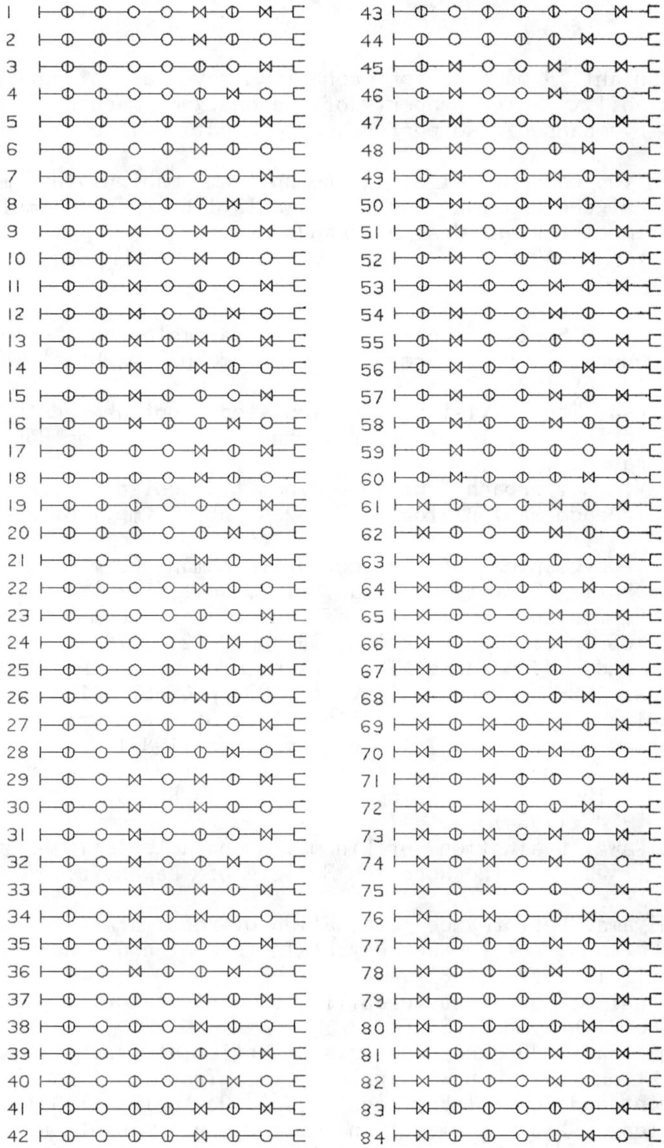

FIG.8. The geometries of the 7R robot manipulators

the variances of W of three types of robot manipulator in the condition of obstacle avoidance. The curve 1 in FIG.7 shows the case of the optimum mechanism proposed by Hollerbach. The curve 2 shows the case of the mechanism signed No.32 in FIG.8. The curve 3 shows the case of the mechanism signed No.54 in FIG.8.

CONCLUSION

(1) Variant 7R manipulator mechanisms have variant workspaces and manipulability. The property of the designed mechanism can be described by means of two performance evaluation index stated in this paper.

(2) In the case of selecting the optimum manipulator mechanisms, a designer can make use of the data library and simulation graphs to meet variant design demands.

REFERENCE

1. Y.C.Tsai, A.H.Soni,"The effect of link parameter on the working space of general 3R robot arms", Mechanism and Machine Theory, Vol.19, No.19/16,(1984).

2. Y.Hanafusa,"The division of articulated robot arms using the geomretric charcteristic of their workspace", J.of the Robbotics Society of Japan, Vol.6, No.6.

3. John. M. Hollerbach,"Optimum kinematic design for a seven degrees of freedom manipulator", The 2nd Int. Symp. of Robotics Research,(1985).

4. E.Nakano,"Development of a direct-drive humanlike manipulator", J. of the Robotics Society of Japan,Vol.5, No.1,(1987).

5. B.Roth,"Performance evaluation of manipulators from a kinematic viewpoit", NBS Special Publication 459, pp39-62,(1976).

6. A.Kumar and K.J. Waidron,"The workspace of a mechanical maipulator",Trans. ASME J.Mech.Design, Vol.103,pp665-672,(1981).

7. P.R.Paul and C.N.Stevension,"Kinematics of robot wrists", Int. J. of Robotics Research, Vol.1, No.2, pp31-38,(1983).

8. T.Yoshikawa,"Analysis and control of robot manipulators with redundancy", The 1st Int. Symp. of Robotics Research, The MIT Press ,pp735-747,(1984).

9. T.Yoshikawa,"Evaluation of kinematic characteristic of robot mechanisms", J. of the Robotics Society of Japan,Vol.5, No.4, (1987).

10. M.Uchiyama,"Performance evaluation of robot arms using the Jacobian", Trans. of the Society of Instrument and Control Engineering, Vol.21, No.2 ,(1985).

11. G.Kinoshita,"The manueverability of manipulator with multi-joints", The 21th Symp.of SICE.

12. H.Hanafusa,"Redundancy analysis of articulated robot arms and its utilization for tasks with priorgity", Trans. of the Society of Instrument and Control Engineering,Vol.19, NO.15, (1983).

13. Gui tong, Xiu daoxi, Yuan peizang,"Analysis of redundant manipulator arms using the general inverse of the Jacobian and its application", The 6th Symp. of the Mechanism Society of China,(1988).

SECTION 9 : DUAL AND MOBILE ROBOTS

Structure of the Minimum-Time Control Law for Two Cooperative Arms Along a Specified Path

Yaobin Chen
Electrical Engineering and Computer Science Department
The George Washington University
Washington, DC 20052

Alan A. Desrochers
Electrical, Computer, and Systems Engineering Department
Rensselaer Polytechnic Institute
Troy, New York 12180-3590

ABSTRACT

The structure of the minimum-time control law for two robot arms moving a common object along a specified path subject to "hard" control constraints is studied using a closed chain dynamic model of the dual-arm system. By using the "Extended Pontryagin's Minimum Principle" and a set of parameterized dynamic equations, it is shown under some reasonable assumption that *one and only one* control torque of all the joints is always in saturation on every finite time interval along its time-optimal trajectory, while the rest of them adjust their torques so that the path constraint is not violated.

INTRODUCTION

It has been widely recognized that multiple robot systems play an important role in many applications, ranging from manufacturing and space exploration. In manufacturing automation, a variety of tasks, such as assembly, large part transfer and so on are performed by multiple cooperative robots. In space exploration, space station assembly, maintenance and servicing require an efficient manipulation of multiple coordinating robot arms. To enhance the overall productivity and the efficiency of system operation, one of important issues to be addressed is the minimization of the total traveling time for the system. The main objective of this paper is to conduct a preliminary investigation into the structure of the minimum-time control (MTC) law for two manipulators cooperatively moving a common object along a pre-specified motion path.

The problem considered herein can be stated as: given the dynamic equations of the two arms which coordinatively move a common object, find the control torques applied to each joint of the two arms that will drive the system from a initial position to any desired final position along a predetermined motion path in minimum time, assuming the control torques are subject to "hard" constraints. Several researchers have studied the MTC problem for single robot systems (See [1-4] as well as references cited there). A detailed survey of this research can be found in [6]. It is worth mentioning that Shin and McKay [1], Bobrow, et al. [2] and Pfeiffer and Johanni [3] independently came to similar solutions by following the same objective and formulation, although their numerical algorithms and motivations are slightly different. Most of previous work concentrated on obtaining the numerical solutions to the problem. It should be noted that Chen and Desrochers [4] have systematically studied the structure of the MTC law for single robot arms along prespecified path based on the dynamics parameterization used by other authors. The theoretical result was verified by numerical simulations.

A recent literature survey of multiple robot systems indicates that much research effort has been devoted to the modeling, motion and force control [6-8]. To the best knowledge of the author, however, little work has been done for the MTC of multiple arms cooperatively handling a common object. The extension of the developed theory for the single robot case to the case of multiple robots is not trivial and requires further research effort to the problem pertinent to the solution structure and numerical algorithms. This paper presents a preliminary investigation into the structure of the MTC law for two robot arms moving a common object along a constrained geometric path. Section 2 presents a closed chain dynamic model using the Lagrange's formulation. In Section 3, the problem is formulated by a dynamics parameterization approach. Then, the necessary conditions for optimality are given as the "Extended Pontryagin's Minimum Principle" (EPMP) for coping with a general MTC problem with state-dependent control constraints. Section 4 presents the main results through two theorems and their proofs. Finally, numerical computation aspects are discussed and conclusions are given.

DYNAMIC MODEL OF TWO COORDINATED ROBOTS

In this paper, we consider the two arms handling a common object as a single mechanical system consisting of kinematic closed chains. For convenience, some assumptions regarding the dynamics of the system are made as follows. Each robot has N-joint serial rigid links with inelastic joints; The two robots firmly hold the common rigid object so that there is no relative movement between its end-effector and the object; All the kinematic and dynamic parameters are known; Only viscous frictions are considered; Finally, the dynamics of the actuators are not included in the model. Therefore, the system can be modeled as a single closed kinematic chain consisting of 2N links and 2N joints.

Kinematic and Dynamic Coupling:

Suppose the two arms are denoted as robot a and robot b, respectively. The object is denoted by **C**. It is apparent that the two robot arms are both kinematically and dynamically coupled when manipulating a common object. The kinematic coupling constraints are determined by expressing the Cartesian velocity of the center of mass of the object **C** in terms of the joint velocities of robots a and b, that is,

$$\dot{x}_c(t) = J_{ac}(q^a)\,\dot{q}^a(t) \tag{1}$$

$$\dot{x}_c(t) = J_{bc}(q^b)\,\dot{q}^b(t) \tag{2}$$

where x_c is the velocity vector of the **C**, q the joint coordinates of each robot, and J the corresponding Jacobian matrices. Combining the two constraints above yields

$$[J_{ac}(q^a)\;,\; -J_{bc}(q^b)]\;\dot{q}(t) = 0 \tag{3}$$

where $q(t) = [(q^a)^T, (q^b)^T]^T$.

The dynamic coupling effects are quantified by the dynamics of the held common object, that is,

$$M_c\,\ddot{x}_c(t) + C_c(x_c,\dot{x}_c) = L_a(q^a)\,f_a + L_b(q^b)\,f_b \tag{4}$$

where M_c is the (6x6) mass matrix of the object, C_c the (6x1) Coriolis and gravity term, L the (6x6) transposed Jacobian matrices which transform the generalized contact forces of robot a (or b) to their equivalent at **C** (See [9] for the detailed).

Dynamic Equations of the Closed Chain System:

The Euler-Lagrange's equations of motions for each robot arm are

$$M_a(q^a)\ddot{q}^a + C_a(q^a, \dot{q}^a) + J_a^T(q^a) f_a = \tau_a \tag{5}$$

$$M_b(q^b)\ddot{q}^b + C_b(q^b, \dot{q}^b) + J_b^T(q^b) f_b = \tau_b \tag{6}$$

where M is the (NxN) inertia matrix, C the (Nx1) vector which includes Coriolis, centrifugal, gravity, and joint viscous friction forces, J the (6xN) Jacobian matrix, and τ is the control torques. Using the kinematic and dynamic constraints (1)-(4) and the robot dynamic equations (5)-(6), a complete dynamic model of closed chain system can be derived as follows (see [9] for detail)

$$M(q)\ddot{q} + \Delta(q)C(q,\dot{q}) + J^T(q)[J(q)M^{-1}(q)J^T(q)]^{-1}\dot{J}(q)\dot{q} = \Delta(q)\tau \tag{7}$$

where

$$\Delta = I_{2Nx2N} - J^T[JM^{-1}J^T]^{-1} JM^{-1}$$

$$J(q) = [-L_b^T L_a^{-T} J_a, J_b]$$

M is the so-called pseudo-inertia matrix (real symmetric positive definite), and C consists of all the coupling and nonlinear terms. If D(q) is nonsingular, (7) can be written as

$$M_q(q)\ddot{q}(t) + C_q(q,\dot{q}) = \tau \tag{8}$$

which gives a compact expression of the closed chain dynamic model.

FORMULATION OF THE EQUIVALENT MTC PROBLEM

Parameterization of the Robot Dynamics

Now let us consider that the motion of the held object is constrained to a fixed path in joint space, and that the path can be expressed as a parameterized curve. Using the path parameter $s(t)\in \mathbf{R}^1$ (e.g. the path length or distance), the generalized coordinate q(t) can be implicitly written as a function of s(t), i.e.

$$q(t) = \varphi(s(t)) \qquad s(t) \in [0, s_f] \tag{9}$$

where s_f is the final value of s(t), $\varphi(s(t))$: $[0, s_f]\rightarrow\mathbf{R}^{2N}$ and each component $\varphi_i(s(t))$ is continuous with respect to s(t). Taking the first and second time derivatives for both sides in (9) yields

$$\dot{q}(t) = \frac{d\varphi(s)}{ds(t)}\frac{ds(t)}{dt} = \frac{d\varphi(s)}{ds(t)}\dot{s}(t)$$

$$\ddot{q}(t) = \frac{d\varphi}{ds}\ddot{s} + \frac{d_2\varphi}{ds^2}(\dot{s})^2$$

Plugging into (8), we obtain a parameterized form of the closed chain dynamic equations

$$D(s(t))\ddot{s}(t) + H(s(t), \dot{s}(t)) = \tau(t) \tag{10}$$

where

$$D(s) = M_q(\varphi(s))\frac{d\varphi(s)}{ds} \in \mathbf{R}^{2N}$$

$$H(s,\dot{s}) = M_q(\varphi(s))\frac{d^2\varphi(s)}{ds^2}(\dot{s})^2 + C_q(s,\dot{s}) \in \mathbf{R}^{2N}$$

Note that if s(t) represents the distance or length of the desired path, then $\dot{s}(t)$, $\ddot{s}(t)$ are the path velocity and acceleration, respectively. Therefore, we call them the *pseudo-velocity* and *pseudo-acceleration* . The i-th equation of (10) is written as

$$d_i(s(t))\,\ddot{s}(t) + h_i(s(t), \dot{s}(t)) = \tau_i(t) \tag{11}$$

With the parameterization described above, we define the path parameter and its pseudo-velocity as new state variables, i.e. let

$$x(t) = [x_1(t), x_2(t)]^T = [s(t), \dot{s}(t)]^T$$

Then the following 2N+1 state equations are obtained

$$\dot{x}_1(t) = x_2(t)$$

$$d_i(x_1(t))\,\dot{x}_2(t) + h_i(x(t)) = \tau_i(t) \qquad \text{for all } i=1,2,\ldots, 2N \tag{12}$$

with the initial and final conditions: $x(0)=x_0$, $x(t_f)=x_f$ and the constraints on the control torques:

$$\tau_{imin} \leq \tau_i(t) \leq \tau_{imax} \qquad \text{for all } i=1,2,\ldots, 2N, \quad t\in [0, t_f]$$

From Eq.(12), one sees that given the current state x(t), the quantity $\dot{x}_2(t)$, i.e. the pseudo-acceleration, if known, determines uniquely the control torques for all the joints of each robot, so that the manipulation of this scalar variable s(t) (or $x_1(t)$) can replace the manipulation of 2N-vector control torques and a set of constraints (the path parameterization equations). It is this observation that we would like to use to transform the control torque constraints into a set of constraints on the $\dot{x}_2(t)$.

Define $\dot{x}_2(t)$=u(t). Using (12) and the control constraints, after simple algebraic manipulations, we have

$$u_{min}(x) \leq u(t) \leq u_{max}(x) \tag{13}$$

where

$$u_{min}(x) = \max_i\{L_i(x)\}, \quad u_{max}(x) = \min_i \{U_i(x)\}$$

$$L_i(x) = \frac{\tau_{imin}\langle d_i > 0 \rangle + \tau_{imax}\langle d_i < 0 \rangle - h_i(x)}{d_i(x_1)}$$

$$U_i(x) = \frac{\tau_{imax}\langle d_i > 0 \rangle + \tau_{imin}\langle d_i < 0 \rangle - h_i(x)}{d_i(x_1)}$$

The function $\langle\,\rangle$ is defined as

$$\langle \text{Cond.} \rangle = \begin{cases} 1 & \text{if Cond. is true} \\ 0 & \text{otherwise} \end{cases} \tag{14}$$

<u>Statement of the Equivalent MTC Problem:</u>

Given the initial state x(0), determine the control variable u(t) if it exists, which will drive the robot arm from x(0) to any desired final state x_f in minimum time, where u(t) is subject to (13) for $t \in [0, t_f]$. Mathematically,

$$\min_{u(t)\in U(x)} \quad J(u(t)) = \int_0^{t_f} 1\, dt$$

Subject to

$$\begin{aligned} \dot{x}_1(t) &= x_2(t) \\ \dot{x}_2(t) &= u(t) \\ x(0) &= x_0\,,\ x(t_f) = x_f \end{aligned} \tag{15}$$

with the state-dependent control constraints **U** given in (13).

Thus, the original MTC problem with 2N-nonlinear differential equations subject to inequality constraints and equality constraints is transformed into an equivalent double integrator system subject to a set of state-dependent control constraints. Analytically, it will be seen that this transformation greatly simplifies the study of the MTC structure.

Necessary Conditions for Optimality: Extended Pontryagin's Minimum Principle:

With reference to [10], the main theoretical results on the necessary conditions for optimality with state-dependent control constraints are summarized, allowing the derivation to be omitted here.

THEOREM 1 (Extended Pontryagin's Minimum Principle or EPMP) For the problem stated in (15), if the control $u^*(t)$ is time-optimal at x_0 and generates the optimal trajectory corresponding to the solution $x^*(t)$ of the state equation in (15), then there exists a nontrivial solution p(t) of the costate equation defined as

$$\dot{p}(t) = -\left(\frac{\partial f(x(t),u(t))}{\partial x}\right)^T p(t) + \left(\frac{\partial S}{\partial x}\right)^T \mu(t) \tag{16}$$

where the Hamiltonian function H is defined as

$$H(x, p, u) = 1 + p^T(t)\, f(x, u) \tag{17}$$

such that

(a) $H(x^*(t), p^*(t), u^*(t)) \leq H(x^*(t), p^*(t), u(t))$

(b) $H(x^*(t), p^*(t), u^*(t)) \equiv 0, \quad \forall\, t \in [0, t_f]$

where u(t) is piecewise continuous, S and μ are determined by the state-dependent constraints (see [9]). Δ

STRUCTURE OF THE MTC LAW

The structure of the pseudo-acceleration trajectory is obtained directly by using the EPMP in Theorem 1 for the equivalent MTC problem.

THEOREM 2 (The structure of the pseudo-acceleration trajectory) For the equivalent MTC problem stated in (15), the optimal control (i.e. the pseudo-acceleration of the given path) is always saturated at either the upper bound $u_{max}(x)$ or the lower bound $u_{min}(x)$, depending upon the sign of the switching function $p_2^*(t)$.

Proof: See [9].

Theorem 2 gives only the structure of the optimal pseudo-acceleration not that of individual actuators. We first make an assumption before giving the main result.

Assumption 1: We assume

$$\frac{\tau_{imax}\langle d_i>0\rangle+\tau_{imin}\langle d_i<0\rangle-h_i}{d_i} \neq \frac{\tau_{jmax}\langle d_j>0\rangle+\tau_{jmin}\langle d_j<0\rangle-h_j}{d_j}, \quad \forall t\in [t_1, t_2],\ i\neq j$$

THEOREM 3: Under Assumption 1, the structure of the MTC law for two cooperative arms moving a common object along a prespecified motion path subject to the constant control torque constraints requires that *one and only one* actuator is always saturated. Mathematically, for i, j∈ {1,2,..., 2N}, i≠j, and t ∈ $[t_1, t_2]$ c $[0, t_f]$,

$$\{\tau_i(t) \mid \tau_i(t)=\tau_{imin} \text{ or } \tau_{imax}\} \cap \{\tau_j(t) \mid \tau_j(t)=\tau_{jmin} \text{ or } \tau_{jmax}\} = \phi$$

Proof : See [9].

Remark: Without Assumption 1, the structure of the MTC trajectory is in general that at least one of the control torques is always saturated along the optimal trajectory.

DISCUSSION AND CONCLUSIONS

The structure of the MTC law for the two cooperative arms transferring a common object along a specified path was investigated by using the extended Pontryagin's minimum principle and the parameterization method. It has been shown that under some reasonable assumption one and only one actuator is always in saturation, while the rest of them adjust their torques so that the prespecified path constraint on the motion will not be violated. To solve the problem numerically, efficient computational algorithms are required. The algorithms given in [1,2,11] for the single arm case can be readily extended to the dual-arm case. To validate the theoretical result on the MTC structure, a computer simulation for two 3-link planar robots handling a rigid object is being conducted. To further enhance computational efficiency, the reduced-order dynamic model should be employed. For the purpose of on-line implementation, new efficient algorithms need to be developed which is currently under investigation. As a result, the structure of the MTC solution should be able to provide new insight into the dynamic characteristics of two cooperative robot arms moving a common object, pertaining to both path planning and design specifications.

REFERENCES

1. K.G. Shin and N.D.McKay, "Minimum-Time Control of Robotic Manipulators with Geometric Constraints", IEEE Trans. on AC, Vol. AC-30, pp.531-541, 1985.
2. J.E. Bobrow, et al., "Time-Optimal Control of Robotic Manipulators Along Specified Paths", Int. J. Robotics Research, Vol.4(3), pp.3-17, 1985.
3. F. Pfeifferand R. Johanni, "A Concept for Manipulator Trajectory Planning", IEEE J. Robotics and Automation, Vol. RA-3(2), April, pp.115-123, 1987.
4. Y. Chen and A. A. Desrochers, "Structure of the Minimum-Time Control Law for Robotic Manipulators with Constrained Paths," Proc. of the IEEE Int. Conf. on Robo. and Autom., (Philadelphia, PA), pp.1210-1215, April 1988.
5. Y. Chen,"Minimum-Time Control of Robotic Manipulators", Ph.D. Dissertation, Rensselaer Polytechnic Institute, Troy, NY, 1988.
6. G. Rodriguez, "Recursive Forward Dynamics for Multiple Robot Arms Moving a Common Task Object," IEEE Trans. R&A, Vol.RA-5, pp. 510-521, Aug. 1989.
7. T.J. Tarn, et al., "New Nonlinear Control Algorithms for Multiple Robot Arms," IEEE Trans. Aerospace and Elec. Sys., Vol.24(5), pp.571-183, Sept.1988.
8. J.T. Wen and K. Kreutz, "Motion and Force Control for Multiple Cooperative Manipulators," Proc. IEEE Conf. Robot. and Autom. (Phoenix, AZ), pp.1246-1251, May 1989.
9. Y. Chen, "Structure of the Minimum-Time Control Law for Two Cooperative Arms Along a Specified Path," Submitted to the 29th CDC, Feb. 1990.
10. G. Leitmann, The Calculus of Variations and Optimal Control, An Introduction, Plenum Press, New York, 1981.
11. J.-J.E. Slotine and H.S. Yang, "Improving the Efficiency of Time-Optimal Path-Following Algorithms," IEEE Trans. R&A, Vol.5, pp.118-124, Feb. 1989.

SINGLE-MASTER MULTI-SLAVE MANIPULATOR SYSTEM BASED ON VIRTUAL INTERNAL MODEL

KAZUHIRO KOSUGE
Visiting Scientist, CIDMS, Massachusetts Institute of Technology
Cambridge, MA 02139, USA

JUN ISHIKAWA and KATSUHISA FURUTA
Dept. of Control Engineering, Tokyo Institute of Technology
2-12-1 Oh-okayama, Meguro-ku, Tokyo 152, JAPAN

MASARU SAKAI
Nachi-Fujikoshi Corporation
20 Ishigane, Toyama 930, JAPAN

ABSTRACT

A single-master multi-slave manipulator system is proposed in this paper. The system has a master arm with six degrees of freedom and two or more slave arms, each of which has six or more degrees of freedom. Different from conventional master-slave manipulator systems, an operator maneuveres a master arm to control variables related to tasks. We introduce the concept of task-oriented control which tailors the relation between the motion of a master arm and the variables related to tasks using a virtual internal model [3]. An experimental system, which has two joysticks as a master arm and two six-degrees-of-freedom industrial manipulators as slave arms, is developed and illustrates the idea.

INTRODUCTION

In general, a master slave manipulator system consists of a master arm maneuvered by a human operator and a slave arm used for real operations, that is, a single slave manipulator is maneuvered by an operator using a single master arm. Advanced applications sometimes require several slave arms manipulated in coordination; the coordination of slave arms increases the load capacity, rigidity and dexterity of the system.

In the framework of conventional single-master single-slave manipulator, slave arms in coordination are to be manipulated by an operator using master arms, each of which is corresponding to one of the slave arms. It is difficult for an operator to manipulate master arms simultaneously, because the operator has to coordinate the motions of the arms and control not only positions of the arms but also forces/moments applied to the manipulated object. In this paper, we propose a single-master multi-slave manipulator system based on the concept of "task-oriented control"[2]. In this system, an operator is required to manipulate a single master arm with six degrees of freedom whose motion is related to the task they want to do.

In the following sections, we introduce the idea of the task-oriented control of a single-master multi-slave manipulator system, and propose a control architecture of the system. An experimental single-master multi-slave manipulator system is then developed. This system has two joysticks as a master, each of which has three degrees of freedom, and two slave arms, each of which has six degrees of freedom. Several experiments illustrate the concept of the system.

SINGLE-MASTER MULTI-SLAVE MANIPULATOR

As shown in fig.1, a single-master multi-slave manipulator consists of a master arm manipulated by an operator and a set of slave arms used for a task. A master arm is a master device by which an operator can specify motions of six degrees of freedom. In general, the coordination of slave arms increases the load capacity, the rigidity and the dexterity of the system. Fig.2 shows typical examples for these three cases. Let's consider how we can use a single-master to control slave arms in these examples.

In fig.2 (a), two arms manipulating a heavy object that a single slave arm cannot handle. In this example, what an operator of the system wants to do is not to specify the motion of two arms but to specify the motion of the object with six degrees of freedom. Apparently, the motion of a single-master can specify the motion of the object. In fig.2 (b), two arms, in coordination, are using a tool having interactions with its environment. In this example, what the operator wants to do is to specify the position/orientation of the tool and the force/moment that the tool apllies to its environment. A single-master can specify the position/orientation of the tool and the force/moment applied to the environment by employing a bilateral feedback. In fig.2 (c), two arms are assembling two parts, a bolt and a nut. In this case, what the operator has to do is to control the difference in position and orientation of these two parts and the interface force/moment between these two parts. A single-master with six degrees of freedom can specify the difference in position and orientation as well as the interface force/moment by employing a bilateral feedback. Thus, in a single-master multi-slave manipulator systems, we do not specify motions of slave arms using a master arm, but we use a master arm to specify task-oriented variables necessary for the execution of a task.

In some cases, such as fig.2 (a) and fig.2 (b), the information given by a master arm is not sufficient for the control of slave arms; take fig.2 (a) and fig.2 (b) for examples, we have to specify the internal force/moment applied to the manipulated object which is not given by the motion of a single-master. Usually, the necessary information, which is not given by a master arm, can be determined by appropriate knowledge about the task.

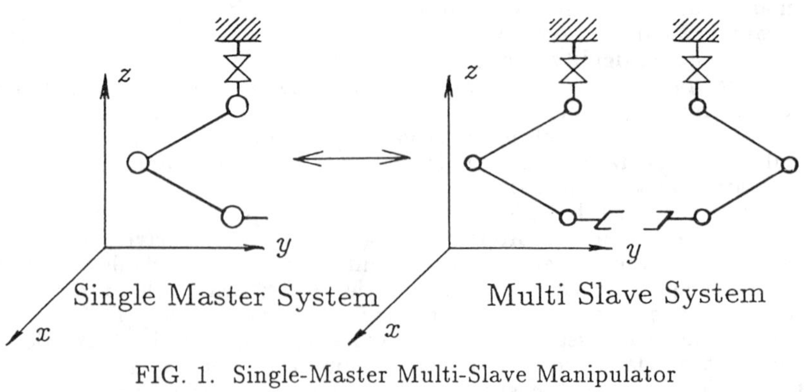

FIG. 1. Single-Master Multi-Slave Manipulator

TASK-ORIENTED CONTROL

Concept of Task-Oriented Control

Authors have proposed task-oriented control of single-master single-slave manipulator systems [2]. In this section, we review and extend the idea to the control problem of single-master multi-slave manipulators.

Motions of an object handled by manipulators to do a task, in general, consist of two types of motions, gross motions and fine motions. The gross motion is a preparatory motion for actual work, and the fine motion is a motion which comprises the work itself [5][6]. In the conventional bilateral feedback of master-slave manipulators, the motion of a slave arm is controlled so as to follow the motion of a master arm and the motion of a master arm is controlled so as to feedback to the operator the interface force/moment between slave arm and its working environment. To do a task by means of such a force reflective master-slave manipulator, the operator has to specify both gross motions and fine motions. Other control algorithms have been developed to increase the realism with which an operator can generate fine motions necessary for tasks [8], etc. How to assist the operator to generate fine motions has been a key issue in such a framework of master-slave manipulator control.

The task-oriented control, as a new paradigm of master-slave manipulator control, generates task-oriented fine motions by itself instead of assisting an operator to generate them. Fine motions necessary for execution of a task are designed a priori, generated automatically based on the interface force/moment and superimposed on gross motions specified by an operator through a master arm. Other sensory information in addition to interface force/moment may be necessary and used for generating fine motions. In a single-master multi-slave manipulator system, task-oriented variables are controlled by a master arm. The task-oriented variables are concerned with the motion of an object, such as motion of a tool, the difference in position/orientation of parts, etc. Instead of specifying the task-oriented variables directly by a master arm, task-oriented control specifies fine motions for a task. How to design fine motions is an important issue for this control scheme. Several research has been done concerned with the analysis and the design of fine motions [5][6][7]. It should be noted that the fine motions are not necessarily the mimic motions of a human operator. We can use artificial fine motions analyzed and designed for manipulators.

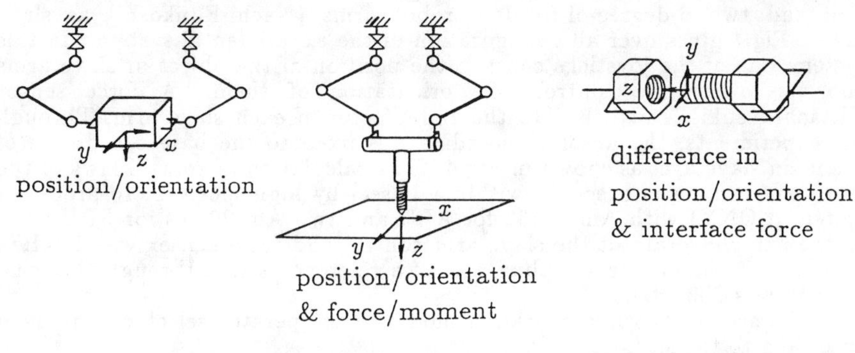

(a) Load Capacity (b) Rigidity (c) Dexterity

FIG. 2. Manipulation by Slave Arms

Control Architecture of Task-Oriented Control Using VIM

Some of the authors have proposed the control architecture of a master-slave manipulator system using a virtual internal model (VIM) [1]. A VIM is a reference model driven by sensory information [3], and was used to realize task-oriented functions which assist an operator to execute a task. It describes a dynamic relation between the motion a master arm and that of a slave arm [1] for the conventional single-master single-slave manipulator system. We extend the idea to the task-oriented control of a single-master multi-slave manipulator system. In single-master multi-slave manipulator systems, the motion of a master arm specifies task-oriented variables. We use a VIM to describe the relation between the motion of a master arm and that of the task-oriented variables. The slave arms are controlled based on the task-oriented variables generated by a VIM. Many control algorithms have been proposed for the control problem of coordinating robot arms [9]-[16]. Here we apply a method proposed in [9] and [4] by authors. Please refer to [9] for more detail of the control algorithm of slave arms. Fig.3 shows the control architecture of the system.

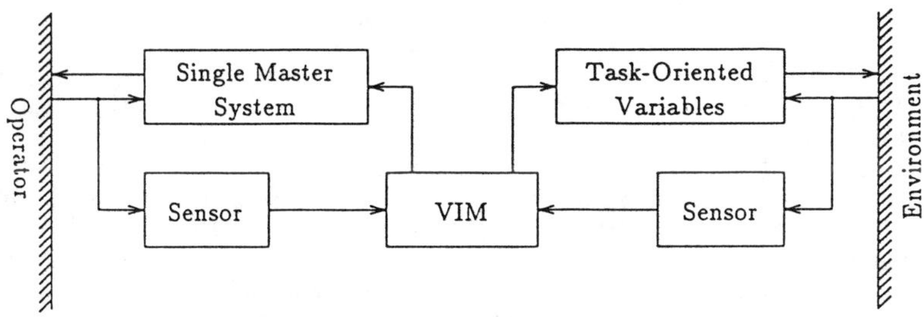

FIG. 3. Control Architecture

EXPERIMENTAL SYSTEM

In order to illustrate the concept of the single-master multi-slave manipulator system, we have developed an experimental system with two slave arms. The system consists of two 3-degree-of-freedom joysticks as a master arm and two 6-degree-of-freedom robot arms (Nachi-Fujikoshi) as slave arms. Fig.4 gives over all configuration of the experimental system. In this system, one of the joysticks controls the position of the object or slave arms and the other one controls the orientation of them. A force sensor (Hitachi-Kenki) is attached to the end-effector of each slave arm. Through the experiments, the absolute coordinate is fixed to the base coordinate of the main slave arm as shown in Fig.4. The calculation of control laws of the two slave arms was executed within 1 [msec] by high-speed micro-processor system (SONY) with Am 29332 for ALU and two Am 2935's for FPU, that is, the sampling rate of the slave arm control and force sensor was 1 [kHz], on the other hand, the sampling rate of a reference signal through the master arm was 200 [Hz].

The system has five operation modes. The operator select one of them according to the situation.

(1) Home Position Mode

This mode is to have each slave arm move quickly to its home position. The operator can modify the home position using a key-board.

(2) Single Mode

In this mode, the master arm controls a selected single slave arm.

(3) Parallel Mode

This mode is used to have all the slave arms move in parallel. The position and orientation of master arm correspond to those of each slave arm.

(4) Symmetric Motion Mode

In this mode, the operator controls a selected slave arm through the master arm and the other slave arm moves symmetrically to the object. For example as the selected slave arm approaches the object, the other slave arm approaches the object at the same rate. This mode is convenient to grip the object by the slave arms.

(5) Coordinated Motion Mode

In this mode, the operator can controls the position and orientation of the object through the master arm with task-oriented fine motions.

This experimental system does not have the bilateral function yet. So, what we can control using a master arm is limited to six degrees of freedom.

Fig.5 shows a palletizing task using this system. In this case, the task-oriented variables are the position/orientation of the manipulated box. The operator specified the gross motions of the box and fine motions were superimposed on them by the task-oriented control.

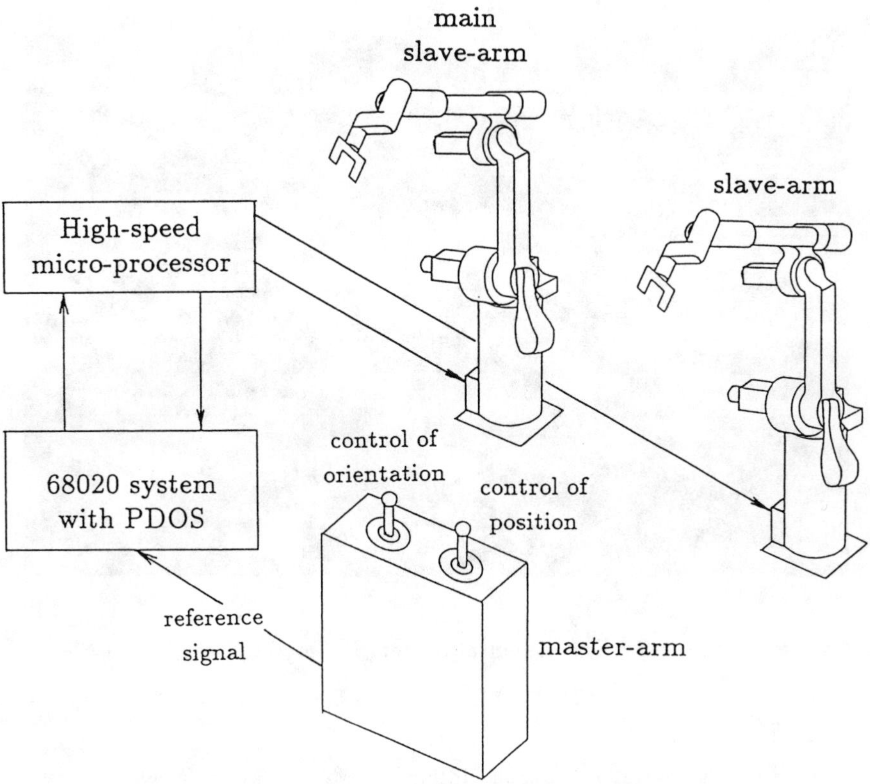

FIG. 4. Experimental System

CONCLUSIONS

Some tasks require several slave arms operated in coordination. In the framework of the conventional master-slave manipulator systems, slave arms are to be controlled by master arms each of which corresponds to a slave arm. In this paper, we have proposed a single-master multi-slave manipulator system and its task-oriented control. In the proposed system, an operator controls task-oriented variables using a master arm. Slave arms are controlled based on the task-oriented variables generated by a VIM according to the motion of a master arm and sensory signals such as interface force/moment between a manipulated object and its environment. With the proposed system, the operator no longer has to consider how to control the slave arms in coordination and can concentrate on their task during the operation.

We have developed an experimental single-master multi-slave manipulator system with two slave arms, each of which has six degrees of freedom, and illustrated the concept of our system. With the task-oriented control, we could execute a palletizing task, although the experimental system dose not have a bilateral function and a master arm can specify the task-oriented variable up to six degrees of freedom. The use of the bilateral feedback as well as task-oriented functions will enable the system to deal with more complex tasks.

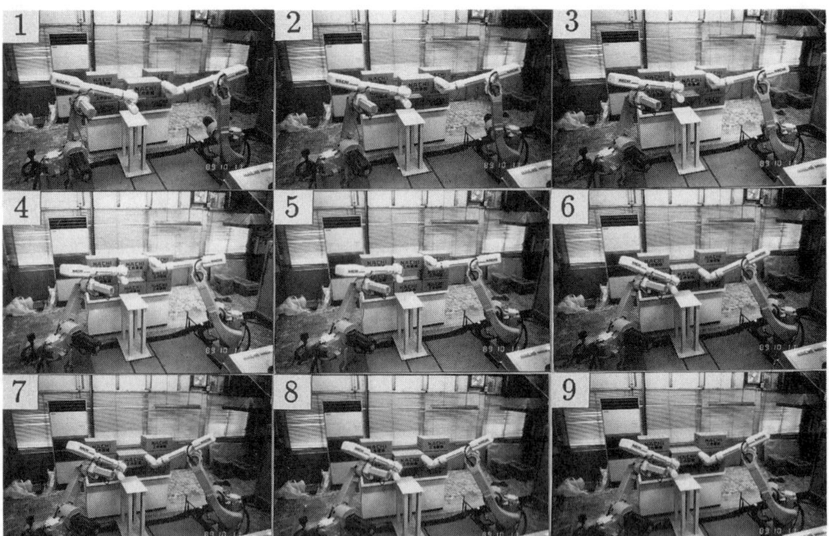

FIG. 5. Experimental System Palletizing Boxes

REFERENCES

[1] K.Furuta, K.Kosuge, Y.Shiote and H.Hatano, "Master-Slave Manipulator Based on Virtual Internal Model Following Control Concept", Proc, of 1987 IEEE International Conference on Robotics and Automation, pp.567-572, 1987

[2] K.Kosuge, A.Sato and K.Furuta, "Task-Oriented Control of Master-Slave Manipulator", Proc. of 1990 Japan-USA Symposium on Flexible Automation, to appear, Kyoto, 1990

[3] K.Kosuge, K.Furuta and T.Yokoyama, "Virtual Internal Model Following Control of robot Arms", Proc. of 1987 IEEE International Conference on Robotics and Automation, pp.1549-1554, 1987

[4] K.Kosuge, J.Ishikawa, K.Furuta and M.Sakai, "Control of Single-Master Multi- Slave Manipulator System Using VIM", Proc. of 1990 IEEE International Conference on Robotics and Automation, 1990

[5] D.E.Whitney, "A survey of Manipulation and Assembly", Chapter 8, Robotics Science, edited by Michael Brady, MIT Press, Cambridge, 1989

[6] D.E.Whitney, "Quasi-Static Assembly of Compliantly Supported Rigid Parts", Trans. of ASME, Journal of Dynamic System, Measurement and Control, Vol. 104, No.1, pp.65-77, 1982

[7] H.Asada and Y.Kakumoto, "The Dynamic RCC Hand for High-Speed Assembly", Proc. of 1988 IEEE International Conference on Robotics and Automation, pp.120-125, 1988

[8] Blake Hannaford, "A Design Framework for Teleoperators with Kinesthetic Feedback", IEEE Trans. on Robotics and Automation, Vol.RA-5, No.4, pp.426-434, 1989

[9] K.Kosuge, M.Koga, K.Furuta, K.Nosaki, "Coordinated Motion Control of Robot Arm based on Virtual Internal Model", Proc. of 1989 IEEE International conference on Robotics and Automation, pp.1097-1102, 1989

[10] E.Nakano, S.Ozaki, T.Ishida and I.Kato, "Cooperational Control of the Anthropomorphous Manipulator 'MELARM'", Proc.of the 4th International Symposium on Industrial Robots, pp.251-260, Tokyo 1974

[11] S.Kurono, "Coordinated Computer Control of a Pair of Artificial Arms", Biomechanism 3, The Tokyo University Press, pp.182-193, 1975

[12] Y.Nakamura, K.Nagai and T.Yoshikawa, "Mechanics of Coordinative Manipulation by Multiple robotic Mechanisms", Proc. of 1987 IEEE Conference on Robotics and Automation, pp.991-998, 1987

[13] M.Uchiyama and P.Dauchez, "A Symmetric Hybrid Position/Force Control Scheme for the Coordination of Two Robots", Proc.of 1988 IEEE International Conference on Robotics and Automation, pp.350-356, 1988

[14] S.Hayati, "Hybrid Position/Force Control of Multi-Arm Cooperating Robots", Proc.of 1986 IEEE International Conference on Robotics and Automation, pp.82-89, 1986

[15] T.J.Tarn, A.K.Bejczy, and X.Yun, "Design of Dynamic Control of Two Cooperating Robot Arms; Closed Chain Formulation", Proc.1987 IEEE International Conference on Robotics and Automation, pp.7-13, 1987

[16] Y.F.Zheng and J.Y.S.Luh, "Control of Two Coordinated Robots in Motion", Proc. 24th IEEE Conference on Decision and Control, vol.1, pp.1761-1765, 1985

MOTION CONTROL LIBRARY FOR KINEMATIC COORDINATION OF MULTIPLE ROBOTIC DEVICES

RICK GUPTILL *, SHAHEEN AHMAD and YUNGTAT LAM

Real-time Robot Control Laboratory, School of Electrical Engineering, Purdue University, West Lafayette, IN 47907, USA

ABSTRACT

In this paper we describe a software package for kinematic coordination of multiple robotic devices using homogeneous transforms. The multi-robot trajectory generation scheme is described in this paper, the computational load placed on a 68020 microprocessor during the library execution is also presented.

1. INTRODUCTION AND MOTION SPECIFICATION

There are a large number of software packages for single robot coordination [1] [2] [3] [4] using homogeneous transforms. In the multi-robot area there has been considerable amount of work done in their coordination and control see [4]. In this paper we describe a software package for multi-robot cooperation.

Let T_n represents the homogeneous transform describing the position and orientation of the last link the the robotic device. We will assume a solution exists to find T_n given the vector of joint variables ($\vec{\theta}$) for the device. We refer to this as the *forward kinematics* solution, or, $T_n = K(\vec{\theta})$. In general the *inverse kinematic* operation will need a configuration variable to choose an arm solution and thus, $\vec{\theta} = K^{-1}(T_n, \text{arm–configuration})$. By including the T_n transform of a device in a concatenation of transforms describing intermediate coordinate frames in a work space, we form a *position equation*. For example, a position equation instructing a robotic device to move to a block on a table could be specified with the following transforms describing the task. P_1 : [base][T_n][gripper] = [table][block] ; TOOL: [gripper]. (Note throughout this paper we will use ";" for new line when it is used to illustrate software code.)

Note P_1 will be used to denote the position at which this transform equation is satisfied. The transform specified as the TOOL, in this case the [gripper] transform, is the coordinate frame in which straight line movement will take place if it is specified. The internal representation of a position equation is a doubly linked list [2].

To move the desired device to a position we have specified, we pass the position pointer to the move command: move(P_1).

The move command puts the position equation onto the *motion queue* of the device listed in the position equation. The motion queue is a "first in first out" queue which keeps the order of positions for each device to move to. To specify how the motion is to take place, a motion structure is created to store information

*Now with International Robotics Inc., Vancouver, BC, Canada

pertaining to the move such as velocity, Cartesian mode or joint mode, segment travel time, and acceleration time. Our motion structure contains these motion information as well as a pointer to the position to which the motion refers.

Tasks involving coordinated movement between two or more robotic devices always have one device in the role of leader or master. This device requires a position specification which does not depend on another device. The follower or slave device uses the T_n transform of the leader to solve for its position and orientation every sample period. Therefore, we make the restriction that only one device may appear in a single position equation without loosing any flexibility in task specification.

2. TRAJECTORY GENERATION

Trajectory generation for multiple devices essentially consists of trajectory generation for each device separately, treating the position and orientations of other devices as constant transforms each sample time. It is necessary, therefore, to first solve the device which does not depend on the position of the others. To simplify the software structure, we will use the same trajectory generator for every device we wish to control. We start by deriving the equations of motion for a point in α-space which we will call α_{out}. α-space can be a point in joint space or Cartesian space. To move our point from point α_B to point α_C we can use the equation $\alpha_{out} = r(\alpha_B - \alpha_C) + \alpha_C$ and let r = 1 - h, as h varies from 0 to 1. Motion between positions is divided into two regions, a transition region and a linear region, see Figure 1. The transition region, which starts at $-t_{acc}$ in time before the current point (position) and extends to t_{acc} after the point, is further divided into the sum of two motions. The first motion is that of a point starting from rest at point α_B at time $-t_{acc}$ and transitioning to constant velocity at t_{acc} after α_B. The second motion is that of a point on the previous path at $-t_{acc}$ moving toward α_B coming to a stop at α_B at time = t_{acc}. The linear region from $\alpha_B(t_{acc})$ to $\alpha_C(t_{seg} - t_{acc})$ is simple to compute and is not divided into separate motions. The time t_{seg} is the travel time from the point α_B to α_C.

The boundary conditions on the point moving from α_B to α_C is used to find a matching seventh order polynomial for the transition region and a linear polynomial for the linear velocity region. As we vary h from 0 to 1, α_{out} will move from rest at α_B to α_C. At time $-t_{acc}$ before α_C a new transition process will repeat for the next position α_D.

For the second motion during the transition, we again use a seventh order polynomial and derive the coefficients using the boundary conditions. Because this motion is to be added to the transition motion described above, we derived this second transition such that it is the difference between the present position and the next position, with this difference going to zero at t_{acc}. The subroutine which computes the coefficients of the matching polynominal is called Λ. Now that we have both motions derived, we can add them to obtain the resultant motion in α-space from $\alpha_{out}(-t_{acc})$ transitioning past α_B and moving toward α_C. The transition past point α_C occurs if another position is specified, and stops at point α_C if no further points are specified. To come to a complete stop using the system just derived, we move to the last position twice.

3. STRAIGHT LINE MOTION

To move from P_B to P_C in Cartesian space, we use a drive transform [2] into the position equation. The drive transform $D(r) \mid r \in [0,1]$, represents a translational and an orientation change from P_B to P_C, both proportional to r. In between P_B , and P_C the robot device position is calculated from the transform equation, [base_of_device][device][gripper] = [reference][pointB][D(r)]. For the linear portion of the motion between positions, we will use the drive transform described in [5]. It consists of a rotation ψ around a unit vector u, vector u is defined in terms of the first two Euler angles, ϕ and θ, and u =(cos(ϕ)sin(θ), sin(ϕ)sin(θ), cos(θ)) , where $0 \le \theta < \pi$. The translation is defined to be (rx,ry,rz). To move in Cartesian mode we use the equation $T_n = \Gamma\ (L_C(t)\ D(r)\ R_C(t))$, the L_C is the left hand side of the transform equation involving D(r), and R_C, is the right hand side of D(r) when solving for T_n, as given in the above. Here we let r vary from 1 to 0. To generate the position vector in joint space, we call the inverse kinematics for the device.

$\vec{\theta}_{out}$ = device→K^{-1} ($\Gamma(L_C(t)\ D(r)\ R_C(t))$, arm_configuration)

When we transition from one Cartesian motion path segment to another, transitions have to be calculated in cartesian space. To determine the boundary conditions in Cartesian space, we need to calculate the drive parameters and find the difference between two successive drive parameters to obtain the velocities. To compute the secondary drive parameters we use the subroutine Δ_s which computes the Euler angles given T_n, and the concatenated transforms to the left and right of the drive transform.

Because P_B might involve a functionally defined transform in its definition, we form a new position equation, P_T, used only for Cartesian transitions. $P_T : T_n = \Gamma(L_T\ D_s(p)\ R_T)$, where $D_s(p)$ is the secondary drive transform and $\rho\ \varepsilon$ [0,1]. To use this drive transform requires that all drive parameters are in the same form. For this reason we will not use the rotation around a vector, but will instead use the drive parameters as they are computed, in the form of Euler angles. The form of the drive parameter calculation subroutine has less computations required than the primary drive parameter subroutine described above.

During the transition, the drive parameters are computed using the seventh order polynomial whose coefficients were found using the drive parameter boundary conditions.

If we evaluate the transition position equation we obtain the straight line path which drives the robotic device from point $P_B(-t_{acc})$ to $P_B(t_{acc})$. But we actually need is the difference in Cartesian space which will be added to the first equation of motion. This amount which matches the difference in Cartesian space is given by the relationship: $[T_{matching}]^{-1}[T_B(-t_{acc})]$ where $T_{matching} = \Gamma(L_T\ D_s(p)\ R_T)$

4. Summary of Trajectory Generation Scheme: By deriving the equations of motion as two motions added together, we obtain a very simple method to smoothly transition between functionally defined positions. The two types of motions, joint and Cartesian, result in four types of transition possibilities. To take advantage of the transition motion being divided into two motions and then added together, we divide the trajectory generator into two subroutines; *a main trajectory generator* which computes the joint vector during the linear region of movement, and a *matching trajectory generator* which provides the matching joint vector during transitions. In doing this, we provide a natural way to parallelize the trajectory generator.

5. PERFORMANCE ANALYSIS

The computation times on a 25MHz 68020 processor with cache enabled with a 20MHz 68881 co-processor are also listed in Table 1. The Γ function computations assume five transform multiplies. The misc entry combines joints to encoders conversion, joint limit checks, and calculations involving the **r** parameter. Table 1 shows the trajectory generator running on one processor controlling one six axes PUMA robot requires 13 milliseconds. The sample periods with the largest amount of computation occur during the Cartesian mode to Cartesian mode transition. Given a uniform sample time and one processor for each the main and matching trajectory subroutines, the point at $t_{seg} - t_{acc}$ during a Cartesian to Cartesian transition requires the most computation and defines the sample time, here it is approximately 8 milliseconds for the PUMA example. During this sample period, one Γ () function, one K^{-1} (), one primary drive transform and one drive parameter calculation are required by the main trajectory subroutine.

6. SOFTWARE INTEGRATION AND PROGRAMMING EXAMPLE

The motion control library is divided into two types of subroutines, those related to the trajectory generator, and those called by the user to specify the robot task in a *user_task* (). The number of *user_task* ()s written correspond to how the user wants to break up the communication coordination of the task. Global variables in the user file allows semaphores to be set up for intertask communication. To illustrate motion coordination, we describe the programming of a robotic task which performs non-destructive testing of composite materials using two five axes motion devices sharing a common axis, and a one axis table, see Figure 2. The master robot holds an ultrasonic transmitter and the slave holds the receiver. The table orients the part so that the current scanning location is normal to both the receiver and the transmitter. Variations in the ultrasonic signal produce a internal map of the part including any defects as the part is scanned.

We will assume we are scanning boat hulls to locate any defects in the hull material. The boat hull will be mounted on the table. The [surface] transform will be functionally defined and will return the next scanning surface point each sample time. The position equation for the table is:

P_{table} : [table] = [table_orientor] ; TOOL : [table]

where [table_orientor] actually evaluates [surface] and extracts the information needed to align the table. The position equation to control the master_device is

$P_{transmit}$: [base_of_master][master_device][transmitter]

= [table→T_n][boat_hull][surface] ;

TOOL : [transmitter]. And the equation for the slave_device is

$P_{receive}$: [base_of_slave] [slave_device][receiver] = [base_of_master]

[master_device→T_n] [transmitter] [thickness_of_boat_hull] ; TOOL : [receiver].

By first moving the table, the master and then the slave manipulators into their starting locations:

move(P_{table_start});
move(P_{master_start});
move(P_{slave_start});

we can accomplish the test with three calls to move:

move(P_{table});
move($P_{transmit}$);
move($P_{receive}$);

By expanding the non-destructive testing example to include a gantry robot to supply the scanning robots with new boat hulls, we can show communication coordination. Using two user_task () subroutines, we can easily separate the functionality of the robotic tasks while providing shared variable communication coordination. The gantry can be removed from the device task list while the other two robots are scanning the part to reduce the sample time. The partial C program below outlines the two concurrent tasks.

```
/* An example of two user-written tasks which provide communication
coordination between three robotic devices while dividing the functionality
of the task for ease of programming.
static int in_position = FALSE;
static int testing_done = FALSE;
static int all_done = FALSE;
    user_task1 ()
    { DEVICE_PTR gantry; int i;

        gantry = place_on_task_device_list ("GANTRY");
        /* have the gantry supply boat hulls */
        for (i = 0; i < number_of_boat_hulls; i++)
            { in_position = FALSE;
            pick_up_new_boat_hull (gantry);
            place_on_bench (gantry);
            in_position = TRUE;
            while (testing_done == FALSE) /* wait */ ;
            remove_boat_hull (gantry);
            }
        all_done = TRUE;
        }

    user_task2 ()
    { DEVICE_PTR table, master_device, slave_device; int i;
        table = place_on_task_device_list ("TABLE");
        master_device = place_on_task_device_list ("MASTER");
        slave_device = place_on_task_device_list ("SLAVE");
        while (all_done == FALSE)
            { /* scan the boat hull */
            while (in_position == FALSE)
                /* wait */ ;
            .
            move (P_table_start);
            move (P_master_start);
            move (P_slave_start);
            .
            move (P_master_start);
            move (P_transmit);
            move (P_receive);
            .
            testing_done = TRUE;
            }
    }
```

7. SUMMARY

In this paper we have outlined the mathematical basis for coordinated kinematic control of multiple robotic devices. We have also discussed a software implementation consisting of a device independent trajectory generator which is divided into a main subroutine and a matching subroutine which can be designed to run on a multiprocessor system. The position specification method and the trajectory generator allow motion coordination while the multiple user tasks allow communication coordination and programming flexibility.

REFERENCES

1. Shaheen Ahmad, Colin Besant, "Robot Level Programming Languages and the SRIL-90 Language," Proceedings of the IEEE Compsac Conference, Nov. 1984, Chicago, IL.

2. Paul, R. P. (1981a). Robot Manipulators: Mathematics, Programming, and Control , Cambridge: MIT Press.

3. Hayward, V., (1983). "RCCL User's Manual, Version 1.0", TR-EE 83-46, School of Electrical Engineering, Purdue University.

4. Shaheen Ahmad, Heng Guo, "Dynamic Coordination of Dual-Arm Robotic Systems with Joint Flexibility," Proceedings of the IEEE Conference on Robotics and Automation, April 1989, Philadelphia, PA.

5. Paul, R. P., & Zhang, H., (1985). "Robot Motion Trajectory Specification and Generation", Robotics Research: The Second International Symposium, MIT-Press Series in Artificial Intelligence.

TABLE 1 Trajectory Generator Subroutine Timing for the PUMA robot. Each entry in this table is a six tuple, they are as follows: number of multiplies, adds, sines/cosines atan2, square roots and the last entry is execution time in milliseconds.

$\Gamma(1) = \{165,120,0,0,0,(2.5)\}$, $\Delta = \{119,88,2,3,2,(1.8)\}$, $\Delta_s = \{104,88,0,3,2,(1.5)\}$,

K^{-1} (puma) $= \{22,14,2,5,1,(2.0)\}$, K (puma) $= \{24,13,4,0,0,(1.2)\}$, D(r) $= \{19,9,1,0,0,(0.4)\}$,

$D_s(p) = \{64,40,3,0,0,(0.8)\}$, $\Lambda = \{54,36,0,0,0,(0.7)\}$, misc $= \{45,62,0,0,0,(1.0)\}$

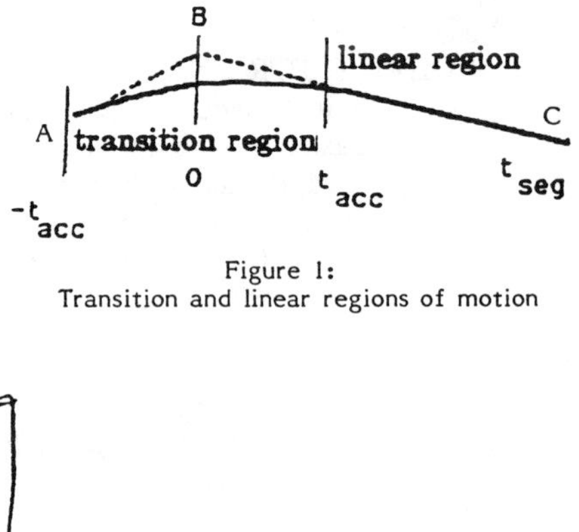

Figure 1:
Transition and linear regions of motion

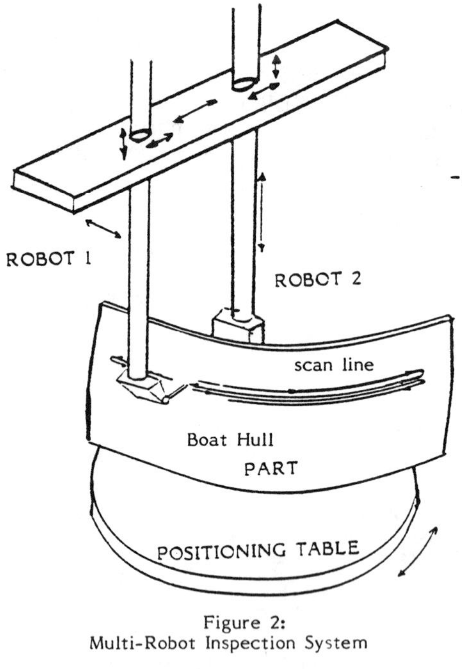

Figure 2:
Multi-Robot Inspection System

ADAPTIVE LOAD SHARING CONTROL FOR COOPERATING TWO MANIPULATORS TO HOLD OBJECTS ROBUSTLY

Masaru Uchiyama and Toshiaki Yamashita
Department of Precision Engineering
Tohoku University
Aramaki Aza-Aoba, Aoba-ku, Sendai 980, Japan

1 Introduction

Recently the importance of dual-armed, or more generally, multi-armed robots has been noted because the application of robots to unstructured environments requires the robots to carry out sophisticated tasks that cannot be achieved only by a single-armed robot. Robotics for space exploration is a good example of such application [1]. Stimulated by the application, research on multi-armed robots has grown up, and modeling and control of multi-armed mechanisms has become an interesting target of academic research [2]. Moreover, since the mechanism of multi-armed robots has a similarity with that of multi-fingered hands, communication with the research on multi-fingered hands [3] may possibly create an interesting fusion of research fields.

Keeping this in mind, our group has been involved in research on force control of two manipulators that hold a common object to carry or to manipulate [4, 5,6,7,8,9]. Results obtained so far include static and kinematic equations to define task vectors [5,6], formulation of a hybrid position/force control scheme using the task vectors [4,6], experimental verification of the scheme [7], and, recently, extension of the definition of the task vectors to implement load sharing capability [8]. The most recent paper [8], however, presents only a framework for load sharing control. To find an effective load sharing algorithm remains unsolved although an attempt to find it was made [9]. The main point of the attempt was to realize robust holding of the object by adjusting the sharing of loads adaptively between the two arms. Focusing on the same topic, what this paper is going to present, is an adaptive load sharing algorithm which, recently, has been obtained, and its experimental verification.

To derive the load sharing algorithm, first, contact conditions between the end-effectors and the object are considered. A set of applicable forces and moments on the contact point is defined as a subset of the all forces and moments. Next, the load sharing matrix defined in [8] that satisfies the conditions is searched for. In case there are many candidates for the matrix, the best one should be chosen. Therefore, an algorithm to find the best one is then discussed. This paper, however, does not present an optimal algorithm. Instead, it presents a heuristic algorithm based on theoretical consideration.

The effectiveness of the algorithm is verified experimentally. The robot system used in the experiment is an industrial robot system consisting of two four degrees-of-freedom Cartesian-type arms with six-axis force/torque sensors at each wrist.

2 Framework for Adaptive Load Sharing

The framework we use in this paper for adaptive load sharing is the theory presented in [8]. According to the theory, task vectors to describe the coordinated task: for example, to carry an object by holding it, is defined by using a load sharing matrix, which is a set of coefficients to determine the load distribution between the two arms. Although details of the derivation, which can be found in [8], are omitted in this paper, only results are presented in the following.

Let denote force/moment vectors applied by the arms to the object as $\boldsymbol{f}_{b1}$, $\boldsymbol{f}_{b2}$, velocity vectors of the arms as $\boldsymbol{s}_{b1}$, $\boldsymbol{s}_{b2}$, and homogeneous transformation matrices to represent positions and orientations of the arms as

$$\boldsymbol{H}_{bi} = \begin{bmatrix} \boldsymbol{n}_{bi} & \boldsymbol{o}_{bi} & \boldsymbol{a}_{bi} & \boldsymbol{p}_{bi} \\ 0 & 0 & 0 & 1 \end{bmatrix}, \tag{1}$$

respectively, where $i = 1$ and 2. It is noted that all vectors represent corresponding quantities at a unique representative point of the object. Using the above notations, the task vectors are defined as follows:

$$\boldsymbol{f}_a = \boldsymbol{f}_{b1} + \boldsymbol{f}_{b2} \tag{2}$$

$$\boldsymbol{f}_r = (\boldsymbol{I}_6 - \boldsymbol{K})\boldsymbol{f}_{b1} - \boldsymbol{K}\boldsymbol{f}_{b2} \tag{3}$$

$$\boldsymbol{s}_a = \boldsymbol{K}^T \boldsymbol{s}_{b1} + (\boldsymbol{I}_6 - \boldsymbol{K}^T)\boldsymbol{s}_{b2} \tag{4}$$

$$\Delta \boldsymbol{s}_r = \boldsymbol{s}_{b_1} - \boldsymbol{s}_{b_2} \tag{5}$$

$$\boldsymbol{p}_a = \boldsymbol{K}_1^T \boldsymbol{p}_{b1} + (\boldsymbol{I}_3 - \boldsymbol{K}_1^T)\,\boldsymbol{p}_{b2} \tag{6}$$

$$\boldsymbol{n}_a = \boldsymbol{K}_2^T \boldsymbol{n}_{b1} + (\boldsymbol{I}_3 - \boldsymbol{K}_2^T)\,\boldsymbol{n}_{b2} \tag{7}$$

$$\boldsymbol{o}_a = \boldsymbol{K}_2^T \boldsymbol{o}_{b1} + (\boldsymbol{I}_3 - \boldsymbol{K}_2^T)\,\boldsymbol{o}_{b2} \tag{8}$$

$$\boldsymbol{a}_a = \boldsymbol{K}_2^T \boldsymbol{a}_{b1} + (\boldsymbol{I}_3 - \boldsymbol{K}_2^T)\,\boldsymbol{a}_{b2} \tag{9}$$

$$\boldsymbol{K} = \mathrm{diag}\,[\boldsymbol{K}_1\ \boldsymbol{K}_2] \tag{10}$$

$$\Delta \boldsymbol{p}_r = \boldsymbol{p}_{b_1} - \boldsymbol{p}_{b_2} \tag{11}$$

$$\Delta \boldsymbol{\Omega}_r = \frac{(\boldsymbol{n}_{b_2} \times \boldsymbol{n}_{b_1} + \boldsymbol{o}_{b_2} \times \boldsymbol{o}_{b_1} + \boldsymbol{a}_{b_2} \times \boldsymbol{a}_{b_1})}{2} \tag{12}$$

where $\boldsymbol{f}_a$, $\boldsymbol{s}_a$ and

$$\boldsymbol{H}_a = \begin{bmatrix} \boldsymbol{n}_a & \boldsymbol{o}_a & \boldsymbol{a}_a & \boldsymbol{p}_a \\ 0 & 0 & 0 & 1 \end{bmatrix} \tag{13}$$

are the external force/moment, velocity and position/orientation of the object, respectively, and $\boldsymbol{f}_r$, $\Delta \boldsymbol{s}_r$, $\Delta \boldsymbol{p}_r$ and $\Delta \Omega_r$ are the internal force/moment, velocity, position and orientation of the object, respectively.

It should be noted that the 6×6 matrix $\boldsymbol{K}$ in the above equations determines the sharing of loads between the two arms. This can be seen from (2) and (3).

3 Adaptive Load Sharing for Robust Holding

3.1 Conditions to Keep Holding

Robust holding means that the robot keeps holding the object in spite of disturbance forces or moments on the object. This can be achieved, of course, by holding the object tightly. Holding the object tightly, however, may cause breaking of the object. In order to avoid this, and to achieve the robust holding, sharing the disturbance forces or moments adaptively between the two arms would be, intuitively, an alternative solution. In this paper we choose the alternative solution to derive an adaptive algorithm for the load sharing.

We consider the situation depicted in Figure 1. In order for the robot to hold the object, the hand in the figure must, first, keep contact with the object and, secondly, not slip on the object. Conditions for this, of course, depend on geometrical and physical conditions of the hand and the object.

Here, to illustrate those conditions, we suppose that the geometry of the hand is as shown in Figure 2, and that the friction on the surfaces is that of Coulomb. On this assumption, the conditions to keep contact and not to slip are given, respectively, as follows:

Conditions to keep contact:

$$F_{zhi} \leq 0 \tag{14}$$

$$| N_{xhi} | \leq r_{yi} F_{zhi} \tag{15}$$

$$| N_{yhi} | \leq r_{xi} F_{zhi} \tag{16}$$

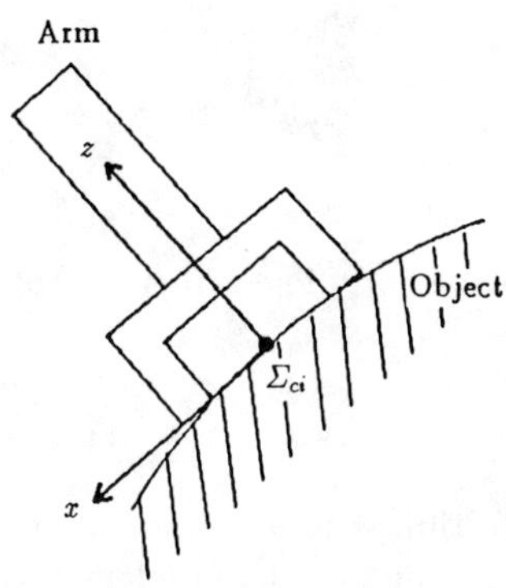

Figure 1: A robot hand on the surface of an object.

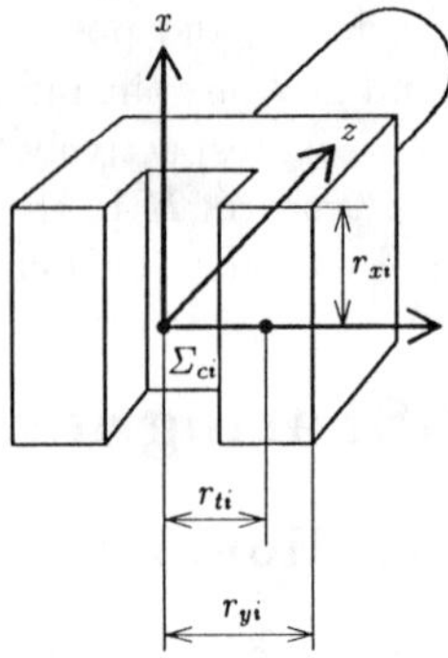

Figure 2: Geometry of a hand.

Conditions not to slip:

$$|F_{xhi}| \leq \mu F_{zhi} \tag{17}$$

$$|F_{yhi}| \leq \mu F_{zhi} \tag{18}$$

$$|N_{zhi}| \leq \mu r_{ti} F_{zhi} \tag{19}$$

where F_{xhi}, F_{yhi}, and F_{zhi} are the x, y, and z components of the contact forces and N_{xhi}, N_{yhi}, and N_{zhi} are the x, y, and z components of the moments. The suffix hi denotes that forces or moments are at the contact points of each hand. μ is the frictional coefficient. Geometrical parameters can be found in Figure 2.

The forces and moments in (14) to (19) are calculated from the forces and moments at the representative point of the object. Therefore, substituting (2) and (3) into (14) through (19) to eliminate F_{xhi}, F_{yhi}, F_{zhi}, N_{xhi}, N_{yhi}, N_{zhi}, and $\boldsymbol{f}_{bi}$, we have a set of linear inequalities regarding the components of the load sharing matrix $\boldsymbol{K}$. The inequalities include $\boldsymbol{f}_a$ and $\boldsymbol{f}_r$ as parameters. In this paper we assume that $\boldsymbol{K}$ is a diagonal matrix:

$$\boldsymbol{K} = \text{diag}[\alpha_j] \tag{20}$$

where

$$0 \leq \alpha_j \leq 1. \tag{21}$$

Then, we have a linear vector inequality to express the conditions to keep holding as follows:

$$\boldsymbol{c}_1\alpha_1 + \boldsymbol{c}_2\alpha_2 + \cdots + \boldsymbol{c}_6\alpha_6 \leq \boldsymbol{c}_0 \tag{22}$$

where $\boldsymbol{c}_j$ is a coefficient vector. The dimension of $\boldsymbol{K}$ has been assumed to be 6×6.

3.2 An Algorithm for Robust Holding

The mission of the algorithm is to determine α_j to satisfy (22). Since (22) is a set of linear inequalities, to find α_j is a problem of a kind of linear programming.

In this paper, before solving it generally, we analyze a special case to present a heuristic algorithm.

The special case to be considered is based on our experimental setup. Details of the setup is found in [4,7]. The industrial robot used in the setup has, as shown in Figure 3, two four degrees-of-freedom Cartesian type arms. Using this robot we consider, as an example, the task shown in Figure 4: to carry the object when collision between the object and the obstacle happens. In this case, the dimension of $\boldsymbol{K}$ is 4×4. Therefore, only α_1 to α_4 have to be determined.

To illustrate the region of α_1 to α_4 that satisfy (22), we calculate the region assuming that the external and internal forces in the direction of z axis are zero and the external moment in the direction of z axis is zero, when only α_1 and α_2 for the external forces in the direction of x and y axes, respectively, are to be considered. Figure 5 illustrate an example of the region for this case.

Based on such numerical analysis for many cases, we have derived the following algorithm to determine α_j:

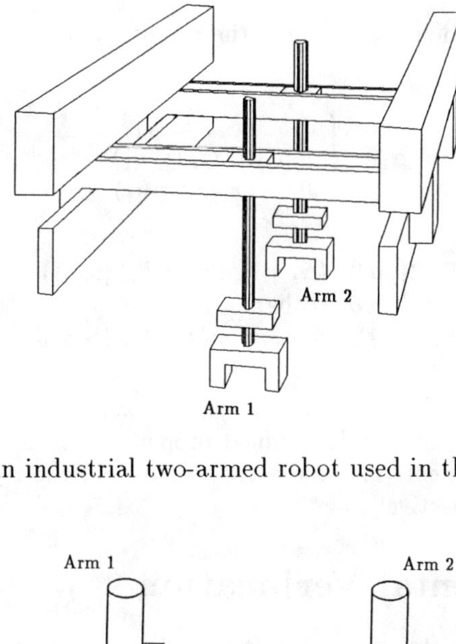

Figure 3: An industrial two-armed robot used in the experiment.

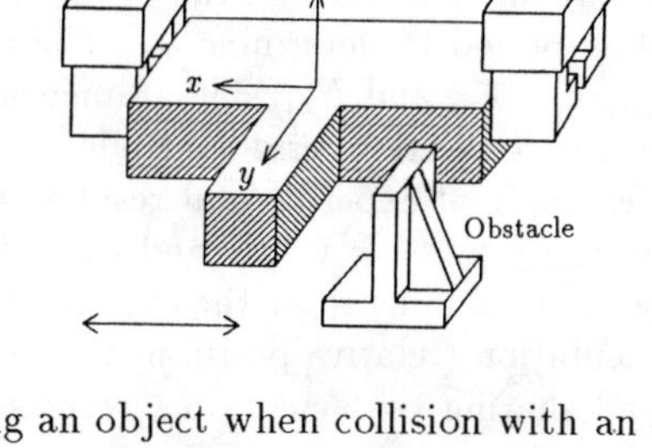

Figure 4: Carrying an object when collision with an obstacle happens.

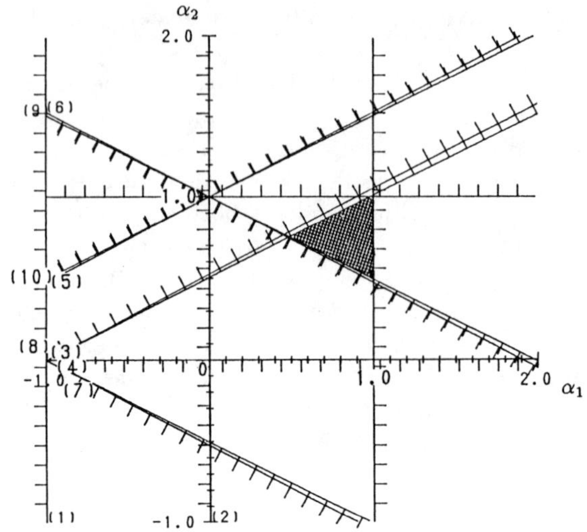

Figure 5: Region that satisfies the conditions to keep holding.

$$\alpha_1 = \begin{cases} 1 & (F_{xa} > a_1) \\ 0.5 & (\mid F_{xa} \mid \leq a_1) \\ 0 & (F_{xa} < -a_1) \end{cases} \tag{23}$$

and

$$\alpha_j = \begin{cases} 0.8 & (\alpha_1 = 1 \text{ and } \mid F_{ja} \mid > a_j) \\ 0.5 & (\text{other case}) \\ 0.2 & (\alpha_1 = 0 \text{ and } \mid F_{ja} \mid > a_j) \end{cases} \tag{24}$$

where F_{ja} represents F_{ya}, F_{za} and N_{za} for j = 2, 3 and 4, respectively, and a_j represents thresholds to be determined properly. F_{xa}, F_{ya}, F_{za} and N_{za} are the x, y and z components of the external forces and the z component of the external moment, respectively.

4 Experimental Verification

The load sharing algorithm derived in the previous section was implemented on the industrial robot. In the implementation, the equation (23) through (24) are calculated every sampling period to determine α_j. The calculation is based on the measurement of F_{xa}, F_{ya}, F_{za} and N_{za}. The framework of the control is the same as in [8,9]; the task vectors are hybrid controlled.

Figure 6 shows an example of experimental results. Conditions for the experiment are a_1 = 1.5 N, $a_2 = a_3$ = 0.1 N and a_4 = 0.2 Nm. The data at the left, center and the right columns show the external forces and moment, internal positions and orientation (relative positions and orientation between the two arms), and the load sharing coefficients α_j, respectively. The data show

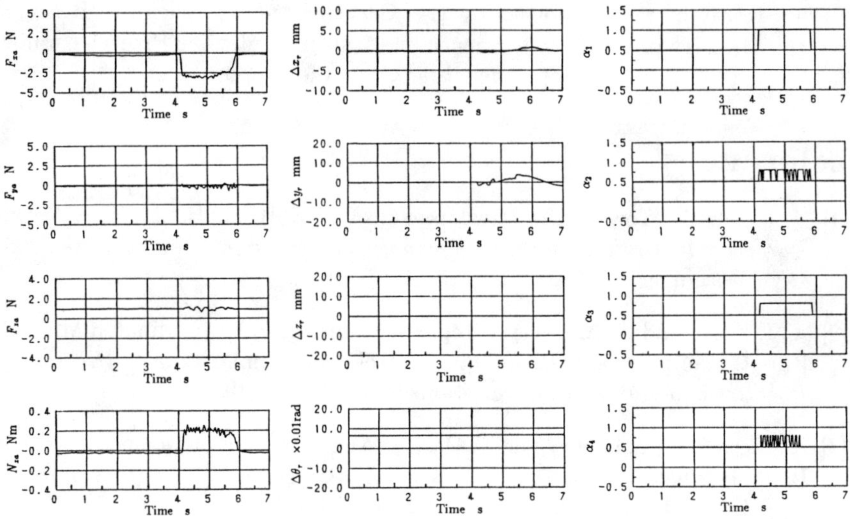

Figure 6: An example of results for experiments on robust holding.

that α_j are adaptively determined from the measurement of the external forces and moment, and that the internal positions and orientation are kept almost constant. This means that the object is held firmly by the arms in spite of the external forces and moment caused by the collision.

Through experiments on the same task on different conditions and other tasks such as sliding an object on another object being pressed on it, we have verified that the algorithm realizes the robust holding. Comparison of the result with results without adaptive load sharing have made clear the effectiveness of the algorithm. Results of the comparison, however, are not reported in this paper.

5 Concluding Remarks

In this paper we discussed the problem of robust holding of an object by a cooperating two-armed robot. The approach we took was to realize it by adjusting the distribution of the forces on the object between the two arms. First, we analyzed contact conditions and formulated the problem as that of linear programming. And, before solving it for a general case, we analyzed a special case numerically for an experimental setup, and derived a heuristic algorithm to be verified experimentally.

Future research will be directed to deriving an algorithm for a general case, and its combination with other approaches such as a hold-tightly approach will yield an important part of an integrated skill system for an intelligent robot in future. In this sense a knowledge-based approach to the problem would be also promising.

This research was supported by the Grant-in-Aid for Scientific Research from the Ministry of Education, Science and Culture under the Project Number 63550178. The authors express their thanks for the support.

References

[1] Y. Umetani, "Toward the Development of COSMO-LAB," in *Proc. Second European In-Orbit Operations Technology Symposium*, Toulouse, pp. xxi–xxvi, December 1989.

[2] A. J. Koivo and G. A. Bekey, "Report of Workshop on Coordinated Multiple Robot Manipulators: Planning, Control, and Applications," *IEEE J. of Robotics and Automation*, vol. 4, no. 1, pp. 91–93, 1988.

[3] A. Bicci, J. K. Salisbury and P. Dario, "Augmentation of Grasp Robustness Using Intrinsic Tactile Sensing," in *Proc. 1989 IEEE Int. Conf. on Robotics and Automation*, Scottsdale, pp. 302–307, May 1989.

[4] M. Uchiyama, N. Iwasawa, and K. Hakomori, "Hybrid Position/Force Control for Coordination of a Two-Arm Robot," in *Proc. of 1987 IEEE Int. Conf. on Robotics and Automation*, Raleigh, pp. 1242–1247, April 1987.

[5] P. Dauchez and M. Uchiyama, "Kinematic Formulation for Two Force Controlled Cooperating Robots," in *Proc. '87 Int. Conf. Advanced Robotics*, Versailles, pp. 457–467, October 1987.

[6] M. Uchiyama and P. Dauchez, "A Symmetric Hybrid Position/Force Control Scheme for the Coordination of Two Robots," in *Proc. 1988 IEEE Int. Conf. on Robotics and Automation*, Philadelphia, pp. 350–356, April 1988.

[7] M. Uchiyama and Y. Nakamura, "Symmetric Hybrid Position/Force Control of Two Cooperating Robot Manipulators," in *Proc. 1988 IEEE Int. Workshop on Intelligent Robots and Systems*, Tokyo, pp. 515–520, November 1988.

[8] M. Uchiyama, "A Unified Approach to Load Sharing, Motion Decomposing, and Force Sensing of Dual Arm Robots," in *Prepr. 5th Int. Symp. of Robotics Research*, Tokyo, pp. 41–48, August 1989.

[9] M. Uchiyama and T. Yamashita, "Asymmetric Hybrid Control of Positions and Forces of a Dual Arm Robot to Share Loads," *First Int. Symp. on Experimental Robotics*, Montréal, June 1989.

A Multi-Arm Tele/Autonomous Executive System

Kam S. Tso, Paul G. Backes, Thomas S. Lee, Samad Hayati
Jet Propulsion Laboratory
California Institute of Technology
Pasadena, CA 91109

Abstract

The Manipulator Control and Mechanization (MCM) subsystem provides task execution capability to the higher level subsystems of the JPL Telerobot Testbed. The MCM System Executive described in this paper is the interface to the higher level subsystems and controls task execution in the MCM subsystem. Robot task macros including guarded motion, compliant motion, grasp, and force reflecting teleoperation are callable via the executive with input parameters. Traded control is achieved by alternately executing autonomous and teleoperation tasks from the higher level subsystems. The same code can also be run on graphical simulation to preview the motions. The subsystem, arm, and sensor status information are reported periodically to other subsystems. The system is operational in the JPL Telerobot Testbed demonstrating robot tasks such as orbital replacement unit (ORU) changeout utilizing a combination of autonomy and teleoperation.

1 Introduction

The JPL/NASA telerobot program is exploring the feasibility of using robots in space for on-orbit assembly, maintenance, and repair operations. The JPL Telerobot Testbed consists of five subsystems: Manipulator Control and Mechanization (MCM), Task Planning and Reasoning (TPR), Run Time Control (RTC), Operator Control Station (OCS), and Sensing and Perception (S&P) [1]. Fig. 1 shows a schematic drawing of the overall architecture.

On the autonomous side, the TPR subsystem will develop task scripts from requests made by the operator. The scripts are passed to RTC which breaks them down to a number of primitive operations that can be executed by MCM. RTC is also responsible for generating collision-free path and maintaining the world model. S&P provides object designation, acquisition and tracking capabilities. MCM provides robot task execution capability in the form of macros. On the teleoperation side, the control is based on the operator generating commands by physically moving two six DOF force reflecting hand controllers with the remote site MCM teleop macro responding to these commands.

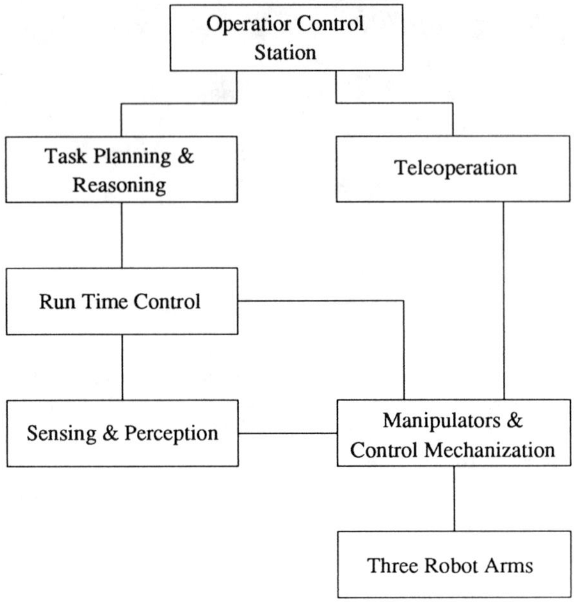

Figure 1: JPL/NASA Telerobot Testbed Architecture

2 MCM Architecture

The MCM subsystem is a real-time system that controls robots, grippers, and hand controllers, acquires sensor data for control, and communicates with other telerobot subsystems. Fig. 2 shows a block diagram of the MCM subsystem. The system is physically divided into two sites: *local* and *remote*. In the local site where the operator is stationed, there are two identical systems each consisting of a Force Reflecting Hand Controller (FRHC), a VME chassis with 68020 processors for computations of kinematics and force reflection, an RGB monitor for force/torque display, and a JPL-built Universal Motor Controller (UMC) for control of the hand controllers. Also there is a SUN-3 computer for the operator interface. The remote site consists of a SUN-4 computer to run the enhanced version of multi-arm RCCL [3] which is capable of generating trajectories for multiple coordinated manipulators. This robot software is used to run two PUMA 560's. A MicroVax II runs the same enhanced RCCL to move a third PUMA 560 for positioning a pair of mounted cameras for the S&P subsystem. An additional pair of side-mounted color cameras is used for stereo display of the remote site for the operator. The reason for not using the SUN-4 to run the third arm is the unavailability of an additional UMC rather than any limitation in the SUN-4 based RCCL software. Each arm is equipped with a Lord wrist force/torque sensor and servoed gripper from TRI Inc. Signals from hand controllers, force/torque sensors, servoed grippers, and teach pendants are read and passed to the SUN-4 through two MOPER chassis which are equipped with four 68020-processors and several parallel and serial communication cards each. Additional details of the MCM subsystem can be found in [2].

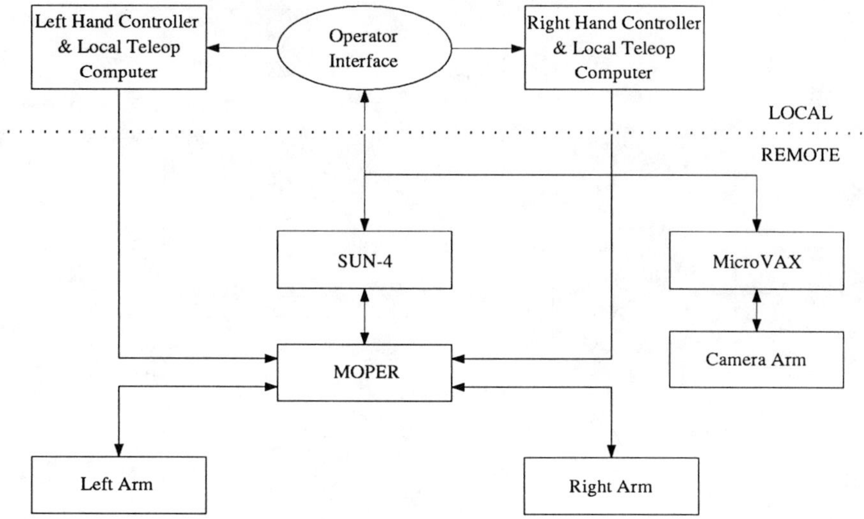

Figure 2: High-Level MCM Architecture

3 Components of MCM Executive

The MCM Executive consists of five software modules distributed on three computers, as shown in Fig. 3. The NIP Interface Module runs on the MicroVAX/VMS gateway computer. The other four components run on both the MicroVAX/UNIX and SUN4/UNIX robot control computers. The followings describe the function of each module.

3.1 NIP Interface Module

All subsystem communication is done using a common software library called the Network Interface Package (NIP), which was developed at the Stanford Research Institute (SRI). NIP is a set of software modules which allows application tasks on separate computers to exchange data over an Ethernet. It enforces a set of valid record protocols which defines the legal record types and record exchange sequences. It also logs and monitors network activities. Currently NIP is implemented on VAX computers running the VMS operating system and uses Digital Equipment Corporation's DECNET protocol for data transfer. Since the MCM software runs on Unix computers, a gateway is needed to interface MCM to other subsystems. The NIP Interface Module acts as such a gateway. It runs on a VMS MicroVAX computer and functions to establish and terminate NIP communication links, forward commands and reports between MCM and the other subsystems. A standard NIP message consists of three components: a header which has a fixed length contains information such as the sender, the responder, protocol, message ID, message type, and interface ID, the length of the comment and data fields, followed by a variable length comment and data field.

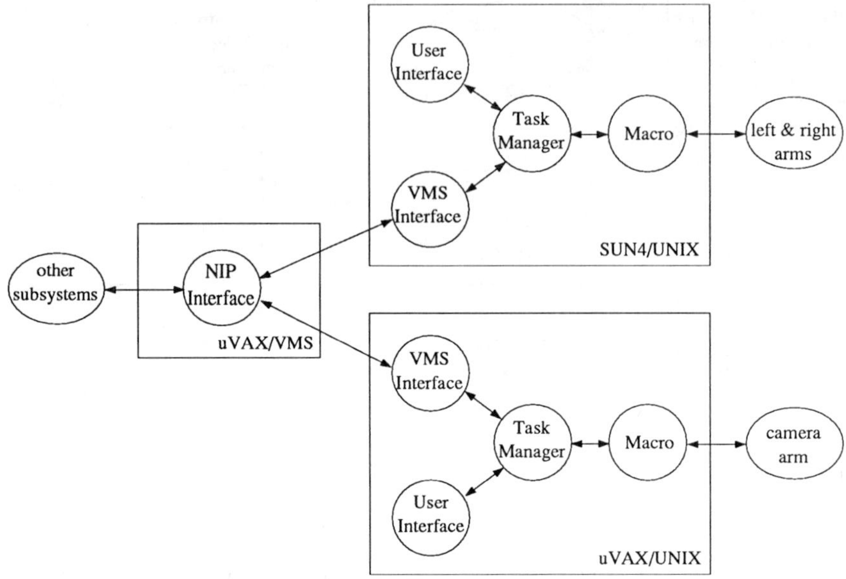

Figure 3: Components of MCM Executive

3.2 VMS Interface Module

The VMS Interface Module acts as an interface for the UNIX robot control computers to the VMS gateway computer. The communication is done by using the SunLink DECNET Network Interface (DNI) package provided by SUN. Since MicroVAXes and SUN workstations use different byte order and floating point number representation, the VMS Interface converts the data field to the right format for incoming and outgoing messages. It also logs all the messages into a file which is very useful for debugging at the subsystem level by verifying that correct data has been sent.

3.3 User Interface Module

Besides accepting commands from other subsystems through the NIP gateway and VMS Interface channel, MCM Executive can also accept commands from a user interface. The input to this interactive interface is very similar to MATLAB and is an extension to ARC (A Robot Calculator) supported by RCCL. ARC is an interactive scalar and matrix calculator which has access to RCCL's robot specific library functions (such as Jacobian, forward kinematics and so on). The extension to ARC allows it to invoke the macros so an operator can interactively perform analysis and then try the results on the robots without leaving the analysis environment.

3.4 Task Manager Module

The Task Manager is the heart of the MCM Executive. It starts up the VMS Interface and User Interface modules and sets up communication links with the two modules using the UNIX pipes. The VMS Interface in turn starts up the NIP Interface. After the startup, the Task Manager accepts commands from either the high level subsystems or the user interface. Once a command has arrived, it checks to see if the requested resources, such as robots and grippers, are currently available. If they are, the corresponding macro will be executed with the input data provided in the command. Otherwise, the command will be placed to the end of a First-In-First-Out (FIFO) queue. Some exceptional commands such as CANCEL-MACRO, SHUTDOWN-MCM will get executed immediately without waiting. The Task Manager detects failures of other components of the MCM Executive. If the VMS Interface or the User Interface aborts due to an internal software error or a communication problem, it will be restarted. If a macro task aborts, a report will be sent to the requesting subsystem informing it that a failure has occurred during the execution of the macro.

3.5 Macros

Robot task execution capability in the MCM subsystem is provided as a set of macros. They can be separated as four categories: guarded motion, compliant motion, grasp, and teleoperation.

Guarded Motion Macros. The guarded motion macros are Cartesian-Guarded-Move, Joint-Guarded-Move, and Move-to-Touch. The Cartesian- and Joint-Guarded-Move move through via points specified as Cartesian frames and joint angles respectively. During the motion, they actively check the values of the force/torque sensor and stop upon a sensed collision. The Move-to-Touch macro moves the arm until contact and backs off toward the initial position until the contact force is reduced to zero.

Compliant Motion Macro. A Generalized-Compliant-Motion macro is implemented which can be used for general compliant motion tasks, such as door opening, crank turning, pin insertion, bolt turning, and leveling by specifying values of a parameter set such as the destination frame, compliance frame, tool frame, and compliance gains.

Grasp Macro. The Grasp macro provides both free and compliant grasps using either a pneumatic or servoed gripper.

Teleoperation Macro. The teleoperation macro uses hand-controller inputs to move the arm. When the macro starts, it connects to a menu which allows an operator to select either Cartesian or joint modes of operation, and with or without force reflection [4].

All the macros use a common set of functions to receive and send data. Without modification or recompilation, a macro can be run as a standalone program invoked by a user or as a module executed by the Task Manager. In the standalone mode, input is read either from the keyboard interactively or from a prepared data file. In the integrated mode, input is supplied by the Task Manager through a UNIX pipe.

4 Status Report

A comprehensive report on the status of the MCM subsystem is sent upon completion of each macro, or per requested by a *status* command. Intermediate reports may also be requested by specifying a reporting rate so that the higher level subsystems can monitor the execution of the macros. Currently the maximum reporting rate is about once every second. The rate is limited both by the CPU cycles left in the computer and the communication overhead to the other subsystems.

All reports have the same format. Each contains a *system status code* for the general status of the MCM subsystem, e.g. error connecting to robot, communication error. A *command status code* for the status of the command, e.g. command in progress, completed successfully, or completed with error. A completion report also contains a *command result code* which returns the result of the completed command. Each macro has its own set of possible results, e.g. stop on position reached, stop on force threshold, and stop on timeout. The report may then be followed by the arm states. If it is a single-arm macro, only the state of the requested arm is sent. A dual-arm macro will return both arm states. In an erroneous situation, such as error connecting to robot, no arm state is sent. An arm state always contains the observed joint angles, T6 frame and configuration, force/torque values, and gripper separation.

5 Database

A database is used during task execution by the macros. This database does not contain a world model which would be useful for task planning. Rather, it provides an execution environment for the macros. The database contains numerous entries, each of which has three fields: object type, name, and value. Commonly used entries include the base frame of the arm, force/torque safety limit, and mass property of the currently grasped object. Commands can be sent by the higher level subsystems to change the values of the database entries.

6 Simulation Capability

Single and dual-arm simulation capability exists via RCCL. The code that runs the arms can be used in simulation without any modification to the code itself. MCM has developed a 3-D graphics package which is interfaced to the simulation software for display on a SUN computer screen[5]. This package is written using the SunCORE graphics library. Several features exist such as adjusting the viewing window size and changing view angles.

7 Future Enhancements

The MCM subsystem we have described uses a set of macros to provide robot task execution capability to higher level subsystems. It has a number of advantages. First, the effort of developing and debugging the software is simplified by minimizing the size of each macro and making them independent of each other. Second, it makes the execution more robust. Since each macro process terminates after completing the task, a problem in one macro is localized to itself. Also, it eliminates cumulative errors, such as memory being allocated but not released by one macro which would eventually cause the system to run out of memory. However, it also has some disadvantages. First, it takes one to two seconds each time a macro process is spawn. Consequently the MCM subsystem cannot respond to a command as fast as would be desirable. Second, the macros take up a large amount of disk space (about one to two megabytes each) since each contains the RCCL libraries. Since we would like to put the software into ROM, the size would make it infeasible to use the system in space. Third, each macro has it own specific monitor functions, reflexes, and control resulting in a large but limited set of possible tasks and reflexes. Also, task execution parameters cannot be modified by the higher level subsystems during execution. A future plan is under way to integrate the task execution capabilities by modularizing monitoring, reflexing, and sensor-based motion. The higher level subsystems can then specify a unique task by utilizing any number of the available modules. New capabilities can be easily added to the system by simply adding the associated module and parameters[6].

8 Conclusions

This paper has described the functional architecture of the MCM subsystem of the JPL/NASA Telerobot project and the MCM System Executive in particular. MCM is a complete autonomous and teleoperation telerobot system and operational demonstrating robot tasks such as orbital replacement unit (ORU) changeout using a combination of autonomy and teleoperation.

Acknowledgments

The research described in this document was performed at the Jet Propulsion Laboratory, California Institute of Technology, under contract with National Aeronautics and Space Administration.

References

[1] Matijevic, J., Zimmerman, W., Dolinsky, S., "The NASA/OAST Telerobot Testbed Architecture," *Proceedings of NASA Conference on Space Telerobotics*, Pasadena, California, Jan. 31 - Feb. 2, 1989.

[2] Hayati, S., Lee, T.S., Tso, K.S., Backes, P.G., and Lloyd J., "A testbed for a unified teleoperated-autonomous dual-arm robotic system," *Proceedings of the 1990*

IEEE International Conference on Robotics and Automation, Cincinnati, Ohio, May 1990.

[3] Lloyd, J., Parker, M., and McClain, R., "Extending the RCCL programming environment to multiple robots and processors," *Proceedings of the 1988 IEEE International Conference on Robotics and Automation*, Philadelphia, PA, April 1988, pp. 465-469.

[4] Lee, T.S., Hayati, S., Tso, K.S., and Backes, P.G., "Dual-arm voice-controlled force-reflecting teleoperation system with a time delay feature," *Proceedings of the Third International Symposium on Robotics and Manufacturing*, British Columbia, Canada, July 1990.

[5] Backes, P., "The MCM/RCCL robot graphics simulator," Vers. 1.2, JPL Inter-Office Memorandum No. 347-89-392 (internal document), May 15, 1989.

[6] Backes, P.G., Tso, K.S., Hayati, S., and Lee, T.S., "A modular telerobotic task execution system," *Proceedings of 1990 IEEE Conference on Systems Engineering*, Pittsburgh, PA, August, 1990.

A Dynamic Decoupling Control Scheme for Two-Arm Coordination via State Feedback

by

Paul I. Ro, Hyeung S. Choi, Gordon Lee
Mechanical and Aerospace Engineering
North Carolina State University

ABSTRACT

This paper presents a new control scheme for decoupling the dynamics of two coordinating robot manipulators. A simple full-state feedback scheme with configuration dependent gains can be devised to decouple the system dynamics such that the dynamics of each arm and that of an object held by the two arms are independent of one another. As an example, the proposed control scheme is applied to coordinate the motion of a candidate two-arm system; a parallel-connected system composed of two motor systems whose shafts are coupled together by a spring-and-damper coupling. The advantage of the proposed scheme is that the same control scheme can be applied both for the closed kinematic chain (object-grasping) case and open kinematic chain (no object-grasping) case.

INTRODUCTION

Many tasks arise in assembly, repair and inspection that require multiple robot manipulators to perform in a coordinated manner. A multitude of challenging research issues arise from multi-arm coordinated control [1-5]. One of the fundamental problems that control designers face is the fact that a dual-arm robotic system manipulating a common load is described by a closed kinematic chain, resulting in system dynamic constraints and a reduction in the degrees of freedom [6]. In [7], Ahmad and Luo described a technique for coordinated motion control of multi-arm manipulators for welding applications. There, a redundant manipulator with seven degrees of freedom (equivalent to two non-redundant arms) is required to weld on specific trajectory along a table. Constraints on singular conditions and motion limits are incorporated into a performance measure to be optimized. The approach requires off-line path planning and, in effect, uses a master/slave control scheme. Carignan and Akin [8] transformed the dual-arm problem to a hierarchical control structure whereby a complete minimization is performed on a reduced-order model of the system in order to construct the payload trajectory; then a parameter minimization is done to find the force distribution of the arms on the payload. This approach yields a suboptimal solution but incorporates the dynamics and control issues into nonconflicting performance measures.

Seraji [9] develops an adaptive position/force control approach to the dual-arm problem. By employing an adaptive PID structure, knowledge of the mathematical model of the system is not required. The coupling effects between the manipulators, through the common payload, and modelled as disturbances in the position and force equations which are then compensated for in the adaptation rule. In [11], Ro and Youcef-Toumi present a leader-follower control approach, but with a reference model structure. The leader manipulator is directed according to a prescribed reference model system while the follower arm follows via interacting force feedback. Robustness issues of the control scheme in the presence of actuator nonlinearities and model uncertainties as well as bounded disturbances are presented in [12].

In this paper, an issue of dynamic decoupling between two cooperating robot arms manipulating a common object is addressed. The object is assumed to be rigid and rigidly held by two robot arms. Depending on the arm configuration and the speed with which the object is manipulated, the dynamic coupling between two robot arms and that of the object can be negligibly small or constitute a significant portion of the overall dynamics. In this paper, a new control scheme is introduced which incorporates a decoupling condition into the two-arm coordination problem. Stability of the approach in a linear sense is guaranteed, while the robustness of the approach can be obtained in a manner similar to what is shown in [12].

TWO-ARM DYNAMICS

The equations of motion for two robot arms grasping an object can be expressed as the following:

$$\mathbf{H}_1\ddot{\mathbf{q}}_1 + \mathbf{C}_1(\mathbf{q}_1,\dot{\mathbf{q}}_1) = \mathbf{T}_1 + \mathbf{J}_1^T\,\mathbf{F}_1 \tag{1a}$$

$$\mathbf{H}_2\ddot{\mathbf{q}}_2 + \mathbf{C}_2(\mathbf{q}_2,\dot{\mathbf{q}}_2) = \mathbf{T}_2 + \mathbf{J}_2^T\,\mathbf{F}_2 \tag{1b}$$

$$\mathbf{M}_o\ddot{\mathbf{x}}_o + \mathbf{C}_o(\mathbf{x}_o,\dot{\mathbf{x}}_o) = -\,\mathbf{L}_1^T\,\mathbf{F}_1 - \mathbf{L}_2^T\,\mathbf{F}_2 \tag{1c}$$

where $\mathbf{q}_1$ and $\mathbf{q}_2$ are *nx1* joint angle vectors for arms 1 and 2; $\mathbf{x}_o$ is the *nx1* vector representing the position and orientation or the object center in the inertial space; $\mathbf{T}_1$ and $\mathbf{T}_2$ are the *nx1* joint torque vectors for arms 1 and 2; $\mathbf{H}_i$ are mass matrices of size *nxn* associated with arms 1 and 2; $\mathbf{M}_o$ is the mass matrix associated with the object; $\mathbf{C}_i$ are nonlinear force vectors of size *nx1*, respectively; and $\mathbf{J}_i$ are the *nxn* Jacobian matrices of arms 1 and 2. Forces $\mathbf{F}_1$ ($\mathbf{F}_2$) represent *nx1* vectors of forces and moments at the interaction between arm 1 (arm 2) and the object, and *nxn* matrices $\mathbf{L}_i$ represent transformations associated with finite lengths between the center of the object and the interaction points. Similar expressions for the dynamics or two-arm systems have been used by [8], [15], and others. Above expression can be rewritten in a matrix form as

$$\mathbf{M}(\mathbf{x})\,\ddot{\mathbf{x}} = \mathbf{u} - \mathbf{C}(\mathbf{x},\dot{\mathbf{x}}) + \mathbf{G}(\mathbf{x})\,\mathbf{f} \tag{2}$$

where

$$\mathbf{x}=\begin{bmatrix}\mathbf{q}_1\\ \mathbf{q}_2\\ \mathbf{x}_o\end{bmatrix},\ \mathbf{u}=\begin{bmatrix}\mathbf{T}_1\\ \mathbf{T}_2\\ 0\end{bmatrix},\ \mathbf{C}=\begin{bmatrix}\mathbf{C}_1\\ \mathbf{C}_2\\ \mathbf{C}_o\end{bmatrix},\ \mathbf{f}=\begin{bmatrix}\mathbf{F}_1\\ \mathbf{F}_2\end{bmatrix},\ \mathbf{M}=\begin{bmatrix}\mathbf{H}_1 & 0 & 0\\ 0 & \mathbf{H}_2 & 0\\ 0 & 0 & \mathbf{M}_o\end{bmatrix},\ \mathbf{G}=\begin{bmatrix}\mathbf{J}_1^T & 0\\ 0 & \mathbf{J}_2^T\\ -\mathbf{L}_1^T & -\mathbf{L}_2^T\end{bmatrix}$$

Eq. (2) states that there are $5n$ unknowns, $\mathbf{x}$ and $\mathbf{f}$, with $3n$ equations. However, $2nx1$ vector, $\mathbf{f}$, can be expressed as functions of other $3n$ unknowns using the kinematic constraints imposed on the system due to rigid grasping, as shown in [8]. The kinematic constraints due to the closed-kinematic chain formed by grasping can be expressed by $2n$ algebraic equations

$$\mathbf{x}_o = \mathbf{r}_1(\mathbf{q}_1) = \mathbf{r}_2(\mathbf{q}_2) \tag{3}$$

where $\mathbf{r}_1$ and $\mathbf{r}_2$ represent the object position and orientation in arm 1 and 2 coordinates, respectively. Equation (3) can be rewritten as

$$\alpha(\mathbf{x}) \equiv \begin{bmatrix}\mathbf{r}(\mathbf{q}_1) - \mathbf{x}_o\\ \mathbf{r}(\mathbf{q}_2) - \mathbf{x}_o\end{bmatrix} = 0 \tag{4}$$

where α represents the closed kinematic chain, and is always equal to *2nx1* null vector for all time. The force, f, can be expressed as a function of known variables by taking the second derivative of constraints (4) with respect to time and substituting eq. (2) to the resulting expression and solving for **f**. The resulting expression of two-arm dynamics is

$$\ddot{\mathbf{x}} = \mathbf{P}(\mathbf{u} - \mathbf{C}(\mathbf{x},\dot{\mathbf{x}})) - \mathbf{Q}\dot{\mathbf{x}} \tag{5}$$

where

$$\mathbf{P} = \mathbf{M}^{-1}[\mathbf{I} - \mathbf{G}(\alpha_x\mathbf{M}^{-1}\mathbf{G})^{-1}\alpha_x\mathbf{M}^{-1}]$$

$$\mathbf{Q} = \mathbf{M}^{-1}\mathbf{G}(\alpha_x\mathbf{M}^{-1}\mathbf{G})^{-1}\dot{\alpha}_x$$

and α_x and $\dot{\alpha}_x$ are *2nx3n* partials of constraint (4) with respect to **x** and that with respect to time, respectively. The above equations of motion holds whenever $(\alpha_x\mathbf{M}^{-1}\mathbf{G})^{-1}$ exists.

A DECOUPLING CONTROL SCHEME VIA STATE FEEDBACK

Here, we propose a state feedback control scheme that will try to "decouple" as much as possible the dynamics of each arm from the other as well as from that of the object. Going back to eq. (5), in order to decouple the dynamics of arm 1 from arm 2 and the object we need to diagonalize two *3nx3n* matrices, namely, **P** and **Q**. For this purpose, we consider

$$\mathbf{u} = -\mathbf{K}_1(\mathbf{x})\,\dot{\mathbf{x}} - \mathbf{K}_2(\mathbf{x})\,\mathbf{x} + \hat{\mathbf{C}} \tag{6}$$

where $\mathbf{K}_1$ and $\mathbf{K}_2$ are feedback gain matrices and feedforward term, $\hat{\mathbf{C}}$ are defined as

$$\mathbf{K}_1 = \begin{bmatrix} k_{11}(q_1) & k_{12}(q_2) & k_{13}(x_o) \\ k_{21}(q_1) & k_{22}(q_2) & k_{23}(x_o) \\ 0 & 0 & 0 \end{bmatrix}, \quad \mathbf{K}_2 = \begin{bmatrix} g_{11}(q_1) & g_{12}(q_2) & g_{13}(x_o) \\ g_{21}(q_1) & g_{22}(q_2) & g_{23}(x_o) \\ 0 & 0 & 0 \end{bmatrix}, \quad \hat{\mathbf{C}} = \begin{bmatrix} \hat{C}_1 \\ \hat{C}_2 \\ 0 \end{bmatrix}$$

The feedforward term represents the estimates of the nonlinear Coriolis, centrifugal, and gravity forces. We note that the bottom row has to be zero because there is no control available for the object. Each element, $\mathbf{k}_{ij}$ or $\mathbf{g}_{ij}$, is an *nxn* matrix of feedback gains that is dependent on the configuration of the arms. This is because the matrices, **P** and **Q**, are configuration dependent and therefore functions of Jacobian matrices. If we apply the control action shown in eq. (6) to the two-arm system of (5), the resulting equations of motion, assuming $\mathbf{C} = \hat{\mathbf{C}}$ become

$$\ddot{\mathbf{x}} + [\mathbf{P}\mathbf{K}_1(\mathbf{x}) + \mathbf{Q}]\,\dot{\mathbf{x}} + \mathbf{P}\mathbf{K}_2(\mathbf{x})\,\mathbf{x} = 0 \tag{7}$$

Looking at the above expression, the decoupling of the dynamics of each arm can be achieved if we can diagonalize the coefficient matrices associated with the velocity vector and the position vector terms. Of our prime concern is to decouple the dynamics between the two arms. This can be achieved by choosing the off-diagonal gain elements $\mathbf{k}_{12}$, $\mathbf{k}_{21}$, $\mathbf{g}_{12}$, and $\mathbf{g}_{21}$ such that the 12-th element and 21-th element of the coefficient matrices become null, that is,

$$(\mathbf{P}\mathbf{K}_1 + \mathbf{Q})_{12} = (\mathbf{P}\mathbf{K}_1 + \mathbf{Q})_{21} = 0 \quad \text{and} \quad (\mathbf{P}\mathbf{K}_2)_{12} = (\mathbf{P}\mathbf{K}_2)_{21} = 0\,. \tag{8}$$

For this particular case, the following choice of gain elements will satisfy the condition shown in eq. (8):

$$k_{12} = -\,p_{11}^{-1}[p_{12}k_{22} + q_{12}]\,,\; k_{21} = -\,p_{22}^{-1}[p_{21}k_{11} + q_{21}]$$

$$g_{12} = -\,p_{11}^{-1}p_{12}k_{22}\,,\text{ and } g_{21} = -\,p_{22}^{-1}p_{21}k_{11} \tag{9}$$

where p_{ij} and q_{ij} represent the ij-th element of $\mathbf{P}$ and $\mathbf{Q}$, respectively. Similarly, the decoupling of the object dynamics from the first arm dynamics can be realized by choosing k_{13}, k_{31}, g_{13}, and g_{31} elements such that

$$(\mathbf{PK_1}+\mathbf{Q})_{13} = (\mathbf{PK_1}+\mathbf{Q})_{31} = \mathbf{0} \quad \text{and} \quad (\mathbf{PK_2})_{13} = (\mathbf{PK_2})_{31} = \mathbf{0} \; . \qquad (10)$$

The conditions in (8) and (10) can not be simultaneously satisfied because there is no control associated with the object; this is represented by the null row vectors for the last rows of $\mathbf{K_1}$ and $\mathbf{K_2}$. The stability of the overall system as well as the desired performance of each arm depend on choosing appropriate gains for diagonal elements, k_{ii} and g_{ii}, such that the individual second order matrix equations have stable coefficients. One complication with the above approach can result because of the fact that these gains are configuration dependent. This issue is being addressed as an on-going research topic, and the result will be presented in a near future.

The approach shown above is useful in that we can define the desired dynamics of each arm independent of the other and of the object. Also, it is particularly useful in that the control scheme can be readily adapted for controlling the arms separately in the case of open kinematic chain (no object-grasping mode). In case the robot arms are maneuvering in space independently, the gains for the off-diagonal elements of $\mathbf{K_1}$ can be set to zero. If at some point an object is detected and the arms start manipulating the object, the gains of off-diagonal elements can be obtained according to (8). This simple but very efficient control feature can be essential in space assembly and repair, or even in factory assembly, where the operating mode of dual-arm may have to change frequently from the "object-grasping" mode to "no object-grasping" mode.

A SIMPLE EXAMPLE

The above control scheme is applied to coordinate the motion of two simple mass-spring-damper systems (representing simple motor systems) coupled by massless object whose dynamics is characterized by a spring and a damper (see Fig. 1). It is quite obvious to see that the dynamics of this dual-mass system can be described by

$$\dot{\mathbf{x}}_1 = \mathbf{A}_1\mathbf{x}_1 + \mathbf{B}_1\mathbf{u}_1 - \mathbf{h}_1\mathbf{f}_c$$

$$\dot{\mathbf{x}}_2 = \mathbf{A}_2\mathbf{x}_2 + \mathbf{B}_2\mathbf{u}_2 + \mathbf{h}_2\mathbf{f}_c \qquad (11)$$

where $\mathbf{x}_i^T = [x_i, \dot{x}_i]$ and $\mathbf{f}_c$ is the coupling force and is defined as $\mathbf{f}_c = \mathbf{K}_c(\mathbf{x}_1\text{-}\mathbf{x}_2)$ where the coupling matrix $\mathbf{K}_c = \mathrm{diag}(k_2, b_2)$. We can rewrite (11) by incorporating the coupling force relation,

$$\mathbf{x} = \mathbf{A}\mathbf{x} + \mathbf{B}\mathbf{u} \qquad (12)$$

where

$$\mathbf{A} = \begin{bmatrix} \mathbf{A}_1-\mathbf{h}_1\mathbf{K}_c & \mathbf{h}_1\mathbf{K}_c \\ \mathbf{h}_2\mathbf{K}_c & \mathbf{A}_2-\mathbf{h}_2\mathbf{K}_c \end{bmatrix}, \quad \mathbf{B} = \begin{bmatrix} \mathbf{B}_1 & \mathbf{0} \\ \mathbf{0} & \mathbf{B}_2 \end{bmatrix}, \quad \mathbf{x} = \begin{bmatrix} \mathbf{x}_1 \\ \mathbf{x}_2 \end{bmatrix}, \quad \mathbf{u} = \begin{bmatrix} \mathbf{u}_1 \\ \mathbf{u}_2 \end{bmatrix}$$

According to the proposed control scheme, we choose the control

$$\mathbf{u} = -\,[\mathbf{g}_1 \; \mathbf{g}_2 \; \mathbf{g}_3 \; \mathbf{g}_4]\;\mathbf{x} \qquad (13)$$

such that the resulting overall equations of motion become

$$\dot{\mathbf{x}} = \begin{bmatrix} \dot{\mathbf{x}}_1 \\ \dot{\mathbf{x}}_2 \end{bmatrix} = \begin{bmatrix} \mathbf{A}_1-\mathbf{h}_1\mathbf{K}_c-\mathbf{B}_1\mathbf{g}_1 & \mathbf{h}_1\mathbf{K}_c-\mathbf{B}_1\mathbf{g}_2 \\ \mathbf{h}_2\mathbf{K}_c-\mathbf{B}_2\mathbf{g}_3 & \mathbf{A}_2-\mathbf{h}_2\mathbf{K}_c-\mathbf{B}_2\mathbf{g}_4 \end{bmatrix} \begin{bmatrix} \mathbf{x}_1 \\ \mathbf{x}_2 \end{bmatrix} \qquad (14)$$

Then, the decoupling condition becomes obvious in that the off-diagonal terms have to go to zero, or

$$g_2 = B_1^* h_1 K_c \text{ and } g_3 = B_2^* h_2 K_c \tag{15}$$

where * represents a pseudo-inverse. Finally, the decoupled dynamics of mass system 1 (or 2) can be shaped into a desired form by choosing the gains for g_1 (or g_2). The stability of each system can also be guaranteed whenever $(A_i - h_i K_c, B_i)$ for i = 1,2 is controllable.

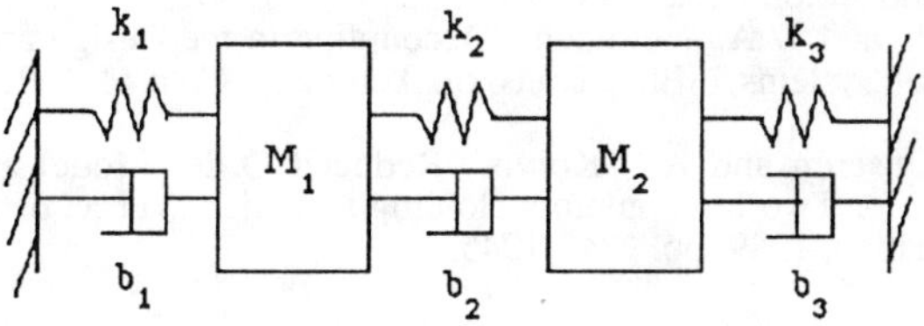

Figure 1. A One-dimensional Coordination

CONCLUSION

A decoupling control scheme for dual-arm coordination is devised. The scheme decouples the dynamics of each arm from the other and that of the object, and utilizes a straight forward full-state feedback with configuration dependent gains. A simple example of a linear one-dimensional coordination is illustrated to demonstrate the usefulness of the control scheme. In the actual implementation, a some form of realizing the configuration dependent gains have to be further investigated along with the issues of stability and performance robustness.

REFERENCES

[1] E. Nakavo, S. Ozaki, T. Ishida and I. Kata, "Cooperational Control of the Anthropomorphous Manipulator MELANM", Proc. of Fourth Int'l Conf. on Architectural Robots, 1974, pp. 251-260.

[2] S. Fujii and S. Kurono, "Coordinated Computer Control of a Pair of Manipulators," Proc. of Fourth World Congress on Theory of Machines and Mechanisms, 1975, pp. 411-417.

[3] J.Y.S. Luh and Y.F. Zheng, "Computation of Input Generalized Forces for Robots with Closed Kinematic Chain Mechanisms," IEEE Journal of Robotics and Automation, Vol. RA-1, No. 2, pp. 95-103.

[4] S. Hayati, "Hybrid Position Force Control of Mult-arm Cooperating Robots," Proc. of IEEE Int'l Conf. on Robotics and Automation, 1986, pp. 82-89.

[5] T.J. Tarn, A.K. Bejczy, and X. Yun, "Coordinated Control of Two Robot Arms," Proc. of IEEE Int'l Conf. on Robotics and Automation, 1989, pp. 1193-1202.

[6] A.J. Koivo and G.A. Bekey, "Report of Workshop on Coordinated Multiple Robot Manipulators: Planning, Control and Applications, IEEE Journal of Robotics and Automation, Vol. RA-4, No.1, 1988, pp. 91-93.

[7] A. Ahmad and S. Luo, "Coordinated Motion Control of Multiple Robotic Devices for Welding and Redundancy Coordination through Constrained Optimization in Cartesian space," IEEE Journal of Robotics and Automation, Vol. RA-5, No. 4, 1989, pp. 409-417.

[8] C.R. Carignan and D.L. Akin, "Cooperative Control of Two Arms in the Transport of an Inertial Load in Zero Gravity," IEEE Journal of Robotics and Automation, Vol. RA-4, No. 4, 1988, pp.414-418.

[9] H. Seraji, "Coordinated Adaptive Position/Force Control of Dual-Arm Robots,

IASTED Int'l Journal of Robotics and Automation, Vol. 3, No. 3, 1988, pp.140-149.

[10] C.O. Alford and S.M. Belyen, "Coordinated Contro of Two Robot Arms, Proc. of IEEE Int'l Conf. on Robotics and Automation, 1984, pp. 468-473.

[11] P. Ro and K. Youcef-Toumi, "A Reference Model Following Control for Motion Coordination of Two Dynamically Interacting Systems," Proc. of ASME WAM89, San Francisco, Dec. 1989.

[12] P. Ro and K. Youcef-Toumi, "A Robust Motion Coordination of Two Dynamically Interacting Systems with Model Uncertainties and Bounded Disturbances," Proc. of 28th IEEE Conf. on Decision and Control, Tampa, Dec. 1989.

[13] M. Cotsaftis and C. Vibet, "Decoupled Control for 2-D N-Link Biped Walking System," Robotics and Autonomous Systems 5, pp. 97-107, 1989.

[14] P.L. Falb and W.A. Wolovich, "Decoupling in the Design and Synthesis of Multivariable Control Systems," IEEE Trans. on Automatic Control, Vol. AC-12, No. 6, Dec. 1967.

[15] M.A. Unseren and A.J. Koivo, "Reduced Order Model and Decoupled Control Architecture for Two Manipulators Holding an Object," Proc. of IEEE Conf. on Robotics and Automation, 1989, pp. 1240-1245.

EXPERIMENTAL VALIDATION OF BI-ARM OPERATIONS FOR SPACE-RELATED ASSEMBLY TASKS

H. BRUHM
Technical University of Darmstadt
Department of Control Engineering
Control Systems Theory and Robotics Group
Schloßgraben 1, 6100 Darmstadt (F.R. of Germany)

A. SEYFFER and **S. WIENAND**
ISRA Systemtechnik GmbH
Mornewegstr. 45a, 6100 Darmstadt (F.R. of Germany)

ABSTRACT

This paper summarizes the experience gained from laboratory demonstrations that were performed in support of the Bi-Arm-Servicer (BIAS) Study [1] under contract from the European Space Agency. In order to cover a reasonable range of the required operational capabilities, the following two reference tasks were defined for the study: The first one is a structural assembly task where a long metal tube has to be inserted into a node element under narrow clearance conditions, while the second one is concerned with electrical connections and requires mating of the two parts of a connector which are attached to the loose ends of the cables to be connected. The paper describes the task elements, hardware and control structure of the laboratory system, the considerations leading to the solution strategies finally employed, and the results of the experiments performed. A major point of interest in the experiments was to determine the type, amount and quality of sensory information required in the operations, as well as suitable control strategies for its exploitation. The set-up used included a vision system and a wrist force/torque sensor in one arm. The two sensors were found to be complementary in the sense that the vision system can quickly provide the rough position information needed in the approach phase of an assembly motion while the slow fine positioning motions in the insertion phase are guided by the force/torque sensor.

INTRODUCTION

In recent studies, the European and American space agencies have developed concepts of (tele-)robotic systems with cooperating arms which promise increased versatility and superior performance over comparable single arm systems for the in-orbit operations envisaged in connection with construction, operation and maintenance of the International Space Station and its accompanying unmanned elements [2,3]. In view of the very limited experience with multi-arm systems in terrestrial applications, early ground-based feasibility demonstration of certain key operations is an important element in these activities.

The objective of the experiments described here was to explore the applicability and the value of a dual-arm system for automatic execution of typical tasks envisaged in the frame of the described scenario. A "low-cost" system equipped with just one wrist force/torque sensor and a camera was used for this purpose in a first step. Two tasks were selected for the experiments where cooperative action of the two arms was expected to play an important role in some, but not necessarily all execution steps: a structural assembly task requiring the manipulation of a large object, and mating of the two parts of an electrical connector, e.g. in the course of a repair task.

The study objective as formulated above implied the clarification of the following questions, in particular:

- Identification of the steps in a task execution sequence where cooperative action of two arms is required or advantageous.

- Identification of the sensor information requirements of the tasks and development of suitable control strategies exploiting this information.

- Development of task solution strategies of minimum complexity, making use of independent and coordinated operation of the two arms and of sensor information and control functions as appropriate.

- Analysis of the complexity added by the cooperative use of two arms.

- Analysis of typical problems that occurred during task execution.

- Analysis of duration and reliability of task execution.

As far as the last question is concerned, the results are valid only for our specific system, due to the highly implementation-dependent nature of the properties considered. They are not discussed in this paper.

TASK DESCRIPTION AND SOLUTION STRATEGIES

In the first task the two parts of a space-qualified electrical connector (see fig. 1) had to be mated. The type of connector used is designed for manual operation by an astronaut in EVA, with space gloves on. No provisions for automatic operation have been incorporated in the design. As a result, suitable gripping surfaces for the robots were missing and there was no visual clue indicating correct rotational alignment of the connector parts. The correct orientation had to be determined by the robot from tactile

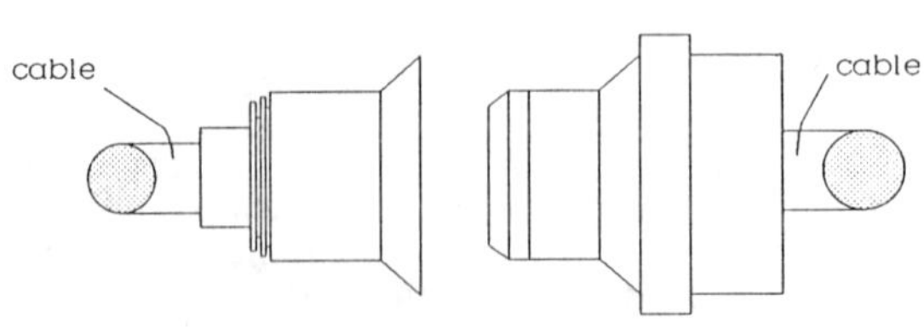

Fig. 1: Connector

information (as by the human operator in manual operation). This and the high mating force of 60 N required to engage the connector pins were additional difficulties of this task as compared to a similar one performed by Dauchez and Kokkinis [4].

The solution strategy developed for this task employed the arm without sensory equipment as an "active fixture" which would grasp the left part of the connector and present it to the other arm in the most convenient position and orientation, thereby allowing to achieve maximum dexterity (see [5] for the definition of dexterity measures) for the subsequent sensor-controlled insertion of the right part of the connector by the other arm. The initial situation for this task, characterized by the accuracy with which position and orientation of the two connector parts were known to the system, was so well-defined that all steps up to the insertion could be performed without sensory support. The gripper fingers were designed in such a way that closure of the grippers would further reduce uncertainties, so that the relative positioning error of the two parts in the approach position for the insertion could be guaranteed to be smaller than the width of the connector chamfer.

In the second task a structural beam element of 1.2m length had to be inserted with one end into a receptacle under narrow clearance conditions (g6/H7, diameter: 32 mm). For safety reasons and in view of the size of the object it was required that two points of the beam should always be kept in a well-defined position, i.e. it had to be transported from its initial position to the approach position near the receptacle with a coordinated bi-arm motion. The insertion itself, however, had to be done by the sensor-equipped arm alone, as the information provided by a single force/torque sensor in the wrist of one arm is not sufficient to determine interaction forces with the environment as long as the other arm is still in contact with the load. A solution strategy was found which reconciled these conflicting requirements through the definition of an approach position where the tip of the beam was already intruding a little distance into the chamfer of the receptacle. In this position the beam could be released by the arm having held it near its front end without a risk of losing the positioning accuracy, as the position of the tip of the beam would already be constrained by the receptacle.

The initial situation in the structural assembly task was distinguished by the assumption of a higher degree of uncertainty, with the position of the receptacle being only approximately known. The exact location of the approach point had to be determined before the beam was grasped. For this purpose the arm with the camera was moved to a position above the scene, pointing it towards the receptacle. An image processing system was used to identify the receptacle and extract the coordinates of the approach point from the visual information.

The choice of the gripping position of the sensor-equipped arm on the beam was a compromise between the following two requirements:

(1) minimization of the lever arm from the sensor to the tip of the beam in order to avoid excessive torques on the sensor during the mating operation

(2) maintaining good control over the orientation of the beam in the mating phase where it is held by only one arm, which required the gripping point to be some minimum distance away from the tip.

A gripping position 400 mm away from the tip was selected. The position of the other arm was at 100 mm from the tip.

SYSTEM STRUCTURE AND CONTROL CONCEPT

The bi-arm system with which the experiments were performed was built up using only commercially available hardware components, in the configuration shown in fig. 2. The two robot arms - a manutec r3 and a KUKA 163/3 - were controlled by Siemens RCM 3 robot control systems. The RCMs were connected to a central PC system - accommodating the coordination, sensory control and user interface functions - via ISRA-PRCI (PC Robot Control Interface) dual-port memory interfaces. Communication with the ISRA-Vision image processing system and the force/torque sensor system (measurement range: 500 N / 50 Nm) was based on RS232 interfaces.

The control system structure as shown in fig. 3 is characterized by independent motion execution modules for each arm including path interpolation, sensor supervision and sensor control functions. Despite the local nature of their function the motion execution modules are resident on the central computer. They generate cartesian position commands which are transmitted to the RCM control units via the PRCI interface with a 32 ms cycle time. The RCMs perform coordinate transformations and joint level control functions.

Coordination of the two arms is the responsibility of the central planning module and is achieved through the generation of coordinated task execution plans for both sides. A task execution plan specifies a sequence of actions (arm motions, gripper action, activation of sensor functions etc.) and the associated parameters such as control mode, goal position, speed and acceleration profile of a motion, set-points for the force control loops and threshold values for the sensor supervision functions.

Coordinated and non-coordinated control modes had to be implemented for the execution of the experiments. Independent operations of the two arms, which constituted the major part of the task execution plans, were performed under the following non-coordinated control modes which could be activated on a single arm only or on both arms simultaneously:

(a) Pure position control: The arm follows its nominal path as described by the task execution plan.

(b) Position control with sensory supervision: The arm again follows a nominal path, but with the possibility of the motion being interrupted by a pre-defined sensor event (e.g. force/torque sensor readings exceeding a given threshold).

(c) Hybrid position/force control: The arm performs a nominal motion with continuous corrections from the sensor control loop(s).

(d) Hybrid position/force control with sensory supervision: Combination of (b) and (c). The motion is both sensor-controlled and sensor-supervised.

In the coordinated control modes, constraints on the tip positions of the two arms relative to each other had to be taken into account continuously during

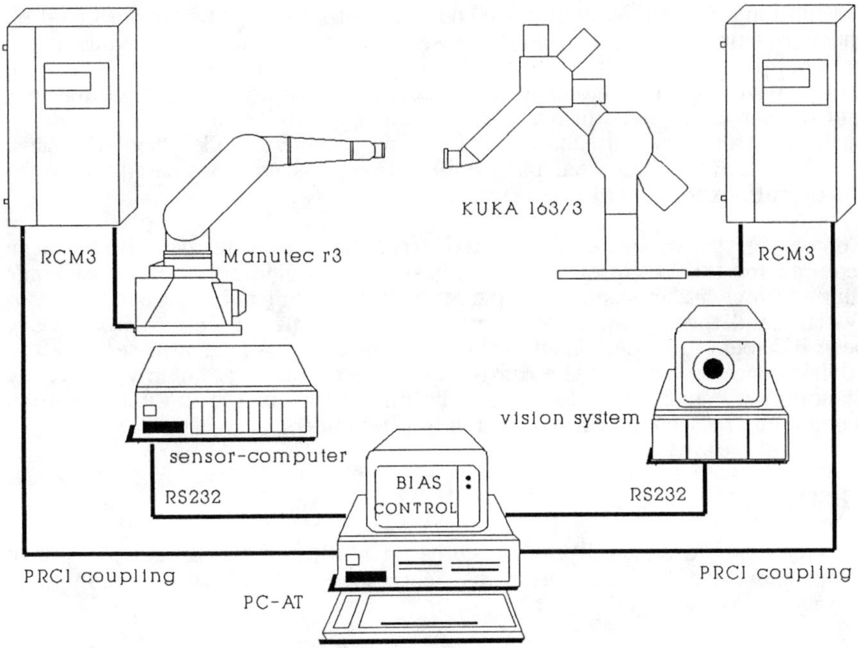

Fig. 2: Hardware configuration of the experimental system.

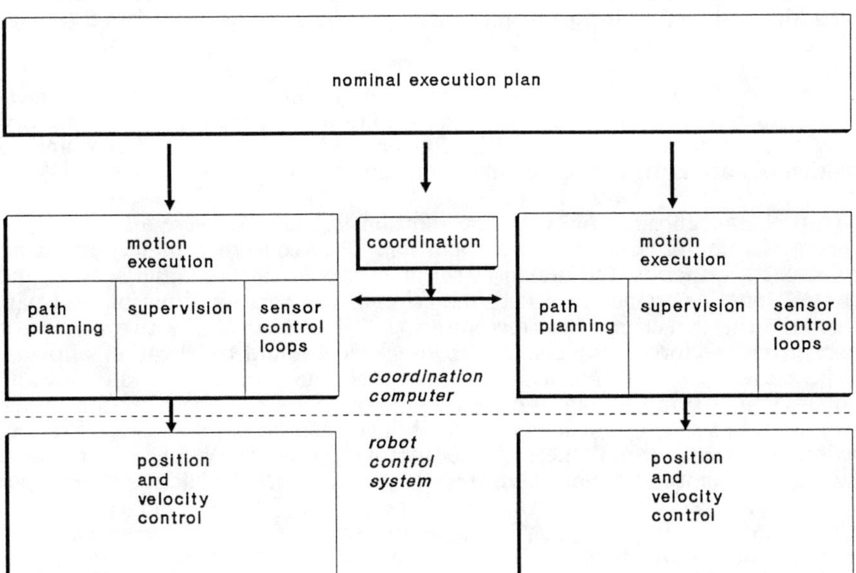

Fig. 3: Control structure of the experimental system.

the motion. The use of symmetric coordination strategies being precluded by the hardware constraints (only one sensor), the master/slave mode (with nominal path exchange) was the only coordinated control mode implemented on the system. In this mode both arms execute coordinated nominal motions and the slave arm is additionally force/torque controlled around the nominal path in all degrees of freedom. A "full" master/slave mode where the slave arm does not have nominal path information, was not considered useful for the operations to be performed.

Comparing this hardware and control structure with what would have been required for the execution of similar tasks by a single arm we can conclude that the increase in system complexity was less than two-fold as the sensor systems and the PC implementing the sensory control functions would have been indispensable components also in a comparable single-arm set-up. The additional complexity in the control system structure was mainly caused by planning and supervision functions; the duplication of the motion execution module for the second arm required only little effort.

RESULTS

As a general statement it can be said that neither of the tasks was easy to perform, with only a few problems being specific to the bi-arm approach, however.

The main advantage of using two arms for the connector mating task was their ability to perform preparatory activities (grasping of the two connector parts and moving them to the approach positions) in parallel and the free orientability of the part held by the fixture arm. Also the fact that a single arm would not have been able to perform the task in the given set-up without an additional fixture for one connector part should be mentioned as a (not at all trivial) advantage.

It was not necessary for the execution of the connector mating task to have two arms with the same dexterity and abilities, one of them being needed only as a fixture for one part of the connector. No advantages of coordinated motion of both arms during the insertion could be identified.

The problems encountered with the solution of this task were all <u>not</u> due to the application of two arms in cooperation. They concerned mainly mechanical aspects - such as the design of suitable gripper fingers - and sensing and control functions, where the determination of the contact situation and particularly the detection of correct rotational alignment of the two connector parts from the force/torque sensor readings were found to be rather difficult. (The correct rotational position was marked mechanically by an indexing bolt). A high translational and rotational compliance in the direction of the two lateral axes of the task frame would have been desirable in this situation, paired with low compliance in the mating direction. With the (passive) mechanical compliance not exhibiting these characteristics, active compliance was the only way to achieve the desired behaviour. This was only possible to a certain extent, however, as the active implementation of high compliance in a configuration which is mechanically very stiff poses critical requirements with respect to the bandwidth of the force control loop, the positioning accuracy of the arm and the resolution of the sensor which must be able to measure the high mating forces as well as the low forces that the compliance controller has to deal with.

The beam handling experiment confirmed the anticipated advantages of better positioning accuracy for the task object and reduced wrist torques in the bi-arm configuration. Simultaneous execution of different subtasks was not possible in this case. Master/slave control with nominal path exchange was found to be an appropriate control mode for the coordinated transfer motion, which included both translation and re-orientation of the beam. The successful execution of this operation was partly due also to good kinematic calibration of the system; it was possible, in fact, with the given mechanical compliance of the system, to execute this motion even without force control, using only nominal path exchange.

Execution of the coordinated motion required two fully articulated arms. Having a force/torque sensor on one arm was found to be sufficient for unconstrained motion of the task object. In the case of external contact of the object (with the receptacle) problems in the resolution of forces and torques at the contact point arose as expected. The strategy of retracting one arm from the object and performing the mating operation with the other arm alone worked satisfactorily, despite the relatively long lever arm that the external contact forces had with respect to the sensor. A shorter lever arm would have been more appropriate in view of the given ratio of force and torque measurement ranges of the sensor. Alternatively, the mating operation could have been performed with a symmetric system (with sensors in both wrists) in a coordinated bi-arm operation where the torque measurement range of the sensors is less important, given the possibility to generate torques by pairs of forces with lines of action having a lever arm with respect to each other..

The relatively high stiffness of the system did again limit the achievable gain in the force control loops. Higher mechanical compliance would have relieved this constraint and would of course have meant also higher tolerance to kinematic errors in the case where the coordinated motion was performed with pure nominal path exchange.

CONCLUSIONS

It could be demonstrated by the experiments that rather complex operations can be performed with a low-cost version of a bi-arm system using just one force/torque sensor. This is especially true for cases where one arm performs only auxiliary functions such as workpiece presentation. Manipulation of large objects in free motion by coordinated parallel action of two arms is another class of operations where such a system can be used with advantage. The master/slave control mode with nominal path exchange or - under very opportune circumstances - pure nominal path exchange were shown to be applicable in these cases. Operations requiring control of external forces on the payload did confirm the known limitations of the master/slave strategies, however. Such operations could only be performed in a single-arm mode.

Lower mechanical stiffness would have allowed for higher gains in the force control loops, with corresponding improvements in performance. These could also be achieved by a reduced sampling time and/or dead time of the force controller. Sampling times of 5-10 ms are considered much more appropriate for the operations performed here than the 32 ms dictated by the RCM control units. The dead time in the present experimental system was approximately 32 ms and could be reduced to ca. 10 ms.

The resolution of the force/torque sensor(s) and a ratio of the force and torque measurement ranges which is matched to the task are two other factors affecting the performance of active compliance. The measurement range for the torques being too small compared to that for the forces is a problem which is typically encountered in the manipulation of large objects under external motion constraints, as for example in the case of the beam insertion task. Bi-arm systems with only one sensor cannot provide a solution to this problem; other sensor designs or the use of a symmetric multi-arm system and corresponding control modes [6-8] are required here.

The information provided by the camera was complementary to that from the force torque sensor. The combination of the two sensors allowed a significantly higher speed of execution for the beam mating task than if only the force/torque sensor had been used, in which case it would have been necessary to determine the location of the hole in the receptacle by means of a sensor-supervised and sensor-controlled search motion. Besides, we are not sure if it would have been possible to detect the hole with this motion being performed as a coordinated bi-arm operation in a (sensor-supervised) master/slave control mode. Particular advantages of a dual-arm system in connection with the use of a camera may be achieved if one arm is used to position the camera and point it towards the area of interest while the other one performs an operation to be supervised through the camera or presents an object to be inspected in the optimal position and orientation with respect to lighting conditions.

Our recommendation concerning the design of bi- and multi-arm systems is to use relatively stiff arms partly or fully equipped with flexible force/torque sensors which constitute instrumented mechanical compliance devices, implementing passive compliance together with a possibility of measuring the deflections. Compensation of these deflections through the control system allows robustness of the end-effector-related positioning accuracy with respect to external forces and torques to be maintained while the stability boundary of the force control loops is shifted to higher gain values. Using a single force/torque sensor and a master/slave strategy for the control of coordinated motions is often sufficient for a bi-arm system, except when a particular force distribution between the arms is to be realized [9-11] or if contact forces between payload and environment are to be controlled.

REFERENCES

1. R.H. Bentall, Bi-Arm Servicer (Work Statement), Appendix 1 to ITT A0/1-1946/87/NL/MAC. ESA/ESTEC, Noordwijk (The Netherlands), July 1987
2. R.H. Bentall and D. Kassing "Man-tended options for European space robotics", 1st European In-Orbit Operations Technology Symposium, Darmstadt (F.R. of Germany), 7-9 Sept. 1987, 475-483
3. P.S. Schenker, "Program objectives and technology outreach", Space Telerobotics Workshop, JPL, Pasadena (CA, USA), 20-22 January 1987, vol. I, 3-17

4. P. Dauchez and T. Kokkinis, "Assembly 'in space' with a two-arm robot", IASTED Conf. on Robotics & Automation, Santa Barbara (CA, USA), Nov. 1989, 5-9
5. C.A. Klein, "Use of redundancy in the design of robotic systems", 2nd Int. Symp. on Robotics Research, 1985, 207-214
6. M. Uchiyama and P. Dauchez, "A symmetric hybrid position/force control scheme for the coordination of two robots", IEEE Int. Conf. on Robotics and Automation, Philadelphia (PA, USA), 24-29 April 1988, 350-356.
7. T.E. Alberts and D.I. Soloway, "Force-control of a multi-arm robot system", IEEE Int. Conf. on Robotics and Automation, Philadelphia (PA), 24-29 April 1988, 1490-1496
8. H. Bruhm, J. Deisenroth and P. Schädler: "On the design and simulation-based validation of an active compliance law for multi-arm robots", Int. J. of Robotics and Autonomous Systems 5 (1989), 307-321
9. D.E. Orin and S.Y. Oh, "Control of force distribution in robotic mechanisms containing closed kinematic chains", J. of Dyn. Systems, Measurement and Control, 102 (1981), 134-141
10. H. Bruhm, "Coordinated motion planning and optimal force distribution for robots with multiple cooperating arms", IFAC Symposium on Robot Control, Barcelona (Spain), 6-8 Nov. 1985, 271-276
11. Y.F. Zheng and J.Y.S. Luh, "Optimal load distribution for two industrial robots handling a single object", IEEE Int. Conf. on Robotics and Automation, Philadelphia (PA, USA), 24-29 April 1988, 344-349.

ACKNOWLEDGEMENT

The work reported in this paper was performed for the European Space Agency (ESA) under contract no. 7406/87/NL/MAC.

Mechanical Design of Quadruped Walking Machine

Yong-San Yoon, Hyung-Joo Hong, Woong-Hee Shon and Zeung-Nam Bien

Korea Advanced Institute of Science and Technology

Abstract

The design procedure of a quadruped walking machine is described with much emphasis on the leg design. It includes the specification of the walking robot, type synthesis of the leg mechanism, walking volume, dimensional synthesis of the pantograph mechanism, stiffness, and selection of motors and power transmission systems. An experimental quadruped walking robot named KAISER II was constructed based on this design procedure. Currently it can walk at 0.10 m/s in static gait on level ground with duty factor of 0.8 and climb up the stairs of 50% grade. The body weight of the robot is 80 Kg including payload of 10 Kg.

1. Introduction

Since the overall motion of the walking robot of concern is mostly based on the leg motion, the performance of the robot is much dependent upon the leg design. The design procedure is much influenced from the problems found in the previous test model. The test model weighs 75 kilograms and walk at 630 mm/min on the level ground with duty factor of 0.9. The most severe problems encountered with that test model was the frictional resistance of the transmission system and structural deformation making it impossible to realize the ideal walking of duty factor 3/4.

In this study, the static level walking speed of 0.15 m/s with duty factor of 0.8, body weight of about 70 kilograms including 10 kilogram payload, and the stairs of 50% grade to climb up were set as the target of the walking robot design.

2. Mechanical Design

The design of the leg influences the performance of the walking robot in mobility, energy efficiency, walking speed, control method, etc. It includes type synthesis to determine the type and dof of the mechanism and dimensional synthesis based on energy efficiency, walking speed, walking volume and stiffness of the leg. Detailed design procedures are to be explained in the following sections.

2. 1 Design Specification

The representative mechanical characteristics of the walking robot are walking speed, obstacle crossing capability, body weight including payload, and deflections. Even though the design specification should be determined from some given tasks, we opted that non-practical specifications previously mentioned because of the experimental nature of our project and experience with the test model. The walking speed of 0.15 m/s is 15 times the speed of our test model. In selecting an obstacle, we chose a regular stairs of 50% grade. Since each leg of the walking robot should carry its body weight and payload during walking causing large deflection at the leg and consuming a lot of energy in gait, we tried to keep the body weight as light as possible. The maximum deflection of the body during gait is set to be 5 mm to achieve the duty factor of 0.8 and to get a modest appearance.

2.2 Type Synthesis of Leg Mechanism

The number of legs of walking robots may assume any number. However biped, quadruped and hexapod are the most popular ones currently. During static gait, some legs are in swing and some legs are in stance. It is desirable that each leg in swing should have three dof to reach any arbitrary point and the supporting legs should provide six dof to move the body in that instance. For a biped robot, one leg may be in swing and the other leg may be in stance thus requiring six dof for each leg. For a quadruped robot, one or two legs may be in swing and two or three legs may be in stance thus requiring three dof for each leg. For a robot with more than four legs, it requires more than 12 dof because each leg requires minimum 3 dof. Thus it seems to be that 12 dof is the minimum for a general static gait motion and the quadruped is the most statically stable with that minimum dof.

The leg mechanism of the walking robot should satisfy the following conditions:

i) Three degrees of freedom
ii) Minimal potential and kinetic energy loss during gait
iii) Structural strength and stiffness
iv) Proper walking volume with moderate leg size
v) No interference and smooth run during operation
iv) Simple calculation for control

In the test model, we used a 3-D pantograph mechanism as shown in Figure 1. That mechanism meets most of the conditions cited above but the stiffness requirement due to the x-y table to provide lateral motion. Thus we switched to the 4-bar mechanism as shown in the Figure 1 for the lateral motion because it can provide a fairly horizontal lateral motion of the foot as shown in Figure 2 with adequate amount of structural stiffness. The vertical structural deflection was the major concern during gait and it could be reduced easily with this new model because the forces acting on the four bar mechanism is much smaller than that on the x-y table of the 3-D pantograph mechanism and the vertical motion of the coupler of the four bar mechanism is not magnified as in the case of the other. For the A-P(Anterior-posterior) and vertical motions, a planar pantograph mechanism is chosen as other walking robots in Figure 1[1,3].

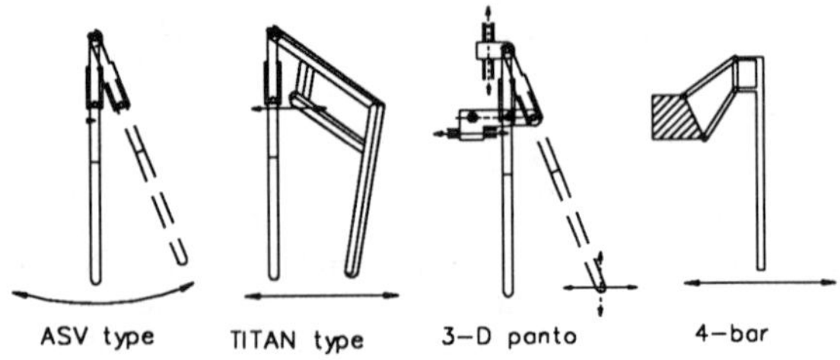

Figure 1. Application of Pantograph For Side Walk

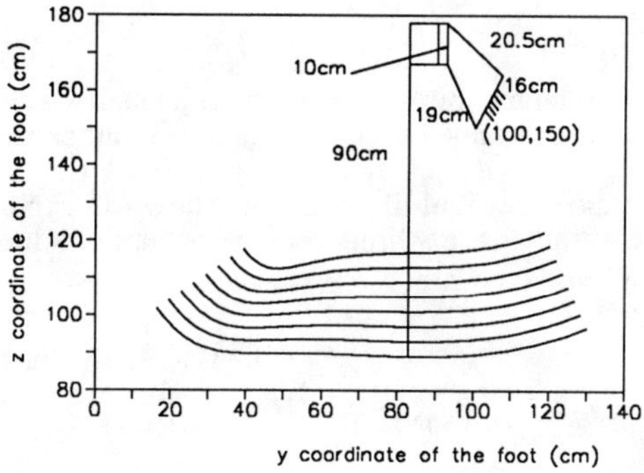

Figure 2 Straightness of Foot Trajectory

2.3 Optimal Design of Pantograph For Speed Improvement

In the case of static level walking for a quadruped walking robot, the walking speed is determined from the forward swing velocity of the foot because forward swing foot requires speed three times faster than stance foot. Accordingly the problem of the walking speed improvement may be summarized as the one that maximize the forward swing velocity of the foot for a given motor capacity by changing the leg geometry and the driving system.

The design parameters of the optimization problem are the leg geometry represented by the four links, stroke, body height and the gear ratio of the power transmission system. About 50% of speed improvement was obtained through the design change and the specific resistance was reduced by 38 %. The optimally designed leg shows greater magnification ratio of the pantograph and larger stroke as shown in Figure 3. This optimal design result is not directly used in the final design but used as a guide to get a faster leg.

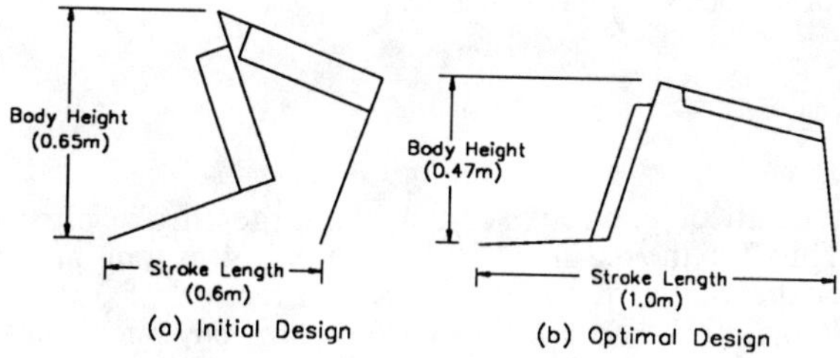

Figure 3. Optimized Leg Configuration[4]

2.4 Walking Volume

The maximum manageable obstacles such as stairs, slopes, ditches, etc are determined by the walking volume of the robot and its relative position with respect to the body. When the point P is driven along the x-axis(A-P direction) and the point Q is driven along the z-axis(Vertical direction) as shown in Figure 6, the foot of the pantograph mechanism may have a limited range of motion due to the limitation of transmission angle as follows:

$$\frac{l_1^2+l_2^2-2\cdot l_1\cdot l_2\cos\psi}{(R+1)^2} \le \frac{X^2}{(R+1)^2}+\frac{Z^2}{R^2} \le \frac{l_1^2+l_2^2+2\cdot l_1\cdot l_2\cos\psi}{(R+1)^2}, \quad 0 \le \psi \le \frac{\pi}{2} \tag{1}$$

The gait cycle for a leg may be divided ideally by the foot trajectory as shown in Figure 4. The swing phase begins by moving the foot up from A and swings forward with the maximum forward velocity of V_{max} and finally touch down to B. The stance phase begins from the foot strike at B moving the foot backward with the ideal foot velocity of $(1-\beta)V_{max}/\beta$ until the foot-off at A.

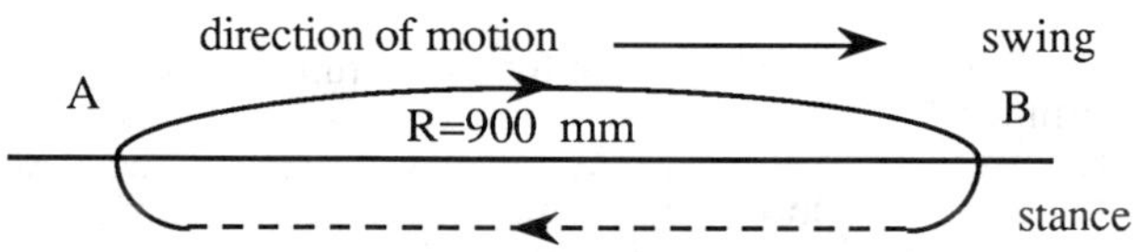

Figure 4. Foot Trajectory with respect to Body

Then to achieve the walking velocity of 0.15 m/s, the stroke R of 0.9 m and duty factor β of 0.8 with maximum leg swing velocity of 1 m/s in x-axis and 0.1 m/s in z-axis can be considered. In this case, the gait cycle takes

$$T_{tot} = T_{AB} + T_{BA} = \frac{R}{V_{max}} + \frac{R\cdot\beta}{(1-\beta)V_{max}} \tag{2}$$

And the velocity of the body is represented by

$$V_{body} = \frac{R}{\beta\cdot T_{tot}} \tag{3}$$

As shown in the result of previous section, the larger stroke provides the faster walking speed. Thus for the assumed body size of 1.4 m long and 1.0 m high at standing posture, the range of motion in A-P direction of 1.0 m seems to be reasonable. The height of the walking volume and the body pitch can be determined from the requirement that it should be able to climb up the stairs of 50% grade. To climb up the stairs, the height of the walking volume should not be less than 650 mm as shown in Figure 5. To provide maximum stability margin in the wave gait of a quadruped walking machine, the relation between the stoke R and the body pitch P should be as follows [2]:

$$\frac{R}{P} = \frac{\beta}{3\beta - 2} \tag{4}$$

In most cases, the value of R/P for maximum stability margin is greater than 1. It means that the walking volume should be overlapped to increase the stability margin in the wave gait but it causes the problem of interference between fore and hind legs. As the ratio is not much greater than 1, the body pitch and the step length are taken to be the same in this design.

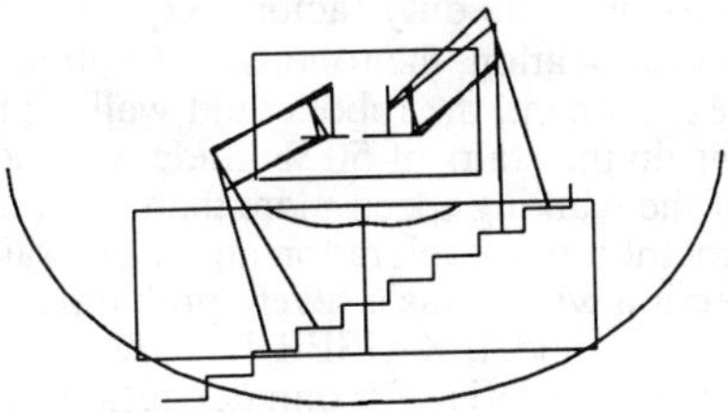

Figure 5. Stair Climbing

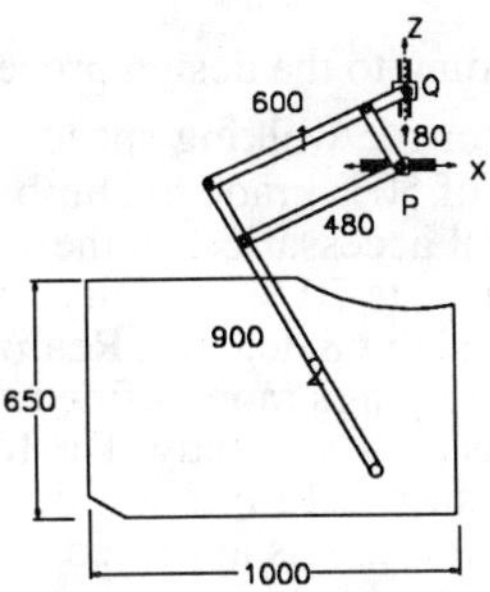

Figure 6. Selected Walking Volume

Then the walking space in x-z plane should be taken as the intersection of the rectangle of 1.0 m by 0.65 m with the annular region bounded by those two ellipses given by Equation (1). The rectangle is related to the limits of slides in x and z axis. Since the upper half of the ellipse can not be used because of constraints on the driving axis, the selection should be in lower ellipse. With the constraint that the foot position during stair walking should be within the walking space as shown in Figure 5, it is determined as shown in Figure 6.

2.5 Stiffness Design

As explained in the previous section, the large amount of the deflection of the foot with fair amount of body weight may severely deteriorate the walking speed. Increasing the stiffness of the leg, however, will demand additional weight increase of the system. So the maximum deflection of the foot is limited as 5 mm when a external load of 70 kg is applied to the foot. The deflection is checked with a finite element analysis for the assumed leg system including driving system. The vertical deflection of foot of the pantograph is mostly affected by the link 1 and link 2 and it is suggested to use a composite material for those links.

2.6 Selection of Motors and Power Transmission Systems

The driving force for A-P motion is mainly due to the inertia force and the frictional force in static level gait or the work against gravity with the body pitching angle of 10 degrees. The leg mechanism selected here is gravitationally decoupled so that the ground reaction may not effect the A-P axis driving in ideal case. However it may not walk sometimes in perfectly horizontal posture and the body pitching angle of 10 degrees is taken into account for the selection of the motors for the A-P axis driving. The maximum force on a foot during static gait may be limited to the half of the body weight as long as the overall posture is in horizontal, which is hard to maintain. Also there are some dynamic effect on the force. Thus it is assumed that maximum force on a foot may reach to the body weight. Frictional force is not easily predictable because of unclear friction mechanism. So we made a mock-up of the leg system to measure the friction force under the realistic loading condition. For the vertical axis drive, speed is not seemed to be critical and thus a large gear reduction is introduced. This is the case also for the lateral axis driving.

3. Results and Discussions

According to the design procedure described here, a quadruped walking robot is designed for the walking speed of 0.15 m/s with the duty factor β equals to 0.8 and the stairs of 50% grade to climb up. After fabrication, the robot weighs about 80 kg with several accessaries. In the walking experiment, the robot could walk as fast as 0.1 m/s using its 70 % capacity and climbs up the stairs of 50 % grade with body pitch angle of about 8 degrees. Realization of the walking speed more than 0.1 m/s in the static gait requires more refined control method for acceleration and deceleration and remains as a further study. The foot deflection which was a severe problem in the test model is successfully eliminated in the new model, KAISER II. However the vibration about roll axis appeared during gait. The criteria of 5 mm deflection at the foot looks reasonable because such amount of deflection does not much disturb the walking in static gait.

Acknowledgements

The support of the Korean Ministry of Science and Technology through generous research grant is gratefully acknowledged.

Reference

[1] Song, S.M., Kinematic Optimal Design of A Six-Legged Walking Machine, Ph.D Dissertation, The Ohio State University, 1984

[2] Song, S.M. and Waldron, K.J., "Geometric Design of a Walking Machine for Optimal Mobility," J. of Mechanisms and Transmissions and Automation in Design. Vol.109, pp.21-28, 1987

[3] Hirose S., Masui, T. and Kikuchi, H., "TITAN III:A Quadruped Walking Vehicle - The Structure and Basic Characteristics," Proc. of the 2nd Int'l Symp. of Robotics Research, MIT press, pp325-331, 1984

[4] Yoon, Y.S. and Hong, H.J., "A Mechanical Design of a Quadruped Walking Machine for the Fastest Speed in Static Level Walking," to be published on Trans. of SME,1990.

The Mobile Robot Path Planning Using Embedding Methodes

Henry Tominaga and Behnam Bavarian
Department of Electrical and Computer Engineering
University of California, Irvine
Irvine, California 92717

Absract

Robot path planning problem in a two-dimensional Euclidean workspace is formulated as a variational problem. Given the dynamic characteristics of the robot and a set of obstacle constraints, the objective is to solve the problem of bringing the robot from some starting point to a desired destination while optimizing a cost function. In general we can show the existence of many local solutions to the optimization problem. we will take advantage of the fact that the topological space in a higher dimensional Euclidean space is simply-connected even when obstacles ("holes") are added, and embed the problem in a three-dimensional space to obtain the unique global optimum solution. Several simulations are carried out to find minimum time solution through obstacles with the mobile robot defined by simple dynamics.

1. Introduction

The goal of this paper is to systematically formulate the problem of mobile robot path planning as an optimal control problem. In general there has been two approaches to the problem of mobile robot path planning in an environment which is cluttered with obstacles. One approach is to use simplified rules based on the Euclidean geometry and straight line path segements or simply rule-based approach. The second approach is to formulated the problem analytically and try to solve the problem for the "best" solution or path. For implementation the rule-based approach shall be the choice, since for a particular mobile robot with limited sensors and computational capability sophisticated algorithems and the analytical model-based approach is not practical. However the analytical approach provides a rigorous way to formulated the problem in a generic sense. Hence, with an analytical generic model we are not limited in exploring various aspects of the problem. For example for many obstacles or few ; or moving obstacles or stationary we only need to add more constraints to the same basic model. Furthermore with this approach more general cost functions can be defined and optimized to find a solution such as minimum time , minimum energy, minimum distance and any combination of them.

The problem of mobile robot navigation can be classified into a hierarchy with increasing level of difficulty in finding a solution depending on the environment and the information available to the robot. A simple classification is the following:

- Navigation in known, static environment;
- Navigation in unknown, static environment;
- Navigation in known, dynamic environment; and

- Navigation in unknown, dynamic environment;

with increasing level of difficulty down the list. Here the class of environments in which the obstacles are moving(non-moving) with respect to an inertial coordinate are called dynamic(static) environments, and the class of environments in which the location and orientation of the obstacles are known(unknown) to the robot are called known(unknown) environments.

In this paper we investigate the first class, navigation in static known environment with the analytical model-based approach. The navigation in static known environment is called the *find path* problem [1] which has been investigated extensively in the past. Lozano-Perez[2] used a configuration space approach by forming the visibility graph of the modified obstacles and the graph is searched for a shortest path between the source and destination. Similar approaches and improvents are reported in[3],[4], [5]. Another approach to the *find path* problem is to model the open space environment using medial axis transfroms[7] or voronoi diagrams[6]. In all the above solutions the basic problem is simplified to a list or a graph representation and simple rules and heuristics are developed to come up with an answer.

Our approach is to formulate the *find path* problem as an optimal control problem by defining simple dynamics for the mobile robot and treating the obstacles as constraints in the state space. It is easy to show that in general the two dimensional Euclidean workspace with obstacles has many local minimum solutions. So we introduce an additional dimension and solve the problem in the new three-dimensional workspace which results in finding the global optimum path.

In the next section the problem formulation is presentedand and section 3 provides the results of several simulations. Section 4 summarizes the paper contributions and the current and future direction of this work.

2. Problem Formulation

The *find path* problem can be formulated as a variational constrained optimization problem as follows. Given the dynamic characteristics of the robot

$$\ddot{\mathbf{x}}(t) = \mathbf{f}(\mathbf{x}(t), \dot{\mathbf{x}}(t), \mathbf{u}(t)) \tag{1}$$

and N obstacles defined by

$$\boldsymbol{\Omega}_i(\mathbf{x}) \qquad i = 1, 2, ..., N \tag{2}$$

find a control history $\mathbf{u}(t)$ that minimizes the cost functional

$$J(\mathbf{u}) = \int_{t_0}^{t_f} \mathbf{g}(\mathbf{x}, \dot{\mathbf{x}}, \mathbf{u}, t)dt \tag{3}$$

subject to the constraints

$$\mathbf{x}(t_0) = \mathbf{x}_0 \; ; \;\; \mathbf{x}(t_f) = \mathbf{x}_f; \tag{4}$$

and the obstacle constraints

$$\mathbf{x}(t) \notin \boldsymbol{\Omega}_i(\mathbf{x}) \;\;\; for \; i = 1, 2, ..., N; \;\;\; and \; \forall t \in [t_0, t_f]. \tag{5}$$

Here $\mathbf{x}$ and $\dot{\mathbf{x}} \in \Re^n$ representing the state variables; and $\mathbf{u} \in \Re^m$ representing the m control variables.

To complete the problem formulation the state equations, Eq.1, The constraints, Eq.2, and the cost functional, Eq.3, should be defined explicitly. We will select the state equations to represent a "point" robot with two degrees of freedom, and two inputs in the two cartesian directions. The optimization criterion is selected to minimize the time, so the final time t_f is necessarily unspecified and the values of the control inputs must be bounded. In order to avoid bang-bang solutions we will implement the control inequality constraints as penalty functions. Finally we need to mathemetically express the admissible regions which are the complements to the area covered by the obstacles. As the first step in formulating the problem we will select to model the obstacles as circular regions. If we have N such obstacles the admissible regions can be expressed as N inequality constraints

$$\begin{aligned} \omega_1(\mathbf{x}(t)) &= (x_1(t) - c_{1x})^2 + (x_3(t) - c_{1y})^2 - {r_1}^2 \geq 0 \\ &\vdots \\ \omega_N(\mathbf{x}(t)) &= (x_1(t) - c_{Nx})^2 + (x_3(t) - c_{Ny})^2 - {r_N}^2 \geq 0 \end{aligned} \tag{6}$$

where c_{ix}, c_{iy} and r_i denote the center coordinates and the radius for the ith obstacles and x_1 and x_3 are the states representing the xy cartesian coordinates.

We will assume that the robot starts at time $t = 0$ at some point other than the origin with zero velocity. The desired destination point is the origin and the robot should have zero velocity at that point. The explicit problem formulation is stated as follows:

Given the state equations

$$\begin{aligned} \dot{x}_1(t) &= x_2(t) \\ \dot{x}_2(t) &= u_1(t) \\ \dot{x}_3(t) &= x_4(t) \\ \dot{x}_4(t) &= u_2(t) \\ \dot{x}_5(t) &= \sum_{i=1}^{N} [\omega_i(\mathbf{x}(t))]^2 \mathcal{U}(-\omega_i) \end{aligned} \tag{7}$$

with the initial conditions

$$x_1(t_0) = x_{10} \; ; \; x_2(t_0) = 0 \; ; \; x_3(t_0) = x_{30} \; ; \; x_4(t_0) = 0 \; ; \; x_5(t_0) = 0. \tag{8}$$

and the x_5 state implements the N inequality constraints with the $\mathcal{U}(-\omega_i)$ being a step function. The state $x_5(t)$ is nondecreasing with time; since the boundary conditions at the initial and final times are zero, $x_5(t)$ equals zero for all t. This forces the constraint $\omega_i(\mathbf{x}(t)) \geq 0$ to be satisfied for all time.

Find the control history $\mathbf{u}(t)$ that minimizes the cost functional

$$J(\mathbf{u}) = \frac{1}{2}\mathbf{x}(t_f)^T \mathbf{G} \mathbf{x}(t_f) + \int_{t_0}^{t_f} [1 + B_1(u_1(t)^2 - u_{max})^2 + B_2(u_2(t)^2 - u_{max})^2] dt \tag{9}$$

where B_1 and B_2 are constant weighting factors and u_{max} is the bound on the magnitude of the control variables.

and two obstacles in the workspace. Figure 3 shows the same workspace with the embedded approach. The initial guess is over the plane and the final trajectory is the global solution. We have carried many more simulations with similar result. However as it is common in any optimization problem the numerical steepest descent solution near the final answer converges slowly. Also the parameters should be properly selected so that the conditions set for the embedding approach to work.

4. Conclusions

We have presented a general approach to the robot navigation *find path* problem. by using the minimum time criterion we have also generalized the traditional minimum distance problem. The technique converges to the global solution without any rule-based heuristics. Currently we are working on the rigorous mathematical formalism for the proof of global convergence for the embedding approach. The formalism can be extended to navigation in three-dimensional spaces by embedding the problem into a four-dimesional space. Furthermore by having the parametric equations of the obstacles be time varying we are extending the result for the solution of the path planning in the dynamic environment. The numerical implementation can also be improved via other approaches such as simulated annealing and parallel distributed processing implementation with neural networks.

References

[1] B. J. Oommen, S. S. Iyengar, N. S. V. Rao, and R. L. Kashyap, "Robot Navigation in Unknown Terrain Using Learned Visibility Graphs. Part I: The Disjoint Convex Obstacles Case," *IEEE J. Robotics and Automation*, Vol. RA-3, no. 6, Dec. 1987.

[2] T. Lozano-Perez, "Spatial Planning: A Configuration Space Approach," *IEEE Trans. Computers*, Vol. C-32, Feb. 1983, pp 108-120.

[3] R. A. Brooks, and T. Lozano-Perez, "A subdivision Algorithm in Configuration Space for Path with Rotation," *IEEE Trans. Syst., Man, Cybern.*, Vol. SMC-15, no. 2, pp.224–233, Mar./ Apr. 1985.

[4] L. Gouzenes, "Strategies for Solving Collision-Free Trajectories Problems for Mobile and Manipulator Robots," *Int. J. Robotics Res.*,Vol. 3, no. 4, pp. 51-65, 1984.

[5] H. T. Moravec, "The Stanford Cart and the CMU Rover," *Proceedings of the IEEE*, Vol. 71, no. 7, pp 827–884, 1983.

[6] F. Mortezaie and B. Bavarian, "Application of the Voronoi Diagram Technique for Mobile Robot Navigation," *Proceedings of the 2nd Int. Symp. on Robotics and Manufacturing*,ASME press, Nov. 1988, pp. 451–457.

[7] R. A. Brooks, "Solving the Find-Path Problem by Good Representation of Free-Space,"*IEEE Trans. System, Mann and Cybern.*, Vol. SMC-13, no. 3, Mar. 1983.

[8] H. N. Tominaga, *A Mathematical Solution to the Two Dimensional Robot Path Planning Problem: Solving a Global Optimization Problem Using Embedding Methods,* MS Thesis, Dept. of Elec. and Comp. Eng., University of California, Irvine, 1988.

The necessary conditions for a locally optimal control can be stated by forming the Hamiltonian which combines the state equations and the cost functional. The details of the formulation is provided in [8]. In summary the conditions are:

1. Satisfying the state equations(7);
2. Satisfying the costate equations from the definition of a Hamiltonian;
3. Satisfying the transversality conditions that give the boundary conditions for the costates;
4. Satisfying the Pontryagin's minimum principle by setting the partial derivative of the Hamiltonian with respect to the control inputs to zero;
5. Satisfying the transversality condition for the final time.

Figure 1 shows a typical solution for two-obstacle problem. The solution depends on the initial guess for the control history. Furthermore the numerical solution converges only locally to the final path and there are in general many local solutions[8]. We propose embedding the two-dimensional problem into a three-dimensional Euclidean space by representing the circular obstacles with spheres and adding penalty function for the third dimensions to the cost functional so that the final path will not have any part outside the plain. Hence the solution will converge to the global minimum. The justification and analysis of this extension are provided in [8]. The changes to the problem formulation will be the following. Two new states representing the z axis in the cartesian coordinate system are added to the state equation (7);

$$\begin{aligned} \dot{x}_6 &= x_7 \\ \dot{x}_7 &= u_3 \end{aligned}$$

with initial conditions, $x_6(t_0) = x_{60}$ and $x_7(t_0) = 0$, and the new cost functional is defined by

$$J(\mathbf{u}) = \frac{1}{2}\mathbf{x}(t_f)^T\mathbf{G}\mathbf{x}(t_f)+\int_{t_0}^{t_f}[1+\alpha(x_6(t))^2+B_1(u_1(t)^2-u_{max})^2+B_2(u_2(t)^2-u_{max})^2]dt \tag{10}$$

where α is a constant and other constants are as defined before. The value of the control $u_3(t)$ is not constrained since it is a fictitious control variable. However the final control history should show $u_3(t) = 0$ for all t. In the next section we will show several simulations for both the two and three dimentional formulation.

3. Simulations

The necessary conditions can be solved numerically using the steepest descent algorithm. The results show that the solution is dependent on the initial trajectory that is specified by the user as the first guess to the solution. This numerically verifies that there are multiple local minimum in our problem and that the global minimum is computed with a "proper" choice of the initial trajectory. Figures 1 and 2 show the two simulations with different initial trajectories for the two dimensional workspace

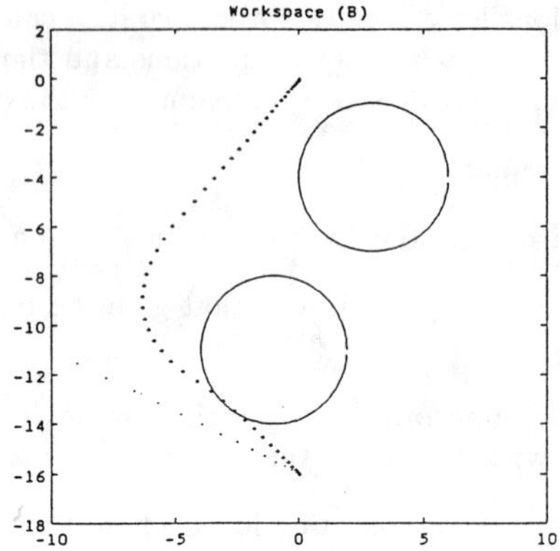

Figure 1: Two Dimensional Formulation; Initial Trajectory to the Left.

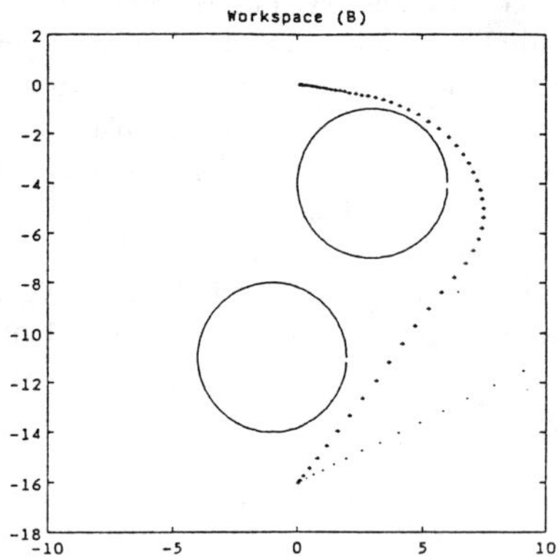

Figure 2: Two Dimensional Formulation; Initial Trajectory to the Right.

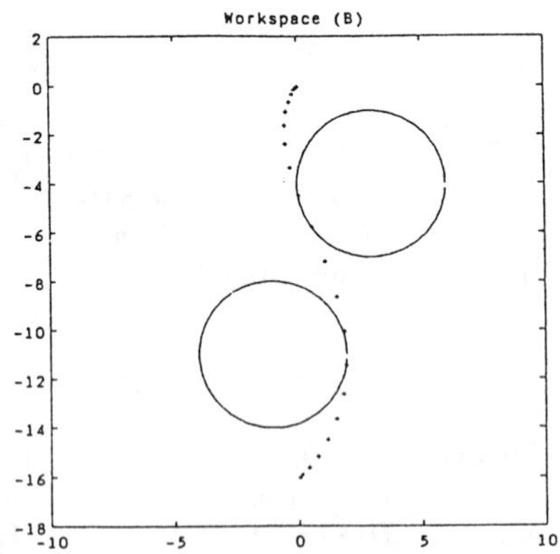

Figure 3: Embedded Formulation; Initial Trajectory over the Plane.

Control Architecture Of An Autonomous Surveillance Mobile Robot For Unstructured Environments

Vivek P. Burhanpurkar, Colin Hinz,
Cyberworks Inc., 6 Kitchener St., Orillia, Ont L3V 6Z9, CANADA.

Ravi Lingarkar and Naresh K. Sinha
Robotics and Intelligent Control Group, Department of Electrical and Computer Engineering, McMaster University, Hamilton, Ont L8S 4L7, CANADA.

ABSTRACT
This paper describes the prototype autonomous mobile robot developed at Cyberworks for the purpose of security and surveillance. Since the robot operates in a dynamic unstructured indoor environment, it has the capability to navigate in the presence of transient objects like people, trash cans, chairs, etc.. The domain to be patrolled is stored as a connected graph, which basically consists of a sequence of digitized waypoints. The robot navigates by deadreckoning between two adjacent waypoints and visual landmark recognition.

Autonomous navigation consists of two components, a global path finding algorithm which uses the connected graph (map) to determine the shortest path between two nodes, and a local collision avoidance scheme to navigate around transient objects lying in the path between two adjacent nodes. The local collision avoidance scheme is a heuristics based navigational algorithm that takes advantage of the knowledge of the robot's immediate surroundings as seen by the onboard multi-frequency rotating ultrasonic array.

The control architecture consists of an onboard distributed computer system to support the parallelism required by the robot to execute all its functions in real-time.

I. INTRODUCTION

The term 'autonomous mobile robot' is applied to a vehicle that is capable of executing a pre-defined generalized task (such as patrolling the corridors of a building), without being pre-programmed for specific details of the task. This further implies that the robot should be capable of reacting to unpredictable dynamic events which may impede the successful execution of the task. To achieve this level of "human scale performance", the robot must self-define its operating domain and reliably locate objects within that domain regardless of size, shape, material composition, etc..

It must collect and process data so as to generate navigational ideas, optimal for the specific situation in which it finds itself at every moment. It must perform these rather "abstract" functions efficiently, quickly, economically and without human intervention either prior to or during the execution of the task. In the case of building sentry, such an ideal robot should transmit video images to a base station while it silently patrols a random pattern through a building without the aid of embedded lines.

Since this research is aimed at practical solutions to some of the problems facing autonomous mobile robots, five key problem areas have been addressed in developing the prototype mobile robot.

1. **Simplifying the way sensor data is processed**

Local data are gathered in real-time, utilizing short range sensors. Data compression and filtering are executed in real-time to decompose the local data into a series of vectors and symbols such that the character of the local domain is retained without the retention of the original vast set of cartesian points. The set of symbolic local data are consolidated into a single global descriptor, which is dynamically optimized into a further reduced set of vectors, and symbols encompassing not only the original object placement data but additionally, navigational assistance data for navigational path planning through the domain.

As the environment is scanned, the data is software pre-processed and salient data is stored in RAM. New salient data is then analyzed in order to reorganize and possibly discard portions of past salient data. Inferences are then made based on this process of data reorganization. The inferences are further verified or modified based on new salient data collected. This process operates in real-time and is the basis upon which the guidance system steers the robot. What data is considered salient and what is not, is determined by a lengthy topological rule hierarchy expert system which acts as kind of data filter.

2. Computational Data Compression

To improve computational efficiencies, new types of data set representations are used. In general, rather than storing many hundred thousand bytes of data representing the cartesian map, the amount of data is reduced in real-time to navigational descriptors which define the nature of free space in the domain through which the robot travels. This results in a reduction of stored and manipulated map data by two to three orders of magnitude. A correspondingly large reduction in data manipulation and interpretation time results.

3. Non-Ideal Data Rationalization

Generic slippage, non-ideal angular visual resolution, random noise errors and other non-idealities corrupt map data. To solve this problem, we have associated probable error "vectors" with highly critical data sets. These vectors combine to reduce or normalize the degree of error in the system and to cancel its effects to some degree.

4. Dynamic Event Rationalization

Dynamic elements in the domain are dealt with by dividing the observable space into regions which are time multiplexed. Comparisons can then be made to filter dynamic from static elements. This approach has had limited success but is significant in its ability to ignore various types of transient dynamic events such as moving people and, to some extent, even noise.

5. Sensing

Camera-based vision systems suffer from cost and computational problems. Ultrasonic systems suffer from ray geometry, angular resolution, signal absorption and response time problems. Laser based rangefinders suffer from signal absorption, specular reflection, ambient light and cost problems. We have developed an ultrasonic system which overcomes some of the non-idealities listed above and takes benefit of the cost advantage of ultrasound. The result is a multi-transducer, multi-frequency, multiplexed, rotating array which is capable of detecting all types of objects regardless of size, shape, geometry, material, etc.. However, the angular resolution of the system still requires optimization.

All the above features have been incorporated in the prototype autonomous surveillance robot. In the following sections we will examine the robot construction, the onboard computer system and the control software for surveillance.

II. ROBOT HARDWARE DESCRIPTION

The mobile unit is a three-wheeled vehicle with two direct drive front wheels (10 inch diameter) providing vehicle propulsion and steering, while a swivel caster gives the necessary support at the rear. The drive wheels are driven by DC motors. The drive motors are independently controlled through pulse-width modulation (PWM). High resolution optical encoders are placed on the wheel periphery to provide odometric information. The encoders provide feedback for every 0.1mm motion of the wheel, thus allowing very precise displacement and velocity information to be used for deadreckoning. The side elevation of the mobile unit is shown in figure 1.

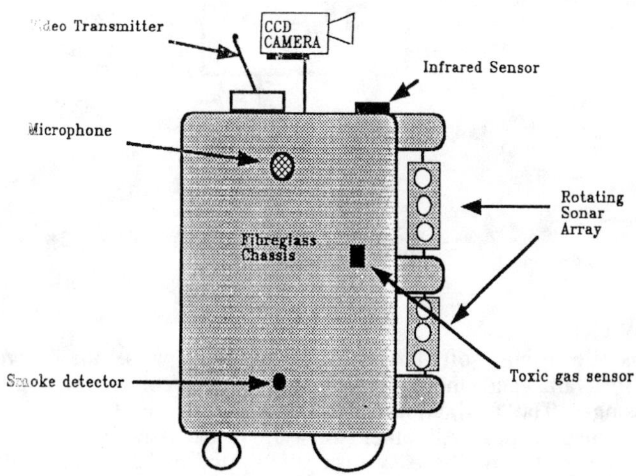

Figure 1. Side view of the surveillance mobile robot.

A twelve element multi-frequency, multiplexed stepper motor controlled rotating ultrasonic matrix array is placed at the front for collision avoidance. Other sensors can be placed as shown in the figure to detect smoke, toxic gases, etc.. The approximate dimensions of the robot are 85 × 47 × 60 cms and it weighs about 350 pounds. The robot is enclosed in a one piece fiberglass shell. The DC motors, stepper motor and the interface electronics are driven by nine 28 AH gel cell deep cycle batteries. These batteries are placed close to the robot's center of rotation to reduce the rotational inertia of the robot in the horizontal plane.

The mobile robot is controlled by two microprocessors arranged in a master/slave configuration as shown in figure 2. For the aforementioned simplification of sensor data manipulation and functionality of the surveillance robot, it was found that two microprocessors were enough, but if necessary, additional CPU's may be connected to the master CPU through the expansion connectors provided. The onboard computer communicates with the base station operator or a more powerful high-level planner through a 4800 baud RF link.

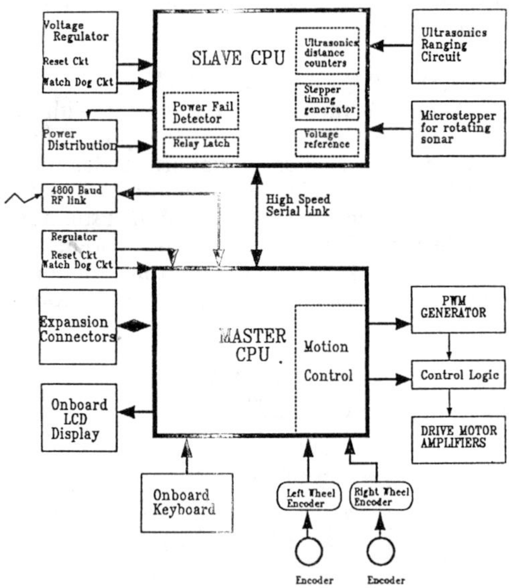

Figure 2. Block Diagram of the onboard distributed computer system

III. SOFTWARE

The mobile robot software consists of two parts, the surveillance navigation software, and the software for motion control and ultrasonic data processing. The navigation software is written in C and then cross compiled to object code. All other software is partly written in C and in assembler language to provide easy manipulation of register data in the CPU and interface chips. In addition, when rapid operation of interrupt routines is required, the interrupt service routines are written in assembly language to ensure that the code has optimum efficiency.

Surveillance Navigation software

Many algorithms have been developed, proposed and are still being developed for the navigation of autonomous mobile robots [BRO83] [JOR86] [BUR87]. In the case of a security robot, key points in the region under surveillance are digitized into a sequence of waypoints, by taking the robot through the region with the help of a joystick by the operator at the base station. This sequence of waypoints forms an undirected connected graph or the map, and each waypoint is a node in the graph. The surveillance program uses this map to navigate between different nodes in the region under surveillance.

The surveillance program operates in three modes, namely: go to, patrol and wander modes. In each mode, the robot navigates between adjacent nodes via deadreckoning and visual landmark recognition. If an obstacle is detected between adjacent nodes, a heuristics based procedure [MIT84] is called, and it waits for 30 seconds for the object to clear from its path. If not, the robot tries to navigate around the obstacle. If the robot fails to navigate around the obstacle, it returns to the node and attempts to find an alternate path to the destination. Only if no alternate paths exists, or if all paths are blocked, will the robot stop and indicate failure to reach its destination. A 30 second wait when a obstacle is detected prevents the robot from unnecessarily navigating around moving objects such as people, moving machines, or other mobile robots. The navigation software is written such that the motion of the robot is continuous until it reaches turns. The robot can also recalibrates its odometric parameters with its infrared sensors and reflective tape placed at irregular intervals on the ceiling of building.

Due to the nature of robotic surveillance, some user preprogramming is required in order to define locations or events which may be considered significant by specific users. This limits the degree of "autonomy" which can be incorporated into the system design. However the same principle described above have been implemented on a totally autonomous cleaning robot. The "CyberVac" was tested by a non-technical group of users in Europe for over a year under "real conditions" of undefined, unstructured environments. CyberVac was able to be placed in any complex room and navigate around furniture in such a way that the entire floor surface of the room was methodically and efficiently cleaned without preprogramming and without an inefficient "first pass" to "learn" a map of the room. If the objects in the room were reorganized in a random manner or moved unexpectedly, CyberVac covered the new floor surfaces without reprogramming. This demonstrated the ability of the principles described above to deliver "human-scale-autonomy" in a task specific application in terms of enabling a mobile robot to "understand" and intelligently service any complex area without human assistance or preprogramming regardless of shape or size of the room or object characterization within the room.

IV CONCLUSIONS

This research has stressed the development of novel sensory input and data processing methodologies which have been implemented for a mobile robot to execute the autonomous and semi-autonomous functions of building cleaning and security. Further a material handling robot and research platform robot have also been implemented for third party research. The ultimate aim of the research is to achieve human scale performance levels for certain specific applications while maintaining realistic price/performance ratios.

ACKNOWLEDGEMENTS

We would like to acknowledge the financial and technical support of the National Research Council of Canada and the Department of Industry Science and Technology of Canada.

REFERENCES

[BRO83] Brooks, R., "Solving the Find-path problem by good representation of free space, **IEEE transactions on Systems, Man and Cybernetics,** Vol. SMC-13, No. 3, pp. 190-197, 1983.

[JOR86] Jorgensen, C. C., Hamel, W. R., and Weisbin, C. R., "Exploring Autonomous Navigation", **BYTE** pp. 223-235, January, 1986.

[BUR87] Burks, R. L, et al., "Autonomous Navigation, Exploration and Recognition", **IEEE Expert**, pp 18-27, Winter 1987.

[MIT84] Mitchell, J. S. B., "An autonomous vehicle navigation algorithm", **SPIE,** Vol. 485, Applications of Artificial Intelligence, pp. 153-158, 1984.

ROAD FOLLOWING AND OBSTACLE DETECTION FOR THE MOBILE ROBOT VISUAL NAVIGATION

Wang Jiangang, Zhu Feng, Wu Jixian, Xu Xinping

Shenyang Institute of Automation
Academia Sinica
Shenyang 110003, P.R.China

ABSTRACT

A video-based road following and obstacle detection algorithm for mobile robot is presented. We use adaptive color segmentation to classify the scene as "road" and "non-road" region first. The "non-road" regions inside the road region are considered as obstacle candidate regions. Then we transform and compare the same obstacle candidates from two cameras to determine it is true or false obstacle. Experiment results of road segmentation and obstacle detection are presented.

INTRODUCTION

In order to travel autonomously, a mobile robot must have vision system for the road following and obstacle avoidance.

Currently, two classes of road segmentation were used for the mobile robot vision system. The first class is based on edge detection to track border lines. Typical methods are developed by the West Germany[1] and University of Maryland[2]. Another class of segmentation is based on color region analysis, including the works from Martin Marietta[3], FMC[4], CMU[5] and Yamanashi University[6]. We have developed an adaptive color region segmentation algorithm[7]. Both of the road sample statistics and the classification thresholds are updated. So the algorithm can adapt the gradual changes of road image.

Another fundamental requirement of mobile robot is obstacle avoidance. Active method can obtain the range image directly, and the obstacles can be detected by the range data processing[5]. The passive approach to obtain range data using stereo images has to solve the complex matching problem. For the obstacle detection, it is not necessary to calculate the description of whole 3-D scene. It only needs to identify the obstacle candidate, the non-road regions inside the road region after road segmentation, as true obstacle or false obstacle.

We use two cameras fixed vertically, like the way of Yamanashi University[6]. The first one is used for road segmentation. If there are obstacle candidates in the first image, we take the second image to transform and compare with the first one. If it is true obstacle, the difference will be detected. This approach is simple and efficient.

The proposed vision system architecture for the road segmentation and obstacle detection is shown in FIG.1.

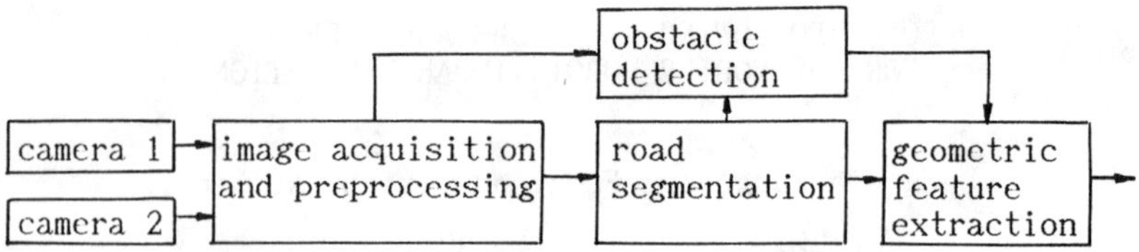

FIG.1. Block diagram of the system

ROAD SEGMENTATION ALGORITHM

Adaptive Color Region Segmentation Algorithm

The generality and adaptation problem can be resolved better by using adaptive color region segmentation algorithm.

First of all, the color image is decoded into Red, Green and Blue components. And the R, G, B images are digitized separately. Each pixel consists of a 3-triple of red, green and blue intensity values. The effects of brightness for road segmentation can be reduced by color normalization. Consequently the problem of less dark shadow can be solved.

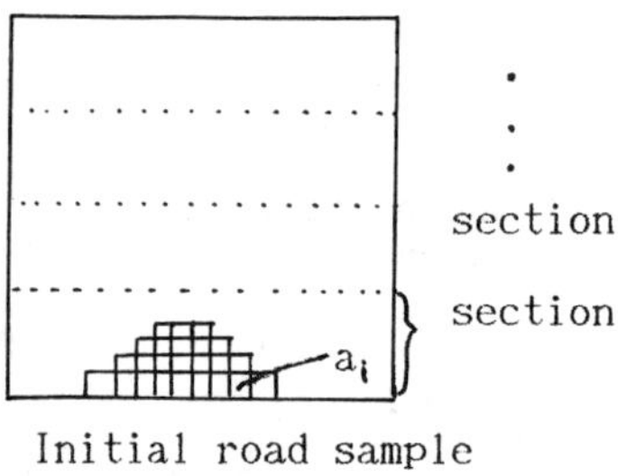

FIG.2. Image section and road sample

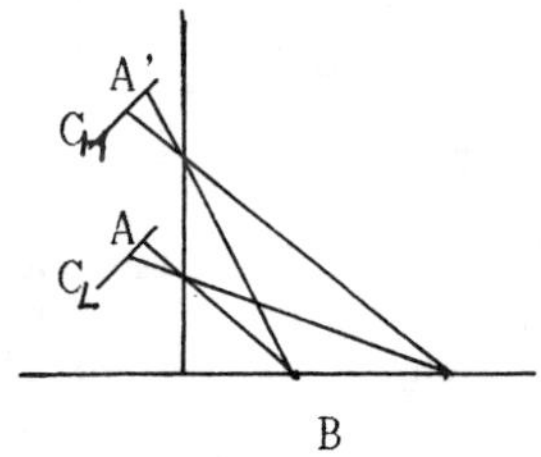

FIG.4. Transform A to A'

Initial road sample is a trapezoid region at the bottom of the image. We first calculate the statistical values of road sample and determine the thresholds. Next step is the classification line-by-line from bottom to up and from left to right. If the statistical values of pixel are less than corresponding thresholds, the pixel will be classified as road class, else the pixel belongs to non-road class.

In the end of the classification of one line, we update the sample statistics which are used to calculate statistical values of pixels of next line. And the road sample is updated also in the end of every restrict image section. Successive image sections are processed as before until termination conditions are satisfied.

The last image can be also used as road sample for the current image when overlap part of two successive images is large enough.

The road boundary extraction will provide boundaries and obstacle candidate regions for obstacle detection algorithm and path planning.

Threshold Calculation

This segmentation algorithm only analyses the road part on the image. And the final result is two classes, road and non-road. We develop a threshold calculation method using statistical values of road sample.

The initial road sample pixels are $a_1, a_2, \ldots, a_i, \ldots, a_n$ (FIG.2). m, v and cv are the mean, variance and covariance of sample region (for r, g and b).

Let's assume:

$$M_v^{(j)} = \mathop{MAX}_{i=1}^{n}(a_i - m_j)^2 \qquad M_{cv}^{(jk)} = MAX \mid (a_i - m_j)(a_i - m_k) \mid$$

where $j,k \in \{r,g,b\}$ and $j \neq k$.

Let's define:

$$V^{(a)} = v_r + v_g + v_b \qquad CV^{(a)} = cv_{rg} + cv_{rb} + cv_{gb}$$

The threshold is calculated as follows:

$$\alpha_j = f(M_v^{(j)}, v_j, V^{(a)}) \qquad \alpha_{kn} = h(M_{cv}^{(kn)}, cv_{kn}, CV^{(a)})$$

where $j,k,n \in \{r,g,b\}$ and $k \neq n$, f and h are non-linear experience function.

Classification Line-by-line

We take a line as a unit in classifying to adapt the gradual change of road part. So the statistical value which is used to calculate pixels statistical values is updated line-by-line.

Let's assume a_{ij} is a pixel at row i and column j.

If a_{ij} satisfies

$$(a_{ijk} - m_{sk})^2 < \alpha_k \quad \text{and} \quad \mid (a_{ijn} - m_{sn})(a_{ijp} - m_{sp}) \mid < \alpha_{np}$$

then a_{ij} belongs to road class, else a_{ij} belongs to non-road class. Where $k,n,p \in \{r,g,b\}$ and $n \neq p$. m_s is sample statistical value.

Geometric Reasoning

The geometric reasoning can correct the segmentation effectively. It also can speed up the segmentation by avoidance of much unnecessary computation and judgement. The boundary-following can provide road shape (straight,winding,cross,etc.) information for path planning.

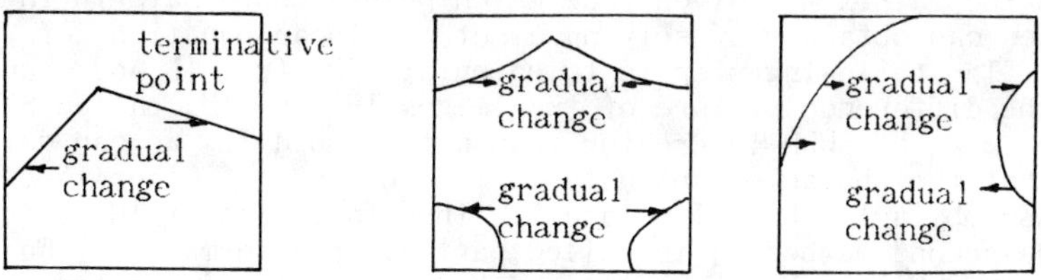

FIG.3. Road models

The road shape, the boundary continuity constraint and the boundary smoothness constraint are used to construct the road models. Several common models we have constructed are shown in FIG.3.

OBSTACLE DETECTION

CH and CL are two cameras fixed vertically (FIG.4). The images of them are IH and IL separately. If there is a probable obstacle region A, called obstacle candidate, in the road region of IL we first assume that it is a false obstacle. Now using the parameters of camera CL, the region B on the ground can be calculated from A. Then we can obtain the region A' which is the projection of B in CH. This image is written as IH'. Next we compare the IH' with the actual image IH. If they are the same, it indicates that the original assumption is right and the obstacle candidate is false obstacle. Otherwise, it is true obstacle. FIG.5 illustrates this method.

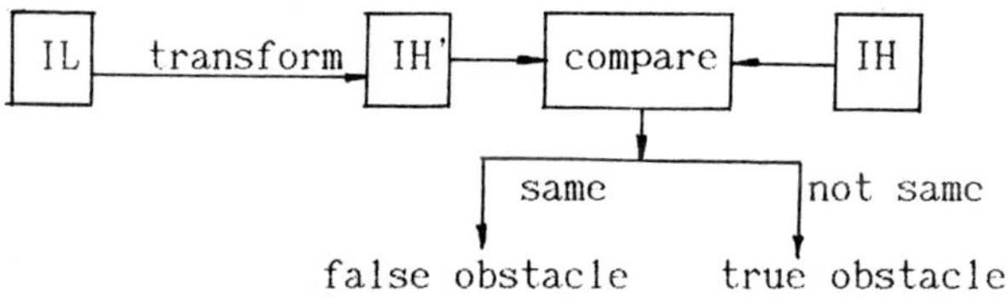

FIG.5. Block diagram of obstacle detection

In order to compare IH' and IH, we have designed two methods.

The first is to transform all the obstacle candidate region to IH'. This transformation includes two steps, the first step is inverse perspective transformation from image plane IL(ul,vl) to the world plane (x,y,z) and the second step transforms the world plane (x,y,z) to another image plane IH (uh,vh). However, if we use the homogeneous coordinate to represent the IH and IL plane, we can prove in theory that the transformation can be written as the following form

$$(ul, vl, 1)*M=(Uh, Vh, t)$$

where M is a 3X3 matrix and uh=Uh/t, vh=Vh/t. The problem is how to get the matrix M. Given four match points coordinate in IH and IL, we can obtain M by solving a set of linear equation of 8th order. The following step is to calculate IH' from IL and then to make the difference picture of two images IH' and IH, called SUBH. If the mean of SUBH is greater than a threshold T, we can determine that the obstacle candidate is true obstacle. Otherwise, it is false obstacle. Here T relates to the statistics of IH.

The second method[7] is called partial transformation. We can obtain the Y_{maxH}, Y_{minH}, Y_{maxL} and Y_{minL} from the segmented image IH and IL, where Y_{maxH} (Y_{maxL}) and Y_{minH} (Y_{minL}) are the maximum and minimum row coordinates of the obstacle candidate region in IH(IL). Then we transform Y_{maxL}, Y_{minL} to Y_{maxH}' and Y_{minH}'. Set $h=Y_{maxH}-Y_{minH}$ and $h'=Y_{maxH}'-Y_{minH}'$, if $|h-h'|>Th$, it is true obstacle. Otherwise, it is false obstacle. The threshold Th depends on the minimum height of the obstacle that the mobile robot can travel. This method needs to have a perfect segmentation, but the advantage of this method is that the exact calibration of cameras is unnecessary.

EXPERIMENT RESULT

The experiment is performed on the VICOM Image Processing System. FIG.6 shows the road segmentation results for a sequence of road images from a CCD camera. The travel speed of observer is about 10km/h and the mean processing time of each image is 5 seconds. The road boundaries and the central line of the road are given.

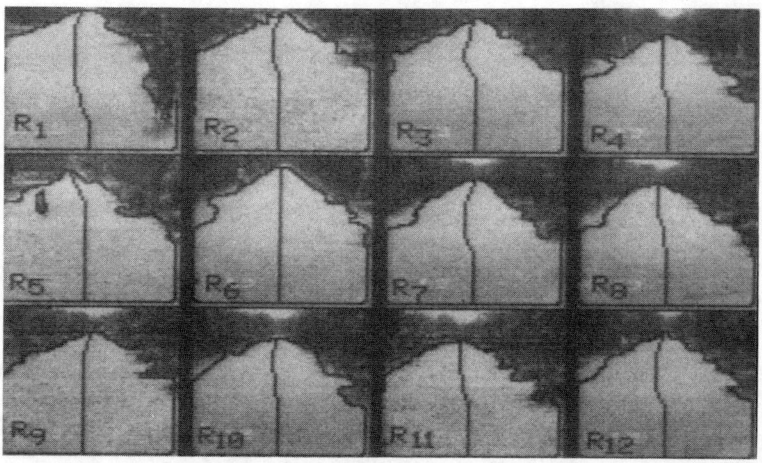

FIG.6. The road segmentation results for a sequence of images

FIG.7 shows the experiment results of obstacle detection algorithm. The labels 1,2,3 and 4 in (a) (b) are four points to calibrate the two cameras, 5 is a true obstacle and 6 is a false one. We can see that, the IH and IH' of true obstacle 5 are different, and the corresponding region in SUBH has a large mean value. Meanwhile, IH and IH' of false obstacle 6 are almost same and the corresponding region in SUBH has a small mean. We also see that it has much more difference of h and h' for the true obstacle 5 and little difference for the false obstacle 6.

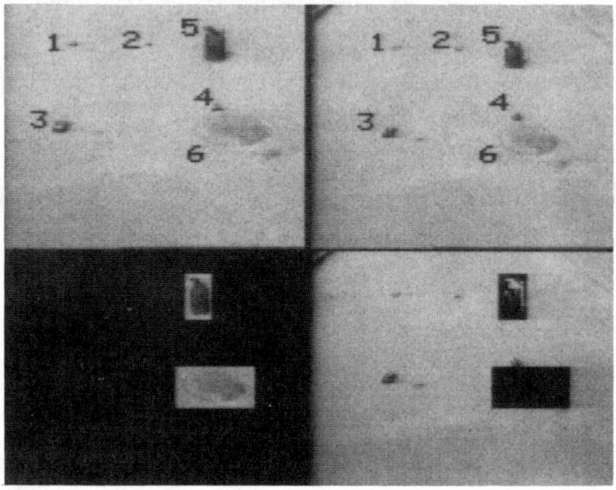

IL	IH
IH'	SUBH

FIG.7. Experiment result of obstacle detection

REFERENCES

1. K.Kuhnera, "A vision system for real time road and object recognition for vehicle guidance", Proc. SPIE Mobile Robot Conf. Cambridge, M.A, 269-272 (Oct. 1986).

2. A.M.Waxman, J.J.Lemoigne, L.S.Davis, B.Svinivasan, T.R.Kushner, E.Liang, T.Siddalingaiah, "A visual navigation system for autonomous land vehicles", IEEE J. of Robotics and Automation, Vol RA-3, No.2, 124-141 (April 1987).

3. M.A.Turk, D.G.Morgenthaler, K.D.Gremborn, M.Mana, "VITS-A vision system for autonomous land vehicle navigation", IEEE Trans. on PAMI, Vol 10, No.3, 342-361 (May 1988).

4. D.Kuan, G.Phipps, A.C.Hsueh, "Autonomous robotic vehicle road following", IEEE Trans. on PAMI,Vol 10, No.5, 648-658 (Sept. 1988).

5. C.Thorpe, M.H.Hebert, T.Kanade, S.A.Shafer, "Vision and navigation for the Carnegie-Mellon Navlab", IEEE trans. on PAMI, Vol.10, No.3, 362-373 (May 1988).

6. H.Mori, S.Katani, H.Ishiguro, S.YasuTomi, Y.Chino, "Self-contained mobile robot in campus road", JRSJ Vol 5, No.5, 29-42. (Oct. 1987)

7. Zhu Tieshan, Shao Zhuxian, Wu Jixian, "A rapid algorithm for color region based road following & obstacle detection of autonomous vision navigation", Proc. 2nd Chinese robotics conf., Shenyang, China, 340-344 (Sept. 1989).

LEGGED LOCOMOTION AT CONSTANT BODY VELOCITY

Y. S. BAEK , E. F. FICHTER
Mechanical Engineering Dept.

B. L. FICHTER
Entomology Dept.
Oregon State University; Corvallis, Oregon 97331

ABSTRACT

Constraints developed for legged locomotion are based on four dimensionless numbers, three defining ratios between actual and extreme parameter values and the fourth relating foot acceleration, leg stroke and body velocity. Figures and equations form a basis for analyzing and designing walking sequences when body moves at constant velocity.

INTRODUCTION

Legged locomotion is a method for manipulating a body through a very large range of motion. It also differs from other manipulation methods because body motion is not controlled directly: body motion is a consequence of direct motion control of feet. Thus when analyzing walking, designing a walking machine, or designing a stepping sequence for an existing walking machine, it is important to understand constraints placed on body motion by leg motion.

This paper presents kinematic constraints on horizontal components of velocity of body and feet. We assume that the horizontal component of body velocity is constant and lies in body yz-plane and that body x-axis has constant direction perpendicular to global z-axis (Fig. 1). The global z-axis is in the direction of gravity but has opposite sense. These restrictions exclude turning but do not exclude travel on sloping or uneven surfaces. We also assume a symmetrical trapezoidal velocity profile (Fig. 2) for feet since this

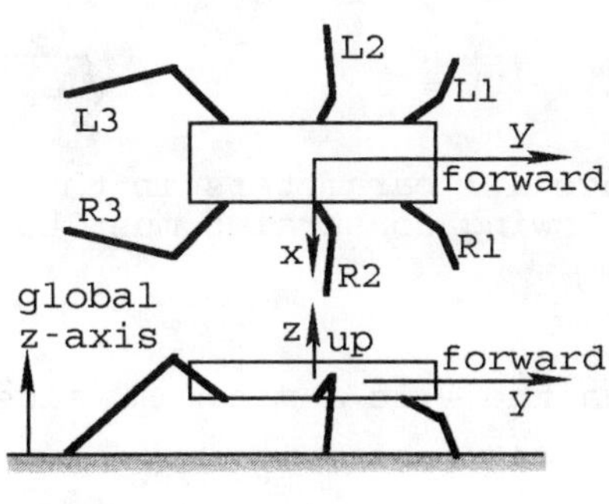

Fig. 1. Local and global coordinate systems.

is simple and realistic [1][2]. Duty factor, β, and step time, t_s, are held constant for simplicity. These assumptions allow all legs to be treated identically.

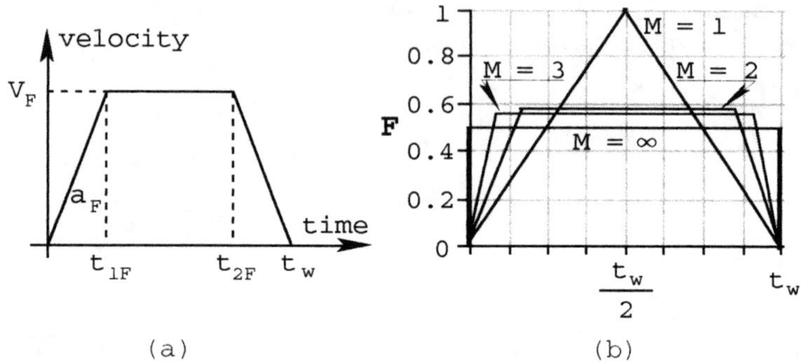

Fig.2. (a). Foot velocity profile relative to ground.
(b). Change in foot velocity profile with shape factor M.

FOOT VELOCITY RELATIONSHIPS

Consider first the algebraic relationship implied by Fig. 2a. The area under the trapezoid, step length d_{sF}, is given by

$$d_{sF} = \left(t_w - 2\,\frac{V_F}{a_F}\right) V_F + \left(\frac{V_F}{a_F}\right) V_F \tag{1}$$

where a_F is foot acceleration, t_w is swing time and V_F is maximum foot velocity. Solve for V_F

$$V_F = \frac{1}{2}\left[\, a_F\, t_w \pm \sqrt{(a_F\, t_w)^2 - 4\, a_F\, d_{sF}}\,\right] \tag{2}$$

Since all parameters in this eq'n are positive, the following constraint must hold.

$$a_F\, t_w^2 - 4\, d_{sF} \geq 0 \tag{3}$$

Thus there is a minimum value of a_F

$$(a_F)_{min} = \frac{4\, d_{sF}}{t_w^2} \tag{4}$$

Other foot acceleration is defined using a multiplier, M

$$a_F = M\,(a_F)_{min} \tag{5}$$

This multiplier, the first dimensionless number we will use, is defined as

$$M = \frac{a_F \, t_w^{\,2}}{4 \, d_{sF}} \tag{6}$$

Foot acceleration can now be eliminated from eq'n (2)

$$V_F = \frac{2 \, d_{sF}}{t_w} \left[M \pm \sqrt{M \, (M - 1)} \right] \tag{7}$$

When foot acceleration is minimum (M = 1)

$$V_F = \frac{2 \, d_{sF}}{t_w} \tag{8}$$

When a_F is increased with constant d_{sF} and t_w, V_F must decrease. Thus only the negative sign in eq'n (7) is valid.

$$V_F = \frac{2 \, d_{sF}}{t_w} \left[M - \sqrt{M \, (M - 1)} \right] \tag{9}$$

It is convenient to define another dimensionless number, the ratio of V_F to $(V_F)_{max}$, from eq'n (8)

$$F = \frac{V_F \, t_w}{2 \, d_{sF}} \tag{10}$$

Fig. 2b illustrates the relationship between M and F.

RELATIONSHIPS BETWEEN VELOCITIES OF BODY AND FEET

When body moves with constant velocity and feet move in a periodic gait, distance traveled by a foot, d_{sF} must equal distance traveled by body, d_{sB} (Fig. 3). In eq'ns (6) through (10), d_{sB} can be substituted for d_{sF}. Also from Fig. 3

$$d_{sB} = V_B \, t_s \tag{11}$$

To relate V_B to V_F, eliminate dsF from eq'n (9).

$$V_F = \frac{2 \, V_B \, t_s}{t_w} \left[M - \sqrt{M \, (M - 1)} \right] \tag{12}$$

Since β is constant, t_s and t_w are related as $t_s \, (1 - \beta) = t_w$; t_s and t_w are eliminated from eq'n (12) (Fig. 4).

$$\frac{V_B}{V_F} = \frac{1 - \beta}{2\left[M - \sqrt{M(M-1)}\right]} \tag{13}$$

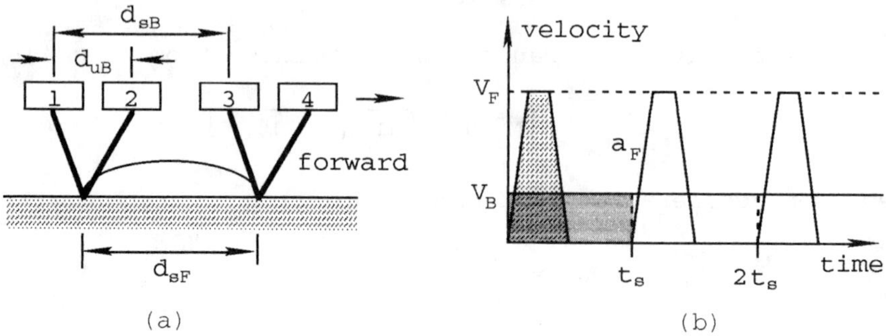

Fig. 3. Relationship between body and foot step length. (a) From side view (b). From velocity profiles.

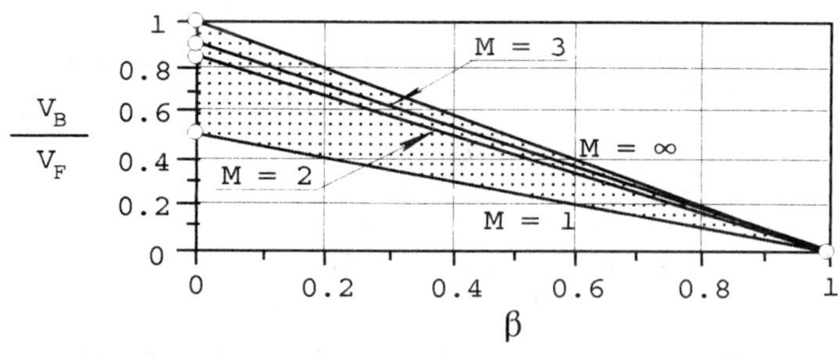

Fig. 4. Duty factor and M limit ratio of body to foot velocities. Shaded region shows allowed values.

LIMIT ON USABLE RANGE OF MOTION OF FOOT

Another limit on motion can be deduced from Fig. 3a; distance body travels while foot is on substrate, d_{uB}, is limited by range of motion of foot relative to body. There is, in addition, a more subtle limitation on d_{uB} illustrated in Fig. 5. At the beginning of the swing, foot velocity is zero so foot lags increasingly far behind body until it accelerates to body velocity. This means that when the swing begins, the foot cannot be at the limit of its range of motion. Furthermore, since the foot must be raised to swing forward, this limit on d_{uB} is not based on range of motion at substrate height. Converting

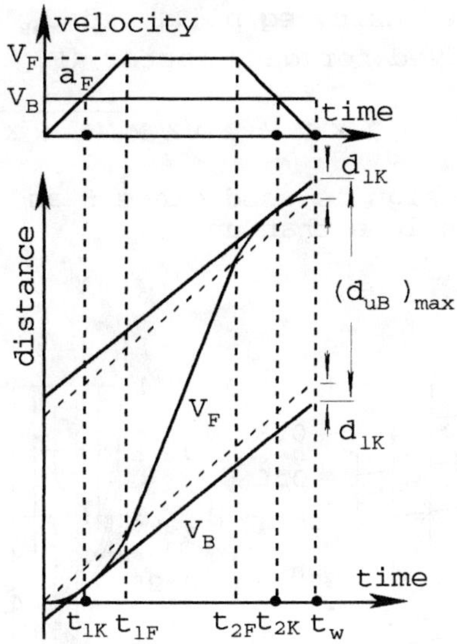

Fig. 5. Body velocity constraint on leg workspace.

these concepts to algebra, minimum distance from foot position to limit of foot range of motion is the area bounded by the foot velocity profile before time t_{1K}

$$d_{1K} = \frac{V_B^2}{2\ a_F} \tag{14}$$

If range of motion of foot relative to body at the required height is $(d_{uB})_{max}$ then maximum leg stroke is

$$d_{uB} \leq (d_{uB})_{max} - \frac{V_B^2}{a_F} \tag{15}$$

since d_{1K} must be added at each end of the stroke.

Eq'n (15) can be rearranged into an inequality between dimensionless ratios K and G

$$K = \frac{V_B^2}{a_F\ d_{uB}} \leq G = \frac{(d_{uB})_{max} - d_{uB}}{d_{uB}} \tag{16}$$

where G is the ratio of unused to used leg stroke.

Eliminating foot acceleration in eq'n (6) and (16) and substituting d_{sB} for d_{sF} gives

$$K = \frac{t_w^2\ V_B^2}{4\ M\ d_{sB}\ d_{uB}} \tag{17}$$

which is simplified using eq'n (11), $\beta\, d_{sB} = d_{uB}$ and $t_w = t_s\,(1 - \beta)$ and solved for duty factor (Fig. 6)

$$\beta = 1 + 2\,K\,M - 2\,\sqrt{K\,M\,(1 + K\,M)} \qquad (18)$$

Only the negative sign is used since K and M are positive and duty factor is less than unity.

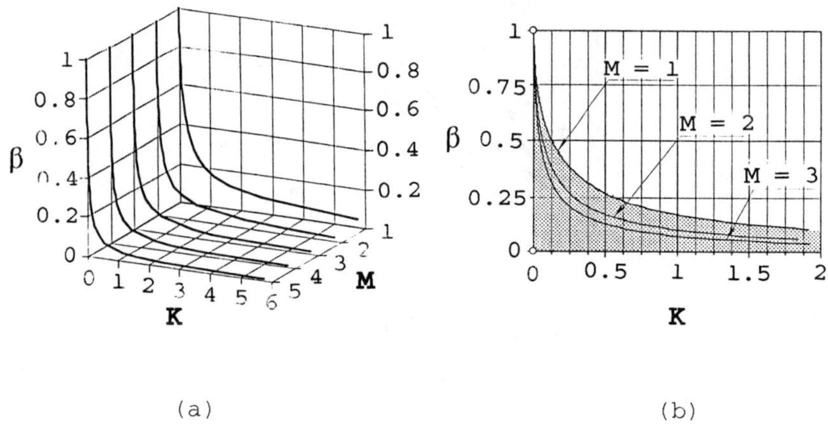

Fig. 6. (a). Locus of possible values forms surface.
(b). Shaded region shows allowed values.

DISCUSSION

Principal parameters of legged locomotion at constant velocity are β, t_s, d_{uB}, V_B, V_F and a_F. Four dimensionless numbers clarify relationships among parameters. Use of these relationships for analysis and design depends on application and will be discussed in a future paper.

REFERENCES

1. Y.S. Baek, Kinematic Analysis of Legged System Locomotion on Smooth Horizontal Surfaces, Ph.D. dissertation, Oregon State University (1990).

2. M. Kaneko, S. Tachi, K. Tanie and M. Abe, "Basic Study on Similarity in Walking Machine from a Point of Energetic Efficiency", IEEE J. of Robotics and Automation, RA-3(1), 19-30 (1987).

Co-operant Mobile Robots

D P Barnes, B Bury, J O Gray.
Control & Instrumentation Group,
Department of Electronic and Electrical Engineering,
University of Salford,
Salford, United Kingdom,
M5 4WT.

ABSTRACT

Members of the Control & Instrumentation Group at the University of Salford have recently established a major new research initiative in the area of co-operant mobile robots. The aim is to demonstrate complex mobile robot task achievement via the co-operative behaviour of many relatively simple mobile robot devices. Many biological systems are a testament to the success of this approach. The work complements existing mobile robot research and will therefore serve to extend the application base of current mobile robot technology. The work programme is concerned with the generic areas of cost effective sensory systems, behavioural control architectures, inter-mobile robot communication and co-operant mobile robot behaviour. Programme funding has allowed work to commence and this paper seeks to detail our thinking and progress in these generic areas. To focus the work, a number of manufacturing applications have been identified and a set of research demonstrators proposed. These are also presented in the paper.

INTRODUCTION

There was a tendency for early robotic systems to be regarded as the centre of their manufacturing universe. Other essential processes being seen as mere peripheral entities under their ultimate control. Over recent years, the move has been away from this omnipotent control approach and has led to a control devolution resulting in the *islands of automation* concept [1]. For economic success, 'trade' is essential and consequently the trend has been to develop distributed, co-operating manufacturing systems. This successful approach has not stopped at the macro 'island' level, it is being developed further and early research has begun in areas such as co-operating robot manipulators and co-operating expert systems [2,3]. Automated manufacturing plant must now be seen as a 'society' of manufacturing 'craftsmen' and 'experts' where co-operative task achievement is the key to success. Co-operating devices are likely to prove invaluable in tomorrow's manufacturing arena and it is to the area of co-operating mobile robots that our research is dedicated.

A long term aim of the group is to establish a comprehensive applications portfolio for co-operant mobile robots that will include manufacturing and construction. The first of these applications has recently been established with the other application forming a natural evolution in complexity. Funding from the Science and Engineering Research Council, ACME Directorate has allowed a three project to commence in the area of co-operant mobile robots for advanced manufacturing and material handling applications. This paper describes our thinking and progress in this area.

MANUFACTURING APPLICATIONS

To ensure the manufacturing relevance of the research, the work is being focused upon a number of specific industrial problems. The research group includes three industrial companies who have specific manufacturing applications for the project deliverables. The applications include : warehouse material handling, product pick and place operations using co-operant mobile automata, NDT of fabricated submarine hulls using multiple mobile robots and hazardous material handling that requires sophisticated co-operant mobile systems in order to fully automate the manufacturing process. Information on the industrial applications has allowed a set of research demonstrators to be defined. Their purpose is to show the effectiveness of the generic research by using two developed mobile robots to demonstrate selected aspects of these industrial problems. Work has initially focussed upon the material handling application that involves moving palleted objects from one location to another.

THE SALFORD APPROACH

The majority of mobile robot research [4,5] has been concerned with the problem of how to endow mobile robotic devices with greater perception and intelligence. Some applications require considerable computing power, high resolution sensory imaging devices, sensor data fusion and world modeling and planning techniques. This approach can be regarded as using the *solitary vertebrate* model for mobile robot research, where researchers are attempting to mimic many of the capabilities found within some vertebrate species.Work has been conducted for many years using this model with varying degrees of success. Clearly, there are applications that require this approach and with the advent of recent computing hardware and techniques, realisation of this robotic solitary vertebrate has begun. An alternative approach, for which there has been little or no research over the past years, is the *social invertebrate* model. Here the goal is to mimic many of the capabilities found within some invertebrate species and in particular those that complete a given task via co-operative behaviour. Their success is not largely due to their cognitive powers, as in the vertebrate case, but rather it is due to the fact that these 'technically' simpler organisms are able to communicate and co-operate. The resultant *superorganism* exhibits an holistic behaviour that stems in a straightforward way from the relatively crude repertories of individual society members. The term *swarm intelligence* [6] aptly describes this holistic phenomenon. The Salford mobile robot research approach is firmly committed to the social invertebrate model and furthermore the authors believed that a sensor-driven or behavioural control architecture [7] is ideal for its realisation.

EXPERIMENTAL APPARATUS

Two mobile robots are currently under construction and constitute the experimental apparatus for the SERC funded research programme. These two devices, affectionately known as 'Fred' and 'Ginger' are shown in Fig. 1. The majority of work to date has involved the careful engineering of these devices so that they not only satisfy our immediate research needs but also provide sufficient extensibility for investigating future ideas.

FIG. 1. 'Fred' and 'Ginger' under early construction

Each mobile robot is based upon a Real World Interface B12 mobile platform. These omni-directional devices, capable of speeds up to 2 m/sec, possess an on-mobile primary control microprocessor together with a number of internal state sensors. An RS232 interface allows communication with an external computer. To enhance the capabilities of the B12 platforms, superstructures have been built that are capable of housing additional sensors, data acquisition, communications and secondary control hardware. These superstructures have three different levels ranging from the B12 platform at Level 0 through to navigation systems at Level 3. Attached to Level 0 is a 'sensor belt' that currently houses a number of Polaroid ultra-sonic transducers. The final sensor belt will contain additional optical and inductive sensors for object proximity detection. Levels 1 and 2 are at an advanced stage of construction with Level 1 requiring some ten different electronic systems ranging from microcontrollers, that act as an interface with the secondary control system, to an advanced optical communications system. Level 2 contains the transputer based secondary controller. This consists of a double extended eurocard transputer motherboard capable of holding up to 16 individual transputer modules. Running through the centre of each superstructure is a common level interconnection 'core'. This can allow up to 24 electronic p.c.b.'s to be connected at each level and has been engineered for maximum flexibility. The superstructures can be regarded as an instrumentation and control 'pod' that can be easily disconnected from a B12 platform for ease of development and maintenance. Similarly, the different superstructure levels can be disconnected from one another and replaced with different levels containing application specific electronics.

Fig. 2 shows a schematic diagram of the hardware system needed for each mobile robot . An external PC transputer development system, which is connected to a SUN workstation network, provides remote communication with the on-mobile secondary controllers for development purposes. The on-mobile transputer network controls the B12 platform by issuing a variety of translate and rotate motion commands. Various B12 internal state sensors can also be interrogated to obtain for

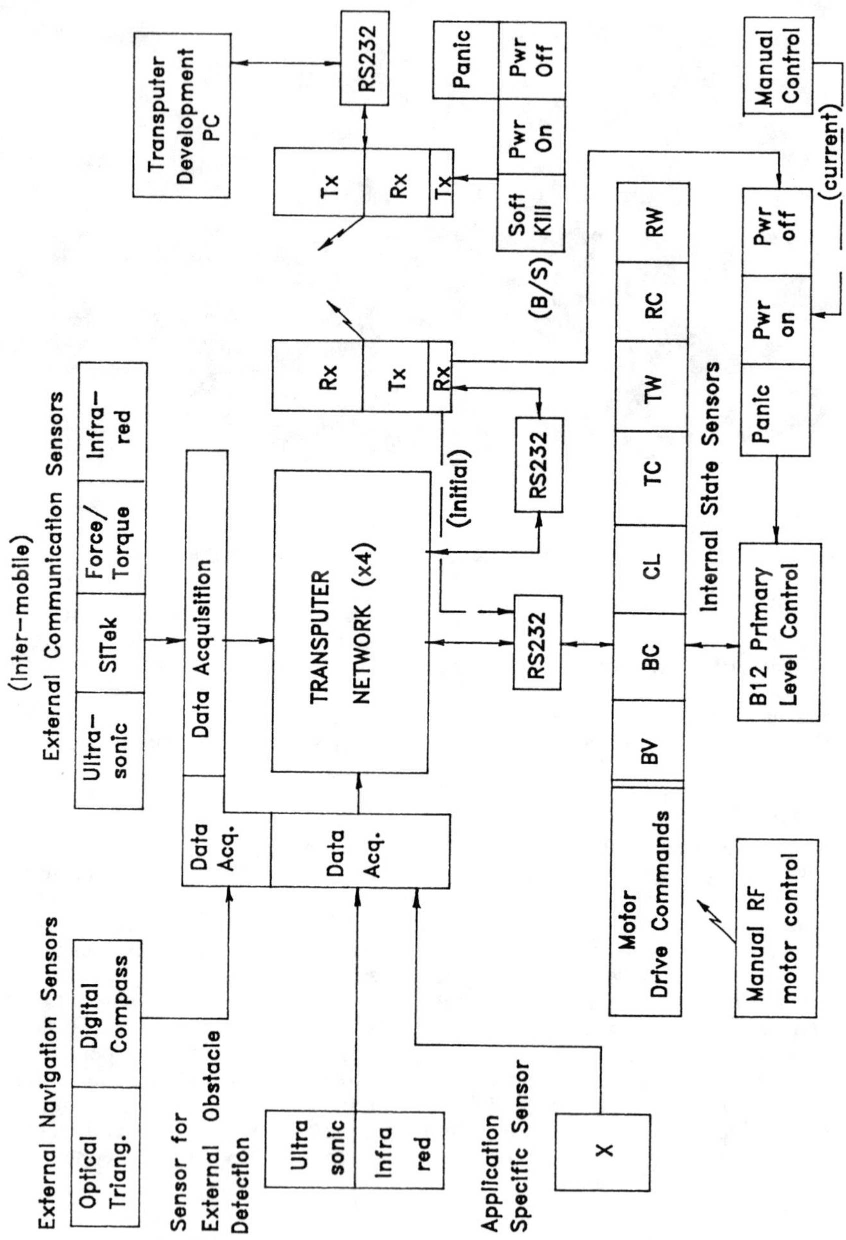

FIG. 2. A schematic diagram of the hardware needed for each mobile robot

example battery voltages and actuator encoder values. The on-mobile transputer network is interfaced to the B12 primary controller and additional external sensors via a number of microcontroller boards capable of performing basic communication and data acquisition functions.

EXPERIMENTATION PROGRAMME

A number of generic research areas are under investigation, our thinking and progress in these areas is described as follows :

Cost effective sensory systems :- A sensor-driven control approach requires a wide variety of transducers and sensing techniques and their choice permeates all aspects of the research. Because of the largely 'world model-less' approach, high resolution imaging sensors are being avoided. The authors believe that an adequate level of competence can be achieved by using relatively inexpensive sensory systems. Above all, the authors intend to avoid vision research which has all too often consumed vast amounts of resources at the expense of performing comprehensive m*obile* robot experiments. The following sensory systems are currently under investigation :

a) O*bject Proximity Detection*... a multi-modal approach is being developed that uses three different types of sensor - an infrared differential reflectivity sensor; a frequency swept ultrasonic ranging sensor and an object presence inductive-coupled oscillatory sensor. Each of these have particular strengths and weaknesses and the research is concerned with obtaining a 'good mix' so that reliable object proximity information can be obtained.

b) N*avigation*... optical beacon navigation sensors are being investigated that use a pulsed infrared light source that can be detected using a 2D phase sensitive transducer. By using two of these devices and suitable aspheric optics, triangulation techniques can be used to calculate a mobile's position relative to the beacon source.

c) *Active Objects*... coupled with the navigation sensors, the concept of active objects is being investigated. In the material handling application there is a clear need to identify the position of an object that requires relocating. Rather than take the vision approach to this problem, the objects will be electronically tagged and effectively transmit their location in a similar fashion to a homing beacon. Electronic tagging is becoming more widely used within manufacturing industry and while it is currently cost prohibitive for large batch production, it does provide a convenient means of 'leapfrogging' what has hitherto been a difficult object sensing problem. Given an advance in electronic tagging technology, then this approach does become economically viable for its wider use within manufacturing industry.

Behavioural control architectures :- Building upon current work in this area, this research is concerned with developing a library of p*rimal behaviours* both for the purpose of low level survivalistic control and the higher levels of purposeful communication and co-operation. Given the heuristic nature of current behavioural controllers and the lessons learnt from other s*hallow knowledge* systems, this research is initially examing the use of c*ertainty grids* and v*irtual force field* methods [8] for the real-time control of our two mobile robots. These techniques can be implemented using a sensor-driven control approach that effectively produces a motion vector output for each behaviour at each level. Adaptive salience can be attributed to different levels and the motion vectors projected onto an actuator command plane that produces a resultant translation and rotation for each mobile robot. The net effect of this is to produce a system that is analogue in nature as opposed to the current discretised behavioural controllers. It is believed that such a system will aid the building of large complex behavioural systems and yield greater control stability.

Co-operant mobile robot communication and behaviour :- Inter-mobile communication and co-operant control are major research areas and current ideas are being based upon studies into animal group behaviour modeling [9]. These areas are a rich source of information and serve to highlight the different internal, environment and neighbour forces that exist within these multiple animal systems. The ultimate aim of this research is to synthesise complex behaviour from the relatively crude competence repertories of our individual mobile robots.

CONCLUSION

Research into co-operant mobile robots for advanced manufacturing and material handling applications has been under way now for some six months. During this time considerable effort has been expended on engineering the two mobile robots. Research has begun in the areas of cost effective sensory systems, behavioural control architectures, inter-mobile communication and co-operant mobile robot behaviour and our early ideas in these areas have been described. The research is being focused by a number of industrial applications which has led to the definition of a set of research demonstrators designed for 'proof of principle' purposes. The use of co-operant mobile robots is likely to prove invaluable in tomorrow's manufacturing arena and it is hoped that more researchers will become involved in this challenging new robotics area.

REFERENCES

1. Russell G T, "A Flexible Manufacturing Assembly Strategy upon Multi-Robot Multi-Sensor Islands of Automation", ACME Grant GR/D 60778.

2. Chimes P R, "Multiple-arm Robot Control Systems", Robotics Age 7, 5-10 Oct. 1985.

3. Freund E, "On the Design of Multi-robot Systems", Proc. of the International Conference on Robotics, IEEE Computer Society Press, Silver Spring, MD, 1984.

4. Brady J M, Durrant-Whyte H F, "A Sensing Autonomous Guided Vehicle for Advanced Manufacturing", ACME Grant GR/E 42419.

5. Rembold I U, "The Karlsruhe Autonomous Mobile Assembly Robot", Proc. IEEE Int. Conf. on Robotics and Automation, 1989, Vol 1, pp 598-603.

6. Beni W, Wang J, "Swarm Intelligence in Cellular Robotic Systems", Center for Robotic Systems, University of California, Santa Barbara, 93106.

7. Brooks R A, "A Robust Layered Control System for a Mobile Robot", IEEE Journal of Robotics and Automation, RA-2, 1986.

8. Borenstein J, "Real-Time Obstacle Avoidance for Fast Mobile Robots", IEEE Transactions on Systems, Man, And Cybernetics, Vol 19, No 5, Sept/Oct 1989.

9. Sannomiya N, Matuda K, "A Mathematical Model of Fish Behavior in a Water Tank", IEEE Transactions on Systems, Man, and Cybernitics Vol SMC 14 No1 84.

WIDE ANGLE TRACKING SYSTEM FOR MOBILE ROBOTS

Ernest L. Hall
University of Cincinnati

Bettie C. Hall
International Computer Robotics Corporation
Cincinnati, Ohio

ABSTRACT

A technique for mobile robot navigation using a wide angle lens to capture images from specified targets in a hemispherical field of view has been investigated. The sensed position information provides a basis for controlling the motion of a mobile robot by comparing the current position to the desired position to determine range and bearing correction signals for the robot motor controllers. These positions may be along a predetermined or computed path. While the basic algorithms for this omnidirectional vision controller have been previously studied, these studies were limited to 30 frame/second update rates. The goal of the present research was to concentrate on methods for improving the speed by an order of magnitude to 300 frames/second to permit the method to be used not only for the autonomous control of slow moving vehicles, but for higher speed vehicles such as automobiles, tanks, ships, aircraft and spacecraft.

INTRODUCTION

An omnidirectional vision navigation system, a global guidance system developed for a mobile robot, is used to determine global position and thus indicate the location of the robot relative to predefined targets. The use of a wide angle or fisheye lens that can see a 2π steradian field of view fits many robotic applications, eliminating rotary scanning and reducing the number of moving components in the system. The omnidirectional vision system has two major characteristics which are inherent in the design of a wide angle lens: first, an object and its image feature are on the same azimuth angle (azimuth angle invariability); second, there is a linear relationship between the zenith angle (the angle between the vertical axis and a target location) and the radial distance of the image location. The sensing system consists of a sensor and a vision processor. The prototype sensor consists of three components: a wide angle lens with image reducing adaptor, a video camera, and a gimbal mount.

The visualization and recognition of an image using an omnidirectional vision system and the use of digital image restoration has been studied for many years.[9] Instantaneous viewing through an extremely wide angle lens provides useful information from a hemispherical field of view, which, in spite of distorted geometric properties, maintains the physical relations between the object and the image field.

Investigations have been made into the use of an omnidirectional vision system to guide a mobile robot, such as a robot lawn mower and a robot cleaning machine, with the information on the basic design and the theoretical analysis provided in references.[9] A recent analysis with experimental verification of the dynamic performance of the sensor that indicates the usefulness of the guidance system for general autonomous mobile robots has also been performed.[10] The linear relationship between the zenith angle of a target and the radial location of the image feature, the error due to relative motion, and the image processing time related to the number of target blobs were studied experimentally. These analyses and experiments indicated that although the basic characteristics of the omnidirectional vision sensor seemed well defined, the question of speed of operation had not been investigated.[10] This led to research to determine if the system could work for high speed mobile robots as described in this report. After investigating the hardware components of the high speed system and analyzing the operation, special simulation software was developed which permitted controlled experiments.

The technique called dynamic, omnidirectional vision guidance has several unique features. A wide angle lens provides up to a 360 degree field of view. A video camera is used to capture the image at a rate of 30 full or 60 half frames/second. A high speed A/D converter captures a frame of image data, which is then stored in memory. An image processing algorithm is used to detect, recognize, and measure the image of a target. Another processing algorithm is then used with calibration parameters to determine the three dimensional position of the target. From two or more target locations, the three dimensional position of the mobile robot is determined. The measured position of the robot may then be compared to a desired position to implement path correction or path-following rules.

Previous research and experimentation indicates that this approach is feasible for slow-moving mobile robots, and some demonstrations have been conducted. However, the reason for the restriction to slow-moving robots is that the vision update and processing time has been 1/30 second or slower, due to the use of standard vision processing methods and equipment.

The 30 frames/second TV rate is more a function of the response of the human visual system and equipment designed for human viewing than a physical limitation. Swift [11] reports that, with high speed cameras and electronics, it is possible to achieve frame rates of one million frames/second. Of course, the costs of the camera and processor increase with speed.

We performed a feasibility study focusing on the research design required to obtain wide angle tracking at rates of 300 frames/second in a system that could be made available at a competitive cost for use in mobile robots. In particular, higher speed cameras, parallel processing, and simple tracking algorithms which could be implemented with digital hardware were investigated with the goal of achieving an order of magnitude speed improvement, which should permit automatic visual guidance of vehicles travelling at speeds of 30 miles/hour rather than the three miles/hour permitted by current systems.

A growing number of technical papers address the problems of robot control. Representative of this work are those of Whitney[12], Nevins and Whitney[13], Salisbury[14] and Raibert and Craig[15].

General Robot Control Problem

Based on the problem requirements, the design of the robot servo control was approached as a general nonlinear system. The problem of servo control of a flexible, multi-degree of freedom kinematic chain, or flexible manipulator must be treated as a nonlinear control problem. That is, the control of such a manipulator is a controlled process, the dynamics of which has the following form:

$$\dot{\underline{x}} = f(\underline{x},\underline{u},t) \tag{1}$$

where $\underline{u}$ is the vector <u>input</u> to the system, $\underline{x}$ is the <u>state</u> vector of the system, f is a nonlinear and possibly time variant function, and "t" represents time. If a performance criteria having the general form:

$$S = \int_{T_1}^{T_2} G\,(\underline{c}(t),\underline{u}(t),\underline{r}(t),t)\,dt \tag{2}$$

and a reference signal, $\underline{r}(t)$, which signifies the desired output response, $\underline{c}(t)$ the actual output response and G represents a loss function or measure of instantaneous change from ideal performance, is given, then an optimal control problem is specified. The problem of optimal control is the determination of the control input, $\underline{u}$, that minimizes the performance criteria, S, subject to certain constraints on $\underline{u}$ and $\underline{x}$.

Vision Servo Control

Navigation and vision guidance using a wide angle lens, correction equations, methods for target detection and recognition, and path control algorithms for mobile robots have been developed. The technique called dynamic, omnidirectional vision guidance has several unique features. A wide angle lens provides up to a 360 degree field of view. A video camera is used to capture the image at a rate of 30 full or 60 half frames/second. A high speed A/D converter captures a frame of image data which is stored in memory. An image processing algorithm is used to detect, recognize, and measure the image of a target. Another processing algorithm is then used with calibration parameters to determine the three dimensional position of the target. From two or more target locations, the three dimensional position of the mobile robot can then be determined. The measured position of the robot may then be compared to a desired position to implement path correction and path following rules.

The operation of the vision servo control system includes five major steps: calibration; target tracking; path correction; control updating; and region filling.

Stability Analysis

Stability is such an important characteristic of a robot control system that one may consider an unstable system useless. In general, one can attempt a system analysis either analytically or experimentally or with a combination of both methods. The analytical method requires exact and detailed knowledge of the system parameters. In this case, the primary unknown information is the robot vehicle. This basically precludes the use of a totally analytical approach. The experimental method can be used for all or part of the system, but is most appropriate for the robot vehicle. Therefore, a combination approach will be described with analytical models used when known, and simulated data for the robot vehicle dynamics. The simulation data was selected to model simple vehicles and would need to be verified for more realistic models before actual use.

A Bode stability analysis was used, since it provides for ease of combination of analytical and experimental data. With this method, we examine the open loop transfer function, $G(s)H(s)$, and examine the frequency response to determine locations at which this function would equal negative one, i.e., $1+G(s)H(s) = 0$. A threshold for phase margin and gain margin is used to ensure practical stability. The phase margin is the amount of phase present at the open loop crossover frequency, plus 180 degrees. The gain margin is the magnitude of the system gain, measured at the frequency for which the open loop phase is -180 degrees. By necessity, only design examples can be presented; however, the constants were selected to correspond to actual components which could be built into a prototype. The main single parameter whose effect is of primary interest is the vision servo rate.

The following open loop design goals should lead to a stable closed loop system: 1) a phase margin of at least 40 degrees
2) a gain margin of at least -10dB.

The overall response of the system can be determined from the component responses by multiplying the magnitude responses and adding the phase responses. For example, a magnitude response starts with a gain greater than 40 dB. This high DC gain will affect the speed of response of the system. However, since the phase response starts at less than -90 degrees, there is an instability indicated at the 0 dB crossover point where the phase is greater than -180 degrees, a condition of instability. The magnitude response also indicates the peak due to the system dynamics. Such resonant modes are characteristic of robot vehicles. The phase response is very complicated since each ZOH and delay add phase to the system response. This makes it possible to also have instability points at high frequencies where the system has modes and the phase is near -180 degrees. A decrease of gain of the system provides the desired stable response. The gain margin and phase margin indicate that the system is stable; however, other parameters could lead to an unstable system. For example, if the computation time is increased from 1/30 to 1/10 second, the gain margin decreases and the phase margin increases. If the phase margin increases to more than 40 degrees, instabilities could be expected. For a more realistic load with several resonant modes, the system would need to be tested carefully for stability.

CONCLUSIONS

This research study and simulation results indicate that the proposed objectives of wide angle tracking at rates of 300 frames/second is feasible with some modifications of off the shelve components and simple algorithms. In addition to the unique high speed camera, a high speed tracker is also required. Several possible variations of the design were considered and it is possible that several different approaches could be successfully demonstrated.

Concepts not considered in this study which could be explored in future research include the use of color or multispectral target acquisition, multiple target tracking, and more sophisticated target recognition algorithms.

Acknowledgement. This material is based upon work supported by the National Science Foundation under award number ISI-8860314.

REFERENCES

1. R.L. Anderson, N. Alvertos, and E.L. Hall, "Omnidirectional real time imaging using digital restoration," 15th Int. Cong. on High Speed Photography and Photonics, Proc. SPIE, Vol. 348, pp. 807-816, 1983.
2. M. Ehtashami, S.J. Oh, and E.L. Hall, "Omnidirectional position location for mobile robots," Proc. Int. Conf. on Intel. Robots and Computer Vision, Cambridge, MA, pp. 62-73, Nov. 1984.
3. Z.L. Cao, S.J. Oh, and E.L. Hall, "Dynamic omnidirectional vision for mobile robots," J. of Robotic Systems, Wiley, 3:1, pp. 5-17, 1986.
4. Z.L. Cao, S.J. Oh, and E.L. Hall, "Dynamic Omnidirectional Vision for Mobile Robots," Proc. of SPIE Intelligent Robotics and Computer Vision Symposium, Cambridge, MA, Sept. 1985.
5. Y.Y. Huang, Z.L. Cao, and E.L. Hall, "Region Filling Operations for Mobile Robots Using Computer Graphics," Proc. IEEE Conf. on Robotics and Automation, San Francisco, CA, April, 1986.
6. S.J. Oh, E.A. Kattan, Z.L. Cao, Y.Y. Huang, and E.L. Hall, "Experience with a Robot Lawnmower," Robots 10 Conf. Proc., April, 1986.
7. Z.L. Cao and E.L. Hall, "Omnidirectional Dynamic Guidance for Mobile Robots," 1986 ASME Computers in Engineering Conference, Chicago, July, 1986.
8. Z.L. Cao, Y.Y. Huang, and E.L. Hall, "Region Filling Operations with Random Obstacle Avoidance for Mobile Robots," Journal of Robotics Systems, 5(2), 1988, pp. 87-102.
9. S.J. Oh, D.J. Ahn, and E.L. Hall, "A Wide Angle Vision System for Mobile Robots," Proc. of IEEE Control Conf., 1988.
10. S.J. Oh and E.L. Hall, "Calibration of an omnidirectional vision navigation system using an industrial robot," Optical Engineering, Sep. 1989.
11. H.F. Swift, "Current and future activities in high speed photography and photonics," SPIE High Speed Photography, Vol. 348, San Diego, CA. pp. 8-14, 1982.
12. D.E. Whitney, "Force Feedback Control of Manipulator Fine Motions," ASME J Dynamic Systems, Measurement and Control, June, 1977, pp. 91-97.
13. J.L. Nevins and D.E. Whitney, "What is Remote Center Compliance and What Can It Do?", Proceedings of 9th International Symposium on Industrial Robots, Washington, March 1979.
14. J.K. Salisbury, "Active Stiffness Control of a Manipulator in Cartesian Coordinates," Proc. of 19th IEEE Conference on Decision and Control, Dec. 1980.
15. M.H. Raibert and J.J. Craig, "Hybrid Position/Force Control of Manipulators," Trans. ASME Journal of Dynamic System Measurement and Control, Vol. 102, June 1982, pp. 126-133.

A VISION GUIDED MULTI-MODULES CONTROL SYSTEM FOR AN AUTONOMOUS MOBILE ROBOT

Valérie SCHAEFFER (*), Claude BAUER (*)

Pierre NOVIKOFF (**)

Jean-Marie ALLÉE (*), Jean-Pierre BÉNÉJAM (*),

(*) *Laboratoire d'Informatique Fondamentale, Université Paris 6*[1]
(**) *Laboratoire de Robotique de Paris, Université Paris 6*

ABSTRACT

A mobile robot must be able to move around "by itself" in an evolutive environment. Our aim, in this paper, is to briefly present our proposals for a conceptual vision-based modelling of the robot environment and for an architecture of the robot control system.
Such a conceptual modelling permits a high level conceptual planning. It also allows to use quasi-conceptual algorithms for the pilotage of the robot, and keeps us from using a CAD modelling.

INTRODUCTION

One condition for the autonomy of a mobile robot is its reasoning capacity about its environment knowledge. Our case study is an autonomous mobile robot evolving in a man-made factory-like environment. The traditional solutions for environment modelling, based on geometry, seem to be not very well suited for a synergistic environment. Thus we propose a more conceptual world modelling based on recognition capacities of vision.

The choices made for this modelling straight affect, in a low manner, the perception and environment recognition modules, and, in a high one, the planning-replanning and global control modules.

We first rapidly introduce our environment modelling, then we present the architecture of the control system managing the classical loop perception-decision-action [1]. In the two last paragraphs we dwell on driving and perception modules.

---

[1] Université Paris 6
4, place JUSSIEU
75252 PARIS Cedex 05, FRANCE

1. ENVIRONMENT MODELLING

This modelling is based on the topography and the functionality (semantic) of the different places. We introduce several levels of topographic modelling . The first concerns the relations between the different world units, the decomposition of a world unit in free space and objects-in-it. These objects are the landmarks for any one who wants to move about in this world unit, this moving appears in the free space and the path is located in relation to some of the objects situated in the world unit.

Thus, we introduce [2] three graphs representing the different levels:
- the global graph expressing the connections between the different world units (nodes of the graph) via the transitions (arcs of the graph) ;
- the navigational graph whose arcs indicate the main roads and whose nodes are the points of view and are located by a well determined configuration of the beacons[2] ;
- the last graph, called inter-reachibility graph, joins each beacon of a world unit to all the others the robot is able to reach from it.

Such a modelling is guided by our will to be as near as possible to the internal representation of the world that a human is able to built and to abstract ourselves all the more of a purely geometrical and metrical representation of the environment. Our representation brings the following improvements:
- less information to memorize ;
- the man-machine communication is easier: there is no need of a translation phase because the human and the machine will use the same concepts and the same words to refer to ;
- the description is made in a natural or pseudo-natural language so it makes more easy a transformation from one environment to another ;
- it is not robot-dependent and allows any robot equipped with an appropriate interface module.

Moreover, such a hierarchical modelling of the environment, using different levels of abstraction allows the production of overall or local resolutions and thus optimizes the planning activity of the robot.

2. CONTROL SYSTEM ARCHITECTURE

Our control system possesses characteristics from intelligent control systems as they are described in the Intelligent Control Theory discussed by A. MEYSTEL in [3]. This theory is opposed to conventional control theory generally based on mathematics, and tends to utilize unconventional approaches to provide systems with elements of recognition, learning and decision.

Our control system for an autonomous mobile robot embodies several specialized modules managed by a supervisor which co-ordinates perception, reasoning and action. These different modules work on a common data base representing the environment modelling introduced in the first paragraph. They use a shared data structure to communicate, which is organized like a blackboard. We briefly present each of these modules.

---

[2] A beacon is a feature easily recognizable by the vision

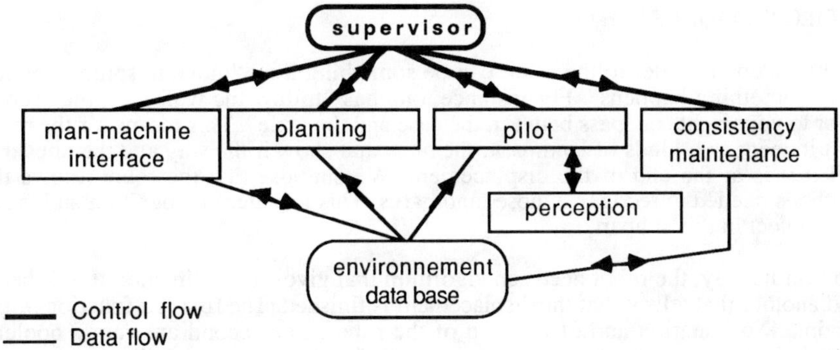

In our architecture, recognition and action are taken into account by two modules. The perception module entails vision and ultrasonic emitter/receivers managing. The pilot module commands the perception module, and its task is to carry out plans computed by the planning module.
If plan execution in real world fails, the pilot module calls basis procedures capable of handling with unimportant estimated error, or with failure with a pre-established answer. In case of major discrepancy between the expected environment state and the perceived one, the pilot module alerts the control which calls the replanning module.
These two modules, perception and pilot modules, will be discussed furthermore in the next paragraphs.

The planning module works in two manners. Off-line planning takes as input one or many orders from the operator to the robot, and then computes and arranges a sequence of actions, the actions size is the one needed by the pilot module to carry these planned actions out.
When plan execution fails in a serious way, replanning is in charge of modifying locally or globally the computed plan, and takes into account the failure cause: a difference between the current world and the model, or a trajectory loss.
The planning module integrates a temporal reasoning, managing the action length, the action temporal relations with its effects and with other actions, and external temporal constraints from the environment or the operator.

The system has to deal with the possible evolution of the robot environment. The consistency maintenance module works on the environment data base, and integrates the environment modifications noticed by the perception module, into the model on which the system reasons. This update is in three parts:
- translation of the perceived information into the model formalism,
- integration of the new information in the existing data base,
- checking of the consistency of the updated data base.

The dialogue between the operator and the control must be possible at every moment of the robot activity. The man-machine interface module allows the introduction of the environment description by the operator, more or less precise requests to the robot, with the start of a question-answer process if the request is ill expressed, or if the given informations are insufficient. This module manages also the system explanations about its activity. When a modification occurs in the environment, the system starts a dialogue with the operator in order to precise, if it is necessary, the environment data base updating.

3. PILOT MODULE

A displacement order to the robot can be something like: “move in some direction until something happens”. For instance, one has “follow the wall until the second door to your right" or "pass between the case and the table”. This means for the robot that it needs two kinds of landmark: the ones that show it the way and the ones that tell it it is at the end of the displacement. We suppose that the robot knows the methods needed to recognize those landmarks. This is pattern recognition and is not our subject in this subpart.

To find its way, the robot needs an algorithm that gives it the direction it will head, and another that tells it that the displacement is finished. The first is a function of the landmark orientation and of the step of the robot. The second answers a boolean signal. This signal tells the ownership to a set of points solutions to linear equations of possible mesures made by the robot. This set of points is a characteristic zone. For example, if we are trying to find the places were the difference between two distance mesures is constant, then, we will find an hyperbole: it is the set of points solutions of the linear equation of the two mesures of distance. This hyperbole is an example of zone that the robot must reach.

The displacement of the robot is then the following: recognize landmarks, choose the direction to be headed, verify if the goal is not reached. But the trajectory followed can cross an obstacle. For instance, in a corridor, a door can be open and the robot will have to avoid it. So we need to integrate the obstacle avoidance in the pilot module. This is done using an algorithm not far from the LUMELSKY's bug 2 algorithm [5]. It says that one can move forward in a straight line until an obstacle is reached. In this case, one has to follow the obstacle until the straight line is crossed again at a point nearer to the goal. We had to adapt this algorithm because the vision module brings a map of the nearby ground where there are three kinds of regions: the free ground, the unfree ground and the unknown ground.

We have tested the navigation module on three experiments: the first is a simulation of a robot on a computer with different missions, the second is a real mission that is for to robot to pass between two landmarks in a room where there was no obstacle. The last is for the robot to reach a landmark avoiding obstacles on its way. The navigation module is implemented on a SUN computer which is radio linked with our mobile robot.

4. VISION MODULE

The vision module is divided in two parts. The first subpart is concerned with pattern recognition. It has to recognize landmarks. As one cannot already recognize a door or a table with efficiency, we use artificial landmarks. The choice of the pattern is made under the following ideas. Considering the geometric description the pattern must be a unique object of the analysed scene which verifies a particular set of geometrical relations. The pattern chosen must be insensitive to noise introduced throughout the detection process and variant with the robot position and the ligthting conditions. Eventually, the pattern must be as simple as possible so that the robot can detect the pattern quickly. In our experiments we chose three black bars placed on a white flat background.

The second subpart of the vision module is concerned with the avoidance of obstacles. We designed an algorithm that can detect an obstacle from a single shot of a monocular camera [6]. This algorithm detects the ground in the image. Then, it extracts the boundaries of the ground and selects the ones that are real and the ones that occluding. Then, using classical geometry, it finds the 3D position of the real boundaries. The knowledge used to classify the boundaries is declarative knowledge. It describes the world in general terms. For instance, “all objects in the scene have vertical boundaries connected to there horizontal ones” or, “most of the objects can have only two horizontal boundaries on the ground seen from any points” and so on.

We have implemented all these algorithms (i.e. the vision module) on a vision processor which is carried by a turret placed onto our mobile robot. The vision module is controlled from the same SUN that the one the pilot module uses.

CONCLUSION

Our research confirms that the conceptual modelling presented in this paper is well adapted to the two levels of planning we involve, pilotage planning and off-line planning.
The first level is largely implemented in the vision and pilot modules. The high level planning module which purpose is to organize the tasks of the robot and to allow robot moving around in an uncertain world will be one next step. In that step, we will also take into account the temporal bearing of such a planning and the consistency maintenance of the environment data base.

REFERENCES:

1. R.Alami, R.Chatila, M.Ghallab, C.Laugier, J.P. Laumond, "Robotique et Intelligence Artificielle", Actes des 3èmes journées nationales du PRC-GDR IA (HERMES, Paris, 1990).
2. C.Bauer, V.Schaeffer, P.Novikoff, J.M.Allée, "Modélisation conceptuelle de l'environnement d'un robot mobile autonome", Actes de la Convention IA 90 (HERMES, Paris, 1990).
3. A. Meystel, "Intelligent Control: A Sketch of the Theory", Journal of Intelligent and Robotic Systems, 2:97-107, 1989.
4. R Chatila, G. Giralt, "Task and Path Planning for Mobile Robot", NATO ASI Series, Vol.F33, Machine Intelligence and Knowledge Engineering for Robotic Applications, Springer-Verlag, 1987.
5. Lumelsky, "Algorithmic and Complexity Issues of Robot Motion in an Uncertain Environment", Journal of Complexity 3:146-182, 1987.
6. Novikoff, Chellali, Milgram, "Navigation of a mobile robot with obstacle and landmark recognition by monocular vision", Proceedings of the ICARCV'90, Singapore, 1990.

SECTION 10 : EDUCATION AND RESEARCH

CIM IN HIGHER EDUCATION OVERVIEW
INDUSTRY PERSPECTIVE

Vincent J. Vitagliano, Eng.Sc.D., P.E.
University Consultant,
Academic Information Systems
IBM Corporation

INTRODUCTION

Manufacturing programs in higher education have existed for many years. In general, with few exceptions, these programs were limited. The interest for such programs generally existed in industrial engineering departments or as special options within mechanical engineering programs.

Today the situation has changed dramatically. A wide variety of manufacturing programs are being implemented in schools of engineering, in schools of technology, in community colleges, in technical schools and in schools of business. The subject is attracting great interest and support not only from within higher education but also from industry and government.

This presentation will review the developments in computer based academic manufacturing programs from the decade of the seventies to the present.

CAD/CAM IN INDUSTRY

In the seventies the CAD/CAM revolution took place in industry. It began with the development of computer aided drafting (CAD) programs that were implemented as part of turn-key systems. In the right environment, such systems, though expensive, were both cost effective and productivity improvers. Once the drawing information was captured in the computing system, it became a simple matter to use this information to generate the paper tape that could control numerically controlled milling machines. And CAM, computer aided manufacturing, came in to being.

With the above developments, engineers then realized that the potential of the computer existed to expand the horizons of CAD to the area of design. Similarly the horizons for CAM began to extend into other areas of manufacturing. In the eighties, designing for manufacturing began to receive significant attention and the concepts of CIM, computer integrated manufacturing, were introduced to expand the limits of both CAD and CAM.

CAD/CAM IN EDUCATION

In the late seventies, engineering school administrators, recognizing what was happening in industry, began to assess the potential impact of CAD/CAM within their own environments. They were confronted with several concerns: was CAD/CAM in industry here to stay? or was it just a passing fad? was it affordable in higher education? A number of educational institutions pioneered in this endeavor in the late seventies and early eighties, and realized very positive results. Yet, the affordability issue was very serious.

In this same time frame, the U.S. began to experience setbacks within the international economy. Other nations were realizing productivity gains that far exceeded what was happening in the U.S. As a result, U.S. Industry, sensing the industrial economic problem, its dependency on graduates with technological backgrounds, and the need for engineering and technology schools to update their academic programs, initiated some decisive steps to effect remedial actions. These actions resulted in providing significant support to higher education to update their academic programs with computer based education as well as to enhance existing or initiate new manufacturing programs. As a result much has happened and continues to happen that will yield significant benefits to U.S. Industry and to the U.S. economy.

Examples of IBM's support of higher education are:

1 the CAD/CAM Initiative Competition and the Manufacturing Systems Engineering Curriculum Development Programs in 1982 and 1983, which were valued in excess of fifty million dollars; and

2 the current CIM in Higher Education Program, initiated in 1988, which is valued in excess of thirty million dollars.

The CAD/CAM Initiative was announced in September 1982 and was directed to engineering schools. It was conceived by Lewis Branscomb, who at the time was IBM's Chief Scientist. Dr. Branscomb, sensing the falling behind of higher education in its ability to stay current with technological developments in industry, recognized that engineering schools were having difficulty commanding resources to procure adequate computing equipment. To highlight an industry requirement that engineering graduates had to be exposed to the latest technological developments, a competition was announced whereby IBM would grant twenty CAD/CAM computing systems of hardware and software each valued at two million dollars. 115 schools submitted proposals from which 20 were selected for support. The intent of these grants was to expand and/or initiate strong academic programs in both instruction and research.

As a complementary part of this CAD/CAM Initiative, and under the direction of Erich Bloch, who at that time was an IBM Vice-President (and who is now the Director of the National Science Foundation), a manufacturing curriculum development competition was also announced. 170 schools participated in this competition. The proposals received defined programs in Manufacturing Systems Engineering that would lead to the granting of masters degrees. From these proposals five schools were selected each to receive two million dollars in cash to implement their proposals.

What was most interesting was the fact that both of the above competitions proved to be catalysts that incented other computing companies and industrial organizations to support related activities. In addition, institutions of higher education reset their priorities and began directing their own resources towards supporting computer based engineering education. One very positive result of this activity was that within a year there were about sixty engineering schools that had expressed interests in initiating similar masters level programs in manufacturing systems engineering.

In 1987 it became apparent that still more could be done by industry to spur on added activities that would enhance manufacturing programs in higher education. This resulted in the CIM in Higher Education Program which was announced in December of 1988. This program is not a grant program. Rather it is directed to both two and four year institutions of higher education with engineering, technology and business programs which already had manufacturing programs, and which were poised to expand their programs toward strong CIM implementations. The intent is to better prepare students in these programs to enter the manufacturing environment and be able to face the multiplicity of challenges that exist therein. CIM related hardware and software is provided the institutions in return for their initiating CIM instructional and demonstration programs that are directed to supporting industrial organizations that they might serve. Though new, this program has received a very positive response for its pragmatic directions, the inclusion of two year schools and schools with technology programs, and for fostering the merging of engineering/technology and business school programs.

To date seventy institutions are benefitting directly from these programs, with many other institutions having received indirect benefits.

SUMMARY

The results of this Industry/Higher Education interaction are most positive, providing very significant benefits to both parties. The receptivity and cooperation of higher education to work closely with industry is proving to be an interesting phenomenon. Most significantly will be the enhancement of CIM programs now taking place, which have the potential to better prepare individuals to both effect and move into tomorrow's computer integrated enterprises.

CIM STARTUP IN A LOW MANUFACTURING DENSITY, HIGH-TECH ENVIRONMENT

JOE H. MULLINS
University of New Mexico, Albuquerque, NM 87131

ABSTRACT

The IBM CIM Alliance program is now starting at the University of New Mexico. Albuquerque has a small but growing manufacturing sector, primarily of small to medium sized companies. Owing to the strong influence of Federal Labs, the environment is very technology intensive, and creates special situations for applications of CIM. Some of these are discussed in this paper.

INTRODUCTION

The need to address the small or the medium sized manufacturing sector has been pointed out in a number of places. The number of jobs created in the industrial sector has historically been largely concentrated in the "foundation firms",as they have been called [1], that is, the medium sized manufacturers employing 20-499 employees. They employ 55% of the people in U.S. manufacturing, and have provided the bulk of the new jobs in the last decade.

Yet, relative to the large manufacturers (>499 employees), they have been slipping in productivity over the last decade or two [2]. The situation is even worse for companies smaller than 20 employees.

There are a number of reasons for this disturbing trend, but at least one reason is that these small and medium sized companies have been much slower to adopt modern computer-based technologies than have their large counterparts [1] [2]. No doubt this is caused by a lack of funds, advice, and expertise in the smaller companies compared to the larger ones.

CIM, as a general concept, covers the manufacturing enterprise from design to delivery, and includes a computer hierarchy from large business oriented systems to real time automation. For a large, high volume manufacturer, the situation is fairly clear: all parts of CIM may be profitably applied.

For the small manufacturer, especially one whose product is low volume and probably high-tech, the situation is something else again. Cost of the system is of paramount importance, and such elaborations as automated storage and retrieval, automated guided vehicles, and elaborate MRP-II packages are probably not appropriate.

Helpful but low-cost solutions are needed, preferably "off the shelf". As the pace and volume of the enterprise picks up, it should be possible to add parts of the CIM system, so thought should be given to modularity and upgradability. But small to medium manufacturers need help.

CATEGORIES TO BE CONSIDERED

Once having determined that a single CIM package approach is not feasible for small manufacturers, it becomes reasonable to look at categories of things that could be applicable, constitute a useful stand-alone facility, and lend themselves to later incorporation into an integrated system.

Some general areas that suggest themselves are

1. CAD packages

2. PC or workstation based inventory or MRP, MRP-II and production planning systems

3. Off-line computer based modeling and simulation, CAPP (computer aided process planning)

4. CNC, CAM or other software driven manufacturing processes or machinery

5. Automation and/or robotics

SOME CHOICES

Keeping in mind the cost and training constraints, we concluded that item 5) was impractical in all but a few cases, so have not seriously addressed it. Item 4) is applicable only at the low end in cost and machine complexity, and we have given it some attention, but it is not a very high priority.

Item 3) is within reach of most manufacturers who have PCs, and thus presents some intriguing possibilities. The problem here, of course, is that small manufacturers tend to have small, simple operations, and computer simulation has definite limits.

Item 2) has potential, but only for relatively simple and inexpensive packages, operating on PC's. We do intend, however, to offer training in these areas, up through and including AS/400 based systems such as MAPICS.

Item 1), CAD, is the only area which has unambiguous benefit to most small manufacturers, especially considering how economical these packages have become. This is an area which will occupy the majority of attention in the early phases of our CIM program. AutoCAD, IBM PC CAD, CADAM, and perhaps others will be available.

Primarily for purposes of instruction, limited facilities will be available to allow students to download designs developed on the PC equipment to software driven machining centers. Operations are limited to aluminum, plastics, or other soft materials, but serve our needs in that they involve equipment of limited cost and complexity.

INSTRUCTIONAL PLANS

We felt we would like to extend our reach into the CIM area as a service to the industrial community, even though our equipment choices are limited. To this end, we have designed a comprehensive CIM curriculum modeled upon a highly successful TQM series that is now underway at the University of New Mexico (UNM).

This plan employs instructional TV (ITV) techniques very extensively, as does the TQM series. This has proven to be an excellent way to reach a geographically dispersed audience such as we have in New Mexico.

The Total Quality Management (TQM) Series

This series began in June, 1989, with a one-day video broadcast on the topic with four speakers. This conference was broadcast only locally over the UNM Instructional TV ITFS network. It was very well received, so was followed up with a local three-day workshop in October, not video broadcast.

The reception of the workshop was outstanding,so an entire video conference was designed to be broadcast nationwide over satellite. This began in January,1990, and is continuing for 14 modules, of approximately four hours each, on the 2nd and 4th Thursdays of each month.

Currently the broadcast has 22 sites, spread widely over the U.S. and Canada, with an estimated audience of 1000. As a result of the favorable reception of this series, we have also instituted a series of workshops, each of approximately three days duration, which we give in Albuquerque and in other locations (Dallas, San Diego, Washington, St. Louis, Boston, for example).

We are convinced that part of the success of this series is due to a careful blending of engineering and management in the series. It is jointly sponsored by the College of Engineering and the Anderson Schools of Management, and we have endeavored to maintain a balance between the disciplines. It is this model we intend to follow in the upcoming CIM series.

The Planned CIM Series

In planning a series of courses that would meet our goal of helping small to medium manufacturers while not neglecting large manufacturers we came to the conclusion that the engineering/management balance is critical.

Since small manufacturers have difficulty absorbing new systems and equipment that are large and expensive, it is especially important to leave them with some procedural and management approaches that can be applied without significant capital investment. And, as has been frequently observed[3][4] , introduction of CIM or other automation systems without the concomitant human systems is futile.

We expect to begin the series with an executive overview CIM seminar in July, 1990, which will be approximately 1/2 day (morning) followed by an afternoon hands-on demonstration. This session will not be video broadcast.

Following this will be a 1/2 day executive overview video CIM conference in the fall, with a much wider audience, and a three day local workshop in the fall or early spring. The three day event will cover the following topics.

- Introduction to CIM
- CIM technology options
- Human resources options
- Management implications of CIM
- CAD/CAM
- Computers in manufacturing
- Justification for CIM

Assuming that the sessions have been well received, we will then plan a major video conference series for 1991, modeled after the successful TQM series. It is expected that the CIM series will cover the above topics, plus an additional topic, "Quality considerations in CIM".

Finally, a series of hands-on workshops will be initiated covering these topics, also to begin in 1991 if the need appears to be there. All of these initiatives will require an adequate student body, since these are professional and continuing education courses, which must operate on a full-cost recovery basis. Academic funding is not available to support them.

CONCLUSIONS AND SUMMARY

Because of a limited resource base and a low density of manufacturing in New Mexico, special choices need to be made with respect to initiating CIM options. Although the technological environment in New Mexico is impressive, it is still necessary to keep expenses for the (mostly small) manufacturers at a low level, while introducing productivity improvements.

Our choice is to concentrate on workstation and PC based systems as much as possible, to provide hands-on training for small manufacturers in such areas as CAD and inventory control systems.

In addition, however, we are planning an extensive series of video conferences in CIM to reach a more geographically dispersed audience, and a series of hands-on workshops that can be given locally or as a "road show", to reach the manufacturers directly, whether locally or throughout the state. Both the video conferences and the workshops will address issues in CIM that affect both the engineering and the management aspects, which we have found to be a critical feature.

REFERENCES

1. Don Luria, "The Case for the Base", Modern Michigan 2 (1), 4-8 (Fall 1989)

2. Don Luria and Carey Treado, "What Does Technology Do?", Modern Michigan 2 (2), 4-7 (Winter 1990)

3. Toward a New Era in U.S. Manufacturing, The Need for a National Vision, Ch 2, 31-47, National Academy Press, Washington D.C. (1986)

4. Dertouzos, Lester, Solow, Made in America, Regaining the Productive Edge, Ch 9, 117-126, The MIT Press, Cambridge, Mass (1989)

A PARTNERSHIP SOLUTION IN A GLOBAL MARKETPLACE

Dr. Charles W. Branch, President,
Trident Technical College

We have gathered here to a large extent because of our faith in the promise of a recovery effort: that of American industry in today's global economy. We are here because the economic playing field was transformed while much of U.S. industry had turned its head. We are here because we have the privilege of experiencing--and the challenge of responding to--what is becoming the second industrial revolution.

It can be described as the third phase of the Industrial Revolution, according to Professor Tom Storier of Bradford University in Oxford, England. In the first phase, machines extended the range of human muscle. In the second, the human nervous system was extended through the invention of the telephone, radio, and television. And here in the third, computers are being used to extend the capabilities of the human brain.

Computer Integrated Manufacturing, despite its ominously cold and mechanical-sounding name to the uninitiated, is not a method designed to eliminate the need for human resources. It is a way to bring those people together to pursue a common goal. Much like a classroom environment, CIM presents the same information to those on the job, and they assimilate that information to produce the desired results.

CIM is an investment for a manufacturer in today's global economic environment, and the CIM in Higher Education alliances developed by IBM across the United States function as the warranty to protect and diversify that investment.

Most importantly, we have gathered here because we understand and appreciate the potential of partnerships between education and industry as a solution for improving America's productivity and attention to quality in a global marketplace. In this respect, "partnership" is synonymous with "progress." Regardless of the type of partnership, however, the key ingredient is the people needed to make things happen.

Partnership, progress, and people. It is an unbelievably powerful combination which can propel a given entity from commonness to greatness while creating the same sort of transformation for those who have a stake in the future of that entity. Trident Technical

College provides my favorite example. In 1985, the people of the Charleston, S.C. had a specific expectation level of Trident Technical College. We were established as an adjunct activity to industrial development in a three-county service area of South Carolina. For more than 20 years, the College reacted well to its environment. An industry would locate or re-locate, and we would respond with the necessary start-up training for the incumbent workforce.

But as positive as that sounds, it soon was not good enough, and the evolvement of Charleston's second largest industry provides an illustration. Charleston is a tourist's mecca and well-known vacation destination. Tourism and hospitality-related services generate 20,000 jobs and $800 million in revenue each year for Charleston County. Yet it was not until four years ago that we began to develop programs of study in hotel, motel, and restaurant management. We may have reacted well but we should have been pro-active.

So we began a turnaround at Trident in 1986 that reminds me a lot of the CIM in Higher Education Alliance. Instead of waiting to be called upon, we began looking for unmet needs. All we had to do was ask. Whether through partnerships or technical advising, we discovered needs that we could satisfy.

Evidence that our mission statement was not etched in granite came quickly. In one instance, local industry sought assistance from Trident, but this time it was not start-up training. It was re-training--in basic skills remediation. The facts were simple: Mobay Chemical Corp. was poised to land lucrative contracts, yet its employees could not comprehend equipment manuals so that plant production efficiency could be improved. Our resulting partnership with Mobay had five components:

1. Assessment of the workforce. We took surveys of Mobay employees to determine basic skill levels.

2. Curriculum development. We custom-designed basic skills courses to match survey results.

3. On site training. This provided a minimum of disruption to plant operation and employee schedules.

4. Flexible scheduling. Again, a minimum of distraction/disruption to established work schedules.

5. Accountability. Mobay needed a promise that we could meet their needs. Our confidence level in our people allowed us to guarantee that if we did not raise the reading levels of Mobay employees by two grades after 25 hours of instruction, they didn't owe us a dime.

Partnerships imply an equality of sharing. We helped Mobay meet an objective. But that partnership also demonstrated to us the theory of "Use what you sell; sell what you use." The courses taught at Mobay are commonly taught in the credit side of the house; now we realized it could be packaged for business and industry.

The local chamber of commerce offered another opportunity to view the same principle. They were in a quandry over the best way to automate their office operations. What type of computer hardware and software? What capabilities? How can we develop a system that can be added to later? These were all questions we helped them answer by freeing up a computer whiz to study their operations and make recommendations.

The local travel and convention bureau had been conducting a tourism profile of visitors to Charleston, which was a service for local restaurateurs and hoteliers. It was a data base to guide them in their marketing strategies. About the time their source of volunteer surveyors dried up, we were there with students from our business programs, who collected that data and compiled a tourism profile.

Along with these partnerships, we systematically began development of:

* Business-education alliances
* Revitalized academic advisory committees
* Active relationships with economic developers
* Internal and external task forces to tap the pulse of the community
* Inventories of faculty and staff talents

In fulfilling these unmet needs, we demonstrated--especially to those that had that certain expectation level--that we are more than a community college. We are a community resource.

One of the dividends that resulted was the wealth of untapped human resources we found in our own back yard. The fact is, entering partnerships and fulfilling unmet needs brings out the best in people and fills them with unbridled enthusiasm for their professional abilities, while unveiling talents never imagined.

At Trident, it turned a former small-time high school drafting teacher into a nationally-recognized leader in CAD/CAM instruction. It transformed a student counselor into a prolific academic program developer and specialist in problem-solving. The computer whiz mentioned earlier was a librarian only a few years ago.

The new attitude offered the right environment for an accomplished and nationally-recognized two-year college teacher to develop the proposal which led to our CIM partnership with IBM. What that alliance has done for Trident in less than three years is a tribute to the vision it has instilled.

On the heels of the 1988 announcement of our CIM in Higher Education partnership with IBM came some more exciting news. We were named to direct field operations for a multi-million dollar effort aimed at restoring competitiveness to metal working plants throughout the Southeast.

Through the Consortium for Manufacturing Competitiveness, that effort holds promises for thousands of industries throughout the Southeast and the people who are employed in those industries. Our portion of that proposal--written by the same individual who submitted our CIM in Higher Ed proposal--was the determining factor in convincing the federal government that we could pull this project off. The good news didn't stop there, however.

The local IBM office is well aware of the calibre of personnel we have at Trident. We work very closely with them, and it would have been impossible to get where we are today without their guidance. Certainly their confidence in the abilities of our people went a long way toward Trident's designation in June of 1989 as one of only six IBM CIM in Higher Education Resource Centers in the United States.

And as a kick-off for the fiscal year 1989, our partnership with IBM got a home. Trident's Center for Productivity and Technology Transfer opened in July 1989. It is a fitting tribute to the combined efforts of many people--ours and those at IBM. It is a showpiece of hardware and software, but most importantly, it is a functioning resource for students, faculty, local industry representatives, visitors from colleges and universities in the region we serve as a resource, and to IBM representatives.

All that's needed is to review a weekly schedule of CPTT activities: a CAD to Management Database seminar, an IBM Lumber Seminar, an Expert Systems Demo, a tour by visitors from an out-of-state university, a desk top publishing demo, an ITT meeting, a Machine Tool Technology Advisory Board meeting, and a meeting of the local Economic Development Board are just a few.

Word travels quickly when you are thrust into

a leadership role. It was gratifying to watch the state's other two-year colleges beat a path to our door as the IBM partnership put Trident at the forefront of computerization across academic disciplines. Yet it was another terrific morale booster to see the parade of out-of-state institutions who have paid visits to this Center for Productivity and Technology Transfer.

We have taken the time to determine the monetary value of all the support, training, hardware, and software we have and will receive as a result of our partnership with IBM through the CIM in Higher Education Alliance. By this time, it's well in excess of $5 million.

Impressive as it is, however, it can't hold a candle to the value gained by exposing students, faculty, staff, and local industrial representatives to technology at their fingertips. You'll find Trident Technical College at the forefront of technology transfer today, a scant two years after we announced our partnership. You'll find participants from across the nation attending robotics seminars at Trident. You'll find Trident in leadership roles in local, statewide, and regional think tanks designed to apply the concepts of automated manufacturing to reap gains in efficiency, quality, and competitiveness for industry.

You'll find one of the most prolific examples of academic computer integration education in a two-year college environment. We are teaching upper level courses that have previously been impossible to teach locally, such as CICS Assembly and Data Base Design.

We are involved in computer-aided instruction using portable PCs. MAPICS is now woven into the fabric of our Accounting and Industrial Management programs. IBM's Professional Office System software has become a welcomed high-tech addition to our Office Information Technology program and our Office Automation Center. And through IBM's ACIS program, we are set to become a demonstration center for WISCWARE.

All this generates a contagious enthusiasm by those who can visualize the future. That is where the ultimate value of the CIM in Higher Education Alliance--the IBM partnership--can show its true colors for any educational institution.

IBM is recognized as an educational problem solver. The partnerships IBM establishes with colleges and universities provide the wherewithal to pump enthusiasm into those who work at those institutions.

The partnerships allows us to become problem solvers and economic development leaders for the areas we serve.

The CIM in Higher Education Alliance offers a golden opportunity. An opportunity to take the concept of Computer Integrated Manufacturing and mix it with the right people, a little risk and a lot of initiative..the end product is a Computer Integrated Enterprise. For American business and industry that means a partnership solution for a global economy.

RESEARCH LABORATORY FOR FISH PROCESSING AUTOMATION

Clarence W. de Silva
NSERC Professor of Industrial Automation
Department of Mechanical Engineering
The University of British Columbia
Vancouver, B.C.

ABSTRACT

This paper outlines the organization, resources, and research and development activities of a newly established laboratory for industrial automation at the University of British Columbia. The laboratory has been established in the Department of Mechanical Engineering primarily to support the research and development activities associated with the NSERC Chair of Industrial Automation. The research is focused on the development of advanced and low-cost technology for flexible automation of the fish processing industry. The main objective is to upgrade the technology used in the mechanical processing of fish, thereby reducing wastage of the primary product, improving efficiency, and making the local industry more competitive in world export markets. Establishment of an infrastructure in industrial automation within the university and locally train/retrain control engineers for local industries are related objectives. As a specific task, an experimental fish processing cell is being developed in the laboratory. The theme of the activities of the laboratory is the integration of advanced control, high-level computer vision, and manipulation, for application in the area of fish processing.

INTRODUCTION

There are many process industries in Canada. Virtually all these industries make use of process control techniques. For example, fish processing plants, lumber mills, pulp and paper plants, mining equipment, wineries and machine tools commonly employ servocontrol. There is no doubt that the performance (e.g. speed, accuracy, product quality) of a process plant can be improved through better technology of automation. Several obstacles are present for achieving performance improvement in this manner. High cost of modernizing existing process plants (for example $10 million was spent by Crown Forest Industries in 1987 to upgrade sawmills and plywood operation in Okanagan, B.C.) and lack of qualified manpower can be cited as the two main obstacles in this context. The situation has been aggravated due to increased dependence on foreign technology products. Natural-resource industries have suffered further due to growing claim by other sectors (e.g. manufacturing) for a major share of the qualified technical manpower.

A slight improvement in process performance could lead to a major increase in revenues. For example in the fish processing industry in the province of British Columbia, a one-percent improvement in raw product recovery (say, by introducing an improved butchering process) could increase the present revenue by as much as $5 million. Research, development, and implementation of advanced technology of automation are crucial to achieving performance improvement in industrial processes. Furthermore, by developing the necessary technology and hardware locally, we will be able to avoid unnecessary delays in implementation while feeding the cost of development back into the local economy.

Three steps can be identified for improving industrial automation within the province in particular and throughout Canada in general: establish within our universities an adequate infrastructure in industrial automation and develop advanced technology that is appropriate for local industries; rapidly transfer low-cost techniques of advanced automation to local industries; and locally train (retrain) control engineers for local industries through university/industry collaboration and interaction. A primary goal of the newly-established Laboratory for Industrial Automation is to make significant contributions towards all three steps.

The research and development activities of the laboratory are focused on the integration of advanced control, high-level computer vision, and sophisticated mechanical tooling and manipulation, and on the application of this technology for improving the performance (speed, accuracy, yield, product quality, etc.) of process plants. A specific project which is carried out in the laboratory is the development of a flexible fish processing cell. The cell controller will be available on a low-cost target machine and will not necessarily replace existing low-level servocontrollers but rather supplement them by serving as a very high speed "human in the loop", performing intelligent monitoring and tuning functions. A high-level computer vision system that uses model-based knowledge, reasoning through past experience, common knowledge, and learning is being developed to guide the process (e.g., butchering or filleting operations). Design and development of a robotic cutting system consisting of a customized cutter and sophisticated fixtures and handling devices, has been undertaken.

RESEARCH ACTIVITIES

Research projects in a variety of areas that are relevant to industrial automation are being carried out in the laboratory. Specific projects include knowledge-based control, model-based vision, coordinated manipulation and control, grasping and handling, and dynamics of fish processing. In knowledge-based control, the previous work of de Silva and MacFarlane [1] is extended to a fish processing cell. The controller consists of a three-level hierarchial structures. The lowest level consists of hard controllers including servo controllers. The intermediate level is an intelligent observer which will monitor the lower level and advise the top level, the intelligent tuner. The intelligent observer is a set of rule-based knowledge sources that are attached to the individual controllers in the lowest level. The rules carry the knowledge needed to evaluate the performance of a particular controller using sensory data and specifications such as error tolerance, oscillation amplitude tolerance, and acceptable rate of error convergence. The intelligent tuner is based on fuzzy reasoning. To develop this, first, expert knowledge on controller tuning is expressed as a set of linguistic statements. Next, membership functions are established for the fuzzy quantities within the fuzzy rules. Rule matching is accomplished using the compositional rule of inference.

Robotics and computer vision have been combined to produce useful results in a variety of applications [2]. One of our goals is to adopt medium to high-level vision into fish processing automation. Vision guidance could be applied in at least two ways here. For example, the cutting contour could be generated first and then it could be used to generate the reference drive signal for the fish cutter. Alternatively, cameras mounted on the cutter could be used to guide the cutter in real time. Our research activities are concentrated on the first approach. The second approach is a developmental activity which might involve the use of alternative sensing techniques such as proximity, tactile, and ultrasonic sensors [3]. In the present application, the complexity of the vision problem is eased to some extent due to the fact that the object (fish) is already recognized, at least in the generic sense. Hence the objective will be to generate the cutting contour sufficiently fast, to meet the process speed requirements. Low to medium-level computer-vision techniques involving image transformation, enhancement, and analysis of basic features such as edges, regions, shapes, colors and texture, are reasonably well established and widely applied. Unfortunately, application of these techniques to directly generate, say a cutting contour for an arbitrary fish would be computationally slow and unacceptable in terms of process speed requirements (less than 300ms per fish). Our research effort is directed to expediting vision-based generation of cutting contour, through the use of model-based vision techniques. A fast method of cutting contour generation is being developed in the laboratory. Consider the system shown in Figure 1. Suppose that a series of models representing all possible variations of fish that are normally expected at a processing machine is generated. Each model will consist of a cutting contour and a set of attribute values which identifies the particular model. A set of attribute values is processed at the vision station during real-time operation of the process. These attribute values, when transmitted into the local database of the rule-based system, will fire a rule which in turn will identify a model of fish. The model will provide the necessary cutting contour to the robotic cutter. The time taken to compute the characteristic attribute values is generally shorter than what would be needed to compute a cutting contour from an image of a fish. The generation of the model base and the rule base are necessarily carried out off line. Hence computer overhead does not present a serious obstacle here. The main steps involved are as follows : Record static and video images of fish; read images into a vision workstation; enhance and analyze the images; identify a cutting contour for each image in terms of pixel coordinates; make dimensional measurements (which will include possible representations of attributes of each image in pixel coordinates and include a scaling measurement); move the reference frame of contour data to one end of the contour; compute the chord length of each cutting contour and nondimensionalize the contour coordinates with respect to this value; curve fit each contour; pick one contour and match it with the remaining contours through a process of

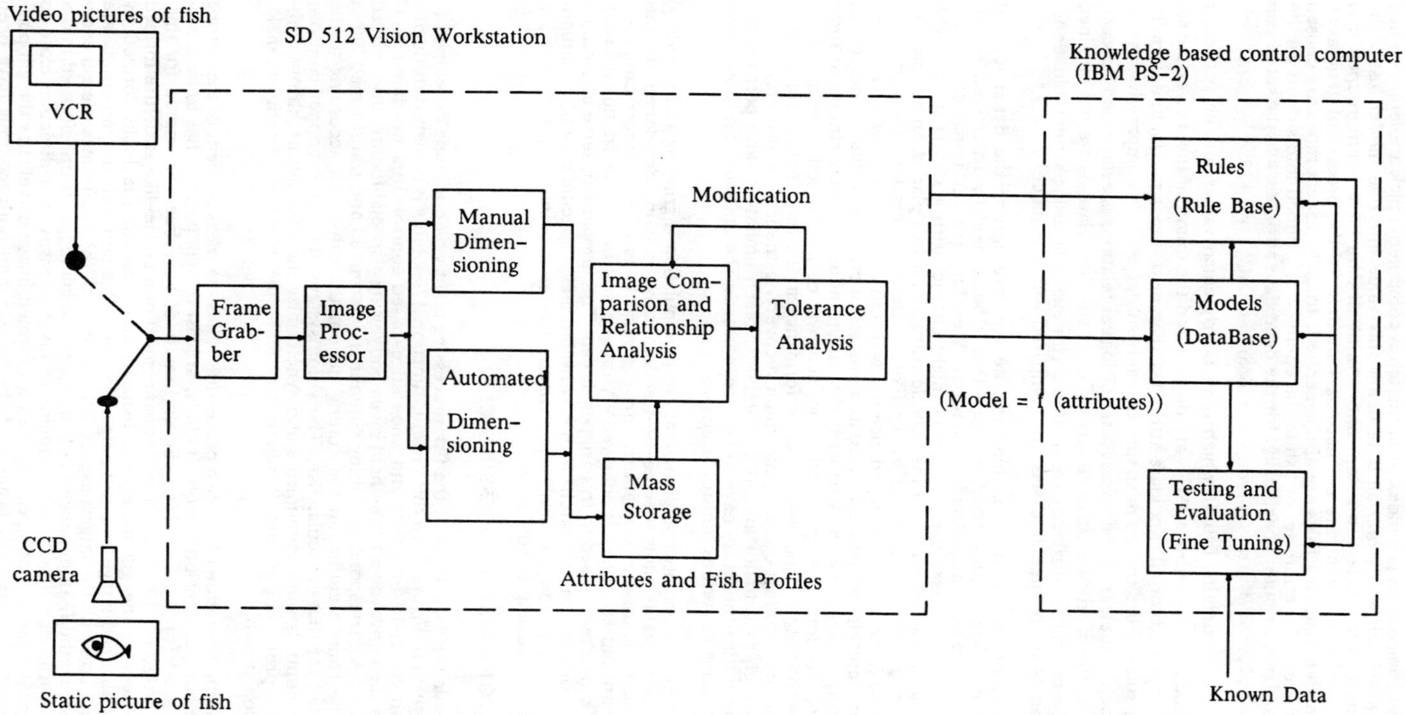

Figure 1. The System for the Off-Line Development of a Model Base for Salmon

frame rotation and comparison (an index of mismatch is computed); pick a tolerance value for the index of mismatch and group all data below this value into one model; using the remaining contours follow the last two steps above until all data have been grouped into models. This procedure will generate the necessary models. In the process of generating the models, the model–matching rules are also generated. Initially, aspect ratios are chosen as candidate attributes. The entire set of fish in the database is grouped into subsets on the basis of each candidate attribute. Specifically, for the candidate attribute ac_i ,the set of fish data $F = \{ f_j : j = 1, 2, \cdots, n \}$ is grouped into subsets AC_{ir}, $r = 1, 2 \cdots, m$ such that each subset AC_{ir} contains fish f_j which have nearly equal values for the particular attribute (ac_i). Note that the subsets AC_{ir} are disjoint and the union of them gives the set F. This process is repeated for all candidate attributes. The next step is crucial to determining the matching rules. First the membership of each subset AC_{ir} is compared with the membership of each model. In this manner, a set of logical relations is established between model sets M_k and the candidate attribute sets AC_{ir} . If suitable relations are not obtained in this manner, the candidate attributes are modified (for example, using set combinations) and the memberships of the modified sets are compared with the models.

In the area of coordinated manipulation we are investigating the dynamics and control problems associated with a flexible object that is being held by an actively reconfigurable fixture and being manipulated by a robot. We allow for additional constraints from the environment. For example, the object might interact with a sliding surface with friction. A particular situation would be the robotic cutting of a fish placed on a table top, while being held by a robotic gripper or active fixture. During operation, the cutter will encounter a variable impedance; the impedance is low in soft flesh and high when cutting a bone. Furthermore, cutting can be interpreted as a process of creating additional degrees of freedom in the body, and gripping may be treated as an introduction of constraints to some existing degrees of freedom. An optimization problem will be formulated in terms of dynamics of the overall system. The associated performance index may include representations for task time and product quality (maximum recovery, esthetic features, etc). Additional constraints such as actuator and tool limitations, reach limitations and perhaps impedance along the cutting trajectory have to be incorporated. The optimization should lead to design modifications as well as control strategies.

Another research activity complementary to manipulation is gripping and handling. In particular, a conformable multi–degree–of–freedom hand is being developed. In this hand a single actuator is used to sequentially drive several links of each finger until conformable closure around an object is achieved. A unique and innovative frictional switch augmented by a flexi–screw is used in this gripper design. This gripper can serve as a reconfigurable fixture in our robotics research. Fish is considered as a specific case of a manipulated object. To aid this study, a nonlinear dynamic model is being developed for fish under a variety of processing conditions.

THE LABORATORY FISH PROCESSING CELL

A schematic representation of the fish processing plant which is being developed in the laboratory is given in Figure 2. In this system, functional flexibility is achieved through a robotic cutter with vision–based driver and knowledge–based control. In a heading operation, each fish on the conveyor is firmly held by its body using a special fixture, and imaged at the vision station. A corresponding cutting contour is determined and transformed into a drive signal for the robotic cutter. At the cutting station, the fish is cut according to the trajectory generated for that particular fish. The vision system, the cutter/gripper controls, and the conveyor controls will communicate and synchronize the activities of various components in the fish processing system. A supervisory real–time operating system will make this communication possible.

Crucial to the operation of the fish processing cell is the measurement of the cutting position using the computer vision station. Initially, at least the gill position has to be measured at a rapid speed (order of 10 Hz.). Several methods have been attempted for this. The methods are based on the isolation and enhancement of an edge–like structure formed by the shadow effect of the gill plate of a fish. In one method, the blobs formed by thresholding the enhanced image and grouping edges through a connectivity procedure, are boxed. The longest box with the largest aspect ratio (length/width) determines the gill coordinates. In another method, an enhanced 2–D image of the gill region is transformed into a 1–D representation through directional averaging of pixel intensity in a direction close to the edge direction. A characteristic peak in the transformed image establishes the gill position. The latter method is generally faster than the former, and both methods have been

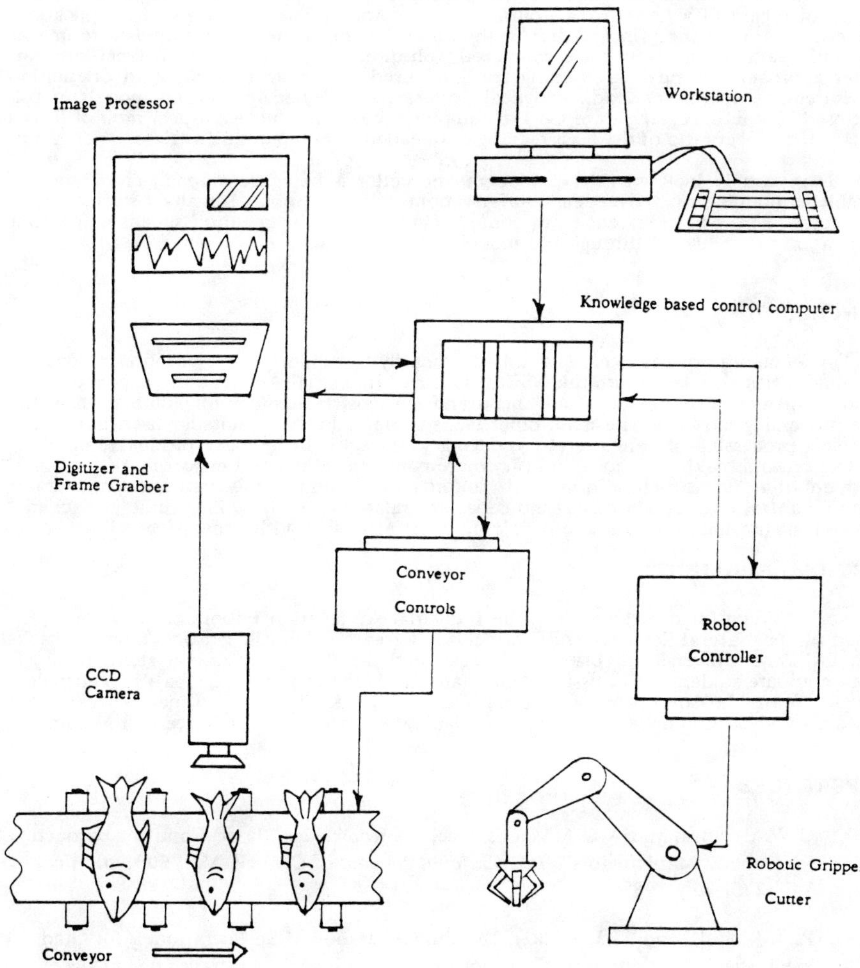

Figure 2 Schematic Representation of the Laboratory Fish Processing Cell

implemented on a vision workstation which contains a multi–tasking operating system with a real–time kernel.

The orientation of fish on the conveyor is also important for the accurate positioning of the cutter. Several vision–based techniques are being developed to rapidly measure the orientation of a fish. In this context, the concept of orientation is being carefully examined. We feel that instead of the widely–used representation of orientation by a principal axis of the second moment of area, the major axis of the primary ellipse that partially fits to the head region of a fish, in the Fourier sense, is appropriate in the present application. Nevertheless, both the principal axis method and the ellipse method are applied to images of pacific salmon and results are compared with direct measurements of orientation. Another approach extends the "boxing method" used for gill measurement to orientation measurement. In this method, analytical expressions obtained for various models of fish enclosed by a least rectangle are used to compute the orientation. An aspect ratio of fish as well as the aspect ratio of the least rectangle is needed to compute the angle of orientation.

A doubly–inclined (V–type) pneumatic cutter is being developed. The cutter assembly is mounted on a five–degree–of–freedom industrial robot. Initially a straight cut is attempted. This will be extended to contour cutting once the required speeds of contour generation are achieved through the model–based approach.

CONCLUSION

A laboratory has been established primarily to support the research and development activities associated with the NSERC Chair of Industrial Automation. The research is focused on the development of advanced and low–cost technology for automation of the fish processing industry. The main objective is to upgrade the technology used in the mechanical processing of fish, thereby reducing wastage of the primary product, improving efficiency, and making the local industry more competitive in world export markets. Establishment of an infrastructure in industrial automation within the university and locally train/retrain control engineers for local industries are related objectives. A specific project carried out in the Laboratory is the development of a flexible fish processing cell.

ACKNOWLEDGMENT

Funding for our research in the Industrial Automation Laboratory has been provided by the Natural Sciences and Engineering Research Council, British Columbia Packers, Ltd., the University of British Columbia, and the B.C. Advanced Systems Institute. Four graduate students (F. Bussani, L.B. Gamage, J. Peng, and M. Saliba), two engineers (N. Riahi and J. Yang), and two visiting scholars (H. Kenig and Y. Kiperwas) carry out research and development activities in the laboratory under the guidance of the author.

REFERENCES

1. C.W. de Silva, and A.G.J. MacFarlane, "A Knowledge–Based Control Approach for Robotic Manipulators", *International Journal of Control,* Vol. 50, No. 1, pp. 249–273, July 1989.

2. P.D. Kovesi, and G.J. Walker, "Vision Generation of Surface Models for Guiding a Sheep Shearing Robot", *Proc. 19th ISIR,* Sydney, Australia, pp. 417–426, November, 1988.

3. C.W. de Silva, *Control Sensors and Actuators,* Prentice–Hall, Englewood Cliffs, NJ, 1989.

IMPLEMENTATION OF SEMI-AUTONOMOUS MOBILE ROBOTS IN SEMI-STRUCTURED ENVIRONMENTS*

Robert W. Albrecht
Professor
Departments of Electrical Engineering and Nuclear Engineering
University of Washington
Seattle, Washington 98195

ABSTRACT

A general plan supported by hardware, software shells, senior/graduate courses and graduate research is in place that facilitates the evolution of autonomous mobile robots capable of navigating and performing prototypic tasks in semi-structured environments. A series of projects is designed to enhance the repertoire of skills of the robots to robustly accomplish a range of tasks from simple to complex. The skills are built within a structure that is designed to be cumulative. The result is forecast to be systematic improvement in a generalized knowledge base that is identified as the key to implementation of semi-autonomous mobile robots in semi-structured environments.

BACKGROUND

1. The Plan

A framework has been adopted that is implementable in the short term and has the potential to be a practical approach to the deployment of semi-autonomous mobile robots in the longer term. The plan is based upon a group of faculty and students imagining that they are a future mobile robot supplier. The hypothetical customer is asked to provide three pieces of information:

(a) A CAD model of the environment for robot operation,

(b) A task description of the expected performance of robots in the environment,

(c) A specification of limitations on modifying the environment for navigation, maneuvering, and manipulating.

Based on these three pieces of information, a special knowledge base is instantiated for the customer's application. This knowledge base is a particular instance of a more general knowledge base that is known to lead to satisfactory robot behavior in similar cases. The knowledge base will permit the hypothetical robot company to simulate the operation of a robot with

*This work supported by an equipment grant from NSF, matching funds from the UW, and generous equipment discounts from Hewlett Packard, GENSYM, and Denning.

appropriate constraints and goals. The simulation is the result of an expert system with suitable inferencing power interacting with the specialized knowledge base. The illustration of this plan is shown in Fig. 1.

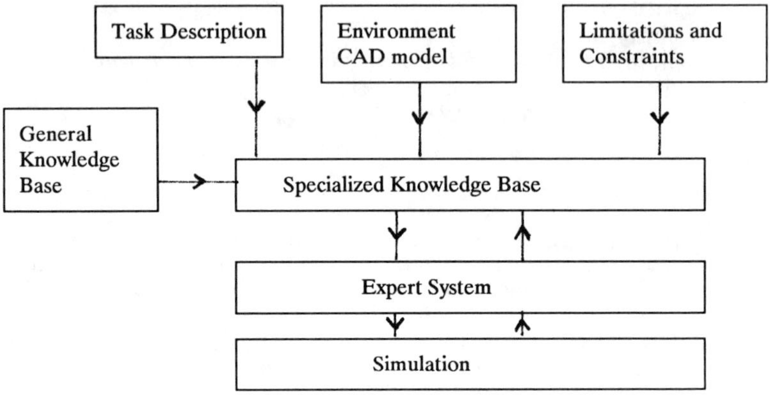

Figure 1. Overall Plan of Mobile Robots Development

After a complete simulation, the hypothetical mobile robot company configures a robot to meet the customer's requirements, tests it, and ships the robot with an appropriate knowledge base to the customer.

The key to this plan is the creation of the general knowledge base that permits specialization to the customer's application. The evolution of this general knowledge base is the central focus of our program.

2. Creating the General Knowledge Base

Our procedure for creating the general knowledge base is as follows:

(a) Create detailed CAD models of real environments that present mobile robots with a variety of challenges (This will also include CAD models of the equipment with which the mobile robot is expected to interact such as docking ports and fixed robot manipulators.),

(b) Create representative task descriptions and constraint sets for the above real environments,

(c) Build a knowledge base within an expert system shell that can interface with real robots in real environments,

(d) Operate real robots in real environments using a real-time expert system and gradually improve performance and versatility,

(e) Test the proposition that new environments with new tasks and new constraints can be adequately simulated by instantiating specialized knowledge bases, performing the simulation, and then checking the validity of the simulation by actually configuring a real space that represents the modelled space,

Repeat steps (d) and (e) until confident that a general knowledge base has been built.

CURRENT STATUS

The infrastructure for carrying out this plan has been built over several years. The key components are:

1. Hardware

Robots:

Two identical mobile robots (named Hercule Poirot and Miss Marple) were acquired from Denning Mobile Robots Inc. from our specifications. The robots are equipped with TV cameras, 24 sonar sensors each, infra-red beacon sensors, laser reflectometers, tactile bumpers, radio links, motor encoders, and on-board computer plus multiple sub-processors.

Work Stations:

Six networked workstations (5 HP 340 C+, 1 HP 370 turbo SRX) dedicated to mobile robots are in operation. Network access to 30 additional work stations in the general EE instructional area is provided.

Peripherals:

Hardware to structure environments includes infra-red coded beacons, laser reflectors, and custom obstacles. Auxillary equipment for instruction and research includes an assortment of printers and plotters including a very large format plotter.

2. Laboratories

Two laboratories are used by the mobile robots program. The robot development laboratory consists of two adjacent rooms in the EE building designated the "roam-room" and the "control room". The robot proving grounds consists of a large bay of the former Nuclear Reactor Building with an overlooking control room and a briefing room. This proving grounds is in the process of being reconfigured as a "mobile robot challenge environment" to simulate the many potential challenges that may face such devices.

3. Software

For environmental CAD modelling the Unigraphics CAD system has been adopted. This system is capable of producing true three dimensional CAD models. The CAD models are produced in IGES files that form the basis for the environmental definitions of both real environments and simulated environments.

For real-time expert system control of the robots and for knowledge base development the Gensym G2 expert system has been put into operation. The system is used for creating knowledge representations for mobile robot operation in real and modelled environments.

For interfacing knowledge bases to both IGES files from Unigraphics and for interfacing to the robots in real time the GSI interface program has been implemented. The interface to robots is facilitated by a special software module called the dispatcher.

For dynamic simulation, both the G2 simulator and MATLAB have been adopted.

4. Courses

Three senior/graduate courses that are taught each year have been developed. Mobile Robots I introduces environmental modelling, properties of robotic vehicles, and special sensors related to robots. Mobile Robots II introduces elements of intelligent control, artificial intelligence, and computer vision. Mobile Robots III gives students the opportunity to carry out experiments to improve the repertoire of skills of robots in actual operation. Enrollment each year is in the range 15 - 30 students.

5. Research

Research at the MS and PhD levels is built upon the infrastructure described above. This is opposite to the customary procedure of building undergraduate courses based on an infrastructure of graduate research.

GENERALIZED KNOWLEDGE BASE

1. CAD Modelling

Detailed CAD models have been created for eight regions in the robot proving grounds. These were produced as student assignments in Robots I. Future activities in Robots I will produce CAD models for the robots development lab and for other spaces such as hallways, elevators, outside regions, etc.

2. Surveyor

The link between CAD modelling and a knowledge base representation of the environment is the responsibility of the surveyor (termed Hipparchus). The surveyor analyzes the CAD model and produces a set of objects with attributes that become the knowledge elements of a knowledge base.

3. Cartographer

A cartography facility (termed Mercator) has been developed that organizes environments (such as robot development lab) and regions (such as roam room) and catalogs the relevant maps into an atlas (a compendium of maps, indices, and useful information). Logical maps (properly termed iconised cartograms with geometric attributes) have been created that have real and virtual walls surrounding each defined region in a logical ordering that permits "reasoning" and walls surrounding internal forbidden space that also have ordered properties permitting reasoning. In addition to walls, many objects have been defined with appropriate attributes. These include coordinate origins, paths, doors, furniture, equipment, scenes, beacons, floor features, etc. All of the components of every region are cataloged and indexed by Mercator for use in reasoning. In addition, Mercator tests for illogical constructions in maps and alerts either the operator or an autonomous robot so that such illogical constructions can be repaired.

4. Robots

Robots and their essential attributes have been logically defined in the knowledge base. These logical definitions are instantiated for various applications. Many of the properties of robots are defined in terms of fuzzy logic with the anticipation that reasoning will increment and decrement probabilities instead of yielding absolute truth or falsity.

5. Procedures

A set of efficient and robust procedures for such tasks as triangulation, path planning, beacon-following, reflector navigation, obstacle avoidance, and dead reckoning have been written and are callable by the expert system.

6. Dispatcher

Software to provide a general interface between workstations and robots has been completed. The software provides a "socket" architecture to simplify interfacing to the expert system and it provides "governors" to filter commands and avoid dangerous robot actions. The dispatcher provides a standard interface with either real robots in real environments or simulated robots in simulated environments.

7. Controller

A multifaceted controller has been developed. One component sets and reads motor and sensor parameters and reports faults. Another controller panel provides control inputs to drive and steer motors and monitors robot motion. A third controller panel provides sensor commands to the robots and graphically displays sensor data. These first three controller components are built within the expert system shell. A fourth controller is a highly graphic tele-operation interface. This is built as a separate module and interfaced to the expert system and the dispatcher. The tele-operation mode is extremely valuable as a guide for rule-creation in the expert system.

8. Simulation

Simulation facilities are under development that provide a means to study robot behavior in environments. This requires the simulation of the robot vehicle, its sensors, and the attributes of environments including navigational aids such as beacons and reflectors. The first primitive simulator has been completed and more advanced simulation is in progress.

9. Intelligence Hierarchy

The schematic of the intelligence hierarchy is shown in Fig 2. The controller that is shown is interfaced to the dispatcher and then to the robots and to the simulator. Rules for the pilot level of the hierarchy involve such skills as negotiating doorways, avoiding transient obstacles, and following environmental cues. Rules for the navigator level relate to using maps, beacons, reflectors, and other information provided by the Cartography facility. The planner level of the hierarchy is more symbolic and less geometric than navigator. Planner deals with analyzing and optimizing travel between regions and environments.

All attributes of this hierarchy are under continuing development. For example, in the Mobile Robots III class each small working group writes rules appropriate to a specific aspect of the hierarchy.

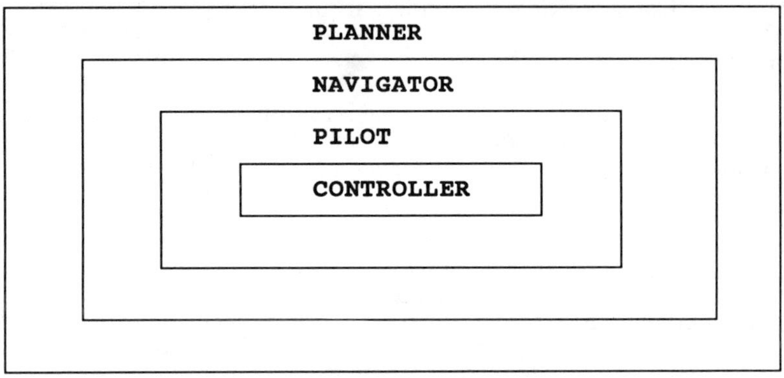

Figure 2: Intelligence Hierarchy Schematic

SUMMARY

A comprehensive approach to the development of practical semi-autonomous mobile robots is in place. A conscientious attempt has been made to structure the program and facilities in such a way that an incremental development can be carried out by a sequence of student projects. Of particular interest is the focus on a general knowledge base and the use of real robots in real environments to gain confidence in simulation as a technique for producing robots that are tuned to their jobs and their environments.

ACKNOWLEDGEMENT

The assistance of faculty colleagues, students in the robots classes, and graduate student researchers has made this work possible.

PARTNERSHIPS - AN INTEGRAL PART OF TECHNOLOGY TRANSFER

CAROL J. SPENCER
Milwaukee Area Technical College

ABSTRACT

Several faulty assumptions about the transfer of technology are made by educators and industrialists. These assumptions are challenged and alternative points of view are presented. In addition, this paper outlines three elements for consideration when undertaking the task of technology transfer.

INTRODUCTION

Frequently, technology transfer is inhibited by oversimplification of the process involved in moving technology from a research and development to a work environment. This oversimplification often includes several faulty assumptions or value sets that should be addressed.

However, before addressing assumptions, it might be helpful to offer some definitions to assist in building common understandings. These definitions are offered as working definitions rather than as dictionary definitions.

Technology: techniques used to do work efficiently and precisely. These technologies can range from the use of a simple pulley device to highly complex computer driven devices.

Advanced Technology: technology that uses computerization and electronics to assist in the performance of work tasks.

Technology Transfer: the activity of moving technology, generally advanced technology, from a research and development environment to an application environment. It's important to note that transfer of technology includes both the transfer of physical aspects (machines, techniques, etc.) and the human aspects of the technology (confidence, understandings, and skills).

Technical Literacy: the basic understanding of any technology in use at the work site by the general work force. This is based on the assumption that foundation levels of technical understandings must be in place before we can successfully transfer more advanced technologies to the workplace.

Commercialization: making a new technology in a research environment into a profitable proposition for the private sector. This term is often incorrectly used as an exact synonym for technology transfer.

In addition to thinking about the "what" of technology transfer, it is important for us to think about the "who." The transfer transaction includes three basic sets of players; originators, receivers, and transferrers. Originators are generally those individuals who develop new technologies--research and development types. The receivers can be a broader group of individuals including any one who may have use for new technologies. This group might include CEO's, cost accountants, middle managers, plant floor workers, and technical educators. The last group, transferrers are those people who assist in moving the technology between the originators and the receivers. This group includes those who go by names such as educators, disseminators, facilitators, consultants, and trainers.

With the above general understandings about the what and the who of technology transfer, let's examine three faulty assumptions about technology transfer. Most of us who have worked with technology are guilty, at least to some degree of making and voicing these assumptions at one time or another.

<u>Faulty Assumption #1</u>

Advanced technology inventions which work in a research and development environment are immediately adaptable to an application environment.

There are several factors that can be offered to prove this assumption false. Research and development, by their very nature are conducted in a controlled environment. The workplace introduces a broad range of variables that are not usually introduced in the research phase. To name a few, we might consider low skilled, poorly motivated workers. The work environment also includes high repetitions and higher speeds in less than ideal physical settings. In addition we might cite tighter resources, production pressures, and the profit motive.

Probably the key words in the faulty assumption are "immediately adaptable." Efforts to transfer any new technology from the lab to the work place need to include the **intermediate steps** as part of the transfer process.

<u>**Faulty Assumption #2**</u>

Individuals who are the originators of new technologies are the most logical individuals to be central to the transfer process.

The primary focus of the developers of new technologies are their skills in a research and developmental environment. Training and education are, at best, secondary skills of these individuals. Often these individuals are completely removed from the application or work environment and are located in an academic or research organization. In some cases these individuals have no recent application or workplace experiences.

Even when the researcher/developer has recent applications experience, that person may carry an academic aura that makes it difficult to establish relationships of trust with plant supervisors and workers who have a more limited academic background.

<u>**Faulty Assumption #3**</u>

Technical and community colleges do not have the understanding of or the access to advanced technology and, therefore, do not have the technical competence to transfer new technologies to application settings.

Many, if not most, technical and community colleges differ from a public image that categorizes those schools as a traditional "vocational school" of decades ago. Two-year post-secondary educations are still perceived by many as second-rate places where we send folks who are good with their hands but maybe not too bright. Unfortunately, this is an extension of the image that many still have of the individual who works on the plant floor with the new technology of today.

Most two-year colleges today have access to more, and more current equipment and technology than most baccalaureate granting institutions. The two-year college mission has forced this group of colleges to be more responsive to local communities, which include, in growing numbers, our industrial communities. Occupational programs are developed and operated under the watchful eye of advisory committees comprised of employers, employees, vendors, and former students. Although the academic credentials of faculty may vary from the traditionally accepted, the technical competence and understanding of technical applications of the faculty members in two-year colleges would be hard to match in any university of research environment.

Addressing the Technology Transfer Task

If we accept that the three assumptions outlined are indeed faulty, then we need to find a way to talk about technology transfer that is positive and proactive. One way to do that is to attack the translation task by acknowledging that the **what** and **how** of technology transfer are different for different receivers.

Three elements must be considered as we undertake the transfer of a specific technology. First, we must consider content. When considering the transfer of a given technology, the transferrer needs to understand the **information need** of the receiver. For example, job demands and responsibilities will dictate what content needs to be shared. A CEO may need to know how a new technology impacts productivity, work environment, quality, and profitability. The mid-manager will need to know how the technology fits into the broader operation, what maintenance requirements, and what training requirements will come with the new technology. The plant floor worker needs to know how to operate the new technology in the daily production routine and how it changes job demands.

The second element to consider is the conceptual level of the presentation. The receivers of the technology will vary greatly not only in what they need to know but in their ability to absorb and understand abstract concepts. The level of formal education and the literacy level of the receiver and how well that is matched by the transferrer's skills greatly impacts upon the success or failure of the transfer of any given technology.

The third primary element to consider in the transfer of technology is the **ability of the transferrer to gain confidence** of the receivers. It is human nature that we generally are more comfortable with those who are more alike than different than we. When planning the transfer of technology we must try to match shared cultures and shared backgrounds and experiences between the transferrer and receiver. This match will facilitate and speed the transfer process.

Partnerships - The Critical Link

As technical and community colleges consider their role as transferrers of technology, partnerships with industry can be the most critical link in addressing the **Faulty Assumptions.** Partnership activities can offer the researcher in industry a "prove out" facility that is close to industry conditions. Partnership activities build competency for communicating about technology in both the researcher and in the college faculty and staff. Partnerships enhance the access to and understanding of technology by the technical and community college faculty and staff.

What is partnership? For the purpose of this discussion, it can be described as a mutually beneficial undertaking where both industry and education partners contribute to and benefit from that undertaking--the transfer of technology to a broader public.

Partnerships should not be taken lightly or without considerable forethought. When a college and an industry begin to contemplate some type of partnership or cooperative venture, a number of questions and concerns soon surface. It is to everyone's advantage to anticipate and plan for difficulties, because these relationships will not survive and thrive unless a mutual understanding of responsibilities is developed prior to the formalization of a partnership.

MATC operates a CIM Development Center that contains an educational facility that serves technical education, graduate engineering education, continuing education for journeymen, and career education for K-12 students. This CIM Center was developed with the assistance of several corporate partners, including Allen Bradley, Rexnord, Kennemetal, Kearney and Trecker, Brown and Sharpe, Bridgeport, and IBM. Our corporate partners share with us the need to transfer new technologies to our industrial community. These corporate partners have made significant contributions of equipment, staff members, and time in the development and ongoing operation of the Center. We at MATC believe that we are assisting our corporate partners in meeting some of their corporate goals as they assist us in our educational goal.

The partnership between Milwaukee Area Technical College (MATC) and IBM through the IBM CIM in Higher Education activity is one that can be pointed to as an example of a carefully developed partnership that enhances the transfer of technology.

IBM has made significant contribution to our institution through hardware, software, and personnel expertise. In return, we work to provide customer briefings, demonstrations of IBM technology in a manufacturing setting, and a facility to prove-out prototypes in a working environment.

We believe that partnerships such as this are critical to our ability as a college to remain current with new technologies. We believe that by combining the technical expertise of our corporate partners with our expertise in teaching in the business and industry environment, together we can enhance the productivity and competitiveness of American industry.

ROBOTICS RESEARCH AT SANDIA NATIONAL LABORATORIES

James J. Wiczer
Sandia National Laboratories[1]
Albuquerque, New Mexico

Abstract

Research at Sandia National Laboratories in technologies underlying robotic systems will be described. This overview will briefly summarize recent research results on several topics, including controls, vision, sensors, motion planning, and automated assembly planning. The activities described here are currently being investigated and will be presented in the context of specific robotic applications.

Introduction
In support of the U.S. Department of Energy (US DOE) operations, Sandia National Laboratories has established a robotics research and advanced development program. It is anticipated that the US DOE will require increasing levels of robotics and automation technology to meet the needs associated with environmental restoration, waste management, waste minimization, and small batch manufacturing in a period of increasing concern for worker safety and environmental awareness. In response to these needs, Sandia has become involved in a broad range of robotics and automation related research and advanced development activities including work in the areas of controls, vision, sensors, motion planning, automated assembly, and systems integration.

This work is organized into several robotics research laboratories in which each lab focuses on an aspect of robotics and automation technology. This paper will briefly describe some of the current technical activities, recent results, and future directions for these robotics research laboratories.

Controls Research
In the area of controls, three main activities are being explored: the control of flexible robotic structures, torque control of direct drive robots actuated by variable reluctance motors, and impedance control for in-contact operations. All of these controls research activities are currently being explored in the context of specific applications.

Flexible Link Control
The control of flexible structures is being investigated for two potential applications -

---

[1]This work was performed at Sandia National Laboratories and supported by the U. S. Department of Energy under Contract DE-AC04-76DP00789.

controlling large aspect ratio robots that must handle relatively heavy payloads compared to the mass of the robot and controlling high speed light weight robots (J. T. Feddema[1]). Commercially available pedestal arm robots are typically able to precisely move payloads that are only 3% - 7% of the robot's mass. As robots are made to move faster with lighter weight links and less massive joints, dynamic effects characterized by flexing motions will occur during translations and payload disturbances; the robot can no longer be controlled as a rigid link, ideal joint structure. Controlling and minimizing these unwanted flexing motions is the goal of the two link flexible robot arm project.

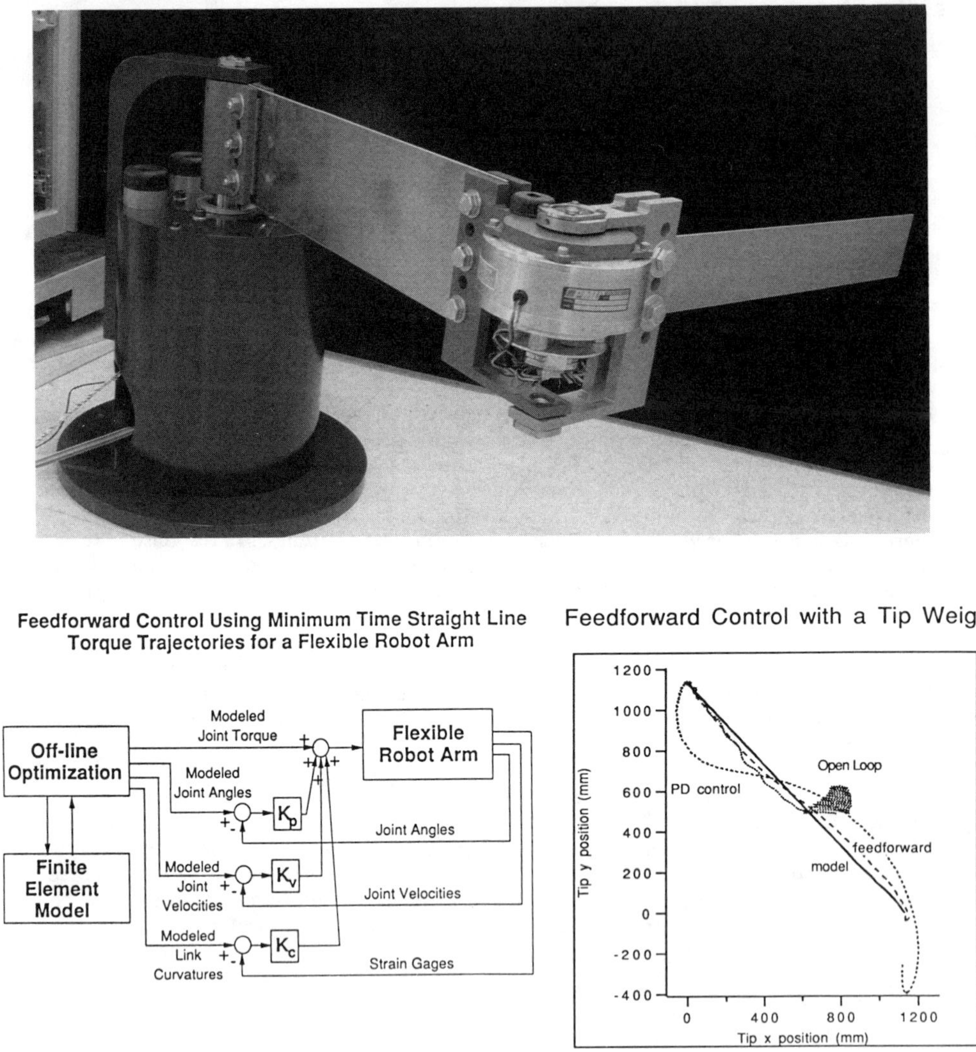

FIG. 1. Flexible link robot arm (a) used with indicated control strategy (b) for feedforward control resulting in displayed position profile (c) for various strategies.

Model-based and sensor-based control strategies have been developed and experimentally evaluated on this planar robot arm structure. An off-line, recursive quadratic optimization algorithm has been used to determine a minimum time, torque profile for the actuators while maintaining the robot arm tip along the pre-selected trajectory and limiting the required torque to the actual actuator output torque limits. An efficient, finite-element model of the robot dynamics for an n-link, m-joint system with bending only in the horizontal plane was used in the optimization and in real-time control strategies. Control schemes tested include an open loop torque profile, a closed loop proportional-derivative feedback joint controller, and a feedforward controller utilizing sensor inputs from an array of strain gages, accelerometers, and joint encoders. The finite element model has been tuned to complete a set of dynamic state calculations in 11 milliseconds on a VME-based single transputer, (Inmos T-800) auxiliary computer board allowing for the implementation of real-time model based control.

A feedforward controller using the torque profile generated by the optimization scheme, proportional-derivative feedback from the joint angles, and proportional feedback from the strain gages to correct for modelling errors has been implemented. The performance of this controller when compared to a simple proportional-derivative controller has shown a factor of 10 reduction in residual oscillation amplitude at the end of the planned trajectory and significantly less excursion from the desired trajectory during the motion. Figure 1 shows a) a photograph of the two link arm, b) a block diagram of the feedforward control strategy, and c) results comparing the various control strategies implemented to date during this study. This work is being extended beyond two-link planar robot arms to include complex robot arm structures moving in 3-dimensional space.

Torque Control of Direct Drive Robots

Torque control of a direct drive robot arms is being explored to enable high performance impedance control of light machining and other in-contact operations (G. Starr & C. W. Wilson [2]). The torque commanded controller is based on several characterization experiments to determine the output torque of variable reluctance motors as a function of rotor position and drive pulse power (P. F. Muir[3]). This information has been loaded

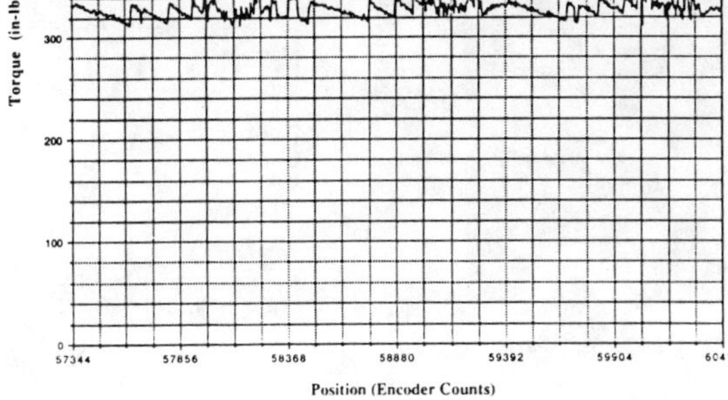

FIG. 2. Actuator torque (vertical axis) as a function of joint angle (horizontal axis) with less than 15% ripple.

into a specially designed microprocessor based, dedicated controller board and used to provide access to low-level torque or position control of joints one and two in an AdeptTwo[4], commercially available robot. Experiments on the AdeptTwo robot arm have shown that this torque controller has achieved less than 15% torque ripple where torque ripple here refers to a variation in the output torque as a function of joint position. Results from experimental investigations of the kinematic and dynamic characteristics of the AdeptTwo are shown in Figure 2. This work is currently being extended to include all four actuators and improve the torque control algorithms to further reduce torque ripple effects.

Impedance Control for In-Contact Operations

Force-position control techniques are currently implemented on an AdeptOne[4] robot as part of a light machining development workcell. The primary focus of this laboratory is robotic edge and surface finishing; these finishing operations are important manufacturing processes for many machined parts. Since these processes are relatively labor intensive and can represent 30% of the total cost of some machined parts, there is significant interest in automating these operations. Use of conventional position control strategies with robot arms has not been successful for these machining operations due to tool wear, variations in physical dimensions due to fabrication processes used to create the workpiece (casting in our application), and uncertainties in the exact workpiece location. For this application, tight control of tool forces is required to accomplish the required machining operations. An impedance control strategy using existing hardware implementations demonstrated in the Robotic Edge Finishing Laboratory have shown that chamfer operations can be performed within typical tolerances of 0.25mm +/- 0.05mm on a surface with significant position uncertainties (C. S. Loucks[5]). This work has demonstrated edge chamfering operations on a sample of 304 stainless steel to create a 0.33 mm chamfer with 2.2 newtons of tool force applied by the robot arm at a material feed rate of 20 mm/sec. Figure 3 shows the Robotic Edge Finishing Laboratory with recent data indicating tool force applied to the workpiece as a function of position along the edge.

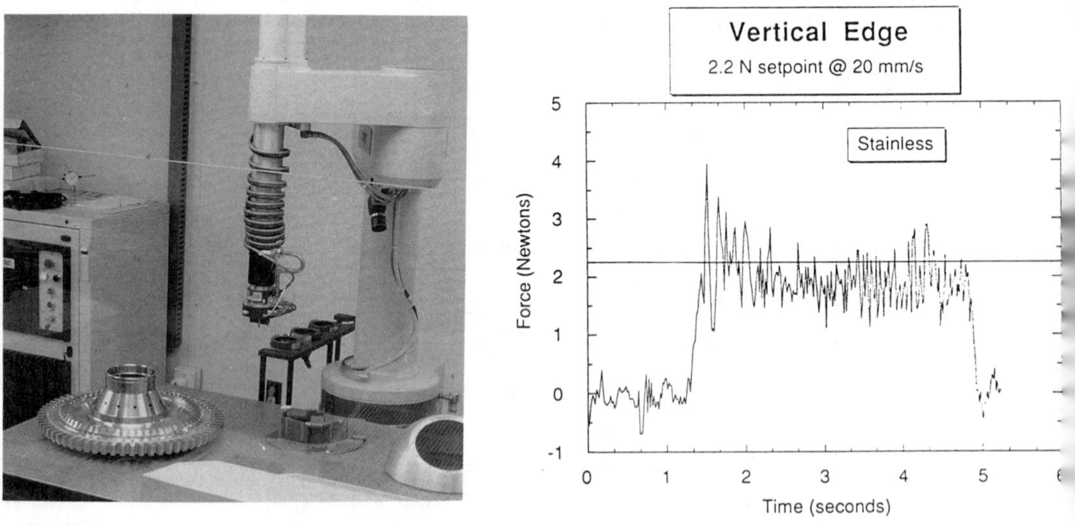

FIG. 3. Robotic Edge Finishing Laboratory (a) used to perform edge chamfering operations with a specified force (b).

Vision

Robotics vision research activities at Sandia have utilized a model-based approach with multiple monocular sensors to determine object positions and orientation in 3-dimensions (L.P. Ray[6]). This research is motivated by vision problems in factory environments in which only a relatively small number of relevant objects (1 - 20) may appear in the field of view and need to be located. The initial identity and location of objects in the field of view are assumed to be known. It is also assumed that the vision system contains geometrical representations of each object of interest for multiple views. Scene changes from the initial conditions typically occur as work in an operating robotic workcell proceeds. Linear features are extracted from each sensor image and are interpreted as observed features corresponding to model features. A search interpretation tree combined with prior position estimates is used to correlate model and sensor features and new object positions are determined by maximum likelihood estimation. As the number of camera views used in the system increases, the uncertainty and error in position and orientation is reduced. Implementations of this vision system have demonstrated the ability to locate multiple moving objects among a cluttered background with partial object occlusion utilizing two camera views. Objects were located to approximately 0.1 inches in translation and 3 degrees in rotation. This work is being extended to perform curved feature extraction and model correlation and to reduce the computation time required for each image analysis. Currently a dedicated Sun 4/260 requires 10 - 25 seconds to perform the needed calculations.

Sensors

Sandia researchers are exploring a new spatially resolved force sensor that simultaneously measures normal and tangential forces[7]. This capacitance based sensor is being applied to tactile sensing in robot end effector applications to predict the onset of slip. In essence, the sensor measures small changes in capacitance due to electrode position changes induced by force applied to parallel plate capacitors fabricated with deformable dielectric media. Currently, this sensor resolves less than 0.1 newtons of normal and tangential force with 10 mm center-to-center spacing between adjacent sensing elements.

Other sensor research includes two types of burr sensors. These sensors are being integrated into automated edge finishing operations on spatially complex machined parts using preview control strategies. Laboratory experiments with this sensor technology have demonstrated sensor resolution of burrs with features smaller than 10 microns in height above the parent material and smaller than 100 microns in lateral extent. This high resolution capacitance imaging sensor has been developed jointly by Novak and Wiczer and is being investigated for real-time closed loop control and inspection of robotic deburring and chamfering processes. As shown in Figure 4, this sensor has different resolution in the vertical and horizontal directions but is well suited for many manufacturing processes.

Motion Planning

Motion planning research has focused on developing algorithms to automatically plan complex robot motions in a cluttered environment without interfering with obstacles present in the environment. This work has concentrated on planning robot motions in "near" real time to accommodate dynamic workspace environments with multiple moving obstacles. Recent work has resulted in a method to obtain boundary equations of configuration obstacles for manipulators in three dimensional environments (Hwang[8]). Configuration obstacles refers to the map of a physical obstacle's Cartesian space coordinates into a robot's joint space representation. In this work, obstacles are assumed to be polyhedral in shape and are approximated by a collection of planar triangular

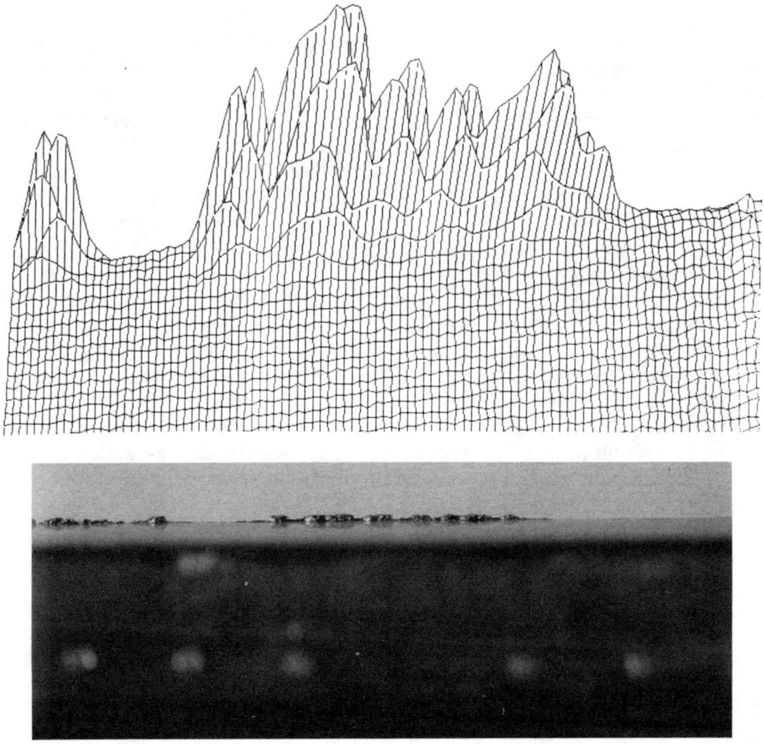

FIG. 4. Computer generated image (a) of output from new proximity sensor used to measure burrs on steel bar sample (b).

segments. The boundary equations for the configuration obstacle is derived from the intersection conditions of a line segment and a triangular patch. Implementations of this algorithm requires robot links to be approximated as cylinders and then these cylinders are reduced to their axes as obstacles are grown by the size of the cylinder radii. These expressions can be used to compute unobstructed robot trajectories or to analyze the geometry of contacts between manipulators and obstacles. This work is being extended to analyze more computationally efficient motion planning techniques that will allow real-time motion planning for robots functioning in dynamic environments.

Automated Mechanical Assembly

Automated mechanical assembly planning work at Sandia has resulted in the capability to create an assembly plan automatically for a broad class of mechanisms using the Archimedes system (D. Strip, A. Maciejewski[9]) and a CAD representation of the assembly. This system has demonstrated the planning and physical assembly of a 13 piece mechanism - consisting of mostly pins and cogs - from a CAD representation using a robot arm, a vision system, and a wrist force-torque sensor. Archimedes deduces geometric assembly constraints and generates assembly plans compatible with these constraints. Archimedes also generates robot instructions to implement the assembly plan. This work is being extended to plan for a wider range of 2.5 D and true 3D mechanisms.

Future Research
Future research activities required to support the next generation of Sandia robotic systems will be focused on the topics described above - controls, vision, sensors, and planning. Current robotic systems at Sandia are being applied to two major areas: robotic systems for environmental restoration and waste management activities and robotic systems for manufacturing processes. New technology requirements for near future systems are anticipated to be in these same areas. The issues associated with environmental restoration and waste management activities are particularly challenging since operational environments are typically severe - often including high levels of radioactivity. Robust error recovery will be increasingly important since errors during the robotic handling of radioactive waste can result in significant negative consequences. Issues such as remote robot maintenance and automated robot recovery will also gain interest.

Technology requirements in support of manufacturing processes will continue to focus on high-speed automation for small batch operations with an emphasis on the capability to rapidly switch between processes for a variety of product types.

1. J. T. Feddema, G. R. Eisler, and D. J. Segalman, "Integration of Model-Based and Sensor-Based Control for a Two-Link Flexible Robot Arm," IEEE Int'l Conf. on Systems Engineering, Pittsburgh, Pa, Aug. 9-11, 1990.

2. G. Starr and C. W. Wilson, "Sensor Driven Robot Control and Mobility", SAND 87-7150, Sandia National Laboratory report, Albuquerque, N.M., May, 1989, pp 30-53.

3. P. Muir, "A Virtual Sensor Approach to Robot Kinematic Identification: Theory and Experiment", IEEE Int'l Conference on Systems Engineering, Pittsburgh, Pa., Aug. 9-11, 1990.

4. Available from Adept Technology Inc., 150 Rose Orchard Way, San Jose, Ca. 95134.

5. C. S. Loucks, C. B. Selleck, S. Thunborg "CAD-Directed Robotic Edge Finishing", SME Technical Paper MR 89-114, Presented at Deburring & Surface Conditioning '89, San Diego, Feb 13-16 1989.

6. L. P. Ray, "Estimation of Modeled Object Pose from Monocular Images", Proc. IEEE Int'l Conf. on Robotics and Automation, Cincinnati, May 13-18 1990, pp 408-413.

7. J. L. Novak, " SCINNS " Int'l Conf on Robotics and Automation, Scottsdale, AZ, June 1989

8. Y. K. Hwang, "Boundary Equations of Configuration Obstacles for Manipulators", Proc. IEEE Int'l Conf. on Robotics and Automation, Cincinnati, May, 1990.

9. D. Strip and A. A. Maciejewski, "Archimedes: An Experiment in Automating Mechanical Assembly", Intn'l Symposium on Robotics and Automation, July 18, 1990, Vancouver, BC.

THE CHALLENGE OF INTER-ENTERPRISE CIM

David McCluney
Director of Marketing
The CIM Institute, UK

Peter Sackett
Director of Academic Programmes
The CIM Institute, UK
Professor of Integrated Manufacturing CIT, UK

ABSTRACT

The role of information technology in the successful pan national manufacturing environment demands a new response to the requirement to provision the key integrated business resource - labour.

The CIM Institute has defined and built a teaching environment which focuses attention on the externalisation of business processes. Projects and research are undertaken which study the increasing complexity of information management in supply chains. Networking with other universities and companies is establishing an environment close to the leading edge of industrial need.

The paper presented describes both the current work of The CIM Institute and its direction in delivering both labour and inter enterprise process development.

INTRODUCTION

The CIM Institute was established in 1986 in response to IBM's request to UK universities to competitively bid for funding to set up an independent educational charity based within an environment of academic excellence. The winning tender came from Cranfield Institute of Technology, Bedford, England. The Head of the College of Manufacturing, invested his staff and expertise to put together the business plan which secured and has sustained this new enterprise.

The mission statement of The CIM Institute sets the challenge:

> 'To be the leading European centre, in association with Cranfield Institute of Technology and others, for education, research, consultancy and advanced development projects, in pursuit of integrated business operations.'

This objective, set down in 1986, arose from concerns parallel to those expressed in the US National Science Foundation report 1988 (1). In both cases Computer Integrated Manufacturing embraces the total business, engineering and manufacturing processes to derive an holistic competitive edge. This leads to concern for organisations and enterprise structure alongside the technical issues of information technology.

There is a misunderstanding to be bridged between business management and people working within effective business processes. In particular where manufactures are concerned the gap has become a chasm. Awareness of this gap and the economic value of bridging it is growing. The modern manufacturing enterprise is a highly dynamic socio-technical system. Success in this environment requires high calibre personnel with skill zones which are in short supply.

In stimulating the universities to address these issues, it was anticipated that a substantial volume of intellectual effort would be fed into the industrial marketplace. Effort to be delivered by people who have: the perspective to evaluate the changing business environment, the perception to select the techniques and technologies that will advance competitive position, and the commitment to manage the process of change. Experience indicates that a vast effort is required to generate awareness of these problems, let alone build the solutions to them.

THE PROCESS

The CIM Institute has addressed these areas. The Masters degree course is currently placing 30 second degree students a year in the field. 60% to 70% of whom are engineers, 15% to 20% women and all are members of the best qualified entry stream on a campus of more than one thousand post graduate students. The course achieved UK Science and Engineering Research Council approval and is already, we believe, the largest single subject Masters course in the UK. The Academic Programmes Director focuses the course on two major area of manufacturing business. The management of people and their skills through the inevitable process of continuous change and the management of information through increasingly capable technologies. Organisational development, total quality management, team building and commitment to education form the background to the 'people' theme whilst Advanced Manufacturing Technologies are explored throughout the course in the context of a rich computing resource environment.

On management, Aleksander (2) notes that ... 'information technology acts as a sort of magnifying glass rather than an improver. If one does things badly it can make them worse ...' However it does enable you to do things better if you think about the business objectives rather than the technology. This in essence is our approach.

Masters students are required to present both Group Projects and Individual Theses based on industrial projects commercially funded. PhD students too focus on industries research requirements. In 1989 four doctorates were awarded to CIM Institute members. The current objective is to have ten research projects of this type active at all times.

The CIM Institute's success with the student community is due in no small part to the outstanding facilities available to each of them. A working environment closely modelled on IBM's own best office/laboratory practice computing facilities is not the 'normal' campus resource. There can be little doubt that this quality contributes directly to the high calibre of the students and their achievements.

The strength of The Institute is built upon this academic base and from this business opportunities arise which have enabled business viability to be achieved.

Industrial education and consultancy are the added value of the Institute's activity. The challenge most frequently addressed can be defined readily as the management of cultural change. Manufacturing Engineers from across Europe pass through The Institute and are given the opportunity to stand back, to see their company as a whole. In re-evaluating their individual roles and contributions, Boards of Directors and Senior Executives are challenged to review their competitive position and evaluate manufacturing technology offerings which will enhance business processes to sustain and increase profitability.

CIM consultancy projects cover shop floor data collection to corporate strategy formulation. The Institute's partnership with Cranfield is particularly valuable here since the community of expertise available means resources can be imported to respond to each specific requirement. The demand for consultancy is strong, particularly in the area of Total Quality Management where a reputation for top down implementation with total management commitment has been established.

The underlying strength of a university based Institute is that focus may be placed on the direction which research should pursue in order to assist industries' long term aims. Cranfield is 75% industry-funded as a result of its unique abilities and structure. There is a culture of competitiveness and entrepreneurial struggle throughout the Cranfield campus. It is from this environment that cases for leading edge activity are made and through it that new technologies and solutions are delivered. Inter-Enterprise CIM is one key area for us. In part this arises because we see it as crucial to European manufacturing in the 90's. In part it arises by design. We planned to make IECIM work from the start and we have a very good local reason to do so. By design we are a manufacturing research establishment with none of our own manufacturing hardware.

IECIM

Inter-enterprise CIM encompasses the compression of concept to customer leadtime, working with Just in Time supply chains and logistic support throughout product life. Against a background of accelerating specialisation and the visual disintegration of previously integrated businesses, these trends drive the requirements for electronic data interchange (EDI) to new levels of complexity.

Competitive edge will be provided by access to creative information processing centres such as specialist design or manufacturing houses. The evolution of such systems is already apparent in niche product areas; the European Airbus, Tornado, European Fighter (EFA) and NH-90 Helicopter projects are good examples which typically handle 100,000 parts per aircraft (3). This technology will rapidly become appropriate, affordable and desirable for a broad spread of business enterprises as the dramatically declining IT:Labour cost ratio is maintained (4). Even the communications costs may be insignificant. A client of ours in Bedford UK, connects his PC to his printer, a distance of 3 metres, via the Company's networked mainframe - this is in Texas, USA. In '2061 Odyssey Three', Arthur C Clarke notes that the historic decision to eliminate all differential charges between local and long distance telephone calls on 31 December 2000 had significant sociological impact. The same is true for manufacturing organisations. In fact the challenge of IECIM is not really a technical one, far more a challenge to management. Power via conventional hierarchy is seriously weakned by the flat structure of the inter enterprise CIM environment (5).

The need to address this environmental change is pressured by the forces of competition and directed by international standards. The United States Department of Defense initiative 'CALS' - Computer- aided Acquisition and Logistics Support - will impact many UK and European companies. It is likely that, together with the European initiatives, this will set the standards and the pace for inter- organisation technical communication.

The perception of the 1990's to which manufacturing integration researchers and their facilities must respond includes:

- Business processes which increasingly cross enterprise boundaries to interface with functional areas in <u>other</u> companies:

    - Product design and definition
    - Manufacturing process definition
    - Manufacturing facilities

- Supplier/customer integration (people and processes) through interchange of commercial/technical data

- Ability to function effectively as links in unbuffered supply/ distribution chains

  The ability to integrate the activities of a number of entities to produce and sell manufactured products profitably will depend on the relationship of these entities and the communication that passes between them. We are accustomed to thinking about this in the context of a single enterprise with different departments, Sales, Design, Engineering, Manufacturing, Distribution. However, within a large marketplace, entities from many different enterprises, or entities which in themselves are independent enterprises, relate via a single product to produce a designed result.

An example might be a Merchandising entity recognising a business opportunity and requesting :

- a Design entity to design it
- a Manufacturing entity to build it
- a Distribution entity to distribute it

and

- a Marketing entity to sell it.

The implication of such an example is that all the entities can be considered as 'flexible' or 'programmable' within their expertise envelope.

In setting out to research the issues which will arise in the 1990's environment the broad spectrum of industry and current activity must be addressed. This diversity encompasses:

- Sub-industry mix
- Types of manufacture
- Establishment size
- Products/processes/technologies
- Functional areas
- Component/material variety
- Sourcing alternatives
- Dynamic state-of-the-art technologies
- Industrial concentration trends

For the UK and many other countries the exposure is extreme due to :

- o International Aspects
    - Widening Horizons (Eastern and Central Europe)
    - EEC and 1992
    - European initiatives (ESPRIT and CIM OSA)
    - Future relations with US/Pacific RIM
- o Conflicting standards and cultures
    - Within UK sub-industries
      Different procedures/protocols/nomenclatures/ numbering systems/corporate cultures
    - With other countries
      Wide divergency in part numbers/designs and drawing standards/structure conventions/product representation/syntax/sourcing practices

IECIM places new prospectives and demands on classical manufacturing business decisions on :

- Preferred methods and techniques
- Advanced materials technology
- Sequential inter-functional dialogue in parallel
- Legal and Financial commitment before financial benefit of production is achieved
- Design for manufactureability and supportability

Action is therefore required to ensure that Industry, Academia and Government together provide a vehicle which will deliver practical guidance and support across a broad range of industry sectors. The facility to effectively research, develop, educate and implement solutions demands a new academic environment.

CONCLUSION

The CIM Institute has developed an infrastructure (6) and an IECIM Demonstrator which enables a number of roles to be defined :

- The provision of an enabling system/test bed
- Research collaboration
- Coordination and leadership (through 'Operational Group')
- National/International focus for advanced teaching and research in CIM
- The provision of an information service and awareness of state-of-the-art
- Promotion of shared efforts across international boundaries

The scope of effort required is huge and it is as well to remember that 'the distance from the objective is not a matter for despair, but an opportunity'.

REFERENCES

(1) NSF report, 'A Research Agenda for CIM - Information Technology', National Research Council, National Academy Press, USA, 1988

(2) Aleksander I, 'Critical Factors in High Technology Management', Proceedings 1st Industrial Summit New Manufacturing Imperatives, 12-15 Jan 1988, London, UK, pp 91-103

(3) Meijer A H, 'Electronic data interchange for product structure data', Int J Computer Integrated Manufacturing, Vol 2, No 4, July-Aug 1989, pp 220-228.

(4) Scott Morton M S, 'The Impact of Information Technology on Organisations in the 1990s', Siemens Review, 3/89, pp 36-41

(5) Kanter R M, 'The New Managerial Work', Harvard Business Review, Nov/Dec 1989, pp 85-92

(6) Sackett P J 'CIM Status and Perspectives', ATV-SEMAPP, CIM Now and in the Future, Technical University of Denmark, Copenhagen, 13/3, 1990

SECTION 11 : NEURAL NETWORKS AND LEARNING

ADAPTIVE NEURAL NET CONTROLLER FOR ROBOTIC MANIPULATORS

J. Yuh and R. Lakshmi
Department of Mechanical Engineering
University of Hawaii, Honolulu, Hawaii 96822, *USA*

S.H. Lee
Department of Electrical Engineering Systems
University of Southern California, LA, CA 90089, *USA*

J.E. Oh
Department of Mechanical Engineering
Han Yang University, Seoul, Korea

ABSTRACT

This paper describes the recent study of the application of neural-net to robot control. An on-line learning algorithm for the robot controller is described by using the error back propagation method. The proposed control scheme does not require a priori knowledge of the physical parameters of the manipulator, its dynamic structure or its configuration, except the upper bound of inertia terms. Results of the study show the promising feature of the neural net controller which learns and adapts to changes in robot system.

INTRODUCTION

As robot systems are used for more sophisticated purposes, the demands placed upon their performance, in terms of accuracy and speed, are increased. High performance requires more intelligence in the control system which has the capability of learning and adapting to changes in the system and its environment. During the past decade, the advanced robot control schemes have been studied. One approach involves using a dynamic model of the robot to calculate the joint drive torques for the specified trajectory. These computed torques can then be used as the reference inputs to a feedback controller [1,2]. This technique, known as the nonlinear model-based scheme (or the computed torque method), can provide the desired performance if the full dynamic model and accurate parameter values of the manipulator are known. In practice, however, it is quite difficult to obtain the accurate manipulator model. Some of the parameters are time-varying and unknown such as payload. Unmodelled dynamics and parameter uncertainty may result in poor performance or even instability of the control system [3]. More recently, the application of adaptive control techniques such as the model reference adaptive control scheme and self-tuning control scheme to the robot control system have been studied [4-9]. These adaptive control techniques generally do not require a priori knowledge of the parameters of the manipulator but require a priori knowledge of its structure. The conventional adaptive control system might exhibit instability due to the excitation of unmodeled dynamics or high frequency noise [10].

Recent developments in the area of neural networks have introduced new applications for architectures which employ massive parallelism in computation. Some of the popular models have been discussed by Rumelhart et al. [11]. In recent years, possibility of the use of neural networks in control system design have been also discussed in the literature [12-17]. Most of the approaches suggested in the above were evaluated with extremely simple examples such as the nondynamic system. Furthermore, they require a long learning period before implementation. Therefore, the neural net controller must be retrained whenever there is a significant change in plant dynamics. It is mainly because the corresponding input to the desired output of the plant is unknown, and the controller cannot be directly trained. Since robot dynamics include time-varying parameters as well as unknown parameters such as payload, the approaches discussed in the literature are not suitable for robot control. According to their approaches, the controller must be retrained before implementation, when the variance of the parameters is wide and different payloads are handled by the manipulator. Otherwise poor performance of the robot system would be expected.

The objective of this paper is to investigate the on-line application of a multilayer neural network to robot control. Our control scheme does not require any information about the robot dynamics except the upper bound of inertia terms, and the lengthy training period before implementation. For every time step, the controller is directly adjusted and implemented to control the robot. Kawato et al. [14] also proposed an on-line neural-net control approach. However, their network has two layers (i.e. no hidden layer) and they use nonlinear terms from dynamic equations of the robot system for nonlinear functions of the network. Their approach could be interpreted as one of the well-known conventional adaptive robot control scheme, *nonlinear model-based adaptive control* [4] with the desired trajectory input to the nonlinear controller. In the following sections, we describe an adaptive control algorithm using the neural net, and discuss the efficiency and effectiveness of the control algorithm with a two d.o.f. manipulator.

CONTROL SYSTEM

The dynamic equations of the manipulator can be described in the joint coordinate system as follows:

$$D(x)\ddot{x} = Q(\dot{x},x) + T_d + U \qquad (1)$$

where x, $\dot{x}$, and $\ddot{x}$ are n-dimensional vectors of joint position, velocity and acceleration, respectively, D is the nxn-dimensional inertia matrix, Q is a n-dimensional torque vector due to Coriolis, centrifugal, gravity, and friction forces, T_d is a n-dimensional torque vector representing unknown signals due to unmodeled dynamics and external disturbances, and U is a n-dimensional joint torque vector. By considering the small perturbations in the variables about the desired values corresponding to the desired robot path, we can obtain the linearized equation for joint s (s = 1 ... n) as follows:

$$J_s(t)\ddot{x}_s(t) = K1_s(t)\dot{x}_s(t) + K0_s(t) + u_s(t) \qquad (2)$$

where J_s is the effective inertia sensed by the motor of joint s, $K1_s$ corresponds to the effective damping coefficient, and $K0_s$ describes an effective disturbance torque, and u_s is a torque applied to joint s. These coefficients depend on the positions, velocities, and accelerations of other joints. This linearized model has been used for robot control system design by several researchers [6,18]. By assuming that the changes in J_s, $K1_s$, and $K0_s$ are small over one sampling period, the Eq. (2) can be discretized by setting $\dot{x}_s=v_s(k)$ and approximating $\ddot{x}_s=[v_s(k+1)-v_s(k)]/T$ as follows:

$$v_s(k+1) = A1_s(k)v_s(k) + A0_s(k) + B_s(k)u_s(k) \qquad (3)$$

where k means the k-th time step, and T is the sampling time period. For each joint, we propose to use a neural network joint control module which has n layers including input, hidden, and output layers. The input layer has three neurons which input the velocity error signal about the desired velocity v_{sd}, the position error signal about the desired position x_{sd}, and constant d. Each hidden layer has multiple neurons. One of them also has the constant input d while others receive the signals from the neurons in the previous layer. The output layer has one neuron which outputs the joint control signal u_s which is computed by the following equation:

$$u_j = q_j^n \qquad (4)$$

$$q_j^{m-1} = f_j^{m-1}[p_j^{m-1}] \qquad (5)$$

$$q_j^1 = 1 \quad \text{for } j=1 \text{ , } = (v_{sd}-v_s) \text{ for } j=2, \text{ and } = (x_{sd}-x_s) \text{ for } j=3 \qquad (6)$$

where superscript and subscript represent layer number and neuron number in that layer, repectively, (i.e. q_j^{m-1} is the value of q at the j-th neuron of the (m-1)-th layer), and

$$p_j^{m-1} = \Sigma_k w_{jk}^{m-1} q_k^{m-2} \qquad (7)$$

where w_{jk}^{m-1} represents the weight between the j-th neuron of the m-th layer and the k-th neuron of the (m-1)-th layer. The activation function f[.] in Eq. (5) can be any function which is differentiable and nondecreasing. In this study, we consider the following function for f[.]: $f[x] = [1-\exp(-x)]/[1+\exp(-x)]$ which is bounded by ± 1, for neurons in the hidden layer, and $f[x] = x$ for neurons in the output layer. It should be noted that if we do not use any hidden layer in the network, this model will be similar to the conventional PD controller.

The main problem of the neural net application to control systems is to find a proper way to train the controller (i.e. to adjust the weight w) which generates the control signals u_s^* which provide the desired performance of the controlled system. We propose to use the error back propagation method, a widely used algorithm for training neural networks. Error back propagation attempts to minimize the squared error for training sample input-output pairs. For each time step, it modifies each weight using the partial derivative of the error with respect to that weight. Since error back propagation adjusts weight in the local direction of greatest error reduction, it is a gradient algorithm. The error back propagation method is described in detail by Rumelhart et al. [11]. Based on this method, the following equation is used to adjust the weight:

$$w_{ij}^m(k+1) = w_{ij}^m(k) + \rho\delta_i^m q_j^{m-1} \qquad (8)$$

where ρ is a learning rate constant, and

$$\delta_i^m = (u_s^* - u_s) \quad \text{for the output layer}$$

$$f'^m_i[p_i^m]\Sigma_k \delta_k^{m+1} w_{ik}^{m+1} \quad \text{for all other layers} \qquad (9)$$

where f'[.] is the derivative of f[.], and u_s^* is the network training signal. Since the value of u_s^* is not known in advance, a direct controller training using the error back propagation method has not been possible. However, if we can estimate the error signal $(u_s^*-u_s)$ without knowing the value of u_s^* explicitly, we can implement the above algorithm for a direct controller training. We describe how to obtain the estimate of the error signal $(u_s^*-u_s)$ in the following. From Eq. (3), since the robot system is assumed to be controllable, the control signal u_s^* exists such that

$$v_{sd}^*(k+1) = A1_s(k)v_s(k) + A0_s(k) + B_s(k)u_s^*(k) \qquad (10)$$

where v_{sd}^* is the desired value. Therefore, we can estimate the error between u_s^* and u_s from Eqs. (3) and (10):

$$\Delta u_s = u_s^* - u_s = C[v_{sd}^* - v_s] \qquad (11)$$

where C is a positive constant. Even though the signal u_s^* is not known, the estimate of Δu_s can be obtained from Eq. (11) by using measurements of the joint outputs if the estimate of C is known. As can be seen from Eqs. (2), (3) and (10), C is dependent on T and J_s. Since the learning rate of the output layer also depends on the sampling time step, from Eqs. (8), (9) and (11), one can combine C and $\rho_{output\ layer}$ such that the modified C: C_{mod} is a function of J_s and T and $\rho_{output\ layer}=1$. To determine C_{mod}, an experiment was carried out. For simplicity of analysis, the presented appproach was tested for a single-link manipulator described by the following equation: $J\ddot{x}=-c\dot{x}-mgL\sin x+u$ where x, $\dot{x}$ and $\ddot{x}$ are joint angle, velocity and acceleration, respectively, u is the input torque, $J=mL^2$, the point mass m is 1 kg, link length L is 1 m, damping coefficient c is 2.6 Nsec/m, and gravitational acceleration g is 9.8 m/sec^2. In the experiment, we used the three layered network which has the one input layer, one hidden layer and one output layer. A hidden layer has three neurons. We also used a repeated motion with a frequency of 1 sec. for the desired joint angle. Each cycle consists of acceleration, constant velocity and deceleration, and constant velocity periods. A constant input to network d=0.1, and all the initial weights $w_{ijk}=0.1$. Results of the experiment carried out with the different sets of (T, C_{mod}) give the following equation:

$$C_{mod} = (10^3)J_{max}T \qquad \text{for } \rho_{output\ layer} = 1 \qquad (12)$$

where J_{max} is the maximum value of J. The above equation can be interpreted as $C = J_{max}/T$ for $\rho_{output\ layer} = T^2x10^3$. The sampling interval varies from one robot to another. It depends on the mechanical time constant. Studies [1] show that the min. sampling interval lies between 5 and 100 ms. (many robot systems use 16 ms.), and a no-load to full-load variation of inertia of 10:1 is expected for a well designed robot. A change in inertia will change the natural structural frequency. A sampling rate of 10 times the structural frequency is recommended. In addition to this experiment, we investigated the effect of the number of layers by implementing the presented approach with two, three, and five layers. As stated earlier, the two layered neural network means that there is no hidden layer, and it is a PD controller with adapation algorithm. We used the same value of (T,C_{mod}) = (0.001,10) for each case. For repeated motion, the results become identical regardless of number of layers. However, the case of two layers (no hidden layer) shows the relatively large output error comparing to other cases during the first three cycles. Therefore, a simple conventional PD controller structure is discouraged to use, and the use of hidden layer is recommended. We also investigated the effect of different number of neurons. We chose the three layer neural network which has one hidden layer, and used (T,C_{mod})=(0.001,10). We tested the neural networks with two, three, and four of neurons in the hidden layer, respectively. Two neurons in the hidden layer means that there is only one neuron which receives the signals from the input layer. It is recommended from the result of this test to use the same number of neurons of the input layer or more in the hidden layer. Due to the limited space, all the results of the experiment cannot be included in this paper.

CASE STUDY

Based on the result of the experiment, the presented approach was applied to control a two d.o.f. planar manipulator similar to one shown in [4]. This manipulator operates in a vertical plane under the gravitational force. It consists of two rigid links and point mass at the end of each link. We assume that payload varies between 0.1 kg and 1 Kg. For each joint, we use the three layered identical neural net controller which has three neurons in the input layer, three neurons in the hidden layer and one neuron in the output layer. The desired repeated motion (Fig. 1) was used for each joint. $v_{sd}^*(k+1)$ in Eq. (11) is defined as $v_{sd}(k+1)+0.01(x_{sd}(k)-x_s(k))/T$ where the second term accounts for the correction in the error of the angular position x_s. To investigate the robustness of the control system with respect to payload uncertainty, payload 1 kg, 0.1 kg, and 0.5 kg were used for the first three cycles, the next three cycles, and the last three cycles, respectively, without any change in the control system. Results (Fig. 2) show that even though the value of payload was changed, each joint was well controlled to follow the desired output.

CONCLUSIONS

In this paper, we have described a neural network joint control module for robotic manipulators. Effect of the different number of layers and effect of the different number of neurons were investigated. The presented control approach does not require any information about the robot system. The result of case study shows that the control system has the adapting and learning capability to changes in the robot system. This simple control module is used to control each joint of the manipulator, and therefore n modules are required for n joints. However, since they can be implemented in parallel, more modules doesn't mean more computation time. The current research task is an experimental investigation on the neural-net controller with a single-link manipulator.

REFERENCES

[1] R.P. Paul, *Robotic Manipulators Mathematics, Programming and Control*, MIT Press, 1981

[2] C.S.G. Lee, et al., "On the Control of Mechanical Manipulators," Proc. 6th IFAC Conf. on Estim. and Param. Ident., Washington D.C., 1982, pp. 1454-1459

[3] O. Egeland, "On the Robustness of the Computed Torque Technique in Manipulator Control," Proc. IEEE Int'l. Conf. on Robotics and Automation, San Fran., 1986, pp. 1203-1208

[4] J.J. Craig, *Adaptive Control of Mechanical Manipulators*, Addison-Wesley

[5] S. Dubowsky, and D.T. DesForges, "The Application of Model Reference Adaptive Control to Robotic Manipulators," ASME J. of Dynamic Sys., Meas. and Control, Vol. 101, 1979, pp.193-200

[6] A.J. Koivo and T.H. Guo, "Adaptive Linear Controller for Robotic Manipulators," IEEE Trans. on Automatic Control, Vol. 28, 1983, pp. 162-171

[7] H. Seraji, "A New Approach to Adaptive Control of Manipulators," Journal of Dynamic Systems, Measurement, and Control, Vol. 109, 1987, pp. 193-202

[8] M. Tomizuka and R. Horowitz, "Model Reference Adaptive Control of Mechanical Manipulators", Proc. IFAC Adaptive Sys. in Control and Signal Processing, San Fran., CA, 1987, pp. 27-32

[9] J. Yuh and W.E. Holley, "Application of Discrete-time Model Reference Adaptive Control to Industrial Robots: A Computer Simulation," Journal of Manufacturing Systems, Vol. 7, No.1, 1988, pp. 47-56

[10] C.E. Rohrs, et al., "Robustness of Continuous-time Adaptive COntrol Algorithms in the Presence of Unmodeled Dynamics," IEEE Trans. on Auto. Control, No. 9, Sept., 1985, pp. 881-889

[11] D.E. Rumelhart et al.,, *Parallel Distributed Processing*, MIT Press, 1986

[12] M.S. Lan, "Adaptive Control of Unknown Dynmaical Systems via Neural Net-work Approach," Proc. of the 1989 ACC, PA, Vol. 1, 1989, pp. 910-915

[13] A.G. Barto, "Adaptive Neural Networks for Learning Control: Some Computational Experiments," Proc. IEEE Workshop on Intelligent Control, RPI, Troy, NY, August, 1985, 1986, pp. 170-175

[14] M. Kawato, et al., "A Hierarchical Neural-Network Model for Control and Learning of Voluntary Movement," J. of Biol. Cybern., Vol. 57, 1987, pp.169-185
[15] Proceedings of Int'l Joint Conf. on Neural Networks, Washington, D.C., June, 1989, Vol. 2, pp. 299-356
[16] D. Psaltis, et al., "A Multilayer Neural Network Controller," Proc. 1987 IEEE Int'l Conf. on Neural Networks, San Diego, CA, June
[17] B. Bavarian, "Introduction to Neural Networks for Intelligent Control," IEEE Control System Magazine, April, 1988, pp. 3-7
[18] J. Yuh, and D.K. Tissue, "Adaptive Control for Mechancial Manipulators Having a Joint Compliance," Proc. 1990 IEEE Int'l. Conf. on Robotics and Automation, May, Cincinnati, OH

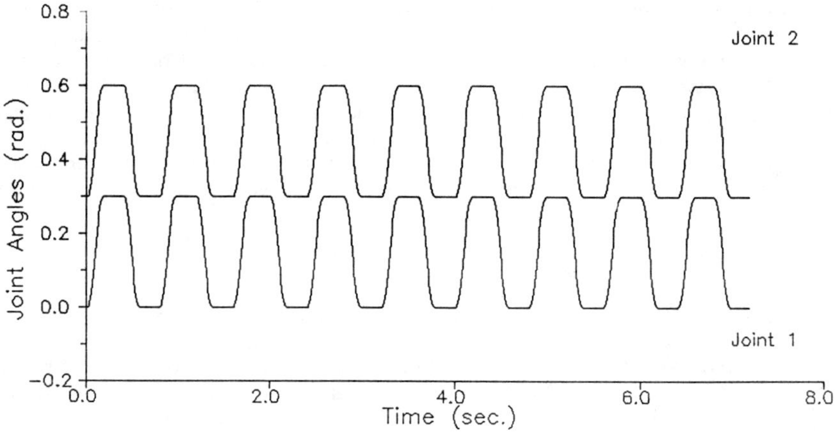

Fig. 1: Desired Joint Angles

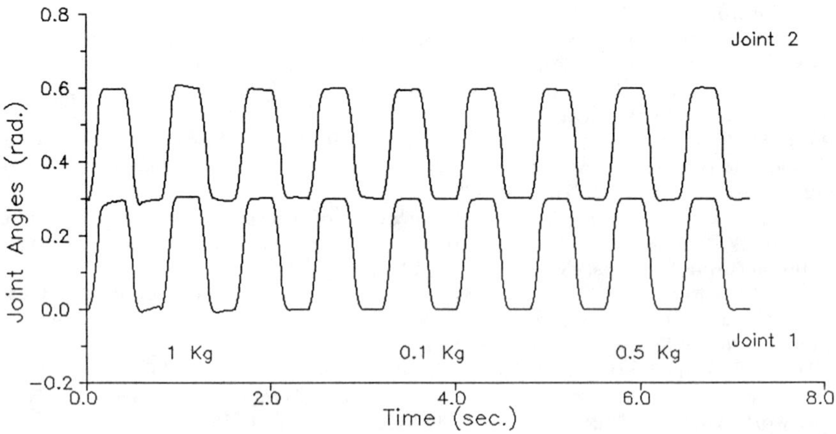

Fig. 2: Case Study -- Joint Angles

TWO DIMENSIONAL ALLOCATION OF IRREGULAR PATTERNS: A SIMULATED ANNEALING APPROACH

Cihan H. Dagli and Alireza Hajakbari

Department of Engineering Management
University of Missouri-Rolla
Rolla, MO 65401

ABSTRACT

In this study the use of simulated annealing approach for finding local optimum solution of stock cutting problem is studied. The conceptual model proposed uses an energy function that measures the area of the rectangle enclosing all the figures, level of similarity between pattern pairs and the degree of overlap among patterns in evaluating various pattern configurations generated by the simulated annealing algorithm. Details of the model are described with an example.

INTRODUCTION

The problem of allocating rectangular and/or irregular patterns arises frequently in applications where it has to be determined how a set of two-dimensional shapes will fit onto a large stock sheet of finite dimensions in such a way that the resulting scrap will be minimized. This problem is common to many industries such as aerospace, shipbuilding, steel construction, shoe manufacturing, clothing, flat glass and furniture.

Since 1950's, the cutting stock problem has received considerable attention as the computers have appeared as fast and economic information processors and the optimization techniques have developed. Solution of this problem is of considerable practical significance. Because of the diversity of the structures of real world stock cutting problems, there exists no general standard method for solving them. Over the years, two main approaches are emerged namely; heuristics and approaches based on linear programming relaxations.

One dimensional stock cutting problems wherein the other two dimensions of the items being cut are assumed constant are studied initially. Eisemann [1] and Haessler [2] proposed algorithms for the solution of these problems. Experience gained from the solution of one-dimensional single plate cutting stock problems is used for solving two dimensional multi-plate problems [3]. Dynamic, integer, and linear programing are used extensively as solution method. Optimization approaches proposed have restrictions in application due to NP-complete nature of the problem. To overcome this difficulty various heuristic approaches are proposed. Albano and Sapuppo [4], Dagli [5, 6], and Dagli and Tatoglu [7] proposed heuristic algorithms which can process both rectangular and irregular patterns. The algorithm of Albano and Sapuppo [4] assumes only line segments for each side, whereas the latter two assume both line and arc segments for the sides of a pattern. Dagli and Tatoglu [7] proposed a two-stage hierarchical approach that deals with the two-dimensional allocation problems of irregular patterns in the multi-plate context. In the first stage of the procedure, initial allocation of patterns to the plates is made through mathematical programming; then based on this initial allocation detailed two-dimensional allocation is made through heuristic algorithms in the second stage.

Heuristics play an important role, since they are flexible enough to take into account various additional restrictions and objectives appearing in practice. The quality of heuristics is generally problem specific, and they can identify a pattern which is "good" for the particular problem in question. On the other hand, Linear Programming algorithms first solve the linear programming relaxation of the cutting pattern oriented model and search for an integer solution using complex methods or simply rounding up to the next integer. In almost all linear programming models proposed, a two stage approach is used. Initially various cutting patterns are generated based on given stock sheet dimensions and small pattern shapes, then a decision variable is assigned for each cutting pattern and a linear programming model is formulated. Generally, linear programming approaches are not widely used in practice, and most of the time, heuristics are preferred for selecting patterns. This is basically due to the fact that the large number of possible patterns combinations (in the order of hundred millions) cannot be represented in linear programming formulations. In this study the feasibility of using simulated annealing approach to overcome solution difficulties posed by mathematical programing model is examined.

SIMULATED ANNEALING METHOD

Simulated annealing is a stochastic technique that has recently attracted significant attention as a suitable technique for optimization problems of very large scale. It has effectively solved the traveling salesman problem. The method has also been successfully used for designing complex integrated circuits.

Metropolis Algorithm [8] forms the basis of the simulated annealing method. In each step of the algorithm an atom is given a small random displacement and the resulting change in the energy of the system is computed. If energy change is less than zero the displacement is accepted, and the configuration with the displaced atom is used as the starting point for the next step. If the change in energy is greater than zero then the displacement is treated probabilistically. The probability that the configuration is accepted for a given energy change (ΔE) is equal to $e^{-\Delta E/KT}$, where K is the Boltzmann constant, and T is the temperature.

There is an analogy between simulated annealing and thermodynamics, specifically with the way that liquids freeze and crystalize, or metals cool and anneal. At high temperatures, the molecules of the liquid move freely with respect to one another. If the liquid is cooled slowly, the thermal mobility is lost. The atoms are often able to line themselves up and form a pure crystal that is completely ordered over a distance up to billions of times the size of an individual atoms in all directions. This crystal is the state of minimum energy for this system. For slowly cooled systems nature is able to find this minimum energy state. On the other hand if a liquid metal is cooled quickly it does not reach this state. So the essence of the process is slow cooling allowing ample time for redistribution of the atoms as they lose mobility.

The simulated annealing applied to optimization problems involves the preparatory steps. First, one must identify the analogues of the physics concepts in the optimization problem itself: the energy function becomes the objective function, the configurations of particles become the configurations of parameter values, finding a low energy configuration becomes seeking a near optimal solution, and temperature becomes the control parameter for the process. Secondly, one must select an annealing schedule consisting of a decreasing set of temperatures together with the amount of time to spend at each temperature.

Third, one must have a way of generating and selecting new configurations.

Annealing schedule

Annealing schedule forms the basis of simulated annealing approach. Hence a good annealing schedule for the problem on hand is the prospect for a promising result. The annealing schedule consists of; initial temperature, temperature decrement, equilibrium condition, and stopping criteria.

First question in the cooling schedule is to determine, how high must the system be heated before annealing begins. Second basic question concerns the minimum temperature the system must reach to achieve the optimum solution. There are a number of different proposals in the literature that have dealt with the above fundamental questions [9, 10].

According to Arts [11], in initial value of the temperature, virtually all transitions are accepted (since $e^{-(\Delta E/T_0)} \simeq 1$ for all configurations). Beyond this temperature, there is a infinite-temperature limit, in which, any move will be accepted regardless of whether it increases or decreases the value of the energy function. In contrast, at any point below the final temperature, system's energy will not decrease, therefore, when the system is reached to its minimum energy, lowering temperature further will not improve the solution.

White [10] studied the behavior of the annealing curve, where energy versus temperature is graphed. His approach is based on the configuration density function $\omega(E)$ which contains all of the information pertaining to equilibrium properties of the system at all temperatures. Assuming that $\omega(E)$ is Gaussian with mean E, White argues that the characteristics of high temperature T_F can be identified with temperature at which E is just within thermal noise (one standard deviation) of E(mean), that is $T = \sigma_\infty$. This can be viewed as a system of $\simeq N$ patterns, that are randomly distributed on the grid. If the grid size is L*L, then the distribution of energies is that of an N-step random walk with step size $\approx L$, and T_0 can be approximated to $T_0 = \sigma_\infty \simeq \sqrt{N} * L$.

There is a simple way to calculate the minimum temperature [10]. Suppose that a configuration has energy E_1, and that exactly one move can take the system to the ground energy level E_F. In two dimensional stock cutting problems, there are $2L^2$ possible moves that can be originated from state E_1. Thus there are $2L^2$ moves, but only one of them goes downhill to E_F. Since the smallest non zero change in reducing scrap is 1, therefore:

$$e^{-(E_1-E_F)/T} \leq 1/2L^2$$

$$T \leq (E_1 - E_F)/2 \ln L$$

$$T_F = (E_1 - E_F)/2\ln L \simeq 1/(2 \ln L).$$

In the above model, the qualitative behavior of E(T) is completely characterized by four quantities; $E_0 \simeq$ E(mean), $T_F = \sigma_\infty$, E_F, and $(E_1-E_F)/2\ln L$. Temperature decrement should be small enough for the system to establish equilibrium. A frequently used decrement value is given by $T_{k+1} = \alpha * T_k$, K = 0,1,2,3,....

where α is a constant smaller than but close to 1. Kirkpatrick [12] first proposed this rule with $\alpha = 0.95$. This rule is extensively used by other researchers [13] with values ranging from 0.5 to 0.99.

SIMULATED ANNEALING APPROACH FOR STOCK CUTTING

Energy function

The energy function for simulated annealing has three main components to map a given configuration of patterns into a positive real number. They are; area of the rectangle that encloses all the patterns, weighted sum of distances between patterns within the configurations, area of overlapping patterns. This can be represented in functional form as:

$$E = af_1+bf_2+cf_3$$

where f_1 denotes the area of the enveloping rectangle for all patterns, and f_2 represent the amount of overlap in a given placement and finally f_3 is a weighted sum distances of patterns within the configuration.

$$f_3 = {}_{i=1}\Sigma^n \; {}_{j=i+1}\Sigma^n \, (w_{ij}d_{ij})$$

Where d_{ij} is the distance between the patterns i and j in a given configuration and w_{ij} represents the tendency of the pattern i to attract pattern j. The values for this index for each pattern pair need to be calculated prior to the simulated annealing approach. The contribution of each component to total energy is controlled through the parameters a, b, c that change based on the patterns allocated. Hence, it is possible to create different set of energy functions based on the irregularity of the shapes under consideration.

Pattern representation

Pattern representation is an important issue in solving stock cutting problems. Choice of efficient representation methods is important, especially for algorithms that require extensive computation time. In the proposed approach matrix representation is used. Each pattern is represented by a matrix of size K by L which is the minimum rectangular enclosure of the irregular shape. The value of K and L is determined based on the desired accuracy of computations.

Attractiveness weight generator

Simulated annealing algorithm selected requires calculation of attractiveness weights for pattern pairs. In the sample problem patterns are restricted to rotate in multiples of 90 degrees. Attractiveness weights of each pattern pair is calculated based on this restriction by evaluating all possible combinations. Rectangular enclosure is selected for each combination generated for pattern pairs. The one that gives the maximum packing density is selected as a weight to be used in the simulated annealing algorithm. The patterns used include arcs, both convex and concave, and straight lines. These patterns are mapped on a grid, the corresponding grid representation of each pattern is used for allocation procedure.

Figure 1 describes the structure of the proposed patterns used to allocate ten irregular patterns. To demonstrate the workings of the simulated annealing algorithm three allocation configurations are generated. They are given in Figures 2, 3, and 4. The value of the energy function is computed for each configuration. Using the procedures mentioned earlier, T_0 and T_F for current problem size of L= 10 and grid size of N=40 are calculated. This gave an initial temperature value of 126.49 and final temperature value of 0.217.

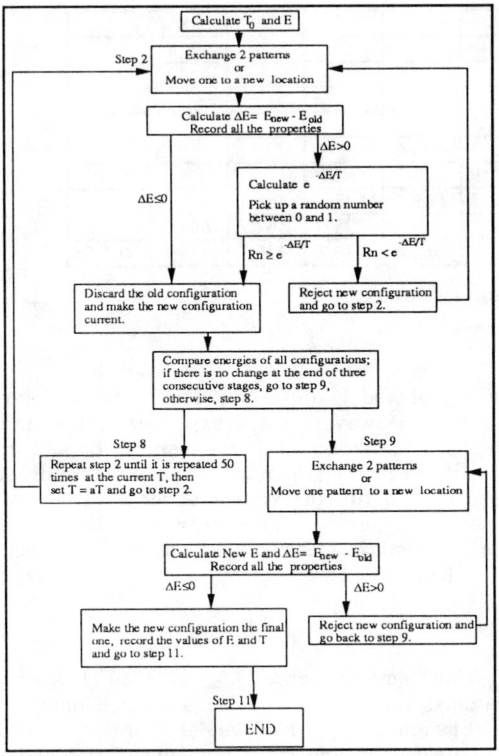

Figure 1. Program Execution Structure

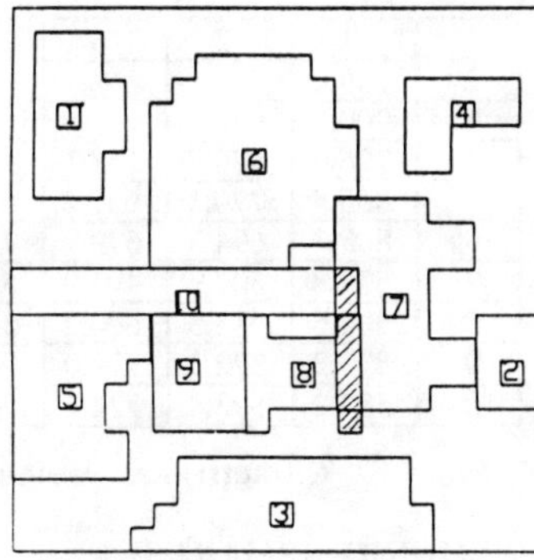

Figure 2:
First allocation configuration
E=529.00(a)+0(b)+292968.2(c)

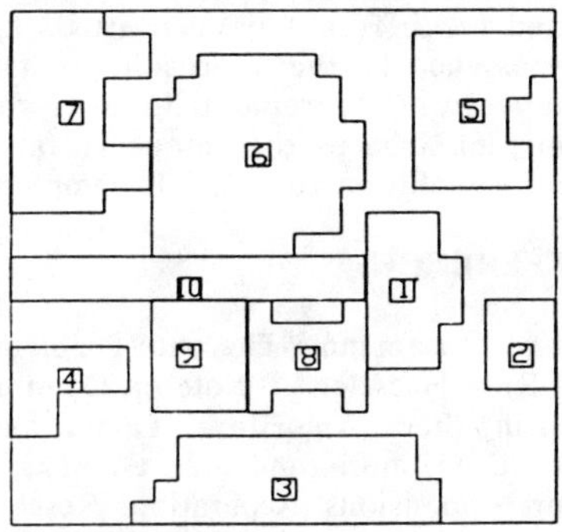

Figure 3:
Second allocation configuration
E=529.00(a)+6(b)+304214.1(c)

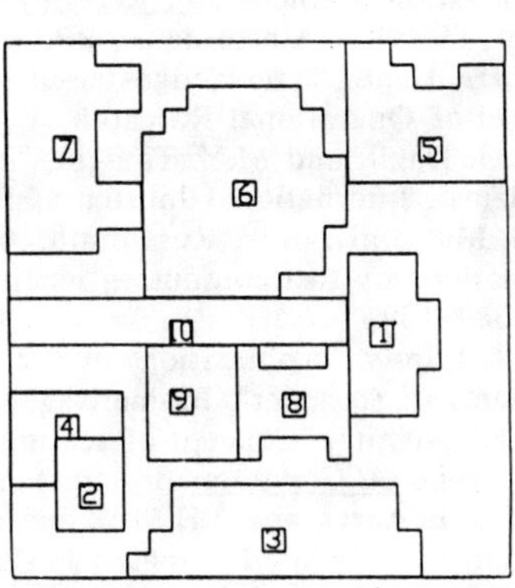

Figure 4:
Third allocation configuration
E=508.00(a)+0(b)+279145.9(c)

Parameters a, b, c are selected based on set of patterns to be allocated. The value of these parameters will be the same for each configuration. The second configuration generated by the simulated annealing algorithm provides a higher energy value then the first configuration. At this stage of the algorithm probability of energy change is computed using Boltzmann machine and compared with a probability value generated from a uniform distribution, this configuration is not accepted as the random number drawn was smaller than the computed probability, thus creating the third configuration. Computer code that will allocate fifty irregular patterns using simulated annealing algorithm is being developed.

1	1	2	3	4	5	6	7	8	9
2	759.0								
3	900.5	640.8							
4	726.0	878.8	558.1						
5	865.5	770.3	637.2	601.0					
6	808.3	694.3	579.9	752.0	743.6				
7	539.8	507.9	491.0	561.1	623.7	652.0			
8	604.6	613.3	600.2	537.7	642.6	638.1	456.4		
9	825.2	872.3	774.0	769.2	839.1	725.6	562.7	694.2	
10	584.2	593.8	722.7	506.1	463.4	539.2	436.8	358.3	491.6

Table 1. Attractiveness weight of Patterns

CONCLUDING REMARKS

The simulated annealing approach proposed in this study provided promising results for the problem on hand so far. However, more experimentation and theoretical development must be sought in selection of parameters a, b, and c, based on different pattern attributes and annealing schedule. The slow cooling is the essence of the approach, so it is useful to run the system under a variety of temperature decrement to analyze the behavior of the energy function. The approach insures the rescue from the local minima entrapment and hence a near optimal solution for the NP-complete problem will be obtained.

REFERENCES

[1] K. Eisemann, "The trim Problem", Management Science, 3,3, 279-284 (1957)
[2] R. W.Haessler,"A Note on Computational Modifications to the Gilmore-Gomory Cutting Stock Algorithm", Operations Research, 28, 4, 1001-1005 (1980).
[3] P.C. Gilmore, and R.E. Gomory, "Multistage cutting-stock problems of two and more dimensions", Operations Research, 13, 94-120 (1965).
[4] A. Albano, and G. Sapuppo, "Optimal allocation of two-dimensional irregular shapes using heuristic search methods" IEEE Transactions on Systems, Man and Cybernetics, 10, 242-248 (1980).
[5] C. H. Dagli, "Cutting Stock Problem: Combined Use of Heuristics and Optimization Methods", In Recent Developments in Production Research, A. Mital Editor, (Elsvier, Amsterdam), PP. 500-506 (1988).
[6] C. H. Dagli, "Knowledge-based systems for cutting stock problems", European Journal of Operational Research, Vol. 44, 160-166 (1990).
[7] C.H. Dagli, and M.Y. Tatoglu, "An approach to two-dimensional cutting stock problems", International Journal of Production Research, 25, 175-190 (1987).
[8] N. Metroplis, A.W. Rosenbuth, A.H. Teller, E. Teller, "Equations of state calculations by fast computing machines", Journal of Chemistry and physics, 21, pp. 1087-1091.
[9] M. Lundy, "Applications of the Annealing Algorithm to Combinatorial Problems in Statistics", Biometrika, 72 191-198 (1985).
[10] S.R. White, "Concept of scaling in simulated annealing", IEEE Int. Conference on Computer design, Portchester, 1984.
[11] E.H.L Aarts, and J.H.M. Korst and P.J.M. van Laarhoven, "A Quantative Analysis of Simulated Annealing Algorithm: A Case Study for the Traveling Salesman Problem", J. Statis. Physics.,50, 189-206 (1988).
[12] S. Kirkpatrick, C.D. Gelatt Jr., M.P. Vecchi, "Optimization by Simulated Annealing", Science 220, pp. 671-680(1983).
[13] E. Bonomi, and J.L. Lutton, "The N-city Traveling Salesman Problem: Statistical Mechanics and the Metropolis Algorithm", SIAM Rev., 26 551-568 (1984).

A HOPFIELD-TANK NETWORK TARGET TRACKER

R. M. Iñigo
University of Virginia

C. Narathong
University of Wisconsin

ABSTRACT

Real-time visual tracking is a difficult problem requiring high speed processing. We have previously reported a fast algorithm capable of estimating displacement for a sequence of images using a conventional rectangular sensor. When used with algoarithmic-spiral sensor, changes of scale can also be estimated. Although the algorithm can be implemented using sequential or parallel digital processing, a Hopfield-Tank (HT) network implementation is potentially simpler and faster.

INTRODUCTION

The purpose of a target tracker is to keep a target centered in the Field of View (FOV) at all times. The tracker uses a sequence of image frames as an input. It then produces the displacements of points (pixels) on target surfaces between successive frames. The output displacements can be mapped directly to the input domain of the control system which in turn performs necessary operations in order to center the target in the FOV. To accomplish the task, a real-time high speed operation is required. Thus, a fast tracking algorithm as well as a special purpose hardware are necessary. In this paper, a highly parallel tracking algorithm to determine target displacements is developed and implemented using a highly parallel neural network.

LINE CORRELATOR TRACKER

The LCT was developed for real time tracking applications as part of a system with biological vision features. The system has two sensors, a conventional uniform rectangular grid and a grid of exponentially increasing rings, conformally mapped to a *computation plane* by the natural log function. The LCT uses the rectangular grid for translation in image plane and the log-spiral grid for scaling and rotation about the optical axis. Details of the architecture and results of simulations are given in [2,3]. For the rectangular grid, displacements in the x direction are given by

$$\Delta x_{lk} = \frac{\sum_{j=1}^{m} R_{lj}^2 - \sum_{j=1}^{m} R_{lj} T_{(l+k)j}}{\sum_{j=1}^{m} \frac{\partial I}{\partial x_{(l+k)j}} R_{lj}} \qquad (1)$$

where l and k are indices within a tracking window of size $n \times (2m+1)$. R and T are two successive image frames. The overall horizontal and vertical displacements are given by

$$\Delta X = \Delta \bar{x}_k \tag{2}$$

where

$$\Delta \bar{x}_k = \frac{\sum_{l=0}^{n} \Delta x_{lk}}{n}$$

and

$$\Delta Y = k \tag{3}$$

respectively. The index k is calculated according to equations below.

$$\sigma_{n_k} = min\,(\sigma_{n_{-k_{max}}}, \ldots, \sigma_{n_{+k_{max}}}) \tag{4}$$

$$\begin{aligned} f^{-1}(\sigma_{n_k}) &= f^{-1}(min\,(\sigma_{n_{-k_{max}}}, \ldots, \sigma_{n_{+k_{max}}})) \\ &= k \end{aligned} \tag{5}$$

where σ_{n_k} is the standard deviation of n estimates for the kth column and f is a one to one function which maps the σ_{n_k} to its index k.

The above algorithm utilizes only point-to-point correlation, not full correlation. It was shown in [1] that computation time of 40%, or more, can be saved over the conventional correlation algorithm. Furthermore, simulation results with real images have shown that the algorithm possesses good noise immunity and it can be applied directly to a tracking window without segmentation.

THE NETWORK ARCHITECTURE

The network is a two-layer HT network [3]. The first layer consists of $(2m + 1)$ planes. Each plane contains n HT networks. There are $(2D + 1)$ neurons in each network, where D is the maximum row displacement. The row displacements are computed by each of these networks. The output of these networks are fed directly to the second layer. This layer computes the overall displacement based on the outputs of the first layer. There is a single plane in the second layer. The plane contains $(2m + 1)\times(2D + 1)$ neurons. The outputs from the second layer are the horizontal and vertical displacements. The following subsections describe the energy functions of the two layers.

The model in the first layer contains binary neurons representing the row displacement between the two images. We use $n\times(2m + 1)\times(2D + 1)$ neurons. For implementation, we discretize the row displacement by letting $-D \leq j \leq +D$. We also let $[V_j]_{lk}$ represent the state of the jth neuron of lth row and kth plane. When the neuron $[V_j]_{lk}$ is 1, this means that the row displacement for the lth row in the kth plane is j. If subpixel accuracy is desired, one can simply increase the number of neurons within the tracking window. For this model to function, an energy function must be developed such that only one neuron within each row is turned on when the network reaches stable state. The energy function for the lth row of kth plane is given below.

$$E_{lk} = \frac{A}{2}\left(\sum_i\sum_{j\neq i}\left[V_i\right]_{lk}\left[V_j\right]_{lk} - \sum_i\left[V_i\right]_{lk}\right) + \frac{B}{2}\left(\sum_j\left[V_j\right]_{lk} - 1\right)^2$$

$$+ C'\sum_j\left(S_{lk} - Q_{lk}i\right)^2\left[V_j\right]_{lk} \tag{6}$$

where

$$S_{lk} = \sum_j R_{lj}^2 - \sum_j R_{lj}T_{(l+k)j}$$

$$Q_{lk} = \sum_j \frac{\partial I}{\partial x_{(l+k)j}} R_{lj}$$

$$C' = \frac{C}{(i+\varepsilon)^2}$$

The first and the second terms in (6) provide row inhibition and global inhibition, respectively. These two terms assure that there is one and only one neuron "on" for the lth row of the kth plane when the network reaches the stable state. These two terms are also known as the constraint terms. The last term is the data term or the objective term. Without this term! a neuron will be on randomly. The presence of this term will force the neuron, which corresponds to the row displacement, turn on. Notice that the third term is taken directly from (1), with i parameter corresponding to Δx_{lk}. Due to the poor scaling property of HT network, the constant C is scaled to sensitize the data to row displacement i. ε $(0 < \varepsilon < 1)$ is added to the scaling term to avoid dividing by zero in case of zero motion. Rearranging (6), the first term is written

$$= -\frac{1}{2}\sum_i\sum_{j\neq i}(-A(1-\delta_{ij}))\left[V_i\right]_{lk}\left[V_j\right]_{lk} - \sum_i\frac{A}{2}\left[V_i\right]_{lk} \tag{7}$$

the second term,

$$= -\frac{1}{2}\sum_i\sum_j(-B)\left[V_i\right]_{lk}\left[V_j\right]_{lk} - \sum_i B\left[V_i\right]_{lk} + 1 \tag{8}$$

and the third term,

$$= -\sum_j(-C')\left(S_{lk} - Q_{lk}\,i\right)^2\left[V_j\right]_{lk} \tag{9}$$

where δ_{ij} denotes Kronecker delta. The general energy function of HT network [9] can be written as

$$E_{lk} = -\frac{1}{2}\sum_i\sum_j\left[T_{ij}\right]_{lk}\left[V_i\right]_{lk}\left[V_j\right]_{lk} - \sum_i\left[I_i\right]_{lk}\left[V_i\right]_{lk} \tag{10}$$

By comparing the terms in (7), (8), and (9) with the corresponding terms in (10), we determine the interconnection strengths (connection matrix, $\left[T_{ij}\right]_{lk}$) and the bias inputs (excitation term, $\left[I_i\right]_{lk}$) as

$$\left[T_{ij}\right]_{lk} = -A(1-\delta_{ij}) - B$$
$$\left[I_i\right]_{lk} = \frac{A}{2} + B - C'\left[S_{lk} - Q_{lk}\, i\right]^2 \qquad (11)$$

Notice that the quadratic terms in the energy function define a connection matrix and the linear terms define input bias current.

The second layer contains binary neurons representing the overall row and column displacements between the two images and consists of $(2m + 1)\times(2D + 1)$ neurons. The horizontal displacement is discretized by letting $-D \le j \le +D$. The vertical displacement is represented by index k. V_{kj} represents the state of the kjth neuron. When V_{kj} is 1, the horizontal displacement is j and the vertical displacement is k. When the network reaches stable state, there should only one neuron on for the entire layer. Since this is a 2D problem, a 4-dimensional energy function is required. Using the x and y subscripts as the row indices and the i and j as the column indices, such an energy function is given by

$$E = \frac{A}{2}\sum_x\sum_i\sum_y\sum_j V_{xi}V_{y_j}\left[1-\delta_{xi,yj}\right] + \frac{B}{2}\left[\sum_x\sum_i V_{xi} - 1\right]^2 - C\sum_x\sum_i\left[\sum_l V'_{li}\right]_x V_{xi} \qquad (12)$$

The first term provides row and column inhibitions, and the second term global inhibition, thus assuring that only one neuron is turned on for the entire layer. The last term forces the neuron corresponding to the object translation to turn on. The V'_l is the neuron output at the lth row of the first layer.

$$E = -\frac{1}{2}\sum_x\sum_y\sum_i\sum_j T_{xi,yj}V_{xi}V_{y_j} - \sum_x\sum_i I_{xi}V_{xi} \qquad (13)$$

Through similar analysis as in the previous section, the weight connection matrix and the bias input are derived.

$$T_{xi,yj} = -A\left[1-\delta_{xi,yj}\right] - B$$
$$I_{xi} = B + C\left[\sum_l V'_{li}\right]_x \qquad (14)$$

AN EXAMPLE OF NEURAL NETWORK TARGET TRACKER USING A TWO-LAYER HOPFIELD-TANK NETWORK

The example presented here illustrates a small, but completed neural network target tracker. In this example, there are three planes in the first layer, Fig. 1. Each plane has three rows of HT neurons. Within each row, there are five neurons, each with negative feedback. The row neurons computes row displacements. the weight connections (T_{ij}) is constant and can be set according to the equation below.

$$T_{ij} = -A(1-\delta_{ij}) - B \qquad (15)$$

where A = 500 and B = 200. The network receives input images from the previous stages. The input images combined with the constants yield the current input term (I_i) which is given by

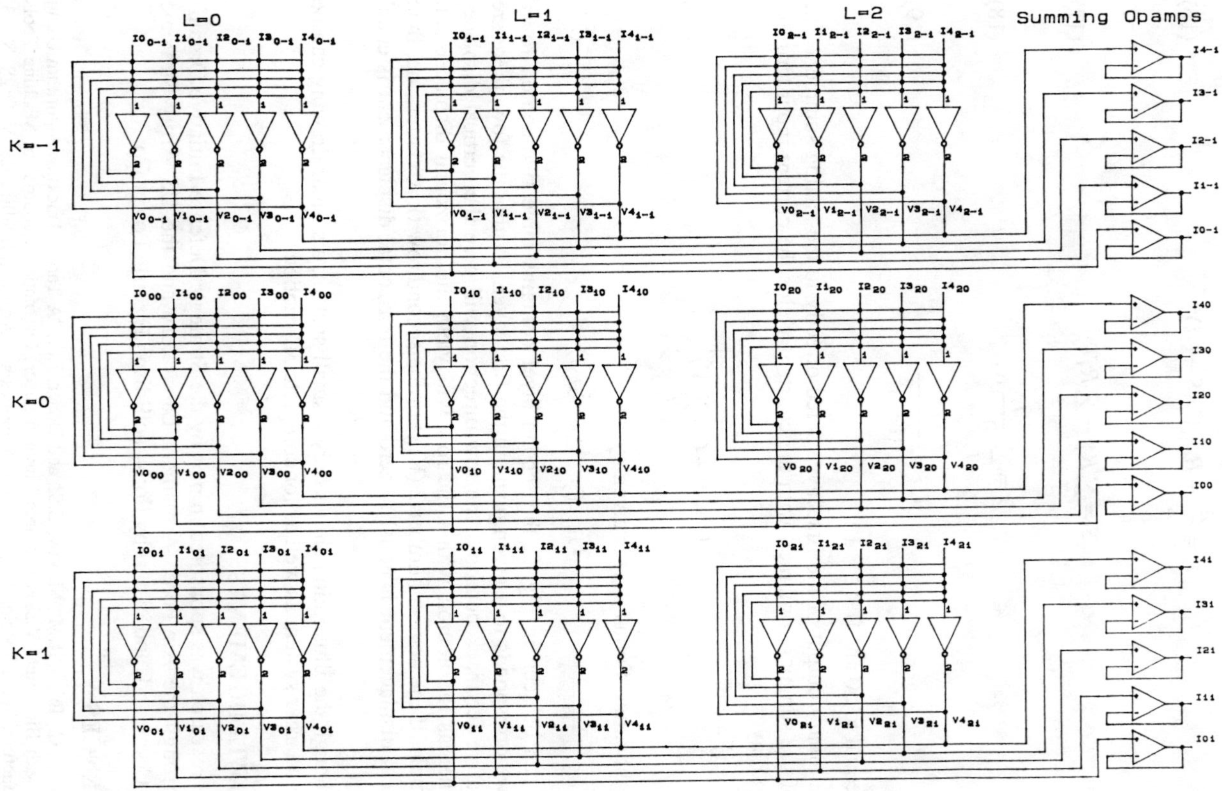

Fig. 1. First Layer HT Network.

$$I_i = \frac{A}{2} + B - C'(S - Qi)^2 \tag{16}$$

where

$$S = \sum_j R_j^2 - \sum_j R_j T_j \tag{17}$$

$$Q = \sum_j \frac{\partial I}{\partial x_j} R_j \tag{18}$$

$$C' = \frac{C}{(i + \varepsilon)^2} \tag{19}$$

C is set equal to 50 and $0 < \varepsilon < 1$.

The binary output of the first layer is fed directly to the analog summer. The output of the summer combined with the constant yields the current input term for the second layer.

$$I_{x_i} = B + C \left[\sum_l V'_{li} \right]_x \tag{20}$$

The weight connection ($T_{xi,yi}$) is given by

$$T_{xi,yj} = -A\,(1 - \delta_{xi,yi}) - B \tag{21}$$

As can be seen in Fig. 2, the second layer contains three five-neuron rows. Each row corresponds to a given plane in the first layer. Within the fow, there are five negative feedback neurons. the row direction will give the vertical displacement, whereas the neuron within the row will yield the horizontal displacement. For example, if the neuron number 2 (V_{10}) of the second row (k=0) is turn on, then the vertical displacement is equal to zero and the horizontal displacement is equal to one.

Although the illustration shows only a small system, a similar drawing can be made for a larger system as well as for complete 3D motion.

ACKNOWLEDGEMENTS

The research was sponsored in part by the Strategic Defense Initiative Organization through the Air Force Armament Laboratory and the US Army Strategic Defense Command, and in part by Potomac Edison Electric Power Co.

REFERENCES

1. Narathong, C., Iñigo, R.M., McVey, E.S., and Doner, J.F., "A Target Tracking Algorithm Using A Sensor with Biological Vision Features," invited paper, International Journal of Machine Tools and Manufacture. in the Special Supplement on Robotics and Artificial Intelligence, Vol. 28, No. 3, (1987).

2. Narathong, C., Iñigo, R.M., McVey, E.S., and Doner, J.F., Real-Time Algorithm and Architecture for Motion Prediction, Proceedings of IEEE Conf. on Computer Vision and Pattern Recognition, Ann Arbor, MI, (June 1988).

3. Hopfield, J.J., and Tank, D.W., "Neural Computation of Decisions in Optimization Problems," Biological Cybernetics, Vol. 52, 141-152 (1985).

4. Iñigo, R. M. and Narathong, C., "A Two-Layer Hopfield-Tank Network for Motion Estimation," Int. Joint Conf. Neural Nets, Conf. Proceedings, (Washington, DC, January 1990).

5. Iñigo, R. M. and Narathong, C., "A Neural Network Target Tracker," SPIE Tech. Symposium on Aerospace Sensing, Conf. Proc., (Orlando, FL, April 1990).

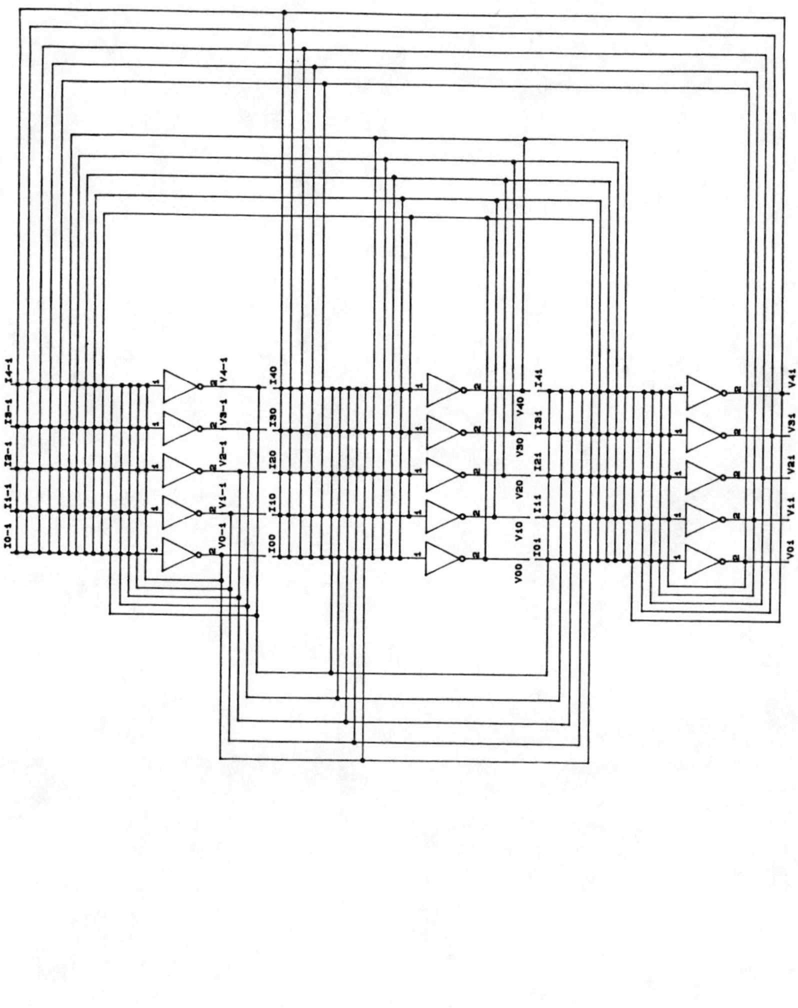

Fig. 2. Second Layer HT Network.

Model Following Using Multi-Layer Perceptrons

D. Hush, C. Abdallah, and B. Horne
Department of Electrical Engineering and Computer Engineering
University of New Mexico
Albuquerque, NM, 87131 USA.

Abstract

A model following controller is proposed for discrete nonlinear systems using a recursive multi-layer perceptron (MLP). The MLP network contains dynamics and is able to minimize the error between the plant and a desired model in many cases.

I. Introduction

In general, the control of nonlinear systems requires a compensator which performs some nonlinear mapping [1]. In practice, however, an analytical form of the compensator is difficult to find and implement. On the other hand, neural networks have been billed as massive computational networks which can implement nonlinear mappings from one linear vector space to another [3]. The proposed scheme attempts to minimize the error between the model and plant outputs by using an MLP neural network as a feedback compensator (see Figure 1). If the plant is known, the weights of the MLP can be adaptively updated using a standard gradient search technique (back-propagation through the plant). This approach has been recently suggested in many control applications [4-6]. By using the classic MLP algorithm [7] however, the application of these controllers was limited to the case of static feedback laws. For output feedback, such a restriction will undeniably prevent us from dealing with many interesting problems. Recently, recurrent networks which contain dynamics have also been suggested [8] but have not caught on in the control community. In [5] and [6], a state was reconstructed by delaying the outputs and the inputs to the system so that the MLP could be used to implement a static mapping from a state space to an input space. The present work uses a similar idea but differs from [6] because it includes the controller directly in the feedback structure and has a different implementation of the dynamic backpropagation algorithm than the one in [5].

In section II, we describe the nonlinear control problem and present the neural network algorithm. An example is given in section III and our conclusions are given in section IV.

II. Recursive Multi-Layer Perceptrons

Our efforts in this research have concentrated on including the controller as an integral part of the update algorithm for a model-following control problem, rather than identifying the system or its inverse (compare to direct adaptive control). This was motivated by the fact that the estimated nonlinear system or its inverse, unlike their linear counterparts, are mathematical models only. They do not give any useful information when the actual system is subjected to different inputs, except when the training input is white noise [5]. The model-following problem was selected for its generality. For example, by replacing the model with a unity gain one recovers the inverse identification problem [2,5]. Many control objectives may be achieved by selecting an appropriate model and trying to minimize the error between the plant/controller output and that of the model. The block diagram of Figure 1 represents the general structure of the problem discussed and the controller is described next.

The controller in Figure 1 consists of two neural networks, a recursive net and a nonrecursive one. The outputs of these two networks are summed to produce the controller output,

$$u(k) = v_N(k) + v_R(k) \tag{1}$$

Each of these networks is a multi-layer perceptron. If we let $\mathbf{v}_i(k)$ represent the output vector of the *ith* layer then signals are propagated through the networks according to

$$\mathbf{v}_i(k) = \mathbf{f}(\, \mathbf{W}_i \mathbf{v}_{i-1}(k) + \mathbf{w}_{i0}) \tag{2}$$

$\mathbf{W}_i$ represents the weight matrix connecting the outputs of layer $i-1$ to the nodes in layer i, and $\mathbf{w}_{i0}$ are the bias weights for layer i. $\mathbf{f}(\cdot)$ is a vector function of a vector which applies the standard sigmoid function, $f(\gamma) = (1+e^{-\gamma})^{-1} - 0.5$, to each component.

The input vector to the nonrecursive net consists of current and delayed values of the input signal, that is for the nonrecursive net $\mathbf{v}_0(k) = \mathbf{d}(k)$, where

$$\mathbf{d}(k)^T = \left[d(k),\, d(k-1),\ldots,\, d(k-o_n) \right] \tag{3}$$

The input to the recursive net consists of previous net outputs, that is for the recursive net $\mathbf{v}_0(k) = \mathbf{u}(k-1)$ where

$$\mathbf{u}(k-1)^T = \left[u(k-1),\, u(k-2),\ldots,\, u(k-o_r) \right] \tag{4}$$

o_n and o_r are the orders of the nonrecursive and recursive net respectively. Note, both nets have only one output node so that $\mathbf{W}_L$ is a row vector and w_{L0} is a scalar. In addition the output node is linear for both nets, that is $f(\gamma) = \gamma$ for the output node.

Our goal here is to find the set of network weights that minimize the total error,

$$E = \frac{1}{2}\sum_{k=0}^{N} e^2(k) = \sum_{k=0}^{N} (r(k)-y(k))^2 \tag{5}$$

where $r(k)$ is the output of the model and $y(k)$ is the output of the plant. The plant is in general nonlinear and its output is a function of previous outputs as well as the current and previous inputs,

$$y(k) = g(\, u(k), u(k-1), \ldots, u(k-p_n), y(k-1), y(k-2), \ldots, y(k-p_r)\,) \tag{6}$$

where p_n and p_r are the nonrecursive and recursive orders of the plant respectively. The weights are found using a gradient search. As such the update equation for weight j is

$$\begin{aligned} w_j(m+1) &= w_j(m) - \mu \frac{\partial E}{\partial w_j} \\ &= w_j(m) + \mu \sum_{k=0}^{N} e(k)\beta_j(k) \end{aligned} \tag{7}$$

where

$$\beta_j(k) = \frac{\partial y(k)}{\partial w_j} = \mathbf{q}_u^T(k)\mathbf{a}_j^{p_n}(k) + \mathbf{q}_y^T(k-1)\mathbf{b}_j(k-1) \tag{8}$$

and

$$\mathbf{q}_u^T(k) = \left[\frac{\partial g(\cdot)}{\partial u(k)}, \frac{\partial g(\cdot)}{\partial u(k-1)}, \cdots, \frac{\partial g(\cdot)}{\partial u(k-p_n)} \right] \tag{9}$$

$$\mathbf{q}_y^T(k-1) = \left[\frac{\partial g(\cdot)}{\partial y(k-1)}, \frac{\partial g(\cdot)}{\partial y(k-2)}, \cdots, \frac{\partial g(\cdot)}{\partial y(k-p_r)} \right] \tag{10}$$

$$\mathbf{a}_j^n(k)^T = \left[\alpha_j(k), \alpha_j(k-1), \cdots, \alpha_j(k-n) \right] \tag{11}$$

$$\mathbf{b}_j(k-1)^T = \left[\beta_j(k-1), \beta_j(k-2), \cdots, \beta_j(k-p_r) \right] \tag{12}$$

The $\mathbf{q}$ vectors above represent partials of the plant with respect to its inputs and outputs. When the plant is unknown then these must be estimated. The $\alpha_j(k)$ term in Eq. 11 is given by

$$\alpha_j(k) = \frac{\partial u(k)}{\partial w_j} = \frac{\partial v_{NR}(k)}{\partial w_j} + \frac{\partial v_R(k)}{\partial w_j} \tag{13}$$

It can be shown that

$$\alpha_j(k) = BP_j(k) + FF(\mathbf{a}_j^{o_r-1}(k-1)) \tag{14}$$

where $BP_j(k)$ are the standard backpropagation equations and $FF(\mathbf{a}_j^{o_r-1}(k-1))$ represents a recursive term due to the feedback. The standard backpropagation term for the nth weight in the mth node of layer l can be expressed

$$BP(k) = v_{l-1,n}(k)h_{lm}(k)\mathbf{w}_{l+1,m}^T \mathbf{H}_{l+1}(k) \cdots \mathbf{W}_{L-1}^T \mathbf{H}_{L-1}(k)\, \mathbf{W}_L^T \mathbf{1} \tag{15}$$

where $\mathbf{1}$ is a vector of 1s, and

$$\mathbf{H}_i(k) = diag\left[h_{i1}(k),\ h_{i2}(k),\ \cdots,\ h_{in}(k)\right] \quad (16)$$

where

$$h_{im}(k) = v_{im}(k)[1-v_{im}(k)] \quad (17)$$

The recursive term in Eq. (14) can be shown to take on the form

$$FF(\mathbf{a}_j^{o,-1}(k-1)) = \mathbf{W}_L^r\mathbf{H}_{L-1}(k)\,\mathbf{W}_{L-1}^r \cdots \mathbf{H}_1(k)\,\mathbf{W}_1^r\mathbf{a}_j^{o,-1}(k-1) \quad (18)$$

In this operation the $\mathbf{a}_j^{o,-1}(k-1)$ vector is fed *forward* through the recursive net in a fashion which is almost the reverse of the $BP(k)$ operation.

In general, the stability of the gradient approach in a feedback loop is not guaranteed even in the linear case. As might be suspected, certain initial conditions and design parameters will lead to unstable behavior of the closed-loop system. This will be an area of further research. On the other hand, we were able to solve many model-following problems, with different external inputs and many different models. A particular model-following example is given next.

III. Example

Let the plant be given by

$$y(k) = \frac{y(k-1)}{1+y^2(k-1)} + u^3(k).$$

and the reference input be

$$d(k) = \sin(0.05\pi k)$$

Let the reference model be the unity gain so that an inverse model is sought. The algorithm was run for 50,000 iterations with white noise training input, after which, the weights were fixed and the results of Figure 2 were obtained. As can be seen, the recursive neural network has learned the inverse of the plant for the particular input d(k).

IV. Conclusions

A recursive neural network was derived and used for the model-following control of a discrete-time nonlinear systems. The network is updated in the feedback loop directly and was able to minimize the model-following error for a number of plants. Many issues however, remain to be resolved, such as the stability of the closed-loop system, the choice of the number of delays and other design parameters. It is felt that the recursive MLP approach presents an alternative to the hard problem of controlling a nonlinear system.

REFERENCES

[1] A. Isidori, *Nonlinear Control Systems: An Introduction,* Lecture Notes in Control and Information Sciences, M. Thoma, Ed., Springer Verlag, Berlin, 1985.

[2] R.H. Hirschorn,"Invertibility of Multivariable Nonlinear Control Systems," *IEEE trans. Automat. Contr.*, Vol. AC-24, No.6, pp.855-865, Dec. 1979.

[3] Funahashi, K., "On the approximate realization of continuous mappings by neural networks," *Neural Networks,* Vol. 2, pp. 183-192, 1989.

[4] Special Section on Neural Networks For Control Systems, *IEEE Control Systems Magazine,* Vol.9, No.3, pp.25-59, April 1989.

[5] K.S. Narendra, and K. Parthasarathy, "Identification and Control of Dynamical Systems Using Neural Networks," *IEEE Trans. on Neural Networks,* Vol.1, No.1, pp.4-27, March 1990.

[6] W. Li, and J.J.E. Slotine, "Neural Network Control of Unknown Nonlinear Systems," *Proc. IEEE Auto. Contr. Conf.*, pp. 1136-1141, Pittsburgh, PA 1989.

[7] Rumelhart, D.E., McClelland, J.L., & Williams, R.J., "Learning internal representations by error propagation," In D.E. Rumelhart & J.L. McClelland (Eds.), *Parallel Distributed Processing.* Cambridge, MA: MIT Press, 1986.

[8] Pineda, F.J., "Generalization of backpropagation to recurrent neural networks," *Physical Review Letters,* Vol. 18, pp. 2229-2232, 1987.

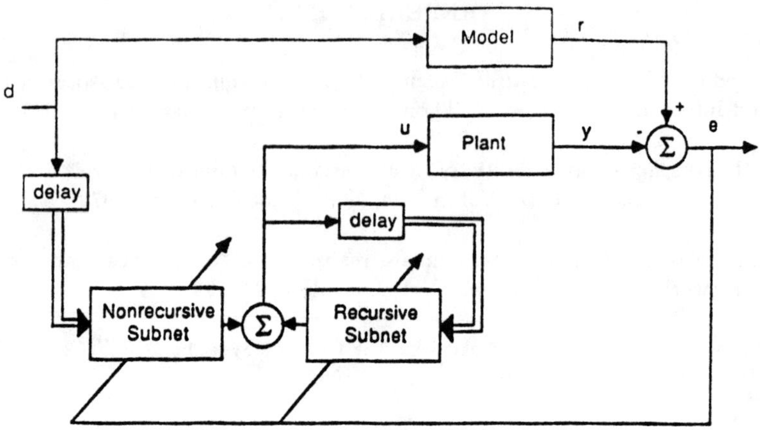

Figure 1

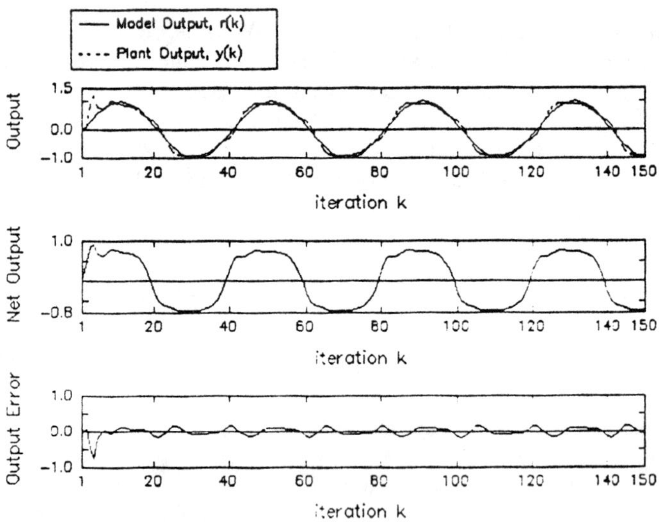

Figure 2

A Neural Network Based Strategy for the Inverse Kinematics Problem in Robotics †

By L. Nguyen and R.V. Patel
Department of Electrical and Computer Engineering
Concordia University, Montreal, Quebec
CANADA H3G 1M8

ABSTRACT

The inverse kinematics problem plays an important role in robot manipulator analysis and control. In this paper, a general neural network architecture with on-line learning capability is presented. First, a multilayer feedforward network is trained off-line to provide an approximate solution to the robot inverse kinematics problem. Then a neural network compensator with on-line learning capability is used to provide a solution within some specified tolerances. Simulations show that this approach is suitable for solving the inverse kinematics problem for a robot manipulator in real-time.

1 Introduction

A kinematics equation provides the relationship between joint displacements and the resulting end-effector positions and orientations. The problem of finding the end-effector position and orientation for a given set of joint displacements is referred to as the forward (direct) kinematics problem. That is, the forward kinematics problem allows one to specify in a unique manner the relationship between the (n x 1) joint vector θ and the (m x 1) Cartesian vector P as

$$P = F(\theta)$$

where F is a continuous non-linear function that depends on the structure and physical parameters of the manipulator. It associates with each θ a unique P. The inverse mapping is given by:

$$\theta = F^{-1}(P)$$

which, in general may have several values of θ corresponding to one value of P. There are two main classes of solution methods for the inverse kinematics problem: those that give closed-form (explicit) solutions and those that give numerical solutions. Closed-form solutions are only possible for certain types of robot manipulators. For instance, it is known that all six-axes, wrist-partitioned manipulators allow closed-form solutions. However, apart from these and some very special cases, the problem of computing manipulator inverse kinematics requires numerical solutions. Such solutions are also used for redundant manipulators. Because of their iterative nature, numerical solution methods are generally much slower than the corresponding closed-form solution methods (when they exist). The use of numerical solution methods raises the problem of real-time implementation of manipulator kinematics control.

Neural networks provide a fast way of learning to produce a set of desired output states, given a set of input states. There are several advantages of using neural networks in robotics and other applications: No programming is required; the time required to

†This research was supported by the Natural Sciences and Engineering Council of Canada under grant A1345

obtain a solution is independent of the number of degrees of freedom; neural networks have an inherent fault tolerant property and they provide the capability of real-time operation.

Recently, Guez and Ahmad [1] have proposed a solution to the inverse kinematics problem using an error back propagation network. They have shown that the inverse kinematics problem can be solved by training a three-layer perceptron network. However, their approach has two main drawbacks: The solutions obtained are of limited accuracy and the time required to train the network is rather large. In addition, the neural network is not able to provide a robust solution when the input differs greatly from the training input. Therefore, an on-line learning capability is also required.

In this paper, a neural network architecture is proposed. It achieves accurate solutions and reduces the training time required by incorporating an on-line learning capability in the network.

2 The Neural Network Architecture

Figure 1 presents the overall architecture of a neural network with on-line learning capability for the robot inverse kinematics problem. The system is composed of two neural network controllers: an approximate inverse kinematic controller and a neural network compensator.

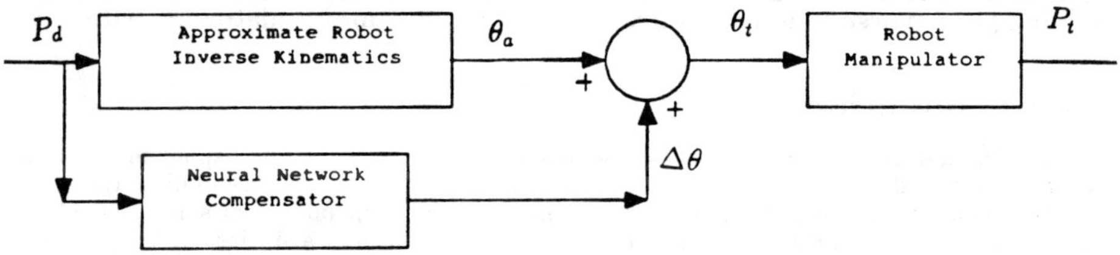

Figure 1. A Neural Network with on-line learning capability.

The approximate inverse kinematic controller is trained off-line to provide an approximate solution to the robot inverse kinematics problem . This network is then used to train a neural network compensator that reduces the error to within some desired tolerance. The advantage of this approach is that no prior knowledge of the robot kinematics is required. In addition, because on-line training is available, the system will be robust to uncertainty in the robot kinematics parameters.

3 The Multilayer Feedforward Network

3.1 Design overview

The feedforward controller is designed to approximate the inverse kinematics of a manipulator. This controller must convert desired Cartesian space positions into joint space variables which can be used as inputs for the robot dynamic controller to drive the robot along the specified path.

In our approach, the manipulator is considered as a black box; only its inputs and outputs are observed. Our objective is to build a feedforward controller to reproduce the inverse kinematics relationship, giving us the required joint angles. The controller must be taught this relationship solely by observation of the manipulator's inputs and outputs.

3.2 The Training Strategy

Psaltis et al [2] have proposed an on-line training scheme for a multilayer neural network controller. The method is shown in Figure 2. They considered the robot as an additional unmodifiable layer. Therefore, the error signal at the system output can be propagated back through the robot using the derivatives of the robot joint variables at the operating point. In the case of the inverse kinematics problem, such a scheme would require the inverse Jacobian to be evaluated at every iteration. To overcome the difficulty where the calculation of manipulator inverse Jacobian is expensive, Chen and Pao [3] proposed using an inverse transfer relationship to approximate the inverse Jacobian at the operating point. The advantage of this approach is that it provides an on-line learning capability, making the system adaptable to changes in manipulator kinematics. However, due to the closed-loop structure, the system is subject to stability problems.

In our approach, the feedforward controller is designed to approximate the inverse kinematics only. The general learning scheme used is shown in Figure 3.

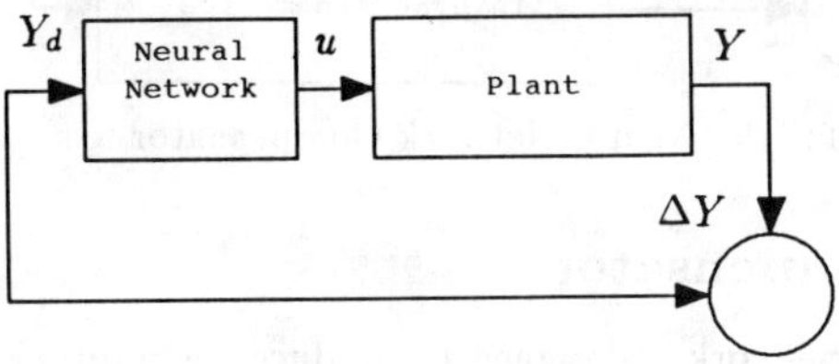

Figure 2. A Specialized Learning Configuration

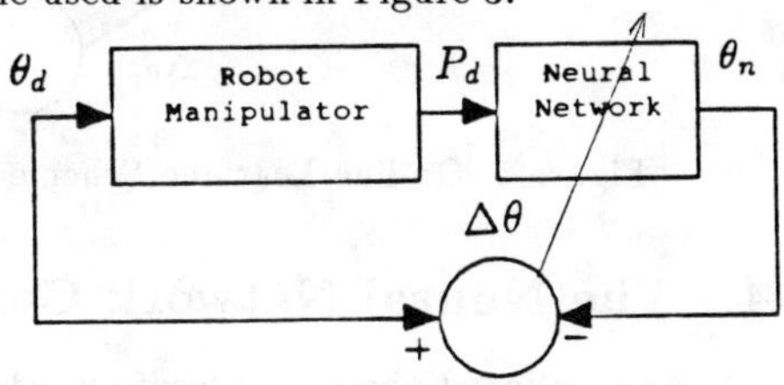

Figure 3. A Generalized Learning Configuration

The drawback of this approach is that it is an off-line procedure. Since the sets of inputs and outputs used for training are stable during the training time, training the network presents no stability problems. The manipulator is considered as a black box. Therefore we cannot train the network to correspond correctly in a specific regions of the manipulator work space. The set of training data must be uniformly distributed over the manipulator work space so that the network can interpolate the intermediate points.

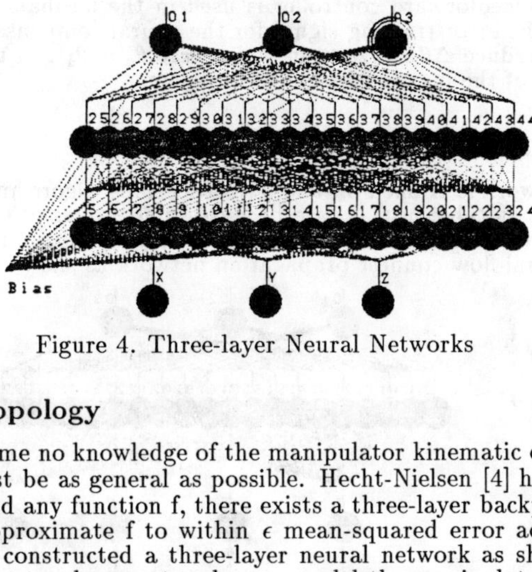

Figure 4. Three-layer Neural Networks

3.3 Network Topology

Because we assume no knowledge of the manipulator kinematic or dynamic parameters, our model must be as general as possible. Hecht-Nielsen [4] has recently showed that for any $\epsilon > 0$ and any function f, there exists a three-layer backpropagation neural network that can approximate f to within ϵ mean-squared error accuracy. Based on this result, we have constructed a three-layer neural network as showed in Figure 4. Simulations show that two-layer networks can model the manipulator inverse kinematics adequately in most cases. However, the three-layer network of Figure 4 is used for the sake of generality. Each node of the network performs the sigmoid function. The learning procedure follows the generalized delta rule for a feedforward network. The purpose of this algorithm is to minimize the mean-squared error between the desired joint variables and the actual joint variables.

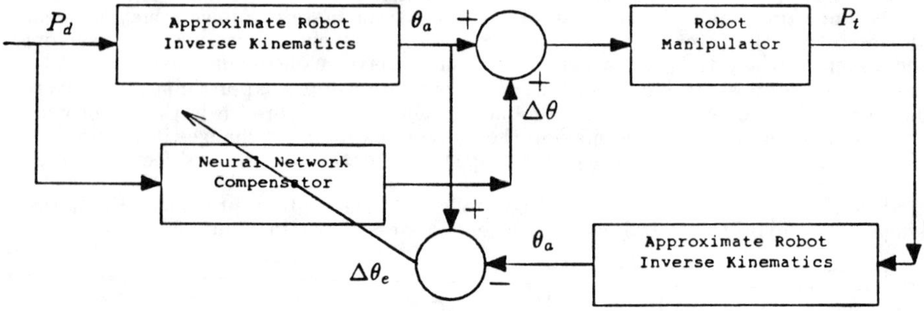

Figure 5. On-line Learning Scheme for the Neural Network Compensator

4 The Neural Network Compensator

As mentioned above, the feedforward network is designed to produce the approximate inverse kinematics of the manipulator. Therefore, the error in this network will be large for inputs which are different from the training inputs. The neural compensator is designed to reduce these errors to within some desired tolerances in the task-space region of interest. The training of the neural compensator is therefore an on-line procedure.

To train the neural compensator, we propose a feedback learning scheme as shown in Figure 5. The system is organized in a closed-loop structure. A desired Cartesian position P_d is fed into the feedforward controller. The output θ_a of the feedforward controller is then combined with the output of the compensator, $\Delta\theta$, to form the manipulator joint angle input θ_t. This is fed into the manipulator to get the actual manipulator position P_t. A copy of the feedforward controller is used in the feedback path. The difference $\Delta\theta_e = \theta_a - \theta_a$ is the error training signal for the neural compensator. The objective of the training is to reduce $\Delta\theta_e$ to zero, in which case $\theta_a = \theta_a$, so that $P_t = P_d$ and θ_t is the exact solution of the inverse kinematics problem.

4.1 Training Strategy and Network Topology

Since the network is trained on-line, a fast learning algorithm must be used. Back propagation is not suitable in this case because small values of the learning coefficient are needed to ensure convergence, resulting in a very slow training process. In our approach, we use a uni-flow counter propagation network as shown in Figure 6.

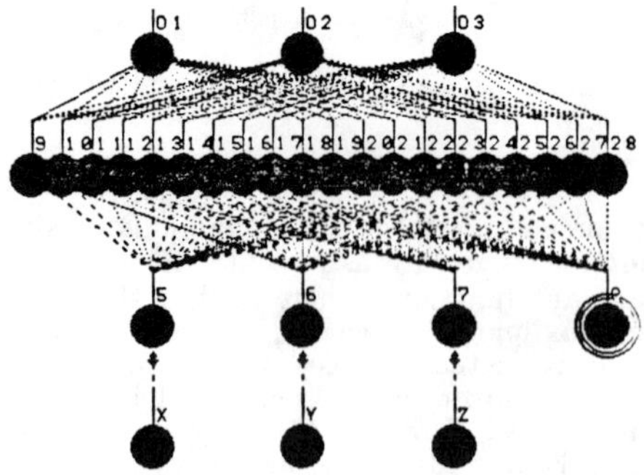

Figure 6. Uni-flow Counter Propagation Network

The first hidden layer is the normalizing layer to normalize the input vector into a unit sphere. The second hidden layer is the Kohonen layer that acts as a nearest neighbour classifier. The Kohonen layer measures the cosine of the angle between the input vector X and the weight vector W. The node with the closest weight vector wins the competition. The learning procedure of this layer is the Kohonen learning rule. The output layer is the Grossberg Outstar which essentially learns to produce a certain output when a particular input is applied. The output layer follows the Widrow-Hoff learning algorithm. This counter propagation network works as a lookup table where the closest input is found and its equivalent output is read. It basically performs a piece-wise linear approximation of the outputs. To achieve a better estimation of the outputs, we use a modified version of the Kohonen layer where two nodes with the closest weight vectors to the input vector win the competition instead of one node. This allows a better approximation of the outputs compared to the standard counter-propagation network [5].

5 Simulation Results

To illustrate our approach, we simulated the inverse kinematics of a PUMA 560 robot. The PUMA 560 is a wrist-partitioned manipulator with six revolute joints. The first three joints achieve desired end-effector positions in its 3-D task space while the last three joints achieve desired orientations of the end-effector. For illustration purposes, only the inverse kinematics corresponding to the positioning of the end-effector were simulated. The orientation of the end-effector was ignored. The reason for choosing this manipulator is that its inverse kinematics can be obtained in closed form. This eases the problem of data generation and verification. The equations of the PUMA 560 inverse kinematics are given in [6,7]. The values of the parameters used in our simulations were obtained from [7].

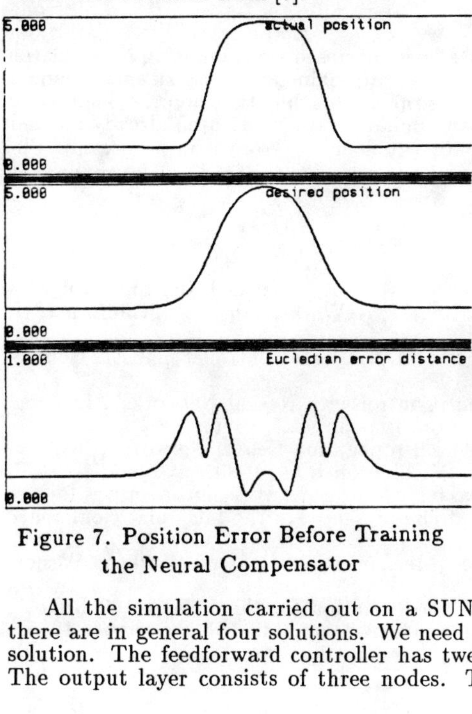

Figure 7. Position Error Before Training the Neural Compensator

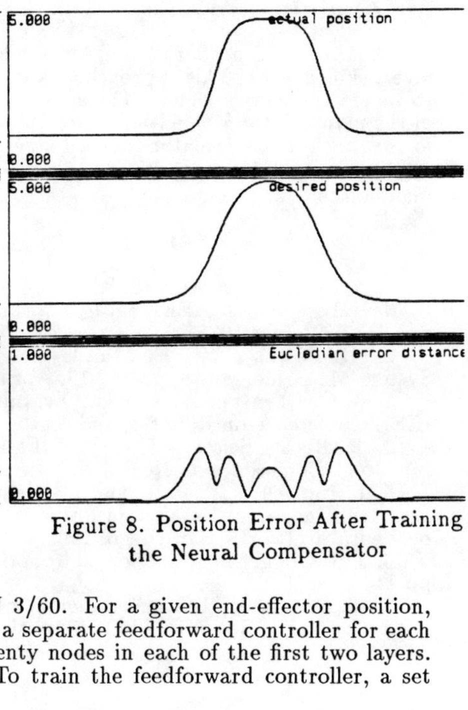

Figure 8. Position Error After Training the Neural Compensator

All the simulation carried out on a SUN 3/60. For a given end-effector position, there are in general four solutions. We need a separate feedforward controller for each solution. The feedforward controller has twenty nodes in each of the first two layers. The output layer consists of three nodes. To train the feedforward controller, a set of inputs and outputs were obtained by measurement. All the weights were randomly initialized to values between -0.5 and 0.5. In the output layer, all the weights were initialized to zero. This is to avoid the case in which a local minimum predominates when the generalized delta algorithm starts. To speed up the convergence rate, we set a high learning rate at the first layer and a low learning rate at the second layer. In our case, we used 0.4 and 0.1 respectively. To avoid being stuck in a local minimum during learning time, a momentum term was used. The training was done for 200 presentations of the data set of 100 samples.

The feedforward controller was then implemented and a copy of it was placed in the feedback path to generate the error training signal for the neural compensator. The normalizing layer contained four nodes. The Kohonen layer consisted of twenty nodes. All the weights of the neural compensator were randomly initialized to values between -0.5 and +0.5.

To train the neural compensator, a cubic spline trajectory was used. Figure 7 shows the errors of the network before training the neural compensator. In training the neural compensator, fifty samples of the cubic spline were used. The result obtained after twenty presentations is shown in Figure 8.

To obtain a more accurate solution to the robot inverse kinematics problem, the number of nodes in the Kohonen layer of the neural compensator needs to be increased. However, this increases the training time. In our simulation, twenty five nodes were found to be sufficient. The number of nodes in the Kohonen layer depends on the number of samples being used to train the neural compensator. Experiments show that the number of nodes in the Kohonen layer should be at least one half of the number of training samples. To improve the training time of the networks, a smaller sub-network should be construct for each output. This will result in faster convergence because the training for each output is independent of that for the others.[5]

6 Conclusions

A new neural network based scheme has been proposed for computing manipulator inverse kinematics. This approach makes use of an approximate inverse kinematic model to generate the error signal. The advantage of the approach is that no prior knowledge of the manipulator kinematics is required. A similar strategy has been already applied for solving the manipulator forward kinematic problem [5]. Work is in progress on obtaining conservative (cyclic) solutions for the inverse kinematics problem for redundant manipulator.

REFERENCES

1. A. Guez and Z. Ahmad, "Solution to the Inverse Kinematics Problem in Robotics by Neural Networks", IEEE Int'l Conf. on Neural NetWorks, San Diego, California, II, pp. 617-621, July 1988.

2. D. Psaltis et al, "A Multilayered Neural Network Controller", IEEE Control System Magazine, vol. 8, pp. 17-19, April 1988.

3. V.C. Chen and Y.H. Pao, "Learning Control with Neural Networks", In Proc. IEEE Conference on Robotics and Automation, Phoenix, AZ, 1989.

4. R. Hecht-Nielsen, "Theory of the Back Propagation Neural Network", In Proc. Int. Joint Conference on Neural Networks, Washington D.C., 1989.

5. L. Nguyen, R.V. Patel and K. Khorasani, "Neural Network Architectures for the Forward Kinematics Problem in Robotics", to be presented at the Int. Joint Conference on Neural Networks, San Diego, 1990.

6. J. J. Craig, "Introduction to Robotics: Mechanics and Control", Addison-Wesley, pp 115-118, 1986.

7. S. Elgazzar, "Efficient Kinematic Transformation for the PUMA 560 Robot", IEEE Journal of Robotics and Automation, vol. RA-1, pp. 142-151, September 1985.

A Modified Hopfield Neural Network for Optimization of Manufacturing Task Scheduling

Zhen-Ping Lo and B. Bavarian
Department of Electrical and Computer Engineering
University of California, Irvine
Irvine, CA 92717

Abstract

The crossbar Hopfield network is applied to solve the Multiple Task Scheduling on multiple servers. The complete formulation of the problem is presented. This includes the neural network model developed for the scheduleing task, and the definition of the energy function for the network which are used to find the characteristic equations for each neuron. The simulation result for a sample run is also presented.

1 Introduction

The scheduling problem has been stimulated by the increasingly competitive world markets for efficiently manufactured goods. Better schedules provide a competitive advantage. There is a great need for better scheduling algorithms and heuristics. The optimum task scheduling for manufacturing processes is in general an NP-complete problem for single server or manufacturing cells. The problem is more severe for scheduling many tasks with precedence constraints among them, timing requirements, set up costs and completion deadlines to several manufacturing cells.

The machine scheduling problem has been studied extensively over the past several decades[1][2]. The problem may simply be stated as follows : n jobs are to be processed on m servers. The scheduling problem is to find the sequence and timing of each job on each machine such that some given performance criterion is optimized, e.g., minimizing the time required to complete all of the jobs.

The classical heuristic approach is to find the next operation or task to be processed based upon some easily computed parameters of the job or machines. Examples of these parameters are processing times, due dates, costs, set up times, arrival times and machine loading[3]. Some of the rules applied to select the next operation are FIFO(First In, First Out), SPT(Shortest Processing Time), and EDD(Earlist Due Date). A complete survey of the rules used in scheduling is presented by Panwalkar and Iskander[4], and S. French[5].

In this paper we provide a new approach to formulate the problem of scheduling n jobs on m machines with neural networks. Recent advances in the study of artificial neural networks provides the basis to solve NP-complete optimization problems such as the scheduling problem with a tractable time complexity. The key approach is to formulate the problem with many first order differential or difference equations and solve them in a parallel distributed fashion. The work of Hopfield[6] was the original attempt to define a general neural network architecture which can be used to solve optimization problems. Hopfield and Tank[7] formulated and solved the Travelling Saleman Problem which is an NP-complete optimization problem with a two dimensional neural network. The approach was further extended for solving the multiple travelling salesmen problem[8] by Wacholder,et.al. In both cases valid suboptimal solutions are found with an order one time complexity of convergence. These works showed the potential of the approach to the optimum decision making problems, which has been pursued for solving the load balancing problem in a hypercube multi-computer system[9] and the deterministic single server[10] scheduling problem.

Neural network architectures for computing the optimization problems are based on two approaches, simulated annealing method[11], and deterministic differential gradient descent method. Simulated annealing is a stochastic strategy for selecting the minimal configuration state, appropriate for problems with multiple extrema. Recently, Hsu[12] proposed a fast simulated annealing in which a D-dimensional Cauchy probability distribution makes occasional long jumps during local sampling of the phase space, thus resulting in a faster convergence than classical simulated annealing. The projected gradient method is an iterative procedure for global minimization of quadratic functions. A starting point is chosen in the interior of the cube and iteration steps are with respect to the energy function. A minima point could be found by applying the method. In this paper we are going to extend a deterministic single server scheduling problem into a multi-server scheduling problem by the gradient descent method.

Section 2 defines the problem of multi-server scheduling. The energy(or cost) function for the problem is defined in section 3, and in section 4 the results of a simulation run is shown. Section 5 contains the concluding remarks and our current and future work on the problem.

2 A Muti-server Problem

Suppose that n jobs $J_i(i=1,2,...n)$ have to be processed on m machines $M_j(j=1,2, ...,m)$. Each job has unequal lengths. We assume that each machine can process at most one job at a time . Also we assume each machine has different ability. The problem is to determine a sequence of jobs on each machine which minimizes the total set up cost and the time required to complete all the jobs. Each task is associated with a due date, denoted by d_i by which the job must be completed or else a penality is incurred for the tardiness in job completion. Also we want to make sure every machine will work on at least one job.

The solution to the m machine n task problem consists of an ordered list of n jobs on m machines. To map the problem onto the neural network, we extended the two dimensional Hopfield network for the TSP problem to a three dimensional neural network box,called Neuro-Box Network or NBN, where the neurons operate in high-gain limit to converge to digital(1, and 0) output. The z-axis represents the jobs, the x-axis represents the machines and the y-axis represents the time. All the three axes are in discrete units, each unit representing the corresponding job, machine and the time it is started. The units of time depend on the problem at hand ; it may be a day, a week or one hour, for n jobs, m machines to be done in k units. Each cube is a neuron in the network and it has excitatory and inhibitory connections to all other neurons in the box. If the state of neurons v_{ijl} is one ,then job J_j is assigned to machine M_i at lth time slot.

For this problem representation, then in the xy planes for each job , there should be only one neuron at state 1 and the rest zero, since that particular job need to be assigned to a machine at a given starting time. In the xz, planes there may be at most m neuron high with the constraint that only one neuron in a row or column can be high, since at a given time job J_j can be assigned to only one machine(row constraint) and two jobs can not be assigned to a machine(column constraint).

J_1	0	0	0	0	0	0
J_2	1	0	0	0	0	0
J_3	0	0	0	1	0	0
J_4	0	0	0	0	0	0

Machine 1 schedule

J_1	0	0	1	0	0	0
J_2	0	0	0	0	0	0
J_3	0	0	0	0	0	0
J_4	1	0	0	0	0	0

Machine 2 schedule

Figure 1: An example of job assignment table for two machines and 4 jobs

Furthermore, in yz plane which shows the schedule of a machine, at most only one neuron can be high in the lth row, since the job is assigned once(row constraints) and two jobs can not start at the same time(column constraint). Since we want each machine to do at least one job, at least one neuron should be high in zy plane. Also if there is more than one job assigned to this machine the difference in their position in y-axis should be greater than the length of time it takes for the first job to be finished at this machine. The total number of neurons in the box with "1" output must equal to n the number of jobs. Fig. 1 shows an example of a 2 machine, 4 jobs assignment.

3 Derivation of Energy Fuction of the Neural Network

Our network will be defined by an energy function. The lowest energy state of the network corresponds to the best scheduling. The functon consists of two parts: the physical objective function and the constraint functions.

The E_t, the objective function, which minimizes the time required to complete all the jobs,

$$E_t = \frac{1}{2}\sum_{j=1}^{m}\sum_{i=1}^{n}\sum_{l=1}^{k} v_{ijl}\left(\frac{l+T_{ij}-1}{C_k}\right) \tag{1}$$

The energy functions for row and column constraints are given by

$$E_1 = \frac{1}{2}\sum_{i=1}^{n}\sum_{l=1}^{k}\sum_{j=1}^{m}\sum_{l_1\neq l}^{k}\sum_{j_1\neq j}^{m} v_{ijl}(v_{ij_1l}+v_{ijl_1}+v_{ij_1l_1}) \tag{2}$$

$$E_2 = \frac{1}{2}\sum_{j=1}^{m}\sum_{l=1}^{k}\sum_{i=1}^{n}\sum_{i_1\neq i}^{n} v_{ijl}v_{i_1jl} \tag{3}$$

$$E_3 = \frac{1}{2}\left(\sum_{i=1}^{n}\sum_{j=1}^{m}\sum_{l=1}^{k} v_{ijl}-n\right)^2 \tag{4}$$

$$E_4 = \frac{1}{2}\sum_{j=1}^{m} f_j^2 H(f_j) \tag{5}$$

$$E_5 = \frac{1}{2}\left(\sum_{i=1}^{n}\sum_{j=1}^{m} v_{ij1}-m\right)^2 \tag{6}$$

$$E_6 = \frac{1}{2}\sum_{j=1}^{m}\sum_{i=1}^{n}\sum_{l=1}^{k} v_{ijl}g_{ijl}^2 H(g_{ijl}) \tag{7}$$

$$E_7 = \frac{1}{2}\sum_{j=1}^{m}\sum_{i=1}^{n}\sum_{l=1}^{k}\sum_{i_1\neq i}^{n} v_{ijl}(h_1^2 H(h_1)\sum_{l_1=l}^{k} v_{i_1jl_1} + h_2^2 H(h_2)\sum_{l_2\neq l}^{l-1} v_{i_1jl_2}) \tag{8}$$

Where
$f_j = 1 - \sum_{i=1}^{n}\sum_{l=1}^{k} v_{ijl}$
$g_{ijl} = l - 1 + T_{ij} - d_i \quad T_{ij} = \frac{L_i}{m_j}$
$h_1 = (T_{ij} + l) - \sum_{l_1=l}^{k} l_1 v_{i_1jl_1}$
$h_2 = \sum_{l_2\neq l}^{l-1}(T_{i_1j} + l_2)v_{i_1jl_2} - l$

$i, j, l, i_1, j_1, l_1, l_2$ are indices. C_k is constant for scaling; v_{ijl} is output of neuron; m_j is work rate of the jth machine; L_i is the ith job length; T_{ij} is the time required by jth machine to complete the ith job; d_i is the deadline of ith job; and $H(f)$ is the heavyside function with value equal to one if $f > 0$, otherwise zero. We are going to use high gain sigmod function to approximate the $H(f)$, i.e.,

$$H(f) = \frac{1}{1 + e^{-\gamma f}} \tag{9}$$

where γ is a constant.

The constraints (2) (3) (4) (5) are required to form M permutation matrices: They ensure that each job is only assigned to one machine and every machine at least does one job in the whole schedule sequence. The constraint (6) ensures every machine to be assigned one job at time 0. The constraint (7) is to enforce the deadline requirements. The constraint (8) means if ith job is assigned to the jth machine at time l, then no other jobs can be assigned to the machine till after $l + T_{ij}$ and also job i_1th can not be assigned before $l_2 + T_{i_1j}$.

A constrained optimization problem can be converted to an unconstrained one by introducing Lagrangian multipliers, λ_α. As suggested in the MTSP[8] solution, if we also let λ_α be the states of seven additional neurons, then the object is to minimize the total cost function:

$$E = E_t + \sum_{\alpha=1}^{7}\lambda_\alpha E_\alpha \tag{10}$$

Applying the gradient technique, we obtain a system of ordinary differential equations

$$\frac{du_{ijl}}{dt} = -(\frac{\partial E_t}{\partial v_{ijl}} + \sum_{\alpha=1}^{7}\lambda_\alpha\frac{\partial E_\alpha}{\partial v_{ijl}}) \tag{11}$$

$$\frac{d\lambda_\alpha}{dt} = -E_\alpha \quad for\ \alpha = 1, 2, \ldots, 7 \tag{12}$$

In order to make the system converge to a stable equiliburium point, equations (11) and (12) are modified according to the suggestions in the MTSP[8] which incoporates a damping term in (11) and solves (12) forward in time, i.e.,

$$\frac{du_{ijl}}{dt} = -\frac{u_{ijl}}{\tau} - (\frac{\partial E_t}{\partial v_{ijl}} + \sum_{\alpha=1}^{7}\lambda_\alpha\frac{\partial E_\alpha}{\partial v_{ijl}}) \tag{13}$$

$$\frac{d\lambda_\alpha}{dt} = E_\alpha \quad for\ \alpha = 1, 2, \ldots, 7 \tag{14}$$

By differentiating E_t and $E_1 \ldots E_7$, we obtain the necessary neural network equations which will be simulated next.

Table 1. Table of machine 1 schedule

Sequence	Starting T	Complete T	Finishing T	Deadline
J5	1	2	2	4
J10	3	3	5	6
J7	6	1	6	9
J8	7	2	8	10

Table 2. Table of machine 2 schedule

Sequence	Starting T	Complete T	Finishing T	Deadline
J2	1	5	5	6
J8	6	1	6	8
J3	7	2	8	9

Table 3. Table of machine 3 schedule

Sequence	Starting T	Complete T	Finishing T	Deadline
J9	1	1	1	2
J4	2	3	4	5
J1	7	4	10*	9

Figure 2: schedule of 3 machines and 10 jobs

4 Simulations

We used the first order Euler method to solve the nonlinear differential equations of the neurons. The $u_{ijl}(0)$ and $\lambda_\alpha(0)$, initial conditions are chosen arbitrarily within the range suggested in [7] and [8]. It turns out that the convergence is very sensitive to selection of the starting points. Many simulations were carried out with varying degrees of success of convergence to valid scheduling sequences. The energy function consists of many competing terms with local minima some of which do not satisfy the constraints. Fig.2 shows the table for one example of assigning 10 jobs to 3 machines. The jobs have deadlines from 2 to 10 time units and different lengths. The schedule meets all the constraints except for the job one which is finished one time unit after it was due. The selection of job lengths and deadline in general does not need to be hard limits i.e., by allowing soft constraint, some of the problems in

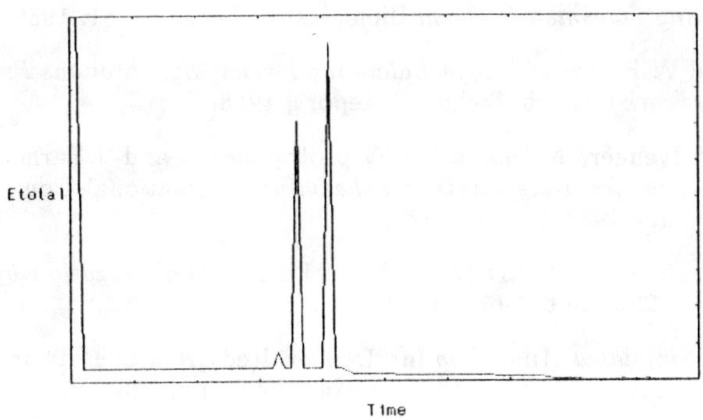

Figure 3: Energy functions vs number of iterations

constrains satisfaction can be relaxed.

5 Conclusion

We have formulated the scheduling problem of n job assignment to m machine with a novel 3-dimensional Hopfield like neural network. The approach is flexible. It incorporates different machine work capacities, different job lengths, and hard deadlines. It is generic and the Neuron-Box Network(NBN) concept allows expension in any of the three dimensions, i.e., it can readily be adopted for any number of jobs, any number of machines and one week planning or one year planning.

Currently we are investigating the stability and convergence issures of the NBN. Also the approach of simulated annealing is being developed in an alternative structure for realizing the Neuron-Box Network operation.

References

[1] K.Baker *Introduction to Sequening and Scheduling* New York: Wiley, 1974.

[2] E.G. Coffman, Jr., Ed., *Computer and Job shop scheduling* New York: Wiley, 1976.

[3] R.W.Conway, W.L.Maxwell, and L.W. Miller *Theory of Scheduling* Reading, MA: Addison-Wesley, 1967.

[4] S.S Panwalkear and W.Iskander *A Survey of Scheduling Rules* Oper. Res., Vol.25, No.1, pp. 45-61, Jan.-Feb.,1977.

[5] S.French *Sequencing and Scheduling: An Introduction to the Mathematics of Job-shop* New York: wiley, 1988.

[6] J.J Hopfield *Neural Networks and Physical Systems with Emergent Collective Computational Abilities* Proc. Nat. Acad. Sci. USA, Vol.79, 2554-2558,1982.

[7] J.J Hopfield and D.W. Tank *Neural Computation of Decision in Optimization Problem* Biological Cybernetics 52. 1985. pp.141-152

[8] E.Wacholder, J.Han and R.C.Momn *A Neural Network Algorithm for the Multiple Travelling Salesmen Problem* Biological Cybernetics 61. 1989. pp.11-19

[9] G.C.Fox and W.Furmanski *Load Balancing Loosely Synchronous Problems with a Neural Network* Caltech Technical Report, 1988.

[10] S.Gulati S.S.Iyenger, N.Toomarian, V.protopopescu and J.Barhen *Nonlinear Neural Networks for Deterministic Scheduling* International Conf. On Neural Network IV June,1987. pp 745-752.

[11] Kirkpatrick, S., Gelatt, C.D.Jr., and Vecchi, M.P. *Optimization by Simulated Annealing* Sci. 220, pp.671-680 ,1983.

[12] H.Szu *Fast Simulated Annealing* In: Denker J(ed) Neural Networks for Computing: American Institute of Physics, New York, pp. 420-425.

TWO-DIMENSIONAL CONTROL THEORY FOR ROBOTIC MANIPULATORS

A. Zilouchian and A. Gautam
Department of Electrical Engineering
Florida Atlantic University
Boca Raton, Florida 33431

ABSTRACT

In this paper a novel two-dimensional (2-D) learning scheme for robotic manipulators is proposed. The scheme is based on an iterative structure which utilizes available data from previous operation of a specified trajectory. It is shown that the fast convergence of the actual trajectory to the desired one is guaranteed upon proper selection of a weighting matrix. The proposed learning control technique is applied to a trajectory tracking problem for a two-link robot manipulator. Convergence of the scheme for various weighting matrices is shown.

I. INTRODUCTION

During the last decade many control schemes have been proposed for robot manipulators. These include nonlinear cancellation method, adaptive control, resolved motion and acceleration control, time-optimal control and learning control. Although adaptive control techniques such as [11], [12] do not require a precise description of the mathematical model of the system and can face parameter variation, they still require a complicated structure for the regulator due to the complexity of the adaptive mechanism. On the other hand, learning control techniques have been recently proposed by different researchers including Arimoto, et al. [1], [2], Craig [7] and Casolino et al. [5], [6] , Gang and Jamshidi [3] which are based on the use of repeated trails of a preassigned trajectory.

In general, human being have the the ability to learn from experience. People with experience can complete a given task in a better way since the experience acts as a sort of feedback information which is used to modify the control action [1]-[2]. Although humans and mechanical robots can perform similar tasks, there are sharp distinctions between the robots and human control strategy. For example human control is largely acquired through learning whereas the operation of a robot is specified by a pre-written algorithm.

In this paper a learning control model for robot manipulators is proposed. The technique utilizes a two-dimensional system model with two independent variables . One variable reflects an instant in time history and the other variable reflects the number of learning execution cycles. The difference between the system output and the desired output is utilized to produce new input sequences such that the desired trajectory can be achieved after a number of iterations.

II. PROBLEM APPROACH

Consider the dynamic model of a serial-link manipulator with n degrees of freedom [10].

$$R(q)\ddot{q}+K(q,\dot{q})+g(q)=\tau \tag{1}$$

where $q=(q_1,q_2,...,q_n)$ denotes the joint angle coordinates, $R(q)$ is a positive definite inertia matrix, $K(q,\dot{q})$ consists of the centrifugal, coriolis forces as well as the viscous frictional forces and $g(q)$ term comes from the potential energy of the manipulator.

Given a desired trajectory with a time interval $i \in [0,N]$, the robot trajectory tracking problem is to obtain the ideal input sequences $u(i)$ $(i=1,...,N)$ such that the robot position (velocity) is coincident with the desired one over the given time interval. To determine such ideal inputs, in this paper a two stage control strategy is proposed. In the first step the following feedback control law [4] is applied to the manipulator described by (1).

$$\tau = u + g(q) + \bar{A}(q_d - q) + \bar{B}(\dot{q}_d - \dot{q}) \tag{2}$$

where A and B are constant matrices and u is an additional control input to further improve the motion of the manipulator. Substituting (2) into (1):

$$R(q)\ddot{q} + K(q,\dot{q}) + \bar{A}(q - q_d) + \bar{B}(\dot{q} - \dot{q}_d) = u \tag{3}$$

Let us define

$$\begin{bmatrix} q \\ \dot{q} \end{bmatrix} \triangleq \begin{bmatrix} x_1 \\ x_2 \end{bmatrix} \tag{4}$$

thus, (3) can be written as

$$R(x_1)\dot{x}_2 + K(x_1,x_2) + \bar{A}(x_1 - x_{1d}) + \bar{B}(x_2 - x_{2d}) = u \tag{5}$$

Therefore the dynamic model of the serial link manipulator in state-space form can be described by:

$$\dot{x} = f(x) + Bu \tag{6}$$

$$y = [0 \quad I]x = Cx \tag{7}$$

where

$$\begin{bmatrix} x_1 \\ x_2 \end{bmatrix} = \begin{bmatrix} q \\ \dot{q} \end{bmatrix} \qquad B = \begin{bmatrix} 0 \\ R(x_1)^{-1} \end{bmatrix}$$

$$f(x) = \begin{bmatrix} x_2 \\ -R(x_1)^{-1}\{K(x_1,x_2) + \bar{A}(x_2 - x_{2d}) + \bar{B}(x_1 - x_{1d})\} \end{bmatrix}$$

If the geometry of the robot was known precisely, classical control schemes could be utilized to obtain the desired trajectory. However, since it is usually impossible to obtain the full description of the system dynamics, a pre-designed input sequence cannot always generate the desired system output. In this paper, we propose a learning control process for generation of sequences $u(i)$ that eliminates the difference between the actual and desired trajectories. However, as the learning process is done in iterative cycle, a new variable "j" can be introduced. Therefore, the desired trajectory tracking problem for the robot can be formulated as

$$x(i+1,j) = f[x(i,j)] + Bu(i,j) \tag{8}$$

$$y(i,j) = Cx(i,j) \tag{9}$$

where the variable "i" represents an instant in time and the variable "j" represents the number of learning execution cycles. Equations (8) and (9) are the general form of a model which has been considered in [3], [13].

III. THE TWO-DIMENSIONAL LEARNING CONTROL SCHEME

Consider the trajectory tracking problem using a 2-D system description. The error between the actual system output y(i, j) and the desired output $y_d(i, j)$ can be defined as:

$$e(i,j)=y_d(i,j)-y(i,j)$$

The purpose of a learning process is to generate a new control sequence u(i, j) such that it can reduce or eliminate the error e(i, j) as the number of iterations increase. This learning process is different from the ordinary feedback employed to control a system. For an ordinary feedback control loop, the instantaneous error between the desired output and system output is utilized to generate a new control input during each iteration. Therefore, in this type of feedback control a new input is provided to the actuators every instant in time, i.e., the input changes from time i to time i+1 and so on. Whereas in the case of the learning control scheme, the input u(i, j) for $i \in [0,N]$ is stored in the memory and the entire error sequence e(i, j) is used to produce a new input sequence u(i, j+1) for the next (j+1)-th iteration. To describe the above idea, the following equation can be utilized:

$$u(i,j+1)=u(i,j)+De(i+1,j) \quad (10)$$

where D is a mathematical operator which maps the error information into the control input modification. The operator D can be chosen to be either a gain or a differentiator or an integrator or any combination of the above. The stability and convergence rate of the learning control process is highly affected by the proper selection of D.

Using the control scheme described by equation (10), the dynamical behavior of the learning control system can be formulated by the following set of equations:

$$x(i+1,j)=f[x(i,j)]+Bu(i,j) \quad (11)$$

$$y(i,j)=Cx(i,j) \quad (12)$$

$$e(i,j)=y_d(i,j)-y(i,j) \quad (13)$$

$$u(i,j+1)=u(i,j)+De(i+1,j) \quad (14)$$

Substituting for e(i + 1, j), equation (14) may be written as:

$$u(i,j+1)=-DCf[x(i,j)]+[I_r-CBD]u(i,j)+Dy_d(i+1,j) \quad (15)$$

A detailed block diagram description of the learning control scheme is given in Figure 1.

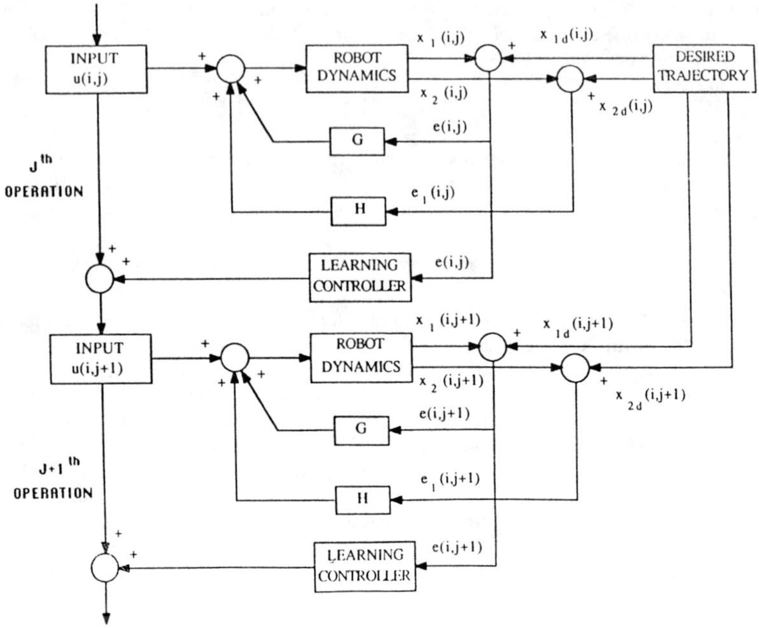

Figure 1.

IV. CONVERGENCE OF THE CONTROL SCHEME

To guarantee the convergence of the proposed technique the system output $y(i, j)$ should approach the desired output $y_d(i, j)$ as the learning process progresses, i.e.,

$$y(i,j) \rightarrow y_d(i,j) \qquad \text{for } i \in [0,N] \text{ as } j \rightarrow \infty$$

or

$$e(i,j) \rightarrow 0 \qquad \text{for } \in [0,N] \text{ as } j \rightarrow \infty$$

The 2-D dynamical equation of error is given as follows:

$$e(i+1,j)=y_d(i+1,j)-y(i+1,j)$$

$$e(i+1,j)=y_d(i,j)-Cx(i+1,j) \tag{16}$$

substituting for $x(i+1, j)$,

$$e(i+1,j)=y_d(i+1,j)-C f[x(i,j)]-CBu(i,j) \tag{17}$$

Similarly,

$$e(i+1,j+1)=y_d(i+1,j+1)-C\{f[x(i,j+1)]\}-CBu(i,j+1)$$

substituting for $u(i, j+1)$ from equation (14), we get

$$e(i+1,j+1)=y_d(i+1,j+1)-C\{f[x(i,j+1)]\}-CBu(i,j+1)-CBDe(i+1,j) \tag{18}$$

subtracting equation (17) from equation (18)

$$e(i+1,j+1)=[I_r-CBD]e(i+1,j)-C\{f[x(i,j+1)]-f[x(i,j)]\} \tag{19}$$

Equation (19) describes the two-dimensional dynamic characteristic of the error function.

Herein it is shown that the error $e(i, j)$ reduces to zero with the progression of the learning algorithm.

Theorem 1: For a dynamical model of a serial manipulator with n degrees of freedom described by equations (11), (12) and (14), if

1.

$$\| I_r - CBD \|_2 < 1 \tag{20}$$

2. The initial errors in each learning process are zero.

$$e(0,j)=0 \qquad \text{for } j=1,2,...,N.$$

Which means that the manipulator returns to its home position after each learning iteration.

3. The vector valued function $f(.)$ satisfies the condition

$$\| f[x_1(i,j)] - f[x_2(i,j)] \|_2 \le a \| x_1(i,j) - x_2(i,j) \|_2$$

where a is a positive number and $\|.\|_2$ is the x_2 norm.

then the error $e(i, j)$ will approach zero as the number of learning iterations increases $(j \rightarrow \infty)$.

In other words,

$$\lim_{j\rightarrow\infty} e(i,j)=0 \qquad \text{for } i=1,2,3,...,N.$$

Proof [10]

As we have already pointed out, the proper choice of the matrix D is essential for the convergence of the proposeed scheme. The weighting matrix should be chosen subject to the restrictions imposed by the condition $\| [I_r - CBD] \|_2 \le 1$.

Herein three lemmas for the convergence of the scheme for a general n-degree of freedom robot are given:

Lemma 1: If the weighting matrix is selected as $D=R(q)$, then the convergence of the control algorithm is always guaranteed.

Proof: [10]

If the value of the inertia matrix is exactly known in advance, we can set the weighting matrix to be equal to the inertia matrix $R(q)$ at each instant and be guaranteed of a fast convergence of the process. However, the inertia matrix is not know exactly and therefore the choice of D will not be exact. The following results are based on the assumption that there exists a certain amount of uncertainty in the value of inertia matrix of a robot manipulator.

Lemma 2: If the weighting matrix is chosen as $D=R(q)+L$, and $\sigma(L) \le \sigma(R)$ then the convergence is guaranteed where σ and σ are the largest and smallest singular values of the corresponding matrices.

Proof:

Choose $D=R(q)+L$

From equation (21) the condition for convergence is given

$$\|[I_r - CBD]\|_2 \leq 1$$

As it has been shown earlier,

$$[I_r - CBD] = [I_r - R(q)^{-1}\{R(q) + L\}] \tag{21}$$

$$\|[I_r - CBD]\|_2 = \|I_r - R(q)^{-1}L\|_2 \leq 1 \tag{22}$$

It is well known that this inequality will hold if

$$\|-R(q)^{-1}\|_2 \|L\|_2 \leq 1$$

$$\bar{\sigma}(R(q)^{-1})\|L\|_2 = \frac{1}{\underline{\sigma}(R(q)} \cdot \|L\|_2 \leq 1$$

$$\|L\|_2 \leq \underline{\sigma}(R(q))$$

$$\bar{\sigma}(L) \leq \underline{\sigma}(R)q)$$

Q.E.D.

However, the bound in lemma 2 is pessimistic. Lemma 3 gives another criterian for the convergence of the scheme.

Lemma 3: If the weighting matrix is chosen as D = R(q) + L, the region of convergence can be given by:

$$-R_{ij}(q) \leq L_{ij} \leq R_{ij}(q) \tag{23}$$

Proof: [10]

VI. SIMULATION RESULTS

The learning control scheme proposed in this paper was used to run simulations on a two-link as well as a three-link Microbot robot [11]. The simulation results for two link robots are shown in Figures 2-5. The output chosen for simulation purposes was the position of the 2nd link for the 2-link robot manipulator. The error plotted is the difference between the actual output and the desired one.

An Example

For a two-link robot the Lagrange-Euler equations of motion are given as:

$$\begin{bmatrix} \tau_1 \\ \tau_2 \end{bmatrix} = \begin{bmatrix} 1/3m_1\ell^2 + 4/3m_2\ell^2 + m_2C_2\ell^2 & 1/2m_2\ell^2 + 1/2m_2\ell^2C_2 \\ 1/2m_2\ell^2 + 1/2m_2\ell^2C_2 & 1/3m_2\ell^2 \end{bmatrix} \begin{bmatrix} \ddot{q}_1 \\ \ddot{q}_2 \end{bmatrix}$$

$$+ \begin{bmatrix} -1/2m_2S_2\ell^2\dot{q}_2^2 - m_2S_2\ell^2\dot{q}_1\dot{q}_2 \\ 1/2m_2S_2\ell^2q_1^2 \end{bmatrix}$$

$$+ \begin{bmatrix} -1/2m_1gC_1 + 1/2m_2g\ell C_{12} + m_2g\ell C_1 \\ 1/2m_2g\ell C_{12} \end{bmatrix} \tag{24}$$

where $C_i \equiv \cos\theta_i$; $S_i \equiv \sin\theta_i$; $c_{ij} \equiv \cos(\theta_i + \theta_j)$
Writing R(q) in simplied form as

$$R(q) = \begin{bmatrix} R_1(q) & r_2(q) \\ r_2(q) & R_2(q) \end{bmatrix} \tag{25}$$

the system equations of the manipulator can be written as

$$B = \begin{bmatrix} 0 \\ R(q)^{-1} \end{bmatrix}_{4x2} \qquad C = \begin{bmatrix} 0\ 0\ 1\ 0 \\ 0\ 0\ 0\ 1 \end{bmatrix}$$

The values for the various parameters used for the purpose of simulation were $\ell_1 = 1$ meter, $m_1 = m_1 = 1$ kg. A table showing the r.m.s. error for various choices of the weighting matrix "D" for a 2-link robot is given in Figure 4.

VII. CONCLUSIONS

In this paper we have proposed a 2-D model for a betterment process for nonlinear robot dynamics that improves the performance of the next trail by using updated data from the previous operation. The convergence of the control scheme for a general n-degree of freedom manipulator under certain conditions was proven. Simulation results for a two-link robotic manipulator using the 2-D learning scheme were also presented.

VIII. REFERENCES

1. Arimoto, S., S. Kawamura, and F. Miyazaki, "Bettering Operation of Robots by Learning", Journal of Robotic Systems, 1-2, 124-140 (1984).
2. Arimoto, S., S. Kawamura, and F. Miyazaki, "Bettering Operation of Dynamic Systems by Learning; A New Control Theory for Servomechanisms or Mechatronic Systems", Proceedings of the 23rd IEEE Conference on Decision and Control, 1064-1069 (December 1984).
3. Geng, S. and M. Jamshidi, "Two-Dimensional System Models for Learning Control Systems", Proceedings of the Second International Symposium on Robotics and Manufacturing, 273-280 (November 1988).
4. Takegaki, M. and S. Arimoto, "A New Feedback Method for Dynamic Control of Manipulators", Transactions ASME, Journal Dynamic System Measurements Control, 103, 113-125 (1981).
5. Bondi, P., G. Casalino and L. Gambardella, "On the Iterative Learning Control Theory for Robotic Manipulators", IEEE Journal of Robotics and Automation, 4, 14-23 (February 1988).
6. Casalino, G. and L. Gambardella, "Learning of Movements in Robotic Manipulators", Proceedings of IEEE Conference on Robotics and Automation (April 1986).
7. Craig, J. J., "Adaptive Control of Manipulators through Repeated Trails", Proceedings of American Control Conference (June 1984).
8. Dubowski and D. T. DesForges, "The Apprlication of Model Referenced Adaptive Control to Robotic Manipulators", ASME Journal Dynamic System, Measurement and Control (September 1979).
9. J. J. Slotine and W. Li, "Adaptive Robot Control: A New Perspective", Proceedings of the 26th Conference on Decision and Control, 192-198 (1987).
10. A. Gautam, "2-D Approximation and Learning Control of Robot Manipulators". Masters Thesis, Florida Atlantic University, December 1989.

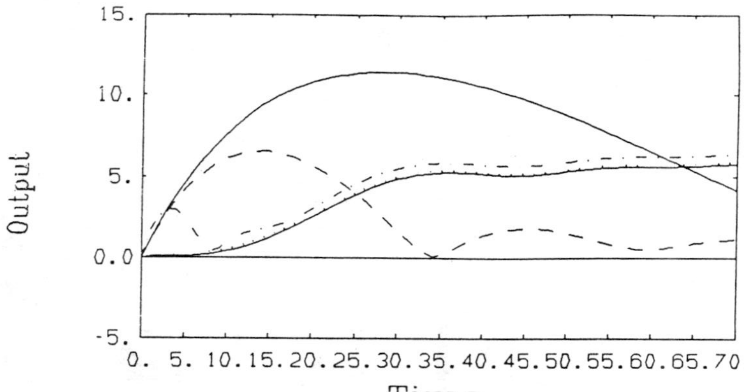

Figure 2.

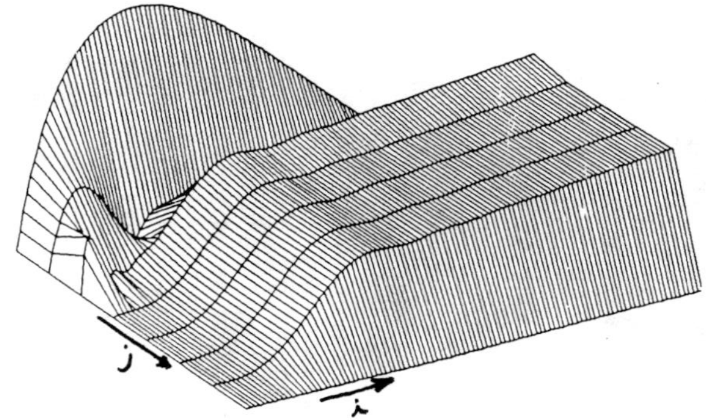

Figure 3 : Output for the 2-link robot with $D_1 = 1$ and $D_2 = 5$

Iter.	$D_2 = R(q)$		$D_2 = 5$		$D_2 = 4$		$D_2 = 3$	
No.	RMS	L_∞	RMS	L_∞	RMS	L_∞	RMS	L_∞
1	710.25	109.86	710.25	109.86	710.25	109.86	710.25	109.86
2	61.021	22.201	133.30	40.767	157.22	46.361	266.99	53.356
3	6.5893	2.9029	6.7639	3.0103	7.1210	3.6582	8.9412	4.0983
4	0.8399	0.2210	0.8943	0.2493	0.9978	0.3056	2.4724	0.3835
5	0.4205	0.1085	0.4693	0.1166	0.5469	0.1272	1.1306	0.1411

Table 1 Table of errors for the 2-link Robot

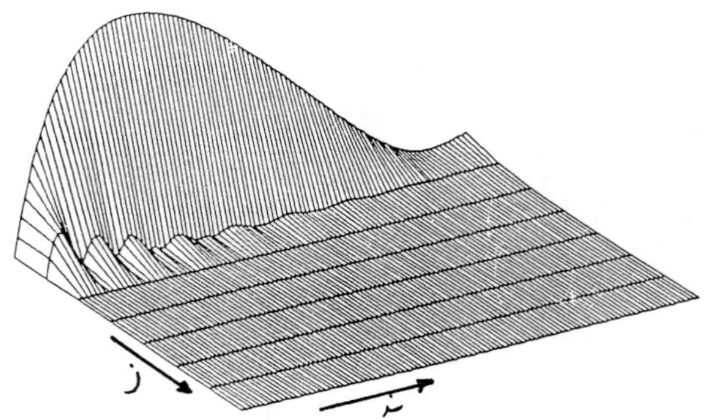

Figure 4 : Error for the 2-link robot with $D_1 = 1$ and $D_2 = R(q)$

SUBJECT INDEX

D

E

F

L

M

N

O

P

Q

R

S

T

AUTHOR INDEX

X

Y

Z